PRINCIPLES OF FORENSIC PATHOLOGY

PRINCIPLES OF FORENSIC PATHOLOGY

FROM INVESTIGATION TO CERTIFICATION

Edited by

SUSAN F. ELY

JAMES R. GILL

ELSEVIER

ACADEMIC PRESS

An imprint of Elsevier

Academic Press is an imprint of Elsevier
125 London Wall, London EC2Y 5AS, United Kingdom
525 B Street, Suite 1650, San Diego, CA 92101, United States
50 Hampshire Street, 5th Floor, Cambridge, MA 02139, United States
The Boulevard, Langford Lane, Kidlington, Oxford OX5 1GB, United Kingdom

Notices

Knowledge and best practice in this field are constantly changing. As new research and experience broaden our understanding, changes in research methods, professional practices, or medical treatment may become necessary.

Practitioners and researchers must always rely on their own experience and knowledge in evaluating and using any information, methods, compounds, or experiments described herein. In using such information or methods they should be mindful of their own safety and the safety of others, including parties for whom they have a professional responsibility.

To the fullest extent of the law, neither the Publisher nor the authors, contributors, or editors, assume any liability for any injury and/or damage to persons or property as a matter of products liability, negligence or otherwise, or from any use or operation of any methods, products, instructions, or ideas contained in the material herein.

ISBN: 978-0-323-91796-4

For information on all Academic Press publications visit our website at https://www.elsevier.com/books-and-journals

Publisher: Stacy Masucci
Acquisitions Editor: Patricia Osborne
Editorial Project Manager: Kristi Anderson
Production Project Manager: Selvaraj Raviraj
Cover Designer: Mark Rogers

Typeset by TNQ Technologies

Dr. Charles S. Hirsch

In honor of the immeasurable value of timeless mentorship and in loving memory of Dr. Charles S. Hirsch.

We also dedicate this book to each of our spouses, Scott and Raffaella, and to our families, for their tireless support and encouragement, not to mention their endless patience in enduring many late nights and cancelled dinners throughout this long, but ultimately very rewarding process.

SFE and JRG

April, 2022

Watercolor by Dr. Cricket Coleman

Contents

10. Approach to blunt, sharp, and transportation deaths

SUSAN F. ELY AND JAMES R. GILL

11. Approach to asphyxial deaths

SUSAN F. ELY AND JAMES R. GILL

12. Approach to burns, blast, and radiation injury

SUSAN F. ELY AND JAMES R. GILL

13. Approach to environmental deaths

SUSAN F. ELY AND JAMES R. GILL

14. Approach to toxicological deaths

THOMAS GILSON, DAVID DOLINAK, SUSAN F. ELY, AND JAMES R. GILL

15. Special topics: in-custody deaths, physical altercations, neglect (adult), approach to unexpected hemorrhage, and delayed and concealed homicides

SUSAN F. ELY AND JAMES R. GILL

16. Communication

SUSAN F. ELY, JAMES R. GILL, JONATHAN HAYES, AND JONATHAN LUCAS

17. The forensic pathologist's public health role

ERIN G. BROOKS

List of contributors

Bradley Adams, PhD, Office of Chief Medical Examiner, Department of Anthropology, New York, NY, United States

Erin G. Brooks, MD, Department of Pathology and Laboratory Medicine, University of Wisconsin Hospital and Clinics and School of Medicine and Public Health, Madison, WI, United States

David Dolinak, MD, Cuyahoga County Medical Examiner's Office, Regional Forensic Sciences Laboratory, Cleveland, OH, United States

Jamie Elifritz, MD, New Mexico Office of the Medical Investigator, Forensic Imaging, University of New Mexico, Albuquerque, NM, United States

Susan F. Ely, MD, MPHTM, Office of Chief Medical Examiner, New York, NY, United States; New York University Grossman School of Medicine, Department of Forensic Medicine, New York, NY, United States *(The views expressed in this book do not necessarily represent the views of the city of New York, the OCME, or New York University)*

James R. Gill, MD, State of Connecticut Office of the Chief Medical Examiner, Farmington, CT, United States; Yale University, New Heaven, CT, United States; University of Connecticut, Storrs, CT, United States; Quinnipiac University, Hamden, CT, United States

Thomas Gilson, MD, Cuyahoga County Medical Examiner's Office, Regional Forensic Sciences Laboratory, Cleveland, OH, United States

Jonathan Hayes, MD, Office of Chief Medical Examiner, New York, NY, United States; New York University Grossman School of Medicine, Department of Forensic Medicine, New York, NY, United States

Heather S. Jarrell, MD, New Mexico Office of the Medical Investigator, University of New Mexico, Albuquerque, NM, United States

Kelly Keyes, D-ABMDI, RTI Center for Forensic Sciences, Research, Triangle Park, NC, United States

Kristen Landi, MD, Office of Chief Medical Examiner, New York, NY, United States; New York University Grossman School of Medicine, Department of Forensic Medicine, New York, NY, United States

Jonathan Lucas, MD, Department of Medical Examiner-Coroner, Los Angeles, CA, United States; David Geffen School of Medicine, University of California Los Angeles, Los Angeles, CA, United States

Michelle N. Stram, MD, ScM, Office of Chief Medical Examiner, New York, NY, United States; Department of Forensic Medicine, New York University Grossman School of Medicine, New York, NY, United States

In memoriam

The inspiration for this book is born out of years of training and of being trained, out of hours of sitting around a conference table or in random hallways, engaged in animated, sometimes collegially contentious, but always interesting fat chewing, out of the same quotidian issues, concepts, questions, lost sleep, and hand wringing we endure as forensic pathologists, particularly during the training of our fellows, day after day, year after year. It is born out of the recognition of the importance and often far-reaching repercussions of our decisions, out of our mindfulness of the survivors we serve. We firmly believe that few agonize more over getting it right than do forensic pathologists. It is in that spirit and with a nod to the challenges we face that we present the foregoing concepts in forensic pathology practice, memorialized in one bound text.

Before doing so, it would be disingenuous not to credit the late Charles S. Hirsch (CSH) for instilling in the editors of this book an unapologetic love and sense of duty to this singular profession. Without him, the words on these pages would have considerably fewer guideposts and less inspiration. In our practice, our commitment to maintaining integrity, our perspectives, and all manners of communication, verbal and written, his impeccable influence is not only felt, it is central. Never does a day go by that we don't ask ourselves, "What would Dr. Hirsch do? What might he say?", at which point we would draw from the seemingly bottomless reservoir of what we fondly refer to as "Hirschisms," and regale anyone willing to listen with the perfect turn of phrase to fit the moment. At a ceremony honoring and celebrating both the man and his career on the momentous occasion of his retirement, he offered the following words:

> Day in, day out, we serve people during the worst and most stressful experiences of their lives—the aftermath of the unexpected death of a friend or relative. The public is minimally aware of this, and the media are mainly concerned with sex and violence. The only people with a real understanding of our service are those of us who do it and the unfortunates who have unexpectedly lost a loved one. What we do every day makes a difference, and we do it without public acclaim. Herein lies the virtue and true gratification of our work. That feeling of gratification must come from our routine rather than the isolated achievement of a lofty goal. In this context, the mundane trumps the spectacular.

This encapsulates both his gift of oratory and principles perfectly. He was a man whose deft use of the English language to succinctly convey meaning was as close to perfect as we have ever encountered, his mind having the capacity to cut through the pork of any matter with exquisite precision, ease, and humor. Needless to say, our better angels and elevated discourse are his legacy. He is our muse.

To be fair though, due credit should be extended to Drs. Lester Adelson and Alan Moritz, Dr. Hirsch's medicolegal father and grandfather, respectively. As Dr. Hirsch was to us, so too were these two men to him. Without these giants, pioneers, and role models to countless forensic pathologists, past and present, Dr. Hirsch would likely have neither discovered forensic pathology nor his specific compass and professional inspiration. Their combined contributions to the development of forensic pathology and its core training principles as we know them today are immeasurable. In his tribute to Dr. Adelson in 1989, Dr. Hirsch aptly described Dr. Adelson's contributions as thus, "Lester Adelson's intellectual legacy is unequaled by any other forensic pathologist who has touched ink to paper."[1] Along with other seminal writings, most notably his treatise, *The Pathology of Homicide*, his should be required reading for all trainees; we can think of no better way to begin a career in forensic pathology than to enter the headspace of Lester Adelson.[2] The work of these three men, Moritz, Adelson, and Hirsch, whose careers overlapped in Cleveland at the Cuyahoga County Coroner's Office and Case Western Reserve University, are inextricably linked in their influence on so many. [3] For this reason, it is more accurate to concede that, together, they are responsible for the inspiration and intended tenor, tone, and overall goals of this text, despite the fact that the authors never met the elder two of this trio Figs. 1 and 2.

Currently, there are many practicing former trainees of Dr. Hirsch scattered around the world; but to dispel what may be a misperception, I would offer this: most do not maintain a rigid adherence to specific conclusions, opinions, or dogma, per se. After all, it was CSH who encouraged robust, if collegial, debate, always reminding us, "Let us agree to disagree, but let's not be disagreeable" when the discussion threatened to go south and become contentious or unprofessional. He, of all people, recognized the reality of reasonable, thoughtful, and well-trained individuals having disparate opinions, while also recognizing that we must make room for those differences as respectable and dignified physicians and scientists, to use the discussion as the fuel that might enhance the incisiveness of our thought processes. Instead of a strict adherence to final conclusions, per se, his primary focus in all matters and his perennial and consistent message centered on *how* we arrived at our conclusions, not as much *what* those conclusions were (not at all to diminish the latter, about which he had many unambiguous opinions). His focus, however, was on curating and perfecting that *process* into one that would invariably lead to the most accurate, defensible, and internally consistent opinions. He would ask what answers we were seeking and why, and how those inquiries might ultimately support sound forensic conclusions. He forced us to defend our conclusions by first defending the process by which we arrived at them. In the end, this focus was predicated on his belief that sound and consistent thought processes and practices would inevitably and naturally lead to the most accurate answers, to the truth of any matter. To wit, he would ask if we were being internally consistent, evidence-based, letting common sense prevail, not allowing an adherence to objectivity deny reality, making decisions with integrity, maintaining defensible decisions, and, most importantly and at all cost, avoiding being arbitrary and capricious, the most grievous of errors!

FIGURE 1 Lester Adelson, MD. *Portrait made available by his family.*

The beauty of his central tenets is that integrity of practice is something we can all get behind in these difficult and sometimes fraught times for forensic pathology, something everyone can agree upon as a profession, and, as such, elevate the work. His are a set of values more than dogma. While as a subspecialty, we strive to find common ground by putting standards of practice into place, we must also recognize the reality and inevitability of reasonable differences of opinion between reasonable people. Moreover, forensic pathology practice, perhaps more so than most other areas of medicine, is challenged by the unavoidable and sometimes vast differences in governmental budgets and municipal fiscal support of our work, local governments and politics, geographical limitations, the disparate population compositions and sizes we serve, and the variability in practice patterns set in motion during the early years of training in a variety of reputable, but different programs. While that recognition is important, it need not render hopeless our joint attempts to find common ground in our collective practices. Rather than throwing up our hands in defeat, working together to find and set scientifically and forensically sound principles and practices, regardless of these variables, is not only likely to enhance the respect that we as professionals might enjoy, it is incumbent upon all of us to work toward.

FIGURE 2 Alan Richards Moritz. *Photographer unknown., "Photograph of Alan Richards Moritz, circa April 1949.," OnView: Digital Collections & Exhibits, accessed March 27, 2022, https://collections.countway.harvard.edu/onview/items/show/17807.*

May we find common ground not by where we work or where we trained, but rather by science, common sense, and above all, the unvarnished story the human body tells.

References

1. Hirsch CS. Lester Adelson, MD. *Am J Forensic Med Pathol*. 1989;10(3):261–263.
2. Adelson L. *Pathology of Homicide*. 1st ed. Springfield: Charles C Thomas; 1974.
3. Adelson L. Alan Richards Moritz, MD: An appreciation. *Am J Forensic Med Pathol*. 1981;2(4):297–298.

Biography

SUSAN F. ELY

Dr. Ely is currently the Fellowship Training Program Director and a senior medical examiner for the Office of Chief Medical Examiner of the City of New York (OCME). She is also a Clinical Associate Professor in the Department of Forensic Medicine at the New York University School of Medicine. Prior to becoming program director, Dr. Ely served as Deputy Chief Medical Examiner for Bronx County. She received her medical degree and masters of public health and tropical medicine at Tulane University in New Orleans and completed internships in pediatrics and internal medicine, followed by a residency in anatomic and clinical pathology and a fellowship in forensic pathology, the latter at the New York City Office of Chief Medical Examiner. She is a peer reviewer for forensic pathology journals, has published in both forensic and clinical journals, and is a member of the American Academy of Forensic Sciences and the National Association of Medical Examiners. She has extensive experience in expert testimony and has lectured on forensic topics for varied audiences both in the United States and abroad.

The views expressed in this book do not necessarily represent the views of the city of New York, the OCME, or New York University.

Biography

JAMES R. GILL

Dr. Gill is the Chief Medical Examiner of Connecticut and has clinical medical school faculty appointments in the Departments of Pathology at Yale, the University of Connecticut, and Quinnipiac University. He is a past President of the National Association of Medical Examiners (NAME) and the Chair for the Forensic Pathology Committee of the College of American Pathologists. He did his pathology training at Yale and Memorial Sloan-Kettering Cancer Center and his forensic pathology fellowship at the New York City Office of Chief Medical Examiner. Prior to Connecticut, he was the Deputy Chief Medical Examiner for Bronx County in New York City.

He has published over 100 scientific articles and book chapters on a variety of forensic pathology and public health topics. He was on the American Board of Pathology's test-writing committee for the forensic pathology boards for 10 years. He has testified as an expert witness in forensic pathology over 450 times and given hundreds of lectures in forensic pathology around the country and the world, including Spain and New Zealand.

Foreword

This new textbook is different from other forensic pathology textbooks. Dr. Gill and Dr. Ely have produced a comprehensive guide for the day-to-day practice of forensic pathology by adroitly emphasizing and delineating the proper approach to death investigation and the common sense thinking necessary for the competent forensic pathologist to interpret the scene investigation, the autopsy findings, and the ancillary testing in order to best fulfill the role of the medical examiner in modern death investigation.

Fifty years ago, I started my forensic pathology fellowship and since then, I have been a reader and collector of most of the better known forensic pathology textbooks, including, among others: Taylor's *Medical Jurisprudence*; Sydney Smith's *Forensic Medicine*; Gonzales, Vance, Helpern, and Umberger's *Legal Medicine, Pathology and Toxicology*; Adelson's *The Pathology of Homicide*; Spitz and Fisher's *Medicolegal Investigation of Death*; DiMaio's *Forensic Pathology*; and Dolinak's *Forensic Pathology*. All have a place on my bookshelf, and each has proved valuable to me in many ways.

Gill and Ely's new comprehensive textbook follows that impressive list and is a welcome and important addition to those classic textbooks. Those textbooks concentrate heavily on the physical aspects of forensic pathology: the scene, the gross autopsy findings, the microscopic features, and the toxicologic and ancillary testing results in individual cases and/or groups of related cases. This textbook not only addresses those important issues in detail but also additionally looks at them through the eyes of how an experienced forensic pathologist would use them to adequately certify cause and manner of death, and best answer questions raised by the medicolegal investigation. This detailed approach to the philosophy of the modern forensic pathologist promotes the understanding of the uniqueness of each case.

The format of the textbook is a logical approach to the science and philosophy of forensic pathology, starting with introductory chapters defining the role of the forensic pathologist and his/her place in our modern society. In particular, Chapter 2, on thinking forensically, is the key to the substrate of philosophy running throughout the book.

The textbook covers areas that the older textbooks could not because of advances in the field. The chapters on postmortem imaging and 3D printing and on technological advancements in forensic pathology practice are important additions needed for a modern textbook. Other sections, such as those on therapeutic complications and in-custody deaths are also topics valuable for any current textbook on forensic pathology.

Gill and Ely bring excellent scholarly credentials to this comprehensive task. They have also recruited other highly qualified forensic scientists for specialized chapters. Throughout the book, their approach to forensic pathology is solidly based on their training and experience and the guiding principles of their longstanding mentor and mine, Dr. Charles Hirsch, the former Chief Medical Examiner of New York City.

This textbook has something for everyone associated with forensic pathology. Practicing forensic pathologists will use it as a resource and refresher. Forensic pathology fellows and residents in pathology will rely on it for training and how to think forensically. Practicing physicians will consult it for understanding how to interact with families. It is a valuable resource for attorneys to understand how forensic pathologists approach cases to come to conclusions of cause and manner of death. Public officials will benefit from reading this textbook to understand how forensic pathologists certify deaths with a logical approach, and with important implications for public health and the statistical analysis of deaths.

The medical field of forensic pathology has long waited for a comprehensive textbook of this scope and depth, emphasizing the philosophical and common sense approach to death investigation. That wait is over.

Ross Zumwalt, M.D.
Anatomic Pathologist and Senior Associate Consultant of Laboratory Medicine and Pathology, Mayo Clinic, Rochester, MN

Preface

"If government has an interest in the lives of its citizens, it must also have an interest in their deaths." -Dr. Charles S. Hirsch

"The right to search for the truth implies also a duty one must not conceal any part of what one has recognized to be true." -Albert Einstein

In forensic pathology (FP), there are few textbooks geared toward training, let alone any specifically aimed at the *conceptual* approach to forensic practice. Medicolegal death investigation is predicated on scientific, investigative, and physical evidence, but it also relies upon an immeasurable degree of nuance and context that thread together its component parts; indeed, these are the features that distinguish FP from other forensic sciences. FP is not primarily a comparative or quantitative discipline—we do not match bullets, compare tire tread patterns, or calibrate mass spectrometers to determine drug concentrations. FP is a medical subspecialty built on science, training, experience, judgment, and ultimately, opinion. We obtain histories to generate hypotheses about which diseases or injuries may have caused death. We confirm or refute these hypotheses with the use of the physical examination (the autopsy), the circumstances, and diagnostic studies in order to formulate medical diagnoses—ultimately, to render opinions regarding the cause and manner of death. It is this understanding of the complexity lying at the intersection of findings and context that hones forensic thinking, ultimately distinguishing good from great forensic pathologists; it constitutes the very art of FP practice.

Current FP textbooks focus primarily on autopsy findings, laboratory studies, and their associated case-based interpretations. This is important foundational knowledge to learn in order to enhance the expertise of the autopsy and investigative components of FP. The thought processes, however, that form the basis of the interpretation of these physical and investigative findings, in a structured and consistent manner, are what distinguish a technician from a physician. Unfortunately, there is a dearth of textbooks that focus specifically on laying out a systematic approach to forensic thinking and practice, one that encourages the consistent, accurate, detailed, and comprehensive investigation of varied and specific types of death circumstances with the goal of arriving at opinions and certifications in a way that is both reproducible and defensible.

This book is *not* intended to be a one-size-fits-all prescription for precisely how to certify all causes and manners of death, nor is it intended to specify word choice in the cause of death statements (although we will offer many examples). We recognize that there are different schools of thought, particularly with regard to manners of death. Rather, this book is intended to offer a framework within which FP is practiced, offering an approach for evidence-based forensic thinking that will most accurately lead to conclusions that are mechanism-based and ever mindful of the importance of consistency to ensure that we meet our burdens of proof. We aim to lay the groundwork that, with conscious effort, specifically avoids the trap of being arbitrary and capricious and minimizes confirmation bias. We will also emphasize the importance of nimbleness, of maintaining the delicate balance between making allowances for case-by-case differences in decision-making, while operating from within a specific construct of thought processes and unchanging forensic concepts.

Further, we fully recognize that no textbook or even fellowship training period can possibly present every imaginable contingency to trainees in both experience and case breadth. It is precisely with that in mind that we have focused on a presentation of foundational concepts that we hope will set the stage for the forensic pathologist to approach myriad medicolegal death scenarios and meet any challenge by applying the same concepts, standards, and pathophysiologic principles to arrive at the most accurate conclusions with the greatest frequency. That said, while the *concepts* outlined in this book are broadly relevant to FP in general, the *practice* suggestions we present are more with the forensic pathologist in training and the junior forensic pathologist in mind. By this, we mean that many practice instructions we offer err on the side of doing more than less, of encouraging the relatively more comprehensive, if still thoughtful and selective, approach that is in keeping with the foundational training period and our training standards, while mindful of the trap of dogmatism. Over the many postgraduate years spent in FP practice, we recognize that the development of experience will allow some of these general practice instructions to become more case-specific, by virtue of an increasingly seasoned and nimble pathologist applying judgment and necessary discretion, while also allowing for jurisdictional differences in approach. We emphasize a concept that will be reiterated in this text: there is no one approach that has the capacity to specifically and individually accommodate *all* decedents or medicolegal death investigations. As Dr. Hirsch would say, "Routine is never a substitute for judgement."

This book will focus on "forensic thinking": what questions need to be asked, how do we answer them, and how do we use those answers in the context of a death investigation to reach correct conclusions? As just one example, we will suggest that the determination of medicolegal opinions should *begin* with the identification of the mechanism(s) of death and proceed from there. Among other things, this book will include scenarios that are encountered both frequently and less frequently, offering mental check lists to guide a consistent approach that is tailored to each type of death circumstance, thus maximizing the FP's ability to meet their burdens of proof systematically.

To that end, we strive to identify and promulgate sound forensic practice concepts rather than granular and prescriptive dictums or standards. Our goal is to present a systematic approach for making and asserting reproducible forensic decisions and opinions, but to do so in a way that seeks to find common ground among all practicing forensic pathologists, to put down in a deliberate fashion what "thinking forensically" means, and how we might better flesh out its underpinnings and incorporate them into our daily practice. The overarching spirit of this book is best encapsulated by two aphorisms from the late Charles S. Hirsch: "When your standards go, your principles are not far behind" and "Let common sense prevail."

Acknowledgments

There were many colleagues who not only taught us forensic pathology but also generously shared teaching images and many pearls. These include our forensic pathology co-fellows, Margaret Prial and Eric Peters, plus Carolyn Kappen, our forensic neuropathology co-fellow at the time. Yvonne Milewski, Jonathan Hayes, Monica Smiddy, Mark Flomenbaum, Corrine Ambrosi, Luz Alandy, Hernando Mena, Vernon Armbrustmacher, and Marina Stajic deserve special mention, as they all taught us so much. There are many colleagues who have shared their material with us over the years and who may recognize some of these images. We are not able to accurately or adequately acknowledge them individually, but we thank them here for their trust, generosity, friendship, and courtesy.

In addition, we would like to both acknowledge and personally thank Drs. Kristin Roman, Rebecca Folkerth, and Yingying Tang for their invaluable editorial assistance in the areas of maternal mortality, neuropathology, and molecular genetics, respectively.

Beyond his substantial involvement as a contributing author, we are indebted to Dr. Bradley Adams for his flexibility and willingness to avail himself to us in providing ongoing editorial advice and high-quality images. His writing is incorporated into Chapters 4, 15, and 19.

Most especially, we wish to thank Dr. Andrea (Cricket) Coleman for creating all of the medical illustrations published in this text, including the cover art, as well as for her invaluable assistance in developing the cover design and layout. Also, as a representation of her considerable talents, a copy of her portrait of Dr. Hirsch is displayed in the beginning of this book (Dedication); the original painting currently hangs in the lobby of the "DNA building" at the NYC OCME. Her biographical profile is highlighted below.

About the Artist (cover art and all medical illustrations):

Andrea "Cricket" Coleman, MD was born in Chicago, and grew up in Los Angeles, Cambridge MA, and New York. She got her BA in Fine Art from Vassar, majoring in figurative painting, but the example of her grandmother (Eppie Lederer, known to millions of Americans looking for advice and reassurance as "Ann Landers") led her back to a career oriented toward service. While pursuing her post-baccalaureate at Columbia, she interned in the Forensic Biology department of the Office of the Chief Medical Examiner in New York City (OCME), which led to a keen fascination with forensics. After graduating in medicine from SUNY Downstate, Dr. Coleman did an anatomic pathology residency at Mt. Sinai in New York. She returned the OCME as a fellow in forensic pathology, spending a second year doing subspecialty training in forensic neuropathology and cardiac pathology. After her fellowship years, she joined the office permanently, working primarily out of the Brooklyn office before being recruited to the Manhattan office to help train the forensic fellows. Throughout her medical career, she has kept up a prolific output in fine art. While she is perhaps best known for her watercolors, she is medium-agnostic, equally adept in oils, ceramics, sculpture, and fiber work. The waiting list for her portrait work, which ranges from children to cats, dogs and horses, is very long indeed.

CHAPTER

1

Forensic pathology: overview

Susan F. Ely and James R. Gill

Medicine, to produce health, has to examine disease. -***Plutarch***

Nothing can equal the thoughtful medicolegal autopsy as a teacher of the amazing variability of human responses to disease and injury.
- ***Dr. Charles S. Hirsch***

Introduction: who we are and our unique role as physicians

By the most basic of definitions, pathology is the study of disease and injury. While building upon a foundation of general pathology, **forensic pathology** resides at the center of medicolegal matters and, in the United States, requires further specialized training beyond medical residency. As a discipline, *it stands at the crossroads of medicine and law* and is inextricably linked to the arenas of public health and safety, criminal justice, quality assurance in healthcare, law enforcement, and public service. Dr. Lester Adelson offered the broadest definition of forensic medicine as, "The application of medical knowledge, medical 'know-how', and medical 'know-why' to aid in the administration of justice,"[1] similar to Smith's definition of the same: "Forensic medicine (is)... that body of medical and paramedical scientific knowledge that may be used for the purposes of administration of the Law."[2]

Entering into this far-reaching career path might be viewed as equal parts noble and daunting. At their most elemental, Dr. Adelson aptly described forensic pathologists (FPs) as "family physician(s) to the bereaved."[3] He went on to say the following,

> Death in all its forms and all its guises is the professional realm of the forensic pathologist. From this standpoint, it is reasonable to regard him as the last generalist in medicine. He is constantly exposed to the age-old truth that tells him that although there is only one way in which to be born, there are many ways in which to die.
>
> ... Even under the best of circumstances, death is almost always a traumatic experience to the survivors. The emotional impact of death is magnified enormously when it comes unexpectedly.
>
> Interpreting autopsy and associated laboratory findings falls logically within the expertise of the 'attending postmortem physician', an appellation descriptive of the forensic pathologist's role at this juncture.

To his point, our obligation to survivors cannot be understated. Forensic pathologists (FPs) are uniquely positioned to listen to those who most need a voice during a time of tremendous loss and confusion. Dr. Hirsch recalled hearing Dr. Adelson "explain the inexplicable to bereaved families and thereby help them accept a reality that previously seemed impossible," thus supporting their effort to "put grief and guilt into perspective."[4] Put simply, forensic pathologists have the capacity to be considerably more than grief counselors. They have the singular authority and opportunity to exact lasting good, to impart helpful and credible medical and scientific information, perspective, knowledge, and understanding, while also offering compassion, honesty, and comfort in a way that few others can, thereby giving people at the worst times in their lives closure and a path forward. We are the physicians who serve as the primary medical link between a sudden and unforeseen loss and those who are left to make sense of it. This is not only very much a part of the job, but also it is a task and responsibility not for the faint of heart and that should be viewed as a sacred oath of service.

Principles of Forensic Pathology
https://doi.org/10.1016/B978-0-323-91796-4.00012-X

FPs are also the only physicians specifically trained in **injury pattern recognition** in both the dead and the living (the latter, referred to as clinical forensic medicine, is commonly a part of training and practice outside the United States). While trauma surgeons are trained to manage and save the lives of victims of diverse and severe forms of injury, they are not specifically trained to recognize all injury patterns, per se. In order to resuscitate the victims of firearms injuries, for example, one does not need to be able to necessarily differentiate entrance from exit gunshot wounds nor to determine range of fire. In fact, when asked to distinguish exit from entrance wounds, clinicians have been shown to be correct only half the time.[5] Rather, this skill set falls squarely into the wheelhouse of FPs. At its core, forensic pathology as a science is descriptive and observational, using physical findings and circumstances to provide expert opinions. As Dr. Hirsch astutely observed in describing his discovery of forensic pathology practice,

> Then came the introduction to the complexities of trauma. We quickly learned ..., that the only people who think that trauma is simple are those who haven't seen it or those who haven't thought about what they've seen.[4]

Also, unlike most physicians, many FPs work as medical examiners, outside of hospitals, either for a governmental/municipal agency or in private practice. There are some medical examiner practices that are part of an academic center, which offers another option for those who are academically inclined. Over the course of a career, a FP will be asked on several occasions to define, under oath, many terms and concepts, not the least of which is "forensic pathology." As such, it is imperative to have an accurate understanding of its practice and a full-throated definition at hand. Among the many tasks falling under their purview, including the handling of chemical and select natural deaths, "injury pattern recognition" is a skill that should be clearly stated as specific and central to this field. This sets the stage in simple terms and makes clear the importance of the highly specialized and advanced level of training on which our expertise depends and uniquely qualifies us as specialists, set apart from all other physicians.

In addition to the above, our involvement in *the investigation of sudden and unexpected deaths* of all types and/or in those with *public health or safety* at issue constitutes other large areas falling under our purview. Moreover, FPs are often involved in the *education* of not only just students, residents, and fellows, but also of scores of other professionals, both medical and legal, as well as families, jurors, governmental professionals, and the public at large. Additionally, an appreciable amount of a forensic pathologist's professional time is spent *testifying as an expert witness* in various legal forums, another task fairly unique to forensic pathology in contrast to most other medical specialties.

Beyond having a clear understanding of the specificity and extent of our expertise as a profession, there is the importance of recognizing just how few of us there are. To date, there are approximately 500–600, full-time practicing, board-certified FPs in the United States, despite the fact that it has been determined that over 1200 are needed to adequately and competently attend to the nation's medicolegal death investigation burden.[6] Consequently, there is a focused, robust, and urgent movement to address this shortage by boosting and recruiting trainees as early as high school and college. To support this initiative, a National Institute of Justice grant supports forensic pathology fellowship training programs and educational loan forgiveness. That is all to say that while this may someday change, there has existed for some time and still exists an immediate and intense need for physicians with the specific expertise only afforded by forensic pathology training. Regardless of this shortfall of FPs, there is evidence to suggest that the specific societal need for this field of medicine will always exist and will weather any storm. During the COVID-19 pandemic, hospitalists, obstetricians, and FPs were the only medical specialties that did not experience a decrease in work. This shortage has helped increase salaries in some jurisdictions (although salaries are still largely and woefully inadequate) and provided ample opportunities for positions in varied geographic areas. There are numerous ways that FPs may ameliorate their income, including through private consulting, locum tenens work, and teaching.

In sum, the practice of forensic pathology differs greatly from hospital-based and clinical pathology. It requires not only a specific mindset and approach as described above, but also it involves service to those in need, education of both medical and nonmedical professionals, expert testimony, and multiple public health roles, all predicated on the rendering of opinions born out of an organized, systematic, observational, contextual, and scientific approach. By design, the unique framework within which we operate and make consequential decisions should aim to be defensible, reproducible, internally consistent, evidence-based, adherent to standards of integrity and well-defined, time-honored forensic principles, and, at no time, be arbitrary or capricious, in appearance or reality.

Historical perspective

The word "forensic" is rooted in the Latin word "*forensis*," referring to a "forum" for the public. Discussions around cause and mechanism of death date as far back as 44BC and proceed from there, including a mechanistic thesis for crucifixion in 33AD.[7] By the 600s, coroners were mentioned in English law, but it wasn't until 1194 that the Articles of Eyre established the Coronial system. In 1250AD, Hsi Yuan Lu published the *China Handbook on Autopsy*.[8] The first formal lectures in forensics were given at the University of Leipzig by Michaelis and Bohn, and the first crime laboratory, Locard, was established in Lyon, France, in 1910.[2]

One of the first recorded autopsies in the "New World" occurred in 1662 in Connecticut.[9] In a series of firsts in the United States, the *physician* coroner was first established in Maryland in 1868 and Massachusetts was the first state to institute a (noncentralized) medical examiner system (1877).[10] But, it wasn't until 1918 that New York City established the first chief medical examiner system[11,12] with the power of autopsy, ushering in the modern era of medicolegal death investigation, followed in 1939 by the state of Maryland establishing the first *statewide* chief medical examiner system. This was preceded by the first domestic crime laboratory, which was developed in Los Angeles in 1932.

The primary underpinnings of academic forensic pathology as a recognized medical specialty in the United States took root most significantly in 1937, a year in which two critical milestones were achieved. The first was the establishment of the Department of Legal Medicine at Harvard Medical School, singular at the time in its autonomous status beyond merely a division. **Dr. Alan Moritz** served as its first chair, ushering in an academic acceptance and respectability that would erect institutional scaffolding within which forensic medicine would begin to develop in earnest, beyond only a governmental function.[13] Not long after its establishment, two important academic programs were developed: *Seminars in Homicide Investigation* and *Rockefeller Fellowships in Legal Medicine*, both serving as templates for similar scholarly activities in the future. After Harvard, Dr. Moritz went to Cuyahoga County (Cleveland) to chair the Department of Pathology at Case Western Reserve University and was able to attract Drs. Adelson (forensic pathology) and Sunshine (toxicology) as faculty. The rest, as is said, is history.

The second important event in 1937 was the publication of the first edition of the seminal text, "*Legal Medicine and Toxicology*," written by then leaders of this nascent field, Drs. Gonzalez, Vance, and Helpern in the Office of Chief Medical Examiner in New York City, and considered the "bible" for the field at the time.[11] As a result, and in conjunction with the Harvard development, a seismic shift from relative obscurity to legitimacy took hold and with it, the birth of "Legal Medicine," a sanctioned new field of medicine and research.[1,2,10,14] *It would not be until 1959, however, that the designation of "legal medicine" would be largely replaced by the field of "forensic pathology," as a result the American Board of Pathology instituting its board certification for the first time,* a recognition of both symbolic and actual significance. As a result, at the time of Dr. Adelson's 1989 telling of the evolution of forensic pathology, the number of board-certified FPs ballooned from 30 to 639 in the span of only 30 years, from 1959 to 1987, more than any other pathology subspecialty at the time. Interestingly, from that time to the present, the growth curve has roughly flattened. Between 1959 and 2020, there have been 1842 physicians who have become board-certified in forensic pathology (data from the American Board of Pathology).

Coroner versus medical examiner systems in the United States and abroad

State and county statutes dictate how medicolegal death investigations are performed, with either a medical examiner or a coroner (ME/C) as the official given the authority to head the investigations. Most systems are organized at a county level, while still others are run by state, regional, or district (including city) jurisdictions. With some exceptions, coroners are *elected* lay persons or physicians who consult with pathologists who have varying degrees of training to assist in their investigations, including the performance of autopsies. By contrast, most chief medical examiners are *appointed* physicians who have been trained in forensic pathology.[10,15,16]

The coroner system that still exists in many parts of the United States today was brought to this country from England, where inquests were overseen by the crown (or "crowner"), hence its name. A trend to replace coroners with nonelected medical examiners in the United States began in earnest in 1877, peaking in the mid-1950s and continuing until 1990, after which few have been converted. The 1950s spike is thought to have been at least partly in response to the publication of the **1954 Model Postmortem Examination Act**, which outlined legislation for the development of medical examiner death investigation systems.[10,15,16] Many of those that were converted from coroner to medical examiner systems had already involved physicians in primary roles of the death investigations, making most of these transitions more natural and seamless.

According to Hanzlick's 1998 JAMA survey-based analysis of both systems, 38 states had at least one "statute-established medical examiner system in place at some level of government (county, district, or state)" and 22 in which no coroners were involved at all.[10] The latter served roughly 48% of the of the US population at the time. Still, among the seven types of systems he outlined, the most prevalent, representing 18 states, was the state medical examiner system, with no autonomous county jurisdictions and no coroner involvement. While these state systems are scattered across the county, their densest concentration was/is in the Northeast. The second most common type involved a mixed county system, where some were medical examiners and others were coroner jurisdictions within the same state (11 states). Among the latter, most are *not* overseen by law enforcement (Sheriff-Coroner systems).[10]

Outside of the United States, a variety of medicolegal death investigation systems exist.[17–19] To a greater or lesser degree, coroner systems are in place in England, Wales, Ireland, Canada, and several countries that were part of the British Empire. Elsewhere in continental Europe, the investigative responsibility primarily falls to the judiciary. In either instance, there is involvement and administrative oversight by both law enforcement and the legal system, and the decision to perform autopsies does not generally fall under the purview of the FP. Historically, more coroner-involved autopsies have been performed in England, Wales, and Ireland than in continental Europe, where sudden, natural, and noncriminal traumatic deaths have been of less interest to authorities, although these practices have been changing over more recent years.[20]

Medicolegal death investigations are done for the public interest. The public interest includes service for the public health and criminal justice communities. Three key features of a properly empowered medicolegal death investigation system are subpoena power, the ability to perform autopsies without permission, and the authority to take charge of the body. As the ME/C may investigate and certify deaths that are emotionally and politically charged, it is important to have an independent office. Ideally, an office would be both free of oversight by law enforcement, as well as an independent agency with its own budget, and the chief medical examiner would not be a political appointee who serves at will to an executive authority.

What constitutes "medical examiner/coroner jurisdiction"?

Ever mindful of substantial jurisdictional variability, below is a list of categories of death commonly requiring ME/C involvement by local law:

- All violent/unnatural deaths (caused by mechanical injury in a variety of circumstances, suicidal, accidental, and homicidal)
- All deaths resulting from the effects of illicit or prescription medications or poisons
- All sudden, unexpected, and/or unexplained deaths of otherwise healthy persons (adults and children/infants)
- Deaths occurring in public view or unattended by a physician (in homes or public places)
- All deaths caused by or occurring in the custody of police or detention/correctional facilities and/or when the decedent is a ward of the state
- Fetal demise if born in the absence of a physician or midwife
- Hospital stillbirths if there is a likely causal relationship to maternal trauma or drug use
- Deaths occurring during or from complications of elective therapeutic or diagnostic procedures (not all jurisdictions)
- Deaths resulting from employment and/or occurring at a place of work (with/without OSHA involvement)
- Certain deaths in which there are time-sensitive questions of public health and safety
- Unusual or suspicious deaths not otherwise falling into one of the above categories

The current NAME Forensic Autopsy Performance Standards[21] *state:*

The forensic pathologist or representative shall investigate all:
A2.1 deaths due to violence.
A2.2 known or suspected nonnatural deaths.
A2.3 unexpected or unexplained deaths when in apparent good health.
A2.4 unexpected or unexplained deaths of infants and children.
A2.5 deaths occurring under unusual or suspicious circumstances.
A2.6 deaths of persons in custody.
A2.7 deaths known or suspected to be caused by diseases constituting a threat to public health.
A2.8 deaths of persons not under the care of a physician.

More commonly than not, decedents meeting the definition of one or more of the above categories fall under ME/C purview for death certification purposes, but the body may or may not be brought to the morgue for physical examination in all cases. If it is, there exists fairly wide jurisdictional variability as to whether an autopsy or only an external examination is deemed necessary. For example, in some offices, persons dying from obvious self-inflicted/suicidal injury (e.g., gunshot wound of head, hanging) would not undergo autopsy given the compelling circumstantial and anatomical findings; some may even be certified and released from the scene or hospital directly, with more limited forensic pathology involvement beyond death certification. These variabilities are predicated on considerable differences between the specific features of the case, as well as on staffing, geographic operational and transportation challenges, and fiscal limitations.[21]

The current NAME Forensic Autopsy Performance Standards[21] state:

The forensic pathologist shall perform a forensic autopsy when:
B3.1 the death is known or suspected to have been caused by apparent criminal violence.
B3.2 the death is unexpected and unexplained in an infant or child.
B3.3 the death is associated with police action.
B3.4 the death is apparently nonnatural and in the custody of a local, state, or federal institution.
B3.5 the death is due to acute workplace injury.[1]
B3.6 the death is caused by apparent electrocution.[1]
B3.7 the death is by apparent intoxication by alcohol, drugs, or poison, unless a significant interval has passed and the medical findings and absence of trauma are well documented.
B3.8 the death is caused by an unwitnessed or suspected drowning.[1]
B3.9 the body is unidentified and the autopsy may aid in identification.
B3.10 the body is skeletonized.
B3.11 the body is charred.
B3.12 the forensic pathologist deems a forensic autopsy is necessary to determine cause or manner of death, or document injuries/disease, or collect evidence.
B3.13 the deceased is involved in a motor vehicle incident, and an autopsy is necessary to document injuries and/or determine the cause of death.

The forensic autopsy (see also Chapter 5)

When a forensic autopsy *is* deemed necessary, it may provide multiple data points for the FP to consider, both anatomic and through ancillary testing, most notably toxicological, from which several questions may be answered over time. The questions a forensic autopsy may answer commonly extend far beyond merely cause and manner of death, particularly in criminal matters. Many observations made during the external exam alone take on a much more focused importance. The absence of certain findings in some deaths may be equally as important as the presence of findings.

Differing substantially from the typical hospital autopsy, the forensic autopsy not only includes the assessment and specific description and interpretation of injuries, when present, but also is performed and viewed in the context of a complete and sometimes multiagency investigation, drawing upon information extending beyond the autopsy suite. A meticulous external examination and toxicology testing are two features that differentiate these two settings. Forensic autopsy practice starkly contrasts with most hospital autopsy practices, where the autopsy is performed as more of a laboratory test, the findings cataloged and reported, but typically with *less* of a need for an investigative and autopsy-based synthesis and ultimate opining as to the cause and manner of death. Hospital pathologists are rarely involved in the death certification process. In fact, most hospital death certificates are completed by clinicians, in the absence of an autopsy.[22–24]

As will be described in much more detail in proceeding chapters, death is a functional event, not an anatomic one. **The power of the autopsy lies not only in the identification of a potential anatomic substrate to explain death (e.g., an obstructive thrombus in the left anterior descending artery), but also in its exclusion of other causes of death (e.g., injury).** In fact, the so-called "negative autopsy" can be a very powerful determination in certifying certain deaths. Ancillary studies performed after the collection of body fluids and tissues constitute another important part of the forensic autopsy, but such studies and the need for them are case/context-dependent and based on the sound judgment of the forensic pathologist. There is no "one-size-fits-all" formula. One of the most common analyses performed, however, is toxicological; this may have bearing on opinions of both cause and/or manner or on unrelated issues of interest by other stakeholders (e.g., was the driver intoxicated at the time of the accident?).

[1] Unless sufficient antemortem medical evaluation has adequately documented findings and issues of concern that would otherwise have required autopsy performance.

Equally important, however, is that FPs as the finders of fact, as well as the large cadre of professionals and family members they commonly encounter, understand the limitations of the autopsy and of science, in general. Further, in our approach to any autopsy, we must first ask, "Why am I doing an autopsy? What questions need to be answered? (e.g., is there occult trauma?). In addition, we must also try to anticipate what questions may *later* arise by police, families, attorneys, other physicians, and myriad agencies. The autopsy is our one chance to achieve that end. A "complete" autopsy is one that answers all subsequent questions. We would rather explain "why we did something than why we did not." It is better to do the extra examination than to not do it. An analogous clinical scenario is exemplified by a decision point in pediatrics, when physicians are taught that if you think a child could have meningitis, then you go ahead and perform the lumbar puncture. See Chapter 3 for further elaboration on this topic.

In the foregoing vein, below is a list of some of the benefits and issues that might be addressed through the performance of an autopsy:

1. Establish cause of death
2. Assist in determining manner of death
3. Monitor public health
   - Produce accurate vital statistics
   - Identify emerging and changing diseases
4. Assist in assessing the quality of medical practice (quality assurance)
   - Evaluate effectiveness and/or safety of new diagnostic or therapeutic modalities
   - Compare and contrast antemortem versus postmortem findings
5. Inform survivors and assist in providing information for postmortem matters, such as life insurance claims
6. Identify occult disease that may have heritable relevance to family members
7. Assist in answering questions as expert witnesses in various courts of law, criminal, civil, and family
8. Address the many questions of stakeholders, some unanticipated/not necessarily related to cause or manner of death.[25]

Death certification (see also Chapter 3)

More than any other medical specialist, FPs are acutely aware of the importance and complexity of death certification and of the myriad repercussions that may ensue as a result of it. Beyond the important field of palliative care and its focus on *dying* and dignified end of life care, medical training in the United States is not geared toward understanding *death* and its repercussions, per se. Arguably, there may be an unconscious institutional aversion to focusing on what many physicians might perceive as the ultimate failure of medical efforts. Some countries, such as Sweden, historically have much higher autopsy rates (>80%), but even there, the rates have been decreasing.[26] Regardless of the underpinnings of this educational oversight, and to no fault of their own, few clinicians are specifically and competently trained in death certification, much less trained in a comprehensive understanding of the pathophysiology of death (also Chapter 3). Most physicians learned to complete a death certificate during their internship from the second-year resident who learned how to do it the previous year from their second-year resident. Predictably, the same mistakes are repeated.[27–44]

Proper death certification is a physician's last and most solemn responsibility to a patient, a responsibility that does not cease with cardiac activity and the moment of death. Despite this, more times than not, and largely for practical reasons, the responsibility of teaching proper death certification and the management of its consequences fall into the hands of the FP. Predicated on the lack of emphasis and training on competent death certification in medical school and clinical medicine in the United States, it is probably most prudent and beneficial to all involved to face the reality of the system and appropriately designate the FP as the holder of the reins, given the medicolegal backdrop. Additionally, in many jurisdictions, it is against the law for a physician other than a ME/C to knowingly certify any death other than those caused exclusively by natural disease, often a misdemeanor that is, admittedly, rarely, if ever, prosecuted.

The far-reaching effects of incompetent or inaccurate death certification cannot be understated, hence the critical, if not urgent importance of addressing the national FP shortage. Mishandling of the death certification process can initiate a cascade of events that potentially leaves the survivors dealing with months or years of a morass of economic and/or legal complications, some difficult to reverse. These often substantial issues are layered over the acute and ongoing fallout of the sudden loss of a family member. We must not lose sight of the living behind the dead, the people gutted by an unexpected loss and in desperate need of time sensitive answers. They are often dealing with intense grief while simultaneously handling attendant and often complex financial matters.

As such, poorly executed death certification constitutes a failure of a physician's responsibility at the worst times in people's lives. The FP's obligation to the dead and their potential service to the survivors is not insignificant. We are at the helm of not only answering questions that may facilitate closure and lessen the emotional and financial upheaval of families, but also have the solemn obligation and unique position to settle such matters as insurance claims and otherwise provide necessary and objective opinions to those managing myriad financial and legal matters during the typically byzantine probate and estate settling process, as well as in various civil matters, including litigation.

In the criminal realm, our duties are no less sober, or central. We are not and should not cast ourselves as judge, jury, and executioner; rather, we are one cog in the complex machinery of criminal investigations and their adjudication. That said, our role is critical, sometimes central, more in some cases than in others, but is enhanced by our neutrality, by the fact that we are not attached to the outcomes of matters of criminal and civil litigation. We are typically afforded the privilege by the courts, based on our extensive training and experience, of being qualified as expert witnesses, a designation conferring both gravity and significant responsibility. That recognition, along with a clear-eyed understanding of our burden of proof in criminal matters, facilitates the checks and balances that should attend any death certification discussion and classification, making us ever mindful of avoiding walking headlong into the territory of "arbitrary and capricious." As Dr. Hirsch cautioned, "Consistency of classification is crucial if the medicolegal official is to be an unbiased participant in the criminal justice process"[45] and, very importantly, "He who accuses must prove."

Planning for a career in forensic pathology: advice to future trainees

In the experience of the authors, FPs tend to discover this field in one of two ways: early and methodically, predicated on varied and formative life experiences, or late and sometimes serendipitously, after training or practicing in other medical specialties. The two primary authors of this text represent both paths. Neither is necessarily preferable or better; rather, it is the final common pathway of finding FP that matters.

For those in the United States who are clear about their desired career path prior to residency or even medical school, pathology residency training may be viewed as largely a means to an end. In fact, it is not uncommon for many residents who are laser-focused on forensic pathology to enter into a three-year anatomic pathology (AP) only residency training track. Others may enter the combined anatomic and clinical (AP/CP) track. Both tracks have merit and relevance to FP practice.

Most important for either path is the real-life exposure to forensic pathology afforded by spending significant time with practicing FPs, typically in a nearby ME/C system. Shadowing practicing FPs in the autopsy room, conference room, courtroom, and beyond is both necessary and eye-opening. Critical in making an informed career choice is the firsthand observation of the day in, day out challenges and rewards, the fascinating and sometimes humbling discovery of the intricacy of human disease and injury, a realistic sense of the political, financial, and administrative frustrations and limitations, and the varied professional and collaborative relationships of the field. But, equally, if not more important than all of that is recognizing and embracing the fact that, at the heart of forensic pathology practice, one must systematically form and ultimately *defend* expert opinions. If this notion makes you uneasy, take note and reflect on whether you believe you could ultimately gain comfort with that responsibility. Feeling daunted by it early on is natural, but realizing that proper training and a stringent, evidence-based practice will naturally lead you to defensible opinions should help ease that anxiety. Also, of course, finding a mentor is key.

All that said, while charting a path (AP/CP vs. AP only, etc.) is important, one needs to overcome many hurdles to become a FP. Focus should be placed on each of those hurdles, one at a time. A strong foundation in general pathology is important. The better general pathologist one is and the better one learns clinical medicine, the better FP one will be. Don't get ahead of yourself; there will be time to learn all of the nuances of FP. By all means, "keeping a hand" in forensic pathology early on is fine, even advisable, as long as it doesn't intrude on other training. That said, knowing that you want to practice forensic pathology earlier in your professional career can help you to focus on certain pathology topics (e.g., death certification) or engage in research projects or case reports during medical school or residency training. These kinds of academic pursuits are easier to initiate when FP relationships are forged and you begin to attend national meetings as a student or trainee. During pathology residency, we would advise doing as many autopsies (including eviscerations) as possible. If the residency program only offers a forensics rotation in the second or third year, try to spend at least a day or so with the medical examiner during the first year.

There are wide variations in training structure between the United States and countries abroad, so much so that a granular discussion exceeds the scope of this book. Suffice to say, however, that the medical school/residency/

fellowship paradigm, so familiar to physicians in the United States, does not necessarily exist in that precise form abroad. In much of continental Europe, England, Wales, and Ireland, for example, forensic pathology training falls largely to the universities, with studies focused between both undergraduate and postgraduate curricula, depending on the country. In their 2002 article calling for the harmonization of current forensic practice, Mavroforou and Michelodimitrakis emphasized the importance of the global standardization of FP practice and training to the extent possible, highlighting the caveats inherent in the lack of autonomy of practicing FPs in Greece.[20]

When searching for a training program, "goodness of fit" becomes critical. This may be based on many practical variables, not the least of which is geography, proximity to extended family, school districts, cost of living, and so on. But, of equal importance to consider and reflect upon is the type of program you seek and the people and mentors with whom you wish to work, side by side, for a year. The size of the program and the number of fellows may have importance for some and little for others. Ditto regarding academic appointments and the availability of scholarly activity opportunities. Clearly, this is a highly subjective and personal proposition.

Fellowship training presents a critical window in one's overall professional path.[46–49] It is during this period that the previously "undifferentiated" pathologist takes form substantially. Regardless of where one lands, it is important that you dive into fellowship and take full advantage of the singular opportunities that this limited time affords. It will be equal parts busy, stimulating, and exhausting. In fact, forensic pathology training could easily fill two years, given the breadth of training it requires and the challenges presented by attempting to maximize the exposure to varied cases and myriad forensic experiences in such a short period of time. But, one year *is* a sufficient time to have the groundwork laid and to absorb the necessary and foundational concepts that will become central to one's practice over the course of a career. The relationships forged during the training year will also prove foundational, the mentors who guided you being only a phone call away for the duration of your career, and your co-fellows, if any, becoming the equivalent of professional siblings. Whatever the details, most remember their training year with fondness and appreciation, recognizing that the FP one becomes is largely owed to that intensive and formative twelve months

References

1. Adelson L. Forensic pathology then and now. Retrospect and reflections. *Am J Forensic Med Pathol*. 1989;10(3):251–260.
2. Smith S. The history and development of forensic medicine. *Br Med J*. 1951;1(4707):599–607.
3. Adelson L. The forensic pathologist. "Family physician" to the bereaved. *JAMA*. 1977;237(15):1585–1588.
4. Hirsch CS. Lester Adelson, MD. *Am J Forensic Med Pathol*. 1989;10(3):261–263.
5. Collins KA, Lantz PE. Interpretation of fatal, multiple, and exiting gunshot wounds by trauma specialists. *J Forensic Sci*. 1994;39(1):94–99.
6. National Research Council. *Strengthening Forensic Science in the United States: A Path Forward*. Washington, DC: The National Academies Press; 2009.
7. Byard RW. Forensic and historical aspects of crucifixion. *Forensic Sci Med Pathol*. 2016;12(2):206–208.
8. Peng Z, Pounder DJ. Forensic medicine in China. *Am J Forensic Med Pathol*. 1998;19(4):368–371.
9. Flanagin A, Sullivan-Fowler M. The first postmortem in the country. *JAMA*. 1993;270(16).
10. Hanzlick R, Combs D. Medical examiner and coroner systems: history and trends. *JAMA*. 1998;279(11):870–874.
11. Eckert WG. Charles Norris (1868–1935) and Thomas A. Gonzales (1878–1956), New York's forensic pioneers. *Am J Forensic Med Pathol*. 1987; 8(4):350–353.
12. Eckert WG. Medicolegal investigation in New York City. History and activities 1918–1978. *Am J Forensic Med Pathol*. 1983;4(1):33–54.
13. Adelson L. Alan Richards Moritz, M.D. an appreciation. *Am J Forensic Med Pathol*. 1981;2(4):297–298.
14. Hanzlick R. History of the national association of medical examiners and its meetings, 1966–93. *Am J Forensic Med Pathol*. 1995;16(4):278–313.
15. Hanzlick R, Combs D, Parrish RG, Ing RT. Death investigation in the United States, 1990: a survey of statutes, systems, and educational requirements. *J Forensic Sci*. 1993;38(3):628–632.
16. Hanzlick R, Parrish RG, Combs D. Standard language in death investigation laws. *J Forensic Sci*. 1994;39(3):637–643.
17. Hiss J, Kahana T, Arensburg B. Forensic medicine in Israel. *Am J Forensic Med Pathol*. 1997;18(2):154–157.
18. Castiglione AG, Lomi A. Forensic autopsies in Italy. *J Forensic Sci*. 1993;38(3):622–627.
19. Chang L, Zhang B, Yan P, Fowler D, Li L. Current medico-legal death investigation system in China. *J Forensic Sci*. 2011;56(4):930–933.
20. Mavroforou A, Michalodimitrakis E. Forensic pathology on the threshold of the 21st century and the need for harmonization of current practice and training: the Greek concept. *Am J Forensic Med Pathol*. 2002;23(1):19–25.
21. Peterson GF, Clark SC. Forensic autopsy performance standards. *Am J Forensic Med Pathol*. 2006;27(3):200–225.
22. Weedn VW, Menendez MJ. Reclaiming the autopsy as the practice of medicine: a pathway to remediation of the forensic pathology workforce shortage? *Am J Forensic Med Pathol*. 2020;41:242–248.
23. Turnbull A, Osborn M, Nicholas N. Hospital autopsy: endangered or extinct? *J Clin Pathol*. 2015;68:601–604.
24. Roberts WC. The autopsy: its decline and a suggestion for its revival. *N Engl J Med*. 1978;299(7):332–338.
25. Hanzlick RL. The "value-added" forensic autopsy: public health, other uses, and relevance to forensic pathology's future. *Acad Forensic Pathol*. 2015;5(2):177–185.
26. Lindström P, Janzon L, Sternby N. Declining autopsy rate in Sweden: a study of causes and consequences in Malmö, Sweden. *J Intern Med*. 1997;242(2):157–165.

27. Gobbato F, Vecchiet F, Barbierato D, Melato M, Manconi R. Inaccuracy of death certificate diagnoses in malignancy: an analysis of 1,405 autopsied cases. *Hum Pathol.* 1982;13(11):1036–1038.
28. Zumwalt RE, Ritter MR. Incorrect death certification. An invitation to obfuscation. *Postgrad Med.* 1987;81(8):245–247, 250, 253–244.
29. Pritt BS, Hardin NJ, Richmond JA, Shapiro SL. Death certification errors at an academic institution. *Arch Pathol Lab Med.* 2005;129(11): 1476–1479.
30. Gill JR, Goldfeder LB, Hirsch CS. Use of "therapeutic complication" as a manner of death. *J Forensic Sci.* 2006;51(5):1127–1133.
31. Lu TH, Shih TP, Lee MC, Chou MC, Lin CK. Diversity in death certification: a case vignette approach. *J Clin Epidemiol.* 2001;54(11):1086–1093.
32. Ravakhah K. Death certificates are not reliable: revivification of the autopsy. *South Med J.* 2006;99(7):728–733.
33. Cina SJ, Selby DM, Clark B. Accuracy of death certification in two tertiary care military hospitals. *Mil Med.* 1999;164(12) 897–899.
34. Maudsley G, Williams EM. "Inaccuracy" in death certification—where are we now? *J Publ Health Med.* 1996;18(1):59–66.
35. Slater DN. Certifying the cause of death: an audit of wording inaccuracies. *J Clin Pathol.* 1993;46(3):232–234.
36. Smith Sehdev AE, Hutchins GM. Problems with proper completion and accuracy of the cause-of-death statement. *Arch Intern Med.* 2001;161(2): 277–284.
37. Jordan JM, Bass MJ. Errors in death certificate completion in a teaching hospital. *Clin Invest Med.* 1993;16(4):249–255.
38. Fischtein D, Cina SJ. Errors on death certificates requiring amendments: the Broward County experience. *Am J Forensic Med Pathol.* 2011;32(2): 146–148.
39. Cambridge B, Cina SJ. The accuracy of death certificate completion in a suburban community. *Am J Forensic Med Pathol.* 2010;31(3):232–235.
40. Brooks EG, Reed KD. Principles and pitfalls: a guide to death certification. *Clin Med Res.* 2015;13(2):74–82.
41. McGivern L, Shulman L, Carney JK, Shapiro S, Bundock E. Death certification errors and the effect on mortality statistics. *Publ Health Rep.* 2017; 132(6):669–675.
42. Randall B. Death certification: a primer, Part IV—problems in death certification. *S D Med.* 2014;67(8):315–317, 319.
43. Myers KA, Farquhar DR. Improving the accuracy of death certification. *CMAJ (Can Med Assoc J).* 1998;158(10):1317–1323.
44. Hanzlick R. Quality assurance review of death certificates: a pilot study. *Am J Forensic Med Pathol.* 2005;26(1):63–65.
45. Adams VI, Flomenbaum MA, Hirsch CS. Trauma and disease. In: Spitz WU, ed. *Spitz and Fisher's Medicolegal Investigation of Death.* 4th ed. Springfield, IL: Charles C Thomas; 2006:436–459.
46. Hanzlick R, Haden-Pinneri K. Forensic pathology fellowship training positions and subsequent forensic pathology work effort of past forensic pathology fellows. *Acad Forensic Pathol.* 2011;1(3):322–327.
47. Hanzlick R. A concise summary of approved/accredited forensic pathology fellowship training programs in the United States: 1961–2011. *Acad Forensic Pathol.* 2011;1(1):112–121.
48. Nine JS, Zumwalt RE. Integrating the Accreditation Council for Graduate Medical Education general competencies into forensic pathology fellowship training. *Am J Forensic Med Pathol.* 2005;26(4):334–339.
49. Pinckard JK, Hunsaker D, Weedn VW. A comprehensive analysis of forensic science training in forensic pathology fellowship programs. *J Forensic Sci.* 2003;48(2):362–367.

CHAPTER

2

Thinking forensically: planning, anticipating, observing, documenting, inquiring, synthesizing, opining, and communicating, in that order

Susan F. Ely and James R. Gill

Our science and art are useful tools in the larger patchwork of investigation that ultimately is stitched into a quilt of truth. -*Dr. Charles S. Hirsch*

The importance of context

In 1956, Dr. Alan R. Moritz, one of the founding fathers of forensic pathology (FP), gave the Ward Burdick Award address at the 35th annual meeting of the American Society of Clinical Pathologists in Chicago. His address, entitled "Classical Mistakes in Forensic Pathology," was subsequently published and remains a seminal work in the field of forensic pathology. In it, he describes errors in autopsy practice, interpretations, and forensic reasoning. He starts by emphasizing the importance of being aware of the objectives and differences between medicolegal and hospital autopsies.[1] In their 2012 paper entitled, "Classical Mistakes Made In Forensic Toxicology by Forensic Pathologists," a toxicological rendering of the Moritz original, Gill and Stajic asserted the following, "One of the greatest mistakes that a forensic pathologist can make with toxicology is opining the cause of death based solely on a drug concentration, without consideration of the entire case."[2]

The late Joe Davis similarly cautioned forensic pathologists to require all information that is available in every case, so that we can then "make meaningful forensic correlations based on the entirety of the case." He wisely noted, "The need for circumstantial details bears an inverse relationship to the severity of disease alterations revealed by autopsy."[3] Stated differently, unless a finding is incompatible with continued life, it must be contextualized. In their treatise on child abuse, Hirsch and Zumwalt offered the following:

> We strongly emphasize the need for investigative teamwork and the imperative to interpret autopsy findings in the light of the entire case study. Narrowly focused tunnel vision, restricted exclusively to autopsy findings taken out of context of the investigative universe, will see only part of the truth. Errors of interpretation resulting from such defective perception will leave some instances of fatal maltreatment unrecognized and, in other instances, will not provide truthful, non-sinister explanations for injuries. None of us can rest comfortably with the notion that we may have overlooked a homicide or misinterpreted homicidal injuries as an accident. However, any discomfort resulting from such false conclusions is luxury compared to our dread of the converse.[4]

In view of the preceding sage words, a theme clearly comes into stark relief. Analogous to the real estate mantra, "location, location, location," FP practice is predicated on "context, context, context," representing a foundational tenet and the central difference between it, hospital pathology, and most other forensic science disciplines. The best FP practice relies upon multiple departments and a collaboration with both multidisciplinary and multiagency specialists. The most precise puzzle solving, yielding the most accurate answers for cause and manner of death, is born out of contextualizing not only carefully observed findings, both physical and chemical, but also from recognizing the company they keep and the setting in which they arose. To wit, the same concentration of cocaine in a

Principles of Forensic Pathology
https://doi.org/10.1016/B978-0-323-91796-4.00019-2

previously healthy individual found with a needle in their arm carries altogether different pathophysiologic importance than it does in a person who has suffered a gunshot wound of the head, when that injury is determined to be unequivocally fatal. The latter renders the cocaine irrelevant, at least in terms of any pathophysiologic role it may have played in the death, the decedent dying *with* the drug, not *from* it. In that vein, many, if not most, accidental illicit drug intoxications are not necessarily "overdoses," per se, as many lay people may presume, a misconception that the cause is easily diagnosed by a cursory glance at a number on a toxicology report, analogous to a hemoglobin or cholesterol concentration, having clearly defined reference values. Rather, fatal intoxications represent the most reasonable explanations for individuals' deaths in which the circumstances and the exclusion of more compelling causes lead to evidence-based and scientific opinions. "There is an infinite gap between the lab and the street." The so-called "fatal concentration" of an illicit or prescribed substance, in a vacuum, is often more illusion than reality (see Chapter 14). "Toxicology gives us information, but not conclusions."

Medicolegal autopsies diverge substantially from hospital autopsies in many ways, including, but not limited to standards of practice for each, given their widely disparate objectives. For one, the *external* examination is a singularly important and potentially rich source of contextual information, based on observations of injuries, patterned scars, clothing, degrees of hygiene, nutritional and hydrational status, and even edentulousness. Much more so than in the hospital setting, these may constitute important data points that have bearing on our framing and ultimate impressions of the case. Regardless, the devil is in the details, details which are carefully recorded in a way that is distinctly different than they would be in hospital autopsy practice, where an individual's clothing would not even be available for examination, much less be of likely interest or relevance to the pathologist. Issues surrounding time of death, the recognition of advanced postmortem changes, postmortem animal predation, and the collection of gross and trace evidence, with the attendant and stringent documentation, written and photographic, represent just a few of the areas that have little to no place in routine hospital pathology practice. The establishment and burden of decedent identity alone is not relevant to the hospital pathologist. In his tribute to Dr. Lester Adelson in 1989, Charles Hirsch wrote the following about his discovery of and realizations about forensic pathology:

> ... it was startling to realize that the pathology of the hospital and the clinic is only one profile of human illness. The flip side is, of course, the pathology of the street and the home as observed in medicolegal practice. It's a different pathology and its interpretation requires the application of medicolegal logic, as well as scientific medical reasoning.[5]

Given the scope of most investigations and the purview and tasks of the typical medical examiner/coroner (ME/C), the forensic autopsy itself serves as an important, but perhaps not final "lab test" in the context of an entire death investigation, depending on the case. Dr. Hirsch addressed the importance of mental flexibility and context with the below assertion:

> Rigid criteria and inflexible rules simplify life because they absolve us of the responsibility to think and to doubt. When we yield to such temptations in our professional life, two unfortunate consequences arise. First, in a limited sense, we violate a cardinal principle of death investigation by failing to evaluate fatalities on their individual merits. Second, in a general sense, when rules override judgment, we sacrifice a measure of our intellectual credibility.[6]

The role of inference: releasing the grip on exclusively anatomic findings

Forensic pathologists must have a solid and practical understanding of the pathophysiology of death and dying, much of which is based on functional, not necessarily anatomic pathology (see Chapter 3). "The fundamental difference between living and dead organisms is functional, not structural; therefore, the pathologist who focuses exclusively on structurally demonstrable, anatomic causes of death is doomed to failure." This notion serves as an important underpinning of the conceptual transition from hospital to FP practice.

A 66-year-old man with advanced heart disease suffers a witnessed sudden death and is found to have marked three-vessel coronary atherosclerosis, cardiac hypertrophy, and a negative rapid urine toxicology screen at autopsy. The overwhelming probability, by virtue of a known pathophysiologic substrate, the compelling precipitousness of the death, and the exclusion of other causes, is that he died from a fatal cardiac arrhythmia (*mechanism)* due to atherosclerotic and hypertensive cardiovascular disease (*proximate cause)*. We can conclude this with confidence not because we have special postmortem electrocardiographic capabilities, but because no other explanation makes sense. In our training and experience as forensic pathologists, we know that there exists a short list of natural diseases having the capacity to cause so-called "instantaneous death" (fatal cardiac arrhythmias, saddle pulmonary thromboembolus, certain ruptured aneurysms, and pontine strokes, to name a few), and with a complete autopsy able to exclude all of those, this leaves the forensic pathologist only one choice: to believe what is observed and draw the most, perhaps the only reasonable functional mechanistic conclusion.

It also is important in these deaths to understand that one may not find a coronary artery thrombus or evidence of an acute myocardial infarct (MI). These are seen in approximately half of sudden deaths due to atherosclerotic cardiovascular disease. The remaining instances have fixed atherosclerotic disease and myocardial fibrosis. A study from Spain examined the causes of sudden cardiac death (SCD) in 216 adults aged 35–49 years who underwent forensic autopsy.[7] Ischemic heart disease was diagnosed in 140 (65%) decedents, of which 74 had a coronary thrombosis or acute MI. The remaining had fixed coronary artery atherosclerotic stenosis and/or myocardial fibrosis.

SCD due to coronary artery atherosclerosis is not always associated with a structurally demonstrable coronary artery thrombus or an acute MI; often, there is no acute lesion at autopsy. The detection of an acute MI or a coronary artery thrombosis is not required to make a diagnosis of SCD due to atherosclerosis. MI is not synonymous with SCD. Some of these deaths occur so suddenly that there is no time for overt pathologic changes to develop. Morphologic changes of MIs are first detectable by light microscopy from 4 to 12 h after the onset of marked ischemia. Coronary artery atherosclerosis may cause a lethal cardiac arrhythmia by different pathways. Some of these pathways leave no demonstrable evidence of it having occurred, but these purely electrical deaths are commonly brought to the attention of forensic pathologists. The likely pathogenesis is myocyte ischemia (due to an imbalance of myocardial oxygen supply and demand seen in the setting of fixed stenoses) that causes a ventricular arrhythmia.[3] Ischemia, like hypoxemia, is arrhythmogenic.

In a similar fashion, the resistance to diagnose hypertensive cardiovascular disease at autopsy, predicated on the absence of a clinical history of hypertension, turns a blind eye to common sense and the diagnostic sensitivity of the autopsy as a medical procedure. We suggest that the combination of cardiac hypertrophy and fine granularity of the kidneys (a gross manifestation of arteriolosclerosis/small vessel disease) equates to hypertensive cardiovascular disease, with or without any reasonable contributions from chronic stimulant drug misuse or any other etiologically specific cause of cardiac hypertrophy. To deny this diminishes our role as diagnosticians and defies deductive reasoning and logic in the name of objectivity.

Objectivity, neutrality, and observational acumen: being autoptic

Forensic pathologists are neutral arbiters of the facts. As such, they need to approach every case with care, precision, and a mental nimbleness for the surprises the body may present, but in doing so, maintain a healthy degree of circumspection. They must strike an important balance between taking context into account while remaining open to the myriad functional and anatomic possibilities they may encounter at the table. Thankfully, the body does not lie; it can be trusted, but it is limited by the pathologist's thoroughness, experience, and powers of observation. If forensic pathologists approach deaths with tunnel vision, assuming the findings will be one thing, but the body surprises them with another, it is the body rather than the anticipated diagnosis that is right, underscoring its capacity to "slay a beautiful theory with an ugly fact." As Dr. Davis said, "It is the holy duty of a forensic pathologist to accept a more logical explanation regardless of the source, feelings, sides, etc."

> There are none so blind as those who refuse to see.

We should aim to be autoptic (literally, "see for yourself"), to believe what we see, and not just what we are looking to find. This minimizes potential distortions in perceptions and interpretations of gross and microscopic findings to comport with preconceived theories (i.e., cognitive bias). Ideally, our hypotheses will be flexible to what the body ultimately tells us. If the findings do not fit the story, it's the story that is flawed, not the body, and this should prompt further investigation. If a story does not make sense relative to the autopsy findings, there is usually a good reason, a question that has not yet been asked, a scene photo not yet taken or perused, a body position unknown, a half-truth told, or a rock that has otherwise not yet been overturned. Be wary of physicians who have an unwavering attachment to their diagnoses ("The only time they let loose is to get a tighter grip."). Simply put, we must be aware of and try to avoid confirmation bias in our approach to autopsies. To that end, forensic pathologists can use consensus conferences and quality assurance reviews to weed out biases that are inherent to human nature.

Moreover, while a good forensic pathologist will be observant, taking note *that* a particular finding exists, a more astute forensic pathologist will take it a step further and seek an explanation, especially when a finding is unexpected and/or deviates from a preconceived theory. An astute FP will be circumspect and inquisitive. Why is the gunshot entrance wound atypical? They will notice not merely *that* a large caliber, jacketed bullet did not exit the body, but they will ask *why* it didn't exit the body? They will ask *why* the kidneys are pale, the heart is empty, and the body is starting to putrefy, despite the death having occurred recently. This real-time inquiry and search for reasonable explanations dovetail the autoptic approach and is where the forensic synthesis, understanding, and mechanistic evidence building of the case begin in earnest. It is where the rubber meets the road. If one misses the opportunity to start that thought process at the table, it makes it all the more difficult to do so at a later time, when one's memory of the details of the case has dimmed; an important window of opportunity for the most accurate dot connecting will have been missed.

The invaluable balance and nuance born out of taking all available contextual information into consideration while recognizing that the autopsy findings may reveal "holes" in the story is central to the practice of FP, and to all medical disciplines, for that matter. That said, to throw out investigative information and dogmatically perform all autopsies in a blinded fashion is no more forensically sound, responsible, or rational than tightly adhering to a preconceived theory regardless of what the physical examination reveals. Operating in the extremes is likely to yield less accurate and evidenced-based opinions. As Dr. Hirsch reminded us, "Refusal to consider circumstantial information can lead to a denial of reality in the name of objectivity."

Further, the discovery of incongruities between the autopsy findings and the story presented before the examination of the body are critical to share with those for whom the inconsistencies may have import: clinicians, prosecutors, families, child protective services, police, and so on. Potentially all of these stakeholders have much to gain by receiving the most accurate information born out of careful scientific inquiry and discovery and comparing those to the existing investigative narrative. In most instances, there is very little downside for full disclosure of the physical findings, relayed in a timely fashion. Any effort having the capacity of bringing the investigation closer to an accurate conclusion is a worthwhile endeavor. This constitutes the quality assurance that forensic pathologists provide as professionals, having potentially far-reaching value for many, not just for law enforcement officials.

Admittedly, balanced and measured thinking and practice take effort and time, but they are also more likely to encourage the most reproducible and defensible approach and associated conclusions. More is better when it comes to the information to consider in any investigation. Striking a balance and taking all data points, contextual, historical, anatomical, clinical, and physiological, into account enhances our ability to arrive at the most accurate answers in a way that is neither arbitrary nor capricious, thus maximizing the likelihood of revealing the truth. It affords us the ability of being the expert voices of reason and science both in the courtroom and the autopsy room, a necessary, privileged, effortful, and ultimately humbling position to maintain.

> We render a diagnosis on everything we see and not on isolated findings.

Finally, at the center of maintaining perspective in our role is remembering, at all times, that as a profession, we "have no dog in the fight" and that our classifications are not necessarily the final word for adjudicative purposes. While we should be accommodating and sensitive to the needs of families wherever possible and within the confines and high standards of our practice, we must ultimately "call them as we see them" and let the chips fall where they may. If the public sees that we can be trusted with the promise of neutrality, come what may, everyone wins in the end. As such, we should aim to have no emotional attachment to the outcomes of our opinions, hence our impartiality as a cog in the larger machinery. Once we bow to outside pressures, monetary, political, or otherwise, or attach ourselves emotionally as advocates in our certification decisions, our objectivity is threatened and our opinions and credibility potentially undercut, in reality or in the perception of reality.

> When emotion collides with intellect, intellect always loses.

The bare truth is that we represent only one piece in the investigative puzzle. Classifying as accident the death of a pedestrian victim of a motor vehicle collision involving a "hit and run," for example, does not preclude authorities from prosecuting the death criminally, any more than our certification of a death as a homicide forces them to seek an indictment. By design, each profession maintains different roles, responsibilities, and standards of practice. Those involved in law enforcement should be reminded that our death certification burdens, definitions, standards, nomenclature, and classifications are rooted in FP and death certification practices, not the letter of the law, per se. Ours is primarily a public health function that happens to also be used by the legal system (see Chapter 17). This is why we have a manner designation of "homicide," not "murder." With rare exceptions, our burden to prove intent is primarily reserved for suicide classifications.

> Our science and art are useful tools in the larger patchwork of investigation that ultimately is stitched into a quilt of truth.

Anticipating, planning, observing, documenting, inquiring, synthesizing, opining, and communicating, in that order

With the foregoing thesis in mind, we propose ten tenets of forensic thinking and practice:

Craft a plan before and after an autopsy, but remain flexible and loosely attached to the initial differential diagnoses. "Routine is never a substitute for judgment."

"If you think of it, do it," because, "I'd rather explain why I did something than why I did not." Consciously aim to minimize errors of omission.[1]

Be carefully observational and "autoptic"; see for yourself and "describe what you see." Document all that you see, even if you do not fully recognize its significance or even its etiology at the moment of observation ("That's an excellent example of whatever it is.").

Step back from the table to allow a broader perspective. By distancing yourself from the trees, you are more likely to recognize the forest (i.e., look for patterns).

"Never leave the autopsy room without at least considering the mechanism of death." Consider mechanism, and then cause. This will, in turn, lead you to manner. Anything else places the cart before the horse.

Wherever possible, seek primary sources of information and obtain "the news, not the editorial." In a suspected choking death, ask that investigators speak directly to the witness(es) who saw the event, not just the medical resident who reported the death.

Recognize and admit to the limitations of the autopsy and of scientific analysis, in general (e.g., time of death estimates, time of injury estimates, undetermined causes of death). "Honesty is the only stigma that attaches to a pathologist who leaves an occasional case undetermined." Don't be afraid of saying, "I don't know." Honesty lends more credibility than a disingenuous guess.

In all matters, but especially in your professional opinions, avoid being arbitrary and capricious; remain internally consistent, thoughtful, and evidence-based in your practice and conclusions. "Hunches are not evidence" and suspicions are not proof.

Our words matter; choose them carefully to convey a specific meaning. "We are in the business of communication." We must communicate well and clearly in both verbal and written forms, ever mindful of a diverse audience with often disparate opinions, needs, agendas, and levels of forensic understanding. Moreover, in public releases, "Distribution of partial information guarantees complete misunderstanding."

In your communications with the bereaved, remember that you are a human being before you are a physician. Our role in providing helpful, clear, and definitive scientific conclusions can be just as important to survivors as that of our ability to provide closure and solace. Honesty, sensitivity, and good listening skills will go a long way.

Preautopsy analysis and planning

Autopsies are likely to have a much higher "rate of return" in yielding incisive, useful, and valid conclusions when even brief, but intentional and thoughtful consideration and analysis takes place before, during, and immediately afterward. Based on the initial investigative information, the forensic pathologist should ask what question(s) must be answered from the autopsy. In some instances, the inquiry is very broad, such as, what is the cause of death in this previously healthy 12-year-old boy? Other times, it may be more focused, such as, did that fall cause the death of this 75-year-old man or was it a terminal collapse from known heart disease?

Enter the autopsy suite with a loose attachment to a differential diagnostic list that is constructed from your forensic knowledge and experience, dovetailed with what is known at that point in time about the case. As Lester Adelson wisely observed, "The most important part of the autopsy is what you do with your head, not with your hands," and, "A question is halfway to wisdom."[8]

This intentional approach has the potential of focusing and modifying dissection methods, testing, and observational skills. If one anticipates shock of any kind (hypovolemic, cardiogenic, neurogenic, or septic) as the mechanism of death, for example, the forensic pathologist should look for and be more acutely vigilant about identifying evidence to support or disprove that thesis (the presence or absence of poorly formed or barely discernible lividity, marked visceral and muscular pallor, a nearly bloodless autopsy or empty heart, etc.) and then photographically document the results of that search, negative or positive.

A little contextual forethought may lead the forensic pathologist to collect sexual assault evidence before the body is rinsed and the first incision is made and it is too late, take an inferior vena cava blood culture before handling and removing the heart, remove the spinal cord, or check for a cardiac air embolism. Some procedures can be done at a later time, after further consideration and a return to the body; others cannot. Certain decisions are made in real time as a result of an unexpected autopsy finding. An obstructive saddle pulmonary thromboembolus (PTE) is a common example of such a discovery. Once this Class I finding is identified, its source is best sought and the lower extremities explored, a dissection that is not routinely performed in all autopsies. In addition, attention would need to be turned to identifying the proximate cause of the deep venous thrombosis, since that will inform the manner of death (unnatural versus natural). If unclear, primary malignancies must be carefully excluded, organ by organ. Additional testing might be called for, such as a molecular thrombophilia analysis, a toxicological analysis, or electrolyte testing to exclude dehydration, if no structurally demonstrable or historical cause is identified. For many hospital pathologists, the "investigation" ends with the detection of the pulmonary thromboembolus; for the forensic pathologist, it is where it begins.

A foolish consistency is the hobgoblin of little minds. ***Ralph Waldo Emerson***

Anticipating questions and future issues

While every autopsy should be thorough, the forensic pathologist is well advised to attempt to anticipate pointed questions or concerns that may arise in specific cases, such as those, for example, with an elevated degree of public interest and likely attendant review and scrutiny, or in which multiple stakeholders may have differing, often adversarial interests and opinions. This is when the forensic pathologist's role as "neutral arbiter of the facts" becomes both relevant and important and may lead to a more prudent and exhaustive photographic documentation of pertinent positive *and* negative findings at the autopsy table, findings that can be made available for any interested party to review at a future time. As a general rule, "If you think of it, do it." As Adelson noted, a perfect autopsy is one that "anticipates and answers all the later questions."[8]

Potential questions ultimately born out of civil, criminal, and familial matters, anticipated to the extent that they can be, are best considered at the autopsy table. This might direct both ancillary testing and photographic documentation. For example, later questions may be focused on whether a given injury is compatible with consciousness or not, on potential survival intervals (if unknown), and on "pain and suffering" relative to specific identified trauma. An opinion about possible consciousness is important in that, without it, there is no pain and suffering, since these constitute experiences that require perceptive functioning, by definition. This is an issue at the heart of many civil suits and may or may not be ultimately answerable by the forensic pathologist alone. It should nonetheless at least be considered during the autopsy and postautopsy analysis, where potentially relevant. *Autopsies can and often do provide many answers beyond just those surrounding cause and manner of death.* This has been called the "value-added" forensic autopsy, in which the autopsy provides much more, going beyond answering only basic questions of certification.[9]

Consider the following example: A woman is struck and run over by a motor vehicle. As a result, her torso is partially run over, causing severe crush-type injury, but no head trauma. At autopsy, a 1500 mL hemoperitoneum from a multiply lacerated liver is discovered. The hemorrhage indicates continued circulation/life for at least some minutes, and the absence of significant head trauma opens up the likelihood of consciousness during at least some part of that survival interval. Without consciousness, there would be no pain and suffering. Scientifically ascertaining the likelihood not just of circulation/life, but that of consciousness relative to the identified injuries is central to opinions around pain and suffering. With a shattered skull or transected brainstem (complete disruption not just of the vital centers, but also of the reticular activating system that maintains consciousness), pain and suffering are irrelevant from the moment that injury is sustained. Without demonstrable head injury, on the other hand, consciousness is possible, perhaps likely, but concussive injury cannot be ruled out in many instances. The latter scenario represents the gray area we grapple with as expert witnesses.

Postautopsy analysis and planning

Just as important as the preautopsy plan, **the postautopsy plan** is born out of the progression from preautopsy theories to demonstrable autopsy findings, or lack thereof. By the end of the autopsy, the pathologist rules certain entities out, allowing others to rise to the top of the differential diagnostic list. If not clarified by the autopsy findings alone, the postautopsy analysis should aim to create a discrete plan for necessary tests, investigative questions, and/or phone calls (through police, operative notes, discussions with next of kin or clinicians, return scene visits, etc.) that will allow the FP to amend the pending death certificate later, as well as sort out what other data points are needed, not necessarily for the cause and manner of death, but to finalize the autopsy report. When death certification is made **"pending,"** the FP should always consider and define the specific endpoint of the pathophysiologic, laboratory, and/or investigative inquiry, one that is predicated on specific mechanistic, causal, or circumstantial questions that require answering in order to arrive at a defensible certification conclusion, not a thoughtlessly constructed, unfocused laundry list of all possible or irrelevant questions that might be answered (see the below discussion regarding decisions to make initial death certificates "pending").

As an example, an unwitnessed death of a man is found with drugs and drug paraphernalia and has a positive rapid toxicology test for illicit drugs at the autopsy table. Even though he has a history of epilepsy, he need not undergo a full postmortem neuropathologic work up (at least for death certification purposes; there may be other compelling reasons). Seizure-related fatalities are diagnoses of exclusion. A neuropathologic examination may find a reputed seizure focus or evidence of prior seizures, but it cannot diagnose a terminal seizure having caused an unwitnessed death (an "electrical" event) and does not determine the ultimate certification alone. When a drug intoxication will likely serve as an **efficient intervening cause** of death, investigating the circumstances and pathogenesis of the epilepsy may constitute a rabbit hole that utilizes unnecessary resources, a "shotgun" approach to diagnostic testing that may distract the pathologist, particularly one in training, from the opportunity to "think forensically" and stay focused on the mechanism of death. An analogous example is a death due to a suicidal hanging in an individual who happens to have been shot in the torso years earlier. Other than removing any remaining ballistics from the body at autopsy (for completeness), no further investigative time or energy need be directed toward the remote firearms event if it in no way contributes mechanistically to the death (see discussion on mechanism in the next section).

Observing and documenting

Step back from the table to allow a broader perspective. *By distancing yourself from the trees, you are more likely to recognize the forest (patterns).*

"Never leave the autopsy room without at least considering the mechanism of death." *Anything else places the cart before the horse.*

Documentation

Documentation is a critical element of FP practice. Because of the interface between legal matters and forensic and clinical findings, the FP has a central role as the spokesperson for the dead, with the intention of drawing conclusions having forensic veracity and meeting specific burdens of proof.

To that end, it is incumbent upon the forensic pathologist to anticipate and be mindful of the responsibility they have in carefully contributing to the data bank of evidence in a given medicolegal death investigation, one that can be objectively evaluated long after the autopsy concludes. Depending on the case, there may be interest and review of every aspect of the death, interest that extends well beyond the autopsy suite and the day of autopsy, sometimes for years to come. One aspect of our job, then, is to memorialize our observations in thoroughly and carefully written form (notes and reports; see Chapter 16), further corroborated with high quality photographic and, at times, radiographic images.

That said, while documentation is important, it is only as valuable as are the powers of observation of the forensic pathologist. The preceding illustrative example of the documentation of physical findings supporting a mechanism of "shock" is only relevant if there is the forethought of looking for, and then recognizing and documenting those signs. Stated differently, documentation is greatly enhanced by planning and the honing of observational skills. "You can't see what you can't believe" or haven't considered. And, it follows, then, that you can't document what you fail to notice.

The value of pertinent negatives

Equally important among the finely honed observational skills of forensic pathologists is the ability to consider, notice, and document that which we do *not* see, that is, the **pertinent negatives**. The autopsy is a very important tool in its capacity to rule out diseases or injuries that may have been part of the differential diagnostic considerations or suspicions (ours or others') before we made the first incision. Upon completing a full autopsy, it is just as important to ask ourselves what we have succeeded in *excluding* as in what we have found. Many of the answers we provide to relatives and friends are born out of their concerns around injury, potential foul play, or suffering. One of the primary reasons to autopsy an alcoholic who frequents bars and has a history of repeated falls is to exclude fatal trauma, accidental or inflicted. Not infrequently, what we exclude allows us to promote another disease having lethal potential to the fore, making a cause and manner of death determination on the day of autopsy as a direct result of the autopsy. Advanced heart disease in an alcoholic with a significant and space-occupying/fatal subdural hematoma is demoted to a disease he died *with*, rather than *from*. In the absence of that head injury, however, that same heart disease would likely take center stage on the death certificate and preclude the need for a police investigation.

Considering disease and injury lethality

Generally speaking, the forensic pathologist is tasked with recognizing disease and injury along a spectrum of severity and then making determinations as to the likelihood of their lethality. *Findings* due to natural disease can be categorized into **Classes I through V.**[10–12] Class I represents the most advanced findings that are generally considered to be incompatible with continued life (e.g., a saddle pulmonary thromboembolus or a ruptured aortic dissection that rapidly produces three liters of acute left hemothorax). This is discussed in detail in the next chapter. While there is no such classification system for injury findings, the forensic pathologist still views injuries similarly with regard to assessing their severity and determining their capacity and likelihood to cause or contribute to death. An analogous example of a "Class I" *injury* finding would be a gunshot wound that perforates the brainstem. Injuries having a lower lethal potential may contribute to death not at all or just to a lesser extent, justifying the "demotion" of the latter to contributory cause status.

Injury pattern recognition

Actively cultivating abilities in **injury pattern recognition** is another important component of the professional arc of the forensic pathologist. One such skill involves developing an understanding of **"patterned injuries"** versus **"patterns of injury."**

A **patterned injury** is an individual injury that possesses specific qualities in size, morphology, and/or fine detail that reflect the particular object or surface that caused it. Sometimes, it is only clear to the forensic pathologist *that* a patterned injury has been identified, not necessarily *what* surface or object caused it, or *how* (a blow vs. passive contact). "That's an excellent example of whatever it is" is sometimes the only intellectually honest assessment one can make in the initial discovery of such an injury. Other times, the pattern is classic and immediately identifiable (e.g., tire tread contusions, "grab marks," seatbelt contusions, dicing injuries in victims of motor vehicle accidents, or scalp lacerations and associated skull fractures from blows with a hammer).

Regardless of whether the putative object or surface having created the patterned injury is known at the time of autopsy or not, documenting it before the body is released is the important and time-sensitive task to undertake from an evidentiary and investigative standpoint. At a later time, objects and surfaces that are discovered during an ongoing investigation can always be presented to the finder of fact and compared to photographs, allowing decisions and interpretations to be made about the likelihood of their causal association to the injury. Importantly, when describing patterned injuries, it is not enough and is, in fact, woefully inadequate to document in a report that, "there is a patterned injury," without further qualification and a detailed written description to justify that impression. The devil is in the details, and the description is only as helpful as it is specific. All such injuries should also be carefully measured, described, and photographed.

In contrast to patterned injuries, recognition of **patterns of injury** often requires a wider-angle lens, with a stepping away from the body to get the bigger picture. *Examples of patterns of injury* include, but are not limited to, planar injury of the face and/or prominences of one side of the body from a fall onto a flat, resistant surface in a person who is already dead or unconscious, a tight focus of a multiplicity of wounds on one side or region of the body, indicating the likely body position at the time of injury infliction, nonfatal wounds on the dorsal or volar surfaces of the upper extremities or legs in a victim of multiple inflicted sharp force trauma, specific body region patterns of inflicted vs. self-inflicted firearms or sharp force injuries, etc.). As an example, tightly clustered, similarly oriented stab or incised wounds of the neck, volar wrist, or left chest, having the appearance of control and intent, particularly with "hesitation marks," and without penetration of the overlying clothing, are classic for self-inflicted injury, whereas multiple haphazardly located, gaping and/or swallow-tail type wounds, broadly distributed on the body, including the back and dorsal arms, more likely indicate homicidal injury that involved relative movement of the knife to the body during a violent struggle. Similarly, when the face, breasts, and/or genitalia are injured, particularly excessively, the likelihood of the injury being self-inflicted approaches zero; it is also highly *unlikely*, although possible in such circumstances, that the decedent and the perpetrator were strangers to one another. Injuries sustained in a domestic violence setting versus a "street fight" may also take on disparate patterns in the quality and quantity of wounding. If you don't look for these patterns, you will in all likelihood not see them.[13,14] The term "overkill" denotes a typically emotional and/or psychological and excessive component to an assault. By contrast, the burglar who is surprised by an awakened resident is not likely to stab the decedent 80 times. This type of information may help guide the police in terms of whether they should attempt to locate and interview former or current intimate contacts of the decedent. See Chapter 10 for a more in-depth discussion.

Also important is noting **"the company injuries keep."** As an example, a skin defect on the abdomen, ambiguous for a sharp versus blunt injury (laceration), in a sea of immediately surrounding clear-cut stab wounds, is highly unlikely to represent a laceration, given not only the company the wound keeps but also because of its location on the body (it is very unusual to sustain a laceration of the abdomen, its soft tissue not being stretched over bone). "Let common sense prevail." Some findings, particularly external injuries, can be glaring, but ultimately irrelevant to the cause of death when one considers precisely how that injury might mechanistically relate to the fatal outcome. If a decedent is noted to have 42 closely clustered, barely hemorrhagic, superficial stab wounds penetrating only into the subcutaneous tissues of the back, but also a contact perforating gunshot wound to the back of the head, the former injuries should not be considered as contributory to the cause or mechanism of death when the gunshot wound injury is incompatible with continued life. Be aware of the "distraction of the dramatic." That said, a thorough description of the sharp force injury *does* belong in both the body of the autopsy report and as part of its final diagnoses. The death certificate, however, is not a substitute for a complete investigation and thorough autopsy. Without some mechanistic link, no disease or injury should appear as a primary or contributory cause on the death certificate (see further discussion below and in Chapter 3).

Consider the mechanism of death before leaving the autopsy room

When there are competing potential causes of death, identifying the mechanism of death will help point the forensic pathologist toward the correct primary cause. **A forensic pathologist should never walk away from the autopsy table without at least considering the mechanism of death.** Mechanism informs cause, which, in turn, informs manner. Skipping over the mechanism and jumping straight to the cause not only places the cart before the horse, but also presents a considerable risk of arriving at a conclusion that is both incorrect and baseless in the context of the entirety of the case. As FPs, our professional lives are centered around the defense of our opinions, opinions that we have extensive training and take great care in forming. Of all the detours that most forensic pathologists try to avoid, arriving at indefensible conclusions sits very high on the list. Sometimes, the mechanism is complex (more than one), but, regardless, it is best to start with a defensible, evidence-based mechanism or mechanisms and go from there with the thought process.

This brings us to the important discussion of **how a forensic pathologist best determines the mechanism(s) of death in an evidence-based and nonarbitrary fashion.** Whether the proximate cause is a disease or injury, there are pathophysiologic and/or biochemical alteration(s) that link that cause of death to the moment of death. To identify those "links" (immediate causes and mechanisms), the forensic pathologist begins the inquiry at the autopsy table, ultimately dovetailing those observations with investigative data points. Known or anticipated **sequelae** of the identified disease or injury should be considered, sought, and documented. Examples of such sequelae (immediate causes and mechanisms) include hemorrhage (sometimes frank internal or external exsanguination) with associated degrees of tissue pallor, hemoaspiration patterns, various other kinds of shock that cause visceral and muscular pallor, amniotic fluid or air emboli, sepsis, DIC (see discussion below), space-occupying intracranial hemorrhage, and pulmonary thromboemboli, to name only a few.

The identification of some mechanisms may require further investigative inquiry, laboratory testing, witness statements, or medical records (i.e., must be inferred because they represent *functional derangements*, e.g., arrhythmias or seizures, as opposed to structurally demonstrable findings). When the death is reliably *witnessed* and its circumstances clearly defined, for instance (e.g., the death is sudden and preceded by alertness vs. a gradual mental status deterioration, followed by obtundation), the list of potential mechanisms shortens considerably. A classic example is demonstrated by a death that may be due to a prescription narcotic intoxication, but where other advanced natural diseases offer potential competing causes. If such an individual has a witnessed death that is preceded by agitation and acute dyspnea and tachypnea, and he is observed to be distressed, awake, alert, and panicked by his hypoxemia, then a narcotic-induced respiratory depression becomes a much less likely consideration. This presentation does not comport with the typical mechanism by which narcotics and depressant drugs generally kill (classically preceded by hypoventilation and obtundation/unconsciousness/coma, sometimes with audible "snoring"). If the history as relayed is credible, it should be considered in the interpretation of anatomic and toxicological findings, particularly when more than one kind of potentially fatal alternate respiratory disease, such as bronchial asthma or emphysema, exist.

Importantly, while explicitly memorializing sometimes complex mechanism(s) on the death certificate is neither always necessary nor necessarily advisable, it is no less important for the forensic pathologist to recognize them in order to ultimately arrive at the most accurate proximate cause. To illustrate, we return to shock. If visceral pallor indicating shock is noted, its etiology must be sought. Shock is a nonspecific clinical diagnosis; visceral and muscular pallor are often the associated pathologic sequelae, but that observation is not specific to its etiology, per se. Shock may be cardiogenic, septic, hypovolemic/hemorrhagic, or neurogenic. It falls squarely within the wheelhouse of the forensic pathologist to determine which one of these modalities best explains that terminal outcome, based upon both the autopsy findings (the type of disease or injury identified) and the available clinical and/or investigative information.

For example, the findings of marked visceral pallor in a young man who was shot through major vasculature of the anterior neck, most of his blood spilling onto the sidewalk and his death following within minutes of the shooting, support the conclusion that he died from hypovolemic shock. Alternatively, the man who was rendered quadriplegic and bed-ridden from a remote gunshot wound through the cervical spinal cord, having suffered terminal septic complications from paralysis-associated infected decubitus ulcers, may also have signs of shock (pallor), but his are of a different pathogenesis (i.e., sepsis). Both may have similar *anatomic* sequelae of shock, but having arisen through disparate pathways. In these instances, the autopsy alone will likely be inadequate and may render inaccurate conclusions of mechanism. Context, whether clinical or investigative, is often central to the veracity of opinions of mechanism. The devil is in the details, both at the autopsy table and in the pre- and postautopsy inquiries.

Inquiring

Wherever possible, seek primary sources of information and obtain "the news, not the editorial."

As a part of forensic pathologists' attempts to interpret findings and make pathologic correlations, they might need to consider further questions that would be helpful to ask and have answered both during the autopsy, when there may be more time-sensitivity in interpreting findings and directing autopsy decisions (e.g., knowing body positions at scenes of death, whether cricoid pressure was applied during resuscitative maneuvers that might explain soft tissue hemorrhage of the neck, whether a bullet was removed during surgical intervention, etc.) and after the autopsy. Some postautopsy questions might be most easily posed during follow up conversations with the next of kin, who may have witnessed a benign fall at home, are knowledgeable about antemortem symptoms, a drug habit, suicidal ideations, or past medical history, or can simply provide phone numbers of physicians. Other questions are best directed to medical professionals, including physicians and emergency services personnel (how much blood was evacuated, was a gunshot wound defect incorporated into a thoracotomy incision, were forceps used to remove a foreign body firmly lodged in the larynx, etc.), and for still others, to the police or prosecutor's offices. Have a low threshold for picking up the phone.

> We solve more cases with the telephone than with the microscope.

When dealing with any outside informant, however, including law enforcement, families, lawyers, or other physicians, it is advisable to **obtain objective investigative data points, rather than opinions and conclusions without further qualification.** When presented with others' impressions, inquire further to gather the specific facts that formed the basis of their conclusions ("Seek the news, not the editorial."). While considering all incoming information, in the end, the forensic pathologist must draw final conclusions based on investigative and physical *facts*, not on assumptions.

It is also very important to **seek primary sources of information.** Inserting one or more messengers between the forensic pathologist and the primary observer or gatherer of facts increases the likelihood of inaccuracy and miscommunication with each added layer. Like the game of operator, there always exists a high likelihood of important and/or nuanced information being lost in translation as the information and communication gulf widens. It also robs the FP of the opportunity of having a dialogue, a give and take about the details and issues at the center of a case, with the primary investigator, caretaker, or witness. Having a direct conversation with an assigned law enforcement officer or detective, for example, leaves little to no room for confusion and guesswork and affords the forensic pathologist the opportunity to specifically and clearly delineate their questions and needs relative to the death investigation. Expectations are set and a relationship is started, setting the stage for many follow up discussions, if needed.

All that said, it is important that during an active criminal investigation, the FP can and should speak directly to the police and prosecutor, but they should _not_ speak directly with potential suspects or the family and friends of suspects. In doing so, the forensic pathologist may compromise the investigation prior to the convening of a grand jury and also potentially risks becoming both an **expert witness** *and* a **fact witness** in future legal proceedings. In suspicious or clearly criminal matters, it is best to follow protocol and leave the interrogations and detective work to law enforcement officials. See Chapter 4 for a further discussion on this topic.

Interpreting and synthesizing

Recognize and admit to the limitations of the autopsy and of scientific analysis, in general (e.g., *time of death estimates, undetermined causes of death). "Honesty is the only stigma that attaches to a pathologist who leaves an occasional case undetermined." Don't be afraid of saying, "I don't know." Honesty lends more credibility than a disingenuous guess.*

Once a careful and thorough autopsy has been performed, all findings have been documented, and any further investigative inquiries have been made, it is time to interpret all of the gathered data and synthesize it into either a definitive conclusion/opinion to be communicated on a death certificate *or* develop a clear plan for what is needed to do so at a later time. Before that synthesis can be accomplished, however, there must be a clear understanding of the foundational definitions, concepts, and limitations upon which FP is predicated and practiced. "Someone who has all the answers all the time is either a scoundrel or a fool."

By design, interpretation and synthesis precedes the formation of opinions. *This is best achieved by having basic knowledge of the following concepts and their role in death certification (see also Chapter 3 for a more granular discussion of these concepts):*

Mechanism of death: As already advised earlier in this chapter, when interpreting findings and trying to piece together the component parts of a person's death, start with the mechanism of death, and then go from there. When faced with multiple potential underlying causes of death, identifying the mechanism of death within the context of the entire case will most accurately point to the correct cause. "*A* cause of death is not necessarily *the* cause of death."

Cause of death: The disease, injury, or *combination* responsible for the death.

Proximate (underlying) cause of death: The etiologically specific disease and/or injury that, in a natural and continuous sequence, unbroken by any **efficient intervening cause,** produces the fatality and without which the end result would not have occurred.

The importance of etiologic specificity: Why? Requisite to competent death certification is a clear understanding of the concept of **etiologic specificity.** A saddle pulmonary thromboembolus (PTE) is a nonspecific *immediate* cause of death. Depending on its *underlying/proximate* cause, however, it may represent a natural death, a TC, an accident, a suicide, or even a homicide. Obesity or metastatic ovarian cancer constitute competent (natural) proximate causes for PTE, while a thromboembolus caused by immobility due to quadriplegia complicating a remote inflicted gunshot wound of the cervical spine would be classified as a homicide. The identification of the "bottom line" etiology potentially takes the case down vastly different paths. It is for this reason that when a death certificate cites only a PTE alone as the cause of death (immediate, not proximate), the certification is woefully incomplete, and a reviewing medical examiner's office should reject any requests for cremation until the proximate cause has been identified. If injurious, most ME/C offices take jurisdiction over the certification of the death, even if it is not deemed necessary to transport the body to the morgue, per se (see Chapter 3 for examples of common mechanisms and immediate causes).

Disseminated intravascular coagulopathy (DIC) and **sepsis** are two common mechanisms encountered in patients who have endured a critical care hospital stay for any number of reasons.

These nonetiologically specific mechanisms are largely clinical and/or clinicopathological diagnoses, and, as such, must be evidence-based conclusions when opined by the medical examiner/coroner.

When one collects a **blood culture** at autopsy, a positive result requires contextualization and case correlation, analogous to a toxicology result. If *pathologic* bacteria are identified by blood culture (as opposed to the polymicrobial growth of rare bacteria that are usually indicative of postmortem overgrowth rather than antemortem/pathologic growth), the most one can say is that there was "bacteremia" at the moment of death. Unless there is clinical and/or physical/autopsy evidence to support the sequela of "sepsis," such as pallor indicative of septic shock, *not* explained by other types of shock, and/or medical documentation of clinical "septic shock" (i.e., the clinician's impressions based on severe hypotension, often requiring pressor therapy, with leukocytosis or leukopenia, +/− fever, an infectious source, and/or positive hospital blood cultures), the term "sepsis" should not be used.

The same rigor should be applied to the diagnosis of DIC. While only a mechanism, it is still *not* a finding typically based on autopsy observations alone; it must make sense contextually/clinically and have evidence to support it. It is advisable to not opine on such a diagnosis predicated on "possibilities," but rather on evidence-based clinical and/or anatomic data points (i.e., hospital laboratory values consistent with DIC and/or a clinical history or physical/autopsy findings of diffuse and/or poorly explained/spontaneous mucosal, soft tissue, cavity, or visceral hemorrhage, IV site oozing, etc., in a setting that correlates well with DIC).

In short, it is best to resist the urge to speculate and draw a conclusion of "sepsis" or "DIC" without evidence to support such conclusions (clinical/laboratory and/or anatomic findings). As a general rule, speculation is a dirty word in FP! Once DIC or sepsis *are* reasonably diagnosed, however, the next and most important step is to identify their proximate cause. *Like PTE, sepsis and DIC may be the mechanisms or sequelae of natural deaths, homicides, and everything in between.*

Appropriate uses of contributory cause(s)/"part 2": These are *conditions* contributing to death, but not resulting in the underlying cause of death in part 1. This is neither required for death certification nor to be used to list every diagnosis from the autopsy report.

For *natural deaths*, it is appropriate and common practice to place other mechanistically related natural diseases/comorbidities in part 2 (e.g., diabetes mellitus, obesity, emphysema, etc.), if they reasonably contribute to the death given the circumstances. But, part 2 should not be littered with a laundry list of "interesting" or "unexpected" entities that have no contributory lethal potential. They can, however, be nonetiologically specific entities/findings; *etiologic specificity is not a requirement for part 2 findings* (examples: thrombocytopenia, cirrhosis, atrial fibrillation, etc., without further qualification). That said, *if* the etiology of the entity *is* known and would change the manner from natural to a therapeutic complication or other unnatural manner, the etiology of the part 2 entity is best specified and the manner changed accordingly. Remember that, in most jurisdictions, in order for a natural death to be classified as such, it must be *exclusively* (100%!) due to natural causes. Under those practice guidelines, even a minor contribution from an unnatural cause would change the manner. In other words, the default practice for manner classification in those jurisdictions is: injury (even if a minor contribution) trumps natural disease.

This brings up another point with regard to contributory causes. If adding one or more diagnoses in part 2 will change the manner of death from natural, it is best to make sure that the diagnosis is necessary to invoke, that the injury or therapy, in all likelihood, contributed to death (cannot be reasonably ignored). In other words, the primary cause in these fatalities is *not* adequately explained *without* invoking the contributory cause(s) (e.g., a natural disease that would normally/reasonably not be fatal from hemorrhagic complications *but for* the risk conferred by the part 2 entity of *over*-anticoagulation therapy).

The converse is also true for listing natural disease in part 2 when the proximate cause of death is an injury. If the death can be reasonably explained purely by the injury, there is no need to invoke additional natural disease in part 2. Inappropriate uses of part 2 may result in a variety of unintended ramifications. For example, some accidental insurance policies will only pay out a benefit if injury is the sole cause of death. Adding a renal cell carcinoma as a part 2 disease in someone who dies from an aortic transection from a high-speed motor vehicle collision is both unwarranted and counter-scientific. On the other hand, for the person who dies from smoke inhalation with a carboxyhemoglobin saturation of only 30%, the addition of pulmonary emphysema or atherosclerotic cardiovascular disease in part 2 may be appropriate and necessary. This pattern of thinking leads us naturally to the following concept.

Ask the "but for" question with regard to causes (parts 1 or 2): *"But for this disease/injury, would the death have likely occurred?" (On this day?)*

Manner of death (how the cause arose): This is informed by the proximate cause. Burdens of proof vary depending on the manner, but the minimum requirement is **probability** (>50% likely), not possibility or absolute/100% certainty (see the below discussion on **burdens of proof**). Ideally, however, forensic pathologists aim to achieve more than just a mere 51% likelihood in most of their unnatural certifications, but, for homicides, the burden is "a reasonable degree of medical/scientific certainty" (i.e., implicit is that determination is that no other *reasonable* contingencies/explanations exist). For many, a very high degree of probability is also the standard for the certification of suicides. Finally, and very importantly, in terms of manner, **there is no statute of limitations dictating the amount of time that can elapse between injury occurrence and its ultimate lethal consequence.** It is for this reason that we are able to certify "delayed" homicides, suicides, and accidents. If a person rendered quadriplegic by being shot in the neck dies from its direct complications 40 years later, his death is no less a homicide then than it would have been had he died on the day of the shooting, the duration of time elapsed between the injury and the ultimate death having no bearing on the injury's culpability. See Chapter 15.

The basic pathophysiology of death and dying: This is critical in determining the mechanism. Ideally, an FP should understand degrees of injury and disease severity, potential survival intervals, and mechanisms of death for specific diseases and injuries (death in seconds vs. minutes vs. hours, or beyond); respiratory (minutes) versus cardiac (seconds) deaths; irreversible brain damage and death by neurologic criteria (so-called "brain death"); and the immediate incapacitation/"stopping power" of certain injuries or natural processes. With regard to injuries, bleeding and asphyxial deaths more commonly take their toll over the course of minutes, while cardiac deaths and brain stem injuries typically do so in seconds/often instantaneously (see Chapter 3 for further discussion).

Classes I–V definitions of natural findings (recall that Class I is incompatible with continued life); see Chapter 3.

Is a given finding a cause or a consequence of death? Examples include: thermal injury sustained in bed while smoking cigarettes—does this represent a fire victim or a natural death with postmortem charring?; elevated or markedly reduced postmortem body temperatures, a cause or a consequence of death/postmortem equilibration?; the question of possible positional asphyxia (i.e., dying *with* or *from* a found position of potential respiratory compromise); submersion in a bathtub—cause or consequence?; "choking"—passive, resuscitative artifact, or causal—death *from* eating or *while* eating; the significance of minor terminal head trauma, especially when other potentially/definitely lethal injury or disease has been identified, such as a pulmonary thromboembolus or a gunshot wound of the head. In the circumstances of positional asphyxia and bathtub drownings (see Chapter 11), the FP is left with the burden of explaining *why* the decedent would not remove themself from these compromising circumstances. Identifying the reason reveals the proximate cause of death. If no such explanation exists, the basis for the opinion that they died from these injuries is likely faulty, at best. As Dr. Hirsch was known to say, "Neurologically competent, nonintoxicated persons do not drown in bathtubs." Such certifications demand an explanation.

A **"fatal concussion"** as a mechanism of death, often suspected based on structurally demonstrable, but sublethal head trauma, can be reasonably excluded by reports of a Glasgow Coma Scale of 15 or a video tape showing the decedent standing up and walking off or otherwise maintaining consciousness after sustaining the injury, only to collapse an hour later. Alternatively, the video may show the decedent clutching his chest, after which he appears to lose consciousness, *then* fall to the ground or down a staircase, which raises the specter of him having suffered a fatal dysrhythmia (due to natural causes) *before* his head made contact with the resistant surface. This is all to say that some investigative data points fly squarely in the face of the mechanism of a fatal concussion, or at least confer less veracity to that conclusion. The functional nature of this mechanism explains why these cases can be quite challenging in the absence of eyewitnesses, video surveillance footage, or written emergency services neurologic assessments.

The question is not *can* an injury or disease kill, but *did* it? Asked differently, **did an individual die *with* or *from* the identified disease or injury?** Science and evidence need to be lined up in order to answer this question accurately. *"Sequence doesn't (necessarily) make a consequence."*

Cause or consequence of death–vital reactions: Occasionally, a "unifying theory" bubbles to the top and threads most of the autopsy findings together nicely. Regardless, all findings must be individually and collectively assessed for their importance within the context of the entire case, but also with regard to whether they appear to have formed before or after death. Assessing for vital reactions through careful gross and/or microscopic observations allows the FP to determine whether certain findings should be considered *at all* in terms of the cause of death.

A common example of a *gross* intravital reaction (i.e., evidence of circulation and therefore of the disease or injury having occurred during life) is **a substantial hemorrhagic response** to the pathology (when expected). Minimal to no hemorrhage in association with an injury, such as a fracture or laceration, for example, is often indicative of lack of circulation (death) when it was sustained. Such injuries often occur with a terminal/perimortal collapse due to an unrelated cause or with postmortem movement or transport of a dead body.

As another example, the absence of airway hyperemia and soot deposition makes it highly unlikely that an individual was breathing/alive during a fire. Conversely, the presence of both indicates smoke inhalational injury, even in the absence of an elevated carboxyhemoglobin saturation (the latter scenario can be encountered in flash fires[6,15]). You have to believe what you see. Similarly, while **frothy pulmonary edema** is a nonspecific finding, its presence represents a compelling intravital marker, regardless of cause. While it may be cardiogenic, neurogenic, inflammatory, or injurious in nature, when "frothy," it likely indicates air movement over the mucus membranes and alveolar passages. In other words, it strongly suggests a respiratory effort in a living and breathing, if even unconscious, person, and therefore, represents a vital reaction. In a potential drowning victim, for example, the observation of frothy pulmonary edema, including a "foam cone," constitutes a distinctly different observation than just (nonfrothy) fluid in the lungs or stomach, the latter which may have gained access to the body passively during the postmortem interval. Pulmonary edema is not diagnostic of drowning, per se, but, it strongly supports an intravital response when grappling with the question of, "Did this person die *in* the water (cardiac arrest before submersion) or *from* the water?" In this setting and in the absence of other causes, the presence of pulmonary edema is a compelling indicator in support of the latter.

A common *microscopic* vital reaction is **inflammation in response to disease or injury,** although its absence in an acute injury would be expected. The takeaway is that a significant hemorrhagic response (with or without inflammation), hyperemia, frothy pulmonary edema, and/or inflammation in response to disease or injury are all consistent with circulation and respiration, at least to some extent (i.e., the disease or injury having occurred during life), regardless of their causal relationship to death. "It takes circulation to get inflammation." Importantly, however, is that **circulation and respiration,** per se**, are not reliably predictive of consciousness** and these two intravital variables should not be conflated with consciousness. Even more important to recognize, however, is the variability of the human body from person to person (host variability), injury to injury (minor to marked, cutaneous to soft tissue to visceral, etc.), and from circumstance to circumstance. *"In reality, pathologic processes seldom cooperate fully with attempts to date or age them with precise reliability."*

Is a given finding a sequela or a primary cause? For example, a *terminal seizure* following a gunshot wound of the head, a MI, or diabetic ketoacidosis does not represent a primary seizure-related death, but rather a nonspecific sequela of an unrelated proximate cause; *anoxic-ischemic encephalopathy* is a common final pathway following cardiac arrests due to myriad primary causes, from MIs to drug intoxications to mechanical injuries, as opposed to a primary neuropathologic process. *"Do not confuse correlation with causation."*

What was the time of death? When was the fatal injury sustained relative to death? These are commonly asked questions, the former from friends and family, and both as part of certain criminal investigations and their adjudication. Absent video surveillance, eyewitnesses, or the involvement of medical professionals, neither is typically answered with specificity, based on the examination of the body alone. Rather, in most death investigations, the answers to both should include a disclaimer with respect to **scientific limitations** and the importance of dovetailing physical evidence with investigative data points. Generally speaking, the further in time from the moment of death the discovery of a body occurs, the less specific will be the estimates of the postmortem interval. As such, forensic pathologists are more likely to speak in terms of *units and ranges* rather than specific numbers. Estimates lean toward the qualitative more than the quantitative. As a result, and not uncommonly, we are instead asked under oath whether a specific hypothetical point in time is "consistent" (*reasonably* possible) or "inconsistent" (*not reasonably* possible) with a suggested time of injury occurrence or death, and we can answer with "yes," "no," or, "it depends," offering further elaboration and explanation, if necessary. See Chapter 4 for further discussion on this topic.

Opining

In all matters, but especially in your professional opinions, avoid being arbitrary and capricious; *remain internally consistent, thoughtful, and evidence-based in your practice and conclusions. "Hunches are not evidence."*

"He who accuses must prove."

Opining is the "end game" of the forensic pathologist's task list in any medicolegal death investigation. It is predicated on our training and experience and ultimately represents the primary responsibility born out of our expertise. Everything we do is done in the service of achieving the goal of arriving at not only defensible conclusions, but also correct ones. That said, we must remain open to new information and not view our opinions as having been written in stone. We make them given all available information at the time of their rendering. A rigid attachment to them, leading to a refusal to consider new information that may come forward, detracts from our neutrality. Often, such new information is not relevant and therefore does not change the conclusions we initially drew, but a fair and measured consideration is a reasonable expectation in our role as neutral participants in any investigation. We consider all information that is presented to us, regardless of when that occurs.

The formation of opinions in forensic medicine is often consequential. As such, they are determined within a construct of specific **burdens of proof** (Table 2.1). Burdens of proof are defined by **degrees of certainty**. Burdens of proof depend on the manner of death, a concept uniquely relevant to physicians who are forensic pathologists (see also Chapter 3). A clear understanding of these dovetailed concepts is requisite for both competent death certification and expert testimony. Both are outlined below.

TABLE 2.1 Certainty
Burdens of proof and degrees of certainty: Possible versus probable.

Degrees of certainty		Definition/needed for:
Possible	>0%	Does not defy laws of science
Probable	>50%	Natural causes of death
Reasonable degree of medical certainty	>>50%	Homicides
Certainty beyond a possible doubt (impossible)	100%	Never required
"Consistent with"	>>0%	A *reasonable* possibility

Possibilities

Possible (>0%): An event is possible if it does not defy the laws of nature. There may be reasonable and unreasonable possibilities. Homicide classifications do not allow for other *reasonable* possible causes.

Probabilities

Probable (>50%): Preponderance of evidence (more likely than not) permits reasonable doubt.

This is the civil court burden, the degree of certainty needed for opinions on death certifications in natural, certain accidental, and therapeutic complication deaths. It is important to note the difference from scientific probability (0.05), which is the likelihood that the difference between two sets of data due to chance is 5%. This is not used in medicolegal investigations because such investigations involve examining only a single datum.

Reasonable degree of medical/scientific certainty**

This is defined as certainty beyond a reasonable doubt (no exact %, but far exceeds 50%). The assertion of this degree of certainty assumes that all other *reasonable* explanations and contingencies have been excluded. This is the degree of certainty when there is no good reason to believe otherwise. It is one that you would require to make the most important decisions in your life or the degree of assurance that a reasonable person relies upon in his or her most important business dealings. Certainty beyond a reasonable doubt is used when considering homicide or, for many (accident vs.), suicide classifications. If there is another reasonable possibility, a reasonable degree of certainty burden is not met. When presented, however, with an *unreasonable* possibility as an alternate theory to a homicidal death, our burden *is* still met, even if we cannot absolutely exclude that possibility. As Dr. Hirsch once replied when presented with such an unreasonable theory, "While I can't say that's impossible, it's so highly unlikely that it's standing shoulder to shoulder with impossibility."

**When our opinions are used in criminal legal proceedings, a higher degree of certainty than probability is used, i.e., reasonable degree of medical/scientific certainty.

Certainty beyond a possible doubt (100% certainty)

This burden is **not** required to uphold any medical opinion because it embodies the premise that all other hypotheses are impossible (e.g., you cannot scratch your right elbow with your right hand). This is not even the burden of prosecutors in adjudicating homicides.

"Everything is consistent; nothing is inconsistent."

With the above in mind, determination of the final cause and manner of death will most reliably meet the necessary burdens with both reproducibility and veracity if the foregoing principles described in this chapter are consistently incorporated into the decision-making process. Further, in communicating and memorializing those written opinions in a legal document (i.e., the death certificate), word choice matters (see "Communicating" below, as well as

Chapter 16). Clear, deliberate, and cogent death certifications that convey the necessary proximate cause, manner, and, where appropriate, a description of circumstances, without conjecture or the creation of more questions than answers, will most frequently hit the mark. Immediate causes and mechanisms, while central to the formation of the final death certification and classification, may be part of the cause of death statement, depending on whether they enhance its communicative value, but they are not required to appear on paper, per se.

That said, the **death certificate** is not intended to be a substitute for the entire case and investigative file. Ideally, **it should succinctly represent the sound, clear, and evidence-based conclusions relating only to cause and manner of death**, along with an objective mention of any circumstances or factors necessary to clarify or further explain either, depending on the case. For example, while the toxicology results in a homicidal shooting or the presence or absence of nonfatal injuries indicative of a struggle in a knife encounter may have substantial bearing on future questions and legal matters relating to the death, those details would have no place on the death certificate, per se, if they played no role whatsoever in the *mechanism* of death or were unnecessary to cite in explaining the manner determination.

As an illustrative example, if, as a result of a motor vehicle collision with another car, blunt impact injuries kill a driver while he is intoxicated with alcohol, a description of how the accidental collision occurred should remain a neutral statement of an indisputable fact: e.g., "driver involved in collision with other motor vehicle," rather than, "driver involved in collision with other motor vehicle while under the influence of alcohol." The latter not only is unnecessary to invoke in order to meet the burden of explaining to the registrar how the accidental injury arose, but it is also based on a certain degree of both judgment and speculation. It presupposes the effects of a specific ethanol concentration on an individual level (highly variable) and also assumes that other behavioral, environmental, physiological, or circumstantial variables were *not* at play to an equal or greater extent in causing the collision (fault or errors of other drivers, mechanical vehicular factors, weather, road conditions, glare, texting, distractions by other occupants of the vehicle or other cars, bad news on the radio, dizziness, fatigue, a seizure, and so on). As such, making such a judgment, unnecessary for the purposes of competent death certification, would risk entry in the territory of arbitrary and capricious. Fortunately, it is neither necessary nor reasonably possible for medical examiners to opine and report as to the "why" of most roadway collisions in order to certify such deaths, but rather to merely document *that* they accidentally occurred. Questions relating to the cause(s) of the collision may likely hold importance for many stakeholders, but these would be most appropriately addressed by the traffic scene investigators/law enforcement. Medical examiner/coroners may be presented with theories from those who investigate and reconstruct such incidents, but, with rare exception, and for the purposes of death certification, those specific variables are neither necessary nor appropriate to include on a death certificate. One should avoid "speculation built on a foundation of assumption." Notable exceptions to this rule would be scenarios involving vehicular homicides or suicides, where the determination of *intent* bears a different burden of explanation, requiring investigative evidence of why and how the event arose. See Chapters 3 and 10 for further discussions on the topic.

"Descriptive" causes of death represent a unique situation presented in death certification in which the cause of death does not neatly fit into a discrete mechanistic or etiologically specific disease or injury paradigm and/or is incomplete with the disease or injury standing alone without further qualification. In such instances, a description of the death circumstances is incorporated into the cause of death statement, along with which disease and/or injury processes were, at least to some degree, at play. This approach might be employed in circumstances such as the sudden death of a person with advanced heart disease during a physical altercation or armed robbery, where no structurally demonstrable fatal injuries, per se, are identified, but the event is well-documented and the injurious mechanism is determined to be *functional* (i.e., stress-induced cardiac dysrhythmia). Because of the homicidal nature of such deaths, certification simply as "atherosclerotic and hypertensive cardiovascular disease" would only serve to confuse, and not adequately represent the nature of the case or explain the manner determination. Objectively disclosing the way in which such a sudden cardiac event arose in either the cause of death statement (the so-called "descriptive" cause of death) or in the "how the injury arose" section of the death certificate would provide a justification and further clarification of the manner opinion. This approach might also serve as a helpful approach in a *surgical death* in which the discrete mechanism is unclear, despite a comprehensive medical record and pathological review. In such situations, the use of a descriptive cause allows the pathologist to communicate that "but for" the surgery and/or anesthesia, the death would not have reasonably occurred on this day, while acknowledging that the precise way (mechanism) by which the procedure led to death exceeds the pathologist's diagnostic capabilities. When such a descriptive cause of death is utilized for an unnatural/injurious death, "see above" is usually sufficient in the **"how injury occurred"** field. See Chapters 3 and 15 for further discussions.

Making "pending" death certification decisions

Determining whether to "pend" a case is usually based more on substance than style. That said, there is some degree of nuance, as well as reasonable differences in comfort levels among forensic pathologists. Some are more "aggressive" in certifying a case on the day of autopsy, while others are more conservative, wanting more information or documentation before making a final determination. Neither is necessarily the "right" or standard way of doing things, per se, but in any practice, the following principles might be kept in mind:

As explained previously, the death certificate is a legal document that reflects a *probability* opinion (>50%), not one of absolute *certainty (100%)*. We would like to achieve a healthy and reasonable balance between the urge to "pend" every case when evidence-based probability has *already* been achieved on the day of autopsy versus certifying cases prematurely and based on assumptions, before all the evidence has been lined up and reasonable and necessary questions have been asked and answered. That said, becoming paralyzed by the fear of rendering what most would view as a reasonable and defensible opinion is to be avoided. One learns with experience over the course of years when a *higher* burden than mere probability (>50%) is necessary (suicides, homicides, certain accidents, in-custody deaths, complex cases, etc.) and how to become increasingly confident in one's opinions. The best way to enhance the development of this confidence, however, is by applying the same concepts and standards of death certification decision-making to every case.

Aim to have a discrete justification and plan for any pending cases, whether waiting for medical records, further police investigation, a histologic determination of the likely proximate cause, toxicological analysis, a neuropathologic evaluation, a scene visit, or some combination thereof, it is advisable to think it through and know *why* you are deferring the final certification. Having a plan and an endpoint as to what, *specifically*, is needed to render a final opinion is also helpful in that it can help in the management of a voluminous caseload, a common burden of trainees. While there is absolutely no shame in a pending death certificate, knowing *why* we are doing so will likely organize our workflow and enhance the consistency of our forensic practice and the veracity of our final opinions.

Avoid thoughtlessly and reflexively "pending" cases as a matter of practice or principle, treating all cases as the same. It is helpful to remain mindful of the fact that pending a death certificate may have immediate repercussions on the next of kin (e.g., life insurance claims) and, while we do pend more cases than not (e.g., intoxication deaths), it is important to not do so without cause. There are survivors behind our work, people who are often gutted by a sudden loss and also dealing with the associated financial fall out. With that in mind, one is advised not to "pend" a case because one *can*, but rather because one *should*.

All that said, on rare occasion, FPs may defer the certification of a death to allow a family time to come to terms with the shock of a suicide, for example, to create a buffer between their acceptance and that piece of paper. They can be truthfully told that the toxicology will not realistically change the final manner determination, but that you will do them the service of waiting for it before memorializing your final opinion on paper. This gives them some perhaps much needed time and space to come to terms with the reality of what has occurred. This is the art of FP practice. There are the nuanced ways in which we might serve grieving families, where the occasional and intentional decision to "pend" is done in the service of more good than harm. There is no "one size fits all" approach.

Communicating

Our words matter; choose them carefully to convey specific meaning. *"We are in the business of communication." We must communicate well in both verbal and written forms, ever mindful of a diverse audience with often disparate opinions, needs, agendas, and levels of forensic understanding.* **In your communications with the bereaved, remember that you are a human being before you are a physician.** *Our role in providing helpful, clear, and definitive scientific conclusions can be just as important to survivors as that of our ability to provide closure and solace. Honesty, sensitivity, and good listening skills will go a long way.*[16,17]

We are at the final stage of the forensic pathologist's task list. The case has been thoroughly investigated, the decedent examined, physical evidence documented, inquiries made, and opinions rendered. Now comes the time to communicate our conclusions. The audience will vary, necessitating a nimbleness and flexibility in how we will need to convey sometimes sensitive information. The common thread between these various conversations, however, will be truthfulness. Additionally, while operating within the investigative universe, we strive to seek the news, not the editorial, but the converse is more appropriate when we are communicating our final conclusions. Start by communicating the editorial, not necessarily the news, unless necessary or specifically requested to do so. Definitive conclusions offer comfort. Also, the amount of detail certain audiences will desire or be able to bear may be minimal. It is important to be sensitive to varied needs.

This topic is so important that the authors dedicated an entire chapter to it (see Chapter 16). As such, the proceeding discussion represents only a few salient points to consider in relation to our communications in forensic practice.

As forensic pathologists, our communications in both written and verbal forms are central to our work. Generally speaking, ME/C communication can be categorized into one of four areas: **written documents** (death certificates, autopsy reports, press releases, emails, case notes, etc.), **direct conversations** (in person or telephone discussions with families, lawyers, law enforcement, other governmental agencies, physicians, etc.), **expert testimony/legal proceedings**, and **teaching/public speaking.**[18]

Death certification has been addressed above and will be further discussed in detail in Chapter 3.

In our **written reports**, a cogent summary of our diagnostic conclusions should be clear, and the overall organization of the report deliberate in its thoughtful and communicative construction. There are many ways to achieve this, demonstrated by the variability of report formatting among ME/C offices. While further follow-up clarification may be necessary to nonmedical readers, the specific crafting of the *final diagnoses* on the first (or last) page of the autopsy report should nonetheless clearly reflect the case synthesis and the cause and manner of death (*the editorial*), whereas the body of the report would ideally include a well-organized, clear, objective, and detailed recitation of the findings and observations that support those conclusions (*the "news"*).[1] A guiding principle in all written forensic communications is to be succinct, but thorough, detailed, and clear, declarative statements usually being preferred over the passive voice, unless otherwise necessary. In criminal matters, all written documents are potentially *discoverable*, including emails. As such, waxing philosophic or having substantive discussions in email communications regarding a homicide or otherwise suspicious death is strongly discouraged.

In our direct conversations with myriad individuals, **professionalism** should always set the tone and guide the exchange. In dealing with grieving friends and family, exhibiting humanity and authenticity, recognizing the delicate nature of the situation and the varied ways grief manifests, and demonstrating reflective listening skills will go a long way, as will the delivery of clear, conclusive, and timely information and opinions, sensitively conveyed. Wherever possible, making a grieving family wait days for an initial phone call should be avoided.

With regard to communicating "preliminary" findings to the **public/press**, "distribution of partial information guarantees complete misunderstanding." On the other hand, "information void is the mother of paranoia," so managing expectations with a full disclosure as to when complete and final conclusions might be revealed and an honest explanation as to why more investigative time is needed and beneficial to all involved will likely prove helpful in minimizing frustration and the creation of conspiracy theories. People deserve and need honest process explanations, even if not immediate answers.[1] A more granular discussion specifically addressing press communications, particularly with regard to relatively highly charged or high-profile deaths will be covered in Chapter 16, as will guidance, clarity, and expectations relevant to expert testimony in various legal forums and advice related to public speaking in academic and other settings.

References

1. Moritz AR. Classical mistakes in forensic path. *Am J Clin Pathol.* 1956;26(12):1383–1397.
2. Gill J, Stajic M. Classical mistakes made in forensic toxicology by forensic pathologists. *Acad For Path.* 2012;2(2):228–234.
3. Davis JH, Wright RK. The very sudden cardiac death syndrome – a conceptual model for pathologists. *Hum Pathol.* 1980;11(2):117–121.
4. Angermeier P, Zumwalt RE, Hirsch CS. No Anatomic Cause of Death No. FP 85-4 (FP-141). *Am Soc Clin Pathol Foren Pathol Check Sample.* 1985; 27(4):1–3. FP 85-84.
5. Hirsch CS, Lester Adelson MD. *Am J Forensic Med Pathol.* 1989;10(3):261–263.
6. Walter JE, Hirsch CS, Zumwalt RE. Never say never. Negligible carboxyhemoglobin in the victim of a smoldering mattress fire. *Am J Forensic Med Pathol.* 1984;5(3):239–244.
7. Morentin B, Audicana C. Population-based study of out-of-hospital sudden cardiovascular death: incidence and causes of death in middle-aged adults. *Rev Esp Cardiol.* 2010;64(1):28–34.
8. Adelson L. *Pathology of Homicide.* 1st ed. Springfield: Charles C Thomas; 1974.
9. Hanzlick RL. The "value-added" forensic autopsy: public health, other uses, and relevance to forensic pathology's future. *Acad Foren Pathol.* 2015;5(2):177–185.
10. Adams VI, Flomenbaum MA, Hirsch CS. Trauma and disease. In: WU S, ed. *Spitz and Fisher's Medicolegal Investigation of Death.* 4th ed. Springfield, IL: Charles C Thomas; 2006:436–459.
11. Hirsch CS, Zumwalt RE. Anderson's pathology. In: Damjanov I, Linder J, eds. *Anderson's Pathology.* 10th ed. Mosby; 1995:80–109.
12. Adelson L. No anatomic cause of death; the enigma of the forensic pathologist. *Conn Med.* 1954;18(9):732–738.
13. Oliver WR, Fang X. Forensic pathologist consensus in the interpretation of photographs of patterned injuries of the skin. *J Forensic Sci.* 2016; 61(4):972–978.

14. Oliver WR. Effect of history and context on forensic pathologist interpretation of photographs of patterned injury of the skin. *J Forensic Sci.* 2017;62(6):1500–1505.
15. Hirsch CS, Bost RO, Gerber SR, Cowan ME, Adelson L, Sunshine I. Carboxyhemoglobin concentrations in flash fire victims: report of six simultaneous fire fatalities without elevated carboxyhemoglobin. *Am J Clin Pathol.* 1977;68(3):317–320.
16. Hirsch CS. Talking to the family after an autopsy. *Arch Pathol Lab Med.* 1984;108(6):513–514.
17. Adelson L. The forensic pathologist. "Family physician" to the bereaved. *JAMA.* 1977;237(15):1585–1588.
18. Adelson L. The pathologist takes the witness stand. *Am J Clin Pathol.* 1964;41:484–486.

CHAPTER

3

The pathophysiology of death and death certification

Susan F. Ely and James R. Gill

Death is a functional event, not an anatomical event. The pathologist who searches for recent pathologic alterations to explain every death is doomed to failure. *-Drs. Vernon Adams and Charles S. Hirsch*

Common things are common because they are the things that occur with the greatest frequency. *-Dr. Charles S. Hirsch*

The pathophysiology of death

There are two legal **definitions of death:** one involves the cardiopulmonary system and the other involves the brain; both contain the word irreversible.[1] If a sudden cardiac arrest was synonymous with death, no one would ever be able to be successfully resuscitated. Death is defined as the irreversible cessation of function of the heart and lungs or the entire brain, including the brainstem. The latter group of deaths has implications for organ donation (the preferred term is "donation," not "harvesting") and for the medical examiner/coroner (ME/C), who is often in the position of deciding whether to restrict donation once death has been pronounced by neurological criteria (also indelicately referred to as "brain death"). In these instances, the patient is usually on a ventilator and the heart is still beating. Since the brain is not as resilient in response to oxygen deprivation as are other organs, including the heart, it is not unusual for a person to be "successfully" resuscitated (e.g., defibrillation of the heart), despite the brain already having suffered irreversible damage during the ischemic period.

Three conditions are needed to diagnose **death by neurological criteria:** persistent coma, the absence of brainstem reflexes, and the inability to breathe independently. Coma is diagnosed when there is no response (i.e., no eye opening, verbal response, or limb movement) to a painful stimulus.[1] Brainstem function is assessed by testing pupillary responsiveness to light and the presence or absence of a gag/cough reflex with throat suctioning. The third criterion is assessed through an apnea test, which involves temporarily removing the patient from mechanical ventilation and observing for spontaneous breathing. If no breathing is witnessed for 10 minutes and the blood carbon dioxide level increases by 20 mm of mercury or more, the patient has met the criteria for neurologic death. Other radiographic tests, such as radionuclide studies, transcranial Doppler ultrasound, or cerebral angiography may be used as an alternative to the apnea test.[1]

One pitfall encountered by many inexperienced autopsy pathologists is the expectation of finding an anatomic "smoking gun" diagnosis to explain every death. This becomes particularly problematic when considering sudden cardiac deaths. The belief that one must find an acute thrombus or an acute myocardial infarct is common, if misguided. A sudden death in a person with fixed advanced atherosclerotic stenosis, with or without myocardial fibrosis, is due to atherosclerotic cardiovascular disease until proven otherwise (by toxicology, microscopic exam, history, etc.). A study in Spain reviewed cardiac autopsy findings in sudden deaths. They[2] found that only about half had a thrombus or acute myocardial infarct; the rest had fixed disease and/or myocardial fibrosis. The inexperienced autopsy pathologist also may underestimate the ability of hypertensive cardiovascular disease to cause death. Substrates for arrhythmogenicity, such as a clinical history of hypertension, elevated cardiac mass, and small vessel disease, should not be disregarded, but the absence of a history of hypertension should not necessarily preclude the postmortem diagnosis when morphologic evidence exists. We must not "deny reality for the sake of objectivity."

https://doi.org/10.1016/B978-0-323-91796-4.00015-5

Blood pressure cannot be taken after death. The prosector makes a diagnosis of hypertensive cardiovascular disease based on the history (when present) and/or cardiac hypertrophy in conjunction with gross and/or microscopic evidence of small vessel disease. It should be noted that a hypertensive may not always have cardiac hypertrophy, but when identified, small vessel disease is one way to distinguish cardiac hypertrophy due to chronic abuse of cocaine or other stimulants (no small vessel disease) from hypertensive cardiovascular disease. If the decedent is old enough, the development of gross fine granularity of the kidneys or microscopic (renal) arteriolar medial hypertrophy supports the latter. Aortic stenosis also may cause cardiac hypertrophy, as can chronic alcohol misuse and other less common natural diseases.[3–10]

There are often no *acute* structurally demonstrable gross or microscopic findings in **sudden cardiac deaths** simply because the individual died too quickly for the cells to develop the morphological changes seen with the naked eye or the microscope. Survival intervals and medical interventions are two variables that are familiar to both hospital and forensic pathologists. But, it is the forensic pathologist (because of the statutory nature of their work) who is more likely to investigate deaths without survival intervals and with diseases that are untouched by medical hands. Where else but in a medical examiner/coroner office is one likely to see a death due to an untreated acute appendicitis?

In order to conceptualize the lethal capacity of various autopsy findings, a classification system was created.[11,12] This **four-tiered classification system (Classes I–IV)** grades autopsy findings of *natural disease* by their degree of certainty in explaining a death.[11,12] A **Class I** autopsy finding is incompatible with continued life (e.g., ruptured aortic dissection, saddle pulmonary thromboembolism [PTE], pontine hematoma, Fig. 3.1,

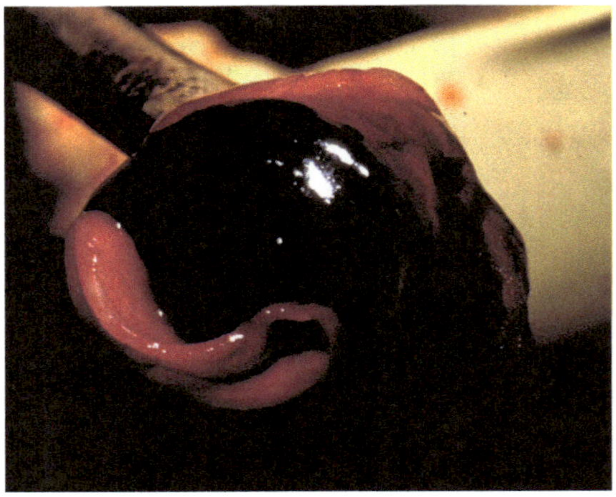

FIGURE 3.1 **Aortic dissection.**

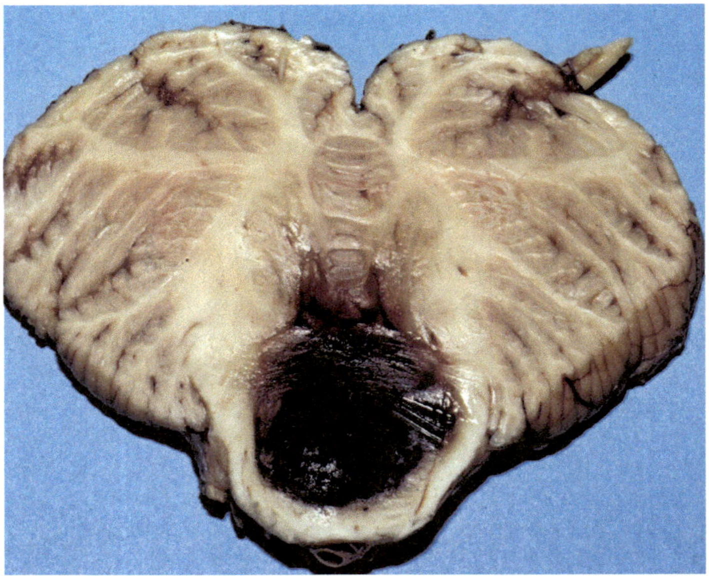

FIGURE 3.2 **Hypertensive pontine stroke.**

Fig. 3.2); a **Class II** finding is a disease with lethal potential sufficiently advanced to be recognized as capable of causing death, but compatible with continued life (e.g., marked coronary artery atherosclerosis, gastrointestinal hemorrhage); **Class III** includes findings of marginal pathology (e.g., slight cardiac hypertrophy, hepatic steatosis) that are usually insufficient by themselves to explain death; and **Class IV** represents conditions with clinical history, but no structurally demonstrable lethal lesion (e.g., epilepsy). Some use a **Class V** designation for an undetermined cause (neither identifiable pathology nor history of disease). A similar approach may be used for traumatic injury.

These classification systems are important in that they lay the groundwork for a conceptual understanding of reasonably expected sequelae and clinical presentations of a myriad of injuries and diseases. A clear and scientifically based understanding of the pathophysiology of disease and injury allows the forensic pathologist to frame reasonable and detailed descriptions of the events that may have led to death. Stated differently, this knowledge sets the stage for the forensic pathologist to opine about **potential survival intervals** relating to a variety of disease/injury categories and severities, which in turn, may play a role in their responses to questions in legal proceedings that extend beyond mere cause and manner.

Milton Helpern once said that every death is due to a "breakdown in the body's oxygen cycle."[13] There are multiple steps in the oxygen cycle in which this breakdown may occur. When considering the pathophysiology of death, the location of the breakdown in the cycle may be critical in determining both the underlying cause and the manner of death. Where the specific oxygen interruption occurred will likely lead one to the "how" of the breakdown, working back through mechanisms and immediate causes to reach the proximate or underlying cause of death and ultimately, the manner. Possible mechanisms by which the oxygen cycle becomes deranged include:

1. Lack of environmental oxygen available for the body to inhale.
2. Inability of oxygen to reach the alveoli.
3. Inability of the oxygen to cross the alveolar wall into the blood.
4. Inability of the hemoglobin to carry the oxygen.
5. Inability of the blood to circulate the oxygen.
6. Inability to unload the oxygen.
7. Inability of the cells to utilize oxygen.

For each of these locations, there are diseases, injuries, or processes that can potentially interfere with the oxygen cycle; the manner of death is predicated on whether the identified underlying cause is a disease or an injury (mechanical or chemical). Derangements of environmental oxygen may be brought about by a drowning or entrapment in airless spaces. The inability of oxygen to reach the alveoli may be caused by an upper airway obstruction (choking on a bolus of food or laryngeal edema due to anaphylaxis, for example). The crossing of oxygen into the pulmonary circulation may be hampered by bronchopneumonia, diffuse alveolar damage (DAD), or amyloidosis. There also may exist an inability of the blood to carry oxygen (e.g., hemoglobinopathies, anemia, methemoglobinemia due to a nitrite ingestion, or carbon monoxide poisoning due to a fire) or to circulate it (e.g., cardiogenic, hypovolemic, or septic shock). Cells poisoned by cyanide may be unable to utilize the oxygen.[14–20] Even the dramatic decapitation death ultimately results from a breakdown in the delivery of oxygen to the brain from the rest of the body.

Due to its common mechanistic role in traumatic deaths, **hypovolemia (exsanguination)** is a common mechanism of death encountered by the forensic pathologist. Generally speaking, to die primarily or even exclusively from exsanguination, an otherwise healthy person typically *rapidly loses 50% of their total blood volume*, not slowly over days or weeks.[21–23] Almost no one will survive a rapid loss of one half of their blood volume, regardless of their degree of health. So, there are *two* factors that must be considered in a suspected exsanguination death: *volume* of loss and the *rate* of loss. The blood volume of an average adult is approximately 65 (female)–75 (male) mL/kg body mass; for children, it is 70–75 mL/kg, for infants, 75–80, and for neonates, 85–90 mL/kg, peaking up to as high as 105 mL/kg by the end of the first month of life, after which it returns to the infant range over the ensuing months.[24,25] If a person slowly loses blood (e.g., an ulcerated colonic cancer), the hematocrit may gradually decrease from 45% to 22%, causing them to be perhaps at higher risk for suffering weakness or a myocardial infarct, but they will likely remain relatively hemodynamically stable, despite the substantial anemia. This is because the body has time to compensate with fluid shifts. Interestingly, a person who dies from losing half of their blood volume quickly would be expected to have a normal hematocrit immediately following the loss. This is because the body had not yet had a chance to compensate with fluid shifts that ultimately "dilute" the red blood cell mass in order to maintain intravascular volume. It also should be noted that not all persons who bleed to death have empty hearts (see below).[26]

All that said, *persons with advanced age and/or* **significant comorbidities** may die following the loss of much lower blood volumes because they have a lower physiologic tolerance. For example, a person with advanced coronary artery disease may depend upon a normal hematocrit or blood pressure to maintain an adequate supply of oxygen to the heart. This is why calculating the total blood volume and determining how much blood was actually lost may be helpful in death certification. If a person rapidly loses 50% of their blood volume, there would not be a need to invoke a comorbidity in part 2. True exsanguination ("bleeding to death") is incompatible with continued life, and it constitutes the injury equivalent of a Class I autopsy finding. If, on the other hand, an individual dies after having lost only 20% of their total blood volume, one may consider including one or more part 2 diagnoses in order to better/more adequately explain the fatal outcome following the relatively smaller blood loss. To be clear, when these persons succumb to loss of lower blood volumes, their deaths would not be classified as "exsanguinations," per se, but rather "hemorrhagic complications" of the culpable injury, with contributions from the natural comorbidities listed in part 2. Keep in mind also that when blood loss volumes cannot be easily measured (e.g., when blood loss is primarily external, into the environment, not into the body), moderate to marked visceral and muscular pallor is a competent and accurate marker of exsanguination-related shock. "The body doesn't lie."

Another question that commonly arises is how to determine whether someone was *alive* at the time of injury (i.e., had circulation). The amount of hemorrhage in response to a given injury is often an important indicator in making that determination. One does not necessarily need to observe a large amount of hemorrhage to call an injury fatal, but one should also not necessarily call any small amount of hemorrhage evidence of circulation; the amount seen should be both contextualized and also sufficient to indicate blood pressure/a vital response (i.e., the injury having occurred "intravitally"). Even a dead body retains a modicum of passive "blood pressure" exerted by the weight of the surrounding tissues and gravitational forces on the intravascular blood. However, it takes a fair amount of "pressure" to push a large amount of blood through tissues or against gravity (i.e., into the head from the heart). Tardieu spots are an example of postmortem extravasated/extravascular passive blood seepage. They demonstrate that "hemorrhage"/blood seepage may occur after death, even into tissues, but are typically in a relatively small quantity. In a person who sustains blunt injury after death (e.g., a dead body is thrown out of a fifth floor window), one would *not* expect to find a large hemothorax, extensive soft tissue hemorrhage surrounding fractures, visceral pallor, or a shrunken/wrinkled spleen. This constellation of reactions constitutes evidence of circulation and indicates an intravital response to the injury. Of note, attempts at cardiopulmonary resuscitation (and even embalming) may cause further postmortem extravasation of blood into tissues and cavities in areas with previously injured blood vessels.

Functional neurologic mechanisms of deaths include **spinal shock** and **fatal concussion/diffuse axonal injury.** One autopsy finding that may point to a subtle/nonstructurally demonstrable fatal head injury is the **"empty heart" sign.**[26] It has been estimated that approximately 15% of deaths due to instantly lethal, structural *or* functional central nervous system trauma have an empty or nearly empty heart at autopsy. If, in a nondecomposed body, the quantity of blood in the pericardial sac after removal of the heart is less than 15 mL (most are associated with 5–10 mL), the heart is classified as "empty." Empty hearts are most typically seen in blunt injury deaths, including motor vehicle collisions and descents from heights, but they also are observed with perforating gunshot wounds of the head or neck. This phenomenon is thought to be a function of an immediate neurogenic shunting of blood to the periphery that occurs contemporaneously with the injury event. As such, exsanguination from vascular injuries with external or internal blood loss precludes a true "empty heart" designation.[27] A bloodless or nearly bloodless heart in those deaths, if observed, is merely a function of the limited intravascular blood volume, rather than a neurogenically mediated shunting of blood, per se.

So, where does the blood go? Autopsy observations indicate unequivocally that the blood is not in the thoracic or abdominal viscera or cavities. Similar to the heart, the major arteries and veins are also typically partially collapsed and nearly empty.[26] Consequently, the autopsy is nearly bloodless and only a small amount of blood pools in the body cavities with evisceration. Also, the blood is not in the skin or subcutaneous tissues because these decedents typically have poorly developed or even barely discernible livor mortis. The blood likely resides in the skeletal muscles and/or skeleton, as the traumatic neurologic death is thought to result in a concomitant and precipitous neurogenic loss of the sympathetic arteriolar tone, one that normally governs circulation in the skeletal muscles and bones.

The detection of an empty heart should focus one's attention on subtle injuries of the brainstem or high spinal cord, if obvious and structurally demonstrable neurotrauma is not otherwise observed (e.g., shattered skull). The documentation of this finding is useful in helping to determine whether concomitant heart disease contributed to the death in a person who has sustained head injury with minimal grossly detectable lethal sequelae. Consider, for example, an unwitnessed accident, in which a 45-year-old man is found dead with some head injury (laceration of the scalp and patchy nonspace-occupying subarachnoid hemorrhage), after having fallen from a ladder. He also has marked arteriosclerotic heart disease. The finding of an empty heart in such a scenario points to a neurogenic rather than a cardiogenic mechanism, allowing the death to be ultimately attributed solely to the head injury.[26]

Competing possible causes of death may present a challenge. In these instances, looking for the **mechanism of death** (often, the clinical history may help) can point to the proximate cause of death. For example, a person with chronic obstructive pulmonary disease (COPD) and coronary artery atherosclerosis presents to the emergency department complaining of shortness of breath, increased sputum production, and wheezing. While being examined, the patient arrests. At autopsy, emphysema and chronic bronchitis are detected, along with advanced three-vessel coronary artery atherosclerosis. How should the death be certified? Alternatively, consider the same patient presenting with crushing substernal chest pain, diaphoresis, and pain in the left arm that started while shoveling snow. While being examined, the patient arrests. At autopsy, emphysema and chronic bronchitis are detected, along with advanced three-vessel coronary artery atherosclerosis. How should the death be certified? In the first instance, based on the history, the lung disease deserves part 1/primary cause status. The heart disease may be included in part 2, as an acute COPD exacerbation may further stress a compromised heart (hypoxemia is arrhythmogenic). Because the lung disease is fueling the most acute process, its placement in part 1 is supported by the dominant mechanism of the death. In the second instance, the history describes a typical sudden cardiac death and the primary cause would be best listed as atherosclerotic coronary artery disease. Chronic bronchitis/pulmonary emphysema, however, may reasonably be included as a (part 2) contributing condition since limited oxygen reserve may play a role in the oxygen supply versus demand mismatch (with its attendant arrhythmogenicity) that may enhance the likelihood of a cardiac event. Admittedly, these distinctions are not critical, per se, in these specific examples of natural fatalities, but the development of a consistent scientific and mechanistic approach to opinion formation will lay the groundwork for habits of forensic reasoning that will likely serve the forensic pathologist (FP) well when later faced with a particularly challenging death certification having much higher stakes, including matters affecting a person's civil liberties.

The above two examples demonstrate *the importance of the clinical history* in identifying mechanisms that will more reasonably and scientifically lead to the primary cause. In both, but for the contextual backdrop, the determinations would not have been possible. One should not do an autopsy in a vacuum and should especially avoid certifying deaths based purely on the autopsy findings, when not obviously lethal, that is. There are some causes of death that are quite self-evident (based on *inclusionary* Class I findings), requiring less in terms of history, while others are based predominantly, if not entirely on history and the *exclusion* of other reasonable causes at autopsy (a Class IV death). As Joseph Davis famously said, "The need for circumstantial details bears an inverse relationship to the severity of disease alterations revealed by autopsy."[28]

Further, it is important to point out the central role the ME/C may play in helping to dispel common misconceptions around the **cause of death** versus **the failure to resuscitate,** as well as in **interpreting the veracity and significance of purported resuscitative artifacts.** The latter is based on the totality of the case context and the constellation of findings. Failure to resuscitate does not represent a cause, but rather the inability to stop a disease or injury having initiated the lethal sequence of events from continuing to do so. This distinction is an important one, as it not uncommonly causes confusion on the part of family members. A delayed ambulance response or an esophageal intubation neither provide causal explanations of the death nor bear culpability for it, per se. Advanced life support measures are emergency procedures that are instituted to either revive a person who is already in cardiac arrest and would remain so without medical or surgical intervention, or who is clearly on a rapidly lethal path unless further measures are instituted (e.g., someone who is rapidly exsanguinating from a ruptured aortic aneurysm).

Along the same lines, but viewed conversely, *assigning blame to resuscitative measures in producing visceral injuries in persons with no other reasonable cause of death can be equally problematic and counter logical.* If, for example, an otherwise healthy child, with no natural disease at autopsy, suffers inflicted injuries, including abdominal blunt trauma that results in a hemodynamically significant hemoperitoneum from a lacerated liver, the suggestion that these findings represent resuscitative artifacts (from chest compressions) begs the question of cause of death. Specifically, if the child did not die from this hemorrhage, for what reason was he in cardiac arrest and requiring said resuscitation in the first place? That is not to say that resuscitative artifacts are not commonly encountered by forensic pathologists, but they also can serve as potential distractions when their significance is conflated or distorted, or their very existence wrongly diagnosed.[29–47]

Although vital statistic bureaus require a single cause on the death certificate, this is not always possible. Deaths are often multifactorial, with contributions from a variety of disease processes and even injuries. Therefore, one must not compromise the proper certification of death to meet an epidemiological request. Vital statistics bureaus employ a variety of codes and measures to accommodate and accurately reflect the multiple causes and conditions that may be included on a death certificate. That said, parts 1 and 2 of the cause of death statement must have scientific/mechanistic veracity and are not to be used as a provisional anatomic diagnosis list, representing an exhaustive list of all identified gross or historical conditions, regardless of their causal association with death.

Vital statistics bureaus only have access to the information on the death certificate. Therefore, it is important that the certificates are properly and competently completed from the start. Unfortunately, there is much room for improvement in the quality of death certification in the United States. For example, cardiopulmonary arrest is not a cause of death, yet it is still the sole "cause" listed on countless death certificates. It does not belong on a death certificate alone, as it is a synonym for death and medical jargon that simply means that the heart and lungs have stopped. The heart and lungs ultimately stop in every death. The question is, what caused the heart and lungs to cease functioning? There is rarely a benefit in including "cardiopulmonary arrest" in the cause of death statement since it is assumed that if a death certificate is being completed, the person is indeed dead. Because it can be placed on every death certificate, it adds nothing in explaining *why* the person died.

Death certification

The temptation to make an epicrisis of the death certificate must be resisted.

Completing a proper death certificate is both a skill and an art. It requires logic, patience, all necessary medical studies, and, of course, a proximate cause of death. A cause of death should not represent a comprehensive recitation of all diseases and injuries (i.e., an epicrisis), but rather a concise description of why a person died. There is much to learn about the science and art of death certification. There are an infinite number of permutations and variations of deaths; however, over 90% have a fairly straightforward cause. The other 10% require much more thought and consideration. Even after decades of practice, we still get those cases that make us go, "hmm" and scratch our heads. Having a solid foundation and consistent approach allows one to work through those unusual puzzles. It is also important to keep in mind the needs of the varied audiences (vital records, families, treating physicians, public health and child and adult protective service agencies, insurance companies, lawyers, and law enforcement) who will be reading and interpreting our death certificates. Likely fewer than 10% result in energetic discussions and differences of opinion among competent forensic pathologists; this is even more common in discussions of manner of death.

Remember, both the cause and manner of death are opinions based on evidence-based probability. Every case is unique and must be examined on its own merits, using the principles described here. In addition, the forensic pathologist in training would be well advised to read and reread the 1995 ASCP Check Sample, *Problem-Solving in Death Certification*, by Charles Hirsch and Mark Flomenbaum.[48] This constitutes essential reading in death certification practices, setting the stage for many important concepts that will be repeatedly encountered and should ideally be introduced early in fellowship training (see Appendix D for the required reading list for trainees).

The death certificate is a legal document that memorializes a medical opinion. Although it is not legally binding (just ask an insurance company), when signing a death certificate, one should consider doing so signing "under oath," as it is an administrative medicolegal opinion that may have far reaching effects. Approximately 1% of the population dies per year.[49] In 2018, 1,317,730 deaths were reported to ME/C offices and 604,700 (45.9%) were accepted for further investigation.[50] Therefore, in a given population of 1,000,000, for example, there will be 10,000 deaths per year, and roughly 4500 will be accepted for ME/C investigation; approximately one third of those will undergo a postmortem examination (external examinations and autopsies combined).[50] One national group (Scientific Working Group for Medicolegal Death Investigation) discussed a goal of one autopsy per thousand population annually.[51] In fact, most offices are operating below that level, yet may still be meeting service expectations. A goal of 0.70 autopsies per 1000 population may be more realistic.[51,52] Therefore, as an example, for a population of 1,000,000, the ideal number of autopsies to be performed would be 1000; however, 700 may suffice.

For most deaths, the burden of proof for death certification is a "probability" (>50%, or more likely than not). But for some, namely homicides, the burden is substantially higher: probability beyond a reasonable doubt (so-called "reasonable degree of medical or scientific certainty"). This degree of certainty indicates that all other *reasonable* causes/manners of death have been excluded. This stringent burden is used to protect the civil liberties of those accused. These certifications may result in trials that put people in prison for the rest of their lives or may result in the death penalty. That said, the burden is never "certainty beyond a shadow of a doubt (100% certainty)," even in homicides, as this would present an insurmountable hurdle that is not required in any jurisdiction or court in the United States. See Chapter 2 for a further burden-of-proof discussion.

Although a person's freedom or life may not be hanging in the balance with a suicide determination, many forensic pathologists require a higher burden of proof in certifying suicides (beyond just a probability). Suicides can be equally, and sometimes more contentious and emotionally charged than homicides, often having substantial financial and potential personal fall out (fueled by insurance claims, religious beliefs, guilt, and denial). This may be so because they are the only manner of death requiring the demonstration of intent, a burden which is not usually required even for a homicide determination. Forensic pathologists rarely get "sued," but classifications of suicide are one designation for which litigation may occasionally ensue.[53]

Unfortunately, however, death certification policies and practices are not uniform in the United States or within those states not having a single, centralized state medical examiner agency having the ability to maintain uniformity. For example, in some coroner jurisdictions, a forensic pathologist may be relied upon to provide the cause of death, while the coroner, who has the ultimate responsibility of actually overseeing the investigation and of completing the death certificate, may or may not be a physician, depending on the jurisdiction, and is responsible for the manner. This has led to criticisms of the coroner system in some parts of the country. There are many deaths in which a medical understanding of the cause of death will impact the manner of death. Two such examples include drug intoxications and infant head injuries. The interpretation of postmortem toxicology results (see Chapter 14) requires scientific expertise, owing to the variety of pitfalls that may be encountered and the importance of context (anatomic, physiologic, and circumstantial). One must understand how the postmortem drug concentration may be affected by the postmortem interval, the sample tested (postmortem redistribution), and survival intervals, as well as issues of physiologic tolerance, bioaccumulation (e.g., in cirrhotics), potentiation by multiple drugs, the caveats of interpreting gastric and vitreous humor concentrations, the concept of efficient intervening cause, and view the results through the lens of context, such as the circumstances just preceding death (consciousness? obtunded?).[54–78] Interpretation of infant head injuries largely relies upon a correlation between the reputed method and timing of the injury compared to the developmental age, as well as the pathophysiology reflected by the autopsy findings.[79–89]

Some differences in death certification are matters of style, over which we express preferences, but are not dogmatic. Other differences, however, are matters of substance and we will explain why that may challenge the forensic pathology fellow. Common mistakes in death certification will also be addressed in later chapters, which delve more deeply into death certification regarding the cause and manner of death in a variety of circumstances encountered by the forensic pathologist.

Cause of death

The cause of death section on the death certificate has two areas (Fig. 3.3). **Part 1** is for the cause of death and **part 2** is reserved for contributing conditions not resulting in the cause listed in part 1.[90] There may not always be a part 2

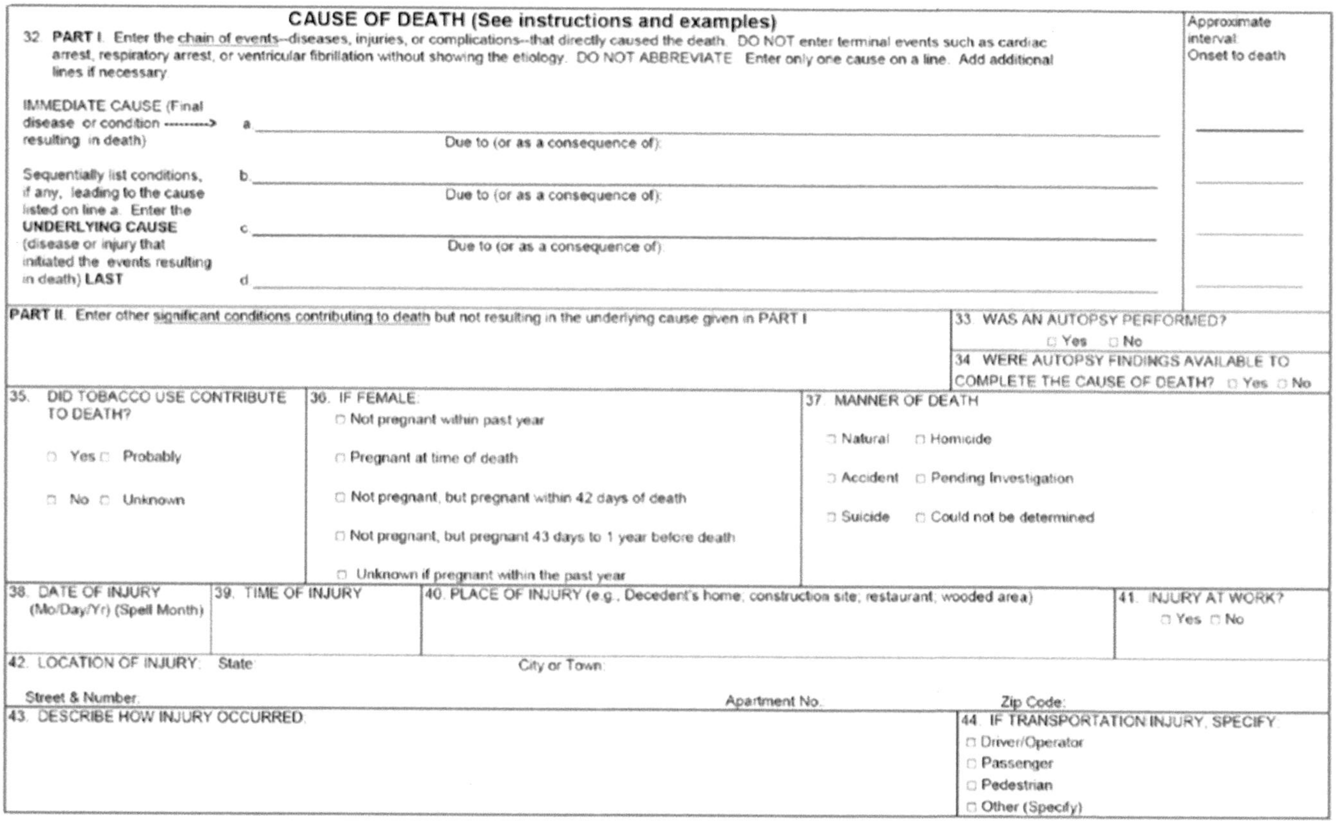
CAUSE OF DEATH (See instructions and examples)

32. PART I. Enter the chain of events--diseases, injuries, or complications--that directly caused the death. DO NOT enter terminal events such as cardiac arrest, respiratory arrest, or ventricular fibrillation without showing the etiology. DO NOT ABBREVIATE. Enter only one cause on a line. Add additional lines if necessary.

Approximate interval: Onset to death

IMMEDIATE CAUSE (Final disease or condition ---------> resulting in death) a. ______ Due to (or as a consequence of):

Sequentially list conditions, if any, leading to the cause listed on line a. Enter the UNDERLYING CAUSE (disease or injury that initiated the events resulting in death) LAST b. ______ Due to (or as a consequence of): c. ______ Due to (or as a consequence of): d. ______

PART II. Enter other significant conditions contributing to death but not resulting in the underlying cause given in PART I

33. WAS AN AUTOPSY PERFORMED? □ Yes □ No

34. WERE AUTOPSY FINDINGS AVAILABLE TO COMPLETE THE CAUSE OF DEATH? □ Yes □ No

35. DID TOBACCO USE CONTRIBUTE TO DEATH? □ Yes □ Probably □ No □ Unknown

36. IF FEMALE: □ Not pregnant within past year □ Pregnant at time of death □ Not pregnant, but pregnant within 42 days of death □ Not pregnant, but pregnant 43 days to 1 year before death □ Unknown if pregnant within the past year

37. MANNER OF DEATH □ Natural □ Homicide □ Accident □ Pending Investigation □ Suicide □ Could not be determined

38. DATE OF INJURY (Mo/Day/Yr) (Spell Month)

39. TIME OF INJURY

40. PLACE OF INJURY (e.g., Decedent's home; construction site; restaurant; wooded area)

41. INJURY AT WORK? □ Yes □ No

42. LOCATION OF INJURY: State: City or Town: Street & Number: Apartment No.: Zip Code:

43. DESCRIBE HOW INJURY OCCURRED:

44. IF TRANSPORTATION INJURY, SPECIFY: □ Driver/Operator □ Passenger □ Pedestrian □ Other (Specify)

FIGURE 3.3 **Standard US death certificate.**

TABLE 3.1 Definitions: causes, mechanism, manner.

Cause of death: The disease, injury, or combination of disease and injury responsible for the fatality.
Proximate cause of death (underlying): The disease or injury that in a natural and continuous sequence, unbroken by any efficient intervening cause, produces the death and without which the death would not have occurred; it must be an etiologically specific disease or injury.
Immediate causes of death: The complications of the underlying cause interposed between proximate cause and death. There may be one or more immediate causes, which may occur over a brief or prolonged interval, but none absolves the underlying cause of its ultimate responsibility. Immediate causes usually are not etiologically specific and therefore cannot stand by themselves on the death certificate.
Mechanisms of death: The alterations of physiology and biochemistry whereby the causes exert their lethal effects (e.g., DIC, sepsis, cardiac arrhythmias, congestive heart failure, asphyxiation, and exsanguination). They may due to a wide variety of diseases and injuries and are never etiologically specific.
Manner of death: The manner of death is an explanation of how the cause arose, either natural or unnatural, and is based largely on the circumstances, but is informed by the proximate cause.

(neither necessary nor relevant in many deaths). The **cause of death** is **the disease and/or injury responsible for the lethal chain of events that resulted in death** (Table 3.1). It may be broken into three components: mechanisms, immediate causes, and the underlying (proximate) cause of death. A little known "secret" that many people are not aware of is that the underlying cause (proximate cause) of death is all that is required in part 1 of the death certificate; the other two components are included only if known or particularly relevant or informative in the certification. The coders of vital records, however, tend to want more rather than less information. Regardless, and despite a suggestion to the contrary by virtue of the construction of part 1, *one is not required to write something on each line of the death certificate;* the proximate cause may be put on the top line, with nothing else listed (Table 3.2).

So, how do we define these three components of the cause of death? **Mechanisms** are **nonspecific physiologic or biochemical alterations that link the cause of death with the moment of death**. They include such derangements as exsanguination, hypoxia, asphyxia, metabolic acidosis, DIC, sepsis, heart failure, and arrhythmias. They are nonspecific and therefore cannot stand alone since a variety of diseases and injuries may cause death through these mechanisms. Stated differently, they are not *etiologically specific*. For example, a person may exsanguinate from a stab wound of the heart or a duodenal ulcer that has eroded into the gastroduodenal artery. To reach the proximate cause, one must ask what caused the exsanguination (i.e., what is its proximate cause?).

Some deaths have a clear mechanism of death, and this can be useful when there are competing causes of death (Table 3.3). For example, consider a person who has advanced emphysema, lung adenocarcinoma, and marked coronary artery disease at autopsy. Is the death from the emphysema, the lung cancer, the heart disease, or some combination thereof? The mechanism of death either identified at autopsy or inferred from the clinical history can help guide the forensic pathologist in selecting the most accurate underlying/primary cause of death. For example, if the lung cancer resulted in a postobstructive bronchopneumonia and sepsis, the cancer would be the proximate cause. If, however, the patient presented with chest pain, diaphoresis, and electrocardiogram changes caused by an acute myocardial infarct, with no infection at autopsy, the underlying cause of death would be atherosclerotic cardiovascular disease, with the emphysema as a contributing condition. Depending upon the extent of the malignancy and its interference with pulmonary function, the lung cancer also might be included as a contributing condition. On the other hand, a small peripheral adenocarcinoma in the lung may represent merely an incidental finding and should not be included on the death certificate. That said, some jurisdictions do have a separate box for reporting cancer for vital statistical purposes, regardless of whether it played a role in the death. While thought processes such as the ones demonstrated above form the foundation for identifying the appropriate proximate cause, this exercise is relatively academic when it does not affect the manner of death. When the choice of the proximate cause changes the manner, on the other hand, these conceptual considerations take on greater importance.

TABLE 3.2 Common proximate causes of death.

Common Causes of Death
Natural
Hypertensive cardiovascular disease
Arteriosclerotic cardiovascular disease
Hypertensive and atherosclerotic cardiovascular disease
Asthmatic bronchitis/acute and chronic bronchial asthma
Chronic bronchitis/pulmonary emphysema
Lobar pneumonia (NEVER bronchopneumonia alone)
Hyperglycemia and ketoacidosis due to diabetes mellitus/diabetic ketoacidosis
Complications of malnutrition due to inanition in a person with cognitive disability (uunspecified)
Obesity
Ruptured cerebral artery aneurysm
Anomalous origin of the left coronary artery.
Pulmonary thromboembolism due to deep venous thrombosis complicating … (disease or injury)
Complications of senile dementia of the alzheimer-type (without neuropathology diagnosis)
Fire, ice, and air
Hanging
Asphyxiation due to covering of head with fastened plastic bag
Asphyxia due to choking on bolus of food due to (multiple sclerosis, ethanol intoxication, etc.)
Drowning by submersion in bathtub due to epilepsy/drowning due to epilepsy ("found submerged in bathtub" in how injury occurred field)
Thermal injuries of approximately 35% body surface area and inhalation of products of combustion
Hypothermia due to environmental exposure to cold
Substance use/toxicological causes
Acute intoxication by to the combined effects of …
Bacterial endocarditis due to chronic intravenous drug use disorder
Acute and chronic alcoholism
Additive effects of/adverse reaction to … for treatment of …
Blunt injury
Fractures and visceral injuries due to blunt impacts(s) of head, trunk, and extremities
Complications of persistent vegetative state due to remote blunt impact injury of head
Cardiac arrhythmia due to blunt impact of chest (commotio cordis)
Concussion of cervical spinal cord due to craniocervical dislocation due to blunt impact to head

Continued

TABLE 3.2 Common proximate causes of death.—cont'd

Guns, knives, and other homicides
Stab wound(s) of _ w/perforations/penetrations of …
Gunshot wound of head with perforations/penetrations of …
Sudden death while bound and restrained during robbery due to arteriosclerotic heart disease
Children
SUID (undetermined)
Fatal child abuse syndrome with …

TABLE 3.3 Examples of mechanisms, immediate causes, and proximate (underlying) causes.

Mechanism (not always listed on the death certificate)	Immediate cause	Proximate cause
Gastrointestinal hemorrhage	Cirrhosis with ruptured esophageal varices	Chronic alcohol use disorder
Metabolic acidosis	Acute pancreatitis	Cholelithiasis with impacted gallstone
Pulmonary thromboembolism	Deep venous thrombosis	Pancreatic carcinoma
Respiratory insufficiency	Bronchopneumonia	Novel COVID-19 respiratory viral infection
Cardiac arrhythmia	Aortic stenosis	Rheumatic heart disease
Cardiac arrhythmia	Coronary artery thrombosis	Atherosclerotic coronary artery disease
Uremia	Kidney failure	Diabetes mellitus
Anaphylaxis	Treatment with penicillin	*Streptococcal* pharyngitis
Cardiac arrhythmia	Stress related to physical altercation	Hypertensive and atherosclerotic cardiovascular disease

Mechanisms are *not* required on the death certificate, but, when structurally demonstrable, they can sometimes be informative for families and clinicians; so, in certain instances, they are best included to enhance communication. For example, a chronic alcohol user with cirrhosis of the liver dies in the hospital from an upper gastrointestinal hemorrhage due to ruptured esophageal varices. The mechanism would be exsanguination, and the immediate cause of death would be the ruptured varices. While it would be acceptable and technically accurate to list "chronic alcohol use disorder" by itself as the competent proximate cause of death on the certificate, it would not be optimally communicative. This certificate would not be as informative as one listing the other clearly lethal and structurally demonstrable complications of the chronic alcohol use. It also might cause confusion for the family, who do not understand the pathophysiologic link of the mechanism to the cause and believe the death was simply due to "bleeding from the stomach." On the other hand, when the mechanism is reasonably deduced, but is functional, not structural in nature (e.g., an arrhythmia), it is less common practice to specifically list it on the death certificate unless necessary in the context of the case, allowing the proximate cause to stand alone (e.g., when a witnessed sudden death is caused by hypertensive and atherosclerotic disease).

In the above case, the immediate cause is ruptured esophageal varices. **Immediate causes** of death are **complications of the underlying cause interposed between the proximate cause and the fatal result (mechanism).** Not every death has an identifiable immediate cause. For death certification purposes, *immediate causes and mechanisms are essentially interchangeable*, but neither should be a substitute for an etiologically specific underlying cause. An example of all three is exemplified in the following cause of death statement: hemopericardium with cardiac tamponade (mechanism) due to ruptured myocardial infarct (immediate cause) due to arteriosclerotic heart disease (underlying/proximate cause).

The **proximate (or underlying) cause of death** is defined as **that which in a natural and continuous sequence, unbroken by any efficient intervening cause, produces the death and without which the death would not have occurred.** The underlying cause of death *must be an etiologically specific disease or injury.* The latter can be either *structurally demonstrable/anatomic* (i.e., blunt impact injury with a subdural hematoma) or *functional and descriptive* (e.g., an arrhythmia due to hypertensive and atherosclerotic disease in the setting of the pathophysiologic stress resulting from a physical altercation). Nearly all death certificates require a competent (etiologically specific) proximate cause. **There are six deaths, however, without a proximate cause of death** (Table 3.4). The first is a complication due to elective cosmetic plastic surgery that is performed to remedy no identifiable disease or injury. Wrinkles due to aging or a large nose are not codable diseases. The second is a maternal death resulting from a complication, such as one related to anesthesia, for example, occurring during the labor and delivery or elective Cesarean section in an otherwise healthy mother (a healthy pregnancy is not a disease). A third is a complication associated with elective organ donation in which the living donor dies from a complication of the surgery to remove a healthy organ. The fourth is a complication from a screening procedure (e.g., colonoscopy). The last two involve preventive medicine: complications of birth control methods and of vaccinations. That said, these death certifications may include (part 2) contributing conditions. There are also instances in which an underlying cause of death likely exists, but cannot be specifically determined. These represent examples of the limitations of our scientific knowledge and diagnostic capabilities and not investigative failings. Usually, these deaths are certified as "undetermined" or "sudden death of undetermined etiology," with a *natural* manner. Every ME/C office should have some undetermined deaths (usually fewer than 5% of the deaths investigated). As Hirsch and Zumwalt noted, "Honesty is the only stigma that attaches to a pathologist who leaves an occasional case undetermined."[91]

Part 2 of the death certificate is reserved for conditions that contribute to death, but do not result in the underlying cause listed in part 1. These also are known as *"other significant conditions."* They do not need to have etiologic specificity, however, if they are an injury or the result of an injury, the manner of death should reflect this, no matter how minor the contribution (natural deaths must be 100%, exclusively natural). Part 2 should not be used for interesting or unusual findings, unless they make a causal contribution to death, no matter how minor. Often, and appropriately, part 2 is left blank (i.e., it is not required for competent death certification).

In natural deaths, it is very common practice to place other natural diseases/comorbidities/conditions in part 2 (diabetes mellitus [DM], obesity, emphysema, thrombocytopenia, cirrhosis, atrial fibrillation, etc.), without further qualification. That said, part 2 is *not* intended to be blithely populated with a laundry list of entities having no

TABLE 3.4 Six deaths that do not have a proximate cause.

Type	Example of cause of death statement
Cosmetic surgery (nonreconstructive)	Pulmonary thromboembolism due to deep venous thrombosis complicating cosmetic rhytidectomy and rhinoplasty
Maternal death during childbirth	Amniotic fluid embolism complicating (normal) full-term pregnancy; complication of anesthesia for elective c-section
Living-organ donor complications	Hemorrhagic complications during nephrectomy for living-organ donation
Screening colonoscopy	Peritonitis due to perforation of nondiseased large bowel during screening colonoscopy
Birth control complications	Sepsis complicating intrauterine device insertion
Vaccination complications	Anaphylaxis following administration of influenza vaccine

reasonable contributory lethal potential, particularly if including a condition would change the manner from a natural death to a therapeutic complication (TC) or an unnatural manner. See also Chapter 2 for additional discussion on this topic.

Finally, two additional concepts in death certification deserve special mention. On a daily or near daily basis, it seems, the issue of **"style versus substance"** arises in conversations and invariably surrounds differences in the approach to certifying a particular death. Differences of opinion are as inevitable as they are unavoidable, particularly within the practice of forensic pathology. Dr. Hirsch was fond of recognizing this inevitability by suggesting, "Let us agree to disagree, but let's not be disagreeable."

With regard to our trainees, however, for whom we are responsible for laying the groundwork of certain unyielding practice principles, our focus must be overwhelmingly on *substance*—a constant drumbeat of foundational and nonnegotiable death certification concepts: the importance of etiologic specificity, the requisite proximate cause (with rare exceptions—see above), the importance of thinking mechanistically, of understanding burdens of proof by manner of death, and of sticking to the facts and evidence as we know them. On the flip side of that are stylistic differences, an expected manifestation born out of our inherently different training, personalities, and preferences. How Dr. X might turn a phrase relative to Dr. Y (e.g., one may use as few words as are required and the other certifier with as many words strung together as possible) is not as important as the elemental meaning of their words, the specific facts, and conclusions that form the basis of their opinions. That is to say, if the death is "coded" the same way by vital records, then one is dealing with style over substance. There are minimalists, there are lumpers, and there are, of course, splitters. In all likelihood, by the end of one's training year, after ambiently taking in a variety of styles, a fellow will emerge, having found a stylistic voice that makes sense to them, hopefully and importantly, within the greater backdrop of forensic substance, the latter allowing much less flexibility, by design.

What is critically important, however, is that we know the difference between the two. Stylistic differences are allowable, substantive ones less so. In our experience, however, forensic pathologists may sometimes unwittingly ascribe a strong opinion to their "style," when, if fact, it represents more a matter of substance. The better we are at defining the parameters and importance of the substance of forensic pathology practice and its underpinnings, the less tempted we may all be to hide behind "personal preferences" that are actually consequential. In the end, being grounded in basic forensic principles should help us to avoid the inconsistencies that serve as fertile soil for arbitrary and capricious decision making.

The last point focuses on making an exception to a death certification "rule" ("If you know the rules well enough, you are allowed to break them occasionally"). While we assiduously try to avoid redundancy in both death certification and report writing, we occasionally must bend the rules in order to achieve a more important goal. The death certificate is a legal document that broadcasts an opinion based, by definition, on *probability*. As such, to include **the word "probable"** in a cause of death statement seems patently redundant, in perhaps the same way that including "cardiac arrest" may seem similarly unnecessary. That said, there are specific scenarios in which each of these phrases might, by necessity, have a place on the death certificate.

On occasion, in a death involving an apparent chemical intoxication, for example, we are limited by the lack of available blood to submit for toxicologic analysis (due to a prolonged survival interval, sample deterioration, or because the method of the testing we wish to perform is otherwise limited). That said, the investigative and autopsy findings (e.g., sodium nitrite and a suicide note at the death scene, in combination with gray-green lividity at autopsy) are so compelling for a specific cause and manner of death, combined with the exclusion of other causes, that we feel confident in asserting an opinion, despite the limitation of not having confirmatory toxicological evidence. We choose to not "deny reality in the name of objectivity." Despite our confidence in making the call, however, we may wish to deliberately "flag" the death certificate, not at all to convey any ambivalence, per se, but rather to show that while we are aware that no toxicologic testing was performed, we are no less sanguine in our opinion. A mechanism for this disclosure is achieved by use of the word "probable" before "sodium nitrite poisoning." That one word serves to highlight our acknowledgment of the limitations of the investigation to anyone who may wonder if we noticed, but also indicates that, despite knowledge of that limitation, the death could still be competently certified. "Probable" serves as a semantic nod, a code that conveys that, yes, we are aware, but we are letting common sense prevail.

"Pending" death certificates

In ME/C practice, it is common to issue a death certificate that has a cause and manner of death listed as "pending" or "pending further study." Most require additional laboratory and/or investigative information (e.g., toxicology testing, histology, medical record review, a police investigation, etc.) before the death can be certified. These additional investigations may take weeks to even months to complete. In order to allow the family to proceed with timely funeral arrangements, however, the body may be released with a "pending" certificate. This allows the body to be cremated or buried because the autopsy provided the opportunity to collect any necessary fluid or tissue samples, removing the need to retain it further.

Once the additional testing and/or investigating is completed, a final/amended death certificate may be issued, listing the ultimate cause and manner of death opinion. In the interim, a pending death certificate may cause some inconvenience for the family with regard to insurance claims, but it does serve as a proof of death, so can be used for matters involving banking, social security, etc. Some insurance companies will not pay a benefit on a claim until they have the final certificate listing the definitive cause and manner of death. This may be related to the need to exclude a homicide involving the beneficiary as a suspect, a suicide clause, or a double indemnity accidental clause. Many insurance policies have exclusions for a suicide having occurred within the first 6 or 12 months of the purchase of the policy. In some instances, the insurance company just needs to know that there is no concern for homicidal or otherwise suspicious/criminal activity on the part of the beneficiary. As such, depending on the policy, a letter or verbal response from the ME/C stating, "there is no concern for foul play" (the so-called **"no foul play" letter**) may be enough to allow the insurance company to process the claim and pay the benefit. See the previous Chapter 2 for further discussion on this topic.

Common death certificate mistakes

Common mistakes in death certification include: not listing a proximate cause, the "cart before the horse" error, misuse of part 2, listing more than one etiologically specific cause of death, a hodgepodge/laundry list of numerous unrelated or noncausal diagnoses, and the inclusion of terminology relating to how an injury occurred in the cause of death statement (Table 3.5). Numerous studies have demonstrated a considerable number of errors on routine hospital death certificates in both public and academic medical centers.[92–110] These errors range from not including a proximate cause of death, the misuse of part 2, to logic errors.

A common example of *not including an underlying cause of death* is the isolated listing of bronchopneumonia. **Bronchopneumonia**, except in a young infant or those of advanced age, is invariably a complication of another disease process. It is *not etiologically specific*, as many deaths from a variety of both natural and unnatural proximate causes can ultimately result in a terminal bronchopneumonia (immediate cause). The question is, what caused the bronchopneumonia? Failing to ask this question may lead to the grievous oversight of a traumatic proximate cause and manner. Common causes of bronchopneumonia include viral influenza, pulmonary emphysema, lung cancer, postoperative complications, dementia, and relative immobilization/being bedbound from a variety of diseases or injuries. By contrast, **lobar pneumonia** *is a different disease process and may stand alone on the death certificate.* An

TABLE 3.5 Common death certificate mistakes.

Error	Example (correction)
No proximate cause	Chronic renal failure (due to diabetes mellitus)
Incorrect order	Atherosclerosis due to myocardial infarct (order should be reversed)
Hodgepodge/laundry list	Hypercalcemia due to pulmonary embolism due to hypertensive cardiovascular disease due to pancreatic carcinoma (pulmonary thromboembolism due to pancreatic carcinoma)
Addition of injury information in cause	Blunt head injury due to motor vehicle collision (blunt head injury)
Use of death synonyms	Cardiopulmonary arrest (arteriosclerotic cardiovascular disease)

example of a typical lobar pneumonia is pneumococcal pneumonia, associated with a consolidated lobe that anyone may acquire. It is unusual to die from it, but when it happens, lobar pneumonia does not require an underlying cause on the death certificate.

The proximate cause is often forgotten in therapeutic complication **(TC) deaths.** For example, a person may die from anaphylaxis due to penicillin therapy, but penicillin therapy is not a codable disease or injury and therefore does not qualify as a proximate cause. The underlying cause is the disease or injury that the penicillin was administered to treat (i.e., the *indication* for the therapy). *On the death certificate,* **TC deaths require three data points***: the complication(s), the therapy or diagnostic procedure, and the underlying disease for which the treatment or diagnostic procedure was administered.* Deaths from complications of therapy involving an injury as the proximate cause should have a manner of death that corresponds to the circumstances of the *injury.* That is, even if the therapeutic event is listed in the cause of death statement, the injury manner will "trump" TC as the manner in these deaths (i.e., such fatalities would not be classified as TC deaths, per se). For jurisdictions that do not have the TC option, TC deaths are typically certified with a natural or accidental manner. The problem with an accidental manner is that it may inadvertently raise the specter of medical malpractice, which may or may not be relevant. The problem with certifying the death as natural, on the other hand, is that it may give the appearance of minimizing or ignoring the complication. Natural deaths are typically due 100% to disease or old age, so, if an external contributing factor plays a role in the death, the manner is not certified as natural.

Since the proximate cause must be a codable disease or injury, not a procedure or nonetiologically specific complication of that procedure, a TC cause of death is typically framed as follows:

(Complication) of/during/following
(the procedure) for the treatment/evaluation of
(proximate cause: etiologically specific **disease**).
For example,

Complications of small bowel obstruction due to adhesions related to remote total abdominal hysterectomy with bilateral salpingo-oophorectomy for the treatment of menometrorrhagia due to uterine leiomyomata

Alternatively, if the underlying cause for the remote hysterectomy was blunt force trauma, which had led to a uterine laceration (how injury occurred: motor vehicle collision), the manner would be accident rather than TC. Proximate cause informs manner and **the proximate cause must be a disease in order to classify a death as a TC. In addition, the therapy/procedure listed on the death certificate of a TC must be <u>elective</u>,** *not* an emergent and potentially life-saving one, wherein death would have overwhelmingly resulted regardless of the intervention (see Chapter 8 for a more detailed discussion of therapy-related deaths).

The cart before the horse error involves listing the proximate cause first, followed by the immediate cause. An example would be "atherosclerotic coronary artery disease due to myocardial infarct." Some clinicians who fill out death certificates mistakenly think that the most important diagnosis should be listed first. In fact, the layout of the death certificate has "due to or as a consequence of" between each line. So, the death certificate in this instance should be certified as, "myocardial infarct due to atherosclerotic coronary artery disease," the bottom line containing the proximate/underlying cause.

Another common error involves *the misuse of part 2.* For example, a person dies from a gunshot wound of the head. At autopsy, the prosector discovers a large renal cell carcinoma that had previously gone undiagnosed. What better way to share your discovery with the world than to put it on the death certificate? But this is an improper use of part 2. The renal cell carcinoma should be described in the body of the autopsy report and memorialized in the final diagnoses; however, since it did not cause or contribute to the death in any way, it does not belong on the death certificate.

<u>*Injurious or therapy-related conditions listed in part 2 have the potential to change the manner of death, so must be included only if they need to be invoked, after thoughtful consideration.*</u> Thoughtlessly (and unnecessarily) adding *noncontributory* part 2 diseases or injuries in homicide cases and/or when doing so changes the manner of death is frowned upon. In such circumstances, part 2 is best used only after thoughtful and evidence-based consideration and *only* if it contributes to death *and* when the primary cause of death is insufficient by itself to explain the death. Avoid being arbitrary and capricious. That said, adding various medical conditions (i.e., natural diseases and conditions) that "aren't helping" and are reasonably mechanistically related to the fatal outcome is a common and acceptable practice in *natural* deaths (common examples: DM, obesity, emphysema, etc.). Finally, part 2 conditions need not be etiologically specific (e.g., "cirrhosis," or "anticoagulation therapy" can stand alone in part 2 *without* the qualifier "for the treatment

of atrial fibrillation"), although, admittedly, the latter would change a natural manner to a TC in jurisdictions where that designation is an option, so thoughtful consideration is still in order.

The general rule is that in part 1, there is *one (proximate) cause of death per person*. A commonly encountered exception to that rule, however, involves the designation of **"hypertensive and atherosclerotic cardiovascular disease"** in a clear arrhythmic death without an acute thrombus or myocardial infarct. This recognizes the fact that both disease processes are capable of causing a sudden death and that, more often than not, these diseases coexist as dovetailed pathophysiologic processes. Beyond that common pairing, however, the rule is usually one cause of death per person. An example of an error of this type would be a death certificate listing, "myocardial infarct due to DM and atherosclerosis." A better certification might be "myocardial infarct due to atherosclerotic cardiovascular disease," with DM listed in part 2.

Careful examination of the criteria for part 2 may indicate that a rule has been broken by listing diabetes mellitus as a contributory factor since the definition for part 2 states, "other significant conditions contributing to death, but not resulting in the underlying cause." We know that DM is a risk factor for atherosclerosis so, in that sense, it does contribute to atherosclerosis. Regardless, it is fairly common practice to include DM as a part 2 condition in deaths due to cardiovascular causes. Occasionally, such as in deaths related to diabetic ketoacidosis, DM is the primary, stand-alone proximate cause (see Chapter 7 for a more in-depth discussion of DM as a cause of death).

The so-called *hodgepodge death certificate* is characterized by a random listing of a variety of clinical diagnosis without any logical pathophysiological connection. Essentially, the certifier is merely regurgitating on paper a problems list without further consideration of their lethal potential, the causal relationships between the entities, or the mechanism of death. For example, a hodgepodge certification might be, "hypercalcemia due to pancreatic carcinoma due to hypertensive cardiovascular disease," with a contributing (part 2) condition is listed as "pulmonary embolism." Clearly, this is a listing of conditions with some degree of connection, but in no logical sequence. Hypercalcemia and PTE may be complications of pancreatic carcinoma, but hypertensive cardiovascular disease does not cause pancreatic carcinoma. This death certificate could easily be reworked to indicate a death due to PTE due to deep venous thrombosis (immediate cause) due to pancreatic adenocarcinoma (proximate cause). Part 2 could reasonably include the hypertensive cardiovascular disease, but "hypercalcemia" should be removed altogether since, while it may accurately reflect one of the patients' conditions at the moment of death, it likely had no mechanistic relationship to the development of a pulmonary embolism or to his demise, in general.

The **Virchow triad** describes the risk factors for thromboembolic disease. The first is a **hypercoagulable state**, which may result from malignancies, such as pancreatic carcinoma. The second is **injury of the vessel**. The third is **stasis**, which may be seen with obesity, relatively inactivity or immobility following injury or surgery, and/or diseases involving the heart, in which there is poor cardiac output. So, in this instance, the pancreatic carcinoma would be the dominant cause and risk factor for the hypercoagulable state. The hypertensive cardiovascular disease, depending on its extent, may have a contributing role in stasis and so it can be reasonably included in part 2. Note that although the part 2 condition of hypertensive cardiovascular disease may be a contributing factor to the immediate cause (deep vein thrombosis), it is not contributory to the proximate cause (pancreatic adenocarcinoma), and so does not defy the part 2 rule that excludes conditions that directly result in the underlying cause. *Note also the importance of identifying the proximate cause in PTE deaths since, depending on the underlying cause, the manner of death may be unnatural.* For example, a death due to PTE in a bedbound, quadriplegic man who had been rendered so after having been shot in the cervical spine 20 years before his death, would be certified as a homicide, even if there were other factors, such as obesity, that contributed to his thrombotic risk. We will elaborate further on this specific discussion in Chapter 15, but suffice to say that we cannot emphasize more strongly that proximate cause informs manner and that the mechanism informs the proximate cause.

One must be careful in *not letting the tail wag the dog*, that is, not listing the terminal problem in part 1 and the major underlying substrate in part 2. For example, a person with advanced heart disease sustains a hip fracture from an accidental fall. Following a successful surgery, the patient goes into atrial fibrillation three days postoperatively and is transferred to the medicine service, and a few days later, goes into congestive heart failure, develops a bronchopneumonia, and dies. In this instance, the heart disease is best listed in part 1 and the hip fracture, in part 2. Some may use a descriptive cause of death and include it all in part 1 (e.g., "cardiac complications following surgery for femur fracture due to blunt injury in patient with atherosclerotic cardiovascular disease"). In either instance, the manner of death would be accident. A similar scenario arises when there is a complication of a blood transfusion reaction that most healthy people would survive, but the patient has marked coronary disease and end-stage renal disease. In this example, the transfusion might be demoted to part 2 and the ASCVD to part 1. The manner in these transfusion-related deaths would, of course, be predicated by the proximate cause/the etiologically specific indication for the blood replacement therapy.

Another common death certification error involves *listing how an injury occurred in the cause of death statement.* For example, consider a person who dies in a motor vehicle collision, with extensive blunt injury identified at autopsy. The cause of death is blunt impact injuries, period; however, some may erroneously list "blunt injuries *due to motor vehicle collision*" as the cause. A motor vehicle collision is an *event*, a circumstance; it is neither a disease nor an injury, so it should not be listed as a cause of death of any sort, immediate or proximate. Rather, when a death results from a mechanical or chemical injury, there is a separate space on the death certificate for documentation of **"how injury occurred"** and that is where a description of this type belongs. For example, if a person who dies from injuries sustained in a motor vehicle collision, the "how injury occurred" field might state, "Driver involved in motor vehicle collision," "Driver in motor vehicle collision with tree," "Pedestrian struck by motor vehicle," "Pedestrian struck by motor vehicle that left scene," or "Passenger in motor vehicle collision," depending on the details of the accident. For public health and vital statistical purposes, it is important to indicate the driver/passenger/pedestrian information and any other pertinent circumstantial information in this dedicated location on the death certificate. Completion of this section of the death certificate is intended to ensure that there will be no confusion about why a specific manner was chosen (see ahead).

Manners of death

Natural deaths

Natural deaths are defined as those caused exclusively (100%) by disease or old age (Table 3.6). If an injury (physical, chemical, or functional—see below) contributes to a death, no matter how minor, the fatality is no longer 100% natural. Violent deaths are subclassified as accidents, homicides, or suicides. Also, one has the option to leave the manner of death undetermined when the circumstances and findings leave substantial or *reasonable* doubt about the classification. Some deaths due to advanced natural disease may have an injury that contributes to the death, but the natural disease is the driving force. In such instances, the disease maybe listed in part 1 and the injury or intoxication may be listed in part 2. Some jurisdictions base the manner of death certification only on what is listed in part 1. In such instances, even with an injury listed in part 2, the manner would be certified as natural. The potential problem with this method of certification is that it puts the pathologist on a slippery slope. There can be some subjectivity as to which process is listed in part 1 and which is listed in part 2. In addition, some certifiers may use a "descriptive" cause of death which includes both the disease and the injury in part 1.

An example of the split part 1/part 2 natural and (chemical) injury certification is exemplified by a typical cocaine death.

Example 1: A 30-year-old man was found dead in his secured apartment. He had a history of cocaine use. The autopsy was negative for disease or injury. Toxicology testing detected 0.4 mg/mL of cocaine in the peripheral blood. The death is certified as an acute cocaine intoxication (accident).

TABLE 3.6 Manner of death definitions.

Natural	Death 100% due to disease or old age
Therapeutic complication	Death due to a predictable complication of appropriate treatment
Accident	Violent death that is *not* the result of an intentional or criminal act of another (inadvertent)
Suicide	Violent death caused by an act of the decedent with the intent to kill oneself
Homicide	Violent death at the hand of another person or deaths due to the hostile or illegal acts of another person
Undetermined	No probable manner can be determined

Example 2: A 55-year-old man was found dead in his secured apartment. He had a history of cocaine use. The autopsy revealed advanced three vessel coronary artery disease with scarring of the left ventricle and no acute coronary arterial thrombosis. Toxicology testing was negative for cocaine and other drugs of misuse. His death is certified as atherosclerotic coronary artery disease (natural).

In each of these deaths, the autopsy and toxicology findings are sufficient to render an unequivocal determination of the cause of death. But, now consider example 3.

Example 3: A 55-year-old man was found dead in his secured apartment. He had a history of cocaine use. The autopsy detected advanced three vessel coronary artery disease with scarring of the left ventricle and no acute coronary arterial thrombosis. Toxicology testing detected 0.4 mg/mL of cocaine in the peripheral blood. How should his death be certified? There are two findings that, individually, would explain the death. Therefore, both could be included on the death certificate. Some may choose only to include the cocaine intoxication, as it is the cause that will drive the manner determination; others may feel that it would be arbitrary to pick only one, as both are capable of causing death and, in combination, likely have potentiating mechanistic effects. Therefore, there are four options for certification:

1. Part 1: Atherosclerotic cardiovascular disease. Part 2: Acute cocaine intoxication
2. Part 1: Acute cocaine intoxication. Part 2: Atherosclerotic cardiovascular disease
3. Part 1: Acute cocaine intoxication complicating atherosclerotic cardiovascular disease
4. Acute cocaine intoxication

While slightly different in style, the first three of these statements accurately communicate the cause of death in substance. The fourth has a substantive difference. The third certification includes two etiologic specific causes, which is not typical of death certification constructs, but is used on occasion by some. Regardless, in our opinion, all should be certified as *accidents*. Per a NAME position paper on cocaine deaths (now sunsetted), the detection of cocaine associated with a fixed coronary lesion is generally considered accidental.[56]

Therapeutic complications

TCs are deaths that result from *predictable* complications of *appropriate* treatment (or diagnostic procedures).[95,111–113] As deaths from complications of diagnostic and therapeutic procedures are not exclusively due to disease, they do not meet the definitional criteria for a natural manner. It may also be perceived as unduly inflammatory to certify these deaths as accidents. Therefore, it is useful and appropriate to employ the more neutral TC designation in these deaths. The term is neither accusatory nor a synonym for malpractice. Additional benefits of using TC as a manner include improved consistency of classification, the avoidance of arbitrariness in selecting natural versus accident, and the ease of tracking these deaths.

The use of "TC" as a manner of death was initiated in Cuyahoga County (metropolitan Cleveland), Ohio in the 1960s. The practice has been introduced in other jurisdictions (New York City, Connecticut, Massachusetts, and Suffolk County, New York) by explaining to the local registrar of vital statistics the reasons for the terminology. Although the registrar accepts these "TCs" as the manner of death, they ultimately code the manner for their purposes, as required by national and international coding systems.

In some jurisdictions, the terms *Medical Misadventure* or *Therapeutic Misadventure* are used in these instances. The term *misadventure* is potentially inflammatory and implies an accident. Medical "accidents" are usually due to *inadvertent* acts or complications resulting from *inappropriate* therapy (i.e., errors). TCs constitute known, predictable, even if relatively rare complications of appropriate medical treatment or diagnostic procedures, not unfortunate accidents.

Some forensic pathologists will not use nonstandard or novel terminology for causes and manners of death (i.e., "if it is not in the ICD, it is not usable"). We consider the systems of classification and coding as our servants, not our masters. If a new or nonstandard term better expresses our professional opinion, we encourage a discussion with the local vital statistics bureau to explain the reasoning, given our professional expertise. Following a collegial and hopefully convincing discussion, the new term may be accepted by the registrar, and the nosologist can find the ICD coding to best fit the new designation.

Accidents

Accidents are violent deaths that do not result from an intentional or criminal act of another. Typical drug intoxication deaths due to misuse and most motor vehicle collisions deaths are common examples. Other examples include injuries from most residential fires, choking on a food bolus, many drownings, construction mishaps, inadvertent intoxications, or injury related to errors in medical therapy, among many others.

Other common accidental deaths involve falls in the elderly. These may cause hip fractures or subdural hematomas. Deaths resulting from *pathologic* hip fractures are natural. These typically result from marked osteoporosis or malignancies involving bone. In these instances, the fall is precipitated by the fracture. If, on the other hand, the fracture is due to a blunt impact from a trip and fall, the death would be classified as an accident, since the fracture is a consequence of the fall, not the cause of it. The circumstances and history are key to sorting out these two manners of death.

The elderly are at particular risk for sustaining subdural hematomas. Due to some degree of cerebral atrophy commonly associated with aging, there tends to be more "space" within which the brain can move, thus increasing the risk of the tearing of bridging veins with a sudden acceleration–deceleration injury. In addition, anticoagulation therapy and liver disease can be risk factors. The forensic pathologist must be wary of the so-called "spontaneous" subdural hemorrhage. Most of these are a consequence of occult injury, which may not be evident as a gross impact site of the scapular soft tissues or because the patient is unable to offer a history (unconscious, dementia, dead). Keep in mind also that direct head impact is not necessarily requisite in the mechanics of the tearing of bridging veins. Rather, enough shearing forces from a fall may lead to tearing of these vessels in the absence of a direct head impact, particularly in a person with advanced age and brain atrophy, in whom the bridging veins are placed under increased tension as a result. These become even more challenging when there is an acute and chronic or subacute component to the hematoma, broadening the time interval for the injury occurrence.

There are well-described "spontaneous" subdural hematomas that may be associated with end-stage renal disease. These are felt to be nontraumatic in origin. Anticoagulation during hemodialysis combined with the platelet dysfunction related to advanced renal disease may play a role. Of course, occult minor trauma can also be the cause. When presented with a hospital death involving a "spontaneous" subdural hematoma, the ME/C is advised to take jurisdiction, as it may later prove to be traumatic in origin and even a homicide.

Chronic alcohol users are at increased risk for subdural hematomas. This is related to liver disease (the liver is responsible for producing most clotting factors), thrombocytopenia, and/or platelet dysfunction caused by chronic alcohol ingestion, combined with their vulnerability for falling, becoming involved in altercations, and/or being menaced while intoxicated. This is one reason why one should have a relatively lower threshold for performing autopsies in chronic alcohol users, particularly when there is external, even if not apparently fatal, evidence of blunt injury. In a retrospective review of 1000 decedents with known or suspected acute or chronic alcohol use, 388 (38.8%) had an external injury. While 115 (11.5%) had an internal injury at autopsy and 29 (2.9%) had an injury that caused or contributed to the death, only one (0.1%) had an internal injury that caused death *without* external injury somewhere on the body. The study demonstrated the rarity of occult lethal injury diagnosed at autopsy in persons with acute or chronic alcohol use.[114] That said, those suffering from alcohol misuse may have a subdural hematoma without evidence of an external impact site of the *head*, per se. In addition, an autopsy may not only be performed in this group solely to exclude trauma. There is a relatively high incidence of substance misuse in this cohort, and internal findings (e.g., occult esophageal varix rupture complicating cirrhosis) may help determine whether the death is from chronic alcohol use or another disease process.[115,116] Perhaps most importantly, unless the fall is witnessed, one should resist the urge to assume that a subdural hematoma in a chronic alcohol user is always from an accidental fall, easily explained by an intoxication. Finally, it is imperative that we help our law enforcement colleagues understand the important concept of the potential **"lucid interval"** when investigating deaths caused by (traumatic) subdural hemorrhages (see Chapters 10 and 15 for further discussion on this topic).

Suicides

Suicides are injurious deaths (mechanical or chemical) caused by an act of the decedent with the intent to kill oneself. These represent one of the more challenging categories of deaths to certify because the ME/C must demonstrate volition and *intent* (i.e., the person intended to take their own life). There are suicide operational criteria that can be used to help make these determinations.[117] **Intent** may be indicated directly with documentation of clear verbal or written suicidal ideation and plans and/or by the inherently volitional nature of the act (most hangings). It should be noted, however, that fewer than half of the people who commit suicide leave a note. Therefore, the absence of a note does not exclude the death representing a suicidal act. Indirect evidence from a combination of one or several data points also may help to support this designation. Indirect evidence may include: expressions of farewell or hopelessness, previous suicide attempts/threats, recent stressful events, significant losses, or depression, efforts to learn about means of committing suicide, rehearsal, or precautions to avoid rescue, and unexpected preparations for death. Unexpected preparations may include putting one's will and life insurance documents in a prominent position (e.g., on the kitchen table), putting out extra food for pets, or bringing a pet to a shelter.[117]

Most traumatic suicides (e.g., hangings, gunshot wounds, descents from height) show clear intent by the very act itself. Loading a gun, pressing the barrel against the temple, and pulling the trigger are deliberate and demonstrable acts. Climbing over a balcony railing or constructing and then placing a noose over one's head are also examples of volitional acts demonstrating cogent and linear activity, by definition. Suspected suicidal intoxication deaths may be more challenging (see Chapter 14). As such, care must be taken in sample collection and in assessing postmortem drug concentrations. Suicidal ingestions by prescription medications or by illicit drugs in a person with substance use issues may be more difficult, but not impossible to interpret; markedly elevated drug concentrations with a supportive scene investigation may allow for a diagnosis. For example, a decedent may have a combination of a variety of prescribed medications detected in postmortem toxicology. Several are found to be elevated, but not by an order of magnitude. If the scene investigation finds inappropriately empty (recently prescribed) pill bottles, this may represent important information. Pill counts at the death scene, specifically noting the numbers and dosages of pills prescribed, the daily dosing instructions, and the date the prescriptions were filled, are important and relevant data points in these cases. The name of the prescribing physician also is important to document should there be a need for further follow up.

Examination of the gastric contents also may help in demonstrating intent. Pills are meant to dissolve in the stomach and so it is uncommon to find actual whole or nearly whole pills there. One may find congealed clumps which may represent a bezoar of pills or a chalky, granular pill sediment that becomes apparent after the gastric contents have been left to stand in a container for several minutes as the autopsy proceeds. Pills may be white, red, green, yellow, or pink; some also may collect in the proximal small bowel, as the stomach contents may empty during the antemortem period. Collection and testing of gastric contents (ensuring that they are well homogenized) for toxicological testing may be useful in equivocal cases. That said, the absence of an elevated amount of drugs in the stomach does not exclude an intentional intoxication death, but may help to diagnose one when present, so it is important to document the total volume of the gastric contents at autopsy, even if only a representative sample is submitted. This allows for a more accurate calculation of the total amount of gastric drugs ingested to be later extrapolated from the blood concentrations quantitated in the laboratory. The toxicology laboratory reports a drug concentration for the gastric contents. This concentration (which may seem impressively elevated) is used to determine the actual amount of drug in the stomach. The actual total amount of residual drug in the stomach contents is calculated by multiplying the gastric concentration by the total volume of gastric contents measured at autopsy, if not submitted entirely. This calculation will provide the actual weight ("mg") of the putative drug(s) still in the stomach at the time of death. In all of this computing, however, it is very important to remember that the gastric medication concentrations and total gastric drug volume do not represent pathophysiologic evidence for toxicity, per se, since the contents of the stomach still constitute the external environment, preabsorption into the blood. Rather, the residual drug amount in the stomach may be used to support *intent* in the appropriate context and when substantially elevated. When not elevated, it may mean either that an intentional/suicidal ingestion has not occurred *or* that it has, but that most or all of the gastric contents had been absorbed by the time death ensued. The postmortem blood concentrations within the broader context of the entire case investigation should determine which of these two scenarios has occurred.

For example, toxicological analysis of the gastric contents detects an acetaminophen concentration of 11,712 mg/kg (1,206 mg in 103 gm). If 103 g constitutes the total gastric contents, then the total residual gastric volume of drug equates to about four 325 mg tabs of acetaminophen. The concentration will also vary depending upon how much "food/fluid" is in the stomach. If the total gastric contents are 1000 gm, then the person ingested over 12,000 mg (or over 37 (325 mg) tablets). The latter would demonstrate intent while the former example may not.

A medical and social history is often important in these deaths. It sometimes is helpful to speak with the treating psychiatrist, but one should not expect a psychiatrist to state that they felt their patient was suicidal, as this might be perceived as an admission of malpractice. It may be more useful to seek objective facts, such as what the psychiatric diagnosis was or if there were any prior suicidal attempts. When was the last visit? Had the patient stopped coming? Had the patient stopped getting prescription refills for antidepressants? Was there a recent change in medication? All of these variables may go into the "mix" of data points when investigating such deaths.

Finally, the designation of suicide requires reasonable proof of intent to *fatally* harm oneself, the act being reasonably viewed as life-ending. Some acts are volitional and potentially self-harming, or even reckless, but would not reasonably or necessarily be expected to lead to death (e.g., "cutting," fire starting, or head banging). If any of these acts *inadvertently* leads to death, they would not likely meet the definition of suicidal acts, per se. This is an important distinction. Self-harming and/or self-destructive behaviors, per se, are not always viewed as potentially and intentionally life-threatening.

Homicides

At every *homicide there is one accurate witness without self-serving statements. The pathologist interrogates that witness.*

A homicide is a violent death at the hand of another person or a death due to the hostile or illegal acts of another person. There are notable exceptions, such as most fatal contact sports-related injuries, benign (noncriminal) physical play or wrestling, such as that between two siblings or friends, and, by death certification convention, most motor vehicle collision deaths. Also, most medical complications, even when malpractice, also are not certified as homicides, even though the hand of another played a role. If, by contrast, in a medical-related or motor vehicular fatality, the death was caused by an *intentional* act to cause harm (e.g., injecting someone purposely with 10 times the prescribed dose of insulin to cause harm or using a motor vehicle as a weapon to run someone over), then these would be appropriately certified as homicides.

Deaths occurring in a delayed fashion relative to the injury may present more of a challenge. This may occur with accidents, suicides, and homicides. If a person survives a *life-threatening* injury, but later succumbs to a complication of that injury, the cause and manner of death would be dictated by the circumstances of the initial injury. One must be able to demonstrate a direct pathophysiological link between the injury and the death, answering in the affirmative that "but for" the injury, this death would not have occurred, that this determination is as clear now as it would have been had the decedent died on the day that the injury had been sustained. In these circumstances, the passage of time does not relieve the original injury of its culpability in the ultimate fatal outcome. Sometimes, the only link between a remote firearms injury and the consequent death years later is the residual ballistics evidence in the body. If police reports and medical records have been expunged and the only story the FP has been given for the shooting is the family's, having ballistics evidence and associated injuries helps close the gap. In the absence of this, one might have more difficulty in meeting the burden necessary to call the death a homicide (see also Chapter 15).

The legal tests of causality applied to death certification practices are particularly helpful in these **remote injury deaths**. The critical duty in certifying such a death is to determine if the injury is the proximate (underlying) cause of death. For example, the wheelchair-bound paraplegic (due to a remote gunshot wound of the spinal cord) dies from sepsis due to infected decubitus ulcers. In this example, there are at least three conditions that meet the "but for" test (the gunshot wound, the paraplegia, and the decubitus ulcers), but only the gunshot wound meets the proximate cause test. The search for the proximate cause ends with the etiologically specific disease or injury that initiated the lethal chain of events.[118]

Certifying a fatality as a homicide after an interval of many years between injury and death may put a difficult or impossible burden on the police and prosecutor (in some instances, the original crime had been previously adjudicated). Such challenges, however, have no bearing on the death certification. Death certificates are an administrative medicolegal classification based on established standards and definitions. There is no statute of limitations for death certification; as such, the forensic pathologist need not be hampered by the amount of time that has passed between the original life-threatening injury and its ultimate fatal consequence. Rather, the burden for the FP lies only with the demonstration of evidence that supports the unequivocal link between the two, by history and/or clinicopathological correlation. Beyond that, what the police and prosecuting attorney do in response to any such death certification is their responsibility.[48] Based on our autonomous classifications, they are neither obligated to prosecute nor limited in their adjudicative decisions.

"Homicide by heart attack" is a description of a death due to the hostile or illegal acts of another that cause death through the involvement of underlying cardiac disease. That said, similar certifications may be made when the lethal outcome involves a hypertensive cerebrovascular event or a ruptured saccular cerebral aneurysm resulting from the stress of similar criminal activity. These susceptible, sometimes elderly people are subjected to the stress of a robbery, assault, or physical altercation, but do not sustain demonstrably lethal visceral injuries or fractures. Instead, *these deaths result from the combined effects of the* **underlying disease** *with the superimposed* **stress** *of a criminal act, the former, the proximate cause, and the latter, the (functional) immediate cause.* The stress of the criminal act may manifest in such ways as an arrhythmia-induced sudden death, with or without a myocardial infarct (the most common manifestation), a hypertensive stroke, an aortic dissection, or a ruptured berry aneurysm. The manner is certified as homicide because the death occurs at the hand of another person and is the direct result of their illegal act (robbery, assault, restraint, physical altercation, etc.). The specific susceptibility or vulnerability of the victim does not absolve the assailant of criminal responsibility. "The Constitution of the United States does not guarantee an assailant the right to a healthy victim"; one takes their victim as they find them.[48]

In some of these instances, a traditional death certificate format does not effectively represent the substance of the death circumstance. Instead, the most accurate and communicative avenue to take may involve the use of what is called a **"descriptive" death certification**, one that simply states what happened. For example, a person with advanced coronary atherosclerosis is susceptible to sudden death at any time. This person may suddenly die while being subjected to the stressful ordeal of an armed robbery. The temporal proximity to the event and the suddenness of this death is best explained through the mechanism of a cardiac arrhythmia, which is directly related to the *criminal activity (defined as unlawful acts falling outside the normal stresses and activities of daily living)*. It is impossible to measure stress at autopsy, and in the absence of an implantable cardiac device, it is also not possible to take a retrospective postmortem electrocardiogram. That said, the temporal relationship of the criminal events to the cardiac arrest cannot be reasonably ignored. Therefore, the death is certified descriptively: "Sudden Death during Armed Robbery and Threat with Gun due to Arteriosclerotic Heart Disease." The heart disease is the underlying/proximate cause of death, but the criminal menace and related stress-induced arrhythmia (deduced) constitutes a **functional injury** and the **immediate cause**.[48] Recall that if an injury of any type is listed in parts 1 or 2 of the death certificate, the natural manner is trumped by the manner predicated by the injury circumstance—in this case, a homicide.

Note also that not only can criminal acts lead to homicidal deaths in the absence of structurally demonstrable fatal injuries, they can also do so in the absence of physical contact. Examples of the latter include criminal menace—behavior reasonably perceived as threatening to life or limb, such as threats of severe physical harm, criminal pursuit, or the wielding and threat of a deadly weapon. In death certifications of a "descriptive" nature, it usually suffices and makes the most sense to let the cause of death statement speak to both the cause and the circumstances in one place, indicating simply, "see above" in the "how injury occurred" section of the death certificate.

The hostile or illegal acts also may trigger other natural disease processes. These circumstances typically involve stress and exertion (e.g., fleeing from an armed pursuer), resulting in an acute asthma exacerbation[119] or a ruptured cerebral artery aneurysm due to the associated hypertensive surge from the external stress (Fig. 3.4, Fig. 3.5).[120] When a sudden death is precipitated by a physical altercation, such as any battery (defined as a physical act that results in harmful or offensive contact with another person, but without that person's consent), even an angry physical "scuffle" between a father and son, the manner of death is homicide. In the absence of physical contact between the father and son, however, their argument is merely categorized as a normal stress of daily living and does not rise

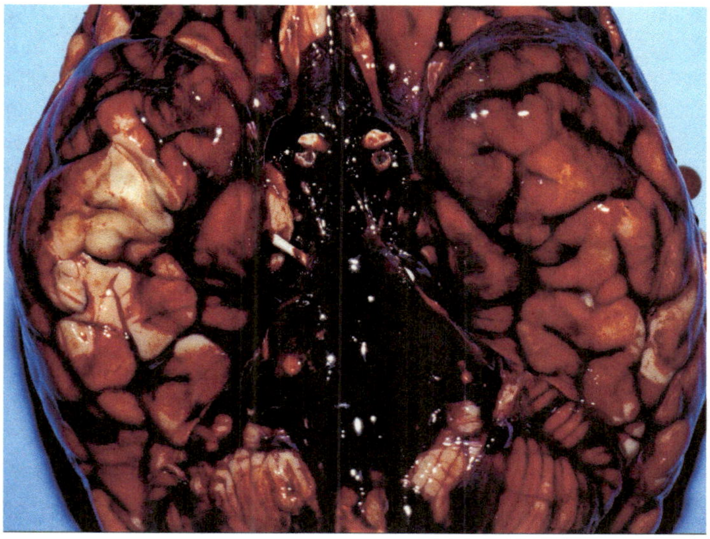

FIGURE 3.4 **Ruptured Saccular (berry) Aneurysm.** Subarachnoid hemorrhage at the base of the brain.

FIGURE 3.5 **Cerebral artery aneurysm (formalin-fixed brain).**

to the level of criminal activity (assuming no weapons are brandished). In this circumstance, the son is no more responsible for his father's death than would be the wife if the father had died while having sexual intercourse with her. The manner of death in those scenarios would be natural.

As previously stated, a cause of death statement *must* contain in it a disease *and/or* injury. The latter can be either structurally demonstrable/anatomic (e.g., a subdural hemorrhage) or functional and descriptive (e.g., the pathophysiologic stress of a physical altercation). If unsure whether a circumstance rises to the level of **a functional injury** (i.e., whether it clearly falls outside the activities/stresses of daily living), then, it does not belong on the death certificate. In fact, if unclear about where is falls, the most intellectually honest thing to do would be to call the cause (and, by necessity, the manner) "undetermined." If, on the other hand, sexual intercourse likely contributes to a sudden cardiac death in a person with advanced heart disease (this is not unusual), it doesn't constitute a functional injury because it does not fall outside the activities and stresses of daily living (not criminal in nature). As such, the possibility of the likely association between sexual activity and the sudden death need not, indeed, should not be memorialized on the death certificate. The 62-year-old-man with advanced heart disease who dies during sexual intercourse with his wife (and no drugs) has died a natural death.

If a **physical contact/altercation/scuffle** *is* invoked in the cause of death statement as the immediate cause, however, it constitutes a (functional) *injury*, by definition, because it **falls outside the activities and stresses of daily living**. Conversely, if the certifier has determined, for whatever reason, that the putative activity *does* fall within the activities and stresses of daily life, then the physical contact has no place on the death certificate since, by definition, it does not constitute an injury (according to the certifier). Further, while including "physical altercation" in the cause of death statement constitutes a decision and an opinion (to a reasonable degree of medical certainty) that something outside the activities of daily living has occurred, one needs not know nor opine about culpability (i.e., about who was more "aggressive" or who "started the fight," etc.) to arrive at that determination. Death certification standards and burdens differ from those of law enforcement and prosecutors. We are not judge, jury, and executioner.

Some forensic pathologists may attempt to quantitate the extent of involvement of the "altercation" in light of other potential causes of death, thus averting a homicidal determination. For example, there is a physical altercation, but the certifier determines, for whatever reason, that it did not, in and of itself, cause or contribute to death, but still adds a phrase such as "following physical altercation" or "temporally associated with physical altercation" to the cause of death statement, despite assigning a nonhomicidal certification. Instead, they may base the manner of death on the underlying cause (e.g., cocaine intoxication, as an accidental death, or atherosclerotic and hypertensive cardiovascular disease as a natural death, etc.), regardless of the physical altercation being invoked in the cause of death statement.[121–123] But, the analysis and decision about whether physical contact is part of the cause of death should be made *before* its inclusion on the death certificate. If the death is being viewed as a natural death, then the inclusion of functional injury becomes confusing. In other words, the natural classification would indicate that the physical contact is not being viewed as rising to the level of an injury/criminality. Once the injury has been made part of the death certification, however, an opinion about its contribution to death appears to have been effectively made. Ignoring it and checking the "natural" manner checkbox is not only confusing in terms of the definition of natural deaths and its incompatibility with death certification conventions, it also results in a legal document that seems to be contradicting itself.

Not all physical contact constitutes "death at the hand of another" (criminal contact) and common sense should apply. If someone dies from the stress of the physical contact that occurs during a lawful police encounter (such as, handcuffing following the capture of an armed bank robber, without excessive, nonprocedural restraint) or during a sanctioned sporting event (such as a check during an ice hockey game) that is played by the rules, these are not homicides. If, on the other hand, two hockey players throw off their gloves, a fist fight breaks out, and one dies from a blunt head injury, that would constitute a homicide.

Not uncommonly, there may be involvement of *both* a physical altercation *and* an intoxication or disease and, depending on the case, both may appear on the same death certificate. In some jurisdictions, if the stress of any "physical altercation" contributed to death, no matter how small its contribution, even with five other stimulant drugs, the death will still be classified as a homicide. An acute intoxication, however, should not be arbitrarily listed as a contributing condition unless it plays a physiologic/mechanistic role in death and/or if the injury does not make sense without invoking the intoxication. Deaths from a gunshot wound of the head, an incised wound of the carotid artery, or an expanding and space-occupying epidural hematoma due to a blunt impact and fracture of the temporal bone, should not have an intoxication included on the death certificate. That said, when the death is due to the stress of criminal activity, it may seem arbitrary to selectively exclude **stimulant drug intoxications** as contributory. For this reason, it is not unusual that the final certification of many apparent "homicide by heart attack" (stress-related) deaths are deferred until after toxicological testing is completed. The manner will be no less a homicide, but waiting makes allowances for a possible toxicological contribution.

The final nuance of some homicide determinations is the concept of an **efficient intervening cause**. Just because a person may be quadriplegic from a remote gunshot injury does not mean that whenever they die, the death will automatically be certified a homicide. If that person dies from a metastatic small cell carcinoma of the lung, the manner of death would be natural, the remote gunshot wound and resultant quadriplegia being causally *unrelated* to the lung cancer. A person who dies from a TC from treatment of a nonlife-threatening injury caused by another, also will not likely meet the burden or definition of a homicide (see Chapter 8). For example, consider a 48-year-old woman who sprained her upper arm and shoulder during an assault. Despite conservative medical treatments for 12 months, she continued to experience pain in the arm and shoulder and was diagnosed with a posttraumatic sympathetic dystrophy. During an elective cervical sympathectomy, she arrested following the induction of general anesthesia. Autopsy diagnosed marked atherosclerotic stenoses in two of her major coronary arteries. The complication of this elective (not lifesaving, even if necessary) therapy constitutes an "efficient intervening cause." Neither her nerve pain nor the injury that caused it were ever life-threatening, at the time of the injury occurrence or at the time of her surgery, so the TC becomes dominant and relieves the original injury of its causal responsibility (see Chapters 8 and 15 for further discussions on special topics in homicide investigations and certification).[48]

Undetermined cause and/or manner

> Honesty is the only stigma that attaches to a pathologist who leaves an occasional case undetermined. ***Drs. Charles S. Hirsch and Ross Zumwalt***

There will be cases in which no cause of death is found despite a thorough medicolegal death investigation. This is simply a recognition of the limitations of our science and the inability of a morphologic study to demonstrate functional derangements. If one is able to exclude trauma and intoxication, the manner may at least be certified as natural.

That said, there will also be individuals whose *manner* of death also remains undetermined. An office having no undetermined deaths would defy reasonable statistical predictions and should cause a raised eyebrow. "Someone who has all the answers all the time is either a scoundrel or a fool."

By way of an example, an unidentified decedent without injuries is found floating in a river. Without knowledge of the identity of the individual and/or the circumstances of the death or psychosocial history, the manner of death is best certified as undetermined, even if the cause is determined to be drowning. Other examples include intoxicated persons who sustain apparently unwitnessed fatal blunt impact injuries, especially head injuries with space-occupying subdural hemorrhages. Absent other information, the temptation to assume that an intoxication-related accidental fall led to the injury must be resisted. Just because no one saw or reported anything suspicious does not mean that nothing suspicious happened ("Absence of evidence is not evidence of absence"; "a negative is never proof."). In these situations, the manner of death would best be classified as "undetermined" unless the pattern of injuries is inconsistent with an accident, as when the physical findings are clearly indicative with an assault/beating, and therefore a homicide, regardless of the fruit born out of a police investigation.

Sometimes, both the cause and manner of death must be left undetermined. An undetermined manner classification should be used only when necessary, but it is not synonym for "uncomfortable." The undetermined manner should not serve as a shield that protects the ME/C from making a difficult or unpopular determination when all necessary investigative facts are known and available. Usually, it is when there are *unknowable variables* that we are forced down the undetermined path.

With rare exceptions, if the cause of death is traumatic, an undetermined manner should not be assigned until there is a police investigation. If no such police investigation has been started by the time of autopsy, it is commonly requested by the ME/C. Regardless, the investigation is best served by providing very specific questions and requests to the case detective. Do not assume that they will know the specific investigative information you seek or necessarily know about or understand the concept of lucid interval, the understanding of which may expand their investigation in time and space (see also Chapter 15).

Undetermined classifications assume that all investigative and scientific avenues have been exhausted before coming to this unsatisfying conclusion, or that the investigation is limited by legal mandate ("pursuant to law" religious objections). In many instances, an undetermined classification requires more, not less work, particularly in cases in which there is an autopsy, to assure ourselves and anyone else reviewing the case that we looked under every available rock.

A fatality in which the cause of death is an injury (chemical or mechanical) and the manner is "undetermined" must still have the injury section completed. For example, in an acute mixed intoxication from prescription medications with an undetermined manner (suicide vs. accident), the injury section is still completed to the extent that the details (time and place of occurrence) are known. For the "how injury occurred" section, however, one may state, for example, "Excessive ingestion of multiple prescription medications with unclear intent" when the circumstances are inconclusive. Similarly, for deaths involving a descent form height, the cause is invariably blunt injury. But, if the circumstances are unclear, the manner may be classified as "undetermined" and the "how injury occurred" section should explain the reasoning for that manner designation by stating something like, "descent from height, circumstances unclear (or unknown)." When both the cause *and* the manner are undetermined ("U-squared, U^2"), then the injury section of the death certificate need not be completed because, by definition, since the list of potential (unknown) causes includes both those that are natural and those that are unnatural (i.e., the injury section is not completed in the former).

Forensic pathologists occasionally encounter "suspicious" deaths in which the ultimate certification gets distilled down to homicide versus accident. These are the deaths that have the capacity of making us uncomfortably unsure as to whether an inflicted injury directly caused or hastened death. An example is the sometimes difficult interpretation of ambiguous or minor neck "injury" under quasi-suspicious and/or unclear circumstances involving a potentially vulnerable person, such as a prostitute who uses illicit drugs. There may be only a focal abrasion of the neck or slight hemorrhage in one strap muscle while the toxicology testing detects a concentration of cocaine falling within or below the range of a typical accidental drug intoxication fatality. The forensic pathologist may even think, "In my gut, I believe this is a homicide," a truthful statement of fact and experience, but not necessarily what is required by law to memorialize that opinion on a legal document. The latter requires a conclusion reached to a reasonable degree of medical and scientific certainty (i.e., excludes all other reasonable contingencies) and is based on solid scientific and investigative evidence. As Dr. Hirsch was known to say, "Hunches are not evidence." Such a case may be best and most accurately left undetermined.

It is incumbent upon us as medical examiners to meet our burdens of proof, the highest being in homicide determinations. Someone's civil liberties may be at stake and one's "hunch," even if correct, will not and should not hold up in court. We must view our role as expert witnesses as the privilege it is and exercise it with the utmost responsibility, based on science, training, and experience. Our most challenging cases are often best shared with trusted colleagues in a "consensus" case conference setting to see if there is general agreement. Following this, if there are still unresolved questions in a death that is likely a homicide, but in which an alternate theoretical, but *reasonable* explanation also exists, the most responsible, intellectually honest, and sound decision might be to classify the death as "Undetermined." As unsettling as this sometimes feels, it is best to under call a homicide rather than to overcall one. This is simply a recognition of the limitations of our work.

Importantly, the only time that an undetermined cause and manner should be used in death certification is either in a case in which an autopsy has been performed, OR, if no autopsy has been done, one in which the external examination (with or without toxicological testing) was performed "pursuant to law" (i.e., religious objection to autopsy). In other words, if honoring a nonreligious/personal objection means that the cause and manner of death will be left as undetermined, then the FP is best advised to *not* honor the objection, but rather to perform an autopsy (or an external examination with toxicology, if applicable) to meet their medicolegal obligation. That said, honoring/not overriding a religious objection assumes that the forensic pathologist is confident that they have reasonably ruled out concern for potential criminality and have no defensible cause to push further. If the FP *does* have reasonable concerns, it is best to discuss this with the family in the hopes that they will withdraw their objection. If they don't, however, there may exist the need for them to present their objection before a judge.

Finally, understand also that not all "pursuant to law" objections are certified as "undetermined" cause and manner. Many can be reasonably certified with a cause and/or manner, without the necessity of an autopsy. Where mechanical and chemical causes have been reasonably excluded, but the (natural) cause remains unclear, an undetermined cause with a natural manner is a perfectly reasonable conclusion and a competent, even if unsatisfying certification. Moreover, where fatal injuries have been reasonably documented in hospital and the circumstances investigated, the performance of an autopsy may be wholly unnecessary for death certification purposes, unless a criminal investigation or concerns dictate otherwise.

War injury

War injury is a singular *manner of death* that may be (rarely) used in some jurisdictions. Its intended purpose is for the death certification of persons who die from the sequelae of remote injuries that were sustained in combat during active military service. Not dissimilar to the approach to delayed homicide deaths, the "but for" question is asked with regard to the direct mechanistic link between mechanical or chemical injuries sustained in combat or as a prisoner of war and the death some years later. Examples include complications from remote Agent Orange warfare or longstanding neurologic, gastrointestinal, pulmonary, or immobility sequelae of remote shrapnel injuries. By contrast, when a fatality involves personnel who die in or more proximate to combat, the military typically handles the certification, using a military death certificate that indicates "killed in action" rather than the homicidal manner that is used on civilian death certificates.

Other death certificate fields and miscellaneous issues

Injury information

The "injury" information section of the death certificate is required for all homicides, accidents, suicides, and any injurious deaths (chemical, mechanical, or functional) having an undetermined manner. Since, by definition, there is no injury in natural deaths and TCs, the injury section is left blank for these deaths. This section is also left blank when both the cause *and* manner have been deemed "undetermined." In many deaths, including some homicides, the time of injury (and sometimes the date) is not known. When a body is decomposed, for example, both the time and date of injury occurrence will likely not be precisely known for death certification purposes (therefore, left "unknown" in the injury section), although, in conjunction with investigative data points, may be reasonably estimated in later criminal proceedings. If there is evidence that a body was moved after death from the location of injury occurrence to another location, the place of injury may be unknown as well.

While there is plenty of room for *stylistic* differences among forensic pathologists, the **"how injury occurred"** statement is very important and, ideally, will be completed thoughtfully, objectively, consistently, and without conjecture. For example, a blunt head injury (the proximate cause of death) can occur in many ways (e.g., a fall from height, struck by another with an object, struck by a motor vehicle, etc.), the various ways in which the injury arose informing the manner. The cause of death indicates what the injury is while the "how injury occurred" section explains the circumstances in which the injury was sustained; the goal is that the explanation in this section makes clear the reason a specific manner box was checked. Some jurisdictions may combine these (e.g., "Blunt head injury due to motor vehicle accident"), but the authors do not. The intent of the vital records bureau is clear, in that they do not want deaths certified in this way. If they did, there would be no reason to have separate lines. The person is dying from the blunt trauma, not the motor vehicle collision. As described above, cause of death statements contain codable diseases and/or injuries; a motor vehicle collision is neither. Rather, it is a circumstance.

It follows, then, that, unlike the cause of death statement, the "how injury occurred" field does not typically include codable diseases or injuries (exceptions exist). If, for example, a depressed individual with recent suicidal ideations is witnessed to intentionally jump from the roof of a building to their death and the manner is deemed a suicide, the proximate cause of death would be blunt impact injuries and the "how injury occurred" section would state, "jumped from height" or a similar description that conveys the same information. While "history of depression" may be included in the investigative report and even listed as a historical finding in the final diagnoses section of the autopsy report, as it contributes to the contextual backdrop of the final fatal event, it is not typically included in the "how injury occurred" section, per se (nor in the cause of death statement), as it is neither a mechanism of the death nor is it necessary in the explanation of how/under what specific circumstances the decedent descended from the roof on that day. Including the word "jumped" in the "how injury occurred" section, along with the suicidal manner designation, is sufficient explanation because the intentionality of the event is inferred, without the necessity of further describing *why* the decedent was suicidal. Separate and apart from longstanding clinical depression, an individual in a similar suicidal descent circumstance might alternatively have just lost a spouse, been diagnosed with terminal cancer, or suffered a devastating financial blow. These investigative and contextual data points, while important in the overall evaluation and understanding of the case, are not usually included on the death certificate, their inclusion being as unnecessary as the inclusion of "depression."

Similarly, if an investigation indicates that someone's injury arose from a (nondescent from height) accidental fall (to the floor, for example), it is not our burden or place to explain *why* they fell, unless it is abundantly clear and well-documented and/or uniquely important to the case to qualify. In most instances, "fell" or "fell to floor" in the "how injury occurred" section is enough in cases of benign falls at home or elsewhere. Our burden is to determine, through a thorough investigation and autopsy *that* an individual fell (i.e., the use of the word "fell" indicating that we have reasonably determined that they were not "pushed" or otherwise menaced), not necessarily *why* they fell. Often, accurately identifying the explanation(s) for the fall is not reasonably possible and will remain unknowable despite a thorough investigation, but this should not hamper most certifications; requiring this data point for certification purposes would designate many death certificates as incomplete or undetermined. Whether an elderly person sustaining a fatal fall at home (causing a fatal hip fracture or a subdural hemorrhage, for example), fell because of Parkinson-related imbalance, medication or cardiac-related dizziness, large or cumbersome slippers, or a fold in the rug may never be clearly known, but, as long as the *inadvertent* nature of the fall is documented and supported by the investigative facts, the certification is competent. Saying "fell while intoxicated," "motor vehicle collision while intoxicated" is as unnecessary as it is presumptuous and arbitrarily ignores all other potential and myriad variables (clumsiness, faulty sidewalks, dangerous road conditions, dizziness, syncope, slipping on a banana peel, texting, distraction from passengers, glare, and on and on). Moreover, autopsy toxicology results are not available on the day of autopsy, but that should not prevent certification in cases *not* requiring a further explanation of the death circumstances through toxicology testing (e.g., most mechanical homicides, accidental falls, and clear-cut accidents, such as those from motor vehicle collisions).

By contrast, *if the injury does not make sense without the invocation of an intoxication, it may well need to be included as the underlying cause of death or the "how injury occurred" section* (e.g., an adult drowning in a bathtub or choking on a food bolus). It also may be appropriate and necessary to include, for example, in an injurious death with an undetermined manner, in order to further clarify the reasoning behind choosing the undetermined designation. For example, an ME/C may be having difficulty distinguishing between an accidental versus a suicidal manner of death in a fatal descent from height involving a person who has a history of depression, but who is also intoxicated with phencyclidine or psilocybin at the moment of death. In this instance, an undetermined manner with a "how injury occurred" section stating, "descent from height while intoxicated with phencyclidine (or psilocybin)" may help to explain the decision to classify the manner as undetermined (or as an accident, as the case may be).

Crafting an accurate and clarifying "how injury occurred" statement, which is devoid of unnecessary judgment, requires careful deliberation and adherence to certain principles. That said, relative to other parts of the death certificate, this section also has the potential to create the most confusion and consternation, for experienced forensic pathologists and trainees alike. Understanding the goals and limitations of the information conveyed in this section and recognizing its significance will likely enhance both its accuracy and its communicative value. There are several points of potential confusion and/or misuse.

The importance of "how injury occurred": Do's and Don'ts

What is common is common, what is rare is rare.[124]

The following summarizes important concepts, common mistakes, and traps to avoid:

1. **Unless otherwise necessary to explain the manner, avoid listing codable diseases and injuries in this section.**
2. **Only describe the circumstances of the injury or injuries that are listed in the cause of death statement.** These injuries are often incompatible with continued life. As an example, if the decedent was hogtied and bound, but the cause is "gunshot wound of head," the "how injury occurred" line should say something like, "shot by other," and not include a description of the binding and hogtying, despite the fact that this may be quite important in the investigation and subsequent adjudication. If the decedent had sustained multiple nonfatal blunt injuries (contusions on various surfaced of the body) in a violent struggle that preceded a fatal incised wound of the neck and only the incised wound is listed in the cause of death statement, then only "cut by other" (or something similar) would appear in the how injury occurred field, excluding an explanation of the other signs of the struggle. A detailed description of *all* other nonfatal injuries, of course, *does* belong in both the final diagnoses and body of the autopsy report, but the death certificate should not to be used a substitute for the entire case file, with a full explanation of *all* noncausal injuries in the "how injury occurred" section.
3. **Postmortem inflicted injury** (dismemberment, burning of the body, etc.) demand a thorough description within the autopsy report (including the final diagnoses), but should not be otherwise circumstantially explained in the "how injury occurred" section. No matter how dramatic, if an injury is sustained in the postmortem period, there is no place for its description on the death certificate.
4. **Adding "while intoxicated (with)" at the end of a "how injury occurred" statement is primarily reserved for situations where it is necessary to explain a manner in a death due to mechanical injury** (e.g., drowning in a bathtub) that otherwise wouldn't make sense without it.
5. **Unless well documented (video, witnesses) and/or otherwise important in the context of the case (e.g., work-related), it is not necessary to identify the reason or reasons why an individual fell to the floor or ground from a standing position, just *that* they fell** (i.e., were not pushed or otherwise menaced). Whether the fall was caused by dizziness, inattentiveness, external factors, intoxication, or some combination thereof may never be fully understood, but this is fine for death certification purposes. *Remain mindful of the trap of speculation.*
6. **Attempt to maintain balance and accuracy: where possible, try to avoid unnecessary judgments, sticking to neutral and objective statements of fact** (Table 3.7 and Table 3.8). That said, try also not to hide behind "objectivity" when all available evidence points to only one reasonable conclusion. "Don't deny reality for the sake of objectivity." Tables 3.3–3.7 offer many examples of stylistic differences (e.g., "shot by another," without further qualification and "shot by another," with more circumstantial detail; "hanged self" vs. "hanged self with rope"). Using more objective detail may be a stylistic choice made by some FPs as a matter of practice, while for others, it might only come into play when they feel it is necessary to clearly explain the chosen manner (e.g., an accidental vs. homicidal shooting death).

TABLE 3.7 Common unnatural causes, manners, and how injury occurred.

Cause	Manner	How injury occurred?
Blunt injury of head	Accident	Driver in motor vehicle collision
Acute cocaine intoxication	Accident	Substance use
Drowning	Accident	Submerged in pool
Gunshot wound of head	Suicide	Shot self
Gunshot wound of head	Homicide	Shot by another
Incised wounds of wrists	Suicide	Cut self
Hanging	Suicide	Hanged self (with rope)
Thermal injuries and smoke inhalation	Accident	Residential fire
Sudden cardiac arrest during physical altercation in person with arteriosclerotic cardiovascular disease	Homicide	Physical altercation (or, see above)
Multiple blunt impact injuries	Suicide	Descent (or jumped) from height

TABLE 3.8 Examples of statements in "how injury occurred."

Shot by another
Shot by another who was violating safety rules of shooting range
Shot by another during unsupervised handling of gun
Shot by hunter reportedly shooting at unseen game
Shot by police
Intentionally provoked police to shoot
Shot by poacher using illegal weapon
Shot self while playing with gun
Shot self
Playing Russian roulette (with revolver)
Hanged self/hanged self with ...
Placed plastic bag around head and tied at neck/covered head with plastic bag and fastened at neck
Aspirated packets of cocaine wrapped in plastic bag during encounter with police
Fire due to careless discard of smoking materials
Fire due to faulty wiring
Incendiary fire
Residential fire (with/without a further description of the genesis of the fire determined by the fire marshal)
Smoking while using supplemental oxygen
Inadvertently scalded self in bathtub
Inadvertently touched 1400-volt power line
Firefighter injured in the line of duty
Firefighter with self-contained breathing apparatus that ran out air during residential fire
Occupational needle stick injury
Broke through 10th floor window during psychotic episode
Assaulted
Jumped/fell from height
Descent from height/(during agitated psychosis/psychotic episode/following use of phencyclidine)
Pedestrian struck by truck
Crushed in trash compactor
Pedestrian struck by motor vehicle (that left scene of accident)
Driver/passenger in motor vehicle collision
Car passenger/hit pole/ejected
Jumped in front of train while experiencing acute bipolar disorder episode (vs. intentionally jumped in front of train)
Unnatural exposure to anthrax
Acute and chronic substance use
Direct (or assisted) intravenous administration of above listed agents
Overuse of prescription medications

TABLE 3.8 Examples of statements in "how injury occurred."—cont'd

Inadvertent strangulation during autoerotic activity
Intentional inhalation of car exhaust
Intentional ingestion of excessive prescription medications
Smuggling drugs in gastrointestinal tract
Police officer shot in the line of duty
Injury inflicted by another
Punched and fell to the ground, striking head
Struck multiple times on head with object
Struck by motor vehicle following physical altercation
Sharing sleep surfaces with two adults
Medical, nutritional, and hydrational neglect of child/infant/dependent adult
Inadvertently cut self with broken bottle (while intoxicated)—if intoxication necessary to explain outcome

Time of death/found/pronounced and miscellaneous information

The investigator must obtain this date and the time of death/pronouncement from emergency services or hospital physicians, whomever made the pronouncement. "Unknown" or "not available" is unacceptable, as this constitutes information that is required on the death certificate. Every death that we investigate has such a date and time. If the body remains at the scene and emergency services technicians (EMTs) never "pronounced," then the time of death will be the time the body was found. EMTs who decide not to bring a body to the hospital have effectively pronounced the person even though there may not have been a formal declaration. If the EMTs believed the person was alive or potentially could have been resuscitated, they would have initiated treatment.

Tobacco use and pregnancy information are usually included on the death certificate, but requirements vary by jurisdiction. If prompted, the certifier needs to report if tobacco use contributed to the death or not. The choices are yes, no, probably, and unknown. For most deaths, this is unknown, but for deaths listing COPD as a primary or contributory cause, it would be illogical to check the "unknown" box. For pregnancy, the information needed may include:

1. Not pregnant within one year of death
2. Pregnant at time of death
3. Not pregnant at death, but pregnant within 42 days of death
4. Not pregnant at death, but pregnant 43 days to one year before death
5. Unknown if pregnant within one year of death

If the decedent was pregnant within the prior one year, the following questions may also be asked:

1. Live birth
2. Spontaneous termination/ectopic
3. Induced termination
4. None
5. Date of outcome

Assigning gender on death certificates

Most death certificates have been changed to accommodate gender nonbinary individuals. The "Sex" fields may indicate the decedent's gender identity at the time of death (M, F, X, Undetermined, Unknown), while a field such as, "Sex Assigned at Birth" will reflect the decedent's sex/anatomic sex at birth. When included on the death certificate, both fields *must* be completed by the ME/C. The ME/C makes this determination to the best their ability to honor the wishes of the decedent, to best consider the needs of the family, and to meet vital records requirements.

Injury at work

There is a checkbox for "at work" on the death certificate.[90,125] The Department of Labor is one stakeholder that is interested in these data.[126,127] Per CDC recommendations, this field should be completed if anything other than natural disease is included in parts 1 or 2 of the death certification, including homicides, suicides, and accidents. The "Yes" selection should be checked if the injury occurred at work. Otherwise, enter "No." An injury may occur at work regardless of whether the injury occurred in the course of the decedent's "usual" occupation or whether the injury, per se, was work-related.

Examples of injury at work and injury not at work follow:

YES

Injury while working or in vocational training on the job premises.
Injury while on break or at lunch or in the parking lot <u>on</u> the job premises.
Injury while working for pay or compensation, including at home.
Injury while working as a volunteer law enforcement official, etc.
Injury while traveling on business, including to/from business contacts.
Work activity where a vehicle is the work environment, such as a taxi or a truck.

NO

Injury while engaged in personal recreational activity on the job premises.
Injury while a visitor (not on official work business) to job premises.
Homemaker working at homemaking activities.
Student in school.
Working for self for no profit (mowing yard, repairing own roof, hobby).
Commuting to or from work.

Death certificates in disasters/mass fatality incidents (see also Chapter 4)[128]

One should include the disaster by name on the death certificate, when possible. This may be indicated in the "how injury occurred" box, however, when the manner of death is natural, this wouldn't be possible, causing these natural deaths to be lost to vital record statistics. To remedy this, the CDC recommends that part 2 be used to communicate the involvement of the disaster when the resultant deaths are natural, such as, part 1: Arteriosclerotic cardiovascular disease; part 2: Exertion shoveling snow during the February 2, 2022 snowstorm (natural). Examples of "how injury occurred" descriptions in unnatural disaster-related deaths include: Car drove into flooded roadway after Hurricane Elsa, or hypoxemia due to nonoperational oxygen generator during power outage caused by 2022 New Orleans tornado. The manner determination of a death related to a disaster depends on a variety of factors, including when the particular "exposure" occurred. Typical causes are listed in Table 3.9.

Key death certification tips

1. **Always identify an underlying (etiologically specific) cause of death**; if none is identifiable, designate the cause as either "undetermined" or list the appropriate mechanism (e.g., cardiac arrhythmia) followed by "of undetermined etiology" to memorialize your recognition of the absence of a proximate cause.
2. **Use part 2 for conditions that contribute to death, but do not result in the underlying cause listed in part 1.**
3. **One cause of death per person**, however, the needs of the vital statistics bureau must be subordinate to the criminal justice system, so there may be exceptions.
4. **Injury anywhere on the death certificate trumps natural (or TCs)** in informing manner. If the immediate cause of death is an injury (drowning), it trumps the proximate cause (epilepsy) if the proximate cause is natural (the manner would be accident in this example, to reflect the drowning).

TABLE 3.9 Considerations with certification of disaster deaths.

Mass terrororist events (e.g., WTC/9-11)
For unidentified bodies: "Physical injury"
Injuries or complications (e.g., sarcoid) stemming from the acute event (homicides)
Complications (pulmonary fibrosis) from chronic exposure during clean-up/recovery (natural)
Crushed by debris (manner depends on whether this occurred from the acute event or during clean-up/recovery—homicide vs. accident)
Wildfire death considerations
Deaths directly resulting from the fire
Deaths occurring during evacuation
Deaths occurring while evacuated
CDC terminology for disasters
Direct cause: A death is directly attributable to the forces of the disaster or a direct consequence of these forces (structural collapse, drowning, fire, radiation exposure)
Indirect: Not from a direct impact but due to unsafe or unhealthy conditions around the time of the disaster including: while preparing (fall while boarding up windows)
Responding to (car crash while escaping)
During recovery from (lack of access to hemodialysis. CO poisoning from generator)
Partially attributable: a death that cannot be tied definitively to the disaster, but in which the disaster has more likely than not played a contributing role in the death (e.g., stroke in person with underlying disease, suicide in person with mental illness). "BUT FOR" principle.

5. **The proximate cause trumps the immediate cause.** If there are two different injuries, from two separate events, one as the immediate cause injury (drowning), and the other as the proximate cause injury (posttraumatic seizure disorder due to remote gunshot wound of head), *the injury in the proximate cause trumps that in the immediate cause in informing the manner,* so long as the one in the proximate cause was *potentially* fatal at the time of the event (essentially, any gunshot wound of the head is *potentially* fatal, even if it ends up not causing death at the time).
6. If you have competing causes of death, ask, "What is the likely mechanism?" **The mechanism of death will point you toward the most accurate cause (and manner) of death.**

References

1. Seifi A, Lacci JV, Godoy DA. Incidence of brain death in the United States. *Clin Neurol Neurosurg*. 2020;195:105885.
2. Morentin B, Audicana C. Population-based study of out-of-hospital sudden cardiovascular death: incidence and causes of death in middle-aged adults. *Rev Esp Cardiol*. 2010;64(1):28–34.
3. Roberts WC, Siegel RJ, McManus BM. Idiopathic dilated cardiomyopathy: analysis of 152 necropsy patients. *Am J Cardiol*. 1987;60(16): 1340–1355.
4. Basso C, Thiene G, Corrado D, Buja G, Melacini P, Nava A. Hypertrophic cardiomyopathy and sudden death in the young: pathologic evidence of myocardial ischemia [see comments]. *Hum Pathol*. 2000;31(8):988–998.
5. Kannel WB, Gordon T, Offutt D. Left ventricular hypertrophy by electrocardiogram. Prevalence, incidence, and mortality in the Framingham study. *Ann Intern Med*. 1969;71(1):89–105.
6. Davies MJ, Popple A. Sudden unexpected cardiac death–a practical approach to the forensic problem. *Histopathology*. 1979;3(4):255–277.
7. Lucena J, Blanco M, Jurado C, et al. Cocaine-related sudden death: a prospective investigation in south-west Spain. *Eur Heart J*. 2010;31(3): 318–329.

8. Pelliccia A, Maron BJ, Spataro A, Proschan MA, Spirito P. The upper limit of physiologic cardiac hypertrophy in highly trained elite athletes. *N Engl J Med*. 1991;324(5):295–301.
9. Suarez-Mier MP, Aguilera B, Mosquera RM, Sanchez-de-Leon MS. Pathology of sudden death during recreational sports in Spain. *Forensic Sci Int*. 2013;226(1–3):188–196.
10. Basso C, Michaud K, d'Amati G, et al. Cardiac hypertrophy at autopsy. *Virchows Arch*. 2021;479(1):79–94.
11. Adelson L. No anatomic cause of death; the enigma of the forensic pathologist. *Conn Med*. 1954;18(9):732–738.
12. Adams VI, Flomenbaum MA, Hirsch CS. Trauma and disease. In: WU S, ed. *Spitz and Fisher's Medicolegal Investigation of Death*. 4th ed. Springfield, IL: Charles C Thomas; 2006:436–459.
13. Nuland S. *How We Die: Reflections of Life's Final Chapter. Vintage*. 1995.
14. Padwell A. Cyanide Poisoning. Case studies of one homicide and two suicides. *Am J Forensic Med Pathol*. 1997;18(2):185–188.
15. Winek CL, Fusia E, Collom WD, Shanor SP. Cyanide poisoning as a mode of suicide. *Forensic Sci*. 1978;11(1):51–55.
16. Vogel SN, Sultan TR, Ten Eyck RP. Cyanide poisoning. *Clin Toxicol*. 1981;18(3):367–383.
17. Fernando GC, Busuttil A. Cyanide ingestion. Case studies of four suicides. *Am J Forensic Med Pathol*. 1991;12(3):241–246.
18. Chikasue F, Yashiki M, Kojima T, et al. Cyanide distribution in five fatal cyanide poisonings and the effect of storage conditions on cyanide concentration in tissue. *Forensic Sci Int*. 1988;38(3–4):173–183.
19. Cina SJ, Raso DS, Conradi SE. Suicidal cyanide ingestion as detailed in Final Exit. *J Forensic Sci*. 1994;39(6):1568–1570.
20. Gill JR, Marker E, Stajic M. Suicide by cyanide: 17 deaths. *J Forensic Sci*. 2004;49(4):826–828.
21. Byard RW. The 'break enter and die' syndrome may involve significant injury to major neck vessels. *J Forensic Sci*. 2011;56(Suppl 1): S252–S254.
22. Gill JR, Storck K, Kelly S. Fatal exsanguination from hemodialysis vascular access sites. *Forensic Sci Med Pathol*. 2012;8(3):259–262.
23. Handlos P, Marecova K, Smatanova M, Dvoracek I, Dobias M. Fatal hemorrhage from an arteriovenous fistula. *J Forensic Sci*. 2018;63(5): 1577–1581.
24. Nadler SB, Hidalgo JH, Bloch T. Prediction of blood volume in normal human adults. *Surgery*. 1962;51(2):224–232.
25. Sharma R, Sharma S. *Physiology, Blood Volume*. StatPearls. *Treasure Island (FL)*. StatPearls Publishing LLC.; 2022.
26. Hirsch CH, Zumwalt RE. The "empty heart" sign. *Am J Forensic Med Pathol*. 1986;7(2):112–114.
27. Contostavlos DL. Exsanguination from impact head trauma; the explanation for the "empty heart" sign. *Forensic Sci Int*. 1998;95(3):201–212.
28. Davis JH, Wright RK. The very sudden cardiac death syndrome – a conceptual model for pathologists. *Hum Pathol*. 1980;11(2):117–121.
29. Hood I, Ryan D, Spitz WU. Resuscitation and petechiae. *Am J Forensic Med Pathol*. 1988;9(1):35–37.
30. Matshes EW, Lew EO. Two-handed cardiopulmonary resuscitation can cause rib fractures in infants. *Am J Forensic Med Pathol*. 2010;31(4): 303–307.
31. Hart AP, Azar VJ, Hart KR, Stephens BG. Autopsy artifact created by the Revivant AutoPulse resuscitation device. *J Forensic Sci*. 2005;50(1): 164–168.
32. Ziegenfuss MD, Mullany DV. Traumatic liver injury complicating cardio-pulmonary resuscitation. The value of a major intensive care facility: a report of two cases. *Crit Care Resusc*. 2004;6(2):102–104.
33. Meron G, Kurkciyan I, Sterz F, et al. Cardiopulmonary resuscitation-associated major liver injury. *Resuscitation*. 2007;75(3):445–453.
34. Krischer JP, Fine EG, Davis JH, Nagel EL. Complications of cardiac resuscitation. *Chest*. 1987;92(2):287–291.
35. Krischer JP. Complications of cardiac resuscitation. *CHEST J*. 1987;92(2):287.
36. Hlavaty L, Sung L. Resuscitation and prevalence of external facial, neck, and chest injuries in infants. *Am J Forensic Med Pathol*. 2015;36(4): 301–304.
37. Maxeiner H, Jekat R. Resuscitation and conjunctival petechial hemorrhages. *J Forensic Leg Med*. 2010;17(2):87–91.
38. Pham H, Enzenauer RW, Elder JE, Levin AV. Retinal hemorrhage after cardiopulmonary resuscitation with chest compressions. *Am J Forensic Med Pathol*. 2013;34(2):122–124.
39. Thompson M, Langlois NEI, Byard RW. Flail chest following failed cardiopulmonary resuscitation. *J Forensic Sci*. 2017;62(5):1220–1222.
40. Dupont V, Rouge-Maillart C, Gaudin A, Jeanneteau A, Jousset N, Malbranque S. Left diaphragm laceration due to cardiopulmonary resuscitation. *J Forensic Sci*. 2016;61(4):1135–1138.
41. Ondruschka B, Baier C, Bernhard M, et al. Frequency and intensity of pulmonary bone marrow and fat embolism due to manual or automated chest compressions during cardiopulmonary resuscitation. *Forensic Sci Med Pathol*. 2019;15(1):48–55.
42. Leadbeatter S, Knight B. Resuscitation artefact. *Med Sci Law*. 1988;28(3):200–204.
43. Buschmann CT, Tsokos M. Frequent and rare complications of resuscitation attempts. *Intensive Care Med*. 2009;35(3):397–404.
44. Raven KP, Reay DT, Harruff RC. Artifactual injuries of the larynx produced by resuscitative intubation. *Am J Forensic Med Pathol*. 1999;20(1): 31–36.
45. Garland J, Tse R. Neck injury and conjunctival petechiae in a woman who underwent cardiopulmonary resuscitation and subsequently died from meningococcal sepsis. *Forensic Sci Med Pathol*. 2017;13(4):432–435.
46. Milling L, Leth PM, Astrup BS. Life-threatening and suspicious lesions caused by mechanical cardiopulmonary resuscitation. *Am J Forensic Med Pathol*. 2017;38(3):219–221.
47. Pollanen MS. Pitfalls and artifacts in the neck at autopsy. *Acad Forensic Pathol*. 2016;6(1):45–62.
48. Hirsch CS, Flomenbaum M. Problem-solving in death certification. *Am Soc Clin Pathol Forensic Pathol Check Sample*. 1995;95–1:1–31. FP.
49. National Research Council. *Strengthening Forensic Science in the United States: A Path Forward*. Washington, DC: The National Academies Press; 2009. https://doi.org/10.17226/12589.
50. Brooks C. Medical examiner and coroner offices. In: *U.S. Department of Justice OoJP, Bureau of Justice Statistics*. 2018:ed2021.
51. Scientific-Working-Group-on-Medicolegal-Death-Investigation-(SWGMDI). *Regional Medicolegal Autopsy and Death Investigation Centers-Construction, Staffing, and Costs-A Report and Recommendations*; 2013, 2021 http://www.swgmdi.org/images/si6.facilityconstruction.published.9-17-13.pdf.
52. Gill JR. TheNational association of medical examiners ad hoc state medical examiner committee, state medical examiner systems, 2013: staffing, autopsies, strengths, limitations, and needs. *Acad For Path*. 2014;4(1):24–31.
53. Hanzlick R. Lawsuits against medical examiners or coroners arising from death certificates. *Am J Forensic Med Pathol*. 1997;18(2):119–123.

54. Gettler A. The historical development of toxicology. *JFS*. 1956;1:3–25.
55. Wetli CV. Investigation of drug-related deaths. An overview. *Am J Forensic Med Pathol*. 1984;5(2):111–120.
56. Stephens BG, Jentzen JM, Karch S, Wetli CV, Mash DC. National Association of Medical Examiners position paper on the certification of cocaine-related deaths. *Am J Forensic Med Pathol*. 2004;25(1):11–13.
57. Gill JR. Practical toxicology for the forensic pathologist. In: Tsokos M, ed. *Forensic Pathology Reviews*. Vol 2. Totowa, NJ: Humana Press; 2005: 243–272.
58. Skopp G. Postmortem toxicology. *Forensic Sci Med Pathol*. 2010;6(4):314–325.
59. Manini AF, Nelson LS, Olsen D, Vlahov D, Hoffman RS. Medical examiner and medical toxicologist agreement on cause of death. *Forensic Sci Int*. 2011;206(1–3):71–76.
60. Andresen H, Gullans A, Veselinovic M, et al. Fentanyl: toxic or therapeutic? Postmortem and antemortem blood concentrations after transdermal fentanyl application. *J Anal Toxicol*. 2012;36(3):182–194.
61. Drummer OH. Post-mortem toxicology. *Forensic Sci Int*. 2007;165(2–3):199–203.
62. Drummer OH, Gerostamoulos J. Postmortem drug analysis: analytical and toxicological aspects. *Ther Drug Monit*. 2002;24(2):199–209.
63. Flanagan RJ, Connally G. Interpretation of analytical toxicology results in life and at postmortem. *Toxicol Rev*. 2005;24(1):51–62.
64. Flanagan RJ, Connally G, Evans JM. Analytical toxicology: guidelines for sample collection postmortem. *Toxicol Rev*. 2005;24(1):63–71.
65. Merves M, Goldberger B. *Forensic Toxicology. Principles of Addictions and the Law-Applications in Forensic, Mental Health, and Medical Practice*. Elsevier; 2010:193–201.
66. Leikin JB, Watson WA. Post-mortem toxicology: what the dead can and cannot tell us. *J Toxicol Clin Toxicol*. 2003;41(1):47–56.
67. Richardson T. Pitfalls in forensic toxicology. *Ann Clin Biochem*. 2000;37(Pt 1):20–44.
68. Drummer OH. Postmortem toxicology of drugs of abuse. *Forensic Sci Int*. 2004;142(2–3):101–113.
69. Flanagan RJ. Interpretation of postmortem toxicology: more complicated than it might seem-part 1. *Adv Drug React Bullet*. 2008;249:955–958.
70. Flanagan RJ. Interpretation of postmortem toxicology: more complicated than it might seem-part 2. *Adv Drug React Bullet*. 2008;250:959–962.
71. Anderson WH, Prouty RW. Postmortem redistribution of drugs. In: Baselt RC, ed. Chicago: YearBook Medical; 1989:70–102. Baselt RC, ed. Advances in Analytical Toxicology; Vol 2.
72. Byard RW, Butzbach DM. Issues in the interpretation of postmortem toxicology. *Forensic Sci Med Pathol*. 2012;8(3):205–207.
73. Gill JR, Lin PT, Nelson L. Reliability of postmortem fentanyl concentrations in determining the cause of death. *J Med Toxicol*. 2013;9(1):34–41.
74. Labay LM, Caruso JL, Gilson TP, et al. Synthetic cannabinoid drug use as a cause or contributory cause of death. *Forensic Sci Int*. 2016;260: 31–39.
75. Gill JR. From death to death certificate: what do the dead say? *J Med Toxicol*. 2017;13(1):111–116.
76. deRoux SJ, Dunn WA. "Bath salts" the New York city medical examiner experience: a 3-year retrospective review. *J Forensic Sci*. 2017;62(3): 695–699.
77. Alper KR, Stajic M, Gill JR. Fatalities temporally associated with the ingestion of ibogaine. *J Forensic Sci*. 2012;57(2):398–412.
78. Gill J, Stajic M. Classical mistakes made in forensic toxicology by forensic pathologists. *Acad For Path*. 2012;2(2):228–234.
79. Jenny C, Hymel KP, Ritzen A, Reinert SE, Hay TC. Analysis of missed cases of abusive head trauma. *JAMA*. 1999;281(7):621–626.
80. Billmire ME, Myers PA. Serious head injury in infants: accident or abuse? *Pediatrics*. 1985;75(2):340–342.
81. Hall JR, Reyes HM, Horvat M, Meller JL, Stein R. The mortality of childhood falls. *J Trauma*. 1989;29(9):1273–1275
82. Tarantino CA, Dowd MD, Murdock TC. Short vertical falls in infants. *Pediatr Emerg Care*. 1999;15(1):5–8.
83. Case ME, Graham MA, Handy TC, Jentzen JM, Monteleone JA. Position paper on fatal abusive head injuries in infants and young children. National association of medical examiners ad hoc committee on shaken baby syndrome. *Am J Forensic Med Pathol*. 2001;22(2):112–122.
84. Plunkett J. Fatal pediatric head injuries caused by short-distance falls. *Am J Forensic Med Pathol*. 2001;22(1):1–12.
85. Schaber B, Hart AP, Armbrustmacher V, Hirsch CS. Fatal pediatric head injuries caused by short distance falls. *Am J Forensic Med Pathol*. 2002; 23(1):101–103. author reply 103–105.
86. Geddes JF, Hackshaw AK, Vowles GH, Nickols CD, Whitwell HL. Neuropathology of inflicted head injury in children. I. Patterns of brain damage. *Brain*. 2001;124(Pt 7):1290–1298.
87. Dolinak D, Reichard R. An overview of inflicted head injury in infants and young children, with a review of beta-amyloid precursor protein immunohistochemistry. *Arch Pathol Lab Med*. 2006;130(5):712–717.
88. Duhaime AC, Alario AJ, Lewander WJ, et al. Head injury in very young children: mechanisms, injury types, and ophthalmologic findings in 100 hospitalized patients younger than 2 years of age. *Pediatrics*. 1992;90(2 Pt 1):179–185.
89. Gill JR, Goldfeder LB, Armbrustmacher V, Coleman A, Mena H, Hirsch CS. Fatal head injury in children younger than 2 years in New York City and an overview of the shaken baby syndrome. *Arch Pathol Lab Med*. 2009;133(4):619–627.
90. Nashelsky M. Death certification-revisiting the primary CDC and NAME literature. *Acad For Path*. 2012;2(2):120–125.
91. Zumwalt RE, Hirsch CS. Subtle fatal child abuse. *Hum Pathol*. 1980;11(2):167–174.
92. Gobbato F, Vecchiet F, Barbierato D, Melato M, Manconi R. Inaccuracy of death certificate diagnoses in malignancy: an analysis of 1,405 autopsied cases. *Hum Pathol*. 1982;13(11):1036–1038.
93. Zumwalt RE, Ritter MR. Incorrect death certification. An invitation to obfuscation. *Postgrad Med*. 1987;81(8):245–247, 250, 253-244.
94. Pritt BS, Hardin NJ, Richmond JA, Shapiro SL. Death certification errors at an academic institution. *Arch Pathol Lab Med*. 2005;129(11): 1476–1479.
95. Gill JR, Goldfeder LB, Hirsch CS. Use of "therapeutic complication" as a manner of death. *J Forensic Sci*. 2006;51(5):1127–1133.
96. Ravakhah K. Death certificates are not reliable: revivification of the autopsy. *South Med J*. 2006;99(7):728–733.
97. Cina SJ, Selby DM, Clark B. Accuracy of death certification in two tertiary care military hospitals. *Mil Med*. 1999;164(12):897–899.
98. Maudsley G, Williams EM. "Inaccuracy" in death certification–where are we now? *J Public Health Med*. 1996;18(1):59–66.
99. Slater DN. Certifying the cause of death: an audit of wording inaccuracies. *J Clin Pathol*. 1993;46(3):232–234.
100. Smith Sehdev AE, Hutchins GM. Problems with proper completion and accuracy of the cause-of-death statement. *Arch Intern Med*. 2001; 161(2):277–284.
101. Jordan JM, Bass MJ. Errors in death certificate completion in a teaching hospital. *Clin Invest Med*. 1993;16(4):249–255.

102. Fischtein D, Cina SJ. Errors on death certificates requiring amendments: the Broward County experience. *Am J Forensic Med Pathol*. 2011;32(2): 146–148.
103. Cambridge B, Cina SJ. The accuracy of death certificate completion in a suburban community. *Am J Forensic Med Pathol*. 2010;31(3):232–235.
104. Brooks EG, Reed KD. Principles and pitfalls: a guide to death certification. *Clin Med Res*. 2015;13(2):74–82.
105. McGivern L, Shulman L, Carney JK, Shapiro S, Bundock E. Death certification errors and the effect on mortality statistics. *Public Health Rep*. 2017;132(6):669–675.
106. Randall B. Death certification: a primer, Part IV–problems in death certification. *S D Med*. 2014;67(8):315–317, 319.
107. Myers KA, Farquhar DR. Improving the accuracy of death certification. *CMAJ*. 1998;158(10):1317–1323.
108. Hanzlick R. Quality assurance review of death certificates: a pilot study. *Am J Forensic Med Pathol*. 2005;26(1):63–65.
109. Goldstein RD, Blair PS, Sens MA, et al. Inconsistent classification of unexplained sudden deaths in infants and children hinders surveillance, prevention and research: recommendations from the 3rd International Congress on Sudden Infant and Child Death. *Forensic Sci Med Pathol*. 2019;15(4):622–628.
110. Tatsumi K, Shapiro S, Bundock E. Death certificate surveillance: a component of death investigation. *Acad Forensic Pathol*. 2013;3(1):99–104.
111. Gill J, Maloney K, Hirsch C. The consistency and advantage of therapeutic complication as a manner of death. *Acad For Path*. 2012;2(2): 176–181.
112. Gill JR, Ely SF, Toriello A, Hirsch CS. Adverse medical complications: an under-reported contributory cause of death in New York City. *Public Health*. 2014;128(4):325–331.
113. Krehbiel K, Matshes E, Lathrop S, Pinckard J. Medical therapy-related deaths and the medical examiner. *Acad For Path*. 2015;5(3):421–435.
114. Stahl-Herz J, Pasquale-Styles M. Autopsy-Diagnosed Injury Deaths in Persons with Acute or Chronic Alcohol Use: A Review of 1000 Deaths with History of Alcohol Use (4.7 Abstract). Paper Presented at: National Association of Medical Examiners Annual Meeting2021; West Palm Beach, FL.
115. Riddick L, Luke J. Alcohol-associated deaths in the District of Columbia–a postmortem study. *J Forensic Sci*. 1978:493–501.
116. Rehm J, Gmel G, Sempos CT, Trevisan M. Alcohol-related morbidity and mortality. *Alcoh Res Health*. 2003;27(1):39–51.
117. Rosenberg ML, Davidson LE, Smith JC, et al. Operational criteria for the determination of suicide. *J Forensic Sci*. 1988;33(6):1445–1456.
118. Lin P, Gill JR. Delayed homicides and the proximate cause. *Am J Forensic Med Pathol*. 2009;30(4):354–357.
119. Gill JR, McCubbin KR, Landi KK. Homicide by asthma. *Acad For Path*. 2011;1(1):122–127.
120. Prahlow JA. Homicidal cerebral artery aneurysm rupture. *J Forensic Sci*. 2004;49(5):1082–1085.
121. Graham MA. Investigation of deaths temporally associated with law enforcement apprehension. *Acad For Path*. 2014;4(3):366–389.
122. Mitchell Jr RA, Diaz F, Goldfogel GA, et al. National association of medical examiners position paper: recommendations for the definition, investigation, postmortem examination, and reporting of deaths in custody. *Acad Forensic Pathol*. 2017;7(4):604–618.
123. Hirsch CS. Restraint asphyxiation [letter]. *Am J Forensic Med Pathol*. 1994;15(3):266.
124. Davis JH. Bodies found in the water. An investigative approach. *Am J Forensic Med Pathol*. 1986;7(4):291–297.
125. Instructions for completing the cause-of-death section of the death certificate for injury and poisoning. In: *Statistics CfDCaPNCfH*. Hyattsville, MD: U.S. Department of Health and Human Services; 2004:1–2.
126. Scott DF. A study of logger fatalities from 1992–2000. *Inj Prev*. 2004;10(4):239–243.
127. Centers for Disease Control and Prevention. Nonoccupational logging fatalities–Vermont, 1997–2007. *MMWR Morb Mortal Wkly Rep*. 2008; 57(10):260–262.
128. Rocha LA, Fromknecht CQ, Redman SD, Brady JE, Hodge SE, Noe RS. Medicolegal death scene investigations after natural disaster- and weather-related events: a review of the literature. *Acad Forensic Pathol*. 2017;7(2):221–239.

CHAPTER

4

The scene investigation, postmortem changes, and time of death

Susan F. Ely, Kelly Keyes and James R. Gill

The need for circumstantial details bears an inverse relationship to the severity of disease alterations revealed by autopsy. ***Dr. Joseph Davis***

People do die the way they live; death investigation would be chaos if the situation were otherwise. ***Dr. James Luke***[1]

Determination of the cause and manner of death should always take into account all available information. This includes the circumstances of the death, environment at the scene, social and medical history, autopsy findings, and toxicologic results. Reliance upon any one factor is fraught with error. ***Drs. Roger Mittleman and Charles Wetli***[2]

The scene investigation represents one of the most important decision points in forensic practice. Should an investigator go to the scene? Should the body be brought into the medical examiner/coroner (ME/C) office for an examination? If you drop the ball here, the opportunity for accurate death certification may be lost. The collection and/or documentation of important evidence in support of the facts of the case also begins here. Observations of the body with respect to its position, its specific location, and to the postmortem interval (lividity, rigor and algor mortis, degree of putrefaction, etc.), before further time passes and refrigeration interferes, cannot be as accurately reproduced at a later time. The scene investigation offers exactly one opportunity at a specific place and moment in time for the documentation of physical evidence while the body is still at the death scene. Documentation of clues relating to the time of death (and many other determinations; see below), both on the body and in the environment (e.g., date on the last newspaper that was brought into the house) is more important at the death scene than anywhere else. Crime scenes and deaths in public view have their own challenges and requirements, as do mass fatality incidents (MFIs) with comingling and devastating injuries that may impair visual identification.

The written narrative

Reporting a chronological narrative of what happened (and what did not happen) constitutes the backbone of a quality death investigation report. What happened before death, specifically? Was the decedent sick or complaining of anything in particular just prior to death or were they in their usual state of health? Were they short of breath or were they complaining of chest pain? Did they suddenly collapse or were they heard snoring and later found dead? Were they behaving erratically and incoherently or were they cogent and purposeful, even if clearly distraught? What were the objective observations at the scene? How and when was the body discovered? By whom? Were attempts at resuscitation made? Was there a suicide note, pills strewn on the bed with inappropriately empty pill bottles nearby, a dripping blood trail from the kitchen sink to the bathroom sink to the bed, or an indication of a struggle with blood on the floor and multiple surfaces, including cast-off bloodstains of the ceiling (n.b., always look at the ceiling)?

Providing this context and degree of detail is critical in facilitating the ME/C in the proper certification of the death and in answering other corollary questions that may be asked of them at a later time. Beyond that, key information that is standard to ask for includes the last time the decedent was reliably known to be alive (seen or spoken to), the name and contact information of next of kin, the name of the primary care medical provider in (natural) deaths at home (this can also be obtained from the prescription bottles or medical alert bracelets), and, for traumatic

Principles of Forensic Pathology
https://doi.org/10.1016/B978-0-323-91796-4.00030-1

deaths, the time (it can be approximate), date, and place of the injury (police documents may later be required to corroborate this information, if necessary, but this information may be adequate in providing a start point). Depending on the death circumstances, certain demographic information, such as country of origin, recent travel, or ethnicity may have bearing on public health matters or considerations of certain disease entities.

It is always preferable to *extract information from* **primary sources** *rather than from secondary or tertiary sources, wherever possible,* and to seek and report "the news, not the editorial." Later forensic conclusions must be drawn from credible facts, not from others' opinions, per se, particularly from those who are not forensically trained. For example, some emergency medical technicians (EMTs) are willing to opine about the cause of death and the postmortem interval. These opinions or observations can be problematic if inaccurate and have the capacity to sway or confuse friends or family at the death scene. Not uncommonly, it may become necessary to later clarify with them that while EMTs are medically trained, they are not usually forensically trained, per se. Documenting an opinion of the decedent's cardiologist about the extent of heart disease, on the other hand, constitutes a credible and perhaps helpful opinion from a trained expert about something falling within the breadth of their training and experience. The opinion of a neighbor about what they believe *may* have happened can be noted, but should be viewed through a very different lens. Care must be taken when having a discussion with a psychiatrist about their patient who just committed suicide. One should not necessarily expect the psychiatrist to offer the opinion that their patient was suicidal. If they do, they are potentially admitting to malpractice because they would have had a duty to act. Unfortunately, the reality is that psychiatrists may not have known that their patient was suicidal. For this reason, even though the psychiatrist may indeed be speaking truthfully, one mustn't stringently rely upon this information to determine whether or not the death was a suicide.

Questions from the **medicolegal death investigator (MDI)** should instead be geared toward the specific observations or facts that support an individual's opinion rather than toward the opinion itself. If a neighbor asserts that she thinks the decedent was "really sick" or "really depressed," for example, the best follow-up questions probe into how and why they came to that conclusion, on what specific observations they made (e.g., substantial weight loss, new onset reclusiveness, giving away their cat, constant coughing, complaints of dizziness, etc.). If they were a direct witness to the death, they should be asked to describe in detail what they saw rather than what they think probably occurred or caused the death. The police or emergency medical service (EMS) telling an MDI that "there wasn't enough water in the tub for the decedent to drown" is much less informative and has less veracity than offering objective facts and observations (specific water level and position of the body/face relative to water, as observed by the first person to discover it). Seek the news, not the editorial.

The foregoing mental checklists present potentially pertinent questions and issues that are best asked and considered by an MDI, by phone or at the scene, in an effort to assist the ME/C in most accurately evaluating the case and certifying the death. They are organized *alphabetically* and by disease/injury type, scene type, or special topic/subject category. It is important to emphasize, however, that no two deaths are identical, regardless of the general categories within which they fall. As such, "checklists" should serve only as potentially helpful guideposts and reminders, not as prescriptions. They should be used to the extent necessary and as deemed appropriate and relevant in the judgment of the investigator.

General tips

- *Secured location* (e.g., chain locked), versus bolt or just slam-locked?
- Why was person found? Odor? Texts containing suicidal ideations?
- Is it a single room occupancy housing facility? If so, for geriatric residents, formerly homeless residents, or those with a substance use history? Was it a psychiatric patient residence? A studio apartment? Unkept or hoarder conditions? (Fig. 4.1)
- When was the decedent last seen alive?
- Were there any recent complaints, acute illnesses, falls/traumatic events, or were they in their usual state of good health? Any past medical and/or social histories?
- Are there any prescription medications, alcohol, or drugs on scene? If so, where are they found in relation to the decedent? Is the alcohol covered in dust and in a cabinet or out and open, with many empty bottles or cans found in the trash?

Anaphylaxis and food

What is the manner?—this informs when jurisdiction should be taken:

- If *food or pollen* causes anaphylaxis, then the manner is **Natural** (food is necessary to live and does not constitute a toxin).

FIGURE 4.1 Hoarder kitchen.

- If caused by a *medication*, the manner would be **Therapeutic Complication, Natural, or Accident** (depending on jurisdictional choices), assuming it is not for the treatment of an injury.
- A *bee sting* is an envenomation, so is **Accident.**
- *Fish toxin poison* (Ciguatera, Scombroid) is **Accident.**
- *Wild mushroom poisoning* is **Accident.**
- *Escherichia coli/Staphylococcus/Salmonella food poisoning* is **Natural.**
- Food bolus choking while swallowing (see choking below) is **Accident.**

Asthma/emphysema

- Did the decedent have recent shortness of breath? Wheezing? Stridor?
- Prior hospitalization for asthma? Intubated? If so, how many times? When was last hospitalization?
- Home oxygen or nebulizer? If so, were they in place on the decedent and turned on, or only nearby?
- Recent increase in the use of inhalers (Fig. 4.2)? Steroid dependency?
- Smoker?

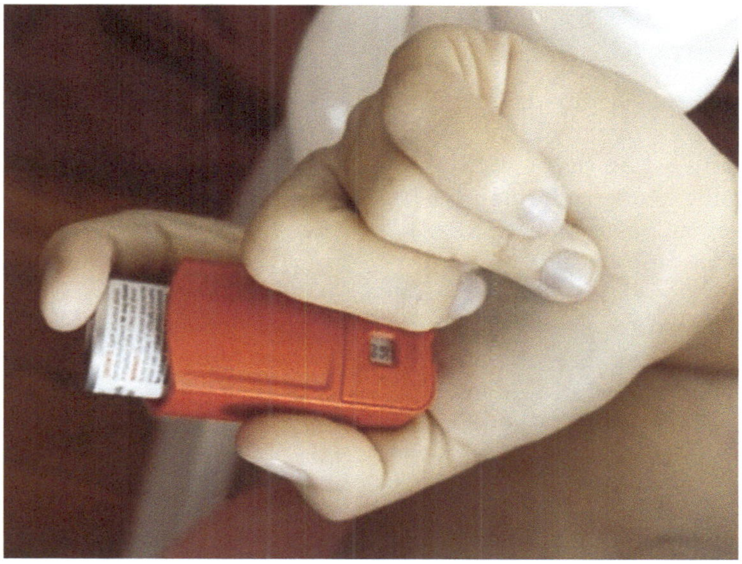

FIGURE 4.2 Asthma inhaler.

Bathtub deaths (see also Chapters 6 and 11)

- Was the body fully in the tub, or partially in and out? Appeared to have fallen into the tub?
- Appeared to have been taking a bath or a shower?
- Clothed or naked?
- Head/face above or below the water line?
- Head toward the faucet end of the tub?
- Water running? Were/are toys or other objects in the tub? If so, what were/are they, and where? At the time of discovery, was the drain inadvertently obstructed by washcloths or other objects or was it intentionally plugged?
- What was the water temperature at the time of discovery?
- Was the decedent known to take baths regularly or occasionally?
- If the tub has been drained, ask the above questions of the person who *first* discovered the body and look for a water/soap line.
- Electric appliances nearby?
- Always empty the tub (especially if water is bloody or cloudy) to see if anything else is in the tub with the decedent (weapon, pill bottle, etc.)
- Medical and drug histories: Heart disease, Epilepsy? Stroke? Multiple Sclerosis? Alcoholism? Substance use history (illicit or prescription)? Were there pills, illicit drugs, or alcohol nearby?
- What parts of the body were wet (including scalp hair)? Was there submersion artifact of the palms and/or soles? Was there skin slippage? If so, in what distribution? These observations become particularly important if the body has been removed from the tub prior to the arrival of the MDI.

Cancer

- Especially with potential external exams, the ME/C must know the status of the cancer (e.g., terminal, remission/remote, cured, recent therapy?). If a person found dead at home has a history of heart disease and breast cancer, more information is needed to accurately certify the death, especially in the absence of an autopsy. Is the cancer in remission or is it an active problem? For example, if the patient had a low-grade cancer that was treated with radiation and has had no recurrences for 10 years, it would probably be ignored for the purposes of death certification and the ME/C would certify the death with the heart disease. On the other hand, if the patient had lung metastases and was recently discharged from the hospital to home for comfort care, the cancer would take center stage. Specifically asking how the patient had been in the days or hours leading up to their death is sometimes very informative. Blood at the scene may relate to vascular ports, inadvertently or intentionally removed, or complications of gastrointestinal cancers. Inquiring about the decedent's mental health/suicidality may be pertinent as well, depending on the case. The ME/C is often called about home hospice deaths, but, if deemed natural, these fatalities are often made "no cases," allowing the hospice physician to certify the death.

Carbon monoxide (nonfire) (see also Chapter 14)

- If the body is found in a car, was it running? If the motor was off, was the ignition in the on position? Was the gas tank full? Was the battery dead? What was the make and year of the car? Does the death appear to be a suicide? If yes, how? Was an apparatus set up (Fig. 4.3)?
- Carbon monoxide poisoning is often considered when there is more than one fatality in the same residence (see below discussion on "companion cases"). There may be a need to alert authorities so that they can exclude a ventilation duct blockage or faulty heating system.
- Deaths occurring in homes during power outages should include a search for gas-powered generators. They may be set up in the basement, with the deaths occurring in the upper floors.
- Suicides may involve carbon monoxide in a variety of ways. These include use of car exhaust, as described above, or from a small charcoal grill in an enclosed location or use of chemicals to generate CO by mixing formic and sulfuric acid ($H_2SO_4 + HCOOH \rightarrow CO + H_2SO_4 \bullet H_2O$). When inhaled, methylene chloride (found in paint strippers) can be converted to CO by the liver.[3,4]

Choking on food (see also Chapter 11)

It is rare for a neurologically intact, nonintoxicated adult to choke to death on food. In this context, choking is defined as an internal occlusion of the airway. Adults chew and swallow boluses of food thousands of times in their

FIGURE 4.3 **Duct from tail pipe to passenger compartment.**

life and yet this type of death is rare. When it does happen, there is usually an underlying reason that would otherwise explain a chewing, swallowing, or gag reflex impairment (e.g., remote stroke, multiple sclerosis, psychiatric illness, alcohol and/or depressant drug intoxication, dementia, etc.). The forensic pathologist (FP) needs to know what the risk factor(s) is/are because it may constitute the underlying cause of death. Conversely, choking deaths in young children do not necessarily require an underlying cause.

- It is important to speak directly to any witnesses (obtain the news, not the editorial); do not rely solely upon secondhand information.
- Was the person eating just before they collapsed?
- What were they eating?
- Were they drinking alcohol?
- Was the person witnessed to collapse?
- Was it sudden or was there a seeming struggle or extended pause/alert, but silent period before unconsciousness? If so, what, specifically, was seen and how long did it last? Did they grab their neck, point to their throat? Could they vocalize?
- Was the person speaking just before they collapsed? The "café coronary" is the mistaken diagnosis of a "heart attack" while eating. These deaths may be chokings.[5]
- What specifically did the EMTs find when they attempted to intubate? Was anything lodged in his airway? If yes, what and where, loosely in the oropharynx or wedged in larynx or trachea? Were forceps needed to extract it? Speaking directly with first responders can be very helpful, if possible, as this level of detail may not be found in their written reports.
- Remember there is a difference between terminal aspiration (agonal aspiration of partially digested/regurgitated gastric contents because the person is dying from something else) versus choking on a bolus of food. Emergency responders may mistake terminal aspiration for the cause of death (i.e., someone dying *while or after* eating vs. dying *from* eating); for that reason, they need to give you the facts, not their opinion. After death, sphincters relax and the contents of a full stomach may decompress through the esophagus.
- Were there any neuropsychiatric problems that may have limited mastication or impaired swallowing function or the gag reflex (e.g., severe behavioral or psychiatric disability, edentulousness, multiple sclerosis)?

Communicable disease (e.g., tuberculosis, meningitis, CJD, etc.; see chapter 17)

Tuberculosis and meningitis

Both are reportable communicable diseases to the Department of Health.[6–10] If a patient has known tuberculosis (TB) and has been in the hospital receiving treatment for several weeks before death, this death generally does not need to be made a ME/C case. If TB is suspected, but not diagnosed (for example, the patient died shortly after admission), then the death may be made an ME/C case in many jurisdictions. Prisoners with TB are typically ME/C cases, regardless of the clinical scenario. The bottom line is that not all TB deaths need to be certified/autopsied by a ME/C; case by case judgment, jurisdictional policies, and common sense must prevail.

A similar approach applies for meningitis. If the diagnosis has been made at the hospital, there is no need to pursue further. That said, if meningitis is reasonably suspected, but has not been diagnostically worked up, particularly in a younger individual, for whom the cause of death is unclear, ME/C involvement becomes very important. If **acute bacterial meningitis** is diagnosed at autopsy, particularly if caused by **neisseria meningitidis**, it constitutes an urgent public health matter, as time-sensitive prophylaxis of first responders and other close contacts may be in order.[6]

AIDS/HIV

- Risk factor(s)? Sexually transmitted (not an underling disease), intravenous drug use (the underlying/ proximate cause), transfusion-related (why transfused? Injury? Assault?). If an injury, the circumstances will inform the manner when human immunodeficiency virus/acquired immunodeficiency syndrome (HIV/AIDS) is involved with the death.[7,11–14]
- Is the history of HIV (+) or full-blown AIDS?
- Any opportunistic infections?
- Treatment? CD4 count, viral load?
- Recent illness in the past few days or usual state of good health?
- Were they expected to die at this time in relation to the HIV infection?

Creutzfeldt–Jakob disease

With regard to spongiform encephalopathies (Creutzfeldt–Jakob disease [CJD])[15,16], there are two instances in which the ME/C is likely to accept jurisdiction of the case:

- The **department of public health** is already aware of the suspected diagnosis or death and is **specifically requesting ME/C to complete an autopsy.**
- The **patient is suspected to have the rapid-onset type of CJD** (symptoms and death occur in a matter of a few weeks/months rather than the chronic, senile type that occurs over many months to years) **AND the diagnosis has not been confirmed by biopsy AND the hospital is unwilling or unable to perform the autopsy.**
- The purpose of taking jurisdiction is to make sure that tissue is collected to send for diagnostic confirmation. The hospital can do this, but may be unwilling or unable to, and then it becomes the responsibility of the ME/C to make sure it gets done.
- The rapid onset type can be highly infectious and is of particular public health interest. The chronic, senile type does not constitute a public health concern.
- If the hospital is going to perform the autopsy, the case (a natural death) does not need ME/C involvement, but a record of the report and its disposition should be made.
- If the diagnosis has already been made by BIOPSY and reported to the Department of Health, then the case (a natural death) does not need ME/C involvement, but a record of the report and its disposition should be made.
- If it is the chronic (symptoms for years) type of suspected CJD, then it is not the highly infectious type of public health concern, and the case (a natural death) does not need ME/C involvement, but a record of the report and its disposition should be made.

Because of contamination hazards, many ME/C offices do NOT perform a complete autopsy in suspected rapid-onset CJD deaths. Rather, they may remove the brain only, involve limited personnel, and send the tissue to a lab that specializes in confirmation.

Novel coronavirus (COVID-19)

Most COVID-19 deaths occur in the hospital, but may be reportable to the ME/C because of the threat to public safety and/or when deaths occur at home. If it is a presumptive or definitive diagnosis, some offices will decline jurisdiction and allow the treating physician to certify the death.[17–23]

Delayed hospital deaths following recent or remote injury (see also Chapter 15)

- The death certificate requires the *date, time, and location of the injury occurrence* (type of place and address).

- If the hospital does not have this information, the police or next of kin may know (if a noncriminal/accidental injury was sustained at home, without police involvement, the family is typically the best source of information). This should be part of the initial investigative information gathering.
- If outside the home, extract the location and date of injury from the EMS sheet—this is needed to locate police reports.
- What were the initial descriptions of the injuries? Obtain radiology reports, as needed.
- After a hospital survival period ranging from hours to days, the earliest/admission blood samples are best reflexively requested (see above); the ME/C can later decide on the need to have them analyzed, depending on the case.
- *Hospital course:* Unconscious from the beginning? Surgery? Amount of blood loss? Blood product replacement? Thrombolytic therapy?
- Any hospital toxicology results?
- Expected to die?
- Medical records: For a protracted admission (weeks to months), it is often best to request only selected documents, if needed at all, depending on the details of the case (e.g., admission history and physical, imaging, microbiology, urine toxicology screen, operative notes, etc.).
- Access to *electronic health records* at the time of the death investigation **triage has improved the efficiency and quality of clinical information.**

Descent (fall versus jump) from height

- How high and deep was the barrier (wall/window ledge/balcony/roof top parapet) that the decedent had to scale? This helps address the "intent" issue needed for suicide (Fig. 4.4).[24–45]
- Were shoes left behind?
- Was there any video surveillance?
- If from a fire escape or roof top, was the decedent known to spend any time there, smoking cigarettes, etc.?
- Was the decedent known to use potentially hallucinogenic drugs? Was there broken glass and/or a broken window and/or evidence of disarray at the scene?
- Was there any rope near the window or extending from the roof, indicating a possible accidental fall while attempting to gain access through a window?

FIGURE 4.4 Decent from height.

Diabetes mellitus

- Was the decedent insulin-dependent? Was there a glucometer[46] at the scene? Any glucose readings? Sugar diary? Check the refrigerator for insulin.
- Was there recent malaise or an intercurrent illnesses (setting the stage for diabetic ketoacidosis [DKA])?
- Was the decedent prescribed any *atypical/second generation antipsychotic medications* (clozapine, olanzapine, quetiapine, and risperidone); if so, when started? Any recent dosage changes or weight gain? Keep in mind that atypical antipsychotic medications can cause new onset diabetes mellitus (DM), with a telescoped course, sometimes even presenting with DKA.

Drowning (see also Chapter 13)

- What happened? Witnesses? Any evidence of struggle, splashing, yelling? Or just a quiet, sudden submersion while swimming?
- If found in a pool, did they appear to have been swimming (i.e., wearing a bathing suit) or to have fallen in inadvertently? If the latter, based on what evidence? Was any other object in the pool with them?
- Anyone else in the water? Age(s)?
- How deep was the water (to be later compared to the height of the decedent)? Rip tides? Calm water? Weather conditions?
- Did the person know how to swim?
- When was the person last seen alive?
- Any history of suicidal ideations or attempts? Domestic violence?
- What was the last thing the person ate or drank?
- *If a child (bathtub, pool, other body of water):*
    - Where was the person who was supposed to be supervising the child? What were they doing and for how long? Was the child capable of getting into the pool, opening a gate, etc.?
    - Any other children at home? Age(s)?
    - Has either parent previously had a child who died?
    - Any prior child protective services involvement?
    - Was there a fence or gate around pool? If so, did it have a functional lock mechanism? Was there a ladder if an above ground pool?
- Was the person pushed in?
- Was anyone holding the person under water?
- Any past medical problems (surgeries, alcohol, drug use, heart disease)?
- Was the person trying to see how long he could hold his breath under water (hyperventilation syndrome)?
- Was there a diving board or was the person seen to dive into the shallow end (cervical vertebral fracture?)
- If found floating in a public waterway sometime after having entered the water from an unknown location, identity is the first order of business in the investigation. Once identified, ask when were they last seen alive? What were they doing? What was their mental state? Did they have their keys and wallet with them or uncharacteristically leave them at home? Is there a body of water that is close to their residence? Are they known to go there for any reason (walks, fishing, etc.)? Can they swim? Were they clothed? Were they wearing a bathing suit or other swim gear?

Drugs and ethanol/substance misuse deaths (see also Chapter 14)

- Was the decedent enrolled in a methadone maintenance treatment program? If so, which one and for how long, and on what dosage regimen?
- Were there alcohol bottles (Fig. 4.5)?
- What was the decedent last known to be doing and when relative to their discovery? Were they heard snoring prior to being found dead? Is this normal for them?
- In what specific position was the decedent found? Was there purge? A "foam cone"? Track marks?
- What medications are present and/or prescribed? What other medications are available in the home?
- If the death was pronounced in a hospital, was a hospital toxicology screen performed? if yes, what were the results? And, what battery of drugs did they test for, specifically?

FIGURE 4.5 **Alcohol "nips."**

- Are hospital admission bloods available to request? (see below; only relevant if there was a survival interval of hours or more following a return of circulation after resuscitative efforts).
- Was any therapeutic fentanyl, morphine, benzodiazepines, or other depressant medication given during the hospital stay? If so, when, specifically, relative to the time and date stamp on the obtained hospital admission bloods?
- Was a blood alcohol concentration obtained upon admission?
- Were drugs or drug paraphernalia found at the death scene? If so, what were they, specifically? Needles? Drug pipe? Spoon? Copper/wire scrubbing pads? Glassine envelopes?
- If there is a sudden surge of overdose deaths in a specific area, there may be "bad"/highly potent drugs on the street. Make a note of any markings on the drug packet and consider having drugs sent to the toxicology lab for examination.[47,48]
- Many ME/C offices will decline jurisdiction on most hospital deaths of patients who have infectious (e.g., AIDS, hepatitis, endocarditis) or anoxic–ischemic complications of IVDA. That said, many/most ME/C offices will take jurisdiction of the death of an IVDA if the answer is "yes" to any of the following questions:
  1. Was the patient in the custody of police or a correctional facility?
  2. Did any physical injury cause or contribute to death?
  3. Is the death a sequela of an overdose (e.g., anoxic encephalopathy)?*
  4. Did an acute drug intoxication cause or contribute to death?*

Communication with families (see also Chapter 16)

- In the investigative report, include the telephone number(s) and relationships for the next of kin. Note any language barriers or contact preferences.
- What are the family wishes regarding autopsy? Were any personal, religious, or cultural objections raised? (See below)
- The MDI should attempt to make contact with the family unless precluded by the time of day (e.g., 4 am). At the very least, make sure that the contact number is in the report so the ME/C can try to call them in the morning or later in the day, if needed.
- If a hospital death, ask the reporting physician if the family is: (1) aware of the death, and (2) aware of the ME/C involvement and plan. If the family is unavailable to speak to the MDI, the medical staff should provide the family's contact information and the MDI might also leave their contact information for the doctor to provide to the family.

* Depending on the jurisdiction and staffing capabilities, the latter two categories may only require certification by the ME/C, but not necessarily transport into the office.

Family preferences for and objections to autopsy

- When a family specifically wants an autopsy to be done, ask why? What is their specific concern or question? Having this information will allow the ME/C to communicate to them before the examination, often answering questions and meeting their needs, while obviating the need for an autopsy. They can also manage their expectations regarding the limitations of the autopsy and what it can and cannot determine (specific time of death for the latter, for example). If the family has no preference regarding autopsy, this too should be documented in the narrative. On cases in which the MDI judges that an autopsy is unlikely to be performed, they should explain that to the family. If a preference for an autopsy is expressed, that too should be documented in the MDI report, while communicating to them that the decision about whether to perform an autopsy or not will ultimately fall within the discretion of the ME/C.
- On the opposite side of the spectrum are **objections** to autopsy, sometimes adamant. These are even more important to clearly document in the investigative report and elsewhere since, unlike an external exam reverting to an autopsy exam, once an autopsy has been done, it cannot be undone.
- Objections can be based on **personal or cultural preferences** or **religious beliefs or tenets.** In some states, there is a specific law defining religious objections to autopsy that prescribes the rights of those asserting them, as well as the rules and legal exceptions to the protection of those rights. That said, a careful, considerate, and communicative approach should be the norm. *The investigator's job is to clearly document that an objection has been expressed and what the basis of that objection is.* **Ideally, such documentation should be placed in bold in the first sentence of their report.** *The name, relationship, and phone number of the individual asserting the objection should also be clear.* Best practices would also suggest that the investigator assures the family member/executor/friend registering their objection that this information will be shared with the ME/C, and that the ultimate decision will be communicated to them before an autopsy is performed, but that it will be at the discretion of the ME/C.

Final exit suicides (see also Chapter 11)

- Was the plastic bag completely over head when first found?
- Was the bag tightly secured around neck? Was it ripped open or removed by EMS or others?
- Note and photograph any condensation within the bag
- Medications: Pill counts? A pill slurry or any alcohol nearby?
- If a helium or nitrogen tank was used, was the valve open? Is any helium/nitrogen left in tank?
- Was there a suicide note, instructions, a journal, or an instructional suicide book in view?

Fire (see also Chapter 12)

Hospital deaths: Remember to ask for the **admission carboxyhemoglobin saturation**, if measured (and/or *request admission bloods* for ME/C testing of same) and inquire about the duration of resuscitative efforts by EMS and hospital staff (specifically involving oxygen therapy). If there was a prolonged survival interval in a burn unit, document the thermal injury as calculated by the **percent body surface area involvement** that was initially assessed prior to subsequent debridement surgeries and healing. Ask if there was inhalation injury?

Gastric bypass surgery (see also Chapter 8)

Obtain details about the procedure, such as the operative note.[49–55] What exactly did they do surgically? Did they place an inferior vena cava filter? How many days after surgery was the death? Signs of infection? Gastrointestinal signs and symptoms? Suspected pulmonary thromboembolism? Excess pain medication?

Gunshot wounds (also see Chapter 9)

- If a bullet was recovered during surgery, find out from where in the body it was recovered (try to get the operative report). What did it perforate? By the time an autopsy is performed, the anatomy will have been distorted by the surgery and possible healing and it may be difficult to trace the path of the bullet without this clinical information. With gunshot wounds (GSWs) deaths in the hospital, the MDI still should touch base with the police. In particular, inquire about whether it was a police-involved shooting or anything otherwise unusual about it. For hospital deaths, the physician should be asked specifically if they altered any wounds (such as, incising a GSW during emergency surgeries or thoracotomies or using a GSW defect as an entry point for a thoracostomy tube, etc.).

- If there is a scene investigation and a gun is present, note its specific location relative to the decedent, the decedent's experience with and habits around using and/or cleaning the gun, and whether clothing was worn and perforated.
- If a child, how was the gun normally stored?

Head injury (also see Chapter 10)

- After the incident, was the decedent conscious at any time?

(Knowing that someone was unconscious immediately after a traumatic incident and that they remained unconscious is a very useful data point in a head trauma investigation, as this may represent a sign of diffuse axonal injury (DAI); having a conscious/lucid interval excludes DAI/"fatal concussions")

- Were any statements about the incident made to rescue/police/medical personnel by the decedent or others? Are any witness statements or phone numbers documented in the 911 call report or the ambulance call report?
- Request a copy of the neurosurgical report, if applicable, and of all head computed tomography (CT) scan reports, the initial/presurgical report being the most important.
- If the clinician uses the phrase, "intracranial" hemorrhage, ask them to specify what that means—its size/significance relative to the neurologic compromise (herniation?) and where it is/was in the cranial cavity: Subdural? Intracerebral? Basal ganglia? Contusion hematoma?
- Is/was there any facial/scalpular/soft tissue injury on exam or by initial imaging? This is particularly important in religious autopsy objections (i.e., when an autopsy will not be performed) or when there was a prolonged survival interval.
- Ask the hospital to send the radiology report if you have any doubt; hospital physicians may not appreciate the important difference between a subdural hematoma and a subarachnoid hemorrhage.
- Obtain a medical history, if known: Alcoholic? Hypertensive? Taking anticoagulants? Known to have an unsteady gait?
- Ask for the admission blood alcohol concentration, if performed.
- If "found down," how were they found and where specifically (e.g., in relation to a staircase, in the street, partially under a car?). Were there any witnesses or video surveillance?
- Are the police already involved in the investigation?
- When did the head injury occur relative to the death?

Deaths following heart surgery

- Request a copy of the dictated operative report or cardiac catheterization report.
- Heart valve replacement: ask for the etiologically specific disease/indication, not just aortic stenosis; rather, rheumatic? degenerative? bicuspid aortic valve?
- Heart transplants: Why? What was the underlying disease? Any rejection issues?

Hip fractures

- *Natural (pathologic) versus accident:* If the fracture is due to osteoporosis or metastatic tumor, the manner of death may not be accidental (due to injury), but rather natural. One way to sort this out is by history. If the fracture caused the fall, it is pathologic. In general, if the fall caused the fracture, it is an injury, even if the person has osteoporosis.
- Even if the death occurred during surgery or postoperatively, the manner is still determined by the etiology and circumstances of the fracture. If in the postoperative period, when did the death occur relative to the surgery?
- If it is an injury, obtain the date, time, location, and how the injury occurred (fell?)?
- Anticipate and document objections when speaking to families (this is common in the elderly, following benign falls at home).

Hospital bloods (see also Chapter 14)

- Try to get the earliest blood sample (i.e., drawn in the emergency department/upon admission) that is available and indicate in the investigative report when the sample was drawn. Avoid requesting "hospital blood" without further qualification; rather, ensure that the earliest remaining sample is obtained. If the chemistry/hematology lab does not have any remaining tubes of blood drawn upon admission, it is best to also check with the **blood bank.** The blood bank is a good resource for admission bloods because they are usually required to retain the sample for at least seven days. If the blood bank has nothing, then a chemistry/hematology sample that was drawn a day or two after admission may be acceptable enough, depending on the details of the case.
- These samples are collected for toxicology testing in people who have had prolonged survival intervals in the hospital. Since the autopsy blood may not reflect the state of intoxication at the time of the chemical or mechanical injury/collapse, these samples supply the concentrations that best represent what was in the decedent's system closest in time to that event. It is also good practice to routinely request admission bloods on "delayed" homicides (homicides in which there is some delay—up to a week or so, for the purposes of blood specimen request—between the injury occurrence and the death), motor vehicle, and drug intoxication deaths. Delayed in this setting generally means a hospital survival interval of greater than 2 h.

Hyperthermia/hypothermia (see also Chapter 13)

- In suspected cases, obtain a core body temperature at the scene or ask the hospital if they obtained one?
- If the death was pronounced in a hospital, document any intravital activity to ascertain whether the measured body temperature was a cause or a consequence of death (intravital evidence can be as minor as a brief return of any heart rhythm or electrical activity, even pulseless electrical activity (PEA), seizure activity, etc.). Even the absence of rigor mortis is helpful in ascertaining whether the temperature was achieved before or after death.
- Was the decedent wearing appropriate clothing for the weather? Any evidence of paradoxical undressing (some hypothermia deaths)?
- Note recent temperatures/weather
- Risk factors: Drug or medication use (depressant drugs, cocaine, alcohol, amphetamines, antidepressants, diuretics), Diabetes? Obesity? Dementia? Psychiatric illness?
- If the decedent was found indoors, how was scene environment being cooled or heated, if at all (e.g., Was there functioning air conditioning, fans, or heating?)
- Inquire with the initial responders what the temperature of the scene felt like when they first arrived (before altering it by adjusting windows, etc.)? Was it stifling hot, cool, frigid?

Infants/newborns (see also Chapter 6)

- Request the PLACENTA on all newborn/stillborn cases.
- Expedite the Sudden Unexplained Infant Death work—once the initial investigation is completed, an investigator then performs a scene investigation/reenactment.
- **"Cerebral palsy"** is not etiologically specific; if the infant did not have documented neurologic compromise from birth (antepartum or intrapartum complication), more of an explanation is needed (e.g., might this be a remote case of abusive head trauma in which the child survived with neurologic impairment?). "Cerebral palsy" is usually best used to describe the sequelae of a prenatal or perinatal hypoxic–ischemic event.[56–61]
- Including in the investigative narrative the events of the day or two prior to the infant's death can be very helpful in giving the ME/C a better sense of the case. Establish a timeline of what and how the infant was doing for the previous days to week: Ill? Vomiting? Cold symptoms? Fever (tactile vs. measured)?
- What developmental milestones had been met? Delayed?
- Last meal: what and when?
- Last seen alive?
- Who found the baby and who was home when the baby was found?
- When was the baby/child last in a state of normal health? How so (describe)? By who's observations? (Caregiver's? Pediatrician's? Neighbor's?)
- Where was baby found (bed, crib, etc.) and in what position (back, side, or face down)? What else, specifically, was in the sleep environment with the infant?

FIGURE 4.6 **Blood at scene of gunshot wound.**

- Who or what else was in bed with or on the infant? If there were other persons, how were their bodies positioned relative to the infant's? What were their ages and sizes?
- Any history of trauma?
- Core temperature on arrival to emergency department?
- Infant deaths during normal child birth (e.g., dystocia, nuchal cord) are natural deaths if there is documented life following delivery.
- Birth history? Premature? Neonatal intensive care unit stay? Prenatal issues or diagnoses?

Injuries in general

- It is best to maintain objectivity and be descriptive, avoiding opinion-based proclamations in the investigative narrative when describing injuries (Fig. 4.6). Doing so precludes the possibility of the ME/C having to later contradict investigator conclusions that conflict with theirs. Using language such as "defects consistent with sharp force trauma" or "...possible sharp force trauma" (if less clear), "multiple defects on the face, back, and left lower extremity" instead of opining about entrance versus exit wounds or a brow laceration from a terminal fall in firearms deaths. It is best to give the news, not the editorial.

Kidney disease

- Hemodialysis and renal failure should always have an underlying diagnosis (proximate cause). Hypertensive disease? Lupus? DM? HIV? Nonsteroidal anti-inflammatory drugs?

Liver disease

- Cirrhosis is not an etiologically specific disease. What caused it (alcohol, virus, medication)?[62,63]
- If viral hepatitis, what was the risk factor for transmission: transfusion (indication? injury?—this informs manner if cirrhosis is part of the cause), IVDA?
- Acute, fulminant liver failure cause (seek and document a potential cause: alcohol, medication, viral, poisoning, acetaminophen ingestion?)
- In potential acetaminophen toxicity deaths: inquire about hospital blood concentrations (may need to get admission bloods); inquire about circumstances—intentional, inadvertent, cointoxication/use or overuse with alcohol or related to preexistent liver disease?

Medications at scene

- Record medication information: medication, dose, directions, date filled, pills remaining relative to what would be expected, pharmacy (with pharmacy telephone number), to whom the medication was prescribed, and name and contact information of the physician.
- An image of the label of the prescription bottle and the pill contents poured on a table are an easy way to document the pill count (if necessary; see below).
- If potentially relevant, call the pharmacy and get the physician's office number to obtain further medical history. If a suspected fatal ingestion, perform pill counts (ideally, with photographic documentation).
- Check the refrigerator for insulin.

Motor vehicle collisions (see also Chapter 10)

- Who died?: driver, pedestrian, passenger?
- The phrase "motor vehicle *collision*" is preferable to "motor vehicle *accident*."
- Did the collision occur during the commission of another nonvehicular crime (e.g., bank robbers fleeing from the scene of the crime or from the police?). If yes, this death might constitute a vehicular homicide.
- Did the vehicle leave the scene (*not* a vehicular homicide for death certification purposes, even if being handled by authorities as a crime); in some jurisdictions, the fact that the vehicle left the scene is indicated in the "how injury occurred" section of the death certificate.
- Any evidence of intent to cause injury (to self or to another)? If the vehicle was intentionally used as a weapon to kill another person or persons, this too constitutes a vehicular homicide.
- What were the roadway, weather, and lighting (time of day) conditions?
- If a pedestrian, is it known why and how they were in the roadway? This is particularly important when there is a pedestrian in an unusual area, such as a highway. A suicidal intent or disorientation might be considerations under these circumstances. Is it known what they were wearing or if they appeared intoxicated or wounded prior to impact? Were they just standing in the middle of the roadway?

Multiple medical diseases/potential causes of death

- When there is a history of multiple medical problems or multiple possible causes of death at autopsy, the mechanism or immediate cause of death can be very helpful in leading the ME/C to the proximate cause of death. If there is an autopsy objection, then the circumstances of death become even more important. For example, if a person with heart disease and emphysema dies at home, what is the cause of death? If that person was complaining of chest pain and suddenly collapsed, the heart disease would make the most sense in part 1 and the emphysema, in part 2. If on the other hand, the wife said that he was having a chronic obstructive pulmonary disease exacerbation for two days, with increased sputum production and shortness of breath, refused to go to the hospital the night before, and was found dead in bed the next morning, the respiratory disease would most appropriately be placed in part 1 and the heart disease in part 2. *Mechanistic data gathering of this type can be immensely helpful in accurate death certification.*

"No scene"/public view/direct to ME/C deaths

- All/most of the information required with a scene visit (the front page of the investigator's report) should be included in deaths where no scene investigation is possible. This includes: date, time, location of injury, and family wishes regarding autopsy. See above for hospital cases.
- Consideration should be given to performing scene visits even if the body was already removed (e.g., deaths pronounced in hospital or, rarely, those swiftly removed from public view). Such cases may include deaths in which the scene examination may prove vital for proper death certification, such as may be the case in accidental versus suicidal descents from a height.

Obesity

Obesity is a competent proximate cause of death.[64–68]

Organ donation

- There is a difference between "*organ*" and "tissue" donation. In order to qualify as a potential organ donor, the person must be declared dead by neurologic criteria, but still have respiratory and circulatory function (typically assisted by a ventilator and even pressors). Tissue donation (cornea, long bone, saphenous vein, heart valves) may occur after cessation of circulatory and respiratory function.[69–72]
- Once a person is declared dead and has been accepted as a ME/C case, the organ/tissue donation staff cannot procure anything without the express permission of the ME/C. If the organ donor staff are in the process of locating the next of kin to get permission, many offices will allow them to hold the body at the hospital until this is resolved.

Occupational safety and health administration—OSHA (see also Chapter 3)

- Is the occupational safety and health administration (OSHA) involved? If not and the death occurred in the workplace (nonclerical), have a low threshold for notifying OSHA.[73–75]
- Was the decedent working near or possibly using any electrical devices?

Pregnancy (see also Chapter 3)

Depending on the state, the death certificate may specify one or more of the following[76]:

1. Not pregnant within one year of death
2. Pregnant at time of death
3. Not pregnant at death, but pregnant within 42 days of death
4. Not pregnant at death, but pregnant 43 days to one year before death
5. Unknown if pregnant within one year of death

If pregnant within one year of death:

1. Live birth? Date of outcome
2. Spontaneous termination/ectopic
3. Induced termination
4. None

If a pregnant or recently pregnant woman is found dead or witnessed to die at home with an unclear cause (not with drugs and drug paraphernalia), inquire about the date of delivery, type of delivery, delivery outcome (live birth?), and any complications (if applicable), versus the gestational age, if still pregnant at the time of death, as well as recent complaints, signs/symptoms, recent trauma, and prenatal/postnatal and/or longstanding medical issues, including postpartum depression.

Seizures (see also Chapter 7)

- Cause/Etiology? Alcohol, childhood onset, terminal, posttraumatic (if the latter, what, when, where, which hospital?).[77–81]
- Neurologic work-up completed or presumptive diagnosis?
- Obtain the neurologist's name and phone number.
- Rough date of onset (childhood, adulthood, when relative to head trauma, if applicable)? Duration? Frequency? Last one? Grand mal?
- Treatment: Which medications were prescribed? Compliant with medications? Were there any "breakthrough" seizures?
- How was the body found? (Prone? In bed? Apparently sleeping, not just ON the bed? In a bathtub?). Was the decedent in a position that may have compromised the airway?

Sharp injury (see also Chapter 10)

- For hospital deaths, it is important to specifically ask the physician if they altered any wounds (such as extending a stab wound, placing a chest tube in a stab wound, or incorporating a stab wound into a thoracotomy incision).[82–91]

Stillbirths (versus live births) (see also Chapter 6)

- Take jurisdiction if:
  1) Delivered in the absence of a physician (out of hospital)
  2) Illegal abortion
  3) Maternal trauma or drug use-related
  4) Therapeutic complication during delivery
  5) Some other unusual or suspicious circumstance
- Obtain the PLACENTA (and/or the placental pathology report, if already examined)
- Clinical/prenatal and parturition history (maternal and fetal): Gestational DM? Preeclampsia/Eclampsia? Infection? Prolonged rupture of membranes? Abruption? Placenta previa? Accreta/Increta/Percreta? Last fetal movements or documentation of life (ultrasound)?

In many jurisdictions, a "fetal" demise/termination of pregnancy certificate is completed unless:[76]

There was a heartbeat, tone, movements (voluntary or involuntary), vocalizations of any sort, spontaneous breathing, swallowing, or gasping, even momentary, after birth (obtain and document the Apgar score).

If answers to the above are NO (Apgar score of 0/0), then:

1) **Spontaneous:** the ME/C may need to complete a certificate of fetal *demise* for a spontaneous termination of pregnancy, depending on the case details.
2) **Induced:** the ME/C may or may not be required to complete a certificate of fetal death for induced termination of pregnancy, depending on the jurisdiction.

If the fetus is:

<20 weeks gestation (e.g., weight <300 gm, or more in some jurisdictions**) or crown-rump length is < 20 cm, then the fetus is assumed to be under 20 weeks). May be disposed of as other anatomic tissue.

>20 weeks: funeral home director must dispose of the remains.

Suicides

- When a decedent has reportedly committed suicide, but is brought to the hospital from the scene, it often is still important for a scene investigation (with photographic documentation) to be initiated if the hospital death/pronouncement occurs within a reasonable amount of time of the putative injury occurrence.
- Important observations and evidence that might aid in the ultimate death certification (suicide vs. accident, for example) may be waiting at the scene (a suicide note, an open window with a chair next to it, a will and testament intentionally left out in plain view, etc.).
- For a descent from height death, the need to document the height and width of the traversed barrier is no less important when the decedent happens to have been brought into the hospital before pronouncement. This is the same for pill counts in suspected suicidal ingestions. Indicating and documenting the specific putative drugs or chemicals that were potentially ingested may prove helpful in later guiding toxicology testing.[24,25,28,30,40] Pill counts may also be necessary, depending on the case.
- Our burden of proof is very high in suicides, given our obligation to prove intent. As such, sidelining the scene investigation in these deaths just because the body happens to be in the hospital is rarely prudent. That said, on occasion, too much time has elapsed between the injury occurrence and the death (delayed suicides) for a scene investigation to be informative, or even possible. In those instances, the ME/C is left with the police investigation and whatever other historical documentation (usually through the family), they may be able to gather to make their final determination.

Sudden death

- If a sudden "cardiac arrest," inquire about the circumstances[78,92–120] of death:
- Witnessed? Speaking, and then suddenly slumped over? Eating? Seizure activity before versus following the collapse? Found dead? Drugs or drug paraphernalia nearby (particularly stimulants)? Sudden dyspnea, followed by a cardiac arrest (more in keeping with a pulmonary thromboembolus)? Inhalers nearby?

** These criteria are jurisdiction-dependent.

Surgeries and therapeutic complications[121,122] *(see also Chapter 8)*

- Date, type of anesthesia, type of surgery, findings.
- Did the patient recover from anesthesia?
- Document the name of the surgeon and the office telephone number. In a hospital or ambulatory care facility? Dentist?
- Why was the organ/tissue resected? What disease (proximate cause) did they find?
- Requesting the surgical pathology report may be helpful.
- Was the surgery elective or emergent (this may affect the manner of death: therapeutic complication vs. natural, respectively)?
- Did the patient reach his/her baseline postoperatively in terms of pain, ambulation, nutrition, etc. (important for cardiac and pulmonary thromboembolism deaths)?

Time of death/found/pronounced

- The investigator must get this date and time and include it in the report. "Unknown" or "not available" is unacceptable. This is information that is required for the proper completion of the death certificate and every death that the ME/C investigates must have such a date and time. If the body remains at the scene and EMS never "pronounced," then the time of death will be the time the body was found.

Postmortem changes and establishing the basis for postmortem interval estimates

Like so many determinations in the practice of forensic pathology, **time of death/postmortem interval estimates** are born out of multiple biological and environmental variables that are viewed through a contextual prism.[121–140] Nowhere in our practice, however, is the television crime scene investigation expectation, or "CSI effect," more troublesome in its setting up of unreasonable expectations than it is with those surrounding our ability to determine an *exact* time of death. To the dissatisfaction of family members and attorneys alike, who expect us to produce answers about this on a daily basis, we nearly never are able to determine the moment of death with the kind of precision they hope for, absent an eyewitness to the death. Moreover, the further in time the discovery of a body is from its death, the less accurate the time of death estimate becomes. That said, we must do our best to consider all data points and answer questions as accurately as possible, but usually only in terms of **likelihoods**. The same approach applies to estimating the time of injury relative to a death, if not contemporaneous with it and when it is not known through credible witnesses, video surveillance, or a hospitalization. The latter is referred to as the potential **survival interval**, which is determined by the type and severity of the injury and how it would be expected to manifest pathophysiologically and to kill mechanistically. This topic is discussed further in Chapters 2 and 3.

Opinions generated at the scene and sometimes even at autopsy are investigative opinions. These differ from opinions offered in court, which are typically to a reasonable degree of scientific certainty (when involving criminal matters). During the onset of the investigation, one may discuss possibilities in order to assist in the initial stages of the investigation. These investigative opinions may be presented later at court, especially if they turn out to be incorrect. At trial, one should acknowledge that those opinions were initial investigative opinions based on the information available at that time. As new information is learned, those opinions may change.

Before the scientific analysis even begins, the first question that should be asked is, "When was the person last (reliably) seen alive or spoken to?" That is, when were they last definitively known to be alive? An accurate answer to that question alone should provide a reasonable start point in the estimation of time of death. Beyond that, when estimating the postmortem interval, the FP will speak in terms of **ranges** (hours, days, weeks, months) more than in discrete *numbers*, per se. Similarly, forensic pathologists can more easily answer questions posed to them about whether a specific hypothetical time of death is **"consistent"** or **"inconsistent"** with the forensic evidence, physical and circumstantial. These questions are usually asked in criminal/homicide investigations, not uncommonly while the FP is under oath during grand jury inquiries or trials. In these settings, answering in the affirmative that something is "consistent with" a proposed hypothetical simply means that it is *reasonably possible* (but, not necessarily the only possibility). Conversely, answering that something is "inconsistent" means that it is *not reasonably possible*. Because criminal trials are predicated on reasonable doubt, this terminology has significance and relevance to the triers of fact.

At the scene of death, the MDI is tasked with observing myriad things, not the least of which are factors in the environment and on the body, many of which will be taken into consideration by the forensic pathologist in their subsequent determination of the postmortem interval/time of death. These include, but are not limited to:

Environmental variables

- Ambient temperature
- The position of the windows (open or closed)
- The presence or absence of functional air conditioning or heat
- When the mail or last newspaper had been brought inside or the last cellphone call made?
- The death of a pet
- Is the body covered/insulated by anything (blankets—how thick? What type of fabric? Multiple layers of clothing, etc.)?

Bodily variables

- Livor mortis (dependent intravascular postmortem blood pooling/hypostasis), fixed (nonblanchable) versus unfixed (blanchable)
- Algor mortis (postmortem cooling), as measured by a thermometer, when indicated.
- Rigor mortis (postmortem rigidity/stiffening of the body), easily broken/breakable/moderate/strong/passed through
- Extent of putrefaction, if any (early, moderate, advanced)
- Extent of mummification, if any
- Body habitus (body mass index)

These observations are most helpful at the scene stage of the investigation; as such, forensic pathologists rely heavily upon the MDI. Once transported from the death scene to the morgue, possibly hours later, and placed in a refrigerator overnight, the conditions of death will have been substantially altered and the postmortem changes noted the following day by the ME/C will have much less scientific importance and relevance in the subsequent time of death estimates. It is for this reason that the investigator's death scene observations are so critical, but particularly so for time of death estimates. In addition to assistance with postmortem interval, the distribution of livor and rigor mortis may aid is detecting whether a body has been moved after death (i.e., an inappropriate rigor or livor mortis pattern relative to the position found).

Generally speaking, **livor mortis (lividity)** becomes fixed 8 h following death, *on average.*[141,142] As stated above, whether its pattern is consistent or inconsistent with the body's found position is also important for the MDI to observe and note, but this is not relevant to the time of death estimate, per se. When describing lividity, it is best to do so using *quantitative modifiers: scant/barely discernible versus well-formed and blanching versus fixed/nonblanching*. Also, the color of the lividity may be a clue to the cause of death. Lividity is often "cherry red" with elevated carboxyhemoglobin saturations.[143] When death is accompanied by elevated saturations of methemoglobin, livor is usually gray-blue to brown-purple (as seen with sodium nitrite use).[144–146] Methylene chloride, fluoroacetate, early putrefaction, embalming, and refrigeration also may imbue a cherry-red appearance.[141,142]

In terms of **algor mortis**, assuming the ambient environmental temperature is cooler than the body's temperature during life (the latter normally roughly 98.6°F/37°C), the body temperature will drop to equilibrate with the environmental temperature. During a heat wave, particularly if the body is found outside, the body temperature may even increase, if the ambient temperature exceeds that of the body's core temperature at the time of death.

A number of formulae have been developed using temperature to estimate the postmortem interval.[147–151] Historically, Moritz reported a formula for the number of elapsed hours equating to approximately 1.5°F per hour (or 98.6 minus body temperature °F/1.5).[152] Other formulae consider "average environmental conditions" and note the clothed body will cool in air at a rate of two and one-half degrees per hour for the first six hours, averaging a loss of 1–2°F per hour for the first 12."[153] Others believe that at an environmental temperature of about 60°F, a well-developed, clothed adult requires 24 to 30 hours to reach the temperature of the surroundings and cooling occurs at the rate of between 1 and 1.5°F per hour for the first 12 hours and at the rate of about 1°F per hour for the next 12 to 18 hours.[154]

Other studies demonstrate that apparently healthy persons will maintain normal body temperatures for the first four to five hours after death, giving rise to a postmortem *temperature plateau*. In effect, this observation contradicts the preceding formulae in the early postmortem period.[121,123,128–130,134,155–160] Many variables do affect the body's cooperation with any algorithm. These include the degree of **obesity** (a fat pannus provides endogenous insulation, protecting the body from heat loss very effectively, even after death; the endomorph retains heat for a much longer time than the ectomorph), **fever** at the moment of death (not only would this be expected to accelerate the onset of putrefaction, but also it would increase the amount of time it would take for the body to achieve the ambient environmental temperature), **stimulant drugs** (cocaine and others, such as methamphetamine) not uncommonly elevate the body's temperature, a form of an exogenous "fever," having a similar effect on algor mortis, **clinical hyperthermia** (as from heat stroke, environmental exposure, neuroleptic malignant syndrome, etc.), and **exogenous insulation** (e.g., blankets or multiple layer of clothing).

The development of rigor mortis also adheres to an average temporal algorithm, with the usual exceptions, caveats, and disclaimers, as is the case with all postmortem processes. Immediately following death, the body becomes flaccid. Progressively, the body stiffens due to a biochemical process at the cellular skeletal muscle filament level.[161–171] Very generally speaking, a commonly used time frame is that rigidity develops and increases over the first 12 postmortem hours until it peaks, maintains this peak for an additional 12 h, and then begins to decrease over the ensuing 12 h, in theory, returning the body to its original limp condition by the end of that period. Following this formula, one would expect a body to have entered and *passed through* the entire arc of rigor mortis by roughly 36 h after death. But, as one might expect, many variables may come into play and supervene, affecting this predictable course of events. Examples of such variables include strenuous physical exertion or exercise, terminal status epilepticus, and/or an elevated body temperature just prior to death, all of which have the capacity to affect (usually telescope) the rigor mortis trajectory.[172,173] Also, when describing rigidity, *we suggest using adjectives that are evocative of the degree/strength of the muscular involvement (i.e., easily broken/breakable, moderate, strong)* and its anatomic symmetry, or lack thereof. Once rigor is broken, it will not return. Rigor starts in all muscles at the same time, but because of their small size, the jaw (masseter) muscles may appear to develop rigor "first."

In some instances, there may be moderate rigor mortis, raising the question as to whether the rigor is waxing or waning. The body temperature and clarity of the cornea may help distinguish between the two. For example, moderate rigor with clear cornea/irides would favor rigor that is on the upslope of its development.

Decompositional changes (putrefaction, mummification, and adipocere formation) are fueled by the body's endogenous bacterial flora, taking place over time and with great variability, depending on the host and environmental factors (Fig. 4.7).[137,174–192] Putrefaction is enhanced by a warm, humid environment and mummification, by a warm, arid environment. As such, an elevated body temperature of any cause (obesity, fever, stimulants, clinical hyperthermia, layered insulation) accelerates the onset of decomposition, thus telescoping its course, as does intense heat exposure (body in a hot desert, for example). **Early visceral putrefaction** is the earliest *gross* sign of putrefaction, so would not be visible to the MDI at the scene; it is characterized by a focal green discoloration of the bowel and/or an orange-red discoloration of organs and tissues, particularly the aortic intima. Within an average of 24 h postmortem, however, a subtle green discoloration of the skin of the abdomen often appears (most commonly, over the right lower quadrant), this being better detected in lighter skinned individuals. Following this, varying degrees of focal skin slippage, marbling (a prominent vascular pattern seen on the skin's surface), bloating, diffuse green skin discoloration, a foul odor, vitreous humor volume loss/displacement, and purged blood from the orifices and within body cavities, particularly the thoraces, develop over time.[164,192–202] By the time all of these features have appeared, days to even a week or more have usually passed. Typically, decompositional changes are best described with *temporal modifiers: early, moderate, and advanced.*

Mummification proceeds over a similar time period in dry environments, but tends to affect the face and distal extremities (hands and feet) more readily (Fig. 4.8).[203–206]

Not uncommonly, the affected body parts take on a parchment-like and leathery quality, making fingerprinting difficult, if not impossible in some cases, but in the absence of putrefactive facial bloating, visual identification is sometimes still possible. **Adipocere formation** constitutes a specific decompositional process that develops over a

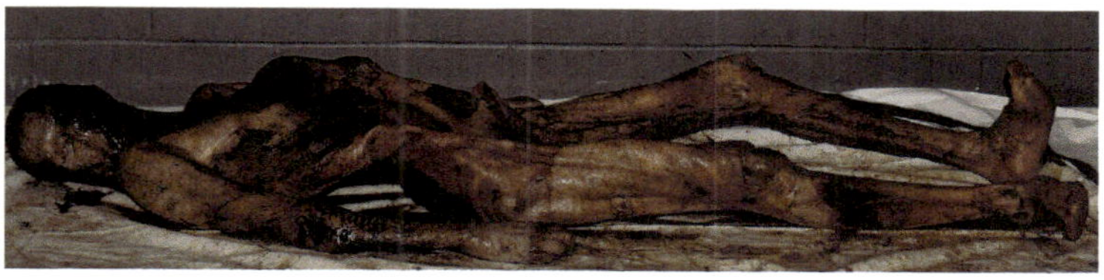

FIGURE 4.7 Decomposed body.

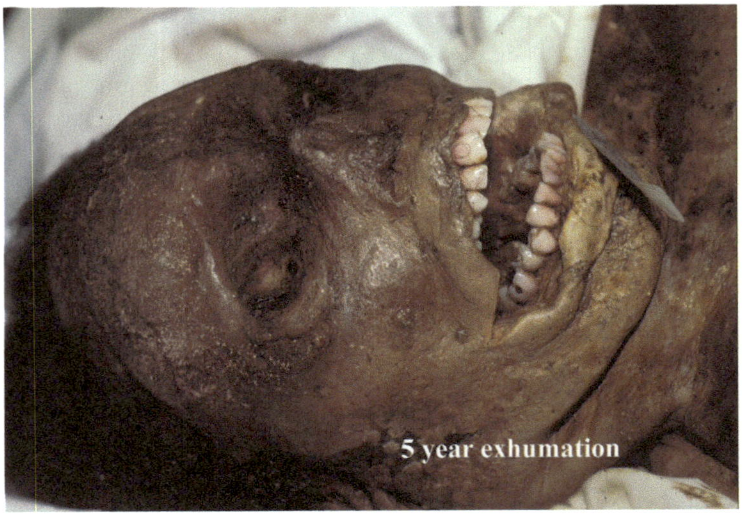

FIGURE 4.8 **Mummified remains.**

more protracted period and under moist conditions.[207] Also known as grave wax, it may be seen in exhumed bodies or in those who have been submerged in water. It typically takes weeks to months to develop. The earliest that has been described is six weeks.[192,207–224]

While not postmortem changes, per se, artifacts/injuries associated with **animal predation/scavenging** typically occur in the postmortem period (Fig. 4.9).[137,140,225–230] When indoors, particularly in a residential setting, the putative animals are most commonly *dogs, cats, and rodents*. In public waterways, they would be *marine animals*. This is not

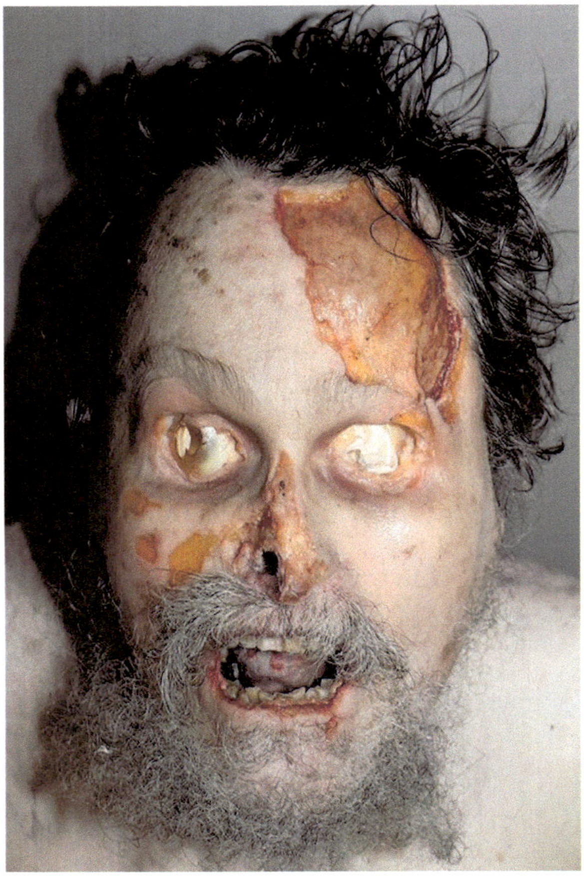

FIGURE 4.9 **Postmortem animal predation.**

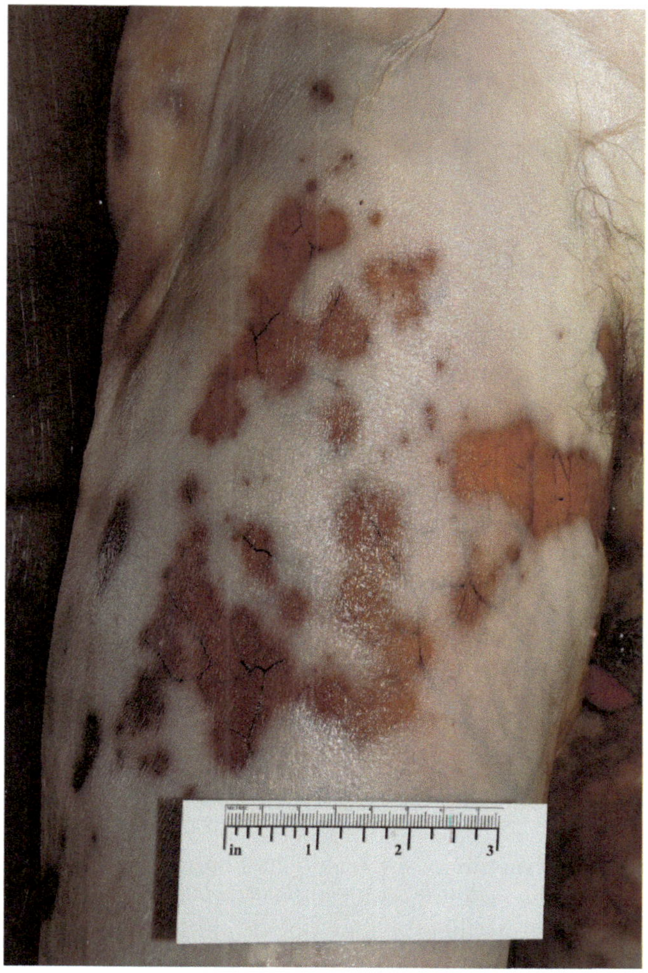

FIGURE 4.10 **Postmortem insect activity.**

to say that deaths caused by dog mauling and occasional shark attacks don't occur, but for the purposes of this chapter, we will restrict the discussion to animal activity taking place in the postmortem period (see Chapter 12 for a discussion of the former).

Depending upon the length of the postmortem interval and the size (and breed) of the animal, these injuries range from minor nibbling on the edges of the face and distal extremities to complete eviscerations, with partial skeletonization. They can be quite dramatic in appearance to the inexperienced eye, particularly when there is blood transfer and smearing that appears suspicious. If necessary and otherwise unclear, a dog's stool can even be examined and X-rayed for human bones.

Rodent and cat predation/scavenging tend to be less dramatic. Scalloped edges are often observed around the periphery of the wounds. In addition to their fairly characteristic morphology and distribution (soft/exposed areas like ears, eyelids, etc.), the bloodlessness of these wounds is noteworthy and helpful in ascertaining their etiology.[137,225–230] A pet at the scene, sometimes still with blood on its paws or face, is, of course, helpful.

Insect activity (anthropophagy) is a component of decomposition in that it can affect its rate of progression.[184,231–236] **Roach activity/roach bites** usually appear as clusters of nonspecific dried, small, orange-red abrasions rather than the defleshing observed from larger animal feeding (Fig. 4.10). The feeding action of **ants** causes irregular, serpiginous, scalloped areas of superficial skin loss, and small punctate and scratch-type lesions. Usually these injuries are orange-pink to yellow and diffusely scattered over the skin. They consist of small and rather shallow gnawed holes that can be easily misinterpreted as antemortem abrasions or burns. They are usually not associated with bleeding, but hemorrhage can sometimes be seen in congested parts of the body. Ant bites frequently give rise to suspicion, especially if located on the neck, potentially mimicking antemortem injuries. They

FIGURE 4.11 **Ritual skull.**

are also occasionally misinterpreted as a patterned abrasion caused by the imprinting of a blunt or other putative object.[237,238] Both roach and ant bites can also occur during life, in an insect-infested home.

Time of death versus time of pronouncement: reports and testimony

For a variety of reasons, many family members and friends not only desperately want to know the time of death of a loved one but are also confused by the seemingly definitive "time of death" listed on the death certificate. What they fail to notice is the fine print indicating that the time noted refers to the "time of death or the *time of pronouncement*." In fact, in the vast majority of out of hospital deaths, it is that latter to which that time refers. The most dramatic example of this is illustrated by the official time of death registered for a decedent found decomposed. Clearly, this individual died long before the time stamp that appears on the death certificate, the actual moment of death not ever likely to be determined with precision. It is for this reason that forensic pathologists spend considerable time and thought focused on the estimation of the time of death/the postmortem interval, as discussed above.

In that vein, before a forensic pathologist testifies in criminal court in a homicide trial and asserts an opinion surrounding the time of death, it's important for them to spend a good deal of time reviewing all of the details of a case in order to give consideration to what is reasonable (consistent) or unreasonable (inconsistent), if that issue is relevant to the case (often, it is not), as they anticipate testimony and define the limits of their opinion. For that reason, it is important for the MDI to recognize the importance of an excellent scene investigation that has paid exquisite attention to detail. The ME/C will have confidence in the assertions they will base on the scene observations if they know that they can trust the veracity of the investigation. Once in court, the FP would be well advised to resist the urge to play the "numbers game," getting ensnared by specificity when speaking about the time of death. But, knowing one's opinion and the forensic basis for it before taking the stand makes defending it much easier and clearer to the jurors. This applies to both forensic pathologists and investigators alike, the latter, not uncommonly called upon to testify about their scene observations as they may relate to questions asked of the forensic pathologist regarding the putative postmortem interval.

Similarly, when speaking to police, child protective services, prosecutors, and juries, MDIs and FPs should also exercise caution in how they characterize the "time of death," particularly in situations where death by neurologic criteria was established days after the individual presented in cardiac arrest, virtually near or at death's door. It is best to be descriptive to avoid being misunderstood on record. Asserting in a report or on the stand that a person "died" on "x" date, some days following their presentation to the hospital, while technically true in terms of their pronouncement time, may convey an inaccurate picture. In many infants subjected to abusive head trauma, for example, they present essentially dead on arrival, but are resuscitated to the extent that they are able to be maintained on advanced life support for days or longer, during which time brain death protocols are completed, by medical necessity, before they are ultimately pronounced dead. In other words, it's quite possible, even likely, that many of these babies were essentially dead on arrival, and but for advanced critical care measures, would have been declared as such had the required documentative clinical procedures to confirm the irreversible brain damage not taken some number of days to complete. The return of circulation resulting from resuscitative efforts does not necessarily mean a return of brain function. In fact, when brain death criteria are met over the course of a few days after presentation, it essentially means that only circulatory function had been temporarily restored with the institution of advanced life support measures, not neurologic (i.e., the brain had already ceased to function). To better document the reality of this pathophysiology then, we believe it's more communicative, objective, and accurate to say (or write in an investigative narrative) when the infant was **"pronounced dead by neurologic criteria,"** rather than saying when the baby "died." As Gertrude Stein said, "A difference, to be a difference, must make a difference." The authors would categorize assisted circulation as opposed to neurologic viability important entities to distinguish.

The trauma scene

Scenes of death due to mechanical injury present various challenges and may trigger different mental checklists depending on the trauma type and the circumstances of death. If a homicide, a crime scene response is the norm and includes careful trace evidence collection and photographic documentation. If there are open wounds, particularly if blunt or sharp, the quantity and pattern of blood distribution are usually carefully observed and documented. Is there cast-off on the ceiling, indicating inflicted blows that breached the scalp? Is there abundant contact transfer and signs of a struggle, handprints, or dragging as evidenced by smears through pools of blood? How high up on the wall is the blood seen? Or, instead, is there a more benign pattern, indicating an accidental injury sustained while the decedent was conscious, with subsequent blood drops around the home, including in the bathroom and kitchen sinks, with scattered bloodstained tissues or paper towels carefully placed on the bedside table? Accidental cuts in the kitchen, fatal gastrointestinal bleeding (always check the toilet for blood, as well as the distribution of blood on the body), or defective and bleeding arteriovenous shunts (used for hemodialysis) can produce very bloody scenes that, without careful inspection, may appear quite alarming to the inexperienced eye.

In general terms, the **quantity of blood** is commonly *overestimated* at scenes by nonforensically trained persons.[239,240] Few other situations more aptly present the "distraction of the dramatic" than do blood-covered scenes. Important to remember is that if even a pint of liquid blood of normal viscosity is poured out onto a nonabsorptive surface, such as the floor, it will spread out to roughly a millimeter in thickness, any slopes in the floor notwithstanding, covering a visually impressive area, despite only being a pint of blood. That said, blood loss might also be *underestimated* if the body is left to bleed to death primarily on thick, absorptive surfaces, such as a bed. External blood loss might also be largely gone if the perpetrator attempted to "clean up." For this reason, vigilance in identifying and documenting evidence of such attempts is important (e.g., the police and/or the MDI should look for and document any bloody or wet mops, rags, and/or sponges or mop wipe patterns left on the floor). This is all important to document as it may have bearing on the ultimate determination of *mechanism of death*. Fortunately, "the body doesn't lie," so even if the scene has been modified or is otherwise difficult to interpret, the body will likely reliably indicate whether exsanguination was the mechanism through such gross findings as visceral pallor and/or poorly formed lividity, unless advanced postmortem change or skeletonization precludes this determination. Finally, caution is advised in attempts at distinguishing purged blood emanating from putrefied remains from actual antemortem/perimortem blood loss. It would be unusual for a pool of purged blood to contain large blood clots or have a spatter pattern indicating hemoptysis/expectoration, for example.

Beyond evaluations of the quality and quantity of blood at scenes, the hands often need to be securely covered ("bagged") before transport from the scene in suspicious or clearly homicidal deaths (see below). Other scene activities include ballistics analysis, conducted by crime scene/law enforcement personnel in firearms deaths, which details the locations and numbers of casings and bullets, documentation of the location of a gun or knife relative to the body in apparent suicides, before they are secured by law enforcement, the search for suicide notes or other evidence of suicide planning, and the size of a window opening and the distance from the base of the window to the floor in descent from height deaths, along with possible hand or shoe prints on the dust or dirt-covered windowsill or roof parapet, among many other tasks (see below).

Crime scenes and deaths in public view

Crime scenes

Homicides reflexively trigger crime scene investigations by law enforcement authorities, whether the victim died/was pronounced in the hospital or is still at the scene of death. When the victim of a homicide or possible homicide is *not* brought to the hospital, a crime scene is also established, but includes both the body and the involvement of an MDI. Typically, the body is neither moved nor touched until a representative of the ME/C office arrives and has a chance to conduct their examination of the decedent. This type of ME/C scene investigation is usually conducted in coordination with the crime scene unit of the local law enforcement agency. Members of such a unit have special training in trace and other evidence collection at scenes of possible crimes. On occasion, however, an MDI will arrive at a scene that is thought to be nonsuspicious, but after making an initial assessment, decides to request a crime scene response because of something obviously or subtly concerning that is found on the body and/or in the environment. In these situations, the MDI takes relevant photographs of the body and the scene, as may law enforcement personnel, independently. More often than not, bags are placed over the hands and secured at the wrist to protect potential trace evidence during transport to the ME/C office. If the crime scene detectives are unable or unwilling to respond to such a scene, the MDI may still "bag" the hands and ask the police to secure the scene pending the autopsy findings (which often are not known until the next day).

Deaths in public view

Deaths occurring in view of the public present a specific challenge to both police and death investigators. The overarching goal is typically to discretely remove the body from the glare of onlookers, who may be equal parts traumatized and morbidly curious, sometimes even angry. But, depending on the nature of the case (e.g., a homicide), a longer scene investigation, one that includes the body remaining, may be necessary and unavoidable, dragging on longer than is ideal. Also, depending on the time of day and the nature of the death (e.g., retaliatory gang violence), an unruly mob occasionally gathers, potentially putting both the scene integrity and the responders at risk. Moreover, the longer the body remains at the death scene, the higher the likelihood that members of the press will eventually arrive, adding an additional layer of complexity to an already sensitive situation.

For a shooting, ballistics evidence is documented and collected, along with any other trace evidence that may be necessary to gather. If the scene involves a noncriminal descent from height, a pedestrian who was struck by a vehicle, or a train strike victim, for example, the body may often be removed to a more hidden location while awaiting the transport vehicle, that is, if not brought to the hospital by EMS and pronounced there. Regardless, when called out on such scenes, the MDI should make them a top priority; they should take precedence over all other cases for all of the reasons stated above. That said, it should go without saying that any responder to the scene (MDI and ME) should always dress and behave professionally (even be careful about smiling) and have an official identification/shield in hand. You never know when you will unwittingly end up on the front page of a local or international newspaper or, perhaps worse, be captured on a video that is later shared on the internet. If that happens, you'll be glad you were the consummate professional.

Field challenges in the recognition and identification of human skeletal remains

Another challenge that MDIs face on occasion is when they must respond to the scene of skeletonized or partially skeletonized remains. In situations where the remains are clearly human, the task is more about maintaining their integrity, as well as protecting potential trace evidence, particularly when suspicious. On other occasions, an ME/C office is notified of "human" or otherwise suspicious skeletal remains. When the MDI responds to such scenes, it may be quite obvious that the remains they are presented with are nonhuman. Other times, it may be more equivocal and require the assessment of a forensic anthropologist (see "Scenes and the Forensic Anthropology Consultant" sidebar, Box 4.1)[181,227,241–287].

BOX 4.1

Scenes and the Forensic Anthropology Consultant

Bradley Adams, Ph.D.

Forensic anthropologists have expertise with skeletal anatomy, human variation, and the identification of skeletal pathology. These skills can assist many aspects of medicolegal investigations, ranging from scene recoveries to laboratory analyses.

Some types of **scenes** that may benefit from anthropological involvement include: *buried bodies, dispersed remains, fire scenes, and mass fatality events.* Forensic anthropologists often have training in archaeological recovery methods. The same principles used for excavation of a prehistoric site also apply to the recovery and documentation of remains from a forensic scene investigation. For the recovery of buried remains, it is important to use proper excavation techniques and to understand soil composition and soil disturbance. This ensures that buried remains are not inadvertently damaged during exposure and removal and that their context is properly recorded. For dispersed remains on the ground surface (e.g., animal scavenged), it is also important to use proper search methods so that human remains and associated evidence are not missed.

An understanding of **taphonomy** (changes that occur to the body from the time of death to the time of discovery) is also important for scene recovery. An example would be the recovery of burned remains from a fire scene. To an untrained individual, fragmentary burned bones or charred body parts that have undergone thermal changes may blend in with the fire debris and be overlooked. An understanding of taphonomy can also assist during laboratory analysis, such as differentiating marks on bones caused by postmortem animal scavenging or root-etching from perimortem sharp force trauma. In addition, an anthropologist may use their understanding of taphonomic changes to assist with a determination of time-since-death, especially for extended periods of time involving weeks, months, or years.

In the laboratory, anthropologists can determine whether or not skeletal remains are human, assess medicolegal significance, analyze skeletal trauma, assist in the resolution of commingling, and provide details to facilitate decedent identification.

When isolated skeletal remains are discovered and reported to law enforcement, it is imperative to quickly determine if they are, in fact, human. If a cranium is present, this is generally straightforward, but for unassociated bones and bone fragments, it can be more challenging. It is also worth noting that skeletal remains from infants and children look very different from fully formed adult bones since growth and development, including epiphyseal fusion, has not yet occurred and they may be misidentified by the untrained observer. Since a forensic anthropologist is an expert in skeletal anatomy, a visual determination of whether or not bones are human is usually easy, even with fragmented bones, based on species variation in bone morphology. Quite often, this can be accomplished remotely via digital photographs that include a scale, so that time and effort is not wasted with a scene visit for **nonhuman bones**. A prompt determination of whether or not bones are human will allow for the appropriate escalation or deescalation of personnel and resources.

Just because bones are determined to be human does not necessarily mean that they have medicolegal significance. Input from an anthropologist can assist with this determination.[288] Consider an example where a skull is found on a shelf by law enforcement personnel while serving a warrant. Evaluation of the skull may reveal features indicating that it is an *anatomical teaching specimen* and, as such, likely lacks medicolegal significance. Perhaps more common are instances where bones are unexpectedly discovered during construction work. A determination must be made as to whether the bones have medicolegal significance (e.g., *clandestine burial*) or if they lack medicolegal significance (e.g., *disturbed cemetery*). Through taphonomic indicators, such as the state of preservation of the remains and contextual clues found at the scene, it is usually possible for a forensic anthropologist to make this determination. In the case of historic/prehistoric graves, there will generally be other agencies that would be notified of the discovery and take jurisdiction, such as the State Historic Preservation Office and Native American tribes. The Native American Graves Protection and Repatriation Act (NAGPRA), established in 1990, is a federal law created to protect the cultural and biological remains of Native Americans and their ancestors. It outlines a requirement and process to return certain Native American cultural items (including human remains) to lineal descendants, culturally affiliated Indian tribes, or Native Hawaiian organizations. Federal agencies and museums, universities, state agencies, local governments, or any institution that receives Federal funds must comply with NAGPRA.[289]

Anthropologists can provide important information for decedent identification, especially in cases where the body

Continued

BOX 4.1 *(cont'd)*

is badly decomposed or skeletonized. Through an analysis of specific skeletal features throughout the skeleton, a **biological profile** can be developed. This refers to the decedent's *age at death, their biological sex, their population affinity (i.e., ancestry), and their living height.* Each of these components can help narrow the pool of potential missing individuals. In general, age is estimated in subadults from assessment of dental development and epiphyseal fusion. Age assessment in adults is less precise and focuses on degenerative changes observed in the skeleton. Biological sex can be accurately determined in postpubescent individuals through skeletal variation, especially features present on the pelvic bones. It is important to remember that biological sex is not synonymous with gender and the terms are not interchangeable. Anthropologists can estimate population affinity from analysis of morphological and metric variation on the skull. Stature estimation is usually calculated from regression models that use long bone measurements (e.g., maximum length of the femur).

Often times, forensic anthropologists are critical members of an ME/C office's *multidisciplinary disaster response team,* assisting with both field response and morgue operations (see "Mass Fatality Incidents").[298] In some mass fatality events, there may be **commingling** of remains. This refers to the intermixing of fragmentary remains from more than one individual. An anthropologist can assist with segregation and reassociation of commingled remains back to specific individuals. This can be accomplished by conjoining fragmented remains along a fracture site or by using other techniques such as articulation and osteometrics. For very fragmentary incidents, there will be a heavy reliance on DNA for reassociation and decedent identification.

Bones (both human and nonhuman) also may be used in various rituals (e.g., **Santeria** and **Palo Mayombe,** Fig. 4.11).[288] The forensic anthropologist can assist in determining their type and potential provenance (e.g., previously buried).[290–296] In some instances, mercury is used in the ritual which may cause concern for exposure.[288,291,297]

Attempts at postmortem concealment of bodies in homicides (dismemberment, incineration, and burial)

The approach to this singular category of death investigation and autopsy from the perspective of the FP is described in great detail in Chapter 15. That said, for the MDI, a host of challenges may arise at their entry point into the investigation, some similar to those encountered in the investigation of skeletonized and/or partial remains, but these deaths nearly always involve a crime scene investigation. The scenes the MDI may be presented with will be highly varied, depending on the type of postmortem concealment attempt and the location of the disposal. As always, a partnering with law enforcement authorities in the containment and handling of the scene and a stringent and careful adherence to protocols in the gathering and documentation of physical and other evidence will keep the work less overwhelming.

Depending on the condition of the remains, an anthropologist/anthropology team might need to respond to the scene, if available for that jurisdiction, particularly when burial is involved (see sidebar). On occasion, animals have discovered such remains and may have ingested and/or relocated some of the bones or other body parts to a slightly more distant location. Anthropologists are particularly adept at handling and mapping out this type of complex scene. Also, in a dismemberment case, it is not unusual for the investigator to be faced with more than one crime scene investigation over time, depending on where and when the victim's body parts were deposited. The case numbers for each may ultimately be merged into one, so the similarities in the circumstances and even the proximity of these *linked "cases"* and the probability of their relatedness should be so indicated in the MDI narrative and/or clearly indexed somewhere in the case file. The use of cadaver dogs, when available, may also be useful to locate remains.[299]

Also, importantly, while the police investigation begins with the crime scene involving these unidentified remains, the second stage of that investigation can only begin in earnest once the decedent is identified. At that point, the police may have an address, family and friends to interview, an apartment to examine, and a neighborhood to canvas, and the investigation consequently tends to take a turn. Identification, in many ways, represents an investigative reset for law enforcement personnel in these cases; the investigation can only go so far in the absence of an identification, hence its critical importance in crime solving.

Finally, it is important to consider the handling of trace evidence, such as bindings of various sorts, trash bags, and duct tape in these and other complex homicidal circumstances. It is best if the MDI focuses as much as possible on maintaining the integrity of the remains and associated gross and trace evidence prior to their transport to the morgue. Unwinding tape, for example, is to be avoided, where possible, as its adhesive surface may be of evidentiary value for trace and fingerprint analysis. Ideally, even once in the morgue, it can be carefully cut off the body and secured by the FP after examination and photographic documentation, as opposed to unwound.

Mass fatality incidents (MFI)

MFIs or events, sometimes called multifatality incidents, present a host of unique challenges to both the medicolegal death investigator and the forensic pathologist alike.[283,298,300–308] Many articles and books are devoted to this important and complex topic, most written by professionals who have spent much of their careers dedicated to studying and enhancing readiness for such events, man-made or natural, most requiring a response self-limited by days to weeks, but others (rarer), requiring one that is protracted over years. The overarching theme that emerges from what has been written by those with the most extensive experience is that forethought and preparedness are critical, including drills and exercises with multiagency collaboration and planning. All that said, a comprehensive discussion of this topic exceeds the scope of this book. The foregoing, however, will attempt to highlight some central concepts for the FP to keep in mind.

In many jurisdictions, the ME/C is given express authority by local or state governmental bodies and laws to be the lead agency in managing the responses to such events. MFIs are typically defined as events in which enough fatalities have occurred that multiple agencies and special operations are needed.[283,298,300,306–311] Depending on the ME/C office, they are characterized by one or more of the following:

- A discrete incident resulting in more decedents to be recovered and examined than can be managed by the local ME/C jurisdiction (>5–10, depending on the jurisdiction). The centers for disease control and prevention (CDC) defines a disaster as a serious disruption of the functioning of society, causing widespread human, material, or environmental losses that exceeds the local capacity to respond.
- An event or ongoing situation that requires a protracted recovery and fatality management effort (e.g., the events of and following the collapse of the twin towers on 9/11/01 or those of a protracted pandemic, such as that resulting from COVID-19).[300,312,313]
- An incident requiring a multiagency response.
- Multiple fatalities involving potentially contaminated remains (biologic, radioactive, explosive, chemical, or nuclear)

Depending on the size of the office, a dedicated and specialized response team may be in place and ready, along with specific related policies and procedures that support a swift mobilization when needed. Regardless of the planning and process designed for each office and depending on the magnitude of the incident, the response will often require the involvement of every departmental level and specialty available within the agency, including, but not limited to: FPs, MDIs, forensic anthropologists (see sidebar), identification staff, many who may be involved in the establishment of a family assistance center (FAC), if the incident is of substantial proportions, forensic biology (DNA) staff (also typically part of the FAC), mortuary and special operations staff, photographers, radiology staff, evidence staff, transportation staff, clerical staff, IT staff, and administrative staff. Typically, an MFI triggers an "all hands on deck" response within an ME/C agency, often temporarily siphoning some part of the work force away from the routine work that they would normally be managing.

That said, the regular and essential workflow must continue, even if it is temporarily modified; the homicides, suicides, accidents, and natural deaths don't take a convenient pause in response to such multifatality events and still need attention like all other fatalities. This balancing act can present many challenges for both the leaders and the workforce of the agency and requires intense organization and cooperation. Sometimes, policies and procedures must be modified by necessity to both preserve and maximize the workforce as the incident unfolds; these may involve temporarily changing guidelines, including those dictating which kinds of cases should be brought to the office and, of those, which need to be autopsied. Again, this constitutes a separate balancing act in deciding what is considered essential under new and extraordinary conditions and what can be handled differently, all without substantially compromising standards and practice principles. At no time is team work more important than when an agency's operational needs are stretched nearly to their elastic limit by an unexpected and large fatality incident.

Examples of events that might lead to a potential MFI in any jurisdiction, large or small, include, but are not limited to:

- Large building fires, residential or commercial
- Building collapses (from a variety of causes)
- Transportation: aircraft, boat, train, or multimotor vehicular road crash incidents
- Mass or multiple shootings
- Crowd stampedes (such as at concerts)
- Carbon monoxide incidents
- Pandemic-related fatalities, which may be protracted
- Terrorist attacks (nuclear, biologic, explosive, aircraft, etc.)
- Natural disasters (flash flooding, hurricanes, tornadoes, tsunamis, wildfires, etc.)

From an investigative standpoint, personal safety of staff comes first, followed by the assiduous documentation of the remains, personal effects, identification of next of kin, if known, evidence collection, if any, and the overall scene description as a start point to the written and photographic narrative. Rarely, there is a decontamination protocol that must be followed before full scene processing can proceed, so having specific dedicated staff with hazmat training could prove very helpful. Sometimes, regional MFI systems are in place, allowing neighboring jurisdictions to coordinate and assist one another under these extreme circumstances. Regional MFI drills and exercises prove very fruitful in these coordination efforts. On occasion, external support, such as Disaster Mortuary Operational Response Teams, may need to be put in place, if the size and/or duration of the incident outpaces available staff and resources.

The MFI cases are typically entered into the agency's database as **companion cases**, even if initially unidentified, or are otherwise tagged as related to the same event. The investigator may be joined at the scene by an anthropologist or an **anthropology team**, depending on the scale of the incident and the condition of the remains. In aircraft crash events, for example, there may be extensive **dispersal and commingling** of remains, as well as widespread fire damage. The accurate identification and cataloging of all remains and their personal effects will become critical. Some will require later reunification, with merging of more than one case number into one, after the completion of an anthropologic, radiologic, contextual, and/or DNA evaluation and confirmation. On occasion, radiologists and forensic dentists will be employed as consultants in the effort. In some jurisdictions, the availability of **rapid DNA field testing** makes these efforts much easier in terms of delineating and identifying fragmented remains and in general, providing answers to families more quickly (see Chapter 19).[314,315]

For a larger scale MFI, the setting up of a discrete and dedicated **FAC** and/or outreach operation becomes imperative. Using a centralized database, shared by multiple agencies, streamlines this effort.[300,301,312,313,316] Also, when the death certificates for such fatalities are completed, there should be a coordinated effort toward consistency, particularly if more than one forensic pathologist is involved in the certifications. While the cause of death statements may differ slightly (e.g., smoke inhalation vs. thermal injuries in a mass fatality fire incident), the manners, time and place of injury occurrence, and the "how injury occurred" section are typically identical, if one discrete event is responsible for all fatalities.

While not qualifying as a "multifatality incident," per se, *the occasional discovery of two decedents at the same location* presents a specific challenge to the MDI and FP alike and deserves specific mention as part of this discussion. These deaths are viewed as **"companion cases"** and should reflexively trigger a mental checklist. A typical scenario involves two decedents (sometimes more) found together at home or in some kind of residential environment, such as a homeless encampment. In all such scenes, careful attention should be paid to the postmortem changes of each decedent, with emphasis on any disparities. The contemporaneous or closely associated deaths of two individuals usually result from one of the following scenarios:

- **Two natural deaths** that have occurred *in sequence* as the result of a caregiver dying from natural (or other) causes, followed by the natural death of a dependent (a bedbound and/or severely demented adult, a mentally ill or developmentally disabled child or adult, an infant, etc.) as a result of inanition, infection, etc. The caregiver in this scenario usually has physical signs of having died first (i.e., of there having been a longer postmortem interval than that of the dependent).
- **Two suicidal deaths** (various modalities), usually with scene evidence of a suicide pact. There is typically no disparity in the postmortem changes between the two decedents.[317]

- **Two accidental, drug intoxication deaths**, particularly when involving either "laced" or highly potent illicit substances, such as fentanyl and/or its analogues.[318–320] Most commonly, the decedents have similar postmortem changes.
- **Deaths from environmental derangements**, such as a carbon monoxide leak.[142,321–328] The decedents have usually undergone similar postmortem changes.
- **Double homicide** (usually clear by the scene investigation and wounding patterns); the degree of postmortem change is often/usually similar between the decedents.
- **Homicide–Suicide** (again, often clearly indicated by both the scene investigation and the wounding patterns). The postmortem intervals may be similar or disparate. When the latter, the perpetrator of the homicide will have findings indicating a longer postmortem interval.[329–342]

Avoiding the top 10 investigative oversights

1. Document your discussion with the family about the potential need for an autopsy (and their preferences).
2. *Proximate cause:* Before hanging up with the clinician, ask yourself if there is enough information to certify the death. In some instances, an autopsy will be needed to determine the proximate cause, but in many deaths, the proximate cause is known. In those instances, however, you may need to inquire about the proximate cause because clinicians tend to focus on immediate causes and mechanisms. Therefore, ask: "Why was the aortic valve replaced?" When they say "Due to aortic stenosis", reply: "From what? Rheumatic disease? Bicuspid valve? Etc."
3. *Cancer* and competing causes of death: If there is a prior history of cancer, ask if it was in remission or an active medical issue.
4. *Scene photographs:* Don't just take a picture of the body; also take some wider shots of the scene (e.g., room); for close up photographs, use a scale. Photograph medication vials if not excessive in number.
5. For *homicide* victims pronounced at the hospital, inquire about the following:
   a. Specifically ask if they cut through or put any tubes in a wound.
   b. Recovered any ballistic material.
   c. If there was surgery, what did they find and do?
   d. Call the law enforcement agency and get information about the death (e.g., is this a police-involved shooting?).
   e. Request hospital admission bloods if there was a prolonged (>2-hour) survival interval.
6. *Family information:* At a scene, have any family fill out the identification form, if available in your jurisdiction, and obtain any medical history. If an injury, acute or remote, may have or definitely caused or contributed to the death, get the time/date and place of occurrence information from the family up front, if known (often necessary if the injury was sustained at home or otherwise had no police involvement) or from the police. Without this information, the death certificate cannot be completed.
7. *Medical scene:* Get the name and number of the family physician/primary care physician. Document prescriptions, recording the name of the doctor and the pharmacy phone number. Call the pharmacy and ask for the physician's phone number. If it is a suspected pill ingestion, count the remaining pills, and document the milligram dosage per pill and dosage instructions, the date filled, and the total number of pills prescribed.
8. *Methadone:* In a suspected substance use death, ask if the decedent was in a designated methadone maintenance or substance use treatment program, and if so, what their dosage and frequency of clinic visits were (daily or weekly supply?). See Chapter 14.
9. *Head injury deaths with a survival interval:* Get copies of head CT/magnetic resonance imaging reports. Was the person ever known to be conscious after the injury? Was any neurosurgery performed? What did the surgeon do/find?
10. In apparent *hyperthermia or hypothermia deaths* pronounced in hospitals: What was the initial body temperature upon arrival? Was it intravital? (Indicated by any movement, nonpurposeful or purposeful, seizure activity, electrical cardiac activity of any type?)

References

1. Luke J. Strangulation as a method of homicide. Study (1965-1966). In: *New York City. Archives of Pathology.* Vol 83. 1967:64–70 (1).
2. Mittleman RE, Wetli CV. Cocaine and sudden "natural" death. *J Forensic Sci.* 1987;32(1):11–19.

3. Fagin J, Bradley J, Williams D. Carbon monoxide poisoning secondary to inhaling methylene chloride. *Br Med J*. 1980;281(6253):1461.
4. Rague J, Grush J, Buchanan J. A case series of chemical dermal injury requiring operative intervention after prolonged dermal methylene chloride exposure. *J Med Toxicol*. 2021;17(2):222–226.
5. Mittleman RE, Wetli CV. The fatal cafe coronary. Foreign-body airway obstruction. *JAMA*. 1982;247(9):1285–1288.
6. Welsby PD, Golledge CL. Meningococcal meningitis. *BMJ*. 1990;300(6733):1150–1151.
7. Friedman W, Singh F. Tuberculosis, AIDS, and Death among substance abusers on welfare in New York City. *NEJM*. 1996;334:828–833.
8. Dada MA, Lazarus NG, Kharsany AB, Sturm AW. Sudden death caused by myocardial tuberculosis: case report and review of the literature. *Am J Forensic Med Pathol*. 2000;21(4):385–388.
9. Ozsoy S, Demirel B, Albay A, Kisa O, Dinc AH, Safali M. Tuberculosis prevalence in forensic autopsies. *Am J Forensic Med Pathol*. 2010;31(1): 55–57.
10. Thu M, Winskog C, Byard RW. Tuberculosis and sudden death. *Forensic Sci Med Pathol*. 2014;10(2):266–268.
11. Klatt, Nichols, Noguchi. Evolving trends revealed by autopsies of patients with AIDS. *Arch Pathol Lab Med*. 1994;118:884–890.
12. Craemer. Postmortem viability of HIV. *NEJM*. 1994;331:1315.
13. Palell F, Delaney K, Moorman A, et al. Declining Morbidity and mortality among patients with advanced HIV infection. *NEJM*. 1998;338: 853–860.
14. Havlir D, Barnes P. Tuberculosis in patients with HIV infection. *N Engl J Med*. 1999;340(5):367–373.
15. Manuelidis L. The dimensions of Creutzfeldt-Jakob disease. *Transfusion*. 1994;34(10):915–928.
16. Mittenzwei R, Dibernardo L, Hulette C, Harrison WT. Creutzfeldt-jakob disease mimicking neurosyphilis: a complex presentation, histopathological findings, and special precautions for autopsy. *Am J Forensic Med Pathol*. 2019;40(4):371–375.
17. Hanley B, Lucas SB, Youd E, Swift B, Osborn M. Autopsy in suspected COVID-19 cases. *J Clin Pathol*. 2020;73(5):239–242. https://doi.org/10.1136/jclinpath-2020-206522. In this issue.
18. Xu Z, Shi L, Wang Y, et al. Pathological findings of COVID-19 associated with acute respiratory distress syndrome. *Lancet Respir Med*. 2020.
19. Ruan Q, Yang K, Wang W, Jiang L, Song J. *Clinical Predictors of Mortality Due to COVID-19 Based on an Analysis of Data of 150 Patients from Wuhan*. China: Intensive Care Med; 2020.
20. Barton LM, Duval EJ, Stroberg E, Ghosh S, Mukhopadhyay S. COVID-19 autopsies, Oklahoma, USA. *Am J Clin Pathol*. 2020.
21. Richardson S, Hirsch JS, Narasimhan M, et al. Presenting characteristics, comorbidities, and outcomes among 5700 patients hospitalized with COVID-19 in the New York City area. *JAMA*. 2020.
22. National Center for Health Statistics. *Guidance for Certifying Deaths Due to Coronavirus Disease 2019 (COVID–19). Hyattsville, MD*. 2020.
23. Gill JR, DeJoseph ME. The importance of proper death certification during the COVID-19 pandemic. *JAMA*. 2020;324(1):27–28.
24. Gill JR. Fatal descent from height in New York City. *J Forensic Sci*. 2001;46(5):1132–1137.
25. Shaw KP, Hsu SY. Horizontal distance and height determining falling pattern. *J Forensic Sci*. 1998;43(4):765–771.
26. Bergner L, Mayer S, Harris D. Falls from heights: a childhood epidemic in an urban area. *Am J Publ Health*. 1971;61(1):90–96.
27. Lehman D, Schonfeld N. Falls from heights: a problem not just in the Northeast. *Pediatrics*. 1993;92(1):121–124.
28. Scalea T, Goldstein A, Phillips T, et al. An analysis of 161 falls from a height: the 'jumper syndrome. *J Trauma*. 1986;26(8):706–712.
29. Barlow B, Niemirska M, Gandhi RP, Leblanc W. Ten years of experience with falls from a height in children. *J Pediatr Surg*. 1983;18(4):509–511.
30. Goonetilleke UK. Injuries caused by falls from heights. *Med Sci Law*. 1980;20(4):262–275.
31. Simonsen J. Fall from height into the water–pattern of lesions. A preliminary report of 10 cases. *Acta Med Leg Soc*. 1983;33(2):1119–1124.
32. Risser D, Bonsch A, Schneider B, Bauer G. Risk of dying after a free fall from height. *Forensic Sci Int*. 1996;78(3):187–191.
33. Li L, Smialek JE. The investigation of fatal falls and jumps from heights in Maryland (1987-1992). *Am J Forensic Med Pathol*. 1994;15(4):295–299.
34. Perret G, Flomenbaum M, La Harpe R. Suicides by fall from height in Geneva, Switzerland, from 1991 to 2000. *J Forensic Sci*. 2003;48(4): 821–826.
35. Turk EE, Tsokos M. Pathologic features of fatal falls from height. *Am J Forensic Med Pathol*. 2004;25(3):194–199.
36. Plattner T, Kopp A, Bolliger S, Zollinger U. External injuries to the neck after free fall from great height. *Am J Forensic Med Pathol*. 2004;25(4): 285–287.
37. Cross R. Fatal falls from a height: two case studies. *J Forensic Sci*. 2006;51(1):93–99.
38. Atanasijevic TC, Savic SN, Nikolic SD, Djoki VM. Frequency and severity of injuries in correlation with the height of fall. *J Forensic Sci*. 2005; 50(3):608–612.
39. Fanton L, Bevalot F, Schoendorff P, Lalliard S, Jdeed K, Malicier D. Toxicologic aspects of deaths due to falls from height. *Am J Forensic Med Pathol*. 2007;28(3):262–266.
40. Gill JR. Falls from height. *Phys Find Adult*. 2016:496–500.
41. Ferguson C, Sutherland T. Murder by pushing: an exploratory analysis of homicidal falls from a height. *Am J Forensic Med Pathol*. 2018;39(3): 192–200.
42. Zivkovic V, Nikolic S, Babic D, Djonic D, Atanasijevic T, Djuric M. Pontomedullary lacerations in falls from a height–a retrospective autopsy study. *J Forensic Sci*. 2012;57(3):654–657.
43. Petaros A, Slaus M, Coklo M, Sosa I, Cengija M, Bosnar A. Retrospective analysis of free-fall fractures with regard to height and cause of fall. *Forensic Sci Int*. 2013;226(1–3):290–295.
44. Milanowicz M, Kedzior K. Active numerical model of human body for reconstruction of falls from height. *Forensic Sci Int*. 2017;270:223–231.
45. Casali MB, Blandino A, Grignaschi S, Florio EM, Travaini G, Genovese UR. The pathological diagnosis of the height of fatal falls: a mathematical approach. *Forensic Sci Int*. 2019;302:109883.
46. DeJoseph M, Hoppa E. Death investigation of Diabetes mellitus: scene investigation and interrogation of technology. *Acad Forensic Pathol*. 2016;6(2):164–173.
47. Lozano JG, Healy NL, Kimberley Molina D. The prevalence of paraphernalia found at the scene of drug-related deaths. *J Forensic Sci*. 2016; 62(3):691–694.
48. Davis GG, Cadwallader AB, Fligner CL, et al. Position paper: recommendations for the investigation, diagnosis, and certification of deaths related to opioid and other drugs. *Am J Forensic Med Pathol*. 2020;41(3):152–159.

49. Melinek J, Livingston E, Cortina G, Fishbein MC. Autopsy findings following gastric bypass surgery for morbid obesity. *Arch Pathol Lab Med.* 2002;126(9):1091–1095.
50. Deitel M. Overview of operations for morbid obesity. *World J Surg.* 1998;22(9):913–918.
51. Eriksson S, Backman L, Ljungstrom KG. The incidence of clinical postoperative thrombosis after gastric surgery for obesity during 16 years. *Obes Surg.* 1997;7(4):332–335.
52. Pace WG, Martin Jr EW, Tetirick T, Fabri PJ, Carey LC. Gastric partitioning for morbid obesity. *Ann Surg.* 1979;190(3):392–400.
53. Steinbrook R. Surgery for severe obesity. *N Engl J Med.* 2004;350(11):1075–1079.
54. Schauer PR, Ikramuddin S, Gourash W, Ramanathan R, Luketich J. Outcomes after laparoscopic Roux-en-Y gastric bypass for morbid obesity. *Ann Surg.* 2000;232(4):515–529.
55. Hall JC, Watts JM, O'Brien PE, et al. Gastric surgery for morbid obesity. The Adelaide Study. *Ann Surg.* 1990;211(4):419–427.
56. MacDonald D. Cerebral palsy and intrapartum fetal monitoring. *N Engl J Med.* 1996;334(10):659–660.
57. Rosen RS, Armbrustmacher V, Sampson BA. Mortality in cerebral palsy (CP): the importance of the cause of CP on the manner of death. *J Forensic Sci.* 2003;48(5):1144–1147.
58. Evans PM, Alberman E. Certified cause of death in children and young adults with cerebral palsy. *Arch Dis Child.* 1991;66(3):325–329.
59. Gill JR, Morotti RA, Tranchida V, Morhaime J, Mena H. Delayed homicides due to infant head injury initially reported as natural (cerebral palsy) deaths. *Pediatr Dev Pathol.* 2008;11(1):39–45.
60. Folkerth RD. Neuropathologic substrate of cerebral palsy. *J Child Neurol.* 2005;20(12):940–949.
61. Dudley M, Chapman BM. Death of a 6-year-old boy with mental retardation: accident versus child abuse. *J Forensic Sci.* 2012;57(4):1124–1128.
62. Antia FP. Ethyl alcohol: physiologic, pathologic and clinical aspects. *J Assoc Phys India.* 1972;20(7):511–515.
63. Altschule M. Cirrhosis due to plant toxins. *Med Sci.* 1963;14(3):38.
64. Hunsaker D, Hunsaker J. Obesity epidemic in the United States. In: Tosokos M, ed. *Forensic Pathology Reviews.* Vol 2. Totowa, NC: Humana Press Inc.; 2004:61–100.
65. Kral JG. Morbid obesity and related health risks. *Ann Intern Med.* 1985;103(6 (Pt 2)):1043–1047.
66. Duflou J, Virmani R, Rabin I, Burke A, Farb A, Smialek S. Sudden death as a result of heart disease in morbid obesity. *Am Heart J.* 1995;130: 306–313.
67. Byard RW. The complex spectrum of forensic issues arising from obesity. *Forensic Sci Med Pathol.* 2012;8(4):402–413
68. Maloney KF, Schoppe CH. Obesity and non-atherosclerotic cardiovascular disease. *Acad Foren Pathol.* 2013;3(1):8–12.
69. Sanner M. A comparison of public attitudes toward autopsy, organ donation, and anatomic dissection. A Swedish survey. *JAMA.* 1994;271(4): 284–288.
70. Wolf DA, Derrick SM. Undetermined cause and manner of death after organ/tissue donation. *Am J Forensic Med Pathol.* 2010;31(2):113–116.
71. The End of Life: Guidelines for Health Professionals Concerning, Death Certificates, Autopsies, and Organ and Tissue Donations. Vol 6th. New York: New York Academy of Medicine; 1994.
72. Jason D. The role of the medical examiner/coroner in organ and tissue procurement for transplantation. *Am J Forensic Med Pathol.* 1994;15(3): 192–202.
73. Davis GG, Brissie RM. A review of crane deaths in Jefferson County, Alabama. *J Forensic Sci.* 2000;45(2):392–396.
74. OSHA hazard bulletin: tree care work: falls and falling object hazards. In: *Labor-Department.* 2014. Washington, DC.
75. Cannon D, Taylor A. *Occupational Safety and Health Administration (OSHA) Investigation of an Occupational Diving Death (Platform Presentation). Paper Presented at: Annual NAME Meeting; October 18, 2021.* 2021. West Palm, FL).
76. Hoyert DL, Minino AM, Anderson RN. *Physicians Handbook of Medical Certification of Death.* 2 ed. Hyattsville: Department of Health and Human Services, Centers for Disease Control and Prevention; 2003.
77. Annegers JF, Hauser WA, Coan SP, Rocca WA. A population-based study of seizures after traumatic brain injuries. *N Engl J Med.* 1998;338(1): 20–24.
78. Nashef L, Fish DR, Garner S, Sander JW, Shorvon SD. Sudden death in epilepsy: a study of incidence in a young cohort with epilepsy and learning difficulty. *Epilepsia.* 1995;36(12):1187–1194.
79. Nashef L, Sander J. Sudden unexpected deaths in epilepsy-where are we now? *Seizure.* 1996;5:235–238.
80. Shields LB, Hunsaker DM, Hunsaker 3rd JC, Parker Jr JC. Sudden unexpected death in epilepsy: neuropathologic findings. *Am J Forensic Med Pathol.* 2002;23(4):307–314.
81. Lear-Kaul KC, Coughlin L, Dobersen MJ. Sudden unexpected death in epilepsy: a retrospective study. *Am J Forensic Med Pathol.* 2005;26(1): 11–17.
82. Rogde S, Hougen HP, Poulsen K. Homicide by sharp force in two Scandinavian capitals. *Forensic Sci Int.* 2000;109(2):135–145.
83. Thoresen SO, Rognum TO. Survival time and acting capability after fatal injury by sharp weapons. *Forensic Sci Int.* 1986;31(3):181–187.
84. Prahlow JA, Ross KF, Lene WJ, Kirby DB. Accidental sharp force injury fatalities. *Am J Forensic Med Pathol.* 2001;22(4):358–366.
85. Gill JR, Catanese C. Sharp injury fatalities in New York City. *J Forensic Sci.* 2002;47(3):554–557.
86. Spitz W, Petty C, Fisher R. Physical Activity until collapse following fatal injury by firearms and sharp pointed weapons. *J Forensic Sci.* 1961; 6(3):290–300.
87. Karakasi MV, Nastoulis E, Kapetanakis S, Vasilikos E, Kyropoulos G, Pavlidis P. Hesitation wounds and sharp force injuries in forensic pathology and psychiatry: multidisciplinary review of the literature and study of two cases. *J Forensic Sci.* 2016;61(6):1515–1523.
88. Schmidt U, Pollak S. Sharp force injuries in clinical forensic medicine–findings in victims and perpetrators. *Forensic Sci Int.* 2006;159(2–3): 113–118.
89. Thomsen AH, Hougen HP, Villesen P, Brink O, Leth PM. Sharp force homicide in Denmark 1992-2016. *J Forensic Sci.* 2020;65(3):833–839.
90. De-Giorgio F, Lodise M, Quaranta G, et al. Suicidal or homicidal sharp force injuries? A review and critical analysis of the heterogeneity in the forensic literature. *J Forensic Sci.* 2015;60(Suppl 1):S97–S107.
91. Franchi A, Kolopp M, Coudane H, Martrille L. Precise survival time and physical activity after fatal left ventricle injury from sharp pointed weapon: a case report and a review of the literature. *Int J Leg Med.* 2016;130(5):1299–1301.
92. Liberthson RR. Sudden death from cardiac causes in children and young adults. *N Engl J Med.* 1996;334(16):1039–1044.

93. Bharati S, Lev M. Sudden death in athletes - conduction system: practica1 approach to dissection and pertinent pathology. *Cardvasc Pathol.* 1994;3(2):117–127.
94. Raval MP, Welti CV. Sudden death from cocaine induced excited delirium: 45 cases (abstract). *Am J Clin Pathol.* 1995;104:329.
95. Corrado D, Thiene G, Nava A, Rossi L, Pennelli N. Sudden death in young competitive athletes: clinicopathologic correlations in 22 cases. *Am J Med.* 1990;89(5):588–596.
96. Irwin J, Cohle S. Sudden death due to diabetic ketoacidosis. *Am J Forensic Med Pathol.* 1988;9:119–121.
97. Multiple. Sudden death due to low-energy chest wall impact (commotio cordis). *N Engl J Med.* 1998;339:1398–1399.
98. Chute D, Grove C, Rajasekhara B, Smialek JE. Schizophrenia and sudden death: a medical examiner case study. *Am J Forensic Med Pathol.* 1999; 20(2):131–135.
99. Thogmartin JR. Sudden death in police pursuit. *J Forensic Sci.* 1998;43(6):1228–1231.
100. Adelson L. Sudden death from coronary disease. *JAMA.* 1961;176(2):131–135.
101. Kerwin AJ. Sudden death while driving. *Can Med Assoc J.* 1984;131(4):312–314.
102. Haugen RK. The Caf'e Coronary. Sudden Deaths in Restaurants. *JAMA.* 1963;186:142–143.
103. Tough SC, Green FH, Paul JE, Wigle DT, Butt JC. Sudden death from asthma in 108 children and young adults. *J Asthma.* 1996;33(3):179–188.
104. Thiene G, Basso C, Corrado D. Cardiovascular causes of sudden death. In: Silver MD, Gotlieb AI, Schoen FJ, eds. *Cardiovascular Pathology.* New York: Churchhill Livingstone; 2001:326–374.
105. Lecomte D, Fornes P, Nicolas G. Stressful events as a trigger of sudden death: a study of 43 medico-legal autopsy cases. *Forensic Sci Int.* 1996; 79(1):1–10.
106. Beh P, Byard RW. Cardiac tumors and sudden death. *Forensic Sci Med Pathol.* 2014;10(2):269–271.
107. Bierton C, Cashman K, Langlois NE. Is sudden death random or is it in the weather? *Forensic Sci Med Pathol.* 2013;9(1):31–35.
108. Kasuda S, Kudo R, Yuui K, Nakata M, Hatake K. Sudden death due to bilateral pulmonary thromboembolism following laparoscopic cholecystectomy. *Am J Forensic Med Pathol.* 2016;37(3):198–200.
109. Cummings P, Trekla D, Springer K. *Sudden Death.* Atlas of Forensic Histopathology. Cambridge: Cambridge University Press; 2011.
110. Atkinson. Pathobiology of sudden death: coronary causes. *Cardvasc Path.* 1994;3(2):105–115.
111. Randall B. Fatty liver and sudden death. A review. *Hum Pathol.* 1980;11(2):147–153.
112. Hugar BS, Praveen S, Kainoor SK, Shetty AR. Sudden death in Marfan syndrome. *J Forensic Sci.* 2014;59(4):1126–1128.
113. Toni C, Iannaccone F, Chella P, Basolo F, Pucci A. Sudden death in a case of recurrent Takotsubo syndrome. *Forensic Sci Med Pathol.* 2019;15(4): 595–597.
114. Schroder AS, Edler C, Sperhake JP. Sudden death from acute epiglottitis in a toddler. *Forensic Sci Med Pathol.* 2018;14(4):555–557.
115. Tomcik MA, Gerig NR, Prahlow JA. Sudden death from ruptured intracranial vascular malformation. *Forensic Sci Med Pathol.* 2011;7(2): 185–191.
116. Kitulwatte ID, Kim PJ, Pollanen MS. Sudden death related myocarditis: a study of 56 cases. *Forensic Sci Med Pathol.* 2010;6(1):13–19.
117. Scott T. Sudden death in emergency care: responding to bereaved relatives. *Emerg Nurse.* 2013;21(8).
118. Mitchell BL. Sickle cell trait and sudden death. *Sports Med Open.* 2018;4(1):19.
119. Harris KM, Haas TS, Eichner ER, Maron BJ. Sickle cell trait associated with sudden death in competitive athletes. *Am J Cardiol.* 2012;110(8): 1185–1188.
120. Browne RJ. Sickle cell trait and sudden death. *Sports Med.* 1994;18(6):373–374.
121. Marshall TK. The use of body temperature in estimating the time of death and its limitations. *Med Sci Law.* 1969;9(3):178–182.
122. Jacoby MR. The determination of the time interval following death by morphological, physical and chemical observation. *Chicago Med Sch Q.* 1970;29(4):203–209.
123. Brown A, Marshall TK. Body temperature as a means of estimating the time of death. *Forensic Sci.* 1974;4(2):125–133.
124. Burton JF. Fallacies in the signs of death. *J Forensic Sci.* 1974;19(3):529–534.
125. Burton JF. The estimated time of death. *Leg Med Annu.* 1977;1976:31–35.
126. Suzutani T, Ishibashi H, Takatori T. Studies on the estimation of the postmortem interval. 3. Rigor mortis (author's transl). *Hokkaido Igaku Zasshi.* 1978;52(6):269–283.
127. Henry JB, Smith FA. Estimation of the postmortem interval by chemical means. *Am J Forensic Med Pathol.* 1980;1(4):341–347.
128. Brown A, Hicks B, Knight B, Nokes LD. Determination of time since death using the double exponential cooling model. *Med Sci Law.* 1985; 25(3):223–227.
129. Green MA, Wright JC. Postmortem interval estimation from body temperature data only. *Forensic Sci Int.* 1985;28(1):35–46.
130. Hutchins GM. Body temperature is elevated in the early postmortem period. *Hum Pathol.* 1985;16(6):560–561.
131. Biddinger PW. Postmortem wound dehiscence. A report of three cases. *Am J Forensic Med Pathol.* 1987;8(2):120–122.
132. Stephens RJ, Richards RG. Vitreous humor chemistry: the use of potassium concentration for the prediction of the postmortem interval. *J Forensic Sci.* 1987;32(2):503–509.
133. Henssge C, Madea B, Gallenkemper E. Death time estimation in case work. II. Integration of different methods. *Forensic Sci Int.* 1988;39(1): 77–87.
134. Knight B. The evolution of methods for estimating the time of death from body temperature. *Forensic Sci Int.* 1988;36(1–2):47–55.
135. Jaffe FA. Stomach contents and the time of death. Reexamination of a persistent question. *Am J Forensic Med Pathol.* 1989;10(1):37–41.
136. Sparks DL, Oeltgen PR, Kryscio RJ, Hunsaker 3rd JC. Comparison of chemical methods for determining postmortem interval. *J Forensic Sci.* 1989;34(1):197–206.
137. Mann RW, Bass WM, Meadows L. Time since death and decomposition of the human body: variables and observations in case and experimental field studies. *J Forensic Sci.* 1990;35(1):103–111.
138. Cina SJ. Flow cytometric evaluation of DNA degradation: a predictor of postmortem interval? *Am J Forensic Med Pathol.* 1994;15(4):300–302.
139. Lange N, Swearer S, Sturner WQ. Human postmortem interval estimation from vitreous potassium: an analysis of original data from six different studies. *Forensic Sci Int.* 1994;66(3):159–174.
140. Byard RW, James RA, Gilbert JD. Diagnostic problems associated with cadaveric trauma from animal activity. *Am J Forensic Med Pathol.* 2002; 23(3):238–244.

141. Jaffe FA. Pathogenicity of carbon monoxide. *Am J Forensic Med Pathol*. 1997;18(4):406–410.
142. Gill JR, Goldfeder LB, Stajic M. The Happy Land homicides: 87 deaths due to smoke inhalation. *J Forensic Sci*. 2003;48(1):161–163.
143. Nielsen PR, Gheorghe A, Lynnerup N. Forensic aspects of carbon monoxide poisoning by charcoal burning in Denmark, 2008-2012: an autopsy based study. *Forensic Sci Med Pathol*. 2014;10(3):390–394.
144. Hideyuki Nushida MN. An autopsy case of fatal methemoglobinemia due to ingestion of sodium nitrite. *J Forensic Res*. 2014;06(01).
145. Mudan A, Repplinger D, Lebin J, Lewis J, Vohra R, Smollin C. Severe methemoglobinemia and death from intentional sodium nitrite ingestions. *J Emerg Med*. 2020.
146. Varlet V, Ryser E, Augsburger M, Palmiere C. Stability of postmortem methemoglobin: artifactual changes caused by storage conditions. *Forensic Sci Int*. 2018;283:21–28.
147. Nelson EL. Estimation of short-term postmortem interval utilizing core body temperature: a new algorithm. *Forensic Sci Int*. 2000;109(1): 31–38.
148. Rutty GN. The estimation of the time since death using temperatures recorded from the external auditory canal : Part I: can a temperature be recorded and interpreted from this site? *Forensic Sci Med Pathol*. 2005;1(1):41–51.
149. Wardak KS, Cina SJ. Algor mortis: an erroneous measurement following postmortem refrigeration. *J Forensic Sci*. 2011;56(5):1219–1221.
150. Andersson MG, Ceciliason AS, Sandler H, Mostad P. Application of the Bayesian framework for forensic interpretation to casework involving postmortem interval estimates of decomposed human remains. *Forensic Sci Int*. 2019;301:402–414.
151. Eden RE, Thomas B. *Algor Mortis*. StatPearls. *Treasure Island (FL)*. 2020.
152. Moritz A. *Pathology of Trauma*. 2nd ed. Philadelphia: Lea and Febiger; 1954.
153. Rampini SK, Schneemann M, Rentsch K, Bachli EB. Camphor intoxication after cao gio (coin rubbing). *JAMA*. 2002;288(1):45.
154. Rhodes B, Gordon I, Turner R. *Medical Jurisprudence: A South African Handbook*. South Africa: Stewart Printing Co; 1942.
155. Marshall TK. Temperature methods of estimating the time of death. *Med Sci Law* 1965;5(4):224–232.
156. Shapiro HA. The post-mortem temperature plateau. *J Forensic Med*. 1965;12(4):137–141.
157. Nokes LD, Hicks B, Knight BH. The post-mortem temperature plateau–fact or fiction? *Med Sci Law*. 1985;25(4):263–264.
158. Nokes LD, Hicks B, Knight B. The use of trachea temperature as a means of determining the post-mortem period. *Med Sci Law*. 1986;26(3): 199–202.
159. Simonsen J, Voigt J, Jeppesen N. Determination of the time of death by continuous post-mortem temperature measurements. *Med Sci Law*. 1977;17(2):112–122.
160. James WR, Knight BH. Errors in estimating time since death. *Med Sci Law*. 1965;5:111–116.
161. Kobayashi M, Ikegaya H, Takase I, Hatanaka K, Sakurada K, Iwase H. Development of rigor mortis is not affected by muscle volume. *Forensic Sci Int*. 2001;117(3):213–219.
162. Varetto L, Curto O. Long persistence of rigor mortis at constant low temperature. *Forensic Sci Int*. 2005;147(1):31–34.
163. Chakravarthy M. "Rigor mortis" in a live patient. *Am J Forensic Med Pathol*. 2010;31(1):87–88.
164. Gill JR, Landi K. Putrefactive rigor: apparent rigor mortis due to gas distension. *Am J Forensic Med Pathol*. 2010.
165. Pillay VV, D'Souza DH. Rigor mortis or something else? *Am J Forensic Med Pathol*. 2010;31(4):e2.
166. Sauvageau A, Viner S, Davies J. Muscle stiffness in critical illness is not a form of rigor mortis. *Am J Forensic Med Pathol*. 2010;31(4):e5.
167. D'Souza DH, Harish S, Rajesh M, Kiran J. Rigor mortis in an unusual position: forensic considerations. *Int J Appl Basic Med Res*. 2011;1(2): 120–122.
168. Tsokos M, Byard RW. Putrefactive "rigor mortis". *Forensic Sci Med Pathol*. 2012;8:200–201.
169. Anders S, Kunz M, Gehl A, Sehner S, Raupach T, Beck-Bornholdt HP. Estimation of the time since death–reconsidering the re-establishment of rigor mortis. *Int J Leg Med*. 2013;127(1):127–130.
170. Esnault P, Prunet B, Lacroix G, D'Aranda E, Gaillard Y, Boret H. Instantaneous rigor after fatal pholcodine intoxication. *Br J Clin Pharmacol*. 2014;77(3):578–579.
171. Mesri M, Behzadnia M, Dorooshi G. Accelerated rigor mortis: a case letter. *J Res Med Sci*. 2017;22(123):126.
172. Krompecher T, Bergerioux C, Brandt-Casadevall C, Gujer HR. Experimental evaluation of rigor mortis. VI. Effect of various causes of death on the evolution of rigor mortis. *Forensic Sci Int*. 1983;22(1):1–9.
173. Moreva EV. Pharmacological analysis of rigor mortis of the skeletal muscles; effect of sodium azide and fluoride on rate of rigor mortis. *Biull Eksp Biol Med*. 1954;38(10):54–56.
174. Simmons T, Cross PA, Adlam RE, Moffatt C. The influence of insects on decomposition rate in buried and surface remains. *J Forensic Sci*. 2010; 55(4):889–892.
175. Jirón LF, Cartín VM. Insect succession in the decomposition of a mammal in Costa Rica. *J N Y Entomol Soc*. 1981:158–165.
176. Rodriguez B. Decomposition of buried bodies and methods that may aid in their location. *JFS*. 1985;30:836–852.
177. Komar DA. Decay rates in a cold climate region: a review of cases involving advanced decomposition from the Medical Examiner's Office in Edmonton, Alberta. *J Forensic Sci*. 1998;43(1):57–61.
178. Pfeiffer S, Milne S, Stevenson RM. The natural decomposition of adipocere. *J Forensic Sci*. 1998;43(2):368–370.
179. Shalaby OA, deCarvalho LM, Goff ML. Comparison of patterns of decomposition in a hanging carcass and a carcass in contact with soil in a xerophytic habitat on the Island of Oahu, Hawaii. *J Forensic Sci*. 2000;45(6):1267–1273.
180. Vass AA, Barshick SA, Sega G, et al. Decomposition chemistry of human remains: a new methodology for determining the postmortem interval. *J Forensic Sci*. 2002;47(3):542–553.
181. Decomposition: Pathology, Entomology, and Anthropology. *USCAP 2004 Forensic Pathology Evening Lecture*. 2004.
182. Vass AA, Smith RR, Thompson CV, et al. Decompositional odor analysis database. *J Forensic Sci*. 2004;49(4):760–769.
183. Voss SC, Forbes SL, Dadour IR. Decomposition and insect succession on cadavers inside a vehicle environment. *Forensic Sci Med Pathol*. 2008; 4(1):22–32.
184. Campobasso CP, Marchetti D, Introna F, Colonna MF. Postmortem artifacts made by ants and the effect of ant activity on decompositional rates. *Am J Forensic Med Pathol*. 2009;30(1):84–87.
185. Lee Goff M. Early post-mortem changes and stages of decomposition in exposed cadavers. *Exp Appl Acarol*. 2009;49(1–2):21–36.
186. Cross P, Simmons T. The influence of penetrative trauma on the rate of decomposition. *J Forensic Sci*. 2010;55(2):295–301.

187. Anderson GS. Comparison of decomposition rates and faunal colonization of carrion in indoor and outdoor environments. *J Forensic Sci*. 2011;56(1):136–142.
188. Parks CL. A study of the human decomposition sequence in central Texas. *J Forensic Sci*. 2011;56(1):19–22.
189. Gruenthal A, Moffatt C, Simmons T. Differential decomposition patterns in charred versus un-charred remains. *J Forensic Sci*. 2012;57(1):12–18.
190. Sutherland A, Myburgh J, Steyn M, Becker PJ. The effect of body size on the rate of decomposition in a temperate region of South Africa. *Forensic Sci Int*. 2013;231(1–3):257–262.
191. Cockle DL, Bell LS. Human decomposition and the reliability of a 'Universal' model for post mortem interval estimations. *Forensic Sci Int*. 2015;253:136 e131–139.
192. Caruso JL. Decomposition changes in bodies recovered from water. *Acad Forensic Pathol*. 2016;6(1):19–27.
193. Byard RW, Gehl A, Anders S, Tsokos M. Putrefaction and wound dehiscence: a potentially confusing postmortem phenomenon. *Am J Forensic Med Pathol*. 2006;27(1):61–63.
194. Brondum N, Simonsen J. Postmortem red coloration of teeth. A retrospective investigation of 26 cases. *Am J Forensic Med Pathol*. 1987;8(2):127–130.
195. James R, Gilbert J, Byard RW. Posttraumatic diaphragmatic hernia and death–etiologic factor or putrefactive artifact? *Am J Forensic Med Pathol*. 1999;20(1):66–69.
196. Sims DN, Lokan RJ, James RA, et al. Putrefactive pleural effusions as an alternative sample for drug quantification. *Am J Forensic Med Pathol*. 1999;20(4):343–346.
197. Prieto JL, Magana C, Ubelaker DH. Interpretation of postmortem change in cadavers in Spain. *J Forensic Sci*. 2004;49(5):918–923.
198. Singh MK, O'Donnell C, Woodford NW. Progressive gas formation in a deceased person during mortuary storage demonstrated on computed tomography. *Forensic Sci Med Pathol*. 2009;5(3):236–242.
199. O'Donnell CJ, Baker MA. Postmortem CT findings of gastromalacia: a trap for the radiologist with forensic interest. *Forensic Sci Med Pathol*. 2010;6(4):293–297.
200. Fernando T, Winskog C, Byard RW. An unusual pattern of decomposition associated with suicidal electrocution in a bath. *J Forensic Sci*. 2013;58(4):1085–1087.
201. MacKenzie JM. Examining the decomposed brain. *Am J Forensic Med Pathol*. 2014;35(4):265–270.
202. Maujean G, Vacher P, Bagur J, Guinet T, Malicier D. Forensic autopsy of human decomposed bodies as a valuable tool for prevention: a French regional study. *Am J Forensic Med Pathol*. 2016;37(4):270–274.
203. Campobasso CP, Falamingo R, Grattagliano I, Vinci F. The mummified corpse in a domestic setting. *Am J Forensic Med Pathol*. 2009;30(3):307–310.
204. Harding BE, Wolf BC. The phenomenon of the urban mummy. *J Forensic Sci*. 2015;60(6):1654–1657.
205. Chen CC, Yang CK, Chen CY, Lee HC, Wang SM. Comparison of rehydration techniques for fingerprinting the deceased after mummification. *J Forensic Sci*. 2017;62(1):205–208.
206. Finaughty DA, Morris AG. Precocious natural mummification in a temperate climate (Western Cape, South Africa). *Forensic Sci Int*. 2019;303:109948.
207. Forbes SL, Stuart BH, Dadour IR, Dent BB. A preliminary investigation of the stages of adipocere formation. *J Forensic Sci*. 2004;49(3):566–574.
208. Kahana T, Almog J, Levy J, Shmeltzer E, Spier Y, Hiss J. Marine taphonomy: adipocere formation in a series of bodies recovered from a single shipwreck. *J Forensic Sci*. 1999;44(5):897–901.
209. Yan F, McNally R, Kontanis EJ, Sadik OA. Preliminary quantitative investigation of postmortem adipocere formation. *J Forensic Sci*. 2001;46(3):609–614.
210. Forbes SL, Stuart BH, Dent BB, Fenwick-Mulcahy S. Characterization of adipocere formation in animal species. *J Forensic Sci*. 2005;50(3):633–640.
211. Jackowski C, Thali M, Sonnenschein M, Aghayev E, Yen K, Dirnhofer R. Adipocere in postmortem imaging using multislice computed tomography (MSCT) and magnetic resonance imaging (MRI). *Am J Forensic Med Pathol*. 2005;26(4):360–364.
212. Stuart BH, Craft L, Forbes SL, Dent BB. Studies of adipocere using attenuated total reflectance infrared spectroscopy. *Forensic Sci Med Pathol*. 2005;1(3):197–201.
213. O'Brien TG, Kuehner AC. Waxing grave about adipocere: soft tissue change in an aquatic context. *J Forensic Sci*. 2007;52(2):294–301.
214. Dumser TK, Turkay M. Postmortem changes of human bodies on the bathyal sea floor–two cases of aircraft accidents above the open sea. *J Forensic Sci*. 2008;53(5):1049–1052.
215. Frund HC, Schoenen D. Quantification of adipocere degradation with and without access to oxygen and to the living soil. *Forensic Sci Int*. 2009;188(1–3):18–22.
216. Notter SJ, Stuart BH, Rowe R, Langlois N. The initial changes of fat deposits during the decomposition of human and pig remains. *J Forensic Sci*. 2009;54(1):195–201.
217. Duraes N, Cortez D, Algarra M, et al. Comparison of adipocere formation in four soil types of the Porto (Portugal) district. *Forensic Sci Int*. 2010;195(1–3):168 e161–166.
218. Ubelaker DH, Zarenko KM. Adipocere: what is known after over two centuries of research. *Forensic Sci Int*. 2011;208(1–3):167–172.
219. Moses RJ. Experimental adipocere formation: implications for adipocere formation on buried bone. *J Forensic Sci*. 2012;57(3):589–595.
220. Widya M, Moffatt C, Simmons T. The formation of early stage adipocere in submerged remains: a preliminary experimental study. *J Forensic Sci*. 2012;57(2):328–333.
221. Gibelli D, Di Giancamillo A, Taborelli A, et al. Burial of piglet carcasses in cement: a study of macroscopic and microscopic alterations on an animal model. *Am J Forensic Med Pathol*. 2013;34(1):43–49.
222. Schoenen D, Schoenen H. Adipocere formation–the result of insufficient microbial degradation. *Forensic Sci Int*. 2013;226(1–3):301 e301–306.
223. Fiedler S, Berns AE, Schwark L, Woelk AT, Graw M. The chemistry of death–Adipocere degradation in modern graveyards. *Forensic Sci Int*. 2015;257:320–328.
224. Byard RW. Adipocere-the fat of graveyards. *Am J Forensic Med Pathol*. 2016;37(3):208–210.
225. Carson EA, Stefan VH, Powell JF. Skeletal manifestations of bear scavenging. *J Forensic Sci*. 2000;45(3):515–526.

226. Koszyca B, Gilbert JD, Byard RW. Antemortem trauma from rodent activity the popiel phenomenon. *Forensic Sci Med Pathol*. 2006;2(4): 269–272.
227. Gapert R, Tsokos M. Anthropological analysis of extensive rodent gnaw marks on a human skull using post-mortem multislice computed tomography (pmMSCT). *Forensic Sci Med Pathol*. 2013;9(3):441–445.
228. Young A, Stillman R, Smith MJ, Korstjens AH. An experimental study of vertebrate scavenging behavior in a Northwest European woodland context. *J Forensic Sci*. 2014;59(5):1333–1342.
229. Pokines JT, Sussman R, Gough M, et al. Taphonomic analysis of rodentia and lagomorpha bone gnawing based upon incisor size. *J Forensic Sci*. 2017;62(1):50–66.
230. Toledo VA, Fonseca GM, Gonzalez PA, Ibarra L, Torres FJ, Saez PL. A morphological and morphometric study of bite marks caused by mice (Mus Musculus) on different baits for forensic purposes. *J Forensic Sci*. 2017;62(2):462–467.
231. Benecke M. Six forensic entomology cases: description and commentary. *J Forensic Sci*. 1998;43(4):797–805.
232. Amendt J, Krettek R, Zehner R. Forensic entomology. *Naturwissenschaften*. 2004;91(2):51–65.
233. Matuszewski S, Bajerlein D, Konwerski S, Szpila K. Insect succession and carrion decomposition in selected forests of Central Europe. Part 1: pattern and rate of decomposition. *Forensic Sci Int*. 2010;194(1–3):85–93.
234. Matuszewski S, Szafalowicz M. Temperature-dependent appearance of forensically useful beetles on carcasses. *Forensic Sci Int*. 2013;229(1–3): 92–99.
235. Lindgren NK, Bucheli SR, Archambeault AD, Bytheway JA. Exclusion of forensically important flies due to burying behavior by the red imported fire ant (*Solenopsis invicta*) in southeast Texas. *Forensic Sci Int*. 2011;204(1–3):e1–3.
236. Byard RW. Autopsy problems associated with postmortem ant activity. *Forensic Sci Med Pathol*. 2005;1(1):37–40.
237. Krinsky CS, Lathrop SL, Brown P, Nolte KB. Drugs, detention, and death: a study of the mortality of recently released prisoners. *Am J Forensic Med Pathol*. 2009;30(1):6–9.
238. Heath KJ, Byard RW. Ant activity as a source of postmortem bleeding. *Forensic Sci Med Pathol*. 2014;10(3):472–474.
239. Peschel O, Kunz SN, Rothschild MA, Mutzel E. Blood stain pattern analysis. *Forensic Sci Med Pathol*. 2011;7(3):257–270.
240. Frank M, Schmucker U, Stengel D, et al. Proper estimation of blood loss on scene of trauma: tool or tale? *J Trauma*. 2010;69(5):1191–1195.
241. Crowder CM, Wiersema JM, Adams BJ, Austin DE, Love JC. The utility of forensic anthropology in the medical examiner's office. *Acad Forensic Pathol*. 2016;6(3):349–360.
242. Falsetti AB. A thousand tales of dead men: the forensic anthropology cases of William R. Maples, Ph.D. *J Forensic Sci*. 1999;44(4):682–686.
243. Walsh-Haney HA. Sharp-force trauma analysis and the forensic anthropologist: techniques advocated by William R. Maples, Ph.D. *J Forensic Sci*. 1999;44(4):720–723.
244. Spence MW, Shkrum MJ, Ariss A, Regan J. Craniocervical injuries in judicial hangings: an anthropologic analysis of six cases. *Am J Forensic Med Pathol*. 1999;20(4):309–322.
245. Wescott DJ, Moore-Jansen PH. Metric variation in the human occipital bone: forensic anthropological applications. *J Forensic Sci*. 2001;46(5): 1159–1163.
246. Bartelink EJ, Wiersema JM, Demaree RS. Quantitative analysis of sharp-force trauma: an application of scanning electron microscopy in forensic anthropology. *J Forensic Sci*. 2001;46(6):1288–1293.
247. Komar DA. Twenty-seven years of forensic anthropology casework in New Mexico. *J Forensic Sci*. 2003;48(3):521–524.
248. Steadman DW, Haglund WD. The scope of anthropological contributions to human rights investigations. *J Forensic Sci*. 2005;50(1):23–30.
249. Mundorff AZ, Vidoli G, Melinek J. Anthropological and radiographic comparison of vertebrae for identification of decomposed human remains. *J Forensic Sci*. 2006;51(5):1002–1004.
250. Langley NR. An anthropological analysis of gunshot wounds to the chest. *J Forensic Sci*. 2007;52(3):532–537.
251. Miller E. *A Forensic Anthropological Perspective from Los Angeles County. Paper Presented at*. Los Angeles, CA: National Association of Medical Examiners; 2005.
252. Zelson-Mundorff A, Steadman D. Anthropological perspectives on the forensic response at the World Trade Center disaster. *Gen Anthropol Bull Gen Anthropol Div*. 2003;10(1):1–5.
253. Di Nunno N, Saponetti SS, Scattarella V, et al. DNA extraction: an anthropologic aspect of bone remains from sixth- to seventh-century ad bone remains. *Am J Forensic Med Pathol*. 2007;28(4):333–341.
254. Dedouit F, Guilbeau-Frugier C, Telmon N, et al. Virtual autopsy and forensic anthropology of a mummified fetus: a report of one case. *J Forensic Sci*. 2008;53(1):208–212.
255. Hinkes MJ. Migrant deaths along the California-Mexico border: an anthropological perspective. *J Forensic Sci*. 2008;53(1):16–20.
256. Ubelaker DH. Issues in the global applications of methodology in forensic anthropology. *J Forensic Sci*. 2008;53(3):606–607.
257. Konigsberg LW, Herrmann NP, Wescott DJ, Kimmerle EH. Estimation and evidence in forensic anthropology: age-at-death. *J Forensic Sci*. 2008;53(3):541–557.
258. Grivas CR, Komar DA. Kumho, Daubert, and the nature of scientific inquiry: implications for forensic anthropology. *J Forensic Sci*. 2008;53(4): 771–776.
259. Robinson C, Eisma R, Morgan B, et al. Anthropological measurement of lower limb and foot bones using multi-detector computed tomography. *J Forensic Sci*. 2008;53(6):1289–1295.
260. Christensen AM, Crowder CM. Evidentiary standards for forensic anthropology. *J Forensic Sci*. 2009;54(6):1211–1216
261. Adams BJ, Herrmann NP. Estimation of living stature from selected anthropometric (soft tissue) measurements: applications for forensic anthropology. *J Forensic Sci*. 2009;54(4):753–760.
262. Park DK, Park KH, Ko JS, et al. The role of forensic anthropology in the examination of the Daegu subway disaster (2003, Korea). *J Forensic Sci*. 2009;54(3):513–518.
263. Brogdon BG, Sorg MH, Marden K. Fingering a murderer: a successful anthropological and radiological collaboration. *J Forensic Sci*. 2010; 55(1):248–250.
264. Christensen AM, Passalacqua NV, Schmunk GA, et al. The value and availability of forensic anthropological consultation in medicolegal death investigations. *Forensic Sci Med Pathol*. 2015;11(3):438–441.

265. McNulty JP, Gapert R. Forensic anthropology and radiography in the examination of an unknown mummified hand. *Forensic Sci Med Pathol*. 2013;9(4):602–606.
266. Porta D, Poppa P, Regazzola V, et al. The importance of an anthropological scene of crime investigation in the case of burnt remains in vehicles: 3 case studies. *Am J Forensic Med Pathol*. 2013;34(3):195–200.
267. Obertova Z, Adalian P, Baccino E, et al. The status of forensic anthropology in europe and South Africa: results of the 2016 FASE questionnaire on forensic anthropology. *J Forensic Sci*. 2019;64(4):1017–1025.
268. Carew RM, Morgan RM, Rando C. A preliminary investigation into the accuracy of 3D modeling and 3D printing in forensic anthropology evidence reconstruction. *J Forensic Sci*. 2019;64(2):342–352.
269. Thomas RM, Parks CL, Richard AH. Accuracy rates of ancestry estimation by forensic anthropologists using identified forensic cases. *J Forensic Sci*. 2017;62(4):971–974.
270. Wilson EK. Reanalysis of Korean war anthropological records to support the resolution of cold cases. *J Forensic Sci*. 2017;62(5):1127–1133.
271. Bethard JD. Historical trends in graduate research and training of diplomates of the American board of forensic anthropology. *J Forensic Sci*. 2017;62(1):5–11.
272. Schaffer WC. Total body mass estimation from anthropometric measurements in modern young adult U.S. Populations with healthy body fat percentages (NHANES III). *J Forensic Sci*. 2016;61(6):1431–1439.
273. Machado MP, Simoes MP, Gamba TO, et al. A wormian bone, mimicking an entry gunshot wound of the skull, in an anthropological specimen. *J Forensic Sci*. 2016;61(3):855–857.
274. Lesciotto KM. The impact of Daubert on the admissibility of forensic anthropology expert testimony. *J Forensic Sci*. 2015;60(3):549–555.
275. Shirley NR, Ramirez Montes PA. Age estimation in forensic anthropology: quantification of observer error in phase versus component-based methods. *J Forensic Sci*. 2015;60(1):107–111.
276. Garvin HM, Passalacqua NV. Current practices by forensic anthropologists in adult skeletal age estimation. *J Forensic Sci*. 2012;57(2):427–433.
277. Boyd C, Boyd DC. Theory and the scientific basis for forensic anthropology. *J Forensic Sci*. 2011;56(6):1407–1415.
278. Spradley MK, Jantz RL. Sex estimation in forensic anthropology: skull versus postcranial elements. *J Forensic Sci*. 2011;56(2):289–296.
279. Fourie Z, Damstra J, Gerrits PO, Ren Y. Evaluation of anthropometric accuracy and reliability using different three-dimensional scanning systems. *Forensic Sci Int*. 2011;207(1–3):127–134.
280. Blau S, Briggs CA. The role of forensic anthropology in Disaster Victim Identification (DVI). *Forensic Sci Int*. 2011;205(1–3):29–35.
281. Davy-Jow SL, Lees DM, Russell S. Virtual forensic anthropology: novel applications of anthropometry and technology in a child death case. *Forensic Sci Int*. 2013;224(1–3):e7–10.
282. Calmon M. Forensic anthropology and missing persons: a Brazilian perspective. *Forensic Sci Int*. 2019;298:425 e421–425 e426.
283. Wiersema JM, Woody A. The forensic anthropologist in the mass fatality context. *Acad Forensic Pathol*. 2016;6(3):455–462.
284. Anderson BE, Spradley MK. The role of the anthropologist in the identification of migrant remains in the American southwest. *Acad Forensic Pathol*. 2016;6(3):432–438.
285. Craig E. The role of the anthropologist in the application of NamUs. *Acad Forensic Pathol*. 2016;6(3):424–431.
286. Derrick SM, Figura B. The role of the anthropologist in identification at two urban medical examiner offices: New York city and harris county. *Acad Forensic Pathol*. 2016;6(3):413–423.
287. Wiersema JM. Evolution of forensic anthropological methods of identification. *Acad Forensic Pathol*. 2016;6(3):361–369.
288. Gill JR, Rainwater CW, Adams BJ. Santeria and Palo Mayombe: skulls, mercury, and artifacts. *J Forensic Sci*. 2009;54(6):1458–1462.
289. Rose JC, Green TJ, Green VD. NAGPRA is forever: osteology and the repatriation of skeletons. *Annu Rev Anthropol*. 1996;25(1):81–103.
290. Wetli C, Martinez M. Forensic sciences aspects of Santeria, a religious cult of african origin. *J Forensic Sci*. 1981;26:506–514.
291. Alison Newby C, Riley DM, Leal-Almeraz TO. Mercury use and exposure among Santeria practitioners: religious versus folk practice in Northern New Jersey, USA. *Ethn Health*. 2006;11(3):287–306.
292. Pasquali EA. Santeria. *J Holist Nurs*. 1994;12(4):380–390.
293. Sandoval MS. Santeria. *J Fla Med Assoc*. 1983;70(8):620–628.
294. Perlmutter D. *Syncretic Religions. Investigating Religious Terrorism and Ritualistic Crimes*. 1st ed. Boca Raton: CRC Press; 2003:181–234.
295. Pasquali EA. Santeria: a religion that is a health care system for Long Island Cuban-Americans. *J N Y State Nurses Assoc*. 1986;17(1):12–15.
296. Pokines JT. A Santeria/Palo Mayombe ritual cauldron containing a human skull and multiple artifacts recovered in western Massachusetts, U.S.A. *Forensic Sci Int*. 2015;248:e1–7.
297. Forman J, Moline J, Cernichiari E, et al. A cluster of pediatric metallic mercury exposure cases treated with meso-2,3-dimercaptosuccinic acid (DMSA). *Environ Health Perspect*. 2000;108(6):575–577.
298. Conlon E. Mass Fatality Managment and the Effects of Commingling. In: Adams B, Byrd J, eds. *Commingled Human Remains*. Oxford: Academic Press; 2014.
299. Komar D. The use of cadaver dogs in locating scattered, scavenged human remains: preliminary field test results. *J Forensic Sci*. 1999;44(2): 405–408.
300. Gill JR. 9/11 and the New York City Medical Examiner's Office. *Forensic Sci Med Pathol*. 2006;2(1):29–32.
301. Carroll E, Johnson A, DePaolo F, Adams BJ, Mazone D, Sampson B. Trends in United States Mass Fatality Incidents and Recommendations for Medical Examiners and Coroners. *Acad Forensic Pathol*. 2017;7(3):318–329.
302. Anderson M, Leditschke J, Bassed R, Cordner SM, Drummer OH. Mortuary operations following mass fatality natural disasters: a review. *Forensic Sci Med Pathol*. 2017;13(1):67–77.
303. Eitzen D, Zimmermann A. Setting up an off-site emergency mortuary facility (EMF) to deal with a DVI incident: disaster victim management (DVM). *Forensic Sci Med Pathol*. 2012;8(2):189–193.
304. Lake AW, James H, Berketa JW. Disaster Victim Identification: quality management from an odontology perspective. *Forensic Sci Med Pathol*. 2012;8(2):157–163.
305. Mundorff AZ. Integrating forensic anthropology into Disaster Victim Identification. *Forensic Sci Med Pathol*. 2012;8(2):131–139.
306. Graham EA. Disaster victim identification. *Forensic Sci Med Pathol*. 2006;2(3):203–207.
307. Hayes J. How to Survive a Mass Fatality Incident, Mostly. *Acad Forensic Pathol*. 2017;7(3):502–514.

308. Rocha LA, Fromknecht CQ, Redman SD, Brady JE, Hodge SE, Noe RS. Medicolegal Death Scene Investigations After Natural Disaster- and Weather-Related Events: A Review of the Literature. *Acad Forensic Pathol*. 2017;7(2):221–239.
309. Jordan FB. The role of the medical examiner in mass casualty situations with special reference to the Alfred P. Murrah Building bombing. *J Oklahoma State Med Assoc*. 1999;92(4):159–163.
310. Mundorff AZ. Anthropologist-directed triage: Three distinct mass fatality events involving fragmentation of human remains. In: Adams BJ, Byrd JE, eds. *Recovery, Analysis, and Identification of Commingled Human Remains*. Totowa, NJ: Humana Press; 2008:123–144.
311. Hennessey M. Data management and commingled remains at mass fatality incidents (MFIs). In: Adams BJ, Byrd JE, eds. *Recovery, Analysis, and Identification of Commingled Human Remains*. Totowa, NJ: Humana Press; 2008 337–356.
312. Budimlija ZM, Prinz MK, Zelson-Mundorff A, et al. World Trade Center human identification project: experiences with individual body identification cases. *Croat Med J*. 2003;44(3):259–263.
313. Mundorff AZ, Bartelink EJ, Mar-Cash E. DNA preservation in skeletal elements from the World Trade Center disaster: recommendations for mass fatality management. *J Forensic Sci*. 2009;54(4):739–745.
314. Gin K, Tovar J, Bartelink EJ, et al. The 2018 California Wildfires: Integration of Rapid DNA to Dramatically Accelerate Victim Identification. *J Forensic Sci*. 2020;65(3):791–799.
315. Masters A, Ogden R, Wetton JH, Dawnay N. Defining end user requirements for a field-based molecular detection system for wildlife forensic investigations. *Forensic Sci Int*. 2019;301:231–239.
316. Cash HD, Hoyle JW, Sutton AJ. Development under Extreme Conditions: Forensic Bioinformatics in the Wake of the World Trade Center Disaster. Paper Presented at: Pacific Symposium on Biocomputing 20032003; Singapore.
317. Prat S, Rerolle C, Saint-Martin P. Suicide pacts: six cases and literature review. *J Forensic Sci*. 2013;58(4):1092–1098.
318. Best D, Gossop M, Man LH, Stillwell G, Coomber R, Strang J. Peer overdose resuscitation: multiple intervention strategies and time to response by drug users who witness overdose. *Drug Alcohol Rev*. 2002;21(3):269–274.
319. Rodda LN, Pilgrim JL, Di Rago M, Crump K, Gerostamoulos D, Drummer OH. A Cluster of Fentanyl-Laced Heroin Deaths in 2015 in Melbourne, Australia. *J Anal Toxicol*. 2017;41(4):318–324.
320. Boddiger D. Fentanyl-laced street drugs "kill hundreds". *Lancet*. 2006;368(9535):569–570.
321. Ramos V, Hernandez AF, Villanueva E. Simultaneous death of twins. An environmental hazard or SIDS? *Am J Forensic Med Pathol*. 1997;18(1): 75–78.
322. Ruszkiewicz A, de Boer B, Robertson S. Unusual presentation of death due to carbon monoxide poisoning. A report of two cases. *Am J Forensic Med Pathol*. 1997;18(2):181–184.
323. Stefanidou ME, Maravelias CP, Dona AA, Pistos CM, Spiliopoulou CA, Athanaselis SA. Carbon monoxide-related deaths in Greece: a 23-year survey. *Am J Forensic Med Pathol*. 2012;33(2):128–131.
324. Byard RW. Carbon monoxide - the silent killer. *Forensic Sci Med Pathol*. 2019;15(1):1–2.
325. Sircar K, Clower J, Shin MK, Bailey C, King M, Yip F. Carbon monoxide poisoning deaths in the United States, 1999 to 2012. *Am J Emerg Med*. 2015;33(9):1140–1145.
326. Trichter J, Helpern M. The problem in New York City of accidental carbon monoxide poisoning. *NY Med*. 1952;8(20):30–37.
327. Unintentional non-fire-related carbon monoxide exposures–United States, 2001-2003. *MMWR Morb Mortal Wkly Rep*. 2005;54(2):36–39.
328. Lisbona CF, Hamnett HJ. Epidemiological Study of Carbon Monoxide Deaths in Scotland 2007-2016. *J Forensic Sci*. 2018;63(6):1776–1782.
329. Hannah SG, Turf EE, Fierro MF. Murder-suicide in central Virginia: a descriptive epidemiologic study and empiric validation of the Hanzlick-Koponen typology. *Am J Forensic Med Pathol*. 1998;19(3):275–283.
330. Palermo GB, Smith MB, Jenzten JM, et al. Murder-suicide of the jealous paranoia type: a multicenter statistical pilot study. *Am J Forensic Med Pathol*. 1997;18(4):374–383.
331. Byard RW, Knight D, James RA, Gilbert J. Murder-suicides involving children: a 29-year study. *Am J Forensic Med Pathol*. 1999;20(4):323–327.
332. Campanelli C, Gilson T. Murder-suicide in New Hampshire, 1995-2000. *Am J Forensic Med Pathol*. 2002;23(3):248–251.
333. Rouge-Maillart C, Jousset N, Gaudin A, Bouju B, Penneau M. Women who kill their children. *Am J Forensic Med Pathol*. 2005;26(4):320–326.
334. Marzuk PM, Tardiff K, Hirsch CS. The epidemiology of murder-suicide. *JAMA*. 1992;267(23):3179–3183.
335. Byard RW, Veldhoen D, Kobus H, Heath K. "Murder-suicide" or "murder-accident"? Difficulties with the analysis of cases. *J Forensic Sci*. 2010;55(5):1375–1377.
336. Sautter J, Gapert R, Tsokos M, Oesterhelweg L. Murder-suicide by carbon dioxide (CO_2) poisoning: a family case from Berlin, Germany. *Forensic Sci Med Pathol*. 2014;10(1):97–102.
337. du Plessis M, Hlaise KK. Homicide-suicide (dyadic death): a case study of double hanging. *Am J Forensic Med Pathol*. 2012;33(3):262–264.
338. Regoeczi WC, Gilson T. Homicide-Suicide in Cuyahoga County, Ohio, 1991-2015. *J Forensic Sci*. 2018;63(5):1539–1544.
339. Knoll JL, Hatters-Friedman S. The Homicide-Suicide Phenomenon: Findings of Psychological Autopsies. *J Forensic Sci*. 2015;60(5):1253–1257.
340. Dogan KH, Demirci S, Gunaydin G, Buken B. Homicide-suicide in Konya, Turkey between 2000 and 2007. *J Forensic Sci*. 2010;55(1):110–115.
341. Bridges FS, Lester D. Homicide-suicide in the United States, 1968-1975. *Forensic Sci Int*. 2011;206(1–3):185–189.
342. Panczak R, Geissbuhler M, Zwahlen M, Killias M, Tal K, Egger M. Homicide-suicides compared to homicides and suicides: systematic review and meta-analysis. *Forensic Sci Int*. 2013;233(1–3):28–36.

CHAPTER

5

The forensic autopsy

Susan F. Ely and James R. Gill

The autopsy establishes truth, detects change, provides data, instructs learners, and promotes justice. -*George Lundberg, 1984*

Partial autopsies give partial answers. -*Dr. Charles S. Hirsch*

The big picture

There is no standard definition for a *complete* autopsy. Autopsies are performed to answer the questions at hand, as well as those reasonably foreseeable for the future.[1–63] A "complete" autopsy is one that attempts to anticipate and answer all questions.[64] Partial autopsies have little place in forensic pathology practice.[65,66] *Partial autopsies do give partial answers.* If a surgeon simply wants to know what a liver nodule is (an abscess vs. a hematoma), a limited autopsy might be of use. But, since the autopsy is intended to determine why a person died, as well as answer myriad unforeseen questions separate and apart from the cause and manner of death, a full examination is needed. At a minimum, to the forensic pathologist, an autopsy that does not include a gross dissection of the cranial contents, the anterior neck structures, and the contents of the chest, abdomen, and pelvis is incomplete. On the opposite end of the spectrum are external examinations. There are many decedents who, because of the location and condition of their remains (decomposed at home), must be brought into the medical examiner's office for both identification and death certification purposes. Many who are of advanced age and/or have substantial comorbidities have died, in overwhelming likelihood, from natural causes and usually do not require an autopsy, despite falling under the purview of the medical examiner/coroner (ME/C).

Differences predicated on jurisdictional office philosophy and levels of comfort, standards of practice, staffing/workforce issues, and fiscal limitations will dictate which decedents are viewed (external examination only) and which are autopsied. Variations in practice reasonably exist between ME/C offices. That said, it is important to have the autopsy triaging practices for any given jurisdiction defined, published as a policy or standard operating procedure, and made known to the practicing forensic pathologists (FP), even if it must bend on a case by case or situational basis (e.g., during a pandemic—COVID, or epidemic—opioid). This provides a framework within which expectations are set and case triage decisions can be reasonably made. This is particularly important in a training environment. Before trainees develop experience and mature into seasoned professionals who make increasingly difficult decisions, they must first start with a fairly well-defined, principle-based framework, with discrete guidelines. Embedded within any policy, however, should be allowances for professional discretion, for which FPs will be supported. There is no policy manual in which every possible contingency can be outlined and prescribed, nor should there be. When it comes to guidelines set for professionals who practice in a field rife with nuance, it is advisable to create accommodations through such mechanisms as consensus conferences and make allowances for the expected "gray" areas of practice. If carefully considered, reasonable allowances can be made to accommodate a variety of unforeseen contingencies, without sacrificing principles of practice. "When your standards go, your principles are not far behind."

Equally important to address, particularly for those who are relatively new to the field, is the issue of autopsy dissection speed. This is a common area of insecurity for the inexperienced fellow. Being fast in the autopsy room should *not* be the highest priority early in fellowship, nor is it necessarily a predictive measure of work quality.

Principles of Forensic Pathology
https://doi.org/10.1016/B978-0-323-91796-4.00014-3

While learning how to work efficiently, utilize one's time, and pace oneself are all important skills that develop during the course of the fellowship year and beyond, particularly on busy case days, speed, per se, is not an indicator of a job well done, especially if it reflects cut corners. Much more important is to do the work that will allow the FP to sleep at night. In the inimitable words of Dr. Hirsch, "If you think of it, do it." Fellows can rest assured knowing that by the end of their training year, speed, confidence, and work quality will have converged in a way that they couldn't have anticipated on day one. Everything in good time.

The forensic autopsy: general tips

1. Look critically at the body to be sure that the height and weight reported to you comport with the body that you are examining. If the measurements are not consistent, have the body reweighed and/or remeasured. Carefully check all identification tags/wrist bands.
2. Be meticulous about observing and describing external characteristics.
3. Weigh the organs and measure the volumes of fluids (do not estimate).
4. The thoracic and abdominal organs should be removed prior to the examination of the head, and the brain should be removed prior to the internal examination of the neck; this sequence relieves vascular engorgement of the neck, hence minimizing the risk of artifacts that could be misinterpreted as injury.
5. Following reflection of the scalp and exposure of the skull, the pathologist should examine the surface of the calvarium and subscalpular tissues before the skull is sawed and opened. Additionally, they should always be present to observe the calvarium being removed to immediately document, collect, and measure any extra-axial hemorrhage, if it exists, before it spills or slides out onto the autopsy table.
6. Strip the dura mater from the skull or you risk overlooking fractures; strip the psoas to expose pelvic fractures or hidden abscesses; strip the parietal pleura to identify subtle rib fractures or calluses, especially in pediatric cases. Fractures also can be assessed through rib separation, by incising the intercostal muscles.
7. Do not be afraid of dark places. Open the middle ears of children, drowning victims, and possible victims of lightning. Open and aspirate the sphenoid sinus in drowning victims and in suspected asphyxial deaths. Examine the posterior oropharynx (foreign bodies, infections/abscesses, etc.), behind the ears, the axillae, the anus, and in the gluteal, mammary, and scrotal folds, the latter grouping particularly important when searching for a "missing" firearms defect (entrance or exit).
8. Have a low threshold for removing the spinal cord.
9. If you identify a structurally demonstrable immediate and/or proximate cause of death, all or a representative portion of that finding should be retained in your stock jar, even if you photograph it first. Avoid burying (or cremating) causal evidence.
10. If needed for the ultimate certification of a likely intoxication case, request any necessary hospital admission bloods *before* going down to start an autopsy; most hospitals discard these specimens after 3–5 days and this may happen while you're performing the autopsy! Don't wait.

Photography

Err on the side of photographing both pertinent negative and pertinent positive findings. There is no need to be stingy in photographic documentation. The adage "film is cheap" was never more relevant than it is now in this digital photography age. Also, if a significant finding on the body is being described in the autopsy report, it is best that it also be documented in the case photo file, and vice versa.

All forensic cases should have a full set of "overall" photographs, including all posterior surfaces. All homicides, pediatric, prisoner, police custody, trauma, and surgical/therapeutic complication (TC) cases will in overwhelming likelihood require further photographic documentation, beyond just the external examination and generic views. Any person who dies in police custody or following any period of hospital treatment (for any condition) while in police custody should have full-body photographs *and* radiographs taken. Individual injuries should be photographed in the customary fashion, whether causal to the death or not.

The suggested approach to photography includes the following: take overall photographs of the body as it is received (i.e., with clothing and/or artifacts of therapy/resuscitative efforts) and, ultimately, including the wounds as they are received ("unwashed" injury photos). It is important to note and **document the *degree* and *pattern* of bloodstaining** on the body before having the technician wash the crusted blood away. Is the blood mostly crusted in the scalp hair and on the face and hands or is it concentrated perianally and in and around the mouth?

With respect to range of fire determinations *in firearms injuries, the skin around entrance wounds should be carefully examined for fouling and photographed as to whether it is present or not, before it is wiped clean.* If an entrance wound is bandaged, the skin surface side of the dressing should also be examined for any fouling or gunpowder residue. It is also best to avoid aggressively irrigating entrance wounds with copious soap and water while the body is being prepared for the "clean" photographs. Gunpowder residue and/or foreign bodies that may reside within wound tracks can be washed away with such unnecessary practices. All wound tracks should be dissected with the knowledge that they have not been altered in any way (unless unavoidably so through surgical intervention). Abrasive brushes should also be expressly avoided in the cleaning of any decedent's body.

Following the undressing, cleaning, and drying of the body, the photographs are repeated. *An anatomic landmark should be included in the intermediate view photograph for orientation and future recall and identification.* Photographs should include a scale or ruler, which indicates the ME case number, placed inferior (caudal) to the injury, or whatever the focus of the photograph may be. If not at the anatomically inferior aspect of the photo, the scale should at least always be in the *same* position relative to the subject of the image in all photographs. This way, the orientation is the same, even in extreme close ups in which anatomic landmarks are no longer in view. For wounds, it is best to start with a "context shot," followed by a close up. All photographs are the forensic pathologist's ultimate responsibility and should be taken under their direction and supervision, that is, if not taken by them personally. This way, when later asked on the stand whether they personally took the photographs, they can respond, "no, but they were taken under my direction," a truthful statement of their role.

It is good practice to photograph homicide clothing, whether received with/on the body or later through law enforcement personnel (see below).

Radiography

Radiology is an important adjunct to forensic pathology practice, for both diagnostic purposes, as well as to improve the efficiency of the autopsy procedure (read: bullets). Plain film and full body scanners are commonplace in ME/C facilities, but computed tomography (CT) and magnetic resonance (MR) scanners are gaining increasing popularity and accessibility (see Chapter 18 for a detailed discussion of advanced imaging techniques). For deaths in which visual identification is not possible due to disfigurement by trauma or postmortem change, head, chest, and dental radiographs can be obtained for comparison with antemortem X-rays. In many instances, identification of decedents through radiography provides a straightforward, routine, and scientific modality. Most interpretations fall within the expertise of a board-certified forensic pathologist and should not require the employment of a consultant radiologist or odontologist.

The following suggestions constitute common forensic radiology practice for ***homicides or other special circumstances:***

Perform radiographs on all homicide victims who have sustained penetrating trauma (e.g., firearms, stabbings, bombs, etc.) and a chest radiograph on all homicides, regardless of the site of trauma.

Perform full body radiographs on all infants/most pediatric cases, in cases of known or suspected child or elder abuse, and all deaths in police or prison custody. Cone-down views of the knees and elbows may be useful in suspected child abuse cases.

Perform radiographs of the pelvis <u>and</u> lower extremities on all *pedestrian* motor vehicle collision victims, regardless of whether it involved a "leaving the scene" circumstance.

Radiologic tasks and tips

Remember that the first task with regard to postmortem X-rays is to learn under what circumstances they should be requested at case triage. The second task is just as, if not more important: to *review* the radiographs <u>before</u> autopsy, even when performing them only for identification purposes in a decomposed body.

Suggestions:

If requesting X-rays to document **projectiles**, review the X-rays *before* starting the autopsy. Failing to do so not only may lengthen and confuse the autopsy effort, but also places the cart before the horse. In addition, if the math isn't adding up or you discover wounds during the autopsy that necessitate additional views, don't hesitate to perform them.

If a **sharp injury** case, especially with injuries to the neck involving major vasculature, look for radiologic evidence of a potential cardiac air embolism *before* starting the autopsy. If suspicious for one, you may want to incise the inferior vena cava-cardiac junction under water when removing the heart. If you look at chest X-rays often enough, it enhances the likelihood of noticing a deviation from normal when there is a substantial hyperlucency around the heart.

If a **pedestrian,** the hips and entire lower extremities should be X-rayed. Look for skeletal injuries *before* opening the legs. The purpose of the leg dissections is to identify *any* injury and measure its location above the heel, even if the injury only involves the soft tissues. This exercise adds data points to the investigation, providing a potential correlation to the putative circumstances (vehicle size, brakes applied or not, speed at impact, etc.).

If dealing with a **decomposed** body, look for findings that might be helpful in the identification process (hip or knee prostheses, staples, sternal wires, etc.) and/or for old ("souvenir") bullets that may require removal. If knee hardware is seen on X-ray, for example, letting the identification staff know may streamline efforts (i.e., specifically requesting antemortem comparison X-rays for that body part). Orthopedic hardware does not need to be removed and retained unless there is a case-related reason to do so. **Pacemakers/defibrillators**, however, typically are retrieved and stored during the course of an autopsy but, unlike the incidental discovery of a projectile on X-ray, the presence of a pacemaker or defibrillator does not necessarily reflexively force an autopsy if the case was originally triaged as an external examination only.

When incidentally identified by X-ray, **"old" bullets** are typically removed for evidentiary purposes (regardless of their causal relationship to death); the FP would likely neither have looked for nor found these (no blood trail) in the body, especially a decomposed one, without the benefit of an X-ray tipping them off.

If a **baby/pediatric case**, evaluate the X-rays for trauma, fracture calluses, bone density, etc.

In a fetus in which **live birth** versus **stillbirth** is being ascertained, look for pulmonary air patterns and gastric and other intestinal bubbles/gas patterns. When oxygen therapy has been administered, these observations may become difficult to interpret (see Chapter 6).

For a **thermal injury** death, radiographs may be useful for identification as well the identifications of projectiles. A markedly burned/charred body may obscure cutaneous gunshot wounds. Deaths due to explosive injury deserve radiographs for both identification and to locate potential bomb debris or shrapnel.

Evidence

ME/C offices have various procedures in place for evidence collection and release. Establishing policies regarding a chain of custody for evidence handling is required for later introduction/admissibility in courts of law.

Clothing

Clothing should be routinely photographed and submitted as evidence in the following circumstances: homicides, police custody deaths, pedestrian victims of "hit and run"/leaving the scene motor vehicular collisions, and in otherwise suspicious cases, including certain pediatric deaths.

Clothing should be examined and described as part of the autopsy report, regardless of case type or whether it will be submitted as evidence or not. In homicides and other select cases (see above), be detailed in the description of the condition of the clothing (e.g., bloodstained vs. blood-soaked, soiled, intact, torn, patterned soiling, such as that consistent with tread marks, or previously cut as part of resuscitative efforts). All defects (*not* "wounds") and their locations on the garments should be described and matched/compared with the wounds on the body. If a defect corresponds with a gunshot entrance wound, comment on the presence or absence of "bullet wipe" and/or "grossly discernible" (or grossly indiscernible) gunpowder residue, with a further description and photographic documentation of its breadth and density. Clothing may become a key source of evidence in discerning ambiguous gunshot wounds and/or in determining range of fire. Noting the specific layering of the clothing before the body is undressed (which item was the outermost, intermediate, and innermost) is important, as it may also have bearing on firearms-related laboratory testing or trace evidence interpretations.

For unidentified or visually unrecognizable bodies in water, a granular description and photographic documentation of clothing articles are often important, and should include sizes, brands, colors, and specific design details. This may aid greatly in contributing to the databases used by missing person investigators and identification staff.

The sexual assault kit

Sexual assault evidence kits are collected in suspected or known instances of sexual assault (Fig. 5.1). With the ability of low-copy DNA testing, swab sample collection is not just limited to the oral, genitalia, and anal areas. Swabs of the neck, breasts, and wrists are areas to consider as well, depending on the case. In addition, outturned pockets (evidence of a suspected robbery) have yielded samples of evidentiary value. Since the optimum collection is completed *before* the autopsy, one should have a low threshold for sampling. As Charles Hirsch would say, "I would rather explain why I did something than why I did not." With DNA databases, it has become policy to test all sexual assault kits (SAKs), even if there is no known suspect or even definitive evidence of an assault having occurred. See below for further discussion on this topic.

Suspected weapons

Never allow an exposed suspected murder weapon (e.g., knife, hammer, and gun) into the morgue/autopsy suite. This may negate any subsequent forensic biology findings recovered from the weapon. These items of evidence should be examined in dedicated evidence examination areas that are separate from the autopsy suites or other body/tissue examination areas. Measurements made from images that include a scale are often sufficient for comparison purposes.

Ballistics evidence

The FP plays a central role in the **ballistics chain of custody** in criminal investigations when they retrieve and handle ballistics evidence. Aside from describing the projectiles in the autopsy report, it is also best to individually label and photograph all ballistics evidence removed from the body or found with the remains (sometimes, ballistics evidence is found in the clothing). Often, very small fragments cannot be located and retrieved, but are of

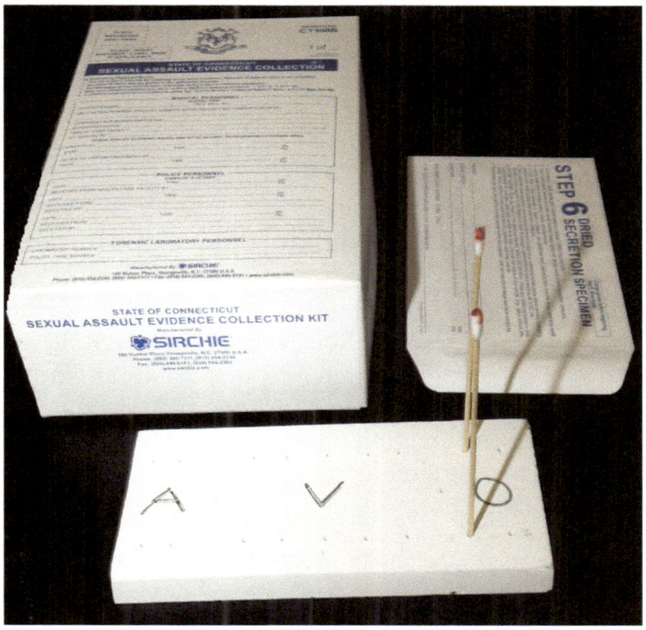

FIGURE 5.1 **Sexual assault collection kit.**

insignificant evidentiary value. Aside from those, however, every effort should be made to retrieve all bullets and their significant fragments and submit them as ballistics evidence, even if they have no relationship to the death. The exception to this rule involves *birdshot or buckshot pellets* from shotguns. Removal of only a representative sampling of such ballistics evidence is common/standard practice given that these projectiles have no rifling and are of little value individually or even in their entirety, in contrast to bullets, even when fragmented. The latter are weighed to assess caliber; this is not relevant to a collection of shot. That said, the presence of shotgun wadding or a cartridge within the wound is of important evidentiary value (i.e., should be submitted) and also has bearing on range of fire estimates. All bullets should be qualitatively described in the autopsy report with respect to their general size (not a caliper measurement, per se; this is reserved for the ballistics expert), their degree of deformity, if any, the location of the deformities (nose, base, flattened side, etc.), whether they are jacketed or coated in any way, and whether the jacket has become separated from the core slug. Unusual projectiles, such as wadcutters, should also be clearly and qualitatively described and photographed.

When performing an autopsy on a gunshot victim, it is important to collect all the blood from the chest and abdominal cavities in a measuring cup or bucket, and to not discard it until all bullets have been accounted for. Bullets are elusive, particularly those of small caliber. They may reside in clots or in ladled liquid blood, and no FP wants to have to pull a drain apart to retrieve one that may have inadvertently slipped down there. Making sure the drain is plugged before starting an autopsy on a firearms victim may save considerable time and frustration. Finally, don't allow the mortuary technicians to discard the body bag until all bullets have been accounted for. See Chapter 9 for a granular discussion on "bullet accounting."

The standard external autopsy examination

In contrast to a hospital autopsy, the external examination of a forensic autopsy is a much more detailed and careful procedure. Standard characteristics (eye color, hair length, the presence of tattoos and scars, the presence of needle puncture or track marks, cachexia, level of hygiene, layered clothing, evidence of prior cutting, etc.) may seem trivial when considering the goal of most autopsies; however, to a family member, they are important. That said, some observations have more diagnositc value than others (e.g., earlobe creases and hirsutism in relation to the likelihood of cardiac disease). Depending on the case, such observations may even have significant bearing on the identification of an unknown decedent or to a criminal investigation. If a pathologist were to mistakenly state that the liver weighed 10 kg, most nonmedical next of kin would not blink an eye. But if you state that the eyes are brown when they are actually blue, the family will question your entire report and even question whether the autopsy was performed on the correct person. Moreover, with homicidal trauma, in particular, the external examination may constitute a much more complex and detail-oriented task than the internal. Cataloging and documenting multiple gunshot wound defects are not only time-consuming, but also it is of substantial evidentiary value, as the attendant photographs that are taken are much more likely to reach the eyes of a juror than one taken of the perforated liver. Chapter 16 (Communication) and Appendix C outline the standard elements of the external autopsy examination.

The standard internal autopsy examination

Each pathologist employs their own style and techniques in the performance of an autopsy. The **Virchow method** is probably the most widely used in forensic practice; however, the en bloc **Rokitansky method** also is still widely used, but primarily in hospital autopsy settings. The Virchow (organ by organ/in situ) method is particularly important in forensic autopsy practice, as it affords contextual discoveries and in situ documentations that are particularly important in ascertaining the relationships of anatomic structures to one another, as well as in the identification and understanding of the sources of hemorrhage, inflammation, infection, and surgical complications, and in following wound tracks, revealing patterns, and searching for projectiles. Chapter 16 (Communication) and Appendix C outline the standard elements of the internal autopsy examination.

Cardiovascular pathology

Forensic pathologists are trained in cardiovascular pathology and, over the course of a career, will examine thousands of hearts. There may be instances, however, in which employing a pathologist who specializes in cardiovascular pathology may be both useful and necessary. This would include cardiac pathologists and pediatric pathologists, both of whom specialize in congenital heart disease, among other things, such as coronary artery anomalies and various cardiomyopathies. That said, *the most important observation a pathologist can make during the course of an autopsy examination with regard to the heart is an accurate recording of its mass relative to the decedent's body mass.* Within the bell curve of cardiac pathology, the interpretation of this single datum will likely produce more pathologic fruit than any other.

After their deactivation, **pacemakers and implantable defibrillators** should be removed from the body, cleaned thoroughly, and stored as evidence. These can be made available to families who may request them at a later time for analysis/interrogation. It is not uncommon for next of kin to feel that a defibrillator was faulty and must have caused or contributed to death. In reality, this (a form of electrocution) is vanishingly rare. That said, these devices are not full proof and may *fail to save* the life of a person with advanced endogenous heart disease, just as an esophageal intubation fails to revive the victim of a fatal shooting. *A procedure that fails to prolong life or fails to interrupt the dying process should not be conflated with the cause of the death.* "Do not confuse correlation with causation."

Beyond the quotidian routine of weighing the heart and assessing the kidneys for small vessel disease (fine granularity), advancements in molecular genetic science and testing have greatly assisted in the diagnosis of channelopathies and cardiomyopathies, having answered many previously unanswered questions, especially when the forensic pathologist is confronted with the sudden death of a person with a grossly normal heart and otherwise negative autopsy. A purple top tube of blood or tissue in a preservation medium (such as RNAlater) is typically all that is needed for such diagnostics. See Chapter 19 for a more in-depth discussion of this important topic.

Neuropathology

Formal neuropathologic consultation is useful and sometimes necessary in select instances. Typically, the brain is first fixed in formalin for approximately two weeks, which allows for thin sections to be sliced and evaluated. While an exhaustive discussion regarding both case selection and diagnoses exceeds the scope of this text, suffice to say that the field of forensic neuropathology is central to the practice of forensic pathology. There are no strict criteria that delineate which brains should be examined by a neuropathologist; however, blunt injury homicides and suspected abusive head trauma deaths represent two examples of such cases that will likely benefit from this consultation.

The following are a few tips to keep in mind when handling the brain at autopsy:

- The selection of cases for neuropathologic examination is made at the discretion of the forensic pathologist. In general, cases in which there may be an important (causal), but nuanced neurological problem are best examined after brains have been fixed in formalin. With the exception of many/most blunt impact homicides, brains with acute head injuries, including those related to firearms injuries, are typically examined fresh at the autopsy table (i.e., not retained as a whole organ).
- **Infant brains** are nearly always better examined after formalin fixation.
- If the brain is retained, the dura mater should be saved with it.
- Presenting specific questions to the neuropathologist prior to or at the brain cutting will facilitate the most efficient examinations with the highest yield.
- If, upon removing the calvarium, an **acute basilar subarachnoid hemorrhage** is encountered (i.e., within and obscuring the circle of Willis), each carotid artery at its entry into the subdural space should be carefully examined (before or after brain removal) and the vessels within the circle of Willis should be explored for ruptured cerebral aneurysms. If no aneurysms are identified, a search for evidence of blunt trauma with vertebral artery injury is the best next step. Occasionally, hyperextension or torsion of the neck or blunt force injury to the cervical spine or lateral head may result in a lacerated vertebral artery at or near its attachment to the dura. In reality, this specific pattern of basilar subarachnoid hemorrhage is most commonly caused by a ruptured saccular (berry) aneurysm located within the circle of Willis. After removal of the brain, the basilar subarachnoid hemorrhage can be gently removed at the autopsy table (using hydrogen peroxide) before a search for the aneurysm is undertaken.
- An acute **subdural hemorrhage** (less than two days) does not usually adhere to the dura and, therefore, will not be demonstrable when the neuropathologist examines the brain after fixation. It is therefore the task of the FP to document its presence at the time of autopsy, with photographs and a description, noting the volume and specific location of the blood. *Saving the liquid subdural blood for toxicology can be valuable*, particularly if there has been a survival interval (after a survival interval, toxicologic testing of the subdural blood will better represent the

substances, if any, in the system of the decedent *at the time of injury occurrence* than will the femoral blood collected at autopsy). If not sawing the skull personally, the FP should always at least be present and standing by when the cranium is removed to expose the brain. If a subdural hemorrhage or hematoma is discovered unexpectedly, this is the time to recognize, document, and quantitate it, before it falls onto the table.

- If cut fresh, **routine sections** for the stock jar might include: the superior-middle frontal gyri, temporal lobes (with amygdaloid nucleus and parahippocampal gyrus), basal ganglia, thalamus, cerebellum, the entire brainstem, and at least a sampling of dura mater. Decisions to remove and retain the **spinal cord** are made on a case-by-case basis, but are fairly routine in infant deaths, instances of acute or remote spinal cord trauma that relates to the death (typically, with some form of a persistent vegetative state or paralysis, or the exclusion thereof), and several natural processes, such as multiple sclerosis.

Forensic anthropology

For ME/C offices fortunate enough to have ready access to on-site or off-site forensic anthropologists as members of the medicolegal team, the benefits are too numerous to count. While the appropriate forensic applications and practices for this specialized consultation service are described in much more detail in Chapters 4, 15, and 19, suffice to say that the anthropology team can be a critical partner to the forensic pathologist in the handling of mass disasters that involve devastating trauma and comingling of remains, deaths involving dismemberment and other attempts at postmortem concealment (incineration, burial, etc.), the evaluation of ambiguous or otherwise vexing skeletal abnormalities and trauma in the "fresh" body, including the skeleton of the neck in cases of compression, in the discrimination of human versus nonhuman remains, and, of course, in assisting with the evaluation and identification of unknown and skeletonized remains, including their likely stature, ancestry, sex, age, and potential clues to their cause and manner of death. If "in house," an anthropologic table side consultation that answers questions in real time, during the performance of the autopsy, can be invaluable.

Toxicological sampling

Toxicological testing is a routine part of any forensic autopsy practice, being performed often/in most cases, irrespective of whether the cause of death is a drug intoxication, per se. Per professional autopsy guidelines, blood (peripheral preferred), vitreous humor, and urine should be routinely collected and stored, if available, with instructions to test being at the discretion of the FP. Given its relatively avascular environment (hence, removing it from the metabolic derangements and cell lysis typical of the postmortem vascular environment), testing of the **vitreous humor** is the primary means by which *electrolyte and glucose analyses are done*, making it an important sample to carefully collect at autopsy. Many ME/C offices will also routinely collect other samples such as gastric contents, liver, bile, and brain. Kidney tissue may be helpful in suspected heavy metal intoxications and hair may be useful for evaluating chronicity of use or administration of substances. In moderately putrefied remains, toxicological analysis is possible, but more limited and less quantitatively accurate. In addition, vitreous humor is unavailable in most bodies having undergone moderate to advanced putrefaction, the fluid having been displaced by bacterial gas. Intravascular blood is also often not retrievable in remains with advanced putrefaction, leaving purged thoracic cavity blood (if available), liver, brain, and bile as alternate sample sources.

When approaching a **suspected suicide by ingestion**, it is best to accurately measure the entire gastric contents. Note its quality (Whole tablets/capsules? How many? Discernible inscriptions on the tablets? Chalky fragments? Granular or chalky sediment after settling? etc.); pouring it through a fine strainer into the measuring cup is often helpful in that regard. Paying close attention to the contents of the small bowel before it is rinsed by the mortuary technician might also prove fruitful. Once measured, the gastric contents can be submitted entirely to the toxicology laboratory, unless voluminous. Regardless, and importantly, knowing the precise total volume will aid in a later calculation of the estimated total quantity of drug(s) ingested (a measure of intent) once the toxicology report has been issued. An internet search of the tablet inscriptions can easily identify pills that may be found. See the toxicology chapter (Chapter 14) for further discussion.

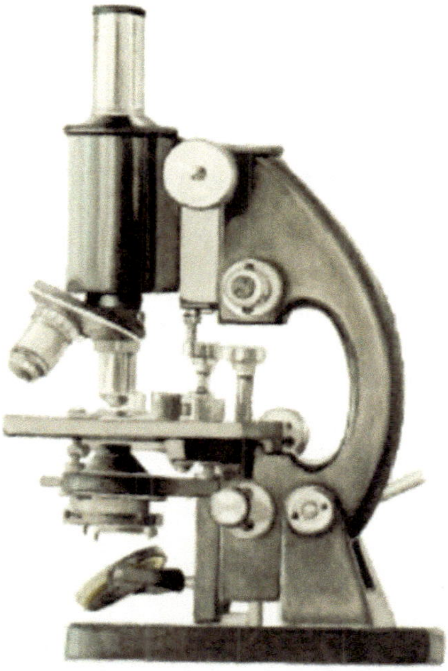

FIGURE 5.2 **Microscope.** *Watercolor by Dr. Cricket Coleman.*

Histology

Aside from infants, who typically require a more extensive histological examination, it is usually up to the forensic pathologist to decide on a case-by-case basis what tissues, if any, need to be examined microscopically (Fig. 5.2). This approach differs considerably from most hospital autopsy practices, in which an extensive histologic evaluation is reflexive rather than focused on a specific question or questions. By contrast, forensic histopathology is particularly important in the "aging" of lesions, both natural and injurious. Whether establishing a reasonable time frame for the formation of a deep venous thrombosis (DVT), of abdominal visceral contusions or lacerations in inflicted pediatric trauma, or in the acuteness of strap muscle hemorrhages in a strangulation, the histologic examination is central in the full work up of *selected* cases. All that said, Charles Hirsch was quick to warn, "In reality, pathologic processes seldom cooperate fully with attempts to date or age them with precise reliability." All that is to say that the microscopic examination is helpful in that it adds an additional data point to the entire constellation of findings, offering mainly units/ranges in timing, rather than quantitative precision. Like most findings in forensic pathology, microscopic findings must be interpreted within the greater context of the case. Additionally, a formalin-filled jar of representative tissue samples is collected and retained on all autopsies so that, if questions later arise, there is an opportunity to perform a further evaluation.

There are some natural (and even unnatural) deaths that may not be diagnosed adequately on gross examination alone or at all, but can be through a microscopic examination. These include such entities as myocarditis, encephalitis, pulmonary hypertension, sickle cell disease, microscopic abnormalities of the cardiac conduction system, and even ethylene glycol poisoning (calcium oxalate crystals in the kidney). Other histologic examinations may be performed for "completeness," especially in deaths that are likely to be widely and comprehensively reviewed, and also as part of the establishment of a diagnosis of exclusion. It is not unusual for a private consultant to criticize an autopsy for not completing a "full" histological examination. These consultants are often hospital-based, nonforensic pathologists who believe an autopsy is incomplete unless a section of autolyzed small bowel has been examined histologically in every case. We disagree.

Microbiology

On occasion, a broad or focused microbiologic work-up during an autopsy is warranted. This is most common in the pediatric setting (see Chapter 6), in which samples for both bacterial and viral studies are frequently submitted. A blood culture is best obtained after some forethought. If decided before the autopsy, a venipuncture with the careful withdrawal of blood is usually easily obtained from the inferior vena cava after reflection of the heart, before its removal; this will yield the most sterile inoculum.

Other possible microbiologic studies include respiratory viral swabs (upper/nasopharyngeal and/or lower/bronchial), a bacterial culture of cerebrospinal fluid (CSF) (obtained by lumbar puncture performed during the external examination), bacterial swabs of tissues or abscesses, mycobacterial cultures, and HIV testing. Following the autopsy, samples for bacterial culture are best stored in an incubator while those for viral analysis can be adequately stored in a refrigerator.

Finally, it is worth noting that the discovery of a *Neisseria meningitidis* infection at autopsy constitutes a **time-sensitive public health emergency** since it involves a highly contagious infectious agent that has the capacity to kill otherwise healthy persons. If suspected, rapid polymerase chain reaction or latex agglutination testing of the brain or CSF is best performed immediately (ideally, on the same day). If positive, communicable disease authorities should be notified so that appropriate prophylaxis of close contacts, including first responders, can be initiated in a timely fashion. If the decedent was a student or prisoner, this has the potential of involving multiple people and turning into a multifatality event. It is with this in mind that we emphasize the importance of having a good working knowledge of a list of reportable communicable diseases mandated by the local department of health and clear protocols for sampling, submission, and testing. See also Chapter 17 for further discussion.

Approach to the trauma autopsy

More than any other part of the forensic autopsy, the introduction of trauma and the ability to systematically and accurately evaluate it constitutes the most abrupt departure from hospital autopsy practice. The forensic autopsy report should be organized in such a way as to maximize its communicative quality. **One key element to achieve this end includes creating a separate and dedicated section for all injuries ("Injuries, External and Internal"), no matter how minor, and resisting the urge to repeat them elsewhere in the report** (See further discussion below and in Chapter 16).

In general, when listing and describing injuries in both the final diagnoses and the body of the report, they should be organized and described by body region (head, neck, trunk, and extremities) and further, from superficial to deep (external followed by associated internal findings). For example, for an injury of the head, one would first describe any skin injury, followed by the associated subjacent subscalpular/subgaleal injury, then, any cranial injury, epidural, subdural, and/or subarachnoid hemorrhages, and finally, intracerebral findings (e.g., contusions, hematomas, perforations, bullet tracks, etc.).

In the United States, for consistency in weapon comparisons and for the best comprehension by the nonforensic audience, including families, jurors, and lawyers who may ultimately read the report, **injuries are best measured in *inches*, both externally and internally**; height and weight are also probably best reported in standard units (inches, feet, and pounds). This, of course, is unnecessary in many, if not most countries outside the United States. That said, by contrast and convention, **measurements for natural disease are typically reported in metric units** (millimeters and centimeters), in keeping with standard medical and surgical reports in the United States and abroad.

Some injuries may be quite extensive (e.g., multiple blunt force trauma in a person who was run over by a train). The goal here is to not get too overwhelmed by the extent of the injuries and their documentation, becoming lost among the trees. As Charles Hirsch would say, "How do you eat an elephant? One bite at a time." In addition, photographs will serve as an important adjunct and quality assurance (QA) in the documentation of injuries. One risk of these types of cases is what is described as, "the distraction of the dramatic." Recognition of this very human tendency is the first step in performing a thorough and appropriate examination and documentation.

Following the above algorithm will make the work both manageable and naturally organized, minimizing the myopic overwhelm. That said, forensic autopsy report writing is equal parts an art and an exercise in complete and accurate documentation. While the latter should not be minimized, "complete and accurate" is not synonymous for an ill-conceived laundry list of findings and excessive verbiage, absent the necessary synthesis and organization; therein lies the art. While no observation is too minor to note, the way in which the observations are interpreted, framed, and ultimately communicated in writing separates a mediocre, even confusing report from an outstanding one. When there is a multiplicity of wounds, the ME/C would be well advised to step back from the autopsy table to gain the wide-angle view of the case from the outset, to notice the "forest." Not only will this allow them to see patterns in wound distribution ("the company injuries keep") but also it will open the door to a more contextual description. If the pattern and distribution of the wounds relative to each other are not highlighted by the FP, it will not likely be noticed or understood by most readers of the report. Clustering of excessive wounds, for example, is sometimes the best and only way to effectively communicate 104 stab and incised wounds. To enumerate and describe them individually in careful and perhaps excessive detail, without further qualification regarding their distribution and relationship to one another is neither an efficient nor necessarily helpful approach. This approach results in a technically complete, although potentially very confusing report, and also represents a lost opportunity in communicating important broader observations and patterns, seen only by a trained eye. Instead, one should so describe if the wounds are all tightly clustered on the lateral left torso (what might this mean with regard to a reconstruction of the events?). Sometimes, less is more in report writing. An excellent autopsy report begins at the autopsy table. See Chapter 16 for a more in-depth discussion on this topic.

Special techniques and examinations

Eye enucleation

This is often performed in a possible infant abuse case or to exclude any subtle trauma in otherwise benign infant deaths. If this is planned before autopsy, the standard *vitreous fluid aspiration should be avoided* to prevent the introduction of artifact. More commonly, in deaths without suspicious circumstances, per se, the forensic pathologist may start by only removing the orbital plates overlying the eyes, unroofing the orbital space, without actually removing the globes. This way, the optic nerve sheaths are exposed and can be evaluated and photographed. If free of hemorrhage/possible trauma, eye enucleation might be avoided. If removed, however, the eyes should be photographed and placed in separate labeled and formalin-filled containers for later evaluation and histologic processing. See also Chapter 6.

Intracardiac air embolism

These are often visible by plain film radiography (a hyperlucent rim around the heart) and/or by employing the bubble test (opening the heart under water, within the anteriorly incised and water filled pericardial sac); although rare, the most common scenario in which this is encountered is with sharp force trauma of the neck involving major veins, such as the jugular vein(s). If available, a CT scan is an excellent imaging modality for detecting air embolism.[67–82]

The layered anterior neck examination

While an anterior neck dissection is a routine component of all forensic autopsy examinations, a layered dissection with careful photographic documentation is typically reserved for circumstances that have prompted concern for inflicted neck compression or other neck injury.

1. Following removal of the brain and thoracic organs, the skin reflection is continued to expose the anterior neck structures.
2. A layered strap muscle neck dissection is performed in instances of suspected or known neck injury, with photographs taken of the layers. The tongue, hyoid bone, larynx, and trachea may also be removed en bloc.
3. The hyoid bone is the superior most skeletal structure of the anterior neck dissection. It is important to palpate it in situ to assess for any displaced fractures. After removal and tissue stripping, it can then be further examined for hemorrhage and fracture (note: if free of fracture in situ, any fracture identified ex situ should be considered as a possible artifact of removal). Further specimen radiographs and forensic anthropological examination may also be useful, depending on the case, but only after necessary soft tissue has been removed and retained for stock or possible histologic sampling.
4. The remaining dissection continues as usual, including a careful inspection of the cervical vertebrae and the prevertebral musculature.
5. In decomposed remains (that is, when an open casket wake is not possible), one may extend the midline trunk incision from the sternum to the chin (an inverted peace sign). This will enhance exposure and decrease the risk of causing artifactual injuries, which may be compounded by the decomposition.

The posterior neck examination[83–85]

This is not performed routinely, however, in *neck compression or otherwise suspicious deaths, motor vehicular or descent from height deaths with minimal demonstrable trauma, deaths immediately following cervical surgeries, and in some drowning deaths*, this procedure may support or identify the cause of death.

Reflect the skin from the base of the midline occiput to the mid back and transversely from posterior shoulder to shoulder (cruciate incision).

Perform a layer-by-layer removal of posterior neck muscles with photographic documentation.

Inspect posterior elements of the vertebral column.

Expose and inspect the atlanto-occipital and atlanto-axial facet joints.

If exposing and removing the cervical spinal cord, remove the posterior elements and saw through the posterior arches at a roughly 45 degree angle, bilaterally.

Remove and inspect the cervical spinal cord and dura together (photograph in situ first); the high cervical spinal cord is best visualized and the least artifactually damaged by performing a posterior approach (vs. an anterior cervical spinal cord removal).

Examine the posterior longitudinal, cruciate, and alar ligaments.

Concerns for sexual assault

Until proven otherwise, assume that a person who has been **strangled** has also been sexually assaulted, and vice versa. Deaths with features of **"overkill"** also should prompt a sexual assault work-up, which includes collection of all trace evidence, not just orifice sampling/semen collection. The collection of a full SAK may prove important in other select cases, where there might have been DNA transfer resulting from intimate/physical contact, with or without sexual activity, per se. For example, if a **grab mark** (a patterned bruise configuration) or **bite mark** is identified, or a strangulation is suspected, appropriate skin swabs should be collected (e.g., including the neck, even in the absence of visible skin injury in cases suspicious for neck compression). A **pelvic exenteration,** following the collection of the SAK, may also be considered in selected suspicious deaths involving suspected sexual assault and/or those with gross genital trauma.

Postmortem sperm harvesting

On occasion, at the request of next of kin, urologists are asked to perform postmortem sperm harvesting within a defined postmortem interval (with appropriate family consent). It is best to allow them to harvest *before* the autopsy; that said, as long as the bladder, rectum, and testes have not been dissected, harvesting can still be accomplished. An ice pack may need to be placed on the scrotum until the urologist is available to perform the harvest.

Pulmonary thromboembolus and posterior lower extremity dissection

Ideally, a death caused by a large pulmonary thromboembolus (PTE) will involve at least a gross examination of the veins and tissues of the pelvis and lower extremities.

A gross examination of the pelvic vasculature is performed during the routine autopsy, whereas that of the lower extremities constitutes an additional autopsy procedure. The latter examination is typically performed by placing the body in the prone position to reveal the posterior lower extremities. A posterior midline incision of each lower limb is made and the skin and subcutaneous soft tissues are reflected to expose the musculature. Either deep longitudinal or cross-sectional incisions of the muscle will then expose the vascular bundle, the incision of which should reveal the presence or absence of thrombi. If present, gross photographs can be helpful for documentation and a subsequent age estimation. Statistically, the most common locations for the formation of DVT are the calves and popliteal fossae.[27,86–98] Some may feel that consideration of whether the deep thrombi are found unilaterally or bilaterally may be of interest and relevance to the case. Sometimes, bilaterality supports a generalized venous stasis mechanism, such as that seen with obesity or immobility due to a variety of causes, whereas, a unilateral DVT may be more be consistent with an ipsilateral lower extremity injury, such as a casted tibial fracture. All that said, this pattern is by no means reliable or diagnostic, per se. Many cases of PTE due to obesity, immobility, cancer, or other noninjurious, systemic causes have only one lower extremity DVT and those with to a limb injury, a thrombotic involvement of both lower extremities. Given the fact that embolic phenomena are somewhat unpredictable, consistency of the laterality of the findings is not reliable.

Histological examination of deep venous thrombi offers the best opportunity to estimate the "age" of the clot and to document the presence or absence of thrombophlebitis (vs. the more common phlebothrombosis). A section of the thromboembolus may help confirm its antemortem origin (i.e., exclude the possibility of it being a postmortem clot), but will shed little light on the initiation and duration of the thrombotic process since it typically represents the most proximal portion of the thrombus, the fragment that was able to break free and embolize (therefore, usually not adherent to the wall), reflecting only the segment that most recently formed. *A more representative indicator of the age of the initiating thrombotic process would reside in the most distal segment of the deep vein thrombus in the lower extremities or pelvis.*

In terms of parlance, remember that the word **"clot"** is more nonspecific and descriptive, not necessarily indicating an intravital pathologic process, so can be used in association with either antemortem or postmortem coagula. *The word **thrombus**, on the other hand, infers an intravital pathologic process,* so its use should be reserved for a clearly antemortem process. This is often fairly clear from the gross examination, but in more equivocal cases, microscopy might be necessary to sort this out. See Chapter 7 for a further discussion on this topic.

Dissections of surgical and injection sites

Recent surgical incisions (e.g., hip replacement) should be dissected to rule out hematomas, infection, or abscess formation. A dissection of a hemodialysis arteriovenous shunt/fistula site in a death from an associated exsanguination should exclude or document infection, injury, or mechanical damage. Occasional dissections of needle puncture and/or "track marks" in a person with intravenous drug use may reveal perivenous hemorrhage. In addition, there may be a foreign-body giant cell reaction identified microscopically, which may support the chronicity of use. This information may be useful when discussing the death with next of kin, who may believe the decedent had never previously used drugs, alleging that someone must have criminally slipped them the drug without their knowledge.

Pedestrian fatalities: special procedures

- Dissect the pelvis and anterior lower extremities to assess for long bone and/or pelvic fractures and/or soft tissue contusions and hematomas, with or without avulsion pockets/shearing injury.
- Measure the precise location or span (if broad) of such injuries from the bottom of the heel. Generally speaking, pelvic and/or femoral fractures are more common from impacts sustained from taller vehicles, such as sport utility vehicles and trucks; tibia/fibular fractures are more common from traditional sedans.
- Ideally, these dissections are performed after the pelvis and lower extremities have been X-rayed and the X-rays have been reviewed.

Dissection of the facial soft tissues and skeleton

On occasion, poorly defined or absent external injuries of the face necessitate a deeper and more thorough dissection. Scenarios that might prompt the need for this special procedure include those in which facial injuries are expected by history (video/witness statements describing a blow or blows to the face and/or documented clinical sequelae), but are not fully apparent on external examination alone. Inapparent injuries might result from a survival/healing interval that has muted their appearance by the time of the autopsy and/or from a blow sufficient to cause the decedent to have fallen to the ground and strike their head, while leaving insufficient evidence (abrasions or contusions) of that contact on the skin's surface. In such circumstances, a deeper dissection of the soft tissue and skeletal structures of the face sometimes reveals hemorrhage(s) and/or fractures that were not visible or palpable on the surface. This is particularly true in darker-skinned individuals.

On balance and to be fair, an alternate outcome of this dissection is the *exclusion* of deeper injury, a negative finding equally important to ascertain and document, potentially indicating the absence of the contact that was believed to have been made. The FP is the neutral arbiter of the facts, ultimately providing both inclusionary *and* exclusionary evidence for the purposes of documenting or excluding injuries that may help in establishing the most accurate conclusions around the cause and manner of death.

The face is a part of the body that the FP will assiduously avoid incising at all cost. Most all are sensitive to the needs and preferences of family members when it comes to performing examinations that preclude subsequent funeral services involving the viewing of the face. That said, it is sometimes necessary to more fully evaluate the face beyond its routine external examination, particularly in criminal investigations. In these situations, the facial dissection allows the skin of the face to be carefully **peeled back** to reveal deeper structures that can be further evaluated, dissected, and photographed, *without* precluding a later viewing during funeral services.

While there is more than one approach in performing this procedure, one way is to extend the inferior aspect of the scalp incision (just behind each ear) somewhat posteriorly (i.e., to the posterolateral neck), then inferiorly and laterally, until it abuts the lateral most aspect of the Y-shaped incision on the anterolateral shoulder. With this technique, a roughly triangular flap is created and can be reflected from the anterolateral shoulders to expose not only the lateral neck structures but also each side of the face. Importantly, this method intentionally leaves undisturbed the attachments of the midline facial structures such that when the bilateral flaps are returned to their original locations, the appearance of the face remains unchanged from its original state.

Full body flay

Certain deaths require a more exhaustive autopsy and postautopsy approach, sometimes involving more than one "special" procedure or dissection, as well as laboratory testing and record review that extends beyond that performed in the majority of routine cases. Most deaths occurring in the *custody of law enforcement personnel or correctional/detention facilities and those that are suspicious for abusive/inflicted trauma involving dependent or otherwise vulnerable persons (infants/children or vulnerable adults)* represent the most common such scenarios.

One of the procedures commonly employed in these deaths, and in other select homicides/potential homicides, is a **full body flay**. In addition to the normal Y-shaped incision of the anterior torso, the flay involves incising the posterior neck and trunk and both the upper and lower extremities in order to fully expose and evaluate the underlying subcutaneous, muscular, and skeletal tissues for any evidence of trauma, documenting both its location and extent. As always, it is best to memorialize this dissection through photographic documentation, regardless of the discovery of negative or positive findings. These can be made available to others for what will invariably be part of an extensive subsequent review.

Autopsy of the pregnant or recently pregnant woman

Maternal deaths during pregnancy, delivery, or in the postpartum period represent a singular type of fatality, both in their rarity and in their uniquely tragic context. In contrast to many other fatalities, these deaths are also more likely to be highly emotionally and legally charged events for both the survivors and clinicians alike. Not uncommonly, these deaths are reviewed by many stakeholders (families, lawyers, clinicians, risk management, etc.), not the least of whom are public health and governmental bodies, such as multidisciplinary city and state maternal mortality review committees. As such, these fatalities are best investigated and certified by the most experienced and impartial medicolegal professionals; in the overwhelming majority of instances, this individual is the ME/C. As they have increased in incidence in this country relative to other countries, public health alarm bells have sounded and the push to have the ME/C involved has increased proportionately.

The foregoing definitions and autopsy guidelines should be considered in the investigations of these deaths:

Pregnancy-related fatalities: These are defined as deaths that *would not have occurred had the decedent not been pregnant.* Examples of causes in such deaths include amniotic fluid emboli, hemorrhagic complications of uterine atony following prolonged, labor, an instrumentation-related perforation of the uterus during an induced termination of pregnancy procedure, a coronary artery dissection, or a suicide directly related to new onset postpartum depression. It is important to note, however, that like delayed deaths from remote injuries, pregnancy-related deaths adhere to no statute of limitations between the initiation of the pregnancy-related disease or complication and the death. To wit, if a woman develops a pregnancy-related cardiomyopathy, but doesn't die from its direct complications until months or even years later, it is no less a pregnancy-related death.

Pregnancy-associated fatalities: These are deaths that occur either while the decedent is pregnant or up to 45 days after the pregnancy ends, irrespective of pregnancy outcome, but *are not a direct result of the pregnancy. These deaths are temporally related to the pregnancy rather than causally related to it.* Examples include deaths related to accidents, such as motor vehicle collisions or illicit drug intoxications, or to homicidal violence, to name only a few. Admittedly, deaths due to certain preexisting natural conditions, such as hypertension, diabetes, and autoimmune disorders may be exacerbated by pregnancy. If this is determined to be the case after a careful medical record review, then these cases are best viewed as pregnancy-related.

Neither "pregnancy-related" nor "pregnancy-associated" designations appear on death certificates, per se, but the relationship of any given maternal fatality to the pregnancy should be clearly conveyed in the cause of death statement if the death is indeed pregnancy-related. This should also be made clear in the synthesis of the final diagnoses in the autopsy report of pregnancy-related deaths. Whether clear or not, the FP will most assuredly be asked by families, clinicians, lawyers, and public health officials about the role the pregnancy played in the death, if any, so these classifications must be carefully considered and appropriately determined.

The following are suggestions in approaching these deaths:

1. Perform a **complete autopsy** wherever possible (i.e., absent religious or adamant personal objections), including *histologic, toxicologic, and, depending on the case, molecular genetic testing.* A full autopsy is likely to answer many more questions that will invariably arise, separate and apart from the cause and manner of death.
2. Have a low threshold for **specialty consultations** depending on the case details (e.g., *neuropathology, cardiac pathology, and anesthesiology).*
3. **Review medical records**, including prenatal, intrapartum/parturition, and the early postpartum periods, depending on when the death occurred. That said, record review may also be important if the initiating complication of pregnancy is remote from the death. For example, if the death is related to a ruptured ectopic pregnancy caused by a mechanical obstruction of the fallopian tube by an adhesion, the etiology of the adhesion must be sought in order to establish the proximate cause and manner of this particular pregnancy outcome (death). If it is discovered that the adhesion resulted from a remote-inflicted gunshot wound and its related emergency exploratory laparotomy, the manner will likely be homicide. If hemorrhagic complications from a placenta accreta are determined to be related to the scar from a remote *elective* Cesarean section, the manner may be TC, in jurisdictions where that manner is available.
4. **Request and examine surgical specimens**, where applicable (emergency hysterectomies, delivered placentas, etc.). If unavailable, be sure to request the pathology reports and recuts of the microscopic tissue sections. Remember that, in many cases, these specimens may be central to establishing the proximate cause of death.
5. **Examine the fetus if it died with the mother before an emergency Cesarean section was possible.** This is not uncommon in violent deaths. The fetus and placenta should be thoroughly examined/autopsied and carefully measured and weighed to accurately determine gestational age and to exclude intrauterine growth retardation. Even if only to reveal the absence of endogenous disease or developmental abnormalities, these examinations are important, particularly in criminal investigations. *Be sure also to retain tissue for possible paternity testing.* While only one death certificate will be issued in these circumstances (the mother's), prosecutorial avenues may be pursued for the loss of two lives. Rather than a death certificate, the fetus is usually (depending on jurisdiction) issued a spontaneous termination of pregnancy certificate, which contains a cause of *demise* (not death), but no manner (for death certification purposes, there is no death unless there is extrauterine life).
6. **Fill out the pregnancy checkbox on the death certificate** (for vital statistical purposes). On occasion, family and friends are unaware of an early pregnancy at the time of the mother's death, so it is advisable to notify them of your discovery when checking that box.
7. **Manners of death in maternal deaths** are informed by the mechanism and the proximate cause of death, as they are in all deaths, and can therefore be classified as natural, TC, or unnatural (homicide, suicide, or accident), depending on the case. In order to be classified as a TC, however, the intervention must *not* have been emergent. Keep in mind also, that unless a pathological process exists, pregnancy-related or not, there may be no proximate cause (disease or injury) in some of these TCs. For example, a normal pregnancy does not constitute a disease or injury. So, if the death of an otherwise healthy woman with a normal, healthy pregnancy results from an anesthetic complication during an elective Cesarean section, the death would be appropriately classified as a TC, with a descriptive cause of death statement and no identifiable proximate cause, per se (see Chapter 3).

Forensic pathologists play a central role in the effort to prevent maternal deaths by identifying and communicating the clinical, psychosocial, and pathophysiologic factors contributing to these untimely deaths. When maternal deaths rise in a technologically advanced and wealthy society, it likely represents a systems failure.

The negative autopsy: the importance of what we exclude

The negative autopsy is both unsatisfying and frustrating for most forensic pathologists. Yet, it is a statistical reality and an important **"diagnosis" (of exclusion)** in many death investigations. For example, the exclusion of demonstrable disease or injury in explaining a death may be necessary in order to make a diagnosis of one due to drowning, epilepsy, or an acute intoxication. In rarer instances, no diagnosis, or only a circumstantial/historical one, such as epilepsy, can be made. Depending on its availability, molecular genetic testing may also play a role in the evaluation of these deaths (see Chapter 19). At a minimum, and depending on the context of the death, the exclusion of chemical and mechanical injury in such "negative" cases still may allow a conclusion to be drawn: cause, Undetermined; manner, Natural. While somewhat unsatisfying, this is not inconsequential. Much of the comfort and closure derived from what we explain to families and friends following an autopsy stems from what we are able to definitively *exclude*. The importance of this cannot be underscored more emphatically.

Organ and tissue donation

There is a difference between organ and tissue donation. Live patients are waiting for whole organ transplantation. In order to qualify as an organ donor, the donor must be declared dead by neurologic criteria, but still have a heartbeat/circulation (typically with ventilator assistance). Tissue procurements (cornea, long bone, saphenous vein, and heart valves), on the other hand, may occur several hours after cessation of circulatory and respiratory function. In either instance, once a person is declared dead, the organ/tissue donation staff cannot proceed without ME/C permission, that is, if the ME/C has taken jurisdiction of the death. That said, the ME/C is commonly able to allow the donations to proceed. Not uncommonly, however, some limitations are made, particularly when the death is suspicious or clearly criminal.[99–102]

In terms of the autopsy report, *it is advisable to create a* **separate heading** *within both the final diagnoses and the body of the report,* detailing this specialized postmortem procedure, with a documentation and itemization of the removed organs and tissues. If a diagnosis such as PTE is suspected, but is made difficult by the postmortem procurement of the heart, it is best to pick up the telephone and ask the involved procurement technician pointed questions about what they may or may not have seen in the pulmonary trunk as they removed the heart. Let us not forget the adage, "We solve more cases with the telephone than with the microscope."

Dissections and the funeral home

The following guidelines are suggested in an effort to allow funeral directors to accomplish their restorative/embalming work. While the ME/C is typically statutorily empowered to perform any necessary dissections needed to appropriately evaluate the body, one should be sensitive to the tasks of the embalmer and the feelings of the family. With the foregoing interests in mind, the following guidelines may be useful:

- Do not make an incision on the face or hands unless there is a compelling necessity.
- If a carotid artery is inadvertently cut high in the neck, place a string/ligature around the artery in the area of the high cut and leave the string long as an aide to the embalmer.
- Removal of the entire cervical vertebral column is unnecessary for the evaluation of most cervical trauma. In the large majority of instances, radiographs coupled with anterior and/or posterior neck dissections and removal of the spinal cord will provide competent, expert evaluation of trauma to the cervical vertebral column. A dissection removing the entire cervical vertebral column in selected instances remains a matter of professional judgment to be exercised by the pathologist.
- All autopsy incisions should be properly sutured and all blood and other stains should be washed from the surface of the body following autopsy. Aside from the common decency and consideration of the foregoing practice, it is an important public health matter.
- Avoid extending the upper limbs of the Y-shaped incision too far posteriorly on the shoulders (they may leak when embalmed) or so far above the clavicles that they risk being seen at the viewing/wake.
- Take care when performing a "face skin reflection" (a relatively rare technique) to keep the midportion of the facial tissue and eyes "anchored" to allow for proper alignment during closing and suturing (see above).
- For bodies that had been embalmed prior to autopsy, one may add a separate section in the autopsy report (something like **"Funeral Home Procedures/Artifacts"**), similar to that described above in **"Postmortem Tissue and Organ Procurement."** Under those headings, one should describe all that applies. The following is an example:

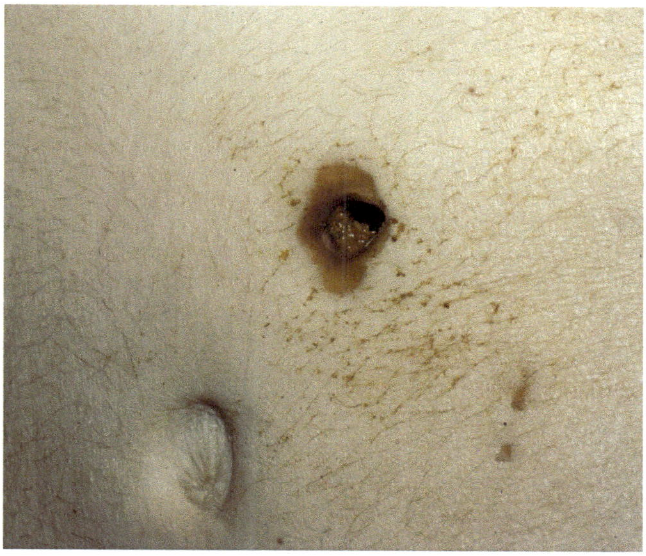

FIGURE 5.3 **Trocar defect of the abdomen.**

The body has been embalmed prior to arrival. Beige foundation make-up has been applied to the face and neck, plastic caps have been placed under the eyelids, the mouth has been sutured closed, and the oral cavity filled with gauze material. Plastic plugs fill trocar defects in the left lower and upper abdomen and there is a sutured incision in the right supraclavicular region. Internally, crystalline material packs the anterior neck region, embalming fluid pools in the thoracic and abdominal cavities; there are multiple trocar (Fig. 5.3) *defects of the viscera, and all organs are moderately fixed by embalming solution.*

If toxicological testing is necessary, it is also important to notify the toxicology laboratory that they will be receiving specimens post-embalming, so that they can interpret and test for volatiles accordingly.

Religious objections to autopsy

A religious objection to the performance of an autopsy may be raised when a decedent was observant of a particular religion during their life, one that holds opposition to autopsy as a tenet of their faith. The most common such faiths include Judaism, Jehovah's Witnesses, Christian Science, and Islam.[103,104] Some jurisdictions must adhere to local laws that prevent a medical examiner/coroner from performing an autopsy over the objection of a surviving relative of the deceased, an objection that asserts that such a procedure is contrary to the decedent's religious beliefs (of any type).[105] Usually, these laws have exceptions in compelling and specific instances, exceptions that allow the ME/C to override such a religious objection. These scenarios typically include homicides, suspicious deaths and/or those involving any criminal investigation, or one potentially posing a threat to public health (e.g., suspected tuberculosis or bacterial meningitis).[105] More commonly than not in these circumstances, a combination of patience and thoughtful communication with the family and/or their clergy results in an understanding and acceptance of the procedure and a withdrawal of the objection, without further legal involvement (see Chapter 16).[106] Sometimes, the offering of a complete, but minimally invasive autopsy quells concerns and demonstrates respect for issues born out of strong religious beliefs. To be clear, however, the authors are not suggesting the offering of a partial autopsy (torso only, head only, etc.). "Partial autopsies give partial answers." Below, however, outlines an autopsy modification that may be possible and that we support; this procedure is most commonly requested by and in keeping with the needs of those of Jewish faith:

- In all instances, every effort should be made to expedite the release of the body.
- The entire autopsy may be performed in a body pouch or shroud without the water running.
- The autopsy procedure should be as minimal as possible without sacrificing the determination of the cause of death.
- Replace all organs in their proper place (e.g., brain in suitable small plastic bag within the skull).

- All instruments are wiped clean with a paper towel and the paper towel placed in the pouch with the body (and sharp instruments in a special sharps container).
- All blood or articles of clothing containing blood that are not needed for pathological or evidentiary purposes should be sent with the remains to the funeral home.
- When possible, the entire body and especially the face and genitalia should be kept covered during the examination.

In the autopsy report, a special statement should precede the body of the report: *This autopsy was modified secondary to religious objections. There has been no modification in the examination of the injuries. Those organs free of injury or natural disease have been externally examined in the least intrusive fashion consistent with the evaluation of the decedent's lethal injuries.*

The forensic team: investigators, technicians, clerical staff

> Our goal always has been to provide technically excellent service in a timely fashion and to do it with compassion and sensitivity for the special needs of bereaved families. Every employee of our agency contributes to the achievement of that goal and deserves a measure of the credit for our success. Every job is important, and failure of any division of the OCME will adversely affect the performance of the entire agency. -***Hirsch C. Retirement Speech. 2013***

Although the forensic pathologist in the 1970s television show, "Quincy, M.E." got all the glory, we all know how important his assistant, Sam, was in the case work. Behind the curtain of every medical examiner/coroner office is a team of professionals responsible for serving people during the worst and most stressful experiences of their lives: the aftermath of the unexpected death of a friend or relative. This team includes the forensic pathologists, the mortuary department, identification staff, forensic imaging (radiography and photography), medicolegal death investigators (MDIs), forensic anthropologists, forensic biologists, medical records staff, information technology staff, public affairs, procurement, clerical support, odontologists (Fig. 5.4), entomologists, transport services, toxicology and histology laboratory personnel, including countless laboratory technicians, human resources, outreach, health and safety, security, facilities, all levels of administration, special operations, and the legal department.[108]

MDIs have one of the best and yet most emotionally challenging jobs in the agency. The best part is that they get to perform the first forensic examination on a decedent. The challenge and thrill of detecting the first clues to the cause of death can be very rewarding. The emotional challenge is interacting, on a daily basis, with families who have just lost a loved one. They see the raw emotion, anger, sadness, disbelief, and tears. It takes a very special person to do this work, day in and day out, while maintaining their professionalism and empathy for families. The **mortuary technicians**, who transport the remains, encounter similar situations, as do **outreach** and **identification staff**. **Criminalists** are the backbone of the laboratories. Without them, the house of cards would surely collapse. Every member of the team plays a vital role; the interplay and cooperation among all departments is vital to the functioning of a successful agency.

Forensic anthropologists are important resources in medicolegal death investigation. Forensic anthropology is a board-certified specialty that requires qualified practitioners to complete intensive and long-term postgraduate

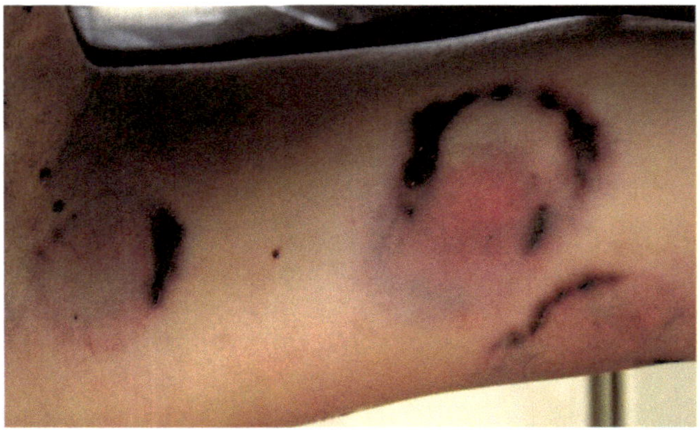

FIGURE 5.4 **Human bites of the arm.** Forensic odontologists may be consulted to evaluate such suspected human bite marks.

study. The American Board of Forensic Anthropology (ABFA) is the certification organization for forensic anthropologists in the United States, but there are other certifying bodies internationally. The ABFA offers certification programs at the MA/MS and Ph.D. levels.[109]

Some offices have **in-house bereavement specialists or victim advocates**. Their presence in an ME/C office allows the family quick access to assistance with grieving, funeral arrangements, the death certification process, and group support options.[110–117]

Quality assurance

Quality assurance (QA) is an important component of forensic pathology practice, just as it is in any medical discipline. Forensic pathology is a medical specialty based on basic science, medicine, training, experience, and judgment, and relies upon history, among other data points, to generate hypotheses about diseases or injuries (the differential diagnosis). The history, physical examination (autopsy), and other diagnostic studies strive to confirm or refute hypotheses or suspicions.[118]

The medicolegal death investigation community is not immune to bias and error. As such, it has implemented ways to minimize it. Medicolegal death investigation offices should be independent of law enforcement and insulated from political control.[119,120] Forensic pathologists provide the greatest benefit and best serve the public and society at large by performing independent, thorough investigations, without relying upon second-hand or filtered information. There are peer-review processes, such as consensus conferences, internet discussion groups (where forensic pathologists share challenging cases with colleagues), and QA reviews by both internal and external experts (the latter via accreditation and inspection).[8,121–129] Standard internal second reviews of a medical examiner's report is routine in many offices. The ultimate QA mechanism for many cases, however, is the justice system. Except for the rare civil action, nonforensic pathologist physicians rarely are subjected to a critical legal review of their medical reports and diagnoses. Forensic pathologists know that any one of their cases (especially the homicides) may become the center of a legal matter. They know that their opinions may be scrutinized to the point of an attorney pointing out a spelling or grammar error. They know that they will have to explain how they arrived at their diagnoses and certifications, as they should. This oversight involves many stakeholders, including forensic investigators, forensic pathologists, criminalists, prosecutors, grand juries, trial juries, judges, appellate courts, defendants, plaintiffs, and other forensic pathology experts.

Cognitive bias is an important QA issue, so it is imperative that the FPs have a clear understanding of what it is. **Cognitive bias** represents a distortion in reasoning, evaluating, remembering, or other cognitive processes that results from holding onto one's beliefs, regardless of contrary information.[130,131] **Confirmational bias** is the tendency to search for or interpret information that corroborates preconceptions, and **contextual bias** occurs when extraneous information influences a decision that may not be supported by objective evidence and/or would otherwise not have been made. Cognitive bias has been studied in forensic decision making.[130–133] A study looking at fingerprint examiner tasks, for example, showed that they may be biased by knowledge of contextual information (e.g., knowing the DNA results).[133]

The consideration of history and circumstances by the MDI has been criticized as enabling cognitive bias. This criticism ignores the fact that the forensic pathologist is a physician, and that the extraction of historical data is an integral and necessary part of all medical disciplines. In fact, the lack of obtaining a relevant history may be viewed by many as malpractice.[118,134] Medical practice is not represented by a single mechanical process, such as that encountered in fingerprint or DNA comparisons. It demands a thorough "history and physical" and the medical specialist is the person with the knowledge and expertise to decide what information is and is not relevant. History and circumstances provide key information to all physicians.[135] In a study of 630 clinical cases, history was found to be the most important part of the diagnosis in 56%. The physical examination was the most important in 17% and the laboratory investigation in 23%.[136] Without history, all physicians are at an increased risk for making an incorrect diagnosis.

Safety in the autopsy environment

There are ways to decrease and prevent autopsy injuries and exposures. Autopsy staff can minimize risk of occupational infections by following standard precautions,[137,138] which include:

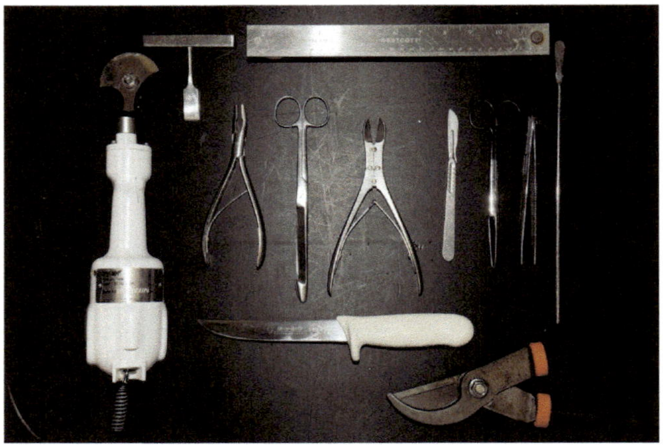

FIGURE 5.5 **Autopsy tools.**

1. All bodies are treated as infectious, regardless of known infectious states, as are all surfaces and equipment used during the autopsy.
2. All fluid and tissue specimens are considered infectious.
3. All postmortem invasive procedures should include the use of complete protective equipment for everybody in the autopsy room who is at risk for fluid or tissue contamination.
4. Instruments and surfaces contaminated during postmortem procedures should be routinely decontaminated with a chemical germicide.
5. All specimens should be collected in a container with a secure lid to prevent leakage.

Universal precautions include proper face shielding (to prevent from being splashed) and the use of **N-95** respirators (high-efficiency particulate air-filtered respirator, with appropriate fit-testing). Dr. Hirsch used to say that bacteria do not have "wings or springs," but they may be aerosolized during an autopsy.[138] These aerosols may contain airborne particles (1–5 um diameter) that can become suspended in air and may be inhaled. There are well-documented instances of tuberculosis transmission at autopsy.[139,140] Water sprayed onto the body and contaminated surfaces, oscillating saws, and fluid aspirators can aerosolize infectious agents (Fig. 5.5).[138] Opening of the chest plate with shears instead of a vibrating saw will decrease the risk of aerosolization of particles. Autopsy rooms should operate at negative pressure with the air expelled directly to the outside of the building. Down-draft autopsy tables, ultraviolet irradiation of the air, and high-efficiency particulate air filtration technology also are available. In one medical report, a patient who did not transmit tuberculosis before death transmitted it during autopsy.[139] Of the healthcare workers caring for this patient for several weeks on a medical ward, not one of the 40 treating staff converted, even though respiratory precautions were not used. All five autopsy staff, however, who attended the three-hour autopsy, did convert. A second study reported an outbreak of tuberculosis in medical students at the University of Sydney.[140] Eight of 35 previously skin-test negative students who attended an autopsy (one hour of exposure) of an immunosuppressed patient with unsuspected active tuberculosis converted.

All staff involved in autopsy work are advised to have periodic purified protein derivative skin testing,[138] an appropriate hepatitis B immunization, and fit-testing for respirators.

Sharp injury risk

To decrease the **risk of sharp injuries**, the wearing of cut-resistant (mesh) gloves (at least on one's nondominant hand) is recommended.[138] Scalpel blades with blunt (as opposed to pointed) tips decrease the risk of a "stab"

FIGURE 5.6 **Scalpel.** *Watercolor by Dr. Cricket Coleman.*

through even a mesh glove (Fig. 5.6). Needles should not be recapped and only one prosector should be doing the evisceration at a time. If a scalpel blade comes off in the body—everything stops until it is located and retrieved.

Radiation risk

A passive radiation detection system at the morgue entrance will offer protection to inadvertent radioactive exposures. Radiation from medical diagnostic or therapeutic procedures are common, and from terrorism/criminal actions, rare, but they do occur. Brachytherapy for prostate cancer involves implanting radioactive (iodine-125 or palladium-103) seeds. Iodine-125 has a half-life of 60 days. ME/C offices should have a radiation protocol for these alerts.[138]

Cardioverter-defibrillator risk

Implantable cardioverter-defibrillators may remain active after death and are potentially hazardous.[141,142] They can be shut off with the use of a magnet and the leads can be disconnected (instead of cut) with a special screwdriver.[138] If a scissors must be used to cut the leads, the handle should be made of rubber or plastic and the cardioverter deactivated beforehand for maximum safety.

References

1. Passarino M. Prevalence of myocarditis at autopsy in turin. *Arch Path Lab Med*. 1997;121:619–622.
2. Lindblad B, Eriksson A, Bergqvist D. Autopsy verified pulmonary embolism in a surgical department 1951-88. *Br J Surg*. 1991;78:849–852.
3. Stevanovic G, Tucakovic G, Dotlic R, Kanjuh V. Correlation of clinical diagnoses with autopsy findings: a retrospective study of 2,145 consecutive autopsies. *Hum Pathol*. 1986;17(12):1225–1230.
4. Yesner R, Robinson M, Goldman L, Reichert C, Engel L. A symposium on the autopsy. *Pathol Annu*. 1984;20:441–477.
5. Jimenez FA. The first autopsy in the New World. *Bull N Y Acad Med*. 1978;54(6):613–619.
6. Sehonanda C, Blum. Changing patterns of autopsy findings among persons with AIDS in an inner city. *Arch Pathol Lab Med*. 1996;120:459–464.
7. Hoda SA, Gerber MA. The value of the autopsy for the 1990s. *Hum Pathol*. 1990;21(9):980–981.
8. Adickes ED, Sims KL. Enhancing autopsy performance and reporting. A system for a 5-day completion time. *Arch Pathol Lab Med*. 1996;120(3):249–253.
9. Valdes-Dapena M. The postautopsy conference with families. *Arch Pathol Lab Med*. 1984;108(6):497–498.
10. Dorsey. A perspective on the autopsy. *Am J Cancer Prev*. 1977;69:217–219.
11. Hutchins GM. Practice guidelines for autopsy pathology. Autopsy reporting. Autopsy Committee of the College of American Pathologists. *Arch Pathol Lab Med*. 1995;119(2):123–130.
12. Sanner M. A comparison of public attitudes toward autopsy, organ donation, and anatomic dissection. A Swedish survey. *JAMA*. 1994;271(4):284–288.
13. Lundberg GD, Voigt GE. Reliability of a presumptive diagnosis in sudden unexpected death in adults. The case for the autopsy. *JAMA*. 1979;242(21):2328–2330.
14. Rosai J. The posthumous analysis (PHA). An alternative to the conventional autopsy. *Am J Clin Pathol*. 1996;106(4 Suppl 1):S15–S17.
15. Sarode VR, Datta BN, Banerjee AK, et al. Autopsy findings and clinical diagnoses: a review of 1,000 cases. *Hum Pathol*. 1993;24(2):194–198.
16. Bauer FW, Robbins SL. An autopsy study of cancer patients. I. Accuracy of the clinical diagnoses (1955 to 1965) Boston City Hospital. *JAMA*. 1972;221(13):1471–1474.
17. Stemmermann GN. Unsuspected cancer in elderly Hawaiian Japanese: an autopsy study. *Hum Pathol*. 1982;13(11):1039–1044.
18. Zarbo R, Baker P, Howanitz P. The autopsy as a performance measurement tool–diagnostic discrepancies and unresolved clinical questions. *Arch Pathol Lab Med*. 1999;123:191–198.
19. Bayer-Garner IB, L MF, Lamps LW. Pathologists in a teaching institution assess the value of the autopsy. *Arch Pathol Lab Med*. 2002;126(4):442–447.
20. Adelson L. Symposium on autopsy and the law. The anatomy of justice. *Bull N Y Acad Med*. 1971;47(7):745–757.
21. Plattner T, Scheurer E, Zollinger U. The response of relatives to medicolegal investigations and forensic autopsy. *Am J Forensic Med Pathol*. 2002;23(4):345–348.
22. Mayo. The metabolic autopsy: postmortem screening in cases of sudden, unexpected death. *Communique*. 2003;28:1–6.
23. DiMaio. *Standardization of Medico-Legal Investigation and Autopsy Procedures*. 2003.
24. O'Grady G. Death of the teaching autopsy. *BMJ*. 2003;327(7418):802–803.
25. Combes A, Mokhtari M, Couvelard A, et al. Clinical and autopsy diagnoses in the intensive care unit: a prospective study. *Arch Intern Med*. 2004;164(4):389–392.
26. Goldman L, Sayson R, Robbins S, Cohn LH, Bettmann M, Weisberg M. The value of the autopsy in three medical eras. *N Engl J Med*. 1983;308(17):1000–1005.
27. Bergqvist D, Lindblad B. A 30-year survey of pulmonary embolism verified at autopsy: an analysis of 1274 surgical patients. *Br J Surg*. 1985;72(2):105–108.
28. Spiliopoulou C, Papadodima S, Kotakidis N, Koutselinis A. Clinical diagnoses and autopsy findings: a retrospective analysis of 252 cases in Greece. *Arch Pathol Lab Med*. 2005;129(2):210–214.

29. Melinek J, Livingston E, Cortina G, Fishbein MC. Autopsy findings following gastric bypass surgery for morbid obesity. *Arch Pathol Lab Med.* 2002;126(9):1091–1095.
30. Underwood J. Resuscitating the teaching autopsy. *BMJ.* 2003;327(7418):803–804.
31. Croft PR, Racz MI, Bloch JD, Palmer CH. Autopsy confirmation of severe pulmonary interstitial fibrosis secondary to Munchausen syndrome presenting as cystic fibrosis. *J Forensic Sci.* 2005;50(5):1194–1198.
32. Peterson GF, Clark SC. Forensic autopsy performance standards. *Am J Forensic Med Pathol.* 2006;27(3):200–225.
33. Christiansen LR, Collins KA. Natural death in the forensic setting: a study and approach to the autopsy. *Am J Forensic Med Pathol.* 2007;28(1): 20–23.
34. Taggart MW, Craver R. Causes of death, determined by autopsy, in previously healthy (or near-healthy) children presenting to a children's hospital. *Arch Pathol Lab Med.* 2006;130(12):1780–1785.
35. Sauvageau A, Racette S. Postmortem changes mistaken for traumatic lesions: a highly prevalent reason for coroner's autopsy request. *Am J Forensic Med Pathol.* 2008;29(2):145–147.
36. Ravakhah K. Death certificates are not reliable: revivification of the autopsy. *South Med J.* 2006;99(7):728–733.
37. Asnaes S, Paaske F. Uncertainty of determining cause of death in medicolegal material without autopsy, an autopsy study. *Forensic Sci Int.* 1980;15(2):103–114.
38. Michiue T, Ishikawa T, Quan L, Zhu BL, Maeda H. Forensic pathological evaluation of injury severity and fatal outcome in traffic accidents: five illustrative autopsy cases of clinically unexpected death. *Forensic Sci Med Pathol.* 2008;4(3):153–158.
39. Spitz WU. The medicolegal autopsy. *Hum Pathol.* 1980;11(2):105–112.
40. Shojania KG, Burton EC. The vanishing nonforensic autopsy. *N Engl J Med.* 2008;358(9):873–875.
41. Horowitz RE, Naritoku WY. The autopsy as a performance measure and teaching tool. *Hum Pathol.* 2007;38(5):688–695.
42. Tester DJ, Ackerman MJ. The role of molecular autopsy in unexplained sudden cardiac death. *Curr Opin Cardiol.* 2006;21(3):166–172.
43. Parham DM, Savell VH, Kokes CP, et al. Incidence of autopsy findings in unexpected deaths of children and adolescents. *Pediatr Dev Pathol.* 2003;6(2):142–155.
44. Milroy CM. The autopsy in cases of unascertained sudden death. *Curr Diagn Pathol.* 2007;13(5):401–409.
45. Turnbull A, Osborn M, Nicholas N. Hospital autopsy: endangered or extinct? *J Clin Pathol.* 2015;68(8):601–604.
46. Gill JR. *Autopsy: Infectious and Serious Communicable Diseases.* 2016:279–284.
47. Ruan X, Chiravuri S, Kaye AD. From death to death certificate: a call for preventing autopsy from dying. *J Med Toxicol.* 2017;13(2):195–196.
48. van den Tweel JG, Wittekind C. The medical autopsy as quality assurance tool in clinical medicine: dreams and realities. *Virchows Arch.* 2016; 468(1):75–81.
49. Takeda A, Hitosugi M, Furukawa S. Autopsy cases of motorcyclists dying of trauma or disease. *Am J Forensic Med Pathol.* 2017;38(3):222–225.
50. Ifteni P, Barabas B, Gavris C, Moga M, Burtea V, Dracea L. Sudden cardiac death: autopsy findings in 7200 cases between 2001 and 2015. *Am J Forensic Med Pathol.* 2017;38(1):49–53.
51. Thiene G, Saffitz JE. Autopsy as a source of discovery in cardiovascular medicine: then and now. *Circulation.* 2018;137(25):2683–2685.
52. Goldman L. Autopsy 2018: still necessary, even if occasionally not sufficient. *Circulation.* 2018;137(25):2686–2688.
53. Humez S, Delteil C, Maurage CA, et al. Does the medical autopsy still have a place in the current diagnostic process? A 6-year retrospective study in two French University hospitals. *Forensic Sci Med Pathol.* 2019;15:564–569. https://doi.org/10.1007/s12024-019-00170-x.
54. CDC. *The Autopsy, Medicine, and Mortality Statistics.* 2001.
55. Hanley B, Lucas SB, Youd E, Swift B, Osborn M. Autopsy in suspected COVID-19 cases. *J Clin Pathol.* 2020;73(5):239–242.
56. Milroy CM, Parai JL. Fat embolism, fat embolism syndrome and the autopsy. *Acad Forensic Pathol.* 2019;9(3–4):136–154.
57. Hansma P. The evolution of the autopsy. *Acad For Path.* 2015;5(4):638–649.
58. Lundberg GD. Now is the time to emphasize the autopsy in quality assurance. *JAMA.* 1988;260(23):3488.
59. Lundberg GD. The archives of pathology and laboratory medicine and the autopsy. *JAMA.* 1984;252(3):390–392.
60. Lundberg GD. Medicine without the autopsy. *Arch Pathol Lab Med.* 1984;108(6):449–454.
61. Craig JR, Johnson L, Lundberg GD, Tatter D, Edmondson HA, McGah S. An autopsy survey of clinical and anatomic diagnoses associated with alcoholism. *Arch Pathol Lab Med.* 1980;104(9):452–455.
62. Krywanczyk A, Amoresano E, Tatsumi K, Mount S. Autopsy service death certificate review an educational experience and public health service. *Arch Pathol Lab Med.* 2020;144:1092–1096.
63. Lynch MJ. The autopsy: legal and ethical principles. *Pathology.* 2002;34(1):67–70.
64. Adams VI, Flomenbaum MA, Hirsch CS. Trauma and disease. In: WU S, ed. *Spitz and Fisher's Medicolegal Investigation of Death.* 4th ed. Springfield, IL: Charles C Thomas; 2006:436–459.
65. Moritz AR. Classical mistakes in forensic path. *Am J Clin Pathol.* 1956;26(12):1383–1397.
66. Oliver WR. Moritz revisited: modern mistakes about how we think about forensic pathology. *Academic Foren Pathol.* 2015;5(2):186–200.
67. Frankel H. Air embolism during posterior spinal fusion. *Can J Anesth.* 1988;35:511–514.
68. Albin K. Venous air embolism during lumbar laminectomy in the prone position. *Anes Analg.* 1991;73:346–349.
69. Bajanowski T, West A, Brinkmann B. Proof of fatal air embolism. *Int J Leg Med.* 1998;111(4):208–211.
70. Adams V, Guidi C. Venous air embolism in homicidal blunt impact head trauma. Case reports. *Am J Forensic Med Pathol.* 2001;22(3):322–326.
71. Jackowski C, Thali M, Sonnenschein M, et al. Visualization and quantification of air embolism structure by processing postmortem MSCT data. *J Forensic Sci.* 2004;49(6):1339–1342.
72. Leth PM. The use of CT scanning in forensic autopsy. *Forensic Sci Med Pathol.* 2007;3(1):65–69.
73. Agarwal SS, Kumar L, Chavali KH, Mestri SC. Fatal venous air embolism following intravenous infusion. *J Forensic Sci.* 2009;54(3):682–684.
74. Byard RW. The 'break enter and die' syndrome may involve significant injury to major neck vessels. *J Forensic Sci.* 2011;56(Suppl 1): S252–S254.
75. Fujioka M, Niino D, Ito M, Matsuoka Y. Fatal paradoxical air embolism diagnosed by postmortem imaging and autopsy. *J Forensic Sci.* 2012; 57(4):1118–1119.
76. Riddick L, Brogdon BG. Fatal air embolism during renal dialysis. *Am J Forensic Med Pathol.* 2012;33(1):110–112.

77. Makino Y, Shimofusa R, Hayakawa M, et al. Massive gas embolism revealed by two consecutive postmortem computed-tomography examinations. *Forensic Sci Int*. 2013;231(1–3):e4–10.
78. Nagai H, Maeda H, Kuroda R, et al. Lethal pulmonary air embolism caused by the removal of a double-lumen hemodialysis catheter. *Am J Forensic Med Pathol*. 2014;35(4):237–238.
79. Zivkovic V, Nikolic S. Pneumopericardium as a consequence of air embolism. *Forensic Sci Med Pathol*. 2015;11(1):122–123.
80. Roquero LP, Camelo-Piragua S, Schmidt C. Cerebral air embolism: a clinical, radiologic and histopathologic correlation. *Am J Forensic Med Pathol*. 2016;37(4):241–244.
81. Edler C, Klein A, Puschel K, Schroder AS. Fatal air embolism in hospital confirmed by autopsy and postmortem computed tomography. *Forensic Sci Med Pathol*. 2018;14(2):251–254.
82. Mercurio I, Capano D, Torre R, et al. A case of fatal cerebral air embolism after blunt lung trauma: postmortem computed tomography and autopsy findings. *Am J Forensic Med Pathol*. 2018;39(1):61–68.
83. Adams VI. Neck injuries: II. Atlantoaxial dislocation–a pathologic study of 14 traffic fatalities. *J Forensic Sci*. 1992;37(2):565–573.
84. Adams VI. Neck injuries: I. Occipitoatlantal dislocation–a pathologic study of twelve traffic fatalities. *J Forensic Sci*. 1992;37(2):556–564.
85. Adams VI. Neck injuries: III. Ligamentous injuries of the craniocervical articulation without occipito-atlantal or atlanto-axial facet dislocation. A pathologic study of 21 traffic fatalities. *J Forensic Sci*. 1993;38(5):1097–1104.
86. Gill JR. The medicolegal evaluation of fatal pulmonary thromboembolism. In: Tsokos M, ed. *Forensic Pathology Reviews*. Vol. 3. Totowa, NJ: Humana Press; 2005:285–304.
87. Fineschi V, Turillazzi E, Neri M, Pomara C, Riezzo I. Histological age determination of venous thrombosis: a neglected forensic task in fatal pulmonary thrombo-embolism. *Forensic Sci Int*. 2009;186(1–3):22–28.
88. Micallef MJ. The autopsy and diagnosis of pulmonary thrombo-embolism. *Forensic Sci Med Pathol*. 2018;14(2):241–243.
89. Dalen JE, Alpert JS. Natural history of pulmonary embolism. *Prog Cardiovasc Dis*. 1975;17(4):259–270.
90. Carson JL, Kelley MA, Duff A, et al. The clinical course of pulmonary embolism. *N Engl J Med*. 1992;326(19):1240–1245.
91. Gillum RF. Pulmonary embolism and thrombophlebitis in the US, 1971-85. *Am Heart J*. 1987;114:1262–1264.
92. Wagenvoort CA. Pathology of pulmonary thromboembolism. *Chest*. 1995;107(1 Suppl):10S–17S.
93. Goldhaber SZ, Elliott CG. Acute pulmonary embolism: part I: epidemiology, pathophysiology, and diagnosis. *Circulation*. 2003;108(22): 2726–2729.
94. Dalen JE. Pulmonary embolism: what have we learned since Virchow?: treatment and prevention. *Chest*. 2002;122(5):1801–1817.
95. White RH. The epidemiology of venous thromboembolism. *Circulation*. 2003;107(23 Suppl 1):I4–I8.
96. Kakkar VV, Howe CT, Flanc C, Clarke MB. Natural history of postoperative deep-vein thrombosis. *Lancet*. 1969;6:230–232.
97. Knight B, Zaini MRS. Pulmonary embolism and venous thrombosis. *Am J Forensic Med Pathol*. 1980;1(3):227–232.
98. Knight B. The dating of pulmonary emboli. *Acta Med Leg Soc*. 1984;34:190–192.
99. Lundberg GD. Attitudes toward autopsy and organ donation in Sweden and the United States. *JAMA*. 1994;271(4):317.
100. Pinckard JK, Graham MA. Heart valve tissue donation does not preclude the diagnosis of clinically significant pediatric cardiac abnormalities. *Am J Forensic Med Pathol*. 2003;24(3):248–253.
101. Jason D. The role of the medical examiner/coroner in organ and tissue procurement for transplantation. *Am J Forensic Med Pathol*. 1994;15(3): 192–202.
102. Wolf DA, Derrick SM. Undetermined cause and manner of death after organ/tissue donation. *Am J Forensic Med Pathol*. 2010;31(2):113–116.
103. Mittleman RE, Davis JH, Kasztl W, Graves Jr WM. Practical approach to investigative ethics and religious objections to the autopsy. *J Forensic Sci*. 1992;37(3):824–829.
104. Gleason DF. Histologic grading of prostate cancer: a perspective. *Hum Pathol*. 1992;23(3):273–279.
105. New York State Public Health Law §4210-c.
106. Davis GJ, Peterson BR. Dilemmas and solutions for the pathologist and clinician encountering religious views of the autopsy. *South Med J*. 1996;89(11):1041–1044.
107. Hirsch C. *Retirement Speech*. 2013.
108. Scientific-Working-Group-on-Medicolegal-Death-Investigation-(SWGMDI). *Regional Medicolegal Autopsy and Death Investigation Centers-Construction, Staffing, and Costs-A Report and Recommendations*; 2013, 2021. http://www.swgmdi.org/images/si6.facilityconstruction.published.9-17-13.pdf.
109. Christensen AM, Passalacqua NV, Schmunk GA, et al. The value and availability of forensic anthropological consultation in medicolegal death investigations. *Forensic Sci Med Pathol*. 2015;11(3):438–441.
110. Stroebe M, Schut H, Stroebe W. Health outcomes of bereavement. *Lancet*. 2007;370(9603):1960–1973.
111. Berry RS, Aurelius MB, Barickman N, Lathrop SL. Utility of a grief services program for medical examiners' offices. *J Forensic Sci*. 2013;58(2): 380–384.
112. Hofmeister U, Navarro S. A psychosocial approach in humanitarian forensic action: the Latin American perspective. *Forensic Sci Int*. 2017;280: 35–43.
113. Cacciatore J. Appropriate bereavement practice after the death of a Native American child. *Fam Soc*. 2009;90(1):46–50.
114. Sandler IN, Ayers TS, Wolchik SA, et al. The family bereavement program: efficacy evaluation of a theory-based prevention program for parentally bereaved children and adolescents. *J Consult Clin Psychol*. 2003;71(3):587.
115. Kõlves K, Zhao Q, Ross V, Hawgood J, Spence SH, De Leo D. Suicide and other sudden death bereavement of immediate family members: an analysis of grief reactions six-months after death. *J Affect Disord*. 2019;243:96–102.
116. Kristensen P, Weisæth L, Heir T. Bereavement and mental health after sudden and violent losses: a review. *Psychiatr Interpers Biol Process*. 2012; 75(1):76–97.
117. Hull MM. Hospice nurses. Caring support for caregiving families. *Cancer Nurs*. 1991;14(2):63–70.
118. Baker AL. History, cognitive bias, incompetence, and corruption are not the same things. Baker, A. Presentation at at the human factors in forensic science: why cognitive bias can lead to flawed expert opinions and testimony, how its influence can Be minimized, and what challenges testifying experts and judges can expect to face. Plenary Session at the Annual Meeting of the American Academy of Forensic Sciences. February 18, 2015 AAFS Plenary Session.

119. National Research Council. *Strengthening Forensic Science in the United States: A Path Forward*. Washington, DC: The National Academies Press; 2009.
120. Melinek JTL, Oliver WR, Schmunk GA, Weedn VW. National association of medical examiners position paper: medical examiner, coroner, and forensic pathologist independence. *Acad For Path*. 2013;3(1):93–98.
121. Zarbo B, Howanitz. Quality Assurance of Autopsy Permit form information, performance time, Prelim report time. *Arch Path med*. 1996;120: 346–352.
122. Anderson RE. The autopsy as an instrument of quality assessment. Classification of premortem and postmortem diagnostic discrepancies. *Arch Pathol Lab Med*. 1984;108(6):490–493.
123. Burke MP, Opeskin K. Audit in forensic pathology. *Am J Forensic Med Pathol*. 2000;21(3):230–236.
124. Sinard JH, Blood DJ. Quality improvement on an academic autopsy service. *Arch Pathol Lab Med*. 2001;125(2):237–245.
125. Cain MD, Brazelton J, Dye DW. Identifying errors in forensic autopsy reports using a novel web-based program. *Acad Forensic Pathol*. 2016; 6(1):103–108.
126. Ersoy G, Akyildiz EU, Korkmaz G, Albek E. Interinstitutional review of slides for forensic pathology: types of inconsistencies. *Am J Forensic Med Pathol*. 2010;31(3):222–226.
127. Hanzlick R. Quality assurance review of death certificates: a pilot study. *Am J Forensic Med Pathol*. 2005;26(1):63–65.
128. Oliver WR, Fang X. Forensic pathologist consensus in the interpretation of photographs of patterned injuries of the skin. *J Forensic Sci*. 2016; 61(4):972–978.
129. Davis GG, Cadwallader AB, Fligner CL, et al. Position paper: recommendations for the investigation, diagnosis, and certification of deaths related to opioid and other drugs. *Am J Forensic Med Pathol*. 2020;41(3):152–159.
130. Cooper GS, Meterko V. Cognitive bias research in forensic science: a systematic review. *Forensic Sci Int*. 2019;297:35–46.
131. Oliver WR. Comment on Cognitive bias research in forensic science: a systematic review. *Forensic Sci Int*. 2019;304:109898.
132. Curley LJ, Munro J, Lages M, MacLean R, Murray J. Assessing cognitive bias in forensic decisions: a review and outlook. *J Forensic Sci*. 2020; 65(2):354–360.
133. Stevenage SV, Bennett A. A biased opinion: demonstration of cognitive bias on a fingerprint matching task through knowledge of DNA test results. *Forensic Sci Int*. 2017;276:93–106.
134. Oliver WR, Fudenberg J, Howe JA, Thomas LC. Cognitive bias in medicolegal death investigation. *Academic Forensic Pathology*. 2015;5(4): 548–560.
135. Leadbeatter S, Knight B. The history and the cause of death. *Med Sci Law*. 1987;27(2):132–135.
136. Sandler G. Costs of unnecessary tests. *Br Med J*. 1979;2(6181):21–24.
137. Lacy JM, Brooks EG, Akers J, et al. Covid-19: postmortem diagnostic and biosafety considerations. *Am J Forensic Med Pathol*. 2020;41(3): 143–151.
138. Nolte KB, Taylor DG, Richmond JY. Biosafety considerations for autopsy. *Am J Forensic Med Pathol*. 2002;23(2):107–122.
139. Templeton GL, Illing LA, Young L, Cave D, Stead WW, Bates JH. The risk for transmission of *Mycobacterium tuberculosis* at the bedside and during autopsy. *Ann Intern Med*. 1995;122(12):922–925.
140. Wilkins D, Woolcock AJ, Cossart YE. Tuberculosis: medical students at risk. *Med J Aust*. 1994;160(7):395–397.
141. Prahlow JA, Guileyardo JM, Barnard JJ. The implantable cardioverter-defibrillator. A potential hazard for autopsy pathologists. *Arch Pathol Lab Med*. 1997;121(10):1076–1080.
142. Bourke M, Green M, Stinson W, Veinot J. Implantable cardioverter-defibrillators and the pathologist: comment and cautionary notes. *J Forensic Sci*. 1998;43(5):969–973.

CHAPTER

6

Approach to forensic perinatal and pediatric pathology

Susan F. Ely, Kristen Landi and James R. Gill

What greater pain can mortals bear than this; to see their children die before their eyes? *-Euripides*

Few deaths receive a more extensive investigation than the untimely deaths of infants and children. The findings are often more subtle than those in adults. As such, they require a more thorough approach to the postmortem examination. While no autopsy should be performed in a vacuum, scene investigation with a reenactment is of particular importance in infant and child investigations and can sometimes provide as much information as does the autopsy. In addition, special studies that are not commonly performed in adult deaths are often performed in pediatric fatalities.

There are three challenging areas in the investigation and certification of infant deaths. The first is determining if a newly delivered "infant" was live born or stillborn. The second involves unexplained infant deaths that fall into the sudden, unexplained infant deaths/sudden infant death syndrome (SIDS)/positional asphyxia groups. The third is the pediatric death due to unnatural causes, including those related to abuse and/or neglect.

The key point to remember in the investigation of live born infants versus stillborn fetuses is that stillborn fetuses are not assigned a manner of death. Therefore, a clearly stillborn fetus who is delivered at home and later found in a dumpster would not be classified as a homicide or any other manner. But, if that same infant was live born, then, at a minimum, due to medical neglect and exposure, a classification of homicide would likely be warranted. As such, the answer to that initial live born versus stillborn question may have considerable legal ramifications (see a more in-depth discussion later in this chapter).

While the work of a medical examiner is often at the crossroads of an unexpected and agonizing time for family members, in pediatric cases, that road is often unpaved and has many subsequent twists and turns. No one expects to bury a child, and while it is important to maintain patience and empathy when dealing with family members in the throes of any death, it is even more important in the setting of these sensitive cases. Although this may not be the first child that the pathologist has autopsied, it is likely the first child that the family has had to bury or cremate.

Finally, because infants and children are either developmentally unable to speak or at least limited in their communication skills, often fearful of saying something that they are not expressly told to say by their parents, or because they are simply shy in the presence of strangers, their ability to express symptoms or complaints is limited at best, even to their parents (if infants), often confounding death investigations. Their dependence on their caregivers extends past mere physical needs and into the realm of a tenacious psychological dependence.

Initial investigation

Once a death has been accepted by a medical examiner/coroner (ME/C), the decedent is now under their jurisdiction. Emergency medical services (EMS) will nearly always transport pediatric patients to the hospital due to the sensitive nature of child deaths and because they are usually discovered shortly after death. Due to the

Principles of Forensic Pathology
https://doi.org/10.1016/B978-0-323-91796-4.00027-1

overwhelming hope by all involved that they can save the child, EMS and emergency department staff will typically go to heroic lengths, often with prolonged resuscitative efforts. As such, there are often *primary and secondary scenes.* The primary scene is where the child was first found unresponsive; the secondary scene is the hospital. Ideally, the investigator visits both.

Once the child is pronounced, they should remain undisturbed by the family, medical staff, and law enforcement personnel until the ME/C investigator has performed their investigation and photographed the child. The investigator is deployed to the hospital as soon as possible after the death is reported. This allows for the examination of the child in a state as close as possible to their initial presentation so that the investigator may document potentially transient findings. Oftentimes, the death is not reported to the ME/C until sometime after pronouncement and, during that time, the family often wants to hold the child. As these deaths are devastating to families, it is normal for parents to want to see and hold their child at this tragic time. It is important to accommodate them, if possible, and the hospital may allow for supervised contact after consulting with the ME/C. A strict "hands off" policy on every infant death is draconian and insensitive. Although this may not be ideal, it is rare that it would have any negative impact on the investigation. For example, in most of these deaths, collection of trace evidence would be of little benefit and therefore concerns for contamination are usually unwarranted. The presence and needs of the family represent additional important reasons why a timely response to the hospital is a necessity.[1,2]

Photographs of the decedent at the hospital should include a picture of the face and full body images, both front and back. This provides an important visual document of the *lividity pattern and any injuries* as observed at that time, before transport to the morgue; these may prove invaluable to the investigation. The investigator is often examining the body a short time after death, generally within hours, and before lividity has become fixed. Therefore, lividity still may shift depending upon the new position of the infant. The autopsy examination often occurs more than 12 hours after the death, and the lividity pattern may shift and/or will likely become fixed by that time. For example, a three-month-old infant who is found unresponsive and face down on an adult bed at 6 p.m., and pronounced at the hospital at 7 p.m., may not have an autopsy until 9 a.m. the following day. The lividity pattern at 7 p.m. may clearly show blanching around the nose and mouth (possibly indicating obstruction from soft bedding or a pillow). Once found and transported to the hospital, the infant lies supine overnight until the autopsy more than 12 h later. This is ample time for the lividity to shift and even become fixed in the "new" gravitationally dependent position. Therefore, photographs taken by the investigator shortly after death may demonstrate a blanching pattern that may or may not be evident at autopsy. This could be one of the data points that helps make the difference between certifying a death as positional asphyxia (accident) versus sudden unexpected infant death (SUID) (natural) or undetermined.

Primary scene

An investigation of the primary scene needs to be coordinated with the investigating police agency and should ideally occur *as soon as possible after the death and prior to autopsy.* Primary scene investigations should be completed *within 24 h of the death.* The investigation should document not only where the decedent was found unresponsive, but also details of the overall environment. The investigator notes the last time the infant was fed, the infant/child's regular diet and sleep habits, any deviations from those habits, who lives in the home, pets in the home, the condition of the home, and the temperature of the sleep environment, among other variables, depending on the age of the child and the questions the circumstances may evoke.

For infants found unresponsive during sleep, particular attention should be paid to the sleep surface and overall environment. Blankets, pillows, stuffed animals, bumpers, and sleep attire should all be documented. If none of these items is apparent in the sleep area, the caretakers should be asked about them and their specific location relative to the infant's face at the time of *discovery,* as the caretaker may have removed them from the area. Additionally, any sheets or blankets in the sleep area should be moved by the investigator, as they may be covering additional padding or cushions that make a seemingly safe sleep surface an unsafe one. All of the bedding that the decedent was found on and the clothing that they were wearing are examined and photographed by the investigator. They may demonstrate bodily fluids (purge) that may indicate the location and position in which the child was found. For example, if the caregiver stated the child died in the crib, but a small blood stain is noted on a sheet in the adult bed, this requires an explanation (perhaps the family or EMS put the child on the bed after discovery, but no assumptions should be made).

Reenactment

A scene reenactment is an integral part of many pediatric investigations and may be as valuable, or sometimes more valuable, than the autopsy itself.[3] The National Association of Medical Examiners, in their 2020 publication on Unexplained Pediatric Deaths, recommends that scene reenactments be performed on all deaths of children aged 24 months or younger, as well as in all those of children with developmental delays or a history of seizures.[4] Scene reenactments involve the use of a doll to demonstrate the position in which a child was last placed (i.e., known to be alive) and the position in which they were found. Given the emotional sensitivity of these fatalities, the idea of a reenactment can seem daunting to caretakers, police, and investigators alike. However, while exercising sensitivity, taking a scientific and matter of fact approach to the process as part of the normal course of an investigation alleviates many potential issues. These reenactments are performed to provide information and insight into the child's death, and are not intended to "trap" or "implicate" a caretaker. Most caretakers want to do anything in their power to help find out what happened to their child, and the reenactment is a powerful way for them to actively participate in the investigation and take a step towards closure.

Reenactment photographs can provide a wealth of information.[3,5] While performing the reenactment, *the investigator should hand the doll to the person, parent or other caretaker,* who last placed the decedent to rest and ask them to position the doll (the use of a nondescript/nonlife-like "doll" may be easier for the family) as they had last placed the child. In other words, the investigator should hand the doll to the caretaker to place, not place the doll as directed by the caretaker. Watching how the caretaker places the doll may also provide valuable information, such as giving clues as to how the infant had been prior to being laid down, sometimes indicating things that had never occurred to the caretaker to mention. For example, a reenactment may lead to the determination that an infant had fallen asleep in the arms of the caretaker (who then nodded off for a short time) and was potentially dead prior to being laid on a safe sleep surface, where they were later found supine. Again, this distinction might constitute the difference between a determination of positional asphyxia (accident) and SUID (natural or undetermined).[6–27]

Medical history

Knowledge of the decedent's medical history and any significant family history can contribute to an understanding of pediatric deaths. The ME/C should review the medical history, including any recent or remote illnesses, operations, immunizations, medications, hospitalizations, and prior injuries, which provide objective data to be compared with third party corroborative or conflicting evidence. A reported history that a child bruises easily, for example, may be refuted by previous normal medical examinations or laboratory tests. Similarly, neonatal and birth records may call into question birth trauma as a cause of injuries or neurologic compromise. The forensic pathologist should make every effort to obtain and review these records, where deemed potentially elucidating. Aside from providing medical information, primary pediatricians who have established a relationship with the caregiver(s) are best positioned to observe patterns of behavior, conscientiousness, familial hardship, repeatedly missed or kept appointments, and behavioral clues, both positive and negative. While a review of the medical record is an important component of many investigations, a direct conversation with a pediatric professional can provide a much more fleshed out picture than that extracted from the words on the pages.

Any prior pediatric deaths in the family should raise the question of child abuse or previously unsuspected familial disease. Birth history and maternal history are reviewed in most instances of infant deaths. This is especially important in cases of prematurity, multiple prior spontaneous abortions, and congenital anomalies. It is always best to err on the side of saving additional tissue in RNAlater for possible future genetic studies in cases of sudden and unexpected pediatric deaths or in those with a familial history of sudden and unexpected death.

Radiology

Prior to the autopsy examination, all pediatric decedents should receive a set of full-body radiographs. Radiographs can document old, recent, and acute injuries, as well as skeletal abnormalities or disease.[28–32] Radiographs are reviewed *prior* to the autopsy. If they are technically inadequate, they should be repeated. In suspected or known child abuse, neglect, or malnourishment cases, separate full (extended) hand and wrist X-rays may be useful. These should be examined for evidence of acute and/or chronic injury (e.g., healed rib fractures). In the evaluation of a live birth versus a stillbirth, pulmonary and gastrointestinal gas patterns are specifically evaluated (see below).

Photography

Photographs should be taken in all pediatric cases. It is the responsibility of the forensic pathologist to decide which photographs to take based upon the details of the case, but it is best to err on the side of taking more photographs. Photographs document injuries and/or neglect, and are often used in court for homicidal or accidental deaths. Photographs of pertinent negatives are equally important (e.g., eyes, mouth, subscalpular soft tissues, genitalia, optic nerve root sheaths, and anus). Many aspects of pediatric cases can only be adequately documented through photographs. No verbal or written description, no matter how eloquent or detailed, can fully portray the graphic impact of the picture of a child's body covered by multiple patterned injuries of different ages and/or the image of a child who has been starved to death.[33–41] In homicides, these photographs will be scrutinized by the courts, both by the prosecution and the defense, and may be used by other experts who may consult on the case. Therefore, quality is extremely important. Work closely with the photographer. Clean tables, instruments, and gloves should be the norm, as well as a clean background for individual specimen images.

The pediatric autopsy

The investigation of infant and child deaths usually requires a more wide-ranging approach than that of adult deaths. As such, there are some variations in the autopsy procedures and the use of additional special studies. These variations are directed by the specific details of the case history and/or the physical findings as the investigation proceeds. The autopsy report follows the same format as that for adults, but with some additional information (see Appendix C). As is the case with adults, it is best to take a systematic approach to the examination so as not to miss any important findings.

It is helpful to include standard reference weights and measurements for the decedent's age; these can be inserted parenthetically after each autopsy weight and measurement, with percentiles for reference. These can be found in appendices of pediatric pathology textbooks.[42] Prior to the external examination portion of the report, an introductory paragraph can detail the foregoing approach for ease of communication. Such a qualifier might state:

> Accepted weights and measurements, where available, are listed parenthetically following the autopsy weights and measurements, correspond to a corrected gestational age of_____, and are taken from ...

Examples of wording:

- *The body is that of a normally developed, 6100 gm (54th percentile) male infant, whose appearance is consistent with the given age of three months. The crown-heel length is 61 cm (52nd percentile).*
- *The heart weighs 39 g (38.4 ± 6.2 g; 50th percentile) and has a normal distribution of right dominant coronary arteries without atherosclerotic stenoses or anomalies.*

External examination

As with adults, it is best to take a head to toe approach, both anteriorly and posteriorly, recording any injuries, scars, identifying features, etc. Note normal and abnormal developmental features (numbers of fingers and toes, palmar creases, low set ears, abnormal facies, cleft palate, epicanthal folds, nasal bridge, micrognathia, microcephaly, etc.). It is helpful to list standard facial measurements (inner and outer canthal distances, interpupillary distance, palpebral fissure length, philtrum length, mouth width, etc.) to document objective measurements supporting the presence or absence of apparent syndromic facies.

Standard Measurements:

1. Crown-heel length
2. Crown-rump length

3. Head circumference
4. Chest circumference
5. Abdominal circumference
6. Hand length (base of palm to the tip of the middle finger)
7. Foot length (heel to the tip of the great toe)

It is fairly standard to comment on the state of nutrition/hydration (e.g., thin vs. well nourished, prominent ribs, sunken eyes, sunken fontanelles, poor skin turgor, prominent cheek bones, scant/thin abdominal pannus), nasal patency (choanal stenosis/atresia), and the state of dentition (edentulous, erupting, deciduous, cavities, thrush, etc.). Pinching the skin is helpful in assessing turgor, an indication of hydrational status (well hydrated skin will be elastic and snap right back, whereas dehydrated skin will remain "tented"). Note if mucus membranes are dry or moist. Note the presence or absence of scalp hair (alopecia; the pattern may suggest natural vs. traumatic causes).

Examine the oral cavity and comment on the appearance of the frenulae, the oral mucosa, and the integrity of the palate. Note any chest, abdominal, or craniospinal deformities or dimpling over the vertebral column, the appearance of the umbilicus, the integrity and patency of the anus, and any luminal contents. Note the presence or absence of a diaper or other rashes and whether they appear treated with creams or powders or whether superinfected and/or crusted. The postmortem finding of anal dilation or an exposed pectinate line in children may raise the concern of sexual abuse.[43–47] A study of 65 pediatric deaths concluded that anal dilatation alone cannot be used as a marker for prior sexual abuse and the exposure of the pectinate line should not be confused with tears or fissures of the anal verge.[43]

Examine the external genitalia and note the following:

Boys: exudates, venereal warts, vesicles, ulcers, bites, circumcision, testicular descent, urethral placement/malformation (hypospadias, epispadias), and sexual ambiguity.

Girls: exudates, discharge, venereal warts, integrity of the hymen and the vaginal mucosa, the presence of foreign objects, sexual ambiguity, and sexual precociousness for age (Tanner stage).[48–55] The latter may be a harbinger of hypersexualized exposures or even chronic sexual molestation.

Postmortem changes and artifacts

Be detailed in your description of the distribution of livor mortis. As described above, an anterior distribution of livor with sparing around the nose and mouth may be the only indication of positional/mechanical asphyxia. That said, it is important to recognize that, depending on the specific circumstances of the case and other findings, this pattern may represent a consequence rather than a cause of death. Be aware of the possibility of postmortem insect activity and pseudofurrows on the neck caused by refrigeration (Fig. 6.1).

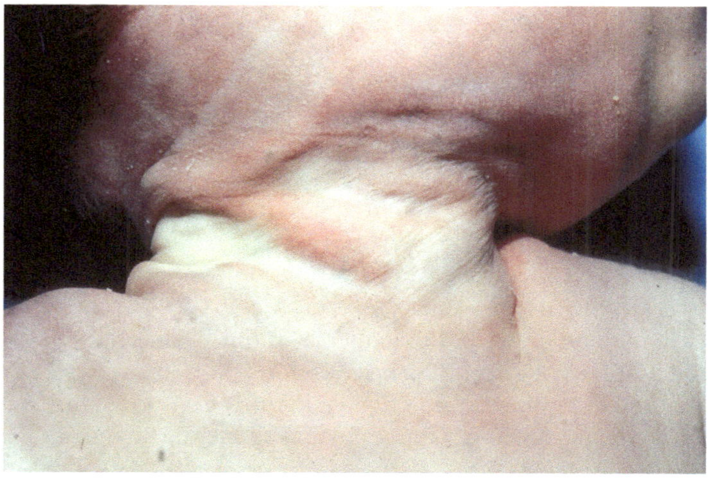

FIGURE 6.1 **Refrigeration artifact of the neck (pseudofurrow).**

Evidence of therapeutic intervention

Make note of the location of any needle puncture sites, catheters, tubes, or ecchymoses caused by resuscitative efforts. To the untrained eye, these findings may be misinterpreted as injuries related to abuse or neglect. In child abuse cases, these findings must not be confused with antemortem injury inflicted by another. Intraosseous catheters can fracture the bone and may be misinterpreted as abusive injury if one does not take note of the iatrogenic puncture overlying the anterior tibia. That said, *substantial visceral and skeletal abdominal and/or chest injuries with evidence of a vital reaction are not adequately explained by terminal resuscitative efforts and should be assiduously investigated if no reasonable or alternate explanation has been provided.*[56,57] See "Pediatric Trauma" below.

Clothing

Describe any clothing received with the child. Comment on whether the garments are clean, intact, and wet/dry. Comment about whether the clothing is age, season, and weather appropriate. If the child is wearing a diaper, comment on whether it is fecal, urine, or bloodstained. A soiled diaper maybe saved for toxicology testing when other specimens are not available and a diaper soaked with urine may support the conclusion that an infant was well-hydrated.

Injuries

Describe and document all injuries (see below and Chapters 9–13).

Internal examination

General

Examine all serosal surfaces for petechiae. Describe all tumors, neoplasms, and congenital abnormalities. Weigh all organs and describe their sizes and configurations.

Body cavities

Comment on the integrity of all body cavity linings and the hemidiaphragms. Comment on any unusual accumulations of fluid (hemorrhage, exudates, fibrin, etc.) and document the volume of any fluid/blood found in the body cavities. Note any adhesions. Comment on the positioning of the organs in relation to each other (normal situs, situs in versus, etc.).

Head and spinal cord

Note the condition of the meninges (cloudy, exudate, hemorrhage) for signs of meningitis. Note any cerebral malformations or hydrocephalus. Examine the middle ears and mastoids. If neural tube defects are suspected, a posterior approach to the spinal column is recommended for the best visualization. The orbital roofs are removed and the eyes and optic nerves are examined (and photographed) in situ for hemorrhage. If hemorrhage is noted, the eyes and optic nerves are best removed for further gross and microscopic examination after fixation, unless deemed unnecessary owing to an adequate explanation for their presence (e.g., motor vehicle collision death). The spinal cord should be examined in all infant deaths, with rare exceptions. The brain and spinal cord should be submitted for neuropathologic examination, where available, in all decedent's aged 12 months or younger, and in most other pediatric deaths without obvious trauma (motor vehicle accidents, gunshot or sharp force injuries that do not impact the spine) or without clearly natural causes (e.g., acute bronchial asthma). An unexplained subdural hemorrhage

(SDH) in infants should raise the suspicion for abusive head trauma (AHT), and the cervical spinal cord should be examined for hemorrhages of the cervical nerve roots. The cervical neck can be dissected in situ or removed en bloc, fixed, decalcified, and sectioned at the levels of the nerve roots to demonstrate the presence or absence of hemorrhage or other injury.[58,59]

Neck

Layer the muscles of the neck anteriorly. Examine the posterior neck musculature if abusive head injury or whiplash injury is suspected. Look for foreign bodies at all levels of the airway, examine the tongue, as well as the upper airway, epiglottis, and the posterior oropharynx for retropharyngeal abscesses/tonsillitis.

Cardiovascular system

Note the distribution of the great vessels and their relationship to the lungs. Often, the best way to examine these relationships is to remove the heart and lungs en bloc and examine them ex situ. If congenital malformations are suspected, an en bloc examination is best to document the changes and anatomic relationships; remove the heart, lungs, and aorta as one bloc. Weigh the heart and lungs together, and then reweigh them individually after their examination. In certain cases, such as when an anomalous pulmonary venous return is suspected, it is beneficial to keep the liver attached to the aorta and remove them with the heart and lung bloc. Cut the aorta "long" (near the diaphragm) to prevent cutting across an aortic coarctation. Examine the aortic arch and the distribution of the major aortic branches. Note any vascular rings and examine the aorta and pulmonary diameter at takeoff from the heart to assess for normal sizes and equal caliber. Note the coronary arterial distribution. Comment on the chambers and whether or not they are hypertrophic or dilated. If they are hypertrophic, note whether it is uniform throughout the chamber. Comment on any necrosis, mottling (suggestive of myocarditis), or hemorrhage of the myocardium. Comment on whether or not the foramen ovale and ductus arteriosus are open, probe-patent, or closed. Note any septal defects. Ensure that the valves are properly formed with the expected number of leaflets and cusps. If deemed necessary, consider a consultation with a cardiac or pediatric pathologist and/or retaining a sample of myocardium in RNAlater (see below). There is little, if any, downside to retaining some myocardial tissue in RNAlater in all pediatric deaths.

Respiratory system

Note the number of lobes (isomerism). Palpate for consolidation (note that grossly discernible findings of bronchopneumonia may be minimal and difficult to appreciate in infants) and look for pulmonary hypoplasia. Bronchopulmonary dysplasia is a microscopic diagnosis, and the findings include large, simplified alveolar structures, a dysmorphic capillary configuration, and variable interstitial cellularity and/or fibroproliferation.[60] Note any true antemortem aspiration of food or foreign bodies, as opposed to perimortem artifact from attempted resuscitation. Examine the trachea and esophagus for a fistulous connection. Look for tracheal stenosis and scarring. Comment on congenital anomalies of the lungs: sequestration, congenital pulmonary airway malformation, etc.

Pediatric pathologists recommend *inflating one or both lungs* prior to sectioning, which allows for a better microscopic examination of the alveoli to assess for pulmonary hypoplasia and interstitial pneumonias, including viral changes. It may, however, introduce artifact, depending on the needs dictated by the case. In suspected stillborn deaths, for example, the float test (see below) can be performed prior to inflation. Inflation can be accomplished by injecting formalin through the main stem bronchus, clamping the bronchus with a hemostat, and submerging the lung in formalin until the end of the autopsy, at which point they can be sectioned and sampled. One sample from each lobe is submitted and, depending upon the size of the lung, each section may be from the hilum to the pleura.

In cases where it is unknown if the subject was live or stillborn, the fresh lungs may be placed in water to see if they float, using a piece of liver as a control. It should be noted, however, that attempted resuscitation may artifactually introduce air into the lungs and decomposition may make the liver float (see below for further discussion).

Digestive system

Note any mucosal hemorrhages and ulcers, rotations and attachments of the intestinal tract, dilitations/points of apparent obstruction, Meckel's diverticulum, internal herniation, inflammation, intussusception, or infection. Describe in detail the contents of the stomach and the intestines (seedy, yellow, meconium, etc.), including the appearance and consistency of the feces. Food in the stomach (formula, breast milk, etc.) is the only pathognomonic finding for a live birth (an infant must be alive to ingest food).

Genitourinary system

Note renal agenesis, dysplasia, obstruction, and congenital absence of the reproductive organs. Note if the testes are descended.

Hematolymphoid system

Note atrophy or agenesis, asplenia or polysplenia, lymph node enlargement, and thymic petechiae and weight. Note that *numerous small mesenteric lymph nodes ("shotty" lymphadenopathy) are typically observed* as a function of the naïve and normally developing immune system during the novel antigen exposures of early childhood. The absence of the thymus in an infant or young child may be an indicator of congenital absence and immunocompromise or the harbinger of *chronic* pathophysiologic and/or psychologic abuse (thymic involution), so must be regarded seriously and investigated accordingly.[61,62]

Endocrine system

Weigh both adrenal glands (most reference tables provide a combined mass) and note if there is hypo- or hyperplasia. Note any adrenal hemorrhage.

Musculoskeletal system

Note evidence of connective tissue disorders or inborn bone diseases (e.g., osteogenesis imperfecta). Look for evidence of nutritional deficiencies (e.g., rickets, scurvy) or congenital syphilis. A radiographic analysis can only enhance this evaluation.

Histology

A full histologic workup is standard in nearly all nontraumatic pediatric deaths, as well as in pediatric homicides, traumatic or chemical. This typically includes sections of the heart, lungs, liver, pancreas, kidneys, spleen, stomach, esophagus, intestine, reproductive organs, and adrenal glands. A section of the trachea at the level of the thyroid gland can demonstrate edema or inflammation of the trachea and proper development of the thyroid.

Generous use of histological examinations will prevent later charges that you ignored medical diseases or relied solely on the gross examination to the exclusion of more subtle features. Even in violent deaths, the histologic examination may reveal indicators of chronic stress, malnutrition, or metabolic disease, in addition to information that may or may not aid in the aging of injuries.

Often, pediatric deaths require a number of histologic sections to adequately evaluate all of the organs. Histologic examination of representative injuries also may separate them by age. In homicidal deaths, it is important to include sections of injuries for diagnosis, estimation of the age of injury, and for further documentation. If you do not submit sections of healing fractures, we suggest at least saving them in the stock jar for later submission as necessary.

Neuropathology

A formal neuropathologic evaluation of the brain and spinal cord following formalin fixation is typically performed in infants and children. It provides the best chance to detect subtle injuries, make sense of brain pathology after a period of hypoxic encephalopathy, detect unusual natural diseases or congenital anomalies, and/or document the anatomical stage of development (such as the degree of myelination). Neuropathology may be of great benefit for either clinical correlation or to document pertinent negatives. Unless there is a compelling reason not to do so, the spinal cord should be removed during the following autopsies:

1. All children under one year of age
2. All suspected child abuse fatalities, regardless of age

Toxicology

The same standard specimens are taken as for adults: peripheral (and cardiac) blood, vitreous humor, and urine (liver, brain, and gastric contents, as needed). Hair and soiled diapers may be tested; hair may give an indication of chronicity.[63–75] **Vitreous humor chemistry** *is indicated in cases of suspected dehydration, neglect, or some major illness.* In most toxicology laboratories, vitreous electrolytes and glucose must be specifically requested on the toxicology submission worksheet. In infants dying unexpectdly, it is best *not to collect vitreous humor until the internal examination of the head and eyes has been completed.* If a SDH is found unexpectedly, the eyes are likely to be much more valuable as pathologic specimens to demonstrate potential retinal hemorrhages (RHs); prior vitreous humor collection may introduce artifacts that can obscure or confuse the gross and microscopic examination of the eyes. See Chapter 14 for further discussion regarding pediatric considerations in toxicological analyses.

Metabolic studies

Metabolic screening may be performed on all newborns and infants by a blood and/or bile spot card, depending on jurisdictional availability.[26,76–82] Additionally, a given state's newborn screening results may be requested and obtained from departments of health.

Microbiologic studies

The symptoms of infectious disease may be subtle in neonates and infants. Postmortem studies for bacteria and viruses should be routinely submitted in most infant deaths.[83–93] At a minimum, nasopharyngeal swabs and blood cultures should be submitted, with rare exception. Depending upon the circumstances, cerebral spinal fluid, lung swabs or tissue, tracheal swabs, meningeal tissue or swabs, etc., may also be submitted. Cultures and swabs are often submitted from pediatric emergency rooms; a review of medical records may prove helpful. If **bacterial meningitis** is suspected by history or gross examination of the brain or cerebrospinal fluid, the department of health must be notified as soon as possible. Diagnoses of infectious diseases using molecular techniques involve target amplification methods (e.g., polymerase chain reaction [PCR]) and a nucleic acid sequence analysis of the amplicon.[83,94–99] Unlike phenotypic characteristics that may be variable, molecular methods examine stable genotypic characteristics. Another major diagnostic advantage of PCR is that it does not require the labor-intensive cultivation of the microbial biomass.[100]

Technologies based upon the **PCR method** will become nearly ubiquitous gold standards of diagnostics in the near future, for several reasons. First, the catalog of infectious agents has grown to the point that virtually all of the significant infectious agents of the human population have been identified. Second, an infectious agent must grow within the human body to cause disease; essentially, it must amplify its own nucleic acids in order to cause a disease. This amplification of nucleic acid in infected tissue offers an opportunity to detect the infectious agent by using PCR. Third, the essential tools for directing PCR (primers) are derived from the genomes of infectious agents, and with time, those genomes will be known, if they are not already. The ability to detect any infectious agent rapidly and specifically by molecular analyses is currently available; the major hurdles are cost and application.[100–102]

Molecular genetics

It is advisable to save a sample of blood in a purple top tube and/or heart, liver, or spleen specimens in a special fixative (RNAlater) for potential genetic studies in *all* cases.[103–116] Molecular genetic laboratories offer a number of panels that test for a variety of hereditary abnormalities including those for cardiac abnormalities (channelopathies, cardiomyopathies) and epilepsy. Paternity issues also may arise. See Chapter 19 for a more in-depth discussion.

Forensic odontology

Suspected bite marks should be swabbed for forensic biology (amylase) and DNA testing. A forensic odontology consultant may be helpful in discriminating between a human bite mark impression from one caused by the family dog (typically, quite different in morphology) or between a young sibling and an adult caretaker.[117–122]

Forensic biology

If a suspected homicide, the collection of fingernail scrapings/clippings/swabs and scalp hair specimens may be indicated. Sexual assault kits also may be performed, as needed. As above, when identified, human bite marks should be swabbed and the swabs submitted to the forensic biology laboratory.

Forensic anthropology

In some ME/C offices, anthropologists may play an important role in child fatalities in which inflicted trauma is suspected. The forensic pathologist has the ultimate responsibility for evaluating all the different pieces of evidence when making their eventual determination of cause and manner of death. For child abuse fatalities, this can often be challenging and may benefit from anthropological input. Through consultation between the ME/C and anthropologist, bones or bone segments may be removed during autopsy for additional evaluation that can be useful for death certification and, in the case of homicides, court testimony. Areas of fracture, suspected fracture, and fracture callus are removed during autopsy for additional studies that may include histology and anthropology. Any skeletal specimens removed at autopsy are best reradiographed, as visualization of fractures and associated healing (if present) will be improved. 3D printing of skeletal injuries based on CT scans may be useful in court (see Chapters 18 and 19). Maceration of specimens should only be completed after histologic sampling. Anthropological interpretation of fractures on the clean bones can provide an additional line of documentation and interpretation of fractures that can be useful to the ME/C and which can be very illustrative in court proceedings. In some instances, additional, more subtle fractures may be observed after maceration that would have otherwise been missed on X-ray evaluation and during autopsy.

Pediatric trauma

Blunt impact injuries

A detailed description and photographic documentation of each injury is vital. Photographs and detailed descriptions may be used as evidence to document and age injuries (acute, recent, and remote), which may demonstrate the chronicity of abuse in suspected child trauma deaths. A detailed description of a patterned injury may suggest an object used to inflict injury. 3D printing (models) of the injuries, particularly when they involve the skull, can be particularly helpful and illustrative in criminal court proceedings. See Chapter 16 for a further discussion.

It is important to specifically mention the following:

1. Size
2. Shape; include whether the edges are straight, curved, angled, etc., and whether the injury hugs/wraps around the body contours or remains only in one plane/on prominences. Think of loop or strap marks versus a rigid object.

3. Color: purple, red, green, or yellow; blurred/distinct margins, scarring
4. Anatomic location and distribution: note whether or not the injury overlies a bony prominence and note symmetry
5. Internally, note substantial hemorrhage, the formation of a fibrinous reaction, purulence, or granulation tissue, each a sign of an intravital injury response

Reexamination of the decedent the day *after* autopsy may help delineate contusions. As the blood drains from the surrounding tissues after death, the contusion may become more apparent/prominent. Remember that the changes occur in the *background, not in the contusion*. This point may be important to explain later at trial. If such changes are noticed, re-photograph them.

Internal *blunt* injury may not be evident externally. This is particularly true of **blunt abdominal trauma**, owing to the absence of bony tissue in the abdominal wall that would otherwise offer resistance against a **focal application of force.** When injury is sustained to the abdomen in circumstances *other* than significant traumatic events, such as motor vehicle collisions or descents from substantial heights (e.g., out of a high window), pediatric abdominal viscera are not normally expected to tear easily, unless there is a focal application of force.[123–125] That is to say that a simple fall to a flat, resistant surface, such as the floor or the ground, at home or at play, does not typically produce enough deceleration forces to inflict the shearing that would be necessary to lacerate viscera. By contrast, a considerable and *protrusive* force applied to the abdominal wall, one that pushes it in and creates a crush type injury of the viscera, as they get trapped between the resistant vertebral column and the application of force, would have a much higher likelihood of causing laceration and shearing. To be clear, a child falling a short distance to the flat ground or floor, per se, without substantial protrusion of a resistant object into the abdominal cavity, such as a bicycle handle, would represent an inadequate explanation for severe visceral blunt trauma, including lacerations.[126–138] The term "short distance" falls has been used to represent falls from beds or other similar furniture.[139–142]

Aside from protrusive blunt objects, another scenario that is more commonly known to create this focal abdominal application in children is **inflicted blunt trauma**, where a fist, knee, elbow, or even a two-handed forceful squeeze, depending on the size of the child, may be sufficient to explain such severe injuries (Fig. 6.2). These focal applications are usually delivered in the form of one or more forceful blows that can produce the substantial protrusive injury we describe here.[126,135,143] Of importance, the structures at the greatest risk for this category of injury are those roughly in the *abdominal midline*, where they are most likely to be crushed against the anterior vertebral bodies (Fig. 6.3).[126,135,138,143] As such, *structures such as the mid to left liver, proximal duodenum, pancreas, and fixed midline mesenteric root are particularly vulnerable to this mechanism of injury.* Moreover, when one or more of these midline structures are injured, their specific location further supports this focal application(s) of force mechanism (Fig. 6.4).

In conjunction with the vulnerable viscera, there also is the seeming incongruity of the often inconspicuous (atraumatic) appearance of the external abdominal wall. Again, this is a reflection of the abdominal wall's lack of skeletal resistance, a resistance that the focal force does not meet until it abuts the rigid vertebral column. This can be confusing to investigators since, unlike facial, chest, and extremity blunt trauma, the abdomen of the child victim of blunt trauma may appear *atraumatic*, despite the child being in visible and severe distress, even at death's door, but for reasons that may elude the inexperienced observer. Never was the aphorism, "absence of evidence is not evidence of absence" more apt than when evaluating fatal inflicted trauma of the abdominal viscera in the

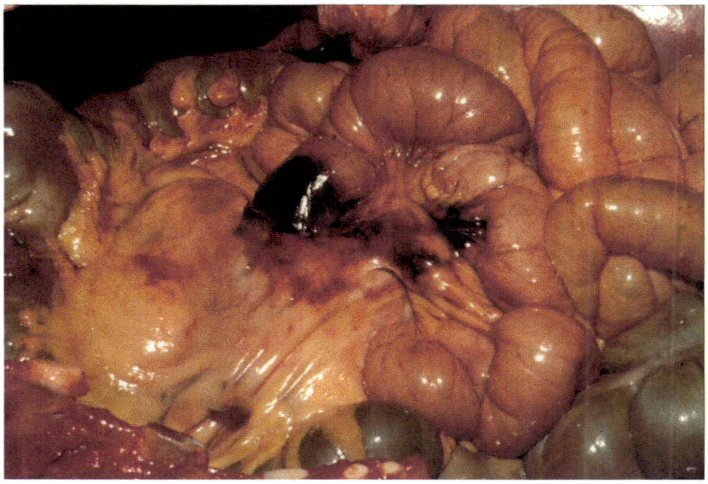

FIGURE 6.2 **Mesenteric hemorrhage at the ligament of Treitz.**

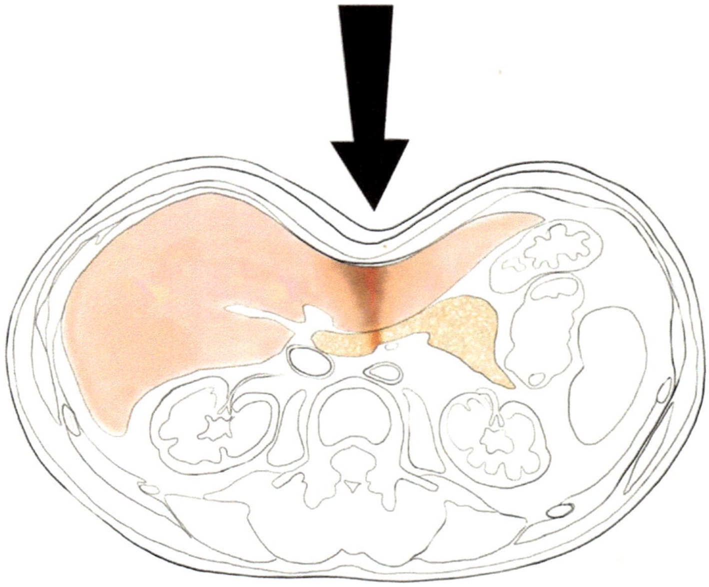

FIGURE 6.3 Schematic of focal application of abdominal force. *Watercolor by Dr. Cricket Coleman.*

absence of visible external (cutaneous/subcutaneous/muscular) abdominal wall injury. The broad front of the soft tissue of the abdominal wall allows it to absorb the trauma, while the deeper tissues become "trapped." Recall that the injury caused by blunt force is a function of the delivered *kinetic energy*, modified by the area of the surface inflicting the injury and the duration of the impact (the duration of the transfer of energy). Both the area (A) and the time (T) inversely affect the extent of injury (see Chapter 10). In addition, due to their blood vessel elasticity, infants and children may require more force to disrupt the vessel wall. This same vascular elasticity may also minimize the formation of petechiae in circumstances where they might be expected.

In all suspicious pediatric deaths, a full body reflection of the skin (flay dissection; see Chapter 5) to examine the subcutaneous, muscular, and skeletal tissues may reveal subtle subcutaneous hemorrhages/contusions or subtle skeletal injury. This dissection would also include a posterior neck dissection, another important component of the overall examination in this setting. Be compulsive about the documentation of details. The autopsy report should include statements detailing the presence or absence of cervical soft tissue hemorrhages, fractures, and spinal epidural or subdural hemorrhages. Pertinent positives and negatives relative to all findings at autopsy are very important to document and require a certain degree of forethought.

Be sure to strip the parietal pleura and incise the intercostal muscles to detect subtle rib fractures. Any acute fracture or fracture callus observed on X-ray or by palpation should be dissected, photographed, and, if necessary,

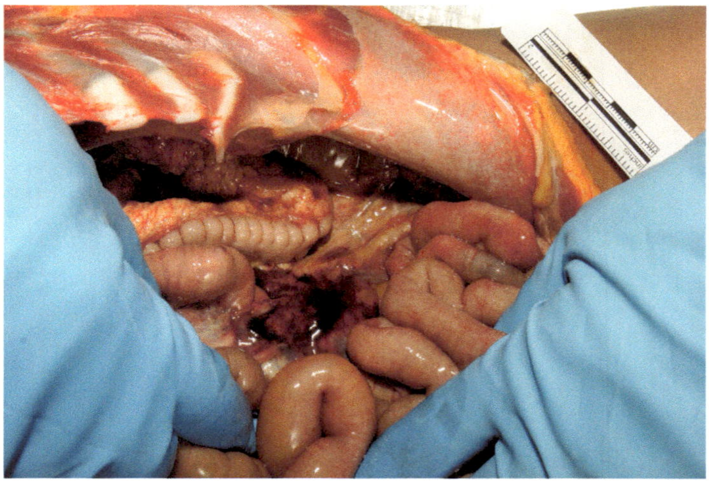

FIGURE 6.4 Mesenteric root hemorrhage.

appropriate sections submitted to histology, with the remainder saved in the stock jar. Anthropology and radiology consultants can also serve as very helpful adjuncts in these investigations.

Fatal child abuse syndrome (aka, battered child syndrome) and abusive head trauma

The absolute dependence of infants and children upon their parents or adult custodians renders them susceptible to a range of fatal maltreatment and neglect that defies the imagination of a thousand nightmares.[144]

... the autopsy findings in this group range from nothing at all to blatantly grotesque. However, regardless of the nature of the autopsy findings, no genus of homicide victims presents a more complex or difficult challenge to the forensic pathologist and other investigative authorities.[144] ***Drs. Ross Zumwalt and Charles S. Hirsch.***

Fatal child abuse syndrome

Some of the most difficult deaths that a forensic pathologist encounters are those due to fatal child abuse syndrome. In these fatalities, the child is typically found to have injuries of varying ages, with different stages of healing, sometimes, if not often, with superimposed evidence of chronic neglect, nutritional, emotional, medical, and/or even educational. The findings tell a graphic tale of repetitive traumatic incidents. The duty of the forensic pathologist in such cases, over and above the routine mission of establishing the cause and manner of death, is to determine whether these injuries were nonaccidental and inflicted over time. This demands an extremely thorough autopsy, with detailed descriptions of the individual injuries and their overall distribution on external and/or internal examination. Additional postmortem studies may aid in estimating the age of an injury, with the caveat noted by Dr. Hirsch that, "In reality, pathologic processes seldom cooperate fully with attempts to date or age them with precise reliability." Instead, it is more important to be able to distinguish injuries of clearly different ages to a reasonable degree of scientific certainty than it is to try to pinpoint the specific age of any one injury.[143]

Beyond the aging of injuries, microscopic tissue examination also can exclude endogenous/natural pathologic processes and can confirm marked thymic involution, a dramatic end-organ manifestation of chronic physical and/or psychological stress that is sometimes observed in children who are chronically abused. Histologic and radiologic examinations will commonly provide important information concerning wound healing (such as fractures) and may allow the differentiation between grossly similar appearing wounds (Fig. 6.5).[31,143,145] Full body radiographs are even more important in these instances and may demonstrate not only acute and/or chronic injuries but may also provide information about bony changes consistent with starvation, neglect, and developmental arrest. They also may exclude such diseases as osteogenesis imperfecta. The mechanism(s) of death in these fatalities can be

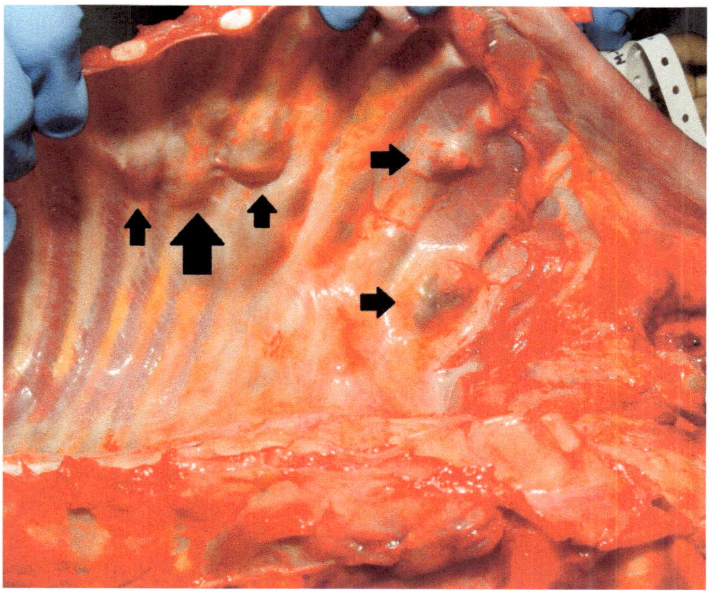

FIGURE 6.5 **Healed rib fractures.** Note multiple fracture calluses (arrows).

complex, even elusive, requiring vigilance and stringent documentation not only at autopsy but also through a thorough medical and sometimes educational record review, multiple ancillary studies, and often, specialty consultations.

Abusive head injury

Abusive head trauma (AHT) is the terminology now more commonly used in lieu of what some formerly called "shaken baby syndrome."[146–155] The diagnostic and nomenclature shift was made due to competing theories about whether or not a point of impact was needed to produce the observed trauma and whether rotational acceleration/deceleration forces were enough to produce the injuries. The so-called triad of findings in these cases consists of SDH, retinal hemorrhages (RHs), and an encephalopathy; however, these findings are not invariable in all AHT deaths and, when present, must be interpreted within the context of the case, including the entire constellation of findings, both physical and investigative.[156–170] The "triad," in and of itself, is not pathognomonic for nonaccidental injury, per se. Victims of AHT may have no external or internal evidence of impact injury or they may have one or more impact sites on internal examination. As such, any SDH in an infant should raise the suspicion of injury. Accidental causes of SDH in an infant, such as motor vehicle collisions or falls from a considerable height, usually can easily be excluded by history and examination. The findings, as with all cases, are considered within the context of the circumstances of the death, including the scientific veracity of the explanations provided by caregivers.

The SDHs in these instances are typically small (nonspace occupying), often appearing as "teaspoons of blood" on the brain/dura, without a mass effect. Because dural bridging veins are torn during the removal of the calvarium at autopsy, the artifactually torn veins may ooze over time, resulting in some accumulation of blood on the brain. This blood may be mistaken for a traumatic SDH. In order to avoid this pitfall and make the proper diagnosis, one is advised to observe and document the surface of the brain immediately after opening of the skull.

The SDH in most instances of AHT is not, in and of itself, fatal. It is merely a marker for the lethal mechanism, that is, an indication of the forces that have passed through the head and brain. Nevertheless, the SDH and the dura may be examined histologically for evidence of acute, recent, or remote injuries that may indicate repeated injury over time or a survival interval. Rotational acceleration/deceleration produces shearing forces of the cerebral axons that can lead to diffuse axonal injury. These injuries can be seen microscopically and are enhanced with amyloid precursor protein (APP) immunohistochemical staining, depending on the survival interval. As such, submission of the brain and spinal cord for neuropathologic consultation is useful.

Although APP is a sensitive marker for axonal injury, it is not specific for mechanical trauma, as APP accumulation also results from hypoxic–ischemic injury and metabolic disorders. APP positivity has been demonstrated in injured axons within 35–45 min following head trauma in the adult human.[171] Various semiquantitative approaches with different operational criteria have been used to evaluate white matter injury at specific regions of interest (i.e., the corpus callosum, internal capsule, brainstem, and cerebellar peduncles).[172–174] Examination of the pattern and distribution of APP staining can distinguish traumatic axonal injury from hypoxic–ischemic injury. Since secondary hypoxic–ischemic injury often accompanies trauma, this distinction is important. Some infants are resuscitated and placed on a ventilator for days or longer, before being declared dead by neurological criteria. This results in extensive hypoxic–ischemic injury, which may complicate the evaluation of traumatic axonal injury. It is in these instances that the APP staining pattern and the clinical history may together help in determining the etiology.[175–178] It has been demonstrated that the microscopic patterns of APP positive axons in traumatic and hypoxic–ischemic etiologies, in conjunction with the examination of hematoxylin and eosin–stained sections, can determine the etiology of the axonal pathology in most instances.[175] With trauma, one typically sees axonal changes within distinct tracks, while with ischemia, they cross different components of the brain (i.e., the latter representing a vascular distribution). See Chapter 10 for a further discussion.

RHs are a frequent manifestation of AHT, characterized by marked acceleration–deceleration forces, with or without blunt head impact (formerly described as shaken baby syndrome).[157,160–162,164,179–183] Approximately 85% of affected children have RHs, with approximately two thirds having extensive (too numerous to count), multilayered hemorrhages that extend to the ora serrata.[167,180,181] The eyes are unroofed and photographed, as described above, removed (without vitreous aspiration), fixed in formalin, and later examined grossly and microscopically for retinal and optic nerve sheath hemorrhages (Fig. 6.6). Iron stains may reveal old hemorrhage.[184] Rather than a simple either/or description, the microscopic number (from few to too numerous to count) and distribution (posterior, multilayered, or extending to the ora serrata) of the hemorrhages are important to document and describe. These observations may be corroborated by an antemortem (hospital) ophthalmologic examination. As noted above,

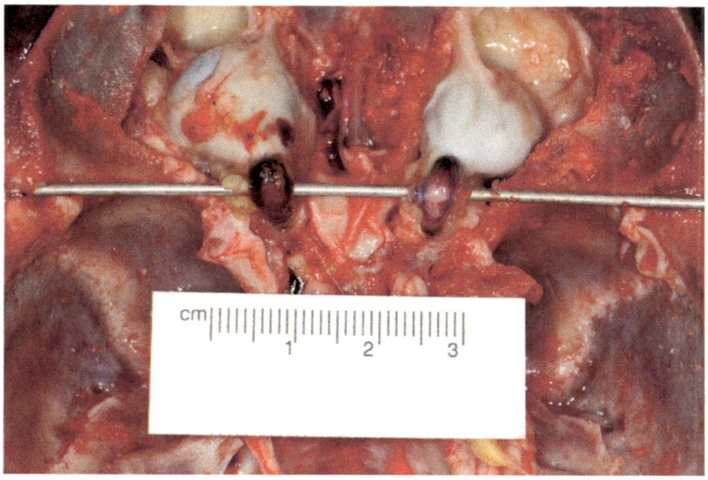

FIGURE 6.5 **Optic nerve sheath hemorrhages.**

vitreous humor should not be withdrawn until orbital and optic nerve sheath hemorrhages are excluded on gross examination.[32,185–189] Birth-related RH occur in one-quarter of normal deliveries and are far more common after deliveries involving instrumentation. Commonly bilateral, they are predominantly intraretinal, posterior, resolve rapidly, and very rarely persist beyond six weeks.[165,190,191]

Subcutaneous hemorrhages of the upper arms, chest wall, and thighs are seen in some instances, as these are sites where the perpetrator may grab the child while inflicting the head injury. A subcutaneous dissection of the skin and underlying soft tissues (flay dissection) may reveal injuries that are not evident externally. Posteromedial rib fractures also may be seen, indicative of a forceful squeezing of the chest while being gripped. Avulsion fractures of the extremities also may result from the limbs flailing about.[29,192–195] Instances involving a demonstrable head impact, with subgaleal hemorrhage and skull fractures, are much easier to recognize and diagnose as trauma (Figs. 6.7 and 6.8).

Because this can be a point of confusion, it is important to address lucid intervals. **Lucid intervals** would *not* be expected in deaths involving marked traumatic axonal injury (TAI) in any setting, adult or pediatric. Rather, lucid intervals (defined as a period of mental clarity) may occur in specific and well-recognized neuropathologic settings, such as when there is an injury of the head that is *insufficient* to cause *direct* diffuse brain impairment (the latter typically occurring with either traumatic axonal injury or hypoxic–ischemic injury), but that instead triggers bleeding (subdural or epidural hemorrhage) or malignant brain edema that expands over time and raises intracranial pressure. Eventually, the elevated pressure from either or both compromises brain function, resulting in unconsciousness (the end of lucidity), and death, if the pressure is not relieved. It follows then that a neurologically devastated infant who dies from a head injury, including a small, noncompressive SDH and a nonswollen brain, did *not* have a lucid

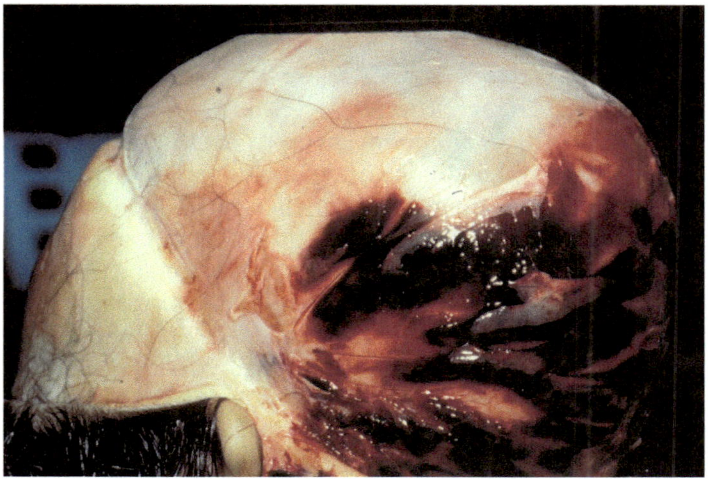

FIGURE 6.7 **Subgaleal hemorrhage with an underlying skull fracture (impact site).**

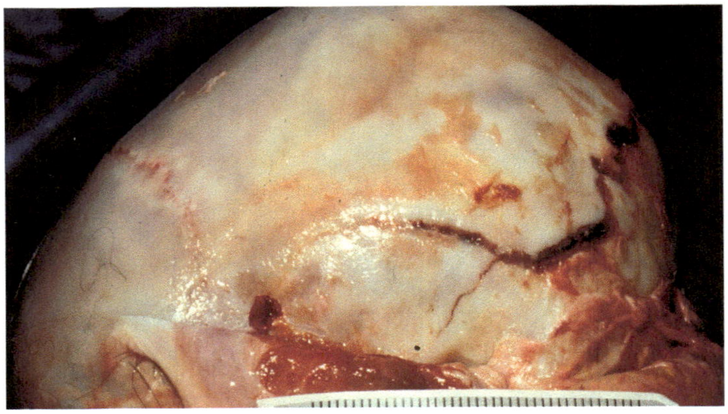

FIGURE 6.8 **Skull fracture.** Skull fracture after the soft tissue is removed (from the infant in Fig. 6.7).

interval. Infants with considerable head injury (including marked TAI) may be able to withdraw or open their eyes, but they would neither appear "normal" nor function normally. It is for this reason that if the infant had remained on life support for some days or more prior to autopsy, the results of the *initial* computed tomography (CT) scan of the head become important to review, as it is typically elucidating in terms of the mechanism of neurologic compromise/death. Specifically, the absence of effacement of the basal cisterns on **admission imaging studies** excludes malignant edema as an explanation for the infant's initial sudden cardiorespiratory arrest, as well as a lucid interval that could have preceded it, and instead supports direct trauma to and impairment of the brain at the moment of injury.[196–199] See Chapter 10 for further discussion.

Patterns of inflicted injury: infants versus toddlers/preschoolers

When comparing and contrasting the psychosocial and physical patterns of inflicted injury in the infant versus toddler/preschooler groups, several observations emerge.

Generally speaking, the head is more commonly the location of fatal inflicted injury in infants, while the torso (mostly the abdomen), more so in toddlers and preschoolers, sometimes the latter in conjunction with the head.[126,134,143,200,201] The relatively large head, which rests on a comparatively weak/poorly controlled neck and body, is disproportionately vulnerable in infants relative to the older cohort.[148,149,202,203]

Based on interviews and confessions, infant deaths are much more likely to be temporally related to a baby's uncontrollable fits of crying, after which an impulsive response may be provoked in the caregiver, sometimes ending with violent shaking.[150,152,204] By contrast, it is more typical that misbehavior or bathroom accidents lead to punitive measures, such as corporal punishment, against the toddler or preschool child by the caregiver. The latter group of children are typically walking and talking, so are capable of "bad" behavior or accidents that create messes, chaos, or otherwise test and provoke adults.

The AHT that typically kills infants rarely presents with a "lucid interval" (i.e., if caused by marked TAI, is typically immediately devastating, if not immediately fatal, producing neurologic compromise at the time of injury occurrence; see above), whereas *the fatal blunt abdominal trauma that is most characteristic of toddler/preschool deaths tends to present with a more protracted and evolving course of symptomatology, characteristically worsening over time* (usually over a day to days).[148,150,196–199,204–209] As such, the investigation is typically more involved for the latter group, requiring a careful cataloging of the child's overall behavior over time, the quality and progression of their movement and play, appetite/development of anorexia, fever, general distress, vomiting, and lethargy. Objective outside observers, ear witnesses and eyewitnesses, such as neighbors, doormen, or store owners, to the child's subsequent activities and behavior, along with video surveillance, might be very helpful in establishing objective data in these criminal investigations. *In the end, the dovetailing of the investigative data points with the gross and microscopic physical injuries helps to reconstruct a reasonable time frame for the injury occurrence relative to the child's death.* Beyond that, it is important to keep in mind that the forensic pathologist (FP) is the expert who can best predict with reasonable medical certainty, the range of symptoms and distress that are the direct result of various injuries found at autopsy, and what one would reasonably or unreasonably expect. Pediatricians, particularly those who specialize in child abuse, are also often involved in these cases and can provide valuable insights and clinical and diagnostic information as well. For these children, as time progresses, the existence of evolving and worsening sequelae of fatal injury over time would predictably lead to the appearance of a clearly and increasingly ill and distressed child, one who does not look "fine" or "normal," even to the inexperienced eye, and who is in need of medical attention. As such, the specter of neglect might also be raised in such cases.[33,34,36,37,201,209–215]

Asphyxia

The findings associated with asphyxial deaths can be subtle, particularly in infants. Typical injuries seen in some *homicidal* asphyxial deaths in older children, teens, or adults result from the victim's struggle with the assailant. Due to the disparity in size and strength between the assailant and the victim, however, a smothered infant may show few, if any, signs of the assault. However, there may be subtle findings. In the setting of heightened suspicion, injury of the frenulum (laceration, contusion) should be considered inflicted trauma until proven otherwise, but must be measured against the type of resuscitative measures used, if any. The same can be said for pinch marks of the nostrils. Force-feeding may lead to aspiration of foreign material. Aspiration deaths may be delayed, with the mechanism of death related to the resultant pneumonia. That said, aspiration of small amounts of gastric contents may occur during the dying process, a consequence of death, not the cause of it.

As noted above, the certification of accidental **positional or mechanical asphyxia** in infants and neonates is usually made based upon the scene investigation and autopsy findings, as well as the additional special studies being negative for a potentially fatal disease and/or injury (i.e., a diagnosis of exclusion and investigative context). The *lividity pattern* may be the only pertinent finding at autopsy that indicates that the child may have died as a result of positional asphyxia. Also important is the inquiry and establishment of the infant's **developmental milestones** at the time of death. Were they able to lift their heads (if so, for how long?), consistently roll over (prone to supine and supine to prone?), sit up, eat solid foods, etc.? **Overlay deaths** constitute a specific category of infant asphyxia in the setting of bed sharing. Occasionally, this mechanism of death is quite clearly demonstrated through the historical narrative and reenactment. Bed surface sharing with an intoxicated adult further complicates the investigation of a suspected overlay death and may result in a manner determination of homicide. Regardless, whether accidental or homicidal, overlay deaths constitute a unique type of positional asphyxia in infant fatalities. As such, the term "overlay" is best specifically listed somewhere on the death certificate to communicate this singular outcome (whether listed in the cause of death statement or in the "how injury occurred" section). It is similar to "autoerotic asphyxia" only in terms of its specificity, so deserves specific mention. Beyond these scenarios, at baseline, young children, given their relatively underdeveloped motor coordination and dexterity and often limited dentition, are at a heightened risk of **choking** on certain foods (e.g., hotdogs, candy, gum), particularly while running or playing. Also, because toddlers and younger children are oral by nature, they are at additional risk for choking on myriad (nonfood) foreign bodies that they happen to place in their mouths (balloons, buttons, etc.).

The criteria for diagnosing asphyxial deaths in infants are described later in this chapter, in the certification section.[4]

Bathtub drownings

Deaths of infants and young children in bathtubs are handled differently than are those of adults and they are also usually of greater concern. These fatalities should be both viewed and investigated with a great deal of circumspection and care owing to children's inherent vulnerability and dependence on caregivers. As a group, infants have a *developmental risk of drowning*, in contrast to the types of risks found in adults, which are typically related to natural disease or intoxication. The death of any infant or young child in a bathtub should always be approached with the burden to exclude child abuse or neglect.

Until they reach the age of full motor and cognitive ability to roll, sit, pull up, stand, walk, and communicate, infants and some young children are at risk of drowning in even a small amount of water (depending on their age and/or their degree of motor and cognitive delay). Given the amount of time it takes to suffer severe to irreversible anoxic brain injury from drowning (minutes), an infant should never be left unattended in a bathtub with water. Leaving an infant with a similarly aged or only slightly older toddler is not viewed as sufficiently safe either. In toddlers, the drowning may be accidental.

In all instances, a thorough scene investigation with photographic documentation and measurements of the tub and interviews with the caretaker(s) are necessary. This information is then correlated with the findings at autopsy and the historical information of the child's neurologic and motor abilities prior to death (Was the child able to stand up?). Always consider the child's height with reference to the tub dimensions and in relation to any injuries.

The myriad circumstances and variables that must be considered and investigated in such deaths exceed the scope of this discussion, but suffice to say that a careful, detailed, thorough, and multidisciplinary approach is in order, including, but not limited to, specific tub and water-level measurements, the latter directly measured by the medicolegal death investigator (MDI) or, if drained, through that documented from the police investigation, photographic documentation of the drain and potentially obstructive objects, such as toys, washcloths, or drain plugs, the wetness of the tub, the surroundings, and the decedent (including the scalp hair), detailed witness statements, including the length of time that the caregiver reported to have left the child unattended, and an understanding of the decedent's developmental milestones (if an infant), medical history (including cardiac disorders, seizures, etc.), medications, and the presence or absence of neurologic compromise or developmental delay. Invariably, law enforcement authorities/police and child protective services will and should be involved.

When drowning is determined to be the mechanism/cause of these deaths, manners are typically classified as accidents or homicides, depending on the details of the case; there is no "one size fits all" classification system. When homicides, however, they are classified as such through evidence of negligence (more common in infants who were left unattended) or as an intentional, volitional act, which is, admittedly, much less common (i.e., a child intentionally being held under water).

As an example of the latter, when a neurologically intact, developmentally normal three-year-old child "drowns" in a bathtub because they were "left alone," having no history of neurologic risk and no autopsy evidence of blunt force head injury to suggest that an accidental fall followed by unconsciousness led to the drowning, the caregiver story becomes inconsistent and suspect. *Neurologically competent, normally developed, nonmedicated, and uninjured three-year-olds should not drown in bathtubs any more than should adults.* It is in these situations, an intentional drowning becomes a concern, particularly if the family has a substantiated history with child protective services and/or of domestic violence.

Regardless, the reasoning for the manner determination, whatever it may be, must be clearly stated in the "how injury occurred" section, since an explanatory proximate cause/risk of a natural or chemical type is less likely to exist in an infant/young child drowning than it is when involving an adult (i.e., "drowning" may stand alone in the cause of death statement in an infant/child, while its explanation is detailed in the "how the injury occurred" field, which should also clearly support the chosen manner—accident vs. homicide). Conversely, in adults, the proximate cause typically serves as the explanation for the drowning (e.g., "drowning due to epilepsy"), with only "found submerged in bathtub," or something similar, stated in the "how injury occurred" section. See Chapter 11 for further discussion on this topic.

Thermal injuries

Aside from clearly accidental house fires, isolated burns on a child are highly suspicious for child abuse, unless there is a reasonable explanation. Steam (e.g., from a broken radiator) can cause extensive thermal skin injury, hyperthermia, or inhalation injury when the exposure is in an enclosed place, particularly when the child is nonambulatory.[216–218] A detailed description of the injury is necessary and should include the following features:

1. Type: flame, scald, steam, contact, or chemical
2. Severity: first, second, or third degree (or partial thickness/full thickness) burns
3. Distribution: body surfaces, confluent areas, spared areas (pressure points vs. skin folds, indicating flexion), stocking/glove distribution (may indicate forceful submersion), streams (indicates running liquid), splash or spatter marks
4. Characteristics: straight, well-demarcated margins (discrete scald lines from submersion in hot water), circular/oval (may indicate cigarettes), patterns related to hot objects (curling iron, heating grate, clothing iron)
5. Effects of healing: scars, pigment changes, exudates; excessive scarring on a young child is cause for concern and further investigation
6. Estimate the percentage of the total body surface area affected: use the pediatric "rule of nines" (different in children and infants as compared to adults, given the disparate proportions of their heads, torsos, and limbs, as compared to each other and to adults, Fig. 6.9).

The question that the forensic pathologist must answer in these instances is whether the pattern of injury is consistent with the caregiver's report of the circumstances. The pattern of the burn may determine how it

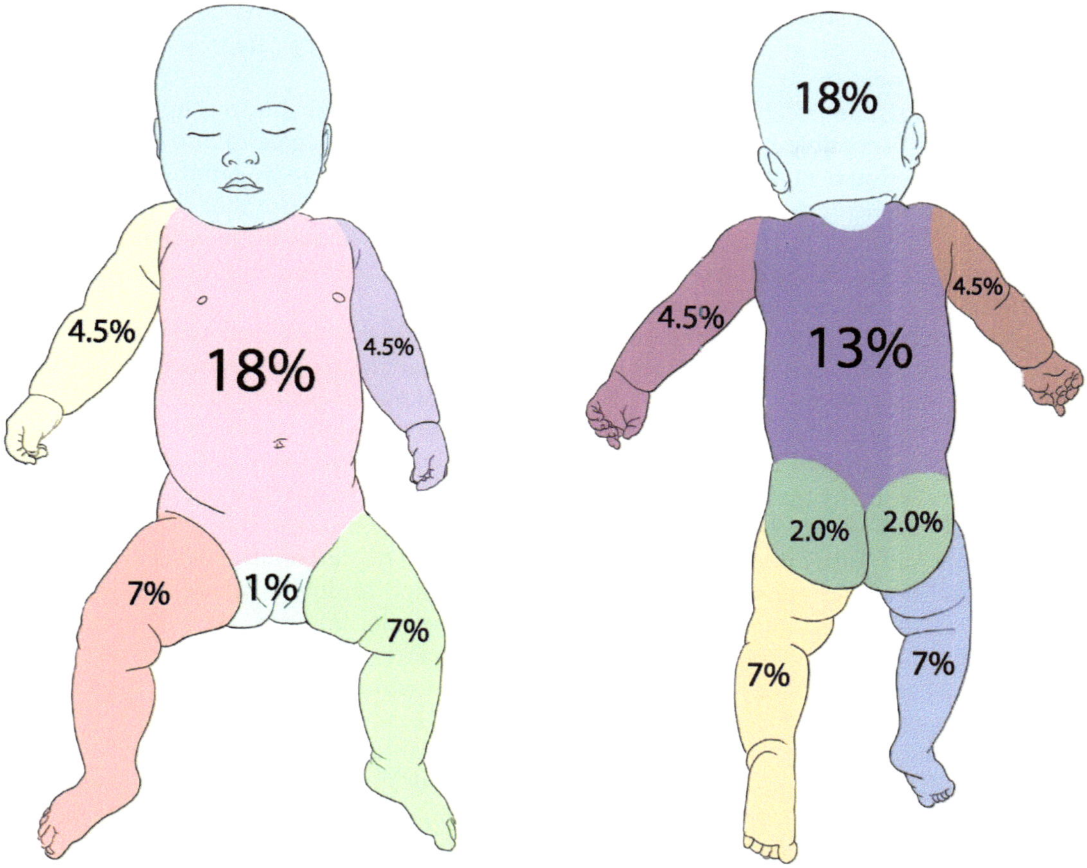

FIGURE 6.9 **Rule of nines for infants.** *Watercolor by Dr. Cricket Coleman.*

occurred (e.g., immersion vs. splash). For example, clear immersion pattern injuries are inconsistent with injuries caused by a pot of boiling water that splashed from the stove. Water at 44°C (111°F) will cause a superficial cutaneous burn in approximately 5 h while water at 60°C (140°F) will cause a full thickness scald burn in approximately 3 s. Checking the maximum temperature of the reputed faucet may help in determining if a given history is consistent or inconsistent with the extent of the burn.[219–223] Toxic epidermal necrolysis, secondary to a viral infection or medication, can cause sloughing of the skin. These patients are treated as burn patients and die from similar complications.

Neglect

All childhood deaths are tragedies, but the loss may be compounded when they were preventable and a parent or caregiver, often unintentionally, contributed to the death. **Environmental neglect** is a term that describes some of these deaths. These include medical neglect, malnutrition, hyperthermia due to being left in cars or other environments,[224–228] ingestion of an accessible drug or poison,[229–236] and unsupervised drownings.[210,211,237–250] Neglect is a component in many of these deaths. Neglect is defined as the failure of a caretaker to provide the goods, services, or care necessary to maintain the health or safety of a child or vulnerable adult. The challenge that ME/C face in some of these deaths is how to certify the manner of death. In the fatalities in which neglect is involved, the intent, quality, and extent of the negligence usually guides the manner certification. Unfortunately, the circumstances surrounding each of these deaths is unique and there are no standard guidelines to help the certifier weigh these varied factors. A household in which excessive amounts of illicit substances, such as heroin and cocaine, are left out in plain view on a coffee table, easily accessible to a toddler to pick up and ingest, sets the stage for clear criminal neglect, supported by both the recklessness of the environmental risk and the involvement of illegal substances. This scenario is wholly different than one in which a toddler stumbles across a bottle of cleaning solution or an elderly relative's cardiac medication and (fatally) ingests it in a few minutes. Regardless, some of these fatalities may be certified as accidents in one jurisdiction while certified as homicides in another. While some are more consistently certified as accidents or homicides, many fall into a gray area. These represent a category of deaths that ME/C often struggle with and agonize over to certify.

As Hirsch and Zumwalt stated: "The absolute dependence of infants and children upon their parents or adult custodians renders them susceptible to a range of fatal maltreatment and neglect that defies the imagination of a 1000 nightmares."[144] **Intentional starvation** and **extreme medical neglect** (e.g., failure to administer lifesaving medication) are well-established examples of deaths with an extent of negligence that clearly warrants a homicide manner determination.[33,34,209,251,252] There is little professional discord about these. Other examples of environmental pediatric neglect are less straight forward and the manner determinations may vary from jurisdiction to jurisdiction based on local customs, policies, and attempts at consistency.

In response to allegations surrounding some infant **malnutrition deaths**, attention may be placed on a rare intestinal malabsorption syndrome, a putative abnormality of the adrenal glands, or a breastfeeding failure.[253–258]

A pediatrician or gastroenterologist may opine about a long list of malabsorption syndromes that may have gone undetected. But, this still begs the question as to why the child was not brought to medical attention. An image of the emaciated child is usually sufficient to demonstrate that, even if there was a rare defect, the child was medically neglected. It would be clear to anyone who sees the image of these children that they demanded immediate medical care. **Pediatric growth charts** are an objective demonstration of malnutrition and they may indicate when (and under whose custody) the child started to lose weight, when they started to "fall off the curve." Evidence that the child was able to gain weight (for example, while in foster care for a period of time or while hospitalized) also helps to exclude an endogenous malabsorption syndrome and points to an external factor (i.e., deprivation of food). With **marasmus** (Gr: withering), in which there is inadequate energy intake in all forms, the child has a shrunken, wasted appearance, with loss of muscle and the subcutaneous fat mass. **Kwashiorkor**, by contrast, is due to marked protein deficiency, but adequate energy intake, resulting in edema and ascites.[33,35–41,234,251,252,259–261]

At autopsy, useful microscopic sections include the thymus, intestines, and bone marrow. **Vitreous electrolytes** may demonstrate dehydration (elevated urea nitrogen and sodium).[262] Externally, one may see tenting of the skin and hyperpigmentation. Other findings may include thymic involution, an empty stomach/small intestine, gastrointestinal ulcers, rickets, bone marrow hypoplasia, depleted liver glycogen stores, decreased organ weights, a distended gallbladder, a colon with hard, dry feces, and bronchopneumonia or pyelonephritis.[261]

Adrenal glands of the human fetus at term are as large as those of adults, weighing 8–10g or more. With congenital hypoplasia, however, the adrenal glands weigh <2 g. The fetal adrenal glands are principally composed of an inner fetal zone that accounts for 85% of the total volume; this zone soon disappears (apoptosis) after birth.[263] As a consequence, the total normal weight of the glands decreases by approximately 50%. A morphometric study in autopsy cases and an ultrasound study during antenatal and neonatal periods indicate that the adrenal glands may shrink more rapidly in infants born at full term, decreasing to their normal infantile size within the first two weeks after birth. The fetal cortex usually disappears by 3–4 weeks of age. These normal changes must be recognized when considering a diagnosis of congenital adrenal hypoplasia.

Congenital adrenal hyperplasia represents a group of autosomal recessive disorders caused by impaired cortisol synthesis. In infants, insufficient aldosterone production causes salt wasting, failure to thrive, and potentially fatal hypovolemic shock.[82,264] The weight and histopathology of the adrenal glands may be a useful diagnostic indicator. In congenital adrenal hypoplasia, the adrenal glands are reduced in size and have a decreased fetal zone in neonates, scattered cytomegalic cells, and decreased cellular lipid content. A similar clinical picture may result from dehydration in inadequately fed neonates and also can be associated with those who are breastfed.[253–257]

Consultations

Have a low threshold for using consultants as needed: neuropathologists, pediatricians, pediatric radiologists, forensic odontologists, pediatric pathologists, and cardiac pathologists.

Child fatality review boards

Most jurisdictions have a Child Fatality Review Panel (CFRP), or similar designation, which is a multidisciplinary committee that includes prosecutors, departments of child and family welfare, law enforcement, pediatricians, social workers, and the ME/C. The CFRP typically reviews all of the pediatric deaths (under the age of 18) in order to determine trends, review the involvement (if any) of other state agencies in the child's life, and to determine ways forward to prevent similar deaths from occurring. In addition, some jurisdictions may also have an office of the Child Advocate that monitors and evaluates public and private agencies that are charged with the protection of children, and reviews state agency policies and procedures to ensure they protect children's rights and promote their best interests.

Mandatory reporting

Most jurisdictions have agencies that investigate child abuse and neglect. If an FP encounters a case in which child abuse or neglect is *suspected* (even if as yet inconclusive), they must report it. Medical examiners, as physicians, are considered mandatory reporters. Although police may already be involved in the investigation by the time the ME/C becomes involved, *do not assume* that the child welfare authority has been made aware of the death. Sometimes, the suspicion of child abuse may not come to the attention of these authorities until weeks or months after the autopsy since the death certification was initially made pending during the investigative period, and the classification of homicide not made until much later. Or, as another possible scenario, an unremarkable autopsy examination of a one-year-old child may point toward a natural cause, but the toxicology returns a month later, positive for fentanyl. For these reasons, as soon as the *suspicion* of abuse, injury, or neglect arises, the notification should be made; other children living in the home may be at imminent risk, so time is of the essence whether only suspicions or final confirmations. The bar is low for reporting, as it can be lifesaving.

Home births: stillbirth versus live birth

The ME/C typically accepts jurisdiction of stillborn fetuses whenever there is a question of maternal trauma, maternal drug use, illegal abortion, or a therapeutic complication.[265–272] If there is uncertainty about accepting jurisdiction or performing an autopsy, it is probably better to be conservative and accept the case. A Certificate of Stillbirth is issued in many jurisdictions for any stillborn fetus that has reached at least 20 weeks of gestation and died before birth/died in utero. If there was a breath, even an isolated gasp, a heartbeat, swallowing, tone, pulsations of the umbilical cord, or movement of voluntary muscles at or after birth, independent of age or size, then the **infant** is considered **live born** and a Certificate of Death must be issued (as well as a Certificate of Birth). If there is no evidence of any, even momentary extrauterine life (see below), the fetus maintains its **"fetal" designation** and its uterine expulsion is classified as a **stillbirth**.[273]

Total body radiographs should be performed, as should toxicological analysis, along with all other routine analyses performed in an infant death, if possible, depending on the age of the fetus. Cocaine and/or benzoylecgonine may be detected in the fetal viscera long after an intrauterine demise. Every effort should be made to obtain and examine the **placenta**, since the placenta is commonly the source of the underlying cause and/or immediate cause/mechanism of death. If the placenta has been examined as a surgical pathology specimen at a hospital, at least the report should be requested, but the actual specimen as well, if not yet discarded.

The forensic pathologist is not uncommonly faced with the death of a fetus or newborn whose birth, witnessed only by the mother, occurred at home/otherwise in the absence of medical professionals, or the fetus and placenta were discovered discarded in public (i.e., in the garbage). A determination as to whether the decedent was born alive is the crucial question to answer in these cases, a question that is often difficult and sometimes even impossible to answer. A comprehensive approach is described below. In short, if live born, even if briefly, radiographs typically identify gas in the gastrointestinal and/or respiratory tracts and/or the partially or fully aerated lungs float in water once removed at autopsy. In the absence of confounding factors, aerated lungs and a gastric air bubble are supportive evidence of live birth. *Food (breastmilk, formula) in the stomach is pathognomonic for live birth with a postnatal survival.* Once the infant is determined to have been live born, the investigation begins in earnest to determine both the cause and the manner of death. The medicolegal investigator and the police should ask questions of the caretakers and witnesses to the birth, if any, to ascertain information concerning movement, breathing, crying, or other activities observed at the time of birth, and may need guidance from the forensic pathologist concerning what questions to ask in order to ascertain this information and how best to pose them. See below.

Approach to the live born versus stillborn infant/fetus and possible neonaticide

When an ME/C is involved in the investigation of a found infant/fetus, discovered under varied circumstances (left outside, in dumpsters, thrown out windows, brought dead to emergency rooms after being delivered at home, etc.), but whose birth was unwitnessed by any medical professional, the following issues should be addressed, not all of which are relevant in every case:

Was the baby stillborn or live born?

Establishing whether the baby was **stillborn** or live born is the first order of business in these cases and is not necessarily an easy question to answer definitively.[266–269,272–284] Ultimately answering it, however, will determine whether a death certificate or a stillbirth certificate will be issued. When issued, the latter (often called a Certificate of Stillbirth) contains a cause of *fetal demise* rather than a cause of death. A stillbirth certificate indicates that the fetus died in utero/was stillborn, having had no evidence of ex utero existence, by either examination and/or investigation/clinical history, if even only for seconds. *If there is no live birth, there is no death, and therefore, no "cause of death," "manner of death," or traditional death certificate.* Instead, **a stillbirth certificate lists the "cause of (intrauterine fetal) demise,"** but ***no manner.*** Making this determination (live birth vs. stillbirth) typically involves connecting several data points, some investigative and others physical/anatomic and radiologic (see further discussion below).

The World Health Organization defines live birth as:

> The complete expulsion or extraction from its mother of a product of conception, irrespective of the duration of the pregnancy, which, after such separation, breathes or shows any other evidence of life—e.g., beating of the heart, pulsation of the umbilical cord or definite movement of voluntarily muscles—whether or not the umbilical cord has been cut or the placenta is attached.[285]

Autopsy: determination of gestational age, live born versus stillborn, and cause of demise or death

A comprehensive autopsy, not dissimilar to the SUDI work up, must be completed, even if there is demonstrable trauma. This includes the following:

Full body plain film radiographs are performed to assess both the likelihood of live birth and the gestational age (aiding in live birth assessment: pulmonary and gastrointestinal air patterns). Realize that a brief postpartum survival in a live born infant may not involve a robust respiratory effort. Sometimes, there is only brief or agonal air swallowing (hence only a gastric bubble being appreciated radiologically). Moreover, assessment of ossification centers and air patterns by a pediatric radiology consultant can aid in both gestational age estimates and the likelihood of live birth.

Careful body, skeletal, and organ **measurements** and observations can aid in **assessing gestational age and the likelihood of an independent ex utero** existence (i.e., the transition from fetus to infant).[265,267,268,273,278,284,286,287] These include body mass, head circumference, crown-rump length, crown-heel length, foot length, and organ weights. Other measurements and observations useful in assessing gestational age and pathologic syndromes are nail development, facies, inner and outer canthal distances, ear placement, palate formation, nasal/choanal and rectal patencies, supernumerary digits, etc. All measurements are compared to appropriate pediatric charts in order to most accurately estimate gestational age.

External and internal **maceration** (sterile tissue breakdown, "autolysis," following in utero demise) is indicative of stillbirth and should be specifically described. Sometimes, this is quite obvious, particularly when the demise occurred fairly long before delivery/expulsion. It is characterized by diffuse skin sloughing with a beefy red discoloration of the skin/dermis and tissues, cranial suture overriding, an airless pulmonary/gastrointestinal radiology pattern, and sinking lungs during the "float test" (see below). These changes should be referred to as "maceration," a word that infers stillbirth, by definition. Usually, it takes about 6 h or more between intrauterine demise and delivery before grossly detected maceration is seen.[268,269]

The word "decomposition" is not typically relevant unless there was an in utero infection, such as chorioamnionitis, or some other maternal infection. Otherwise, by definition, the fetus typically exists in the sterile in utero environment and autolysis, rather than putrefaction, follows the demise, hence the more typical absence of a foul odor, bloating, and green discoloration in most cases. In this setting, the gastric or other gastrointestinal air patterns take on greater significance (i.e., *not* typically artifacts of putrefaction). Acute inflammation of the gastric mucosa may be a sign of chorioamnionitis as a result of swallowed infected amniotic fluid that incites an inflammatory response. Pertinent negatives and positives with regard to maceration should always be a routine part of the autopsy report documentation in these cases. Be cautious in your conclusions, knowing that *while maceration is proof of stillbirth, its absence does not exclude it*, as a fetus can die just before or at the moment of delivery and be stillborn, but have no maceration given the acuteness/timing of the demise. As such, all anatomic/radiologic and investigative data points must be considered before drawing a conclusion.

The flotation test (docimasia), largely of historical interest, is only one data point and must be interpreted within the context of the entire case.[288] It is not conclusive. Submerge the lungs in water and photographically document whether they float or sink. You can also do this lobe by lobe. If the entire lung block is floated, ligate the airway first. Some suggest photographing a submerged organ such as the liver as an internal control. Others suggest ligating and floating the stomach and GI tract before opening them to document air swallowing, particularly if no resuscitative efforts have been performed.

There are two identified determinants of potential *false-positive float tests:* "decomposition" and respiratory resuscitative efforts. Also, testing a section of liver as an internal control may identify putrefaction since a putrefied liver may float while a normal liver will sink. *False negatives* may be seen in cases of extreme prematurity and/or a very weak respiratory effort (in duration or robustness). As pointed out above, not all live born infants have a robust respiratory effort, enough to substantially and uniformly expand the lungs. The float test still may be performed, as one may be critizied for not doing it.

Evidence of circulation (e.g., extensive hemorrhage) *and/or a vital reaction* (e.g., inflammation) *to any* **injuries**, such as those from a substantial descent from height (as may be seen when an infant is thrown out of a window) support life at the time that they were sustained. If present, measure specific volumes of hemorrhage (cavity and in soft tissues) relative to the calculated total body blood volume by age and body mass, photographically document the size and locations of injuries, histologically sample injuries, and document visceral and muscular pallor (intravital evidence of shock). Occasionally, infants die further out from delivery, following a substantial period (hours) of exposure and neglect. That said, many deaths in live born infants occur so close to delivery that an inflammatory response is unlikely; hence, its absence does not exclude a live birth. "Absence of evidence is not evidence of absence."

One should *contextualize any documented injuries against the* **delivery description** *given by the mother*, if the mother is indeed known (see below). Some injuries (e.g., cephalohematomas in vertex deliveries) are more common/less sinister than others. Breech deliveries may cause abdominopelvic visceral injury. Birth-related skull fractures, while not common, tend to be linear and parietal when they do occur and may be related to cephalopelvic disproportion and/or fetal macrosomia. Knowing who the mother is and her general body habitus can very helpful in this regard.

A careful **examination of the placenta and umbilical cord** may demonstrate or exclude pathology or fetal stress and aid in the gestational age estimation.[265,271,277,284,289–295] In addition to other routine observations and measurements, document whether the cord appears to have been torn or cut. A torn umbilical cord can be associated with lethal fetal hemorrhage in live born infants. Positive immunohistochemical staining for tryptase, CD68, and alpha1-antichymotrypsin on umbilical cord tissue may have some utility in helping support the opinion of live birth.[273,292,296–298]

Investigative issues and checklist

1. Determine whether <u>*any*</u> cardiopulmonary resuscitation (CPR) was administered by medical personnel or lay people. This is particularly important with regard to respiratory resuscitative efforts (intubation, bagging, or mouth-to-mouth) so that the radiologic pulmonary and gastrointestinal air patterns can be accurately interpreted. The presence or absence of chest compressions having been administered is also important to know, particularly if blunt force trauma of the torso is observed. Dovetail all of the above with interviews and/or medical records, including the ambulance call report.
2. The police should make every attempt to locate and interview the birth mother; it is best to give specific and granular questions for them to ask her.
   - Specifically, she should be asked if and for how long she knew she was pregnant. It is not uncommon, especially in teen mothers, for the pregnancy to be unwanted and/or otherwise hidden from family members and friends. Make sure they explore and document her neuropsychiatric functioning, if she is in an abusive situation and/or foster care, if a minor, if there is a child protective services history, previous pregnancies or miscarriages, any prenatal history or care, substance use, and past medical history
   - Is the mother obese, making more plausible the assertion that she did not know she was pregnant?

- She should be asked about the onset of labor (specifically, what happened?) and how and where the delivery proceeded. Where was she? Outside? At home? If so, where? On the toilet, in the bathtub, shower, on a bed? How did the baby present (vertex, arm first, etc.)? Was she alone?
- Were there *any* signs of life in the infant and for how long? The latter is less important than the former, but both may provide important data points in the interpretation of findings. It is best for the inquiring investigator to be specific in the questioning: was there eye opening, any tone at all, any movement at all, mouth opening, grimacing, swallowing of air, gasping, breathing, or crying? Was there meconium in the mouth or nose? If there was crying, was it weak, forceful, interrupted, etc.? Any *plausible* signs of life, however brief, indicate a live birth. Compare and contrast her description against the physical and radiographic findings and the observations of others who may have been present.
- After the delivery, what did she think? What did she do next? Was there heavy bleeding from her or the cord? Did she faint? Did she attempt CPR, did she immediately wrap the baby from head to toe thinking it was dead? What did she wrap it in? Did she call 911, a family member, a boyfriend? The police should check the 911 records.

3. If the mother is still unknown and at large (not uncommon), local hospitals should be canvassed for young girls or women of child-bearing age presenting postpartum and/or with vaginal bleeding. If located, specifically ask about the whereabouts of the placenta, if it is not with the fetus. If the mother presented to a hospital and the placenta was delivered there, inquire with the pathology department before is it processed and/or discarded.
4. The police should also interview any other witnesses to the birth or immediate postpartum period, including family, neighbors, medical personnel, etc., asking similarly detailed questions.

Potential etiologically specific causes of neonatal death or fetal demise

- **Prematurity** incompatible with any/significant extrauterine life
- **Congenital or genetic abnormalities** incompatible with life or sufficient to explain the death/demise
- **Birth anoxia** caused by myriad intrapartum issues, maternal or fetal, when the birth occurred outside the attendance of medical personnel.
- **Placental disease/insufficiency/acute processes**, such as abnormal placentation leading to insufficiency or maternal hemorrhage. **Placental abruption** can be caused by hypertensive disease of pregnancy, stimulant drugs, such as cocaine, or trauma.
- **Maternal death of any cause** (natural or unnatural)
- **Maternal chemical injury**: Maternal drug intoxications, maternal smoke inhalation, with or without maternal death.
- [a]**Maternal mechanical injury** leading to fetal/infant demise/death: This includes maternal injuries occurring under varied circumstances (motor vehicle collisions, assaults, etc.), but that lead to fetal demise/stillbirth or induce the delivery of a live born infant who consequently dies. If the infant is live born, the manner of the infant's death will be predicated on the circumstances under which the mother's injuries were sustained.
- **Direct mechanical injury to fetus/infant**: Injury to the infant may occur in utero, independent of any injury to the mother (e.g., fetal head trauma due to mechanical induction of abortion by use of an implement inserted into the vagina by the mother), in utero as part of a constellation of injures to the mother (e.g., motor vehicle collision), or ex utero (intentional, such as smothering, blunt injury, etc., or inadvertent birth-related or postpartum injury).
- **When a neonaticide/homicide**, the cause of death is either through intentionally inflicted injury/commission or by neglect/omission, such as exposure or lack of necessary medical attention following delivery.

More times than not, the definitive death certification or certification of stillbirth for these cases is deferred (made pending) on the day of autopsy.

a As a matter of criminal investigations, understand that even if an infant is stillborn and therefore will have no manner of death determination, if the fetus died in utero as a direct consequence of criminal activity against the mother, the stillbirth may still be prosecuted as part of the larger homicide or assault case, predicated on the crime perpetrated against the mother and the laws of the jurisdiction.

What is SUID, SIDS, SUDC, and SUDP?

There is a difference between SUIDs (sudden *unexpected* infant death) and sudden *unexplained* infant deaths, although they both share the same acronym of SUID and refer to infants under 12 months of age who typically die during sleep/in a sleep environment.[299–304] That said, SUIDs refer to the deaths *prior* to the investigation and *ultimately*, will largely include deaths with undetermined causes (**sudden unexplained infant death** or **unexplained sudden death**), rare natural deaths, and asphyxial deaths (e.g., positional asphyxia). The use of SIDS has fallen out of favor by most, however, some still rarely use it, but with strict criteria.[299–307] Deaths that should not be certified as SIDS include those in which a scene investigation is not or cannot be performed or if an infant dies while bed sharing, sleeping on soft compressible bedding, or during prone sleeping on anything other than a hard mattress designed for infant safety.

Sudden unexplained death in childhood (SUDC) is a category of death in children between the ages of 1 and 18 that remains unexplained after a thorough investigation, including an autopsy. Most often, a seemingly healthy child goes to sleep and never wakes up. At this time, we do not know what causes SUDC, how to predict it, or how to prevent it. There is an SUDC Foundation that supports families and research efforts.[308] Sudden unexpected death in pediatrics (SUDP) is a designation that includes both infants and children.

Death certification

In 2019, the National Association of Medical Examiner's (NAME) panel on SUDP published a review of the investigation, certification, and needs of the family with regard to these deaths.[4] They described the challenges of ICD-10 coding discrepancies born out of the information provided on the death certificate.[4,299] Their goals were to improve and standardize cause of death reporting to further the accuracy of this important epidemiologic data. This also included the use of a synoptic report (**see** Figs. 6.10 and 6.11).[4] Other reporting forms include the **SUID Reporting Form** and the **CDC's scene investigation form** for medico-legal investigators.

It is not unusual to have multiple **questionably causal factors** associated with these deaths. These potentially causal factors may be categorized as either **intrinsic (threats to life arising from inside the body)** or **extrinsic (threats originating outside of the body)**. Examples of concerning intrinsic factors include: low birth weight, prematurity, concurrent nonlethal illness, and a history of febrile seizures. In addition, these would also include natural conditions of unknown significance (e.g., cardiac channelopathy gene variants of unknown significance). The extrinsic factors are conditions in the immediate environment that are a potential threat to life, but have not been deemed as the cause of death. Examples include side or prone sleeping, if unable to roll to a supine position, overbundling without evidence of hyperthermia, objects in the immediate sleep environment, a sleep environment not specifically designed for infant sleep, soft or excessive bedding, sleep surface sharing, or injuries/toxicological findings that are nonlethal or have unknown lethality, as well as circumstances/findings that are concerning for an unnatural death.[4] Their recommended criteria and format for cause of death statements and manners of death are listed in Tables 6.1 and 6.2.[4]

The NAME panel also proposed criteria to strengthen the amount of evidence that is needed to make a diagnosis of an "asphyxial" death (see Table 6.3**)**.[4]

History of SIDS and sudden unexplained death in infancy

Historically, deaths in which the investigations, full autopsies, and laboratory findings were all negative would have met the criteria for SIDS or would have been certified as undetermined/natural. Later, SIDS was no longer used if the deaths involved unsafe sleep conditions.[4,16,309] These unsafe sleep deaths were certified as SUID or undetermined, and a qualifying statement (e.g., placed prone in adult bed) was typically included parenthetically after the undetermined cause of death statement.

Synoptic Report for Sudden Unexpected Pediatric Death

Decedent's Last Name	Decedent's First Name	Decedent's Middle Name	Case Number

Cause of Death	○ Unexplained Sudden Death ○ Other: ______	
Intrinsic Factors	○ NO ○ YES, describe: ______	
Extrinsic Factors	○ NO ○ YES, describe: ______	
Manner of Death	○ Undetermined ○ Natural ○ Accident ○ Homicide ○ Suicide ○ Therapeutic Complications	
Investigation	○ Medicolegal investigation with scene visit and doll reenactment ○ Medicolegal investigation with scene visit, no doll reenactment ○ Medicolegal investigation without scene visit ○ Other, describe: ______	
Medical History	○ Decedent medical history reviewed with no causative, contributory, or risk factors identified ○ Decedent medical history reviewed, minor findings. Describe: ______ ○ Decedent medical history reviewed with potential causative, contributory, or risk factors identified. Describe: ______ ○ Incomplete medical record review ○ No medical record review ○ Medical records unavailable or nonexistent	
Sleep Environment Concerns	*Sleeping Position*	○ Supine sleeping on safe surface (If INFANT: alone, on surface designed for infant sleep, with firm mattress and fitted sheet and absent soft bedding and objects) ○ Prone sleeping on safe surface ○ Supine sleeping on unsafe surface for age ○ Prone sleeping on unsafe surface for age ○ Other sleep position, describe: ______
	Sharing Sleep Surface **(mark all that apply)**	○ Sleep surface sharing with adult(s), how many? ______ ○ Sleep surface sharing with a child/children, how many? ______ ○ Sleep surface with animal(s), how many? ______ ○ N/A, not sharing sleep surface
	Other Sleep Environment Concerns	○ Complex sleep surface sharing (other circumstances not covered above, such as with intoxicated adult or obese individual), describe: ______ ○ Unknown/unconfirmed sleep environment ○ N/A, not sleep-related death
Other Environmental Concerns (mark all that apply)	○ Markedly elevated temperature in home or excess bundling ○ Markedly cold temperature in home ○ Cigarette smoking in or outside home ○ Illicit drug use in or outside home ○ Other environmental concern(s) not listed above, describe: ______	

Excerpt from UNEXPLAINED PEDIATRIC DEATHS: INVESTIGATION, CERTIFICATION AND FAMILY NEEDS
Available for download at sudpeds.com. For additional inquiries contact the SUDC Foundation (800-620-SUDC)

Last modified December 2021

FIGURE 6.10 Synoptic report for sudden unexpected pediatric death. *Excerpt from Bundock E, Corey T, editors. National Association of Medical Examiners Panel on Sudden Unexpected Death in Pediatrics: Unexplained Pediatric Deaths: Investigation, Certification, and Family Needs San Diego: Academic Forensic Pathology International, 2019. Open access on NCBI: https://www.ncbi.nlm.nih.gov/books/NBK577018/.*

Synoptic Report for Sudden Unexpected Pediatric Death

Decedent's Last Name	Decedent's First Name	Decedent's Middle Name	Case Number

Synoptic Report for Sudden Unexpected Pediatric Death (Continued)

Other Objective Concerns (mark all that apply)	◯ Postmortem changes out-of-keeping with stated history ◯ Non-lethal injury or injuries ◯ Non-lethal illness or disease process ◯ Illness or injury of unknown significance ◯ Non-lethal toxicologic finding ◯ Toxicologic finding of unknown significance ◯ Evidence of possible neglect ◯ Other objective concern(s) not listed above, describe: ____________
Autopsy	◯ Complete autopsy (macro-/microscopic exam of thoracic, abdominal and pelvic organs, brain and spinal cord) without subspecialty pathologists consulted (pediatric, cardiac, and/or neuro-pathologists) ◯ Complete autopsy (macro-/microscopic exam of thoracic, abdominal and pelvic organs, brain and spinal cord) with subspecialty pathologists consulted (pediatric, cardiac, and/or neuro-pathologists); Specify consultations: ____________ ◯ Incomplete autopsy; Explain: ____________ ◯ Autopsy not performed; Explain: ____________

Toxicology	◯ Toxicology performed included testing for (check off all that apply)
	    ◯ Alcohol
	    ◯ Illicit drugs
	    ◯ Prescription Drugs
	    ◯ Non-Prescription Drugs
	◯ Toxicology testing not performed due to inadequate or absent specimens
	◯ Toxicology testing not performed

Ancillary studies (mark all that apply)	◯ Bacterial cultures done ◯ Viral cultures / studies done ◯ Vitreous electrolytes done ◯ Genetic cardiac studies done ◯ Metabolic screen done ◯ Other ancillary studies done; Describe: ____________ ◯ Ancillary studies not performed due to inadequate, inappropriate or absent specimens ◯ Ancillary studies not performed

Form completed by:

Name: ____________________

Title: ____________________

Agency: ____________________

Date: ____________________

Printed Name: ____________________ Signature: ____________________

Excerpt from UNEXPLAINED PEDIATRIC DEATHS: INVESTIGATION, CERTIFICATION AND FAMILY NEEDS
Available for download at sudpeds.com. For additional inquiries contact the SUDC Foundation (800-620-SUDC)

Last modified December 2021

FIGURE 6.11 Synoptic report for sudden unexpected pediatric death. *Excerpt from Bundock E, Corey T, editors. National Association of Medical Examiners Panel on Sudden Unexpected Death in Pediatrics: Unexplained Pediatric Deaths: Investigation, Certification, and Family Needs San Diego: Academic Forensic Pathology International, 2019. Open access on NCBI: https://www.ncbi.nlm.nih.gov/books/NBK577018/.*

TABLE 6.1 SUDP cause and manner of death.

Cause of death	Manner of death
Unexplained sudden death (no identified intrinsic or extrinsic factors)	Natural or undetermined
Unexplained sudden death (intrinsic factor(s) identified)	Natural or undetermined
Unexplained sudden death (extrinsic factor(s) identified)	Undetermined
Unexplained sudden death (intrinsic and extrinsic factors identified)	Undetermined
Undetermined (not further specified)	Undetermined
Undetermined (insufficient data)	Undetermined

TABLE 6.2 Criteria with recommended cause of death statements with various factors.

Cause of death: Unexplained sudden death (no identified intrinsic or extrinsic factors)		
Criteria for infant deaths		
1		Infant less than one year of age in apparent good health that dies suddenly and unexpectedly.
2		For sleep-related deaths.
	A	Placed alone, supine, in infant-specific sleep environment (e.g., crib, bassinet, portable crib, play pen) with flat, firm sleep surface, uncluttered by objects, and without potential areas of entrapment.
	B	Found unresponsive or dead, in the same sleep environment, with no obstruction of the nose and/or mouth or compression of neck/chest to cause asphyxia given the developmental abilities of the infant, as described by finder and demonstrated by doll reenactment (unless doll reenactment declined by caregiver).
3		The infant was not overly dressed or bundled for the environmental temperature.
4		Competent caregiver not impaired by drugs or alcohol.
5		Physical findings on body and at scene consistent with history provided by caregiver.
6		Completion of scene investigation and doll reenactment unless caregiver declines.
7		Review of child medical records and family health history.
8		Complete autopsy with histology, comprehensive toxicology testing (including vitreous chemistries if possible), and skeletal survey.
9		No anatomic, metabolic, toxicologic, chemical, historical, or external cause of death identified. Genetic testing is recommended but not required for this certification.
10		No extrinsic or intrinsic risk factors are identified.
Criteria for children greater than 12 months of age		
1		Child greater than 12 months of age in apparent good health that dies suddenly and unexpectedly.
2		For sleep-related deaths
	A	Found unresponsive or dead, with no obstruction of the nose and/or mouth or compression of neck/chest considered sufficient to cause asphyxia given the developmental abilities of the child (prone position without obstruction of nose and/or mouth may be present).
3		Physical findings on body and at scene consistent with history provided by caregiver.
4		Competent caregiver not impaired by drugs or alcohol.
5		Completion of scene investigation and doll reenactment if age or circumstance appropriate and caregiver cooperative.
6		Review of child medical records and family health history.
7		Complete autopsy with histology, comprehensive toxicology testing of blood (including vitreous chemistries if possible).

TABLE 6.2 Criteria with recommended cause of death statements with various factors.—cont'd

8	No anatomic, metabolic, toxicologic, chemical, historical, or external cause of death identified. Genetic testing is recommended but not required for this certification.
9	No extrinsic or intrinsic risk factors are identified.
Cause of death: Unexplained sudden death (intrinsic factor(s) identified):	
	A cause of death cannot be determined and criteria for unexplained sudden death (no identified intrinsic or extrinsic factors) are not met due to intrinsic natural abnormalities that are either known intrinsic risk factors for sudden death (including, but not limited to, low birth weight, preterm birth, small for gestational age, concurrent nonlethal illness, febrile seizures) or of unknown significance (including, but not limited, to mutations of unknown significance). Trauma and other unnatural etiologies are sufficiently excluded.
Cause of death: Unexplained sudden death (extrinsic factors identified)	
	Criteria: A cause of death cannot be determined and criteria for "unexplained sudden death (no identified intrinsic or extrinsic factors)" are not met due to unintentional extrinsic factors that increase risk for unnatural death. For infants this includes side or prone sleep if unable to roll to supine, overheating, objects in sleep environment, infant in sleep environment not specifically designed for infant sleep, soft or excessive bedding, and sleep-surface sharing. At all pediatric ages, this may include, but is not limited to, nonlethal injuries or injuries of unknown significance, nonlethal toxicologic findings of unknown significance, or circumstances otherwise concerning for unnatural death.
Cause of death: Unexplained sudden death (intrinsic and extrinsic factors identified):	
	Criteria: A cause of death cannot be determined and a combination of intrinsic and extrinsic factors are present.
Cause of death: Undetermined (not further specified):	
	Criteria: A cause of death cannot be determined due to circumstances or findings that make the above classifications inapplicable. Examples include inconsistent histories and/or other evidence that raise uncertainty about manner of death, or those which remain undetermined but were not sudden, and those with competing causes.
Cause of death: Undetermined (insufficient data):	
	Criteria: A cause of death cannot be determined because investigation, death scene examination, or autopsy were substantially limited, incomplete, or insufficient. Examples may include legal/religious restrictions, delayed report of death that limits scene investigation, and decomposition.

TABLE 6.3 Criteria for certification of an infant death as being caused by an asphyxial etiology.

1. The child must have a complete/full autopsy.
2. Toxicology, histology, vitreous electrolytes, microbiological cultures, and review of the medical history are performed as necessary and are noncontributory.
3. The infant must have obstruction of both nose and mouth or compression of the neck or chest, that is reliably witnessed or demonstrated by reenactment, or other reliable evidence of overlay or entrapment.
4. Asphyxiation must be probable given the age and stage of development.
5. Other potentially fatal findings or concerning medical conditions are reasonably excluded.

Conclusion

Infant and pediatric deaths are some of the most challenging fatalities that medical examiners encounter, both intellectually and emotionally.[154,308–312] That said, it is always important to remember that the family is going through an incredibly difficult time as well and, although it may not seem so initially, they will be infinitely appreciative of all of the hard work and time ultimately spent on what is, in essence, a physician attending to their loved one. Never hesitate to consult with colleagues in the handling of these cases; no forensic pathologist escapes the difficulty of these deaths, no matter how experienced. A second pair of trusted eyes is nearly always helpful.

References

1. Adelson L. The forensic pathologist. "Family physician" to the bereaved. *JAMA*. 1977;237(15):1585–1588.
2. Hirsch CS. Talking to the family after an autopsy. *Arch Pathol Lab Med*. 1984;108(6):513–514.
3. Pasquale-Styles MA, Tackitt PL, Schmidt CJ. Infant death scene investigation and the assessment of potential risk factors for asphyxia: a review of 209 sudden unexpected infant deaths. *J Forensic Sci*. 2007;52(4):924–929.
4. Bundock E, Corey T, eds. *National Association of Medical Examiners Panel on Sudden Unexpected Death in Pediatrics: Unexplained Pediatric Deaths: Investigation, Certification, and Family Needs*. San Diego: Academic Forensic Pathology International; 2019.
5. Pasquale-Styles MA, Regensburg M, Bao R. Sudden unexpected infant death certification in New York city: intra-agency guideline compliance and variables that may influence death certification. *Acad Forensic Pathol*. 2017;7(4):536–550.
6. Kemp JS, Thach BT. Sudden death in infants sleeping on polystyrene-filled cushions. *N Engl J Med*. 1991;324(26):1858–1864.
7. Mitchell EA, Scragg R, Stewart AW, et al. Results from the first year of the New Zealand cot death study. *N Z Med J*. 1991;104(906):71–76.
8. Ponsonby D, Wang. Factors potentiating the risk of SIDS associated with the prone position. *NEJM*. 1993;329:377–382.
9. Valdes-Dapena. A Pathologists perspective on SIDS-1991. *Pathol Annu*. 1991:133–164.
10. Poets S. Prone sleeping position and SIDS. *NEJM*. 1993;329:425–426.
11. Oyen N, Markestad T, Skaerven R, et al. Combined effects of sleeping position and prenatal risk factors in sudden infant death syndrome: the Nordic Epidemiological SIDS Study. *Pediatrics*. 1997;100(4):613–621.
12. Mallak CT, Milch KS, Horn DF. A deadly anti-SIDS device. *Am J Forensic Med Pathol*. 2000;21(1):79–82.
13. Mitchell E, Krous HF, Donald T, Byard RW. An analysis of the usefulness of specific stages in the pathologic investigation of sudden infant death. *Am J Forensic Med Pathol*. 2000;21(4):395–400.
14. Person TL, Lavezzi WA, Wolf BC. Cosleeping and sudden unexpected death in infancy. *Arch Pathol Lab Med*. 2002;126(3):343–345.
15. Carpenter RG, Irgens LM, Blair PS, et al. Sudden unexplained infant death in 20 regions in Europe: case control study. *Lancet*. 2004;363(9404):185–191.
16. Krous HF, Beckwith JB, Byard RW, et al. Sudden infant death syndrome and unclassified sudden infant deaths: a definitional and diagnostic approach. *Pediatrics*. 2004;114(1):234–238.
17. Alexander RT, Radisch D. Sudden infant death syndrome risk factors with regards to sleep position, sleep surface, and co-sleeping. *J Forensic Sci*. 2005;50(1):147–151.
18. Krous HF, Byard RW. Controversies in pediatric forensic pathology. *Forensic Sci Med Pathol*. 2005;1(1):9–18.
19. Li L, Fowler D, Liu L, Ripple MG, Lambros Z, Smialek JE. Investigation of sudden infant deaths in the State of Maryland (1990–2000). *Forensic Sci Int*. 2005;148(2–3):85–92.
20. McGarvey C, McDonnell M, Hamilton K, O'Regan M, Matthews T. An 8 year study of risk factors for SIDS: bed-sharing versus non-bed-sharing. *Arch Dis Child*. 2006;91(4):318–323.
21. Perrizo K, Pustilnik S. Association between sudden death in infancy and co-sleeping: a look at investigative methods for Galveston County Medical Examiners Office from 1978–2002. *Am J Forensic Med Pathol*. 2006;27(2):169–172.
22. Byard RW, Jensen LL. How reliable is reported sleeping position in cases of unexpected infant death? *J Forensic Sci*. 2008;53(5):1169–1171.
23. Schnitzer PG, Covington TM, Dykstra HK. Sudden unexpected infant deaths: sleep environment and circumstances. *Am J Publ Health*. 2012;102(6):1204–1212.
24. Byard RW, Bright F, Vink R. Why is a prone sleeping position dangerous for certain infants? *Forensic Sci Med Pathol*. 2018;14(1):114–116.
25. Matshes EW, Lew EO. An approach to the classification of apparent asphyxial infant deaths. *Acad Forensic Pathol*. 2017;7(2):200–211.
26. Pinneri K, Matshes EW. Recommendations for the autopsy of an infant who has died suddenly and unexpectedly. *Acad Forensic Pathol*. 2017;7(2):171–181.
27. Duncan JR, Byard RW. *Sudden Infant Death Syndrome: An Overview*. Adelaide (AU): University of Adelaide Press; 2018.
28. Laskey AL, Haberkorn KL, Applegate KE, Catellier MJ. Postmortem skeletal survey practice in pediatric forensic autopsies: a national survey. *J Forensic Sci*. 2009;54(1):189–191.
29. Kleinman PK, Marks SC, Blackbourne B. The metaphyseal lesion in abused infants: a radiologic-histopathologic study. *AJR Am J Roentgenol*. 1986;146(5):895–905.
30. Kepron C, Pollanen MS. Rickets or abuse? A histologic comparison of rickets and child abuse-related fractures. *Forensic Sci Med Pathol*. 2015;11(1):78–87.
31. Malone CA, Sauer NJ, Fenton TW. A radiographic assessment of pediatric fracture healing and time since injury. *J Forensic Sci*. 2011;56(5):1123–1130.
32. Weber MA, Risdon RA, Offiah AC, Malone M, Sebire NJ. Rib fractures identified at post-mortem examination in sudden unexpected deaths in infancy (SUDI). *Forensic Sci Int*. 2009;189(1–3):75–81.
33. Davis JH, Rao VJ, Valdes-Dapena M. A forensic science approach to a starved child. *J Forensic Sci*. 1984;29(2):663–669.
34. Knight LD, Collins KA. A 25-year retrospective review of deaths due to pediatric neglect. *Am J Forensic Med Pathol*. 2005;26(3):221–228.
35. Meade JL, Brissie RM. Infanticide by starvation: calculation of caloric deficit to determine degree of deprivation. *J Forensic Sci*. 1985;30(4):1263–1268.
36. Berkowitz CD. Fatal child neglect. *Adv Pediatr*. 2001;48:331–361.
37. Solarino B, Grattagliano I, Catanesi R, Tsokos M. Child starvation and neglect: a report of two fatal cases. *J Forensic Leg Med*. 2012;19(3):171–174.
38. Kellogg ND, Lukefahr JL. Criminally prosecuted cases of child starvation. *Pediatrics*. 2005;116(6):1309–1316.

39. Hughes EA, Stevens LH, Wilkinson AW. Some aspects of starvation in the newborn baby. *Arch Dis Child*. 1964;39(208) 598–604.
40. Madea B, Ortmann J, Doberentz E. Forensic aspects of starvation. *Forensic Sci Med Pathol*. 2016;12(3):276–298.
41. Neuberger FM, Jopp E, Graw M, Puschel K, Grupe G. Signs of malnutrition and starvation–reconstruction of nutritional life histories by serial isotopic analyses of hair. *Forensic Sci Int*. 2013;226(1–3):22–32.
42. Stocker JT, Dehner LP. *Pediatric Pathology*. Vol 1. Philadelphia: Lippincott Williams & Wilkins; 2002.
43. McCann J, Reay D, Siebert J, Stephens BG, Wirtz S. Postmortem perianal findings in children. *Am J Forensic Med Pathol* 1996;17(4):289–298.
44. Bruni M. Anal findings in sexual abuse of children (a descriptive study). *J Forensic Sci*. 2003;48(6):1343–1346.
45. Byard RW, Donald TG, Rutty GN. Non-traumatic causes of perianal hemorrhage and excoriation in the young. *Forensic Sci Med Pathol*. 2008; 4(3):159–163.
46. Panella M. Disseminated neonatal herpetic infection simulating abusive anal trauma. *J Forensic Sci*. 2011;56(4):1058–1061.
47. Adams JA. Understanding medical findings in child sexual abuse: an update for 2018. *Acad Forensic Pathol*. 2018;8(4):924–937.
48. Boudreaux MC, Lord WD, Dutra RL. Child abduction: aged-based analyses of offender, victim, and offense characteristics in 550 cases of alleged child disappearance. *J Forensic Sci*. 1999;44(3):539–553.
49. Adelson L. Pedicide revisited. The slaughter continues. *Am J Forensic Med Pathol*. 1991;12(1):16–26.
50. Eg MB, Hansen LA, Sabroe S, Charles AV. Hymenal lesions and legal outcome in sexually abused girls with a history of vaginal penetration. *Forensic Sci Int*. 2015;252:163–167.
51. Aprile A, Cesca E, Cecchetto G, Viel G, Mognato G, Gamba P. Partial bowel obstruction in a 2-month-old child. A delayed diagnosis of anal abuse. *Forensic Sci Int*. 2009;192(1–3):e7–9.
52. Goodyear-Smith F, Laidlaw T. What is an intact hymen? A critique of the literature. *Med Sci Law*. 1998;38:289–300.
53. Lauritsen AK, Meldgaard K, Charles AV. Medical examination of sexually abused children: medico-legal value. *J Forensic Sci*. 2000;45(1): 115–117.
54. Sharaf El-Din AA, Elkholy SM, Metwally ES, Farag HA. Pattern of female sexual assault in qalyubia governorate, Egypt, during the period from 2009 to 2013: a retrospective study. *Am J Forensic Med Pathol*. 2015;36(4):276–284.
55. Hansen LA, Mikkelsen SJ, Sabroe S, Charles AV. Medical findings and legal outcomes in sexually abused children. *J Forensic Sci*. 2010;55(1): 104–109.
56. Kohlmeier RE, Dimaio VJ, Sharkey F, Rouse EA, Reeves KE. The timing of histologic changes in liver lacerations. *Am J Forensic Med Pathol*. 2008;29(3):206–207.
57. Byard RW, Wick R, Gilbert JD, Donald T. Histologic dating of bruises in moribund infants and young children. *Forensic Sci Med Pathol*. 2008; 4(3):187–192.
58. Matshes EW, Evans RM, Pinckard JK, Joseph JT, Lew EO. Shaken infants die of neck trauma, not brain trauma. *Acad For Path*. 2011;1(1):82–91.
59. Peterson JE, Love JC, Pinto DC, Wolf DA, Sandberg G. A novel method for removing a spinal cord with attached cervical ganglia from a pediatric decedent. *J Forensic Sci*. 2016;61(1):241–244.
60. Coalson JJ. Pathology of new bronchopulmonary dysplasia. *Semin Neonatol*. 2003;8(1):73–81.
61. Valdes-Dapena M. The pathologist and the sudden infant death syndrome. *Am J Pathol*. 1982;106(1):118–131.
62. Fracasso T, Vennemann M, Pfeiffer H, Bajanowski T. Organ weights in cases of sudden infant death syndrome: a German study. *Am J Forensic Med Pathol*. 2009;30(3):231–234.
63. Wang X, Johansen SS, Nielsen MKK, Linnet K. Segmental hair analysis-interpretation of the time of drug intake in two patients undergoing drug treatment. *J Forensic Sci*. 2019;64(3):950–955.
64. Tournel G, Pollard J, Humbert L, Wiart JF, Hedouin V, Allorge D. Use of hair testing to determine methadone exposure in pediatric deaths. *J Forensic Sci*. 2014;59(5):1436–1440.
65. Chatterton C, Turner K, Klinger N, Etter M, Duez M, Cirimele V. Interpretation of pharmaceutical drug concentrations in young children's head hair. *J Forensic Sci*. 2014;59(1):281–286.
66. Garcia-Bournissen F, Rokach B, Karaskov T, Gareri J, Koren G. Detection of stimulant drugs of abuse in maternal and neonatal hair. *Forensic Sci Med Pathol*. 2007;3(2):115–118.
67. Rust KY, Baumgartner MR, Meggiolaro N, Kraemer T. Detection and validated quantification of 21 benzodiazepines and 3 "z-drugs" in human hair by LC-MS/MS. *Forensic Sci Int*. 2012;215(1–3):64–72.
68. Musshoff F, Kirschbaum KM, Graumann K, Herzfeld C, Sachs H, Madea B. Evaluation of two immunoassay procedures for drug testing in hair samples. *Forensic Sci Int*. 2012;215(1–3):60–63.
69. Zhu KY, Leung KW, Ting AK, et al. The establishment of a highly sensitive method in detecting ketamine and norketamine simultaneously in human hairs by HPLC-Chip-MS/MS. *Forensic Sci Int*. 2011;208(1–3):53–58.
70. Tagliaro F, Bortolotti F, Viel G, Ferrara SD. Caveats against an improper use of hair testing to support the diagnosis of chronic excessive alcohol consumption, following the "Consensus" of the Society of Hair Testing 2009 [Forensic Science International 196 (2010) 2]. *Forensic Sci Int*. 2011;207(1–3):e69–70. author reply e71.
71. Lee S, Han E, In S, Choi H, Chung H, Chung KH. Analysis of pubic hair as an alternative specimen to scalp hair: a contamination issue. *Forensic Sci Int*. 2011;206(1–3):19–21.
72. Lee S, Park Y, Han E, et al. Thebaine in hair as a marker for chronic use of illegal opium poppy substances. *Forensic Sci Int*. 2011;204(1–3): 115–118.
73. Imbert L, Dulaurent S, Mercerolle M, Morichon J, Lachatre G, Gaulier JM. Development and validation of a single LC-MS/MS assay following SPE for simultaneous hair analysis of amphetamines, opiates, cocaine and metabolites. *Forensic Sci Int*. 2014;234:132–138.
74. Wang X, Drummer OH. Review: interpretation of drug presence in the hair of children. *Forensic Sci Int*. 2015;257:458–472.
75. Boumba VA, Di Rago M, Peka M, Drummer OH, Gerostamoulos D. The analysis of 132 novel psychoactive substances in human hair using a single step extraction by tandem LC/MS. *Forensic Sci Int*. 2017;279:192–202.
76. Scaglia F, Longo N. Primary and secondary alterations of neonatal carnitine metabolism. *Semin Perinatol*. 1999;23(2):152–161.
77. Contribution of selected metabolic diseases to early childhood deaths–Virginia, 1996–2001. *MMWR Morb Mortal Wkly Rep*. 2003;52(29): 677–679.

78. Ziadeh R, Hoffman EP, Finegold DN, et al. Medium chain acyl-CoA dehydrogenase deficiency in Pennsylvania: neonatal screening shows high incidence and unexpected mutation frequencies. *Pediatr Res*. 1995;37(5):675–678.
79. Manoukian AA, Ha CE, Seaver LH, Bhagavan NV. A neonatal death due to medium-chain acyl-CoA dehydrogenase deficiency: utilization of the neonatal metabolic screen in a functional approach to sudden unexplained infant death. *Am J Forensic Med Pathol*. 2009;30(3):284–286.
80. Yusupov R, Finegold DN, Naylor EW, Sahai I, Waisbren S, Levy HL. Sudden death in medium chain acyl-coenzyme a dehydrogenase deficiency (MCADD) despite newborn screening. *Mol Genet Metabol*. 2010;101(1):33–39.
81. Pryce JW, Weber MA, Heales S, et al. Postmortem tandem mass spectrometry profiling for detection of infection in unexpected infant death. *Forensic Sci Med Pathol*. 2012;8(3):252–258.
82. Caplan MJ. A practical forensic approach to fatal pediatric endocrinopathies. *Acad Forensic Pathol*. 2016;6(2):258–270.
83. Brooks EG, Gill JR. Members of the national association of medical examiners Ad Hoc committee for B, infectious D. Testing for infectious diseases in sudden unexpected infant death: a survey of medical examiner and coroner offices in the United States. *J Pediatr*. 2015;167(1): 178–182.
84. Davies MW, Mehr S, Garland ST, Morley CJ. Bacterial colonization of toys in neonatal intensive care cots. *Pediatrics*. 2000;106(2):E18.
85. Shivaprakash MR, Rajagopal V, Nagarathna S. Latex agglutination test in the diagnosis of pyogenic meningitis. *J Commun Disord*. 2004;36(2): 127–131.
86. Viswanath G, Praveen, Hanumanthappa AR, Chandrappa NR, Mahesh CB. Bacteriological study of pyogenic meningitis with special reference to latex agglutination. *Indian J Pathol Microbiol*. 2007;50(1):97–100.
87. Sadler DW. The value of a thorough protocol in the investigation of sudden infant deaths. *J Clin Pathol*. 1998;51(9):689–694.
88. Cohen MB. Etiology and mechanisms of acute infectious diarrhea in infants in the United States. *J Pediatr*. 1991;118(4 Pt 2):S34–S39.
89. Chambers HM. The perinatal autopsy: a contemporary approach. *Pathology*. 1992;24(2):45–55.
90. Basinger JM, Fiester SE, Fulcher JW. Mortality from neonatal herpes simplex viremia causing severe hepatitis. *Forensic Sci Med Pathol*. 2019; 15(4):663–666.
91. Astrup BS, Johnsen IB, Engsbro AL. The role of Coxsackievirus A16 in a case of sudden unexplained death in an infant - a SUDI case. *Forensic Sci Int*. 2016;259:e9–13.
92. Oliveira M, Amorim A. Microbial forensics: new breakthroughs and future prospects. *Appl Microbiol Biotechnol*. 2018;102(24):10377–10391.
93. Morris JA, Harrison LM, Partridge SM. Postmortem bacteriology: a re-evaluation. *J Clin Pathol*. 2006;59(1):1–9.
94. Naber. Molecular pathology detection of infectious disease. *NEJM*. 1994;331(18):1212–1215.
95. Olano JP, Walker DH. Diagnosing emerging and reemerging infectious diseases: the pivotal role of the pathologist. *Arch Pathol Lab Med*. 2011; 135(1):83–91.
96. Leggieri N, Rida A, Francois P, Schrenzel J. Molecular diagnosis of bloodstream infections: planning to (physically) reach the bedside. *Curr Opin Infect Dis*. 2010;23(4):311–319.
97. Cathomas G. Molecular diagnostic in infectious disease pathology: an update. *Ann Pathol*. 2009;29(1):S19–S21.
98. Muldrew KL. Molecular diagnostics of infectious diseases. *Curr Opin Pediatr*. 2009;21(1):102–111.
99. Millar BC, Xu J, Moore JE. Molecular diagnostics of medically important bacterial infections. *Curr Issues Mol Biol*. 2007;9(1):21–39.
100. Sibley CD, Peirano G, Church DL. Molecular methods for pathogen and microbial community detection and characterization: current and potential application in diagnostic microbiology. *Infect Genet Evol*. 2012;12(3):505–521.
101. Hosokawa-Muto J, Sakai H, Sassa Y, Fujinami Y, Kishimoto M, Nakahara H. Rapid detection of pathogenic virus genome sequence from throat and nasal swab samples using an exhaustive gene amplification method. *Forensic Sci Med Pathol*. 2019;15(3):399–403.
102. Hensby-Bennett S, Garland J, Philcox W, et al. Rapid Streptococcus pneumonia antigen detection on postmortem urine in a death due to pneumococcal meningitis. *Am J Forensic Med Pathol*. 2019;40(3):269–272.
103. Ackerman MJ, Siu BL, Sturner WQ, et al. Postmortem molecular analysis of SCN5A defects in sudden infant death syndrome. *JAMA*. 2001; 286(18):2264–2269.
104. Di Paolo M, Luchini D, Bloise R, Priori SG. Postmortem molecular analysis in victims of sudden unexplained death. *Am J Forensic Med Pathol*. 2004;25(2):182–184.
105. Oliva A, Pascali VL, Hong K, Brugada R. Molecular autopsy of sudden cardiac death (SCD): the challenge of forensic pathologist to the complexity of genomics. *Am J Forensic Med Pathol*. 2005;26(4):369–370.
106. Basso C, Burke M, Fornes P, et al. Guidelines for autopsy investigation of sudden cardiac death. *Virchows Arch*. 2008;452(1):11–18.
107. Basso C, Carturan E, Pilichou K, Rizzo S, Corrado D, Thiene G. Sudden cardiac death with normal heart: molecular autopsy. *Cardiovasc Pathol*. 2010;19(6):321–325.
108. Campuzano O, Beltran-Alvarez P, Iglesias A, Scornik F, Perez G, Brugada R. Genetics and cardiac channelopathies. *Genet Med*. 2010;12(5): 260–267.
109. Tester DJ, Ackerman MJ. Postmortem long QT syndrome genetic testing for sudden unexplained death in the young. *J Am Coll Cardiol*. 2007; 49(2):240–246.
110. Mazzanti A, Priori SG. Molecular autopsy for sudden unexplained death? Time to discuss pros and cons. *J Cardiovasc Electrophysiol*. 2012; 23(10):1099–1102.
111. Tester DJ, Ackerman MJ. The role of molecular autopsy in unexplained sudden cardiac death. *Curr Opin Cardiol*. 2006;21(3):166–172.
112. Severson E, Timm K. The value of molecular autopsy: genetic testing reveals long QT mutations in an autopsy-negative, postpartum sudden unexpected death. *Acad For Path*. 2014;4(4):552–557.
113. Glengarry JM, Crawford J, Morrow PL, Stables SR, Love DR, Skinner JR. Long QT molecular autopsy in sudden infant death syndrome. *Arch Dis Child*. 2014;99(7):635–640.
114. Shanks GW, Tester DJ, Ackerman JP, et al. Importance of variant interpretation in whole-exome molecular autopsy: population-based case series. *Circulation*. 2018;137(25):2705–2715.
115. Kauferstein S, Kiehne N, Jenewein T, et al. Genetic analysis of sudden unexplained death: a multidisciplinary approach. *Forensic Sci Int*. 2013; 229(1–3):122–127.
116. Cunningham KS. The promise of molecular autopsy in forensic pathology practice. *Acad Forensic Pathol*. 2017;7(4):551–566.

117. Boglioli LR, Taff ML, Turkel SJ, Taylor JV, Peterson CD. Unusual infant death: dog attack or postmortem mutilation after child abuse? *Am J Forensic Med Pathol*. 2000;21(4):389–394.
118. Bernitz H, Owen JH, van Heerden WF, Solheim T. An integrated technique for the analysis of skin bite marks. *J Forensic Sci*. 2008;53(1): 194–198.
119. Bowers CM, Pretty IA. Expert disagreement in bitemark casework. *J Forensic Sci*. 2009;54(4):915–918.
120. Deming JE, Mittleman RE, Wetli CV. Forensic science aspects of fatal sexual assaults on women. *J Forensic Sci*. 1983;28(3):572–576.
121. Tsokos M. Diagnostic criteria for cutaneous injuries in child abuse: classification, findings, and interpretation. *Forensic Sci Med Pathol*. 2015; 11(2):235–242.
122. Page M, Taylor J, Blenkin M. Reality bites–A ten-year retrospective analysis of bitemark casework in Australia. *Forensic Sci Int*. 2012;216(1–3): 82–87.
123. Roshkow JE, Haller JO, Hotson GC, Sclafani SJ, Mezzacappa PM, Rachlin S. Imaging evaluation of children after falls from a height: review of 45 cases. *Radiology*. 1990;175(2):359–363.
124. Michiue T, Ishikawa T, Quan L, Zhu BL, Maeda H. Forensic pathological evaluation of injury severity and fatal outcome in traffic accidents: five illustrative autopsy cases of clinically unexpected death. *Forensic Sci Med Pathol*. 2008;4(3):153–158.
125. Ndiaye A, Chambost M, Chiron M. The fatal injuries of car drivers. *Forensic Sci Int*. 2009;184(1–3):21–27.
126. Callahan K, Knight LD. The pancreas in child abuse. *Acad Forensic Pathol*. 2018;8(2):219–238.
127. Davison AM, Lazda EJ. Small bowel perforation and fatal peritonitis following a fall in a 21-month-old child. *Forensic Sci Med Pathol*. 2008; 4(4):250–254.
128. Hitosugi M, Takatsu A, Shigeta A. Injuries of motorcyclists and bicyclists examined at autopsy. *Am J Forensic Med Pathol*. 1999;20(3):251–255.
129. Jarvis HC, Gill JR. Bicyclist fatalities in New York city. *Acad For Path*. 2015;5(4):667–675.
130. Muggenthaler H, Drobnik S, Hubig M, Fiebig W, Mall G. Fatal abdominal injuries in a bicycle-pedestrian collision - reconstruction using multibody simulation. *Forensic Sci Med Pathol*. 2017;13(2):230–233.
131. Collins KA, Nichols CA. A decade of pediatric homicide: a retrospective study at the Medical University of South Carolina. *Am J Forensic Med Pathol*. 1999;20(2):169–172.
132. deRoux SJ, Prendergast NC. Lacerations of the hepatoduodenal ligament, pancreas and duodenum in a child due to blunt impact. *J Forensic Sci*. 1998;43(1):222–224.
133. Dedouit F, Guilbeau-Frugier C, Capuani C, et al. Child abuse: practical application of autopsy, radiological, and microscopic studies. *J Forensic Sci*. 2008;53(6):1424–1429.
134. Molina DK, Clarkson A, Farley KL, Farley NJ. A review of blunt force injury homicides of children aged 0 to 5 years in Bexar County, Texas, from 1988 to 2009. *Am J Forensic Med Pathol*. 2012;33(4):344–348.
135. Hugar BS, Yajaman GP, Kainoor S, Shetty AR. Death due to isolated jejunal tear following blunt abdominal trauma. *J Forensic Sci*. 2014;59(5): 1420–1422.
136. Erfurt C, Hahn G, Roesner D, Schmidt U. Pediatric radiological diagnostic procedures in cases of suspected child abuse. *Forensic Sci Med Pathol*. 2011;7(1):65–74.
137. Pircher R, Grosse Perdekamp M, Lutz-Bonengel S, Pollak S, Thierauf A. Primarily unrecognized thoracoabdominal impalement in a motorcyclist. *Forensic Sci Int*. 2013;226(1–3):e37–41.
138. Shkrum MJ. Delayed fatal hemorrhage due to small bowel mesenteric laceration. *Am J Forensic Med Pathol*. 2015;36(4):236–238.
139. Levin AV. Fatal pediatric head injuries caused by short distance falls. *Am J Forensic Med Pathol*. 2001;22(4):417.
140. Schaber B, Hart AP, Armbrustmacher V, Hirsch CS. Fatal pediatric head injuries caused by short distance falls. *Am J Forensic Med Pathol*. 2002; 23(1):101–103. author reply 103–105.
141. Root I. Head injuries from short distance falls. *Am J Forensic Med Pathol*. 1992;13(1):85–87.
142. Thompson AK, Bertocci G, Rice W, Pierce MC. Pediatric short-distance household falls: biomechanics and associated injury severity. *Accid Anal Prev*. 2011;43(1):143–150.
143. Dye DW, Peretti FJ, Kokes CP. Histologic evidence of repetitive blunt force abdominal trauma in four pediatric fatalities. *J Forensic Sci*. 2008; 53(6):1430–1433.
144. Zumwalt RE, Hirsch CS. Pathology of fatal child abuse. In: Helfer RE, Kempe RS, eds. *The Battered Child*. 4th ed. Chicago: The University of Chicago Press; 1987:247–285.
145. Baker AL. *Pediatric Fractures Workshop*. AAFS; 2009.
146. Narang S. A Daubert analysis of abusive head trauma/shaken baby syndrome. *Hous J Health L & Pol*. 2011;11:505–632.
147. Shaken baby syndrome: rotational cranial injuries-technical report. *Pediatrics*. 2001;108(1):206–210.
148. Case ME, Graham MA, Handy TC, Jentzen JM, Monteleone JA. Position paper on fatal abusive head injuries in infants and young children. National association of medical examiners Ad Hoc committee on shaken baby syndrome. *Am J Forensic Med Pathol*. 2001;22(2):112–122.
149. Chiesa A, Duhaime AC. Abusive head trauma. *Pediatr Clin*. 2009;56(2):317–331.
150. De Leeuw M, Beuls E, Parizel P, Jorens P, Jacobs W. Confessed abusive blunt head trauma. *Am J Forensic Med Pathol*. 2013;34(2):130–132.
151. Findley K, Barnes P, Moran D, Squier W. *Shaken Baby Syndrome, Abusive Head Trauma, and Actual Innocence: Getting it Right (April 30, 2012)*. University of Wisconsin Legal Studies Research; 2012. Paper Series Paper No. 11952012.
152. Adamsbaum C, Grabar S, Mejean N, Rey-Salmon C. Abusive head trauma: judicial admissions highlight violent and repetitive shaking. *Pediatrics*. 2010;126(3):546–555.
153. Christian CW, Block R. Abusive head trauma in infants and children. *Pediatrics*. 2009;123(5):1409–1411.
154. Case ME. Abusive head injuries in infants and young children. *Leg Med*. 2007;9(2):83–87.
155. Hymel KP, Jenny C, Block RW. Intracranial hemorrhage and rebleeding in suspected victims of abusive head trauma: addressing the forensic controversies. *Child Maltreat*. 2002;7(4):329–348.
156. Shuman MJ, Hutchins KD. Severe retinal hemorrhages with retinoschisis in infants are not pathognomonic for abusive head trauma. *J Forensic Sci*. 2017;62(3):807–811.
157. Binenbaum G, Forbes BJ. The eye in child abuse: key points on retinal hemorrhages and abusive head trauma. *Pediatr Radiol*. 2014;44(Suppl 4): 571–577.

158. Pham H, Enzenauer RW, Elder JE, Levin AV. Retinal hemorrhage after cardiopulmonary resuscitation with chest compressions. *Am J Forensic Med Pathol*. 2013;34(2):122–124.
159. Binenbaum G, Rogers DL, Forbes BJ, et al. Patterns of retinal hemorrhage associated with increased intracranial pressure in children. *Pediatrics*. 2013;132(2):e430–434.
160. Lantz PE, Couture DE. Fatal acute intracranial injury, subdural hematoma, and retinal hemorrhages caused by stairway fall. *J Forensic Sci*. 2011;56(6):1648–1653.
161. Matshes EW. *Retinal and Optic Nerve Sheath Hemorrhages Are Not Pathognomonic of Abusive Head Injury*. Seattle: American Academy of Forensic Sciences; 2010. Paper presented at:.
162. Togioka BM, Arnold MA, Bathurst MA, et al. Retinal hemorrhages and shaken baby syndrome: an evidence-based review. *J Emerg Med*. 2009;37(1):98–106.
163. Ong BB, Gole GA, Robertson T, McGill J, de Lore D, Crawford M. Retinal hemorrhages associated with meningitis in a child with a congenital disorder of glycosylation. *Forensic Sci Med Pathol*. 2009;5(4):307–312.
164. Kivlin JD, Currie ML, Greenbaum VJ, Simons KB, Jentzen J. Retinal hemorrhages in children following fatal motor vehicle crashes: a case series. *Arch Ophthalmol*. 2008;126(6):800–804.
165. Hughes LA, May K, Talbot JF, Parsons MA. Incidence, distribution, and duration of birth-related retinal hemorrhages: a prospective study. *J Am Assoc Pediatr Ophthalmol Strabismus*. 2006;10(2):102–106.
166. Chlebicki MP, Ang B, Barkham T, Laude A. Retinal hemorrhages in 4 patients with dengue fever. *Emerg Infect Dis*. 2005;11(5):770.
167. Schloff S, Mullaney PB, Armstrong DC, et al. Retinal findings in children with intracranial hemorrhage. *Ophthalmology*. 2002;109(8):1472–1476.
168. Pitetti RD, Maffei F, Chang K, Hickey R, Berger R, Pierce MC. Prevalence of retinal hemorrhages and child abuse in children who present with an apparent life-threatening event. *Pediatrics*. 2002;110(3):557–562.
169. Kanter RK. Retinal hemorrhage after cardiopulmonary resuscitation or child abuse. *J Pediatr*. 1986;108(3):430–432.
170. Schoppe CH, Lantz PE. Are peripapillary intrascleral hemorrhages pathognomonic for abusive head trauma? *J Forensic Sci*. 2013;58(1):228–231.
171. Hortobagyi T, Wise S, Hunt N, et al. Traumatic axonal damage in the brain can be detected using beta-APP immunohistochemistry within 35 min after head injury to human adults. *Neuropathol Appl Neurobiol*. 2007;33(2):226–237.
172. Reichard RR, White- III CL, Hladik CL, Dolinak D. Beta-amyloid precursor protein staining of nonaccidental central nervous system injury in pediatric autopsies. *J Neurotrauma*. 2003;20(4):347–355.
173. Dolinak D, Reichard R. An overview of inflicted head injury in infants and young children, with a review of β-amyloid precursor protein immunohistochemistry. *Arch Pathol Lab Med*. 2006;130(5):712–717.
174. Blumbergs PC, Scott G, Vis J, Wainwright H, Simpson D, McLean A. Topography of axonal injury as defined by amyloid precursor protein and the sector scoring method in mild and severe closed head injury. *J Neurotrauma*. 1995;12(4):565–572.
175. Reichard RR, Smith C, Graham DI. The significance of beta-APP immunoreactivity in forensic practice. *Neuropathol Appl Neurobiol*. 2005;31(3):304–313.
176. Gentleman SM, Nash MJ, Sweeting CJ, Graham DI, Roberts GW. Beta-amyloid precursor protein (beta APP) as a marker for axonal injury after head injury. *Neurosci Lett*. 1993;160(2):139–144.
177. Gorrie C, Oakes S, Duflou J, Blumbergs P, Waite PM. Axonal injury in children after motor vehicle crashes: extent, distribution, and size of axonal swellings using beta-APP immunohistochemistry. *J Neurotrauma*. 2002;19(10):1171–1182.
178. Jensen LL, Banner J, Ulhoi BP, Byard RW. beta-Amyloid precursor protein staining of the brain in sudden infant and early childhood death. *Neuropathol Appl Neurobiol*. 2014;40(4):385–397.
179. Rorke-Adams LB. The triad of retinal haemorrhage, subdural haemorrhage and encephalopathy in an infant unassociated with evidence of physical injury is not the result of shaking, but is most likely to have been caused by a natural disease: No. *J Prim Health Care*. 2011;3(2):161–163.
180. Wygnanski-Jaffe T, Morad Y, Levin AV. Pathology of retinal hemorrhage in abusive head trauma. *Forensic Sci Med Pathol*. 2009;5(4):291–297.
181. Levin AV. Retinal hemorrhages: advances in understanding. *Pediatr Clin*. 2009;56(2):333–344.
182. Clark BJ. Retinal hemorrhages: evidence of abuse or abuse of evidence? *Am J Forensic Med Pathol*. 2001;22(4):415–416.
183. Gilliland MG, Luckenbach MW, Chenier TC. Systemic and ocular findings in 169 prospectively studied child deaths: retinal hemorrhages usually mean child abuse. *Forensic Sci Int*. 1994;68(2):117–132.
184. Gilliland MG, Folberg R, Hayreh SS. Age of retinal hemorrhages by iron detection: an animal model. *Am J Forensic Med Pathol*. 2005;26(1):1–4.
185. Matshes EW, Lew EO. Two-handed cardiopulmonary resuscitation can cause rib fractures in infants. *Am J Forensic Med Pathol*. 2010;31(4):303–307.
186. Clouse JR, Lantz PE. Posterior Rib Fractures in infants associated with cardiopulmonary resuscitation (abstract G12). In: *Paper Presented at: 60th American Academy of Forensic Science Annual Meeting; February 2008*. 2008 (Washington, D.C).
187. Dolinak D. Rib fractures in infants due to cardiopulmonary resuscitation efforts. *Am J Forensic Med Pathol*. 2007;28(2):107–110.
188. Maguire S, Mann M, John N, et al. Does cardiopulmonary resuscitation cause rib fractures in children? A systematic review. *Child Abuse Negl*. 2006;30(7):739–751.
189. Love JC, Symes SA. Understanding rib fracture patterns: incomplete and buckle fractures. *J Forensic Sci*. 2004;49(6):1153–1158.
190. Watts P, Maguire S, Kwok T, et al. Newborn retinal hemorrhages: a systematic review. *J AAPOS*. 2013;17(1):70–78.
191. Emerson MV, Pieramici DJ, Stoessel KM, Berreen JP, Gariano RF. Incidence and rate of disappearance of retinal hemorrhage in newborns. *Ophthalmology*. 2001;108(1):36–39.
192. Naqvi A, Raynor E, Freemont AJ. Histological Aging of Fractures in Infants: a practical algorithm for assessing infants suspected of accidental or non-accidental injury. *Histopathology*. 2019;75(1):74–80.
193. Raynor E, Konala P, Freemont A. The detection of significant fractures in suspected infant abuse. *J Forensic Leg Med*. 2018;60:9–14.
194. Lonergan GJ, Baker AM, Morey MK, Boos SC. From the archives of the AFIP. Child abuse: radiologic-pathologic correlation. *Radiographics*. 2003;23(4):811–845.

195. McGraw EP, Pless JE, Pennington DJ, White SJ. Postmortem radiography after unexpected death in neonates, infants, and children: should imaging be routine? *AJR Am J Roentgenol*. 2002;178(6):1517–1521.
196. Case ME. Forensic pathology of child brain trauma. *Brain Pathol*. October 2008;18(4):562.
197. Denton S, Mileusnic D. Delayed sudden death in an infant following an accidental fall: a case report with review of the literature. *Am J Forensic Med Pathol*. 2003;24(4):371–376.
198. Gilliland MG. Interval duration between injury and severe symptoms in nonaccidental head trauma in infants and young children. *J Forensic Sci*. 1998;43(3):723–725.
199. Snoek JW, Minderhoud JM, Wilmink JT. Delayed deterioration following mild head injury in children. *Brain*. 1984;107(Pt 1):15–36.
200. Gill JR, Vincent G, Morotti RA. Toddler fatalities: the terrible twos. *Academic Forensic Pathology*. 2012;2(4):332–337.
201. Ellis PS. The pathology of fatal child abuse. *Pathology*. 1997;29(2):113–121.
202. Ommaya AK, Goldsmith W, Thibault L. Biomechanics and neuropathology of adult and paediatric head injury. *Br J Neurosurg*. 2002;16(3): 220–242.
203. Margulies SS, Thibault KL. Infant skull and suture properties: measurements and implications for mechanisms of pediatric brain injury. *J Biomech Eng*. 2000;122(4):364–371.
204. Vinchon M, de Foort-Dhellemmes S, Desurmont M, Delestret I. Confessed abuse versus witnessed accidents in infants: comparison of clinical, radiological, and ophthalmological data in corroborated cases. *Childs Nerv Syst*. 2010;26(5):637–645.
205. De Leeuw M, Beuls EA, Jorens PG, Parizel PM, Jacobs W. History of an abusive head trauma including a lucid interval and a retinal hemorrhage is most likely false. *Am J Forensic Med Pathol*. 2013;34(3):271–276.
206. Plunkett J. Fatal pediatric head injuries caused by short-distance falls. *Am J Forensic Med Pathol*. 2001;22(1):1–12.
207. Bruce DA, Alavi A, Bilaniuk L, Dolinskas C, Obrist W, Uzzell B. Diffuse cerebral swelling following head injuries in children: the syndrome of "malignant brain edema". *J Neurosurg*. 1981;54(2):170–178.
208. Plunkett J. Lucid interval in infantile fatal head trauma. *Child Abuse Negl*. 1998;22(10):943–946.
209. Case ME. Distinguishing accidental from inflicted head trauma at autopsy. *Pediatr Radiol*. December 2014;44(4):632–640.
210. Lavelle JM, Shaw KN, Seidl T, Ludwig S. Ten-year review of pediatric bathtub near-drownings: evaluation for child abuse and neglect. *Ann Emerg Med*. 1995;25(3):344–348.
211. Feldman KW, Monastersky C, Feldman GK. When is childhood drowning neglect? *Child Abuse Negl*. 1993;17(3):329–336.
212. Alexander R, Crabbe L, Sato Y, Smith W, Bennett T. Serial abuse in children who are shaken. *Am J Dis Child*. 1990;144(1):58–60.
213. Ross AH, Juarez CA. A brief history of fatal child maltreatment and neglect. *Forensic Sci Med Pathol*. 2014;10(3):413–422.
214. Havlik DM, Nolte KB. Fatal "crack" cocaine ingestion in an infant. *Am J Forensic Med Pathol*. 2000;21(3):245–248.
215. Whitehead FJ, Couper RT, Moore L, Bourne AJ, Byard RW. Dehydration deaths in infants and young children. *Am J Forensic Med Pathol*. 1996; 17(1):73–78.
216. Case ME, Graham MA, Handy TC, Jentzen JM, Monteleone JA. National association of medical examiners Ad Hoc committee on shaken baby syndrome. Position paper on fatal abusive head injuries in infants and young children. *Am J Forensic Med Pathol*. June 1, 2001;22(2):112–122.
217. Bhootra BL, Kitinya J. Deaths from accidental steam inhalation during traditional therapy. *J Clin Forensic Med*. 2005;12(4):214–217.
218. Kudchadkar SR, Hamrick JT, Mai CL, Berkowitz I, Tunkel D. The heat is on... thermal epigl ottitis as a late presentation of airway steam injury. *J Emerg Med*. 2014;46(2):e43–46.
219. Feldman KW, Schaller RT, Feldman JA, McMillon M. Tap water scald burns in children. *Pediatrics*. 1978;62(1):1–7.
220. Alexander RT, Fowler DR. Modeling the distribution of scald type burns in a child. *Acad Forensic Pathol*. 2016;6(4):638–656.
221. Moritz AR. Studies of thermal injury: III. The pathology and pathogenesis of cutaneous burns. An experimental study. *Am J Pathol*. 1947;23(6): 915–941.
222. Allasio D, Fischer H. Immersion scald burns and the ability of young children to climb into a bathtub. *Pediatrics*. 2005;115(5):1419–1421.
223. Moritz AR, Henriques FC. Studies of thermal injury: II. The relative importance of time and surface temperature in the causation of cutaneous burns. *Am J Pathol*. 1947;23(5):695–720.
224. Zumwalt R, Petty CS, Holman W. Temperatures in closed automobiles in hot weather. *For Sci Gazette*. 1976;7:7–8.
225. Booth 3rd JN, Davis GG, Waterbor J, McGwin Jr G. Hyperthermia deaths among children in parked vehicles: an analysis of 231 fatalities in the United States, 1999–2007. *Forensic Sci Med Pathol*. 2010;6(2):99–105.
226. Quinton RA. Certification of vehicular hyperthermia deaths in the pediatric population. *Acad Forensic Pathol*. 2016;6(4):657–662.
227. Krous HF, Nadeau JM, Fukumoto RI, Blackbourne BD, Byard RW. Environmental hyperthermic infant and early childhood death: circumstances, pathologic changes, and manner of death. *Am J Forensic Med Pathol*. 2001;22(4):374–382.
228. Grundstein AJ, Duzinski SV, Dolinak D, Null J, Iyer SS. Evaluating infant core temperature response in a hot car using a heat balance model. *Forensic Sci Med Pathol*. 2015;11(1):13–19.
229. Parker KM, Lera Jr TA. Acute isopropanol ingestion: pharmacokinetic parameters in the infant. *Am J Emerg Med*. 1992;10(6):542–544.
230. Baker AM, Johnson DG, Levisky JA, et al. Fatal diphenhydramine intoxication in infants. *J Forensic Sci*. 2003;48(2):425–428.
231. Levine B, Moore KA, Aronica-Pollak P, Fowler DF. Oxycodone intoxication in an infant: accidental or intentional exposure? *J Forensic Sci*. 2004;49(6):1358–1360.
232. Nine JS, Rund CR. Fatality from diphenhydramine monointoxication: a case report and review of the infant, pediatric, and adult literature. *Am J Forensic Med Pathol*. 2006;27(1):36–41.
233. Tardiff K, Marzuk PM, Leon AC, et al. Accidental fatal drug overdoses in New York City: 1990–1992. *Am J Drug Alcohol Abuse*. 1996;22(2): 135–146.
234. Palmiere C, Staub C, La Harpe R, Mangin P. Parental substance abuse and accidental death in children. *J Forensic Sci*. 2010;55(3):819–821.
235. Minera G, Robinson E. Accidental acute alcohol intoxication in infants: review and case report. *J Emerg Med*. 2014;47(5):524–526.
236. Allibe N, Eysseric-Guerin H, Kintz P, et al. Amitriptyline poisoning of a baby: how informative can hair analysis be? *Forensic Sci Int*. 2015;249: 53–58.
237. Somers GR, Chiasson DA, Smith CR. Pediatric drowning: a 20-year review of autopsied cases: III. Bathtub drownings. *Am J Forensic Med Pathol*. 2006;27(2):113–116.

238. Somers GR, Chiasson DA, Smith CR. Pediatric drowning: a 20-year review of autopsied cases: II. Pathologic features. *Am J Forensic Med Pathol*. 2006;27(1):20–24.
239. Somers GR, Chiasson DA, Smith CR. Pediatric drowning: a 20-year review of autopsied cases: I. Demographic features. *Am J Forensic Med Pathol*. 2005;26(4):316–319.
240. Simon HK, Tamura T, Colton K. Reported level of supervision of young children while in the bathtub. *Ambul Pediatr*. 2003;3(2):106–108.
241. Quan L, Cummings P. Characteristics of drowning by different age groups. *Inj Prev*. 2003;9(2):163–168.
242. Alpert B. Bathtub drowning: unintentional, neglect, or abuse. *Med Health R I*. 2003;86(12):385–386.
243. Schmidt P, Madea B. Death in the bathtub involving children. *Forensic Sci Int*. 1995;72(2):147–155.
244. Kemp AM, Mott AM, Sibert JR. Accidents and child abuse in bathtub submersions. *Arch Dis Child*. 1994;70(5):435–438.
245. Mizuta R, Fujita H, Osamura T, Kidowaki T, Kiyosawa N. Childhood drownings and near-drownings in Japan. *Acta Paediatr Jpn*. 1993;35(3): 186–192.
246. Diekema DS, Quan L, Holt VL. Epilepsy as a risk factor for submersion injury in children. *Pediatrics*. 1993;91(3):612–616.
247. Jensen LR, Williams SD, Thurman DJ, Keller PA. Submersion injuries in children younger than 5 years in urban Utah. *West J Med*. 1992;157(6): 641–644.
248. Quan L, Gore EJ, Wentz K, Allen J, Novack AH. Ten-year study of pediatric drownings and near-drownings in King County, Washington: lessons in injury prevention. *Pediatrics*. 1989;83(6):1035–1040.
249. Pearn JH, Brown 3rd J, Wong R, Bart R. Bathtub drownings: report of seven cases. *Pediatrics*. 1979;64(1):68–70.
250. Pearn J, Nixon J. Bathtub immersion accidents involving children. *Med J Aust*. 1977;1(7):211–213.
251. Adelson L. Homicide by starvation: the nutritional variant of the "battered child". *JAMA*. 1963;186(5):458–460.
252. Sarvesvaran ER. Homicide by starvation. *Am J Forensic Med Pathol*. 1992;13(3):264–267.
253. Cooper W, Atherton H, Kahana M, Kotagal U. Increased incidence of severe breastfeeding malnutrition and Hypernatrmeia in a metropolitan area. *Pediatrics*. 1995;96:957–960.
254. Clarke TA, Markarian M, Griswold W, Mendoza S. Hypernatremic dehydration resulting from inadequate breast-feeding. *Pediatrics*. 1979; 63(6):931–932.
255. Evans T, Davies D. Failure to thrive at the breast: an old problem revisited. *Arch Dis Child*. 1977;52:974.
256. Sofer S, Ben-Ezer D, Dagan R. Early severe dehydration in young breast-fed newborn infants. *Isr J Med Sci*. 1993;29(2–3):85–89.
257. Kaplan JA, Siegler RW, Schmunk GA. Fatal hypernatremic dehydration in exclusively breast-fed newborn infants due to maternal lactation failure. *Am J Forensic Med Pathol*. 1998;19(1):19–22.
258. Livingstone VH, Willis CE, Abdel-Wareth LO, Thiessen P, Lockitch G. Neonatal hypernatremic dehydration associated with breast-feeding malnutrition: a retrospective survey. *CMAJ (Can Med Assoc J)*. 2000;162(5):647–652.
259. Madea B. Death as a result of starvation. In: Tsokos M, ed. *Forensic Pathology Reviews; No. 2*. Totowa: Human Press; 2004.
260. Heiger AA. Complacent infantile starvation. *Conn Med*. 1981;45(8):515–516.
261. Gill JR. Pediatric starvation by neglect. *Acad Forens Pathol*. 2013;3(1):46–53.
262. Huser CJ, Smialek JE. Diagnosis of sudden death in infants due to acute dehydration. *Am J Forensic Med Pathol*. 1986;7(4):278–282.
263. Ishimoto H, Jaffe RB. Development and function of the human fetal adrenal cortex: a key component in the feto-placental unit. *Endocr Rev*. 2011;32(3):317–355.
264. Russell MA, Opitz JM, Viseskul C, Gilbert EF, Bargman GJ. Sudden infant death due to congenital adrenal hypoplasia. *Arch Pathol Lab Med*. 1977;101(4):168–169.
265. Rayburn W, Sander C, Barr Jr M, Rygiel R. The stillborn fetus: placental histologic examination in determining a cause. *Obstet Gynecol*. 1985; 65(5):637–641.
266. Bove KE. Practice guidelines for autopsy pathology: the perinatal and pediatric autopsy. Autopsy Committee of the College of American Pathologists. *Arch Pathol Lab Med*. 1997;121(4):368–376.
267. Genest DR, Williams MA, Greene MF. Estimating the time of death in stillborn fetuses: I. Histologic evaluation of fetal organs; an autopsy study of 150 stillborns. *Obstet Gynecol*. 1992;80(4):575–584.
268. Genest DR, Singer DB. Estimating the time of death in stillborn fetuses: III. External fetal examination; a study of 86 stillborns. *Obstet Gynecol*. 1992;80(4):593–600.
269. Sims MA, Collins KA. Fetal death. A 10-year retrospective study. *Am J Forensic Med Pathol*. 2001;22(3):261–265.
270. Pommier S, Adalian P, Gaudart J, Panuel M, Piercecchi-Marti MD, Leonetti G. Fetal age estimation using orbital measurements: 3D CT-scan study including the effects of trisomy 21. *J Forensic Sci*. 2009;54(1):7–12.
271. Silingardi E, Santunione AL, Rivasi F, Gasser B, Zago S, Garagnani L. Unexpected intrauterine fetal death in parvovirus B19 fetal infection. *Am J Forensic Med Pathol*. 2009;30(4):394–397.
272. Bhale CP, Vare A, Gupta A. Fetal autopsy-categories and causes of death at a tertiary care center. *Am J Forensic Med Pathol*. 2021;42(1):12–15.
273. Phillips B, Ong BB. "Was the infant born alive?" A review of postmortem techniques used to determine live birth in cases of suspected neonaticide. *Acad Forensic Pathol*. 2018;8(4):874–893.
274. Lavezzi WA, Keough KM, Der'Ohannesian P, Person TL, Wolf BC. The use of pulmonary interstitial emphysema as an indicator of live birth. *Am J Forensic Med Pathol*. 2003;24(1):87–91.
275. Lavezzi WA, McKenna BJ, Wolf BC. The significance of pulmonary interstitial emphysema in live birth determination. *J Forensic Sci*. 2004; 49(3):546–552.
276. Marchetti D, Belviso M, Marino M, Gaudio R. Evaluation of the placenta in a stillborn fetus to estimate the time of death. *Am J Forensic Med Pathol*. 2007;28(1):38–43.
277. Causes of death among stillbirths. *JAMA*. 2011;306(22):2459–2468.
278. Warren MW. Radiographic determination of developmental age in fetuses and stillborns. *J Forensic Sci*. 1999;44(4):708–712.
279. Bunai Y, Nagai A, Nakamura I, Ohya I. Fetal death from abruptio placentae associated with incorrect use of a seatbelt. *Am J Forensic Med Pathol*. 2000;21(3):207–209.
280. Thureen PJ, Hall DM, Hoffenberg A, Tyson RW. Fatal meconium aspiration in spite of appropriate perinatal airway management: pulmonary and placental evidence of prenatal disease [see comments]. *Am J Obstet Gynecol*. 1997;176(5):967–975.

281. Catanese CA, Gilmore K. Fetal gunshot wound characteristics. *J Forensic Sci*. 2002;47(5):1067–1069.
282. Stenton S, Cohen MC. Assessment of neonaticide in the setting of concealed and denied pregnancies. *Forensic Sci Med Pathol*. 2020;16(2): 226–233.
283. Collins KA, Popek E. Birth injury: birth asphyxia and birth trauma. *Acad Forensic Pathol*. 2018;8(4):788–864.
284. Cohen MC, Scheimberg I. Forensic aspects of perinatal deaths. *Acad Forensic Pathol*. 2018;8(3):452–491.
285. *American Academy of Pediatrics and American College of Obstetricians and Gynecologists, Guidelines for Perinatal Care*. 4th ed. 1997. Elk Grove Village, IL and Washington, DC.
286. Guddat SS, Gapert R, Tsokos M, Oesterhelweg L. Proof of live birth using postmortem multislice computed tomography (pmMSCT) in cases of suspected neonaticide: advantages of diagnostic imaging compared to conventional autopsy. *Forensic Sci Med Pathol*. 2013;9(1):3–12.
287. Yilmaz R, Erkaymaz O, Kara E, Ergen K. Use of autopsy to determine live or stillbirth: new approaches in decision-support systems. *J Forensic Sci*. 2017;62(2):468–472.
288. Mazuchowski EL, Franco DM, Berran PJ, Harcke HT. The virtual Hydrostatic test. *Am J Forensic Med Pathol*. 2017;38(1):24–28.
289. Mittal N, Byard RW, Dahlstrom JE. A practical guide to placental examination for forensic pathologists. *Forensic Sci Med Pathol*. 2020;16(2): 295–312.
290. Ananth CV, Lavery JA, Vintzileos AM, et al. Severe placental abruption: clinical definition and associations with maternal complications. *Am J Obstet Gynecol*. 2016;214(2):272 e271–279.
291. Baber Y, Iles L, O'Donnell C. Determining live birth: an unusual case. *Am J Forensic Med Pathol*. 2015;36(4):262–264.
292. Johnson DJ, Matthies LK, Roberts KA, Yorker BC. Isolation and individualization of conceptus and maternal tissues from abortions and placentas for parentage testing in cases of rape and abandoned newborns. *J Forensic Sci*. 2010;55(6):1430–1436.
293. Oyelese Y, Ananth C. Placental abruption. *Obset Gynecol*. 2006;108:1005–1016.
294. Stallmach T, Hebisch G. Placental pathology: its impact on explaining prenatal and perinatal death. *Virchows Arch*. 2004;445(1):9–16.
295. Bove. Practice guidelines autopsy: perinatal and pediatric autopsy. *Arch Pathol Lab Med*. 1997;121:368–376.
296. Tantius B, Rothschild MA, Valter M, Michael J, Banaschak S. Experimental studies on the tensile properties of human umbilical cords. *Forensic Sci Int*. 2014;236:16–21.
297. Evans SR, Mileusnic-Polchan D. Patterns of breaks in umbilical cords by different mechanisms. *J Forensic Sci*. 2012;57(5):1222–1225.
298. Gheorghe A, Banner J, Hansen SH, Stolborg U, Lynnerup N. Abandonment of newborn infants: a Danish forensic medical survey 1997–2008. *Forensic Sci Med Pathol*. 2011;7(4):317–321.
299. Crandall LG, Reno L, Himes B, Robinson D. The diagnostic shift of SIDS to undetermined: are there unintended consequences? *Acad Forensic Pathol*. 2017;7(2):212–220.
300. Garstang J, Ellis C, Griffiths F, Sidebotham P. Unintentional asphyxia, SIDS, and medically explained deaths: a descriptive study of outcomes of child death review (CDR) investigations following sudden unexpected death in infancy. *Forensic Sci Med Pathol*. 2016;12(4):407–415.
301. Mage DT. An anomaly in U.S. SIDS data reported in the CDC wonder.cdc.gov mortality database. *Forensic Sci Med Pathol*. 2013;9(2):283.
302. Matturri L, Ottaviani G, Benedetti G, Agosta E, Lavezzi AM. Unexpected perinatal death and sudden infant death syndrome (SIDS): Anatomopathologic and legal aspects. *Am J Forensic Med Pathol*. 2005;26(2):155–160.
303. Matthews T. Sudden unexpected infant death: infanticide or SIDS? *Lancet*. 2005;365(9453):3–4.
304. Byard RW, Ranson D, Krous HF, Workshop P. National Australian workshop consensus on the definition of SIDS and initiation of a uniform autopsy approach to unexpected infant and early childhood death. *Forensic Sci Med Pathol*. 2005;1(4):289–292.
305. Nashelsky MB, Pinckard JK. The death of SIDS. *Academic Forensic Pathology*. 2011;1(1):92–99.
306. Sawaguchi T, Nishida H. SIDS doesn't exist. *Am J Forensic Med Pathol*. 2001;22(2):211–212.
307. Gibson AA. Current epidemiology of SIDS. *J Clin Pathol*. 1992;45(11 Suppl):7–10.
308. SUDC Foundation Website. https://sudc.org/#.
309. Corey TS, Hanzlick R, Howard J, Nelson C, Krous H. A functional approach to sudden unexplained infant deaths. *Am J Forensic Med Pathol*. 2007;28(3):271–277.
310. Case ME. Inflicted traumatic brain injury in infants and young children. *Brain Pathol*. October 2008;18(4):571–582.
311. Morad Y, Avni I, Capra L, et al. Shaken baby syndrome without intracranial hemorrhage on initial computed tomography. *J Am Assoc Pediatr Ophthalmol Strabismus*. December 1, 2004;8(6):521–527.
312. Case ME. Accidental traumatic head injury in infants and young children. *Brain Pathol*. October 2008;18(4):583–589.

CHAPTER

7

Approach to natural deaths (adult)

Susan F. Ely and James R. Gill

Common things are common because they are the things that occur with the greatest frequency. -*Dr. Charles S. Hirsch*

Despite the public perception of medical examiner and coroner (ME/C) work, natural deaths comprise a large part of any forensic pathology practice. Many deaths are reported to the ME/C, however, not all are accepted for further investigation. Many persons who likely died from natural causes, except in specific scenarios (e.g., in-custody or unexpected deaths) are not autopsied, and for still others (most dying in hospital), their deaths are not even certified by the ME/C. Some deaths initially may appear suspicious or otherwise unnatural (e.g., an elderly woman on blood thinners found at the bottom of the stairs), but then prove to be a natural death at autopsy. Other deaths may appear natural (e.g., a chronic alcoholic with heart disease found dead in bed), but turn out to be due to occult trauma (e.g., a subdural hematoma).

One indication for an autopsy is to rule out occult fatal trauma (e.g., chronic alcoholics with minor external injury, infants); in many of those instances, trauma is indeed excluded by an autopsy.[1] This is similar to the surgical approach for a suspected acute appendicitis. It is professionally acceptable for surgeons to find some normal appendices when performing surgery for a suspected acute appendicitis. Preoperative diagnostic acumen is not 100%, and the risk of missing one ruptured appendix outweighs the risk of removing a normal one. The same concept and approach apply to forensic pathology.

In addition to in-custody deaths, the natural and medical deaths that are usually investigated by the ME/C include those with a suspected threat to public safety (e.g., suspected, but undiagnosed bacterial meningitis), deaths occurring outside the attendance of the medical professional or medical facility (deaths at home, in the workplace, in public places, etc.), deaths related to medical therapy or diagnostic procedures (see Chapter 8), and those that are unexpected and/or appear to be potentially unnatural (e.g., fall in an elderly patient or a chronic alcoholic with a suspected head injury). The key point in the certification of these deaths is to identify the proximate (underlying) cause. The death certificate of a person who dies of an upper gastrointestinal hemorrhage (GIH) must include the disease that caused the bleeding (e.g., gastric cancer, peptic ulcer disease, ruptured esophageal varix, etc.). ME/C are experts in how to certify all causes and manners of death. Therefore, the forensic pathologist (FP) also may play an important role in guiding clinicians in the proper certification of natural deaths, as there still exist substantial shortcomings in death certifications by some clinicians, largely based on a deficiency in basic medical training in this specific area.[2–29]

The main difference between many of the natural deaths that are ultimately referred to the ME/C and those that are not is either the precipitousness with which the former occur and/or the failure of some patients to seek medical care, resulting in out-of-hospital deaths. These persons may present to the ME/C with such "instantaneously" lethal, Class I findings as a saddle pulmonary thromboembolus, a ruptured aortic dissection, spilling several liters of blood into the thorax within minutes, or a ruptured berry aneurysm in the circle of Willis, causing an immediate "tamponade" of the brainstem, thus paralyzing vital cardiac and respiratory function within seconds. None of these scenarios affords the patient the luxury of time to even call 911, much less receive diagnostic and therapeutic medical attention. Rather, these individuals are found dead at home, at work, or in the street, confounding family, coworkers, and strangers with the dramatic and sudden quality of their death.

These sudden fatal outcomes are augmented by the "silent" nature of fairly common conditions and diseases that also happen to have a high lethal potential. These include, but are not limited to, asymptomatic deep venous thromboses (that may ultimately dislodge and transform into a saddle pulmonary thromboemboli [PTE]), hypertensive cardiovascular disease (HCVD) with consequent cardiac hypertrophy (leading to sudden onset arrhythmias,

https://doi.org/10.1016/B978-0-323-91796-4.00004-0

ruptured aortic dissections, or hemorrhagic basal ganglia strokes, without warning or prodrome), or coronary atherosclerosis, which, until severe enough to become symptomatic, maintains its lethal potential, but is silent. The way in which coronary atherosclerosis usually kills precipitously (i.e., not allowing time to reach medical attention) is through an ischemia-induced cardiac arrhythmia.

The lethal potential of these conditions explains the "sudden" nature of these deaths; the clinical silence prior to death explains their "unexpected" quality. An alternative explanation, however, for why certain individuals die suddenly and out of hospital lies not necessarily in the occult nature of their pathologic conditions, but rather in the behaviors, habits, and social constructs that keep them away from doctors. A high pain tolerance/stoicism, denial of symptoms, fear of disease, "white coat" fear, and/or living in isolation with little help, assistance, or visitation offer just a few explanations. After excluding neurologic, psychiatric, or advanced age-related incapacity, how else does one explain the finding of a ruptured myocardial infarct (Fig. 7.1) in a person having been found "suddenly and unexpectedly" dead, prior to which they were reported to have been in their "usual state of health"? To die with such a Class I lesion, this person would have had to endure days of crushing chest pain, having been unable or unwilling to seek help, or have been self-medicating.

Conditions and diseases having a more protracted and well-defined, but not immediately fatal course have more of a potential for ending in hospital deaths (myriad complications of congestive heart failure, pneumonia, renal failure, or diabetes, for example). Unless a hypertensive cerebral hemorrhage involving deep and critical structures of the basal ganglia or pons, most cerebrovascular events (CVEs), or "strokes," elsewhere in the brain cause more morbidity than sudden mortality, per se. Many of these patients develop sudden onset symptoms that are not immediately fatal, allowing them time to reach medical attention and undergo a diagnostic work up. As such, when fatal, these too tend to fall into the well-documented hospital death category. As a point of clarification, the authors prefer to refer to these cerebral vascular phenomena as CVEs rather than cerebrovascular *accidents* (CVAs), the latter terminology creating potential confusion for the nosologist in vital records.[30] By the time many of these persons die in hospital, after days or weeks or even months of diagnostic and therapeutic medical attention, their natural deaths are neither remotely sudden nor unexpected and the ME/C is never called. Stated differently, there is a selection bias for the quality of lethal natural conditions and characteristics of patients seen by in an average ME/C office.

The sudden and unexpected death

> There is an abrupt, dramatic transition from the glowing activity of life to the lasting stillness of death which requires explanation.
> *Dr. Lester Adelson.*[31]

Truly sudden natural deaths are largely due to a handful of diseases that involve the heart and the brain. The short list includes cardiovascular (e.g., atherosclerosis, hypertension, aortic stenosis, ruptured aortic aneurysms) and neurologic diseases (e.g., ruptured cerebral artery aneurysm, hemorrhagic pontine strokes). William Boyd noted, "Of all the aliments which may blow out life's little candle, heart disease is the chief."[32] Most diseases have the potential for causing a "sudden" death, although there is often an identifiable prodrome, if a careful history can be obtained. Pulmonary emphysema with a ruptured bleb and subsequent pneumothorax can cause a rapid death. A large GIH due to ruptured esophageal varices can also be quite rapidly fatal.

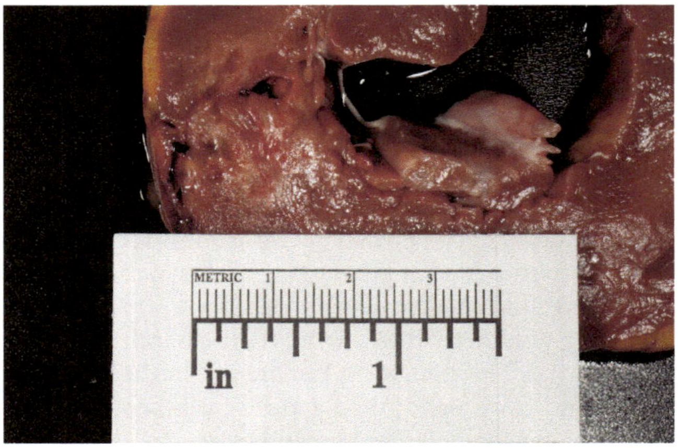

FIGURE 7.1 **Class I finding: ruptured myocardial infarct.** Note the defect in the lateral wall of the left ventricle.

Sudden cardiac death (SCD) is an unexpected death due to cardiac causes that occurs within a limited period (usually within 1 h of the onset of symptoms or without any antecedent symptoms).[33] An important component of the history is that prior to this event, the person appeared in their usual state of health and was performing everyday activities. It has been described as, "like being struck by lightning on a sunny day."[34] This specific set of circumstances causes understandable confusion and heartache for family members. They have a particularly hard time coming to terms with this discordance: the vibrancy of life interrupted without warning and leaving the stillness of death. In the United States, cardiovascular disease accounts for approximately one out of every three deaths.[35] Almost half of these cardiac deaths are SCDs and many are the first, and unfortunately only presentation of this disease.[36]

The national annual incidence of SCD ranges from 180,000 to >450,000 per year.[37] This wide range is due to the various case definitions, ascertainment criteria, data sources, and estimation methods.[37] Hospital-based studies may underestimate them because some patients never make it to the hospital. The use of death certificates also presents challenges, as some clinicians seem to exclusively focus on the cardiovascular system when certifying *any* death because, as they reason, the person died because "the heart stopped." As pointed out in Chapter 2, this logic represents the antithesis of proper death certification and also potentially artificially inflates the reported incidence of SCD.[19,20] The forensic pathologist often becomes involved in the investigation of these deaths since they are sudden and often unexpected, the latter at least in the eyes of family members. The criteria of "sudden and unexpected" are often part of the reporting rules and jurisdictional mandates for medicolegal death investigation systems.[37]

First, we will define each of the components of SCD and how history can be very helpful in determining the mechanism and cause of death, but also over which deaths the ME/C even needs to take jurisdiction.

Sudden

Although death within one hour of symptom onset is typically included in the definition of a SCD, there is no broadly agreed upon interval and some studies use 6 or even 24 h intervals.[38,39] Instead of focusing on a specific interval and the added confusion over the length of resuscitation or agonal survival periods, some prefer the term **"instantaneous"** (like switching off a light) to describe the degree of suddenness that is emblematic of these deaths.[39] For the forensic pathologist, the rapidity of the incapacitating event is the most important component and, absent stimulant drugs, the list of diseases or natural mechanisms that present in this specific fashion is short: a *cardiac arrhythmia* (due to a variety of endogenous cardiac pathologies or toxic-metabolic derangements), a *saddle pulmonary thromboembolism*, and *a ruptured cerebral aneurysm or (usually hemorrhagic) stroke involving or otherwise affecting the brainstem.* A history of recent noncardiac signs or symptoms, such as fever or weight loss, would likely result in further investigation (including perhaps an autopsy, cultures, etc.) since it would be unlikely, but not impossible, to have two fatal mechanisms exerting their effects at the same time.

Rapid medical responses and cardiac defibrillators in public locations may prevent, reverse, or at least delay some of these sudden deaths. Successful cardiac defibrillation may restart the heart, but if the cerebral ischemic interval was long enough, it may all be for naught. It is not unusual to "survive"/have circulation restored and sustained on a respirator for a few days until death by neurological criteria is established. Despite this extended interval, some would consider this a SCD. It would technically, however, be classified as a **sudden cardiac arrest (SCA),** not a SCD.[37] Improvements in critical care and advanced life support measures have set the stage for nomenclature that, by necessity, distinguishes these two categories by virtue of the availability of medical treatment, not necessarily by any substantial differences in their pathophysiology.

As such, **SCA** and **SCD** have different definitions.[37] According to the American Heart Association and the American College of Cardiology guidelines, **SCA** is defined as the abrupt loss of cardiac function *"from an unexpected circulatory arrest, usually due to a cardiac arrhythmia occurring within an hour of the onset of symptoms, in which medical intervention (e.g., defibrillation) reverses the event,"* while **SCD** is defined as *"death from an unexpected circulatory arrest, usually due to a cardiac arrhythmia occurring within an hour of the onset of symptoms."*[40] The difference between these two entities in many cases may be more circumstantial than pathogenic.

Unexpected

Sudden death is not always unexpected. In many jurisdictions, the reportable deaths are those that are *unexpected* rather than sudden. Some clinicians may refuse to certify these deaths because of the absence of a recent medical evaluation or a known potentially lethal disease. Not all sudden or "unexpected" deaths automatically require a medicolegal investigation or an autopsy. If there is a compelling medical history and clear circumstances, an autopsy may not be necessary.[41] For example, a patient with terminal metastatic lung carcinoma may "suddenly" die from a pulmonary thromboembolism. Although sudden, it is not necessarily unexpected. *The opinion of what is considered "unexpected," however, may differ between the death investigator, the treating physician, and the next of kin.* The next of kin understandably and very frequently view the death of a loved one as unexpected, often wanting and desperately needing to know why the person died. They not uncommonly believe that an autopsy is the only means to arrive at those answers with any precision, not understanding, as nonphysicians, the nuances and lethal potentials of cardiac and other diseases. These concerns may be considered by the ME/C, even if these deaths may be reliably certified without an autopsy. But, in the end, the ME/C holds the final discretion over the autopsy decision-making. That said, with careful and detailed discussions, the ME/C can often more than adequately explain natural causes and mechanisms of death in the absence of performing an autopsy, dissipating a family member's sometimes intense concerns and confusion in the process.

Cardiac

There are numerous causes of **sudden unexpected natural death**. Similar to William Boyd's observation above, Charles Hirsch noted, "Coronary artery disease is the captain of the ship of sudden death."[42] But, *sudden, unexpected natural death includes both cardiac and noncardiac* (e.g., ruptured cerebral artery aneurysm) *etiologies*.[38,39,43] Some include pulmonary thromboembolism and aortic disease (e.g., aortic dissection) ac SCDs. We will focus on the primary **cardiac causes of sudden death** since they result in the majority of such deaths. In addition, the proximate causes of the majority of aortic aneurysms and dissections are atherosclerosis and hypertension, respectively. *Forensic dogma expects 60% of medicolegal autopsies to be natural deaths, of which 60% will be due to cardiac disease. Among the cardiac causes,* **90% will be due to atherosclerosis.**[38,43] The other common cardiac causes include: hypertensive cardiovascular disease (HCVD), congenital coronary artery abnormalities, valvular disease (e.g., aortic stenosis), myocarditis, dilated or hypertrophic cardiomyopathies, and myriad genetic causes (e.g., long QT syndrome, Brugada syndrome [BrS], and channelopathies).[38,39,43]

The first step in the understanding of SCD due to coronary artery atherosclerosis is to realize that it is *not* always associated with a coronary artery thrombosis or an acute myocardial infarct.[44] In fact, commonly, there is no *acute* finding at autopsy. The detection of an acute myocardial infarct or coronary artery thrombosis is *not* required to invoke atherosclerosis as the cause of death. Myocardial infarct is not a synonym for SCD. Some of these deaths occur so rapidly that overt pathologic changes do not have time to develop. The likelier mechanism in most instances is myocyte ischemia (due to an imbalance of the myocardial oxygen supply and demand typically seen in the setting of fixed stenoses) that causes a ventricular arrhythmia (fatal tachyarrhythmias, such as ventricular fibrillation, are more common than fatal bradyarrhythmias in the setting of ischemia).[45] In addition, the detection of a lethal, acute cardiac process does not necessarily remove the need to perform toxicological analysis in these deaths, depending on the circumstances. *Acute and/or chronic cocaine and methamphetamine use may contribute* to or directly cause some of these apparently sudden natural cardiac deaths. If that were the case, it would change the manner of death from natural to accident.[46,47]

People can die suddenly from heart disease with a grossly and microscopically normal heart.[45,48] Molecular testing of some of these hearts that are "too good to die" not uncommonly detect molecular genetic explanations for the deaths. Tester and Ackerman studied 49 sudden, unexpected deaths in decedents (1–43 years old) who underwent a medicolegal autopsy and were referred for molecular testing. Molecular analyses of these autopsy-negative, sudden unexplained deaths (SUDs) detected putative cardiac channel mutations (channelopathies) in 35% of the decedents.[49]

Structurally, the heart should be analyzed in four components: coronary arteries, myocardium, cardiac valves, and the conduction system. Common findings/diseases can be categorized within these four categories and one can work down the differential diagnostic list for each while performing the examination and conducting an investigation.

Coronary arteries

While examining the coronary arteries, one should examine them to see if their origins and courses are normal, and if there are atherosclerotic stenoses, dissections, aneurysms, or vasculitidies.[50] The examinations performed by a clinician and an autopsy prosector each have their advantages and disadvantages. If there is a survival interval, clinicians can evaluate functional parameters such as contractility and valve function (echocardiography), electrical alterations (electrocardiogram [EKG]), myocyte damage (serologic markers), and vascular stenoses (catheterization). Pathologists are able to examine the static heart grossly, microscopically, and molecularly, but can only infer functional derangements. In addition to atherosclerosis, there are other coronary artery pathologies, including various types of anomalies (e.g., acute angle take-offs), arteritidies, and coronary artery dissections, classically, if rarely, seen in postpartum women.[38,51–55]

Myocardium

During the autopsy of a person having suffered a sudden death, one looks to the myocardium for acute or remote infarcts, myocytes with hypertrophy or disarray, amyloidosis, and myocarditis (Fig. 7.2), including sarcoidosis. Myocyte hypertrophy is a sign of HCVD or hypertrophic cardiomyopathy (HCM). A number of associated mutations have been identified as hypertrophic cardiomyopathies that primarily involve genes encoding for sarcomeric proteins.[38,56–61] Care must be taken to not overcall disarray in sections of the septum, as this area of the heart may show nonpathological myocyte "disarray" due to the normal intersecting arrangement of myocytes (Fig. 7.3).[38,56,62]

Myocarditis is a histological diagnosis made by the patchy detection of infiltrating lymphocytes and myocyte necrosis. While it has multiple causes, it is most commonly of viral etiology.[63–68] Progressive scarring from myocarditis occurs over time and may result in delayed heart failure and structural changes (a less common chronic phase, if complete resolution does not ensue). These patients often present in congestive heart failure, and this is supported by the pathology that is typically seen by the hospital prosector. By contrast, some of these patients die suddenly during the more active/acute phase of the disease; the mechanism in these deaths is most likely arrhythmogenic in nature. These sudden, fatal arrhythmias are presumed to occur as the myocardium becomes unstable/irritated due to the inflammation, edema, myocyte necrosis, and fibrosis, and often come to the attention of the ME/C predicted on their suddenness and lack of a preexisting diagnosis (unexpectedness). Because this may occur in the early course of the inflammatory process, one must adequately sample the heart (eight to ten sections) so as not to miss these patchy inflammatory foci.

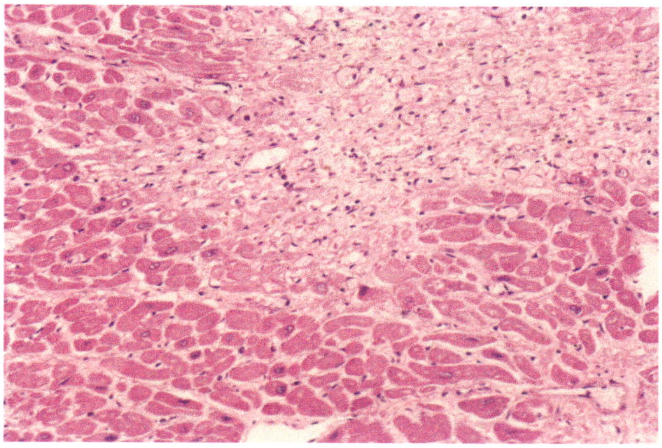

FIGURE 7.2 **Myocarditis.**

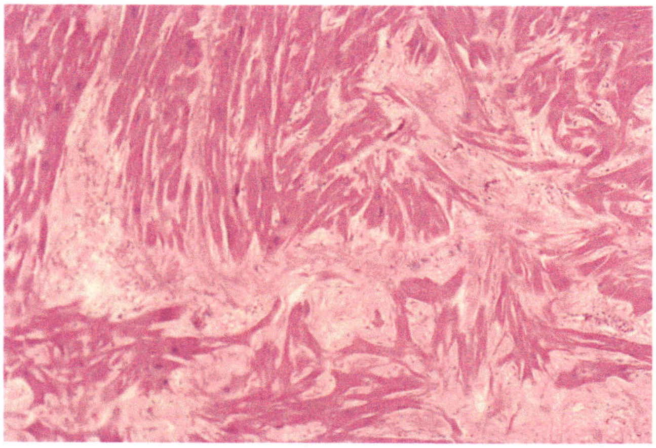

FIGURE 7.3 **Myocyte disarray of HCM.**

Molecular studies have discovered mutations for structural cardiomyopathies (myosin, troponin) and myocyte electrolyte channels (sodium, chloride, calcium) that have further elucidated syndromes that were previously only recognized by clinical and electrocardiographic features (e.g., BrS, long QT syndrome).[49,69–74] These "channelopathies" comprise a group of genetic diseases affecting the electrical system of the heart and are caused by mutations in genes encoding ion channel proteins that are a recognized cause of SCD in the general population.[70,73] Cardiac "channelopathies" should be sought when no structural cardiac abnormality or extra-cardiac cause of death has been identified in the sudden and unexpected death of a previously healthy person with an otherwise negative autopsy (see also the "Molecular Genetics" section in Chapter 19).[49,73,75–80]

Cardiac valves

Sudden death related to valvular disease is commonly due to aortic stenosis or chordal rupture of the mitral valve. Since the etiology of the valvular disease is the proximate cause of death, one must determine if the aortic stenosis is due to a bicuspid congenital anomaly, senile degenerative/dystrophic calcification, a history of rheumatic heart disease, etc. For the flail mitral valve, the proximate cause of death may be mitral valve prolapse (MVP) or a myocardial infarct involving a papillary muscle due to coronary artery atherosclerosis.

Over time, aortic stenosis results in systolic overload and left ventricular hypertrophy. Common prodromal signs and symptoms include exertional angina, syncope, and dyspnea. The risk of sudden death in asymptomatic adults is low, but does occur.[81–83] The mechanism for sudden death is related to cardiac hypertrophy and microscopic myocardial ischemia, with a consequent arrhythmia.[84] At autopsy, subendocardial ischemia (as evidenced by myocytolysis) represents the arrhythmogenic substrate.

MVP is due to myxomatous degeneration of the mitral valve and is one of the most common cardiac abnormalities in the general population. It is usually benign and asymptomatic, but associated arrhythmias and sudden deaths have been rarely described[38,56,85–88] Since MVP is such a common disorder, it is likely that some of these previously reported "MVP" deaths may actually have been caused by an unrecognized channelopathy.

The conduction system

In addition to the myocardium, the sinoatrial and atrioventricular nodes and bundle of His constitute the major components of the conduction system of the heart. Ischemia (nodal artery dysplasia) or myocardial lesions (e.g., sarcoidosis, amyloidosis) may interfere with signal transmission.[53,89–93] In select instances, microscopic examination of the conduction system may be of benefit, but systemic processes (e.g., sarcoidosis, amyloidosis, myocarditis) are unlikely to involve the conduction system alone.[92,94,95]

Clinicians frequently equate mechanisms of death with proximate causes of death. Since there is an ultimate cardiac arrhythmia in virtually all deaths, terminal cardiac events are often used to certify the cause of death. Clinicians may use "myocardial infarct" as the cause of death for any sudden death. This results in the potential for an overcertification of deaths due to "atherosclerotic heart disease."[96]

Death from **HCVD** is an under recognized and underreported cause of death. Most clinicians focus on coronary artery disease or structurally demonstrable *immediate* causes of cardiac death (e.g., aortic dissection) instead of the underlying cause (i.e., HCVD). Moreover, a clinical history of hypertension is not required to make a postmortem diagnosis in the appropriate setting (i.e., findings of cardiac hypertrophy in combination with renal arteriolosclerosis/small vessel disease) at autopsy. See also Chapter 2.

A dissecting aneurysm is not an etiologically specific disease; it is an immediate cause. It may be caused by HCVD, Marfan syndrome, syphilis, or an acute cocaine intoxication. Therefore, to be consistent and to carry out a competent death certification, the underlying cause for the arterial dissection should be sought and indicated. *The majority of aortic dissections and spontaneous intracerebral hemorrhages (typically of the basal ganglia, pons, or cerebellum) are due to HCVD, and this underlying cause should be indicated on the death certificate.* Molecular studies for aortopathies are also available in searches to identify hereditary syndromes (see Chapter 19).

Failure to diagnose HCVD at autopsy in the absence of a clinical history of hypertension is a "manifestation of caution-induced paralysis of judgment." It is not possible to take a blood pressure after death; however, we submit that cardiac hypertrophy in conjunction with renal arteriolosclerosis, with or without additional causes, represents the pathology of chronic hypertension. Other causes of cardiac hypertrophy include aortic stenosis (e.g., bicuspid valve, rheumatic disease, degenerative calcific pathology), chronic alcohol use, or chronic cocaine or other stimulant use; any of these may represent primary or contributory factors in the right setting. *The absence of small vessel disease may help distinguish HCVD from chronic cocaine abuse or other causes.*

When the autopsy findings confirm the foregoing criteria, the diagnosis of HCVD may be made, regardless of the availability of clinical documentation. Since blood pressure is not static, a clinical history of a "normal" blood pressure should not necessarily deter the FP from making a diagnosis of HCVD. Some people with chronic hypertension are normotensive at rest, the labile blood pressure easily fluctuating in response to many variables. The body doesn't lie; we must believe what we see.

The foregoing discussion presents a diagnostic scheme for SCDs involving coronary atherosclerosis and/or HCVD for body weights between 100 lb. (45.5 Kg) and 200 lb. (90.9 Kg). **A common rule of thumb is that the upper limit of normal heart weight is 0.5% of body mass (or 5 gm per kg body mass, 5 gm/kg × 70 kg = 350 gm),**[97] **excluding persons with marked obesity, a heart mass in excess of that upper limit defines cardiac hypertrophy, regardless of the etiology, and results in enhanced arrhythmogenicity that is proportionate to the extent of the increase, even in the absence of left ventricular hypertrophy,** per se. With rare exceptions (athletes) and irrespective of body mass, however, **any heart weighing over 500 g is likely relatively arrhythmogenic.**[98–102]

Beyond an arrhythmogenic mechanism, there are *several other ways that HCVD kills* in structurally unequivocal ways, often leading to Class I findings that are incompatible with continued life. These include **ruptured aortic dissections, hypertensive strokes (basal ganglia, pons, or cerebellum, in decreasing order of frequency), and ruptured saccular (berry) aneurysms,** the latter particularly when in the circle of Willis, adjacent to vital brainstem centers. HCVD is the likely proximate cause in all but the last. *Ruptured berry aneurysms are etiologically specific*, so can "stand alone" on the death certificate. They are congenital vascular abnormalities that tend to enlarge over the course of a lifetime. Still, many die *with* them rather than *from* them, without ever having been the wiser. That said, while HCVD, per se, does not constitute a chronic disease needed to explain the formation of a berry aneurysm, activities that precipitously cause a surge in blood pressure *are* usually implicated. These activities may range from the stress and enjoyment of daily living to criminal activity. If the former, these deaths are classified as natural, with the culpable benign activities *not* appearing on the death certificate, per se. If criminal, however, the functional injury (criminal stress) becomes part of the cause of death statement (see Chapters 3 and 15—"homicide-by-heart attack, or stroke"). Pulmonary hypertension, primary more than secondary, may occasionally cause sudden death. At autopsy, it is largely a microscopic diagnosis; however, one may see gross atherosclerotic plaques of the pulmonary arteries near the hilum (Fig. 7.4).

Finally, excluding criminal menace, the only other scenario that can circumvent HCVD as the sole proximate cause of above blood pressure-related immediate causes are **stimulant drug intoxications** (*exogenous* acute hypertension). If cocaine or another stimulant, such as methamphetamine, is involved, the natural manner is eclipsed by the efficient intervening chemical injury and the death becomes accidental. HCVD (*endogenous* hypertension) can, and likely should, still be included somewhere in the cause of death statement (part 1 or part 2, depending on the case), but because the death is no longer 100% (exclusively) due to natural causes, the natural classification is lost.

By contrast, if coronary atherosclerosis is marked and the heart weight is normal, the diagnosis is atherosclerotic heart disease.[97] If coronary atherosclerosis is slight and heart weight is increased (and paired with renal

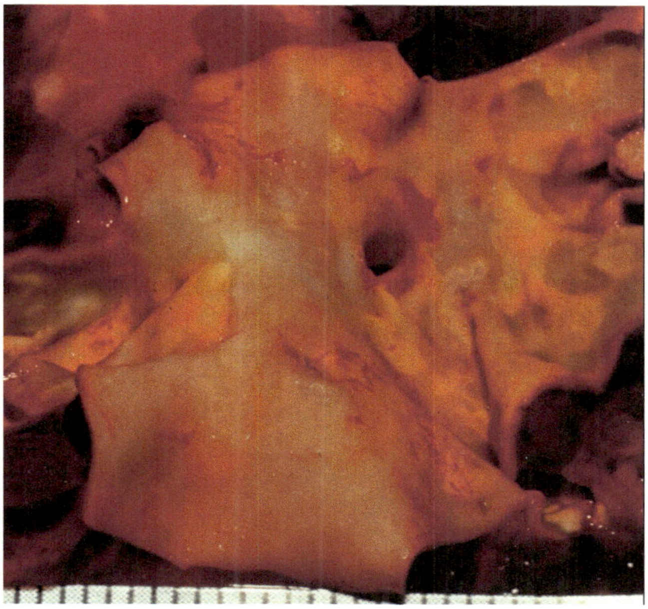

FIGURE 7.4 **Pulmonary hypertension.** Atherosclerotic plaques of pulmonary artery.

arteriolosclerosis), the diagnosis is hypertensive heart disease. If coronary atherosclerosis is moderate or marked and heart weight is increased, the diagnosis is hypertensive and atherosclerotic heart disease. The latter (common) pairing represents one of the rare occasions where we may violate the rule of one proximate cause of death per person. If this combined designation is unacceptable in your practice, list the more impressive of the two in part 1 and the other in part 2.

Diabetes mellitus (DM) is a risk factor for atherosclerosis. Therefore, if death results from any form of arterial disease in a person with diabetes, the vascular disease is the underlying cause of death and DM may be reasonably listed as a contributing cause in part 2 of the death certificate. In some instances, one may use DM as the underlying cause of the atherosclerosis (for example, a 27-year-old patient with brittle diabetes who has marked coronary artery disease).

Specific forensic issues in sudden cardiac death

There are certain SCDs that present unique investigative challenges and are therefore of particular forensic interest. These include SCDs in young athletes, during criminal altercations (homicide-by-heart attack—see Chapters 3 and 15), and those that occur in other hostile environments (e.g., swimming pools, bathtubs, the driver's seat of a motor vehicle, etc.), where there is a potential for trauma.

Sudden death in young athletes[103–109]

Sudden death in a young, previously healthy athlete, although uncommon, is invariably unexpected.[56,62,110,111] The majority are of cardiac etiology, including cardiomyopathies, congenital coronary artery anomalies, myocarditis, and channelopathies. Since some of these are heritable, determining the cause of death is important for the family, not only to know the cause of the sudden loss, but also to determine if genetic counseling and/or screening of the family is warranted.

Despite increasing interest in this topic, there are no large-scale studies examining the prevalence of sudden deaths in athletes. Maron et al. assembled a registry of 1866 sudden athlete deaths in the United States over 27 years, in both high school and college athletes.[110] They found that the incidence of sudden death in athletes may have been previously underestimated and, although still relatively low, occurred at a rate of fewer than 100 per year.

SCD in athletes occurs most commonly during or immediately *after* physical exertion.[56,110] SCD should be distinguished from other causes occurring during sports activities, including complications of heat stroke, sickle cell disease/trait, bronchial asthma, and illicit drug use.[56,110,112–115] In the Maron registry, **56% of sudden deaths were due to cardiac disease** *and 25% were due to trauma, including 3% attributable to commotio cordis* (see below), and another 2% to hyperthermia. Nontraumatic, noncardiac causes of death (e.g., asthma) and undetermined causes accounted for another 8%.[110]

The causes of sudden death in athletes vary by age. In a review of sudden death in athletes, Pigozzi and Rizzo studied "young" (<35 years) and "older" athletes (≥35 years).[56] Not surprisingly, in the "young" group, the majority of deaths were attributed to structural and/or genetic cardiac abnormalities, such as an inherited cardiomyopathies or congenital coronary artery anomalies. In the "older" athletes, acquired causes, such as atherosclerotic cardiovascular disease, were the most common, followed by valvular disease and HCVD.[56]

Causes of SCD in athletes also differ by geographic region. In the United States, the most common causes of SCD in athletes are HCM (36%) and congenital coronary artery anomalies (17%).[110] In Italy, the most common cause is arrhythmogenic cardiomyopathy, followed by coronary artery disease.[56,62,111]

HCM *is the most common cause of SCD in young athletes in the United States, particularly in Black* **men.**[56,62,110,111] HCM must be distinguished from the physiologic hypertrophy resulting from rigorous exercise (the "athlete's heart").[111] In the athlete's heart, the left ventricular hypertrophy is symmetrical and associated with an increased end-diastolic cavity dimension, while the hypertrophy in HCM is more typically asymmetrical and not associated with diastolic dilation of the left ventricular cavity.[56,111,116] Family history may help with the diagnosis, as most athletes who die from HCM have at least one affected first-degree relative.[56,62]

Congenital coronary artery anomalies *are the most common cause of SCD in* **women** *athletes.*[56] Anomalous origins of the left main coronary artery from the right coronary sinus and of the right coronary artery from the left coronary sinus are the most common abnormalities seen at autopsy.[56,62] These anatomic anomalies likely exert their lethal effects through exercise-induced myocardial ischemia; for example, an anomalous left main coronary artery arising from the right coronary sinus may become compressed between the aortic and pulmonary trunks during systole.[111]

Arrhythmogenic cardiomyopathy (also known as arrhythmogenic right ventricular cardiomyopathy) is a rare cause of SCD in athletes in the United States, but is the most common cause of death identified in young athletes in areas of *northern Italy*.[56,111] The regional difference is likely due to a higher prevalence of mutations in the Italian population, although it has been suggested that a lower incidence of SCD from other causes (due to mandatory preparticipation screening of athletes) may be responsible for the differences.[56] A study that compared death rates in Italy (Veneto) and the United States (Minnesota) found that sudden cardiovascular deaths in young competitive athletes have not differed significantly in recent years, despite the use of preparticipation 12-lead screening EKGs in Italy.[117]

Cardiac **"channelopathies"** are considered when no structural cardiac abnormality or extra-cardiac cause of death has been identified in these settings. Sports-related SCD is associated with long QT syndrome and catecholaminergic polymorphic ventricular tachycardia, which results in an exercise-induced ventricular tachycardia.[56]

Commotio cordis (or "cardiac concussion") is one of the most common causes of sudden death in young athletes without underlying heart disease.[56,111] It is defined as a SCA following blunt impact to the chest, which results in ventricular fibrillation.[118] A swine model found that if the chest impact occurred between 30 and 15 msec before the peak of the T-wave on the EKG, ventricular fibrillation occurred in 9 of 10 instances.[119] Accordingly, these deaths are most frequently seen *in sports with high-speed projectiles* (e.g., baseball, lacrosse, or hockey).[56,120] Although these are trauma-related sports deaths, they are included here because they may occur without structurally demonstrable trauma at autopsy. Commotio cordis occurs most commonly in young athletes (under 17 years), which is likely related to the size and compliance of the chest wall.[56] Since commotio cordis deaths are **accidents** (i.e., related to an injury), an autopsy and a thorough investigation is needed for the diagnosis. *The investigation includes interviews with primary sources* (i.e., those who saw the event) in order to document whether there was chest impact seconds before the collapse. One would expect several seconds of consciousness after the impact, despite immediate cessation of normal cardiac activity, because there would be an expected lag between the cardiac arrest and unconsciousness, while blood pressure is transiently maintained, despite the cardiac dysrhythmia; this is temporally similar to what is observed in most electrocutions.

Sudden cardiovascular death in hostile environments

A SCA may occur in unusual locations during a variety of activities. When the event occurs in a filled bathtub or other body of water, the question of whether the person died naturally from heart disease or unnaturally from injury (e.g., drowning) often arises (the perennial question of, "did they die *in* or did they die *from* the water?"). The same question may be considered for a driver found dead in an automobile after a collision. In the latter scenario, the autopsy findings (cardiovascular disease in the absence of lethal trauma), in conjunction with the circumstances, should more than suffice in answering that particular question and in determining the proper certification.

Deaths in the bathtub or larger bodies of water may cause greater diagnostic dilemmas with regard to the cause and manner of death, especially when unwitnessed.[121] Did the person have their sudden cardiac event while in the water or did they drown? Stated differently, was the water submersion a cause or a consequence of death? Since there are no pathognomonic findings to make a diagnosis of drowning, one must rely upon the circumstances of death. If there is an acute and demonstrable cardiac finding, the diagnosis may be clear, particularly when that finding is a Class I lesion (i.e., incompatible with continued life). There are some ME/Cs, however, who still may empirically certify such a death as a drowning, despite detection of a coronary artery thrombosis or an acute myocardial infarct.[122] Due to the high likelihood of death following a cardiac arrest due to an acute coronary thrombosis, however, it is also reasonable to certify these deaths as due to heart disease (natural), despite the subsequent submersion.[122] The likelihood that a person would survive the cardiac arrest is extremely unlikely, whether or not they subsequently submerge in water. There is no need to invoke a drowning to explain the death under these circumstances ("He who accuses must prove."). Even without an acute thrombosis, however, advanced chronic heart disease may be sufficient to certify the death as a natural SCD with the proper investigative circumstances (e.g., a former Olympic swimmer is observed to become suddenly and quietly unresponsive in a health club swimming pool, without an antecedent struggle to suggest a muscle cramp, and is also found to have a compelling/potentially lethal cardiac finding at autopsy, etc.). More importantly, perhaps, is that neurologically competent, nonintoxicated individuals do not drown in bathtubs (see Chapter 9 for a more extensive discussion). The drowning must make sense/have an adequate explanation, depending on the setting.

A sudden natural death "at the wheel" is a well-known and well-described scenario in motor vehicular deaths.[123–128] Although a sudden death may occur so quickly that it has been compared to "turning off a light," there may be a prodrome of a few seconds in which the person is aware of the symptoms. Palpitations, light-headedness, headache, crushing chest pain, or nausea may occur. This likely explains why low-speed collisions (or none at all) are typically seen when these sudden deaths involve the operator of the motor vehicle. The driver may sense something is wrong and try to stop or steer the car to the side of the road,[127] or as symptoms progress, they may start to lose muscle tone and involuntarily release pressure from the gas pedal, thus bringing the vehicle to a gradual stop.

A study of *sudden natural deaths* among automobile drivers in Baltimore found that heart disease was the most frequent cause of death in these drivers.[127] Of 81 deaths, 80 were due to cardiovascular disease effecting either the heart, the aorta, or the brain. Among these 80 decedents, 25 had acute coronary artery thromboses or myocardial infarcts, while 40 had only chronic atherosclerotic and/or hypertensive cardiovascular disease. There was little damage to property and no serious injuries to pedestrians, passengers, or other drivers. More than half of the drivers were apparently able to stop the vehicle before a collision occurred.[127]

The absence of lethal mechanical injuries in conjunction with advanced heart disease is not pathognomonic for a SCD at the wheel. Depending on the details of the collision, the absence of obvious lethal injury in a motor vehicle fatality may warrant further investigation, including, but not limited to a posterior neck dissection to exclude occult trauma, a carbon monoxide analysis, a basic toxicological analysis, and a scene investigation to exclude positional or traumatic asphyxial mechanisms.

SCDs continue to be one of the most common deaths investigated by ME/Cs and therefore these deaths must be properly recognized and certified. Despite the detection of advanced heart disease, one still should consider a proximate or contributing intoxication, particularly involving stimulant drugs, such as cocaine, amphetamines, or synthetic cannabinoids. Although cardiac disease is frequently the cause of sudden death in athletes, one must consider other causes (e.g., commotio cordis, asthma, hyperthermia, sickle cell trait). Homicide-by-heart attack does occur and one must employ strict criteria when making this determination in order to maintain consistency and fairness, while also meeting our homicidal burden of proof (see Chapters 3 and 15 for a detailed discussion). One should not be misled by the initial physical surroundings of the death (i.e., in a motor vehicle collision or swimming pool) and certify a sudden natural death as an accidental death simply by circumstantial association. "Do not confuse correlation with causation."

Pulmonary thromboembolism

Deaths due to pulmonary thromboemboli (not pulmonary "emboli") are commonly encountered by the forensic pathologist. Emboli originate from a variety of sources: fat, air, tumor, amniotic fluid, or thrombus, the last being, far and away, the most common (Figs. 7.5 and 7.6). The detection of a saddle thromboembolus at autopsy does not signify the end of the investigation, but rather the beginning, as the underlying risk factor(s)/proximate cause must be sought. Gross, microscopic, genetic, and even toxicologic evaluations can help define thrombotic risk factors, the age of the thrombus, and underlying etiologies. PTE often have medicolegal repercussions; therefore, a reproducible protocol is needed as part of the investigation. In particular, the time course between the formation of a deep vein thrombosis (DVT) and the resultant thromboembolus may have important medicolegal implications.

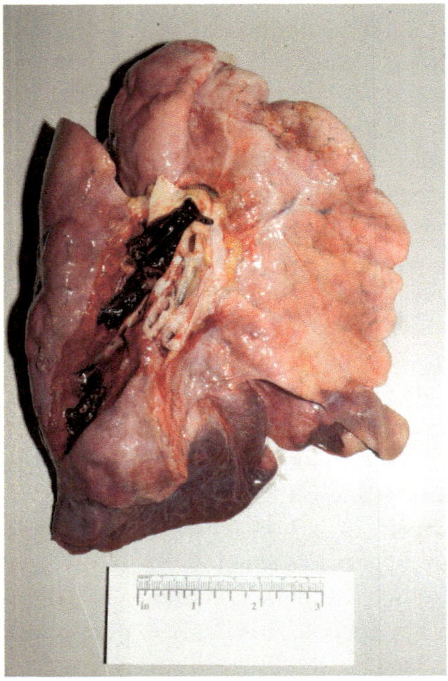

FIGURE 7.5 **Thromboembolus in the proximal pulmonary artery (hilum).**

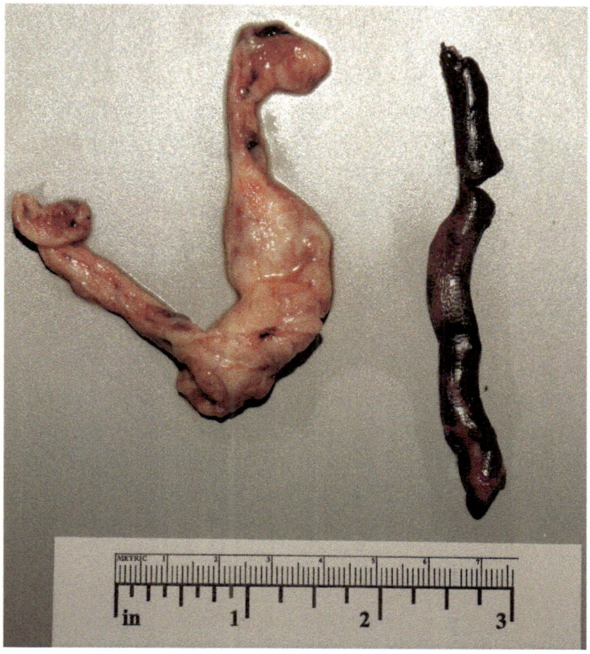

FIGURE 7.6 **Postmortem clot (left) versus a thrombus (antemortem).**

A typical death caused by a PTE is usually reported to the ME/C because of its sudden and unexpected nature, not because "PTE" per se is a reportable ME/C condition. In fact, it typically isn't until its discovery at autopsy that this immediate cause is elucidated. Often, however, it is high on the differential diagnostic list when, if the fatal event is witnessed, its precipitousness nature is described, along with it being preceded by a somewhat panicked, acute onset dyspnea and ominous sense of dread on the part of the decedent, just before the rapid onset of unconsciousness. Not uncommonly, one or more experiences of dizziness or syncope have preceded the fatal event.

Beyond its discovery, identifying its proximate cause is central to the investigation, as it determines whether the medicolegal investigator is dealing with a natural death or an unnatural, even homicidal one. Understanding the pathogenesis and pitfalls of venous thromboembolism (VTE) allows the pathologist to properly certify the proximate cause of death.

Virchow described three risk factors for thrombosis that bear his eponym, **Virchow's triad**. They are *endothelial injury, stasis, and hypercoagulability.* With the advent of genetic testing, heritable thrombophilias (genetic hypercoagulability) may be definitively diagnosed. PTE is an immediate cause of death diagnosed by both the forensic and hospital pathologists. As stated above, these out-of-hospital deaths tend to be fall under the purview of the ME/C predicted on their suddenness. Hospital PTE deaths related to complications of therapy or diagnostic studies also may be reported to the ME/C because of their unexpected nature and temporal relationship to therapy, and although there may be clinical suspicion, the autopsy is the gold standard for the diagnosis. A fatal PTE that complicates a proximate injury will typically also fall under the jurisdiction of the ME/C, primarily because of the history of trauma, acute or delayed, rather than prior knowledge of a PTE, per se. **PTE that cause sudden, unexpected death tend to be large and often saddle**, not the small, peripheral thromboemboli that may go undetected or cause sublethal and less specific symptoms, such as syncope, dizziness, or mild dyspnea. Large PTE diagnosed in a typical ME/C practice are usually **Class I findings** (incompatible with continued life).

The process begins with a thrombus that forms in the deep veins of the lower extremities or pelvis due to endothelial injury, stasis, and/or a hypercoagulable state. The thrombus may gradually enlarge as more fibrin, platelets, and blood continue to layer (lines of Zahn). There are three possible outcomes: lysis, organization with recanalization and resorption, or embolism. Prior to the development of these outcomes, the thrombus typically propagates proximally, which increases the risk of embolism.

Major risk factors *for deep venous thrombosis include: lower extremity fractures, major surgery (especially lower extremity orthopedic procedures), or spinal cord injury.* **Moderate risk factors** *include: hormone therapy, paralysis of any cause, malignancy/chemotherapy, prior venous thrombosis (including hereditary thrombophilic risks), congestive heart failure, and the postpartum period.* **Weak risk factors** *include inactivity due to prolonged sitting, obesity, varicose veins, severe dehydration, and pregnancy.*[129–138] A less common "soft" risk factor, particularly when none other is identified, is **acute and chronic narcotic use**, arguably related to the associated sedentary lifestyle.[139] While a known risk factor of arterial thrombosis, current data supporting cocaine or other stimulants as risk factors for *venous* thrombosis are less convincing.[140]

Unlike narcotics, cocaine alone is also not expected to be associated with a sedentary lifestyle, per se, and in the absence of *chronic* cocaine use resulting in severe heart disease, with or without congestive heart failure, the evidence for its association with DVT/PTE is not as clear and may depend on a variety of factors. In the end, the manner of death for a PTE death is dictated by the underlying disease or injury causing the DVT.

A pulmonary thromboembolism due to a DVT resulting from any type of trauma (e.g., hip fracture) should *not* be certified as a natural death. A DVT complicating *elective* surgery for nontraumatic causes (e.g., hip replacement for degenerative arthritis) may be certified as a therapeutic complication or natural death, depending upon the jurisdiction, whereas a PE/DVT resulting from a broken leg or hip following a fall is accidental. The DVT in the latter scenario may be caused by the stasis brought on by injury-related inactivity (i.e., the DVT may be in both the injured and uninjured leg as a result) or from the injury itself (i.e., the DVT is identified only in the injured leg and is related to the congestion and engorgement related to healing), versus a combination of both.

Even if there is no grossly apparent deep vein thrombus, microscopic examination of the veins can be of benefit. One may see a microscopic thrombus or signs of a recently embolized thrombus (e.g., phlebitis). Gross features that allow the differentiation of true antemortem thrombi from a postmortem clot include the shape, color, size, overall appearance, and consistency of the coagulum, and the extent of its attachment to the blood vessel wall. None of these findings are absolute and one must be mindful of the pitfalls. *Grossly, the classic saddle pulmonary thromboembolus is a tangled, heterogeneous, red-blue-tan, often granular thrombus distending and bulging from the proximal pulmonary arteries/pulmonary arterial trunk and is often slightly adherent to the vessel wall.* Typical postmortem clots slip out of the blood vessel, have a distinctive purple and tan color, a gelatinous consistency (so-called "chicken fat"), and often have the shape (mold) of the blood vessel in which they are found. The word **"clot"** is nonspecific and descriptive; it references neither an antemortem nor a postmortem process, per se. **"Thrombus,"** on the other hand, indicates pathology and, therefore, an intravital process. Acute deep vein thrombi are more likely to embolize than are organizing thrombi that have become adherent to the blood vessel wall from ingrowth of granulation tissue. Microscopic examination is particularly helpful with grossly equivocal thromboemboli (i.e., in determining whether the clot formed antemortem or postmortem).

At autopsy, some emboli may lodge in the right heart and can be missed if they drop out of the vena cava during removal of the heart. More typically, the pulmonary hilum contains unmistakable protruding thromboemboli that are revealed with incision of the pulmonary truck and removal of the heart. After photographing the PTE in situ, the pulmonary arteries should be further opened from the hilum toward the periphery to search for ramifications of the proximal embolus or separate emboli (sometimes more organized) within the lobar and segmental arteries.

Although a residual deep vein thrombus is not always detected at autopsy, a posterior dissection of the lower extremities and deep veins is the preferred method of evaluation for DVTs, rather than simply "milking" the upper leg veins, a technique that is inadequate to include, exclude, or document thrombi (see Chapter 5). We view an autopsy in which a large PTE (immediate cause) has been identified incomplete without dissection of the lower extremity and pelvic vasculature. Adherent deep venous thrombi are best documented with photography and microscopic sections, when deemed necessary. When no DVT source of the PTE is identified, the only reasonable conclusion is that it embolized entirely.

Therapeutic interventions in PTE deaths that may obscure their detection include vigorous resuscitation and/or the use of a thrombolytic agents, which may displace or partially dissolve the thromboembolism, respectively. Histological examination of the lung may demonstrate thromboemboli in small vessels. The possible and inadvertent removal of a saddle pulmonary thromboembolus by a tissue procurement team who have removed the heart for valvular tissue should also be considered. Arterial embalming may dislodge postmortem clots in veins that may then "embolize" to the heart. Microscopic examination can distinguish these "pseudoemboli."

Various criteria have been used to microscopically estimate the age of a thrombus. Typically, they are classified as **acute, organizing, or remote,** but it is not reasonably possible to determine their precise age. The most that can usually be offered is an opinion as to whether the formation of the thrombus is or is not consistent with a specified history. "In reality, pathologic processes seldom cooperate fully with attempts to date or age them with precise reliability." The "dating" of an acute fatal thromboembolus is of little value, as it constitutes the most proximal portion of the DVT and therefore represents the most recently formed segment. For medicolegal purposes, the maximum age of the deep vein thrombus is usually the most relevant to the investigation, so adequate and distal sampling of the DVT is recommended.

DNA testing is costly and the clinical benefits for relatives have been debated. In select instances, this may help in fully elucidating the cause of death. Candidates for these tests include: those persons > 45 years of age, pregnant or recently pregnant decedents, those with a history of recurrent or unexplained stillbirths, oral contraceptive pill use, hormone replacement, or treatment with chemotherapy, those with only weak risk factors (long car ride, flights, drug use), or those with a deep venous thrombosis of undetermined etiology. See Chapter 17 for a further discussion of this topic.

Seizure-related deaths and epilepsy

While most primary seizure episodes are nonfatal, they do have lethal potential; forensic pathologists are the best physicians to investigate and confirm a potentially fatal outcome. *One must distinguish between deaths directly due to a seizure from terminal seizure activity that are a nonspecific manifestation of the dying process.* In medicolegal death investigations, "seizure activity" as a prequel to death is overwhelmingly more common than are primary seizures, per se, as a cause of death. Perimortal seizure activity is often merely a marker of dying, a function of cerebral hypoperfusion with almost any mechanism of death (brain injury from a gunshot wound, exsanguination, cardiac arrhythmia, hypoglycemia, etc.) and, as such, is part of the investigative narrative medical examiner/coroners encounter nearly daily, as opposed to a seizure as the primary cause, which constitutes a small minority of the deaths we investigate.

When it occurs as a primary cause, however, *both the seizure and its fatal mechanism are* **functional in nature**. Even when the product of a structurally demonstrable lesion, such as a tumor or a dural cicatrix from a remote penetrating injury, the seizure itself, the electrical misfiring, analogous to a cardiac arrhythmia, leaves no evidence of having occurred at the moment of death. It is a "diagnosis of exclusion" based on the autopsy (and what the autopsy excludes) and the history and circumstances of the death. Epilepsy/seizure-related fatalities are classic examples of a **"Class IV" exclusionary finding** (see Chapter 3), *characterized by a lethal mechanism which is not structurally demonstrable* (even if the seizure-causing lesion is). Therefore, in such cases, the autopsy, by revealing nothing, raises the degree of certainty for a seizure-related cause, a conclusion reached by history and exclusion of other reasonable causes.[31,141]

For example, an 80-year-old man with chronic obstructive pulmonary disease (COPD), advanced heart disease, and a nontraumatic idiopathic seizure disorder (e.g., "epilepsy") is found dead in bed (i.e., an unwitnessed death, Fig. 7.7). The burden of proof to conclude that a seizure was the primary or even a contributory cause of death will not have been met by exclusionary criteria. Alternatively, if a 20-year-old woman with poorly controlled epilepsy is found dead in bed (also an unwitnessed death), with no other identifiable cause of death (anatomically or chemically) after a complete autopsy, epilepsy would be the probable cause of death (even with therapeutic antiepileptic drug concentrations). These two cases also demonstrate the recognition and use of the statistical likelihood of a particular cause of death. An elderly person with COPD and advanced heart disease is much more likely to have died from those diseases than from a seizure.

When seizures are the primary cause/mechanism of death (not from secondary trauma such as from a seizure-related motor vehicle collision, positional asphyxia, choking, or drowning), they can be so in both the young and old, likely by the deleterious effects of apnea, elevated blood pressure, and/or the autonomic instability commonly experienced during generalized/tonic–clonic seizure episodes, particularly status epilepticus, and less so with partial complex/focal impaired awareness seizures. But, as we now better understand, epileptic deaths are not exclusively related to seizure activity. In 1971, Hirsch and Martin described a small proportion of epileptics who suffered witnessed sudden deaths without an identifiable cause, but in the *absence* of a sustained tonic–clonic seizure at the time of witnessed cardiac arrest, positing that they were related to "acute disruptions of brainstem cardiac or respiratory functions or both as a consequence of seizure discharge, regardless of the type."[142] This theory and description preceded the more recent coining of **Sudden Unexpected Death in Epilepsy (SUDEP).**[143–153] So how does one approach a death which has a reasonable likelihood of being seizure-related?

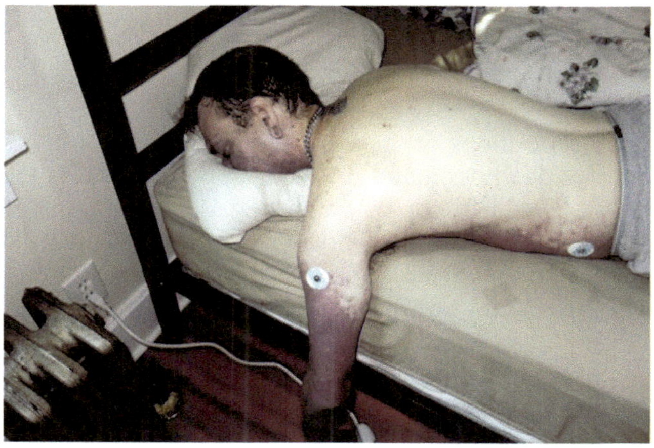

FIGURE 7.7 **Person with epilepsy found prone in bed.**

Seizure investigation

Once it is established that there is a true seizure disorder and not just terminal seizure activity, one must establish the seizure history by speaking to the family and/or by reviewing medical records. A quick phone call to the treating neurologist or primary care physician (contact the prescriber of the antiepileptic medication) is the most efficient means for gathering a rapid treasure trove of important clinical, drug compliance, and etiologic information.

1. Was the seizure witnessed or was the person found dead?
2. If found dead, what was the position of the body at the time of its discovery? A prone, face down in bedding position may raise the specter of seizure-related positional asphyxia
3. What was the type of seizure that was witnessed or what is the typical seizure type by history (e.g., generalized/tonic-clonic versus focal impaired awareness/partial complex)?
4. When was the last known seizure?
5. What was the seizure frequency (once every couple of years will likely be viewed quite differently than seizures having occurred every week)?
6. Were there any known seizure triggers?
7. Were the seizures since birth/childhood or onset later in life?
8. What medical work up had been done (computed tomography, magnetic resonance imaging, electroencephalogram, etc.) and was there a specific diagnosis (e.g., tumor, previous surgery, congenital or acquired metabolic disorders, vascular disorders, infection, alcohol/drug use, primary neurodegenerative/neuropathologic disorders, and, of course, idiopathic)?
9. Are there any other medical problems?
10. Most importantly, was there a remote injury of the head? (if yes, see below)

Posttraumatic epilepsy

Posttraumatic seizures are a known complication of brain injury and are categorized as early onset (within one week of the injury) and late onset (after one week).[154–157] Risk factors for the development of posttraumatic seizures include brain contusions with a subdural hematoma, skull fracture, unconsciousness or amnesia for more than one day, penetrating injuries with a cerebral-dural cicatrix, and age over 65.[154,157] These risks remain for years following the injury; however, *approximately 50% of the seizures have an onset within the first year following the injury.*[157]

1. If there was remote head injury, did the seizures predate the injury or were they documented to begin after the injury? What was the precise nature and extent of the injury? Was there any documentation made by a hospital?
2. What was the timeline of onset relative to the time of injury and was it medically documented?

Medications

Antiseizure medications should be documented at the scene and targeted in the subsequent toxicological analysis. This may require notification of the toxicology laboratory as to what medications one is interested in identifying.

1. Were there any prescribed antiepileptic medications? What were the dosage instructions?
2. If so, was the patient compliant?
3. If compliant with medications, were there any "breakthrough" seizures?
4. Were there any recent changes in the medication(s)?
5. Were the pill counts appropriate?

While subtherapeutic serum concentrations of an antiepileptic medication may prove compelling in the final cause of death determination, seizure deaths can and do occur in persons with "therapeutic" drug concentrations (so-called "breakthrough" seizures). As with all forensic determinations, the cause of death conclusion is ultimately made based on the entire constellation of investigative and autopsy findings, inclusionary and exclusionary. In addition, certain prescription medications and other substances, including stimulants and alcohol (the latter with intoxication or withdrawal), can lower the seizure threshold in epileptics, so may be of value as part of the toxicologic evaluation.

Autopsy and postmortem work-up

As in drownings, there are no pathognomonic autopsy findings in a seizure-related death. *While a "***foam cone,"** **bite marks of the tongue, urinary incontinence, and/or conjunctival petechiae** *may be compelling findings in the right* *setting, they are nonspecific and their absence does not preclude a diagnosis of a seizure-induced death.* Tongue bites may occur with perimortem seizure activity, and they may be absent after a seizure if the teeth do not happen to clamp down on the tongue.

The decision to retain the brain for formal neuropathologic evaluation is made on a case-by-case basis and depends on disparate case goals, ranging from the need to seek the etiology of new onset or previously unevaluated seizure disorders to documenting the pathology of seizure-related injury in a potential homicide. The reflexive practice, however, to save the brains of all persons who exhibit terminal seizure activity, in the absence of a seizure history, per se, should be resisted; these decisions should be made thoughtfully and contextually. While a neuropathologic examination may be important in selected cases, potentially providing support for a seizure disorder (old cerebral contusion, hippocampal abnormalities and variations), in an unwitnessed death, it does *not* have the capacity to definitively tell the FP whether the decedent actually experienced *and* died from a terminal seizure, per se. That determination must come from the forensic pathologist after their careful review of the entirety of the death investigation.

Molecular testing is available to analyze for genes that are associated with both nonsyndromic and syndromic causes of epilepsy (e.g., tuberous sclerosis complex, Angelman syndrome, mucopolysaccharidosis). This testing may be warranted in select "idiopathic" cases. See Chapter 17 for further discussion of this topic.

Death certification

Of all the immediate cause injuries related to seizure deaths/epilepsy, **positional asphyxia** warrants special mention, as it is contextual (i.e., not structurally demonstrable) and can be easily missed, if not at least considered. *It is not unusual for epileptics to have seizures while sleeping, as the passage through sleep/wake stages is relatively epileptogenic.*[158–160] When found prone, face down in a pillow, for example, positional asphyxia is often at least considered by many forensic pathologists. The question that arises is whether their "found" position is a cause or a consequence of death, as asphyxia does not need to be invoked to explain a seizure-related death, per se, given the lethal potential of seizures and/or epilepsy alone. This determination is not necessarily easy to make, so should not be made blithely, as it will change the manner of death. *Finding a thick pool of mucus and foam on a pillow immediately under the nose and mouth, for example, may help,* as it might indicate respiratory compromise from airway obstruction/smothering during the (unconscious) seizure activity in a face down prone position.

These suspected positional asphyxia deaths are best viewed on a case-by-case basis. A designation of "positional asphyxia" as the immediate cause still requires a proximate cause (e.g., epilepsy), but because of invoking an injury as the immediate cause, it results in an **"accidental" manner classification**. The same applies to choking or bathtub drownings related to epilepsy: drownings or chokings are immediate causes, and **epilepsy, the proximate cause.**

There are three scenarios that may be encountered when certifying a seizure death.

1. A seizure, per se, and/or an epilepsy-related sudden death provides the mechanism of death in a person with a seizure disorder.
2. A seizure results in an injury that leads to death (positional asphyxia, drowning, choking).
3. A seizure is the mechanism of death in a person with a seizure disorder, but there may be other comorbidities that contribute (part 2).

Sudden Unexpected Death in Epilepsy (SUDEP) is the current parlance for certification in deaths from an established seizure disorder that is nontraumatic, idiopathic, and otherwise unexplained. While a discussion about the appropriate use of the SUDEP designation exceeds the scope of this text, it is important to recognize that it is typically *applied to the sudden deaths of* ***epileptics*** *in whom other more compelling causes have been excluded and in which status epilepticus has not played a terminal role. The manner is such deaths is* **natural**.

For seizure disorders secondary to a disease, we suggest "seizure disorder due to/complicating (disease/neuropathologic disorder/acute and chronic alcoholism/tumor, etc.)." When the seizure disorder is secondary to an injury, one may certify the death as "posttraumatic seizure disorder following/due to (name etiologically specific remote head injury: blunt, penetrating, etc.)," with the manner of death informed by the circumstances of the remote putative injury. Other examples of causes of death with unnatural manners of death include: "drowning due to epilepsy" or "drowning complicating posttraumatic seizure disorder due to remote gunshot wound of the head." In the gunshot wound example, there are two listed "injuries"; however, the injury that caused the seizure disorder would take precedence as the dominant cause because it is both the proximate injury and is homicidal.

There may be instances of epilepsy deaths with other conditions invoked as comorbidities in part 2 of the death certificate. In these instances, the additional comorbidities may be included to "cover the waterfront." Comorbidities, however, can be challenging when investigating a potential *homicidal* posttraumatic seizure death. The burden of proof is highest in homicides, requiring the exclusion of all other reasonable causes. If the death was not witnessed and there is another disease that may reasonably explain a sudden death, care must be taken in the certification to avoid being arbitrary or capricious. When the stakes are this high, one must be forthcoming with this dilemma to ensure that justice and fairness are not subverted.

Anaphylaxis

Anaphylaxis is a potentially fatal allergic reaction with a rapid onset (minutes to hours) that may manifest as an itchy rash, throat or tongue swelling, acute shortness of breath, sometimes with stridor, laryngeal edema (Fig. 7.8), and/or hypotension/cardiovascular collapse.

Common triggers include insect bites/stings, food, and medications. Other causes include latex exposure and, rarely, vaccinations. On occasion, no allergen is identified. The mechanism involves the release of chemical mediators from hypersensitive inflammatory cells triggered by exposure to an allergen.

Allergic anaphylaxis is usually caused by the interaction of an allergen with specific immunoglobulin E (IgE) antibodies. These sensitized antibodies bind to receptors in tissue mast cells and blood basophils, causing release of inflammatory mediators, such as histamine, tryptase, phospholipid-derived compounds (e.g., prostaglandin D2, leukotrienes, thromboxane A2, and platelet-activating factor), as well as several chemokines and cytokines. Target organs include the skin, mucous membranes, cardiovascular and respiratory systems, and the gastrointestinal (GI) tract.[161–170]

The primary treatment is an epinephrine injection, intravenous fluids, antihistamines, and steroids. Since people prone to anaphylaxis may carry an epinephrine autoinjector, this may represent important scene evidence. *The determination of the cause and* **manner** *in anaphylactic deaths* **will depend upon the type of trigger** *initiating the allergic reaction.*[161,171–187] There is some variation as to the manner of death for some of these deaths. For the anaphylactic death caused by a medication, the cause and manner will depend upon the indication for which the medication was prescribed. In an allergic reaction to penicillin for the treatment of a bacterial pharyngitis, the proper certification would include the infection as the proximate cause (i.e., anaphylaxis due to the administration of penicillin for the treatment of streptococcal pharyngitis). For jurisdictions in which the use of therapeutic complication is an option, TC would be the manner. For those without that option, the death would likely be certified as natural or accident, depending on local practice.[188] If the medication was administered as part of a treatment for a potentially life-threatening injury, on the other hand, the manner would be informed by the circumstances of the injury (e.g., if the reaction was to antibiotics given during the course of treatment for a hip fracture due to blunt injury sustained from a fall, the manner would be accident). If the anaphylaxis was caused by a **bee sting** (which is an injury), the manner would also be **accident**.[167]

Allergies to food that result in anaphylaxis are rare, but do occur. The manner of death classification in these instances may vary between forensic pathologists. Since food is necessary for life, some might view and certify these deaths as natural. Others may note that food represents an exogenous factor, such that related anaphylactic deaths should be certified as accidents. The discussion can become complex when one considers something like a latex

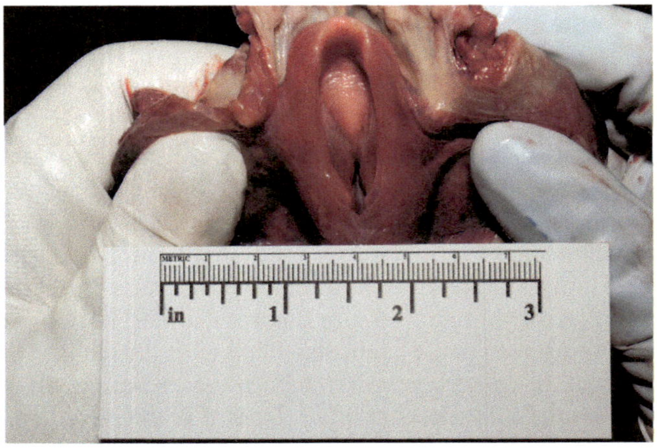

FIGURE 7.8 **Laryngeal edema in anaphylaxis.**

allergy. Is it natural or an accident? **Latex** is an emulsion of polymer microparticles in water that are found in nature, but there is synthetic latex as well. Is the manner of death of anaphylaxis from exposure to a latex glove in a medical setting natural, accidental, or a therapeutic complication? What about the manner of death in anaphylaxis from exposure to a latex condom? These are the circumstances that often initiate robust and sometimes animated discussions among forensic pathologists.

In addition to the history, gross findings (supraglottic/laryngeal and pulmonary edema), and microscopic findings (Giemsa-positive degranulation of mast cells and basophils in laryngeal/supraglottic and/or tongue tissue), one may also measure the **tryptase** concentrations.[161–166,168,170,187,189–204] It should be noted that **peripheral vascular collapse rarely occurs without supraglottic edema.** Mindful of the fact that endogenous/nonpathologic levels of tryptase exist, testing for plasma histamine or serum tryptase concentrations may be helpful in confirming the diagnosis, depending on their specific concentrations and other supportive findings. *Plasma histamine concentrations increase within 10 min of onset and decrease within 30 min.* Total tryptase concentrations represent the total amount in the body, but not the total that was acutely degranulated. *Serum* **mature tryptase** *(previously called beta-tryptase) is a measure of what was degranulated*; concentrations peak 60–90 min after the onset of an episode and may persist for as long as 5 h. The estimated positive predictive value of tryptase elevations in 259 subjects with anesthesia-associated anaphylaxis was 92.6%, and the estimated negative predictive value of normal tryptase levels was 54.3%.

In evaluating a possible anaphylactic reaction, *a ratio of total tryptase (alpha + mature) to mature tryptase* of 20 or greater is consistent with mastocytosis, whereas *a ratio of 10 or less suggests anaphylaxis*. In vitro tests are available to detect serum-specific IgE antibodies. Drug-reactive IgE antibodies have been demonstrated to be involved in anaphylactic reactions.[205]

It's also worth mentioning that **angioedema** has a specific presentation and, not uncommonly, can be elicited by *ACE inhibitors*, particularly in African-American patients. Oddly, this rather dramatic and acute onset upper airway, facial, and tongue swelling can occur many months or even years into established therapy, often without a dosage or brand change.[180,198,206–210] Other (nontherapeutic) allergens may also be implicated in causing angioedema, but when it seems likely after a thorough investigation that an ACE inhibitor is responsible, the death would best be certified as a therapeutic complication, the proximate cause being the medical indication for the therapy (usually HCVD).

In a child with upper airway symptoms, **epiglottitis** (*Hemophilus infiuenza*, type B, Hib) is in the differential diagnosis. Due to routine use of the Hib vaccine in the United States since 1990, the incidence of invasive Hib disease has decreased. When present, there is typically acute suppurative epiglottitis and laryngitis with gram-negative, coccobacilli.

Alcohol use disorder (formerly known as chronic alcoholism) and chronic substance use disorder

Alcohol use disorder

Excessive alcohol use is responsible for more than 95,000 deaths in the United States annually or 261 deaths per day. These deaths shorten the lives of those who die by an average of almost 29 years, for a total of 2.8 million years of potential life lost.[211] In 2021, these deaths were surpassed only by accidental drug intoxication deaths.

Following ingestion, peak blood ethanol concentrations are typically achieved within 30–120 min. Most ethanol, however, is absorbed in the proximal small bowel; therefore, food in the stomach may delay absorption for hours. At autopsy, one may detect a distinct odor following the decedent's ingestion of alcohol. As ethanol is odorless, these smells are derived from the various metabolites and congeners associated with ethanol ingestion and metabolism.

Chronic alcohol misuse can cause death with or without a concurrent acute ethanol intoxication. There are a variety of anatomic and toxic metabolic changes attributable to chronic alcohol use, including myriad complications from *hepatic steatosis and cirrhosis, including acute GIH and hepatic encephalopathy, alcoholic cardiomyopathy, acute and chronic pancreatitis, and hematologic, immunologic, and nutritional derangements.* The effects on the central nervous system include *seizures* (commonly manifesting during withdrawal), Wernicke's encephalopathy, cerebellar atrophy, and the Marchiafava–Bignami syndrome. *Delirium tremens* has a 10% mortality rate. Moreover, this chronic disease not only puts those who misuse alcohol at an enhanced risk of dying from violent means, accidental or inflicted, particularly when intoxicated, it also makes them more susceptible to fatal hemorrhagic complications when they do get injured, given the prevalence of liver pathology in this population of decedents.[211–215] As Lester Adelson astutely pointed out, "The most uncommon commodity, common sense, is soluble in alcohol."

Since ethanol distributes in total body water, its concentration will vary in different body fluids depending upon the water content of that fluid. The standard legal blood alcohol concentration (BAC) is measured in whole blood. Hospitals, however, may quantitate ethanol in plasma, which gives a higher concentration than whole blood calculations because of the greater water content. A plasma concentration of 0.10 gm% is equivalent to 0.08 gm% in whole blood. Ethanol may be tested for in vitreous humor and urine autopsy specimens. Since vitreous has a greater water content than blood and it takes longer for ethanol to reach the vitreous, the blood-vitreous ratio can be used to determine whether the blood ethanol concentration was still rising (preabsorptive) at the time of death. **At equilibrium, the blood ethanol concentration is roughly 75% of that of the vitreous** (BAC = 0.75 VAC or VAC = 1.3 BAC). Only in the absorptive phase (before alcohol has reached its peak) and early postabsorptive phase, before equilibration, the blood ethanol concentration will be higher than the vitreous concentration. If the blood-vitreous ratio is greater than 0.95 (i.e., blood ethanol is greater than vitreous), then death occurred before equilibrium had been achieved.

Caution must be exercised when interpreting **urine ethanol concentrations** *since the bladder is a reservoir that may not have been completely empty when the person started ingesting ethanol and/or may not be emptied for some time after the ingestion stops.* It is, in fact, possible to detect ethanol in the urine while having a zero blood ethanol concentration (e.g., if the person had not urinated for some time after the ingestion stopped) and also it's possible to have a urine ethanol concentration that is lower than the blood concentration (e.g., if the bladder was nearly full when the ingestion began). The typical 1.3–1.4 ratio is technically only for *ureteral* urine, which, in practice, is not recovered at autopsy. It may be recovered in the living if the specimen is collected shortly following a first voiding. Regardless, the extrapolation of blood ethanol from urine ethanol concentrations is not sufficiently accurate in autopsy samples for forensic purposes. The *qualitative* detection of ethanol in the bladder or vitreous, however, may be helpful in interpreting blood ethanol concentrations in putrefied bodies, where endogenous (bacterially produced) ethanol can create interpretive challenges.

Putrefaction is a microorganism-driven process and some bacteria may ferment alcohols. This postmortem production of ethanol may result in the misinterpretation of ethanol detected in putrefied remains. The ethanol may represent **antemortem ingestion (exogenous ethanol)** or **postmortem bacterial production (endogenous ethanol).** *Isopropanol and acetone* are other common byproducts of putrefaction (see below in discussion of DM). Typically, blood ethanol concentrations are reliable for the first 24 h after death and correspond to the antemortem concentration. One study examined ethanol concentrations in 130 putrefied decedents. They found that 23 had *endogenous* ethanol production, with *most concentrations measuring less than 0.07 mg/dL.* Four decedents, however, had ethanol concentrations of 0.11, 0.12, 0.13, and 0.22 mg/dL.[216] How does one reliably distinguish between antemortem ingestion and postmortem (endogenous) production?

One may differentiate endogenous (postmortem) ethanol from exogenous (antemortem) ethanol ingestion by analyzing the urine and/or vitreous. Since the urinary bladder and vitreous body are relatively sequestered sites, there may be delayed, little, or no bacterial intrusion and thus little or no endogenous production of ethanol in these sites. If ethanol is detected in the blood, but not in the urine or vitreous humor, this would support the blood ethanol being endogenous. If, on the other hand, ethanol is detected in all three samples, antemortem (exogenous) ingestion is the most likely explanation. Unfortunately, in markedly decomposed bodies, there is usually no vitreous humor (and sometimes no urine) available to test. In these instances, unless the circumstances are compelling, the interpretation of the blood ethanol should be conservative. When the blood concentration is extremely high, it becomes clearer that at least *some* amount of ethanol was ingested/is exogenous. One other caveat involves ethanol being detected in the urine of decedents with DM. Since these patients may have glycosuria, and a resultant bacterial colonization of the urinary bladder, there may be concomitant production of endogenous ethanol in the urine.

Having the toxicology laboratory reflexively test the vitreous humor for ethanol when ethanol is detected in the blood may be useful for two reasons. The first is to sort out exogenous from endogenous ethanol concentrations; the second is to determine whether the decedent was in the postabsorptive phase, as described above.

By convention, the only causes of deaths that include an acute "intoxication," but are classified with a natural manner are those involving ethanol. Conceptually, an acute intoxication in a chronic alcoholic is viewed as an acute exacerbation of a chronic disease, as opposed to a chemical injury that causes death alone. If, however, on his 18th birthday, a young man, who is relatively naïve to ethanol, rapidly drinks a bottle of whiskey, without the bottle leaving his lips, and then dies, his death would be certified as an accident. In this instance, the young man is not a chronic alcoholic and therefore is not accustomed to such an elevated ethanol concentration (has no tolerance), so he is truly dying from its toxic effects. A foolhardy act without anticipation of the consequences is distinctly different than an acute exacerbation of a chronic disease; the former is an inadvertent poisoning.

When there is a markedly elevated blood ethanol concentration, even in a chronic alcoholic, some may certify the death as an accident. Where the cutoff is drawn between natural and accidental concentrations is perhaps debatable, perhaps even arbitrary, as there are well-documented instances of chronic alcoholics walking into emergency departments with BACs of over 0.6 gm%, such an exceedingly high tolerance representing evidence of their chronic disease.[217–223] That said, many certify as, "acute and chronic alcohol use disorder" any death due to alcoholism in which the decedent is intoxicated with alcohol at the time of death (even as low as 0.05%). One must not ignore the fact that acute alcohol intoxication at lower concentrations still has cardiovascular effects, such as elevations in blood pressure, even if not respiratory depression, per se. Even more important to remember, however, is that listing "acute alcohol intoxication" *alone* on the death certificate, without mention of the chronic component of the disease, is neither relevant nor accurate when certifying alcoholic deaths. The only time that specific wording, characterizing an isolated alcohol intoxication, would be appropriate is in the context of a true alcohol poisoning, as illustrated above with the 18-year-old's intoxication. In an alcoholic dying from the constellation of derangements of the disease, the proximate cause choices are either "acute and chronic alcoholism" or "chronic alcoholism," *not* "acute alcohol intoxication" alone.

If, however, a chronic alcoholic dies as a result of other sedative drugs (benzodiazepines, methadone, barbiturates, etc.), in conjunction with alcohol, the manner changes from natural to **accident,** based on the combined effects of the drugs and alcohol, the combination eliminating a mere acute exacerbation of a chronic disease. If the chronic alcoholic, in the desperation of withdrawal, ingests **methanol or isopropanol**, the manner of his death would also be *accidental* (or suicide if the intent was self-harm). In some deaths due to chronic alcoholism, there is an immediate cause such as cirrhosis with liver failure (with jaundice, ascites, etc.) or an acute gastrointestinal hemorrhage (GIH) due to ruptured esophageal varices. To improve communication with families and clinicians, these structurally demonstrable immediate causes might be best included on the death certificate. If an immediate cause of death is *not* structurally demonstrable (i.e., likely functional), the death can be competently certified simply as "chronic alcohol use disorder" or "complications of chronic alcohol use disorder." In the past, these deaths may have been certified as "fatty metamorphosis of the liver" to avoid mentioning alcohol on the death certificate. But, chronic alcohol use disorder is a disease and should be treated and certified as such.

Studies of sudden, nonviolent deaths in these patients illustrate the not uncommon occurrence of sudden death with autopsy findings limited solely to hepatic steatosis. The exact mechanism(s) of these deaths (many unwitnessed) is unknown, but may include ethanol withdrawal, associated hypoglycemia or hypomagnesemia, or seizures.[224] **Alcoholic ketoacidosis** in the setting of an abrupt withdrawal of ethanol combined with severe nutritional deficiencies is another easily missed mechanism of death related to chronic alcohol use. Often, such a decedent is found surrounded by multiple empty alcohol containers, yet the toxicology testing surprisingly yields a negative ethanol concentration. In these cases, testing for and detecting an elevated beta-hydroxybutyrate (BHB), a fatty acid more specific to the ketosis related to chronic alcohol use, might lead to the more mechanistically clear diagnosis of "alcohol ketoacidosis," rather than the somewhat more vague "chronic alcohol use disorder," without further qualification.[225–227]

Chronic substance use

Historically, the nomenclature used to certify the deaths of persons who chronically suffer from drug dependence has been varied. The cause of death statements in such fatalities has ranged from intravenous drug abuse, "narcotism", substance abuse, and polysubstance abuse to acute and chronic substance abuse. Narcotism was used to refer specifically to a narcotic being the toxic agent of choice. More recently, however, there has been a shift away from the word "abuse" and toward "drug use, misuse, or use disorder."

Whatever the word choice, like alcoholism, chronic drug use/misuse, intravenous or not, is a disease. But, unlike chronic alcohol use, *acute exacerbations (intoxications)* of chronic substance use are conventionally classified as *chemical injuries,* whether the misuse involves illicit drugs and/or prescription medications. As such, the associated intoxication fatalities are not classified as natural deaths. Unless there is evidence, toxicological or investigative, to suggest otherwise, *most of these fatalities are classified as accidents* (i.e., the substances being found in routinely observed concentrations and the circumstances not otherwise suspicious). When the death is caused by a mixed accidental drug intoxication, wording such as, "Acute intoxication by the combined effects of (fentanyl, heroin, cocaine, and ethanol)" is commonplace. When the death results mechanistically from both the drug toxicity and from sequelae of the chronic drug use (such as acute endocarditis or acquired immunodeficiency syndrome [AIDS] pneumonia), then a certification that encompasses both the acute and chronic arms of the disease would be more appropriate and reflective of this complex mechanism (e.g., "Complications of Acute and Chronic Intravenous Substance Use Disorder (fentanyl, heroin, cocaine) with Acute Endocarditis").

The potentially lethal effects of **chronic substance use** are many and varied and all would lead to a natural manner classification. These include, but are not limited to, the viral hepatidities and their sequelae (including cirrhosis and hepatocellular carcinoma), necrotizing fasciitis (when the drugs are used intravenously), immunologic, nutritional, and infectious maladies (bacteremia and endocarditis, human immunodeficiency viral infection/AIDS, bronchopneumonia), a generalized wasting syndrome, and the respiratory complications of injected pharmaceutical agents, to name a few. **"Chronic substance use" is a competent and etiologically specific natural cause of death.** The acute sequelae of acute stimulant use, cocaine being the most prevalent, include those referable to acute surges in blood pressure (ruptured saccular cerebral aneurysms, ruptured aortic dissections, myocardial infarcts, fatal arrhythmias, etc.) and would be appropriately classified as accidents. The chronic sequelae of stimulant use, on the other hand, primarily manifests as cardiac hypertrophy, with its attendant risk of SCD. As such, the latter (chronic) stimulant-related fatalities would be classified as natural deaths, as opposed to accidents.

Pancreatitis

Pancreatitis, acute or chronic, is *not an etiologically specific disease.* The two most common causes of acute pancreatitis are **cholelithiasis** with a gallstone blocking the common bile duct distal to the pancreatic duct and **heavy alcohol use.** Acute pancreatitis has a high lethal potential. Signs and symptoms of acute pancreatitis include pain in the upper abdomen, nausea/vomiting, and fever. Serum lipase and amylase are elevated. In chronic pancreatitis, weight loss, fatty stool, and diarrhea may occur. Complications include infection, hemorrhage, pancreatic insufficiency, and DM. Other less common causes of pancreatitis include *trauma, medications, infections (e.g., mumps, human immunodeficiency virus [HIV]), and malignancies.* Chronic pancreatitis may develop as a result of acute pancreatitis and is usually due to many years of heavy alcohol use. Other causes include high concentrations of triglycerides, hypercalcemia, some medications, and certain genetic disorders, such as cystic fibrosis. Microscopically, one may see necrosis, acute inflammation, and hemorrhage.[228–237]

Gastrointestinal hemorrhage (GIH)

Not uncommonly, scene photos demonstrate a pattern of hemorrhage consistent with a GIH, such as apparent **melena** on the buttocks, posterior legs, clothing, towels, or in the toilet. Melanotic stool is black, tar-like, and sticky. The color change, often accompanied by a strong and foul odor, is caused by digestive enzymes breaking down the blood as it moves through the GI tract. The detection of melena supports an *upper* GI source of the hemorrhage. Upper GIHs also may appear as fresh gastric or proximal small bowel blood or so-called "**coffee ground emesis**" (an effect of the gastric acid on the blood) when the hemorrhage exits the stomach through the "northern portal" (*hematemesis*). *Melena* should not be confused with **hematochezia**, which refers to fresh blood in the feces that imbues them with a maroon or red color. This blood typically comes from the lower GI tract and exits through the "southern portal." Fresh blood per rectum is different still, and typically represents rectoanal trauma or natural disease of the distal bowel (hemorrhoids, rectal tumor, etc.)

Depending upon the circumstances and what is known historically, an external examination alone may suffice in such an investigation. When there is an autopsy, extra care should be taken to look for the bleeding source, as it may be subtle. Assessing its quality, distribution, and quantity will help to isolate the location of the source of the hemorrhage and determine whether it is a primary or contributory cause of death. It is not unusual to see coffee ground fluid in the stomach at autopsy in a person who did *not* die from a GIH. Postmortem autolysis and oozing of blood (purge) may give the stomach contents this appearance as well. Noting that there is no blood at the scene and that the duodenum has tan-yellow chyme are supportive of this as a perimortem artifact.

The medical history or other anatomical diagnoses also will help focus the differential diagnosis. **Cirrhosis** *of the liver may suggest* **ruptured esophageal varices** *or* **gastritis** *associated with chronic alcoholism* (Fig. 7.9, Fig. 7.10, Fig. 7.11). One may see a stomach full of fresh blood that gradually becomes more digested distally, transforming into melena. At a scene, one may observe both fresh blood from hematemesis and melena from bleeding that has passed through the entire GI tract. Being careful to examine the bowel *before* it is "rinsed" will afford the opportunity to see any attached blood clots and both the quality and quantity of the blood-laden feces (important in determining the mechanism).

It is not unusual that the definitive source of the blood loss defies the prosector's search, particularly in chronic alcoholics with cirrhosis. This is described clinically as occult GI bleeding and is most commonly defined as an acute and/or chronic loss of blood, the source of which has not been identified after gastroscopy, colonoscopy, and an upper GI series.[238] Some may be explained by an acute gastritis with the coexistent coagulopathy of alcohol-related cirrhosis and platelet dysfunction. Arteriovenous malformations, especially of the small bowel, represent another difficult to identify source. While not GI in origin, another vexing source of abdominal bleeding in the alcoholic

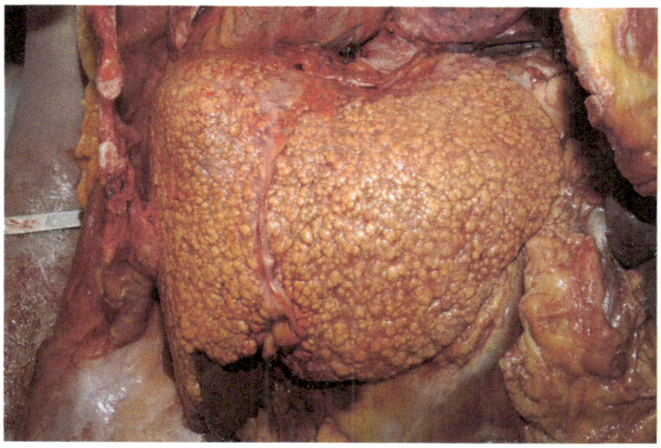

FIGURE 7.9 Cirrhosis of the liver.

FIGURE 7.10 Cross section of a cirrhotic liver.

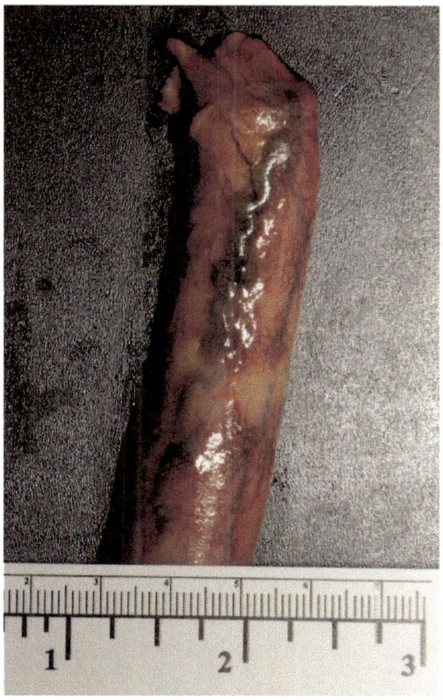

FIGURE 7.11 Esophageal varices; everted esophagus.

is a hemoperitoneum resulting from a ruptured diaphragmatic varix, similar to ruptured esophageal varices in both their pathogenesis and the pathophysiology, but also in the difficulty of their detection once collapsed. Once other natural and traumatic sources have been carefully excluded, you are left with this as a diagnosis by exclusion, but only if you remember to think of it!

After the quality and distribution pattern of the blood has been noted, attention should be turned to its *quantity*. In order to draw conclusions having scientific veracity, the extent of the hemorrhage must be both evaluated and documented. The gastric contents are easily measured and photographed. The intestinal hemorrhage is more difficult to directly measure, so is best described by the extent of the bowel containing frank blood and/or melena. A description such as, "the entire small and large bowel are distended by copious dark red and black, tarry, semisolid contents" would suffice. Pertinent negatives should also be included, descriptions such as "the small intestine contains yellow chyme" and "the large intestine contains semisolid brown stool." Photographic documentation may help memorialize the extent of hemorrhage. Other findings may help support exsanguination as the mechanism of death, evidence that may support subsequent conclusions and opinions. These include visceral (especially the brain and kidneys) and muscular pallor, a shrunken (wrinkled) spleen, and poorly formed lividity (is it scant, patchy, barely discernible?). If you do not think of these signs and specifically look for them, you likely will not notice them. Some GIHs are the primary mechanism of death, others contributory, but not all GIHs are causal in the death; quantity is key, as is the degree of heart disease (which may be necessary to invoke in part II in order to explain why an individual died short of frank exsanguination).

Finally, remember that while substantial acute GIHs can represent immediate causes, they are not etiologically specific and therefore, do not constitute underlying causes of death. Further inquiry is necessary to identify the disease, injury, or treatment that caused the bleeding and initiated the lethal sequence of events. If a bleeding gastric ulcer is discovered at autopsy, for example, its etiology must be sought (is it from peptic ulcer disease, chronic nonsteroidal antiinflammatory drugs, or a gastric cancer?). Moreover, keep in mind that *in hospitalized patients, a GIH may be one of many sequelae in a critically ill patient with stress gastritis, DIC, and/or generalized hypoperfusion-related shock from any number of etiologically specific causes, including a myriad of natural diseases, iatrogenic causes, and trauma.*

Bronchial asthma

Asthma is a chronic inflammatory disease of the pulmonary airways, with recurrent acute exacerbations. The episodes are typically characterized by wheezing, breathlessness, and chest tightness attributable to bronchoconstriction and decreased airflow. The increased airway responsiveness (bronchospasm) may be triggered by a variety of agents and events, including seasonal changes, drugs (e.g., aspirin), respiratory tract infections, inhaled irritants, allergies, exercise, and stress.[113,239–243]

Asthma is a potentially fatal disease that may follow a relatively prolonged clinical course, requiring repeated hospitalization, some with ventilator support.[113,244–246] A well-documented hospital death due to asthma is unlikely to come to the attention of the medical examiner/coroner. Deaths due to bronchial asthma, however, also may occur outside of the hospital. Out-of-hospital deaths may be sudden and occur in young adults, thus raising the question of a drug intoxication. It is not unusual for a person who smokes or insufflates cocaine or other drugs to also have a history of asthma. Therefore, these deaths are often reportable to the ME/C, based on the age of the patient and potential drug involvement alone.

Sudden deaths due to bronchial asthma do occur, particularly in children and young adults.[114,245,247,248] At autopsy, one cannot *see* bronchoconstriction because it is a functional event that cannot be documented grossly or microscopically. Therefore, one must infer it from the history and other autopsy findings. At autopsy, the lungs are often hyperinflated (they classically are seen to touch or come close to touching at the midline when the chest plate is removed, Fig. 7.12) and tenacious mucus, sometimes with frank plugging, may be seen in the smaller pulmonary airways. Microscopically, there is usually some combination of epithelial detachment, mucus gland hyperplasia, inflammatory cell infiltrates of the bronchial wall, with a prominence of eosinophils and mast cells, bronchial smooth muscle hyperplasia/hypertrophy, and deposits along the basement membranes of the bronchial epithelium that give the appearance of a thickened basement membrane. Hypertrophy of the submucosal glands is also commonly seen. If an acute asthma attack with sudden bronchospasm rapidly leads to death, the autopsy findings of an acute exacerbation may be limited. Minimal "asthma" findings at autopsy should not necessarily discourage the FP from making the diagnosis of bronchial asthma, with the appropriate history and the exclusion of other causes.

Homicide-by-heart attack deaths focus on the physiologic stress to the cardiovascular system as it relates to criminal activity that results in a homicidal manner of death determination.[249–252] Stress from a threat or physical

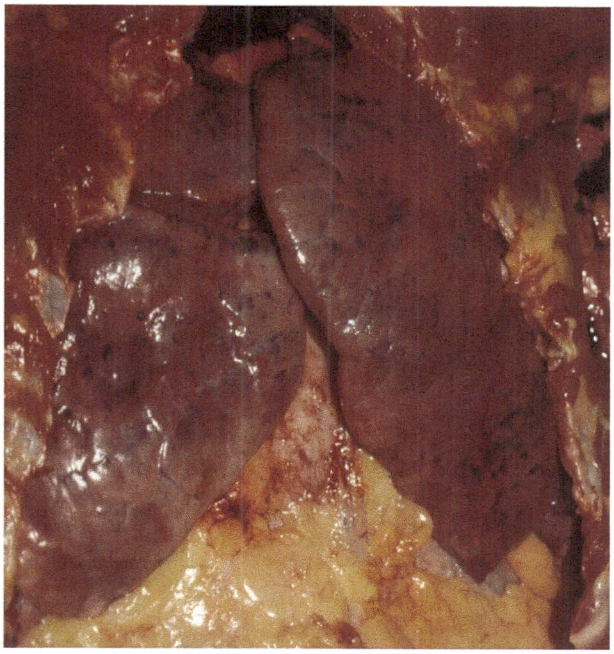

FIGURE 7.12 **Acute bronchial asthma with hyperinflation of the lungs.**

altercation may affect other organ systems as well, including the respiratory system.[249,250,253] In one report, four people with a history of asthma were subjected to the stress of a criminal altercation, but did not sustain functionally significant visceral injuries.[253] The four deaths resulted from the combined effects of underlying bronchial asthma and the superimposed stress. Since operationally, homicide is defined as death at the hand of another person or as resulting from the illegal act of another, these deaths were certified as homicides.[141]

The investigation of an asthma-related homicide should document the underlying medical disease, exclude physical or chemical injury sufficient to explain death, and include a thorough and detailed examination of the circumstances in order to evaluate it within the context of the entire investigation. The causes of death in such scenarios are often certified with a description of the events that triggered the fatality, listed as the immediate cause of death, and bronchial asthma as the underlying cause of death (e.g., respiratory arrest during/immediately following physical altercation with the threat of a baseball bat due to acute and chronic bronchial asthma).[253]

Dementia and old age

Dementia is a potentially fatal, progressive disease.[254] Its mechanisms of death (e.g., bronchopneumonia, inanition) may be incorrectly used as sole causes of death, thus hiding the insidious culprit. Moreover, dementia may serve as the circumstantial explanation and/or proximate cause in some instances of traumatic accidental deaths (choking, hypothermia, and drowning, to name only a few). There also is a risk, however, of "overcalling" dementia as a cause of death. A patient with dementia who dies from a COVID-19 infection is not dying from dementia.[255–257] There are many types of dementia, with varied pathogeneses, and some require an autopsy to make a definitive diagnosis. Without an autopsy, one may certify a death based on the history (e.g., radiographic, neurologic, clinical), using phrases such as "Alzheimer-type Dementia" or "Dementia, Unspecified."

Inanition/progressive inanition is the "exhausted condition that results from lack of food and water" and it may be related to psychiatric (e.g., clinical depression, severe drug dependence) or pathophysiologic conditions (e.g., progressive neurodegenerative disorders, cancer). It is also a common immediate cause of death due to old age or dementia, but is competent to stand alone in part 1 as a proximate cause when not directly caused by or related to any one condition, but commonly, there are one or more conditions listed in part 2.

"Old Age" or "Advanced Age" are competent, etiologically specific causes of death that are often appropriate and typically well-accepted by families, as they are evocative of an inevitable, peaceful, (and blameless) end of life.

Diabetes mellitus

Diabetes mellitus (DM) is a group of metabolic disorders that have the capacity to adversely affect nearly all organ systems, and it is characterized by hyperglycemia over a prolonged duration. Presenting symptoms may include frequent urination and increased thirst (polyuria and polydipsia). The forensic pathologist is likely to see the cardiovascular effects of longstanding diabetes, as well as deaths due to previously undiagnosed, new onset disease. These may present with acute and severe complications, including diabetic ketoacidosis (DKA), hyperosmolar hyperglycemic coma, or sudden death. Type I ("insulin-dependent DM") is more likely to be complicated by DKA. Long-term complications include cardiovascular disease, CVEs, kidney disease, complications of peripheral vascular disease, and damage to the nerves/eyes.

The underlying pathology is due to either a failure of the pancreas to produce enough insulin or a failure of the cells of the body/end organs to respond properly to it. Type 1 diabetes results from an inability of the pancreas to produce enough insulin due to loss of beta cells; this is caused by an **autoimmune** response, but also may occur *secondarily* as a consequence of other disease processes (e.g., **chronic pancreatitis due to chronic alcoholism, cystic fibrosis, other endocrinopathies,** etc.). In addition, some cases may be related to the administration of **atypical antipsychotic medications** (e.g., clozapine, olanzapine, quetiapine, and risperidone), which can cause new onset disease with a telescoped course and, when fatal, may revert the manner of death to a complication of therapy (TC).[258] Moreover, certain illicit drugs, such as cocaine and heroin, can serve as catalysts in tipping the fragile diabetic patient's endocrine balance, triggering a progression into DKA. As such, in diabetic deaths with evidence of both DKA and an **acute cocaine or heroin intoxication**, the intoxication should be included in part 2 and the manner of death classified as an accident. The latter two examples demonstrate that, while DKA can stand alone as a competent, etiologically specific cause of death, other contributory variables may affect the manner designation.

Type 2 diabetes begins with insulin resistance and as it progresses, develops into a lack of insulin production. This was previously referred to as "noninsulin-dependent DM" or "adult-onset diabetes" and is most typically related to obesity and insufficient exercise. There are no pathognomonic autopsy findings that reliably diagnose new onset DM. One may see some pancreatic chronic inflammation or subnuclear vacuolization of the proximal tubules of the kidneys (Armanni–Ebstein lesion), but the best test is vitreous glucose quantitation. Hemoglobin-A1c testing also may be useful, as it gives a temporal overview of glucose control, but not whether there was an acute exacerbation.

Vitreous chemical testing is typically indicated in death investigations that do not have a clinical, gross, or microscopic cause of death. Since uncontrolled DM can result in hyperglycemia and ketoacidosis, *the detection of acetone in the blood (routinely detected during a volatile forensic toxicology screen for alcohols) commonly triggers reflexive vitreous glucose testing in many forensic laboratories.* A quick and cheap test that can be done at the autopsy table is the *glucose/ketone urine dip test;* while not conclusive, this serves as a competent tool in helping to select cases for more definitive glucose testing in the lab. Usually, there is plenty of urine to test, in these deaths, related to the glycosuria.

Vitreous glucose concentrations typically and rapidly decrease after death. It is not unusual to find a barely detectable glucose concentration in postmortem vitreous humor. Therefore, diagnosing hypoglycemia in the postmortem setting based on the vitreous glucose concentrations is not reliable. Vitreous glucose, however, may be useful in diagnosing hyperglycemia. A postmortem vitreous glucose of over 200 mg/dL is nearly always going to represent diabetes, even if not DKA, per se.[259–264] A study of vitreous samples from 27 individuals scheduled for retinal surgery compared the vitreous glucose with the blood glucose.[265] They found that *the mean serum glucose concentrations were approximately double that of the vitreous glucose.*[265]

Ketone bodies include three molecules: **acetoacetate, beta-hydroxybutyrate (BHB), and acetone.** In DKA, a relative insulin deficiency (type I diabetes) or a major increase in insulin resistance (type 2 diabetes) results in severe hyperglycemia, while in alcoholic and starvation ketoacidosis, there is *hypoglycemia* due to reduced nutritional intake. In diabetics, the inability to use glucose due to insulin insufficiency or insulin resistance results in ketone body production. In prolonged fasting/starvation, enhanced ketone body synthesis is the result of glucose unavailability. Ketone bodies are mainly produced in hepatocyte mitochondria and are energy-rich compounds that transport energy from the liver to other tissues. Isopropyl alcohol (isopropanol) is currently considered a marker of ketoacidosis and a product of acetone metabolism in clinical conditions that result in increased ketone bodies. It can be detected in several forensic contexts, beyond direct exposure to isopropyl alcohol itself. These include fatalities involving diabetic and alcoholic ketoacidosis, hypothermia, starvation, and putrefaction. For these reasons, detection of isopropanol does not definitively mean a person has ingested isopropanol.

Several researchers have proposed that **BHB** is a better postmortem indicator of ketoacidosis than acetone.[259] Blood BHB concentrations of up to 500 μmol/L (5.2 mg/dL) have been proposed as normal and over 2500 μmol/L (27.5 mg/dL) as pathological.[266] BHB also may be able to distinguish the ingestion of isopropanol from endogenous production and **aid in** the identification of **alcoholic ketoacidosis.**[267]

Obesity as a cause of death

The understanding of the relationship between obesity and increased morbidity and mortality is as old as medicine.[268] Obesity's deleterious effects have been described in the works of Hippocrates to Virchow. Despite the medical risks of obesity (obesity has been called "medicolegal dynamite" due to its attendant increased risks of complications from medical procedures), it continues to vex clinicians and patients alike. In adults, a body mass index (BMI) of 25 kg/m^2 to 29 kg/m^2 is defined as *overweight* while *obesity* is defined as having a BMI of 30 kg/m^2 or greater.

Due to its multiorgan pathologies (hypertension, cardiac hypertrophy, restrictive lung disease, atherosclerosis, diabetes, liver disease, obstructive sleep apnea, etc.), it is often included as a part 2 diagnosis. In some instances, however, it may be the proximate/primary cause (e.g., saddle pulmonary thromboembolism when obesity is the only or at least the primary risk factor, or a sudden arrhythmic death in a young, morbidly obese person with a 1200-gram heart). Obesity can cause or contribute to death in a variety of ways.[268–276] These mechanisms include the arrhythmogenicity related to cardiac hypertrophy, pulmonary thromboembolism, DM with ketoacidosis, and sleep apnea. The relative inactivity or immobility related to obesity will also likely contribute to cardiac deconditioning, venous stasis dermatitis, which can lead to cellulitis, and an increase risk for deep venous thrombosis and pulmonary thromboembolism.

Bronchopneumonia

While **acute bronchopneumonia** *is a commonly encountered immediate cause of death, it lacks etiologic specificity; as such, it should not stand alone on the death certificate.* Some bronchopneumonias diagnosed at autopsy may be incidental, and not contributory to death. The person was dying *with* (a so-called "nickel's worth of pneumonia"), but not *from* the bronchopneumonia. The forensic pathologist who fails to recognize this may unwittingly also fail to recognize an unnatural manner, including a homicide, by not identifying the proximate cause of the bronchopneumonia when it is, indeed, the immediate cause. Common examples of underlying causes include alcohol use disorder, chronic substance use, advanced cardiovascular disease, DM, any disease, injury, or treatment (including surgery) that puts an individual to bed or otherwise renders them nonambulatory, extremes of age, and respiratory diseases, to name only a few. As such, depending on the proximate cause, a death caused (immediately) by bronchopneumonia has the capacity to be a natural death, an unnatural death (including a homicide), or a therapeutic complication. All this said, true **lobar pneumonia, by contrast,** is etiologically specific and therefore a competent cause of death in itself.

Sudden death and chronic renal disease

This textbook is not intended to represent a primary or comprehensive source of basic pathology. While a detailed discussion of medical kidney disease extends beyond our scope, there are some related areas that the forensic pathologist commonly encounters and that therefore deserve comment.[277,278] The FP's primary focus in the setting of acute and chronic kidney failure is typically to determine its underlying etiology; *"end stage kidney/renal disease" alone is insufficient for death certification purposes.* The most common causes of chronic kidney disease are **DM** and **HCVD**. Other causes include systemic lupus erythematosus, autosomal dominant polycystic kidney disease, acute or chronic drug toxicity (e.g., ibuprofen), paraproteinemia (amyloidosis, multiple myeloma), hypoperfusion of myriad causes, including trauma ("shock kidney"), and sickle cell disease. One should avoid the common medical terminology of "kidney injury," as the word "injury" will set off alarm bells to vital records staff, who are unfamiliar with this common medical usage.

At autopsy, since the concentrations of urea and creatinine are relatively stable in the postmortem vitreous humor, they may help in making the diagnosis. In certain settings, apparent acute tubular necrosis may be seen at autopsy; however, it may be challenging to distinguish this from autolysis. Some causes of kidney failure are related to polycystic kidney disease, which is important to include given its potential hereditary pathogenesis.

Mechanistically, severe kidney disease can primarily cause or contribute to death through the arrhythmogenicity related to electrolyte and toxic-metabolic derangements, exacerbations of fluid overload and congestive heart failure, coagulopathies (uremia-related thrombocytopathy and proteinuria-related deficiencies of clotting factors), and electrolyte, hemorrhagic, infectious, or other complications of hemodialysis and its attendant arteriovenous shunt placement and management. Of note, and most commonly, the latter represent therapeutic complications.

Sudden death from endocrine disorders

Of all the endocrine disorders encountered by the forensic pathologist, DM represents the most prevalent. As such, its discussion warrants a dedicated section (see above). By contrast, the thyroid, adrenal, and pituitary glands are rarely directly involved in causing a sudden death. For completeness, however, a brief summary of each is included.

Both hyperthyroidism and hypothyroidism have the capacity to cause or contribute to death. *Thyrotoxicosis* results from the production of excessive thyroid hormone of any cause. Signs and symptoms may include hyperkinetic movement disorders, irritability, muscle weakness, cardiac arrhythmias (e.g., atrial fibrillation), heat intolerance, hand tremor, and weight loss. *Thyroid storm* is an uncommon complication that is often associated with/precipitated by an infection, leading to worsening of symptoms, a high temperature, and death.[279–287]

Conversely, hypothyroidism may present with poor cold tolerance, fatigue, bradycardia, depression, and weight gain. *Myxedema coma* results from a decompensated form of hypothyroidism that is potentially lethal; the laboratory values may reflect an uncomplicated hypothyroid state, but, somewhat analogous to the underpinnings of DKA, a stressful event (e.g., infection, myocardial infarct, or stroke) can precipitate the myxedema coma. This manifests as an altered mental status and hypothermia, as well as hypoglycemia, hypotension, hyponatremia, hypercapnia/hypoxia, bradycardia, and hypoventilation; despite the name, myxedema is not necessarily seen. When it does occur, myxedema results from a deposition of mucopolysaccharides in the dermis, which, in turn, results in swelling (e.g., pretibial myxedema). It is also a hallmark of Graves' disease, an autoimmune form of hyperthyroidism, and can be seen in Hashimoto's thyroiditis and other chronic forms of hypothyroidism.[288]

The adrenal glands become of interest to the forensic pathologist under three circumstances. The first is when they are hemorrhagic, which may indicate sepsis, particularly fulminant meningococcemia.[289–294] This disease constellation is known as **Waterhouse-Friderichsen syndrome** and, because of its rapidly fatal course, led W.W. Herrick to state that, "no other infection so quickly slays."[295] *Meningococcemia*, with or without meningitis, is a communicable disease and an immediately reportable condition; it constitutes an urgent/emergent public health matter that requires a swift response in order to provide prophylaxis to intimate contacts.[296,297] A *pheochromocytoma* may present as a sudden and unexplained death resulting from cardiac (stress cardiomyopathy, cardiac hypertrophy) and central nervous system (intracerebral hemorrhage) pathology. *Adrenal insufficiency* may be seen in patients on chronic corticosteroid therapy or during infancy. Causes of *acute* adrenal insufficiency include an abrupt withdrawal of long-term corticosteroid medication (a therapeutic complication), Waterhouse–Friderichsen syndrome, and acute stress in those with underlying chronic adrenal insufficiency (Addison disease, tuberculosis, and metastatic disease). Hyponatremia typically occurs with glucocorticoid deficiency.

Sheehan's syndrome, also known as postpartum pituitary gland necrosis, manifests as acute hypopituitarism caused by an infarction complicating hypovolemic shock that occurs during childbirth. The most common initial symptoms are agalactorrhea and amenorrhea or oligomenorrhea after delivery. In some rare instances, the woman with Sheehan's syndrome is relatively asymptomatic and the diagnosis may be delayed for years.[298–300]

Sudden death and psychiatric disease

Schizophrenia is characterized by continuous or relapsing episodes of psychosis. The major symptoms include hallucinations, delusions, paranoia, and disorganized thinking. In 2017, there were an estimated 1.1 million new cases and, in 2019, a total of 20 million cases globally.[301] Schizophrenia is associated with increased mortality that results in a decreased life expectancy by 20–25 years.[302] Although traumatic deaths account for most of this increase, schizophrenic patients also experience increased mortality due to natural causes. Schizophrenia is commonly associated with obesity, poor diet, a sedentary lifestyle, smoking, and an increased rate of suicide.[302] DM carries a higher prevalence among schizophrenics relative to the general population that is independent of psychotropic medical intervention. That said, the adverse effects of atypical antipsychotic agents play an additional role in the development and complications of DM in some.[258,302,303]

There are patients with schizophrenia who die with no recognizable cause of death after a complete autopsy.[304] Altered autonomic physiology might be a factor in these deaths. A report of six such decedents support the opinion that schizophrenia, per se, is a potentially lethal disorder that can present as a sudden death, independent of the above named risk factors.[304] As such, schizophrenia is a competent cause of death in persons found unexpectedly dead and in whom no other cause is identified with a complete autopsy. Recent studies indicate that almost 40% of those with schizophrenia die from complications of cardiovascular disease, including SCD.[305] An additional underlying risk factor may involve BrS, in which calcium channel mutations overlap with those linked with schizophrenia.[305] BrS may also be drug-induced from certain antipsychotic and antidepressant medications.[305] Psychogenic polydipsia is relatively common in patients with schizophrenia (anticholinergic medications can lead to a dry mouth), which may in turn result in hyponatremia.[306,307] A psychotic illness, particularly schizophrenia, may also be the proximate or contributory cause of a fatal hyperactive delerium (see Chapter 15 for a more in-depth discussion).

Sickle cell disease and trait

Sickle cell trait (Hgb AS) rarely causes death in and of itself. Sickling complications among persons with sickle cell trait (eight to 9% of African-Americans and 0.046% of other Americans) are uncommon and typically mild, but serious complications, and rarely death, may manifest in association with extreme physical exertion. Exertion-related sickling is generally related to intensity, not duration. The onset of sickling and related symptoms can occur within minutes of the exertion. Sudden death with sickle cell trait was first described among military recruits and athletes.[112,115,308–391] Some reported deaths associated with a police chase/restraint have also been attributed, at least in part, to sickle cell trait.[392–394] Adverse manifestations and death are even more frequent, of course, in those having sickle cell disease (Hgb SS), with or without exertional stress.

Postmortem findings in these deaths are nonspecific. Sickled cells are typically seen microscopically, and molecular genetic testing or hemoglobin electrophoresis can confirm the diagnosis. A notable caveat is that the mere presence of sickled red blood cells in postmortem microscopic sections is not diagnostic of antemortem sickling, per se, because artifactual sickling can occur postmortem. In the right clinical context (dyspnea, pain crisis, extreme exertion, etc.) and the exclusion of a more compelling cause, however, the finding of "stuffing" and distending of small blood vessels in the pulmonary and glomerular microvasculature and hepatic sinusoids and a patchy distribution of sickled cells may support the diagnosis of (antemortem) sickling.[316] Beyond a sickle crisis, patients with sickle cell disease are also at heightened risk of opioid dependence and misuse resulting from long-term pain management.

Human immunodeficiency virus

Over the past couple of decades, morbidity and mortality related to HIV infection and AIDS have decreased substantially.[395,396] These conditions still bear mentioning mainly because there was a time when HIV/AIDS deaths falling under the purview of an average ME/C office represented a substantial proportion of the caseload, many of such deaths occurring at home and/or related to chronic intravenous substance use.[397,398] Today, largely as a result of highly effective antiretroviral therapy, the AIDS epidemic has largely been eradicated and the handling of these deaths by most ME/C offices is now a rarity, AIDS no longer representing the death sentence it once had.

That said, it is still important to remember key concepts that come into play when considering the placement of HIV and AIDS on the death certificate. To wit, if it is the primary cause of death, *AIDS can stand alone on the death certificate as the proximate (natural) cause of death when it was acquired* **sexually** (recall that sexual intercourse is neither a disease nor an injury, so should not be mentioned anywhere on the death certificate). But, *if the HIV was acquired through* **chronic intravenous drug use**, *HIV would be listed as the immediate cause, with the chronic intravenous substance use serving as the proximate cause,* and the manner still classified as natural. The death would be similarly *certified as natural if the HIV was acquired through* **vertical/maternal transmission.** If, on the other hand, the *HIV was transmitted through a blood transfusion, both the transfusion and the disease or injury requiring the transfusion would need to be listed on the death certificate, and the indication for the transfusion informing the manner.* If the **transfusion**, for example, was given as part of emergency surgery for remote traumatic injury, both would be listed on the death certificate and the manner would be determined by the circumstances of the injury occurrence (accident, suicide, or homicide). If, however, the transfusion was given as part of the treatment for a natural disease, the manner would be classified as a **therapeutic complication** (in jurisdictions where TC is available).

References

1. Stahl-Herz J, Pasquale-Styles M. Autopsy-Diagnosed Injury Deaths in Persons with Acute or Chronic Alcohol Use: A Review of 1000 Deaths with History of Alcohol Use (4.7 Abstract). Paper Presented at: National Association of Medical Examiners Annual Meeting 2021; West Palm Beach, FL.
2. Messite J, Stellman SD. Accuracy of death certificate completion: the need for formalized physician training. *JAMA*. 1996;275(10):794–796.
3. Zumwalt RE, Ritter MR. Incorrect death certification. An invitation to obfuscation. *Postgrad Med*. 1987;81(8):245–247, 250, 253–244.
4. Davis GG. Mind your manners. Part I: history of death certification and manner of death classification. *Am J Forensic Med Pathol*. 1997;18(3): 219–223.
5. Nashelsky MB, Lawrence CH. Accuracy of cause of death determination without forensic autopsy examination. *Am J Forensic Med Pathol*. 2003;24(4):313–319.
6. Pritt BS, Hardin NJ, Richmond JA, Shapiro SL. Death certification errors at an academic institution. *Arch Pathol Lab Med*. 2005;129(11): 1476–1479.
7. Roulson J, Benbow EW, Hasleton PS. Discrepancies between clinical and autopsy diagnosis and the value of post mortem histology; a meta-analysis and review. *Histopathology*. 2005;47(6):551–559.
8. Croft PR, Lathrop SL, Zumwalt RE. Amended cause and manner of death certification: a six-year review of the New Mexico experience. *J Forensic Sci*. 2006;51(3):651–656.
9. Nelson CL, Winston DC. Detection of medical examiner cases from review of cremation requests. *Am J Forensic Med Pathol*. 2006;27(2): 103–105.
10. Cina SJ, Selby DM, Clark B. Accuracy of death certification in two tertiary care military hospitals. *Mil Med*. 1999;164(12):897–899.
11. Maudsley G, Williams EM. 'Inaccuracy' in death certification—where are we now? *J Public Health Med*. 1996;18(1):59–66.
12. Slater DN. Certifying the cause of death: an audit of wording inaccuracies. *J Clin Pathol*. 1993;46(3):232–234.
13. Smith Sehdev AE, Hutchins GM. Problems with proper completion and accuracy of the cause-of-death statement. *Arch Intern Med*. 2001; 161(2):277–284.
14. Jordan JM, Bass MJ. Errors in death certificate completion in a teaching hospital. *Clin Invest Med*. 1993;16(4):249–255.
15. Hanzlick R. Death certificates. The need for further guidance. *Am J Forensic Med Pathol*. 1993;14(3):249–252.
16. Hanzlick R. Death certificates, natural death, and alcohol. The problem of underreporting. *Am J Forensic Med Pathol*. 1988;9(2):149–150.
17. Lin P, Gill JR. Delayed homicides and the proximate cause. *Am J Forensic Med Pathol*. 2009;30(4):354–357.
18. Adams VI, Herrmann MA. The medical examiner. When to report and help with death certificates. *J Fla Med Assoc*. 1995;82(4):255–260.
19. Every NR, Parsons L, Hlatky MA, et al. Use and accuracy of state death certificates for classification of sudden cardiac deaths in high-risk populations. *Am Heart J*. 1997;134(6):1129–1132.
20. Goraya TY, Jacobsen SJ, Belau PG, Weston SA, Kottke TE, Roger VL. Validation of death certificate diagnosis of out-of-hospital coronary heart disease deaths in Olmsted County, Minnesota. *Mayo Clin Proc*. 2000;75(7):681–687.
21. Fischtein D, Cina SJ. Errors on death certificates requiring amendments: the Broward County experience. *Am J Forensic Med Pathol*. 2011;32(2): 146–148.
22. Brooks EG, Reed KD. Principles and pitfalls: a guide to death certification. *Clin Med Res*. 2015;13(2):74–82.
23. Warner M, Paulozzi L, Nolte K, Davis G, Nelson L. State Variation in certifying manner of death and drugs involved 2013. *Acad Forensic Pathol*. 2013;3(2):231–237.
24. McGivern L, Shulman L, Carney JK, Shapiro S, Bundock E. Death certification errors and the effect on mortality statistics. *Public Health Rep*. 2017;132(6):669–675.
25. Randall B. Death certification: a primer, Part IV–problems in death certification. *S D Med*. 2014;67(8):315–317, 319.
26. Myers KA, Farquhar DR. Improving the accuracy of death certification. *CMAJ (Can Med Assoc J)*. 1998;158(10):1317–1323.
27. Mieno MN, Tanaka N, Arai T, et al. Accuracy of death certificates and assessment of factors for misclassification of underlying cause of death. *J Epidemiol*. 2016;26(4):191–198.
28. Hanzlick R. Quality assurance review of death certificates: a pilot study. *Am J Forensic Med Pathol*. 2005;26(1):63–65.
29. Ruiz L, Posey BM, Neuilly MA, Stohr MK, Hemmens C. Certifying death in the United States. *J Forensic Sci*. 2018;63(4):1138–1145.
30. Adelson L. Let's do away with CVA. *JAMA*. 1976;236(21):2390.
31. Adelson L. No anatomic cause of death; the enigma of the forensic pathologist. *Conn Med*. 1954;18(9):732–738.
32. 1934–1936 W. Boyd Canada (Scotland). 2021. www.iapcentral.org/home/hall-of-presidents/1934-1936-william-boyd. Accessed 11/24/2021.
33. Gill JR, Lange RA, Azar OP. What is sudden cardiac death? *Acad Forensic Pathol*. 2011;1(2):176–186.
34. Adelson L. The forensic pathologist. "Family physician" to the bereaved. *JAMA*. 1977;237(15):1585–1588.
35. Roger VL, Go AS, Lloyd-Jones DM, et al. Heart disease and stroke statistics—2011 update: a report from the American Heart Association. *Circulation*. 2011;123(4):e18–e209.
36. Myerburg RJ, Kessler KM, Castellanos A. Sudden cardiac death: epidemiology, transient risk, and intervention assessment. *Ann Intern Med*. 1993;119(12):1187–1197.
37. Davies MJ, Popple A. Sudden unexpected cardiac death–a practical approach to the forensic problem. *Histopathology*. 1979;3(4):255–277.
38. Thiene G, Basso C, Corrado D. Cardiovascular causes of sudden death. In: Silver MD, Gotlieb AI, Schoen FJ, eds. *Cardiovascular Pathology*. New York: Churchhill Livingstone; 2001:326–374.
39. Sampson BA, Adams VI, Hirsch CS. Sudden and unexpected deaths from natural causes in adults. In: WU S, Spitz DJ, eds. *Spitz and Fisher's Medicolegal Investigation of Death*. 4th ed. Springfield: Charles C Thomas; 2006:301–342.
40. Kong MH, Peterson ED, Fonarow GC, et al. Addressing disparities in sudden cardiac arrest care and the underutilization of effective therapies. *Am Heart J*. 2010;160(4):605–618.
41. Gill JR, Scordi-Bello IA. Natural, unexpected deaths: reliability of a presumptive diagnosis. *J Forensic Sci*. 2010;55(1):77–81.
42. Boyd W. *A Textbook of Pathology: Structure and Function in Disease*. Philadelphia: Lea and Febiger; 1970.
43. DiMaio V, DiMaio D. *Forensic Pathology*. 2nd ed. Boca Raton, FL: CRC Press; 2001.

44. Morentin B, Audicana C. Population-based study of out-of-hospital sudden cardiovascular death: incidence and causes of death in middle-aged adults. *Rev Esp Cardiol*. 2010;64(1):28–34.
45. Davis JH, Wright RK. The very sudden cardiac death syndrome—a conceptual model for pathologists. *Hum Pathol*. 1980;11(2):117–121.
46. Mittleman RE, Wetli CV. Cocaine and sudden "natural" death. *J Forensic Sci*. 1987;32(1):11–19.
47. Stephens BG, Jentzen JM, Karch S, Wetli CV, Mash DC. National Association of Medical Examiners position paper on the certification of cocaine-related deaths. *Am J Forensic Med Pathol*. 2004;25(1):11–13.
48. Chugh SS, Kelly KL, Titus JL. Sudden cardiac death with apparently normal heart. *Circulation*. 2000;102(6):649–654.
49. Tester DJ, Ackerman MJ. Postmortem long QT syndrome genetic testing for sudden unexplained death in the young. *J Am Coll Cardiol*. 2007; 49(2):240–246.
50. Basso C, Burke M, Fornes P, et al. Guidelines for autopsy investigation of sudden cardiac death. *Virchows Arch*. 2008;452(1):11–18.
51. Wei JP, Kay D, Fishbein MC. Spontaneous dissection of the distal obtuse marginal coronary artery: a rare cause of sudden death. *Am J Forensic Med Pathol*. 2008;29(2):199–201.
52. Carson HJ, Feickert BL. Coronary arteritis diagnosed at autopsy: three case reports and review of the literature. *Am J Forensic Med Pathol*. 2000; 21(4):349–353.
53. Veinot JP, Johnston B, Acharya V, Healey J. The spectrum of intramyocardial small vessel disease associated with sudden death. *J Forensic Sci*. 2002;47(2):384–388.
54. Alegria JR, Herrmann J, Holmes Jr DR, Lerman A, Rihal CS. Myocardial bridging. *Eur Heart J*. 2005;26(12):1159–1168.
55. Gow RM. Myocardial bridging: does it cause sudden death? *Card Electrophysiol Rev*. 2002;6(1–2):112–114.
56. Pigozzi F, Rizzo M. Sudden death in competitive athletes. *Clin Sports Med*. 2008;27(1):153–181 (ix).
57. Nimura H, Bachinski LL, Sangwatanaroj S. Mutations in the gene for cardiac myosin-binding protein c and late-onset familial hypertrophic cardiomyopathy. *N Engl J Med*. 1998;338:1248–1257.
58. Sutton MSJ, Epstein J. Hypertrophic cardiomyopathy-beyond the sarcomere. *N Engl J Med*. 1998;338:1303–1304.
59. Litovsky SH, Rose AG. Clinicopathologic heterogeneity in hypertrophic cardiomyopathy with regard to age, asymmetric septal hypertrophy, and concentric hypertrophy beyond the pediatric age group. *Arch Pathol Lab Med*. 1998;122(5):434–441.
60. Basso C, Thiene G, Corrado D, Buja G, Melacini P, Nava A. Hypertrophic cardiomyopathy and sudden death in the young: pathologic evidence of myocardial ischemia [see comments]. *Hum Pathol*. 2000;31(8):988–998.
61. Cedrone AJ, Makaryus JN, Catanzaro JN, et al. Sudden cardiac death in a 20-year-old male swimmer. *South Med J*. 2010;103(5):464–466.
62. Maron BJ, Shirani J, Poliac LC, Mathenge R, Roberts WC, Mueller FO. Sudden death in young competitive athletes. Clinical, demographic, and pathological profiles. *JAMA*. 1996;276(3):199–204.
63. Feldman AM, McNamara D. Medical progress: myocarditis. *N Engl J Med*. 2000;343(19):1388–1398.
64. Diaz FJ, Loewe C, Jackson A. Death caused by myocarditis in Wayne County, Michigan: a 9-year retrospective study. *Am J Forensic Med Pathol*. 2006;27(4):300–303.
65. Silingardi E, Rivasi F, Santunione AL, Garagnani L. Sudden death from tubercular myocarditis. *J Forensic Sci*. 2006;51(3):667–669.
66. Burke AP, Saenger J, Mullick F, Virmani R. Hypersensitivity myocarditis. *Arch Pathol Lab Med*. 1991;115(8):764–769.
67. Gore I, Saphir O. Myocarditis: a classification of 1402 cases. *Am Heart J*. 1947;34:827–830.
68. Hardman JM, Earle KM. Myocarditis in 200 fatal meningococcal infections. *Arch Pathol*. 1969;87(3):318–325.
69. Basso C, Carturan E, Pilichou K, Rizzo S, Corrado D, Thiene G. Sudden cardiac death with normal heart: molecular autopsy. *Cardiovasc Pathol*. 2010;19(6):321–325.
70. Campuzano O, Beltran-Alvarez P, Iglesias A, Scornik F, Perez G, Brugada R. Genetics and cardiac channelopathies. *Genet Med*. 2010;12(5): 260–267.
71. Chopra N, Knollmann BC. Genetics of sudden cardiac death syndromes. *Curr Opin Cardiol*. 2011;26(3):196–203.
72. Carturan E, Tester DJ, Brost BC, Basso C, Thiene G, Ackerman MJ. Postmortem genetic testing for conventional autopsy-negative sudden unexplained death: an evaluation of different DNA extraction protocols and the feasibility of mutational analysis from archival paraffin-embedded heart tissue. *Am J Clin Pathol*. 2008;129(3):391–397.
73. Tester DJ, Ackerman MJ. Cardiomyopathic and channelopathic causes of sudden unexplained death in infants and children. *Annu Rev Med*. 2009;60:69–84.
74. Towbin JA. Molecular genetic basis of sudden cardiac death. *Cardiovasc Pathol*. 2001;10(6):283–295.
75. Di Paolo M, Luchini D, Bloise R, Priori SG. Postmortem molecular analysis in victims of sudden unexplained death. *Am J Forensic Med Pathol*. 2004;25(2):182–184.
76. Tester DJ, Ackerman MJ. The role of molecular autopsy in unexplained sudden cardiac death. *Curr Opin Cardiol*. 2006;21(3):166–172.
77. Severson E, Timm K. The value of molecular autopsy: genetic testing reveals long QT mutations in an autopsy-negative, postpartum sudden unexpected death. *Acad Forensic Pathol*. 2014;4(4):552–557.
78. Glengarry JM, Crawford J, Morrow PL, Stables SR, Love DR, Skinner JR. Long QT molecular autopsy in sudden infant death syndrome. *Arch Dis Child*. 2014;99(7):635–640.
79. Wang D, Shah KR, Um SY, et al. Cardiac channelopathy testing in 274 ethnically diverse sudden unexplained deaths. *Forensic Sci Int*. 2014;237: 90–99.
80. Langlois NE. Sudden adult death. *Forensic Sci Med Pathol*. 2009;5(3):210–232.
81. Rosenhek R, Zilberszac R, Schemper M, et al. Natural history of very severe aortic stenosis. *Circulation*. 2010;121(1):151–156.
82. Rosenhek R, Maurer G, Baumgartner H. Should early elective surgery be performed in patients with severe but asymptomatic aortic stenosis? *Eur Heart J*. 2002;23(18):1417–1421.
83. Pellikka PA, Nishimura RA, Bailey KR, Tajik AJ. The natural history of adults with asymptomatic, hemodynamically significant aortic stenosis. *J Am Coll Cardiol*. 1990;15(5):1012–1017.
84. Gould KL, Carabello BA. Why angina in aortic stenosis with normal coronary arteriograms? *Circulation*. 2003;107(25):3121–3123.
85. Virmani R, Atkinson JB, Forman MB, Robinowitz M. Mitral valve prolapse. *Hum Pathol*. 1987;18(6):596–602.
86. Franchitto N, Bounes V, Telmon N, Rouge D. Mitral valve prolapse and out-of-hospital sudden death: a case report and literature review. *Med Sci Law*. 2010;50(3):164–167.

87. Grau JB, Pirelli L, Yu PJ, Galloway AC, Ostrer H. The genetics of mitral valve prolapse. *Clin Genet*. 2007;72(4):288–295.
88. Scala-Barnett DM, Donoghue ER. Sudden death in mitral valve prolapse. *J Forensic Sci*. 1988;33(1):84–91.
89. Burke AP, Virmani R. Intramural coronary artery dysplasia of the ventricular septum and sudden death. *Hum Pathol*. 1998;29(10):1124–1127.
90. Veinot JP, Johnston B. Cardiac sarcoidosis—an occult cause of sudden death: a case report and literature review. *J Forensic Sci*. 1998;43(3):715–717.
91. Jing HL, Hu BJ. Sudden death caused by stricture of the sinus node artery. *Am J Forensic Med Pathol*. 1997;18(4):360–362.
92. Cohle SD, Suarez-Mier MP, Aguilera B. Sudden death resulting from lesions of the cardiac conduction system. *Am J Forensic Med Pathol*. 2002;23(1):83–89.
93. Michaud K, Horisberger B, Romain N, Mangin P. Sudden death following exercise testing related to abnormalities of the cardiac conduction system and coronary artery pathology. *Am J Forensic Med Pathol*. 2004;25(2):161–163.
94. Gulino SP. Examination of the cardiac conduction system: forensic application in cases of sudden cardiac death. *Am J Forensic Med Pathol*. 2003;24(3):227–238.
95. Waller BF, Gering LE, Branyas NA, Slack JD. Anatomy, histology, and pathology of the cardiac conduction system: Part I. *Clin Cardiol*. 1993;16(3):249–252.
96. Fessenden F. *Health Mystery in New York: Heart Disease*. New York Times; 2005, 8/18/2005.
97. Hirsch CS, Flomenbaum M. Problem-solving in death certification. *ASCP Check Sample*. 1995:1–31. FP 95–1.
98. Molina DK, DiMaio VJ. Normal organ weights in men: part I-the heart. *Am J Forensic Med Pathol*. 2012;33(4):362–367.
99. Molina DK, DiMaio VJ. Normal organ weights in women: Part I-the heart. *Am J Forensic Med Pathol*. 2015;36(3):176–181.
100. Philcox W, Garland J, Zaidi F, et al. Higher heart weight in New Zealand Maori and Pacific Islanders. *Am J Forensic Med Pathol*. 2018;39(3):208–212.
101. Tse R, Garland J, Kesha K, et al. The potential diagnostic accuracy of autopsy lung weights, lung-heart ratio, and lung-body ratio in drowning deaths. *Am J Forensic Med Pathol*. 2018;39(3):223–228.
102. Wingren CJ, Ottosson A. Postmortem heart weight modelled using piecewise linear regression in 27,645 medicolegal autopsy cases. *Forensic Sci Int*. 2015;252:157–162.
103. Rich BS. Sudden death screening. *Med Clin North Am*. 1994;78(2):267–288.
104. Maron BJ, Gohman TE, Aeppli D. Prevalence of sudden cardiac death during competitive sports activities in Minnesota high school athletes. *J Am Coll Cardiol*. 1998;32(7):1881–1884.
105. Tabib A, Miras A, Taniere P, Loire R. Undetected cardiac lesions cause unexpected sudden cardiac death during occasional sport activity. A report of 80 cases. *Eur Heart J*. 1999;20(12):900–903.
106. Basso C, Corrado D, Thiene G. Cardiovascular causes of sudden death in young individuals including athletes. *Cardiol Rev*. 1999;7(3):127–135.
107. Thiene G, Carturan E, Corrado D, Basso C. Prevention of sudden cardiac death in the young and in athletes: dream or reality? *Cardiovasc Pathol*. 2010;19(4):207–217.
108. Mozes A, Homoud M, Link M, Weinstock J, Garlitski A, Estes 3rd NA. Screening for sudden death in the athlete. *Cardiovasc Pathol*. 2010;19(6):340–342.
109. Solberg EE, Gjertsen F, Haugstad E, Kolsrud L. Sudden death in sports among young adults in Norway. *Eur J Cardiovasc Prev Rehabil*. 2010;17(3):337–341.
110. Maron BJ, Doerer JJ, Haas TS, Tierney DM, Mueller FO. Sudden deaths in young competitive athletes: analysis of 1866 deaths in the United States, 1980–2006. *Circulation*. 2009;119(8):1085–1092.
111. Maron BJ. Sudden death in young athletes. *N Engl J Med*. 2003;349(11):1064–1075.
112. Eichner ER. Sickle cell trait. *J Sport Rehabil*. 2007;16(3):197–203.
113. DiDario AG, Becker JM. Asthma, sports, and death. *Allergy Asthma Proc*. 2005;26(5):341–344.
114. Becker JM, Rogers J, Rossini G, Mirchandani H, D'Alonzo Jr GE. Asthma deaths during sports: report of a 7-year experience. *J Allergy Clin Immunol*. 2004;113(2):264–267.
115. Luke JL, Helpern M. Sudden unexpected death from natural causes in young adults. A review of 275 consecutive autopsied cases. *Arch Pathol*. 1968;85(1):10–17.
116. Pelliccia A, Maron BJ, Spataro A, Proschan MA, Spirito P. The upper limit of physiologic cardiac hypertrophy in highly trained elite athletes. *N Engl J Med*. 1991;324(5):295–301.
117. Maron BJ, Haas TS, Doerer JJ, Thompson PD, Hodges JS. Comparison of U.S. and Italian experiences with sudden cardiac deaths in young competitive athletes and implications for preparticipation screening strategies. *Am J Cardiol*. 2009;104(2):276–280.
118. Maron BJ, Estes 3rd NA. Commotio cordis. *N Engl J Med*. 2010;362(10):917–927.
119. Link M, Wang P, Pandian N, et al. An experimental model of sudden death due to low-energy chest-wall impact (commotio cordis). *N Engl J Med*. 1998;338:1805–1811.
120. Maron BJ, Gohman TE, Kyle SB, Estes 3rd NA, Link MS. Clinical profile and spectrum of commotio cordis. *JAMA*. 2002;287(9):1142–1146.
121. Geertinger P, Voigt J. Death in the bath. A survey of bathtub deaths in Copenhagen, Denmark, and Gothenburg, Sweden, from 1961 to 1969. *J Forensic Med*. 1970;17(4):136–147.
122. Hanzlick R, Goodin J. Mind your manners. Part III: individual scenario results and discussion of the national association of medical examiners manner of death questionnaire, 1995. *Am J Forensic Med Pathol*. 1997;18(3):228–245.
123. Buttner A, Heimpel M, Eisenmenger W. Sudden natural death 'at the wheel': a retrospective study over a 15-year time period (1982–1996). *Forensic Sci Int*. 1999;103(2):101–112.
124. Cheng LH, Whittington RM. Natural deaths while driving: would screening for risk be ethically justified? *J Med Ethics*. 1998;24(4):248–251.
125. Kerwin AJ. Sudden death while driving. *Can Med Assoc J*. 1984;131(4):312–314.
126. Lau G. Ischaemic heart disease as a natural cause of death in motorists. *Singapore Med J*. 1994;35(5):467–470.
127. Peterson B, Petty C. Sudden natural death among automobile drivers. *J Forensic Sci*. 1962;7(3):274–285.

128. Antecol DH, Roberts WC. Sudden death behind the wheel from natural disease in drivers of four-wheeled motorized vehicles. *Am J Cardiol*. 1990;66(19):1329–1335.
129. Byard RW. Deep venous thrombosis, pulmonary embolism and long-distance flights. *Forensic Sci Med Pathol*. 2019;15(1):122–124.
130. Kasuda S, Kudo R, Yuui K, Nakata M, Hatake K. Sudden death due to bilateral pulmonary thromboembolism following laparoscopic cholecystectomy. *Am J Forensic Med Pathol*. 2016;37(3):198–200.
131. Margiotta G, Carnevali E, Gabbrielli M, Bacci M, Lancia M. Report of a fatal case of pulmonary thromboembolism in a long-distance truck driver. *Am J Forensic Med Pathol*. 2014;35(4):242–245.
132. Rosenfeld HE, Tsokos M, Byard RW. The association between body mass index and pulmonary thromboembolism in an autopsy population. *J Forensic Sci*. 2012;57(5):1336–1338.
133. Fineschi V, Bafunno V, Bello S, et al. Fatal pulmonary thromboembolism. A retrospective autopsy study: searching for genetic thrombophilias (Factor V Leiden (G1691A) and FII (G20210A) gene variants) and dating the thrombus. *Forensic Sci Int*. 2012;214(1–3):152–158.
134. Eom M, Lim SC, Shin Kim Y. Three cases of pulmonary thromboembolism and extensive prayer (invocation) activity as a new possible risk factor. *Am J Forensic Med Pathol*. 2009;30(2):191–194.
135. Gill JR. The medicolegal evaluation of fatal pulmonary thromboembolism. In: Tsokos M, ed. *Forensic Pathology Reviews*. Vol 3. Totowa, NJ: Humana Press; 2005:285–304.
136. Ely SF, Gill JR. Fatal pulmonary thromboembolism and hereditary thrombophilias. *J Forensic Sci*. 2005;50(2):411–418.
137. Kohlmeier RE, Cho CG, Bux RC, et al. Prothrombin gene mutation uncommon in pulmonary embolism. *South Med J*. 2000;93(11):1073–1077.
138. Laffan M. Genetics and pulmonary medicine. 4. Pulmonary embolism. *Thorax*. 1998;53(8):698–702.
139. McColl MD, Tait RC, Greer IA, Walker ID. Injecting drug use is a risk factor for deep vein thrombosis in women in Glasgow. *Br J Haematol*. 2001;112(3):641–643.
140. Wright NM, Martin M, Goff T, Morgan J, Elworthy R, Ghoneim S. Cocaine and thrombosis: a narrative systematic review of clinical and in-vivo studies. *Subst Abuse Treat Prev Policy*. 2007;2:27.
141. Adams VI, Flomenbaum MA, Hirsch CS. Trauma and disease. In: Spitz WU, ed. *Spitz and Fisher's Medicolegal Investigation of Death*. 4th ed. Springfield, IL: Charles C Thomas; 2006:436–459.
142. Hirsch C, Martin D. Unexpected deaths in epileptics. *Neurology*. 1971;21:682–690.
143. Leestma JE. Forensic considerations in sudden unexpected death in epilepsy. *Epilepsia*. 1997;38(11 Suppl):S63–S66.
144. Annegers JF, Coan SP. SUDEP: overview of definitions and review of incidence data. *Seizure*. 1999;8:347.
145. Langan Y, Nashef L, Sander JW. Sudden unexpected death in epilepsy: a series of witnessed deaths. *J Neurol Neurosurg Psychiatry*. 2000;68(2): 211–213.
146. Walczak TS, Leppik IE, D'Amelio M, et al. Incidence and risk factors in sudden unexpected death in epilepsy: a prospective cohort study. *Neurology*. 2001;56(4):519–525.
147. Davis GG, McGwin Jr G. Comparison of heart mass in seizure patients dying of sudden unexplained death in epilepsy to sudden death due to some other cause. *Am J Forensic Med Pathol*. 2004;25(1):23–28.
148. Lear-Kaul KC, Coughlin L, Dobersen MJ. Sudden unexpected death in epilepsy: a retrospective study. *Am J Forensic Med Pathol*. 2005;26(1): 11–17.
149. PCT S, Dalager-Pedersen S, Baandrup U, Dam M, Vesterby-Charles A. Sudden unexpected death in epilepsy: is death by seizures a cardiac disease? *Am J Forensic Med Pathol*. 2005;26(2):99–105.
150. Milroy CM. Sudden unexpected death in epilepsy in childhood. *Forensic Sci Med Pathol*. 2011;7(4):336–340.
151. Atherton DS, Devinsky O, Hesdorffer DC, Wright C, Davis GG. Implications of death certification on sudden unexpected death in epilepsy (SUDEP) research. *Acad Forensic Pathol*. 2016;6(1):96–102.
152. Ellis Jr SP, Szabo CA. Sudden unexpected death in epilepsy: incidence, risk factors, and proposed mechanisms. *Am J Forensic Med Pathol*. 2018; 39(2):98–102.
153. Middleton OL, Atherton DS, Bundock EA, et al. National association of medical examiners position paper: recommendations for the investigation and certification of deaths in people with epilepsy. *Acad Forensic Pathol*. 2018;8(1):119–135.
154. Annegers JF, Hauser WA, Coan SP, Rocca WA. A population-based study of seizures after traumatic brain injuries. *N Engl J Med*. 1998;338(1): 20–24.
155. Hirsch CS, Armbrustmacher V. Trauma of the nervous system. In: Spitz WU, ed. *Spitz and Fisher's Medicolegal Investigation of Death*. 4th ed. Springfield, IL: Charles C Thomas; 2006:994–1077.
156. Annegers JF, Grabow JD, Groover RV, Laws Jr ER, Elveback LR, Kurland LT. Seizures after head trauma: a population study. *Neurology*. 1980; 30(7 Pt 1):683–689.
157. Salazar AM, Jabbari B, Vance SC, Grafman J, Amin D, Dillon JD. Epilepsy after penetrating head injury. I. Clinical correlates: a report of the Vietnam Head Injury Study. *Neurology*. 1985;35(10):1406–1414.
158. Hefti MM, Kinney HC, Cryan JB, et al. Sudden unexpected death in early childhood: general observations in a series of 151 cases: Part 1 of the investigations of the San Diego SUDC Research Project. *Forensic Sci Med Pathol*. 2016;12(1):4–13.
159. Kinney HC, McDonald AG, Minter ME, Berry GT, Poduri A, Goldstein RD. Witnessed sleep-related seizure and sudden unexpected death in infancy: a case report. *Forensic Sci Med Pathol*. 2013;9(3):418–421.
160. McGuone D, Crandall LG, Devinsky O. Sudden unexplained death in childhood: a neuropathology review. *Front Neurol*. 2020;11:582051.
161. Delage C, Irey NS. Anaphylactic deaths: a clinicopathologic study of 43 cases. *J Forensic Sci*. 1972;17(4):525–540.
162. Schwartz LB, Metcalfe DD, Miller JS, Earl H, Sullivan T. Tryptase levels as an indicator of mast-cell activation in systemic anaphylaxis and mastocytosis. *N Engl J Med*. 1987;316(26):1622–1626.
163. Schwartz LB, Yunginger JW, Miller J, Bokhari R, Dull D. Time course of appearance and disappearance of human mast cell tryptase in the circulation after anaphylaxis. *J Clin Investig*. 1989;83(5):1551–1555.
164. Yunginger JW, Nelson DR, Squillace DL, et al. Laboratory investigation of deaths due to anaphylaxis. *J Forensic Sci*. 1991;36(3):857–865.
165. Edston E, van Hage-Hamsten M. beta-Tryptase measurements post-mortem in anaphylactic deaths and in controls. *Forensic Sci Int*. 1998; 93(2–3):135–142.
166. Pumphrey RS, Roberts IS. Postmortem findings after fatal anaphylactic reactions. *J Clin Pathol*. 2000;53(4):273–276.

167. Riches KJ, Gillis D, James RA. An autopsy approach to bee sting-related deaths. *Pathology.* 2002;34(3):257–262.
168. Schwartz LB. Diagnostic value of tryptase in anaphylaxis and mastocytosis. *Immunol Allergy Clin North Am.* 2006;26(3):451–463.
169. Da Broi U, Moreschi C. Post-mortem diagnosis of anaphylaxis: a difficult task in forensic medicine. *Forensic Sci Int.* 2011;204(1–3):1–5.
170. Edston E. Accumulation of eosinophils, mast cells, and basophils in the spleen in anaphylactic deaths. *Forensic Sci Med Pathol.* 2013;9(4):496–500.
171. Prahlow JA, Barnard JJ. Fatal anaphylaxis due to fire ant stings. *Am J Forensic Med Pathol.* 1998;19(2):137–142.
172. Belton AL, Chira T. Fatal anaphylactic reaction to hair dye. *Am J Forensic Med Pathol.* 1997;18(3):290–292.
173. Bennett AT, Collins KA. An unusual case of anaphylaxis. Mold in pancake mix. *Am J Forensic Med Pathol.* 2001;22(3):292–295.
174. Flannagan LM, Wolf BC. Sudden death associated with food and exercise. *J Forensic Sci.* 2004;49(3):543–545.
175. Li M, Goldberger BA, Hopkins C. Fatal case of BOTOX-related anaphylaxis? *J Forensic Sci.* 2005;50(1):169–172.
176. Chiu CY, Lin TY, Hsia SH, Lai SH, Wong KS. Systemic anaphylaxis following local lidocaine administration during a dental procedure. *Pediatr Emerg Care.* 2004;20(3):178–180.
177. Tsujikawa K, Kuwayama K, Miyaguchi H, et al. A fatal case of suspected anaphylaxis with cefoperazone and sulbactam: LC-MS analysis. *J Forensic Sci.* 2008;53(1):226–231.
178. Rubenstein HS. Bee-sting diseases: who is at risk? What is the treatment? *Lancet.* 1982;1(8270):496–499.
179. Barnard JH. Studies of 400 Hymenoptera sting deaths in the United States. *J Allergy Clin Immunol.* 1973;52(5):259–264.
180. Pegues DA, Beck-Sague CM, Woollen SW, et al. Anaphylactoid reactions associated with reuse of hollow-fiber hemodialyzers and ACE inhibitors. *Kidney Int.* 1992;42(5):1232–1237.
181. du Toit-Prinsloo L, Morris NK, Meyer P, Saayman G. Deaths from bee stings: a report of three cases from Pretoria, South Africa. *Forensic Sci Med Pathol.* 2016;12(1):81–85.
182. Byard RW. Death by food. *Forensic Sci Med Pathol.* 2017;14:395–401.
183. Blumenthal R, Scholtz PEP, Shuttleworth JL. Black mamba death: venom versus antivenom? *Am J Forensic Med Pathol.* 2019;40(4):356–360.
184. Hosseini M, Hedjazi A, Bahrami R. Sudden death due to anaphylactic shock in a patient with an intact hepatic hydatid cyst. *Am J Forensic Med Pathol.* 2014;35(4):256–257.
185. Jedidi M, Mlayeh S, Masmoudi T, Souguir MK, Zemni M. Sudden death due to hydatid cyst: thirty-four medicolegal autopsy cases. *Am J Forensic Med Pathol.* 2014;35(1):29–33.
186. Boggs CL, Ripple MG, Ali Z, et al. Anaphylaxis after the injection of buprenorphine. *J Forensic Sci.* 2013;58(5):1381–1383.
187. Esposito M, Montana A, Liberto A, et al. Anaphylactic death: a new forensic workflow for diagnosis. *Healthcare (Basel).* 2021;9(2).
188. Gill J, Maloney K, Hirsch C. The consistency and advantage of therapeutic complication as a manner of death. *Acad Forensic Pathol.* 2012;2(2):176–181.
189. Ansari MQ, Zamora JL, Lipscomb MF. Postmortem diagnosis of acute anaphylaxis by serum tryptase analysis. A case report. *Am J Clin Pathol.* 1993;99(1):101–103.
190. Randall B, Butts J, Halsey JF. Elevated postmortem tryptase in the absence of anaphylaxis. *J Forensic Sci.* 1995;40(2):208–211.
191. Yilmaz R, Yuksekbas O, Erkol Z, Bulut ER, Arslan MN. Postmortem findings after anaphylactic reactions to drugs in Turkey. *Am J Forensic Med Pathol.* 2009;30(4):346–349.
192. Mayer DE, Krauskopf A, Hemmer W, Moritz K, Jarisch R, Reiter C. Usefulness of post mortem determination of serum tryptase, histamine and diamine oxidase in the diagnosis of fatal anaphylaxis. *Forensic Sci Int.* 2011;212(1–3):96–101.
193. Sravan A, Tse R, Cala AD. A decline in 2 consecutive postmortem serum tryptase levels in an anaphylactic death. *Am J Forensic Med Pathol.* 2015;36(4):233–235.
194. Xiao N, Li DR, Wang Q, Zhang F, Yu YG, Wang HJ. Postmortem serum tryptase levels with special regard to acute cardiac deaths. *J Forensic Sci.* 2017;62(5):1336–1338.
195. Byard RW. Anaphylaxis at autopsy. *Forensic Sci Med Pathol.* 2017;13(3):269–271.
196. Kounis NG, Soufras GD, Hahalis G. Accumulation of eosinophils, mast cells, and basophils in the spleen and the coronary arteries in anaphylactic deaths: is the Kounis hypersensitivity associated syndrome present? *Forensic Sci Med Pathol.* 2014;10(1):150–151.
197. Palmiere C. Decline in postmortem serum tryptase levels in anaphylactic deaths. *Am J Forensic Med Pathol.* 2017;38(3):274–275.
198. Martinez-Fernandez P, Vallejo-de-Torres G, Sanchez-de-Leon-Robles MS, et al. Medical and pathologic characteristics of fatal anaphylaxis: a Spanish nationwide 17-year series. *Forensic Sci Med Pathol.* 2019;15(3):369–381.
199. Horn KD, Halsey JF, Zumwalt RE. Utilization of serum tryptase and immunoglobulin e assay in the postmortem diagnosis of anaphylaxis. *Am J Forensic Med Pathol.* 2004;25(1):37–43.
200. Garland J, Philcox W, McCarthy S, et al. Postmortem tryptase level in 120 consecutive nonanaphylactic deaths: establishing a reference range as <23 mug/L. *Am J Forensic Med Pathol.* 2019;40(4):351–355.
201. Tse R, Garland J, Ahn Y. Decline in 2 serial postmortem tryptase measurements beyond 72 hours after death in an antibiotic-related anaphylactic death. *Am J Forensic Med Pathol.* 2018;39(1):14–17.
202. Tse R, Wong CX, Kesha K, et al. Post mortem tryptase cut-off level for anaphylactic death. *Forensic Sci Int.* 2018;284:5–8.
203. Shen Y, Li L, Grant J, et al. Anaphylactic deaths in Maryland (United States) and Shanghai (China): a review of forensic autopsy cases from 2004 to 2006. *Forensic Sci Int.* 2009;186(1–3):1–5.
204. Wang X, Yin C, Su X, Su M. Reliable postmortem molecular diagnosis of anaphylaxis: Co-localization of mast cell degranulation and immunoglobulin E in allergic throat tissues. *Am J Forensic Med Pathol.* 2020;41(4):249–258.
205. Simons FE. Anaphylaxis pathogenesis and treatment. *Allergy.* 2011;66(Suppl 95):31–34.
206. Jason DR. Fatal angioedema associated with captopril. *J Forensic Sci.* 1992;37(5):1418–1421.
207. Roujeau JC, Stern RS. Severe adverse cutaneous reactions to drugs. *N Engl J Med.* 1994;331(19):1272–1285.
208. Dean DE, Schultz DL, Powers RH. Asphyxia due to angiotensin converting enzyme (ACE) inhibitor mediated angioedema of the tongue during the treatment of hypertensive heart disease. *J Forensic Sci.* 2001;46(5):1239–1243.
209. Krizova A, Gardner T, Little DL, Arcieri-Piersanti V, Pollanen MS. Fatal laryngeal angioedema: a case report and a workup of angioedema in a forensic setting. *Forensic Sci Med Pathol.* 2015;11(4):558–563.
210. Gilbert JD, Byard RW. Lethal manifestations of angioedema. *Forensic Sci Med Pathol.* 2019;15(3):494–497.
211. CDC. Deaths from excessive alcohol use in the United States. https://www.cdc.gov/alcohol/features/excessive-alcohol-deaths.html.

212. DiMaio VJ. Sudden, unexpected death due to massive, nontraumatic intra-abdominal hemorrhage in association with cirrhosis of the liver. *Am J Forensic Med Pathol*. 1987;8(3):266–268.
213. Fox L, Crane SA, Bidari C, Jones A. Intra-abdominal hemorrhage from ruptured varices. *Arch Surg*. 1982;117(7):953–956.
214. Hayashi T, Buschmann C, Matejic D, Ingold B, Tsokos M. Fatal intraabdominal bleeding from ectopic varices: report of two autopsy cases. *Forensic Sci Med Pathol*. 2013;9(1):96–99.
215. deRoux SJ, Sgarlato A. Subdural hemorrhage, a retrospective review with emphasis on a cohort of alcoholics. *J Forensic Sci*. 2015;60(5): 1224–1228.
216. Zumwalt RE, Bost RO, Sunshine I. Evaluation of ethanol concentrations in decomposed bodies. *J Forensic Sci*. 1982;27(3):549–554.
217. Sidell FR, Pless JE. Ethyl alcohol: blood levels and performance decrements after oral administration to man. *Psychopharmacologia*. 1971;19(3): 246–261.
218. Roberts JR, Dollard D. Alcohol levels do not accurately predict physical or mental impairment in ethanol-tolerant subjects: relevance to emergency medicine and dram shop laws. *J Med Toxicol*. 2010;6(4):438–442.
219. Dott AB. Blood alcohol levels and intoxication. *JAMA*. 1970;214(12):2196.
220. Perper JA, Twerski A, Wienand JW. Tolerance at high blood alcohol concentrations: a study of 110 cases and review of the literature. *J Forensic Sci*. 1986;31(1):212–221.
221. Johnson H. At what blood levels does alcohol kill? *Med Sci Law*. 1985;25(2):127–130.
222. Hammond KB, Rumack BH, Rodgerson DO. Blood ethanol. A report of unusually high levels in a living patient. *JAMA*. 1973;226(1):63–64.
223. Lindblad B, Olsson R. Unusually high levels of blood alcohol? *JAMA*. 1976;236(14):1600–1602.
224. Randall B. Fatty liver and sudden death. A review. *Hum Pathol*. 1980;11(2):147–153.
225. Olsen Y. What is addiction? History, terminology, and aore concepts. *Med Clin North Am*. 2022;106(1):1–12.
226. Dennis JA, Gittner LS, George AK, Queen CM. Opioid use disorder terminologies and the role of public health in addressing stigma. *Alcohol Treat Q*. 2020;39(1):110–118.
227. Kelly JFEM. Jellinek's disease concept of alcoholism. *Addiction*. 2019;114(3):555–559.
228. Baron TH, Morgan DE. Acute necrotizing pancreatitis. *N Engl J Med*. 1999;340(18):1412–1417.
229. Tsokos M, Braun C. Acute pancreatitis presenting as sudden, unexpected death: an autopsy-based study of 27 cases. *Am J Forensic Med Pathol*. 2007;28(3):267–270.
230. Lin YK, Johnston PC, Arce K, Hatipoglu BA. Chronic pancreatitis and diabetes mellitus. *Curr Treat Options Gastroenterol*. 2015;13(3):319–331.
231. Malka D, Hammel P, Sauvanet A, et al. Risk factors for diabetes mellitus in chronic pancreatitis. *Gastroenterology*. 2000;119(5):1324–1332.
232. Sarles H. Chronic pancreatitis and diabetes. *Bailliere Clin Endocrinol Metabol*. 1992;6(4):745–775.
233. Blamey SL, Imrie CW, O'Neill J, Gilmour WH, Carter DC. Prognostic factors in acute pancreatitis. *Gut*. 1984;25(12):1340–1346.
234. Gill JR. Pancreatitis: a forensic perspective. *Acad Forensic Pathol*. 2016;6(1):237–248.
235. Brown TT, Prahlow JA. Postmortem serum amylase and lipase analysis in the diagnosis of acute pancreatitis. *Acad Forensic Pathol*. 2018;8(2): 311–323.
236. Stoppacher R. Sudden death due to acute pancreatitis. *Acad Forensic Pathol*. 2018;8(2):239–255.
237. Walker AE. The adult pancreas in trauma and disease. *Acad Forensic Pathol*. 2018;8(2):192–218.
238. Mujica VR, Barkin JS. Occult gastrointestinal bleeding. General overview and approach. *Gastrointest Endosc Clin N Am*. 1996;6(4):833–845.
239. Lehrer P, Feldman J, Giardino N, Song HS, Schmaling K. Psychological aspects of asthma. *J Consult Clin Psychol*. 2002:70(3):691–711.
240. Sarafino EP, Dillon JM. Relationships among respiratory infections, triggers of attacks, and asthma severity in children. *J Asthma*. 1998;35(6): 497–504.
241. Loerbroks A, Apfelbacher CJ, Thayer JF, Debling D, Sturmer T. Neuroticism, extraversion, stressful life events and asthma: a cohort study of middle-aged adults. *Allergy*. 2009;64(10):1444–1450.
242. Randolph C. An update on exercise-induced bronchoconstriction with and without asthma. *Curr Allergy Asthma Rep*. 2009;9(6):433–438.
243. Weiler JM. Exercise-induced asthma: a practical guide to definitions, diagnosis, prevalence, and treatment. *Allergy Asthma Proc*. 1996;17(6): 315–325.
244. Matsuse H, Shimoda T, Kohno S, et al. A clinical study of mortality due to asthma. *Ann Allergy Asthma Immunol*. 1995;75(3):267–272.
245. Tough SC, Green FH, Paul JE, Wigle DT, Butt JC. Sudden death from asthma in 108 children and young adults. *J Asthma*. 1996;33(3):179–188.
246. Tatum AM, Greenberger PA, Mileusnic D, Donoghue ER, Lifschultz BD. Clinical, pathologic, and toxicologic findings in asthma deaths in Cook County, Illinois. *Allergy Asthma Proc*. 2001;22(5):285–291.
247. Sur S, Crotty TB, Kephart GM, et al. Sudden-onset fatal asthma. A distinct entity with few eosinophils and relatively more neutrophils in the airway submucosa? *Am Rev Respir Dis*. 1993;148(3):713–719.
248. Champ CS, Byard RW. Sudden death in asthma in childhood. *Forensic Sci Int*. 1994;66(2):117–127.
249. Davis JH. Can sudden cardiac death be murder? *J Forensic Sci*. 1978;23(2):384–387.
250. Turner SA, Barnard JJ, Spotswood SD, Prahlow JA. "Homicide by heart attack" revisited. *J Forensic Sci*. 2004;49(3):598–600.
251. Hlavaty L, Sung L. Applying the principles of homicide by heart attack. *Am J Forensic Med Pathol*. 2016;37(2):112–117
252. Bottari G, Trotta S, Marzullo A, Meliota G, Ciccone MM, Solarino B. Sudden cardiac death after robbery: homicide or natural death? *J Forensic Leg Med*. 2020;75:102057.
253. Gill JR, McCubbin KR, Landi KK. Homicide by asthma. *Acad Forensic Pathol*. 2011;1(1):122–127.
254. Mitchell SL, Teno JM, Kiely DK, et al. The clinical course of advanced dementia. *N Engl J Med*. 2009;361(16):1529–1538.
255. Gill JR, DeJoseph ME. The importance of proper death certification during the COVID-19 Pandemic. *JAMA*. 2020;324(1):27–28.
256. Woolf SH, Chapman DA, Sabo RT, Weinberger DM, Hill L, Taylor DDH. Excess deaths from COVID-19 and other causes, March–July 2020. *JAMA*. 2020;324(15):1562–1564.
257. *National Center for Health Statistics, Guidance for Certifying Deaths Due to Coronavirus Disease 2019 (COVID–19). Hyattsville, MD*. 2020.
258. Ely SF, Neitzel AR, Gill JR. Fatal diabetic ketoacidosis and antipsychotic medication. *J Forensic Sci*. 2013;58(2):398–403.
259. Palmiere C. Postmortem diagnosis of diabetes mellitus and its complications. *Croat Med J*. 2015;56(3):181–193.
260. Coe JI. Use of chemical determinations on vitreous humor in forensic pathology. *J Forensic Sci*. 1972;17(4):541–546.

261. Chaturvedi AK, Botch SR, Canfield DV, Forster EM. Vitreous fluid and/or urine glucose concentrations in 1335 civil aviation accident pilot fatalities. *J Forensic Sci*. 2009;54(3):715–720.
262. Thierauf A, Musshoff F, Madea B. Post-mortem biochemical investigations of vitreous humor. *Forensic Sci Int*. 2009;192(1–3):78–82.
263. Zilg B, Alkass K, Berg S, Druid H. Postmortem identification of hyperglycemia. *Forensic Sci Int*. 2009;185(1–3):89–95.
264. Ali Z, Levine B, Ripple M, Fowler DR. Diabetic ketoacidosis: a silent death. *Am J Forensic Med Pathol*. 2012;33(3):189–193.
265. Kokavec J, Min SH, Tan MH, et al. Biochemical analysis of the living human vitreous. *Clin Exp Ophthalmol*. 2016;44(7):597–609.
266. Iten PX, Meier M. Beta-hydroxybutyric acid—an indicator for an alcoholic ketoacidosis as cause of death in deceased alcohol abusers. *J Forensic Sci*. 2000;45(3):624–632.
267. Petersen TH, Williams T, Nuwayhid N, Harruff R. Postmortem detection of isopropanol in ketoacidosis. *J Forensic Sci*. 2012;57(3):674–678.
268. Maloney K, Schoppe C. Obesity and non-atherosclerotic cardiovascular disease. *Acad Forensic Pathol*. 2013;3(1):8–12.
269. McCleskey BC, Davis GG, Dye DW. Correcting the count: improving vital statistics data regarding deaths related to obesity. *J Forensic Sci*. 2018;63(4):1155–1159.
270. Omond KJ, Langlois NEI, Byard RW. Obesity, body mass index, and homicide. *J Forensic Sci*. 2017;62(4):930–933.
271. Byard RW. The complex spectrum of forensic issues arising from obesity. *Forensic Sci Med Pathol*. 2012;8(4):402–413.
272. Franks PW, Hanson RL, Knowler WC, Sievers ML, Bennett PH, Looker HC. Childhood obesity, other cardiovascular risk factors, and premature death. *N Engl J Med*. 2010;362(6):485–493.
273. Lavie CJ, Milani RV, Ventura HO. Obesity and cardiovascular disease: risk factor, paradox, and impact of weight loss. *J Am Coll Cardiol*. 2009; 53(21):1925–1932.
274. Hunsaker D, Hunsaker J. Obesity epidemic in the United States. In: Tosokos M, ed. *Forensic Pathology Reviews*. Vol 2. Totowa, NC: Humana Press Inc.; 2004:61–100.
275. Yanovski SZ, Yanovski JA. Obesity. *N Engl J Med*. 2002;346(8):591–602.
276. Duflou J, Virmani R, Rabin I, Burke A, Farb A, Smialek S. Sudden death as a result of heart disease in morbid obesity. *Am Heart J*. 1995;130: 306–313.
277. Reynolds MA, Henriksen KJ, Chang A. End-stage kidney disease is overlooked as a proximate cause of death at autopsy. *Am J Clin Pathol*. 2020;153:772–775.
278. Maskell PD, Penney E, Smith PR, Hikin LJ, Morley SR. Post-mortem diagnosis of kidney impairment: an experimental study. *Forensic Sci Int*. 2019;301:271–277.
279. Chiha M, Samarasinghe S, Kabaker AS. Thyroid storm: an updated review. *J Intensive Care Med*. 2015;30(3):131–140.
280. Lynch MJ, Woodford NW. Sudden unexpected death in the setting of undiagnosed Graves' disease. *Forensic Sci Med Pathol*. 2014;10(3): 452–456.
281. Wei D, Yuan X, Yang T, et al. Sudden unexpected death due to Graves' disease during physical altercation. *J Forensic Sci*. 2013;58(5): 1374–1377.
282. Adler J, Colegrove DJ. Contrast induced thyrotoxicosis in a patient with new onset atrial fibrillation: a case report and review. *J Atr Fibrillation*. 2013;6(1):379.
283. Kolesinska Z, Siuda K, Bobkowski W, Niedziela M. Amiodarone-induced thyrotoxicosis with paroxysmal supraventricular (ectopic atrial) tachycardia: case report and review of the literature. *Arch Med Sci*. 2013;9(2):372–376.
284. Eliades M, El-Maouche D, Choudhary C, Zinsmeister B, Burman KD. Takotsubo cardiomyopathy associated with thyrotoxicosis: a case report and review of the literature. *Thyroid*. 2014;24(2):383–389.
285. Al Bannay R, Husain A, Khalaf S. Complete heart block in thyrotoxicosis, is it a manifestation of thyroid storm? A case report and review of the literature. *Case Rep Endocrinol*. 2012;2012:318398.
286. Al Jaber J, Haque S, Noor H, Ibrahim B, Al Suwaidi J. Thyrotoxicosis and coronary artery spasm: case report and review of the literature. *Angiology*. 2010;61(8):807–812.
287. Lim BH, Raman S, Sivanesaratnam V, Ngan A. Thyrotoxicosis in pregnancy–a six year review. *Singapore Med J*. 1989;30(6):539–541.
288. Biondi B, Cappola AR, Cooper DS. Subclinical hypothyroidism: a review. *JAMA*. 2019;322(2):153–160.
289. Garcia NS, Castelo JS, Ramos V, Rezende GS, Pereira FE. Frequency of myocarditis in cases of fatal meningococcal infection in children: observations on 31 cases studied at autopsy. *Rev Soc Bras Med Trop*. 1999;32(5):517–522.
290. Gerber JE, Johnson JE, Scott MA, Madhusudhan KT. Fatal meningitis and encephalitis due to Bartonella henselae bacteria. *J Forensic Sci*. 2002; 47(3):640–644.
291. Hill WR, Kinney TD. The cutaneous lesions in acute meningococcemia; a clinical and pathologic study. *J Am Med Assoc*. 1947;134(6):513–518.
292. Dalldorf FG, Jennette JC. Fatal meningococcal septicemia. *Arch Pathol Lab Med*. 1977;101(1):6–9.
293. Sotto MN, Langer B, Hoshino-Shimizu S, de Brito T. Pathogenesis of cutaneous lesions in acute meningococcemia in humans: light, immunofluorescent, and electron microscopic studies of skin biopsy specimens. *J Infect Dis*. 1976;133(5):506–514.
294. Garland J, Tse R, Cala AD. Neisseria meningitidis isolated in postmortem vitreous humor in a death due to meningococcal sepsis. *Am J Forensic Med Pathol*. 2016;37(4):233–235.
295. Domingo P, Sancho F. Neisseria meningitidis and purpura fulminans in a woman with chronic liver disease. *Lancet Infect Dis*. 2017;17(4):459.
296. Coldiron ME, Assao B, Page A-L, et al. Single-dose oral ciprofloxacin prophylaxis as a response to a meningococcal meningitis epidemic in the African meningitis belt: a 3-arm, open-label, cluster-randomized trial. *PLoS Med*. 2018;15(6):e1002593.
297. Welsby PD, Golledge CL. Meningococcal meningitis. *BMJ*. 1990;300(6733):1150–1151.
298. Marcus BJ, Collins KA. Childhood panhypopituitarism presenting as child abuse: a case report and review of the literature. *Am J Forensic Med Pathol*. 2004;25(3):265–269.
299. Wang AR, Gill JR. The pituitary gland: an infrequent but multifaceted contributor to death. *Acad Forensic Pathol*. 2016;6(2):206–216.
300. Doberentz E, Preuss-Wossner J, Kuchelmeister K, Madea B. Histological examination of the pituitary glands in cases of fatal hypothermia. *Forensic Sci Int*. 2011;207(1–3):46–49.
301. James SL, Abate D, Abate KH, et al. Global, regional, and national incidence, prevalence, and years lived with disability for 354 diseases and injuries for 195 countries and territories, 1990–2017: a systematic analysis for the Global Burden of Disease Study 2017. *Lancet*. 2018; 392(10159):1789–1858.

302. Laursen TM, Nordentoft M, Mortensen PB. Excess early mortality in schizophrenia. *Annu Rev Clin Psychol*. 2014;10:425–448.
303. Meatherall R, Younes J. Fatality from olanzapine induced hyperglycemia. *J Forensic Sci*. 2002;47(4):893–896.
304. Rosh A, Sampson BA, Hirsch CS. Schizophrenia as a cause of death. *J Forensic Sci*. 2003;48(1):164–167.
305. Rastogi A, Viani-Walsh D, Akbari S, Gall N, Gaughran F, Lally J. Pathogenesis and management of Brugada syndrome in schizophrenia: a scoping review. *Gen Hosp Psychiatr*. 2020;67:83–91.
306. Byramji A, Cains G, Gilbert JD, Byard RW. Hyponatremia at autopsy: an analysis of etiologic mechanisms and their possible significance. *Forensic Sci Med Pathol*. 2008;4(3):149–152.
307. DiMaio VJ, DiMaio SJ. Fatal water intoxication in a case of psychogenic polydipsia. *J Forensic Sci*. 1980;25(2):332–335.
308. Vichinsky EP, Styles LA, Colangelo LH, Wright EC, Castro O, Nickerson B. Acute chest syndrome in sickle cell disease: clinical presentation and course. Cooperative Study of Sickle Cell Disease. *Blood*. 1997;89(5):1787–1792.
309. Haupt HM, Moore GW, Bauer TW, Hutchins GM. The lung in sickle cell disease. *Chest*. 1982;81(3):332–337.
310. Vichinsky EP, Neumayr LD, Earles AN, et al. Causes and outcomes of the acute chest syndrome in sickle cell disease. National Acute Chest Syndrome Study Group. *N Engl J Med*. 2000;342(25):1855–1865.
311. Platt OS. The acute chest syndrome of sickle cell disease. *N Engl J Med*. 2000;342(25):1904–1907.
312. Hutchins KD, Ballas SK, Phatak D, Natarajan GA. Sudden unexpected death in a patient with splenic sequestration and sickle cell-beta+-thalassemia syndrome. *J Forensic Sci*. 2001;46(2):412–414.
313. Gladwin MT, Sachdev V, Jison ML, et al. Pulmonary hypertension as a risk factor for death in patients with sickle cell disease. *N Engl J Med*. 2004;350(9):886–895.
314. Hawley DA, McCarthy LJ. Sickle cell disease: two fatalities due to bone marrow emboli in patients with acute chest syndrome. *Am J Forensic Med Pathol*. 2009;30(1):69–71.
315. Thogmartin JR, Wilson CI, Palma NA, Ignacio SS, Pellan WA. Histological diagnosis of sickle cell trait: a blinded analysis. *Am J Forensic Med Pathol*. 2009;30(1):36–39.
316. Scheinin L, Wetli CV. Sudden death and sickle cell trait: medicolegal considerations and implications. *Am J Forensic Med Pathol*. 2009;30(2):204–208.
317. Kark JA, Posey DM, Schumacher HR, Ruehle CJ. Sickle-cell trait as a risk factor for sudden death in physical training. *N Engl J Med*. 1987;317(13):781–787.
318. Thogmartin JR, Wilson CI, Palma NA, Ignacio SS, Shuman MJ, Flannagan LM. Sickle cell trait-associated deaths: a case series with a review of the literature. *J Forensic Sci*. 2011;56(5):1352–1360.
319. Arbefeville EF, Tebbi CK, Chrostowski L, Adams VI. Sudden death after exercise in an adolescent with hemoglobin SE. *Am J Forensic Med Pathol*. 2011;32(4):341–343.
320. Dudley Jr AW, Waddell CC. Crisis in sickle cell trait. *Hum Pathol*. 1991;22(6):616–618.
321. Malowany JI, Butany J. Pathology of sickle cell disease. *Semin Diagn Pathol*. 2012;29(1):49–55.
322. Assanasen C, Quinton R, Buchanan G. Acute myocardial infarction in sickle cell anemia. *J Pediatr Hematol Oncol*. 2003;25(12):978–981.
323. Platt OS, Brambilla DJ, Rosse WF, et al. Mortality in sickle cell disease. Life expectancy and risk factors for early death. *N Engl J Med*. 1994;330(23):1639–1644.
324. Podduturi V, Guileyardo JM. Sickle cell trait as a contributory cause of death in natural disease. *J Forensic Sci*. 2015;60(3):807–811.
325. Mitchell BL. Sickle cell trait and sudden death. *Sports Med Open*. 2018;4(1):19.
326. Wang H, Martone M, Owens ME, et al. NCAA Division I American football players with sickle cell trait have altered hematological responses and hydration status. *Sci Rep*. 2021;11(1):1844.
327. O'Connor FG, Bergeron MF, Cantrell J, et al. ACSM and CHAMP summit on sickle cell trait: mitigating risks for warfighters and athletes. *Med Sci Sports Exerc*. 2012;44(11):2045–2056.
328. Thoreson CK, O'Connor MY, Ricks M, Chung ST, Sumner AE. Sickle cell trait from a metabolic, renal, and vascular perspective: linking history, knowledge, and health. *J Racial Ethn Health Disparities*. 2015;2(3):330–335.
329. Wirthwein D, Spotswood S, Barnard J, Prahlow J. Death due to microvascular occlusion in sickle-cell trait following physical exertion. *J Forensic Sci*. 2001;46(2):399–401.
330. Hyacinth HI, Franceschini N, Seals SR, et al. Association of sickle cell trait with incidence of coronary heart disease among African American individuals. *JAMA Netw Open*. 2021;4(1):e2030435.
331. O'Connor FG, Franzos MA, Nye NS, et al. Summit on exercise collapse associated with sickle cell trait: finding the "way ahead". *Curr Sports Med Rep*. 2021;20(1):47–56.
332. Messonnier LA, Bartolucci P, d'Humieres T, et al. Preventive measures for the critical postexercise period in sickle cell trait and disease. *J Appl Physiol*. 2021;130(2):485–490.
333. Buchanan BK, Siebert DM, Zigman Suchsland ML, et al. Sudden death associated with sickle cell trait before and after mandatory screening. *Sports Health*. 2020;12(3):241–245.
334. Caughey MC, Derebail VK, Key NS, et al. Thirty-year risk of ischemic stroke in individuals with sickle cell trait and modification by chronic kidney disease: the atherosclerosis risk in communities (ARIC) study. *Am J Hematol*. 2019;94(12):1306–1313.
335. Singer DE, Chen L, Shao S, Goldsmith J, Byrne C, Niebuhr DW. The association between sickle cell trait in U.S. Service members with deployment, length of service, and mortality, 1992–2012. *Mil Med*. 2018;183(3–4):e213–e218.
336. Etyang AO, Wandabwa CK, Kapesa S, et al. Blood pressure and arterial stiffness in Kenyan adolescents with the sickle cell trait. *Am J Epidemiol*. 2018;187(2):199–205.
337. Liem RI, Chan C, Vu TT, et al. Association among sickle cell trait, fitness, and cardiovascular risk factors in CARDIA. *Blood*. 2017;129(6):723–728.
338. Xu Z, Zheng Y, Wang X, et al. Stiffening of sickle cell trait red blood cells under simulated strenuous exercise conditions. *Microsyst Nanoeng*. 2016;2:16061.
339. Nelson DA, Deuster PA, Carter 3rd R, Hill OT, Wolcott VL, Kurina LM. Sickle cell trait, rhabdomyolysis, and mortality among U.S. Army soldiers. *N Engl J Med*. 2016;375(5):435–442.
340. Saxena P, Chavarria C, Thurlow J. Rhabdomyolysis in a sickle cell trait positive active duty male soldier. *US Army Med Dep J*. 2016:20–23.

341. Hughes RL, Feig J. Sickle cell trait-related exertional deaths: observations at autopsy and review of the literature. *Mil Med*. 2015;180(8): e929–932.
342. Quattrone RD, Eichner ER, Beutler A, Adams WB, O'Connor FG. Exercise collapse associated with sickle cell trait (ECAST): case report and literature review. *Curr Sports Med Rep*. 2015;14(2):110–116.
343. Key NS, Connes P, Derebail VK. Negative health implications of sickle cell trait in high income countries: from the football field to the laboratory. *Br J Haematol*. 2015;170(1):5–14.
344. Fajardo KA, Tchandja J. Exercise-induced cardiac arrest in a sickle cell trait-positive Air Force recruit: a case report. *Mil Med*. 2015;180(3): e372–374.
345. Abkowitz JL, O'Connor FG, Deuster PA, Thompson AA. Sickle cell trait and safe athletic participation: the way forward. *Curr Sports Med Rep*. 2014;13(3):192–193.
346. Tripette J, Hardy-Dessources MD, Romana M, et al. Exercise-related complications in sickle cell trait. *Clin Hemorheol Microcirc*. 2013;55(1):29–37.
347. Harris KM, Haas TS, Eichner ER, Maron BJ. Sickle cell trait associated with sudden death in competitive athletes. *Am J Cardiol*. 2012;110(8): 1185–1188.
348. Connes P, Harmon KG, Bergeron MF. Pathophysiology of exertional death associated with sickle cell trait: can we make a parallel with vaso-occlusion mechanisms in sickle cell disease? *Br J Sports Med*. 2013;47(4):190.
349. Harmon KG, Drezner JA, Klossner D, Asif IM. Sickle cell trait associated with a RR of death of 37 times in National Collegiate Athletic Association football athletes: a database with 2 million athlete-years as the denominator. *Br J Sports Med*. 2012;46(5):325–330.
350. Lawrie AS, Pizzey A, Trompeter S, et al. Procoagulant activity in patients with sickle cell trait. *Blood Coagul Fibrinolysis*. 2012;23(4):268–270.
351. Ferster K, Eichner ER. Exertional sickling deaths in Army recruits with sickle cell trait. *Mil Med*. 2012;177(1):56–59.
352. Goldsmith JC, Bonham VL, Joiner CH, Kato GJ, Noonan AS, Steinberg MH. Framing the research agenda for sickle cell trait: building on the current understanding of clinical events and their potential implications. *Am J Hematol*. 2012;87(3):340–346.
353. Loosemore M, Walsh SB, Morris E, Stewart G, Porter JB, Montgomery H. Sudden exertional death in sickle cell trait. *Br J Sports Med*. 2012; 46(5):312–314.
354. Key NS, Derebail VK. Sickle-cell trait: novel clinical significance. *Hematology Am Soc Hematol Educ Program*. 2010;2010:418–422.
355. Anzalone ML, Green VS, Buja M, Sanchez LA, Harrykissoon RI, Eichner ER. Sickle cell trait and fatal rhabdomyolysis in football training: a case study. *Med Sci Sports Exerc*. 2010;42(1):3–7.
356. Tsaras G, Owusu-Ansah A, Boateng FO, Amoateng-Adjepong Y. Complications associated with sickle cell trait: a brief narrative review. *Am J Med*. 2009;122(6):507–512.
357. Connes P, Reid H, Hardy-Dessources MD, Morrison E, Hue O. Physiological responses of sickle cell trait carriers during exercise. *Sports Med*. 2008;38(11):931–946.
358. Connes P, Tripette J, Chalabi T, et al. Effects of strenuous exercise on blood coagulation activity in sickle cell trait carriers. *Clin Hemorheol Microcirc*. 2008;38(1):13–21.
359. Makaryus JN, Catanzaro JN, Katona KC. Exertional rhabdomyolysis and renal failure in patients with sickle cell trait: is it time to change our approach? *Hematology*. 2007;12(4):349–352.
360. Mitchell BL. Sickle cell trait and sudden death—bringing it home. *J Natl Med Assoc*. 2007;99(3):300–305.
361. Small E. Sickle cell trait and sudden death. *Pediatr Ann*. 2003;32(11):757–759.
362. Pretzlaff RK. Death of an adolescent athlete with sickle cell trait caused by exertional heat stroke. *Pediatr Crit Care Med*. 2002;3(3):308–310.
363. Franklin QJ, Compeggie M. Splenic syndrome in sickle cell trait: four case presentations and a review of the literature. *Mil Med*. 1999;164(3): 230–233.
364. Holmes PS, Kerle KK, Seto CK. Sickle cell trait and sudden death in athletes. *Am Fam Physician*. 1998;58(8):1760–1761.
365. Lieberman J. Collapse & death associated with sickle cell trait. *Mil Med*. 1997;162(3):ii.
366. Evans P, Murray MJ. Sudden exertional death and sickle cell trait. *Am Fam Physician*. 1997;55(3):784.
367. Kerle KK, Nishimura KD. Exertional collapse and sudden death associated with sickle cell trait. *Mil Med*. 1996;161(12):766–767.
368. Kerle KK, Runkle GP. Sickle cell trait and sudden death in athletes. *JAMA*. 1996;276(18):1472.
369. Murray MJ, Evans P. Sudden exertional death in a soldier with sickle cell trait. *Mil Med*. 1996;161(5):303–305.
370. Harrelson GL, Fincher AL, Robinson JB. Acute exertional rhabdomyolysis and its relationship to sickle cell trait. *J Athl Train*. 1995;30(4): 309–312.
371. Browne RJ. Sickle cell trait and sudden death. *Sports Med*. 1994;18(6):373–374.
372. Brewer GJ. Risks in sickle cell trait. *J Lab Clin Med*. 1993;122(4):354–355.
373. Weisman IM, Zeballos RJ. Sickle cell anemia trait in the military aircrew population. *Aviat Space Environ Med*. 1992;63(5):382–383.
374. Pearson HA. Sickle cell trait and competitive athletics: is there a risk? *Pediatrics*. 1989;83(4):613–614.
375. Davis AM. Sickle-cell trait as a risk factor for sudden death in physical training. *N Engl J Med*. 1988;318(12):787.
376. Sullivan LW. The risks of sickle-cell trait: caution and common sense. *N Engl J Med*. 1987;317(13):830–831.
377. Eichner ER. Sickle cell trait, exercise, and altitude. *Phys Sportsmed*. 1986;14(11):144–157.
378. Sateriale M, Hart P. Unexpected death in a black military recruit with sickle cell trait: case report. *Mil Med*. 1985;150(11):602–605.
379. Foster Jr HW. Maternal death associated with sickle cell trait. *Am J Obstet Gynecol*. 1985;153(3):346–347.
380. Diggs LW. The sickle cell trait in relation to the training and assignment of duties in the armed forces: III. Hyposthenuria, hematuria, sudden death, rhabdomyolysis, and acute tubular necrosis. *Aviat Space Environ Med*. 1984;55(5):358–364.
381. Francis CK, Bleakley DW. The risk of sudden death in sickle cell trait: noninvasive assessmanet of cardiac response to exercise. *Cathet Cardiovasc Diagn*. 1980;6(1):73–80.
382. Sears DA. The morbidity of sickle cell trait: a review of the literature. *Am J Med*. 1978;64(6):1021–1036.
383. Ramirez A, Hartley LH, Rhodes D, Abelmann WH. Morphological features of red blood cells in subjects with sickle cell trait: changes during exercise. *Arch Intern Med*. 1976;136(9):1064–1066.
384. Mease AD, Longo DL, Hakami N. Sicklemia and unexpected death in sickle cell trait: observations of five cases. *Mil Med*. 1976;141(7): 470–473.

385. Ajabor LN, Ezimokhai M. Sickle cell disease and sickle cell trait among subfertile women in Benin city, Nigeria. *Int J Gynaecol Obstet*. 1976; 14(2):182–186.
386. Gibson IH, Love SH. Sickle cell trait: a rare cause of post-operative death in Northern Ireland. *Ulster Med J*. 1974;43(2):143–147.
387. McGarry P, Duncan C. Anesthetic risks in sickle cell trait. *Pediatrics*. 1973;51(3):507–512.
388. Sickle-cell trait and sudden death. *N Engl J Med*. 1970;283(22):1229–1231.
389. Jones SR, Binder RA, Donowho Jr EM. Sudden death in sickle-cell trait. *N Engl J Med*. 1970;282(6):323–325.
390. McCormick WF, Kashgarian M. Abnormal hemoglobins. V. Age at death of patients with sickle cell trait. *Am J Hum Genet*. 1965;17:101–103.
391. Incidence of sickle cell trait—United States, 2010. *MMWR (Morb Mortal Wkly Rep)*. 2014;63(49).
392. Graham MA. Investigation of deaths temporally associated with law enforcement apprehension. *Acad Forensic Pathol*. 2014;4(3):366–389.
393. Thogmartin JR. Sudden death in police pursuit. *J Forensic Sci*. 1998;43(6):1228–1231.
394. Channa Perera SD, Pollanen MS. Sudden death due to sickle cell crisis during law enforcement restraint. *J Forensic Leg Med*. 2007;14(5): 297–300.
395. Quinn TC. HIV epidemiology and the effects of antiviral therapy on long-term consequences. *AIDS (London, England)*. 2008;22(Suppl 3):S7.
396. Ssentongo P, Heilbrunn ES, Ssentongo AE, et al. Epidemiology and outcomes of COVID-19 in HIV-infected individuals a systematic review and meta-analysis. *Sci Rep*. 2021;11(1):1–12.
397. Kibayashi K, Mastri A, Hirsch C. Neuropathology of human immunodeficiency virus infection at different disease stages. *Hum Pathol*. 1996; 27:637–642.
398. Hirsch C. *Maude Abbott Lecture: Forensic Pathology and Public Health. United States and Canadian Academy of Pathology (USCAP); March 4, 1997*. 1997 (Orlando, Florida).

CHAPTER

8

Approach to medical intervention-related deaths

Susan F. Ely and James R. Gill

The right dose differentiates a poison from a remedy. *-Paracelsus.*

In 1999, the Institute of Medicine of the National Academy of Sciences reported that up to 98,000 patients die of preventable medical errors in United States hospitals annually.[1,2] Healthgrades, a healthcare quality rating company, stated that as many as 195,000 people per year may be dying due to preventable errors.[3–5] These reports have focused attention on reducing medical "errors" and subsequent deaths. One recommendation was for mandatory reporting of deaths and serious injuries.[1] Medicolegal death investigation systems are in a position to help. In order to achieve this, however, there first must be a way to clearly identify these deaths to allow them to be classified and tracked consistently and with standardization. The use of "therapeutic complication" (TC) is one such way to accomplish these goals in most instances.[6–10]

Therapeutic complications

In addition to the standard manners of death (natural, homicide, suicide, accident, and undetermined), medical examiners in several jurisdictions have the option to certify a death as TC.[7,9,11,12] Rather than an inadvertent medical *error,* per se, a TC is **a death due to a predictable complication of an appropriate and nonemergent treatment or diagnostic modality.**[7] Most deaths resulting from **elective procedures for the treatment or diagnosis of diseases** (not injuries) are TCs because the patients were not at "death's door" before the surgeries, treatments, or procedures. Such fatalities can be distinguished from natural deaths by employing the *"but for" test.* "But for" the intervention, would the patient likely have died from the disease? If the answer is no, the operation hastened the death and the manner is TC. If, on the other hand, the patient had a ruptured abdominal aortic aneurysm, then died during *emergency* surgery, the operation did not hasten the death, even if it failed to save a dying patient, and the manner would be classified as natural. In the latter instance, the patient would have died in overwhelming likelihood, regardless of the surgery, the lethal sequence of events having started before even reaching medical attention. *When the underlying pathology is incompatible with continued life, therapeutic interventions are better viewed as resuscitative efforts than as "therapies,"* per se. Additionally, it is important to recognize that in the TC paradigm, **the proximate cause, by definition, must be a disease**, not an injury. If the latter, the manner is informed by the circumstances of the injury, irrespective of the intervening therapy (see below for a further discussion).

In jurisdictions not having the option of TC, deaths meeting the "but for" criterion are typically certified as natural or accidental deaths. Most accurately, however, the use of **accident** should be reserved for fatal complications of **inappropriate** or **inadvertent errors in treatment/therapy** (e.g., inadvertently giving a patient 100 times the prescribed dose of insulin). Accidents represent (unintentional) mistakes. When certifying deaths as complications of diagnostic and therapeutic procedures, the death certificate is not complete if it does not indicate the disease for which the medical therapy was administered, as this constitutes the proximate cause of death.

For death certification purposes, deaths occurring during medical care for a disease fall into one of three manner categories: natural, therapy-related, and inadvertent/injuries. In the natural manner group, neither medical therapy nor injury plays a role in causing or contributing to the death. An example of such a scenario would be a death resulting from a saddle pulmonary thromboembolism. A failed attempt at thrombolysis therapy in response would constitute a form of resuscitative effort in a patient already in cardiac arrest/at "death's door." The intervention represents a failure to

https://doi.org/10.1016/B978-0-323-91796-4.00009-X

save the patient rather than a cause of the demise. In other words, without the therapy, the initiator of the lethal sequence of events would undoubtedly have proceeded, unchecked, and with near certainty, caused the death regardless. The second cohort include those that are due to predictable complications of appropriate medical therapies or diagnostic procedures. The third are deaths due to unanticipated complications and/or inappropriate therapies. In general, *these three categories correspond to three manners of death:* **natural, TC, and accident.**

Before assigning a death as a TC, one must have a clear understanding of what is considered a "predictable" complication and what constitutes "appropriate" therapy. One may consider "**predictable**" complications as **those routinely listed in the informed consent statement and sanctioned by a given specialty and/or are reported in the literature (even if rare).**[6,8,13–15] **"Appropriate"** medical therapy is considered as a **treatment or diagnostic procedure, medicinal or procedural, falling within the standard of care and practice, conducted legally (in a sanctioned/licensed clinical facility by a licensed medical professional), and/or sanctioned by the Food and Drug Administration.** Common categories and examples of TCs are listed in Table 8.1.[7] It is also important to understand that, like delayed fatal complications with other manners of death, there is no "statute of limitations" for TCs. As long as there is no identified *efficient intervening cause*, years may pass between the putative therapy and its fatal outcome.

Categories and examples of therapeutic complications

Operative

Postoperative infection: Bronchopneumonia, wound infection, line sepsis, urinary tract infection.
Technical: Dehiscence, bleeding.
Nontechnical/functional: Ileus, cardiac events.
Late/delayed: Adhesions with bowel obstruction, splenectomy associated infection, malabsorption.
Anesthesia-related: Hyperthermia, ventilation complications, anesthesia reactions.
Postoperative: Pulmonary thromboembolism.
Transfusion-related: Transfusion reactions/infections (needed due to surgery).

TABLE 8.1 Common therapeutic complication causes of death.

Complication	Procedure	Proximate cause
Hemorrhagic complications	During elective coronary artery bypass graft surgery	Atherosclerotic coronary artery disease
Pulmonary fibrosis	Methotrexate therapy	Rheumatoid arthritis
Pulmonary thromboembolism due to deep vein thromboses of lower extremities	Elective hip replacement	Osteoarthritis
Bronchopneumonia	Following elective abdominal surgery (hemicolectomy)	Adenocarcinoma of colon
Cardiac arrhythmia	Digoxin toxicity	Idiopathic dilated cardiomyopathy
Sepsis	Cosmetic breast augmentation surgery	(none)
Hypoglycemia	Insulin therapy	Diabetes mellitus
Anaphylaxis	Administration of penicillin	Streptococcal pharyngitis
Transfusion-related acute lung injury (TRALI)	Platelet transfusion	Autoimmune aplastic anemia

Nonoperative

Infection: Line sepsis, urinary tract infections (from indwelling catheters).

Procedural

Technical gastric tubes (leaks/bleeding), bleeding from cardiac catheterization, bronchoscopy, or invasive radiologic procedures.

Nontechnical: Arrhythmia during stress test or catheterization.

Medication: Anaphylaxis, infection during steroid therapy, toxic epidermal necrolysis, gastrointestinal hemorrhage (nonsteroidal antiinflammatory medication, anticoagulant therapies), and pseudomembranous colitis (after antibiotic therapy).

Anticoagulation: Bleeding with anticoagulation therapy.

Hemodialysis: Bleeding, shunt problems, electrolyte disturbances, infected shunt.

Pulmonary embolism: Nonsurgical inactivity, medications (including oral contraceptives/hormone therapies).

Radiation therapy: Fibrosis/infection.

A **medical "accident"** results from an *inadvertent error,* a mistake that falls outside of the predictable complications included on an informed consent form (e.g., the wrong medication or dose inadvertently given to a patient that results in their death). In rare circumstances, the classification of homicide may be used in a clinical/medical environment. These occur when there is *overwhelming evidence of recklessness and/or a flagrant disregard for human life, intentional "mercy" killing (in hospitals or in "assisted suicides"), or when fatal complications occur as the result of an illegal/nonsanctioned procedure, in a facility that is operating in violation of codes of practice or local laws, or any procedure performed by an unlicensed practitioner, even if a physician.*[16–20]

The authors recognize that TC, as a manner of death designation, is not an option for medical examiners/coroners (MEs/Cs) in most jurisdictions in the United States and abroad. It is also not recognized by the World Health Organization (WHO). Therefore, these deaths must be classified as either natural or accidental deaths in these jurisdictions (undetermined is also an option, but should be reserved for deaths with truly unknown circumstances). This binary option results in various jurisdictions using different manners for the same death circumstance, moving further from a standardized death certification practice and accurate vital statistical data gathering.[9] For example, without the ability to use TC, a death due to anaphylaxis related to antibiotic administration may be certified by some offices as a natural death, while in others, as an accident.[9] To use "natural" in such a death ignores the external component, and to designate accident implies a "mistake," which does not accurately represent this specific scenario either. The key in reflecting the nature of the case, regardless of which manner boxes are available to the certifier, is to make sure that the cause of death statement details the specific complication of therapy. In order to properly convey this information to epidemiologists, clinicians, and families requires the inclusion **of three components on the death certificate**. These are: *the putative complication, the causative treatment/procedure or diagnostic modality, and the underlying disease for which the treatment was administered or which the diagnostic procedure was evaluating.* If the TC option is not available, some add "TC" parenthetically at the end of the cause of death statement, with the chosen manner checkbox as natural, by necessity.

Very important to note is that **the manner designation of TC is not synonymous with malpractice.** In fact, more commonly than not, the forensic pathologist (FP) holds no opinion regarding the standard of care per se in a TC death. Such deaths may reflect a broad spectrum of quality of care, but that is neither relevant nor necessary to assert in order for this checkbox to be ticked. Rather, only the criteria supporting the definition of TC must be met. For example, a patient dies from complications following the (emergency) surgical treatment of a small bowel obstruction (SBO) caused by abdominal adhesions complicating remote surgery. The inexperienced investigator may primarily focus on the most recent SBO surgery, rather than on the original cause of the adhesions (the remote surgery). If the adhesions are related to a remote surgery, the proximate cause would be the disease or injury that necessitated that *initial* surgery. Sometimes, the only information that may be known about the original/remote surgery is that it was a "hysterectomy," which is a procedure, not an underlying disease (proximate cause). As such, the indication must still be sought in order to identify the bottom line. If the underlying *indication* for the initial abdominal surgery was a disease, such as uterine leiomyomata, and surgery was elective (i.e., not emergent), then one is dealing with a TC. But, if the hysterectomy had been emergently performed due to excessive bleeding from a placental abruption complicating poorly controlled gestational hypertension (a potentially life-threatening disease), the death would be certified as natural.

In addition, if the disease process itself played a primary, if not exclusive role in the formation of the adhesions, rather than the surgical intervention, per se, one may be dealing with a natural manner. For example, an SBO in a patient with a history of an appendectomy may be a natural or TC, depending on the details of the clinical presentation. To wit, if a normal or even inflamed, but intact appendix had been removed (n.b., due to the risk of underdiagnosing an acute appendicitis—this is not considered malpractice), fatal complications from subsequent adhesions leading to a SBO would constitute a TC. If, on the other hand, the initial surgery was performed following the rupture of an acutely inflamed appendix, the infectious process rather than the surgery would have been primarily culpable for the adhesions, resulting in a natural manner. In addition, surgery in response to a ruptured appendicitis would likely be regarded as emergent, which would further push the death into the natural camp. If in doubt and there is credible evidence for "downgrading" the manner to natural, it is most prudent to do so, choosing the more conservative of the two manners (natural), since "he who accuses must prove."

Prior abdominal trauma adds one more variation to our abdominal adhesion example. Blunt, sharp, and gunshot injuries of the abdomen may result in peritoneal adhesions with or without surgery. Here the forensic pathologist is faced with an entirely different kind of case, one that, at the very least, likely needs an autopsy to exclude an efficient intervening cause and/or contributing conditions. Also, for remote potentially *life-threatening* trauma, the manner is predicated on the circumstances of the original injury (accident, suicide, or homicide). This is a very important concept to understand: injuries "trump" disease in the determination of manner. If a potentially life-threatening injury is identified as the proximate cause (the indication for the therapeutic/surgical intervention), TC is no longer a relevant manner. If a patient dies intraoperatively during the surgical repair of a broken hip (fall at home, not a pathologic fracture), the manner would appropriately be classified as an accident. Moreover, if the proximate cause/indication for a procedure is a disease, but injury of any type is listed in part 2 of the death certificate, the TC manner option again disappears. As in natural deaths, all causes and conditions on the death certificate in TCs, primary or contributory, must be natural diseases.

In rare instances, the treatment of the sequelae of *minor/nonlife threatening injury* (e.g., chronic shoulder pain from a sprain) results in death during a subsequent corrective surgery (e.g., anesthesia complication). In this instance, the appropriate manner would be TC, even if the initiating injurious event that resulted in chronic pain was caused by another person (e.g., an assault). For example, consider the following scenario, similarly described in the 1995 Hirsch—Flomenbaum Check Sample and in Chapter 3: a 48-year-old woman sprained her upper arm and shoulder during an assault. Despite conservative medical treatments for 12 months, she continued to experience pain in the arm and shoulder and was diagnosed with a posttraumatic sympathetic dystrophy. During an elective cervical sympathectomy, she arrested following induction of general anesthesia. An autopsy diagnosed marked atherosclerotic stenoses in two of her major coronary arteries. The complication of this elective (nonlifesaving), even if necessary, therapy constitutes an "efficient intervening cause." Neither her nerve pain nor the injury that caused it were ever life-threatening, at the time of the injury occurrence or at the time of her surgery, so the TC "becomes dominant and relieves the original injury of its causal responsibility."[21]

A helpful way to characterize the original injury as being either "potentially life-threatening" or not is to ask whether the injury "put the patient to bed" at the time, even if temporarily. Generally speaking, injuries severe enough to render a patient bedbound tend to be potentially life-threatening. Also, injuries of the lower extremities have the potential to more readily put patients to bed than do those of the upper extremities, by virtue both of the injured legs and the attendant limited mobility. The latter enhances the potential for mortality in several ways (cardiac deconditioning, infectious complications, such as bronchopneumonia and infected decubitus ulcers, and deep vein thromboses caused by both stasis and in response to injuries). Keep in mind that in the setting of delayed complications of remote life-threatening injuries, the acute threat to life may not remain as such at the time of the subsequent complication in the same way that it had at the time of the injury occurrence, but it may have set the stage for these later complications. Each death must be evaluated based on its own merits and the details of the case (for a further discussion of delayed deaths from injuries, see Chapter 15, "Special Topics").

To summarize, proximate cause informs manner and therefore **a death is only a TC if the underlying/proximate cause is natural, if the procedure that hastened the death was elective, and if there was no contributing traumatic cause.** Absent these three features, the forensic pathologist may be dealing with a natural or an unnatural death, but not a TC. If a surgery is emergent, whether for a disease or a traumatic injury, the underlying cause dictates the manner (natural, accident, homicide, or suicide). When the proximate cause is a potentially life-threatening injury, regardless of whether the procedure was emergent or elective, the manner is predicated on the circumstances of the injury and trumps a TC designation.

Therefore, before accepting jurisdiction, two questions must be asked and answered by the medicolegal death investigator:

(1) What is the *etiologically specific* proximate cause/disease (indication for the culpable surgery)?
(2) Was the procedure/surgery emergent or elective?

Answering these two questions should accurately lead to the proper disposition of a case.

Finally, an "emergent" procedure or treatment is defined as one administered to attempt to save the life of a patient at death's door. That is, but for any medical intervention, there is a high probability of death *Esophageal intubation* during resuscitative efforts, for example, in a person who was already in cardiac standstill, is neither a complication of therapy nor a cause of death; it is simply a failure to reverse the lethal sequence of events already set in motion prior to the intervention. **Failure to save is not a cause of death.** If, on the other hand, that esophageal intubation occurred during an elective hernia repair and the misplacement caused death due to hypoventilatory complications of anesthesia, the death would *not* be natural.

The wording and conceptual approach to the certification of a TC should be framed based on answering three questions that correspond to required components of such a DC, linked by phrases such as, "due to"/"complicating" or "for the treatment (or evaluation) of" (Table 8.1).

1. What is the complication(s)/immediate cause(s)/mechanism(s)? This can be worded as "Complications," "Infectious Complications," "Hemorrhagic Complications," "Respiratory Complications," "Pulmonary Thromboembolus," etc."*...of/during/following...*"?
2. What is the procedure/medical or surgical treatment/drug administration or diagnostic procedure? "for the treatment/evaluation of..."?
3. What is the proximate cause (indication for the medical intervention)?

If the forensic pathologist stops at the procedure, they have not reached the bottom causation line and will have no codable and etiologically specific disease (proximate cause) from which they can determine the basis for their manner determination. That said, **there are six deaths that do not typically have an identifiable or codable proximate cause of death.** These include: *complications of cosmetic surgery, complications of screening procedures (e.g., colonoscopies), fatal maternal complications of surgical child birth (nonemergent cesarean sections) in an otherwise healthy mother, complications of birth control measures, complications of some organ donation surgeries (healthy donors without comorbidities), and complications of vaccinations* (see also Chapter 3).

Deaths related to cosmetic, weight loss, and dental procedures

Liposuction and fat transfer

In 1999, the NYC OCME, in collaboration with clinical toxicologists at the NYC Poison Control Center, published an article in the *New England Journal of Medicine* (*NEJM*), reporting a retrospective review of five deaths due to liposuction.[22] More recently, several additional deaths have been investigated which may reflect the addition of the so-called and widely popular Brazilian butt lift (BBL), also known as fat transfer.[22–37] For a BBL, excess fat is removed with liposuction techniques (typically tumescent) from the hips, abdomen, lower back, or thighs, and a portion of this fat is then strategically injected into the buttocks in order to augment the size and enhance the shape of the buttocks.

"High-volume tumescent liposuction," defined (years ago) as the infusion of greater than 1500 mL of tumescent fluid into the fat, seems to be steadily increasing in practice to excessive volumes (liters) of a lidocaine–epinephrine fluid mixture (tumescent infusate), which is infused into iatrogenically created fat pockets. The timing of complications may help in the investigation. Deaths occurring shortly after the procedure may represent a medication complication (excessive lidocaine) or underlying heart or lung disease. At four to five days postoperatively, infection, fat embolism syndrome, and pulmonary embolism (PE) are the more likely causes. Lidocaine toxicity (to a lesser extent) and fluid overload also should be considered at this later stage, as liters of fluid from the retained tissue infusate continue to be absorbed by the body over time (days). This bears mentioning because one of the otherwise healthy women from the 2000 NEJM series (33 years old with a normal echocardiogram and cardiac enzymes) died from fluid overload >48 h postoperatively. The fluid overload resulted not just in high output heart failure, but also in profound fluid-related hemodilution (hemoglobin of 5.8), exacerbated by a 700 mL surgical estimated blood loss.[22]

While it is prudent to work up fat embolism in non-PE cases, remain circumspect. Avoid over diagnosis of "fat embolism syndrome." This (and PEs) is the presumptive diagnosis of many surgeons, but when a patient's adipose tissue has been substantially traumatized by liposuction and "fat transfer," fat seen in blood vessels should not be surprising/is reasonably expected. The relevant questions, however, should center on the microscopic *extent* dovetailed with the *clinical history* to help distinguish whether the emboli are the cause of death or merely a consequence of the procedure. This requires clinicopathological correlation and a clear understanding of the fat embolism syndrome and its timing, so that the FP can focus questioning of family and doctors when investigating the postoperative period and the patient's signs and symptoms (shortness of breath, confusion, etc.), if any. Fatal fat embolism after penis enlargement by autologous fat transfer also has been reported.[36]

Even though these might be viewed as seemingly innocuous elective procedures, the fatality rate is far from the expected rate of near zero. These investigations are complex and very important to perform thoroughly since otherwise healthy, often young persons are dying unexpectedly. Some of these facilities aggressively advertise and perform upwards of 15–20 procedures per day, with weekend and holiday "specials," little follow up, and no anesthesiologists on site, despite the use of very high doses of infused lidocaine (up to and exceeding 45 mg/kg).[31,38–49] The recent (and relatively untested) addition of BBL adds an additional and potentially confounding and unclear element of risk. An alternative to the traditional BBL has been performed in nonmedical settings by unlicensed "cosmetologists" who are known to inject industrial grade free silicone (not fat) into the buttocks. Silicone injections may result in death via pulmonary embolization of this material (Figs. 8.1 and 8.2).[23,25–27,50] Since these are *unlicensed* providers and the death occurs at the hand of another, they would be justifiably certified as homicides (see Medical homicides section below).

The practical approach to these investigations is multifaceted. For lidocaine testing, one may need to ask the laboratory to quantitate it (while lidocaine is part of most basic screenings, it is not typically quantitated, as, in most deaths, it is viewed as a common resuscitative medication). If a PE is the immediate cause of death, one should consider other risk factors, such as use of oral contraceptives (given the age and sex of many of these patients), documenting the presence or absence of large, compressive uterine leiomyomata, testing for hereditary thrombophilias, calculating the body mass index (BMI), and excluding certain infections that may confer a thrombophilic risk (e.g., COVID-19). These deaths often bring civil law suits (and rarely, criminal charges), and while other variables may not relieve the procedure of its culpability (i.e., but for the procedure, the death would not have occurred at this time), there may be contributory risks that further explain why this patient developed a pulmonary thromboembolus (PTE), while most do not.

The surgical sites should be described in the autopsy report and photographed. Usually, there are hemorrhagic cannula tracks (not large, space-occupying hematomas) within the fat. These penetrations should not breach the deeper muscle or peritoneal cavity. In BBLs, one may only see loose/granular and pulverized fat, which should only be within the fat compartment, not the deeper muscle.

Review of the complete medical and procedure records may guide special toxicology requests (e.g., propofol, etc.) and document the length of the procedure (particularly important in PTE deaths). Some jurisdictions require office-based procedure deaths to be reported to public health oversight agencies to identify patterns that may not be recognized by local ME/C offices due to the small number of cases. These same investigative principles can be applied to other elective and cosmetic procedures.

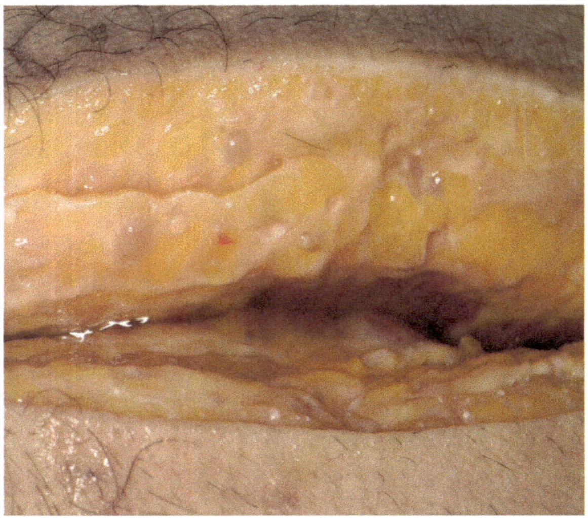

FIGURE 8.1 **Injection sites in the buttock with industrial grade silicone.** At autopsy, injection sites and spheroid collections of silicone are clearly seen in the subcutaneous tissue.

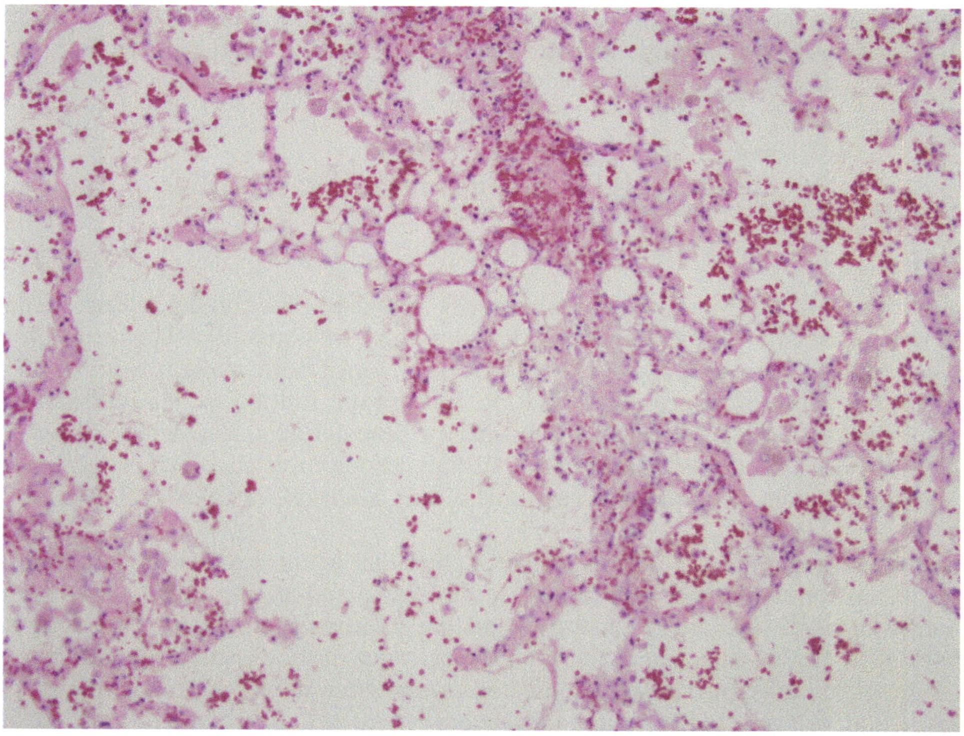

FIGURE 8.2 **Microscopic section of lung.** Silicone emboli in the lung microvasculature.

Bariatric surgery

Because of the substantial health risks associated with obesity, there is a medical indication for bariatric surgery.[51–73] As such, it is not viewed as a cosmetic procedure. As with any surgical procedure, however, there are risks, often heightened by the inherent susceptibility of this patient population.[6,74–86] The proximate cause of death for complications involving bariatric surgery is obesity.[6,51,60,62,65,74,79,81,84–87] "Overweight" and "obesity" are defined as abnormal or excessive fat accumulation that is a health risk.[51,55,61,62] A BMI over 25 is considered overweight, and over 30, obese (see Chapter 7). The mortality rate among patients undergoing bariatric operations is generally cited as being between 0.05% and 2.0%.[6]

A study of 97 deaths due to complications of bariatric surgery found that the most common complication was an anastomotic leak, with subsequent infection.[88] A high percentage of deaths occurred after discharge (40%) and/or >30 days after surgery (37%). The interval between the initial surgery and death ranged from several hours to years. There were eight deaths from complications of bariatric surgery that occurred >1 year after surgery.[83,84,86,87] A study of postoperative deaths in a hospital setting found PE as a common cause.[75] There may also be contributing conditions in these deaths, such as diabetes mellitus or heart disease.

The dental office

Elective dental procedures set the stage for specific, if uncommon, complications. When they cause death, they represent TCs.

The mouth is a rich microbiome, colonized by a variety of commensal bacteria, including both aerobic and anaerobic organisms. Normally, the bacterial flora cause no harm, but in the right setting and if they gain access to the vascular system or the brain, they can wreak havoc. Should an abscess develop following an invasive procedure, the offending bacteria have the capacity to contiguously erode into the basilar skull and brain, through the sinuses or the auditory structures. Once within the cranial vault, a fatal bacterial meningitis can develop quickly, particularly if involving the vital structures of the brainstem. Similarly, bacteremia can quickly develop into septicemia, endocarditis, or bronchopneumonia, rapidly causing death if aggressive organisms, such as anaerobes, are involved.

Patients with prosthetic heart valves should routinely be administered antibiotic prophylaxis prior to dental procedures. Reporting and death certification guidelines are no different for dental procedures than they are for medical and surgical procedures.

Accidents and homicides in medical settings

Medical accidents

Two other manners of death that are much less commonly invoked in the clinical setting are accident and homicide. Medical accidents, while not common, are much less rare than are homicides. The former, by definition, result from "inadvertent errors," not from the predictable, even if rare, complications of appropriate medical therapy that constitute a TC. Examples of **medical accidents** include, but are not limited to, inadvertently giving the wrong medication or dose of medication, perforating a nonpathologic bowel during a screening colonoscopy, inadvertent extubation, etc. Most of these are not just inadvertent, but they also involve chemical or mechanical *injuries*, hence the accidental designation. Fatal accidents in the medical setting are those errors that are serious enough to cause death.

If there is an inadvertent respiratory equipment malfunction that leads to a death (ventilator failure or unplugging, power outage, the endotracheal tube falls out, etc.), it is considered an accidental injury, but only if the underlying/proximate cause of death (necessitating the life-saving equipment and without which the patient would clearly die) is a disease. If, on the other hand, the underlying cause of the ventilator dependence is also an injury, recent or remote, then the manner of death is predicated on the circumstances of that original, potentially lethal injury (accidental, suicidal, or homicidal)/proximate cause.

As an aside, intubation complications are one of the risks involved with the treatment of life-threatening injuries and should not be considered an efficient intervening cause which relieves the underlying injury as the proximate cause. For example, a man sustains a gunshot wound of the cervical spine, rendering him quadriplegic and completely ventilator-dependent for decades. One day, a respiratory therapist inadvertently trips over the ventilator cord, unknowingly unplugging it. The ventilator ceases to function and the patient dies. Because the underlying injury would be lethal without life support, he would not have died due the loss of power but for having been shot several years earlier and kept alive by life support. Tripping over the cord was inadvertent, an accident, but as a result, the initial/original injury was able to exert its lethal effects in the absence of life support. In other words, but for the initial firearms injury, there would be no ventilator dependence and this outcome would not have occurred. The ventilator dependence, its cause, and the most immediate event (power loss) all must be listed in the cause of death statement, but the original injury, the cause of the ventilator dependence, informs the manner in this specific scenario. The proximate circumstance (homicidal injury) trumps the immediate injury circumstance (accidental power loss), but the proximate and immediate causes are directly related and dependent upon one another, so are both listed.

If the same man was crushed by an accidental ceiling collapse while lying in the his hospital bed, while he was still maintained on a functioning ventilator, and the collapse caused acute fatal injuries, such as a shattered skull and avulsed brain, then the death would be classified as an accident, mitigated by a *nonsurvivable* "efficient intervening cause," relieving the first injury of its culpability. The ventilator dependence is moot in the latter scenario, since an avulsed brain would kill anyone, on or off a ventilator. Similarly, if instead, his wife slipped some cyanide into his tube feedings, leading to his death, that too would constitute an efficient intervening cause, unrelated to the firearms injury and capable of killing even healthy persons, but would be certified as a homicide (poisoning by cyanide). If he developed lung cancer and exsanguinated because the tumor invaded the pulmonary artery, the death would be classified as natural. The gunshot wound did not cause the lung cancer; the cancer and its complications constitute an efficient intervening cause that is incompatible with continued life. While these examples are admittedly extreme, they serve to illustrate the burden of identifying mechanism and proximate cause and the importance of recognizing an *efficient intervening cause.*

If a patient is intubated for end-stage emphysema and is of sound mind and alertness, but has suicidal ideations and dies by intentionally pulling out his endotracheal tube (self-extubation) in an attempt to end his life, then some may classify this death as a suicide, while others may classify it as a natural death. Reasonable forensic pathologists may disagree on the appropriate certification of this death. That said, keep in mind that a competent adult can decide what medical treatment they wish to have or not have. Removal of medical treatment is an act of omission, not one of commission. A patient may choose to withdraw potentially life-prolonging chemotherapeutic intervention and enter palliative care instead; this is not generally viewed by FPs as suicidal behavior, but rather an end and quality of life decision. The intentional removal of a breathing tube by a competent adult simply allows nature to take its course. There is no violent act or injury, per se. The definition of suicide is a death due to a violent (or chemical) act initiated by the decedent with the intent to cause their own death. If the death is not certified as an accident following the elective removal of an endotracheal tube from a patient transitioning to hospice, then clearly there is no inadvertency, so no "injury." Accident is invoked in an *inadvertent* extubation because there is an error component to it and it is not a predictable (neither expected nor anticipated) complication. If a similar end-stage chronic obstructive pulmonary disease patient was not of sound mind, but rather was delirious, agitated, and hyperkinetic and were to inadvertently self-extubate while thrashing around in his hospital bed, leading to his death, the manner might reasonably be classified as an accident. These types of deaths often present challenges to the FP and both require and deserve thoughtful, consistent, reproducible, and defensible consideration, using sound forensic principles, while evaluating the totality of each case on its own merits. The cases that give us the most angina as practitioners of responsible death certification present us with the most shades of gray.

The distinction between a TC and an accident may not always be clear. For example, a 45-year-old healthy person has his first screening colonoscopy for cancer. The normal, healthy bowel is perforated during the procedure. This would be certified as an accident. If, however, there was underlying pathology that contributed to the bowel's susceptibility, such as abdominal adhesions or inflammatory bowel disease, the manner might be "downgraded" to a TC, since this outcome would reasonably be viewed as more predictable, given the native pathology of the bowel. In some instances, a discussion with a clinical colleague may be helpful in determining if the complication is viewed as predictable and/or if the therapy was appropriate.

The use of the TC can help capture many, but not all, such complication deaths for public health purposes. They may be missed if the death certificates are not completed properly, with the three components outlined above. In addition, some "medical errors" do not meet the criteria of a TC for death certification purposes, even though the medical intervention, or lack thereof, may have failed to prevent the death. An example of such a scenario would be a person who died from an infection in which the prescribed antibiotic did not provide coverage. This may be ripe for civil litigation, but would extend beyond the realm of the forensic pathologist's investigation and certification of death, an act of omission rather than commission. A second example would be a 62-year-old woman who died from a cecal adenocarcinoma that was missed six months earlier because the gastroenterologist only performed a flexible sigmoidoscopy (not a colonoscopy) for the evaluation of anemia and occult blood in the stool. This death would be certified as natural, despite potential failures in meeting the standard of care for such an evaluation.

Medical homicides

A medical-related homicide is rare; it is defined by intentional acts to harm by extreme medical neglect or by illegal acts.[7,20] Homicides in the clinical setting are predicated on at least one of three findings: a caregiver's complicity (intent) in the death ("assisted suicides" or "angels of mercy" deaths); a clear and reckless (or gross and wanton) disregard for life/safety of the patient; or nonlicensure (physician and/or facility) and/or conducting illegal/nonsanctioned procedures and/or administering non-FDA–approved drugs.[74,89–91]

The underground "practice" that has become increasingly prevalent in this era of BBLs is silicone injections of the buttocks (see "Liposuction and Fat Transfer" above), often purchased on the internet and illegally performed by nonphysician "practitioners." These procedures have claimed the lives of several young women in New York City, Florida, and elsewhere through fatal silicone embolization and may be criminally prosecuted as homicides.[23,25–27,50]

Homicides in the medical setting can and often are predictably complicated and challenging cases to approach, requiring careful consideration and investigation and an objective review of the medical and other records. They must be dealt with on a case-by-case basis, but within a backdrop of an internally consistent approach. Additionally, because the manner is homicide, the certification of these cases requires a higher burden of proof than that required of any other manner classification (reasonable degree of medical/scientific certainty) and may involve collaboration with the prosecutor's office and/or local or state professional oversight agencies.

References

1. Kohn LT, Corrigan JM, Donaldson MS, eds. *To Err Is Human: Building a Safer Health System*. Washington, D.C.: National Academy Press; 2000.
2. Leape LL. Institute of medicine medical error figures are not exaggerated. *JAMA*. 2000;284(1):95–97.
3. Leape LL. Errors are not diseases: they are symptoms of diseases. *Laryngoscope*. 2004;114(8):1320–1321.

4. Leape LL. Making health care safe: are we up to it? *J Pediatr Surg*. 2004;39(3):258–266.
5. Leape LL. Reporting of adverse events. *N Engl J Med*. 2002;347(20):1633–1638.
6. Goldfeder LB, Ren CJ, Gill JR. Fatal complications of bariatric surgery. *Obes Surg*. 2006;16(8):1050–1056.
7. Gill JR, Goldfeder LB, Hirsch CS. Use of "therapeutic complication" as a manner of death. *J Forensic Sci*. 2006;51(5):1127–1133.
8. Gill JR, Storck K, Kelly S. Fatal exsanguination from hemodialysis vascular access sites. *Forensic Sci Med Pathol*. 2012;8(3):259–262. https://doi.org/10.1007/s12024-011-9303-0.
9. Gill J, Maloney K, Hirsch C. The consistency and advantage of therapeutic complication as a manner of death. *Acad For Path*. 2012;2(2):176–181.
10. Gill JR, Ely SF, Toriello A, Hirsch CS. Adverse medical complications: an under-reported contributory cause of death in New York City. *Public Health (Lond)*. 2014;128(4):325–331.
11. Hanzlick R, Hunsaker 3rd J, Davis GJ. Commentary on: Gill JR, Goldfeder LB, Hirsch CS. Use of therapeutic complication as a manner of death. J forensic Sci 2006;51(5):1127–1133. *J Forensic Sci*. 2007;52(2):491. Author reply 492.
12. Krehbiel K, Matshes E, Lathrop S, Pinckard J. Medical therapy-related deaths and the medical examiner. *Acad For Path*. 2015;5(3):421–435.
13. Ely SF, Gill JR. Fatal pulmonary thromboembolism and hereditary thrombophilias. *J Forensic Sci*. 2005;50(2):411–418.
14. Shmuter Z, Frederic FI, Gill JR. Fatal migration of vena caval filters. *Forensic Sci Med Pathol*. 2008;4(2):116–121.
15. Gill JR, Morrow JS, West AB. Fatal hemorrhage following laser hemorrhoidectomy. *J Clin Gastroenterol*. 1994;19(4):343–346.
16. Holbrook J. The criminalisation of fatal medical mistakes. *BMJ*. 2003;327(7424):1118–1119.
17. Filkins JA. "With no evil intent". The criminal prosecution of physicians for medical negligence. *J Leg Med*. 2001;22(4):467–499.
18. Mehlman M. Dishonest medical mistakes. *Vanderbilt Law Rev*. 2006;59(4):1137–1173.
19. Wolf BC, Lavezzi WA. Paths to destruction: the lives and crimes of two serial killers. *J Forensic Sci*. 2007;52(1):199–203.
20. Duncanson E, Richards V, Luce KM, Gill JR. Medical homicide and extreme negligence. *Am J Forensic Med Pathol*. 2009;30(1):18–22.
21. Hirsch CS, Flomenbaum M. Problem-solving in death certification. Check Sample *Am Soc Clin Pathol For Pathol*. 1995;95–1:1–31. FP.
22. Rao RB, Ely SF, Hoffman RS. Deaths related to liposuction. *N Engl J Med*. 1999;340(19):1471–1475.
23. Price EA, Schueler H, Perper JA. Massive systemic silicone embolism: a case report and review of literature. *Am J Forensic Med Pathol*. 2006;27(2):97–102.
24. Bell SK, Gardinali PR. Assessment of silicone polymer composites for environmental forensic applications: a proof of concept study. *J Forensic Sci*. 2010;55(5):1245–1250.
25. Clark RF, Cantrell FL, Pacal A, Chen W, Betten DP. Subcutaneous silicone injection leading to multi-system organ failure. *Clin Toxicol*. 2008;46(9):834–837.
26. Lanzarotta A, Kelley CM. Forensic analysis of human autopsy tissue for the presence of polydimethylsiloxane (silicone) and volatile cyclic Siloxanes using macro FT-IR, FT-IR Spectroscopic imaging and headspace GC-MS. *J Forensic Sci*. 2016;61(3):867–874.
27. Wills S, Byard RW. Lethal silicone embolization syndrome complicating penile, scrotal and breast self-augmentation. *Forensic Sci Med Pathol*. 2019;15(2):304–308.
28. de Jong RH, Grazer FM. "Tumescent" liposuction alert: deaths from lidocaine cardiotoxicity [letter]. *Am J Forensic Med Pathol*. 1999;20(1):101.
29. Platt MS, Kohler LJ, Ruiz R, Cohle SD, Ravichandran P. Deaths associated with liposuction: case reports and review of the literature. *J Forensic Sci*. 2002;47(1):205–207.
30. Matarasso A. Discussion. Analysis of postoperative complications for superficial liposuction: a review of 2398 cases. *Plast Reconstr Surg*. 2011;127(2):872–873.
31. Kim YH, Cha SM, Naidu S, Hwang WJ. Analysis of postoperative complications for superficial liposuction: a review of 2398 cases. *Plast Reconstr Surg*. 2011;127(2):863–871.
32. Terranova C, Sartore D, Snenghi R. Death after liposuction: case report and review of the literature. *Med Sci Law*. 2010;50(3):161–163.
33. Heitmann C, Czermak C, Germann G. Rapidly fatal necrotizing fasciitis after aesthetic liposuction. *Aesthetic Plast Surg*. 2000;24(5):344–347.
34. Laub Jr DR, Laub DR. Fat embolism syndrome after liposuction: a case report and review of the literature. *Ann Plast Surg*. 1990;25(1):48–52.
35. Rapkiewicz AV, Kenerson K, Hutchins KD, Garavan F, Lew EO, Shuman MJ. Fatal complications of aesthetic techniques: the Gluteal region. *J Forensic Sci*. 2018;63(5):1406–1412.
36. Zilg B, Rasten-Almqvist P. Fatal fat embolism after penis enlargement by autologous fat transfer: a case report and review of the literature. *J Forensic Sci*. 2017;62(5):1383–1385.
37. Astarita DC, Scheinin LA, Sathyavagiswaran L. Fat transfer and fatal macroembolization. *J Forensic Sci*. 2015;60(2):509–510.
38. Sterodimas A, Boriani F, Magarakis E, Nicaretta B, Pereira LH, Illouz YG. Thirtyfour years of liposuction: past, present and future. *Eur Rev Med Pharmacol Sci*. 2012;16(3):393–406.
39. Ahmad J, Eaves 3rd FF, Rohrich RJ, Kenkel JM. The American Society for Aesthetic Plastic Surgery (ASAPS) survey: current trends in liposuction. *Aesthetic Surg J*. 2011;31(2):214–224.
40. Berry MG, Davies D. Liposuction: a review of principles and techniques. *J Plast Reconstr Aesthetic Surg*. 2011;64(8):985–992.
41. Pelosi 3rd MA, Pelosi 2nd MA. Liposuction. *Obstet Gynecol Clin N Am*. 2010;37(4):507–519 (viii).
42. Stephan PJ, Kenkel JM. Updates and advances in liposuction. *Aesthetic Surg J*. 2010;30(1):83–97. Quiz 98–100.
43. Ahern RW. The history of liposuction. *Semin Cutan Med Surg*. 2009;28(4):208–211.
44. Triana L, Triana C, Barbato C, Zambrano M. Liposuction: 25 years of experience in 26,259 patients using different devices. *Aesthetic Surg J*. 2009;29(6):509–512.
45. Venkataram J. Tumescent liposuction: a review. *J Cutan Aesthetic Surg*. 2008;1(2):49–57.
46. Kucera IJ, Lambert TJ, Klein JA, Watkins RG, Hoover JM, Kaye AD. Liposuction: contemporary issues for the anesthesiologist. *J Clin Anesth*. 2006;18(5):379–387.
47. Regatieri FL, Mosquera MS. Liposuction anesthesia techniques. *Clin Plast Surg*. 2006;33(1):27–37 (vi).
48. Lawrence N. Current issues in liposuction. *Adv Dermatol*. 2003;19:171–184.
49. Johnson DS, Cook Jr WR. Advanced techniques in liposuction. *Semin Cutan Med Surg*. 1999;18(2):139–148.
50. Crowe JB, Lanzarotta A, Witkowski MR, Andria SE. Analysis of hypodermic needles and syringes for the presence of blood and polydimethylsiloxane (silicone) utilizing microchemical tests and infrared spectroscopy. *J Forensic Sci*. 2015;60(4):1078–1084.

51. Hunsaker D, Hunsaker J. Obesity epidemic in the United States. In: Tosokos M, ed. *Forensic Pathology Reviews*. Vol. 2. Totowa, NC: Humana Press Inc.; 2004:61–100.
52. Nguyen NT, Ho HS, Palmer LS, Wolfe BM. A comparison study of laparoscopic versus open gastric bypass for morbid obesity. *J Am Coll Surg*. 2000;191(2):149–155. Discussion 155-147.
53. Deitel M. Overview of operations for morbid obesity. *World J Surg*. 1998;22(9):913–918.
54. Choban PS, Onyejekwe J, Burge JC, Flancbaum L. A health status assessment of the impact of weight loss following Roux-en-Y gastric bypass for clinically severe obesity. *J Am Coll Surg*. 1999;188(5):491–497.
55. Kral JG. Morbid obesity and related health risks. *Ann Intern Med*. 1985;103(6(Pt 2)):1043–1047.
56. Pace WG, Martin Jr EW, Tetirick T, Fabri PJ, Carey LC. Gastric partitioning for morbid obesity. *Ann Surg*. 1979;190(3):392–400.
57. Rabkin RA. Distal gastric bypass/duodenal switch procedure, Roux-en-Y gastric bypass and biliopancreatic diversion in a community practice. *Obes Surg*. 1998;8(1):53–59.
58. Mognol P, Chosidow D, Marmuse JP. Laparoscopic gastric bypass versus laparoscopic adjustable gastric banding in the super-obese: a comparative study of 290 patients. *Obes Surg*. 2005;15(1):76–81.
59. Kolanowski J. Surgical treatment for morbid obesity. *Br Med Bull*. 1997;53(2):433–444.
60. Santry HP, Gillen DL, Lauderdale DS. Trends in bariatric surgical procedures. *JAMA*. 2005;294(15):1909–1917.
61. Steinbrook R. Surgery for severe obesity. *N Engl J Med*. 2004;350(11):1075–1079.
62. Solomon CG, Dluhy RG. Bariatric surgery–quick fix or long-term solution? *N Engl J Med*. 2004;351(26):2751–2753.
63. Olbers T, Fagevik-Olsen M, Maleckas A, Lonroth H. Randomized clinical trial of laparoscopic Roux-en-Y gastric bypass versus laparoscopic vertical banded gastroplasty for obesity. *Br J Surg*. 2005;92(5):557–562.
64. Livingston EH, Fink AS. Quality of life: cost and future of bariatric surgery. *Arch Surg*. 2003;138(4):383–388.
65. Buchwald H, Avidor Y, Braunwald E, et al. Bariatric surgery: a systematic review and meta-analysis. *JAMA*. 2004;292(14):1724–1737.
66. Anthone GJ, Lord RV, DeMeester TR, Crookes PF. The duodenal switch operation for the treatment of morbid obesity. *Ann Surg*. 2003;238(4): 618–627.
67. Maggard MA, Shugarman LR, Suttorp M, et al. Meta-analysis: surgical treatment of obesity. *Ann Intern Med*. 2005;142(7):547–559.
68. Schauer PR, Ikramuddin S, Gourash W, Ramanathan R, Luketich J. Outcomes after laparoscopic Roux-en-Y gastric bypass for morbid obesity. *Ann Surg*. 2000;232(4):515–529.
69. Hall JC, Watts JM, O'Brien PE, et al. Gastric surgery for morbid obesity. The Adelaide Study. *Ann Surg*. 1990;211(4):419–427.
70. Flum DR, Dellinger EP. Impact of gastric bypass operation on survival: a population-based analysis. *J Am Coll Surg*. 2004;199(4):543–551.
71. Fielding GA, Ren CJ. Laparoscopic adjustable gastric band. *Surg Clin*. 2005;85(1):129–140.
72. Sjostrom L, Narbro K, Sjostrom CD, et al. Effects of bariatric surgery on mortality in Swedish obese subjects. *N Engl J Med*. 2007;357(8):741–752.
73. Buchwald H, Williams SE. Bariatric surgery worldwide 2003. *Obes Surg*. 2004;14(9):1157–1164.
74. Omalu BI, Luckasevic T, Shakir AM, Rozin L, Wecht CH, Kuller LH. Postbariatric surgery deaths, which fall under the jurisdiction of the coroner. *Am J Forensic Med Pathol*. 2004;25(3):237–242.
75. Melinek J, Livingston E, Cortina G, Fishbein MC. Autopsy findings following gastric bypass surgery for morbid obesity. *Arch Pathol Lab Med*. 2002;126(9):1091–1095.
76. Drenick EJ, Fisler JS. Sudden cardiac arrest in morbidly obese surgical patients unexplained after autopsy. *Am J Surg*. 1988;155(6):720–726.
77. Eriksson S, Backman L, Ljungstrom KG. The incidence of clinical postoperative thrombosis after gastric surgery for obesity during 16 years. *Obes Surg*. 1997;7(4):332–335.
78. Podnos YD, Jimenez JC, Wilson SE, Stevens CM, Nguyen NT. Complications after laparoscopic gastric bypass: a review of 3464 cases. *Arch Surg*. 2003;138(9):957–961.
79. Ukleja A, Stone RL. Medical and gastroenterologic management of the post-bariatric surgery patient. *J Clin Gastroenterol*. 2004;38(4):312–321.
80. Flum DR, Salem L, Elrod JA, Dellinger EP, Cheadle A, Chan L. Early mortality among medicare beneficiaries undergoing bariatric surgical procedures. *JAMA*. 2005;294(15):1903–1908.
81. Wolfe BM, Morton JM. Weighing in on bariatric surgery: procedure use, readmission rates, and mortality. *JAMA*. 2005;294(15):1960–1963.
82. Zingmond DS, McGory ML, Ko CY. Hospitalization before and after gastric bypass surgery. *JAMA*. 2005;294(15):1918–1924.
83. Sjostrom L, Lindroos AK, Peltonen M, et al. Lifestyle, diabetes, and cardiovascular risk factors 10 years after bariatric surgery. *N Engl J Med*. 2004;351(26):2683–2693.
84. Alvarez-Leite JI. Nutrient deficiencies secondary to bariatric surgery. *Curr Opin Clin Nutr Metab Care*. 2004;7(5):569–575.
85. Fernandez Jr AZ, DeMaria EJ, Tichansky DS, et al. Experience with over 3,000 open and laparoscopic bariatric procedures: multivariate analysis of factors related to leak and resultant mortality. *Surg Endosc*. 2004;18(2):193–197.
86. Bloomberg RD, Fleishman A, J.E N, Herron JE, Kini S. Nutritional deficiencies following bariatric surgery: what have we learned? *Obes Surg*. 2005;15:145–154.
87. Slater GH, Ren CJ, Siegel N, et al. Serum fat-soluble vitamin deficiency and abnormal calcium metabolism after malabsorptive bariatric surgery. *J Gastrointest Surg*. 2004;8(1):48–55.
88. Baker RS, Foote J, Kemmeter P, Brady R, Vroegop T, Serveld M. The science of stapling and leaks. *Obes Surg*. 2004;14(10):1290–1298.
89. Esmail A. Physician as serial killer–the Shipman case. *N Engl J Med*. 2005;352(18):1843–1844.
90. Fisken RA. Physician as serial killer. *N Engl J Med*. 2005;353(4):430.
91. Taylor GJ. Physician as serial killer. *N Engl J Med*. 2005;353(4):430.

CHAPTER

9

Approach to firearms deaths

Susan F. Ely and James R. Gill

I'd rather autopsy six people shot once than one person shot six times. -*Dr. Charles S. Hirsch.*

Deaths due to trauma

The investigation, autopsy, and certification of traumatic deaths fall within the realm of the forensic pathologist. Forensic pathology is the only medical specialty expressly created to fulfill a governmental need, which is to identify, investigate, and certify deaths that assist the public health and criminal justice systems. The forensic pathologist is the medical expert who examines the living and the dead in order to provide an opinion concerning the cause, mechanism, and manner of disease, injury, or death.[1] Most, if not all, jurisdictions in the United States require a medical examiner/coroner (ME/C) to certify unnatural deaths. Therefore, forensic pathologists see the full spectrum of injury, as well as disease. Most traumatic deaths are due to demonstrable injuries, but a minority fall within the category of *"functional" injuries,* requiring both a description and a context.

In this and the proceeding three chapters dedicated to physical trauma, we intentionally excluded deaths related to environmental and chemical/toxicological injuries, instead including those as part of separate reviews (Chapters 13 and 14). Many mechanical injuries encountered by the ME/C are obvious, but some of the greatest challenges we face in forensic pathology, as in other areas of medicine, involve our aptitude in recognizing an unexpected or subtle finding and the willingness to change an existing diagnosis once another has been discovered. Pediatric injuries are addressed in Chapter 6.

Firearm injury

The standard autopsy findings and descriptions of gunshot wounds are well-detailed in various articles and textbooks, along with the vast array of munitions and firearms, which are beyond the scope of this textbook (Fig. 9.1).[2–5] Instead, we focus on the key qualities of firearm injuries and describe numerous pearls and pitfalls to be aware of when investigating these deaths.

Two sets of terms may cause confusion if not used properly. These are *penetrating versus perforating* and *track versus tract.* A **penetrating injury** is one that involves a projectile or weapon entering a structure, but not exiting it, while a **perforating injury** involves an exit. For example, one may have a penetrating gunshot wound of the head that perforates the brain (but not the skull). A **track** is detectable evidence that something has passed through (the course along which something, such as a bullet, has moved). A **tract** is a system of bodily parts or organs that collectively serve some combined anatomic and physiologic purpose (e.g., gastrointestinal tract). There is a *difference between the use of "to" and "of"* with regard to firearms injuries. **One uses "gunshot to" when referring to the location of the entrance wound, but uses "gunshot wound of" when referring to the entirety of the injury inflicted by the bullet.** Similar formulations are used in blunt and sharp injuries. It is "blunt impact to" versus "blunt injury of" or a "stab to" versus a "stab wound of."

It is best to measure all wounds from the top of the head and from the midline (to the left or to the right), with the exception of the limbs. Given that the upper and lower extremities are potentially mobile relative to the midline of the body (the trunk and head), determining the distance of an extremity wound relative to the midline is not possible

Principles of Forensic Pathology
https://doi.org/10.1016/B978-0-323-91796-4.00024-6

FIGURE 9.1 **Pistol.**

without knowing precisely where in space the limb was at the moment it was shot. For the trunk and head, however, when measuring from the midline, it is best to do so with a straight ruler, not by wrapping soft measuring tape around the body convexity. Determine wound tracks with the viscera in situ, but remember that the anatomical positions of the organs within a body cavity may have been slightly different when the person was shot (or stabbed). If a bullet strikes the vertebral column, do not fail to examine and describe the spinal cord to detect and describe potential segmental softening. Be aware that even if the spinal canal is not penetrated by the projectile, per se, but the vertebral skeleton has been, there can be concussive injury to the spinal cord. If you later learn that the victim suddenly and inexplicably "dropped" after being shot, for example (too quickly to be explained by hypovolemic shock/blood loss), it will have been providential that you considered this mechanism and dissected the canal as a matter of completeness. This is good practice whenever the bullet penetrates any part of the vertebral column.

There are eight things that may come out of a fired gun: (1) gas; (2) soot (the product of the complete combustion of gunpowder, seen as "fouling" if deposited on the skin or clothing); (3) burning and unburnt powder (the products of incomplete combustion of gunpowder, producing "stippling" injury on the body if within the range to do so); (4) primer compounds; (5) copper/nickel vaporized from the cartridge case; (6) metal vaporized from the bullet/jacket; (7) a flame at 1400 F (capable of causing searing); and (8) the bullet.[3] In terms of fouling, its diameter will depend upon the range of fire, the propellant, the angle of the muzzle, the barrel length, the caliber, and the type of firearm (rifle/shotgun vs. handgun), the target material, and any muzzle attachments (silencers/suppressors).

It is important to know what opinions a forensic pathologist may and may not be able to give regarding gunshot wounds (Table 9.1).

In answering each of these questions, there are some key caveats to keep in mind that may affect the opinion asserted. For example, the forensic pathologist must be aware of the fact that contact gunshot wounds, especially over bony areas, may have blowback lacerations that make the wound appear more like a stellate exit wound than a typical circular entrance wound. One must be aware of shored entrances and exits, which may give an exit wound the appearance of an entrance wound and a distant entrance wound the appearance of a contact wound. One must be aware that clothing will likely "filter" gunshot residue (GSR) and so fouling or stippling may not be apparent on the skin with a close-range gunshot wound when it is covered by clothing of various thicknesses. For this reason, it is important to examine the clothing, even if it was cut off and left at the hospital. It is adviasable that investigators ask to obtain the clothing for ME/C examination if it does not accompany the body. See below.

If there is fouling, stippling, a blow-back laceration, or a muzzle stamp, one may be able to give an opinion about the range of fire. If these features are not seen, the wound is either a distant gunshot wound or is of indeterminate range, since something may have filtered out the residue. Measurements of **muzzle imprints** are not necessarily an exact reflection of the gun, but the shape/design/dimensions may be useful for comparisons. One may see false or **pseudostippling** caused by intermediate targets (e.g., tempered glass, jewelry) or from suture defects, insects, or shotgun packing. **Carbon monoxide** from the barrel can cause the immediately underlying muscle to appear

TABLE 9.1 Gunshot wound questions.

1.	Which wound is the entrance wound and which is the exit?
2.	What is the range of fire?
3.	What is the overall direction of the bullet in the body?
4.	What internal tissues/organs were injured?
5.	What is the direction of the graze wound?
6.	How many bullets caused these injuries (re-entrance wounds?)?
7.	What ballistic material was recovered?
8.	How long would the victim have been able to survive after sustaining this wound?
9.	Can you determine the sequence of the gunshot wounds?

pink to bright red in contact gunshot wounds (carboxymyoglobin), but this will fade with exposure to air. The cylinder gap on revolvers allows the sideward escape of gas and burnt and unburnt powder grains, resulting in potential "cylinder-gap stippling," which is distinct from the entrance wound. One may be able to determine the direction of a graze wound by examining the skin tags, the skin tags pointing toward gun.[6,7]

With regard to internal tissue damage, marketing efforts of ammunition/firearm manufacturers focus on size, special designs, and "stopping" power. Ultimately the lethality of a gunshot wound largely depends upon what it hits and how much damage it exacts to that structure. For example, a low-velocity 0.22 caliber bullet fired from a handgun will likely cause death if it perforates the heart, but a 0.40 caliber bullet that strikes the periphery of the lung may be survivable. The wounding ability of a bullet depends upon its *mass and velocity.* Of these two, velocity is supreme. The formula for kinetic energy **(KE) is ½mv²**. Therefore, the KE is doubled when the bullet mass is doubled, but it is quadrupled when the velocity is doubled. High velocity bullets cause very destructive injuries related in part to the **temporary cavity** created by the bullet. Whether a bullet stays in the body, thus expending all of its energy to the body, depends upon several factors. These include the composition, size, and speed of the bullet, as well as what it strikes (both inside the body and through intermediate structures outside of the body).

There are some myths about certain types of ammunition with regard to how much "damage" they produce. One example is the "hollow-point" bullet, which is constructed to deform into a "mushroom"-type configuration with impact. Due to this structural change, the bullet's velocity slows and it is more likely to stay in the body. This is one reason why many police departments use this ammunition. By remaining in the body, there is of less chance of the exiting bullet striking another person. When matched with size and velocity, however, there is little, if any, wounding difference between a hollow-point and a nonhollow-point projectile. That is, one cannot give an opinion that a particular perforating gunshot wound was caused or not caused by a hollow-point bullet based on the extent of wounding.

Overall, the extent of injury a bullet produces depends upon the amount of energy *lost* in the body. *The amount of KE transferred from the bullet to the body depends upon four factors: (1) Amount of KE before impact, (2) Angle of yaw (amount of wobble), (3) The bullet characteristics (caliber, mass, construction, and configuration), and (4) The impacted tissue (density/strength).*[3] **Yaw** is defined as movement around a vertical axis, and the angle of the yaw is measured between the projectile's long axis and its line of flight. Bullets only tumble in the air if they have struck an intermediate target that upsets their stability.[8] Even within tissues, bullets do not tumble. Gelatin studies have demonstrated that pointed bullets may yaw 180° (if their path is long enough) because of their center of gravity, resulting in the base of the bullet shifting to the forward position.[8–11]

Entrance wounds typically are circular (not "round") or oval[12] before skin tension release, but with skin release, they are most typically circular, unless an atypical wound. One may have an **atypical entrance wound** for a variety of reasons.[9,13] If the bullet strikes an **intermediary object or target** before it strikes the body, it may tumble or yaw so that it does not perforate the skin "nose" first. This may be seen with ricochets and in reentrance wounds of the torso following a "through-and-through" wound of the upper extremity or other parts of the extremities. As a result, the torso entrance wound may appear atypical. An atypical entrance wound may be unusually large or have an odd shape (not circular or oval) and/or have an *irregular margin of abrasion.* When describing these wounds, it is important not to confuse the word "asymmetrical" with "atypical." The former should be used in describing the *regular* surrounding margin of abrasion of a gunshot entrance wound that is commonly asymmetrical around the entrance wound defect. Asymmetry does not necessarily denote atypicality; however, asymmetry merely reflects the fact that the nose of the bullet came into contact with the body at an angle, as opposed to at a perfect 90°. By contrast, when the term "atypical" is used to describe a gunshot entrance wound, it is meant to indicate either a misshapen or unusually large perforation and/or an irregular and sometimes oddly large surrounding margin of abrasion. Asymmetry of an entrance wound's margin of abrasion, alone, without irregularity, should not prompt concerns of an intermediary object, per se.

Besides intermediate targets, other causes of atypical wounds are unusual projectiles (faulty ammunition or weapon), target contour variations (e.g., nose and ear), and target texture variations (e.g., eyelids, scrotum, palms, and soles); the latter locations also often associated with **ambiguous wounds** (i.e., difficult to distinguish entrance from exit wound morphology), even if not atypical, per se. Absent GSR, a through-and-through wound of the hand may have entrance and exit wounds that are a challenge to distinguish. A "piggy back" discharge involves a fired bullet striking one that is still lodged in the barrel and may create an odd-appearing (atypical) entrance wound.[14]

Due to the elasticity of the skin, entrance wounds (except contact wounds, see above) are usually smaller than the caliber of the bullet. This is because the skin stretches when first impacted by the bullet. The skin then "recoils" and the resultant defect is smaller than that of the bullet caliber. The diameter of defects in bone or cartilage, on the other hand, will more closely approximate the caliber of the bullet. Some clothing (such as leather) also may stretch with a similar principle applying. Large, gaping defects in clothing should suggest a contact wound, as the gas may cause similar blowback, tearing the clothing in the same way it can the skin.

Charles Hirsch was known to say he would rather "do an autopsy on six people shot once than on one person shot six times." Multiple gunshot wounds can be challenging because of the potential for intersecting tracks, the need to recover multiple projectiles, and to match multiple entrances with their corresponding exit wounds or bullet lodgment sites. In some instances, one may not be able to pair entrance wounds with their corresponding bullets or exit wounds with reliability. When that occurs, it is best to "describe what you see" and conclude only what you are able to conclude with scientific veracity. "If all else fails, tell the truth." If unable to match a given entrance wound to its exit, so state it and explain why. This may happen when entrance wounds are in close proximity to one another. The "bookkeeping" of these wounds becomes paramount.

Graze and tangential gunshot wounds

While usually not contributory to the cause of death, these two firearms injuries must be recognized and carefully described nonetheless. Rarely, tangential wounds of the scalp may generate fatal concussive injury to the brain (see below).

A **graze wound** is caused by the brushing or grazing of the body by a projectile (Fig. 9.2). Most typically, they are oblong abrasions that are sometimes more rounded on one end relative to the other (the former representing the initial projectile contact with the body/skin), but many variations exist depending on the degree of deformity and/or yaw of the bullet at the time of contact. On occasion, there are microlacerations along the long axis of the edges of the wound that may assist in determining the direction the bullet was traveling. Careful observations may reveal that some of these wounds may have been caused by the same bullets that are associated with other wounds, making contact with the skin just before entering or just after exiting the body, based on how the wounds align. Most often, however, they represent stand-alone wounds. Sometimes, a roughly oblong abrasion is suspected to be a graze injury, but it is not specific enough to name it as such. In those instances, it is probably better to describe what is seen under the general blunt injuries section, without committing to it being a graze wound, per se. The possibility that it may represent a graze wound can always be opined later, if asked. Examination of the overlying clothing in these instances may be informative.

A **tangential gunshot wound** is a less ambiguous firearms-related injury, even if not typically contributory to death. It usually consists of a deeper, oblong, lacerated defect, often with one end more rounded than the other. It involves deeper structures than the epidermis alone, usually the subcutis and even the deeper muscular layer, and usually has lacerations along the long margins/edges. As in graze wounds, these indicate bullet directionality, the skin tags pointing in the direction of the firearm (and the microlacerations pointing in the direction of the bullet's trajectory). Rarely, tangential gunshot wounds, particularly those of the head, can involve the skull, creating a **keyhole defect** that may cause or contribute to death.[15,16]

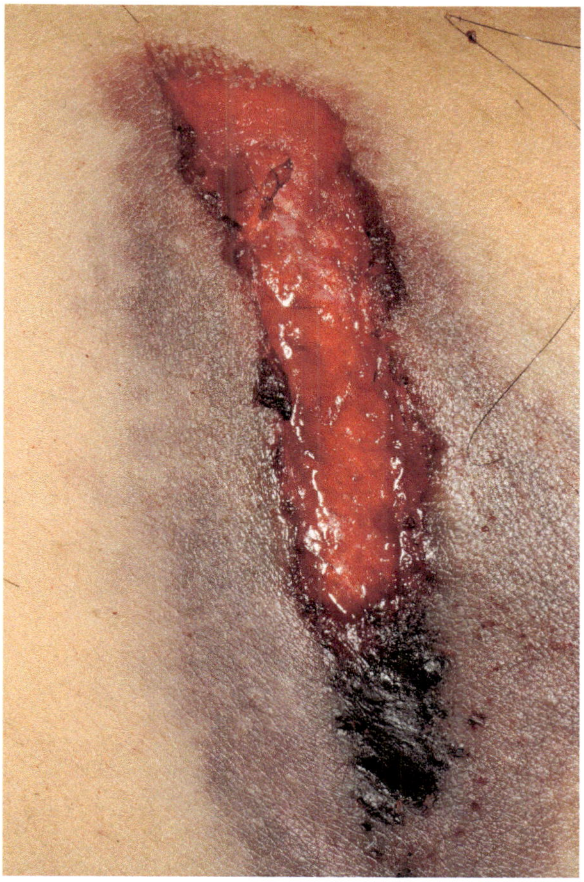

FIGURE 9.2 **Graze wound.**

Retractor artifact

This artifact of resuscitative/therapeutic intervention can be seen with thoracotomies and, in the setting of firearms injuries, can be confused with a tangential gunshot injury. This has the potential for being included in the gunshot wound defect cataloging, thus potentially throwing an unnecessary wrench into the bullet wound accounting equation and witness statements ("The gun was only fired once"). It is characterized by a superficial and somewhat broad skin and subcutaneous tissue defect that sometimes has patterned skin tags or an attached skin flap on one end. It is typically in close proximity to the thoracotomy incision, either in or near the axilla or on the medial upper arm and, more times than not, is anemic-appearing/relatively bloodless. As the name implies, it is inadvertently caused by the use of a retractor (e.g., Finochietto retractor), which is applied to the chest when separating the ribs during a thoracotomy (Fig. 9.3).

Bullet "bookkeeping"

With the exception of deaths involving shotguns or other unusual forms of ballistics, before starting the internal examination, the total number of *entrance* (and *reentrance*) wounds should equal the number of *exit* wounds plus the number of *projectiles* lodged in the body (seen on the preautopsy radiographs). If not, one must ask, why? Has an entrance or exit wound been missed or has a bullet fragmented and created more than one exit wound? There are a variety of reasons for this miscount (see below). Only by complying with this simple calculation can the forensic pathologist be certain that all bullets that entered the body have been accounted for. One should have an answer or a reasonable explanation for the discrepancy (e.g., history of remote gunshot wound with a "souvenir" bullet) (Fig. 9.4).

If you have spent ages looking for a bullet and still cannot find it, take a break for a few minutes and then try again. You might ask for the assistance of a colleague, who may have a fresh perspective on the problem. Nothing (read: blood/clots) should ever go down the drain or be thrown away until all bullets are found; that said, **always make sure the drain is covered** before starting an autopsy of a firearms fatality. Favorite hiding places for bullets

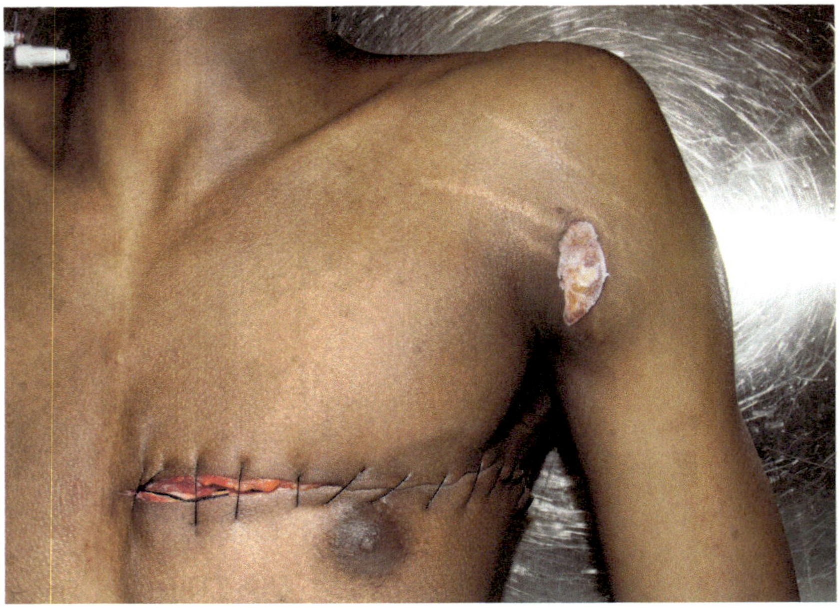

FIGURE 9.3 **Retractor artifact.**

include the subcutaneous fat, the sternum, the vertebral column, the scapulae, and blood clots. Emergency room physicians and surgeons rarely volunteer that they have removed a bullet; they must be asked specifically, but still may not even be aware that a bullet was inadvertently evacuated by suction. Bullets can also embolize. In large bodies, they may be in the lateral soft tissues that are just out of the field of the radiograph plate.

When there is an **extra bullet on X-ray**, *one might consider the following possibilities:*

1. An old ("souvenir") bullet.
2. Partial exit (there is an exit wound, but the main bullet core is still in body).
3. Tandem fired bullet (two bullets create and enter one entrance wound defect, then separate in the body).[14,17–24]
4. "Hidden" entrance wound (e.g., axilla, scrotum, intraoral, intranasal, auditory canal, behind the ear, gluteal folds, and mammary folds).
5. The bullet fragmented or fractured bone resulting in an exit wound from the bullet fragment or fragment of bone.
6. Other metal was carried in with the bullet.
7. The radio-opaque object is not a bullet but a zipper, item of jewelry, or metal tooth crown (if the broken crown is aspirated or swallowed, the subsequent location may cause further confusion).
8. The jacket becomes separated from the bullet core and, to the inexperienced eye, appears as two bullets on the radiographs (be sure to look carefully at the density of each projectile on the radiographs).
9. A gunshot entrance wound has been incised or otherwise incorporated into a thoracotomy, thoracostomy, or other surgical incision.

When there is an apparent **missing bullet**, *one might consider the following possibilities:*

1. Bullet removed during surgery, intentionally or inadvertently.
2. Bullet seen in the radiograph was in the clothing or body bag (thus not found *in* body).
3. Bullet embolism to an area of body not radiographed (more common with smaller caliber bullets).[25–28]
4. Overlapping bullets on radiographs that look like one deformed bullet (get a lateral radiograph or a computed tomography (CT) scan image for a better assessment).
5. Radiographs did not capture a bullet on the edge or periphery of the body (more common in the torso, particularly of obese persons).[29]
6. "Hidden" exit wound (e.g., axilla, scrotum, intraoral, intranasal, auditory canal, behind the ear, gluteal folds, and mammary folds).
7. Misinterpreted wound track because the body was in an unusual position. This should be suspected when a bullet track seems to go nowhere (see below).

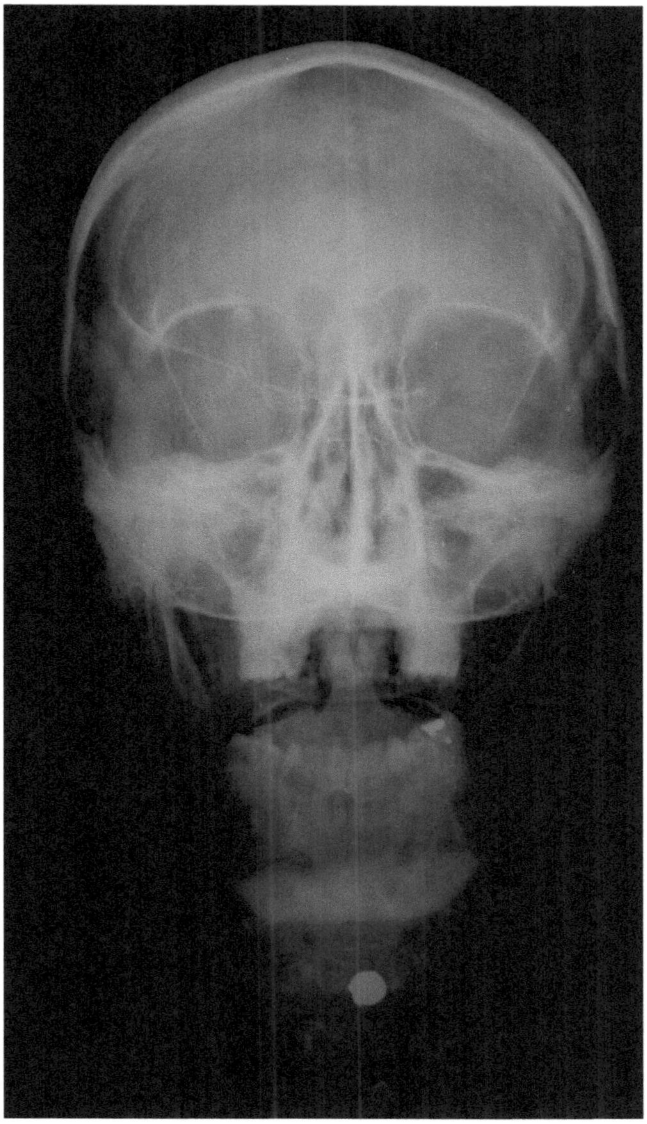

FIGURE 9.4 **"Souvenir" bullet in the neck.**

8. A shored exit wound is misinterpreted as an entrance wound.
9. Postmortem defect, such as an embalming trocar site, misinterpreted as an entrance wound (Fig. 9.5).
10. A gunshot exit wound has been incised or otherwise incorporated into a thoracotomy, thoracostomy, or other surgical incision (Figs. 9.6 and 9.7).

The direction that the bullet travels is given in three overall dimensions (front-back, right-left, and upward-downward), based on the standard anatomic position (standing with arms at sides with palms facing forward). Of course, not everyone gets shot standing in the standard anatomic position, but these are used for standardization, so everyone is using the same references planes. Because autopsies are performed with the body supine and horizontal, one may come across some bullet tracks that do not make sense or that even seem to disappear.

For example, a man sustains a gunshot wound of his buttocks while bent over a fence. The bullet perforates the right pubic bone, which clearly defines the direction (in the standard anatomic position) as back to front and slightly upward and leftward. In the horizontal position, the bullet should have exited the anterior abdomen. There was no exit wound nor any retained bullets on full body examination. However, when the body was positioned with the torso flexed at the waist (in a "bent over" position), it became clear that the track of the bullet went through the right psoas muscle, the subcutaneous tissue of the back, and exited from the back of the head after causing a depressed skull fracture! The head wound initially was attributed to a fall (previously thought to be a terminal impact

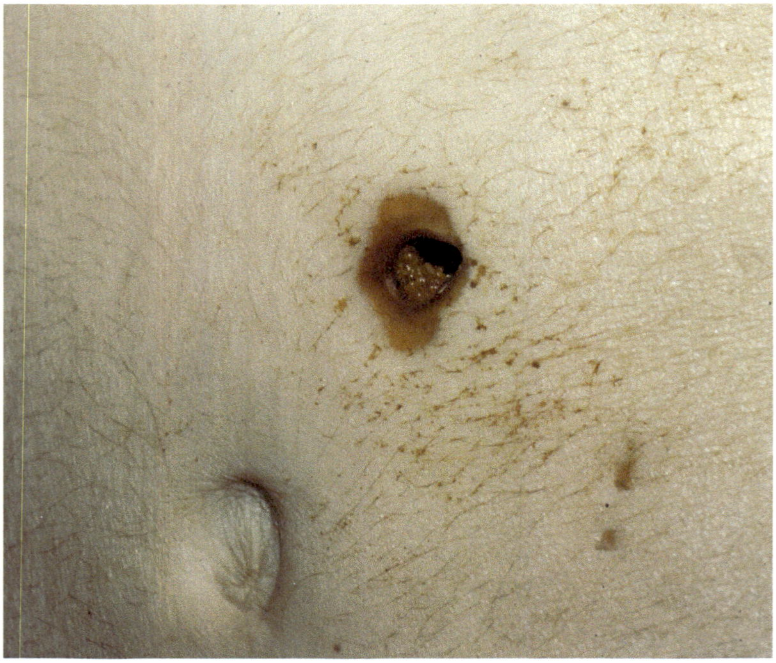

FIGURE 9.5 **Embalming trocar site misinterpreted as an entrance gunshot wound.**

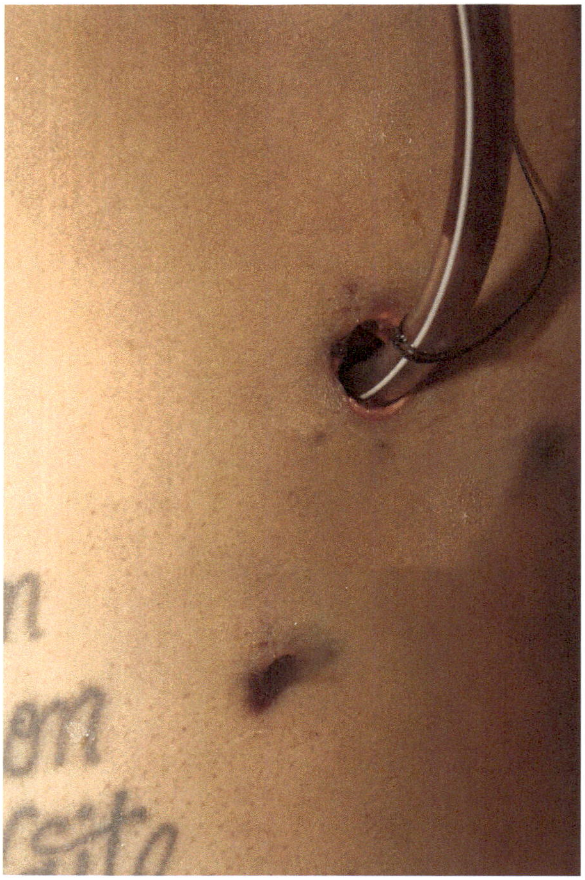

FIGURE 9.6 **Chest tube placed in an entrance gunshot wound.**

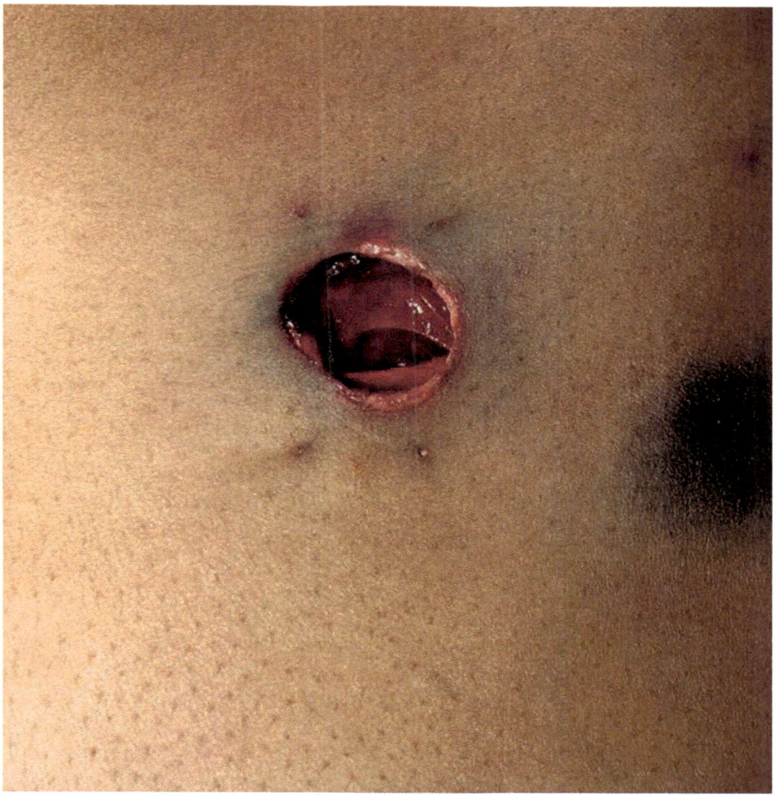

FIGURE 9.7 **Entrance gunshot wound with the chest tube removed.**

laceration). "The only people who think that trauma is simple are those who haven't seen it or those who haven't thought about what they have seen."[30]

Range of fire

The concept of **range of fire** is important to understand, particularly in its limitations. Range of fire is traditionally classified as **contact** (tight and loose), **close**, **intermediate**, and **distant/indeterminate**. Different authors[3] use slightly different terminology and so the best approach is to simply describe the wound features in the body of the autopsy report. "Contact" or "close" range terminology is best not used (just the news, not the editorial). The closest one may come to referring to a range of fire is when describing a muzzle stamp in a tight contact wound. One may add a comment parenthetically, "(comment: consistent with muzzle stamp)." With contact wounds, one may see muzzle stamp abrasions (Figs. 9.8 and 9.9), blowback lacerations (Fig. 9.10), soot within the wound track, and/or gaping defects in the overlying clothing. With intraoral gunshot wounds, one may see lacerations around the mouth or face analogous to blowback lacerations from muzzle gas seen with contact wounds (Fig. 9.11).

Fouling and stippling are the classic features that define and distinguish between close, intermediate, and distant range wounds. The detection of fouling (dust-like, smoky/soot material that washes off the body) *with* stippling (pinpoint *injuries* from the unburnt and partially burnt powder, which do not wash off) defines a **close range** gunshot wound (generally <6″ for most handguns). The detection of stippling, without fouling, defines an **intermediate range** wound (because of their size, the unburnt and partially burnt powder grains travel farther than the soot). Powder grains from a typical handgun may travel anywhere from 18 to 30″ (Fig. 9.12). The absence of both fouling and stippling is designated as **distant range**, which means greater than 30″. Some may prefer the term "indeterminate" for these wounds because it may be that the gun was fired inches away, but an intervening target (e.g., clothing) prevented gunpowder residue from reaching the skin. As the fouling may be washed away, it is important to look for fouling and document its presence or absence with a photograph *before* the body is cleaned. Measurements of its diameter and a description of its density should also be noted, as this may be important for identifying a more specific range (i.e., the tighter and denser the spread, the closer the range). The latter can be done by test-firing the firearm used, if available, using the same ammunition and a target (cotton cloth, pig skin)

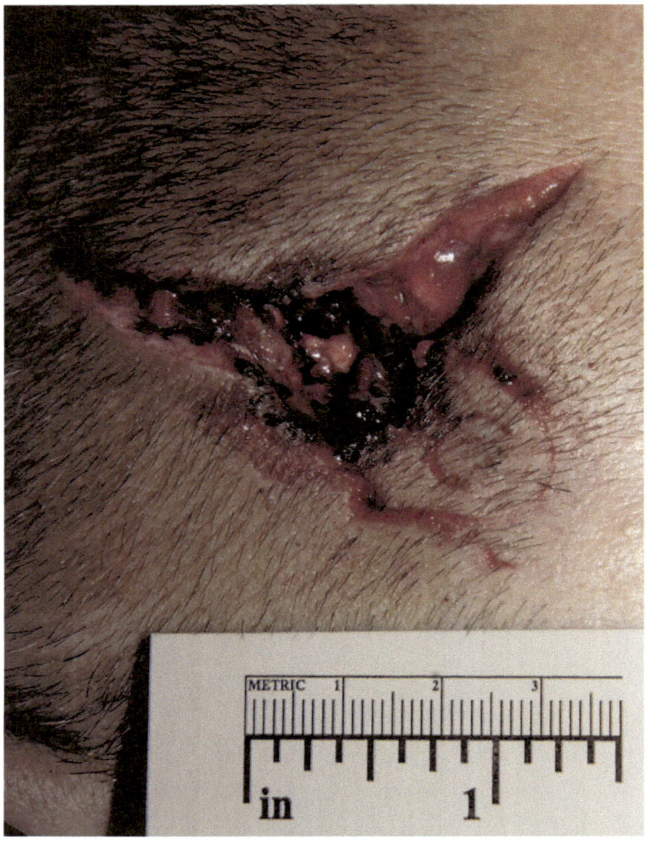

FIGURE 9.8 **Muzzle stamp abrasion with blow back lacerations.**

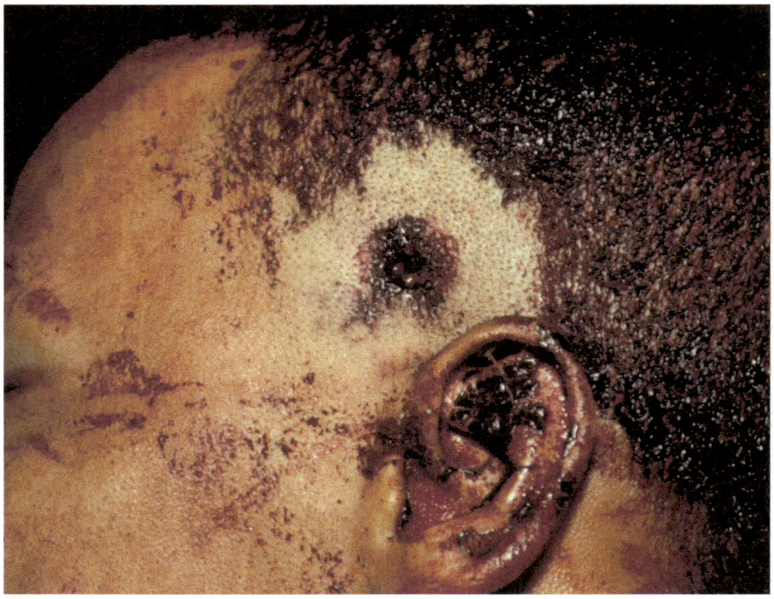

FIGURE 9.9 **Muzzle stamp abrasion at 7:00 position (scalp hair shaved off at autopsy).**

at different ranges, and then comparing the pattern with the autopsy pattern (see also Chapter 5, The Forensic Autopsy). It should be noted that it is both standard and necessary to ultimately wash a body during the examination. This allows for better visualization of the all wounds and results in "cleaner" and clearer images.

When observed, most fouling and stippling surrounds a gunshot entrance wound either symmetrically or asymmetrically and with greater or lesser degrees of density, reflective of the angle of the gunshot relative to the body and

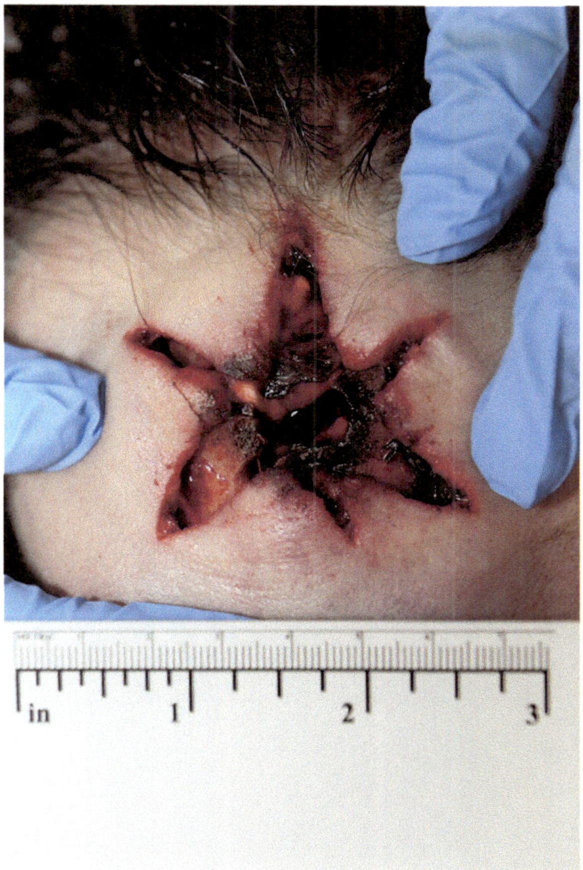

FIGURE 9.10 **Contact gunshot wound with stellate blow back lacerations and singeing.**

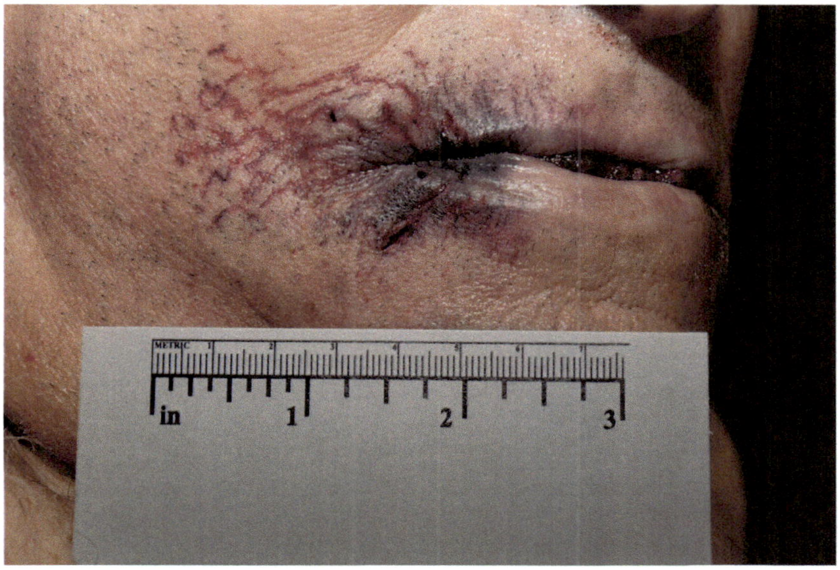

FIGURE 9.11 **Perioral lacerations from an intraoral gunshot wound blowback from a closed mouth.**

of the distance of the muzzle of the gun from the body at the time of its firing, respectively (Fig. 9.13). In tight contact wounds, the soot is blown into the wound, so a careful inspection of the wound track is important; dura and the outer table of the skull are particularly good sources of evidence of gunpowder residue, despite the fact that hemorrhage may obscure or wash away soot within most of the track. On occasion, the pattern of stippling is striking and

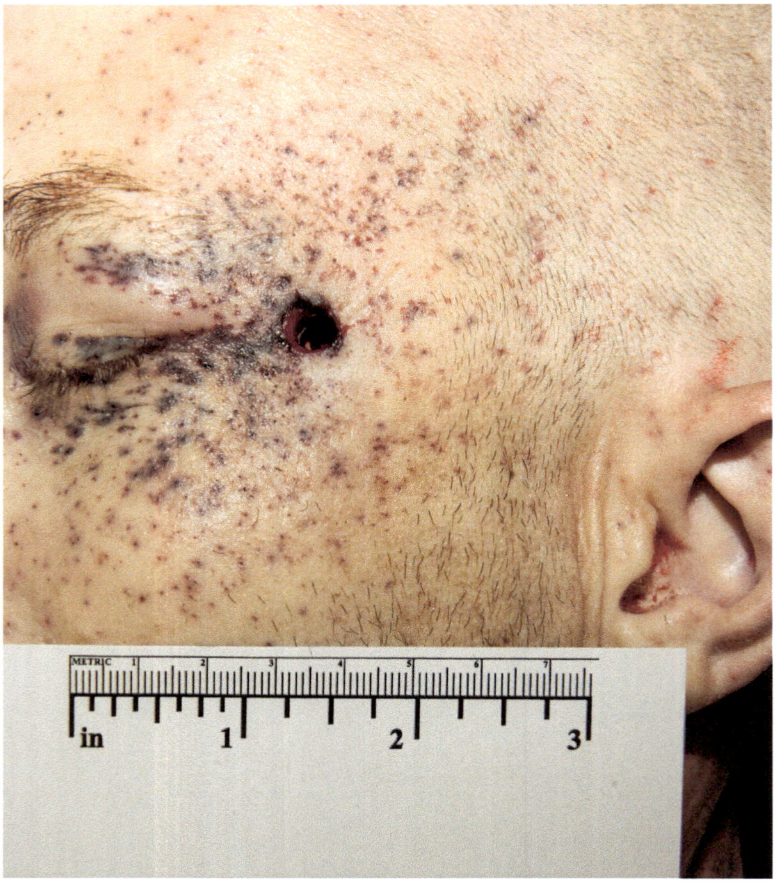

FIGURE 9.12 **Stippling.**

unusual, with discrete and well-demarcated areas of sparing (uninvolved skin where there would have been an expectation for involvement). In these instances, it can be vexing to the forensic pathologist to sort through such patterns as the body lies recumbent on the autopsy table. Usually, however, with some manipulation of the body and inquires made about what the decedent may have been wearing or doing at the time of the shooting, an explanation for these patterns emerges. Examples of explanations for these areas of relative sparing include, but are not limited to, hats/caps, eyeglasses, clothing, flexion of the neck or other body creases, intermediary targets, such as a hand, that partially block an area of skin around the entrance wound, or an irregular body contour, such as that of the face.

While it is usually a boiler plate part of a firearms autopsy report to note the presence or absence of fouling and/or stippling, the forensic pathologist should be well aware of the limitations of interpreting such an observation, as they may well be asked about it at a later time by those less knowledgeable and who may inadvertently misinterpret the words on the page or even wish to deceive or intentionally distort their meaning. When forensic pathologists observe and document the absence of fouling and stippling, they are simply describing what they see on the body alone, without regard to clothing. Even during the warmest months of the year, however, most individuals who are shot *outdoors* are clothed in some fashion (typically, the torso and pelvis, at the very least). Part of a full homicidal firearms forensic pathology autopsy protocol includes a careful examination of the clothing worn at the time of the injury, if still available. The autopsy examination is, of course, a gross one, with photographic documentation, as most crime/forensic laboratories handle the chemical and microscopic residue analysis. That said, the presence or absence of grossly discernible gunpowder residue observed during the forensic pathologist's examination of the clothing is not necessarily the final step in the gathering of range of fire evidence.

The takeaway point is that while the ME/C can clearly make a statement about the absence of fouling or stippling on the *skin* of a firearms victim, they should not have or offer any opinion about the meaning of that observation (in terms of range of fire) until they know which injuries were protected by one or more layers of clothing. More commonly than not, at least the face and, often, the head are unclad, but not always. All of this may change during the coldest months of the year, when nearly every part of the body may be covered by one or more layers. If the victim is pronounced in a hospital and the clothing never makes it to the morgue, it is best to ask about what clothing

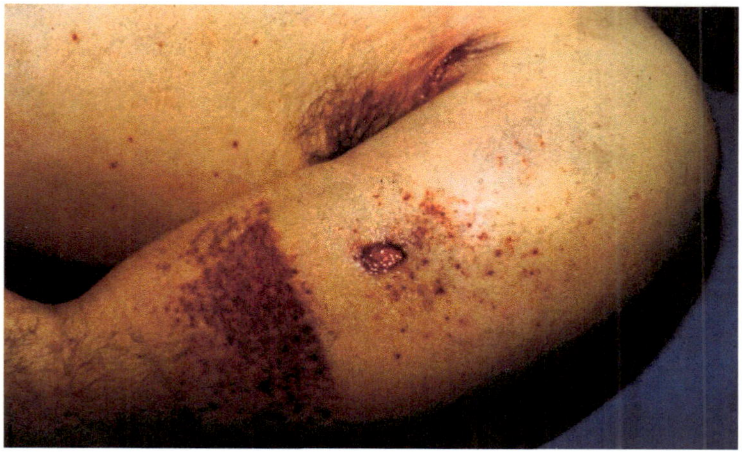

FIGURE 9.13 **Stippling pattern demonstrating the effect of a short-sleeved shirt that prevented the powder grains from striking the covered skin.** Note the sharp line of demarcation where the shirt sleeve spared the skin from stippling injury.

was worn and/or watch a video of the shooting, if available, to observe which layer was the outermost, and request that the clothing be retrieved and delivered to the ME/C office for a careful inspection, photographic documentation, and submission into evidence. Only once the forensic pathologist reliably knows what the person was wearing, and how/where, may they opine about the potential range of fire (with limitations, of course, without a test firing of the actual gun that had been used).

Other helpful information potentially gleaned from homicide clothing:

- Sorting out ambiguous wounds, exit versus entrance, by looking at the size and shape of the defects, gunpowder residue, if present, and **bullet wipe** patterns.
- Documenting the quantity of bloodstaining/soaking [potentially helpful in assessing mechanism(s)].
- Ascertaining which items were likely worn as the outermost layer (if the clothing is received separately); if the decedent is received clad, the outermost to innermost order of the clothing item layering is important to document before the decedent is undressed.

Penetrating gunshot wounds

When a projectile fails to exit the body, always ask yourself why. In doing so, recall the KE equation: MV^2 (mass x velocity^2).[31,32] Essentially whatever decreases the KE will decrease the likelihood that the projectile will have enough energy to exit the body.

The foregoing list offers five potential reasons why bullets may penetrate, but not perforate the body:

Kinetic energy

- **Velocity** (lower velocity results in a lower KE): This is related to a variety of cartridge (e.g., powder load) and firearm factors, as well as intermediate targets that may slow the bullet while in flight.
- **Mass** (lower mass results in lower KE).

Yaw

The greater angle of yaw when the bullet contacts the body, the greater the loss of KE.[31,32]

Bullet construction and configuration

The caliber and shape of the bullet play a role in the "drag" on the bullet by the tissue. **Smaller caliber, pointed nose** (as opposed to blunt) **bullets** will have less "drag" from the tissue and therefore lose less KE.

Unjacketed bullets deform more easily, as jackets tend to hold the bullet together; once a bullet deforms, its velocity slows considerably. Of note, unjacketed bullets are relatively rare ("wadcutters" are one type of such a bullet that is occasionally seen). For example, 0.22 caliber bullets are usually unjacketed, giving them *two* reasons (deformation ability and low mass) for them not to exit the body, yet their small caliber would be a factor that conserves KE thereby increasing an exiting ability (less drag).

Hollow point/hollow nose bullets: These bullets deform easily by design into a "mushroom" configuration, where all of the petal-like leaves of the nose of the bullet peel back. This is the standard ammunition of many police-issued guns, which are designed to penetrate, not perforate, thus minimizing the risk of injury to innocent bystanders. Despite the "hype" associated with the "massive" amount of tissue damage reportedly caused by this bullet, one sees a similar injury pattern with caliber and velocity-matched, nonhollow-point bullets.

Tissue factors (density, elasticity, and strength)

These factors **potentially affect penetration/perforation, particularly** when **the most resistant tissue is involved—bone**. The thicker and denser the bone (e.g., vertebral bodies, base of the skull), the more likely the bullet will deform and therefore be stopped or slowed down (reduced bullet velocity). Sometimes, just a rib or two is enough, depending on the caliber of the bullet.

Intermediary objects/targets

The ultimate effect of intermediary objects on projectiles is a decrease in KE due to the above discussed factors. If there is a penetrating gunshot wound in which, based on the above four factors, one would have expected an exit, one should question, why? As a general rule of thumb in forensic pathology practice, when something does not make sense, there is usually a good explanation, limited only by our inability to identify it. At this point, one should consider an intermediary target which may be an object, surface, or other body part (usually an arm or leg) that the bullet struck (ricocheted off or perforated) *prior* to entering (typically) the head or torso and becoming lodged. The ricochet decreases the velocity, may change the bullet configuration, and increases yaw (wobbling) of the projectile. In order to investigate this hypothesis, the entrance wound of the torso or head is examined for atypicality, as described above. *If you don't think of this, you won't see it.* The exit wound of a through-and-through gunshot wound of the arm also may be "shored" if the arm is the intermediary target and is in contact with the torso at the reentrance site.

Mechanisms of death in firearms fatalities

The mechanisms of death in firearms fatalities are varied and, of course, depend upon the structures injured by the projectiles. Very generally speaking, however, gunshot wounds of the head are much more likely to be immediately incapacitating, if not immediately fatal. A gunshot wound that sends a bullet through the brainstem, for example, will have both immediate "stopping power" and will, in overwhelming likelihood, not be survivable, even if emergency responders are on site very quickly. If captured on video surveillance, the decedent would probably be seen dropping precipitously, with no subsequent *purposeful* movements or consciousness thereafter. A similar drop to the ground, with or without continued consciousness, would likely be witnessed in the victim of a perforating or penetrating cervical spinal cord gunshot wound. For wounds in other parts of the body, many variables come into play in terms of the degree of injury exacted and the duration of a survival interval that might be expected. These variables include such factors as the specific structures perforated (e.g., large arteries, such as the aorta, vs. small veins; peripheral lung vs. the hilum), the caliber and deformity of the bullet, the range of fire, and the health of the victim (Fig. 9.14).

When the mechanism is nonneurologic, most firearms deaths that occur fairly proximate to the injury occurrence (acute deaths) result from hemorrhagic complications. The latter is the most common acute mechanism stemming from wounds of the torso and extremities. As such, most fatalities resulting from exsanguination take minutes (rather than seconds) or even hours to occur. That said, when the aorta is perforated by a bullet, most victims typically remain conscious for less than a minute before hypovolemic shock supervenes from one or both thoraces having rapidly filled with arterial blood under systemic pressure. Further, when the proximal aorta (ascending aorta or aortic arch) is perforated, the conscious interval may be equally brief, but not necessarily because of hypovolemic shock, per se, but rather from a form of cardiovascular shock, the "head of the hose" of systemic blood pressure having been breached and decompressed. Without this pathophysiologic understanding, an inexperienced pathologist might be confused by the absence of large hemothoraces at autopsy, despite identifying a sizable defect in the proximal aorta, along with marked renal and muscular pallor (cardiogenic shock from cardiovascular collapse).

Other less common mechanisms, such as flail chest, pneumothoraces, and hemoaspiration may also occasionally play primary or secondary roles, thereby contributing to a more complex mechanism. When the death is delayed by months or years relative to the injury occurrence, the possibilities for complications expand into those resulting from prolonged immobility, gastrointestinal disturbances, neurologic compromise, or posttraumatic seizures, to name only a few (see discussion in Chapter 15, Special Topics).

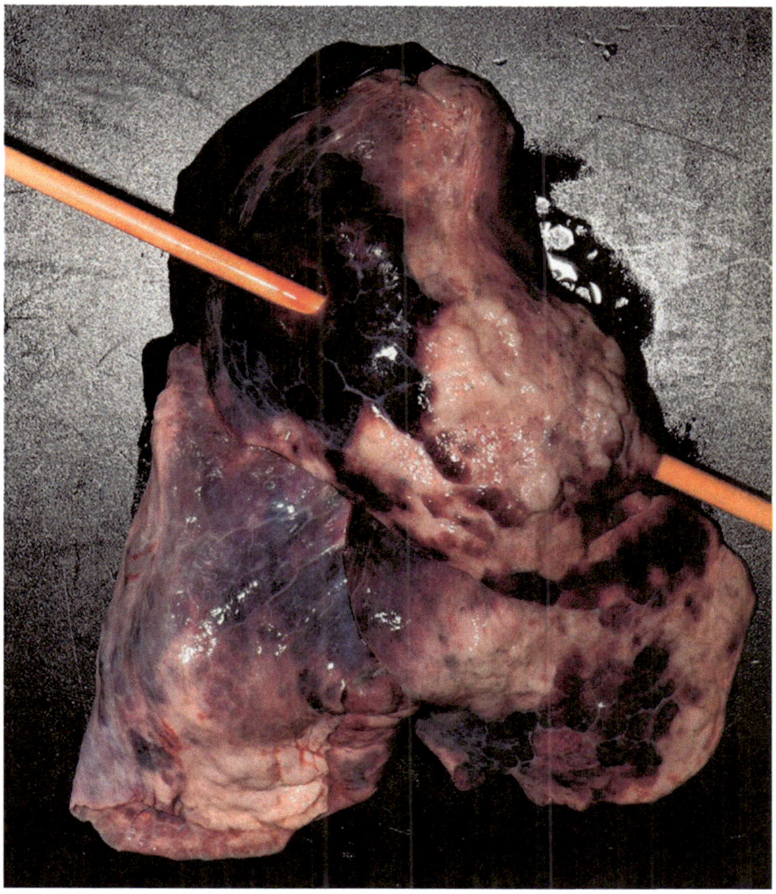

FIGURE 9.14 **Gunshot wound track of the lung.** The use of probes should be limited when discerning the track, but they may be used later for a photographic demonstration of the direction and path.

Manners of death in firearms fatalities: homicides, suicides, and accidents

In most firearms deaths, the manner is self-evident by virtue of either the injury patterns and/or investigative facts (witnesses, video surveillance, and suicide note).[33–35] In other instances, the manner is less clear. For that reason and to enhance the veracity of their conclusions, it is important for forensic pathologists to have a grasp of the general patterns of **suicidal and accidental firearms injuries**. Needless to say, finding a firearm at the scene of death from a reportedly self-inflicted or accidentally inflicted gunshot wound would be a reasonable first step in the movement toward a suicidal or accidental manner determination.

In terms of injury patterns, **the head is overwhelmingly the most common anatomic site for self-inflicted firearms injuries**, with the chest, then the abdomen as far distant second and third locations. On the head, the following locations are chosen by the decedent, in decreasing order of incidence: **right temple, intraoral, forehead, left temple, submental, and back of the head.**[35–38] Contact and close range wounds are much more typical than are intermediate or distant injuries in self-inflicted gunshot wounds.[35,37] Most victims are men, use handguns, and up to 70% are found with the firearm near or on the body, approximately 25% are in the hand, and the remaining, greater than 12 feet from the body.[39] When evidence supports intermediate or distant ranges, accidental and homicidal circumstances become more relevant to consider, depending on the specifics of the case, including the size of the firearm used. It is also distinctly uncommon for victims of suicidal firearms injuries to fire the weapon through their clothing. If the fatal wound is **intraoral**, there may be perioral tearing, depending on whether the mouth was tightly closed at the time of the firing of the gun. In general, it is probably a good idea to use the "intraoral" terminology either in the cause of death statement ("intraoral gunshot wound") or at least somewhere in the final diagnoses of the autopsy report, for vital statistical purposes and tracking of these cases.

Accidental firearms injuries are much less prevalent, but are most typically associated with inadvertent or spontaneous discharges of firearms while cleaning or otherwise handling guns. Differences in manner determinations are not uncommon between jurisdictions when classifying firearms deaths related to Russian roulette, children handling guns, and hunting-related fatalities.[35,40–63]

Gunshot residue testing (GSR)

Although it has gained popularity through television shows, and law enforcement investigators may still request it, testing of the hands for GSR is not as useful as it may appear and is difficult to interpret. A study of 116 suicides by firearms analyzed GSR testing of the hands of the decedents, using scanning electron microscopy with energy-dispersive X-ray spectroscopy (SEM-EDX).[64,65] They reported a 50% false-negative rate. A positive result may be helpful, but it only indicates that a person fired a gun, was near a firing gun, or touched a firearm that had recently been discharged. If a decedent has a contact, close, or intermediate range wound, it is conceivable that GSR can be deposited on their hands regardless of who pulled the trigger. In addition, it is best to avoid the term powder burns, as it is ambiguous to what phenomenon the term is being applied. Some may use "powder burns" to describe searing of the skin from hot gases while others my use it to signify fouling.[3]

Rifles (rifled long guns)

Rifles are long guns with a rifled barrel that are usually fired from the shoulder. A typical rifle has a 30″ barrel, while carbines are shorter (18–24″ barrel). Rifles may fire high velocity bullets (>2000 feet/second), resulting in devastating tissue damage.[66] In high-powered rifle rounds with a semijacketed bullet, the bullet fragments create a lead "snowstorm" effect on radiograph.[66–69] For this reason, the bullet "bookkeeping" described above would not apply.

Shotguns

Shotguns are smooth bore (i.e., no rifling) long guns that typically fire multiple small balls (pellets, or "shot") encased in a plastic tube with wadding (plastic or cardboard) that separates the shot from gun powder (Fig. 9.15).[42,70–79] As there is not one measurable "site of lodgment," a qualitative description of the span of structures the pellets penetrated, with their sequelae, suffices (Figs. 9.16–9.20). *Representative pellets* can be retrieved for evidentiary purposes, as opposed to retrieving every single shot, the latter adding no evidentiary value and constituting an unnecessary and time-consuming exercise in futility, with diminishing returns. Shotguns may also fire single projectiles called shotgun slugs. When found within the wound, the wad is important to recover because it establishes the gauge and maker. Close or contact shotgun wounds of the head are devastating injuries because of the charge of the shot and accompanying gas entering the head.[34]

The "caliber" of a shotgun is classified by its gauge. **Gauge** is defined as the number of lead balls of a certain diameter that, together, weigh one pound. For example, "12 gauge" means that the diameter of the barrel is the same as the diameter of the lead ball of which it would take 12 to weigh a pound (0.410 is an exception, as it uses standard inch caliber). Gauge has nothing to do with the size of the shot in the shotgun cartridge; shells of the same gauge can contain different size shot.

The **shotgun wad**, which separates the shot from the charge, may be seen on X-ray as a faint circle (due to a thin coat of lead). It **can enter the body from a distance of 8–10 feet**, but can travel up to 50′. In general, if the column of shot creates a single entrance wound, then the wad may enter the body. If the shot has spread sufficiently to make multiple single entrance defects, then the wad cannot enter.

Choke is defined as the degree of narrowing of the shotgun barrel designed to control (tighten) the spread of the shot: cylinder bore = no choke (35%), modified (55%–65%), and full (65%–75%). The percent corresponds to the % of shot that fall within a 30″ circle at 40 yards. A shotgun diverter is an attachment that makes the circular shot pattern into a rectangular shape. The term magnum, as it relates to shotguns, means that the cartridge contains more gunpowder than a standard load, resulting in a greater velocity.

There are two size categories of shot: **birdshot and buckshot,** and there are different types of shotgun slugs, including the American Foster and the European Brenneke (with cardboard attached to the base with a screw), as well as an hourglass-shaped **sabot slug**. Slugs rarely exit the body, usually breaking apart into *comma-shaped lead fragments* with penetration.

Any gauge shotgun (except 0.410) will typically create a **round 3/4–1″ skin defect from a distance of up to 2 feet. At a 3-foot range, one sees a defect with scalloped edges; at 4 feet, the pellets begin to separate,** creating satellite defects, and **by 9–10 feet, there are predominantly individual pellet entrance defects.** One may also see **"petal marks"** on the skin from certain plastic shotgun wadding. Because of the *billiard ball effect*, one should not estimate the range of fire based on the pattern ("spread") observed on the radiograph.

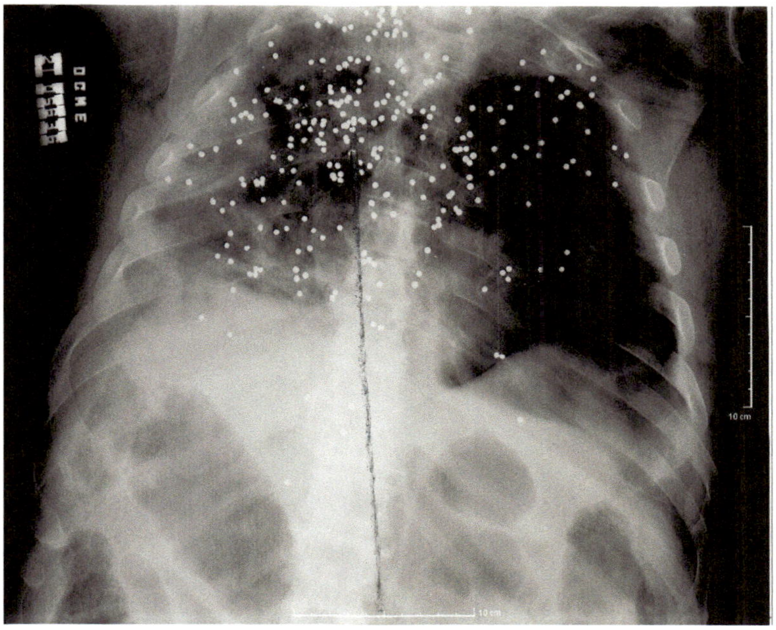

FIGURE 9.15 **Radiograph of shotgun pellets.**

FIGURE 9.16 **Shotgun.**

Ballistics

Ballistics is the science of the motion of projectiles. It includes **interior (in gun), external (in air), and terminal ballistics (penetration of solids, such as wound ballistics).** Although there is some overlap, the firearms examiner is the expert in interior ballistics while *the forensic pathologist is the expert of terminal (wound) ballistics.* For example, the firearm examiner examines the gun for function and for features such as the trigger pull. Typical trigger pulls vary from 12 to 15 lb.[3]

Some firearms do not produce or do not have a reproducible ***rifling pattern.*** These include a revolver without a barrel (the bullet will have a flared base), smooth bore rifles and pistols, a converted starter pistol with a reamed barrel, so-called zip guns (homemade firearms using a car antenna, etc.), a sympathetic discharge (round set off in an adjacent chambered bullet), firearms with a bullet that is of smaller caliber than the barrel (may see "skid" marks), firearms

FIGURE 9.17 **Shotgun cartridge.**

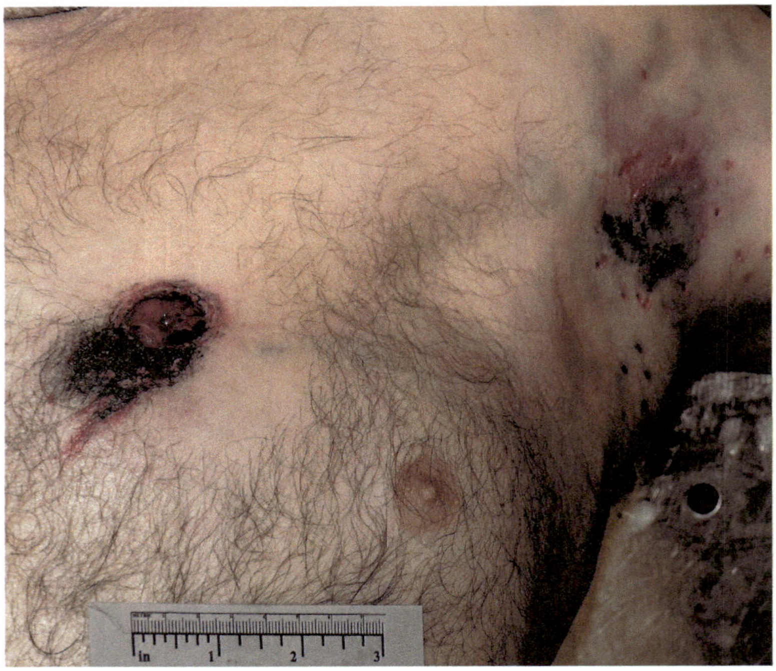

FIGURE 9.18 **Shotgun entrance (chest) and partial exit (axilla).**

with polygonal boring (e.g., Heckler and Koch), and rusted barrels, which will produce unreliable rifling profiles (subsequent firings may have different patterns).[3]

The following is a summary of some more ***unusual ammunition*** *that the forensic pathologist may encounter:*

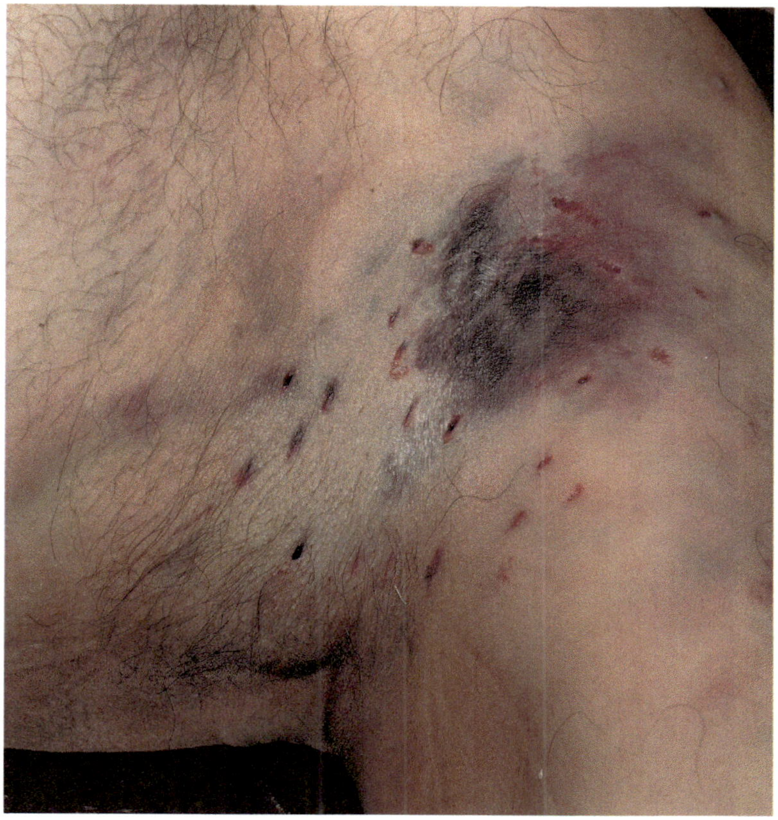

FIGURE 9.19 **Shotgun pellet exit wounds (unusual).**

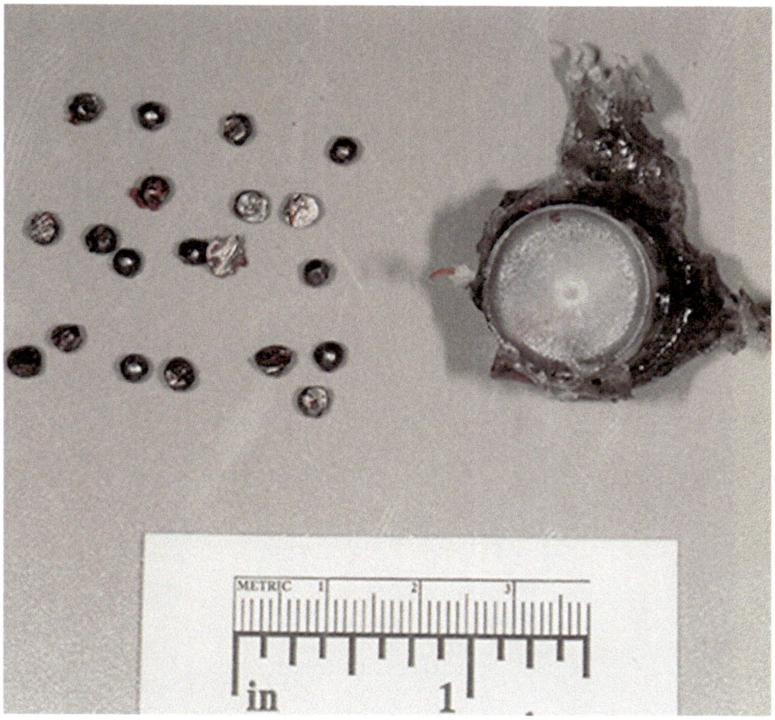

FIGURE 9.20 **Shotgun wadding and pellets (12 gauge).** Recovered from the above shotgun wound.

1. FRANGIBLE BULLETS: Bonded iron or lead pieces; designed to break apart once it hits a target and therefore one cannot discern a rifling pattern.[80,81]
2. RIP: **R**adically **I**nvasive **P**rojectiles are designed to predictably break apart into several metallic shards (or trocars) upon soft tissue penetration.[82] Each trocar then travels independently from the base. The trocars form the nose of the projectile and do not contact the lands and grooves of the barrel and therefore are devoid of a rifling pattern. All the individual ballistic characteristics from the lands and grooves are on the *base* of the projectile.[82,83]
3. HORNADY: Has a polymer plug in the nose with an "H" on it.[84]
4. SPEER SHOT: Rat shot in a handgun (it opens up like shotgun).[85]
5. WINCHESTER EXPANDING POINT 0.25: Has a number 4 steel shot ball in the nose.[84]
6. GLAZER ("Short stop"): A "bullet" with shot that opens on impact. It travels to the target as a single unit, then fragments after impact. It has a copper cup filled with lead shot and capped with Teflon or nylon. Radiographs show pellets and jacket/cup. May also find the plug/cap during autopsy. It is loaded with birdshot of various calibers.[86–88]
7. SHOT SHELL CARTRIDGES: The pellets are separated from the powder charge by the wadding. Shot is kept in place by crimping end of the casing or a plastic cover (N.B. Pellets disperse on discharge at about 4 cm and will produce miniature shotgun pattern on target). No jacketing gets into body and therefore is not seen on radiograph.
8. GOLDEN SABRE, STARFIRE, BLACK TALON: Petals open like claws.
9. KTW: Teflon-coated tungsten or steel (so-called "cop killer" penetrates bullet proof vests and engine blocks) with copper jacket base (copper has rifling).
10. NYCLAD: Nylon-coated lead bullet. The nylon must be recovered for bullet comparison. Often, the nylon is shredded and/or strips off the bullet in the body. The rifling is on the nylon, not the lead.

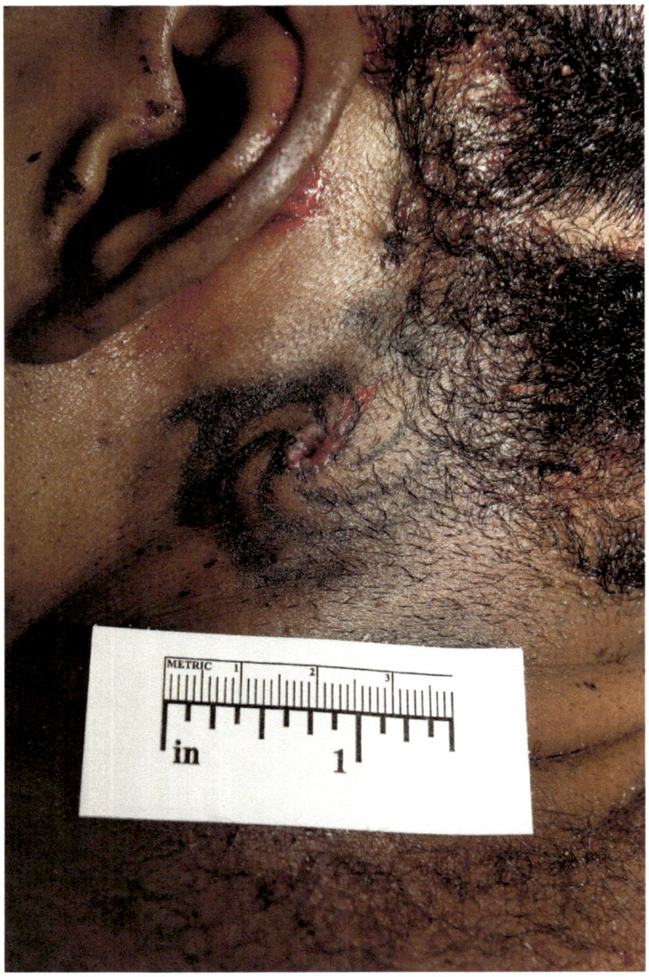

FIGURE 9.21 **Comet-tailing from a plated bullet.**

11. SCORPION/HYDROSHOCK: Half-jacketed hollow-point with a vertical spike.
12. SUBMACHINE GUN/PISTOL: Fully automatic weapon that fires *pistol* ammunition.
13. MACHINE GUN: Fully automatic weapon that fires *rifle* ammunition.
14. PLASTIC BULLETS: Travel at 500 feet/second and may cause a superficial laceration-type wound.
15. RUBBER BULLETS: Also called rubber baton rounds; typically have either a metal core with a rubber coating or a compound mixture that includes rubber. They are a less lethal alternative to metal projectiles, but can still cause serious injuries such as blindness, permanent disability, and death.[86–88]
16. SABOT/ACCELERATOR: Sabot is the sleeve that converts the bullet core to the same caliber as the barrel of the gun. An armor piercing tungsten carbide core sits in a plastic sabot (has slits and five to six petals that fall off bullet after exit from the muzzle, at approximately 14 inches. Sabot can travel 50 feet). The rifling pattern is on the sabot, not the slug.[89]
17. PLATED BULLETS: A unique soot pattern with curved lines (spiral) radiating from the entrance wound defect are due to comet-tailing, an effect that is seen with certain bullets, usually those that are electroplated or plated (Fig. 9.21). This unusual effect occurs when a plated bullet's electroplating material fractures upon firing. Vaporous lead escapes through the fracture site of the plating material. Since the bullet is spinning along its longitudinal axis as it flies through the air, the vaporous lead stream spins as well. Thus, when the bullet enters the target, a spiral-shaped, so-called "comet tail" deposition of vaporous lead can appear on the surface surrounding the entrance site.[90] The destructive power of the gas from the muzzle should not be underestimated. In addition to causing blow back lacerations in intraoral and contact wounds, the gas alone is capable of causing death as evidence by fatalities from discharges with firearms containing blank rounds.[91]

Conclusion

Firearm deaths are a common cause of traumatic injury in the United States. The forensic pathology fellow will have ample opportunity to see a wide variety of firearms injury patterns and manners of death. Suicidal firearm injuries offer an excellent learning opportunity because they involve contact and close-range discharges, and the type of firearm used is usually known.

References

1. Randall B, Fierro M, Froede R. Practice guidelines for forensic pathology. *Arch Pathol Lab Med*. 1998;122:1056–1064.
2. Zumwalt RE, Hirsch CS. Medicolegal interpretation of gunshot wounds. *Am J Emerg Med*. 1987;5(2):133–139.
3. DiMaio VJM. *Gunshot Wounds*. 3rd ed. Boca Raton: CRC Press; 2016.
4. Spitz W, Diaz F. *Spitz and Fisher's Medicolegal Investigation of Death*. 5th ed. Springfield: Charles C Thomas; 2020.
5. Denton JS, Segovia A, Filkins JA. Practical pathology of gunshot wounds. *Arch Pathol Lab Med*. 2006;130(9):1283–1289.
6. Dixon DS. Determination of direction of fire from graze gunshot wounds. *J Forensic Sci*. 1980;25(2):272–279.
7. Heninger MM. An unusual feature of graze gunshot wounds. *Acad Forensic Pathol*. 2016;6(2):291–300.
8. Fackler ML. Wound ballistics. A review of common misconceptions. *JAMA*. 1988;259(18):2730–2736.
9. Haag LC. Wound production by ricocheted and destabilized bullets. *Am J Forensic Med Pathol*. 2007;28(1):4–12.
10. Maiden N. Ballistics reviews: mechanisms of bullet wound trauma. *Forensic Sci Med Pathol*. 2009;5(3):204–209.
11. Haag LC. Base deformation of full metal-jacketed rifle bullets as a measure of impact velocity and range of fire. *Am J Forensic Med Pathol*. 2015; 36(1):16–22.
12. Adelson L. A microscopic study of dermal gunshot wounds. *Am J Clin Pathol*. 1961;35:393–402.
13. Molina DK, Rulon JJ, Wallace EI. The atypical entrance wound: differential diagnosis and discussion of an unusual cause. *Am J Forensic Med Pathol*. 2012;33(3):250–252.
14. Bentley AJ, Busuttil A, Clifton B, Sibbald P. Homicidal tandem bullet wound of the chest. *Am J Forensic Med Pathol*. 1997;18(1):56–59.
15. Dixon DS. Exit keyhole lesion and direction of fire in a gunshot wound of the skull. *J Forensic Sci*. 1984;29(1):336–339.
16. Berryman HE. A systematic approach to the interpretation of gunshot wound trauma to the cranium. *Forensic Sci Int*. 2019;301:306–317.
17. Ellis PS. Fatal gunshot injury caused by an unusual projectile–a barrel-cleaning brush as a tandem bullet. *Am J Forensic Med Pathol*. 1997;18(2): 168–171.
18. Mihailovic Z, Savic S, Atanasijevic T. Firearm suicide committed using an unusual combination of tandem missiles: a bullet, a nail, and a screw. *Am J Forensic Med Pathol*. 2007;28(3):220–222.
19. Tattoli L, Schmid S, Tsokos M. Three rounds as "tandem bullets": unusual findings in a case of a suicidal gunshot to the head. *Forensic Sci Med Pathol*. 2014;10(4):613–618.
20. du Toit-Prinsloo L, Morris NK, Saayman G. Tandem bullet injury: an unusual variant of an unusual injury. *Forensic Sci Med Pathol*. 2014;10(2): 262–265.
21. Khau A, Melinek J. Two cases of tandem bullets-one homicide and one suicide. *Am J Forensic Med Pathol*. 2019;40(3):262–265.
22. Ersoy G, Gurler AS, Ozbay M. Upon a failure to equal entry and exit wounds: a possible case of tandem bullets in view of the literature. *J Forensic Sci*. 2012;57(4):1129–1133.
23. Picoli FF, Mundim-Picoli MBV, Alves AM, Silva M, Franco A, Silva RF. Suicidal tandem bullets to the heart with subsequent embolization - a case report. *Forensic Sci Med Pathol*. 2020;16(3):523–527.

24. Prahlow SP, Wolfenbarger R, Prahlow JA. Tandem bullet homicide. *Acad Forensic Pathol.* 2016;6(1):130–139.
25. Adelson L. Bullet embolism with radiologic documentation. A case report. *Am J Forensic Med Pathol.* 1984;5(3):253–256.
26. Springer J, Newman W, McGoey R. Intravascular bullet embolism to the right atrium. *J Forensic Sci.* 2011;56(Suppl 1):S259–S262.
27. Krispin A, Zaitsev K, Hiss J. The elusive slug: bullet intestinal "embolism". *Forensic Sci Med Pathol.* 2010;6(4):288–292.
28. DiMaio VJ, DiMaio DJ. Bullet embolism: six cases and a review of the literature. *J Forensic Sci.* 1972;17(3):394–398.
29. Thali MJ, Schweitzer W, Yen K, et al. New horizons in forensic radiology: the 60-second digital autopsy-full-body examination of a gunshot victim by multislice computed tomography. *Am J Forensic Med Pathol.* 2003;24(1):22–27.
30. Hirsch CS, Lester Adelson MD. *Am J Forensic Med Pathol.* 1989;10(3):261–263.
31. Stefanopoulos PK, Hadjigeorgiou GF, Filippakis K, Gyftokostas D. Gunshot wounds: a review of ballistics related to penetrating trauma. *J Acute Dis.* 2014;3(3):178–185.
32. French RW, Callender GR. *Ballistic Characteristics of Wounding Agents. Wound Ballistics.* Washington, DC: Office of the surgeon general, Department of the Army; 1962:91–141.
33. Cave R, DiMaio VJ, Molina DK. Homicide or suicide? Gunshot wound interpretation: a Bayesian approach. *Am J Forensic Med Pathol.* 2014;35(2):118–123.
34. Molina DK, Wood LE, DiMaio VJ. Shotgun wounds: a review of range and location as pertaining to manner of death. *Am J Forensic Med Pathol.* 2007;28(2):99–102.
35. Molina DK, DiMaio VJ, Cave R. Handgun wounds: a review of range and location as pertaining to manner of death. *Am J Forensic Med Pathol.* 2013;34(4):342–347.
36. Kohlmeier RE, McMahan CA, DiMaio VJ. Suicide by firearms: a 15-year experience. *Am J Forensic Med Pathol.* 2001;22(4):337–340.
37. Molina DK, DiMaio V, Cave R. Gunshot wounds: a review of firearm type, range, and location as pertaining to manner of death. *Am J Forensic Med Pathol.* 2013;34(4):366–371.
38. Cina SJ, Ward ME, Hopkins MA, Nichols CA. Multifactorial analysis of firearm wounds to the head with attention to anatomic location. *Am J Forensic Med Pathol.* 1999;20(2):109–115.
39. Garavaglia JC, Talkington B. Weapon location following suicidal gunshot wounds. *Am J Forensic Med Pathol.* 1999;20(1):1–5.
40. Cole TB, Patetta MJ. Hunting firearm injuries, North Carolina. *Am J Publ Health.* 1988;78(12):1585–1586.
41. Morrow PL, Hudson P. Accidental firearm fatalities in North Carolina, 1976-80. *Am J Publ Health.* 1986;76(9):1120–1123.
42. Golladay ES, Murphy Jr KE, Wagner CW. Shotgun injuries in pediatric patients. *South Med J.* 1991;84(7):886–888.
43. Baker AM, Keller G, Garcia D. A novel hunting accident. Discharge of a firearm by a hunting dog. *Am J Forensic Med Pathol.* 2001;22(3):285–287.
44. Spagnolia A, Kemp WL. Fatality while bear hunting: a homicide or an accident? *Am J Forensic Med Pathol.* 2019;40(2):144–146.
45. Junuzovic M, Sjoberg A, Eriksson A. Unintentional nonhunting firearm deaths in Sweden, 1983-2012. *J Forensic Sci.* 2016;61(4):966–971.
46. Maujean G, Guinet T, Malicier D. A death involving a fired and deflected bullet: a case report from lyon university institute of forensic science, France. *J Forensic Sci.* 2016;61(1):265–267.
47. Junuzovic M, Eriksson A. Unintentional firearm hunting deaths in Sweden. *Forensic Sci Int.* 2012;216(1–3):12–18.
48. Frazer M, Rosenberg S. Russian roulette with a knife. *J Forensic Sci.* 1983;28(1):268–272.
49. Shields LBE, Hunsaker JC, Stewart D. Russian roulette and risk-taking behavior. *Am J Forensic Med Pathol.* 2008;29:32–39.
50. Marzuk PM, Tardiff K, Smyth D, Stajic M, Leon AC. Cocaine use, risk taking, and fatal Russian roulette. *JAMA.* 1992;267(19):2635–2637.
51. Gill JR, Hirsch CS. Russian roulette deaths. *Am J Forensic Med Pathol.* 2011;32(2):e14.
52. Collins KA. Adolescent Russian roulette deaths. *Am J Forensic Med Pathol.* 2010;31(1):4–6.
53. Fishbain DA, Fletcher JR, Aldrich TE, Davis JH. Relationship between Russian roulette deaths and risk-taking behavior: a controlled study. *Am J Psychiatr.* 1987;144:563–567.
54. Wood Jr NP, Mercy JA. Unintentional firearm-related fatalities, 1970–1984. *Morb Mortal Wkly Rep CDC Surveillance Summ.* 1988;37(1):47–52.
55. Grossman DC, Reay DT, Baker SA. Self-inflicted and unintentional firearm injuries among children and adolescents: the source of the firearm. *Arch Pediatr Adolesc Med.* 1999;153(8):875–878.
56. Johnson MS, Moore M, Mitchell P, Owen P, Pilby J. Serious and fatal firearm injuries among children and adolescents in Alaska: 1991–1997. *Alaska Med.* 2000;42(1):3–10, 27.
57. Galland H. Children, adolescents, and guns in Louisiana: a thought experiment. *J La State Med Soc.* 2000;152(10):523–531.
58. Christoffel KK, Christoffel T. Handguns as a pediatric problem. *Inj Prev.* 1986;5(2):151–156, 1999.
59. Beaver BL, Moore VL, Peclet M, Haller Jr JA, Smialek J, Hill JL. Characteristics of pediatric firearm fatalities. *J Pediatr Surg.* 1990;25(1):97–99. discussion 99–100.
60. Miller M, Azrael D, Hemenway D. Firearm availability and suicide, homicide, and unintentional firearm deaths among women. *J Urban Health.* 2002;79(1):26–38.
61. Rushforth NB, Hirsch CS, Ford AB, Adelson L. Accidental firearm fatalities. *JAMA.* 1977;238(22):2367.
62. Byard RW, Haas E, Marshall DT, Gilbert JD, Krous HF. Characteristic features of pediatric firearm fatalities–comparisons between Australia and the United States. *J Forensic Sci.* 2009;54(5):1093–1096.
63. Prahlow JA. Fatal gunshot wounds in young children. *Acad Forensic Pathol.* 2016;6(4):691–702.
64. Molina DK, Castorena JL, DiMaio VJ. Gunshot residue testing in suicides: Part II: analysis by inductive coupled plasma-atomic emission spectrometry. *Am J Forensic Med Pathol.* 2007;28(3):191–194.
65. Molina DK, Martinez M, Garcia J, DiMaio VJ. Gunshot residue testing in suicides: Part I: analysis by scanning electron microscopy with energy-dispersive X-ray. *Am J Forensic Med Pathol.* 2007;28(3):187–190.
66. Molina DK, DiMaio VJ. Rifle wounds: a review of range and location as pertaining to manner of death. *Am J Forensic Med Pathol.* 2008;29(3):201–205.
67. Fackler ML, Surinchak JS, Malinowski JA, Bowen RE. Bullet fragmentation: a major cause of tissue disruption. *J Trauma.* 1984;24(1):35–39.
68. Straathof D, Bannach BG, Wilson AJ, Dowling GP. Radiography of perforating centerfire rifle wounds of the trunk. *J Forensic Sci.* 2000;45(3):597–601.
69. Haag LC. The forensic aspects of contemporary disintegrating rifle bullets. *Am J Forensic Med Pathol.* 2013;34(1):50–55.
70. Ordog GJ, Wasserberger J, Balasubramaniam S. Shotgun wound ballistics. *J Trauma.* 1988;28(5):624–631.

71. Parks WH. Shotgun wound ballistics. *J Trauma*. 1989;29(2):272.
72. Harruff RC. Comparison of contact shotgun wounds of the head produced by different gauge shotguns. *J Forensic Sci*. 1995;40(5):801–804.
73. Gulmann C, Hougen HP. Entrance, exit, and reentrance of one shot with a shotgun. *Am J Forensic Med Pathol*. 1999;20(1):13–16.
74. Durak D, Fedakar R, Turkmen N. A distant-range, suicidal shotgun wound of the back. *J Forensic Sci*. 2006;51(1):131–133.
75. Linert B, Regnier J, Doyle BW, Prahlow JA. Suicidal shotgun wound employing a shotgun barrel, a shotgun shell, and a BB. *J Forensic Sci*. 2010;55(2):546–548.
76. Cail K, Klatt E. The effect of intermediate clothing targets on shotgun ballistics. *Am J Forensic Med Pathol*. 2013;34(4):348–351.
77. Haag LC. Range of fire determination from the pseudostippling of skin by shotshell buffer material. *Am J Forensic Med Pathol*. 2013;34(1):56–62.
78. Blessing MM, Lin PT. Suicide by shotgun in southeastern Minnesota. *J Forensic Sci*. 2016;61(Suppl 1):S159–S162.
79. Hlavaty L, Amley J, Root K, Avedschmidt S, Jaworski A, Sung L. The difficulties in assessing wounds created by .410 bore ammunition and weapons. *J Forensic Sci*. 2020;65(2):500–507.
80. Kaplan J, Klose R, Fossum R, Di Maio VJ. Centerfire frangible ammunition: wounding potential and other forensic concerns. *Am J Forensic Med Pathol*. 1998;19(4):299–302.
81. Schyma C, Bittner M, Placidi P. The MEN frangible: study of a new bullet in gelatin. Metallwerk Elisenhutte Nassau. *Am J Forensic Med Pathol*. 1997;18(4):325–330.
82. Zhang X, Cain MD, Williams CD, Spears TA, Poulos CK. G2 research radically invasive projectile: the importance of recognizing its imaging and autopsy patterns. *Am J Forensic Med Pathol*. 2021;42(3):248–251.
83. Hakki L, Smith A, Babin J, Hunt J, Duchesne J, Greiffenstein P. Effects of a fragmenting handgun bullet: considerations for trauma care providers. *Injury*. 2019;50(5):1143–1146.
84. Poulos C, Tlomak W. A review of unusual new projectiles. *Acad For Path*. 2013;3(2):144–153.
85. Oommen Z, Pierce SM. Lead-free primer residues: a qualitative characterization of winchester WinClean, remington/UMC LeadLess, federal BallistiClean, and speer lawman CleanFire handgun ammunition. *J Forensic Sci*. 2006;51(3):509–519.
86. Kobayashi M, Mellen PF. Rubber bullet injury: case report with autopsy observation and literature review. *Am J Forensic Med Pathol*. 2009;30(3):262–267.
87. de Freminville H, Prat N, Rongieras F, Voiglio EJ. Less-lethal hybrid ammunition wounds: a forensic assessment introducing bullet-skin-bone entity. *J Forensic Sci*. 2010;55(5):1367–1370.
88. Charlier P, Alvarez JC, Durigon M, de la Grandmaison GL. Unusual death by rubber bullet: should these guns be reclassified as lethal weapons? *Am J Forensic Med Pathol*. 2012;33(1):e4.
89. Ward ME, Nolte KB. Ballistic characterization of the remington premier copper solid sabot shotgun slug. *J Forensic Sci*. 2000;45(6):1259–1266.
90. Prahlow SP, Brown TT, Dye D, Poulos C, Prahlow JA. "Comet-tailing" associated with gunshot entrance wounds. *J Forensic Sci*. 2021;66(3):1154–1160.
91. Zdravkovic M, Milic M, Stojanovic M, Kostov M. Three cases of death caused by shots from blank cartridge. *Am J Forensic Med Pathol*. 2009;30(4):403–406.

CHAPTER

10

Approach to blunt, sharp, and transportation deaths

Susan F. Ely and James R. Gill

You've got to believe what you see.

The body doesn't lie. *-Dr. Charles S. Hirsch*

Blunt injury

There are three general types of blunt injuries: lacerations, abrasions, and contusions. Of the three, lacerations are the result of the greatest amount of force. A laceration is a splitting and tearing of the skin or tissue due to crushing or stretching beyond its elastic limit. Most lacerations have a nonspecific pattern, but some more specifically reflect the instrument or surface having made contact with the body to produce them. Classic features of lacerations include *tissue bridging, undermining (if full thickness and over bone; see below), and a surrounding contusion and/or abrasion* (Figs. 10.1 and 10.2). These features help distinguish them from a sharp injury. An avulsion is a more severe type of laceration injury produced from an oblique/tangential impact that causes a **shearing** or tearing of skin and soft tissues away from the underlying fascia or bone. **Avulsion "pockets"** are characterized by substantially torn tissue creating a typically broad compartment of soft tissue injury (Fig. 10.3) that is filled with blood and/or more gelatinous-appearing soft tissue hemorrhage or hematoma (see "Transportation Injury" below). This typically results from a greater amount of force, with crushing and tearing of tissue due to a moving application of force, such as that produced by a motor vehicle striking a pedestrian or tires running over the body.[1]

In addition to the skin and underlying soft tissues, lacerations may also occur in other tissues and organs, including the aorta, lungs, liver, spleen, pancreas, and brain, to name a few. Aortic lacerations due to deceleration events (e.g., motor vehicle collisions or descents from a considerable height) classically occur at the descending aortic arch, just past the subclavian take-off, where the aorta becomes fixed to the vertebral column. During deceleration, the vertebral column stops with the body while the free aortic arch continues to move; as a result, the greatest stress is at the point of tethering of the aorta to the vertebral column (Fig. 10.4). When describing liver lacerations, one should mention whether or not they involve the more vascular hilum (vs. the periphery), as this has important clinical and injury scaling implications that are of interest to trauma surgeons. Due to its rich blood supply, lacerations of the scalp may be fatal by exsanguination, particularly when multiple.[2]

While lacerations of the viscera are not uncommon in high impact scenarios, such as motor vehicular accidents or descents from heights, they are distinctly uncommon on the skin surfaces in areas of the body without underlying bone (e.g., skin and soft tissues of the abdominal wall). Of all the skin on the body, that of the face and scalp are the most susceptible to laceration injury because of the resistant underlying bone, which reduces the ability of the skin to deform and "absorb" an impact. The scalp/skull unit are a set up for a specific set of injuries, both externally and internally, blunt and firearms injuries alike (see "Craniocervical Trauma" below). The scalp tissue is stretched over the skull in such a way that blunt forces applied to it place it at unique risk for tearing, as the tissues between the application of force and the skull are compressed and crushed. If the forces are great enough, the scalp tissues will stretch and tear under this vice grip scenario, resulting in a laceration that may reflect characteristics of the putative object, often with associated undermining that may assist in determining the directionality of the force.[3] Of note, the

Principles of Forensic Pathology
https://doi.org/10.1016/B978-0-323-91796-4.00025-8

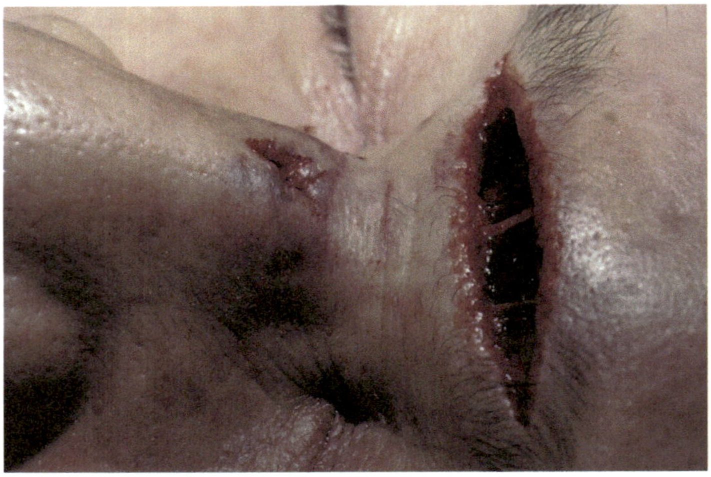

FIGURE 10.1 Laceration with tissue bridging.

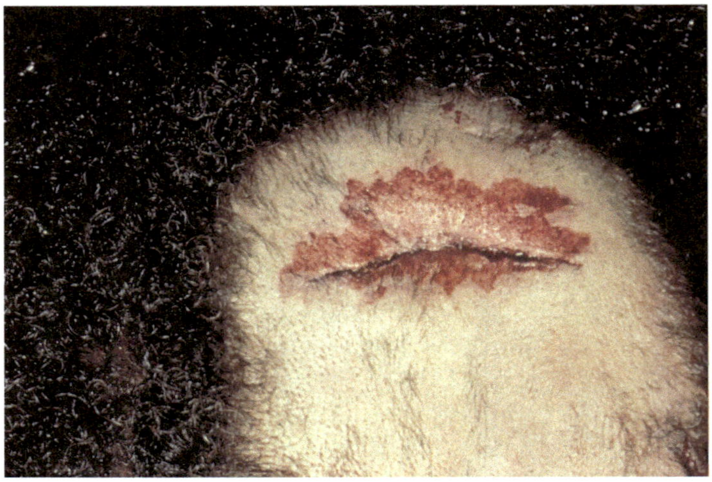

FIGURE 10.2 Laceration with surrounding abrasion (from a fall to a flat, resistant surface).

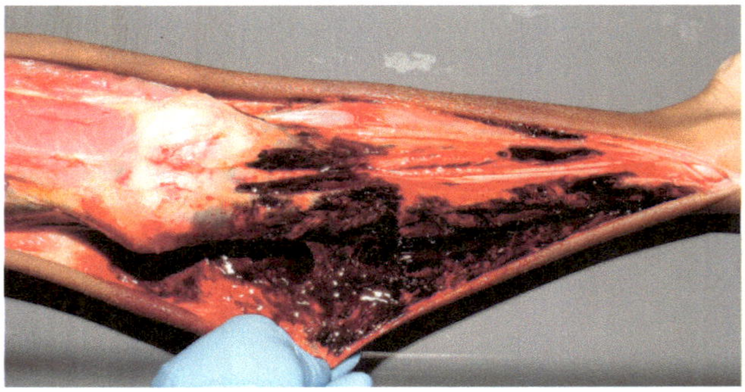

FIGURE 10.3 Blunt (shearing) injury: avulsion pocket.

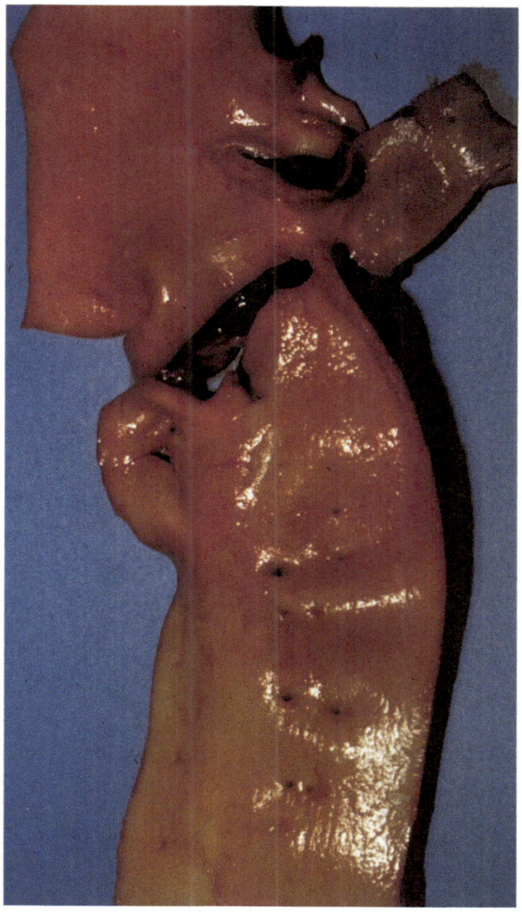

FIGURE 10.4 **Aortic laceration.** Classic location for an aortic laceration in acceleration–deceleration injury (common in descents from heights).

presence of undermining is not a requisite feature of every scalp laceration. It may appear only if the laceration is transscalpular (full thickness) and the force causing it makes contact with the head obliquely or tangentially.

While **fist blows** have the capacity to cause contusions, occasional lacerations, and even fractures of the *face*, depending upon the degree of their force, it is distinctly uncommon for closed *punches* to result in lacerations of the *scalp*, unless something more resistant/dense, such as brass knuckles, is involved. This is a very important concept to understand to the extent that if multiple blunt impacts to the head are evidenced by multiple scalp lacerations (i.e., a bludgeoning), it is important to assert this in the "how injury occurred" field ("struck multiple times with blunt object," as opposed to simply stating, "stuck on head multiple times"). This distinction in not inconsequential and must be documented where there is the scientific veracity to do so, as it may or may not have legal relevance and repercussions. Lacerations of the chest and extremities are distinctly less common than are lacerations of the scalp, and when they occur, are often the result of an underlying displaced fracture, particularly of the long bones or sacrum (i.e., "open" fractures).

As alluded to above, certain features of a laceration may "tell a story" about the likely circumstances of its formation, even before the internal examination is conducted. Both **branching** and **undermining** of a laceration indicate the direction of the application of force that produced it. As an example, if there is a horizontal laceration of the parietal scalp, with undermining on its inferior aspect, the direction of the force that produced the laceration was downward (it is best to document both the location of the undermining relative to the wound, as well as its radius). Similarly, if a vertical laceration on the frontal scalp is Y- or V-shaped, with the branches of the Y/V located inferiorly, the direction of the force is in the direction of the branches and is therefore also in a downward direction. In both instances, and in very general terms, the circumstances of the injury occurrence are more consistent with a blow than from a simple fall with a strike of the head against a resistant object en route to the floor or the ground. In the latter scenario (e.g., falling, with the head striking a dresser in the course of the fall), the forces are directed in an upward direction, incompatible with the injury patterns described above. This is important physical/scientific data that contributes to the body of evidence for the case, even before a police investigation begins.

Finally, separate and apart from the information the individual wounds may offer, the location and distribution of the wound or wounds on the scalp often contribute to an explanation of their genesis, sinister versus benign/innocent. Specifically, a multiplicity of wounds on more than one convexity/plane and wounds above the "brim of the hat" are often more consistent with inflicted rather than accidental injury, in the absence of a descent, motor vehicle collision, or another similar severe wounding scenario (Fig. 10.5).

An **abrasion** is more commonly known as a scrape caused by contact with an abrasive, compressive, or fine, but blunt surface (the latter typified by a scratch from a cat's claw). A typical abrasion involves the removal of the superficial epithelial layer, although there may be deeper abrasions, even down to the muscle. When the force moves in a parallel fashion across the skin (frictional), it often causes a brush-type "burn" pattern that appears pink-red, like a thermal burn, and has a directionality to it. One may determine the specific directionality of the force by looking for heaped up epidermis or a "skin tag" at one edge. Similar to the shoveling of snow, when still attached, the removed epidermis can usually be found at the terminal side of the scrape. A graze gunshot wound is a classic example of such a directional abrasion. As discussed in the last chapter, careful inspection of a graze wound may also reveal microlacerations along the long margins of the wound, leaving behind tiny skin tags that "point" in the direction of the shooter/gun (conversely, the microlacerations "point" in the direction of the traveling bullet). By contrast, if the impact is perpendicular rather than parallel to the skin, it causes more of a compressive injury that leads to a dry, and sometimes parchment-like abrasion. A ligature abrasion or furrow is a classic example of the latter.

Contusions are traumatic extravasations of blood into the tissues due to a mechanical disruption/tearing of blood vessels. Blood pressure is needed to produce a significant extravasation of blood (e.g., hemorrhage and hematomas); however, even after death, due to the effects of gravity and intrinsic residual pressures, one may see lesser (passive) extravasation of blood into the tissues. Anyone who has performed an autopsy can attest to this fact. Distinguishing between the two (antemortem/intravital hemorrhage vs. postmortem blood extravasation) is a function of assessing the degree of hemorrhage. An **ecchymosis** is a more general and descriptive term used to convey any extravasation of blood into the skin or tissues that is greater than 3 mm in size (smaller extravasations are called purpura and petechiae). An ecchymosis is an umbrella term that may represent traumatic, natural, or iatrogenic processes, For example, one may see an ecchymosis surrounding a needle puncture site (a sharp injury, which is either iatrogenic or related to drug use). *All contusions are ecchymoses, but not all ecchymoses are contusions.* That said, if it is unclear whether a soft tissue hemorrhage definitively represents trauma (a contusion), it is better to be descriptive, calling it either an *ecchymosis* or simply a *"soft tissue hemorrhage,"* without further opining as to its pathophysiology. A **hematoma** is a space occupying collection of blood that may be produced by myriad processes, both natural and unnatural.

There are also many individual **patterned blunt injuries** (abrasions, contusions, and lacerations—Fig. 10.6), as well as **patterns of injury** (Figs. 10.7—10.9). *Tram track or railroad track contusions* are an example of the former; they are parallel contusions caused by an impact with a long narrow object, such as a pipe or a bat. Most of the extravasated hemorrhage from the impact occurs along the sides of the wound because that is where there is the

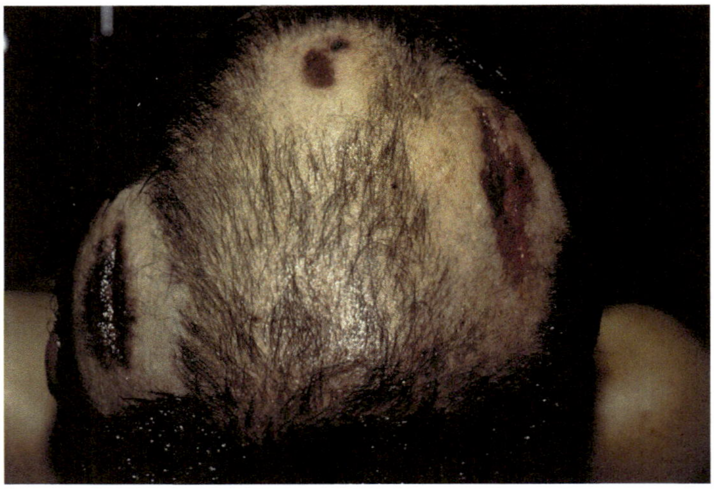

FIGURE 10.5 **Nonplanar impacts due to baseball bat.** Nonplanar impacts of scalp.

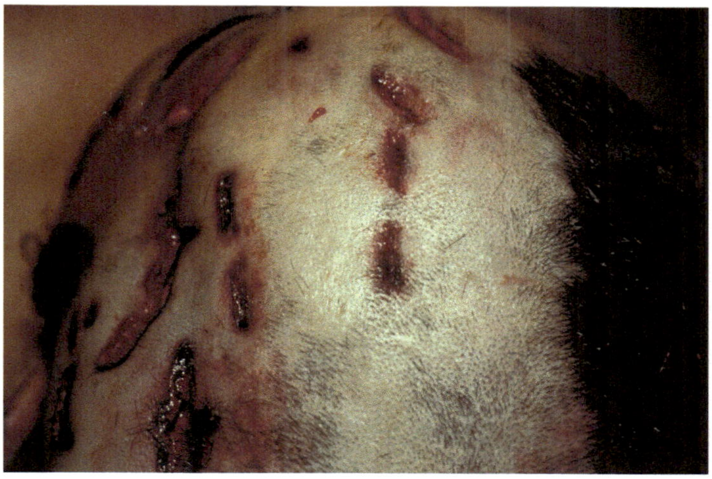

FIGURE 10.6 Patterned lacerations: hammer, claw end.

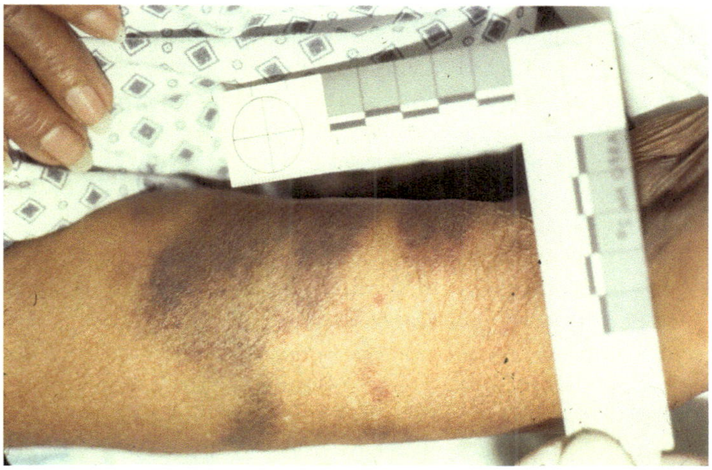

FIGURE 10.7 Patterned contusions: grab mark.

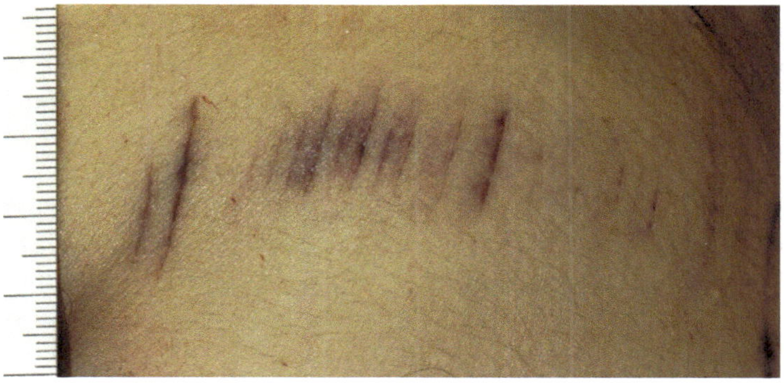

FIGURE 10.8 Patterned abrasions from a barbell.

greatest nonuniform disruption. The term *planar injury*, on the other hand, is used to describe a pattern that involves a group of injuries distributed within a single surface (plane) of the body, and are therefore consistent with a single impact onto a flat, resistant surface, such as the floor or ground. Planar injuries are common in *terminal falls* onto the

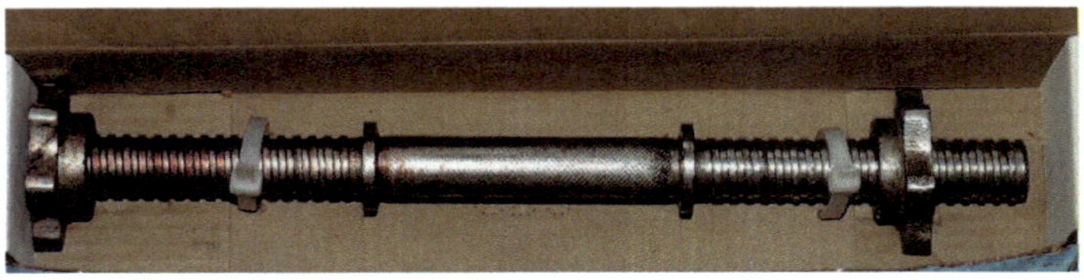

FIGURE 10.9 Barbell used in the assault in Fig. 10.8.

face (especially when the individual is already unconscious/dead), or onto the left or right sides of the body, often leaving ipsilateral markers of the impact on bony prominences (brow, nose, zygoma, chin, shoulder, elbow, hip, knees). As such, planar injury, in and of itself, tends to appear less sinister and sets off fewer alarm bells for the forensic pathologist than does a multiplicity of wounds observed in a different configuration. To wit, a person who has impacts of the same age, on multiple planes of the body, will likely elicit a greater cause for concern of an assault than will planar injury from a simple accidental fall. All that said, the forensic pathologist not uncommonly encounters a clear pattern of blunt injury, but does not have a sense of what object or surface caused it, even after a thorough review of the scene photos. It is no less important to carefully document and describe these patterned injuries in detail, with measurements and qualitative descriptors, for possible comparisons at a later time with weapons or surfaces that are subsequently discovered. The most we can truthfully say when faced with such clearly patterned injury of an as yet unknown source is, "That is an excellent example of whatever it is!"

Contusions will change in color over time and tend to progress from their periphery to their centers. Great care must be taken in attempting to age contusions based on color alone. Acute contusions are generally red and swollen, but the depth of the hemorrhage can affect the color, which eventually becomes blue-purple. Over time, the color will change to yellow-brown and green-yellow due to the breakdown of blood products. Microscopic examination is usually not very helpful except in the potential demonstration of hemosiderin, which may date the contusion at greater than a day or two. Intact red blood cells may be seen for up to several weeks in bluntly injured tissues. People who bruise easily include children, the elderly, and people with cirrhosis, bleeding diatheses, or those who take anticoagulation/antiplatelet medications.

Fractures are essentially lacerations of bone. They may be open, closed, linear, comminuted, spiral, and/or depressed. The pattern of the fracture may be an indication of how the injury occurred. For example, a depressed fracture of the skull is evidence of a focal application of force by an object of considerable density (e.g., a hammer or lead pipe, not a wooden or aluminum broomstick), whereas a linear fracture of the skull is consistent with the skull's impact against a flat, resistant surface, such as the floor or cement, among many other surfaces or objects. Even a few rib fractures in the elderly or those with many comorbidities may be life threatening (Fig. 10.10).

Possible **mechanisms of death due to blunt injuries** are many and varied. They include exsanguination, air embolism, flail chest, asphyxia due to aspiration of blood (commonly associated with some basilar skull and facial fractures, oral trauma, and any other breach of the upper or lower airways), cardiac tamponade, fat embolism, cardiac concussion, rhabdomyolysis with acute renal failure, infection, compartment syndromes, pain-related sudden cardiac death, and cerebral compression from space-occupying hemorrhage (subdural and epidural hematomas), as well as a multiplicity of complex neurotraumatic mechanisms of the head and neck, including concussive injury (see "Craniocervical Trauma" below for a more detailed discussion of the latter two complications). More delayed mechanisms may result from the effects of long-term recumbency/being bedbound or paralyzed (bronchopneumonia, neurogenic bladder and/or bowel, decubitus ulcers with osteomyelitis, and deep venous thrombosis with pulmonary thromboembolism).

Although fat embolism likely occurs in nearly 100% of fractures, it is fatal in only approximately 1%. *Fat embolism syndrome* typically occurs one to four days after the injury occurrence and usually involves displaced or comminuted fractures and/or crush-type injuries of soft tissues. Classically, petechiae of the conjunctiva, skin, and cerebral white matter are seen. For it to be fatal, it is typically systemic (involving the kidneys and brain) and, clinically, a person is expected to develop respiratory distress and confusion within the timeline described above. Fat embolism, even if not fat embolism syndrome, per se, also may be seen with cardiopulmonary resuscitation, sickle cell crisis, and pancreatitis. Like so many entities encountered in forensic practice, the microscopic detection of fat emboli must be viewed through the prism of context, clinical and autopsy, in order to answer the central question, "Is this finding the cause, a coincidence, or a consequence of death?"

Commotio cordis, or cardiac concussion, is a fatal arrhythmia resulting from an impact to the heart through the chest wall. The impact must occur 15–30 ms before the T-wave for it to cause this arrhythmia. Because of the suddenness, there usually is no gross or microscopic injury, other than a possible contusion of the chest wall. Therefore, the

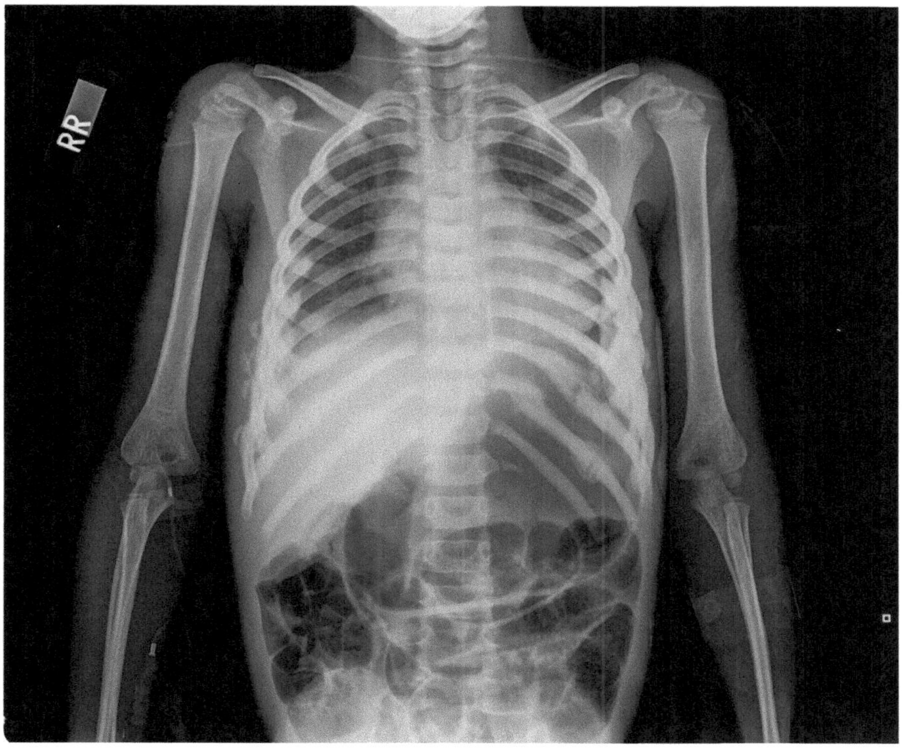

FIGURE 10.10 Fracture calluses of multiple ribs.

clinical history is key in making this diagnosis. It often occurs with athletes and the history is that the person sustains an impact of the chest and then several seconds later, collapses.

The term *rupture* is reserved for natural diseases. Therefore, a cerebral artery aneurysm that bursts is called a ruptured cerebral artery aneurysm, similar to a gastric ulcer or an aortic dissection. By contrast, a tear of the aorta due to a deceleration injury in a motor vehicle collision is a *laceration*, not a ruptured aorta.

The injury caused by blunt force is a function of the delivered **kinetic energy**, modified by the area of the surface inflicting the injury and the duration of the impact (the duration of the transfer of energy). The area (A) and the time (T) inversely affect the extent of injury. This may seem counterintuitive, in that one might intuit that the greater the surface area of the impacting object, the greater the degree of injury, and that the longer a force exerts its effects, also the greater the injury. But, the inverse is actually true. As the surface area of impact increases, the force is distributed over a greater area, dissipating and distributing its effects, thus decreasing the severity of the injury sequelae born by any one specific receiving location. For example, using the same amount of force, the pointed end of the needle will cause more damage than the flat head of the pin. The damage caused by two objects of the same mass, but different size/surface area and density (e.g., a hammer vs. a two by four) will be quite different in severity. As the *duration* of impact increases, the extent of injury also decreases because the body has more time to accommodate to the force. Would you rather be punched with a bare fist or a hand wearing a cushioned boxing glove? If two people jump off a roof and one lands on the concrete and the other lands on a stack of mattresses, who will sustain more injury? The fluffy mattresses (similar to the boxing glove) increase the time (T) of impact, therefore decreasing the extent of the resultant injury. The "formula" is:

$$\textbf{Injury} = \textbf{kinetic energy} \times \mathbf{1/T} \times \mathbf{1/A}.$$

Lastly, **crush injuries**, a subtype of blunt force trauma, bear special mention. Crush injuries occupy a specific category among patterns of blunt force trauma. Not surprisingly, given their descriptive name, they result from considerable compressive forces applied to the body. When severe, as seen in the pedestrian victim who is run over by a motor vehicle, they are striking and self-evident. Devastating skeletal and soft tissue trauma in such circumstances often create deformities that are immediately visible as the forensic pathologist approaches the autopsy table. So severe may be head trauma that a visual (facial) identification of the decedent is precluded. If involving the head, depressed and shatter type fractures may be responsible for externally visible concavities and flattening of the normally round head and extrusion of most of the brain. Sometimes, large and gaping lacerations are also seen, some with associated avulsion injury, especially when caused by the tire(s) of a car, with extensive brain matter left on the roadway. When the compression involves the torso or extremities, similar devastating and avulsive injuries are typically observed. Depending on the location of the crush forces, commonly seen injuries include multiple displaced anterior and posterior rib fractures, "open book," comminuted pelvic fractures, open long bone fractures, with avulsed surrounding soft tissues, and severely lacerated and pulpified viscera, sometimes with traumatic herniation of abdominal viscera into the thoracic cavity through large diaphragmatic lacerations (left more so than right). It is important to recognize that no other mechanism besides severe compression/crush reasonably explains this specific constellation of findings. Sequelae of such devastating injuries include few surprises: an "empty" heart (see below) and visceral pallor are the norm and represent an intravital response to the traumatic shock, an important observation that indicates that the person was not already dead in the roadway before being run over.

Other scenarios that may result in varying degrees of compressive injuries, even if not devastating crush-type, per se, involve, but are not limited to, persons who become pinned under or between one or more large structures (auto mechanics under cars, human stampedes at concerts, etc.), vertically or horizontally, persons within a vehicle during a collision causing compaction of the interior compartment, or with homicidal blunt force violence involving a severe beating by one or more assailants, where stomping with shoes and repeated punches and/or blows with objects/weapons are involved. Looking for patterned injuries consistent with weapons and/or the soles of shoes is particularly important in the latter scenario. An automobile mechanic who is pinned under a car that falls on him due to a faulty lift, for example, may have physical evidence of both crush injuries *and* traumatic asphyxia, the latter with evidence of continued circulation indicated by impairment of venous return from the head to the heart (i.e., usually characterized by marked facial plethora and a florid eruption of facial and conjunctival petechiae and confluent hemorrhages; see Chapter 11). As such, the mechanism of death is these fatalities may be more complex, but the proximate cause still remains blunt force trauma.

Crush injury patterns are especially important to recognize when no witnesses or surveillance videos are available. The body also clearly "tells a story" in a severe beating and stomping, with or without witnesses, the extent, multiplicity, and type of force used (considerable) usually being quite evident. "The body doesn't lie." These findings may also prove important to specifically communicate to law enforcement in "hit and run" fatalities, for example, where the driver, if ultimately located, states that they weren't aware of having struck a pedestrian. The veracity of this statement may later be called into question by authorities, since the body had been run over. For this reason, it is important to document the measurements of the largest (anteroposterior and side to side) dimensions of the torso, if the trunk was included in the compressive injuries, to further clarify and highlight what degree of resistance or "bump" would have been reasonably experienced by the vehicle occupants. Finally, in certifying these deaths, one may feel compelled to include "crush injuries" somewhere in the cause of death statement (or at least within the final diagnoses in the autopsy report), if they are particularly evident and clearly mechanistically important, with or without a component of traumatic asphyxia. A cause such as, "Multiple Blunt Force Injuries of Head and Torso, Including/With Crush Injuries" is an example of such a statement.

Craniocervical trauma

As alluded to above, due to its unique anatomy and tissue composition, blunt and acceleration–deceleration injuries of the head are different than blunt injury of other regions of the body. Head injuries are more complex, with characteristic features that allow the forensic pathologist to offer an opinion as to how an injury may or may not have occurred. Therefore, an understanding of skull fracture patterns, intracranial hemorrhages, cerebral contusion distribution, traumatic axonal injury (TAI), and the likelihood of disturbances of consciousness are important in order to assess and opine about the etiology and clinical behaviors.

Overall, mechanical injuries of the brain are governed by the amount of force applied to the head, the rate of energy transfer, the surface area involved, the anatomic location of the impact, the direction of forces, and the resting or moving status of the head. *Cerebrocortical contusions* start as multifocal punctate hemorrhages with foci of necrosis and injured axons. Large contusion hematomas can develop over a period of hours or days in those who are susceptible (i.e., often coagulopathic). Fracture, herniation, and gliding contusions, as discussed below, are unrelated to the head's movement, while coup and contrecoup contusions are governed by its resting or moving status.

In addition to the autopsy findings, a proper investigation of head injury deaths includes a review of the history and circumstances surrounding the injury, including pertinent medical and surgical treatments and findings, imaging studies (e.g., computed tomography [CT] scans), and toxicological studies. The autopsy examination should not focus exclusively on the brain, but also on other associated injuries, including injuries of the scalp and skull and the presence or absence of an "empty heart" and visceral pallor.[67] Pertinent negative findings should, of course, also be described. As just one example of an important clinicopathological tool, the **Glasgow Coma Scale (GCS)** offers an important, objective, and measurable data point that can be critical in assessing the veracity of functional fatal mechanism determinations, such as the so-called "fatal concussion." The documentation of the absence of a **lucid interval** (see below) can make the difference between the classification of a natural versus an unnatural manner of death.

Skull fractures

Due to its irregular, curved, and "enclosed" structural nature, a variety of skull fractures may occur and potentially provide information about "what happened." The dating of fractures is microscopically unreliable. The speed of fracture formation is faster than that of a speeding bullet because it propagates through a solid (~3000 feet/second).[68] This is demonstrated at the autopsy table with *Puppe's rule* involving skull fractures associated with entrance and exit wounds. In general, fractures form in the direction of force. Skull fractures can be described by morphology and/or location. For example, basilar skull fractures include: *hinge* (horizontal), *ring* (around the foramen magnum), and *longitudinal* (anteroposterior/parasagittal) types. A *diastatic fracture* involves a suture and in children, can be secondary to brain swelling. With linear fractures, *Puppe's rule* says that a later fracture will terminate at (i.e., not cross) an earlier fracture. *Pond fractures* are shallow depressed fractures, while *mosaic* fractures have a spider web-type pattern. *The two major morphological groups, however, are linear and depressed fractures.*

Linear fractures may be simple or complex; however, they are typically seen when there is an impact between the skull and a broad surface (e.g., falling and striking the head on pavement). *Linear fractures of the orbital plates* (the thinnest bones of the skull that overlie the orbital cavities) are common in the setting of blunt impact(s) of the head. Given their delicate nature, they fracture easily relative to other parts of the skull, and not necessarily as a result of forces directly applied to the forehead or eyes, per se, but rather, from considerable forces applied anywhere on the skull, forces that are then transmitted to the thin orbital plates to produce fracture(s). Importantly, these fractures are commonly associated with **"spectacle hemorrhages,"** externally visible soft tissue hemorrhage around the eyes resulting from the settling of blood that has passively seeped through the orbital fracture lines into the facial soft tissues (Fig. 10.11). When bilateral, this pattern of hemorrhage is evocative of eyeglasses (spectacles), with "clean," well-circumscribed margins, limited only by the muscular fascial planes around the eyes. This pattern of periorbital hemorrhage differs from that caused by direct contact injury, the latter described as *contusions*, caused by impact trauma to the eye area rather than passive blood settling. Contusions are more likely to have irregular margins that do not necessarily respect the fascial planes around the eyes. On occasion, there are considerable forces applied to the head, often localized to the frontal region, that cause more severe "blow-out" fractures of the orbital plates. These fractures differ from the simple orbital plate fractures described above in both their morphology and the degree of force necessary to produce them.

Depressed skull fractures differ from linear fractures in that they most typically result from *a focal application of force to the skull by a* **dense object** *having a relatively small surface area* (e.g., a hammer or brick, Fig. 10.12). Basilar skull fractures propagate in the direction of the force. *Longitudinal fractures* are seen with impacts to the front or back of the head. *Hinge (horizontal) fractures* (Fig. 10.13) result from substantial impacts to the sides of the head or to the chin (hyperextension) impacts. *Ring fractures* usually occur in scenarios such as descents from considerable heights, when the brunt of the landing is primarily born by the buttocks or the feet and the forces are transmitted upwards, causing the vertebral column to be driven into the base of the skull.[69] *Comminuted fractures*, sometimes with a

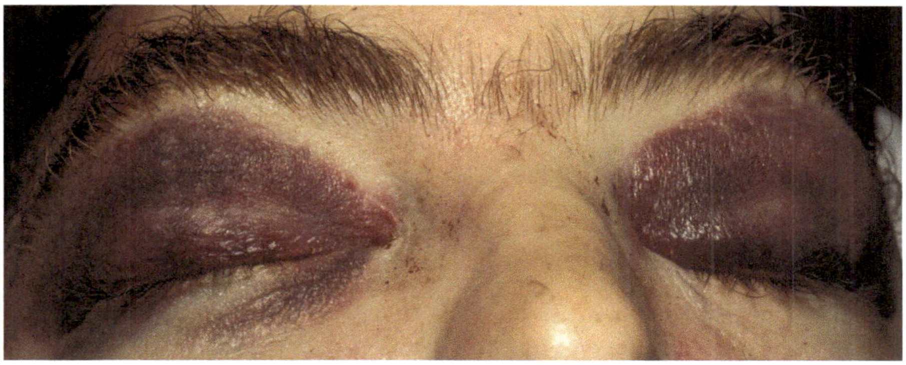

FIGURE 10.11 **Raccoon eye (spectacle) hemorrhages.** These are ecchymoses that are seen with orbital plate fractures and broken noses. Note that the hemorrhage respects the fascial planes because it is arising from the internal injury. If they had been caused by impacts to the periorbital region, one would expect to see irregular contusions involving the bony prominences.

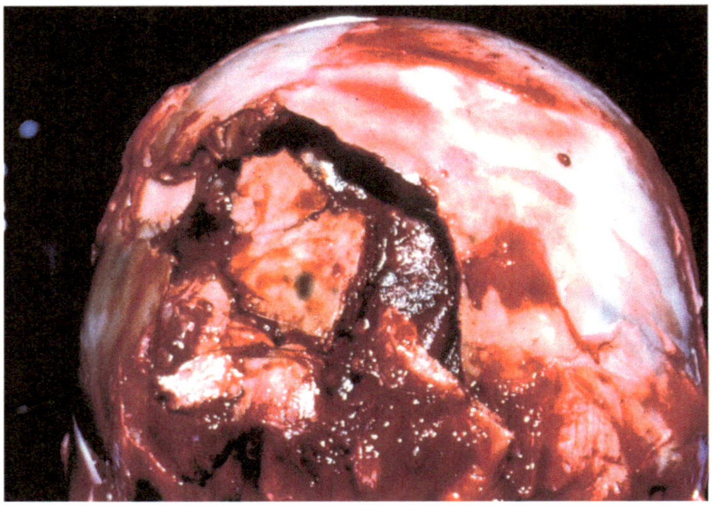

FIGURE 10.12 Depressed skull fracture due to hatchet blow.

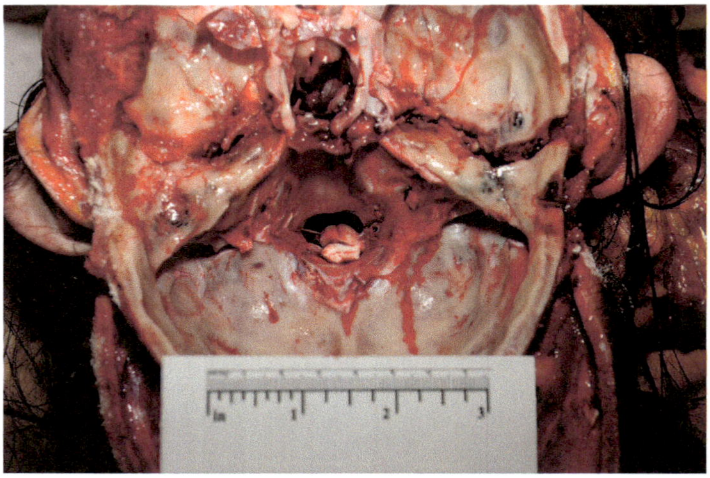

FIGURE 10.13 Hinge fracture of the skull.

complete shattering of the vertex and basilar skull, are largely observed in instances of high velocity impacts involving the head (descents from substantial heights, high velocity motor vehicular collisions, etc.).

Epidural, subdural, subarachnoid, and intracerebral hemorrhages

Hospital clinicians often will report any hemorrhage in the head as an "intracranial" hemorrhage, a parenchymal hemorrhage being the same as a subdural hemorrhage (SDH), which assists the forensic pathologist almost in no way. When faced with this description, one must ask what type of "intracranial hemorrhage" it is. Clinicians (especially residents) may not understand the forensic and death certification importance of the specific type and location of the hemorrhage, extra-axial or intraparenchymal. Regardless of their answer, obtaining a copy of the CT scan report is good practice.

Epidural hemorrhages (EDHs) are due to trauma. As such, deaths due to an EDH must be reported to the medical examiner/coroner (ME/C). Typically, they are caused by a fracture-induced laceration of an artery adjacent to the skull and dura (e.g., a lacerated middle meningeal artery due to a temporal bone fracture). The hemorrhage, which is under arterial pressure, can strip the dura from the inner table of the skull, resulting in a hematoma between the dura and the skull, typically overlying one of the hemispheres. This expanding hematoma raises intracranial pressure (ICP) and often shifts the brain, rapidly resulting in herniation and death if it is not promptly evacuated. Infants and the elderly are more protected from this hematoma because their dura is more firmly attached; prosectors quickly realize this when attempting to strip the dura in these decedents. Rarely, one may see an EDH in the occipital skull due to fracture and laceration of the transverse and occipital sinuses. EDHs are injuries due to contact/impact forces that, in turn, cause the offending fracture, rather than the diffuse shearing forces that are more typical of SDHs.

The majority of **subdural hemorrhages** are also trauma-related (Fig. 10.14). They are often associated with an impact, but an impact per se is neither requisite for their formation nor necessarily adjacent to the location of the SDH. Unlike EDHs, which are contact injuries, with the hemorrhage immediately subjacent to the site of the impact and fracture injury, SDHs may occur without contact or fracture. For example, one may sustain an impact to the left side of the head that results in a SDH over the right hemisphere, the left hemisphere, or both. This is because the hemorrhage is a product of the shearing forces caused by the jolted brain moving relative to the fixed skull, with consequent tearing of the *bridging veins*. There are well-documented subdural hematomas that are caused purely from acceleration—deceleration mechanisms, without an impact per se. These include subdural hematomas caused by roller coaster rides.[70–72]

The neurologic effects of SDHs depend upon several factors. The first is whether there is concomitant diffuse axonal injury (DAI), as determined clinically by a decreased GCS and/or a neuroimaging finding of microhemorrhages in susceptible sites. If there is DAI, neurosurgeons may decline attempts to decompress the hematoma (surgical evacuation) because it will not likely change the prognosis. If the neurologic dysfunction is purely related to the space-occupying hematoma, then prompt evacuation may be lifesaving, that is, if secondary brain injury (herniation) has not already occurred. *The deleterious effects of extra-axial hemorrhage depend upon the location, amount, and rate of its accumulation.* A thin film of SDH has no pathological effect in and of itself. Rather, it is a sign of the forces that have traveled through the head, so may have importance only in that regard. As the volume of blood increases,

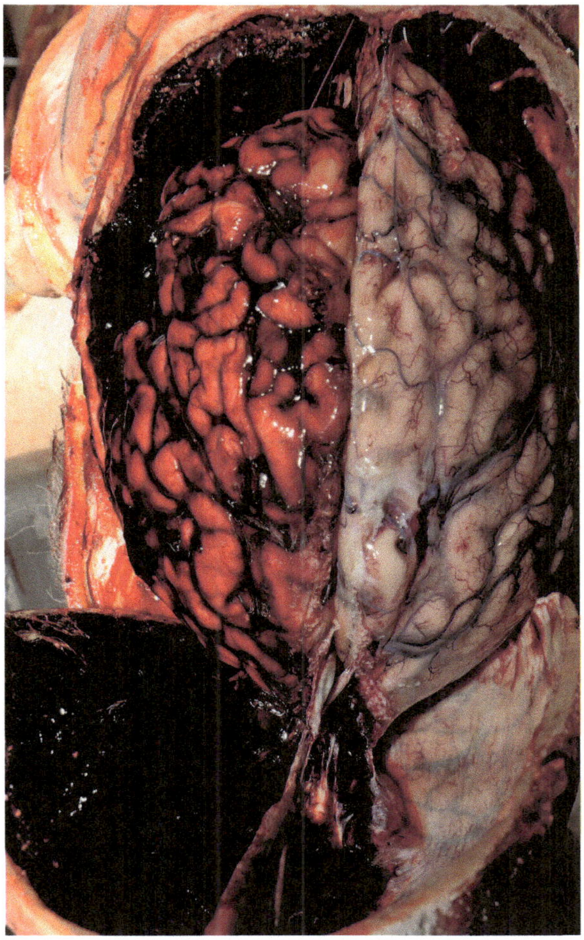

FIGURE 10.14 Subdural hematoma.

it will begin to raise the ICP and displace the brain (herniation). These *secondary effects* depend upon the volume of the blood and the speed with which the hemorrhage accrues. The brain can better accommodate a slowly accumulating volume of blood, resulting in a larger volume of blood needed to tip the scales into neurological compromise. The faster the blood accumulates, however, the less time the brain has to "accommodate" to it. This rate of accumulation in part depends upon the type of vessel that is injured. Most SDHs result from lacerated veins, which operate at lower pressure, while most epidural hematomas are arterial in origin. One expects arterial hemorrhages to accumulate more quickly than the SDHs, but one must remember that the arterial blood also must dissect the dura away from the skull, while the subdural bleeding progresses in an already existent potential space with lower resistance.

For both EDHs and SDHs, however, a **lucid interval** is common. This is the period of consciousness and activity that continues after having sustained the putative injury and before symptoms of *increased ICP* ensue. The latter include such signs as *headache, slurred speech, lethargy, waxing and waning consciousness, and death.* This period is highly variable and depends upon the severity of the vascular injury, the activities of the decedent, and the health of their coagulation system, but can range from hours to days. Because *further* trauma can be sustained during this interval (e.g., an intoxicated individual is punched, falls, sustains head trauma, then gets up and continues activities, falling one or more additional times before becoming unconscious and dying), the investigation may be complex. The more detail the police can gain from witnesses, roommates, family, and video surveillance between the initial injury and death, the better. The wider the interval, however, the more difficult the link may be to make. **Acute on chronic injury** can muddy the waters even further.

Rarely, one may encounter nontraumatic SDHs, or so-called **"spontaneous" SDHs**. For example, typical hypertensive intracerebral hemorrhages (strokes) can dissect through the cerebral parenchyma and arachnoid membrane (a "burst lobe") to produce a secondary SDH.[73–75] In a study by Freytag of 393 fatal hypertensive intracerebral hematomas, 75% broke through the ventricular walls and 6% perforated the arachnoid layer and extended into the subdural space.[74] Anticoagulation therapy, bleeding/clotting disorders, end-stage kidney failure requiring hemodialysis, malignancies, and vascular malformations have also been described as nontraumatic sources. Since most of these may render a person more susceptible to bleeding of any type, some of these are likely to be born out of occult injury that was so minimal that it was not recognized as the causative mechanism.[76–87] Unlike most fatal SDHs, deaths due to these nontraumatic hemorrhages would be certified as natural.[73,78–79,86,88] Because a SDH may occur with little or no evidence of external injury, an autopsy is needed for the diagnosis.[89] Not uncommonly, impact sites, if existent, are only identified with reflection of the scalp. SDHs in infants are discussed in the pediatric chapter (Chapter 6).

Subarachnoid hemorrhage may be seen with both traumatic and natural processes. A classic example of the latter results from a ruptured cerebral artery aneurysm. The putative "bleeding" vessel or vessels that cause SDH and EDH usually are not identified at autopsy, but don't necessarily need to be located to achieve etiologic specificity (based on case context and scalp and/or skull trauma identified at autopsy). With a nontraumatic subarachnoid hemorrhage, on the other hand, the bleeding vessels *are* often found because they usually involve a grossly identifiable structural vascular abnormality or malformation (e.g., aneurysm, arteriovenous malformation); in fact, locating such a pathologic vessel is necessary for the identification of an etiologically specific cause. Cerebral artery aneurysms typically involve the vessels of the circle of Willis and its proximal branches, a saccular ("berry") aneurysm being the most common. In some instances, however, the aneurysm may be located in a more distal segment, giving the appearance on imaging of an "intracerebral" hemorrhage. But, more typically, particularly in an ME/C cohort of decedents, one sees subarachnoid hemorrhage/hematoma at the base of the brain. The amount of hemorrhage is relatively small when compared to an EDH or SDH, but is no less lethal, which is a reflection of its mechanism. Whereas a SDH or EDH typically requires ICP increases and/or hemispheric shifts to be fatal, the location of the aneurysmal subarachnoid hemorrhage at the base of the brain does not. Although the hemorrhage is much smaller, it precipitously collects in a confined vital area where the accommodation of this volume is much less likely than it would be over a cerebral hemisphere. The base of the brain contains key central structures (e.g., cardiorespiratory centers of the brainstem). A sudden expansion of hemorrhage in this area, under high pressure, is enough to disturb those vital structures in a sort of "tamponade" of the brainstem, typically presenting as a sudden death. In addition, subarachnoid blood from any etiology may result in vasospasm, further exacerbating the impairment of central cardiorespiratory centers from cerebral ischemia.

During the autopsy, it is best to remove the fresh subarachnoid blood at the base of the brain (by using hydrogen peroxide) prior to formalin fixation (before it hardens), the latter making the subsequent vascular dissection more difficult. In some instances, a distinct aneurysm is not found. This underscores why it is so important at the time of autopsy to carefully look at the cut common carotid artery stubs in the base of the skull to make sure the aneurysm is not being not left behind in the head.

The other consideration with this basilar pattern of subarachnoid hemorrhage is a *vertebral artery laceration,* which may occur with head impacts, particularly side impacts that create rapid torsion and rotation in accidental or homicidal settings.[71,90–108] This diagnosis requires a more involved dissection of the posterior neck and the extracranial vertebral arteries, as well as a careful examination of the scalp, particularly behind the ears. This is another reason to try to find the aneurysm at the autopsy table. If no aneurysm is found two weeks later during dissection of the formalin-fixed brain, it is too late to dissect the vertebral arteries at that point.[95,97–102,104–108] Rarely, chiropractic manipulations can cause vertebral artery injuries.[101] Such deaths would be certified as therapeutic complications, where this manner is available. Beyond that from vertebral artery injuries, patchy or more confluent hemispheric or cerebellar subarachnoid hemorrhage is a common sequela of blunt force trauma that results from direct impact(s) to the head and/or acceleration–deceleration forces, much in the same way that traumatic SDHs may be produced. These hemorrhages are not typically space-occupying; rather, they represent another marker of the forces that have passed through the head.

Intracerebral hemorrhages may result from trauma or disease. Cortical contusions, parenchymal contusion hematomas, petechiae from fat emboli, and vascular traumatic injury constitute the injury-related etiologies. Nontraumatic intracerebral hemorrhages are caused by elevated blood pressure (e.g., hypertension, acute cocaine intoxication), amyloid angiopathy, vascular malformations, embolic strokes, some ruptured aneurysms, certain infections, and malignancies.[74,75,109–114] The intoxication-related and traumatic causes are of the greatest forensic interest, given their associated unnatural manners.

Finally, while it was briefly addressed above in reference to scalp lacerations, it is important to specifically address **the lethality of being punched in the face/head by a closed fist**. Not uncommonly, the forensic pathologist is faced with the investigation of a death from head trauma involving a substantial subdural hematoma, and accompanied by a story describing a preceding "fight" that reportedly involved one or more facial punches. Several points are important to highlight in the approach to a death accompanied by this narrative.

First, it is important to understand that few punches to the face, unless multiple/excessive, cause death per se. More commonly, one or more punches often send the victim *to the floor or ground*, causing a secondary and considerable impact with a resistant surface, after which there is either immediate unconsciousness, consistent with concussive injury, or a lucid interval.[115–116] Either way, if an acute SDH is subsequently documented in the hospital and/or at autopsy, it is more likely due to the secondary impact with the ground rather than the one punch to the face (the so-called "one punch homicide"). It is no less a homicide, but the mechanics must be understood and made clear to authorities. In these investigations, it is important, of course, to document the extent and pattern of head trauma at autopsy and/or through the medical record, if a substantial survival interval has transpired. Documenting all facial trauma, soft tissue and skeletal, is important. Often, it is self-evident, but when difficult to discern externally, a special dissection of the facial soft issues and skeleton may be necessary to help establish pertinent negative or positive physical evidence (see Chapter 5).

It is also important to point out, however, that the absence of facial injury does not necessarily exclude the possibility of a physical altercation having taken place and being responsible for bringing the decedent to the ground. Some punches barely graze the face. Even thrown, but missed punches, which the decedent manages to dodge, may throw him off balance enough for him to consequently fall to the ground. The physical altercation and the criminal menace would still likely lead to a classification of homicide in such a scenario (see discussions in Chapters 3 and 15). The decedent may even be intoxicated, further throwing him off balance during a physical altercation, but this too would also not necessarily exculpate the encounter (a documented physical altercation) in informing the manner. "The Constitution of the United States does not guarantee a perpetrator the right to a healthy (or unintoxicated) victim." Other important physical evidence would include one or more impacts to the scalp, one of which would likely represent the secondary impact caused by the altercation-induced fall. Both coup and contrecoup cerebral contusions may be observed and must be interpreted through the lens of what may have happened. Detailed witness statements and video surveillance may prove particularly helpful in that regard. The preceding discussion does not apply, of course, to more excessive facial/head trauma inflicted during *a violent beating*, with or without stomping, kicking, and/or punching, which often continues or predominantly occurs on the ground, and that may involve one or more assailants. In this scenario, no secondary fall is necessary to explain the severe functional and structurally demonstrable head trauma and its sequela, sometimes even including crush injuries of the skull or face, and death (see above).

Cerebral cortical contusions

In 1993, Charles Hirsch and Ross Zumwalt taught a short course at the United States and Canadian Academy of Pathology annual meeting. It was entitled *Current Concepts in Forensic Pathology* and addressed the certification of death and a discussion of blunt injuries of the head, including cerebral contusions.[117]

Cerebral cortical contusions grossly appear as punctate and streak-like hemorrhages that are distributed along the crests of the gyri, but may involve the depths of the sulci when numerous (Fig. 10.15). They typically have a cone/wedge-shaped pattern with the apex pointing toward the white matter and the base toward the gyral crest. They may be distinguished microscopically from ischemic lesions by examination of the cortical layers. The subpial layer of the cortex is usually spared with ischemic etiologies, while it is destroyed in contusions. Grossly, hemorrhagic infarcts evenly involve the cortex over the crowns of the gyri and the sulci, while a contusion tends to spare the sulcal cortex. In some patients, large contusion hematomas may develop from these punctate hemorrhages. If a person dies instantly or sustains concomitant injuries of the heart or aorta (e.g., laceration of aorta), cerebral contusions may be inconspicuous and also mechanistically noncontributory, in and of themselves.

Hirsch classified brain contusions into two genera, with three species in each genus. The first genus involves contusions that are independent of the resting or moving status of the head, and the second genus, those that are dependent. *Contusions that are independent of the resting or moving status of the head include:* **fracture contusions, herniation contusions, and gliding contusions.** *Fracture contusions* are caused by an overlying skull fracture. These are distinct from coup contusions (see below). *Herniation contusions* are caused by the transient displacement of the brain against a fixed anatomic aperture. Three common examples include those that result from uncal herniation, with the brain pressing against the free edge of the tentorium, from a herniation of the cingulate gyrus under the falx, and from a cerebellar tonsillar herniation down into the foramen magnum. *Gliding contusions* are linear/oval hemorrhages in the white matter (typically the more frontal gyri, Fig. 10.16), which follow venous drainage and extend toward the lateral ventricles. These are produced with anterior–posterior gliding movements of the brain, which cause venous bleeding. Similar hemorrhages may occur with traumatic DAI.

The second genus of contusions depends upon the resting or moving status of the head. These include **coup and contrecoup contusions**. When a resting, but movable head is struck with a blunt instrument, contusions may form beneath the point of impact. When this occurs without an overlying fracture, it is designated as a *coup contusion*. *Contrecoup contusions* occur when the moving head impacts against a hard, unyielding surface, such as pavement. In these instances, there is no (or only a minor) cortical contusion immediately underlying the site of impact, but rather a conspicuous injury roughly 180° opposite to it. Grossly, one would expect to find subscalpular hemorrhage at the site of impact, but the cerebral cortical contusion would be on the opposite cortex. Contrecoup contusions typically occur in the frontal and temporal lobes as a result of an occipital skull impact, but rarely involve the occipital lobe. Variations in the contours of the skull surface likely play a role in the mechanism of formation of these contusions.

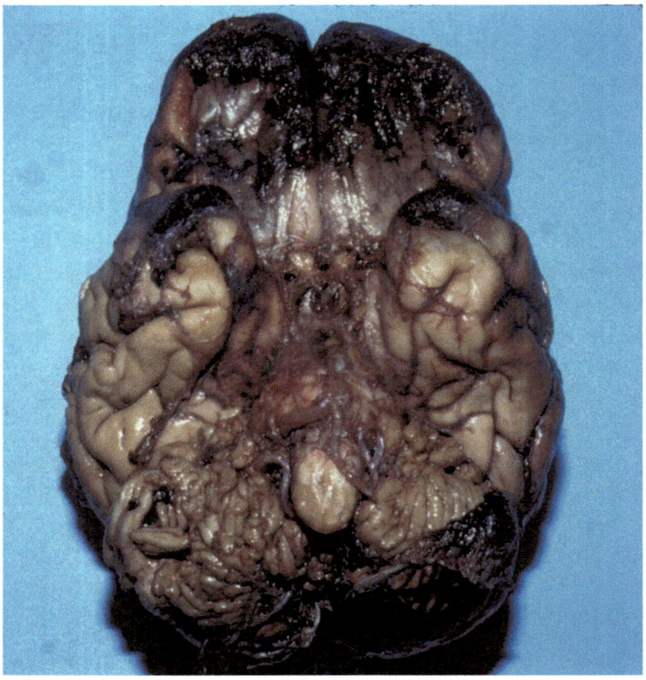

FIGURE 10.15 **Frontal and temporal cerebral cortical contusions.** Contrecoup type, based on an occipital impact site.

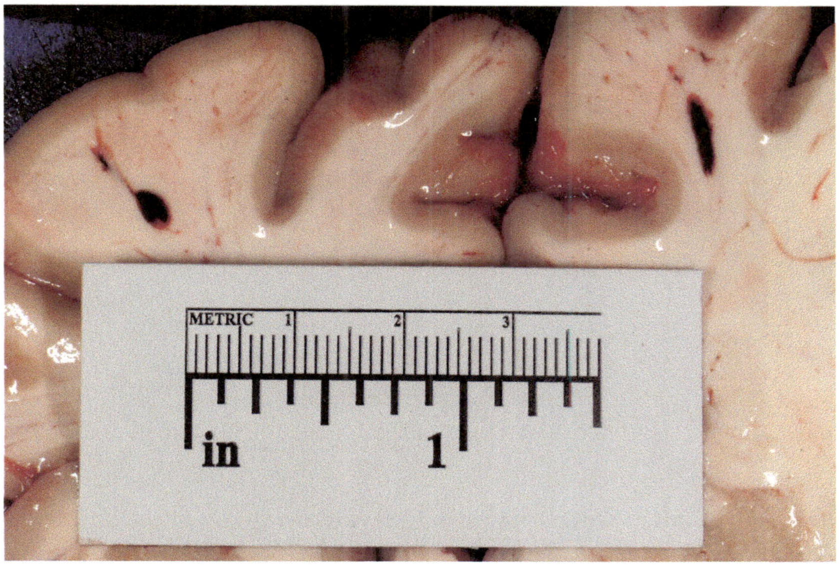

FIGURE 10.16 **Gliding contusions.** Hemorrhagic slits in frontal white matter.

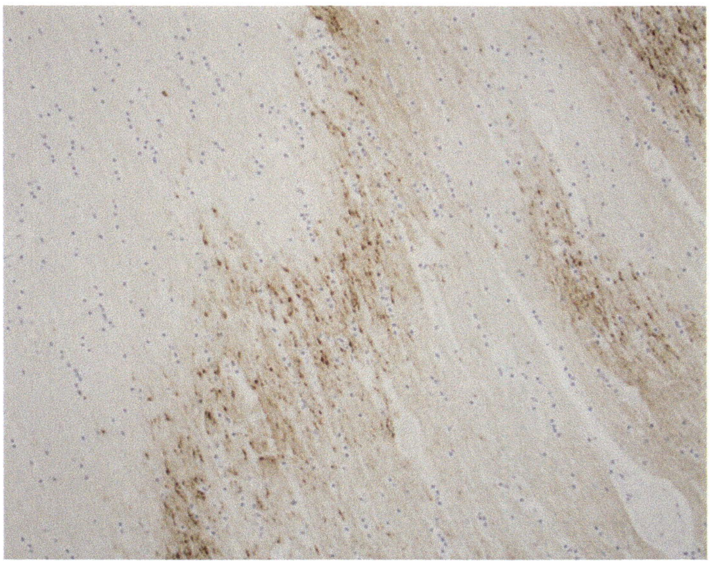

FIGURE 10.17 **Vascular or hypoxic–ischemic pattern in young child on a ventilator.** Amyloid Precursor Protein (APP) immunohistochemical stain demonstrating a "curtain-like" bundled pattern, 100x.

One additional injury scenario involves the immobile head resting on an unyielding surface that is struck with a blunt object. A variety of contusion patterns may occur as a result of this specific scenario. In addition, shattering the skull also may occur primarily in three notable circumstances: descent from a substantial height, crush injury caused by severe compressive forces (such as those produced by being run over by a motor vehicle), and impacts with a dense blunt instrument that causes extensive comminuted fractures on the vault and base of the skull. The fractures in such scenarios may help dissipate the forces applied to the head, and therefore the brains may have cerebral contusions that are either inconspicuous or entirely absent under these extreme circumstances.

Special considerations for infant head injuries are discussed in the pediatric chapter (Chapter 6).

Traumatic axonal injury

Diffuse brain injury involves a spectrum from "mild" concussion to marked DAI, with a permanent comatose state or death (the latter, the so-called "fatal concussion").[118–120] The autopsy manifestations can be subtle and may require a survival interval for them to fully develop. If a person dies very quickly, the brain may appear normal.

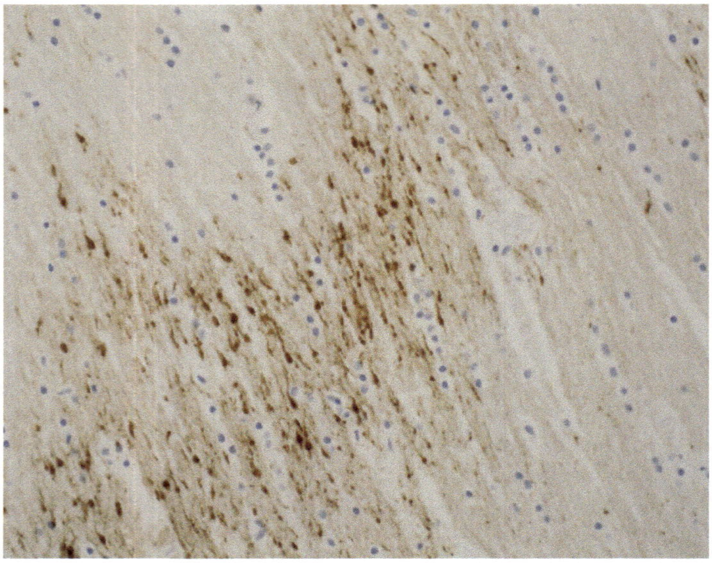

FIGURE 10.18 **Vascular or hypoxic–ischemic pattern in young child on ventilator.** Amyloid Precursor Protein (APP) immunohistochemical stain demonstrating a "curtain-like" bundled pattern, 200x.

FIGURE 10.19 **Young child with abusive head injury.** Traumatic injury patterns in the corpus callosum in a young child dying within hours (i.e., without ischemia) from abusive head trauma with the Amyloid Precursor Protein (APP) immunohistochemical stain revealing scattered single-beaded profiles and axonal bulbs in the corpus callosum, 100x.

FIGURE 10.20 **Young child with abusive head injury.** Traumatic injury patterns in the corpus callosum in a young child dying within hours (i.e., without ischemia) from abusive head trauma with the Amyloid Precursor Protein (APP) immunohistochemical stain revealing scattered single-beaded profiles and axonal bulbs in the corpus callosum, 400x.

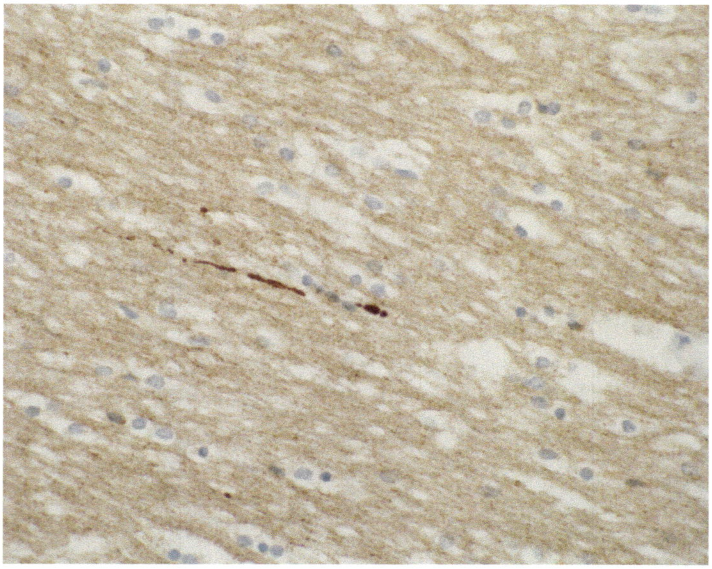

The clinical history (including the GCS), in combination with the autopsy findings, will best allow one to make a diagnosis of a fatal concussion in this scenario. In other instances, there may be a SDH or EDH in conjunction with the concussive injury. Because of the potential for DAI to be caused by the same injury that caused a subdural hematoma, one should not assume that a lucid interval accompanies all subdural or epidural hematomas.

A survival interval is needed to allow for the microscopic diagnosis of DAI with the use of immunohistochemical stains for amyloid precursor protein (APP). The accumulation of APP at axonal swellings/bulbs allows it to be detected with immunohistochemical stains, which is the gold standard for an axonal injury diagnosis.[121–130] APP positivity has been demonstrated in injured axons as early as 35–45 min following head trauma.[128] There are certain specific regions of interest (i.e., the corpus callosum, internal capsule, brainstem, and cerebellar peduncles) and certain patterns of staining that may allow one to identify the etiology of the axonal injury with greater specificity.[122–124] APP is a sensitive marker for axonal injury; however, it is not specific for mechanical trauma (TAI), as it also accumulates with hypoxic–ischemic (vascular axonal injury, VAI) and metabolic disorders. Since secondary hypoxic–ischemic injury often accompanies traumatic injury, it can be a challenge to distinguish between the two. That said, the APP staining pattern, along with the clinical history, may help in determining the etiology.[125,126,129–130] The microscopic distribution of APP-positive axons in TAI and VAI, in conjunction with the examination of hematoxylin and eosin (H&E)-stained sections, allows the cause of the axonal pathology to be determined in most instances.[125] With TAI, one sees axonal changes within distinct tracts, while with ischemia, the changes cross between different components (i.e., vascular territories) of the brain. For example, with ischemia, the axonal injury involves crossing and descending tracts, in contrast to TAI, which follows white matter tracts. One must correlate these findings with the standard H&E stain appearance in the same areas (Figs. 10.17–10.20).

Secondary processes

"Brain swelling" may be *congestive* (increased intravascular blood volume) or *edematous*. Cerebral edema results from an absolute increase in the water content of brain tissue from vascular seepage (edema). *There are four processes that may influence the evolution of a head injury:*

1. Mass effect lesions (usually hematomas: size, speed, and location)
2. Increased ICP (Need a 70 mm hg difference between blood pressure and ICP to perfuse the brain)
3. Cerebral ischemia from decreased perfusion: either local (due to increased ICP) or systemic (due to systemic hypotension)
4. Effects of extracranial complications.

Neck injury

The most important tip for detecting neck injuries is to look for them! This may require a posterior neck examination (see Chapter 5). A posterior neck dissection may be the only way to determine the cause of death in some instances. This is an area in which a postmortem CT scan maybe a very useful adjunct (see Chapter 18).[101,131–136] A posterior neck dissection is performed in suspected neck compression deaths, as well as in blunt injury deaths (e.g., motor vehicle collisions) in which there is not a clear mechanism of death detected during the standard autopsy examination. In these instances, one should consider a posterior neck injury and/or a traumatic asphyxia mechanism (see "Transportation Injury" below). This dissection also might be considered in a blunt injury assault death in which a clear structural mechanism of death is not discovered during autopsy.

Forensic pathologists who routinely perform posterior neck dissections become highly proficient and efficient.[137–139] One should look for soft tissue hemorrhage, ligamentous injury, vertebral artery injury, and vertebral fracture.[90,93,94,96,101,105,106,138–150] There may be hemorrhage around the spinal cord, as well as softening and/or contusive injury of the cord itself. Hospital trauma staff are often appreciative of learning of this diagnosis, and it can serve as an excellent teaching point for junior clinicians. Neurogenic shock manifests with hypotension and bradycardia from loss of sympathetic tone and, in addition to significant head trauma, can also result from cervical and upper thoracic vertebral injuries (Fig. 10.21). Shock without a hypovolemic mechanism is often the clinical manifestation (spinal shock) and helps the clinician answer the question as to why the patient did not respond to therapy.[94,96,98,101,105,106,147,149,151–152] Injury of vertebral arteries (discussed above) is a rare cause of subarachnoid hemorrhage at the base of the brain.[94,96,98,101,105–106,147,149]

As a final point, it is worth noting that a commonly observed and important intravital marker of fatal neurotrauma (head and/or neck) is the **"empty heart" sign**. Hirsch and Zumwalt first coined this phrase when they described this vascular phenomenon in 1986, positing that there is an immediate shift of the core intravascular

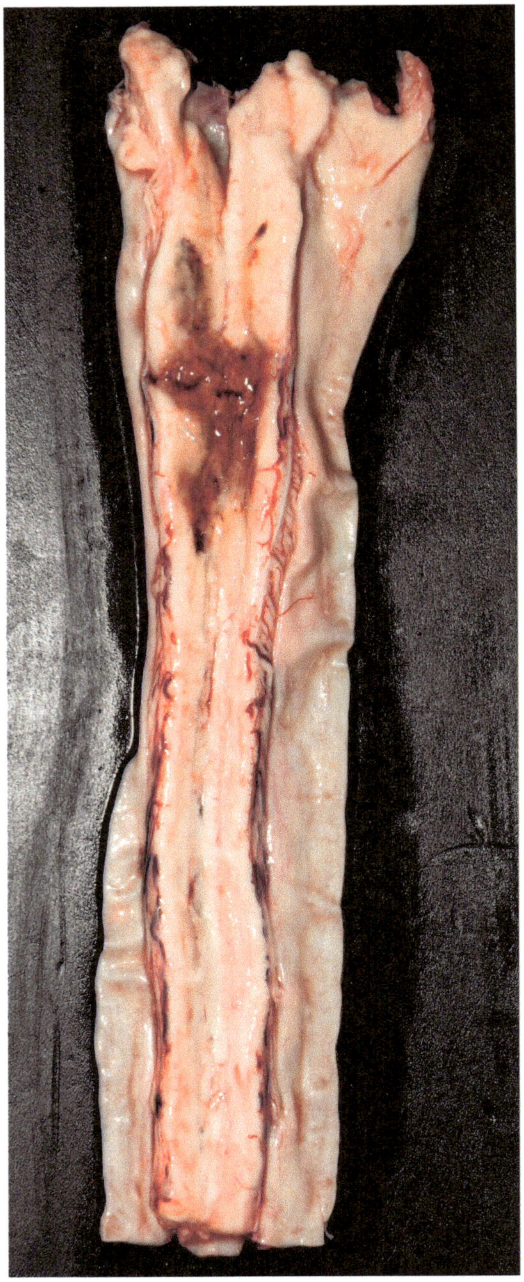

FIGURE 10.21 Contusion of the spinal cord.

blood volume to the periphery (likely to the musculature and bone marrow) in response to a sudden neurogenic cardiovascular collapse (loss of sympathetic tone) and death related to significant neurotrauma.[1] As such, and not surprisingly, it is commonly associated with moderate to marked visceral pallor, similar to that observed in shock due to other etiologies. The benefit of such a sign, when it is observed, rests not only in the fact that it focuses attention to look for subtle injuries of the brainstem and spinal cord, but also, given that it is an intravital phenomenon, in that it indicates that the individual was alive (had circulation, even if not consciousness) at the time of the injury. To wit, if the cervical vertebral column is fractured during postmortem transport of the body, the empty heart sign would *not* be observed at autopsy. The empty heart also can be a useful aid to exclude heart disease as a lethal mechanism in a person with heart disease who also sustains a head impact. An empty heart would point to a neurogenic rather than cardiogenic mechanism of death.[1]

Approach to unwitnessed fatal blunt head and torso trauma

Even if fatal head trauma is well documented in the hospital, until the circumstances are ascertained, the default action is to perform an autopsy and request police involvement when the injury is not known to be witnessed. Some hospitals may describe these as "spontaneous" SDHs, but that represents an editorial, not necessarily the news. Be wary of "spontaneous" subdural hematomas—they should not be assumed to be nontraumatic. Although they may rarely occur (see above), they can also be due to undocumented/unwitnessed trauma with little to no external evidence of injury.

A notable and common exception to the above approach involves deaths following benign falls at home, usually of elderly individuals (assuming the story comports and the family can provide what is needed). Unless otherwise indicated, these circumstances rarely require a police investigation or involvement. Not uncommonly, there are also objections to autopsy in such cases. This does not typically pose an issue since these persons have usually undergone hospital CT scanning and there is also a clear history. The more challenging cases involve a benign fall with a subsequent death, before which imaging was not performed. These may occur in patients in hospice or nursing homes or in deaths that occur following a "lucid interval." The decision to autopsy or not depends upon the clinical history and discretion or concerns of the pathologist.

If someone is brought in "from the street" after having sustained fatal head trauma, such as a large subdural hematoma, seen on hospital imaging studies or documented at autopsy, it is prudent to check the existing police report (if any exist) and/or the 911 call report for witnesses to the injury or video surveillance; the circumstances may be reasonably explained by a significant impact to the head during a witnessed trip and fall. If not reasonably explained, however, the police should investigate.

Depending on the size of the jurisdiction, in order for a specific law enforcement agency or precinct to be assigned the case and begin investigating, the address of the ambulance pick up location may need to be known (the police investigation is not necessarily assigned based on the hospital location). It also is important for the police to understand the potential for a lucid interval after a fatal (significantly space-occupying) subdural hematoma is sustained, thus widening the window of investigation (hours to days or more). If, on the other hand, an individual is "found down" with a cervical vertebral fracture, for example, one that is incompatible with continued movement, the window of investigation is narrowed significantly (i.e., to the location of the occurrence), making any witness statements extremely helpful, if not conclusive. Put another way, an individual with a fatal subdural hematoma may have a terminal "fall"/collapse as a consequence of an expanding hematoma that was sustained hours to days earlier, whereas, no such lucid interval or continued movement would be expected to exist for injuries such as devastating skull fractures, DAI, or cervical vertebral injury with spinal cord trauma.

Finally, it is important to recognize that just because a decedent is intoxicated, one should not assume that a head injury is due to an accidental fall. Intoxication constitutes one of many possible explanations, but not necessarily *the* explanation. Intoxicated people can also be pushed or otherwise menaced, without much, if any, externally discernible scalp injury observed at autopsy. As Dr. Hirsch used to say, "We are not in the business of making assumptions." In such a scenario, a full investigation, with canvassing of potential witnesses, interviews with family, roommates, etc., about the decedent's activities, as well as potential signs and symptoms of head trauma (headache, confusion, lethargy, new onset slurred speech, etc.) of the previous hours to days is in order. Remember that an "ear witness" to an injury occurrence can sometimes be just as valuable as an eyewitness (i.e., a neighbor hears a loud tumble and immediately comes running out of their apartment to find the decedent alone and disoriented and no one running off with their wallet). If, however, despite their best efforts, the police fail to identify ear or eyewitnesses (or video evidence) to further clarify the circumstances of the injury, an undetermined manner is often the most appropriate classification to choose. "Absence of evidence is not evidence of absence" and "A negative is never proof" are two aphorisms that would aptly apply to a police investigation that turns up nothing, but fails to exclude a crime, per se. Arbitrarily certifying the death as an accident would result in the police "closing" the case and the forensic pathologist being on the hot seat when, three months hence, a witness comes forward stating that they saw the decedent pushed down some stairs.

In some instances, a person with such a head injury may survive for minutes or hours and be able to provide a "story" of what happened during this lucid interval. However, this needs to be evaluated carefully. One must consider the reliability of a person's recollection after having sustained a head injury that is of a severity that it ultimately causes death. Add alcohol intoxication to the mix and the reliability of the patients' recollection becomes even more concerning. This is not to say that the decedent's statements, made under this set of circumstances, should be dismissed out of hand, but applying common sense and carefully viewing the statements in the context of the autopsy findings is advisable; remain objective, even if somewhat circumspect.

Somewhat analogous to the lucid interval that often characterizes certain kinds of head trauma (SDHs and EDHs), **fatal peritoneal hemorrhage resulting from lacerations of the spleen or liver** may also take considerable time to manifest as a cardiac arrest due to hypovolemic shock, depending on the size and vascular severity of the injury and the condition of the coagulation system of the victim. As such, these injuries may also be associated with the equivalent of a "lucid interval," a time period during which the injury is occult while the decedent continues with activities. **Delayed splenic "rupture,"** in particular, is typically characterized by a substantial passage of time between the injury occurrence and its lethal outcome. The typical sequence of events is one in which the bleeding associated with the splenic laceration is contained under its capsule for an unspecified amount of time, effectively tamponading the torn vessels for up to days or more, delaying an ultimate exsanguination caused by a fatal hemoperitoneum following a subsequent capsular rupture, sometimes much more distant from the inciting event.[4] As a result, and similar to deaths resulting from SDHs, the investigative window may need to be broader in time and space, extending beyond the moment and location of death (see also Chapter 15 for a more detailed discussion of "Approach to Hemorrhage of Unclear Origin").

Finally, as stated earlier in this chapter, **"intracranial hemorrhage"** is a nonspecific and unhelpful phrase used by many clinicians when reporting deaths to an ME/C office. It could mean an intraparenchymal (nontraumatic) hemorrhagic stroke/cerebrovascular event (e.g., a basal ganglia/hypertensive or even an embolic one), as much as it could mean a traumatic extra-axial subdural/epidural collection that is space occupying *and* fatal, or even a large, compressive intraparenchymal contusion hematoma. That said, when someone, especially someone *elderly*, has fallen, is subsequently imaged, and is reported to have an "intracranial" hemorrhage, always seek clarification as to the location, type, and severity of the hemorrhage within the head before assuming that it is both traumatic *and* fatal (i.e., seeking imaging reports or a credible clinical source is usually clarifying). Put differently, was the fall the cause or the consequence of the hemorrhage? With a bit more investigative probing, it will likely be discovered that many of these fatalities are actually natural deaths (cerebrovascular events) and need not be made ME/C cases.

It's also important to consider that one may have both a hypertensive hemorrhage that causes a fall, but then consequently sustain substantial injuries from that fall, which also contribute to the mechanism of death (skull fractures, SDH), in addition to a sizable (space-occupying) hemorrhage from the stroke. If imaging and/or investigative circumstances cannot sort this out, an autopsy is likely in order. Finally, when a hemorrhage seems out of proportion to the circumstances and/or continues to expand over time in the hospital, it might also be at least worth considering a contribution from anticoagulation therapy, and inquiring as is seen fit.

Sharp injury

Sharp injuries are generally subclassified into stab or incised wounds (the latter also known as cuts, or slashes). The primary difference between the two is that stabs create wounds that are deeper in the body than they are long on the skin, while cuts are typically longer on the skin than they are deep in the body (e.g., surgical incision). The majority of stab and incised wounds are caused by knives, but other objects (e.g., screw drivers, ice picks, scissors, etc.) can produce similar penetrating injuries (Figs. 10.22–10.24). Razor blades, box-cutters, and broken glass are common cutting instruments. By definition, stab wounds are produced by a tip first, forceful and volitional thrusting of the sharp weapon into the body, whereas incised wounds are caused by the long cutting surface of the blade swiping across the body, rather than being thrust into it. As such, the force necessary for the former is greater than that of the latter. The tough human skin will easily deflect the tiny tip of a knife or other sharp weapon, barely causing a superficial flesh wound, but for the forceful thrusting action (see below).[5–8] That said, when asked how much force is necessary to produce a stab wound, an accurate answer would be "considerable," with a qualifier mentioning the volition and the increased force necessary as tissue density increases (e.g., cartilage, bone). By contrast, while incised wounds are often inflicted intentionally, they are also commonly sustained inadvertently during a violent struggle involving a knife or other sharp object, during which there are attempts at deflection, for example.

The manner of death in sharp injury deaths is usually fairly easy to discern, few being certified with an undetermined manner. The number of sharp injuries, per se, is not necessarily a reliable criterion in distinguishing homicides from suicides, in and of itself, whereas the specific *distribution* of the wounds on the body typically offers more salient information in the determination of self-inflicted injury from that inflicted at the hand of another (Figs. 10.25 and 10.26).[6,9–26] While there is some general overlap in body region involvement between homicidal and suicidal sharp force trauma (i.e., left chest, upper extremities, and neck can be seen in both), the *quality of spread*

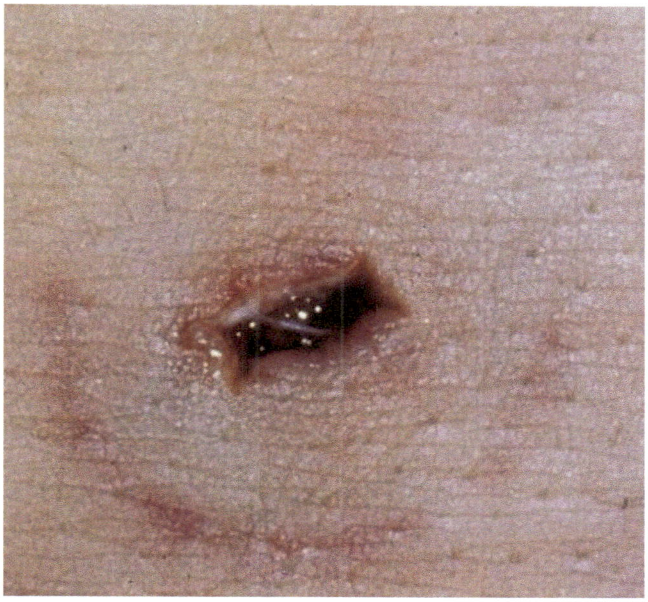

FIGURE 10.22 Flathead screwdriver skin surface wound, with hilt mark.

FIGURE 10.23 Phillips head screwdriver skin surface wound (arrow) in an incinerated victim.

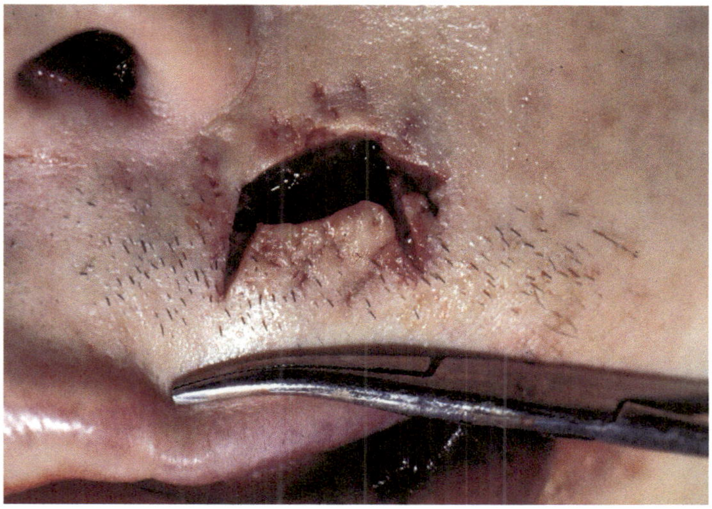

FIGURE 10.24 Broken glass injury (broken bottle).

and the *orientation of the wounds in relation to each other* typically differs considerably depending on the manner. For example, a concentration of multiple colinear wounds of the volar wrist (Fig. 10.27) or tightly clustered superficial stab wounds of the left chest (so-called hesitation wounds, Fig. 10.28) that surround one deeply penetrating fatal wound supports a suicidal manner, while a cut on the palmar surface of the hand (Fig. 10.29) or the dorsal forearms generally speaks more to so-called "defense" injury. A multiplicity of widely distributed wounds on several body surfaces is also more consistent with a pattern of homicidal injury. When the back is injured, it is even clearer. Also, it is exceedingly rare for a person to intentionally cut or stab their own face.[10,14]

A conscious individual who is attacked with a knife will usually instinctively sacrifice their hands/arms or feet/legs (the latter if they happen to be lying on the ground/floor/bed, etc.), using them, often reflexively, to protect their vital areas (face, neck, chest). **Hesitation wounds**[14–16] have been described in 65% of suicides (Figs. 10.27 10.28, and 10.30), while **"defense injuries"** are seen in 49% of homicides (Fig. 10.29).[14] That said, while these injuries may be helpful in the determination of manner, they are not typically mechanistically contributory to death. Important to note here as well are the less common wounds referred to as **"offensive injuries."** When a violent encounter with a knife is particularly bloody and involves a multiplicity of wounds, the handle of the knife used in the attack is more likely to become bloody and slippery. Depending on the size of the knife's guard, the hand holding the knife is at risk of slipping forward, causing inadvertent injury, most commonly to the medial aspect of either the second (index) finger or thumb, depending on how the knife is held.[27,28] That said, it is important for law enforcement to examine the hands of suspects being questioned in connection to such deaths.

When a death is clearly a homicide, but there are no "defense" wounds, there is usually a reasonable explanation if one looks hard enough. Possibilities include a "surprise" attack (e.g., while the victim is sleeping or attacked from behind), an intoxicated victim, developmental or motor limitations of various causes, or an incapacitating wound

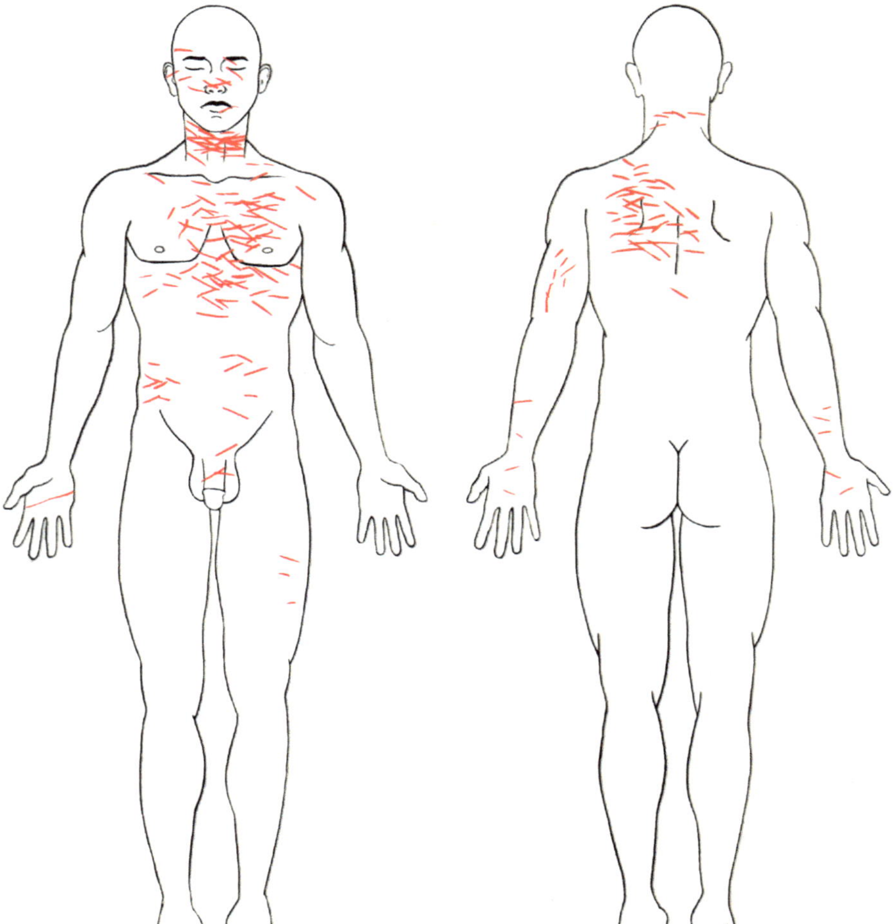

FIGURE 10.25 Distribution of homicidal sharp injuries. *Artwork by Dr. Cricket Coleman.*

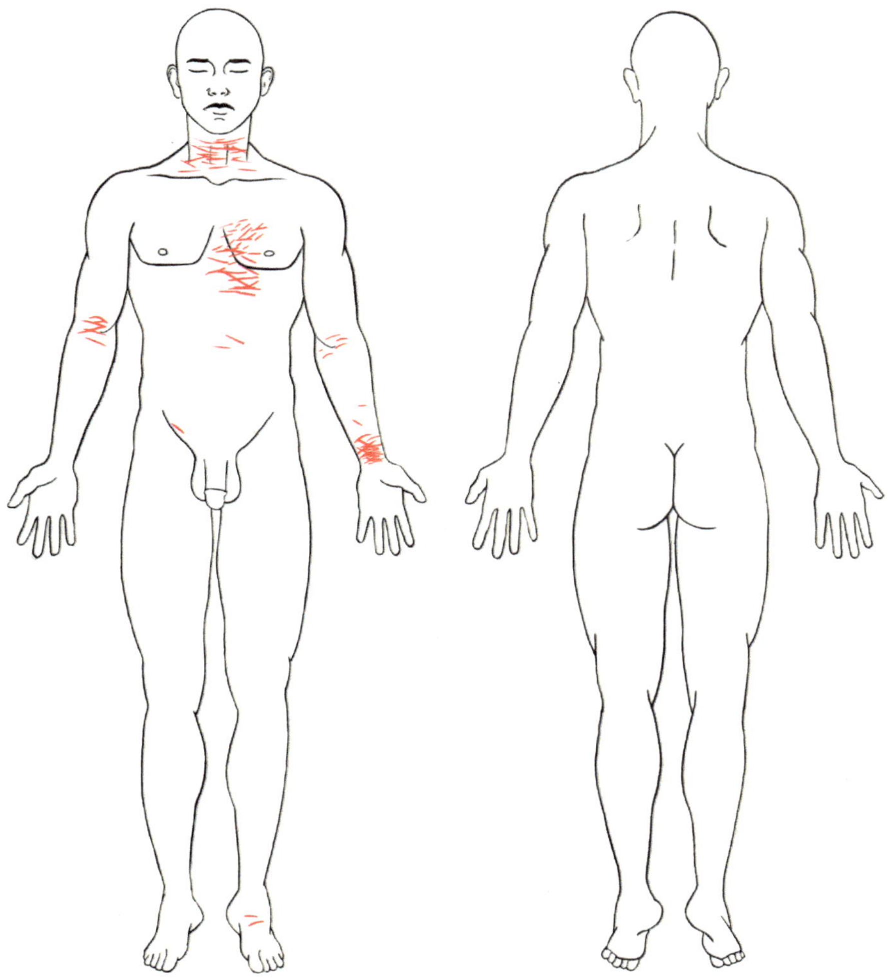

FIGURE 10.26 Distribution of suicidal sharp injuries. *Artwork by Dr. Cricket Coleman.*

inflicted early on in the sequence of wounding. Finally, while it is the anecdotal experience of many that self-inflicted injuries do not typically involve/penetrate through clothing, the data in support of this impression are limited.[29–38] Regardless, it is the view of many that in suicidal sharp injury deaths, the decedent typically either removes their clothing or lifts them before stabbing or cutting themselves, whereas in homicides, the perpetrator does not, an observation that supports a homicidal manner in the latter.

The above paragraph describes what is called a "pattern (distribution) of injuries," but there also exist "patterned injuries" (specific characteristic of individual wounds) that favor homicidal over suicidal sharp force violence. Multiple **gaping wounds**, for example, including some with a **"swallow tail"** or "V-shaped" wound configuration (Fig. 10.31), typically indicate relative movement of the knife to the body, one or both of them moving while the weapon is still penetrating. Such gaping wounds are often haphazardly distributed on multiple surfaces of the body, usually indicative of movement, twisting, and a struggle in a homicidal setting.

A third pattern of injury is one based on the physical injuries and how they reflect the circumstances surrounding the homicidal encounter.[9–15,39] In some instances, for example, one may identify so-called "overkill," which is defined by the infliction of excessive wounds, many more than are necessary to cause death, sometimes with excessive involvement of the face or genitalia and, on occasion, with some even inflicted after death or involving evisceration or avulsion of viscera through gaping abdominal wounds.[40–42] The importance of recognizing this rests with the fact that it may offer important investigative information. **Overkill** often implies rage or jealousy and may cause the police to pursue ex-relations or close contacts of the decedent as part of their investigation.[40,42–44] Regardless, these deaths more commonly involve a fatal encounter between two or more people who know each other well, such as in a domestic context or otherwise in the setting of a so-called "crime of passion," often taking place behind closed doors. By contrast, knife fights that break out in the street or otherwise in public view, as part of an armed

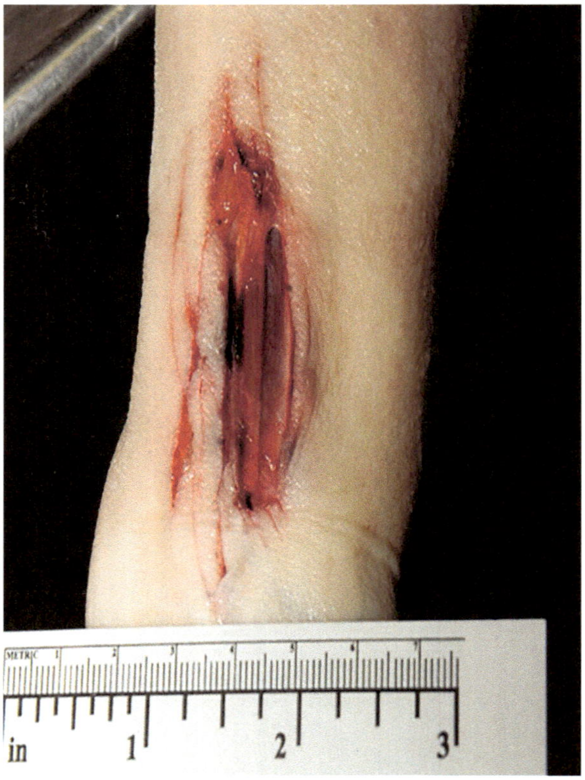

FIGURE 10.27 Incised wounds of wrist (self-inflicted).

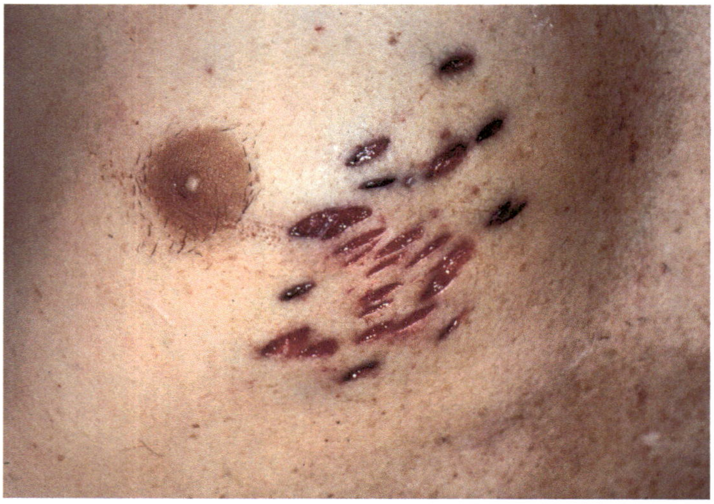

FIGURE 10.28 Hesitation stab wounds of left chest (self-inflicted).

robbery or gang fight, for example, are less likely to have such a pattern of excessive injury and struggle, given the glare of potential witnesses.[21,45–47] Moreover, the weapons used in street encounters are commonly so small as to be concealed in a pocket (literally, a pocket knife). In video surveillance footage, the knife is often difficult, if impossible to see and the violence may even appear as thrown punches rather than stabs. While unwieldy or large kitchen knives may occasionally come into play in street violence, these larger knives, the kind that can be found in an average kitchen, are more commonly used in domestic incidents. Separate and apart from an overkill situation, it is not unusual to see multiple sharp injuries in sharp force homicides, as stab wounds are rarely, if ever, immediately incapacitating. The victim will remain awake until they have lost enough blood to lose consciousness entirely, by a

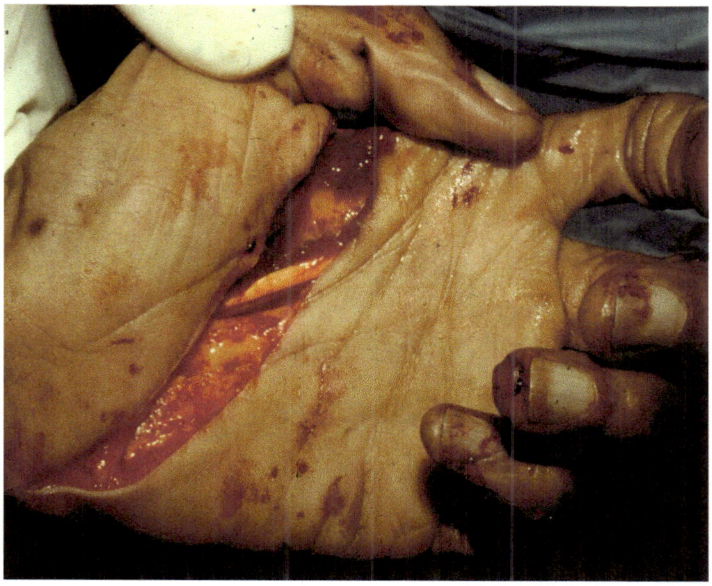

FIGURE 10.29 **"Defense" injury (incised wound) of hand.** Knife grab injury.

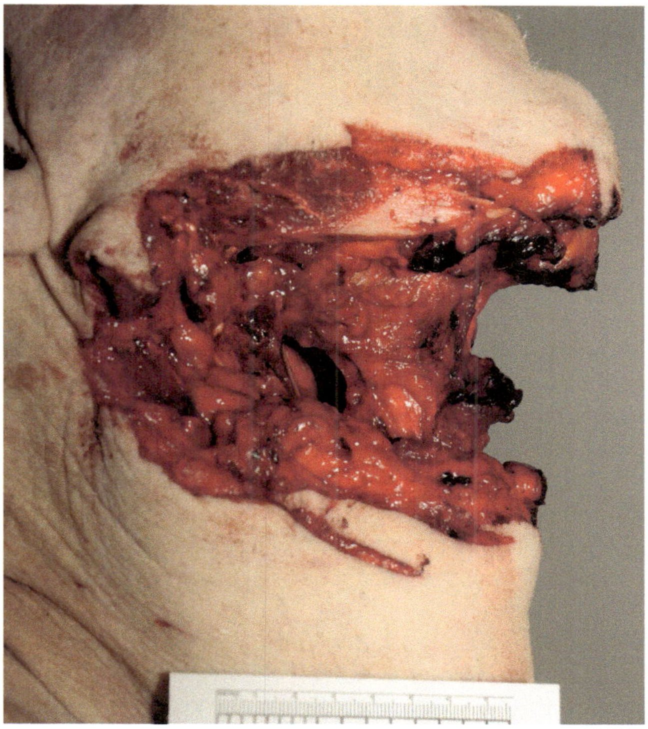

FIGURE 10.30 Circular saw cuts of neck.

hypotensive mechanism (hypovolemic shock), which typically takes at least minutes, sometimes shorter if a major artery is severed, and longer, if only minor veins are penetrated (see below).

Certain questions invariably arise with sharp injury homicides. Which wound was sustained first? Was the perpetrator left or right-handed? What was the size/shape of the knife? Was it serrated? Was the knife found at the scene the murder weapon? How long would the person have been conscious after the injuries?[48–50] Did some of these injuries occur postmortem? How much force was needed to cause this injury?[32,46,51–53] Answers to some of these questions are unknowable from the standpoint of the examination of the body, but some important information can still be gleaned from the autopsy examination of the wounds and their sequelae.

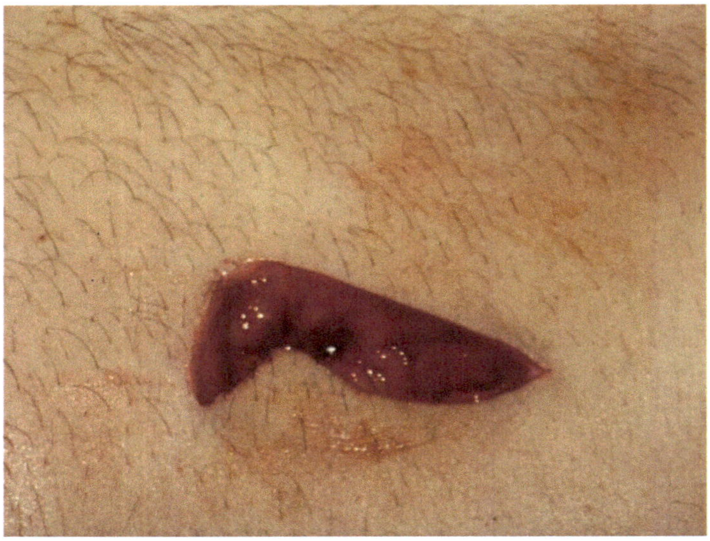

FIGURE 10.31 Swallow tail stab wound.

In the vast majority of instances, the sequence of sharp injuries cannot be discerned. It is also not possible to determine with certainty the handedness of the perpetrator. When confronted with a cluster of stab wounds of the right chest, it is tempting to opine that the perpetrator is left-handed. This first assumes, however, that the assailant and decedent were standing face-to-face and, secondly, that the perpetrator was using a forward thrust. The right-handed perpetrator, however, using a cross-body, backhand attack, could also inflict stab wounds of the right chest. The same wounds might also be reasonably produced were the victim on the ground with the right-handed perpetrator kneeling behind the victim's head and stabbing down at his right chest.

Based on the autopsy findings, there are two definitive ways to be able to state with certainty or near certainty that a specific knife caused a particular stab wound. The first is when the knife is still lodged in the body. The second is if a broken knife tip is lodged in the body. The latter rare finding justifies the common practice of imaging all victims of sharp penetrating trauma. With the help of a tool mark forensic expert, the tip may be linked to a reputed knife. DNA evidence also is very powerful evidence that may link a knife to a body. This is one reason why weapons suspected to have been used in an assault should not be brought to the ME/C office for wound comparison. Wound-knife comparisons can be performed easily through photographs that include an appropriate scale next to the purported weapon. In these purely size–shape comparisons, the forensic pathologist can usually state whether or not a particular knife is capable of causing a particular wound (i.e., is "consistent with"/reasonably possible to be a knife that could have created the wound in question). In the most instances, a knife is consistent with having caused a stab wound if the wound is not *substantially* smaller than the knife is wide at a depth that would correspond to the width of the knife at that level of penetration, *or* if the wound does not have a depth of penetration that is *substantially* deeper than the putative knife is long.

It is because of these specific types of wound–weapon comparisons that it is so important to release skin tension around a stab wound when examining it, either by incising the skin around the wound ("boxing it out") or by manually pushing the skin together to relax the skin around the wound to collapse the elasticity and document its accurate length on the skin surface (Fig. 10.32). Absent this step, the wound is likely to stretch open and gape (Fig. 10.33), more so in areas of high skin tension with limited underlying fatty soft tissue (e.g., neck, skin overlying the pectoralis muscles). While still gaping, it is also less likely that the **blunt and sharp ends (angles)** of the wounds (representing a single-edged knife) will be clearly discernible, another observation that may have relevance to wound–weapon comparisons. For the same reasons, it is similarly important to carefully estimate the depth of the weapon's penetration in the body, allowing for the difficult to quantify body cavity space in the estimate, depending on the wound track. On occasion, owing to drying, therapeutic intervention (suturing wounds, incising and incorporating wounds into a surgical or thoracotomy incision, etc.), or postmortem change, the blunt and sharp angles of a knife wound on the skin surface will not be clearly discernible. For these "ambiguous" wounds, the penetrating defects of deeper structures, such as cartilage, liver, or the myocardium, can be very helpful in elucidating these features.

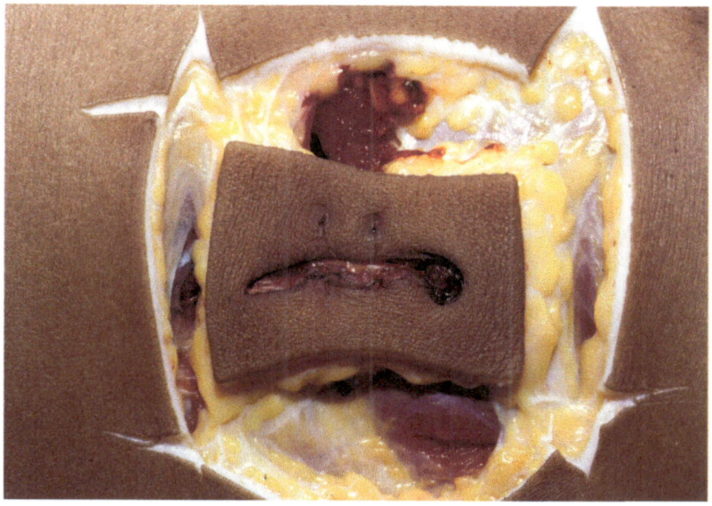

FIGURE 10.32 Stab wound with skin tension release.

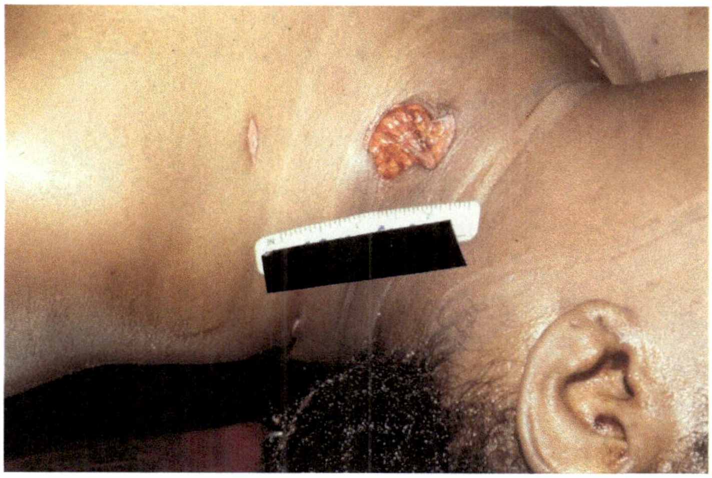

FIGURE 10.33 Gaping stab wound.

Nonfatal patterned superficial incised wounds and scrapes reflecting a knife's **serrations** (Fig. 10.34) are also helpful in the overall wound–weapon comparison analysis. While the characteristics of most stab and incised wounds, per se, do not, in and of themselves, indicate whether the weapon was serrated or not, nearby superficial patterned injuries often do, allowing the forensic pathologist to consider the company the fatal sharp wounds keep. These patterned injuries are usually caused by the swiping of the long cutting surface of a serrated knife's blade across the body during the course of the struggle. Forensic anthropologists may also be able to offer opinions about certain characteristics of the sharp instrument based on the injured bone morphology.[54,55] See Chapter 15 for a further discussion on tool marks and other anthropologic considerations in sharp force trauma deaths.

As previously noted, sharp injuries will rarely, if ever, cause immediate incapacitation. We know that even if one's heart were to stop rather precipitously, as typically occurs with a large **cardiac air embolism** (Figs. 10.35 and 10.36), the brain still has several seconds of consciousness before the drop in blood pressure is "recognized" by the brain and conscious activity ceases. Therefore, the rate of decline leading to death depends upon the specific injury and how rapidly the blood is lost.[49,50,56–61] The transection of a carotid artery, for example, will cause a rapid loss of blood, while an injury of a renal vein will take longer. Even a stab wound of the heart, however, may allow for several minutes of consciousness (Fig. 10.37).[48–50,62] A stab injury of the heart is different than a gunshot wound of the heart because, even though there is an associated defect for each, there is no loss of tissue with a stab wound, as is seen with most firearms wounds. This allows the stab wound defect to functionally "close" better than would a bullet wound, particularly during systole, if the injury involves the ventricle. There is also no blast-type/temporary cavity damage with the former. Regardless, most commonly, **death** from hemorrhage, as is the case in most sharp force

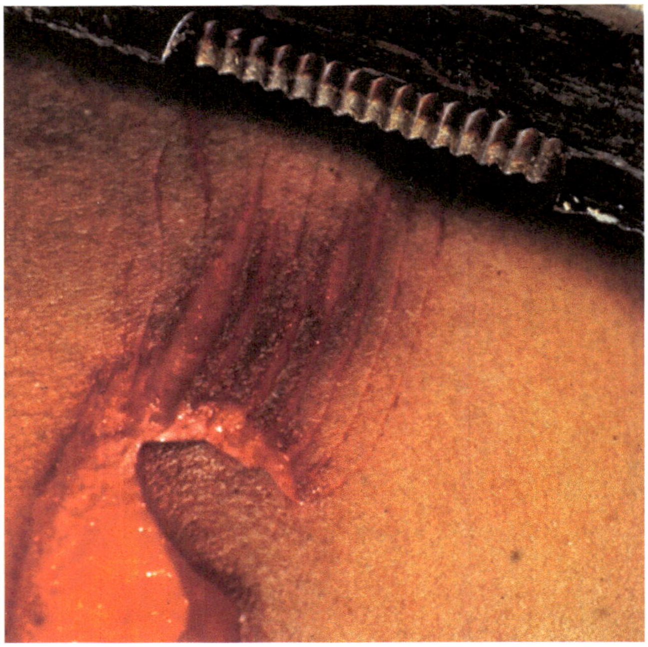

FIGURE 10.34 Serrated knife patterned injury.

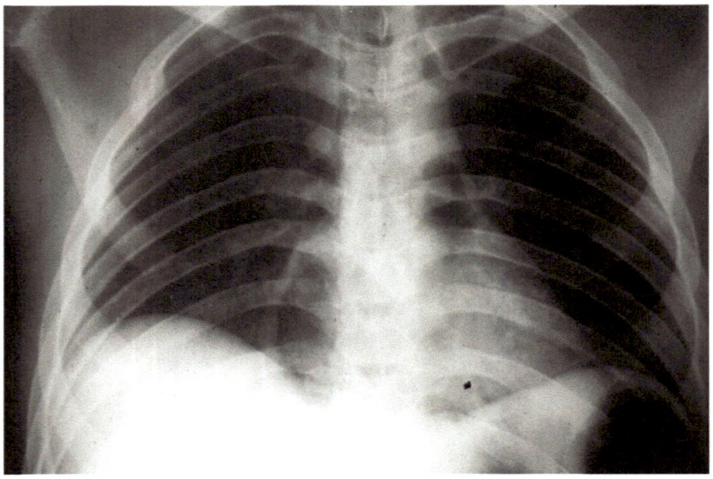

FIGURE 10.35 **Radiograph of chest with "C-sign."** The right heart contains air (lucency) resulting in the right heart border appearing as a "C."

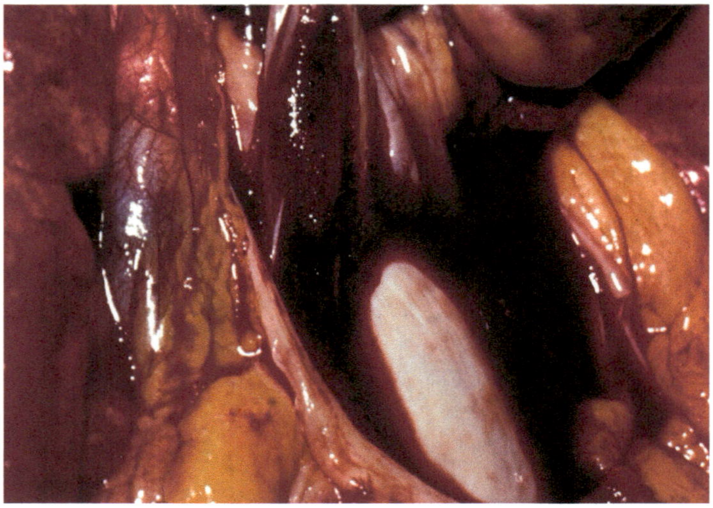

FIGURE 10.36 **Bubbles due to air embolism.** Seen when the IVC/heart junction is incised.

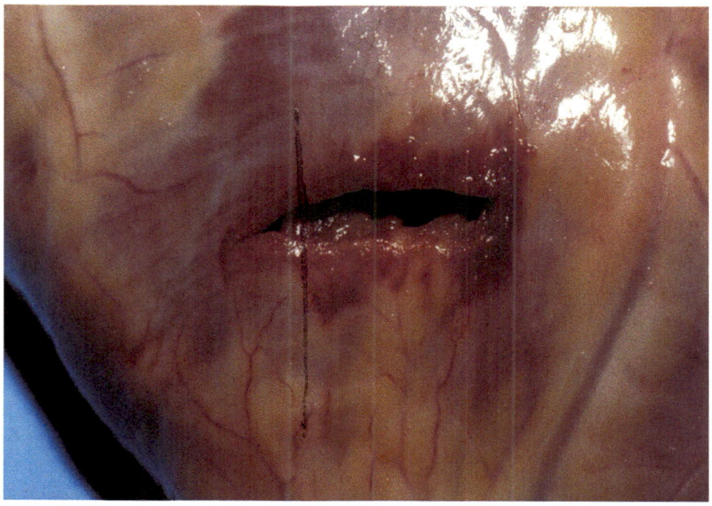

FIGURE 10.37 **Stab wound of heart.** This is helpful if the skin wound is ambiguous for sharp/blunt ends.

trauma fatalities, is **more likely to ensue in minutes rather than in seconds.** *That said, it is important to document physical evidence that supports hypovolemic shock as the mechanism. Specifically, a precise measurement and photographic documentation of the accumulation of cavity blood and/or renal pallor is important, the latter more so for wounds that bleed into the environment, where the volume of blood loss cannot be easily quantitated and photographed (e.g., incised wounds of the neck with major vascular transections).*

The **degree of force necessary** to cause a deeply penetrating stab wound depends upon the pointedness of the tip and sharpness of the knife/sharp object, as well as the structures and tissue types it penetrates. It is not possible to assign a precise number, but it amounts to the degree of force necessary to perforate both clothing (be it a silk blouse or a thick leather jacket) and the skin (the mean force to penetrate human skin has been reported as 11.1 lb (49.5 N)[51]). Because of the elasticity of the skin, once a blade perforates it, it is likely to continue more easily through subsequent soft tissues or organs. Perforation of bone or cartilage, however, or a wound track that is somewhat longer than the blade length speaks to a greater degree of applied force, given the resistance of the tissue penetrated and the force necessary to compress the body enough to create a track longer than the blade length. Having a clear understanding of the physics of sharp force injury is not inconsequential, as questions relating to this topic are not uncommonly asked in criminal proceedings. It is important to be able to form opinions having veracity regarding the reasonable likelihood of a deeply penetrating stab wound being inflicted by accident. A forensic pathologist (FP) might be asked, for example, to opine about the reasonable likelihood of a knife causing a 4-inch in depth stab wound, traveling downward in the chest, through the skin, a rib, and deeper viscera, as a result of the decedent accidentally "falling" onto the blade. The answer to this question should be an unequivocal, "No, that is not a reasonable possibility." The tiny tip of a knife does not have the surface area or density to deeply penetrate the skin and bone without either a forceful thrusting behind it or it being fixed by a stabilizing implement, such as a vice grip. Without one of those two scenarios, the tiny tip would be deflected by the tough human skin, causing, at most, a superficial flesh wound. Moreover, in order to "fall onto" the knife that produces a downward-oriented wound while in a standing position, the decedent would have to fall 4" *up* into the knife, an unreasonable possibility.

The most common **mechanism** of death from sharp injuries is that related to hemorrhage. Hemorrhage may cause death in a variety of ways. *Exsanguination* is loss of enough blood to result in hypovolemic shock that progresses to inadequate perfusion of vital organs and death (generally speaking, the *rapid* loss of 50% of blood volume is fatal). An average adult's **total blood volume is 70 mL/kg x body mass (kg).** So, the blood volume of a 70 kg adult would be roughly 5 L. *Hemorrhage in the restrictive/compressive form* of a hemopericardium/pericardial tamponade typically causes death with a relatively smaller volume of blood (e.g., 450 cc hemopericardium or 150 cc subdural hematoma, rapidly accumulated). A tension pneumothorax or *hemopneumothorax* also may cause death. *Aspirated blood* due to sharp injuries involving the airway, including the oral cavity, may result in an asphyxial mechanism of death that is grossly seen as a geographic pattern (leopard skin-like on cross-sectioning) of the lung(s) (Fig. 10.38). Finally, sharp injuries of the neck that involve the jugular vein, for example, can introduce air into the venous system, resulting in a cardiac air embolism, often causing such a sudden death, if large enough, that very little blood is discovered at the scene of the crime, there absence of visceral pallor, and a surprisingly normal (*not* bloodless) autopsy. Cardiac air

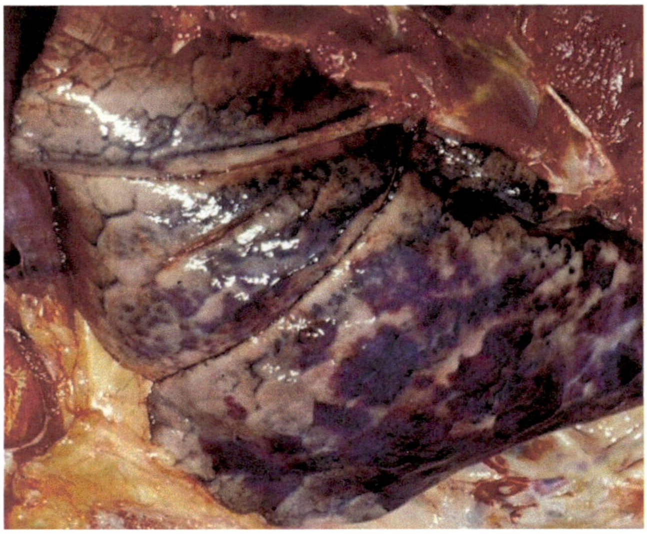

FIGURE 10.38 Lung with blood aspiration pattern.

emboli are best seen on radiographs or by filling the pericardial sac with water at autopsy and observing any escape of air when the right heart-inferior vena cava (IVC) junction is entered while submerged (Fig. 10.36). Finally, when a victim of sharp force trauma dies before they have frankly exsanguinated (i.e., evidenced by their viscera being perhaps mottled, but not markedly pale (Fig. 10.39), as would be expected in frank exsanguination), an explanation should be sought. If it is not explained by a cardiac air embolism mechanism, it is likely explained by the much more common scenario of substantial comorbidities (heart disease, obesity, diabetes, etc.), old age, or additional injury modalities. In these instances, the death may be best and most completely and accurately explained by listing one or more of the comorbid conditions in part 2 of the death certificate or by listing more than one injury type in part 1. Finally, while exsanguination is the most common mechanism of death for sharp force trauma fatalities, the above-listed mechanisms are not mutually exclusive and may act in some combination, resulting in a more complex mechanism of death.

Chop injury

Chop injury represents a unique category of sharp force trauma, although, in reality, its characteristic features often have both blunt and sharp qualities; more accurately, then, chop injuries should be viewed at the intersection of both (Fig. 10.40). Chop injuries are as unmistakable as they are fatal, their dramatic appearance demonstrative of their devastating effects. These injuries are typically caused by heavy/dense, unwieldy weapons with a sharp cutting surface that are expressly used to cause serious physical injury and/or death. Examples of such weapons include axes, meat cleavers, machetes, and hatchets, among others. A typical chop wound is large and gaping, with a broad and deep exposure of the soft tissues and sometimes bone, occasionally with avulsion of muscle. When there are blows to parts of the body covering bone (head, limbs), there are typically underlying crush-type, depressed, and/or comminuted skull fractures that are incompatible with continued life and chips, divots, and sometimes, transections of the long bones (recall the formula, injury = kinetic energy × 1/T × 1/A, where T is time and A is surface area). Also, because of the size of the weapons, a small rim of abrasion is often seen on the skin surrounding the wound (a blunt characteristic). Anthropologic consultation is sometimes helpful in evaluating skeletal injuries in terms of tool mark analysis and the determination of possible weapon types. Patterns of the skin and soft tissue of these wounds may also be observed and helpful with regard to a weapon search (e.g., serrated weapon).

When inflicted on the head and neck, chop wounds are nearly uniformly fatal. Chop injuries to the torso and limbs may also cause death, depending upon the depth of the wounds and the structures penetrated. More commonly, however, as is observed in other forms of inflicted sharp force trauma, chop wounds of the soft tissues and bones of the extremities likely represent defense type injuries more than fatal blows. When chop injuries involve the head and/or neck, they are most commonly immediately incapacitating/rapidly fatal (i.e., often without excessive bleeding), based on both the anatomically devastating sequelae of the injury and the functional neurologic derangements caused by the force of a blow by such a heavy/dense weapon.

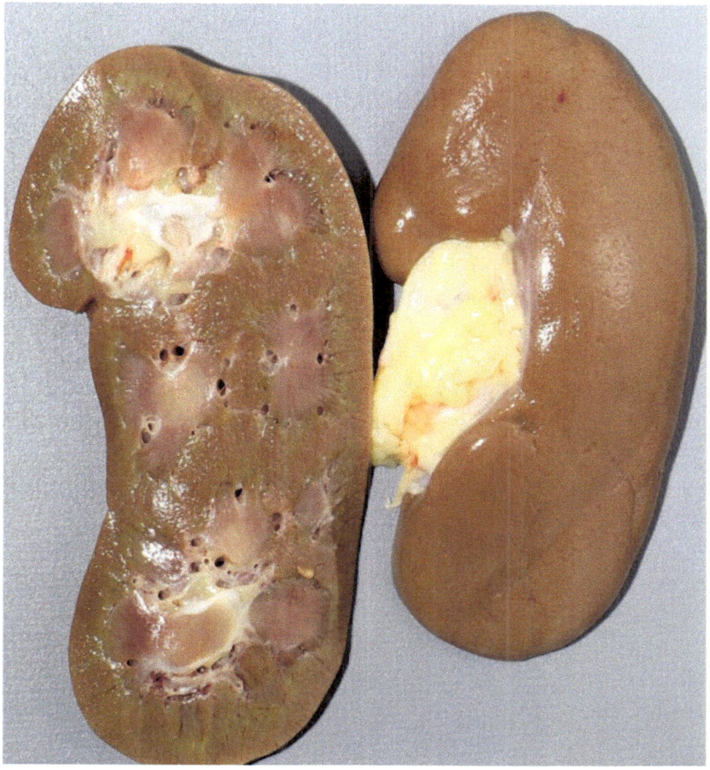

FIGURE 10.39 Pale kidneys due to exsanguination (hypovolemic shock).

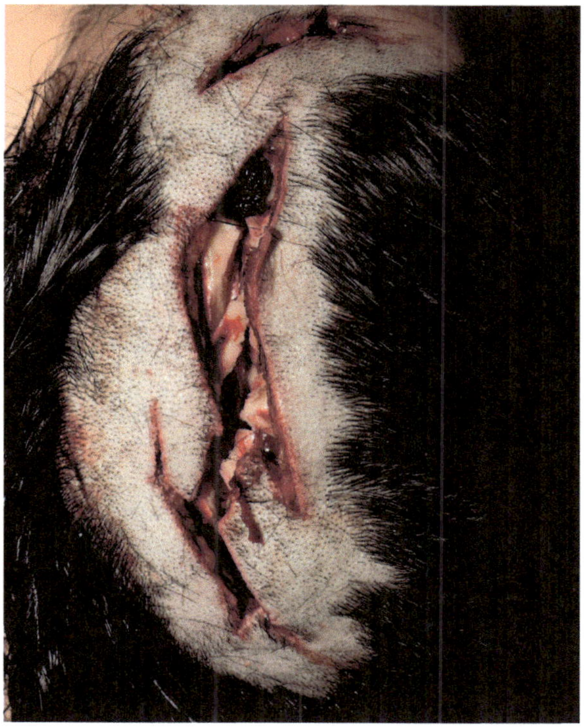

FIGURE 10.40 Chop wounds of scalp/skull (hatchet).

In terms of the circumstances that typically involve chop injuries, while an occasional emotionally disturbed person may randomly wield a chop weapon in public, much more commonly, chop injury involves a victim and assailant who know one another. That said, domestic violence as a backdrop is at the top of the list, and this may or may not involve a history of mental illness on the part of the perpetrator. It is important to remember also that some "chop" weapons may be used in the postmortem dismemberment process (see Chapter 15).

Sharp injury tips

1. Stab wounds can be deeper in the body than the length of the blade if the body region stabbed is compressible (e.g., abdomen), and, of course, stab wounds can be shorter than the length of the blade if the blade is not entirely thrust into the body.
2. The length of the stab wound on the skin may be longer than the width of the blade because there may be cutting along with the stabbing (often with an incised "tail" extending from the wound edge), due to the movement of the blade relative to the body. That said, the reverse does not apply: stab wound skin defects are not expected to be substantially *smaller* than the blade is wide.
3. For incised wounds, the direction of the knife cannot be determined based on the morphology of the cut. Also, both ends of a cut/incised wound are sharp, unlike a stab wound from a single-edged knife, which would be expected to have a blunt and sharp end.
4. All bodies with penetrating injuries will benefit from being radiographed.
5. Think of air embolism with sharp injuries of the neck, especially when there is evidence of limited bleeding in the body (lack of visceral and muscular pallor) and/or at the scene.
6. Look for an intercostal artery injury when faced with a hemothorax without organ injury. Consider an intercostal traumatic aneurysm in delayed traumatic hemothoraces.
7. Unless dragged obliquely across the skin, sharp injuries by serrated knives cannot be distinguished from nonserrated knives when evaluating stab or incised wounds. Examination of injured bone (which can be retained) may offer further information about the sharp implement.[54]
8. Think homicide when there are cuts to the face and/or genitalia.
9. Think of paired instruments (e.g., barbeque fork, scissors) with paired injuries.
10. One cannot reliably determine the "handedness" of a perpetrator based on the distribution of sharp injuries on the body alone.
11. Exercise caution in overcalling stab wounds when they are actually "pseudostab wounds" (dehiscence of old/scarred surgical wounds in a markedly bloated putrefied body); the absence of subjacent tissue penetration will confirm this skin artifact.[63]

Transportation injury

By convention, the deaths of most typical motor vehicle fatality victims (drivers, passengers, and pedestrians) are certified with an accidental manner of death. These even include a pedestrian who is struck by a drunk driver who ran a red light and then fled the scene. As previously noted, an accidental manner of death certification does not prevent a prosecutor from filing charges against a driver. Clear intent in these deaths, however, may affect the manner of death. If there is evidence, for example, that a driver intentionally drove their car into a tree at a high speed, the manner of death would be classified as a suicide. If a driver used the motor vehicle as a weapon in order to run someone over, the manner of death would be classified as a homicide. A second scenario that may result in a homicide determination is when a driver inadvertently strikes a pedestrian or other vehicle while fleeing from the police during a high-speed pursuit following an armed bank robbery, the death occurring during the commission of a felony crime.

On occasion, in an attempt to assess and potentially save the lives of multiple individuals involved in a collision, the occupants are rapidly removed from a vehicle by emergency services personnel, without written or photographic documentation of the victims' locations in the car compartment before removal. In other instances, all unrestrained occupants are ejected, including the driver. Aside from documentation on the death certificate as to whether a decedent was a passenger or the driver, other civil and criminal issues, separate and apart from death certification, may center on the culpability of drivers in their roles in causing collisions, not the least of which may involve toxicological testing. With that in mind, the forensic pathologist has an obligation to look for and document any injuries that may help to delineate an individual's position in the car at the time of the collision. **Dicing injuries and patterned seatbelt contusions** are two examples of such injuries that may be observed (Fig. 10.41). The laterality of each may suggest a driver versus a passenger location, depending on whether there were also rear compartment passengers, which may further complicate the interpretation of findings. Additionally, if there is a flash fire immediately following the collision, such observations might me made virtually impossible if all decedents are severely charred within the compartment. While charred remains may also make the recognition of both the decedents and their injuries more challenging, in most instances, severe/fatal visceral blunt injuries and/or signs

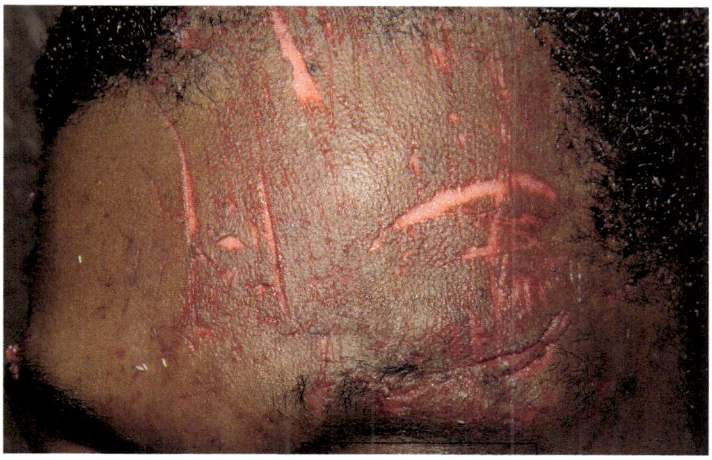

FIGURE 10.41 Dicing injury.

of smoke inhalational injury will still be detectible, even if the **CO-Hgb saturation is negative (flash fire).** On occasion, there is video surveillance of the driver and passengers clearly entering the car as they leave a bar or restaurant, prior to the collision. This compelling investigative evidence, of course, places the physical evidence of dicing and seatbelt injuries in the position of supporting role, rather than main event.

Another vehicular manner of death question occasionally arises when a person dies while driving, but in the absence of a significant collision with another car or fixed object. Rather, the car may slowly decelerate and then comes to rest, after which the driver is found unresponsive or dead. Absent witnesses to the frame-by-frame sequence of events, only an autopsy will definitively unravel the cause of death in this scenario. As alluded to in Chapter 7, most of these fatalities are due to natural or chemical causes, rather than mechanical injury, given the absence of a substantial motor vehicular collision. Causes of such sudden deaths typically include cardiac or cerebrovascular disease. Epilepsy is another natural disease that may incapacitate drivers, with the consequent seizure activity either causing a collision and bodily injury or a SUDEP or seizure-related natural death, leaving the driver flaccid, to slowly veer off the road. If the former, the cause of death will likely be due to blunt force injuries with the epileptic underpinnings of the accident impossible to prove, but not necessary for certification purposes anyway. Even if it were known that the driver had a seizure that led to his loss of control of the car, "epilepsy," per se, would not appear anywhere on the death certificate when the cause was due to blunt injuries. Finally, it is important to at least consider and further investigate a suicidal classification if a pedestrian is struck by a motor vehicle when crossing a highway (there may be other explanations for this behavior, but it is unusual to cross major highways unless intoxicated or otherwise necessary) or when there is a lone motor vehicle that collides with a wall, at high speed, or veers off a roadway that was well known to an unintoxicated driver.

This brings us to an important concept. By convention, in motor vehicle collisions, *the "cause" of the loss of control, of the collision,* per se, *is not typically a point of opinion or documentation asserted on the death certificate, as it is neither necessary nor fully knowable to the degree of probability necessary for death certification purposes.* Stating the reason or reasons for the crash would require some degree of presumption, or even speculation. Many variables may contribute to the sequence of events that end in a vehicular collision. Some of these are mechanical (vehicular failure); others are environmental (road conditions, poor visibility due to weather conditions, glare, etc.); still others are operator-dependent, such as distraction (from texting, talking, pain, daydreaming, scolding children, etc.), intoxication, limitations in hearing or seeing, poor driving ability, or medical conditions; and finally, still other variables involve drivers of other vehicles. The identification of one or more "causes" of the accident not only falls outside the scope of the expertise of medical examiners, but it also may just be plain unknowable. We do not opine about variables that are neither fully defined nor knowable. Most importantly, unlike an adult drowning in a bathtub, which makes no sense without a further explanation, motor vehicle collisions and the deaths that arise from them do not elicit the same confusion that begs for a further explanation. Any further explanation, to the extent that it is known, is contained within the accident investigation police report, and may well have bearing in subsequent civil proceedings, but is not necessary for competent death certification, other than the need for the time and place of occurrence. Also important in the certification of these deaths is the inclusion in the "how injury occurred" field, indicating whether the decedent was a pedestrian, the driver, or a passenger.

As described in Chapter 5, **pedestrian victims** usually prompt additional steps in the autopsy examination in order to fully document and understand the mechanics of the vehicle to body contact, as well as secondary impacts, and to collect gross and trace evidence (scalp hair, paint chips, photographic documentation of a tread pattern on clothing or patterned injuries consistent with a car's grill pattern, depending on the case). This is particularly important when the driver has left the scene, intentionally or obliviously, and law enforcement personnel are actively investigating. An additional **dissection of the lower extremities** is usually performed to rule out/rule in long bone and/or pelvic fractures and soft tissue contusive and shearing injuries, with or without avulsion pockets (Fig. 10.42). It is best to measure these injuries not only in their span and location, but also in their distance from the bottom of the heel so that correlations to the bumper height may be made, if necessary. Pelvic and/or femoral fractures are more common in impacts from tall vehicles, such as sports utility vehicles and trucks, whereas tibial/fibular fractures are more common in impacts from traditional sedans. These dissections are best performed after the pelvis and lower extremities have been X-rayed (see Chapter 5).

Finally, on occasion, no structurally demonstrable fatal injuries are identified during the autopsy of a motor vehicular collision victim, invariably confounding the prosector as the autopsy comes to an end. In these scenarios, it is best to conduct a **careful posterior neck dissection** to document identifiable skeletal injuries and/or soft tissue injuries that are consistent with whiplash injury. It is also important to consider traumatic/compressive or positional asphyxial mechanisms by inquiring about the specific position (e.g., body inverted while unconscious/concussed, but still alive?) in which the decedent was found, the extent and quality of the vehicular damage (e.g., was the interior compartment compressed substantially and/or the steering column pushed up against the decedent's chest and neck?), and whether the decedent's extrication from the vehicle was particularly difficult and of a long duration. Depending on the answers to the foregoing investigative questions, signs of life/circulation during an asphyxial chest or neck compression should be sought and documented (plethora and petechiae of the face and mucous membranes). In the absence of any of the above, the most reasonable conclusion, after exclusion of all other causes (including those that are natural), is a fatal concussion. In such circumstances, a single contusion of the scalp may be the only demonstrable injury at autopsy. This is where clinicopathological correlation becomes even more critical. The documentation of a GCS of 13, for example, should have the FP questioning their certification, since this one clinical data point is incompatible with the diagnosis of a "fatal concussion." In addition to the GCS, the observation of an empty heart would also be very helpful in the analysis of such a mechanism.

Other than motor vehicular fatalities, other less common transportation-related deaths include those involving bicycles, motorcycles, tractors, all-terrain vehicles, subways, trains, and airplanes. Motorized bicycles and scooters have also fairly recently arrived in great numbers in most urban centers, bringing with them an additional layer of risk of mortal injury, civil liability, and attendant investigative complexity. In some major cities, subway fatalities constitute a substantial percentage of a city's transportation deaths, so much so that there is a separate division of the police department dedicated to the investigation of these deaths (and other subway crimes). Subway-related deaths occur on the tracks within both the subway stations (many of these are suicides), as well as deep within the tunnels (many of these are accidents occurring among tunnel dwellers). Homicides, while less frequent, are not uncommon. The causes of deaths in most of these fatalities are grossly self-evident blunt impact injuries (with the occasional electrocution by the third rail), but it is the manner in many of these unwitnessed fatalities that leaves the investigator more challenged.

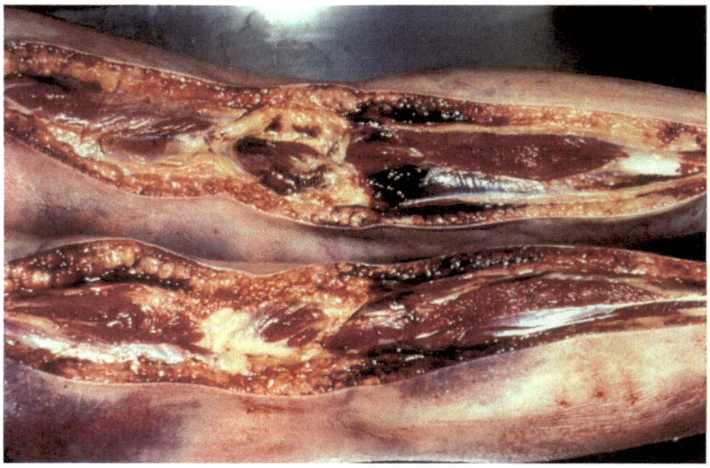

FIGURE 10.42 Pedestrian impact injuries of the legs.

On occasion, an ME/C office is faced with a transportation mass fatality incident. When involving a commercial airplane, for example, the response typically involves multiple agencies and brings with it a singular level of complexity. It requires personnel who are experts in disaster management planning and special operations. By necessity, these events involve the Federal Aviation Association to evaluate the forensic evidence in an attempt to precisely elucidate the cause of the crash. Many stakeholders have an interest in the accuracy of this analysis. From a medical examiner perspective, however, the focus is primarily on the identification of all remains, some potentially comingled and/or devastated by injuries, and the determination of cause of death. It is not uncommon for the ME/C to request anthropologic assistance, if available, to help map out the scene of the crash and catalog the remains. This expertise falls squarely within the wheelhouse of forensic anthropologists, whose field training lends invaluable oversight to these physically challenging and complex investigations (see also Chapter 4).[64–66]

References

1. Tomczak PD, Buikstra JE. Analysis of blunt trauma injuries: vertical deceleration versus horizontal deceleration injuries. *J Forensic Sci*. March 1999;44(2):253–262.
2. Hamilton JR, Sunter JP, Cooper PN. Fatal hemorrhage from simple lacerations of the scalp. *Forensic Sci Med Pathol*. 2005;1(4):267–271.
3. Geisenberger D, Vogt S, Pircher R, Kramer L, Pollak S, Grosse Perdekamp M. Pattern injuries from blows with the muzzle end of a handgun. *Forensic Sci Int*. 2015;257:341–346.
4. Riezzo I, Di Battista B, De Salvia A, et al. Delayed splenic rupture: dating the sub-capsular hemorrhage as a useful task to evaluate causal relationships with trauma. *Forensic Sci Int*. 2014;234:64–71.
5. Bartelink EJ, Wiersema JM, Demaree RS. Quantitative analysis of sharp-force trauma: an application of scanning electron microscopy in forensic anthropology. *J Forensic Sci*. 2001;46(6):1288–1293.
6. Prahlow JA, Ross KF, Lene WJ, Kirby DB. Accidental sharp force injury fatalities. *Am J Forensic Med Pathol*. 2001;22(4):358–366.
7. Ni Annaidh A, Cassidy M, Curtis M, Destrade M, Gilchrist MD. Toward a predictive assessment of stab-penetration forces. *Am J Forensic Med Pathol*. 2015;36(3):162–166.
8. Spagnoli L, Amadasi A, Frustaci M, Mazzarelli D, Porta D, Cattaneo C. Characteristics and time-dependence of cut marks and blunt force fractures on costal cartilages: an experimental study. *Forensic Sci Med Pathol*. 2016;12(1):26–32.
9. Ormstad K, Karlsson T, Enkler L, Law B, Rajs J. Patterns in sharp force fatalities–a comprehensive forensic medical study. *J Forensic Sci*. 1986; 31(2):529–542.
10. Karlsson T, Ormstad K, Rajs J. Patterns in sharp force fatalities—a comprehensive forensic medical study: Part 2. Suicidal sharp force injury in the Stockholm area 1972–1984. *J Forensic Sci*. 1988;33(2):448–461.
11. Karlsson T. Homicidal and suicidal sharp force fatalities in Stockholm, Sweden. Orientation of entrance wounds in stabs gives information in the classification. *Forensic Sci Int*. 1998;93(1):21–32.
12. Karlsson T. Sharp force homicides in the Stockholm area, 1983–1992. *Forensic Sci Int*. 1998;94(1–2):129–139.
13. Rogde S, Hougen HP, Poulsen K. Homicide by sharp force in two Scandinavian capitals. *Forensic Sci Int*. 2000;109(2):135–145.
14. Gill JR, Catanese C. Sharp injury fatalities in New York City. *J Forensic Sci*. 2002;47(3):554–557.
15. Racette S, Kremer C, Desjarlais A, Sauvageau A. Suicidal and homicidal sharp force injury: a 5-year retrospective comparative study of hesitation marks and defense wounds. *Forensic Sci Med Pathol*. 2008;4(4):221–227.
16. Karakasi MV, Nastoulis E, Kapetanakis S, Vasilikos E, Kyropoulos G, Pavlidis P. Hesitation wounds and sharp force injuries in forensic pathology and psychiatry: multidisciplinary review of the literature and study of two cases. *J Forensic Sci*. 2016;61(6):1515–1523.
17. Schaerli S, Schulz R, Gascho D, et al. Injury potential of thrown sharp kitchen and household utensils. *Forensic Sci Med Pathol*. 2018;14(1): 31–41.
18. Nikolic S, Zivkovic V. The possible reconstruction of a sharp injury to a skull. *Forensic Sci Med Pathol*. 2013;9(2):270–273.
19. Valoriani S, Eliopoulos C, Borrini M. Sharp force trauma death in a young individual from medieval gloucester. *Am J Forensic Med Pathol*. 2017; 38(2):111–114.
20. Schmidt U. Sharp force injuries in "clinical" forensic medicine. *Forensic Sci Int*. 2010;195(1–3):1–5.
21. Schmidt U, Pollak S. Sharp force injuries in clinical forensic medicine–findings in victims and perpetrators. *Forensic Sci Int*. 2006;159(2–3): 113–118.
22. Thomsen AH, Hougen HP, Villesen P, Brink O, Leth PM. Sharp force homicide in Denmark 1992–2016. *J Forensic Sci*. 2020;65(3):833–839.
23. Terranova C, Doro L, Zancaner S, et al. Criminological and medico-legal aspects in homicidal and suicidal sharp force fatalities. *J Forensic Sci*. 2020;65(4):1184–1190.
24. Kemal CJ, Patterson T, Molina DK. Deaths due to sharp force injuries in Bexar County, Texas, with respect to manner of death. *Am J Forensic Med Pathol*. 2013;34(3):253–259.
25. De-Giorgio F, Lodise M, Quaranta G, et al. Suicidal or homicidal sharp force injuries? A review and critical analysis of the heterogeneity in the forensic literature. *J Forensic Sci*. 2015;60(Suppl 1):S97–S107.
26. Vassalini M, Verzeletti A, De Ferrari F. Sharp force injury fatalities: a retrospective study (1982–2012) in Brescia (Italy). *J Forensic Sci*. 2014; 59(6):1568–1574.
27. O'Donovan S, Langlois NEI, Byard RW. "Defense" type wounds in suicide. *Forensic Sci Med Pathol*. 2018;14(3):402–405.
28. Schmidt U, Faller-Marquardt M, Tatschner T, Walter K, Pollak S. Cuts to the offender's own hand—unintentional self-infliction in the course of knife attacks. *Int J Leg Med*. 2004;118(6):348–354.
29. Herbst J, Haffner HT. Tentative injuries to exposed skin in a homicide case. *Forensic Sci Int*. 1999;102(2–3):193–196.
30. Vanezis P, West IE. Tentative injuries in self stabbing. *Forensic Sci Int*. 1983;21(1):65–70.

31. Scolan V, Telmon N, Blanc A, Allery JP, Charlet D, Rouge D. Homicide-suicide by stabbing study over 10 years in the toulouse region. *Am J Forensic Med Pathol*. 2004;25(1):33–36.
32. Bolliger SA, Kneubuehl BP, Thali MJ, Eggert S, Siegenthaler L. Stabbing energy and force required for pocket-knives to pierce ribs. *Forensic Sci Med Pathol*. 2016;12(4):394–398.
33. Miller SA, Jones MD. Kinematics of four methods of stabbing: a preliminary study. *Forensic Sci Int*. 1996;82(2):183–190.
34. Nolan G, Hainsworth SV, Rutty GN. Forces required for a knife to penetrate a variety of clothing types. *J Forensic Sci*. 2013;58(2):372–379.
35. Ferllini R. Macroscopic and microscopic analysis of knife stab wounds on fleshed and clothed ribs. *J Forensic Sci*. 2012;57(3):683–690.
36. Burke MP, Baber Y, Cheung Z, Fitzgerald M. Single stab injuries. *Forensic Sci Med Pathol*. 2018;14(3):295–300.
37. Cowper EJ, Carr DJ, Horsfall I, Fergusson SM. The effect of fabric and stabbing variables on severance appearance. *Forensic Sci Int*. 2015;249: 214–224.
38. Kemp SE, Carr DJ, Kieser J, Niven BE, Taylor MC. Forensic evidence in apparel fabrics due to stab events. *Forensic Sci Int*. 2009;191(1–3): 86–96.
39. Chenpanas P, Bir C. Fatal injuries of law enforcement/correctional officers attacked with sharp-edged weapons. *J Forensic Sci*. 2017;62(3): 700–704.
40. Bell MD, Vila RI. Homicide in homosexual victims: a study of 67 cases from the Broward County, Florida, Medical Examiner's office (1982–1992), with special emphasis on "overkill". *Am J Forensic Med Pathol*. 1996;17(1):65–69.
41. Rajs J, Lundstrom M, Broberg M, Lidberg L, Lindquist O. Criminal multilation of the human body in Sweden—a thirty-year medico-legal and forensic psychiatric study. *J Forensic Sci*. 1998;43(3):563–580.
42. Taff ML, Boglioli LK. Gay homicides and "overkill". *Am J Forensic Med Pathol*. 1996;17(4):350–352.
43. Trotta S, Mandarelli G, Ferorelli D, Solarino B. Patricide and overkill: a review of the literature and case report of a murder with Capgras delusion. *Forensic Sci Med Pathol*. 2020;17(2):271–278.
44. Solarino B, Punzi G, Di Vella G, Carabellese F, Catanesi R. A multidisciplinary approach in overkill: analysis of 13 cases and review of the literature. *Forensic Sci Int*. 2019;298:402–407.
45. Humphrey C, Kumaratilake J, Henneberg M. Characteristics of bone injuries resulting from knife wounds incised with different forces. *J Forensic Sci*. 2017;62(6):1445–1451.
46. Annaidh AN, Cassidy M, Curtis M, Destrade M, Gilchrist MD. A combined experimental and numerical study of stab-penetration forces. *Forensic Sci Int*. 2013;233(1–3):7–13.
47. Love JC. Sharp force trauma analysis in bone and cartilage: a literature review. *Forensic Sci Int*. 2019;299:119–127.
48. Thoresen SO, Rognum TO. Survival time and acting capability after fatal injury by sharp weapons. *Forensic Sci Int*. 1986;31(3):181–187.
49. Karger B, Niemeyer J, Brinkmann B. Physical activity following fatal injury from sharp pointed weapons. *Int J Leg Med*. 1999;112(3):188–191.
50. Franchi A, Kolopp M, Coudane H, Martrille L. Precise survival time and physical activity after fatal left ventricle injury from sharp pointed weapon: a case report and a review of the literature. *Int J Leg Med*. 2016;130(5):1299–1301.
51. O'Callaghan PT, Jones MD, James DS, Leadbeatter S, Holt CA, Nokes LD. Dynamics of stab wounds: force required for penetration of various cadaveric human tissues. *Forensic Sci Int*. 1999;104(2–3):173–178.
52. Chadwick EK, Nicol AC, Lane JV, Gray TG. Biomechanics of knife stab attacks. *Forensic Sci Int*. 1999;105(1):35–44.
53. Weber W. Dynamics of stab wounds: force required for penetration of various cadaveric human tissues. *Forensic Sci Int*. 2001;116:229.
54. Crowder C, Rainwater CW, Fridie JS. Microscopic analysis of sharp force trauma in bone and cartilage: a validation study. *J Forensic Sci*. 2013; 58(5):1119–1126.
55. Walsh-Haney HA. Sharp-force trauma analysis and the forensic anthropologist: techniques advocated by William R. Maples, Ph.D. *J Forensic Sci*. 1999;44(4):720–723.
56. Levy V, Rao VJ. Survival time in gunshot and stab wound victims. *Am J Forensic Med Pathol*. 1988;9(3):215–217.
57. Bartoloni G, Trio F, Bartoloni A, Giorlandino A, Pucci A. A fatal stab wound causing selective injury to the left anterior descending coronary artery, myocardial infarction and delayed cardiac tamponade. *Forensic Sci Int*. 2013;229(1–3):e16–18.
58. Charafeddine A. Delayed presentation of cardiac injury after a penetrating stab wound to the chest. *Am Surg*. 2016;82:E31–E32.
59. Harris DG, Janson JT, Van Wyk J, Pretorius J, Rossouw GJ. Delayed pericardial effusion following stab wounds to the chest. *Eur J Cardio Thorac Surg*. 2003;23(4):473–476.
60. Ochi A, Hardikar AA. Delayed presentation of cardiac injury four weeks after stabbing. *Asian Cardiovasc Thorac Ann*. 2020;28(1):62–64.
61. Mayor-Davies J, D'Egidio A, Schein M. Missed stabbed hearts—pitfalls in the diagnosis of penetrating cardiac injuries. Report of 4 cases. *S Afr J Surg*. 1992;30(1):18–19.
62. Spitz W, Petty C, Fisher R. Physical Activity until collapse following fatal injury by firearms and sharp pointed weapons. *J Forensic Sci*. 1961; 6(3):290–300.
63. Gill JR, Cavalli DP, Ely SF. Pseudo-stab wounds: putrefactive dehiscence of remote surgical incisions masquerading as stab wounds. *J Forensic Sci*. 2009;54(5):1152–1154.
64. Byrd JE, Adams BJ. Osteometric sorting of commingled human remains. *J Forensic Sci*. 2003;48(4):717–724.
65. Carroll E, Johnson A, DePaolo F, Adams BJ, Mazone D, Sampson B. Trends in United States mass fatality incidents and recommendations for medical examiners and coroners. *Acad Forensic Pathol*. 2017;7(3):318–329.
66. Conlon E. Mass fatality managment and the effects of commingling. In: Adams B, Byrd J, eds. *Commingled Human Remains*. Oxford: Academic Press; 2014.
67. Hirsch CH, Zumwalt RE. The "empty heart" sign. *Am J Forensic Med Pathol*. 1986;7(2):112–114.
68. Buehler MJ, Abraham FF, Gao H. Hyperelasticity governs dynamic fracture at a critical length scale. *Nature*. 2003;426(6963):141–146.
69. Gill JR. Fatal descent from height in New York City. *J Forensic Sci*. 2001;46(5):1132–1137.
70. Fukutake T, Mine S, Yamakami I, Yamaura A, Hattori T. Roller coaster headache and subdural hematoma. *Neurology*. 2000;54(1):264.
71. Braksiek RJ, Roberts DJ. Amusement park injuries and deaths. *Ann Emerg Med*. 2002;39(1):65–72.
72. Pfister BJ, Chickola L, Smith DH. Head motions while riding roller coasters: implications for brain injury. *Am J Forensic Med Pathol*. 2009;30(4): 339–345.

73. Geisenberger D, Huppertz LM, Buchsel M, Kramer L, Pollak S, Grosse Perdekamp M. Non-traumatic subdural hematoma secondary to septic brain embolism: a rare cause of unexpected death in a drug addict suffering from undiagnosed bacterial endocarditis. *Forensic Sci Int*. 2015; 257:e1–e5.
74. Freytag E. Fatal hypertensive intracerebral haematomas: a survey of the pathological anatomy of 393 cases. *J Neurol Neurosurg Psychiatry*. 1968;31(6):616–620.
75. Patel PV, FitzMaurice E, Nandigam RN, et al. Association of subdural hematoma with increased mortality in lobar intracerebral hemorrhage. *Arch Neurol*. 2009;66(1):79–84.
76. Heninger M. Subdural hematoma occurrence: comparison between ethanol and cocaine use at death. *Am J Forensic Med Pathol*. 2013;34(3): 237–241.
77. Vester ME, Bilo RA, Karst WA, Daams JG, Duijst WL, van Rijn RR. Subdural hematomas: glutaric aciduria type 1 or abusive head trauma? A systematic review. *Forensic Sci Med Pathol*. 2015;11(3):405–415.
78. Sood P, Sinson GP, Cohen EP. Subdural hematomas in chronic dialysis patients: significant and increasing. *Clin J Am Soc Nephrol*. 2007;2(5): 956–959.
79. Bechar M, Lakke JP, van der Hem GK, Beks JW, Penning L. Subdural hematoma during long-term hemodialysis. *Arch Neurol*. 1972;26(6): 513–516.
80. Matsuda M, Matsuda I, Sato M, Handa J. A subdural hematoma following hemorrhagic infarct—report of a case. *Neurol Med Chir*. 1983;23(1): 77–83.
81. Burke MP, O'Donnell C, Opeskin K. Spontaneous acute subdural hematoma complicating arachnoid cyst. *Am J Forensic Med Pathol*. 2010; 31(4):382–384.
82. Pollanen MS. Subdural hemorrhage in infancy: keep an open mind. *Forensic Sci Med Pathol*. 2011;7(3):298–300.
83. Bishop FS, Liu JK, McCall TD, Brockmeyer DL. Glutaric aciduria type 1 presenting as bilateral subdural hematomas mimicking nonaccidental trauma. Case report and review of the literature. *J Neurosurg*. 2007;106(3 Suppl):222–226.
84. Bosma JJ, Miles JB, Shaw MD. Spontaneous chronic and subacute subdural haematoma in young adults. *Acta Neurochir*. 2000;142(11): 1307–1310.
85. Buruma OJ, Sande JJ. Bilateral acute spontaneous subdural hematoma. A case report. *Clin Neurol Neurosurg*. 1976;79(3):211–214.
86. Del Greco F, Krumlovsky F. Subdural haematoma in the course of haemodialysis. *Lancet*. 1969;2(7628):1009–1010.
87. Leonard CD, Weil E, Scribner BH. Subdural haematomas in patients undergoing haemodialysis. *Lancet*. 1969;2(7614):239–240.
88. Brouns R, De Deyn PP. Neurological complications in renal failure: a review. *Clin Neurol Neurosurg*. 2004;107(1):1–16.
89. deRoux SJ, Sgarlato A. Subdural hemorrhage, a retrospective review with emphasis on a cohort of alcoholics. *J Forensic Sci*. 2015;60(5): 1224–1228.
90. Contostavlos DL. Isolated basilar traumatic subarachnoid hemorrhage: an observer's 25 year re-evaluation of the pathogenetic possibilities. *Forensic Sci Int*. 1995;73(1):61–74.
91. Contostavlos DL. Massive subarachnoid hemorrhage due to laceration of the vertebral artery with fracture of the transverse process of the atlas. *J Forensic Sci*. 1971;16:40–56.
92. Opeskin K, Burke MP. Vertebral artery trauma. *Am J Forensic Med Pathol*. 1998;19(3):206–217.
93. Berzlanovich AM, Sim E, Muhm MA. Technique for dissecting the cervical vertebral column. *J Forensic Sci*. 1998;43(1):190–193.
94. Sato Y, Kondo T, Ohshima T. Traumatic tear of the basilar artery associated with vertebral column injuries. *Am J Forensic Med Pathol*. 1997; 18(2):129–134.
95. Reichard RR, Dolinak D, Guileyardo J. Delayed rupture of a vertebral artery as an uncommon cause of fatal subarachnoid hemorrhage: case report. *Am J Forensic Med Pathol*. 2002;23(4):394–397.
96. Collins KA, Sellars K. Vertebral artery laceration mimicking elder abuse. *Am J Forensic Med Pathol*. 2005;26(2):150–154.
97. Ikegaya H, Yajima D, Iwase H. Vertebral artery rupture in a sudden death case after mild trauma. *Am J Forensic Med Pathol*. 2008;29(3): 276–278.
98. Galtes I, Rodriguez-Baeza A, Subirana M, Barberia E, Castella J, Medallo J. A proposed dissection procedure for vertebral arteries in forensic pathology. *J Forensic Sci*. 2012;57(1):212–214.
99. Galtes I, Borondo JC, Cos M, et al. Traumatic bilateral vertebral artery dissection. *Forensic Sci Int*. 2012;214(1–3):e12–e15.
100. Davis JM, Zimmerman RA. Injury of the carotid and vertebral arteries. *Neuroradiology*. 1983;25(2):55–69.
101. Davidson KC, Weiford EC, Dixon GD. Traumatic vertebral artery pseudoaneurysm following chiropractic manipulation. *Radiology*. 1975; 115(3):651–652.
102. DeBehnke DJ, Brady W. Vertebral artery dissection due to minor neck trauma. *J Emerg Med*. 1994;12(1):27–31.
103. Gal J, Herzog W, Kawchuk G, Conway PJ, Zhang YT. Movements of vertebrae during manipulative thrusts to unembalmed human cadavers. *J Manip Physiol Ther*. 1997;20(1):30–40.
104. Lee CK, Gray L, Maguire J. Traumatic vertebral artery injury: detailed clinicopathologic and morphometric analysis of 6 cases. *Am J Forensic Med Pathol*. 2009;30(2):134–136.
105. Medcalf JE, Paul Johnson C, Taktak A, Grabherr S. Variations in the anatomy of the vertebral artery cervical loop segment—a potential predisposing factor for traumatic basal subarachnoid hemorrhage? *Forensic Sci Med Pathol*. 2016;12(4):444–450.
106. Melinek J, Hart AP. Vertebral artery dissection complicating occipital injection of heparin for treatment of thoracic outlet syndrome. *Am J Forensic Med Pathol*. 2012;33(1):76–79.
107. Dudley MH, Garg M. Fatal child abuse presenting with multiple vertebral and vascular trauma. *J Forensic Sci*. 2014;59(2):386–389.
108. Smyk DS, Herath JC. Delayed rupture of a vertebral artery laceration: a case report and challenges for the forensic pathologist. *Forensic Sci Med Pathol*. 2018;14(4):536–540.
109. Nolte KB, Gelman BB. Intracerebral hemorrhage associated with cocaine abuse. *Arch Pathol Lab Med*. 1989;113(7):812–813.

110. Kibayashi K, Mastri A, Hirsch C. Cocaine induced intracerebral hemorrhage: analysis of predisposing factors and mechanisms. *Hum Pathol.* 1995;26:659–663.
111. McIntyre IM, Hamm CE, Sherrard JL, Gary RD, Riley AC, Lucas JR. The analysis of an intracerebral hematoma for drugs of abuse. *J Forensic Sci.* 2003;48(3):680–682.
112. McGee SM, McGee DN, McGee MB. Spontaneous intracerebral hemorrhage related to methamphetamine abuse: autopsy findings and clinical correlation. *Am J Forensic Med Pathol.* 2004;25(4):334–337.
113. Shvartsbeyn M, Phillips DG, Markey MA, Morrison A, DeJong JL, Castellani RJ. Cocaine-induced intracerebral hemorrhage in a patient with cerebral amyloid angiopathy. *J Forensic Sci.* 2010;55(5):1389–1392.
114. Baratham G, Dennyson WG. Delayed traumatic intracerebral haemorrhage. *J Neurol Neurosurg Psychiatry.* 1972;35(5):698–706.
115. Shkrum MJ, Ramsay DA. Death due to concussion and alcohol. *Am J Forensic Med Pathol.* 2000;21(1):94.
116. Molina DK, DiMaio VJ. Head trauma and alcohol: a lethal combination. *Am J Forensic Med Pathol.* 2015;36(4):290–292.
117. Zumwalt R, Hirsch C. *Current Concepts in Forensic Pathology.* USCAP; 1993, 1993.
118. Adams JH, Graham DI, Murray LS, Scott G. Diffuse axonal injury due to nonmissile head injury in humans: an analysis of 45 cases. *Ann Neurol.* 1982;12(6):557–563.
119. Graham DI, Clark JC, Adams JH, Gennarelli TA. Diffuse axonal injury caused by assault. *J Clin Pathol.* 1992;45(9):840–841.
120. Gennarelli TA, Thibault LE, Adams JH, Graham DI, Thompson CJ, Marcincin RP. Diffuse axonal injury and traumatic coma in primates. *Ann Neurol.* 1982;12(6):564–574.
121. Reichard RR, White- III CL, Hladik CL, Dolinak D. Beta-amyloid precursor protein staining in nonhomicidal pediatric medicolegal autopsies. *J Neuropathol Exp Neurol.* 2003;62(3):237–247.
122. Reichard RR, White- III CL, Hladik CL, Dolinak D. Beta-amyloid precursor protein staining of nonaccidental central nervous system injury in pediatric autopsies. *J Neurotrauma.* 2003;20(4):347–355.
123. Dolinak D, Reichard R. An overview of inflicted head injury in infants and young children, with a review of β-amyloid precursor protein immunohistochemistry. *Arch Pathol Lab Med.* 2006;130(5):712–717.
124. Blumbergs PC, Scott G, Vis J, Wainwright H, Simpson D, McLean A. Topography of axonal injury as defined by amyloid precursor protein and the sector scoring method in mild and severe closed head injury. *J Neurotrauma.* 1995;12(4):565–572.
125. Reichard RR, Smith C, Graham DI. The significance of beta-APP immunoreactivity in forensic practice. *Neuropathol Appl Neurobiol.* 2005;31(3): 304–313.
126. Gentleman SM, Nash MJ, Sweeting CJ, Graham DI, Roberts GW. Beta-amyloid precursor protein (beta APP) as a marker for axonal injury after head injury. *Neurosci Lett.* 1993;160(2):139–144.
127. McKenzie KJ, McLellan DR, Gentleman SM, Maxwell WL, Gennarelli TA, Graham DI. Is beta-APP a marker of axonal damage in short-surviving head injury? *Acta Neuropathol.* 1996;92(6):608–613.
128. Hortobagyi T, Wise S, Hunt N, et al. Traumatic axonal damage in the brain can be detected using beta-APP immunohistochemistry within 35 min after head injury to human adults. *Neuropathol Appl Neurobiol.* 2007;33(2):226–237.
129. Gorrie C, Oakes S, Duflou J, Blumbergs P, Waite PM. Axonal injury in children after motor vehicle crashes: extent, distribution, and size of axonal swellings using beta-APP immunohistochemistry. *J Neurotrauma.* 2002;19(10):1171–1182.
130. Jensen LL, Banner J, Ulhoi BP, Byard RW. beta-Amyloid precursor protein staining of the brain in sudden infant and early childhood death. *Neuropathol Appl Neurobiol.* 2014;40(4):385–397.
131. James R, Nasmyth-Jones J. Occurrence of cervical fractures in victims of judicial hanging. *Sci Int.* 1992;54:81–91.
132. Wallace SK, Cohen WA, Stern EJ, Reay DT. Judicial hanging: postmortem radiographic, CT, and MR imaging features with autopsy confirmation. *Radiology.* 1994;193(1):263–267.
133. Uhrenholt L, Nielsen E, Charles AV, Hauge E, Gregersen M. Imaging occult lesions in the cervical spine facet joints. *Am J Forensic Med Pathol.* 2009;30(2):142–147.
134. Molina DK, Nichols JJ, Dimaio VJ. The sensitivity of computed tomography (CT) scans in detecting trauma: are CT scans reliable enough for courtroom testimony? *J Trauma.* 2007;63(3):625–629.
135. Makino Y, Yokota H, Nakatani E, et al. Differences between postmortem CT and autopsy in death investigation of cervical spine injuries. *Forensic Sci Int.* 2017;281:44–51.
136. Maya-Casalprim G, Ortiz J, Tercero A, et al. Cervical spinal cord injury by a low-impact trauma as an unnoticed cause of cardiorespiratory arrest. *Eur Heart J Case Rep.* 2020;4(2):1–6.
137. Adams VI. Neck injuries: III. Ligamentous injuries of the craniocervical articulation without occipito-atlantal or atlanto-axial facet dislocation. A pathologic study of 21 traffic fatalities. *J Forensic Sci.* 1993;38(5):1097–1104.
138. Adams VI. Neck injuries: II. Atlantoaxial dislocation—a pathologic study of 14 traffic fatalities. *J Forensic Sci.* 1992;37(2):565–573.
139. Adams VI. Neck injuries: I. Occipitoatlantal dislocation—a pathologic study of twelve traffic fatalities. *J Forensic Sci.* 1992;37(2):556–564.
140. Davis GG, Glass JM. Case report of sudden death after a blow to the back of the neck. *Am J Forensic Med Pathol.* 2001;22(1):13–18.
141. Viola L, Di Nunno N, Costantinides F, Lombardo S, Mangiatordi S, Di Nunno C. Unusual neck lesion in a pedestrian caused by a road accident. *Am J Forensic Med Pathol.* 2002;23(1):68–71.
142. Bolliger S, Plattner T, Zollinger U. The deadly broomstick: an unusual missile injury to the neck. *Am J Forensic Med Pathol.* 2006;27(4):304–306.
143. Nikolic S, Micic J, Atanasijevic T, Djokic V, Djonic D. Analysis of neck injuries in hanging. *Am J Forensic Med Pathol.* 2003;24(2):179–182.
144. Zivkovic V, Nikolic S, Strajina V, Babic D, Djonic D, Djuric M. Pontomedullary lacerations and concomitant head and neck injuries: their underlying mechanism. A prospective autopsy study. *Forensic Sci Med Pathol.* 2012;8(3):237–242.
145. Ali Z, Fowler DR. En bloc examination of the neck in pediatric homicide cases: a proper way for complete assessment of neck trauma. *Acad Forensic Pathol.* 2016;6(4):622–637.
146. Pollanen MS. Pitfalls and artifacts in the neck at autopsy. *Acad Forensic Pathol.* 2016;6(1):45–62.

147. Koszyca B, Gilbert JD, Blumbergs PC. Traumatic subarachnoid hemorrhage and extracranial vertebral artery injury: a case report and review of the literature. *Am J Forensic Med Pathol*. 2003;24(2):114–118.
148. Matshes EW, Joseph J. Pathologic evaluation of the cervical spine following surgical and chiropractic interventions. *J Forensic Sci*. 2012;57(1): 113–119.
149. Wang M. Death due to extensive cervicomedullary infarction following iatrogenic vertebral artery occlusion. *Forensic Sci Med Pathol*. 2012; 8(3):334–337.
150. Demontis R, d'Aloja E, Manieli C, et al. Case report of sudden death after a gunshot wound to the C2 vertebral bone without direct spinal cord injury: histopathological analysis of spinal-medullary junction. *Forensic Sci Int*. 2019;301:e49–e54.
151. Ditunno JF, Little JW, Tessler A, Burns AS. Spinal shock revisited: a four-phase model. *Spinal Cord*. 2004;42(7):383–395.
152. Ko H-Y, Huh S. *Spinal Shock: Definition and Clinical Implications. Handbook of Spinal Cord Injuries and Related Disorders*. Springer; 2021:165–176.

CHAPTER

11

Approach to asphyxial deaths

Susan F. Ely and James R. Gill

When you have eliminated the impossible, whatever remains, however improbable, must be the truth. *-Sir Arthur Conan Doyle*

Asphyxia

Asphyxial deaths can be some of the most challenging deaths to investigate because the findings may be subtle and have a way of turning up where they are least expected. The textbook images of strangulations represent the easy cases. The challenge arises when there are petechiae with focal strap muscle hemorrhage and equivocal circumstances. The forensic pathologist's determination will likely result in the arrest and trial of a defendant.

Accidental and suicidal asphyxial deaths are usually fairly straightforward; however, they may rarely be staged in an attempt to "cover-up" a homicide. Therefore, one must not fall into the complacency trap and just assume that the "hanging" death is always a suicide. Ask yourself, could this be a homicide? As the old forensic quip reminds us, "think dirty." A similar thought process applies to clinical medicine. Physicians do not want to misdiagnosis a potentially fatal disease with an innocuous benign process. For example, one of the more common causes of back pain in a 60-year-old man is degenerative joint disease. An aortic aneurysm/dissection, however, also can cause back pain.[1] Of course, extensive imaging studies usually are not initially performed on everyone who presents with back pain, but the more life-threatening diagnosis should at least be clinically entertained in the context of the physical exam and the history.[2] Common things are indeed common, but one must always remain open to encountering a zebra.

The main reasons that "asphyxial" deaths can be challenging are based on the fact that all deaths are due to a breakdown in the oxygen cycle and that asphyxia, per se, is not defined by a singular pathognomonic sign that unequivocally confirms the diagnosis. Therefore, any finding that results from a lack of oxygen (e.g., anoxic–ischemic encephalopathy) can theoretically be seen in most, if not all deaths, regardless of how the hypoxemia arose. Most important to understand, however, is the concept that for hypoxemia to be classified as asphyxial in nature, its cause must be *unnatural* (an injury). For that reason, the investigative algorithm in this specific category of deaths involves trying to determine whether and how the asphyxia was brought about (i.e., establishing context). Finding a hotdog lodged in the airway of a three-year-old would constitute the evidence supporting an asphyxial mechanism of death by airway obstruction. Other than the presence of the hotdog, there would be no specific finding at autopsy in this particular case to allow one to make the diagnosis of asphyxia by airway obstruction. *Drowning* represents another category among asphyxial deaths, but it is discussed separately in Chapters 6 and 13.

Asphyxia: accidental and suicidal

Approach to potential choking deaths

It is exceedingly rare for a sober, neurologically intact adult to choke to death on an aspirated bolus of food.[3–5] Therefore, such a death requires a search for an underlying explanation (proximate cause) for the fatal impairment of the swallow and gag reflexes. **Ethanol** *is the most common intoxicant* as the underlying cause of such deaths, and other misused depressant substances (e.g., opioids, sedatives) are occasionally responsible.[5] Neuropsychiatric diseases that are frequently implicated include multiple sclerosis, amyotrophic lateral sclerosis (ALS), Parkinson disease, cerebral vascular episodes, Alzheimer disease and other dementias, and various psychiatric disorders. Recognizing these risk factors may help in both the prevention of and the suspicion for a choking death.[6]

https://doi.org/10.1016/B978-0-323-91796-4.00011-8

A study in San Diego of 133 deaths due to choking found 14% of the decedents with alcohol or other sedatives intoxications and 55% with a documented neurological deficit or anatomic impairment of swallowing.[4] The most common type of food were meat products; 45% occurred at home, 26% at supervised facilities, and 14% at restaurants.[4]

The instruction for part 2 is, "other significant conditions contributing to death, but not resulting in the underlying cause given in part 1." Therefore, it would defy this rule and be both illogical and internally inconsistent to place the intoxication or neurologic disorder in part 2 of the death certificate. Similar to a drowning death due to epilepsy, the manner of choking deaths would not be natural simply because the underlying cause is a disease. The immediate cause of death (aspiration of a bolus of food) is an injury. Therefore, the death does not result exclusively (100%) from disease, so the manner of death is not natural. In general, while death certificates involving traumatic fatalities do objectively describe *how* the injury arose (the "how injury occurred" field), they do not attempt to explain *why* an injury occurred, except in rare instances in which the injury circumstance demands an explanation (see below). For example, a drunk driver who crashes his car dies from the blunt injury (proximate cause). The ethanol may help to explain why the car crash occurred, as could road conditions, texting, glare, and other variables, but none of these contributes physiologically in causing the death, nor is it necessary for certification purposes or even possible to make these judgements with any veracity in our role as a forensic pathologist (FP). Indeed, opining about the *why* of the crash, rather than the *how* of it may well involve conjecture, something forensic pathologists try to avoid at all cost. In rare instances, however, when a traumatic death "does not make sense," it is acceptable to include the "reason," if known, which doubles as the proximate cause. *This approach becomes necessary when an adult drowns in a bathtub, freezes on a park bench, or chokes on a bolus of food.* The fatal outcome of these circumstances demands a further explanation.

On occasion, we conduct an autopsy on a person who has not undergone resuscitative efforts and we unexpectedly discover complete airway obstruction, typically upper airway/laryngeal, by a lodged foreign body (most commonly, a tenacious food bolus, such as a piece of meat), allowing us to conclude that the death resulted from choking (assuming all other more compelling causes, particularly those that are incompatible with continued life, have been excluded). In order to competently certify such deaths, our investigative burden is to identify the etiologically specific proximate cause and/or circumstance that placed the person at risk of choking. Much like drowning in a bathtub, choking is best explained in the cause of death statement and/or in the "how the injury occurred" field. Except in children, it should rarely "stand alone" on the death certificate without further qualification. *It is typically viewed as an immediate cause more than a proximate one.*

In the vast majority of instances, neurologically competent, nonintoxicated persons should not choke to death on food. Exceptions to every rule exist, however. On occasion, a particularly grizzled/tough piece of meat, laughter, or poor mastication unrelated to edentulousness or psychiatric or chemical factors explain the choking by exclusion, but every effort must be made to identify a disease or injury (chemical or mechanical) before reaching that conclusion and crafting a descriptive cause of death statement explaining how the cause arose.

Much more commonly than the above autopsy scenario, we are instead *presented with a diagnosis made by emergency medical service (EMS) or hospital personnel, nursing home staff, or family that an individual "choked to death."* This is the classic situation in which we are being given "the editorial, not the news." These opinions should never be accepted out of hand; instead, some circumspection should be applied and a complete and systematic investigation should be conducted. Many of these deaths, after a thorough investigation, turn out to be natural deaths.

When presented with the initial investigative "diagnosis" of choking, it is very important to ask a series of questions to test the veracity of that claim. Every effort should be made to clarify the sequence of events, timeline, and signs/symptoms, if any, that preceded the loss of consciousness/arrest, and dovetail those with the autopsy findings. For example, was the person eating at the time of arrest? If so, what? Be specific (things like apple sauce and oatmeal are much less concerning from a choking/airway obstruction standpoint than more tenacious and "sticky" foods such as meat, peanut butter, bread, etc.). Was there an indication of a conscious, even if subtle struggle prior to the arrest? Were any signs of airway obstruction evident? *These may include one or more of the following: awake and alert, eyes wide, appearing alarmed, or panicked, but quiet, hands sometimes to the neck/throat area, unable to vocalize because of lack of air movement past the vocal cords, turning blue before losing consciousness, etc.* It is best to specifically ask these questions and not just broadly ask, if there were "any signs of respiratory difficulty?" Remember that during airway obstruction, there is typically a period of silence and/or a conscious struggle to breathe prior to the respiratory arrest. For example, if an elderly hypertensive person is eating breakfast while talking to their spouse and suddenly stops speaking and loses consciousness, in a "lights out" fashion, without any antecedent signs or symptoms of respiratory compromise/distress, or even sudden, if calm aphasia, then, in overwhelming likelihood, they did not choke on food. Rather, they had a cardiac event while they happened to have been eating, supported not only by the precipitousness of the event, but also by the fact that it occurred while they were talking (i.e., moving air across the vocal cords). In this circumstance, the food/eating is most likely *coincident* to the death, not *causal*. By contrast, this careful history taking also helps to identify the so-called "café coronary," which actually is a choking death that is mistaken for a cardiac death/heart attack.[7–9]

If the circumstances sound suspicious for a choking, seek specific information from primary sources, such as first responders (this is best achieved by speaking directly with the emergency medical technician, or at least obtaining a copy of the ambulance call report) and other witnesses (e.g., group or nursing home staff, family/friends). One may inquire if the Heimlich maneuver was applied and what the result was (e.g., did a bolus of food come out of the mouth? If so, what kind and what size?).

One should ask if any attempts at intubation were made, where (in the field or in the emergency department), and what, if anything, was seen or felt during this procedure. Was there resistance in the airway when trying to ventilate the patient? Was there visible food in the airway? If so, where exactly, how lodged/obstructive, what size, and what type of food? Were forceps needed to extract the food? Answers to all these questions usually are not clearly described in the written record, so direct conversations with primary sources may prove important. The reason that primary sources of information are important to seek is because, just like the children's game of operator, the "story" often changes as it is relayed to more people further distanced from the eyewitnesses. Remember that the attending physician that you are speaking with received the history from the emergency medicine resident, who received it from the triage nurse, who received it from the EMT, who received it from a family member. At each step, the history can change and details often become blurred or lost. Also, it is equally important to seek *facts* surrounding the putative choking event, rather than others' opinions/conclusions.

If the food was removed during resuscitative efforts, a subsequent autopsy (absent a Class I finding) will be of limited value, putting the burden on a detailed investigation. There is an important difference between a few crumbs or soft food passively residing in the back of the mouth/posterior oropharynx during the finger sweep of the oral cavity by an EMT and the finding of a firmly lodged food bolus in the larynx that required forceps extraction. This must be specified, as these details matter very much.

It is not unusual for emergency personnel to "overcall" a death as a choking. They may report "gastric contents" in the airway with consequent "aspiration." While this should not be dismissed out of hand, most of these instances represent terminal aspiration of gastric contents regurgitated as a consequence of the agonal and/or resuscitative processes. After death, sphincters in the body relax and with the benefit of gravity or chest compressions, the contents of a stomach may be regurgitated or simply passively relocate. These people are "aspirating" because they are dying rather than dying from aspiration. This is a common agonal phenomenon and is a consequence, not a cause of death. Remain circumspect and ask detailed questions. You might subsequently perform an autopsy in such a case and find a Class I finding such as a ruptured myocardial infarct or a saddle pulmonary thromboembolus, confirming your suspicion that the "aspiration" theory was both premature and inaccurate. Additionally, it is important to be both circumspect and observant at the time of autopsy. With a careful examination, *it is often apparent whether one is looking at masticated/partially masticated food en route to the stomach, on its way "down," so to speak, versus food that has been subjected to the gastric environment and is on its way back up (i.e., passive gastric contents or active emesis, as the case may be).* Most choking events that occur while eating a meal represent the former, so looking for confirmatory evidence of that at autopsy may prove helpful in answering the "cause versus consequence" question. To be clear, aspiration of partially digested gastric contents can and does occur during life and it may also cause morbidity and mortality. But, it more commonly does so by producing an aspiration pneumonia (chemical pneumonitis) that often takes hours to days to become life threatening.

When the inciting event prompting a hospitalization is purported to be a choking, the same aforementioned investigation should ensue and becomes all the more important and central to the final death certification opinion, while the utility/contribution of an autopsy becomes negligible at best. Expected findings at such an autopsy would include various sequelae of anoxic encephalopathy and prolonged immobility, rather than any helpful findings supporting the putative choking event, depending on the length of the hospitalization. The main reason to perform an autopsy is these situations might be to identify a previously undiagnosed Class I finding that would supplant the choking diagnosis (unlikely after a prolonged survival interval) or to identify an otherwise unclear proximate cause (disease or injury) of the choking, if the circumstances and history have not yet provided an explanation. If the investigation, on the other hand, does not support a choking by the criteria used above, but rather is better explained as a natural death, it may well be certified as such, without the need for an autopsy.

As stated above, examples of competent proximate causes for choking include acute intoxications and neurologic diseases (with a resultant effect on the gag reflex), such as multiple sclerosis, an acute or remote cerebrovascular event with residual neurologic deficits, epilepsy/seizure disorders, cerebral palsy (of various etiologies), ALS, and other progressive neurodegenerative diseases. Other considerations include hepatic encephalopathy, Down syndrome (including risks associated with macroglossia), advanced dementia, and Parkinson disease. Psychiatric disease or cognitive delay with attendant behavioral risks (e.g., shoveling large quantities of food into one's mouth, minimal mastication, yelling or thrashing while eating, etc., ingesting/aspirating nonfood foreign bodies, pica) also provides potential explanations. Examining the gastric contents to assess the extent to which the food is masticated may be useful in giving a more objective and granular sense of behavioral and choking risk (including edentulousness). By this reasoning, schizophrenia may be a competent proximate cause for choking, depending on the degree of psychiatric disability and functioning. Obstructive or functional esophageal pathology (tumors, aphasia, strictures, etc.), age, and developmental limitations and risks also can explain choking (e.g., toddlers with limited oral coordination and risk of placing foreign objects into their mouths). If a remote head/neck trauma with residual deficits and/or a posttraumatic seizure disorder is the etiology, one must remember that the manner is predicated on the circumstances of the head or neck injury and these must be sought, in addition to that for the acute choking event. For example, if an inflicted remote gunshot wound leaves an individual quadriplegic, later leading to a choking event that ends his life, the manner would justifiably be certified as homicide (homicide based on the remote firearms event and because the proximate cause trumps the accidental immediate injurious cause—the choking event).

In most choking deaths, the manner is certified as accident; the cause of death statement is based on the circumstances and proximate cause. For example, it may be "upper airway obstruction (food bolus)" or "aspiration of bolus of food" complicating/due to the proximate cause (disease or intoxication or mechanical injury that explains the choking). For example, multiple sclerosis explains how and why a person choked to death on a bolus of food, and therefore a disease is the underlying cause of death. Even though multiple sclerosis is a disease, the immediate cause of death (aspiration of a bolus of food) is an injury. Consequently, the death did not result exclusively from disease and the manner of death is classified as accident.

On rare occasion, the risk or risks leading to a choking event are more complex or sometimes involve circumstantial factors with no identifiable proximate disease or injury. In these instances, it is best to employ a *descriptive approach* to death certification. The foregoing examples illustrate such an approach: "Aspiration of Bolus of Food in Edentulous Person with Schizophrenia." This is a descriptive cause of death statement listing two risk factors: a disease (schizophrenia) and a condition ("edentulousness"), the latter, neither a codable disease nor an injury, per se, despite presenting a choking risk. The "how injury occurred" section in such a death might state, "Choked on food during rapid and excessive oral intake with minimal mastication/chewing." Alternatively, when no identifiable proximate cause is identified, the circumstance that presents the choking risk can also be explained in the "how injury occurred" field of the death certificate (e.g., "Aspirated partially unchewed food while running for bus"). In the bus scenario, the upper airway obstruction can stand alone as the immediate cause since there is no underlying disease or injury (proximate cause); the fatal outcome can be circumstantially explained either in the "how injury occurred" section or embedded within the cause of death statement, parenthetically, after the immediate cause of death, with simply "see above" in the "how injury occurred" section. Occasionally, this descriptive approach is the only logical choice the FP can make when presented with a decedent who is both sober and neurologically and psychiatrically intact.

To summarize, it is astonishingly rare for asphyxia by aspiration of a bolus of food or other foreign bodies (Figs. 11.1 and 11.2) to occur in neurologically intact, sober adults. These adults chew and swallow food thousands upon thousands of times in a lifetime without inadvertently choking to death. Aspiration of food demands an underlying explanation for the fatal impairment of swallowing and gag reflexes. *Ethanol intoxication is the most common underlying cause of such deaths,* but other abused substances (e.g., opioids, benzodiazepines, etc.) are occasionally responsible. It is illogical and internally inconsistent to place the intoxication or neurologic disorder in part 2 of the death certificate because the instruction for part 2 is, "other significant conditions contributing to death but not resulting in the underlying cause given in part 1."

Approach to lack of environmental oxygen

Death due to a lack of environmental oxygen is a diagnosis of exclusion determined after an autopsy and investigation of the circumstances. Typical scenarios include the displacement of oxygen by an inert gas (such as nitrogen or helium) or by an airtight enclosure/airless space (e.g., a plastic bag placed over the head and tightly fastened at the neck or being trapped in an unused refrigerator).[10–29] The cause becomes more complex when the displacing gas is not inert (e.g., carbon dioxide, hydrogen sulfide, carbon monoxide), as such toxic gases are also direct asphyxiants. Deaths due to inhalation of hydrogen sulfide[30–40] or carbon monoxide are pure intoxication deaths through asphyxiation at a cellular level.[41–62] Carbon dioxide inhalation also causes death by intoxication; however, depending upon the circumstances, there may be a component of oxygen displacement as well.[10,16,23,41,63–69] The use of an intoxicant gas in addition to the placement of a plastic bag over the head (e.g., *Final Exit* suicide[70]) may pose challenges in determining the mechanism of death. Is it an intoxication, a suffocation, or some combination thereof?

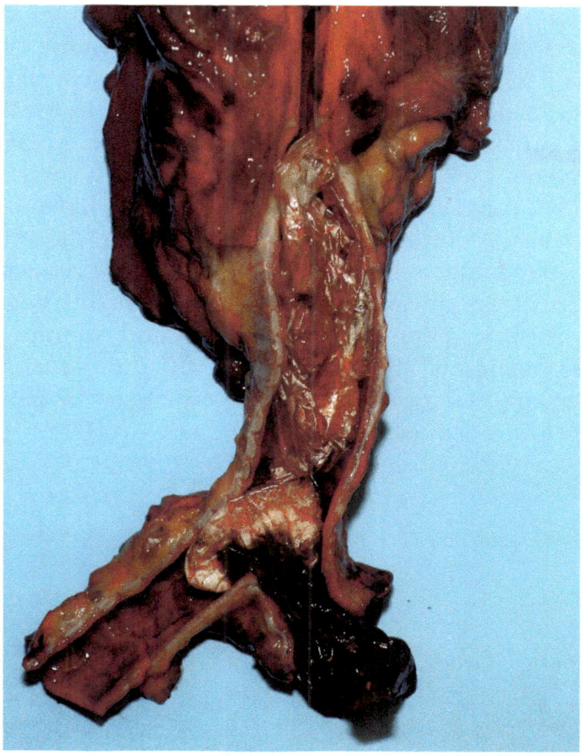

FIGURE 11.1 Plastic bag of cocaine lodged in the trachea.

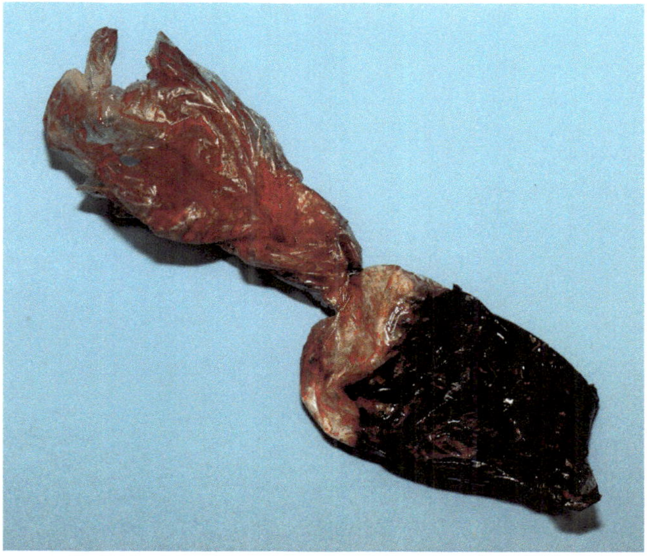

FIGURE 11.2 Bag of cocaine recovered from the trachea.

Approach to final exit (and similar) deaths

Analogous to hanging suicides, these deaths are fairly straight forward in terms of the evidence of intentionality and the ease with which the manner is determined. That said, one should still carefully document the size and placement of the bag over the head and whether and how tightly it is secured around the neck.[71,72] With a tight fastening of the typical small to medium-sized plastic bag, a good seal is created, and consequently an airless space, and the mechanism of death due to asphyxia is through pure anoxia. If, in addition to that, tubing is connected to a tank of an inert, nonintoxicating gas (such as helium, propane, or nitrogen) and inserted into the bag, this will displace the already limited oxygen, inflate the bag (making it more comfortable), and may help displace carbon dioxide (Figs. 11.3 and 11.4). Pure hypoxia (lack of oxygen) is actually a pleasurable feeling. By contrast, the buildup of carbon dioxide in the blood is what causes intense anxiety and the urge to breathe (so-called "air hunger"). The suffering of a patient in palliative care may be mitigated by maintaining ventilatory support (to remove to carbon dioxide), without adding supplemental oxygen, thus allowing nature to take its course.

Documentation of the body and apparatus with the usual "as is" overall photographs, both at the scene of death and at the autopsy table, should clearly document the bag and its fastening "in situ," if still in place after emergency services activities and transport. Taking note of the presence or absence of condensation on the inner surface of the plastic bag and documenting this photographically and in the autopsy report, negatively or positively, is important. The removed bag and fastenings are then photographed separately and submitted as evidence. Condensation supports the conclusions of a sealed fastening of the bag in a breathing individual, its presence being helpful, but its absence, less so, particularly after any manipulation by first responders. If the fastening and/or bag have been altered by police or EMS personnel, this evidence may be lost.

FIGURE 11.3 **Helium tank.** Helium tank used in a *Final Exit* death.

FIGURE 11.4 ***Final Exit* suicide.** Tube connected to a helium tank and inserted into a plastic bag placed over the head.

Noting whether there is an underlying neck impression produced by a tight fastening may rarely implicate a neck compression component, but is much more important in its support of a rapid anoxic mechanism of asphyxia (i.e., a tight seal). When the fastening and bag have been disturbed by first responders, an impression left on the neck and/or the presence of still existent condensation can prove helpful in deducing that the bag had originally been tightly fastened. This is important in how the FP views the case since, with a tight fastening, death ensues in minutes following the placement of the bag, and there usually is no need to defer certification of the cause of death out of a concern for a contributory intoxication. If the person had been conscious enough to place and fasten the bag over the head just minutes before death, no amount of gastric contents containing depressant drugs could be absorbed and exert lethal effects in that short period of time (a few minutes). In other words, a toxicological mechanism is moot in most of these deaths, similar to hanging deaths or those due to gunshot wounds of the head. That said, many of these decedents, at the suggestion of the *Final Exit* prescription, have benzodiazepines and/or other anxiolytic medications in their blood at autopsy, the decedents having taken them in an attempt to quell the anxiety brought on by an otherwise stressful act and to proceed with the suicide. In most instances, however, these medications exert no lethal effects, per se.

Typical cause of death statements include: "Asphyxia by Covering of Head with Fastened Plastic Bag and Displacement of Oxygen by Helium Gas" or "Asphyxia by Covering of Head with Fastened Plastic Bag" (when not connected to a gas tank).

Approach to hangings

A hanging can be viewed as a type of ligature strangulation in which *the compressive forces on the neck are produced by gravity—the weight of the head* (and body, depending on the case), rather than through an *active* squeezing of the neck by any other force, as occurs in strangulations.[74–118]

There is substantial overlap in the physical findings found at autopsy in homicidal ligature strangulations, accidental ligature strangulations, suicidal ligature strangulations, and some hangings. Rather than being recognized through a specific pattern of their ligature injury, ligature compressions are predominantly distinguished by the establishment of the cause of the tightening of the ligature: by self or by other. Just as one cannot tell, based on the wound, who fired the gun that caused a contact gunshot wound of the head, one also cannot necessarily tell from examination of the neck alone, what caused the tightening of the ligature. As such, once again, the scene and circumstances are key. While the pattern and course of ligature abrasions in *hangings* are often distinct from those of most homicidal strangulations, simply looking in a vacuum at the ligature abrasion or furrow in isolation at autopsy may not distinguish between homicidal, accidental, and suicidal ligature *strangulations*.[93,98,99] That said, signs of a struggle are not uncommonly seen in addition to the ligature injury in homicidal strangulations, and, needless to say, this finding often proves conclusive in arriving at a manner determination.

Typical hanging deaths are not truly "asphyxial" deaths because airway compromise often plays a minor mechanistic role. Compression of the cervical vasculature (carotid arteries and jugular veins) results in a local cerebral ischemia (from venous or venous–arterial compression) that causes death. This process is best exemplified by the hanging death of a person with a tracheostomy. Such an individual is able to cause their death by hanging with the ligature superior to the tracheostomy which allows the tracheostomy to bypass any upper airway compression, allowing air to move freely in and out of the lungs.[118] That said, there may be some degree of upper airway obstruction in certain hanging deaths, particularly when the ligature/"noose" is located high in the neck, limited to further slippage upwards only by the protruding angles of the mandible. Rather than the common misconception that there is compression of the rigid larynx, this specific ligature positioning is more likely to displace the base of the soft tongue upwards and posteriorly, potentially causing a functional occlusion/narrowing of the posterior oropharynx. When the ligature is specifically so positioned, the likelihood of airway compression as a mechanism, in addition to the vascular occlusion, is most enhanced by a full suspension hanging, with the feet off the ground/floor (i.e., the forces applied to the neck are a function of the body's full weight, not just that of the head; see below). Given the relatively little difference in compressive forces needed to occlude the jugular veins (2 kg) compared to the carotid arteries (5 kg), this may in part explain why petechiae are not seen in all partial suspension hangings (Fig. 11.5).[99] Moreover, a factor that may explain why petechiae are observed in some fully suspended hangings may be explained by the vertebral artery blood supply.[99] The vertebral arteries are not compressed by a hanging, and even though they are responsible for only about 20% of blood to the brain, they may contribute to the development of petechiae, even with carotid occlusion.

The autopsy of hanging deaths affords the forensic pathologist the opportunity to observe the wide spectrum of neck injury that may or may not be present in a death due to a compression of the neck by a ligature. This mental

FIGURE 11.5 **Hanging (partially supported).**

"database" is important to refer to when investigating a subtle neck compression homicide. A hanging death with a broad sheet, for example, may leave no marks externally or internally, aside from petechiae in some instances. Documentation of the presence or absence of petechiae and their location, along with the size, color, pattern, and upward or vertical configuration of the ligature abrasion or furrow is a common descriptive feature to include in the autopsy report of a hanging victim. A layered neck dissection will demonstrate the distribution of hemorrhage, if any.[119,120]

Many factors contribute to the variability of physical findings in hanging deaths and must be kept in mind, lest the finder of fact mistakenly think that there is a "one size fits all" diagnostic appearance in these deaths. The variability of findings is reflected both in the type of neck injury caused by the ligature itself and in the presence or absence of facial/conjunctival petechiae. These factors include, but are not limited to, the size, rigidity, and degree of abrasiveness of the ligature, the degree of body suspension, the duration of the suspension before the body is discovered, the specific placement of the ligature around the neck (i.e., high or mid-neck), the location of the point of suspension of the ligature relative to the neck opposite to the point of suspension, where the compressive forces are the greatest, and the found position of the body at the scene (e.g., fully suspended and vertical vs. more horizontal and prone, leaning forward, hovering over and supported by the floor, the body roughly parallel to it, with the greatest force exerted on the mid-anterior neck rather than high in the neck at the level of the hyoid bone and tongue). See Table 11.1.

Characteristics of the ligature

The more unyielding, abrasive, and thinner (i.e., smaller surface area, which leads to a more focal application of force) the ligature is, the more pronounced the ligature impression is likely to be (i.e., an abrasion). Ligatures such as telephone cords, belts (Fig. 11.6), computer cables or charger cords, or thin rope, for example, are all likely to leave behind ligature abrasions, and even furrows (trenches), particularly if a full or nearly full suspension is used, given the greater forces applied to the neck in such hangings. By contrast, a ligature constructed from a bedsheet or a piece of clothing, by virtue not only of its surface area but also of its lack of resistance and abrasiveness, will much *less* likely produce an abrasion of any sort, much less a furrow; on occasion, there is a *faint impression* on the skin of the neck from soft ligatures, without demonstrable injury, per se, internally or externally. In addition, the tighter, smaller, and more compressive the ligature, the less likely there will be associated facial/conjunctival petechiae,

TABLE 11.1 Factors that may affect findings in hanging deaths.

Characteristics of ligature: narrow versus broad, padding, abrasive quality
Degree of suspension: fully suspended or a partially supported body
Duration of suspension
Placement of the ligature in the neck's vertical dimension
Location of suspension point

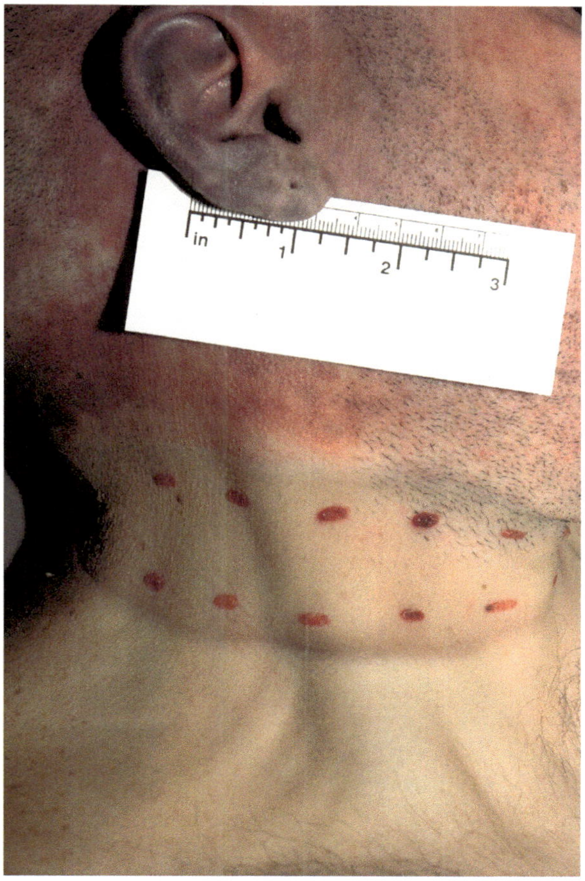

FIGURE 11.6 **Ligature abrasion.** Hanged self with belt; note the pattern created by the belt holes.

predicated on the likelihood of complete occlusion of the carotid arteries with this set of variables (see below). Internal injuries of the neck are also typically commensurate with the degree of compression and the rigidity and surface area of the ligature used.[88,89,91–93,99,106,114,121,122]

Degree of suspension

The greater the degree of suspension (i.e., maximum = feet off the floor), the greater the compressive forces applied to the neck, and the *more* pronounced both the ligature abrasion/furrow and subjacent/internal injuries will likely be and the *less* likely the formation of facial/conjunctival petechiae.

Duration of suspension before body discovery

Not surprisingly, the longer the body is suspended before being discovered and "cut down," the more likely the development of demonstrable injury of any type, externally or internally.[88,89,91–93,99,106,114,121,122]

Placement of the ligature in the neck's vertical dimension

If the ligature and body are positioned in such a way that the greatest compressive forces are applied to the mid-anterior neck, overlying the thyroid and cricoid cartilages, injuries/fractures of those structures are more likely. This is sometimes observed when the body is found suspended in a relatively more horizontal position, leaning forward and hovering over the floor, rather than upright/vertical. If instead, the ligature is placed higher in the neck (more typical), either no skeletal trauma would likely result, or perhaps only the hyoid bone might be injured.[88,89,91–93,99,106,114,121,122]

Location of the point of suspension

The part of the neck adjacent to/underlying the point of suspension of the ligature is typically the area of the neck placed under the least pressure and therefore, sustains the least amount of injury, whereas the area of the neck opposite to the point of suspension is subjected to the greatest compressive forces. For this reason, the most prominent and deepest segment of the ligature abrasion or furrow is expected to be in the latter location. The point of suspension, whether composed of a slip knot or a fixed knot, is most commonly in a posterior position; not uncommonly, no injury, or only an impression of the knot, is observed in this area.[88,89,91–93,99,106,114,121,122]

The manner of death in most hangings is **suicide**. Accidental hangings are inadvertent and typically involve children (entanglements), intoxications, the elderly, and autoerotic practices.[77,94,105,110,123–125] Children may get entangled in drapery cords, high chairs, or other apparently benign structures (e.g., lanyards) that are as rare as they are unusual.[76,77,105,123,126–131] In the older pediatric age group, one must consider various *"asphyxial games,"* in which children apply pressure to the neck.[132] Usually, these are done in groups and involve manual compression that is released following loss of consciousness. If it is done alone and with a ligature, however, there is an increased risk of death. Examination of internet searches and/or interviews with friends may reveal interest in this "game," also known Black Hole, Black Out, Flat-lining, Funky Chicken, Gasp, Knock Out, Rising Sun, Space Cowboy, Space Monkey, Suffocation Roulette, and Tingling. Parents, in their grief and guilt to deny that a death is a suicide, may argue for this diagnosis when not relevant, but it is important to recognize that some deaths that appear to be a suicide may indeed be accidental. This is why a careful investigation must be done.[132]

Intoxicated individuals and the elderly may die from accidental hangings due to their inability to extricate themselves from a precarious positional situation. An intoxicated adult may "pass out" with a necklace getting caught on a suspension point or across a fence or railing.[124,133,134]

Autoerotic practices nearly uniformly involve *men* and if the scene has not been disturbed, they are often fairly obvious (Figs. 11.7 and 11.8).[85,94,135–139] The decedent may be partially naked or wearing usual clothing or bondage wear. *There may be mirrors, lubricant, or pornographic material displayed.* If a family member first discovers the decedent, they may alter the scene. Ligatures often involve some type of *escape-release mechanism* that is intended to relieve the pressure if the person loses consciousness. Regardless of the failure of the fail-safe mechanism in these cases, it should nonetheless be abundantly clear to the forensic investigator that the apparatus could have easily been configured by the decedent and is compatible with such an intent (e.g., in contrast to complex and tight "hog-tying," an example of a binding that is incompatible with a self-created apparatus; Fig. 11.16). There may be *cushioning* between the neck and the ligature for comfort and to prevent injury marks being left on the neck, evidence that the decedent may have wanted to hide from others (Fig. 11.7).

In terms of suicidal hangings, it is very important to recognize that the very act of constructing a noose, attaching it to a fixed structure, placing the noose over one's head, and moving into a dependent head-hanged or body-hanged position, sometimes after having stepped off a ladder, chair, or other intentionally placed elevated object, in and of itself, constitute irrefutable linear, cogent planning and intent, inconsistent with inadvertency or erratic or confused behavior. By definition, absent the mitigating, specific, and rare circumstances described above, the very act of hanging oneself represents evidence of clear intent that is not reasonably interpreted any other way. A suicide note, found in under 50% of suicidal fatalities overall, is of less importance and necessity in the investigation of hanging suicides than in many others.[140–148] Importantly, it must be emphasized that in the certification of "how injury occurred" in hanging deaths of all manners, the wording should be **"hanged self" or "hanged self with rope"** in suicides, along with any further qualifying descriptions for accidents, but that in no instance should "hung self" be memorialized on a death certificate (pictures are hung, people are hanged!). Finally, when there is ligature-associated neck injury, the uses of the words **furrow, abrasion, and impression** (the latter not strictly an injury, per se) should be carefully chosen based on what is seen on the body. Strictly speaking, a "furrow" indicates an abraded trench, if even a shallow one, whereas an abrasion does not. An "impression" indicates visible space-occupying evidence of the ligature, without loss of the epidermis or other demonstrable injury, per se.

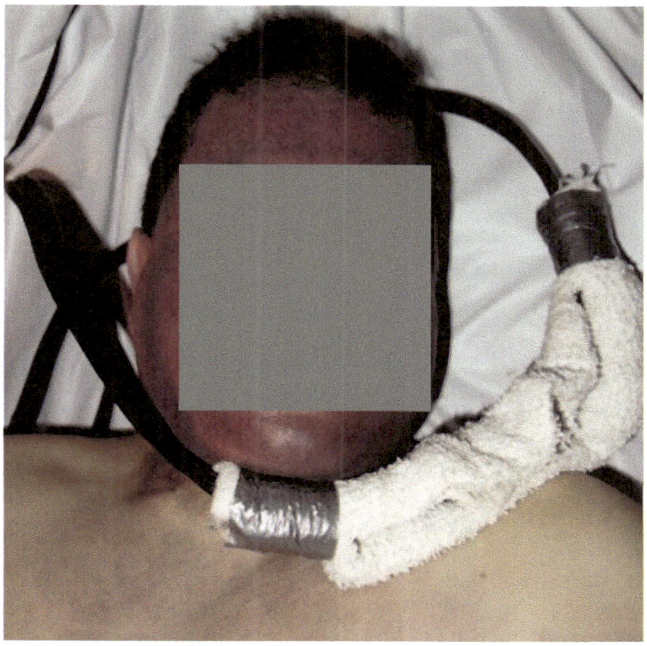

FIGURE 11.7 **Padded ligature in autoerotic asphyxia.**

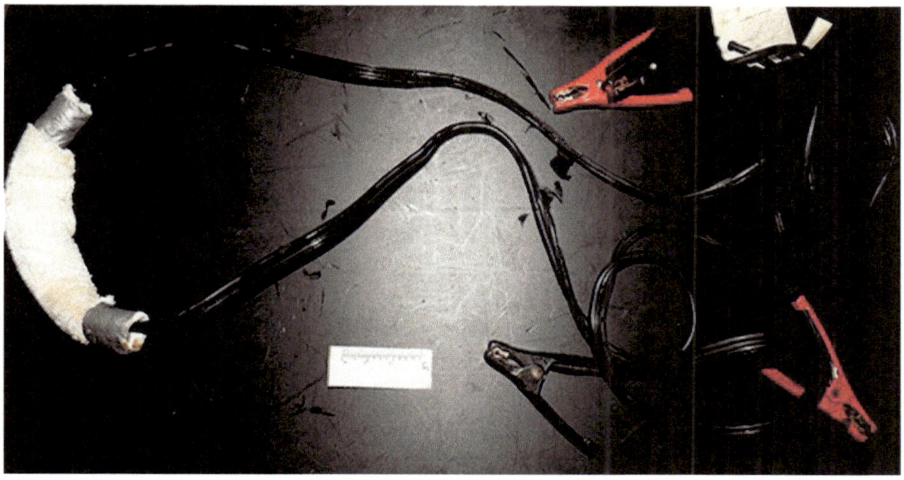

FIGURE 11.8 **Ligature in autoerotic asphyxia.** The ligature was secured with hand-clips which were meant to release with increased weight upon loss of consciousness. They did not release.

Approach to potential positional asphyxia deaths

When you are presented with a decedent in what appears to be a very contorted and/or compromised position at a scene (Fig. 11.9), the temptation is to think or to assume that that position is the primary or a contributory cause of death ("the distraction of the dramatic"). While that may well be true, we must realize that positional asphyxia in one's home is much less common than is dying (due to myriad causes) in odd contortions. That said, to systematically answer this question ("Is it cause or consequence?"), one must first identify *how* and *if* that specific position might cause death (neck or chest compression? smothering? etc.). Is the position potentially lethal, in a vacuum, after excluding other, more likely causes? If so, then, *why*? These questions should be approached in the same way that one would approach an adult found submerged in a bathtub. Seek an explanation for this person's specific risk of injury and death, since sober, neurologically intact individuals typically do not die by positional asphyxia (there are exceptions; if you live on a farm and accidentally fall, then sink into a silo of grain, for example). The point is, one must ask what chemical or neurologic compromise might have reasonably prevented this particular individual from extricating himself from this compromised position. If there are no reasonable explanations, then the likelihood is

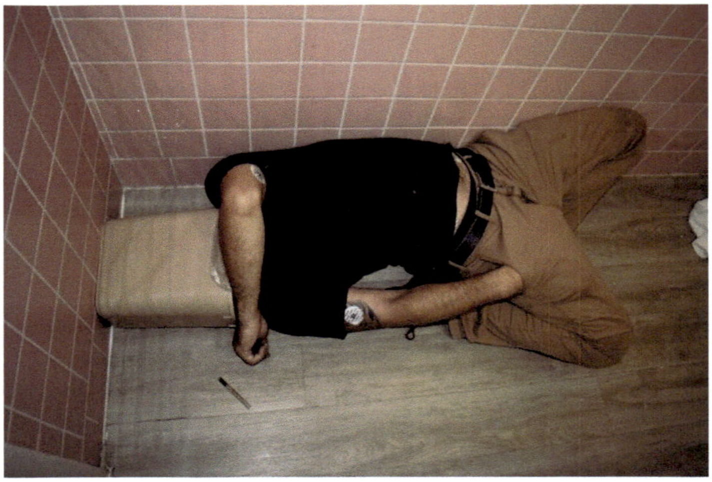

FIGURE 11.9 **Drug intoxication.** Note syringe. Intoxication with head falling into a garbage can.

that their position is a *consequence* of dying, not the *cause* of it. If the position does not appear to be potentially lethal anyway (the "how" question), then asking "why" becomes moot. If positional asphyxia as the mechanism is being seriously considered, then it typically warrants deferring the definitive death certification pending toxicology and/or neuropathology findings to identify the proximate cause. Finally, just as drowning due to multiple sclerosis is an accident, predicated on the injury of drowning despite the proximate cause being natural, positional asphyxia is also an injury, so the manner is unnatural, regardless of the proximate cause. Most are accidents.

Approach to asphyxia: homicides

Apart from police custody deaths and pediatric homicides, suspicious deaths of women or men who may have been strangled to death cause forensic pathologists the most angst. Unlike conclusions drawn from the passing of a bullet through the brainstem, deaths involving compression of the neck and/or torso are often contextualized, and, even with actual neck and/or chest injuries and associated findings, strangulation is often a sort of diagnosis of both inclusion and exclusion, as well as context, particularly when the findings are not striking or they are grossly equivocal. People (women more than men, statistically) can be strangled to death, but they can also survive an attempted strangulation. Both of these victim groups, decedents and survivors, can have similar physical findings (absent the benefit of a thorough internal neck dissection in the living). Both may have facial and conjunctival petechiae and confluent hemorrhages, facial plethora, and contusions and abrasions of the skin of the neck.[149] Living victims sometimes also complain of hoarseness, neck pain, and throbbing headaches immediately after the attack. We know this because, over the years, many forensic pathologists have been asked by prosecutor's offices to opine and testify on the findings and medical records of victims of *attempted* strangulations, giving the forensic examiner the ability to evaluate their injuries and symptoms. So, you might ask, what is the difference between these two groups? One group survived; the other group did not, so only one reasonable conclusion emerges. It is a matter of extent of compression in the latter: a longer time with more force applied, which leads to death.

There are several pearls and pitfalls to consider when faced with a suspected homicidal neck compression death. In suspected strangulations, the FP usually collects trace evidence by using a **sexual assault evidence collection kit.** Similarly, when a death looks like it involved a sexual assault, one should also consider the possibility of an attendant strangulation. Because of the intimate and sometimes impulsive nature of these assaults, they often coexist. Strangulation may be the means chosen to control and quiet a person during an attempted sexual assault.

It is not possible to manually strangle oneself. This is why people in detention centers and prisons must resort to the use of ligatures (hangings or self-strangulations) to accomplish a fatal neck compression. A person may attempt to manually strangle themselves to the point of unconsciousness, but once consciousness is lost, the arms become immediately flaccid and fall away from the neck, interrupting the subsequent sustained and prolonged compression necessary to cause irreversible brain damage and death. If the compression is not sustained after loss of consciousness, the person will gradually regain consciousness again and any further attempts will repeat this futile cycle. On the other hand, a person can use a ligature to strangle themselves, without resorting to a hanging; if they can sufficiently tighten and knot the ligature, it will remain tightly fixed after loss of consciousness. People have been successful at causing death in this way with zip/cable ties that are designed to lock.[150–152]

Facial/Conjunctival petechiae and asphyxia

While there are no pathognomonic findings of asphyxia, facial, conjunctival, and oral petechiae deserve special mention.[108,149,153–157]

Petechiae are pinpoint hemorrhages caused by the rupture of microvasculature in the skin, soft tissues, and organs (Figs. 11.10–11.15). In the setting of asphyxia, several key points should be remembered. First, when considering facial and conjunctival petechiae in asphyxial deaths, it is important to recognize that they are the result of **purely biomechanical vascular phenomena** and not asphyxia/vascular hypoxia or any other type of hypoxemia, per se. This assertion is supported by the fact that they are neither observed in all modalities of asphyxia nor other natural respiratory deaths. Rather, they are produced under a specific set of circumstances, both asphyxial and non-asphyxial alike.[149] Namely, when certain kinds of asphyxial deaths allow arterial blood flow into the head while obstructing venous egress from it, such as is seen in many homicidal neck and/or chest compression deaths and some hangings (see above), this sets up a predictable degree of pressure within the microvasculature of the head that, *if sustained*, often leads to their rupture. The hemorrhages that result are typically tiny (petechiae), reflecting

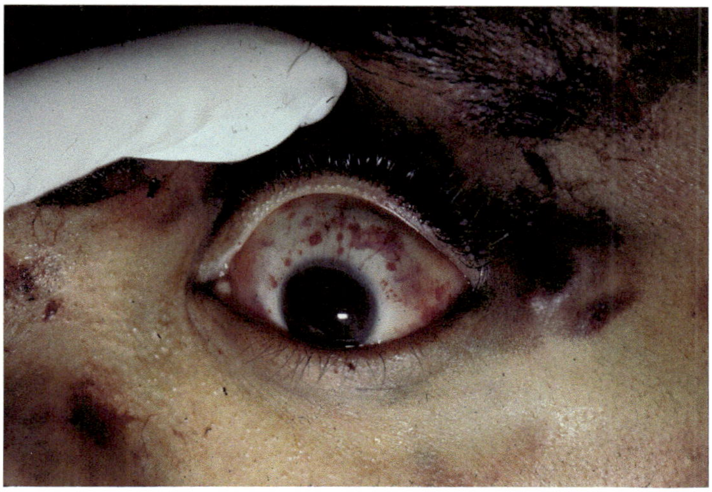

FIGURE 11.10 **Petechiae due to blunt impact (struck by subway train).**

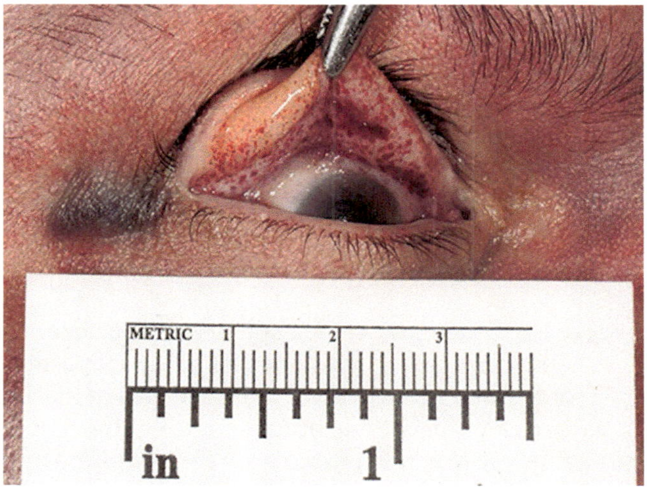

FIGURE 11.11 **Conjunctival petechiae.**

FIGURE 11.12 **Confluent conjunctival hemorrhage.**

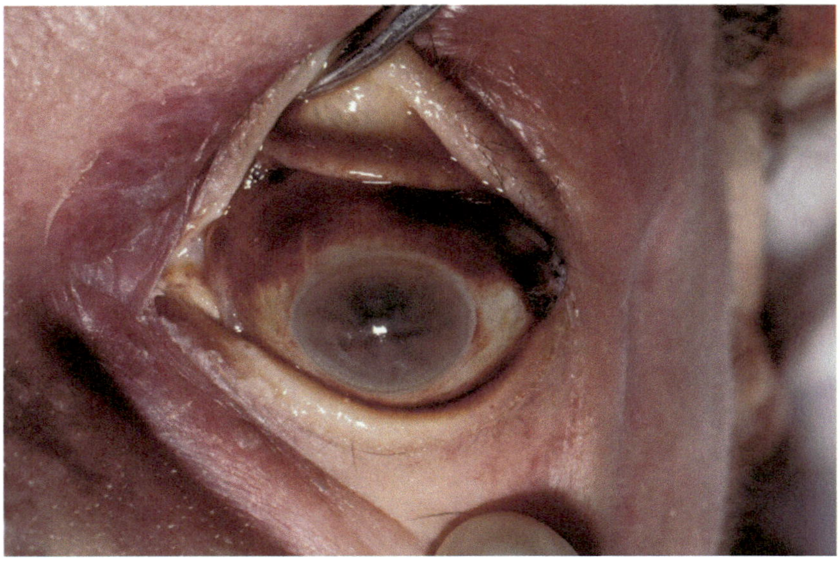

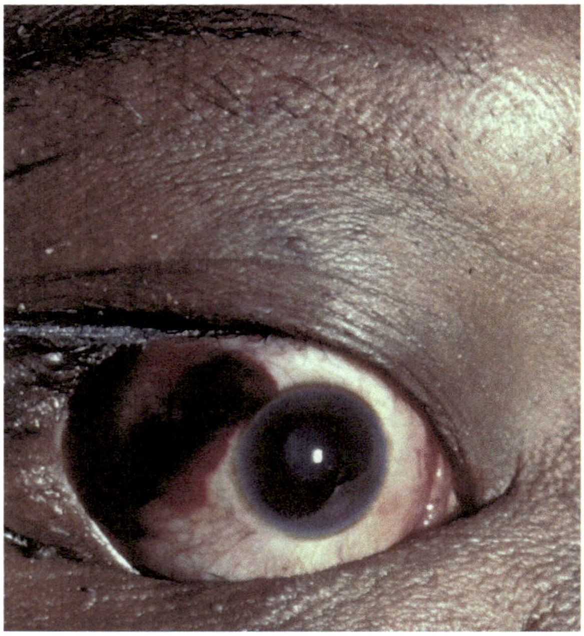

FIGURE 11.13 **Scleral/conjunctival confluent hemorrhage.**

the size of the ruptured vessel. That is not to say that larger facial/conjunctival confluent hemorrhages cannot be seen in this setting; they, in fact, commonly are seen and are merely a reflection of the same biomechanical phenomena that produce petechiae, but perhaps as a result of a longer duration and/or greater force of compression (Fig. 11.12). **Facial plethora** is another finding commonly observed in this setting that further supports the process of sustained venous compression.

The importance of the above lies in the evidentiary value of plethora, petechiae, and confluent hemorrhages of the head when found in the setting of compressive neck and/or chest trauma. Specifically, the petechiae provide supportive physical evidence not only for a beating heart (circulation) but also of the *sustained nature of the pressure* (i.e., minutes, not seconds) applied to the neck and/or chest in order to produce them. It is equally important to point out, however, that, on occasion, considerable forces applied to the neck are evidenced by substantial physical injury of its soft tissues, yet no petechiae or confluent hemorrhages are produced. This combination likely represents even *greater* forces having been applied to the neck, effectively compressing and occluding not only the jugular veins, but also the carotid arteries. This may happen with a particularly strong or forceful assailant, especially when the victim is relatively smaller and/or rendered vulnerable by depressant drugs and/or alcohol and therefore less able to fight back.

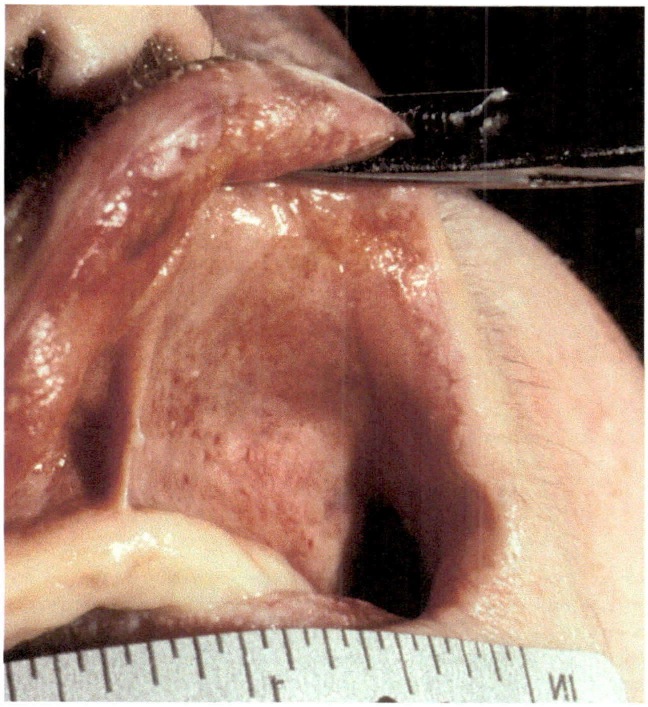

FIGURE 11.14 **Oral mucosal petechiae.**

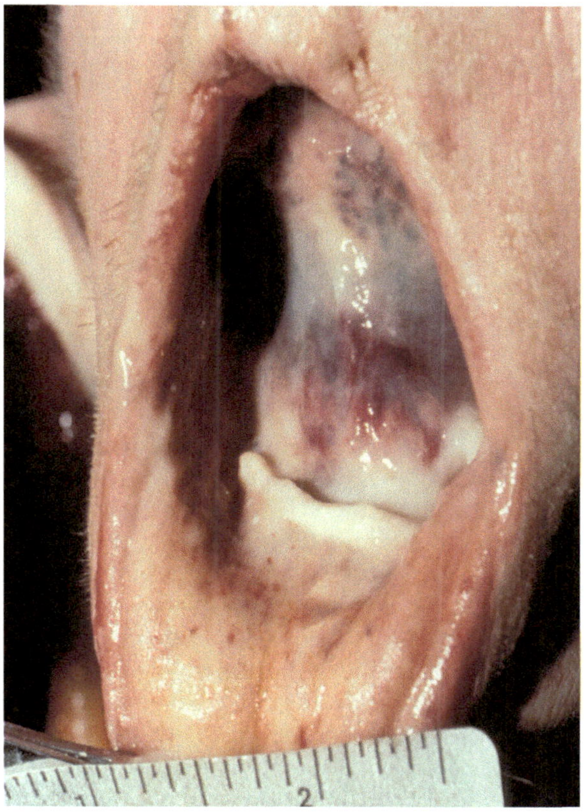

FIGURE 11.15 **Oral mucosal petechiae.**

The important takeaway here is that petechiae are not requisite findings in all homicidal strangulations. Each case must be evaluated on the totality of the constellation of findings and all investigative evidence.

Other nonsinister scenarios in which petechiae may form include any that involve a forceful and/or sudden thoracoabdominal contraction against a closed glottis, as is seen with repeated sneezing or wheezing, coughing paroxysms, status epilepticus, and labor and delivery, to name only a few. The poor venous return experienced with congestive heart failure offers another benign explanation.[108,149,153,155,156,158,159]

In 2010, the late Charles Hirsch, then Chief Medical Examiner of New York City, issued a memo on the interpretation of autopsy findings in homicides due to cervical compression. It follows:

In the adjudication of homicides due to cervical compression (including manual and ligature strangulation and choke holds), medical examiners invariably are asked their opinion about how long it took for the victim to lose consciousness and how long it took thereafter for the victim to die. Answers to those questions are determined by all of the available information about the death, including the autopsy manifestations of injury and the specific characteristics of the victim (e.g., presence or absence of drug and/or alcohol intoxication or heart disease). Therefore, medical examiners must exercise circumspect judgment on a case by case basis in responding to the foregoing questions. Consequently, the parameters that follow in this memorandum must never be regarded as "cookbook" dictum or dogma. Instead, the parameters are offered as general guidelines that describe the average responses of healthy victims who are not intoxicated by ethanol or drugs.

Loss of consciousness

If the carotid arteries are compressed and obstructed, the victim is pale above the area of compression and consciousness is lost in 10–15 s. If the carotid circulation is not interrupted, but venous return is prevented by compression of the jugular veins, the victim is plethoric above the area of compression and has petechiae which appear in approximately 1 min. Depending upon the concomitant degree of airway compromise by displacement and/or compression, consciousness is lost in approximately 1 min.

In any of the foregoing circumstances, if cervical compression is relaxed when consciousness is lost, the victim will awaken because the brain has not yet sustained irreversible injury. This explains why it is impossible to commit suicide by manual strangulation and why a properly applied "carotid sleeper" (lateral compression) choke hold does not cause death if compression is relaxed when the victim goes limp.

Time to death

In order to produce irreversible cerebral injury and death from cerebral ischemia, cerebral obstructive hyperemia, and/or airway compromise, compression of the neck must be maintained for at least a few minutes; i.e., an action that continues after the victim has lost consciousness. If death was contemporaneous with the loss of consciousness, no one would be successfully resuscitated after a potentially lethal cardiac arrhythmia due to any cause.

Carotid sinus and vagal inhibition

During cervical compression, it is theoretically possible for bilateral carotid sinus stimulation to stop the heart. The presence of petechiae above the level of compression excludes cardiac standstill as a mechanism of death because petechiae do not occur in the absence of arterial circulation to the area containing the petechiae. Otherwise, participation of the carotid sinus/vagal stimulation mechanism is speculative, as its occurrence and physiologic consequence cannot be measured in the dead. Furthermore, exclusive focus on the inhibitory limb of the autonomic nervous system is unjustified because it neglects the conflicting, more likely, adrenergic stimulus that arises in response to terrifying, painful injury."

Manner of death

The homicidal burden of proof in deaths of this type can be met time and again, if the investigation is approached with stringency, consistency, and sound forensic practice. When that burden is met, we must not shy away from clearly characterizing these deaths as what they are. Our job as forensic pathologists is to determine, to a reasonable degree of medical certainty, that the neck compression/strangulation not only happened, but that it was fatal. This is accomplished by a thorough investigation and autopsy. If the injury to the neck is obvious, the circumstances reasonably/clearly suspicious, and we are presented with a decedent for whom no other cause is revealed, we will likely determine that strangulation/neck compression is the cause of death, without hesitation (Fig. 11.16). Such cases are relatively more straightforward. Still, there are some medical examiners/coroners (ME/Cs) who might wait for the toxicology, depending on the case, to show that they have considered all evidence, but still, in the end, will be impressed enough by the injuries that they ultimately determine that any drugs detected are incidental to, and therefore not causal in the death, or, in other instances, contributory, but not exculpatory. The bottom line is that FPs should not shy away from diagnosing strangulations when it's clearly supported by all evidence, physical and investigative, simply because, as a group, they are not as uniformly fatal as a bullet that travels through the pons. The likelihood of substantial neck injuries being unrelated to the death of a young woman who happens to be dead without another more compelling cause is exceedingly low, with or without positive toxicology findings. In other words, the death and the substantial injuries are overwhelmingly likely to be related (to a reasonable degree of medical/scientific certainty). We must let common sense prevail in such circumstances and not "deny reality for the sake of objectivity."

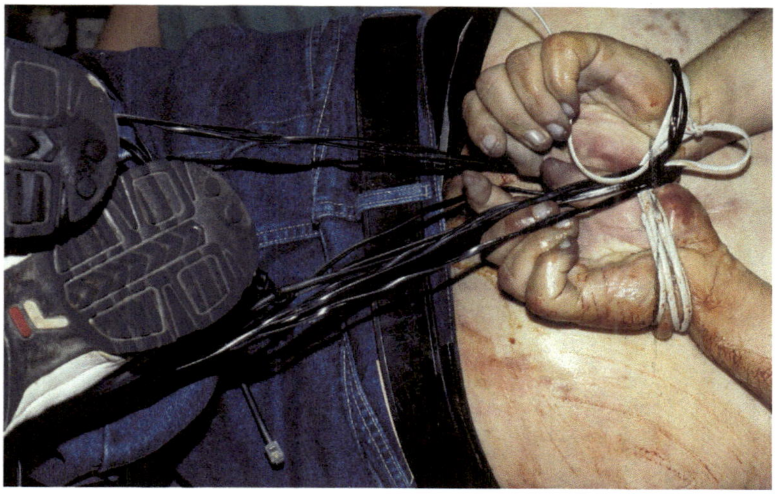

FIGURE 11.16 **Hog-tied.**

That said, the more challenging cases are the ones in which the constellation of injuries are not as impressive or the circumstances not as clearly homicidal, but rather present a *reasonable* concern for another cause. In such cases, positive toxicology results might offer a reasonable alternate explanation for the death. If there is such a *reasonable* potential cause for the death, other than homicidal trauma, and we recognize this, the homicide determination burden has not been met and, as such, the FP will likely be justified in backing away from such a classification. Sometimes, perhaps rarely, undetermined cause and manner may be the necessary determination in these circumstances. Caution should be exercised, though, in the potential overuse of an undetermined classification, a designation not meant to hide behind when unnecessary and when more than sufficient evidence allows us to make the most accurate, defensible, and necessary call.

Because of the expected struggle involved in homicidal neck compression deaths, during which there may be an on again-off again, pulsatile quality to the neck compression, petechiae are much more commonly encountered in homicidal strangulation deaths than they are in hangings and suicidal strangulations, the latter, representing intentional, volitional acts, being devoid of an attendant struggle.[117,160–171] Some homicidal strangulations have features of both manual and ligature forces, and/or a carotid sleeper hold is suspected to have been used, at least in part. For this reason, **"compression of neck"** is a term of art that is sometimes used in the cause of death statements for these fatalities. This broader language encompasses all of the above modalities. **Carotid sleeper** and **bar arm choke holds** are associated with no pathognomonic features, per se. With the *carotid sleeper hold*, the sole finding may be only marked facial plethora and a florid eruption of facial and conjunctival petechiae and confluent hemorrhages without any or with minimal associated neck injury due to the broader application of force (biceps region and forearm applied across the anterolateral aspects of each side of the neck). The *"bar arm"* choke hold applies forearm pressure directly over the airway/larynx, so is likely to involve airway compromise more than vascular compromise; as such, petechiae may be absent. That said, some carotid sleeper hold deaths are associated with a large, broad, and quite prominent strap muscle hemorrhage (Figs. 11.17 and 11.18). This is most commonly observed involving the left sternocleidomastoid muscle, as a right-handed perpetrator applies the choke hold from behind the victim, the assailant's right forearm "landing" and serving as the fulcrum on the left side of the victim's neck, where the greatest compressive forces are applied.[172,173] It is also important to note that some homicidal neck compression deaths involve a concomitant forceful **chest compression**. Such deaths result from the perpetrator lying/sitting on or otherwise compressing the victim's trunk against a resistant surface while also strangling them (i.e., with the compressive forces applied from either the front or from the back side of the decedent's body). In such cases, blunt trauma of the chest wall or back is sometimes observed and plethora, petechiae, and internal tissue hemorrhage might be observed above and/or below a line of demarcation at the level of the torso compression, not just above the level of the neck. If the prosector doesn't look for such a pattern, it can be easily missed. When both are determined to play a role (both asphyxial mechanisms), the cause of death can be easily communicated as **"compression of neck *and* chest,"** or, occasionally, just "compression of chest/torso," or "homicidal asphyxia, including compression of chest (or torso)," when the neck involvement is absent or less clear. *Compression of the neck with smothering* is known as **"burking,"** named after William Burke, who, along with William Hare, used this method in 1828 in Edinburgh to obtain bodies for postmortem anatomical study.

Manual strangulations produce no one set of identical and predictable findings from case to case. There are many ways to forcefully place one's hand(s) on another person's neck to kill them. Two hands may be used while facing the

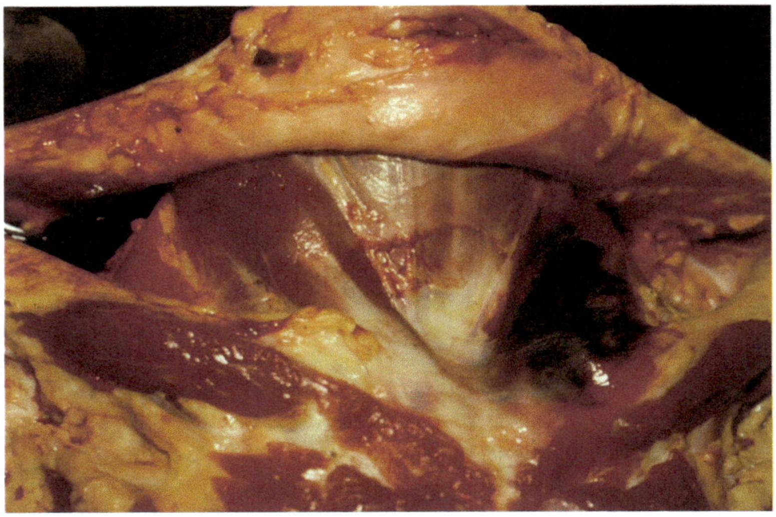

FIGURE 11.17 Left strap muscle hemorrhage in a choke hold neck compression.

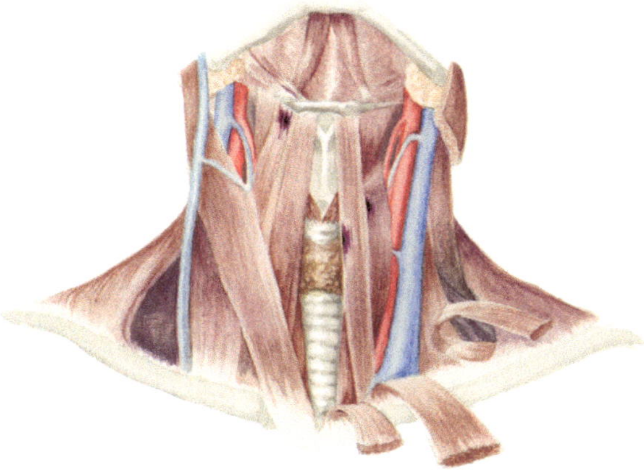

FIGURE 11.18 Strap muscles and vasculature of the neck. *Watercolor by Dr. Cricket Coleman.*

victim, placing eight of the assailant's 10 fingers on the posterior neck and only the thumbs on the anterior neck. An assailant may also strangle from behind, setting up the opposite hand placement, or even from the side. They may use one hand, forcefully compressing high in the neck, while using a resistant surface such as the floor or a wall to close the gap on the compression. This is all to say that the findings from one case to another may be disparate, given the varied strangulation scenarios, but this only highlights *the importance of examining/dissecting the* **entire circumference of the neck** *in all suspected homicidal neck compression deaths* (some do this in all neck compression deaths, regardless of manner); a posterior neck dissection should always be performed in the evaluation of these cases (see Chapter 5). Regardless of these differences, a common thread in most manual strangulations is the typical presence of one or more discrete soft tissue and/or strap muscle hemorrhages that represent the individual pressure points of one or more finger pads, or a broader confluent hemorrhage produced by the palm and/or palmar aspects of the fingers (see Figs. 11.19–11.21). Contrary to popular belief, a fractured hyoid bone in the setting of a manual strangulation is not invariable. While not rare, it is by no means uniformly or even frequently encountered.[174,175] A study by Gonzales of 24 manual strangulations reported 11 with bone/cartilage fractures. Luke studied 25 homicidal strangulations and 12/25 had petechiae, 10/25 had no external injury, and 2/9 of the manual strangulations had hyoid

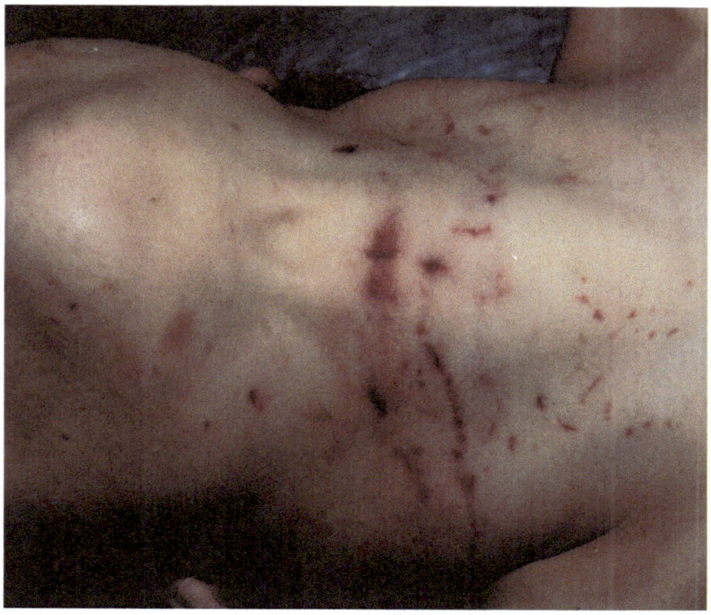

FIGURE 11.19 **Manual strangulation with multiple fingernail abrasions.**

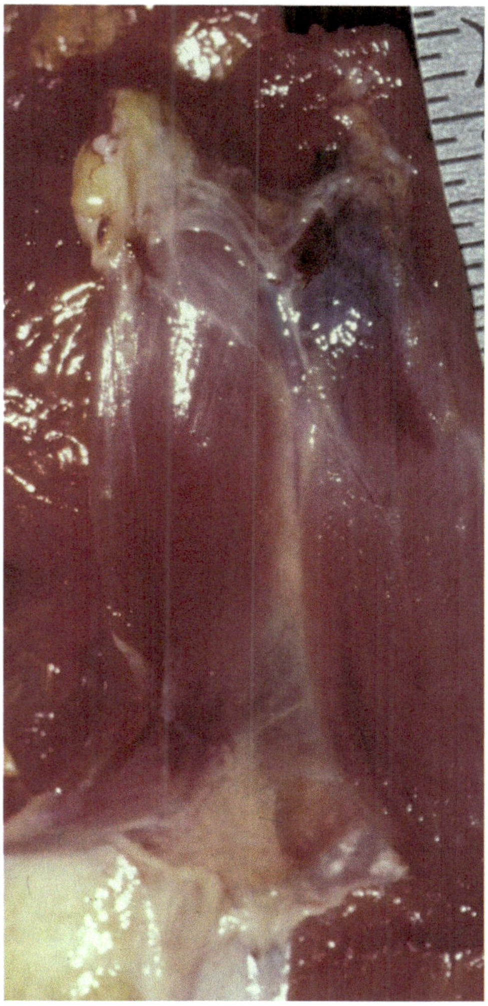

FIGURE 11.20 **Focal strap muscle hemorrhage.**

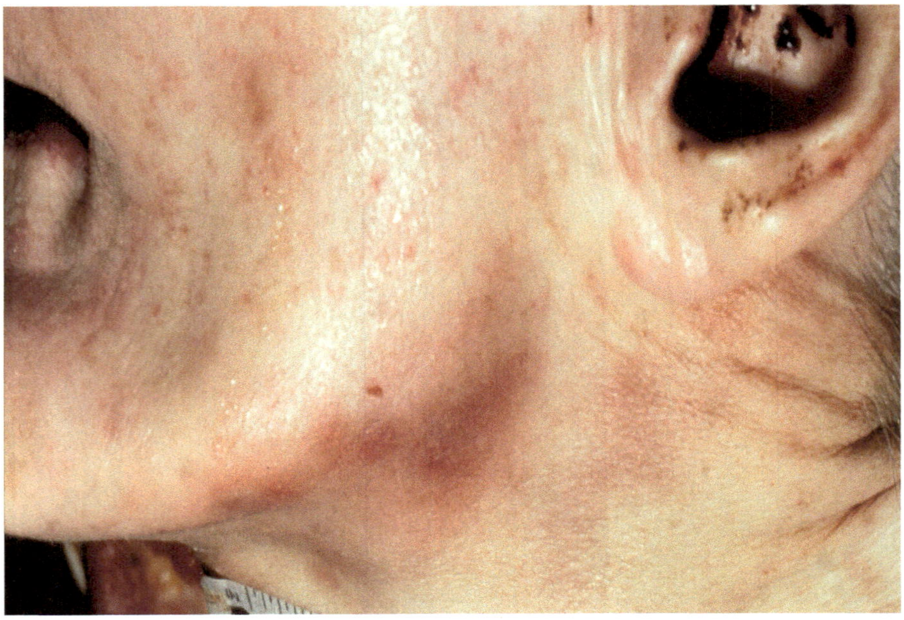

FIGURE 11.21 **Compressive (bilateral) contusion of the angle of the jaw from high neck manual strangulation.**

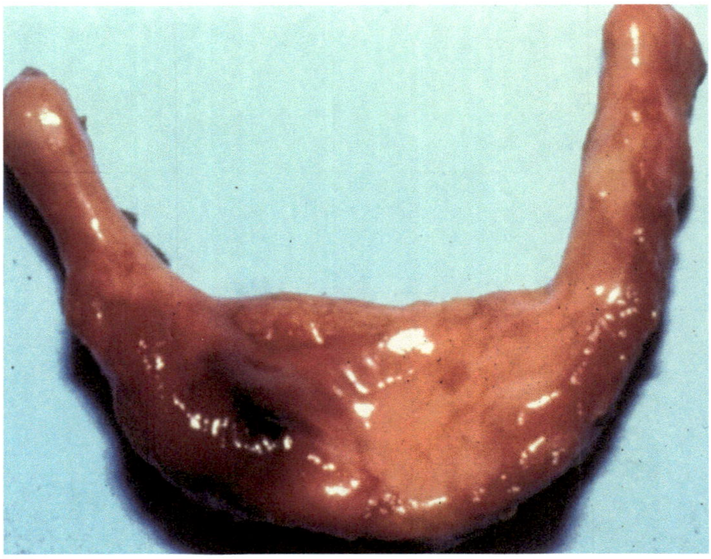

FIGURE 11.22 **Hyoid bone with fracture-related soft tissue hemorrhage.**

fractures (Figs. 11.22 and 11.23). Variables that may favor its incidence include the brittleness of the bone/advanced age of the victim and the precise placement of the hands on the neck (i.e., it would be expected to have a greater frequency with manual strangulations involving a pincer grasp high in the neck).[176–179]

By contrast, in the authors' experience and that of others, with regard to **ligature strangulations**, the soft tissue hemorrhage(s) of the neck, if present, are more likely to be localized subjacent to the ligature injury on the skin (abrasions/furrows), having less vertical breadth of distribution compared to manual strangulations.[150,152,165,180–186]

In any type of homicidal strangulation, *attention should be carefully paid to the* **chin**, a particularly vulnerable anatomic location for injury in a strangulation victim (Fig. 11.24). When the neck of a conscious victim of strangulation is assaulted, a common reflex is to flex the head and neck in an attempt to wedge the chin between the application of force (hands/ligature) and their own neck, to push the hands off, while simultaneously using their own free hands to try to pull the compressive force off their neck. With each of these attempts, victims may inadvertently

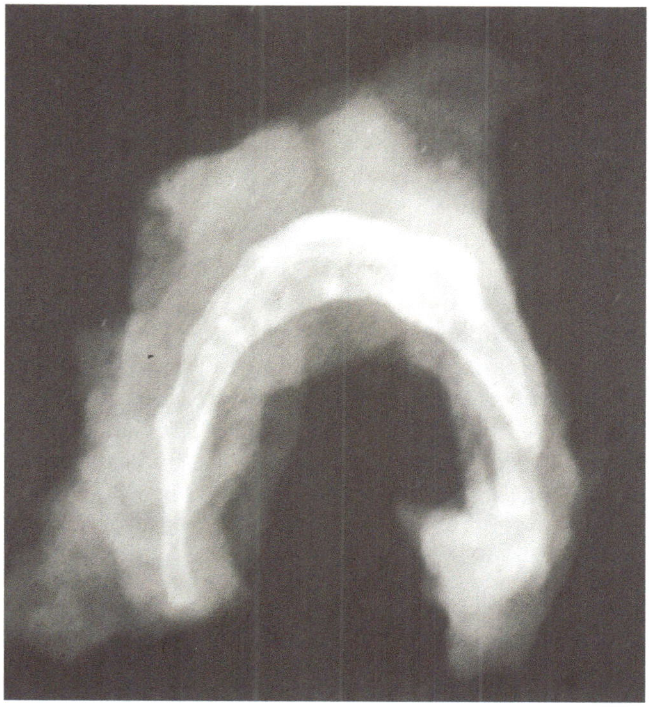

FIGURE 11.23 **Specimen radiograph of fractured hyoid bone.**

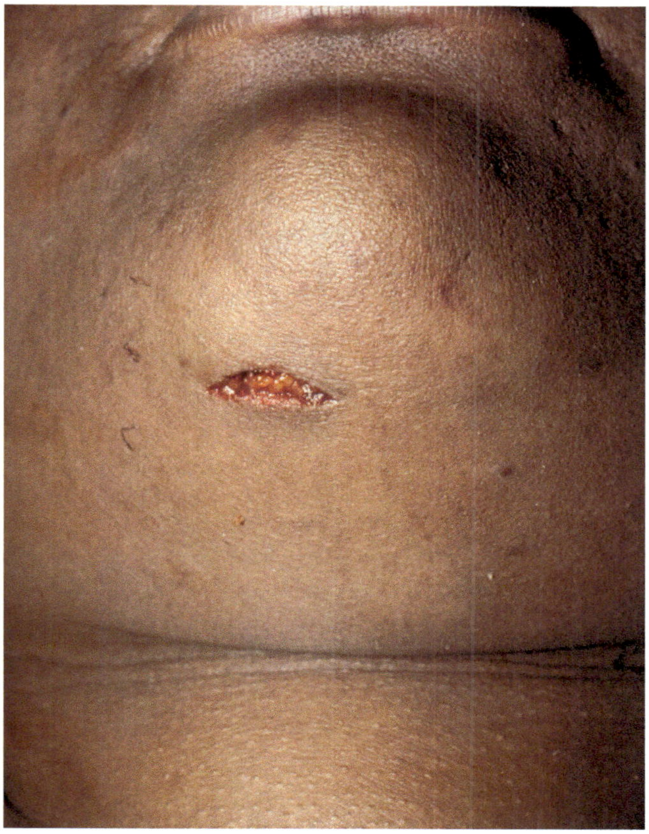

FIGURE 11.24 **Chin injury as a sign of struggle in a strangulation homicide.** Inadvertent puncture wound of the chin (from the decedent's own long nail).

injure their own chin (frictional abrasions, puncture wounds from their own nails, if long, etc.) and also produce abrasions, often linear and vertical, to their own (anterior) necks. For this reason, police may also benefit from documenting any abrasions on a suspect's body that were inflicted by the victim's free hands. In doing so, they may discover vertical linear abrasions on the suspect's back or face, inflicted by the victim in their struggle to break free. This is also why the collection of the victim's nails for DNA analysis is a potentially critical piece of evidence in these cases, particularly when the victim and assailant are not known to be intimate partners, or even to have known one another at all. The importance of noting these injury patterns also lies in their support of the overwhelmingly likely violent struggle that took place until consciousness was lost. Finally, it is also important to submit samples of the soft tissue hemorrhages of the neck for microscopic examination. Documenting the absence of an inflammatory response may have relevance to an alternate theory that might later be presented, one that asserts that the compression occurred, but earlier that day or the day before, not at the moment of death.

Challenges and caveats that sometimes arise in the evaluation and ultimate adjudication of these deaths are many, but include **concomitant drug intoxications** in the setting of a clear strangulation, prone positioning, cardiopulmonary resuscitation (CPR) efforts, and older victims with heart disease, to name only a few.[174,175,187,188] It is important to remember that in a clear-cut case of strangulation, in which the victim is also **intoxicated** by one or more drugs or alcohol, *the strangulation represents an efficient intervening primary cause* in a person who happens to have been intoxicated at the moment of death. Even if one chooses to include the drugs as a contributory cause (this entirely depends upon the case), the intoxication would not exculpate the inflicted neck trauma, when unequivocal, as these two injuries, mechanical and chemical, are not mutually exclusive (see "Manner of Death" above). In the end, these death certification judgments should be made based on the merits of each individual case, viewed through the lens of the training and experience of the ME/C; there is no "one size fits all" formula that can be uniformly applied. **CPR efforts** can be easily evaluated in the context of neck injuries by a simple review of the emergency records, with or without interviews of first responders, where deemed necessary. Absent an attempted venipuncture of the neck (usually self-evident at autopsy) or cricoid pressure applied during intubation efforts, CPR activities, per se, will not likely provide a sufficient or tenable alternate explanation for the combination of substantial compressive injuries of the neck combined with petechiae and hemorrhage of the face and conjunctivae in an otherwise suspicious death. A **prone position** of the body at the time of discovery presents another data point to consider in these deaths. Again, a careful analysis of context and an evaluation of the totality of findings usually resolve any confusion in this setting.

Finally, cause of death statements designating what is sometimes referred to as **"homicidal asphyxia"** and **"homicidal violence"** merit further discussion.[153,189–191] On occasion, a body is discovered under clearly suspicious circumstances (e.g., hog-tied, gagged and/or duct-taped, otherwise concealed in a container or some form of wrapping(s), dismembered, buried, burned, etc.), but the remains are either free of demonstrable fatal injury or have some injury, but are decomposed, burned, or skeletonized to such an extent that it precludes a full examination of all of the soft tissues. Toxicology may or may not detect illicit substances in these instances, but the circumstances are such that a homicidal element is clear after all testing and investigating has been completed. At least two possible options for the cause of death statement are available to the ME/C under these specific circumstances. The first is **homicidal violence of undetermined type**. This designation puts on record that, to a reasonable degree of scientific certainty, the death resulted at the hand of another, but that *the condition of the remains limits the FP's ability to identify the precise mechanism or mechanisms of death.* It constitutes a diagnosis of both exclusion and inclusion. On the other hand, when some degree of injury *is* evident (e.g., sharp force trauma to the skeleton and to what little tissue remains), the forensic pathologist may instead choose to write as the cause, "homicidal violence, including sharp force injuries." This wording recognizes the injury that *was* identified, but without limiting the mechanism of death to that injury alone, given that the examination of the entire remains was not possible. This wording opens the door to a potential multiplicity of wounds, in number and type, including a strangulation that was not possible to identify.

The second cause of death statement that may apply under these circumstances is **"homicidal asphyxia."** This designation is used under similar clearly homicidal circumstances, but the word choice is more specific because, unlike the typically incomplete or otherwise limited remains of homicidal violence deaths, *enough of the remains are available for a complete examination to exclude other nonasphyxial causes (such as penetrating injuries or blunt force trauma).* That said, no structurally demonstrable asphyxial injury is identified. Possibilities include smothering by a volitional covering of the nose and mouth or by placing or wrapping someone in a bag or other coverings while they are still alive, neither of which is necessarily expected to leave demonstrable injuries at autopsy. Another form of asphyxia that might be included under this designation is a manual, ligature, or choke hold strangulation that is too difficult to discern because of advanced putrefaction or postmortem burn injury. Finally, individuals who are hogtied, gagged, bound, duct-taped, wrapped entirely, and/or left inverted/upside down or otherwise positionally compromised, but have no other demonstrable injury, can also be certified with a *descriptive cause*, as opposed to "homicidal asphyxia," instead listing the way in which their body was found (e.g., "Asphyxia by forceful gagging, complex binding of the head and neck, and covering of the of the head with a duct-taped bag"). Because these death certifications are typically determined based on context, a police investigation, and the exclusion of other contributory causes, they are often certified as "pending further investigation" on the day of autopsy.

References

1. Becquemin J-P, Deleuze P, Watelet J, Testard J, Melliere D. Acute and chronic dissections of the abdominal aorta: clinical features and treatment. *J Vasc Surg*. 1990;11(3):397–402.
2. Gonzalez-Urzelai V, Palacio-Elua L, Lopez-de-Munain J. Routine primary care management of acute low back pain: adherence to clinical guidelines. *Eur Spine J*. 2003;12(6):589–594.
3. Coffey A, Pasquale-Styles MA, Gill JR. Fatalities due to choking: internal occlusion of airway. *Acad For Pathol*. 2014;4(1):94–99.
4. Dolkas L, Stanley C, Smith AM, Vilke GM. Deaths associated with choking in San Diego county. *J Forensic Sci*. 2007;52(1):176–179.
5. Nikolic S, Zivkovic V, Dragan B, Jukovic F. Laryngeal choking on food and acute ethanol intoxication in adults–An autopsy study. *J Forensic Sci*. 2010;56(1):128–131.
6. Berzlanovich AM, Fazeny-Dorner B, Waldhoer T, Fasching P, Keil W. Foreign body asphyxia: a preventable cause of death in the elderly. *A J Prevent Med*. 2005;28(1):65–69.
7. Byard RW. Coprophagic cafe coronary. *Am J Forensic Med Pathol*. 2001;22(1):96–99.
8. Haugen RK. The Caf'e Coronary. Sudden deaths in restaurants. *JAMA*. 1963;186:142–143.
9. Mittleman RE, Wetli CV. The fatal cafe coronary. Foreign-body airway obstruction. *JAMA*. 1982;247(9):1285–1288.
10. Gill JR, Ely SF, Hua Z. Environmental gas displacement: three accidental deaths in the workplace. *Am J Forensic Med Pathol*. 2002;23(1):26–30.
11. Ernst A, Zibrak JD. Carbon monoxide poisoning. *N Engl J Med*. 1998;339(22):1603–1608.
12. Jaffe FA. Pathogenicity of carbon monoxide. *Am J Forensic Med Pathol*. 1997;18(4):406–410.
13. James PB, Calder IM. Anoxic asphyxia–a cause of industrial fatalities: a review. *J R Soc Med*. 1991;84(8):493–495.
14. Zugibe FT, Costello JT, Breithaupt MK, Zappi E, Allyn B. The confined space-hypoxia syndrome. *J Forensic Sci*. 1987;32(2):554–560.
15. deJong JL, Adams T. Entrapment in small, enclosed spaces: a case report and points to consider regarding the mechanism of death. *J Forensic Sci*. 2001;46(3):708–713.
16. Halpern P, Raskin Y, Sorkine P, Oganezov A. Exposure to extremely high concentrations of carbon dioxide: a clinical description of a mass casualty incident. *Ann Emerg Med*. 2004;43(2):196–199.
17. Rossi R, De-Giorgio F, Grassi VM, Pascali VL, Lancia M. An unusual suicide: asphyxia by methane gas. *Am J Forensic Med Pathol*. 2013;34(2): 83–85.
18. Manning TJ, Ziminski K, Hyman A, Figueroa G, Lukash L. Methane deaths? Was it the cause? *Am J Forensic Med Pathol*. 1981;2(4):333–336.
19. Gale R, Redner-Carmi R, Gale J. Accumulation of carbon dioxide in oxygen hoods, infant cots, and incubators. *Pediatrics*. 1977;60(4):453–456.
20. Cherian MA, Richmond I. Fatal methane and cyanide poisoning as a result of handling industrial fish: a case report and review of the literature. *J Clin Pathol*. 2000;53(10):794–795.
21. Avis SP, Archibald JT. Asphyxial suicide by propane inhalation and plastic bag suffocation. *J Forensic Sci*. 1994;39(1):253–256.
22. Brighten P. A case of industrial carbon dioxide poisoning. *Anaesthesia*. 1976;31(3):406–407.
23. Cullen DJ, Eger 2nd EI. Cardiovascular effects of carbon dioxide in man. *Anesthesiology*. 1974;41(4):345–349.
24. Caroll GC, Rothenberg DM. Carbon dioxide narcosis. Pathological or "pathillogical"? *Chest*. 1992;102(4):986–988.
25. Downs JC, Conradi SE, Nichols CA. Suicide by environmental hypoxia (forced depletion of oxygen). *Am J Forensic Med Pathol*. 1994;15(3): 216–223.
26. Byard RW, Wilson GW. Death scene gas analysis in suspected methane asphyxia. *Am J Forensic Med Pathol*. 1992;13(1):69–71.
27. Gilson T, Parks BO, Porterfield CM. Suicide with inert gases: addendum to Final Exit. *Am J Forensic Med Pathol*. 2003;24(3):306–308.
28. Gallagher KE, Smith DM, Mellen PF. Suicidal asphyxiation by using pure helium gas: case report, review, and discussion of the influence of the internet. *Am J Forensic Med Pathol*. 2003;24(4):361–363.
29. Weaver LK. Clinical practice. Carbon monoxide poisoning. *N Engl J Med*. 2009;360(12):1217–1225.
30. Adelson L, Sunshine I. Fatal hydrogen sulfide intoxication. Report of three cases occurring in a sewer. *Archiv Pathol*. 1966;81(5):375–380.
31. Kage S, Ito S, Kishida T, Kudo K, Ikeda N. A fatal case of hydrogen sulfide poisoning in a geothermal power plant. *J Forensic Sci*. 1998;43(4): 908–910.
32. Kage S, Kashimura S, Ikeda H, Kudo K, Ikeda N. Fatal and nonfatal poisoning by hydrogen sulfide at an industrial waste site. *J Forensic Sci*. 2002;47(3):652–655.
33. Maebashi K, Iwadate K, Sakai K, et al. Toxicological analysis of 17 autopsy cases of hydrogen sulfide poisoning resulting from the inhalation of intentionally generated hydrogen sulfide gas. *Forensic Sci Int*. 2011;207(1–3):91–95.
34. Ballerino-Regan D, Longmire AW. Hydrogen sulfide exposure as a cause of sudden occupational death. *Arch Pathol Lab Med*. 2010;134(8):1105.
35. Sams RN, Carver 2nd HW, Catanese C, Gilson T. Suicide with hydrogen sulfide. *Am J Forensic Med Pathol*. 2013;34(2):81–82.
36. Smith SS, Cannon DL, Fagan K, Weis CP. Occupational hydrogen sulfide fatalities and thiosulfate levels. *Am J Forensic Med Pathol*. 2017;38(1): 47–48.
37. Adelson L. Fatal hydrogen sulfide intoxication. *Arch Path*. 1966;81:375–380.
38. Bott E, Dodd M. Suicide by hydrogen sulfide inhalation. *Am J Forensic Med Pathol*. 2013;34(1):23–25.
39. Lancia M, Panata L, Tondi V, Carlini L, Bacci M, Rossi R. A fatal work-related poisoning by hydrogen sulfide: report on a case. *Am J Forensic Med Pathol*. 2013;34(4):315–317.
40. Sastre C, Baillif-Couniou V, Kintz P, et al. Fatal accidental hydrogen sulfide poisoning: a domestic case. *J Forensic Sci*. 2013;58(Suppl. 1): S280–S284.

41. Srisont S, Chirachariyavej T, Peonim AV. A carbon dioxide fatality from dry ice. *J Forensic Sci.* 2009;54(4):961–962.
42. Schmitt MW, Williams TL, Woodard KR, Harruff RC. Trends in suicide by carbon monoxide inhalation in King County, Washington: 1996–2009. *J Forensic Sci.* 2011;56(3):652–655.
43. Sheikhazadi A, Saberi Anary SH, Ghadyani MH. Nonfire carbon monoxide-related deaths: a survey in Tehran, Iran (2002–2006). *Am J Forensic Med Pathol.* 2010;31(4):359–363.
44. Shimazu T, Ikeuchi H, Sugimoto H, Goodwin CW, Mason Jr AD, Pruitt Jr BA. Half-life of blood carboxyhemoglobin after short-term and long-term exposure to carbon monoxide. *J Trauma.* 2000;49(1):126–131.
45. Gettler A, Freimuth H. The carbon monoxide content of the blood under various conditions. *Am J Clin Pathol.* 1940;10:603–613.
46. Retzky SS. Carbon monoxide poisoning from Hookah smoking: an emerging public health problem. *J Med Toxicol.* 2017;13(2):193–194.
47. Blackmore D. The determination of carbon monoxide in blood and tissue. *Analyst.* 1970;95:439–458.
48. Kreutz K, Danielsen T, Quinton RA. Suicidal carbon monoxide poisoning by formic and sulfuric acids. *Acad Forensic Pathol.* 2019;9(3–4): 217–224.
49. Byard RW. Carbon monoxide - the silent killer. *Forensic Sci Med Pathol.* 2019;15(1):1–2.
50. Heath K, Byard RW. Lethal carbon monoxide toxicity in a concrete shower unit. *Forensic Sci Med Pathol.* 2019;15(1):133–135.
51. Perera RD. The dangers of carbon monoxide. *N Engl J Med.* 1995;332(13):894.
52. Huff JS, Kardon E. Carbon monoxide toxicity in a man working outdoors with a gasoline-powered hydraulic machine. *N Engl J Med.* 1989; 320(23):1564.
53. Jaffe N. Role of carbon monoxide in coronary disorders. *N Engl J Med.* 1968;279(2):111.
54. Carbon monoxide poisoning - a timely warning. *N Engl J Med.* 1968;278(15):849–850.
55. McBay AJ. Carbon monoxide poisoning. *N Engl J Med.* 1965;272:252–253.
56. Kinoshita H, Turkan H, Vucinic S, et al. Carbon monoxide poisoning. *Toxicol Rep.* 2020;7:169–173.
57. Sircar K, Clower J, Shin MK, Bailey C, King M, Yip F. Carbon monoxide poisoning deaths in the United States, 1999 to 2012. *Am J Emerg Med.* 2015;33(9):1140–1145.
58. Dutra FR. Carbon monoxide from motor vehicles as a cause of illness and death. *Cinci J Med.* 1952;33(7):265–266.
59. Gerostamoulos D, Beyer J, Wong K, Wort C, Drummer OH. Carbon monoxide concentrations in the 2009 Victorian Bushfire disaster victims. *Forensic Sci Int.* 2011;205(1–3):69–72.
60. Trichter J, Helpern M. The problem in New York City of accidental carbon monoxide poisoning. *NY Med.* 1952;8(20):30–37.
61. Martland H. AMA. *Carbon monoxide poisoning.* Vol. 103. 1934:643.
62. Gill JR, Goldfeder LB, Stajic M. The Happy land homicides: 87 deaths due to smoke inhalation. *J Forensic Sci.* 2003;48(1):161–163.
63. Williams HI. Carbon dioxide poisoning: report of eight cases, with two deaths. *Br Med J.* 1958;2(5103):1012–1014.
64. Langford NJ. Carbon dioxide poisoning. *Toxicol Rev.* 2005;24(4):229–235.
65. Brackett Jr NC, Cohen JJ, Schwartz WB. Carbon dioxide titration curve of normal man. Effect of increasing degrees of acute hypercapnia on acid-base equilibrium. *N Engl J Med.* 1965;272:6–12.
66. Thomas WS. Death from carbon dioxide poisoning in a brewery. *Lancet.* 1900;155(3996):928.
67. Rupp WR, Thierauf A, Nadjem H, Vogt S. Suicide by carbon dioxide. *Forensic Sci Int.* 2013;231(1–3):e30–32.
68. Samuel J, Georgievskaya Z. Death in a carbon dioxide therapy bath: a case report and review of the literature. *Acad For Pathol.* 2019;9(1–2): 93–96.
69. Milroy CM. Deaths from environmental hypoxia and raised carbon dioxide. *Acad Forensic Pathol.* 2018;8(1):2–7.
70. Marzuk PM, Tardiff K, Hirsch CS, et al. Increase in suicide by asphyxiation in New York City after the publication of final exit. *N Engl J Med.* 1993;329(20):1508–1510.
71. Haddix TL, Harruff RC, Reay DT, Haglund WD. Asphyxial suicides using plastic bags. *Am J Forensic Med Pathol.* 1996;17(4):308–311.
72. Bullock MJ, Diniz D. Suffocation using plastic bags: a retrospective study of suicides in Ontario, Canada. *J Forensic Sci.* 2000;45(3):608–613.
73. Bowen DA. Hanging–a review. *Forensic Sci Int.* 1982;20(3):247–249.
74. Simonsen J. Patho-anatomic findings in neck structures in asphyxiation due to hanging: survey 100 cases. *FSI.* 1988;38:83–91.
75. Feigin G. Frequency of neck organ fractures in hanging. *Am J Forensic Med Pathol.* 1999;20(2):128–130.
76. Wyatt JP, Wyatt PW, Squires TJ, Busuttil A. Hanging deaths in children. *Am J Forensic Med Pathol.* 1998;19(4):343–346.
77. Denton JS. Fatal accidental hanging from a lanyard key chain in a 10-year-old boy. *J Forensic Sci.* 2002;47(6):1345–1346.
78. Gorniak JM, Sudimack JR, Stanforth JR, Lewis BJ. Hanging deaths with bound hands: what is the manner? *Am J Forensic Med Pathol.* 2007; 28(3):232–234.
79. Guarner J, Hanzlick R. Suicide by hanging. A review of 56 cases. *Am J Forensic Med Pathol.* 1987;8(1):23–26.
80. Gilbert JD, Jensen L, Byard RW. Further observations on the speed of death in hanging. *J Forensic Sci.* 2008;53(5):1204–1205.
81. Sauvageau A, Godin A, Desnoyers S, Kremer C. Six-year retrospective study of suicidal hangings: determination of the pattern of limb lesions induced by body responses to asphyxia by hanging*. *J Forensic Sci.* 2009;54(5):1089–1092.
82. Sauvageau A. Agonal sequences in four filmed hangings: analysis of respiratory and movement responses to asphyxia by hanging. *J Forensic Sci.* 2009;54(1):192–194.
83. Sauvageau A, LaHarpe R, Geberth VJ, Working Group on Human Asphyxia. Agonal sequences in eight filmed hangings: analysis of respiratory and movement responses to asphyxia by hanging. *J Forensic Sci.* 2010;55(5):1278–1281.
84. Clement R, Redpath M, Sauvageau A. Mechanism of death in hanging: a historical review of the evolution of pathophysiological hypotheses. *J Forensic Sci.* 2010;55(5):1268–1271.
85. Sauvageau A, Ambrosi C, Kelly S. Autoerotic nonlethal filmed hangings: a case series and comments on the estimation of the time to irreversibility in hanging. *Am J Forensic Med Pathol.* 2012;33(2):159–162.
86. Kanchan T, Rastogi P, Menezes RG, Nagesh KR. Apparent partial hanging. *Am J Forensic Med Pathol.* 2010;31(4):376–377.
87. Kibayashi K, Mastri A, Hirsch C. Cocaine induced intracerebral hemorrhage: analysis of predisposing factors and mechanisms. *Human Pathology.* 1995;26:659–663.
88. Russo MC, Verzeletti A, Piras M, De Ferrari F. Hanging deaths: a retrospective study regarding 260 cases. *Am J Forensic Med Pathol.* 2016;37(3): 141–145.

89. Samarasekera A, Cooke C. The pathology of hanging deaths in Western Australia. *Pathology.* 1996;28(4):334–338.
90. Nikolic S, Zivkovic V. Cervical spine injuries in suicidal hanging without a long-drop–patterns and possible underlying mechanisms of injury: an autopsy study. *Forensic Sci Med Pathol.* 2014;10(2):193–197.
91. Green H, James RA, Gilbert JD, Byard RW. Fractures of the hyoid bone and laryngeal cartilages in suicidal hanging. *J Clin For Med.* 2000;7: 123–126.
92. Nikolic S, Micic J, Atanasijevic T, Djokic V, Djonic D. Analysis of neck injuries in hanging. *Am J Forensic Med Pathol.* 2003;24(2):179–182.
93. Luke JL. Asphyxial deaths by hanging in New York City, 1964–1965. *J Forensic Sci.* 1967;12(3):359–369.
94. Alongi A, Stassi C, Mondello C, Baldino G, Argo A, Ventura Spagnolo E. Hanging disguised as bondage: accidental or suicidal death? *Forensic Sci Med Pathol.* 2020;16(4):724–727.
95. Gascho D, Heimer J, Tappero C, Schaerli S. Relevant findings on postmortem CT and postmortem MRI in hanging, ligature strangulation and manual strangulation and their additional value compared to autopsy - a systematic review. *Forensic Sci Med Pathol.* 2019;15(1):84–92.
96. Leccia C, Alunni V, Quatrehomme G. Suicidal hanging resulting in decapitation: a case report and review of the literature. *Forensic Sci Int.* 2017;279:e10–e13.
97. Schulz F, Buschmann C, Braun C, Puschel K, Brinkmann B, Tsokos M. Haemorrhages into the back and auxiliary breathing muscles after death by hanging. *Int J Legal Med.* 2011;125(6):863–871.
98. Simonsen J. Patho-anatomic findings in neck structures in asphyxiation due to hanging: a survey of 80 cases. *For Sci Internat.* 1988;38(1–2): 83–91.
99. Luke JL, Reay DT, Eisele JW, Bonnell HJ. Correlation of circumstances with pathological findings in asphyxial deaths by hanging: a prospective study of 61 cases from Seattle, Washington. *J Forensic Sci.* 1985;30:1140–1147.
100. Adair TW, Dobersen MJ. A case of suicidal hanging staged as homicide. *J Forensic Sci.* 1999;44(6):1307–1309.
101. Quinton RA, Dolinak D. Suicidal hangings in jail using telephone cords. *J Forensic Sci.* 2003;48(5):1151–1152.
102. Jordan FB, Schmeckpeper K, Strope M. Jail suicides by hanging. An epidemiological review and recommendations for prevention. *Am J Forensic Med Pathol.* 1987;8(1):27–31.
103. Suarez-Penaranda JM, Alvarez T, Miguens X, et al. Characterization of lesions in hanging deaths. *J Forensic Sci.* 2008;53(3):720–723.
104. Boghossian E, Clement R, Redpath M, Sauvageau A. Respiratory, circulatory, and neurological responses to hanging: a review of animal models. *J Forensic Sci.* 2010;55(5):1272–1277.
105. Dogan KH, Demirci S, Erkol Z, Gulmen MK. Accidental hanging deaths in children in Konya, Turkey between 1998 and 2007. *J Forensic Sci.* 55(3):637–641.
106. Tattoli L, Buschmann CT, Tsokos M. Remarkable findings in suicidal hanging. *Forensic Sci Med Pathol.* 2014;10(4):639–642.
107. Cvetkovic D, Zivkovic V, Nikolic S. Regarding ligature induced blister formation in hanging - the "pinch" effect. *Forensic Sci Med Pathol.* 2018; 14(1):139–141.
108. Sundal JA, Maehle BO, Eide GE, Morild I. Petechial hemorrhages in suicide by hanging: possible contributing variables in petechial development. *Am J Forensic Med Pathol.* 2020;41(2):90–96.
109. Kodikara S. Attempted suicidal hanging: an uncomplicated recovery. *Am J Forensic Med Pathol.* 2012;33(4):317–318.
110. Kodikara S. Accidental hanging: a novel mobile suspension apparatus partially hidden inside the clothes. *Am J Forensic Med Pathol.* 2012;33(4): 311–313.
111. Vapa D, Radosavkic R, Maletin M, Veselinovic I. An unusual suicide case by combination of choking and hanging. *Am J Forensic Med Pathol.* 2012;33(4):305–306.
112. Hejna P, Zatopkova L. Significance of hemorrhages at the origin of the sternocleidomastoid muscles in hanging. *Am J Forensic Med Pathol.* 2012;33(2):124–127.
113. Sauvageau A, Kelly S, Ambrosi C. A filmed hanging without decerebrate and decorticate rigidity: a case report and pathophysiological considerations. *Am J Forensic Med Pathol.* 2012;33(2):176–178.
114. Ambade VN, Kolpe D, Tumram N, Meshram S, Pawar M, Kukde H. Characteristic features of hanging: a study in rural district of Central India. *J Forensic Sci.* 2015;60(5):1216–1223.
115. Hejna P. Amussat's sign in hanging–A prospective autopsy study. *J Forensic Sci.* 2011;56(1):132–135.
116. Hejna P, Rejtarova O. Bleedings into the anterior aspect of the intervertebral disks in the lumbar region of the spine as a diagnostic sign of hanging. *J Forensic Sci.* 2010;55(2):428–431.
117. Geisenberger D, Pollak S, Thierauf-Emberger A. Homicidal strangulation and subsequent hanging of the victim to simulate suicide: delayed elucidation based on reassessment of the autopsy findings. *Forensic Sci Int.* 2019;298:419–423.
118. Spitz W, Diaz F. *Spitz and Fisher's Medicolegal Investigation of Death.* 5th ed. Springfield: Charles C Thomas; 2020.
119. Adams VI. Autopsy technique for neck examination. I. Anterior and lateral compartments and tongue. *Pathol Annual.* 1990;25(2):331–349.
120. Adams VI. Autopsy technique for neck examination. II. Vertebral column and posterior compartment. *Pathol Annu.* 1990;26(Pt 2):211–226.
121. Jayaprakash S, Sreekumari K. Pattern of injuries to neck structures in hanging-an autopsy study. *Am J Forensic Med Pathol.* 2012;33(4):395–399.
122. Andriuskeviciute G, Chmieliauskas S, Jasulaitis A, Laima S, Fomin D, Stasiuniene J. A study of fatal and nonfatal hangings. *J Forensic Sci.* 2016;61(4):984–987.
123. Souheil M, Audrey F, Anny G, Sebastien RJ, Bertrand L. Fatal accidental hanging by a high-chair waist strap in a 2-year-old girl. *J Forensic Sci.* 2011;56(2):534–536.
124. Vadysinghe AN, Sivasubramanium M, Jayasooriya RP. A tree branch instead of a ligature: an unusual accidental hanging. *Forensic Sci Med Pathol.* 2017;13(4):441–443.
125. Kumral B, Ozdes T, Avsar A, Buyuk Y. Accidental deaths by hanging among children in Istanbul, Turkey: retrospective analysis of medico-legal autopsies in 33 years. *Am J Forensic Med Pathol.* 2014;35(4):271–274.
126. Deidiker RD. Accidental ligature strangulation due to a roller-type massage device. *Am J Forensic Med Pathol.* 1999;20(4):354–356.
127. Rauchschwalbe R, Mann NC. Pediatric window-cord strangulations in the United States, 1981–1995. *JAMA.* 1997;277(21):1696–1698.
128. Parchake MB, Kumre V, Kachare RV. Accidental strangulation due to entrapment of saree in crop thrasher machine in an elderly women working at agricultural field. *Am J Forensic Med Pathol.* 2016;37(3):187–189.
129. Osculati A, Fassina G. Two cases of accidental asphyxia by neck compression between bed bars. *Am J Forensic Med Pathol.* 2000;21(3):217–219.

130. Collins KA. Death by overlaying and wedging: a 15-year retrospective study. *Am J Forensic Med Pathol*. 2001;22(2):155–159.
131. Miles SH. Autopsy findings in asphyxia in medical bed rails. *Am J Forensic Med Pathol*. 2009;30(3):256–261.
132. Andrew TA, Fallon KK. Asphyxial games in children and adolescents. *Am J Forensic Med Pathol*. 2007;28(4):303–307.
133. Heath KJ, Cala AD, Byard RW. Metal railing fences and accidental death. *J Forensic Sci*. 2018;63(3):972–975.
134. Byard RW, Gilbert JD. Sleeping accidents in the elderly. *J Forensic Sci*. 2011;56(6):1645–1647.
135. Sauvageau A, Ambrosi C, Kelly S. Three nonlethal ligature strangulations filmed by an autoerotic practitioner: comparison of early agonal responses in strangulation by ligature, hanging, and manual strangulation. *Am J Forensic Med Pathol*. 2012;33(4):339–340.
136. Sauvageau A, Laharpe R, King D, et al. Agonal sequences in 14 filmed hangings with comments on the role of the type of suspension, ischemic habituation, and ethanol intoxication on the timing of agonal responses. *Am J Forensic Med Pathol*. 2011;32(2):104–107.
137. Byard RW, Botterill P. Autoerotic asphyxial death–accident or suicide? *Am J Forensic Med Pathol*. 1998;19(4):377–380.
138. Tournel G, Hubert N, Rouge C, Hedouin V, Gosset D. Complete autoerotic asphyxiation: suicide or accident? *Am J Forensic Med Pathol*. 2001;22(2):180–183.
139. Solarino B, Leonardi S, Grattagliano I, Tattoli L, Di Vella G. An unusual death of a masochist: accident or suicide? *Forensic Sci Int*. 2011;204(1–3):e16–19.
140. Gill JR, Catanese C. Sharp injury fatalities in New York City. *J Forensic Sci*. 2002;47(3):554–557.
141. Demirci S, Dogan KH, Erkol Z, Gunaydin G. Unusual suicide note written on the body: two case reports. *Am J Forensic Med Pathol*. 2009;30(3):276–279.
142. Byard RW. Evidence of premeditation in skin messages in suicide. *J Forensic Sci*. 2016;61(2):566–568.
143. Tumram NK, Ambade VN. Engraved suicide notes: the last note written on body by metallic object. *J Forensic Sci*. 2016;61(Suppl 1):S256–S258.
144. Omalu BI, Macurdy KM, Koehler SA, et al. Forensic pathology and forensic epidemiology of suicides in allegheny county, Pennsylvania: a 10-year retrospective review (1990–1999). *Forensic Sci Med Pathol*. 2005;1(2):125–137.
145. Handelman LD, Lester D. The content of suicide notes from attempters and completers. *Crisis*. 2007;28(2):102–104.
146. Ho T, Yip P, Chiu C, Halliday P. Suicide notes: what do they tell us? *Acta Psychiatr Scand*. 1998;98:467–473.
147. O'Donnell I, Farmer R, Catalan J. Suicide notes. *Br J Psych*. 1993;163:45–48.
148. Ruder TD, Hatch GM, Ampanozi G, Thali MJ, Fischer N. Suicide announcement on Facebook. *Crisis*. 2011;32(5):280–282.
149. Ely SF, Hirsch CS. Asphyxial deaths and petechiae: a review. *J Forensic Sci*. 2000;45(6):1274–1277.
150. Sorokin V, Persechino F, DeRoux SJ, Greenberg MJ. Suicidal ligature strangulation utilizing cable ties: a report of three cases. *Forensic Sci Med Pathol*. 2012;8(1):52–55.
151. Jackson NR, Paul ID. An unusual method of suicide: cable ties. *Am J For Med & Pathol*. 2020;41(3):223–226.
152. Ahmed N, Bauler L, deJong J, Brown T, Prahlow J. *Ligature Strangulation Using Cable Ties: 5 Cases*. 2019.
153. Rao VJ, Wetli CV. The forensic significance of conjunctival petechiae. *Am J Forensic Med Pathol*. 1988;9(1):32–34.
154. Jaffe FA. Petechial hemorrhages. A review of pathogenesis. *Am J Forensic Med Pathol*. 1994;15(3):203–207.
155. Hood I, Ryan D, Spitz WU. Resuscitation and petechiae. *Am J Forensic Med Pathol*. 1988;9(1):35–37.
156. Betz P, Penning R, Keil W. The detection of petechial haemorrhages of the conjunctivae in dependency on the postmortem interval. *Forensic Sci Int*. 1994;64(1):61–67.
157. Ikeda T, Tani N, Aoki Y, et al. Effects of postmortem positional changes on conjunctival petechiae. *Forensic Sci Med Pathol*. 2019;15(1):13–22.
158. Gill JR, Landi K. Traumatic asphyxial deaths due to an uncontrolled crowd. *Am J Forensic Med Pathol*. 2004;25(4):358–361.
159. Maxeiner H, Jekat R. Resuscitation and conjunctival petechial hemorrhages. *J Forensic Leg Med*. 2010;17(2):87–91.
160. Missliwetz J, Vycudilik W. Homicide by strangling or dumping with postmortem injuries after heroin poisoning? *Am J Forensic Med Pathol*. 1997;18(2):211–214.
161. Taff ML, Boglioli LR. Strangulation. A conceptual approach for courtroom presentation. *Am J Forensic Med Pathol*. 1989;10(3):216–220.
162. Ubelaker DH. Hyoid fracture and strangulation. *J Forensic Sci*. 1992;37(5):1216–1222.
163. Pollanen MS. A triad of laryngeal hemorrhages in strangulation: a report of eight cases. *J Forensic Sci*. 2000;45(3):614–618.
164. Di Nunno N, Costantinides F, Conticchio G, Mangiatordi S, Vimercati L, Di Nunno C. Self-strangulation: an uncommon but not unprecedented suicide method. *Am J Forensic Med Pathol*. 2002;23(3):260–263.
165. Pollanen MS, McAuliffe DN. Intra-cartilaginous laryngeal haemorrhages and strangulation. *Forensic Sci Int*. 1998;93(1):13–20.
166. Verma SK, Lal S. Strangulation deaths during 1993–2002 in East Delhi (India). *Leg Med (Tokyo)*. 2006;8(1):1–4.
167. Wahlsten P, Eriksson A. Asphyxia homicides in Finland, 1983–2012. *J Forensic Sci*. 2020;65(5):1548–1556.
168. Pollak S, Thierauf-Emberger A. Homicidal assault to the neck with subsequent simulation of self-hanging. *Forensic Sci Int*. 2015;253:e28–32.
169. Doberentz E, Markwerth P, Madea B. Differentiation of homicidal or suicidal strangulation. *Forensic Sci Int*. 2019;301:e44–e48.
170. Tamsen F, Sturup J, Thiblin I. Homicide injury severity in association with the victim-offender relationship. *Forensic Sci Int*. 2019;300:151–156.
171. Hakkanen-Nyholm H, Repo-Tiihonen E, Lindberg N, Salenius S, Weizmann-Henelius G. Finnish sexual homicides: offence and offender characteristics. *Forensic Sci Int*. 2009;188(1–3):125–130.
172. Kornblum RN. Medical analysis of police choke holds and General Neck trauma (part 1). *Trauma*. 1986;27(5).
173. Kornblum RN. Medical analysis of police choke holds and General Neck trauma (part 2). *Trauma*. 1986;28(1):13–64.
174. Pollanen MS. Subtle fatal manual neck compression. *Med Sci Law*. 2001;41(2):135–140.
175. Pollanen MS. Pitfalls and artifacts in the neck at autopsy. *Acad Forensic Pathol*. 2016;6(1):45–62.
176. Gonzales T. Manual strangulation. *Arch Pathol*. 1933;15:55–66.
177. Rupp JC. Suicidal garrotting and manual self-strangulation. *J Forensic Sci*. 1970;15(1):71–77.
178. Lebreton-Chakour C, Godio-Raboutet Y, Torrents R, et al. Manual strangulation: experimental approach to the genesis of hyoid bone fractures. *Forensic Sci Int*. 2013;228(1–3):47–51.
179. Niort F, Godio-Raboutet Y, Torrents R, et al. Transmission of force to the hyoid bone during manual strangulation: simulation using finite element numerical models. *Forensic Sci Int*. 2015;257:420–424.
180. McMaster AR, Ward EW, Dykeman A, Warman MD. Suicidal ligature strangulation: case report and review of the literature. *J Forensic Sci*. 2001;46(2):386–388.

181. Azmak D. Asphyxial deaths: a retrospective study and review of the literature. *Am J Forensic Med Pathol*. 2006;27(2):134–144.
182. Demirci S, Dogan KH, Erkol Z, Gunaydin G. Suicide by ligature strangulation: three case reports. *Am J Forensic Med Pathol*. 2009;30(4): 369–372.
183. Pramanik P. An unusual method of suicidal ligature strangulation. *J Forensic Sci*. 2016;61(1):274–276.
184. Woodham PC, Fulcher JW, Ward ME. Strangulation by feeding tube in a 23-month-old with down syndrome: case report and review of literature. *Am J Forensic Med Pathol*. 2016;37(4):245–247.
185. Cordner S, Clay FJ, Bassed R, Thomsen AH. Suicidal ligature strangulation: a systematic review of the published literature. *Forensic Sci Med Pathol*. 2020;16(1):123–133.
186. Doberentz E, Geile J, Madea B. Suicidal strangulation with a lashing belt. *Forensic Sci Med Pathol*. 2020;16(3):531–534.
187. Pollanen MS, Perera SD, Clutterbuck DJ. Hemorrhagic lividity of the neck: controlled induction of postmortem hypostatic hemorrhages. *Am J Forensic Med Pathol*. 2009;30(4):322–326.
188. Garland J, Tse R. Neck injury and conjunctival petechiae in a woman who underwent cardiopulmonary resuscitation and subsequently died from meningococcal sepsis. *Forensic Sci Med Pathol*. 2017;13(4):432–435.
189. DiMaio VJ. Homicidal asphyxia. *Am J Forensic Med Pathol*. 2000;21(1):1–4.
190. Gill J, Cavalli D, Ely S, Stahl-Herz J. Homicidal neck compression of females: autopsy and sexual assault findings. *Acad For Path*. 2013;3(4): 454–457.
191. Brown TT, Batalis NI, McClain JL, et al. A retrospective study of the investigation of homicidal childhood asphyxial deaths. *J Forensic Sci*. 2018;63(4):1160–1167.

CHAPTER

12

Approach to burns, blast, and radiation injury

Susan F. Ely and James R. Gill

The right to search for the truth implies also a duty one must not conceal any part of what one has recognized to be true. -*Albert Einstein*

Thermal and inhalational injury

Fire

The overwhelming majority (81%) of the 3600 annual fire-related deaths in the United States occur in the home.[1,2] The manner of death in a thermal burn and/or smoke inhalation fatality from a house fire would depend upon what caused the fire. If it was intentionally set by the occupant to cause their own death, it would be classified as a suicide. If it was caused by the careless discard of a cigarette or a kitchen fire, it would be classified as an accident. If it was intentionally set by another to fraudulently obtain insurance money or as a part of an act of vandalism, it would be classified as a **homicide**. Note that the *intentional setting* of a fatal fire by a cognitively intact person who is developmentally mature enough to understand the inherent danger of fire setting is all that is needed to make a homicide determination. *This burden does not include proof of their intent to kill or otherwise destroy property,* indeed, an intent which may or may not even exist in many such cases. Also, it is important to note that the determination of the intentionality of the fire setting is largely made by investigating law enforcement officials (police and fire marshals [FMs]), upon whom the medical examiner/coroner (ME/C) is dependent for this disposition (the nature/cause of this kind of fire is typically referred to in the official FM report as **"incendiary"**). For death certification purposes, when the manner is *homicide, "incendiary fire"* is the terminology many choose for the "how injury occurred" section, rather than "arson," which is more criminal law terminology.

The approach to the investigation of fire-related fatalities is two-pronged. On the one hand, **the autopsy** must establish that the fire *caused* death, rather than it being a *consequence* of a person dying from a natural or toxicological cause (e.g., cocaine) while holding an inflammatory object, such as a cigarette. If the former, the autopsy aims to further establish the mechanism(s) by which the fire caused the death (thermal burns and/or smoke inhalation, with or without contributory comorbidities; see below). Some combination of the deposition of soot in the larynx (Fig. 12.1) and more distally in the airways (soot in the nares and mouth may deposit ambiently and passively, so is less conclusive), airway mucosal hyperemia, an *elevated carboxyhemoglobin (CO-Hgb)* saturation, frothy pulmonary edema, and so-called "cherry red" discoloration of lividity (Fig. 12.2), tissues and blood constitute typical evidence of a death due to **smoke inhalation**. If, however, the decedent underwent considerable resuscitative efforts that included oxygen therapy and/or had a prolonged survival interval in the hospital, the CO-Hgb testing becomes irrelevant and uninterpretable (unless admission bloods are available to test). If the survival interval was short (minutes to hours), airway findings alone may be sufficient for the determination of smoke inhalation. Also, a prolonged survival interval in a burn unit or intensive care unit provides proof that the decedent was indeed alive at the time of the fire and died as a result of it, rendering less important, if not impossible, the detection and documentation of soot and CO-Hgb at autopsy. In these situations, the ME/C must rely upon the clinical record to make their determination of the cause.

Keep in mind also that chronic *smokers* often live with an elevated CO-Hgb saturation (up to 8%–10%), so this must be considered when interpreting CO-Hgb values.[3,4] **The typical fatal range for CO-Hgb in a *healthy* adult is 40%–60%.**[5–21] Saturations may far exceed this range in the very healthy or the very young (children and infants),

Principles of Forensic Pathology
https://doi.org/10.1016/B978-0-323-91796-4.00026-X

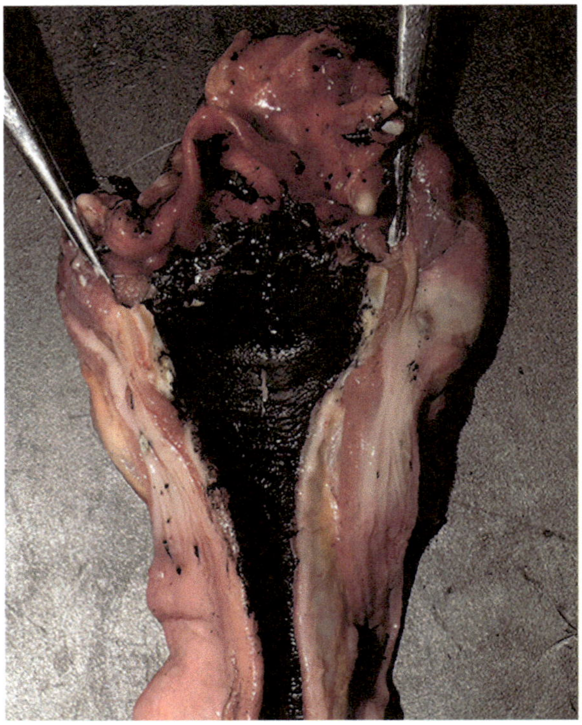

FIGURE 12.1 **Black soot deposition in the airway as evidence of smoke inhalation during life.**

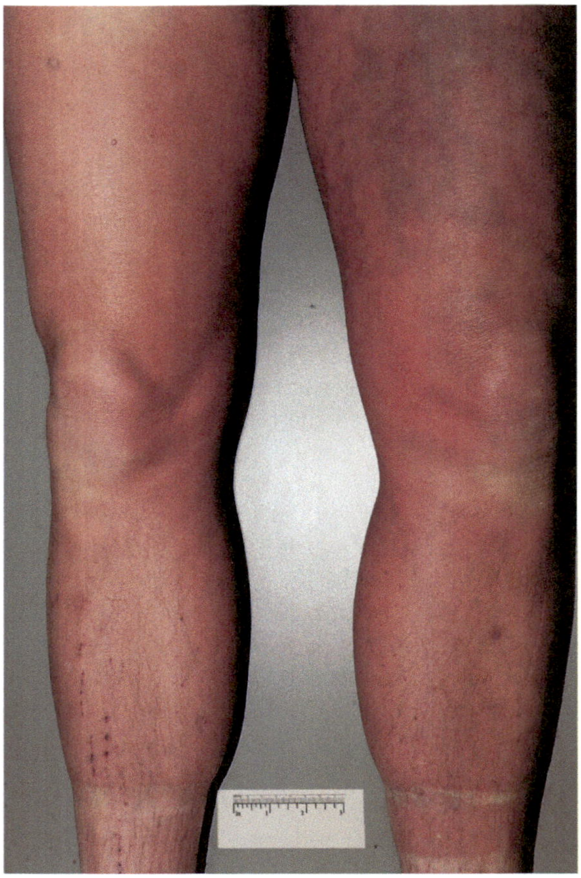

FIGURE 12.2 **Cherry red lividity of carbon monoxide poisoning victim.**

FIGURE 12.3 **Partial and full thickness thermal injury.**

the latter having a higher respiratory rate. *The elderly and/or obese and infirm (those with comorbidities, particularly cardiorespiratory) tend to succumb to the effects of smoke inhalation at lower saturations of CO-Hgb, analogous to traumatic hemorrhagic complications*, where individuals with comorbidities die from a lesser volume of blood loss as compared to their healthier counterparts. In these instances, one or more comorbidities might best be listed in **part 2** to better explain the fatal outcome from the lower Co-Hgb saturation, unless of course the lower level is explained by intervening oxygen therapy. Finally, it is important to recognize that smoke inhalation injury is defined by more than just carbon monoxide poisoning. Depending on the location of the fire, several toxins, such as cyanide and chemicals derived from synthetic materials, are released and aerosolized in a fire. **Extreme heat** alone is very damaging to the airway lining, contributing to the respiratory insult separate and apart from the irritants in smoke and the cellular respiratory toxins.

In addition to smoke inhalation injuries, fire commonly causes **thermal burns**, which may be sustained in conjunction with smoke inhalation, or alone. The standard nomenclature used to describe these injuries categorizes them by the severity in depth of tissue involvement (**partial thickness** vs. **full thickness;** Fig. 12.3) and in body surface area (BSA) involvement ("rule of nines"; see diagram). A **superficial burn** (designated as "first degree" in the old classification system) involves only the epidermis and is characterized by erythema and edema, but no blisters (e.g., a sunburn). A **partial thickness burn** ("second degree") is characterized by blisters (Fig. 12.4), involves both the epidermis and the dermis, does not involve the sweat glands and hair follicles, and also does not typically leave scars. A **full thickness burn** ("third degree") is characterized by a dry and leathery, eschar-like injury that destroys the epidermis and dermis, including the dermal appendages, and typically leaves scarring. Because the nerve endings are destroyed in full thickness burns, the burns are less painful than most partial thickness burns.[22,23] The **rule of nines** is based on the percentage BSA in multiples of nine (Fig. 12.5) that are represented by various anatomic regions of the body, differing between pediatric and adult victims (e.g., infants have a disproportionately larger head relative to their shorter limbs when compared to adults, Fig. 12.6). As a general rule of thumb, *if the age of the victim plus the percentage BSA thermal injury exceeds 110, a fatal outcome is nearly guaranteed.*[24,25] Similarly, *if a victim*

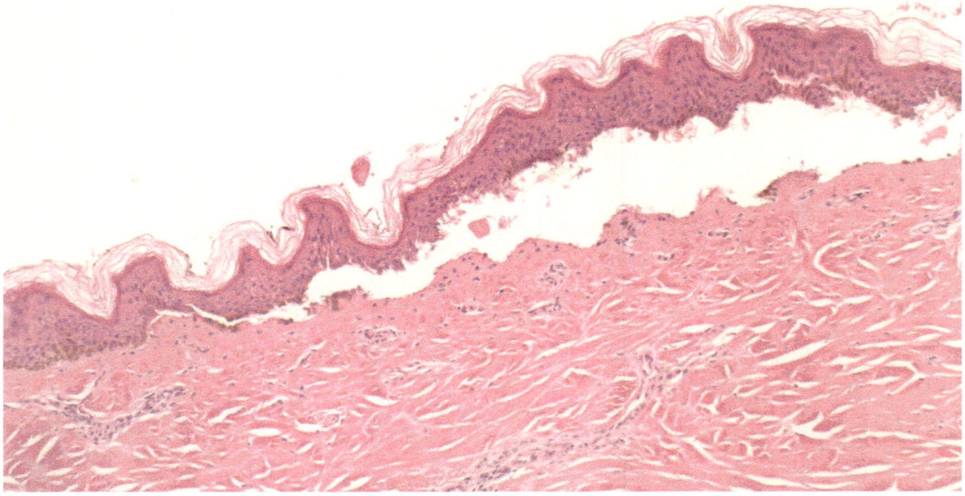

FIGURE 12.4 **Epidermal–dermal split due to thermal injury.**

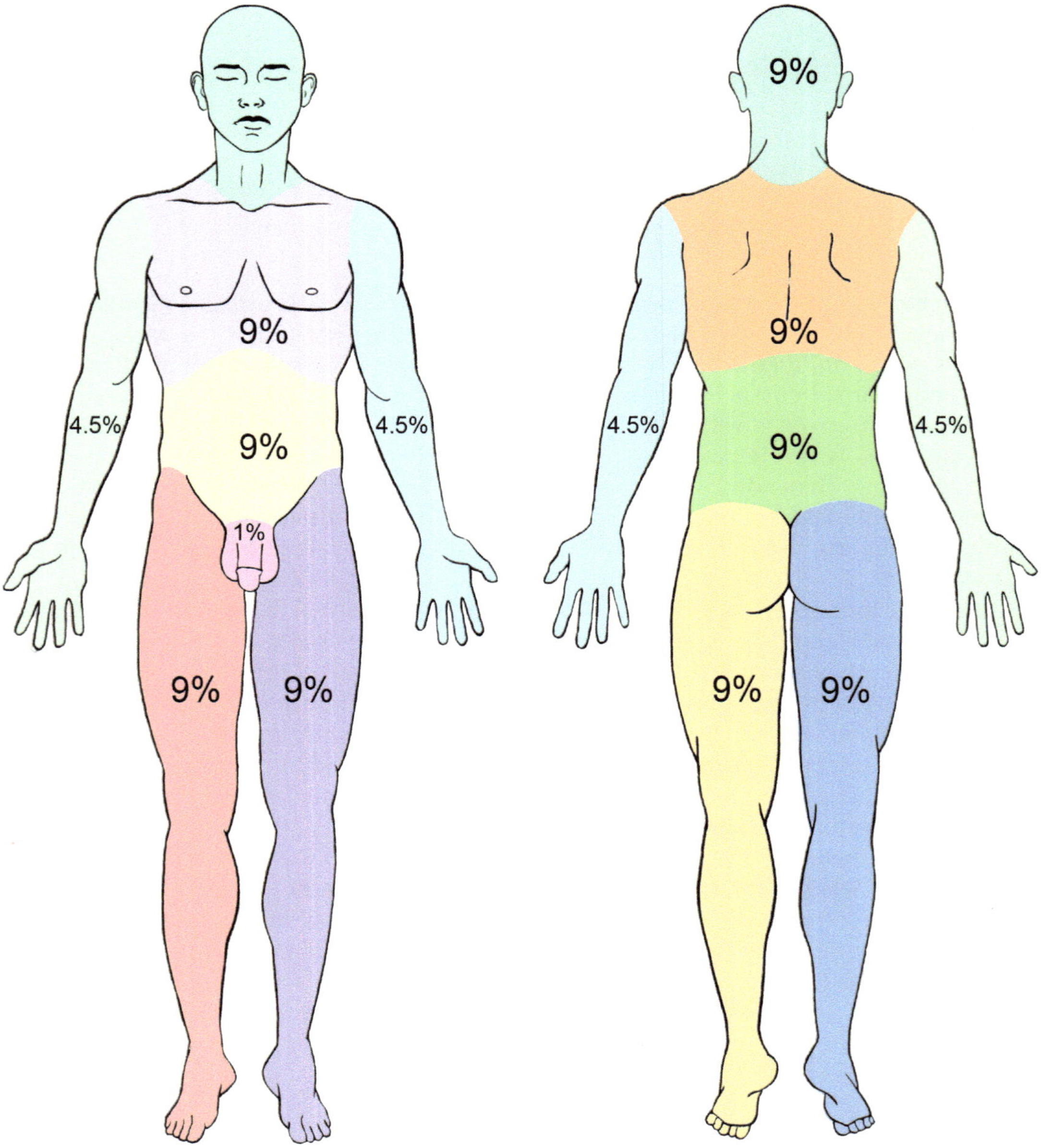

FIGURE 12.5 **Rule of nines (adult).** *Watercolor by Dr. Cricket Coleman.*

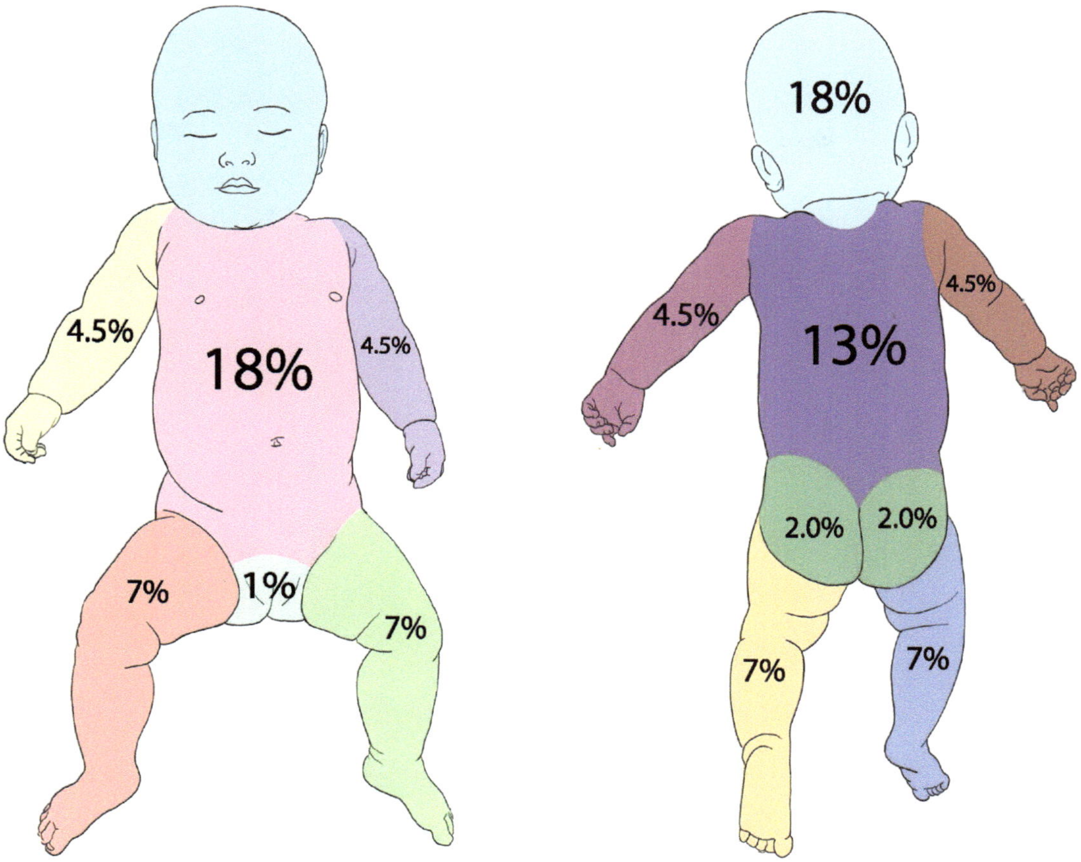

FIGURE 12.6 **Rule of nines (infant).** *Watercolor by Dr. Cricket Coleman.*

is over the age of 60, has suffered >40% BSA thermal injury, and also has smoke inhalational injury, their mortality rate is 90%.[26] For this reason (i.e., for vital statistical purposes), it is good practice to **cite the specific BSA in the cause of death statements** that involve thermal burns.

If a fire victim makes it to medical attention and survives, they are often cared for in an intensive care unit, often one specializing in burn patients. These practitioners are well versed in evaluating burn injury and routinely calculate the patient's percentage BSA involvement upon admission. If such a patient ultimately dies and is brought to the ME/C after a substantial survival interval, it is best to use that *initial clinical BSA* as a reflection of the severity of the original burn. By the time such a decedent is evaluated by the forensic pathologist at the time of autopsy, there is often distortion of the original thermal injury by a substantial degree of healing, debridement, edema, escharotomies, and skin grafting, making an accurate estimation of the original BSA difficult, if not impossible to calculate.

If, on the other hand, the decedent is declared dead on scene with substantial thermal injuries, including char injury (so-called "fourth degree burns," Fig. 12.7), estimating an accurate BSA that represents the moment of death becomes impossible. In other words, if an individual dies in a fire that continues to burn after death, their body will continue to burn. In such instances, the only thing the forensic pathologist can do is document the extent of the thermal injury they observe at the time of discovery, and otherwise focus on the determination of whether the decedent was alive or dead at the time the fire started; they do this by quantitating the CO-Hgb saturation and assessing the airways for evidence of respiration (smoke inhalation injury). Beyond that, the thermal injuries can look quite similar whether sustained before or after death.

Once a person is dead, continued heat and fire often lead to characteristic **thermal artifacts**, including the *pugilistic posture, longitudinal muscle splitting (along the "grain" of the muscle fibers), long bone fractures as the muscles shrink/shorten as they burn, and a dark, granular epidural blood collection* (Fig. 12.8). These must not be confused with antemortem injuries.

The second prong of fire death evaluation is the **investigatory** piece. This involves both the *police* and the *FM*, and a medicolegal death investigator (MDI), if the decedent is pronounced at the scene of the fire, rather than in a hospital. The police conduct a general investigation that aims to establish whether criminality was involved, and how,

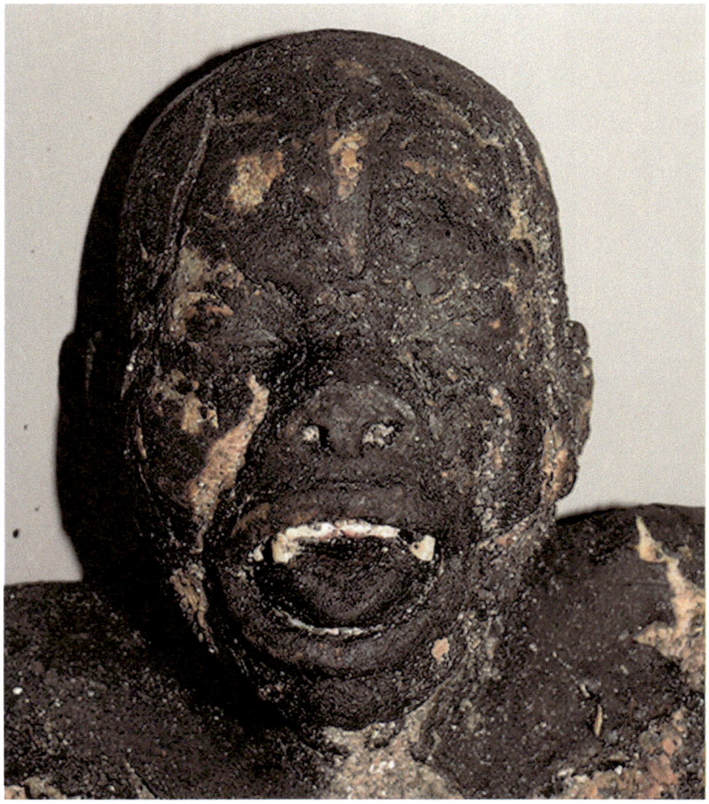

FIGURE 12.7 **Thermal charring ("fourth degree burns").**

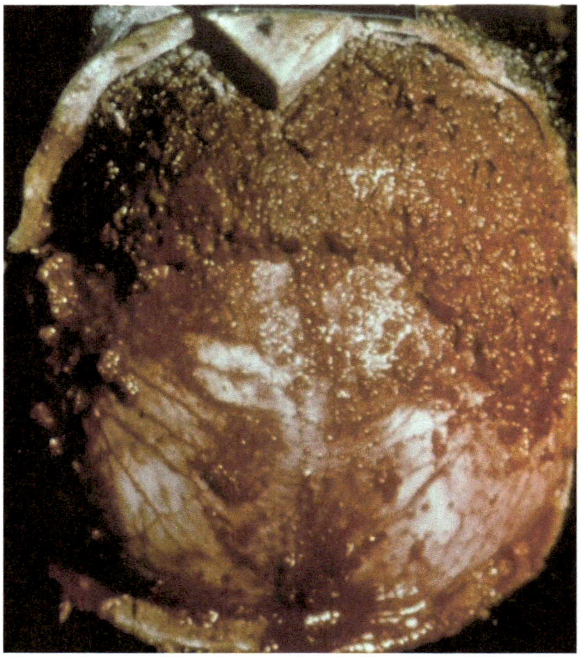

FIGURE 12.8 **Granular epidural blood collection in a fire death.** This is a heat artifact. Note how the epidural blood crosses the midline.

where, and under what circumstances the decedent was found relative to the main location of the fire (e.g., a kitchen fire, bed fire). If a bed fire, finding the decedent some distance from the bed, closer to the door of the apartment, for example, indicates that they were likely alive when the fire started and were attempting to escape when they died. If, on the other hand, they are still found in bed, one must consider the possibilities that they were either dead when the fire started or too intoxicated or otherwise unable (paralyzed, too frail, too young, etc.) to attempt to escape before becoming overcome by the smoke. If intoxication is thought to be the explanation for this failure to escape, it is probably best to await the toxicology results (for both intoxicants and the CO-Hgb saturation), as this may be the only way the decedent's immobility and fatal outcome make sense. If the decedent was bedbound due to paralysis, very young or old age, or comorbidities, then it would make more sense from the beginning, without the need for a toxicological explanation. Rarely, these scenarios represent an intentional act by the decedent (suicide) or another (homicide); see below. To answer which of these theories is applicable (the fire caused the death vs. the death caused the fire), an autopsy must be performed (unless the answer is clear from a survival interval in a hospital).

FMs play an important and specific role in fire investigations, but have a more focused task. They are trained to evaluate the nature and spread of a fire and in doing so, *identify its likely source/cause*. It is the uncovering of this source that ultimately helps to inform the ME/C in the classification of the **manner of death.** To that end, it is premature to establish a manner of death in fire fatalities without first consulting with the FM and police to ensure that their investigation is completed and to be clear on what basis their disposition of the fire was based. Important to point out in these settings, however, is the distinction that must be made regarding the *intent to start a fire*, regardless of malice. By this, we mean, in order for criminality (**arson**) to apply and lead to a FM classification of an "incendiary" fire, intent to harm or kill need not necessarily be present, but an understanding by the fire starter of the inherent recklessness and danger of fire setting should be clear.[27–31] To wit, a fire started by a four-year-old's match playing activity that leads to property damage and loss of life may well *not* be classified as incendiary by the FM, nor the resulting fatalities viewed as homicides by the ME/C. By contrast, this may be viewed very differently if the fire starter is a cognitively normal 14-year-old. While the classification and laws may vary from jurisdiction to jurisdiction, the real question is, in the view of the ME/C, at what age does fire setting change the manner from accident to homicide? There is likely gray area that makes answering this question complex and worthy of an in-depth discussion both among ME/Cs and with prosecutors.

On rare occasions, an **accelerant** is used in intentional fire setting. This is most typically encountered in suicides (**self-immolation deaths**) or as part of **homicidal activity** (either to cause death or in an attempt to destroy the body after a homicide is committed by some other means). When the investigative circumstances heighten concern for accelerant use, samples of any remaining clothing and/or other wrappings can usually be sent for further analysis. Commonly, canisters specially made for such samples are available for their transport. It is sometimes patently clear from the odor of the body alone that an accelerant was used. If it is suspected that the body was set on fire as part of a postmortem concealment attempt, the absence of CO-Hgb in the blood and an examination of the tissues and airway that fails to demonstrate a cherry red discoloration or soot deposition, respectively, suffice in providing evidence in support of this conclusion. A caveat in this instance is presented by a **flash fire**, which *has the capacity to cause death in the absence of an elevated CO-Hgb saturation.*[4,32,33] Flash fires are encountered in thermal explosions, such as those occurring in motor vehicle collisions with engines engulfed in flames, but also may occur with criminal activity that involves the use of an accelerant. This is why the presence or absence of heat injury and soot deposition in the airways is an important additional data point in distinguishing flash fire deaths (*positive* airway findings, despite a *negative* or very low CO-Hgb saturation) from postmortem thermal injury (*negative* airway findings and *negative* CO-Hgb saturation). See Chapter 15 for a further discussion on postmortem concealment efforts involving fire.

Scald and heat burns

Other forms of (nonfire–related) antemortem and postmortem thermal burns include scald injury and injuries from heating sources (radiators, heating pads, etc.). Scald injury can be sustained in an adult or in a child/infant. In *adults*, poorly controlled thermostats in bathroom showers/bathtubs and kitchen spills probably constitute the most common settings for such burns, most being accidental in nature, particularly for those who are frail and/or have motor or cognitive limitations. That said, the more dependent and vulnerable a decedent is, the higher should be the index of suspicion and vigilance in the investigation to exclude neglect-based or volitional/inflicted injuries. A splash pattern would be typical and expected in most benign scenarios. In the *pediatric setting*, on the other hand, a higher index of suspicion and concern is warranted, especially if the scald injury pattern is more typical of submersion (a well-demarcated submersion "line" and relative sparing in the flexion creases of the antecubital fossae, popliteal fossae, and/or groins) and incompatible with a story that indicates an accidental splash mechanism (see Chapter 6). Finally, "scald injury" with epidermal sloughing not uncommonly occurs in the postmortem period when an individual dies from an unrelated (usually sudden cardiac) cause while in the stagnant or running hot water of a bathtub or shower. These injuries should be evaluated within the context of the entire case, including other findings and causes of death. If ultimately determined to have developed in the postmortem period, they should still be described as injuries in the autopsy report, but qualified as being "postmortem" (i.e., a consequence, not a cause of death).

Similarly, **radiator and heating pad contact injuries/burns** may also cause skin damage in the antemortem or postmortem periods, depending on the circumstances, so a careful evaluation and dovetailing of both the context and the autopsy findings is central to their proper interpretation. That said, if interpreted as an intravital injury, an explanation should be sought and provided to explain why the decedent failed to remove the heating element from their body (e.g., intoxication, cerebrovascular event, head trauma, etc.). Finally, on occasion, **smoldering burn injury** is identified during the autopsy examination of a decedent who was brought from a scene where no concerns of a fire were reported. This is sometimes encountered in association with illicit drug deaths (cocaine, heroin) involving a heating element or a "crack pipe," such that, after death (due to a drug intoxication), tissue continues to burn as contact with the heating element persists, at a low temperature and without a flame that would cause damage to the area immediately surrounding the decedent. As such, these injuries can be missed, going completely undetected without a careful inspection of the body at the scene and in the autopsy suite. Often, they are unrelated to the death (i.e., occurred postmortem) and may come as a surprise to the prosector.

Chemical/irritant burns

Chemical burns constitute a distinct category of burn trauma separate and apart from thermal injuries. While encountered much less frequently, they may be seen in the setting of the rare suicide or as part of homicidal violence, including forced ingestions or postmortem criminal activity. Careful attention should be paid to the pattern of injury on the face, particularly in the perioral area, and/or any intraoral trauma, such as mucosal or frenular lacerations and/or contusions that indicate a struggle or force. A close examination of the mucosa of the oral cavity, esophagus, and stomach, taking note of abnormalities in color, damage, or odors, may have relevance in the impressions made or the toxicological testing requested. Bleach, acids, and bases are a few examples of some of the compounds that may be used. The use of pH paper may help distinguish a base from an acid. Bleach may also be used on occasion by perpetrators wanting to "clean up" in the postmortem interval following a homicide, sometimes causing (postmortem) skin damage that is unrelated to the mechanism of death. Other chemicals and toxins used in the suicidal or homicidal settings, such as cyanide and arsenic, will be covered separately in Chapter 14 (Approach to Toxicological Deaths).

Blast/radiation injury

Blast injury

Blast injury is characterized by complex physical trauma caused by the direct or indirect exposure to an explosion. Blast injuries occur in various scenarios ranging from industrial accidents to deliberate attacks, and the response may involve multiple agencies, local and even federal. High-order explosives produce a supersonic overpressure shock wave, while low-order explosives deflagrate (subsonic combustion) and do not produce such a wave. The blast wave caused by an explosion begins with a single pulse of increased air pressure (a few milliseconds) followed by negative pressure (suction). The duration and potential damage of the wave depend upon the explosive material and the distance from detonation. The blast wave progresses as a sphere of compressed and rapidly expanding gases, which displaces an equal volume of air. A person in the path of an explosion will be subjected to excess barometric pressure, as well as pressure from the high velocity wind traveling behind the front of the blast wave. Explosions near or within hard solid surfaces become amplified due to shock wave reflection. *Blast injuries are divided into four types: primary, secondary, tertiary, and quaternary.*

Primary blast injuries are caused by high pressure (over pressure) blast waves. They occur when the person is close to the blast and the entire body is affected. The ears, lungs, and hollow viscus organs are usually the most damaged. There may be a delay for these injuries to result in clinical disability. In fatal instances, the lungs will be markedly hemorrhagic. **Blast lung** refers to extensive pulmonary contusions with damage to alveoli and blood vessels. It is the most common immediate cause of death among people who initially survive an explosion.[34–38]

Secondary blast injuries consist of ballistic trauma caused by shrapnel and other objects propelled by the explosion. *Most deaths are caused by secondary injuries.* **Tertiary blast injuries** are caused by displacement of air by the explosion, which creates a blast wind that can propel people to impact other resistant objects/surfaces. These injuries result from a combination of blunt and penetrating trauma, such as fractures and craniocerebral injuries. **Quaternary**

blast injuries comprise all other injuries not included in the first three, which would include flash burns, crush injury, and inhalation injury.

Blast injuries can cause subtle neurotrauma that may manifest as a loss of memory of the explosion, confusion, headache, an impaired sense of reality, and reduced decision-making ability. Some patients may develop rapid cerebral edema while others develop neuropsychiatric symptoms months or years later. **Blast-induced neurotrauma,** categorized as secondary brain injuries, may include chronic fatigue, headache, and problems with memory, speech, and balance. There is active blast wave posttraumatic stress disorder (**PTSD**) research investigating a cause-and-effect relationship between blast waves and PTSD.

Due to the potential for severe bodily injury and fragmentation/dismemberment, identification of the decedent is a major part of the investigation of these deaths.[39–42]

Radiation injury

Radiation injury may cause *acute radiation syndrome* (also known as radiation sickness or radiation poisoning), which is caused by exposure to high amounts of ionizing radiation in a short period of time.[43,44] Initially, there is nausea, vomiting, a loss of appetite, fever, and skin erythema. Symptoms may start within an hour of exposure and can last for several months, ultimately resulting in recovery or death.

Sources of such radiation include nuclear reactors, cyclotrons, cancer therapy treatments, and nuclear/radiological weapons. There are three organ systems that are usually affected: **the hematopoietic system (primarily the bone marrow), the gastrointestinal tract, and the neurovascular system.** The cells that are most affected by radiation are those that are rapidly dividing and, therefore, sustain the most DNA damage. Effects from the damaged hematopoietic system mainly manifest as a result of *aplastic anemia.* This may lead to infections, bleeding, and anemia. The gastrointestinal syndrome manifests as nausea, vomiting, loss of appetite, and abdominal pain. The neurovascular syndrome typically presents with neurological symptoms such as dizziness, headache, and/or a decreased level of consciousness, occurs within minutes to a few hours, and is usually fatal.

Animal attacks, domestic and in the wild

While an in-depth analysis of fatalities caused by animal attacks exceeds the scope of this book, these deaths may be vanishingly rare or relatively more common, depending on the jurisdiction; as such, they at least deserve mention. In more rural jurisdictions that are within close proximity to national parks, forests, or camping grounds, fatal attacks by *wild animals,* such as bears, may occur with some degree of frequency. In such instances, captured and euthanized animals may require imaging studies and subsequent autopsies to definitively identify and retrieve human remains from their gastrointestinal tracts. Rare shark attacks may be encountered in more southerly resorts, including those involving SCUBA facilities.

In the rest of the country, occasional dog attacks may fall under the purview of the ME/C. While civil litigation alleging negligence may result from such inadvertent attacks, the main situation in which this category of fatality would be certified as anything other than accidental would be in the rare circumstance of a dog being *intentionally* used as a weapon, ordered and/or allowed by its owner or trainer to (fatally) attack another person. That said, the victims of most fatal dog attacks are known to or were the owners of the putative dog. To the inexperienced eye, dog attack injuries (bite and claw marks) may be mistaken for sharp injuries caused by scissors or knives, for example, or identified as blunt force trauma. When involving children, such misdiagnoses can even result in a false allegation of child abuse. For this reason alone, a better understanding and recognition of the patterns seen in these fatalities is strongly encouraged.

Fatal dog attacks must also be distinguished from *postmortem animal predation.* While it is unusual for dogs to feed on cadavers, they will do so if motivated by extreme hunger (see Chapter 4 for an additional discussion). Attack behaviors and canine patterns of fatal injury are somewhat breed-dependent, but typically involve at least some paired "hole and a tear" puncture wounds (a reflection of the two canine teeth) that often involve some degree of pulling and tearing. Claw marks may help in determining the size of the dog based on the estimated paw size. Sometimes, dog fur is found within the wounds. On occasion, veterinarians or odontologists may need to be consulted to discern a species-specific bite injury pattern. *The most prevalent breeds involved in fatal attacks are the unrestrained Pit Bull-type dogs, Rottweilers, and German Shepherds.* The head, throat, and neck are particularly prone to attack once the victim is pushed to the ground, but "defense" injuries are commonly found on the forearms or lower extremities, supporting an antemortem attack, rather than postmortem predation. As such, when part of a fatal attack, the modalities of injury are usually related to exsanguination, asphyxiation, and/or blunt impact head trauma, with skull fractures.

Unlike cats, dogs may eat their prey before they are necessarily dead, a reflection of the speed and frenzy of a dog mauling. Incidents most frequently occur near the victim's home. When the attacking dog is a stray (i.e., unknown to the victim), injuries of the hands and legs are more commonly observed. In a more predatory (rather than defensive) attack, shaking and pulling may result in extensive and disfiguring tissue tearing. Small children are particularly vulnerable to attack based both on their size and their sometimes rapid and erratic movements toward or provocations of the animal. The absence of both a conscious survival interval and witnesses only complicates the investigation. Dogs attacking in packs are, of course, the most dangerous, both in terms of pack behavior and their sheer outnumbering of the victim. While the cause of death in such fatalities may be self-evident, the investigation is important in its elucidating and understanding of the circumstances that provoked the attack (trained attack dog, pregnant female, severe hunger, rabies infection, etc.). Visual identification issues may lead to additional investigative complexity as a result of devastating facial trauma. Trace evidence collection may involve canine-specific swab kits, where available and deemed necessary. Scene photographic documentation, including those of foot and paw prints, as well as an examination of the dog (if available) may be of critical evidentiary value.

In terms of death certification, some may use "complications of multiple puncture wounds" or "multiple (dog) bites/dog bite wounds" in the cause of death statement, with or without blunt injuries, depending on the specifics of the head trauma, and include "dog attack" or "dog mauling" somewhere in the "how injury occurred" field. The manner is most commonly accidental, but of course, depends upon the specifics of the circumstances, as qualified above.[45–55]

References

1. Evarts B. *Fire Loss in the United States during 2018*. October: National Fire Protection Association; 2019, 2019.
2. Istre GR, McCoy MA, Osborn L, Barnard JJ, Bolton A. Deaths and injuries from house fires. *N Engl J Med*. 2001;344(25):1911–1916.
3. Sully CJ, Walker GS, Langlois NEI. Investigation into the potential for post-mortem formation of carboxyhemoglobin in bodies retrieved from fires. *Forensic Sci Med Pathol*. 2018;14(3):342–348.
4. Wirthwein DP, Pless JE. Carboxyhemoglobin levels in a series of automobile fires. Death due to crash or fire? *Am J Forensic Med Pathol*. 1996; 17(2):117–123.
5. Winter PM, Miller JN. Carbon monoxide poisoning. *JAMA*. 1976;236:1502–1504.
6. Ernst A, Zibrak JD. Carbon monoxide poisoning. *N Engl J Med*. 1998;339(22):1603–1608.
7. Carson HJ, Stephens PJ. Four deaths due to carbon monoxide poisoning in car washes. *Am J Forensic Med Pathol*. 1999;20(3):274–276.
8. Jumbelic MI. Open air carbon monoxide poisoning. *J Forensic Sci*. 1998;43(1):228–230.
9. Jaffe FA. Pathogenicity of carbon monoxide. *Am J Forensic Med Pathol*. 1997;18(4):406–410.
10. Gerostamoulos D, Beyer J, Wong K, Wort C, Drummer OH. Carbon monoxide concentrations in the 2009 Victorian Bushfire disaster victims. *Forensic Sci Int*. 2011;205(1–3):69–72.
11. Partovi E, Nunez J, Gill JR. *College of American Pathologists (CAP) Forensic Pathology Resource Committee. 2019 FR-B Forensic Pathology Program: Case 8 – Carbon Monoxide*. Northfield, IL: CAP; 2019.
12. Ruszkiewicz A, de Boer B, Robertson S. Unusual presentation of death due to carbon monoxide poisoning. A report of two cases. *Am J Forensic Med Pathol*. 1997;18(2):181–184.
13. Vreman HJ, Wong RJ, Stevenson DK, et al. Concentration of carbon monoxide (CO) in postmortem human tissues: effect of environmental CO exposure. *J Forensic Sci*. 2006;51(5):1182–1190.
14. Unintentional non-fire-related carbon monoxide exposures–United States, 2001–2003. *MMWR Morb Mortal Wkly Rep*. 2005;54(2):36–39.
15. Weaver LK. Clinical practice. Carbon monoxide poisoning. *N Engl J Med*. 2009;360(12):1217–1225.
16. Schmitt MW, Williams TL, Woodard KR, Harruff RC. Trends in suicide by carbon monoxide inhalation in King County, Washington: 1996–2009. *J Forensic Sci*. 2011;56(3):652–655.
17. Shimazu T, Ikeuchi H, Sugimoto H, Goodwin CW, Mason Jr AD, Pruitt Jr BA. Half-life of blood carboxyhemoglobin after short-term and long-term exposure to carbon monoxide. *J Trauma*. 2000;49(1):126–131.
18. Gettler A, Freimuth H. The carbon monoxide content of the blood under various conditions. *Am J Clin Pathol*. 1940;10:603–613.
19. Retzky SS. Carbon monoxide poisoning from hookah smoking: an emerging public health problem. *J Med Toxicol*. 2017;13(2):193–194.
20. Blackmore D. The determination of carbon monoxide in blood and tissue. *Analyst*. 1970;95:439–458.
21. Lisbona CF, Hamnett HJ. Epidemiological study of carbon monoxide deaths in scotland 2007–2016. *J Forensic Sci*. 2018;63(6):1776–1782.
22. *Studies of Thermal Injury III Path and Pathogenesis of Cutaneous Burns an Experimental Study*. 1947.
23. Nielson CB, Duethman NC, Howard JM, Moncure M, Wood JG. Burns: pathophysiology of systemic complications and current management. *J Burn Care Res*. 2017;38(1):e469–e481.
24. Monafo WW. Initial management of burns. *N Engl J Med*. 1996;335(21):1581–1586.
25. Colohan SM. Predicting prognosis in thermal burns with associated inhalational injury: a systematic review of prognostic factors in adult burn victims. *J Burn Care Res*. 2010;31(4):529–539.
26. Ryan CM, Schoenfeld DA, Thorpe WP, Sheridan RL, Cassem EH, Tompkins RG. Objective estimates of the probability of death from fire burns. *NEJM*. 1998;338:362–366.

27. Ritchie EC, Huff TG. Psychiatric aspects of arsonists. *J Forensic Sci*. 1999;44(4):733–740.
28. Adair TW, Fisher A. Suicide with associated acts of arson: two cases from Colorado. *J Forensic Sci*. 2006;51(4):893–895.
29. Leong GB, Mueller C, Feldsher M. Insane arsonists: an early 21st century sample. *J Forensic Sci*. 2019;64(2):454–459.
30. Adelson L. Role of the pathologist in arson investigation. *J Crim law, Criminol Police Sci*. 1955;45(6):760–768.
31. Lentini JJ. The evolution of fire investigation and its impact on Arson cases. *Crim Just*. 2012;27:12.
32. Hirsch CS, Adelson L. Absence of carboxyhemoglobin in flash fire victims. *JAMA*. 1969;210(12):2279–2280.
33. Hirsch CS, Bost RO, Gerber SR, Cowan ME, Adelson L, Sunshine I. Carboxyhemoglobin concentrations in flash fire victims: report of six simultaneous fire fatalities without elevated carboxyhemoglobin. *Am J Clin Pathol*. 1977;68(3):317–320.
34. DePalma RG, Burris DG, Champion HR, Hodgson MJ. Blast injuries. *N Engl J Med*. 2005;352(13):1335–1342.
35. Tiemensma M, Wadee SA. Blast-induced lacerations - more examples of the diagnostic entity. *Am J Forensic Med Pathol* 2014;35(3):222.
36. Pomara C, D'Errico S, Riezzo I, Perilli G, Volpe U, Fineschi V. Blast overpressure after tire explosion: a fatal case. *Am J Forensic Med Pathol*. 2013; 34(4):306–310.
37. Dussault MC, Smith M, Osselton D. Blast injury and the human skeleton: an important emerging aspect of conflict-related trauma. *J Forensic Sci*. 2014;59(3):606–612.
38. Christensen AM, Smith VA, Ramos V, Shegogue C, Whitworth M. Primary and secondary skeletal blast trauma. *J Forensic Sci*. 2012;57(1):6–11.
39. Budimlija ZM, Prinz MK, Zelson-Mundorff A, et al. World Trade Center human identification project: experiences with individual body identification cases. *Croat Med J*. 2003;44(3):259–263.
40. Mundorff AZ, Bartelink EJ, Mar-Cash E. DNA preservation in skeletal elements from the World Trade Center disaster: recommendations for mass fatality management. *J Forensic Sci*. 2009;54(4):739–745.
41. Warnasch SC. Forensic archaeological recovery of a large-scale mass disaster scene: lessons learned from two complex recovery operations at the world trade center site. *J Forensic Sci*. 2016;61(3):584–593.
42. Gill JR. 9/11 and the New York city medical examiner's office. *Forensic Sci Med Pathol*. 2006;2(1):29–32.
43. Mathes SJ, Alexander J. Radiation injury. *Surg Oncol Clin*. 1996;5(4):809–824.
44. Brown KR, Rzucidlo E. Acute and chronic radiation injury. *J Vasc Surg*. 2011;53(1 Suppl):15S–21S.
45. Avis SP. Dog pack attack: hunting humans. *Am J Forensic Med Pathol*. 1999;20(3):243–246.
46. Chu AY, Ripple MG, Allan CH, Thogmartin JR, Fowler DR. Fatal dog maulings associated with infant swings. *J Forensic Sci*. 2006;51(2): 403–406.
47. Loewe CL, Diaz FJ, Bechinski J. Pitbull mauling deaths in Detroit. *Am J Forensic Med Pathol*. 2007;28(4):356–360.
48. Shields LB, Bernstein ML, Hunsaker 3rd JC, Stewart DM. Dog bite-related fatalities: a 15-year review of Kentucky medical examiner cases. *Am J Forensic Med Pathol*. 2009;30(3):223–230.
49. Stiegler D, Gilbert JD, Warner MS, Byard RW. Fatal dog bite in the absence of significant trauma: capnocytophaga canimorsus infection and unexpected death. *Am J Forensic Med Pathol*. 2010;31(2):198–199.
50. Byard RW. Domestic dogs (*Canis lupus* familiaris) and forensic practice. *Forensic Sci Med Pathol*. 2016;12(3):241–242.
51. Karbeyaz K, Ayranci U. A forensic and medical evaluation of dog bites in a province of Western Turkey. *J Forensic Sci*. 2014;59(2):505–509.
52. Fonseca GM, Palacios R. An unusual case of predation: dog pack or cougar attack? *J Forensic Sci*. 2013;58(1):224–227.
53. Schenk AM, Ragatz LL, Fremouw WJ. Vicious dogs part 2: criminal thinking, callousness, and personality styles of their owners. *J Forensic Sci*. 2012;57(1):152–159.
54. Bury D, Langlois N, Byard RW. Animal-related fatalities–part I: characteristic autopsy findings and variable causes of death associated with blunt and sharp trauma. *J Forensic Sci*. 2012;57(2):370–374.
55. Di Donato S, Ricci P, Panarese F, Turillazzi E. Cane Corso attack: two fatal cases. *Forensic Sci Med Pathol*. 2006;2(2):137–141.

CHAPTER

13

Approach to environmental deaths

Susan F. Ely and James R. Gill

Bacchus hath drowned more men than Neptune. -***Thomas Fuller***

Fatalities related to drownings, electrocutions, hyperthermia, and hypothermia are typically considered environmental deaths and are largely diagnosed with predominant attention focused on the circumstances.

Drowning

Drowning is diagnosed based upon the circumstances and the absence of a disease or injury at autopsy that better explains death.[1–9] The autopsy serves largely to exclude other causes. There may be some "soft" findings that support a diagnosis of drowning, but there is no diagnostic finding.[4,5,7,10–36] Davis described four questions that should be asked when investigating a body found in water.[37] These are:

1. Was the person alive or dead before entry into the water?
2. Did the person drown?
3. Why did the person get into the water?
4. Why was the person unable to survive in the water?

As described by Charles Hirsch, "People die in bathtubs for the same reasons they die in bed or elsewhere in the home. It is untenable to conclude that anybody but a small child has drowned in a bathtub unless disease (e.g., epilepsy), intoxication, or injury renders [them] helpless in such an unfortunate situation."

Adults do not drown in bathtubs or shallow water unless they are unwilling or unable to lift their head above the water. A study in Denmark described common causes and scenarios.[38] Considerations include:[39–43]

1. Intentional (or neglectful) acts: homicide or suicide.[44–48]
2. Mechanical injury: Inadvertent falls with craniocerebral or cervical trauma.
3. Chemical injury (intoxication): Requires rapid incapacitation and failure to awaken from the shock of facial submersion or substantially impairs the individual's motor coordination abilities.
4. Disease: Epilepsy/syncope/amyotrophic lateral sclerosis (ALS)/myasthenia gravis
5. Electrical injury
6. Hypothermia

Adelson said that "One can drown in the Atlantic Ocean or in a child's wading pool, two inches deep."[49] When considering the span of drowning deaths, from an infant in a bathtub to a four-year-old in a swimming pool or a diver in a self-contained underwater breathing apparatus (SCUBA) in the ocean, each face different developmental, social, environmental, and equipment factors.[4,39–43,46,50–86] Neither children nor adults drown in bathtubs unless they are unwilling or unable to lift their heads above water. It does not make sense for adults to drown in a bathtub unless something has incapacitated them in this setting. If they are unwilling to lift their heads above the water, suicide is the appropriate manner of death. If they are unable to lift their head, one must ask, why? If another person held them underwater, the manner of death is homicide. The other explanations (i.e., disease or injury) will largely be certified as accidents (even if the proximate cause is a disease). Put another way, uninjured, neurologically competent, nonintoxicated/unmedicated adults do not/should not accidentally drown in bathtubs (see Chapter 6 for further discussion regarding *pediatric bathtub fatalities*).

Principles of Forensic Pathology
https://doi.org/10.1016/B978-0-323-91796-4.00029-5

The diseases that are most frequently involved with bathtub drownings are those associated with some kind of neurological impairment (e.g., epilepsy, ALS, cerebrovascular events).[9,54,59,68] If epilepsy (a nontraumatic seizure disorder) is the underlying reason for the drowning, it also would be the proximate cause of death, but the manner of death would be accident (informed by the injurious immediate cause—drowning). The circumstances that most frequently cause adults to merely *submerge, rather than drown* in a bathtub are sudden deaths due to heart disease. In these instances, we infer that the person submerged because they died of their heart disease, and the manner of their death is certified as natural. The loss of consciousness caused by a cardiac dysrhythmia results from inadequate perfusion of the brain due to cessation of the heart function. Since death is defined as the irreversible cessation of the heart, these bathtub events would most likely to be fatal regardless of whether the person was in a bathtub or not. If the heart has already stopped, subsequent submersion in water does not contribute to death. There are forensic pathologists who may still certify these deaths as drownings (accident). The reasoning is that the cardiac event happened in a hostile environment and they do not want to ignore it. If, on the other hand, there is a (structurally demonstrable) Class I finding, such as a ruptured myocardial infarct or a saddle pulmonary thromboembolism, this may push some toward a natural manner. There are instances in which the history or extent of disease may be ambiguous. In these instances, one may certify the death as undetermined, or descriptively, such as submersion in water complicating hypertensive cardiovascular disease (or epilepsy). The latter example is a true statement of fact without committing to the conclusion of "drowning," per se. Still, though, once an injury (water submersion) is invoked in the cause of death statement, the manner of death should not be certified as natural.

The difference between the heart disease death in a water environment (natural) versus one with epilepsy (accident) is that, unlike unconsciousness-inducing cardiac arrhythmias, most seizures do not result in death.[39,40,43,87] People may have dozens of seizures in a lifetime and survive each one. Very few people have the opportunity to have more than one episode of ventricular fibrillation (VF) outside of a hospital. In the bathtub scenario, the seizure more likely results in death because of the hostile environment. An analogy can be made with motor vehicle deaths. The death of a person who has a seizure while operating a motor vehicle, causing a collision with a tree, would be certified as an accident if they sustained extensive blunt injury. By contrast, the death of a person who has had an arrhythmic cardiac event while operating a motor vehicle, causing a collision with a tree, would be certified as a natural death, with or without some blunt injury (in this instance, the corroboration of the conclusion that the cardiac arrest preceded the collision would be supported by the lack of a vital reaction to the latter).[88–95]

As forensic pathologists, it is not uncommon to encounter dead adults in bathtubs and showers, both with and without submersion of their heads. Absent facial submersion (directly observed by an investigator or reliably by history), the consideration of drowning as a mechanism of death is usually not relevant. When there is facial submersion, however, the central question is whether it is the cause or a consequence of death.

The conclusion that an adult drowned in a bathtub demands that an explanation be provided in the cause of death statement or in the "how injury occurred" field; otherwise, it makes no sense. As a demographic, elderly persons with one or more comorbidities are disproportionately represented in this group but, more often than not, they have died from natural causes IN bathtubs, not FROM bathtubs (i.e., they slide under the water after death, as a consequence of having had a sudden death from a natural—usually cardiac—cause). In other words, deaths of elderly persons in bathtubs are more common than are bathtub drownings. If, on the other hand, there is reasonable evidence in the history and scene investigation to suggest a drowning, then physical evidence (inclusionary and exclusionary) must be sought through an autopsy and a proximate cause (risk factor for drowning) also sought, identified, then memorialized on the death certificate.

For example, an 80-year-old woman with known atherosclerotic and hypertensive cardiovascular disease, diabetes, and emphysema is found fully submerged in water while obviously taking a bath in a secured apartment. There are no depressant or illicit drugs, no concern or scene evidence to suggest prescription medication or alcohol misuse, no history of depression/suicidality, no evidence of injury or a "foam cone" on examination, and no known neurologic compromise or seizure history. Her death is most likely due to natural causes and the submersion incidental (i.e., consequential, not causal). Statistically and demographically, the likelihood of her death being caused by the water, under this specific set of circumstances, is low, because she has no likely or identifiable drowning risk, but does have considerable risk for a sudden cardiac death. Conversely, if the decedent was an 80-year-old with a chronic alcohol use disorder and a history of alcohol-related seizures and was found similarly, the likelihood of her ultimate immediate cause of death being drowning, neurologically and/or chemically abetted, is reasonable and should be considered in the ensuing work-up.

It is not advisable to render a diagnosis of drowning without the benefit of an autopsy (to identify both exclusionary and Class I inclusionary findings—see below), whereas the determination of a natural death can and often is made in such circumstances, without an autopsy, if the suspicion for drowning is low, as illustrated in the first example above. This is analogous to the classification of similar persons with extensive disease and advanced age found on the floor of their homes and examined externally only, their deaths being certified as natural. Opining instead that they died from blunt force head trauma, without the benefit of an autopsy, based solely on the (likely terminal) small laceration on their brow, would be inappropriate. Analogously, it would equate to concluding that the alcoholic described above died from drowning by circumstance alone, in the absence of an autopsy.

People are also occasionally found partially in bathtubs (legs out of tub in some fashion), partially or fully clothed (or not), or facing the wrong end of the tub (most people who bathe in the tub have their feet toward the faucet/spigot end of the tub). These scenarios have the appearance of a sudden (cardiac) death. The bathtubs are often empty in these scenarios because the person wasn't taking a bath at the moment of death, but rather was in the bathroom when they died suddenly, consequently falling into the tub.

Deaths of individuals found in hot running showers also are often, but not always, sudden in nature and therefore, not typically a drowning concern. And, again, unless there is some reason to expect neurologic or pharmacologic compromise, the "scald injury"/skin sloughing that may be seen on such individuals is likely peri/postmortem (otherwise they would have gotten up!) and therefore, a consequence, not a cause of death. Skin will scald much easier (that is, at a lower temperature and for a shorter duration) after death than during life. This is because during life, circulating blood will pull heat away from the skin and delay the injury (see Chapter 12, "Thermal Injury" section).

Finally, if the death of an adult found submerged in a bathtub is ultimately deemed to be a drowning, the manner is unnatural (accident, suicide, or homicide, depending on the circumstance of the death), regardless of the proximate cause. If the explanation/proximate cause is natural disease (epilepsy, multiple sclerosis, etc.) or related to acute drug/ethanol use (unintentional), then the appropriate manner classification would be accident, this manner constituting the great majority of all adult bathtub drownings.

Note the above framing and thought processes can be applied similarly to the considerations and work up of "positional asphyxia" in an adult.[96–101] Like potential bathtub drownings, there is no place for this consideration in a neurologically intact, unintoxicated adult, without a history of a seizure or other neurologic disorder. If a person with natural comorbidities is found with their face embedded in a pillow, there shouldn't be a consideration about whether they asphyxiated when there is no reason for them to have been unable to extricate themselves from that compromised position. The conclusion should be that they died *in* that position (from natural causes), not *from* that position, for the same reasons we would draw conclusions about dying submerged *in* the bathtub water, not *from* it, under the same circumstances.

There are no pathognomonic autopsy findings in drowning deaths. The diagnosis (including cause and manner) is made based on history, circumstance/investigation, and the exclusion of other more reasonable causes. Despite the fairly common forensic practice of opening and examining the sphenoid sinuses, the finding of fluid is nonspecific and inconclusive and does not create the basis for a drowning diagnosis, per se, any more than does the finding of fluid in the lungs or stomach, areas of the body into which water can passively flow in an already dead, submerged individual.[7,20] The same can be said of the submersion artifact that is often observed on the palms and/or soles; these are not necessarily intravital changes (i.e., do not require respiration or circulation to form), but rather are evidence of prolonged submersion in both the living and the dead.

That said, the finding of ***frothy* pulmonary edema** (exuding from the mouth, within the upper and lower airways, and/or within the pulmonary parenchyma), while nonspecific in and of itself, is an important finding in that it is indicative of an intravital/antemortem respiratory effort, whatever its cause (Fig. 13.1). In other words, it indicates active movement of air through the respiratory tract, hence creating frothing/frothy fluid. And, while nonspecific, in the right setting (e.g., a submerged body), and in the absence of another more reasonable cause, this finding is quite compelling and supportive of a drowning mechanism.

Finally, an autopsy can be very important in ruling out a suspected drowning by identifying a Class I (inclusionary) finding, such as a saddle pulmonary thromboembolism, hemopericardium (with cardiac tamponade) from a ruptured myocardial infarct, or internal exsanguination from an aortic dissection—in other words, entities that are incompatible with continued life in and of themselves.

The approach to bodies in public waterways (pools/oceans/rivers/lakes)

Since these often are unwitnessed events/injuries (i.e., unknown where or under what circumstances the individual entered the water), these cases should be approached with some degree of circumspection, making no assumptions (read: think dirty), similar to our approach to persons who have sustained unwitnessed fatal head trauma, such

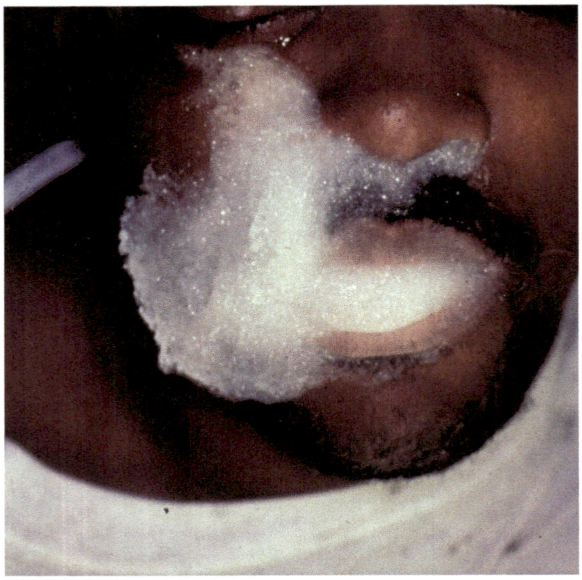

FIGURE 13.1 **Foam in a drowning victim.**

as subdural hematomas.[102] The manner is up for grabs until we know much more, not the least of which is, **Who are you?** This is the first question to ask. This is the critical first order of business and can be hard if the body is badly putrefied or has adipocere and no identification cards. The existence of tattoos in this setting can be very helpful. If the hands have become degloved from submersion, fingerprinting can also be difficult or impossible. Rarely, the bodies are skeletonized; if so, the anthropology team may need to be involved to help with the identification in estimating race, stature, age, and sex.

Until the decedent is definitively identified, the investigation is at a standstill. Once identified, however, the investigation begins in earnest. The police can visit their residence (look for suicide notes, eviction notices, medications, drugs/alcohol), ascertain the proximity of that residence to bodies of water in order to theorize where they may have entered the water, and determine whether there were any barriers to traverse. Most importantly, once identified, the police can interview next of kin, roommates, neighbors, etc., for recent activities, behaviors, or life stressors, find out when they went missing and the circumstances preceding it, whether they left their keys, phone, or wallet at home (often a compelling detail in a potential suicide if this is out of character and deviates from usual habits), inquire about a psychiatric history, if any, including history of recent or remote suicidal ideations or attempts, whether or not they had a habit of walking by the water and, if so, where, whether they had a habit of fishing somewhere nearby, whether they had an alcohol use disorder, and, importantly, whether they could swim. Once police know where they lived, they may even be able to capture them on surveillance video footage (Were they alone? Did they appear intoxicated?).

Remember, where they are *discovered* in the water may be some distance from where they *entered* the water, depending on the length of time they've been in the water and the environmental conditions (currents, storms, etc.). This part of the investigation attempts to answer the second question: **"Why did you end up in the water?"** (Intentional? Accidental? Forceful?) and **"Why didn't you get out?"** Answers to both questions are relevant to the manner of death determination: suicide, accident, homicide, or undetermined. In order to answer the latter question, one must consider both host and environmental variables. If an ocean or river, consider currents/undertows/riptides or very cold conditions (with a hypothermia risk); consider weather patterns/storms; consider swimming ability or intoxication, or natural diseases that might debilitate, conditions that may not uniformly cause death, per se (seizures, strokes, even a muscle cramp or exhaustion, etc.) or debilitating trauma, such as hitting one's head on a rock, causing unconsciousness, that consequently leads to drowning. If there is extensive blunt force trauma, consideration of a descent from height into the water might prompt a search of a nearby elevated bridge. The "why didn't you get out?" question is also important to answer when the water is shallow enough to stand up in, much like our investigations of deaths in bathtubs, situations that require further explanation. In scuba diving deaths, there are also multiple equipment variables to consider, besides the usual dysbarism (adverse effect of gas/air pressures).

Certification of waterway deaths

Cause of Death (How did being in the water kill you?) and Autopsy Findings:

Often, these deaths are drownings (alone or due to an intoxication or trauma), blunt trauma, or some combination of the two. When blunt force trauma is the cause, a descent from a considerable height into the water is often the circumstance and the injuries may have a pattern consistent with that (from a large bridge, for example). Obviously, if an execution style gunshot wound to the back of the head is identified at autopsy, the death is not due to drowning. Rather, the water serves only as the location of attempted postmortem body concealment. Asking about witnesses or video camera surveillance of drowning events, if available (in pools, a witness or video footage is more common), can be extremely helpful. If a struggle to stay afloat is witnessed, it makes the drowning outcome much more straightforward. If, on the other hand, an older man with advanced heart disease is exerting himself swimming laps at a pool and is seen to just suddenly sink under, with no sign of splashing or a struggle, then a sudden and fatal cardiac arrhythmia is more likely the mechanism of death, rather than drowning, particularly if there is no frothy pulmonary edema found at autopsy.

The following suggested investigative checklist might prove helpful in such deaths:

- **Discerning antemortem** versus **postmortem injury** (the latter often from the body making contact with rocks, sand, being struck by boat propellers, etc.) can be challenging in bodies that are putrefied or have adipocere formation, but is easier with recently submerged bodies, where a vital reaction may be clearer.
- **Open the sphenoid sinuses** (this is not conclusive, but part of a thorough drowning work up) and do a qualitative evaluation of the stomach contents.
- Always note whether there is just (liquid) water in the lungs or whether it is **frothy pulmonary edema** (internally and/or exuded in a "foam cone"). As mentioned above, while nonspecific, the latter is more compelling for respiratory distress/effort when drowning (i.e., an intravital response/marker); the former can represent passive postmortem seepage of water into the body, so is less compelling.
- **Document artifacts of submersion:** Look for and learn how to identify *postmortem marine animal feeding* and the wrinkled, pearly white appearance of the palms and soles (Fig. 13.2) indicative of *"submersion artifact."*
- **Be detailed in your description of any clothing** (size, brands, etc.) and make sure to thoroughly check the pockets. All of this may lead to assistance in both the identification process and/or in establishing context for the general investigation if, for example, a wallet or suicide note is discovered! Know also that some or all of the decedent's clothing (with/without personal property) may come off with body movement/currents/animal feeding, depending on how long the person was in the water. So, the absence of shoes, per se, doesn't necessarily mean that the individual went into the water without shoes, but it may.
- **Collect trace (biologic) evidence** depending on the circumstances/level of concern and the condition of the body.

In most places in the world, the highest prevalence for discovering bodies in public waterways is following the cold weather months, in the **early spring**. As the waters begin to warm, the bodies, if having entered the water in the colder months, begin to putrefy and bloat and become more buoyant, floating to the surface and prompting

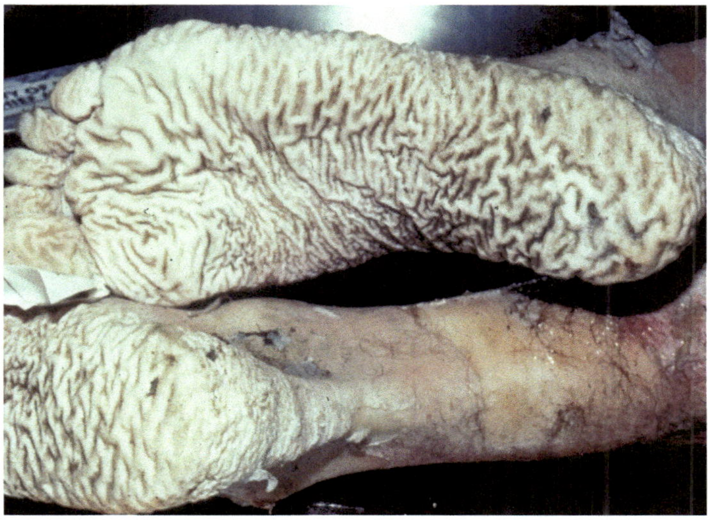

FIGURE 13.2 **Skin submersion artifact (wrinkling).** Typically observed on palms and soles.

discovery. Finally, the certification of these cases is more commonly than not "pending" on the day of autopsy (until the identification, police investigation, and toxicology have been completed).

Scuba (air embolism)

Perhaps surprising to many, the most common cause of deaths involving scuba diving is drowning. In addition to investigating the human and environmental factors in scuba deaths, one should also investigate the equipment, if possible. An "autopsy" of the tank and regulator is important to exclude a malfunction or other issue. Rusted tanks with corrosion may decrease concentrations of oxygen (iron oxide formation), and malfunctioning air compressors may allow carbon monoxide to enter the compressed air. An incorrect helium mix in saturation diving also may occur. Present day scuba monitoring systems have become quite sophisticated, and their review may be helpful in the investigation.[103,104] Obtaining an expert to consult on these deaths can be important. Recreational diving deaths are nearly always litigated.

Human factors that may be involved with scuba deaths include: inexperience (judgment, operator error), panic, risk taking, complacency, disease, and intoxication (alcohol/drugs). Other issues to consider are mechanical trauma (hit by boat), entanglement, and hazardous marine life (unusual).[104–111]

The certification of these deaths involves weighing factors that may weaken or strengthen the presumption of drowning. For example, findings of disease or injury that could have caused death and/or the failure to find evidence that the decedent was alive during some part of the submersion would weaken the presumption of drowning. **Dysbarism** refers to the adverse effects of pressure, including decompression sickness (the bends, Caisson disease).[112] Decreases in pressure are usually the problem, such as when a diver ascends (goes from high to normal pressure). Barotrauma is a type of mechanical trauma due to gas being released into the tissues. At sea level, the body pressure is 1 atm. At 33 feet below sea level, it is 2 atm, and at 66 feet, it is 3 atm. Each 10m (33 ft) of descent into the water increases the pressure on the body by 1 atm.

According to Henry's law, the solubility of a gas in a liquid is proportional to its partial pressure, and to Boyle's law, the volume of compressed gas is inversely related to pressure. For example, assume that you have 1 L of gas at 1 atm of pressure. The volume will decrease to 1/2 L at 2 atm and to 1/8 L at 8 atm. As pressure increases on the body (during descent), more air (which is 80% nitrogen) is therefore inspired through the alveoli, via the blood, to the tissues. If a diver returns to surface too quickly, as the pressure decreases, the volume of gas (nitrogen) increases (Boyle's Law), and the gas in body cavities expands and gas bubbles from the blood may cause air emboli (the "bends"). If the gas is not vented to the exterior, one develops interstitial emphysema, bullae, pneumothoraces, ear drum perforation, and air breaks into pulmonary veins (air embolism). One should always check for a pneumothorax during the autopsy of scuba deaths. Advanced imaging studies also are useful.[113]

Acute decompression sickness may include acute obstruction of small vessels by bubbles (**air embolism**). If it involves the joints, it is known as *the bends*. When it involves the lungs, it is called *the chokes*. Chronic decompression sickness (**Caisson disease**) may lead to ischemic necrosis, aseptic bone necrosis, and spinal cord dorsal and lateral column demyelination. Caissons are watertight enclosures inside of which people can do work underwater (e.g., building the support structures of the Brooklyn Bridge). Pulmonary barotrauma may occur in 2–3 m (~10 feet) of water and may occur during scuba training during a rapid ascent in which expiration was not forceful enough to compensate for the expanding gas.

Saturation diving involves depths greater than 50m and requires a special air mixture, replacing nitrogen with helium, which is less toxic (less narcotism). This is meant to prevent nitrogen narcosis in which the increased pressure of nitrogen forces it into tissue and neurons, resulting in confusion, the so-called "raptures of the depths."

Near-drownings

Occasionally, a person will be found "nearly" drowned in a body of water or bathtub, but rescued by emergency services. Their circulation may be maintained or restored, but they will have suffered anoxic–ischemic encephalopathy to the extent that they are moribund. As such, they may die from complications, particularly respiratory, delayed from the drowning event, after a complicated and prolonged hospital course. These types of deaths should be reported to the medical examiner and are worked up no differently (including a police investigation and a scene visit, if the death is not too remote from the event) than those in which the pronouncement was made at the scene. The immediate cause on the death certificate in such cases might be something like, "Complications following near-drowning due to (proximate cause)" or, if an infant or child, with no stated proximate cause, but rather an explanation of the circumstances in the "how injury occurred" section.

Electricity

The most important issue with electricity is to know that *with an electrocution death, there can be few, if any, external or internal findings.*[114–127] An investigation of the circumstances and a high level of suspicion is key. The importance of recognizing electrocutions lies in the fact that a failure to do so may result in another death. Similar to other special forensic circumstances outlined elsewhere in this book, the identification of these deaths may save lives. There are approximately 1000 fatal electrocutions per year in the United States.

Direct current (DC) is one-directional flow of an electric charge (e.g., battery, lightning). The electrical current flows in a constant direction, distinguishing it from alternating current (AC). AC is an electric current which periodically reverses direction and changes its magnitude continuously with time (typical residential electricity).

Electrical burns

The severity of electrical burns depends upon the voltage, the amount of current, the area of contact, and the duration of contact.[116,125,126,128–130] Not all electrocutions will produce visible burns (Fig. 13.3). Electrocutions are classified as low- and high-voltage events. **High-voltage** *electrocutions involve greater than 1000 volts and virtually all will have visible burns.* For high-voltage electrocutions, direct contact is not needed, as there may be arcing. **Arcing** occurs when an electric current flows through the air from one conductive point to another.[131–133] The heat and energy emitted by the arcing current can cause injury or death. Approximately 5000 V are needed for electricity to arc one - centimeter and 100,000 V are needed for it to travel 35 cm. **Low-voltage** *electrocutions typically involve 110 to 120 V (typical household voltage in the United States) and may or may not be associated with visible cutaneous burns.* It is estimated that 1/3 to 1/2 of low-voltage electrocutions have no associated burns, but direct contact is needed for an electrocution to occur when low voltage. Because of the frequent absence of a burn, the investigator must have a high index of suspicion for an electrocution in these instances.

Electrocutions

Three factors are needed for an electrocution to occur: a *charged source*, a current *pathway* through the body, and a *ground*. For example, birds may be able to perch on high-voltage electric lines without getting electrocuted. The key reason is that there is no ground pathway. A *ground* is defined as a reference point in an electrical circuit from which voltages are measured, a common return path for electric current, or a direct physical connection to the earth. If the bird touches an electrified wire and also touches a grounded object, a path is established that allows the electricity to travel through the bird to the place with no (or lower) voltage (the ground).

The science of electricity

Amperage (I) is the most important factor in an electrocution. Ohm's law states that the current through a conductor between two points is directly proportional to the voltage across the two points ($I=V/R$). Since voltage (V) is usually constant, the variable, resistance (R), will determine how much amperage enters a specific body. Resistance of human skin ranges from 100 (thin, moist skin) to 100,000 Ω (dry, calloused skin). One milliamp (0.001 amp) will give a slight tingle to the skin, while 1–8 milliamps will cause a disconcerting shock. A painful shock occurs

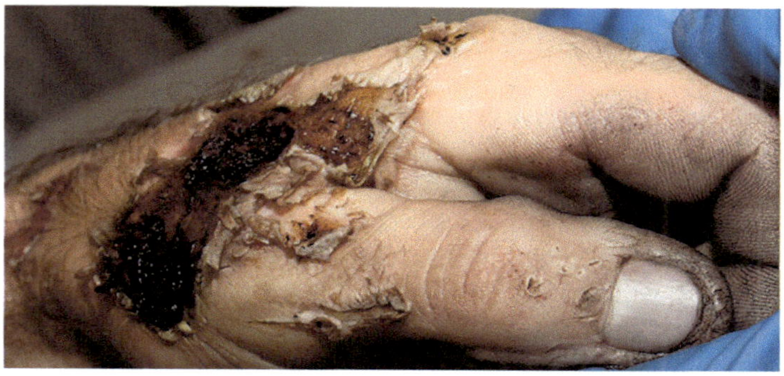

FIGURE 13.3 **Electrical burn of the hand.**

with 10–20 milliamps, and 15–20 milliamps will cause extreme pain and severe muscular contractions. These contractions may result in the inability of the person to release their grip from the electrical device. *VF is seen with 70 to 100 milliamps.* Despite common dogma, typical DC current does not cause asystole.[134] *Asystole and pulseless electrical activity (PEA) are generally not considered inducible except with high electrical stimulation* (e.g., lightning).[134] Studies that have delivered DC current directly to the hearts of patients found that all developed VF, but no patient developed asystole.[135,136] Lightning is a common high-voltage DC pulse that can cause temporary asystole and respiratory arrest (see below). Thus, DC at high-voltage currents can produce asystole, but such shocks also can produce VF.[130,134,136]

Current will follow the shortest pathway with the least resistance from the point of contact to the ground. It will find low-resistance routes in the body as well, typically along vessels and nerves. Blood is an excellent conductor of electricity in the body. This will dictate what internal structures (e.g., heart, brain) the current will pass through. This is important when considering how the electricity causes the death mechanistically.

How does electricity kill?

There are three mechanisms by which electricity causes death. The first is **cardiac**, due to a stoppage of the heart mediated by the electrical interference of the cardiac conduction system. The second is by current passing through the **central nervous system (brain),** causing an electrical disruption of neuronal activity; this will lead to secondary cardiorespiratory effects. The third mechanism is **asphyxial**, due to paralysis of the respiratory muscles.

The speed of incapacitation (unconsciousness) varies among these three mechanisms. For the cardiac mechanism, there may be several seconds of consciousness before the collapse. A person may remain conscious for several seconds after cardiac cessation due to the time it takes for the brainstem to be affected by the drop in blood pressure. This delay is important to remember when investigating suspected electrocutions. *Because the death is not instantaneous, the person may not collapse immediately, next to the electrical source.*

Typical electrocution scenario and investigation

A typical electrocution scenario involves a person sustaining the shock and yelling an expletive. The person then may drop the electrical device or walk away from the electrical source, and several seconds later, collapse. A witness may describe hearing the yell and seeing the person walking toward them and then suddenly collapse. This may appear as a cardiac death if the forensic investigator does not inquire further about what the person was doing shortly before the collapse. The importance of identifying these deaths is for proper certification, particularly if in the workplace, and most importantly, to prevent further deaths from a potentially faulty device or source. For example, if the cause of the electrocution was a faulty electrical device that was not properly grounded, it needs to be identified to prevent another person from meeting the same fate. This is one reason that deaths that may be work-related are usually reportable to the medical examiner/coroner (ME/C). Certain occupations place workers at an increased risk for electrocutions. These include electrical power installers and repairers, earth drillers, electricians, construction workers, ironworkers, roofers, painters, and arborists. The Occupational Safety and Health Administration (OSHA) may be involved in certain work-related deaths, so it is best to have a low threshold for reporting such a work-related event to this agency.

Consultation with an electrician and an "autopsy" of the suspected electrical device may be an important part of the investigation.[124,126,137] An electrician can safely detect voltage differential and current leaks. A frayed wire of an electrical device may result in "electrification" of the device, resulting in a short circuit when used. A short circuit (a "short") is an electrical circuit that allows a current to travel along an unintended path, with no or very low electrical impedance. This results in an excessive current flowing through the circuit. In this instance, the person holding the electrified device becomes the unintended path.

What further complicates these investigations is the frequent absence of external injury. Electrical burns are a function of heat and duration. If contact is quickly broken, one may see no burns, even though a fatal current may have passed through the body. Electrical burns may appear as small, indurated, circumscribed, gray-black injures or areas of skin slippage. The burn depends upon the surface area contacted, so there may not be a burn in an electrocution that occurs in a bathtub with an electrical object that falls into the water. One also may see visible burns that did not cause death. A well-documented example involves a child who bites an electrical cord, causing a local transfer of energy that produces local burns, but no systemic effect.

Examination of the electrical burn usually does not allow one to distinguish the "entrance" from the "exit" by morphology alone. One may see metal fragments at an entrance or exit. Microscopically, one may see intact epidermis with basil palisading and streaming of nuclei. There may also be separation of the epidermis and dermis. The borders are usually distinct, and the tissue damage may be highlighted with alcian blue staining.

One is not able to distinguish burns caused by DC from those of AC. **AC** ranges from 39 to 150 cycles per second and *has the greatest lethality*. In the United States, typical house wiring is 120 V with 60 Hz per second AC. City high-voltage lines are approximately 8000 V and transcontinental high-tension wire lines can carry 100,000 V. The **third rail** of the subway track carries **660 V** and it is a DC, not AC current. The latter produces an unmistakable and patterned electrical injury characterized by a broad (roughly 5–7 inches), band-like, usually depressed, charred burn injury. Finally, electricity does not build up like an intoxicant and so there is no cumulative effect. It either causes cardiac, neurologic, or respiratory compromise or it does not.

Lightning

Lightning causes fewer than 100 deaths per year in the United States.[138] A 1998 study in the Morbidity and Mortality World Report reviewed 1318 deaths from 1980 to 1995, with an average rate of deaths at 82 per year. They reported that approximately *30% of people struck by lightning will die, with 63% of the deaths occurring within one hour.* Nearly 75% struck became disabled in some capacity. Common long-term effects include ocular complications, such as corneal ulcers and cataracts. Florida, Texas, New Mexico, and Arizona were the most common states for victims of lightning strikes and the empire state building in New York City is struck up to 500 times per year. There are over a 100,000 thunderstorms per year in the United States and cloud to ground strikes occur over 30 million times per year.[138] **Keraunopathology** is the study of lightning-induced injury.[139–149] A report of 45 fatal lightning strikes described 41 with cutaneous injuries, 1/3 with ferning, and over 80% with tympanic membrane injury.[150]

Ferning, also known as feathering or Lichtenberg figures, *is pathognomonic for a lightning strike* (Fig. 13.4).[145,151–154] Unfortunately, they are not present in all strikes. Importantly, these are not burns, as they do *fade over time in survivors*, suggesting an erythematous reaction that follows a blood vessel pattern. Despite the impressively high voltage of a lightning bolt (millions of volts), the short duration may leave little burn injury. Clothing is often torn and may even catch on fire. Metal objects (e.g., hair pins) may become magnetized. Hair may be singed and there may focal thermal injuries on the skin in areas in contact with metal objects (such as coins in a pocket).

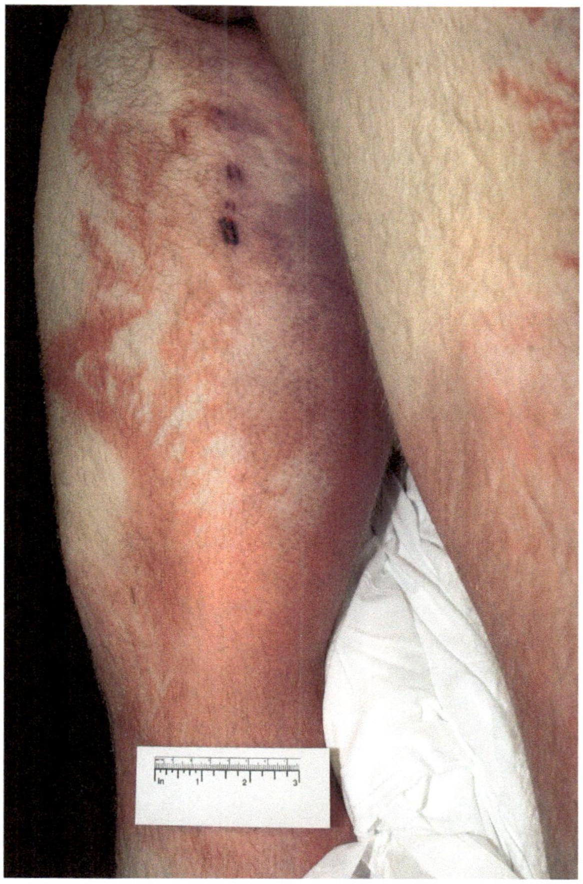

FIGURE 13.4 **Ferning in a lightning strike.**

Apparent sudden deaths outside, during a thunderstorm should raise the suspicion of a lightning related death. In addition to ferning, examination of shoes (for an exit site) and clothing may be revealing (sometimes, the shoe may be blown off the foot).[139–145,151–156]

Lightning is a high-voltage DC pulse that can cause temporary asystole and respiratory arrest. Cardiac activity and spontaneous circulation, however, may return. In some instances, respiratory arrest persists and hypoxia results in asystole or PEA.[157] Prolonged respiratory arrest has been demonstrated in a rodent model, and initial asystole followed by perfusing arrhythmias that end with permanent asystole have been demonstrated in a sheep model.[134] Both asystole and VF have been reported in lightning strikes of humans, with a progression of transient asystole, back to VF, which may deteriorate into asystole over time.[134]

Electromechanical control devices

Electromechanical control devices (EMDs) are battery-powered devices used to temporarily incapacitate a person by a short duration, high voltage, and low amperage electric current that follows low-resistance routes within the body.[158] They are not uncommonly *used during law enforcement apprehension.*[159,160] A study of EMDs by the National Institute of Justice found no conclusive medical evidence of a high risk of serious injury or death from the direct or indirect cardiovascular or metabolic effects of short-term EMD use in healthy, normal, nonstressed, nonintoxicated persons. They concluded that *EMDs do not cause or contribute to death in the large majority of those instances.*[161] Witness statements, videos (police body cameras), and EMD download data may help clarify their role in an in-custody death with regard to the timing of the death in relation to the restraint or discharge of the EMD.[162]

Stun guns are electrical devices meant to incapacitate humans and prod animals (coaxers).[158,163–169] The handheld, battery operated (9 V) stun gun discharges a series of short duration high peak pulses of 10–15 per second that can cause pain and contraction of the muscles. Typical stun guns generate up to 50,000 V, but it is not a continuous discharge and therefore is less likely to cause death. The impulse does not travel between the electrodes, but instead spreads along areas of low resistance. Pathologically, one may see **1/8 to 1/4-inch erythematous paired punctures** and microscopic findings of electrical injury.

Some devices (e.g., **TASERs**) propel darts connected to the electrical device by wires. They resemble a flashlight and shoot two barbs up to 18 feet which can go through two inches of clothing. Electrical flow can occur between the two barbs when the device is discharged, resulting in pain and potential incapacitation.[159–161,170–198]

The spread of the probes on the body is key because the extent of separation will affect the extent of captured nerves and muscle. *At autopsy, one should identify the location of the probes and the distance between them.* Microscopic examination of the skin sites may demonstrate electrical injury, thus, confirming the application of electricity. Most devices, however, record whether there was conduction. It is exceedingly rare for such an electrical device to cause death through the mechanism of electrocution.[161] This is because the thickness of the skin and muscle, along with the probe body locations, are unlikely to be able to affect the cardiac rhythm. Studies, however, have demonstrated that placement of the probes very close to the heart can affect the cardiac rhythm. Studies in swine, in which probes were placed subcutaneously, very close to the heart, demonstrated capture.[170,175,199] Therefore, *it is recommended that these devices be fired at the back of a person, which increases the distance of the probes to the heart.*

If such a device causes death by electrocution, it would be from a cardiac arrhythmia. *Noting the interval between the application of the electrical discharge and unresponsiveness is key.* If the person loses consciousness within seconds of the discharge and the probes are in the vicinity of the heart, a cardiac mechanism is likely. If the person loses consciousness several minutes after the last discharge, however, one can effectively exclude an electrocution, irrespective of probe location. *Because of the subsequent incapacitation, injuries may be incurred from a consequent fall.* If the person is incapacitated at a height (e.g., on the ledge of a building), the subsequent fall may be fatal. Typical law enforcement rules for the use of these devices address the risks of subsequent falls. Some of these devices can be used criminally to torture a person. If this painful, stressful application results in death, in combination with underlying cardiovascular disease (e.g., homicide by heart attack), some of these deaths may be certified as homicides (death at the hand of another), regardless of the legality of their use.

Environmental exposure: hyperthermia

Epidemiologically, there are typically more natural deaths (excess mortality) reported during the hottest months of the year, especially during heat waves and especially for the homeless and persons without air conditioning.[200–207] But not all of those excess deaths are due to hyperthermia. Excessive heat is physiologically stressful, particularly for elderly persons and/or those with comorbid conditions. For example, it may result in dehydration in people on diuretics, trigger asthma/chronic obstructive pulmonary disease exacerbations, tachycardia, etc. These deaths, however, are not classified as accidental "hyperthermia" (an environmental injury), per se, unless evidence supports that the decedent actually achieved an excessively **(typically ≥105°F)** hyperthermic state prior to death, something we typically determine through an *intravital* body temperature measurement.[201,203,205,207–217]

If an apartment is 100°F and the body temperature of the decedent (an ill, elderly person) equilibrates with that environmental temperature after death, there is no way of knowing whether the person died *from* the heat or just *in* the heat. One cannot know what the decedent's temperature was at the moment of death (i.e., the intravital temperature), so the cause of death would not be certified as "hyperthermia." Instead, based on the hot apartment, with a likely equilibrated body temperature, it would be best to err on the conservative side and certify the death as due to natural causes. If, however, an excessively high core temperature is documented intravitally, even if during a brief return of circulation, or if the body temperature exceeds that of the ambient temperature at the scene of death, then, there is a more compelling argument to be made for a hyperthermic death, even in an individual with other comorbid explanations for death. In the end, the determination must be evidence-based and forensically sound, not speculative, as it may have public health ramifications.

Hyperthermia can result from a variety of diseases, intoxications, and other scenarios.[211,212,214,215,218–233] *When it does kill, it is important to understand that hyperthermia can do so alone or as part of stimulant drug intoxications, in certain states of delirium with hyperactivity, or as part of neuroleptic malignant syndrome.* When associated with central nervous system pathology (a so-called "central fever"), the body temperature may be substantially elevated, but these fatalities may or may not be directly related to the hyperthermic state. Rather, the underlying central nervous system (CNS) dysfunction may play a greater role when there is a fatal outcome. Regardless, these deaths are most typically "pending toxicology" or investigation on the day of autopsy. If ultimately deemed a hyperthermia-related death (constituting an environmental injury), those that are inadvertent are certified as accidents. Deaths at the hand of another, even if not intentional (e.g., an infant who dies of hyperthermia from being left in a car on a hot summer day by the parents who went into a restaurant for an hour and a half) may be certified as homicides in many jurisdictions.[231,233]

Environmental exposure: hypothermia

Deaths due to hypothermia are defined as those primarily (or contributorily) caused by a severely reduced core body temperature **(typically <95°F)**.[212,234–271] But, analogous to bathtub drownings, hypothermic deaths demand an explanation, physical or circumstantial, for the fatality to make sense. One should seek an underlying risk factor or factors that explain why the individual did not or could not seek or find a warmer environment. That "risk factor" is also likely to be the proximate cause of death, unless the circumstances alone explain why fatal hypothermia would have reasonably occurred (e.g., a homeless individual unable to locate shelter or an elderly, poorly mobile person living alone in an unheated home). Since not having a home is not a codable disease or injury, it cannot be used as a proximate cause, per se, but can be stated in the "how injury occurred" field or in a descriptive cause of death statement (see below). From a public health perspective, it is important to know how many "undomiciled" people die and from what. Hypothermia deaths often occur in individuals found outside, but can also occur indoors, depending on the debility or dependence of the decedent and/or the availability of heat (which the police and/or investigator should document).

But, the cart must not precede the horse. The primary question that should be asked *before* even that of proximate or contributory causes or circumstantial risk is whether the reported reduced body temperature constitutes a "real" hypothermia death (i.e., Is the body temperature the cause or a consequence of death?). If it is the latter, it is simply a reflection of algor mortis, not antemortem hypothermia. Is the individual dying *from* the cold or dying of something else entirely, *in* the cold? Sometimes, it may be difficult to distinguish the two. Therefore, without compelling evidence of a hypothermic death (e.g., *discrete* **gastric hemorrhages/Wischnewski ulcers**, *paradoxical undressing, documented low core body temperature while alive, no better explanation for death,* etc.), one usually defaults to a nonhypothermic explanation for the death.[216,237,238,240,241,243,245–249,257,259–263,265,266,270–277]

In some instances, a diagnosis of hypothermia, per se, may not affect the manner or the proximate cause of death. For example, consider a person who dies from a fentanyl or heroin intoxication in the basement of an abandoned (unheated) building in the winter. The toxicology and autopsy all support a drug intoxication death. Is this a straight-forward drug overdose or did the person die from hypothermia while they were in an intoxicated state? Absent other findings (e.g., Wischnewski spots), the default would be the intoxication. If on the other hand, there was intravital or autopsy evidence of hypothermia, it would be hypothermia due to environmental exposure complicating acute fentanyl intoxication. The intoxication would still be the underlying cause in either case, but, in the latter instance, the hypothermia would be a sequela that likely played at least some role in the mechanism of death, in combination with the respiratory and CNS-depressant effects of a potent narcotic.

*Hypothermia is best viewed as a **mechanism** of death.* If there is no identifiable proximate cause or causes, natural or unnatural, "hypothermia due to environmental cold exposure" may stand alone on the death certificate, with the explanation for this fatal outcome left to the "how injury occurred" section. It is usually due to environmental cold exposure, even if indoors. **Hypothermia is an injury** and therefore, even when only contributory to death, precludes a natural manner of death, regardless of the underlying reason for the environmental exposure. Hypothermia is one of the deaths due to injury (like an adult drowning in a bathtub, choking on food, or asphyxiating due to a compromised position) that requires further explanation. It does not make sense for an adult to die of hypothermia without further qualification somewhere on the death certificate (part 1, part 2, or in the "how injury occurred" section).

Common explanations (proximate causes) for hypothermia include:

- Intoxication, especially with alcohol misuse, but also with other depressant drug(s)
- Debilitating injury (e.g., subdural hemorrhages, hip fractures, etc.) in individuals found outdoors or indoors
- Natural disease that may debilitate, but not immediately kill (cerebrovascular events, diabetic ketoacidosis, uremia, etc.), inside or outside, depending on the temperature in the home
- Dementia (e.g., with disorientation that leads to wandering outdoors)[250]
- Severe psychiatric debilitation

The most *common circumstantial explanation* involves a homeless individual (with/without superimposed intoxication or advanced natural disease or debility). Because hypothermia is an environmental injury (like hyperthermia), these deaths are typically certified as accidents (unless suicidal, or something criminal/involving neglect). In the foregoing example, the "how injury occurred" field could state "homeless, living outdoors" or simply, "homeless," without further qualification. The same statements could be included parenthetically and descriptively in the cause of death statement, after "Hypothermia due to Environmental Cold Exposure." It is best to avoid writing "*due to* homelessness" in the cause of death statement, however, since homelessness is not a codable disease or injury, so doesn't meet the definition of a proximate cause, per se.

While there are no pathognomonic findings at autopsy, "textbook" signs may include the leopard skin-like pattern of gastric ulcers (Wischnewski ulcers, Fig. 13.5), acute pancreatitis, and paradoxical undressing (rare). The first two, while not a requirement to make the diagnosis, are very helpful because they support an intravital reaction. Another helpful, but nonspecific finding is hyperglycemia (a stress response).[278] Since most decedents have a vitreous glucose of less than 20, postmortem hyperglycemia is compelling in a potential hypothermic setting, particularly if the decedent is not a diabetic.

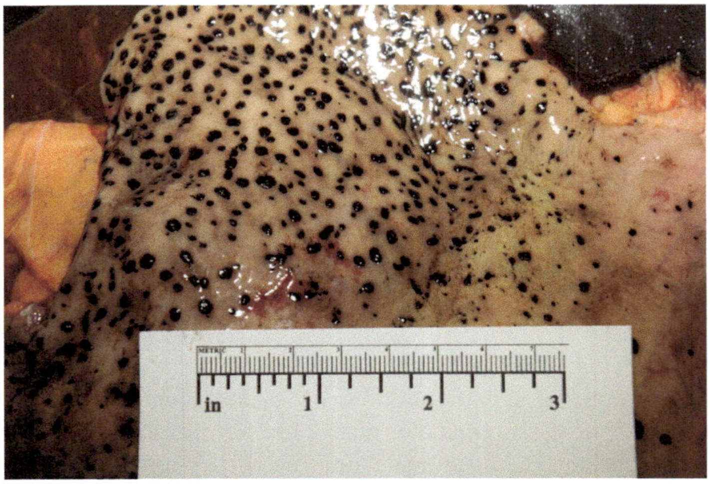

FIGURE 13.5 **Wischnewski ulcers.** Leopard skin-like pattern of gastric ulcers seen in some hypothermia deaths.

Far and away, the most important evidence supporting hypothermia is documentation of a markedly low core body temperature with subsequent return of circulation (no matter how transient or weak), movement, seizure activity, a weak pulse, a dysrhythmia or any cardiac electrical activity (even PEA), any sign of life or documented survival interval whatsoever. Another good clue would be the *absence of* rigor mortis in someone with a very low body temperature. If the temperature was only a function of algor mortis, it would take hours to achieve and the person should be in full rigor. If they aren't, the concern for hypothermia as a mechanism should heighten. If someone is found dead in their apartment with a low temperature (i.e., no medical intervention), it can be a hard call, but not impossible if all physical and investigative findings line up (e.g., a combination of gastric ulcers, a cerebrovascular event diagnosed at autopsy, and an unheated apartment). Hypothermia in a person with dementia who wanders off from an assisted care facility and gets lost in the snow-covered woods would also likely be classified as an accident. That said, many patients dying from hypothermia have low temperatures that are documented by medical professionals as being intravital, as defined by the criteria outlined above, making the diagnosis much easier and leaving the forensic pathologist only with the task of identifying the risk factor(s)/proximate cause and documenting confirmatory physical findings.

Finally, it is important to remember that gastric ulcers that are similar, if not identical, in appearance to those described above can be seen in other situations that present various forms of physiologic stress, separate and apart from hypothermia. In and of themselves, "stress ulcers" are nonspecific. As such, case contextualization leads to a clearer explanation for their development. When found in the setting of a burn victim or other pathologic processes producing marked physiologic stress, including shock, they would be classified as **Curling's ulcers.** When associated with increased intracranial pressure, they might be referred to as **Cushing ulcers**, and with hypothermia, **Wischnewski.**[238,240,273,279–281] Aside from the physiologic stress that all of these have in common, they also all indicate a survival interval during a physiologically stressful period of time, thus representing an intravital process. On occasion, they are identified in an elderly person having died on the floor in their home. If, in such a situation, the autopsy reveals a hip fracture, decubitus ulcers, dehydration, and gastric ulcers, a reasonable conclusion would be that the hip fracture (likely, from a fall) left them on the floor and unable to get up for a period of time, during which time they suffered the physiologic stresses of dehydration, with or without renal insufficiency, pain, and some degree of bleeding. The gastric ulcers in this setting (assuming there was no evidence of concomitant cold exposure) would best be explained by the complex physiologic stress resulting from a survival interval following an injury. A similar outcome might occur from a survival interval following a bleeding scalp injury in an intoxicated person or in one having suffered a debilitating, if not immediately fatal cerebrovascular event.

References

1. Modell JR, Bellefleur M, Davis JH. Drowning without aspiration: is this an appropriate diagnosis? *J Forensic Sci*. 1999;44(6):1119–1123.
2. Wiggins CE, Luke JL. The pathology, diagnosis and medical-legal aspects of death by drowning. *J Oklahoma State Med Assoc*. 1970;63(1):3–7.
3. Szpilman D, Bierens JJ, Handley AJ, Orlowski JP. Drowning. *N Engl J Med*. 2012;366(22):2102–2110.
4. Byard RW. Immersion deaths and drowning: issues arising in the investigation of bodies recovered from water. *Forensic Sci Med Pathol*. 2015; 11(3):323–325.
5. Copeland A. An assessment of lung weights in drowning cases. *Am J For Med Pathol*. 1985;6:301–304.
6. Orlowski JP. Drowning, near-drowning, and ice-water drowning. *J Am Med Assoc*. 1988;260(3):390.
7. Armstrong EJ, Erskine KL. Investigation of drowning deaths: a practical review. *Acad Forensic Pathol*. 2018;8(1):8–43.
8. Bierens JJ, Lunetta P, Tipton M, Warner DS. Physiology of drowning: a review. *Physiology*. 2016;31(2):147–166.
9. Girela-López E, Beltran-Aroca CM, Dye A, Gill JR. Epidemiology and autopsy findings of 500 drowning deaths. *Forensic Sci Int*. 2022;330: 111137.
10. Carter N, Ali F, Green MA. Problems in the interpretation of hemorrhage into neck musculature in cases of drowning. *Am J Forensic Med Pathol*. 1998;19(3):223–225.
11. Fornes P, Pepin G, Heudes D, Lecomte D. Diagnosis of drowning by combined computer-assisted histomorphometry of lungs with blood strontium determination. *J Forensic Sci*. 1998;43(4):772–776.
12. Niles NR. Hemorrhage in the middle-ear and mastoid in drowning. *Am J Clin Pathol*. 1963;40:281–283.
13. Pollanen MS, Cheung C, Chiasson DA. The diagnostic value of the diatom test for drowning, I. Utility: a retrospective analysis of 771 cases of drowning in Ontario, Canada. *J Forensic Sci*. 1997;42(2):281–285.
14. Pollanen MS. The diagnostic value of the diatom test for drowning, II. Validity: analysis of diatoms in bone marrow and drowning medium. *J Forensic Sci*. 1997;42(2):286–290.
15. Alexander RT, Jentzen JM. Neck and scleral hemorrhage in drowning. *J Forensic Sci*. 2011;56(2):522–525.
16. Byard RW, Cains GE, Gilbert JD. Is haemolytic staining of the aortic root a sign of fresh water drowning? *Pathology*. 2005;37(6):551–552.
17. Fucci N, Pascali VL, Puccinelli C, Marcheggiani S, Mancini L, Marchetti D. Evaluation of two methods for the use of diatoms in drowning cases. *Forensic Sci Med Pathol*. 2015;11(4):601–605.
18. Zatopkova L, Hejna P, Janik M. Hemolytic staining of the endocardium of the left heart chambers: a new sign for autopsy diagnosis of freshwater drowning. *Forensic Sci Med Pathol*. 2015;11(1):65–68.

19. Byard RW. Aortic intimal staining in drowning. *Forensic Sci Med Pathol*. 2015;11(3):442–444.
20. Zivkovic V, Babic D, Nikolic S. Svechnikov's sign as an indicator of drowning in immersed bodies changed by decomposition: an autopsy study. *Forensic Sci Med Pathol*. 2013;9(2):177–183.
21. Girela-Lopez E, Ruz-Caracuel I, Beltran C, et al. Histological changes in skeletal muscle during death by drowning: an experimental study. *Am J Forensic Med Pathol*. 2016;37(2):118–126.
22. Haffner HT, Graw M, Erdelkamp J. Spleen findings in drowning. *Forensic Sci Int*. 1994;66(2):95–104.
23. Girela-Lopez E, Beltran-Aroca CM, Jimena I, et al. Structural abnormalities in the human diaphragm in drowning and hanging deaths: preliminary results. *Forensic Sci Med Pathol*. 2020;16(2):265–271.
24. Garland J, Philcox W, Kesha K, et al. Elevated cerebrospinal fluid sodium and chloride levels in a saltwater drowning death. *Am J Forensic Med Pathol*. 2019;40(3):258–261.
25. Paulis MG, Hasan EI. Electrolytes and biochemical changes in cerebrospinal fluid in drowning: experimental rabbit model. *Am J Forensic Med Pathol*. 2018;39(3):236–241.
26. Tse R, Garland J, Kesha K, et al. Elevated postmortem vitreous sodium and chloride level in a salt water drowning death during self-contained underwater breathing Apparatus diving with diving mask in place: case report. *Am J Forensic Med Pathol*. 2018;39(3):247–249.
27. Tse R, Garland J, Kesha K, et al. Combining postmortem vitreous sodium and chloride and lung-body ratio in aiding the diagnosing saltwater drowning. *Am J Forensic Med Pathol*. 2018;39(3):229–235.
28. Tse R, Garland J, Kesha K, et al. The potential diagnostic accuracy of autopsy lung weights, lung-heart ratio, and lung-body ratio in drowning deaths. *Am J Forensic Med Pathol*. 2018;39(3):223–228.
29. Cala AD, Vilain R, Tse R. Elevated postmortem vitreous sodium and chloride levels distinguish saltwater drowning (SWD) deaths from immersion deaths not related to drowning but recovered from saltwater (DNRD). *Am J Forensic Med Pathol*. 2013;34(2):133–138.
30. Zhao J, Liu C, Bardeesi ASA, et al. The diagnostic value of quantitative assessment of diatom test for drowning: an analysis of 128 water-related death cases using microwave digestion-vacuum filtration-automated scanning electron microscopy. *J Forensic Sci*. 2017;62(6):1638–1642.
31. Kakizaki E, Yukawa N. Simple protocol for extracting diatoms from lung tissues of suspected drowning cases within 3h: first practical application. *Forensic Sci Int*. 2015;251:179–185.
32. Barranco R, Castiglioni C, Ventura F, Fracasso T. A comparative digital morphometric study of lung tissue in saltwater and freshwater drowning. *Forensic Sci Int*. 2019;298:157–160.
33. Kakizaki E, Sonoda A, Shinkawa N, Yukawa N. A new enzymatic method for extracting diatoms from organs of suspected drowning cases using papain: optimal digestion and first practical application. *Forensic Sci Int*. 2019;297:204–216.
34. Diaz-Palma PA, Alucema A, Hayashida G, Maidana NI. Development and standardization of a microalgae test for determining deaths by drowning. *Forensic Sci Int*. 2009;184(1–3):37–41.
35. Modell JH, Davis JH, Giammona ST, Moya F, Mann JB. Blood gas and electrolyte changes in human near-drowning victims. *JAMA*. 1968;203(5):337–343.
36. García-Mozo H, Beltran-Aroca CM, Badu IK, Jimena I, Girela-López E. Airborne pollen and spores' deposition in alveolar tissues as a tool in drowning forensic diagnosis. *Aerobiologia*. 2021;37(2):281–291.
37. Davis JH. Bodies found in the water. An investigative approach. *Am J Forensic Med Pathol*. 1986;7(4):291–297.
38. Geertinger P, Voigt J. Death in the bath. A survey of bathtub deaths in Copenhagen, Denmark, and Gothenburg, Sweden, from 1961 to 1969. *J Forensic Med*. 1970;17(4):136–147.
39. Saxena A, Ang LC. Epilepsy and bathtub drowning. Important neuropathological observations. *Am J Forensic Med Pathol*. 1993;14(2):125–129.
40. Budnick LD, Ross DA. Bathtub-related drownings in the United States, 1979-81. *Am J Public Health*. 1985;75(6):630–633.
41. Pearn JH, Brown 3rd J, Wong R, Bart R. Bathtub drownings: report of seven cases. *Pediatrics*. 1979;64(1):68–70.
42. Satoh F, Osawa M, Hasegawa I, Seto Y, Tsuboi A. "Dead in hot bathtub" phenomenon: accidental drowning or natural disease? *Am J Forensic Med Pathol*. 2013;34(2):164–168.
43. Okuda T, Wang Z, Lapan S, Fowler DR. Bathtub drowning: an 11-year retrospective study in the state of Maryland. *Forensic Sci Int*. 2015;253:64–70.
44. Schmidt P, Madea B. Homicide in the bathtub. *Forensic Sci Int*. 1995;72(2):135–146.
45. Mutlu GM, Leikin JB, Oh K, Factor P. An unresponsive biochemistry professor in the bathtub. *Chest*. 2002;122(3):1073–1076.
46. Somers GR, Chiasson DA, Smith CR. Pediatric drowning: a 20-year review of autopsied cases: III. Bathtub drownings. *Am J Forensic Med Pathol*. 2006;27(2):113–116.
47. Yoshioka N, Chiba T, Yamauchi M, Monma T, Yoshizaki K. Forensic consideration of death in the bathtub. *Leg Med*. 2003;5(Suppl 1):S375–S381.
48. Simon HK, Tamura T, Colton K. Reported level of supervision of young children while in the bathtub. *Ambul Pediatr*. 2003;3(2):106–108.
49. Adelson L. *Pathology of Homicide*. 1st ed. Springfield: Charles C Thomas; 1974.
50. Cummings P, Quan L. Trends in unintentional drowning: the role of alcohol and medical care. *JAMA*. 1999;281(23):2198–2202.
51. Davis LG. Suicidal drowning in south Florida. *J Forensic Sci*. 1999;44(5):902–905.
52. Byard RW, Lipsett J. Drowning deaths in toddlers and preambulatory children in South Australia. *Am J Forensic Med Pathol*. 1999;20(4):328–332.
53. Avis SP. Suicidal drowning. *J Forensic Sci*. 1993;38(6):1422–1426.
54. Plueckhahn VD. Alcohol consumption and death by drowning in adults; a 24-year epidemiological analysis. *J Stud Alcohol*. 1982;43(5):445–452.
55. Auer A. Suicide by drowning in Uusimaa province in southern Finland. *Med Sci Law*. 1990;30(2):175–179.
56. Copeland AR. Suicide by drowning. *Am J Forensic Med Pathol*. 1987;8(1):18–22.
57. Byard RW, Houldsworth G, James RA, Gilbert JD. Characteristic features of suicidal drownings: a 20-year study. *Am J Forensic Med Pathol*. 2001;22(2):134–138.
58. Wirthwein DP, Barnard JJ, Prahlow JA. Suicide by drowning: a 20-year review. *J Forensic Sci*. 2002;47(1):131–136.
59. Jumbelic MI, Chambliss M. Accidental toddler drowning in 5-gallon buckets. *JAMA*. 1990;263(14):1952–1953.

60. Accidental drowning of toddlers in buckets. *JAMA*. 1990;264(11):1407.
61. Mann NC, Weller SC, Rauchschwalbe R. Bucket-related drownings in the United States, 1984 through 1990. *Pediatrics*. 1992;89(6 Pt 1): 1068–1071.
62. Somers GR, Smith CR, Wilson GJ, Zielenska M, Tellier R, Taylor GP. Association of drowning and myocarditis in a pediatric population: an autopsy-based study. *Arch Pathol Lab Med*. 2005;129(2):205–209.
63. Somers GR, Chiasson DA, Smith CR. Pediatric drowning: a 20-year review of autopsied cases: II. Pathologic features. *Am J Forensic Med Pathol*. 2006;27(1):20–24.
64. Gorniak JM, Jenkins AJ, Felo JA, Balraj E. Drug prevalence in drowning deaths in Cuyahoga County, Ohio: a ten-year retrospective study. *Am J Forensic Med Pathol*. 2005;26(3):240–243.
65. Somers GR, Chiasson DA, Smith CR. Pediatric drowning: a 20-year review of autopsied cases: I. Demographic features. *Am J Forensic Med Pathol*. 2005;26(4):316–319.
66. Byard R, de Koning C, Blackbourne B, Nadeau J, Krous HF. Shared bathing and drowning in infants and young children. *J Paediatr Child Health*. 2001;37(6):542–544.
67. Capovilla M, Durigon M, de la Grandmaison GL. An original cause of drowning in an industrial environment. *Am J Forensic Med Pathol*. 2007; 28(1):91–93.
68. Quan L, Cummings P. Characteristics of drowning by different age groups. *Inj Prev*. 2003;9(2):163–168.
69. Hwang V, Shofer FS, Durbin DR, Baren JM. Prevalence of traumatic injuries in drowning and near drowning in children and adolescents. *Arch Pediatr Adolesc Med*. 2003;157(1):50–53.
70. Alpert B. Bathtub drowning: unintentional, neglect, or abuse. *Med Health R I*. 2003;86(12):385–386.
71. Stumpp JW, Schneider J, Bar W. Drowning of a girl with anomaly of the bundle of His and the right bundle branch. *Am J Forensic Med Pathol*. 1997;18(2):208–210.
72. Lavelle JM, Shaw KN, Seidl T, Ludwig S. Ten-year review of pediatric bathtub near-drownings: evaluation for child abuse and neglect. *Ann Emerg Med*. 1995;25(3):344–348.
73. Mizuta R, Fujita H, Osamura T, Kidowaki T, Kiyosawa N. Childhood drownings and near-drownings in Japan. *Acta Paediatr Jpn*. 1993;35(3): 186–192.
74. Feldman KW, Monastersky C, Feldman GK. When is childhood drowning neglect? *Child Abuse Negl*. 1993;17(3):329–336.
75. Quan L, Gore EJ, Wentz K, Allen J, Novack AH. Ten-year study of pediatric drownings and near-drownings in King County, Washington: lessons in injury prevention. *Pediatrics*. 1989;83(6):1035–1040.
76. Saberi Anary SH, Sheikhazadi A, Ghadyani MH. Epidemiology of drowning in Mazandaran province, north of Iran. *Am J Forensic Med Pathol*. 2010;31(3):236–242.
77. Stemberga V, Bralic M, Coklo M, Cuculic D, Bosnar A. Suicidal drowning in Southwestern Croatia: a 25-year review. *Am J Forensic Med Pathol*. 2010;31(1):52–54.
78. Byard RW. Drowning deaths in rivers. *Forensic Sci Med Pathol*. 2017;13(3):388–389.
79. Byard RW. Drowning and near drowning - definitions and terminology. *Forensic Sci Med Pathol*. 2017;13(4):529–530.
80. Webber J, Schmidt AC, Sempsrott JR, et al. Fatal and non-fatal drowning in rivers. *Forensic Sci Med Pathol*. 2017;13(4):527–528.
81. Lunetta P, Smith GS, Penttila A, Sajantila A. Undetermined drowning. *Med Sci Law*. 2003;43(3):207–214.
82. Wardak KS, Buchsbaum RM, Walyzada F. The Drowning Index: implementation in drowning, mechanical asphyxia, and acute myocardial infarct cases. *J Forensic Sci*. 2014;59(2):399–403.
83. Leth PM. Homicide by drowning. *Forensic Sci Med Pathol*. 2019;15(2):233–238.
84. Szeremeta M, Janica JR, Niemcunowicz-Janica A, Pepinski W. Fatal drowning as a result of an airplane crash–case report. *Forensic Sci Int*. 2013;226(1–3):e12–15.
85. Byard RW. *Immersion Deaths and Drowning: Issues Arising in the Investigation of Bodies Recovered from Water*. Springer; 2015.
86. Atilgan M, Bulgur-Kirbas D, Akman R, Deveci C. Fatal drowning caused by A swimming pool drainage system. *Am J Forensic Med Pathol*. 2021;42(3):275–277.
87. Hanzlick R, Goodin J. Mind your manners. Part III: individual scenario results and discussion of the national association of medical examiners manner of death questionnaire, 1995. *Am J Forensic Med Pathol*. 1997;18(3):228–245.
88. Motozawa Y, Hitosugi M, Kido M, Kurosu A, Nagai T, Tokudome S. Sudden death while driving a four-wheeled vehicle: an autopsy analysis. *Med Sci Law*. 2008;48(1):64–68.
89. Buttner A, Heimpel M, Eisenmenger W. Sudden natural death 'at the wheel': a retrospective study over a 15-year time period (1982–1996). *Forensic Sci Int*. 1999;103(2):101–112.
90. Antecol DH, Roberts WC. Sudden death behind the wheel from natural disease in drivers of four-wheeled motorized vehicles. *Am J Cardiol*. 1990;66(19):1329–1335.
91. Morild I. Traffic deaths in western Norway. A study from the county of Hordaland 1986-1990. *Forensic Sci Int*. 1994;64(1):9–20.
92. Lau G. Ischaemic heart disease as a natural cause of death in motorists. *Singapore Med J*. 1994;35(5):467–470.
93. Parsons M. Fits and other causes of loss of consciousness while driving. *Q J Med*. 1986;58(227):295–303.
94. Rohrig TP, Nash E, Osawa KA, et al. Fentanyl and driving impairment. *J Anal Toxicol*. 2021;45(4):389–396.
95. Peterson B, Petty C. Sudden natural death among automobile drivers. *J Forensic Sci*. 1962;7(3):274–285.
96. Chaudhari VA, Ghodake DG, Kharat RD. Positional asphyxia: death due to unusual head-down position in a narrow space. *Am J Forensic Med Pathol*. 2016;37(2):51–53.
97. Heath KJ, Cala AD, Byard RW. Metal railing fences and accidental death. *J Forensic Sci*. 2018;63(3):972–975.
98. Straka L, Janik M, Hejna P. Asphyxiated while hunting for treasure: an unusual geocaching fatality. *Forensic Sci Med Pathol*. 2019;15(4): 653–657.
99. Tattoli L, Melloni ND, Di Vella G. Positional asphyxia in a work-related fatality. *Forensic Sci Med Pathol*. 2019;15(4):642–645.
100. Hayashi T, Buschmann C, Correns A, Herre S, Tsokos M. Fatal positional asphyxia. *Forensic Sci Med Pathol*. 2012;8(4):470–472.
101. deJong JL, Lee J, Grande A, Huffman C, Bielby C, Brown T. Positional asphyxia in opioid-related deaths: is it being overlooked? *J Forensic Sci*. 2020;65(6):2008–2012.

102. Davis JH. Bodies in water. Solving the puzzle. *J Fla Med Assoc*. 1992;79(9):630–632.
103. Edmonds C, Caruso J. Recent modifications to the investigation of diving related deaths. *Forensic Sci Med Pathol*. 2014;10(1):83–90.
104. Casadesus JM, Aguirre F, Carrera A, Boadas-Vaello P, Serrando MT, Reina F. Diving-related fatalities: multidisciplinary, experience-based investigation. *Forensic Sci Med Pathol*. 2019;15(2):224–232.
105. Petri NM, Stipancevic H, Sutlovic D, Gojanovic MD. Death of a scuba diver caused by vomiting and panic: a case report. *Am J Forensic Med Pathol*. 2011;32(2):186–189.
106. Byard RW, James RA, Heath KJ. Recovery of human remains after shark attack. *Am J Forensic Med Pathol*. 2006;27(3):256–259.
107. Casadesus JM, Aguirre F, Carrera A, Boadas-Vaello P, Serrando MT, Reina F. Diagnosis of arterial gas embolism in SCUBA diving: modification suggestion of autopsy techniques and experience in eight cases. *Forensic Sci Med Pathol*. 2018;14(1):18–25.
108. Buzzacott P, Nelson C, Hill K, Hires L. Recovery of deceased scuba divers from within flooded subterranean caves. *Forensic Sci Int*. 2017;275: 167–170.
109. Hermans F, Diver RC. *Survey and Analysis of Fatal Accidents in the Commercial Diving Sector*. 2016.
110. Cannon D, Taylor A. Occupational safety and health administration (OSHA) investigation of an occupational diving death (platform presentation). In: *Paper Presented at: Annual NAME Meeting; October 18, 2021*. 2021. West Palm, FL).
111. Papadodima SA, Athanaselis SA, Skliros E, Spiliopoulou CA. Forensic investigation of submersion deaths. *Int J Clin Pract*. 2010;64(1):75–83.
112. Calder IM. Dysbarism. A review. *Forensic Sci Int*. 1986;30(4):237–266.
113. Plattner T, Thali MJ, Yen K, et al. Virtopsy-postmortem multislice computed tomography (MSCT) and magnetic resonance imaging (MRI) in a fatal scuba diving incident. *J Forensic Sci*. 2003;48(6):1347–1355.
114. Klintschar M, Grabuschnigg P, Beham A. Death from electrocution during autoerotic practice: case report and review of the literature. *Am J Forensic Med Pathol*. 1998;19(2):190–193.
115. Marc B, Baudry F, Douceron H, Ghaith A, Wepierre JL, Garnier M. Suicide by electrocution with low-voltage current. *J Forensic Sci*. 2000;45(1): 216–222.
116. Anders S, Matschke J, Tsokos M. Internal current mark in a case of suicide by electrocution. *Am J Forensic Med Pathol*. 2001;22(4):370–373.
117. Schott JC, Davis GJ, Hunsaker 3rd JC. Accidental electrocution during autoeroticism: a shocking case. *Am J Forensic Med Pathol*. 2003;24(1): 92–95.
118. Di Nunno N, Vimercati L, Viola L, Vimercati F. A case of electrocution during illegal fishing activities. *Am J Forensic Med Pathol*. 2003;24(2): 164–167.
119. Bligh-Glover WZ, Miller FP, Balraj EK. Two cases of suicidal electrocution. *Am J Forensic Med Pathol*. 2004;25(3):255–258.
120. Chan P, Duflou J. Suicidal electrocution in Sydney: a 10-year case review. *J Forensic Sci*. 2008;53(2):455–459.
121. Dokov W. Forensic characteristics of suicide by electrocution in Bulgaria. *J Forensic Sci*. 2009;54(3):669–671.
122. Shetty BS, Kanchan T, Shetty M, et al. Fatal electrocution by a support metal wire. *J Forensic Sci*. 2010;55(3):830–831.
123. Sheikhazadi A, Kiani M, Ghadyani MH. Electrocution-related mortality: a survey of 295 deaths in Tehran, Iran between 2002 and 2006. *Am J Forensic Med Pathol*. 2010;31(1):42–45.
124. Rautji R, Rudra A, Behera C, Dogra TD. Electrocution in South Delhi: a retrospective study. *Med Sci Law*. 2003;43(4):350–352.
125. Das S, Patra AP, Shaha KK, Sistla SC, Jena MK. High-voltage suicidal electrocution with multiple exit wounds. *Am J Forensic Med Pathol*. 2013; 34(1):34–37.
126. Wick R, Gilbert J, Simpson E, Byard R. Fatal electrocution in adults – a 30-year study. *Med Sci Law*. 2006;46(2):166–172.
127. Gupta BD, Singh OG, Mehta RA. Death comes through eye: a rare case of electrocution. *Am J Forensic Med Pathol*. 2009;30(3):301–302.
128. Li M, Hamilton W. Review of autopsy findings in judicial electrocutions. *Am J Forensic Med Pathol*. 2005;26(3):261–267.
129. Wick R, Byard RW. *Electrocution and the Autopsy*. 2009;5:53–66.
130. Kroll MW, Fish RM, Lakkireddy D, Luceri RM, Panescu D. Essentials of low-power electrocution: established and speculated mechanisms. *Annu Int Conf IEEE Eng Med Biol Soc*. 2012;2012:5734–5740.
131. Nagesh KR, Kanchan T, Rastogi P, Arun M. Arcing injuries in a fatal electrocution. *Am J Forensic Med Pathol*. 2009;30(2):183–185.
132. Tsokos M. Extensive electrical injuries with opening of the chest wall due to arcing of high-voltage current. *Forensic Sci Med Pathol*. 2012;8(3): 338–339.
133. Bielefeld L, Mierdel K, Pollak S, Perdekamp MG. Electrothermic damage to the nail due to arcing in high-voltage discharge. *Forensic Sci Int*. 2013;233(1–3):149–153.
134. Kroll MW, Andrews CJ, Panescu D. Electrocution: direct-current dogma dies hard. *Am J Forensic Med Pathol*. 2021;42(4):405–406.
135. Sharma AD, Fain E, O'Neill PG, et al. Shock on T versus direct current voltage for induction of ventricular fibrillation: a randomized prospective comparison. *Pacing Clin Electrophysiol*. 2004;27(1):89–94.
136. Lim HS, Flannigan S, Marshall H. Induction by direct current pulse versus 50-Hz pacing on ventricular fibrillation and defibrillation. *J Interv Card Electrophysiol*. 2010;28(3):209–214.
137. Liu S, Yu Y, Huang Q, Luo B, Liao X. Electrocution-related mortality: a review of 71 deaths by low-voltage electrical current in Guangdong, China, 2001–2010. *Am J Forensic Med Pathol*. 2014;35(3):193–196.
138. MMWR. Lightning-associated deaths-US 1980–1995. *MMWR (Morb Mortal Wkly Rep)*. 1998;47:391–393.
139. Cherington M, Kurtzman R, Krider EP, Yarnell PR. Mountain medical mystery. Unwitnessed death of a healthy young man, caused by lightning. *Am J Forensic Med Pathol*. 2001;22(3):296–298.
140. King DEL. Deaths due to lightning strike. *Can Soc Forens Sci J*. 1996;29(4):213–220.
141. Blumenthal R. Lightning fatalities on the South African Highveld: a retrospective descriptive study for the period 1997 to 2000. *Am J Forensic Med Pathol*. 2005;26(1):66–69.
142. Lindstrom R, Bylund PO, Eriksson A. Accidental deaths caused by electricity in Sweden, 1975-2000. *J Forensic Sci*. 2006;51(6):1383–1388.
143. Eriksson A, Ornehult L. Death by lightning. *Am J Forensic Med Pathol*. 1988;9(4):295–300.
144. Murty OP, Kian CK, Ari Husin MH, Nanta Kumar RK, Mohammed Yusuf WY. Fatal lightning strikes in Malaysia. *Am J Forensic Med Pathol*. 2009;30(3):246–251.
145. Byard RW, Thu M, Gilbert JD. Cutaneous manifestations of lightning strike - variability in Lichtenberg figures. *Forensic Sci Med Pathol*. 2017; 13(3):390–393.

146. Ventura F, Barranco R, Bonsignore A, De Stefano F. A unusual lightning death in an indoor setting: a case report. *Am J Forensic Med Pathol.* 2017;38(1):1–4.
147. Pincus JL, Lathrop SL, Briones AJ, Andrews SW, Aurelius MB. Lightning deaths: a retrospective review of New Mexico's cases, 1977-2009. *J Forensic Sci.* 2015;60(1):66–71.
148. Hauser R, Kaliszan M, Basir A, Basir ID. Lightning strike as probable cause of death and determining identity based on the examination of skeletal remains. *J Forensic Sci.* 2013;58(2):527–529.
149. Blumenthal R. Lightning and the forensic pathologist. *Acad Forensic Pathol.* 2018;8(1):98–111.
150. Wetli CV. Keraunopathology. An analysis of 45 fatalities. *Am J Forensic Med Pathol.* 1996;17(2):89–98.
151. Reddy K, Lowenstein EJ. Forensics in dermatology: part II. *J Am Acad Dermatol.* 2011;64(5):811–824. quiz 825–816.
152. Wankhede AG, Sariya DR. Damage due to lightning when it strikes the face. *Forensic Sci Int.* 2013;224(1–3):e1–3.
153. Blumenthal R. The explosive effects of lightning: what are the risks? *Acad Forensic Pathol.* 2016;6(1):89–95.
154. Blumenthal R. Injuries and deaths from lightning. *J Clin Pathol.* 2021;74(5):279–284.
155. Jauhar S. *Out of the Blue, a Lightning Bolt to the Heart.* NY Times; 2004. Feb 10, 2004:F1.
156. Adekoya N, Nolte KB. Struck-by-lightning deaths in the United States. *J Environ Health.* 2005;67(9):45–50, 58.
157. Kagiyama Y, Hill JL, Gettes LS. Interaction of acidosis and increased extracellular potassium on action potential characteristics and conduction in Guinea pig ventricular muscle. *Circ Res.* 1982;51(5):614–623.
158. Robinson MN, Brooks CG, Renshaw GD. Electric shock devices and their effects on the human body. *Med Sci Law.* 1990;30(4):285–300.
159. Pasquier M, Carron PN, Vallotton L, Yersin B. Electronic control device exposure: a review of morbidity and mortality. *Ann Emerg Med.* 2011; 58(2):178–188.
160. Jauchem JR. Deaths in custody: are some due to electronic control devices (including TASER devices) or excited delirium? *J Forensic Leg Med.* 2010;17(1):1–7.
161. *Study of Deaths Following Electro Muscular Disruption: Interim Report.* Washington DC: National Institute of Justice; 2011. In: NIJ, ed2011 http://www.ncjrs.gov/App/publications/abstract.aspx?ID=244890.
162. Ordog GJ, Wasserberger J, Schlater T, Balasubramanium S. Electronic gun (Taser) injuries. *Ann Emerg Med.* 1987;16(1):73–78.
163. Burdett-Smith P. Stun gun injury. *J Accid Emerg Med.* 1997;14(6):402–404.
164. Frechette A, Rimsza ME. Stun gun injury: a new presentation of the battered child syndrome. *Pediatrics.* 1992;89(5 Pt 1):898–901.
165. Fish R. Electric shock, Part III: deliberately applied electric shocks and the treatment of electric injuries. *J Emerg Med.* 1993;11(5):599–603.
166. O'Brien DJ. Electronic weaponry–a question of safety. *Ann Emerg Med.* 1991;20(5):583–587.
167. Walsh J. *Electroshock Torture and the Spread of Stun Technology.* 1997:1247. Vol. 349.
168. Turner MS, Jumbelic ML. Stun gun injuries in the abuse and death of a seven-month-old infant. *J Forensic Sci.* 2003;48(1):180–182.
169. Anders S, Junge M, Schulz F, Puschel K. Cutaneous current marks due to a stun gun injury. *J Forensic Sci.* 2003;48(3):640–642.
170. Dennis AJ, Valentino DJ, Walter RJ, et al. Acute effects of TASER X26 discharges in a swine model. *J Trauma.* 2007;63(3):581–590.
171. Gerber M, Walsh G, Hopmeier M. Sensitivity of TATP to a TASER electrical output. *J Forensic Sci.* 2014;59(6):1638–1641.
172. Vilke GM, Sloane CM, Bouton KD, et al. Physiological effects of a conducted electrical weapon on human subjects. *Ann Emerg Med.* 2007;50(5): 569–575.
173. Bozeman WP, Hauda 2nd WE, Heck JJ, Graham Jr DD, Martin BP, Winslow JE. Safety and injury profile of conducted electrical weapons used by law enforcement officers against criminal suspects. *Ann Emerg Med.* 2009;53(4):480–489.
174. Beason CW, Jauchem JR, Clark 3rd CD, Parker JE, Fines DA. Pulse variations of a conducted energy weapon (similar to the TASER X26 device): effects on muscle contraction and threshold for ventricular fibrillation*. *J Forensic Sci.* 2009;54(5):1113–1118.
175. Jauchem JR. An animal model to investigate effectiveness and safety of conducted energy weapons (including TASER devices). *J Forensic Sci.* 2010;55(2):521–526.
176. Comeaux JA, Jauchem JR, Cox DD, Crane CC, D'Andrea JA. Muscle contraction during electro-muscular incapacitation: a comparison between square-wave pulses and the TASER((R)) X26 Electronic control device. *J Forensic Sci.* 2011;56(Suppl 1):S95–S100.
177. Zipes DP. Sudden cardiac arrest and death following application of shocks from a TASER electronic control device. *Circulation.* 2012;125(20): 2417–2422.
178. Jauchem JR. TASER((R)) conducted electrical weapons: misconceptions in the scientific/medical and other literature. *Forensic Sci Med Pathol.* 2015;11(1):53–64.
179. Criscione JC, Kroll MW. Incapacitation recovery times from a conductive electrical weapon exposure. *Forensic Sci Med Pathol.* 2014;10(2): 203–207.
180. Jauchem JR, Bernhard JA, Cerna CZ, Lim TY, Seaman RL, Tarango M. Effects of a TASER(R) conducted energy weapon on the circulating red-blood-cell population and other factors in *Sus scrofa.* *Forensic Sci Med Pathol.* 2013;9(3):308–320.
181. Dawes DM, Ho JD, Moore JC, Miner JR. An evaluation of two conducted electrical weapons and two probe designs using a swine comparative cardiac safety model. *Forensic Sci Med Pathol.* 2013;9(3):333–342.
182. Havranek S, Neuzil P, Linhart A. Electromuscular incapacitating devices discharge and risk of severe bradycardia. *Am J Forensic Med Pathol.* 2015;36(2):94–98.
183. Becour B. Conducted electrical weapons or stun guns: a review of 46 cases examined in casualty. *Am J Forensic Med Pathol.* 2013;34(2):142–146.
184. Kroll MW, Ritter MB, Guilbault RA, Panescu D. Infection risk from conducted electrical weapon probes: what do we know? *J Forensic Sci.* 2016;61(6):1556–1562.
185. Lucas SR, McGowan JC, Lam TC, Yamaguchi GT, Carver M, Hinz A. Assessment of the TASER XREP blunt impact and penetration injury potential using cadaveric testing. *J Forensic Sci.* 2013;58(Suppl 1):S60–S68.
186. Park EJ, Choi SC, Ahn JH, Min YG. Repetitive TASER X26 discharge resulted in adverse physiologic events with a dose-response relationship related to the duration of discharge in anesthetized swine model. *J Forensic Sci.* 2013;58(1):179–183.
187. Kroll MW, Hail SL, Kroll RM, Wetli CV, Criscione JC. Electrical weapons and excited delirium: shocks, stress, and serum serotonin. *Forensic Sci Med Pathol.* 2018;14(4):478–483.
188. Kunz SN, Calkins H, Adamec J, Kroll MW. Cardiac and skeletal muscle effects of electrical weapons : a review of human and animal studies. *Forensic Sci Med Pathol.* 2018;14(3):358–366.

189. Ho J, Dawes D, Miner J, Kunz S, Nelson R, Sweeney J. Conducted electrical weapon incapacitation during a goal-directed task as a function of probe spread. *Forensic Sci Med Pathol.* 2012;8(4):358–366.
190. Dawes DM, Ho JD, Sweeney JD, Lundin EJ, Kunz SN, Miner JR. The effect of an electronic control device on muscle injury as determined by creatine kinase enzyme. *Forensic Sci Med Pathol.* 2011;7(1):3–8.
191. Jauchem JR. Repeated or long-duration TASER electronic control device exposures: acidemia and lack of respiration. *Forensic Sci Med Pathol.* 2010;6(1):46–53.
192. Jauchem JR, Seaman RL, Klages CM. Physiological effects of the TASER C2 conducted energy weapon. *Forensic Sci Med Pathol.* 2009;5(3): 189–198.
193. Jauchem J, Beason CW, Cook MC. Acute effects of an alternative electronic-control-device waveform in swine. *Forensic Sci Med Pathol.* 2009; 5(1):2–10.
194. Dawes DM, Ho JD, Reardon RF, et al. The respiratory, metabolic, and neuroendocrine effects of a new generation electronic control device. *Forensic Sci Int.* 2011;207(1–3):55–60.
195. Ho JD, Dawes DM, Reardon RF, et al. Human cardiovascular effects of a new generation conducted electrical weapon. *Forensic Sci Int.* 2011; 204(1–3):50–57.
196. Ho JD, Dawes DM, Nystrom PC, et al. Markers of acidosis and stress in a sprint versus a conducted electrical weapon. *Forensic Sci Int.* 2013; 233(1–3):84–89.
197. Ho JD, Dawes DM, Cole JB, Hottinger JC, Overton KG, Miner JR. Lactate and pH evaluation in exhausted humans with prolonged TASER X26 exposure or continued exertion. *Forensic Sci Int.* 2009;190(1–3):80–86.
198. Dawes D, Ho J, Miner J. The neuroendocrine effects of the TASER X26: a brief report. *Forensic Sci Int.* 2009;183(1–3):14–19.
199. Jenkins Jr DM, Murray WB, Kennett MJ, Hughes EL, Werner JR. The effects of continuous application of the TASER X26 waveform on *Sus scrofa*. *J Forensic Sci.* 2013;58(3):684–692.
200. Shen T, Howe HL, Alo C, Moolenaar RL. Toward a broader definition of heat-related death: comparison of mortality estimates from medical examiners' classification with those from total death differentials during the July 1995 heat wave in Chicago, Illinois. *Am J Forensic Med Pathol.* 1998;19(2):113–118.
201. Donoghue ER, Graham MA, Jentzen JM, Lifschultz BD, Luke JL, Mirchandani HG. Criteria for the diagnosis of heat-related deaths: national association of medical examiners. Position paper. National association of medical examiners ad hoc committee on the definition of heat-related fatalities. *Am J Forensic Med Pathol.* 1997;18(1):11–14.
202. Malamud N, Haymaker W, Custer RP. Heat stroke; a clinico-pathologic study of 125 fatal cases. *Mil Surg.* 1946;99(5):397–449.
203. Wolfe MI, Kaiser R, Naughton MP, et al. Heat-Related mortality in selected United States cities, summer 1999. *Am J Forensic Med Pathol.* 2001; 22(4):352–357.
204. Heat-related deaths–four states, July-August 2001, and United States, 1979–1999. *MMWR Morb Mortal Wkly Rep.* 2002;51(26):567–570.
205. Byard RW, Riches KJ. Dehydration and heat-related death: sweat lodge syndrome. *Am J Forensic Med Pathol.* 2005;26(3):236–239.
206. Tek D, Olshaker JS. Heat illness. *Emerg Med Clin.* 1992;10(2):299–310.
207. Semenza W. Heat related deaths during the July 1995 heat wave in Chicago. *NEJM.* 1996;335:84–90.
208. MMWR. Heat-related illnesses and deaths-Missouri, 1998, and United States, 1976–1996. *MMWR (Morb Mortal Wkly Rep).* 1999;248:469–473.
209. Marzuk PM, Tardiff K, Leon AC, et al. Ambient temperature and mortality from unintentional cocaine overdose. *JAMA.* 1998;279(22): 1795–1800.
210. Krous HF, Nadeau JM, Fukumoto RI, Blackbourne BD, Byard RW. Environmental hyperthermic infant and early childhood death: circumstances, pathologic changes, and manner of death. *Am J Forensic Med Pathol.* 2001;22(4):374–382.
211. Simon HB. Hyperthermia. *N Engl J Med.* 1993;329(7):483–487.
212. Nixdorf-Miller A, Hunsaker DM, Hunsaker 3rd JC. Hypothermia and hyperthermia medicolegal investigation of morbidity and mortality from exposure to environmental temperature extremes. *Arch Pathol Lab Med.* 2006;130(9):1297–1304.
213. Zumwalt R, Petty CS, Holman W. Temperatures in closed automobiles in hot weather. *For Sci Gazette.* 1976;7:7–8.
214. Franchi A, Gauchotte G, Gambier N, Raul JS, Martrille L. Postmortem hyperthermia: two case reports and a review of the literature. *Am J Forensic Med Pathol.* 2018;39(4):364–366.
215. Fisher-Hubbard AO, Sung L, Hubbard SA, Hlavaty L. Hyperthermia, thermal injuries, and death from a forced convection heat source: a case report and experimental model. *J Forensic Sci.* 2017;62(3):686–690.
216. Ball CG, Herath JC. Earth, air, water, and fire: histopathology of environmental death. *Acad Forensic Pathol.* 2018;8(3):641–652.
217. Lougee KM, Hess GL, Winston DC. Environmental deaths in southern Arizona. *Acad Forensic Pathol.* 2018;8(1):50–63.
218. Henke AF, Cohle SD, Cottingham SL. Fatal hyperthermia secondary to sunbathing in a patient with multiple sclerosis. *Am J Forensic Med Pathol.* 2000;21(3):204–206.
219. Kohlmeier RE, DiMaio VJ, Kagan-Hallet K. Fatal hyperthermia in hot baths in individuals with multiple sclerosis. *Am J Forensic Med Pathol.* 2000;21(3):201–203.
220. Crandall CG, Vongpatanasin W, Victor RG. Mechanism of cocaine-induced hyperthermia in humans. *Ann Intern Med.* 2002;136(11):785–791.
221. Christiansen LR, Collins KA. Pathologic findings in malignant hyperthermia: a case report and review of literature. *Am J Forensic Med Pathol.* 2004;25(4):327–333.
222. Pamukcoglu T. Sudden death due to malignant hyperthermia. *Am J Forensic Med Pathol.* 1988;9(2):161–162.
223. Kilbane BJ, Mehta S, Backeljauw PF, Shanley TP, Crimmins NA. Approach to management of malignant hyperthermia-like syndrome in pediatric diabetes mellitus. *Pediatr Crit Care Med.* 2006;7(2):169–173.
224. Hollander AS, Olney RC, Blackett PR, Marshall BA. Fatal malignant hyperthermia-like syndrome with rhabdomyolysis complicating the presentation of diabetes mellitus in adolescent males. *Pediatrics.* 2003;111(6 Pt 1):1447–1452.
225. Baluch AR, Oommen SP. Malignant hyperthermia associated with diabetic hyperosmolar hyperglycemic nonketotic state in a young man. *J Clin Anesth.* 2007;19(6):470–472.
226. Rosenberg H, Sambuughin N, Riazi S, Dirksen R. Malignant hyperthermia susceptibility. In: Pagon RA, Adam MP, Ardinger HH, et al., eds. *GeneReviews(R). Seattle (WA).* 1993.
227. Rosenberg H, Pollock N, Schiemann A, Bulger T, Stowell K. Malignant hyperthermia: a review. *Orphanet J Rare Dis.* 2015;10:93.

228. Li W, Zhang L, Liang Y, Tong F, Zhou Y. Sudden death due to malignant hyperthermia with a mutation of RYR1: autopsy, morphology and genetic analysis. *Forensic Sci Med Pathol*. 2017;13(4):444–449.
229. Ramani NS, Stoppacher R, Morani AC, Catanese C. Undiagnosed pheochromocytoma simulating malignant hyperthermia. *Am J Forensic Med Pathol*. 2017;38(3):262–265.
230. Lin H, Zou D, Luo Y, et al. Postmortem diagnosis of fatal hypothermia/hyperthermia by spectrochemical analysis of plasma. *Forensic Sci Med Pathol*. 2019;15(3):332–341.
231. Booth 3rd JN, Davis GG, Waterbor J, McGwin Jr G. Hyperthermia deaths among children in parked vehicles: an analysis of 231 fatalities in the United States, 1999-2007. *Forensic Sci Med Pathol*. 2010;6(2):99–105.
232. Kimura H, Yoshida K, Ohsawa M, Arai T, Tsujimura T. An autopsy case of malignant hyperthermia. *Forensic Sci Int*. 1985;27(1):25–30.
233. Quinton RA. Certification of vehicular hyperthermia deaths in the pediatric population. *Acad Forensic Pathol*. 2016;6(4):657–662.
234. Christensen ED, Lacsina EQ. Mountaineering fatalities on Mount Rainier, Washington, 1977–1997: autopsy and investigative findings. *Am J Forensic Med Pathol*. 1999;20(2):173–179.
235. Ward ME, Cowley AR. Hypothermia: a natural cause of death. *Am J Forensic Med Pathol*. 1999;20(4):383–386.
236. From the Centers for Disease Control and Prevention. Hypothermia-related deaths–Alaska, October 1998–April 1999, and trends in the United States, 1979-1996. *JAMA*. 2000;283(7):878–879.
237. Wedin B, Vanggaard L, Hirvonen J. "Paradoxical undressing" in fatal hypothermia. *J Forensic Sci*. 1979;24(3):543–553.
238. Tsokos M, Rothschild MA, Madea B, Rie M, Sperhake JP. Histological and immunohistochemical study of Wischnewsky spots in fatal hypothermia. *Am J Forensic Med Pathol*. 2006;27(1):70–74.
239. Byard RW, Tsokos M. Forensic issues in cases of Diogenes syndrome. *Am J Forensic Med Pathol*. 2007;28(2):177–181.
240. Birchmeyer MS, Mitchell EK. Wischnewski revisited. The diagnostic value of gastric mucosal ulcers in hypothermic deaths. *Am J Forensic Med Pathol*. 1989;10(1):28–30.
241. Ogata M, Ago K, Ago M, et al. A fatal case of hypothermia associated with hemorrhages of the pectoralis minor, intercostal, and iliopsoas muscles. *Am J Forensic Med Pathol*. 2007;28(4):348–352.
242. Stoppacher R, Yancon AR, Jumbelic MI. Fatalities associated with the termination of electrical services. *Am J Forensic Med Pathol*. 2008;29(3): 231–234.
243. Kettner M, Schnabel A, Ramsthaler F. Suspected paradoxical undressing in a homicide case. *Forensic Sci Med Pathol*. 2012;8(4):426–429.
244. Lim C, Duflou J. Hypothermia fatalities in a temperate climate: sydney, Australia. *Pathology*. 2008;40(1):46–51.
245. Byard RW, Bright FM. Lethal hypothermia - a sometimes elusive diagnosis. *Forensic Sci Med Pathol*. 2017;14(4):421–423.
246. Schweitzer W, Thali M, Giugni G, Winklhofer S. Postmortem pulmonary CT in hypothermia. *Forensic Sci Med Pathol*. 2014;10(4):557–569.
247. Zivkovic V, Nikolic S. The unusual appearance of black esophagus in a case of fatal hypothermia: a possible underlying mechanism. *Forensic Sci Med Pathol*. 2013;9(4):613–614.
248. Bright F, Winskog C, Byard RW. Wischnewski spots and hypothermia: sensitive, specific, or serendipitous? *Forensic Sci Med Pathol*. 2013;9(1): 88–90.
249. Clark KH, Stoppacher R. Gastric mucosal petechial hemorrhages (wischnewsky lesions), hypothermia, and diabetic ketoacidosis. *Am J Forensic Med Pathol*. 2016;37(3):165–169.
250. Byard RW, Langlois NEI. Wandering dementia-A syndrome with forensic implications. *J Forensic Sci*. 2019;64(2):443–445.
251. Perich P, Tuchtan L, Bartoli C, Leonetti G, Piercecchi-Marti MD. Death from hypothermia during a training course under "extreme conditions": related to two cases. *J Forensic Sci*. 2016;61(2):562–565.
252. Morgan M, Schwartz L, Duflou J. Hypothermia secondary to glioblastoma multiforme? Autopsy findings in two cases. *J Forensic Sci*. 2015; 60(2):511–513.
253. Pisani F, Nittis M. An unusual case of hypothermia and paradoxical undressing. *Forensic Sci Med Pathol*. 2020;17(2):338–342.
254. Byard RW. Death in the Arctic - the tragic fate of members of the Franklin expedition (1845). *Forensic Sci Med Pathol*. 2020;17(1):161–166.
255. Byard RW, Bright FM. Lethal hypothermia - a sometimes elusive diagnosis. *Forensic Sci Med Pathol*. 2018;14(4):421–423.
256. Turk EE. Hypothermia. *Forensic Sci Med Pathol*. 2010;6(2):106–115.
257. Turk EE, Sperhake JP, Pueschel K, Tsokos M. An approach to the evaluation of fatal hypothermia. *Forensic Sci Med Pathol*. 2005;1(1):31–35.
258. Dickinson GM, Maya GX, Lo Y, Jarvis HC. Hypothermia-related deaths: a 10-year retrospective study of two major metropolitan cities in the United States. *J Forensic Sci*. 2020;65(6):2013–2018.
259. Hunter WC. Accidental hypothermia. 3. *Northwest Med*. 1968;67(9):837–844.
260. Hunter WC. Accidental hypothermia. II. *Northwest Med*. 1968;67(8):735–739 (contd).
261. Hunter WC. Accidental hypothermia. I. *Northwest Med*. 1968;67(6):569–573 (contd).
262. Doberentz E, Preuss-Wossner J, Kuchelmeister K, Madea B. Histological examination of the pituitary glands in cases of fatal hypothermia. *Forensic Sci Int*. 2011;207(1–3):46–49.
263. Banka K, Teresinski G, Buszewicz G. Free fatty acids as markers of death from hypothermia. *Forensic Sci Int*. 2014;234:79–85.
264. Hyodoh H, Watanabe S, Katada R, Hyodoh K, Matsumoto H. Postmortem computed tomography lung findings in fatal of hypothermia. *Forensic Sci Int*. 2013;231(1–3):190–194.
265. Banka K, Teresinski G, Buszewicz G, Madro R. Glucocorticosteroids as markers of death from hypothermia. *Forensic Sci Int*. 2013;229(1–3): 60–65.
266. Palmiere C, Bardy D, Letovanec I, et al. Biochemical markers of fatal hypothermia. *Forensic Sci Int*. 2013;226(1–3):54–61.
267. Bolte RG, Black PG, Bowers RS, Thorne JK, Corneli HM. The use of extracorporeal rewarming in a child submerged for 66 minutes. *JAMA*. 1988;260(3):377–379.
268. Rothschild M, Schneider V. "Terminal burrowing behaviour"—a phenomenon of lethal hypothermia. *Int J Leg Med*. 1995;107(5):250–256.
269. Rolf CM, Gallagher KE. Hypothermic death in the arctic state. *Acad Forensic Pathol*. 2018;8(1):64–82.
270. Hirvonen J. Necropsy findings in fatal hypothermia cases. *Forensic Sci*. 1976;8(2):155–164.
271. Hirvonen J. Some aspects on death in the cold and concomitant frostbites. *Int J Circumpolar Health*. 2000;59(2):131–136.
272. Devos C, Piette M. Hypothermia and combined post-mortem determination of amylase and isoamylase in the serum and the vitreous humour. *Med Sci Law*. 1989;29(3):218–228.

273. Wolf DA, Aronson JF, Rajaraman S, Veasey 3rd SP. Wischnewski ulcers and acute pancreatitis in two hospitalized patients with cirrhosis, portal vein thrombosis, and hypothermia. *J Forensic Sci*. 1999;44(5):1082–1085.
274. Palmiere C, Teresinski G, Hejna P, Mangin P, Grouzmann E. Diagnostic performance of urinary metanephrines for the postmortem diagnosis of hypothermia. *Forensic Sci Med Pathol*. 2014;10(4):518–525.
275. Morleo B, Teresinski G, Rousseau G, et al. Biomarkers of cerebral damage in fatal hypothermia: preliminary results. *Am J Forensic Med Pathol*. 2019;40(3):242–245.
276. Lin H, Guo X, Luo Y, et al. Postmortem diagnosis of fatal hypothermia by fourier transform infrared spectroscopic analysis of edema fluid in formalin-fixed, paraffin-embedded lung tissues. *J Forensic Sci*. 2020;65(3):846–854.
277. Jakubeniene M, Irnius A, Chaker GA, Paliulis JM, Bechelis A. Post-mortem investigation of calcium content in liver, heart, and skeletal muscle in accidental hypothermia cases. *Forensic Sci Int*. 2009;190(1–3):87–90.
278. Coe JI. Hypothermia: autopsy findings and vitreous glucose. *J Forensic Sci*. 1984;29(2):389–395.
279. Siddiqui AH, Siddiqui F. *Curling Ulcer (Stress-Induced Gastric)*. 2018.
280. Pruitt Jr BA, Foley F, Moncrief JA. Curling's ulcer: a clinical-pathology study of 323 cases. *Ann Surg*. 1970;172(4):523.
281. Gibson M, McDowall G, Turner J, Nahhas M. Clinical experience of a method of continuous intracranial pressure recording in 50 neurosurgical patients. *Intracranial pressure II*. 1975:496–499. Springer.

CHAPTER

14

Approach to toxicological deaths

Thomas Gilson, David Dolinak, Susan F. Ely and James R. Gill

Toxicology testing gives us information, but not conclusions. *-Dr. Charles S. Hirsch*

Drug intoxication deaths are encountered with increasing frequency in medicolegal death investigations. Since 2007, drug overdose has become the most common type of accidental death in the United States.[1] Data from the CDC's National Center for Health Statistics reported over 100,000 drug intoxication deaths in the United States during a 12-month period (ending in April 2021), an increase of 28.5% from the prior period.[2] While many of these deaths may appear "routine," the magnitude of the opioid crisis has not only greatly affected families and the workforce, it has also placed increased scrutiny on accurate postmortem analysis and death certification The proper classification of such deaths requires the integration of toxicology, pharmacology, physiology, pathology, and medicolegal investigative expertise. An understanding of the chemical substances involved in an intoxication, as well as their potential effects, is essential in drawing correlations of their impact on an individual's normal physiologic functioning.

In general, forensic death investigation is directed toward the accurate certification of a cause and manner of death. Like any other area of medicine, the appropriate certification of death (analogous to a clinical diagnosis) rests on the proper synthesis of history, physical examination, and laboratory analysis. It is a fundamental error to overemphasize the importance of the toxicological analysis to the detriment of history/scene investigation and autopsy findings. This is why forensic pathologists, and not forensic toxicologists, certify these deaths.

In order to certify a death as due to a drug intoxication, there are three considerations: (1) the **history and scene** are consistent with a drug intoxication death, (2) the **autopsy** fails to disclose a more compelling disease or physical injury with a severity that is inconsistent with continued life, and (3) the **toxicological concentrations** are in the range typically encountered in such deaths.[3] The general principles of death investigation apply to potential fatal drug intoxication deaths (see Chapter 4), but this chapter will highlight additional factors more specific to the investigation of drug-related deaths.

History and scene investigation

The initial investigation begins with an inquiry focused on any history of alcohol or substance misuse. If present, it is important to know whether the victims of fatal drug intoxications were known to be actively using drugs, in remission, recently incarcerated, or in a treatment facility. Individuals with a history of abstinence, recent incarceration, or a recent stay in a treatment facility are at an increased risk for a fatal relapse due to a loss of tolerance.[4,5] The most common cause of death in the first 18 months following release from prison is a drug intoxication.[5] Depending upon the nature of the substance use, there may have been a history of new onset erratic behavior or activity that suggested drug seeking or withdrawal. Searching a prescription drug monitoring program is helpful in establishing the decedent's history of the use of addictive medications and/or their enrollment in a methadone maintenance treatment program. Other relevant historical information may be more specific to how the decedent obtained the drugs. This might include frequent trips from home at unusual hours, as well as the receipt of unusual packages in the mail (suggesting internet sources for obtaining drugs). In addition, historical information detailing the decedent's behavior prior to discovery may influence a final death determination in an opioid intoxication. For example, an individual may have been awake and alert when they suddenly collapsed, but despite having low to moderate concentrations of opioids in their blood, their death may ultimately be determined *not* to have been caused by opioids (see more below).

Principles of Forensic Pathology
https://doi.org/10.1016/B978-0-323-91796-4.00028-3

Examination of the scene may yield critical information that suggests a drug intoxication (see Chapter 4)). The body position should be noted. *Not infrequently, decedents dying from an opioid intoxication will be found in a kneeling position, with their head down (sometimes referred to as a "prayer" position).* The most obvious findings associated with drug-related deaths include drug paraphernalia, such as drug packaging/residue, syringes, cut straws, burned spoons, razors, glass pipes, and filter material such as copper wiring, which may reflect recent drug use (Fig. 14.1). When documenting the paraphernalia at the scene, one should report the specific items, not just "paraphernalia." Additionally, multiple prescription containers or multiple prescriptions to different individuals at the scene also may be an indication of controlled substance misuse. Burn marks in various locations, including the decedent's fingers or clothing, also may suggest the use of illicit substances via smoking.

A study of over 4000 scene investigations in Bexar County, Texas examined drug paraphernalia at scenes in relation to the cause of death.[6] Drug paraphernalia (defined as needles, caps, burnt spoons, pipes, cookers, foils, small plastic bags, or rolling paper; prescription pill bottles were not included) were found at 247 (6%) scenes. Of these, 194 deaths were drug-related (79%), 48 were not, and five were undetermined. In the 815 deaths that were certified as drug-related, 194 (24%) had drug paraphernalia found at the scene.

Individuals who die from an intoxication also may conceal their drug use; consequently, associated drug paraphernalia may not necessarily be in plain sight. Depending upon the dose and route of administration, an opioid may not result in immediate incapacitation, allowing one time to hide their "works." Unusual packaging should be noted, as this may suggest internet sources for the drugs. It also might prove helpful to seek out cellphones belonging to the decedent, as these may include messages about obtaining and/or distributing drugs. Medications, illicit drugs, and paraphernalia should be secured and transported from the scene with the proper chain of custody, as dictated by local policy. These may prove invaluable for analysis in particularly challenging instances in which toxicology testing is negative despite a high index of suspicion for a drug intoxication. It is usually easier to identify novel drugs from a drug sample as opposed to a biological matrix. These samples also may be critical in tracking the drugs via touch DNA testing. In addition, with delayed deaths that occur in hospitals, testing of drug packets (Fig. 14.2) may specify the actual "opioid" that was detected by the hospital drug screen. The admission blood samples should be sought in these instances; however, sometimes, they have been discarded before death. The blood bank usually is required to keep the "type and cross" sample for seven days, so this laboratory is a good place to check for retained specimens.

Autopsy examination

The autopsy examination is a critical part in the investigation of a potential death due to a drug intoxication. With the recent increase in the number of opioid-related fatalities, some death investigation systems have opted to forego

FIGURE 14.1 Drug paraphernalia: syringe, drug packets, razor blade, and cotton swabs.

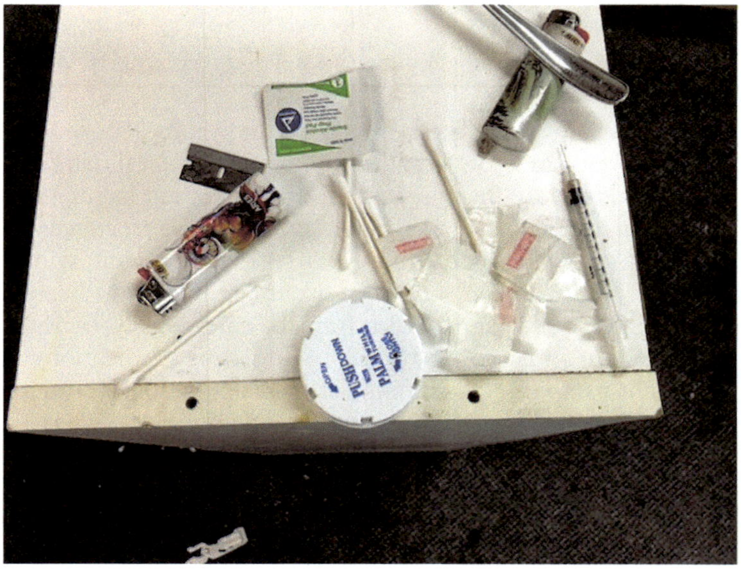

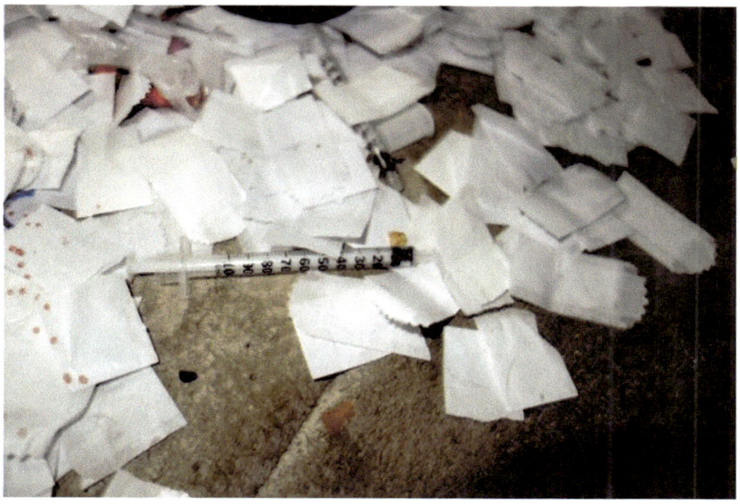

FIGURE 14.2 **Drug packaging: typical glassine drug packets.**

autopsy in select cases for a variety of reasons, including budgetary constraints and personnel shortages.[7] The current recommended best practice is that an autopsy is performed in all instances involving a suspected drug intoxication, except in some delayed hospital deaths.[8,9] An autopsy is used to exclude potential contributing or intervening causes of death, including occult trauma, as this population of decedents has an increased incidence of unnatural death of all manners (accident, suicide, and homicide).[10–12]

The external examination may reveal potential stigmata of drug use. These include puncture sites over superficial veins, perforated nasal septa (Fig. 14.3), cutaneous abscesses and/or irregular scars, chronic burns of finger tips (crack pipes/lighters), a single long finger nail (used for snorting), circular scars in a nonvascular distribution, jaundice, and powder in the nostrils.[13–15] Linear scarring in a distribution overlying superficial veins ("track marks") is evidence of chronic intravenous drug use (Fig. 14.4). Distinct round scars and crusted lesions/ecchymoses in various ages of healing and in a nonvascular distribution (Fig. 14.5) is often seen with "skin-popping." In the latter, instead of an injection into the vein ("mainlining"), the injection is subcutaneous/intramuscular. With needle exchange programs, these scars have become less prevalent (i.e., they are more likely to scar with "dirty" needles).

One of the more predictable findings in an opioid intoxication is frothy white fluid or foam at the mouth or nares. Typically, this "foam cone" is the result of neurogenic pulmonary edema, which has extruded up the airway into the nasopharynx and oral cavity and out the nose and mouth (Fig. 14.6).[9,16–19] This finding is not specific for drug abuse, per se, as it also is scene with congestive heart failure, seizures, spinal shock, and drownings.[17,18,20–32] In the right context, however, it raises one's suspicions. Care should be taken to document the "foam cone" at the scene of death, as this is a fragile physical finding which may be lost in the course of handling of the body and its transport to the death investigation facility.

The internal examination of the potential drug intoxication victim is nonspecific. When involving a depressant drug, particularly opioids, findings often include a distended urinary bladder (urinary retention), pulmonary edema, and, to a lesser extent, cerebral edema (more prominent with a prolonged survival interval).[9,13–15,19,33–39] Internal findings suggestive of chronic drug use include lymphadenopathy (specifically in the porta hepatis area), cirrhosis, and hepatosplenomegaly, and, much less commonly, bacterial endocarditis (Fig. 14.7).[13–15,39] While it has classically been taught that right-sided endocarditis is more diagnostic of chronic drug injections, left-sided heart valves are still more frequently involved with endocarditis caused by intravenous drug use.[40–47] While nonspecific, cardiac hypertrophy can be caused by chronic alcohol and/or stimulant use (e.g., cocaine).[48–52] With the injection of crushed pills, one may see a vasocentric foreign-body giant cell granulomatous reaction in the lung ("shooter's lung," "talkies," "excipient lung disease"), with resultant pulmonary hypertension and cor pulmonale.[53–60]

Histopathologic examination of drug intoxication deaths generally serves to confirm the gross physical findings. Examination of the lungs, especially for polarizable material, may be particularly productive, especially when there is suspicion of intravenous misuse of narcotic pharmaceutical medications.[53,55–58] Crospovidone is an insoluble polymer of N-vinyl-2-pyrrolidone that is used as a disintegrant in pharmaceutical tablets. It can embolize to the lungs when tablet suspensions are injected intravenously.[56] Crospovidone appears as deeply basophilic, coral-like particles (Fig. 14.8) within pulmonary arteries and in extravascular foreign body granulomata. It demonstrates intense staining with mucicarmine, Congo red, and Masson trichrome. Foreign body granulomata and giant cells

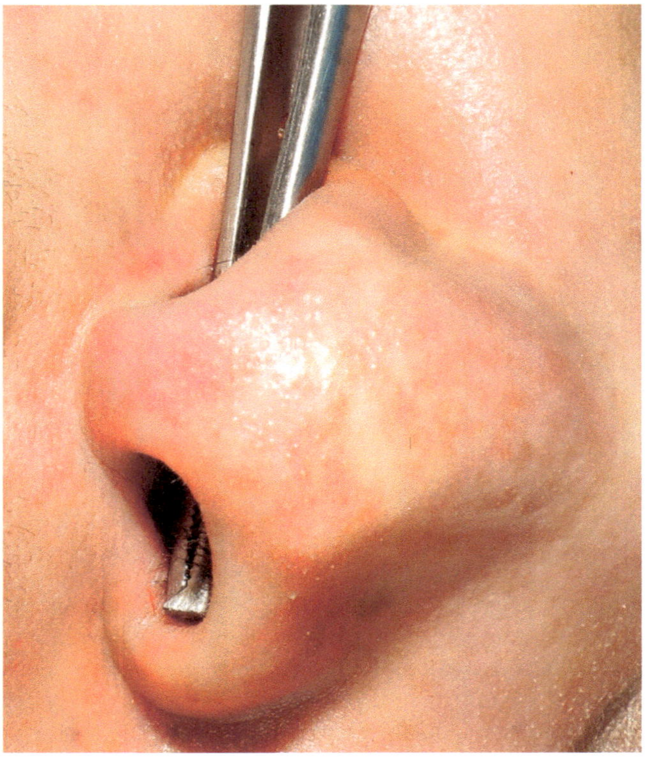

FIGURE 14.3 **Perforated nasal septum in a chronic cocaine user.**

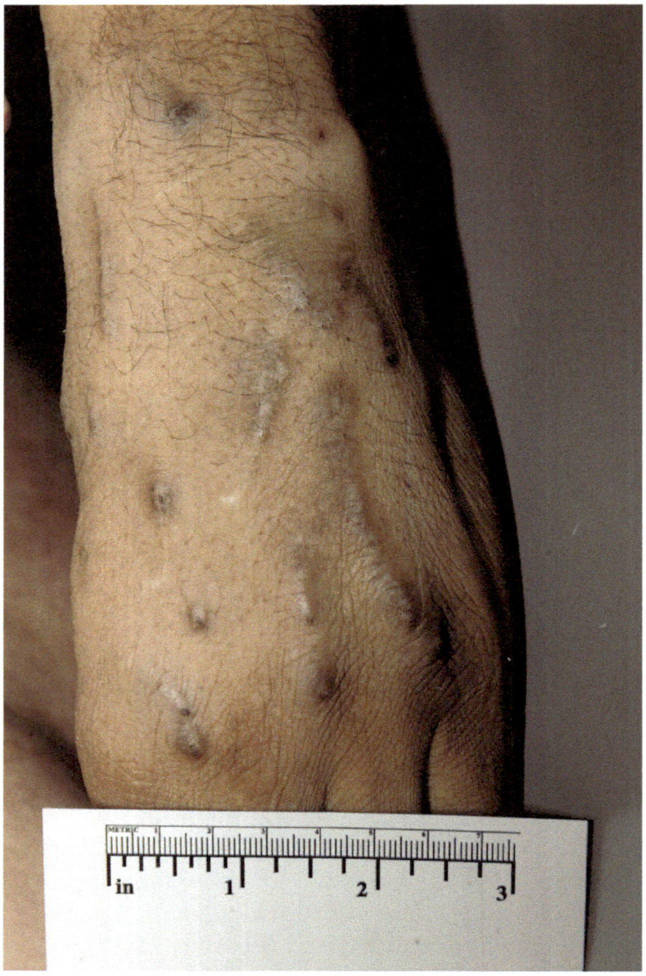

FIGURE 14.4 Cutaneous stigmata of chronic substance use: ("track Marks"): linear scars overlying subcutaneous veins.

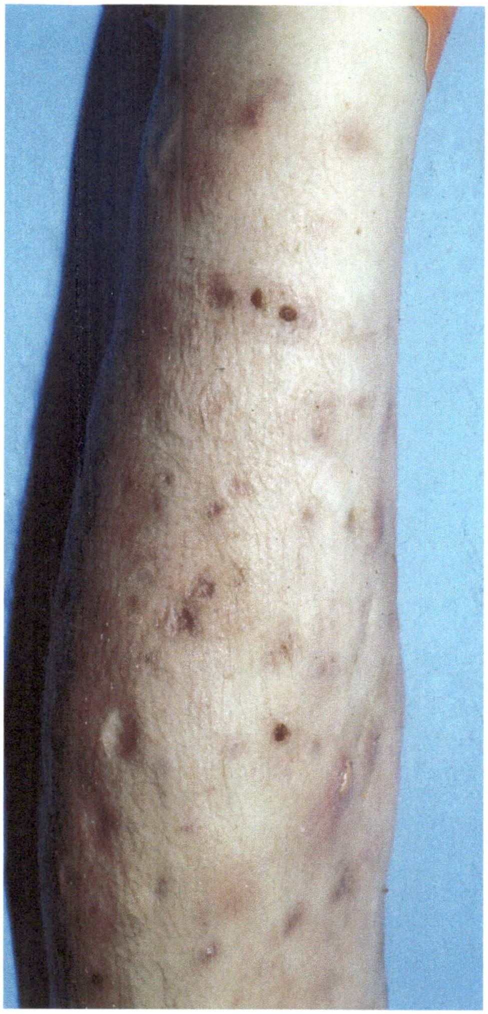

FIGURE 14.5 Skin-popping lesions: recent and remote injection sites in the subcutaneous tissue.

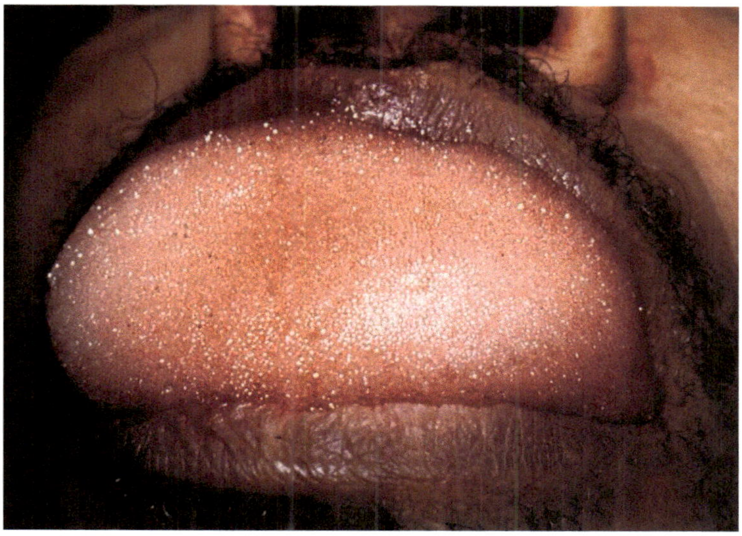

FIGURE 14.6 "Foam Cone": pulmonary edema emanating from the mouth.

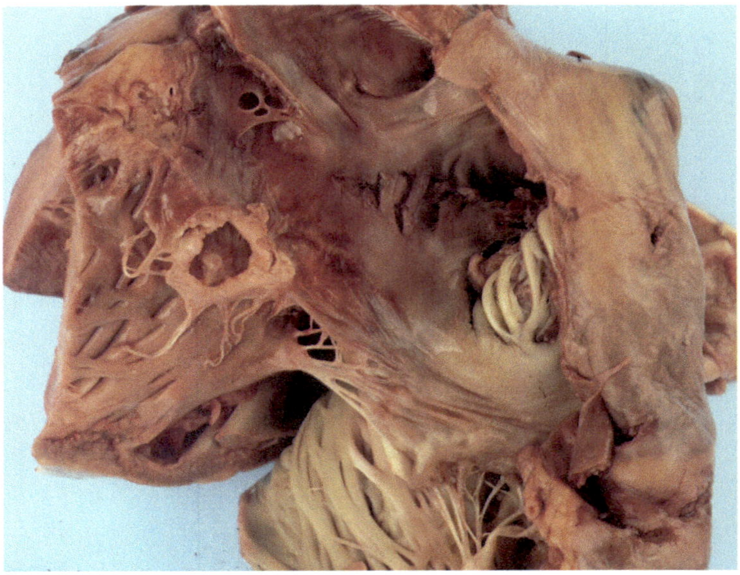

FIGURE 14.7 Bacterial endocarditis: tricuspid valve with vegetation and ulceration.

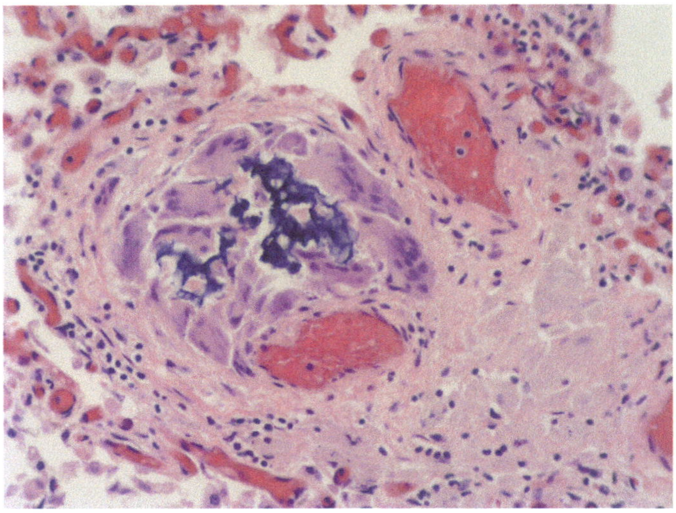

FIGURE 14.8 The basophilic/blue material is Crospovidone. Hematoxyline and Eosin 100X

may be seen in the lung, but also at the injection site. The perivascular distribution of the foreign material distinguishes it from inhalation processes, which would have a more bronchocentric distribution.[61–63]

Routine histological examination of the injection site usually is not necessary; however, it may be useful to reassure concerned family members that the decedent was not surreptitiously poisoned, as there is evidence of prior voluntary use. Microscopy also may reveal patchy acute inflammation ("early bronchopneumonia"), which may indicate a more protracted demise, as may be seen in some opioid deaths.[64] That said, agonal aspiration should not be confused for the cause of death.

Toxicology testing

Autopsy samples

At autopsy, blood (both cardiac and femoral), vitreous humor, and urine are standard toxicology specimens. Blood should be collected with sodium fluoride (NaF) to prevent the in vitro enzymatic breakdown of cocaine

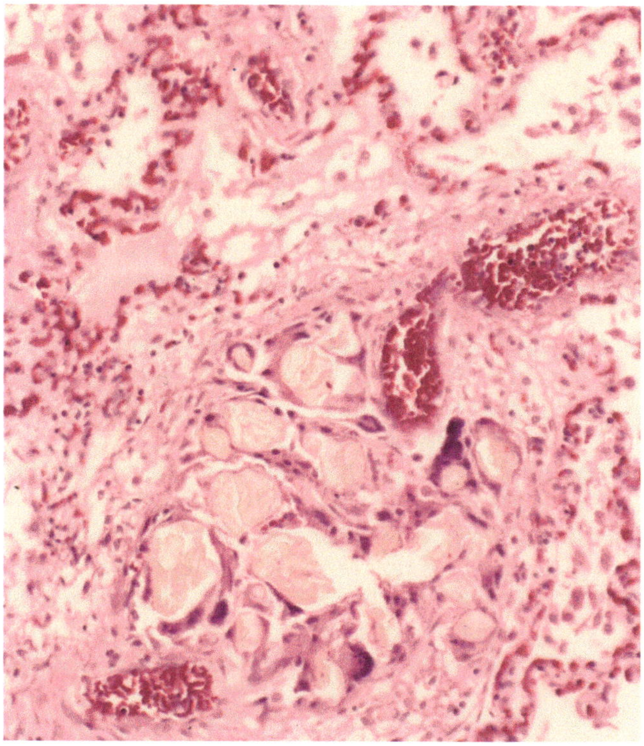

FIGURE 14.9 Excipient lung disease: foreign-body giant cells with polarizable foreign material with a vascular distribution.

and the production of ethanol. A single gray top blood specimen tube provides enough sample to permit a robust analysis of hundreds of licit and illicit drugs and chemicals. Additional specimens are routinely collected to allow for specialized additional testing, if needed. Liver, brain, gastric contents, bile, kidney, subdural blood, and hair each have their own unique applications.[65–97] The scope of toxicological analyses will vary from laboratory to laboratory, and it is imperative that the pathologist be aware of their scope and limitations. The National Association of Medical Examiners has recommended a panel for standard toxicology testing.[9] Additionally, some medical examiner offices stratify toxicology testing with basic and expanded testing, when indicated. The nature of testing is directed by the forensic pathologist and will be, in large measure, driven by the specific circumstances of the death. More comprehensive testing is usually indicated in all manners of suspected intoxication deaths, particularly in younger decedents.

Of the potentially lethal drug classes, the **opioid class** *is currently the most frequently encountered in the United States.* In general, opioids produce analgesia in a dose-related fashion, but cause significant central nervous system (CNS) and respiratory depression, which may show less of a dose-related response. *Typical findings* at an opioid death scene include several of the items mentioned above. In some instances, the scene may have been altered by individuals who had also been using drugs or by family or other concerned individuals who may have attempted to obscure the nature of the decedent's death. Not uncommonly, individuals who were present with the decedent prior to death report that they heard "loud snoring" before later finding them dead.

Other less common drugs may also have deleterious pathological effects.[98,99] For example, **phencyclidine (PCP)** is associated with cardiac dysrhythmias, hypertension, agitation, and seizures. There is some debate within the forensics community as to the lethality of this drug, however, given its arrhythmogenic tendency, it may be included. It also may result in death through a hyperactive delirium.[98,99] PCP blocks dopamine uptake, is lipid soluble, and has a high volume of distribution. It may be detected for days in blood and for many weeks in urine, and therefore, somewhat similar to cannabinoids, its detection indicates that the decedent did at one time take PCP, but the concentration is notoriously difficult to interpret.[98] False positives for PCP may be seen in the urine due to the presence of tramadol, ketamine, diphenhydramine, dextromethorphan, and cathinones.[98,100–103] In drug-related deaths, PCP is most commonly seen in combination with other drugs. When PCP is detected in isolation, comorbidities are likely present.[100]

Death certification

Employing the general investigative framework outlined above, information is generated as part of a comprehensive analysis that allows the forensic pathologist to arrive at the most accurate death certification (see Chapter 3). It is always worth remembering that the detection of licit or illicit substances does not necessarily mean that they caused the death. Reference textbooks and tables that provide potential "toxic" and "lethal" drug concentrations are valuable for interpreting results in suspected drug intoxication deaths,[96,104–108] but they are not a substitute for medical judgement. Survival intervals, postmortem redistribution, tolerance, underlying medical conditions, synergistic effects, and a mechanism that is incompatible with the putative drug, based on a compelling witness information must always be considered. The most important of these variables, however, is the identification of an **efficient intervening cause**, particularly when that cause is incompatible with continued life. The finding of a ruptured myocardial infarct at autopsy, for example, would render the detection of diacetylmorphine (heroin) irrelevant. In fact, given the therapeutic effects morphine is expected to have in treating cardiac ischemia due to occlusive coronary artery disease, it would be arguable that the decedent had not taken enough! If, on the other hand, cocaine had been detected, its likely mechanistic contribution would have skewed the certification in the direction of a drug-related death (accident). That is all to say that exceptions to published tables citing "lethal" concentrations of drugs do exist and need to be considered in the final analysis of any potential drug intoxication death.

There are numerous adverse reactions to drugs that can cause death. Therefore, not all accidental intoxications are "overdoses," per se. The variety of mechanisms by which a drug or a combination of drugs can cause death include: sympathomimetic effects, such as arrhythmias (from stimulant drugs in any amount, for example), myriad known side effects, drug interactions, allergic reactions, idiosyncratic reactions, intolerance (pharmacogenetics), and overdoses.[109,110]

Opioids

There are some drugs that have considerable lethal potential as sole intoxicants (e.g., cocaine or fentanyl), while others are more likely to be contributory or part of a mixed drug intoxication, rather than individually lethal (e.g., benzodiazepines, ethanol). At this point, it is helpful to consider specific drug classes and their lethal potentials. Of the likely individually lethal drugs, the **opioid class** is historically among the most frequently implicated.[1,9,111–120] In general, opioids produce analgesia in a dose-related fashion, but also have the potential to cause life-threatening CNS and respiratory depression. It is important to note that while typical findings at an opioid death scene include those outlined above, it is also possible that drugs used at a different location may be responsible for or contribute to an individual's death.

In general, as opioid use may result in CNS depression, historical information concerning the decedent's behavior prior to discovery should be sought, as it may influence the final death determination. For example, an individual could suddenly collapse with low to moderate concentrations of opioids in their system, but may not necessarily be dying of an "overdose" of opioids. Concomitant natural disease may be given greater weight than the drugs, depending on circumstances that may be inconsistent with an opioid death. The concentrations of opioids, while useful as guidelines, may be markedly low in individuals who die in a lingering fashion, or unusually high in individuals with history of chronic abuse and a high tolerance. This needs to be considered when determining the cause and manner of death.[64]

Stimulants

Another class of *potentially lethal drugs as sole intoxicants* are the stimulants, especially **cocaine and the amphetamine-related drugs**.[3,37,121–123] Mechanistically, these drugs are fundamentally different from the opioids in their mechanism of lethal action in intoxication fatalities. These drugs typically produce death more precipitously, often by cardiovascular disturbances (cardiac dysrhythmia, hypertensive-related hemorrhages) or seizure, and are *not* necessarily dose-dependent in their effect. Even a small amount of cocaine or methamphetamine is, in itself, potentially life-threatening. Cocaine and amphetamines cause sympathomimetic signs and symptoms, such as tachycardia, dysrhythmias, vasospasm, and elevations of blood pressure. These drugs also may have a considerable interplay with natural diseases, in that they are likely to put stress on an already compromised cardiovascular system, enhancing the likelihood of arrhythmias, coronary artery vasospasms, and subsequent myocardial ischemia. Elevations in blood pressure may lead to myocardial and intracerebral infarcts and ruptures of cerebral artery aneurysms or aortic dissections, etc.[26,49–51,124–145]

During analysis of the toxicological results, *one needs to distinguish pharmacologically active substances from inactive metabolites.* This is particularly true of cocaine, which has numerous potential pathophysiologic impacts, but the main metabolite, benzoylecgonine (BE), is considered inactive.[142,146,147] In most instances, the diagnosis of an acute cocaine intoxication should demonstrate *parent* cocaine *in the blood*, not just a metabolite.[142] The half-life of cocaine is about 1 to 1.5 hours, while that of BE is 6 h. Therefore, BE will be detected in the body long after cocaine has been metabolized and is no longer having a pharmacologic effect.

An acute cocaine intoxication diagnosis should not rest exclusively on the detection of parent cocaine in the blood; there are some exceptions. An example would be a *decomposed decedent* in which only BE is detected. As cocaine continues to breakdown after death, one is much less likely to detect the parent compound, even if the person indeed died from its acute effects. This is where the circumstances may be determinative. If the circumstances are compelling (e.g., found dead with a crack pipe in hand), one may certify the death as an acute cocaine intoxication, despite the fact that only BE was detected in the blood. That said, an analysis of brain tissue may prove helpful in some instances where acute cocaine intoxication is suspected, but blood testing is only positive for BE.[82,148]

Another scenario in which parent cocaine may likely be at least partially, if not solely culpable in a death, but no longer detectable in the autopsy blood, involves a decedent who dies from a cocaine-induced pathophysiologic process that started hours to days before death supervened. Such a person may have been witnessed to use cocaine, for example, shortly after which they developed chest pain or a severe headache, only to immediately seek medical attention and be diagnosed with a myocardial infarction, a basal ganglia hemorrhagic stroke, or a ruptured saccular or aortic aneurysm, and then die some hours to days later, after having been sustained on life support. While the survival interval in such deaths would explain the absence of parent cocaine being detected in autopsy toxicological analysis, the circumstantial evidence and temporal proximity of its use to the onset of symptoms would not absolve the cocaine of its overwhelmingly likely role in the fatal outcome. Such determinations allow common sense to prevail.

Such **delayed cocaine deaths** are well described in the literature, but not all are as straightforward as the case illustrated above.[51,52,132,138,141,145,149] BE, without parent cocaine, may be detected in some deaths with a clear cardiovascular mechanism, but with less of a known timeline of cocaine use than was described above. For certification purposes, these deaths can be challenging. Some FPs may ignore the BE altogether and certify the death as a natural cardiovascular death. Others might put "chronic cocaine use" in part 2 (with a natural manner), and still others might include acute cocaine intoxication in parts 1 or 2 (accident). Again, the history can be very important in making these decisions. If there is a history of the onset of chest pain while using cocaine two days prior to death, similar to the above scenario, then invoking an acute intoxication in the pathogenesis of a myocardial infarct is reasonable. Similarly, if the decedent was an otherwise healthy 25-year-old with no alternate causal explanation, an acute intoxication certification would, again, likely be reasonable. The challenge comes with older individuals with extensive underlying cardiovascular disease and an unclear temporal history of symptoms and cocaine use.[131,142] These delayed cocaine deaths are predicated either on advanced life support measures and/or on lethal mechanisms having the capacity to develop over hours or days (e.g., ruptured myocardial infarcts represent a delayed fatal mechanism following the likely initial symptom onset some days earlier, at which time cocaine could have been the instigator, but is now represented only by the residual BE). A similar timeline may be encountered with sentinel hemorrhages in the basal ganglia or berry aneurysms that follow cocaine use, but sluggishly evolve into a larger, ultimately fatal hemorrhages hours to days later, at which time the cocaine has metabolized.

More often than not, **more than one drug** are involved in a lethal intoxication. It is the exceptional case where one may distinguish the pharmacologic effects of only one drug within a given class (such as opioids), and thus exclude another from final certification consideration. Most often, the presence of drug mixtures will necessitate the inclusion of all relevant drugs of significance on the death certificate. It is neither optimally communicative nor standard in these situations to list broad classes of drugs (e.g., opioids, stimulants), rather than listing all the specific drugs involved. This specificity has tremendous practical significance from a public health and epidemiological standpoint.[9,114,116,150–163]

Among the cointoxicating substances that are frequently present are contributory drugs which, in and of themselves, would not likely cause death. First among these in frequency are the **benzodiazepines**, a class of sedative-hypnotic drugs. The safety profile for benzodiazepines is large and these drugs rarely cause death in isolation.[164,165] They do, however, potentiate the depressant effects of opioids and are reasonably considered as contributory to death when both drugs are present. Similar observations may be made for ethanol, which is rarely lethal to the chronic substance user, however, may be toxic to the point of lethality in a novice drinker. Alcohol and benzodiazepines are a well-recognized potentially lethal mixture.[164,166–168] Other drugs in this class of relatively safe compounds, and often seen in association with depressant and stimulant drugs, include antidepressant and antipsychotic medications, muscle relaxants, and sleep aids.[165] To the extent that these have depressant or arrhythmogenic pharmacologic effects, they may be considered additive to the effects of other intoxicants. It is most helpful to adopt a global view of the effects of substances on physiology and pathophysiology in death certification in order to arrive at an accurate and mechanism-based diagnosis.

The cannabinoids are another class of drugs that are frequently detected in the illicit drug-using population. Although **cannabinoids** affect multiple receptors in the brain, they are generally not lethal by themselves, nor are they likely to contribute significantly to lethality in combination with other lethal toxic substances. These must not be confused with the **synthetic cannabinoids**, an entirely different class of drugs that have emerged over the last decade or more, and may have more lethal potential that involves arrhythmogenicity, hypertension, seizure activity, delirium, and/or hyperthermia, to name only a few (see further discussion below).[169–180]

As noted above, drug intoxications involving multiple substances are frequently encountered by the forensic pathologist.[181] The following case scenarios provide a systematic approach to the interpretation of toxicology testing, but a detailed discussion of toxicology testing methodologies exceeds the scope of this chapter. It is assumed in the proceeding case scenarios that a scene investigation, clinical history, and autopsy findings are consistent with an apparent drug intoxication.

Examples of drug intoxications

The majority of drug-related fatalities involve either potent opioids, such as fentanyl or heroin, and/or stimulants such as cocaine or methamphetamine. For the purpose of discussion, the term "opioids" will be used, in general, to include all drugs that act on the opioid mu receptor, therefore including the natural opiates, such as morphine and codeine, the semisynthetics, such as heroin (diacetylmorphine), and the vast array of synthetic opioids, such as fentanyl/fentanyl analogues, oxycodone, and hydrocodone. Opioids can exert potent depressant effects that include sedation and respiratory depression, whereas stimulants can result in cardiovascular complications and seizures. Any of these effects has the potential to be fatal. It is not uncommon for individuals to use opioids and stimulants together, however, increasing the likelihood of a fatal outcome. In fact, as already stated, most drug-related deaths do not involve a single drug toxicity, but instead involve various combinations of drugs.[182–189] Examples of some of the more common drug combinations encountered during medicolegal death investigation follow.

Opioids, benzodiazepines, and other depressants

Potent opioids, such as heroin and fentanyl, can cause death alone via their respiratory depressant and sedative effects. **Benzodiazepines** are a class of widely used (and misused) sedative hypnotics which are considered significantly less lethal, yet are sometimes detected as cointoxicants with various opioids. Benzodiazepines can potentiate the sedative and respiratory depressant effects of opioids. These deaths are appropriately certified as combined intoxications involving particular opioid(s) and benzodiazepine(s). If other drugs are detected and are capable of causing sedation and/or respiratory depression, such as muscle relaxants or sleep medications, it would be appropriate to include them in the cause of death as well. Such drugs may include ethanol, diphenhydramine, gabapentin, and zolpidem. The reason to include *all* relevant drugs in the cause of death statement is to capture the potential *cumulative* effects of the specific drug combination. In addition, it is useful for public health surveillance. In contrast to many single-drug deaths, in multiple intoxicant deaths, the concentrations of the individual drugs may seem relatively unimpressive (at least per standard drug concentration reference publications). When one considers the cumulative effects of the different drugs, however, the overall drug burden on the individual becomes more consequential, as do the potential drug interactions, including synergism. Relatively low drug concentrations will also occasionally be detected in some drug-related fatalities. This may occur if the individual lingered for a prolonged period of time, in a comatose state, while metabolizing drugs to lower concentrations before eventually succumbing to them.[64]

Diphenhydramine is an antihistamine drug which has sedating properties and is commercially available as an over-the-counter treatment for allergies and cold and flu-like symptoms. It also is the active ingredient in some over-the-counter sleep aids. Thus, it may be seen incidentally in suicidal deaths, given that depression may result in insomnia. It also is used as a "cutting agent" in some illicit drug mixtures to dilute the drug and possibly to mitigate the pruritic effects of opioids. While it may be lethal in individuals who ingest it with suicidal intent, the safety margin of diphenhydramine is relatively high. Diphenhydramine would be expected to potentiate the physiologic effects of CNS depressants such as fentanyl, and therefore it would be reasonable to include diphenhydramine in intoxication deaths involving depressant-type drugs.[190] Similarly, **xylazine**, an animal sedative (not an opioid), is an increasingly detected adulterant in illicit drug mixtures and should also be included in the death certification, given its potentiating effects combined with its unresponsiveness to naloxone administration.

Gabapentin is a structurally distinctive medication used in the treatment of seizures, as well as neuralgia. It has become increasingly popular as an off-label medication for the treatment of chronic pain. Gabapentin and related gabapentinoids are capable of potentiating the respiratory and CNS depressive effects of opioids, and as such, should be included in any cause of death statement where it is encountered in this kind of an intoxication.[191–193]

The detection of prescribed legal substances at therapeutic concentrations in postmortem toxicology testing should not be considered, in and of itself, a criterion for their exclusion in a cause of death statement. A certifier is on more defensible ground in drug intoxication deaths to consider the physiologic/mechanistic effects of any substance through the prism of its overall lethal potential and base their decision for the inclusion or exclusion of therapeutic agents on this.

The combination of opioids and stimulants

Opioids, such as **fentanyl and heroin,** primarily act as depressants, whereas cocaine and methamphetamine primarily act as stimulants. These two classes of drugs have very different counteracting physiological effects, yet are not uncommonly detected together. This is in part owing to the widespread use of these two common classes of drugs due to their consumption individually or from the consumption of a mixture of drug material that contains both.[118,181,184,194,195] **Amphetamine** may be detected in licit form or as a result of the metabolism of methamphetamine.

The effects of these two different classes of drugs (opioids and stimulants) should not be considered as mutually exclusive, as they are not uncommonly detected in combination, with potentially cumulative effects. The potentially counteracting effects of these two classes of drugs also do not necessarily render them less toxic. Both have physiologic effects that act in opposition to each other, but in the absence of any convincing clinical history, in practice, it is difficult to distinguish which of these drugs had a more primary toxic/lethal effect on a decedent. This is particularly true of individuals who are found unresponsive with both drugs detected in their postmortem toxicological testing. It may often be speculative and arbitrary to rely on vague historical or nonspecific autopsy findings to discern the relative importance of one substance over the other. In practice, it is reasonable to include both classes of substances on a cause of death statement, certifying these deaths as intoxications by their combined effects.

Methamphetamine occurs in two isomers. *D-methamphetamine* is usually illicit, has significant CNS effects, a high addiction potential, and causes vasoconstriction and stimulant effects on the cardiovascular system. ***L-methamphetamine*** is in some over-the-counter decongestants and has only weak CNS effects. **Amphetamine** is prescribed as Benzedrine or Dexedrine. Methamphetamine is an abused stimulant of choice in some areas of the United States and may be smoked, snorted, or injected.[121,196–199] It has been described in gay communities.[121,196,197,200–202] Other drugs (e.g., amphetaminil, benzphetamine, clobenzorex) may metabolize to methamphetamine, leading to the potential for improper interpretation of the source.[203–205] Methamphetamine metabolizes to amphetamine, generally in the ratio of 10:1. If the concentrations do not approximate this, consider independent exposure to amphetamine or other drugs producing amphetamine and methamphetamine. Consultation with a toxicologist is strongly encouraged with methamphetamine deaths, particularly when there are deviations from the expected 10:1 methamphetamine/amphetamine ratio, when multiple drugs are present, or when findings are not straightforward.[204–206]

Novel (new) psychoactive substances

A novel psychoactive substance (NPS) is a manufactured chemical or a synthetic drug analog that may be used in an abusive manner.[95,102,179,207–241] The more commonly encountered NPSs are the synthetic opioids, which include **fentanyl analogs** such as *fluorofentanyl, acetylfentanyl, and carfentanil.*[195,242–248] The challenge of NPSs stems from the fact that data regarding their effects and interactions are limited because of their lack of widespread use. Additionally, as new compounds come on the illicit market, it may take time for the toxicology laboratories to detect them. It is reasonable for the death investigator to review these data, as well as to extrapolate as to the effects of the NPS based on their structural similarity to other known compounds. An example is fluorofentanyl, which is recognized as pharmacologically and physiologically similar to the base compound of fentanyl. As such, it would be anticipated to have significant CNS and respiratory depressant effects. Therefore, the expectation would be that fluorofentanyl and fentanyl exert cumulative effects in an intoxication setting. Certification of a death in this particular scenario could reasonably be offered as a mixed intoxication due to the combined effects of fentanyl and fluorofentanyl.

Perhaps owing to the quality of various manufacturing processes of illicit fentanyl, other related, physiologically inactive substances may be detected, such as beta-hydroxyfentanyl and propionyl fentanyl (4-ANPP), which may reflect a precursor chemical in the manufacturing process and/or a metabolic product (along with the metabolite norfentanyl). As such, only the main drug, fentanyl, needs to be considered in the cause of death statement.[88,249–251]

Opioids and natural and synthetic cannabinoids

On occasion, a suspected intoxication victim is noted on postmortem toxicology to have both an opioid, such as fentanyl, and cannabinoids. Cannabinoids are among the most used illicit substances and, with their variable degree of legalization, can be expected to be detected with increasing frequency as part of postmortem toxicology testing. Cannabinoids, in and of themselves, are rarely lethal, and their physiologic effects do not potentiate the toxic effects of opioids. The certification of cause of death in this instance would be appropriately attributed to the opioid(s) alone. The inclusion or exclusion of cannabinoids (or any other drug) on the basis of the legal status of the substance rather than its deleterious pharmacologic effects is illogical and best avoided as a practice.

Synthetic cannabinoids are unique, manufactured chemicals that exert their effects through the tetrahydrocannabinol (THC) receptors in the body, causing variable, but potentially more toxic effects than naturally occurring cannabinoids.[169,170,172,173,175,176,179,180,231,252–269] The popularity of synthetic cannabinoids has fluctuated over the last decade, but instances of toxic fatal effects have been reported.[169–180] Most often these toxic effects are attributed to the arrhythmogenicity and/or convulsive potential of these compounds. That said, the toxic effects of synthetic cannabinoids are rare, and it is reasonable in practice to exclude synthetic cannabinoids from a certification in a scenario such as the one illustrated above, given the overwhelming likelihood that the victim's death is due to a significant opioid burden (fentanyl).[169–180]

Methadone

Methadone is a synthetic, long-acting ($T^{1/2} = 15$ h) opioid agonist used in opioid maintenance therapy for opioid dependence, as well as for chronic pain management.[39,185,270–273] Prescribed for daily use, this medication relieves cravings and removes withdrawal symptoms.[274] It may cause QT prolongation and also elicit CYP450 drug interactions.[275] For methadone maintenance therapy, a patient may be started at 20–40 mg per day, typically increasing until the patient "feels" well. Most are maintained on 60–80 mg per day, but some may exceed 200 mg. Tolerance combined with therapeutic effects will drive dosing. A 40 mg dose of methadone may be fatal to an opioid naïve person. Rifampin and ethanol will speed up the metabolism of methadone, while fluconazole (which can inhibit cytochrome P450-mediated drug metabolism) will increase systemic methadone concentrations.[276–278] Some patients also may take amitriptyline in order to potentiate the effects of methadone. Methadone is not detected on routine opiate screens. As such, as part of the program, patients submit to random urine screens. If their urine is "clean," they may be given a week's supply of methadone to take home, instead of having to come in every day to get their dose.

Largely due to tolerance, there is marked overlap between what is considered a "therapeutic" versus a "fatal" methadone concentration.[33,273,275,279–289] Therefore, caution must be applied when considering the role of methadone in a fatality. While there are many variables to consider in the certification of these deaths, there are three broad scenarios that the forensic pathologist will likely encounter. Consider a 56-year-old man with a history of chronic substance use and hypertensive cardiovascular disease, who is found dead in bed. How should the interplay between a methadone maintenance program and the toxicology results affect the certification of the cause and manner of death?

Three scenarios:

1. In a methadone-maintenance program, heart weighs 550 g, Methadone = 1.1 mg/L
2. In a methadone-maintenance program, heart weighs 550 g, Methadone = 1.1 mg/L, BAC 0.25 gm%
3. NOT in a methadone-maintenance program, heart weighs 550 g, Methadone = 1.1 mg/L

Although the methadone in decedent 1 is on the "high" side, we would give him the benefit of the doubt and certify the death as natural due to the heart disease. This opinion is predicated on the assumption that a reasonable degree of tolerance developed from the daily administration of methadone. For decedent 2, the addition of the ethanol would likely move us to an acute intoxication death (accident), the methadone no longer being taken "as prescribed." For decedent 3, the death might be certified as an acute intoxication (accident) with the heart disease in part 2. This represents a general approach; however, other unique factors may result in different certifications. For example, if there was marked pulmonary edema and the daily doses of methadone had already been consumed ahead of schedule, an intoxication certification of death for decedent 1 would be appropriate. Few intoxication deaths require more context than those related to methadone.

The effect of drug metabolism on toxicology results

Drug metabolism can make the identification of the parent drug(s) challenging. For example, the metabolism of some benzodiazepines results in the generation of other "prescribed" drugs in this class, but the certification should be limited, to the best of the certifier's ability, to the parent benzodiazepine that is involved, as opposed to including multiple benzodiazepines on the death certificate, unless of course, there is evidence from the death scene investigation to support their inclusion.[164,290,291] An example would involve the detection of nordiazepam, temazepam, and oxazepam, all of which are known metabolites of diazepam even though temazepam and oxazepam are themselves marketed benzodiazepines.[292] Analyzing the relative concentrations of the various benzodiazepines may also assist in their interpretation (i.e., a substantially higher concentration of the one benzodiazepine relative to much lower concentrations of the others might suggest the latter representing metabolites rather than "coparents").

Another example is **heroin** (diacetylmorphine), which is rarely detected in the postmortem setting due to its short half-life, with a rapid metabolism to inactive 6-monoacetyl morphine (6-MAM), and then to the pharmacologically active morphine, the latter chemically identical to prescription/medicinal morphine.[114,158,293] As **6-MAM** is unique to heroin, in such instances, one can confidently conclude that the detection of 6-MAM (often in conjunction with morphine and codeine) signifies the use of heroin. Although sometimes detected in the blood, 6-MAM, may only be detected in the vitreous fluid and/or urine, owing to its short half-life, but its specific location is not relevant to its role as a marker of heroin.[294] Rather, its presence in any compartment serves its purpose as a tag. Even in the absence of 6-MAM, it is reasonable to consider the decedent with only morphine detected in the postmortem blood, but with heroin at the scene of death, to represent a fatal heroin intoxication. A check of the decedent's prescription drug monitoring site profile also may also help to exclude a morphine prescription, further reinforcing the original impression. Detection of codeine in a concentration below that of morphine, also is indicative of a heroin intoxication. Codeine will metabolize to morphine, but morphine does not metabolize to codeine.[65,114,158,293,295–302] During the extraction and synthesis of heroin, a small amount of codeine may remain. Therefore, if the morphine concentration is greater than the codeine concentration, a heroin intoxication would be favored. If the codeine concentration is greater than that of morphine, it is more likely to represent the use of codeine with a subsequent breakdown to morphine.[114,119,158,303]

Postmortem drug redistribution and breakdown

The detection and quantitation of opioids, benzodiazepines, and other drugs helps determine the potential toxicological burden that a person may have experienced. One must keep in mind, however, that the drug concentrations detected postmortem will vary to some extent from the drug concentrations at the moment of death. This is in part due to the effects of **postmortem drug redistribution**, the process by which drugs can diffuse down concentration gradients in the postmortem period.[94,107,250,304–321] *The extent to which drugs are capable of undergoing postmortem redistribution varies according to the chemical structure of the particular drug.* Fentanyl, for example, has been shown to be capable of undergoing considerable postmortem redistribution.[307,319,322] Also, one should consider that clonazepam and other benzodiazepines can break down rapidly postmortem.[323] Exposure to sunlight of serum containing clonazepam can cause up to 99% loss of the drug within one hour.[324] Sometimes only the inactive drug metabolite will be detected, even though the active parent drug may have been in the decedent's blood at the time of death. Cocaine also breaks down postmortem, highlighting the importance of adding a preservative such as NaF to specimen samples.[325,326] This is important to remember when testing hospital blood samples that may not be preserved with NaF (i.e., in a gray top tube). In these instances, cocaine in blood collected in a non-NaF tube will continue to breakdown and may not be detected by later testing, but cocaine may be detected in the NaF-autopsy sample if cocaine was still in the body at the time of death.

Natural disease in the presence of intoxicants

Since a substantial percentage of victims of drug intoxication are middle aged or older, it is not surprising that a proportion of them will have advanced coexisting disease.[189] As such, the determination of cause and manner of death may be challenging.[327–329] As noted previously, the detection of some drugs may exacerbate the pathophysiology of certain medical conditions.

There are certain anatomical findings that are incompatible with continued life (Class I), and, as such, they are typically viewed as trumping any potential intoxication. For example, a saddle pulmonary thromboembolism is a condition which would render most toxicological findings moot. The detection of such categorically lethal natural conditions in autopsy work is, however, infrequent.[330] Nevertheless, when these findings are encountered, they constitute the cause of death and most toxicological substances can be ignored. The same considerations hold true for trauma-related deaths (e.g., a gunshot wound of the head). All that said, there is one situation in which a Class I finding would directly result from a toxicological cause: when stimulants (most commonly, cocaine) cause such conditions as a ruptured aortic dissection, a ruptured myocardial infarct, a ruptured berry aneurysm, or a basal ganglia or pontine hemorrhage. In these instances, the underlying (or part 2) cause of death would be "Acute Cocaine Intoxication" and the manner, Accident (see below).

In other situations, a review of the clinical history may clarify the role of disease in the setting of the intoxication. For example, an individual with advanced coronary artery atherosclerosis who collapses suddenly, but with opioids detected in their postmortem toxicology, would reasonably be certified as a natural death secondary to ischemic heart disease. It is less reasonable in this setting to ascribe the death to opioids (either as a primary or contributory factor), as the sudden collapse is a typical clinical presentation of a sudden cardiac event, whereas a fatal opioid intoxication does not typically present as a sudden collapse. In addition, opioids (e.g., morphine) are commonly used in the treatment of acute myocardial infarcts. Some may argue that the respiratory effects of opioids could theoretically make a physiologic contribution; however, given the sudden nature of the death, this is unlikely. If, however, in this same scenario, a stimulant, such as cocaine or methamphetamine, is detected, the stimulant would be reasonably included in the cause of death statement, as it potentiates the effects of coronary artery atherosclerosis and myocardial ischemia via vasospasm, tachycardia, and hypertensive changes, all of which affect the heart's supply and demand for oxygen.

In general, stimulants are usually considered as contributory factors in the exacerbation of diseases in which blood pressure elevation or vascular ischemia is an underlying pathophysiologic mechanism. This would include atherosclerotic and hypertensive strokes, ruptured vascular aneurysms, aortic dissections, and cardiac dysrhythmias, both ischemic and nonischemic etiologies. While it is preferable in these situations to identify pharmacologically active stimulant substances, consideration should be given to inactive metabolites such as BE, as the final lethal mechanism in these circumstances may evolve over time (see above).

Another potential interface between a disease and an intoxication involves diabetes mellitus with ketoacidosis. The ketosis-prone diabetic patient is at risk for developing complications secondary to neglect of therapy during an intoxication. In the most general sense, it is reasonable to interpret the lethal pathophysiology of diabetic ketoacidosis in the light of the physiologic effects of the intoxicants involved. Thus, in diabetic ketoacidosis, where stimulants such as cocaine or methamphetamine are detected, it is reasonable to include them as contributory factors in a cause of death statement, as the terminal mechanism of diabetic ketoacidosis involves lethal cardiac arrhythmias, which would be exacerbated by the presence of stimulant drugs. Perhaps more importantly, however, is that, as a little known risk factor for diabetic ketoacidosis (DKA), acute cocaine intoxications, in particular, offset hormonal regulation in such a way as to trigger ketosis, the associated pathophysiologic stress of the intoxication apparently having the potential of creating an endocrinologic tipping point in diabetics. In retrospective reviews of large urban medical centers, those patients with acute or recent cocaine intoxications represented as many as 14%–26% of all DKA admissions.[331–333] It is believed that in addition to associated lifestyle dysfunction that potentially exacerbates medication noncompliance, cocaine also serves as an independent risk factor of DKA. As a stimulant, it augments counter-regulatory hormones, such as catecholamines, that acutely inhibit insulin secretion, thereby increasing hyperglycemia, glucagon, and ketoacid production, even in the absence of a precipitating underlying illness. Arguably, cocaine itself serves as a physiologic stressor mimicking an intercurrent infection that might otherwise precipitate DKA in diabetics. Although to a less dramatic extent, other substances, including heroin, methylenedioxymethylamphetamine (MDMA), and amphetamine all have been reported to influence insulin and glucose metabolism, thus potentially enhancing the risk of DKA.[334–338] Of relevance for certification purposes, however, is that the inclusion of any of these drugs in the cause of death statement would change the manner from natural to accident.

Depressants in the setting of diabetic ketoacidosis may similarly exacerbate the pathophysiology of diabetic ketoacidosis, though by a different mechanism. In diabetic ketoacidosis, or any metabolic acidosis, the compensatory respiratory alkalosis that the body would invoke to offset the endogenous generation of excess lactic acid, would be physiologically blunted by the respiratory depressant effects of opioids. Thus, the opioids would often be considered a contributory factor in death by diabetic ketoacidosis.

Toxicology testing and trauma deaths

In many, if not most, deaths due to mechanical injury, toxicological testing is performed, despite the fact that drugs and alcohol played no role in the mechanism of death nor are necessary for the "how injury occurred" section. Rather, the value of this practice is based on the need for thoroughness in forensic pathology practice, in order to have on hand all data that may (or may not) be useful in addressing future questions that were not necessarily foreseeable at the time of autopsy. "I would rather explain why I did something than why I did not" would apply here. That said, there are some traumatic deaths in which the death doesn't make sense and/or the manner is unclear until toxicology testing is completed. An apparently incoherent, naked man whose behavior is erratic and unpredictable jumps from a substantial height. With or without a psychiatric history, the medical examiner/coroner (ME/C) may feel compelled to defer the certification (primarily the manner determination) in such a case until psychoactive substances are excluded. If a drug such as psilocybin is detected, not only might it inform an accidental (or undetermined, as the case details may dictate) manner, but also it would be listed in the "how injury occurred" section to best communicate why the accidental (or undetermined) checkbox was chosen. Drug intoxications are typically listed in "how injury occurred" *only* when the death certification is incomplete or makes no sense without it. Traumatic suicides, motor vehicular and other transportation deaths, fatal work-related injuries, and homicides represent only a few examples of scenarios in which drug and alcohol involvement are *not* listed on the death certificate (with rare exceptions), but where questions may ultimately arise surrounding the toxicology results, even if not causal or circumstantially necessary to explain the death, per se.

Toxicology testing following hospitalization

In many instances, an intoxicated individual who is found unresponsive is hospitalized and on life support for a considerable period of time (hours, days, or even weeks) before succumbing to the complications and sequelae of a drug toxicity. During this post-intoxication survival period, their body continues to metabolize drugs, and with a prolonged survival period, little or no drug will be detectable in the body at the time of postmortem examination. In many jurisdictions, after such a prolonged hospitalization, an autopsy will not be performed, even if the ME/C takes jurisdiction. In many such instances, an antemortem urine drug screen or other such testing may have been performed during the hospitalization to assist in the antemortem treatment plan. Such test results may prove valuable in identifying which drugs may have been involved in the individual's demise. It is important to recognize, however, that in most **hospital drug testing**, not only is the analysis most typically performed only on *urine* (rather than whole blood, the specimen that most accurately defines an acute intoxication), but also the test is simply a drug screen (without confirmatory and quantitative analyses). As such, it may not distinguish parent drugs from metabolites, does not quantitate the result, is subject to potential false-positive test results, and does not meet forensic reporting standards. As a sample, urine provides an important screening tool, but does not necessarily reflect the serum drug profile. For example, a urine drug screen that is reported as positive for "cocaine" may in reality only reflect the presence of BE. While these tests may prove helpful in supporting the conclusion of a likely drug intoxication death and/or guide further testing, they are limited in specificity. In general, *forensic laboratory analysis* must include confirmation testing, in addition to a positive screen. This may involve the use of a different testing methodology and/or the testing of a second specimen.

In hospital deaths, samples of the decedent's antemortem (admission) blood may be available and procured for postmortem testing. The key is to identify and collect such *admission* blood and/or urine (whatever may be available) from the hospital *promptly*, for postmortem testing, as the specimens may be held for only a short period of time before they are discarded by hospital staff. One should note the date and time stamp of any antemortem specimens and request the "earliest" (closest to the time of admission) bloods available. This is to ensure that the bloods reflect the most acute phase of the intoxication, but also because drugs, such as fentanyl and morphine (a metabolite of heroin), and certain benzodiazepines, such as diazepam, may be therapeutically administered during treatment, and these should not be confused with or mistaken for the drugs that were consumed prior to the hospitalization and that prompted the need for medical attention. In most instances, careful review of the decedent's medical records should help distinguish drugs that were consumed prior to hospitalization from those that were administered during the hospitalization. That said, there may be instances where no admission blood is available and the hospital urine drug screen (along with the investigative and autopsy findings) provides the sole toxicological basis for the death certification. In these instances, it may be best to acknowledge the limited amount of forensic laboratory analysis by using a less specific cause of death opinion that reflects the lack of drug specificity and the possibility of shortcomings, such as certain drugs not being included in the scope of the hospital screen (e.g., "drug toxicity, including opioids").

Interpreting toxicology test results in the decomposed body

Postmortem changes can result in the alteration of various drugs and their concentrations from what was present at the time of death. These changes become gradually more pronounced as the postmortem interval lengthens. As decompositional changes become more pronounced, the identification of drugs becomes more challenging, due primarily to the poorer quality of specimens available for toxicological testing and to the postmortem degradation of drugs. Also, the drug concentrations become more altered due to postmortem drug redistribution, postmortem fluid shifts, and dehydration.[323,325,326,339,340] In general, as decompositional and autolytic processes progress, the quality of toxicological specimens worsens. As a result, with advancing decompositional changes, the value of toxicology results shifts more from quantitative in value to qualitative in value. An example would be *cocaine*, which is susceptible to significant postmortem degradation.[326] As a result, in a decomposed body, it may be that only cocaine metabolites, such as BE (and not parent cocaine), are detected. One should consider the possibility of terminal cocaine toxicity in cases with the appropriate context, even if only the cocaine metabolite(s) is/are detected and parent cocaine is absent (see above).

Microorganisms replicate in the body during decomposition and can generate a significant amount of **(endogenous) ethanol**, which should not be confused with **antemortem (exogenous) ethanol** consumption.[341,342] In some instances, it may be difficult (or impossible) to determine whether the ethanol that is detected is antemortem or postmortem in origin (or a combination of both). This determination can be aided by considering and comparing the ethanol concentrations in blood and vitreous fluid (providing that the specimens are available; vitreous often is absent with moderate to advanced putrefaction). Vitreous fluid is a sequestered fluid, which makes it relatively protected from microorganisms (and hence, postmortem ethanol generation). In instances in which ethanol is generated solely by microorganisms, the vitreous fluid would be expected to have no or very little detected ethanol. Also, the typical maximum ethanol concentration in the blood that is attributed solely to microorganisms has been reported to be about 0.07 gm/dL or so; however, significantly higher concentrations (over 0.2 gm/dL) have been reported.[340–342] As mentioned previously, one should correlate toxicology findings with case investigative information. Regarding instances with cocaine use, the detection of cocaethylene is reflective of antemortem ethanol consumption, as cocaethylene forms via an in vivo enzymatic process, and the generation of cocaethylene is not expected to result from postmortem microorganism-generated ethanol.[343]

Similarly, acetone and isopropanol are often produced in the microorganism-fueled environment of a putrefied body. The presence of these substance must also be contextualized when considering such entities as DKA or alcoholic ketoacidosis. As acetone and isopropanol may undergo conversion to one another, one must be careful to not overdiagnose or underdiagnose isopropanol ingestions.[344,345] B-hydroxybutyrate analysis may help to confirm that the acetone is related to ketoacidosis, as opposed to the result of an isopropanol ingestion (see also Chapter 7).[346–351]

Drug packers ("Mules")

A drug mule or courier is someone who personally smuggles drugs across a border. Some may ingest drug packets for temporary storage in the gastrointestinal tract during their travel before discharging their "cargo" at a safe destination. Some of these drug packets, however, may open in the gastrointestinal tract causing a fatal intoxication.[352–357]

Deaths that occur after recent international travel (e.g., at a motel near an international airport) deserve close scrutiny by the medicolegal investigator. The first step in the autopsy examination of such persons should involve radiology of the abdomen (Fig. 14.10). If drug packets are identified, proper police and evidence staff notification is initiated before the autopsy is started to allow for strict control and documentation of the recovered drug packets during the autopsy. The packets should be counted, weighed, photographed, and a sample given to the toxicology laboratory.

A study of 50 of these deaths in New York City revealed that the majority (42/50) of deaths were caused by acute intoxications due to open or leaking drug packets in the gastrointestinal tract.[355] The number of packets recovered ranged from 1 to 111 (average 46) and the weight of the combined packets ranged from 9.4 to 1200 g. There were five deaths due to intestinal obstructions/bowel perforations.

Suicide versus accident

A suicidal drug ingestion death may be suggested by the decedent's history, which may include depression, suicidal ideation, a witnessed act, a suicide note, and/or by the presence of many pills within the decedent's gastric contents. However, it is important to recognize that the intentional consumption of many pills does not necessarily yield identifiable pills in the gastric contents at autopsy due to the pills dissolving antemortem and/or continuing to dissolve postmortem. Examination of the rest of the intestine may yield relatively intact, or at least recognizable pills,

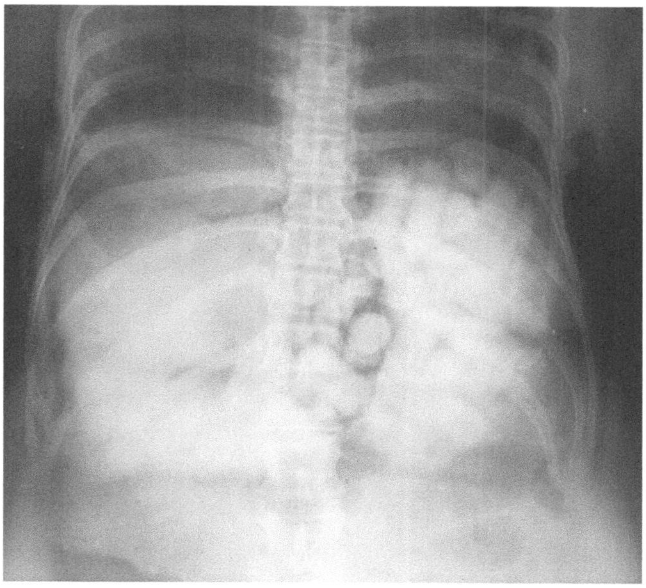

FIGURE 14.10 "Mule"/Drug packer: radiograph of the abdomen with radio-opaque drug packets in the stomach.

particularly if they were coated for sustained enteric release. That said, remember that the contents of the gastrointestinal tract still represent the equivalent of the outside world, or at least of the extravascular environment. As such, their importance lies in their potential *to establish intent* by the examination of their quality and quantity as they relate to pill burden.

In addition, a large parent drug to metabolite ratio may also be reflective of a recent and significant pill ingestion. A suicidal death may also occur with illicit drug consumption. However, in all, absent convincing evidence that the death was due to the consumption of an excessive amount of drug(s) to deliberately end one's life, the default is to list the manner of death as either accident or undetermined, depending on the details of the case.

A common misconception with suicidal ingestions is that the drug concentrations will be astronomically high. In reality, this is not always the case, particularly when one considers the variety of factors that can effect postmortem drug concentrations (e.g., prolonged survival interval, postmortem redistribution, polypharmacy, underlying disease). As in any death, one must consider the circumstances, including pill counts, and the assessment of suicide operational criteria.[358–360] *A common rule of thumb is that if one detects a drug concentration that is at least an order of magnitude higher than expected (e.g., expected 10 mg/L with detected 100 mg/L), one should consider intentional self-harm.*

As the demonstration of intent is needed to certify a death as a suicide, the detection of excess medication in the gastrointestinal tract may be useful. However, there are a few caveats that must be remembered. First, is that an elevated amount of medication in the stomach may support an intestinal suicidal ingestion, but a low amount does not necessarily exclude it. Much of the drug may have been absorbed or passed into the small bowel prior to the moment of death. For this reason, some forensic pathologists collect the small bowel contents as well. Regardless, it is probably prudent to instruct autopsy technicians to refrain from "rinsing" the opened bowel until the forensic pathologist has had a chance to evaluate it. Similarly, the absence of grossly identified pills or pill fragments in the stomach does not exclude a suicidal ingestion. Pills are meant to dissolve and so they may not be seen, or one may just see granular material/sediment (Fig. 14.11) or a clump of amorphous material (some refer to this as "play-doh"). Gastric contents with unusual colors (pink, green) also may suggest an ingestion. Some pills, however, are meant to dissolve slowly (e.g., time-release oxycodone) and may still be visible in the stomach. When whole pills are identified, we suggest that any inscriptions be noted and the pills in their entirety be photographically documented and counted before submission to the toxicology laboratory. Using a fine sieve through which the contents can be poured sometimes makes the identification of individual pills or pill fragments easier. An internet search of the tablet appearance, with or without inscriptions, may also prove helpful in identifying the medication, particularly if unusual; such an identification may also help to guide the toxicology laboratory. As a general rule of thumb, it is a good idea to allow the gastric contents to settle in a measuring cup for a duration of time while the autopsy proceeds, allowing for a later assessment and photographic documentation of a granular or chalky sediment should one develop.

FIGURE 14.11 Granular pill sediment and amorphous white pill paste.

A final caveat is the accurate measurement of the volume of the gastric contents and the appropriate sample submission. The toxicology laboratory usually will report a concentration of drug provided in the sample. Depending upon the amount of "food" present, this concentration may seem impressively high. But that concentration needs to be multiplied by the volume (or weight) of the entire gastric contents in order to extrapolate the actual number of pills (as indicated by the total milligrams) that likely remained in the stomach. One must be sure to send a representative sample for the gastric contents to the lab for testing. Some may choose to submit it entirely, but if the contents are homogeneous, a 100 mL sample may be sufficient. If only a sample is submitted, however, it is all the more important to accurately measure and document the total gastric volume and avoid estimating. Finally, before decanting off the sample, the container holding the sample should be stirred to make sure any pill material that may have settled to the bottom of the container has been resuspended.

Once the toxicology results are later available, a calculation that assists in establishing intent can sometimes be useful in discerning between an accidental and a suicidal drug intoxication death. It is as follows: **putative drug concentration (mg/L) x volume of distribution of putative drug (L/kg) x decedent's body mass (kg).** As a result of this calculation, a ballpark quantity (in milligrams) of pills that were ingested can be estimated (i.e., this calculation attempts to answer the question, "how many pills did the decedent likely take?"). To know how many physical pills this total milligram quantity would equate to, the milligram dosage of the prescribed pills would need to be known. If, for example, 5 mg pills were prescribed for a particular drug, the final milligram total, as calculated above, would need to be divided by 5 (mg) in order to determine the approximate *number of pills* the decedent would had to have ingested to achieve the fatal blood concentration determined at autopsy. To be clear, this calculation is used to help establish *intent* in cases where the manner is otherwise unclear, for any number of reasons, but the calculation has no bearing on the pathophysiology of the putative drug, per se. *This calculation is not intended to produce granular detail, but rather to reveal larger patterns.* Did the decedent likely ingest a handful or more of pills or just 3 or 4?

In nontoxicological suicides, **contextualization** is equally important. The more data points there are, the better, in order to meet our burden of proof. That said, one should not interpret any toxicology report or view any one drug in a vacuum. If, for example, an individual is seen on camera or by credible witnesses to be acting in a clearly cogent manner, demonstrating a linear, purposeful, and coherent act of self-destruction, we would not ignore that because PCP was unexpectedly discovered in toxicology testing. PCP positivity and suicidal manner determinations are not necessarily mutually exclusive. There is no one size fits all. Rather, we must recognize that drug effects are highly variable on an individual basis and that, in this person, the drug did not appear to have had a disorienting effect, even if it may have enhanced impulsivity. If, on the other hand, the individual was exhibiting erratic, incoherent, confused, and clearly delusional behavior on video surveillance or by reliable witness accounts, the PCP would be reasonably viewed as a possible explanation, and most would back off of any claims of suicidality. Absent such witnessed behavioral evidence, drug and alcohol effects are not reliably predicted on an *individual* basis. Rather, all the forensic pathologist can really do in the absence of witnessed antemortem activity is take into consideration the available *aggregate* data pertaining to the range of behavioral and/or pathophysiologic effects that a certain class of drug may have and view it through the prism of the entire context of the death investigation. That said, if there was a finding of PCP in a hanging, many would likely ignore it, given the premeditated act most hanging deaths typically represent. Even without witnesses, a hanging, by definition, indicates intentionality and volition. To construct a noose, tie it to a strong structure that can hold a human body, and then place the noose over one's head and let go requires logical, linear, self-destructive actions (see "Approach to Hanging Deaths" in Chapter 11) This is why, with very rare exception, one does not have to wait for toxicological results in most apparently suicidal hanging deaths.

Still, and finally, it is best to perform toxicological testing in all suicides, even when it will have no bearing whatsoever on the mechanism and cause of death (gunshot wound of head, hanging, etc.) Unforeseeable questions not uncommonly arise in these deaths, sometimes months or years down the line, as families continue to grapple and try come to terms with the reality of what occurred. Invariably and somewhat predictably, many of these questions involve and can be answered by a simple toxicological analysis, performed close to the time of the autopsy. This basic test can put to bed concerns or suspicions that have the capacity to persist, fuel denial, and wreak havoc for years without the benefit of this information.

The apparent drug intoxication death with negative toxicology and no anatomic cause

Although the majority of death investigations in the setting of an apparent drug intoxication are resolved and competently certified following the approach outlined above (complete scene investigation, review of historical information, complete autopsy, and comprehensive toxicology testing), a small number will remain unexplained following the completion of a thorough death investigation. It may be appropriate in these situations to certify the deaths as undetermined cause and manner of death. Prior to issuing this kind of certification, however, it is worthwhile to review the investigation to ensure that it was adequately thorough. Such a review might include a reexamination of all the elements of the investigation to identify potentially overlooked lethal toxic substances that may not have been included in the toxicological laboratory analysis.

The medical history may provide insight into preferred substances of abuse, and it should be confirmed that such substances were actually within the scope of the toxicological testing performed. Any medications recovered at the scene of death should also be reviewed for potential toxicity (e.g., certain antidiarrheal drugs, such as loperamide, may contain narcotics which are potentially fatal in high concentrations).[361–363] Any drug paraphernalia or apparent drug packaging that is recovered at a scene of death should be noted and analyzed. The certifier should vigorously search out the results of the drug testing performed on these items to identify potentially overlooked lethal compounds. Common substances that may not be included in routine toxicological testing include ethylene/propylene glycol, carbon monoxide, cyanide, fentanyl analogues, NPSs, and nitrites.[248,364–386] On occasion, the online purchases of or searches for herbal and black market drugs and chemicals may be born out of a more granular police investigation of the decedent's antemortem activities.

Autopsy and histologic examination should be reviewed for adequacy to exclude less common anatomical causes of death. The medical history should be reviewed for potential physiologic causes of death, such as a seizure disorder or a history of cardiac arrhythmia or a hematologic abnormality. Molecular testing for subtle cardiac disease (e.g., channelopathy) might be considered.[387–389] Vitreous testing for glucose and electrolytes may be useful.[351,390–392] Finally, the scope of toxicological testing should be reviewed and additional testing requested, as indicated by the case details. After such a thorough review, and to the exclusion of anything else more convincing, the certifier may reasonably certify an undetermined cause and manner. Some do not favor the use of a natural manner in the absence of a demonstrable cause of death. Others believe that if one has sufficiently excluded an intoxication or trauma, the use of natural is appropriate, even if the cause remains undetermined.

Legal implications of mixed drug intoxications and other special toxicologic considerations

With the increasing incidence of drug intoxication mortality, there has been a commensurate increase in the frequency of prosecutions of drug-related deaths.[393,394] Traditionally, prosecutions of fatal drug intoxications involved the intentional (homicidal) administration of a drug to a decedent by another person. In the more recent climate, however, the prosecution of drug fatalities has sought to address the culpability of an individual delivering or otherwise distributing a drug to victims of a fatal intoxications. This latter focus may involve particularly lengthy mandatory minimum sentences, particularly at the federal level.

In the scenario involving a prosecution for drug delivery, the cause of death will involve the intoxication. Unless there is evidence of administration of these drugs by another person or the intentional supply of known, predictably lethal amounts of drug (the so called "hot shot"), these deaths are still certified as accidents. This is in keeping with standard forensic practice, but does not preclude the legal prosecution of the death.[9] The designation of a manner of homicide in routine intoxication cases in order to facilitate prosecution is a questionable practice and is best avoided. The manner of death is intended to be used for public health purposes and neither triggers nor prevents prosecution.

The frequency of mixed drug intoxications in these instances may pose certain challenges for the prosecutor, as the distribution of the component drugs listed in the cause of death statement may not all be attributable to one party nor necessarily have equivalent pathophysiologic importance. At this point, statutes may diverge in assigning culpability to the distributor of a drug that *contributes* to death versus the more stringent definition of culpability, which is confined to a distributor of a substance representing a "but for" or "stand alone" cause of death, despite the presence of other drugs cited in the fatal mixture.

A frequently invoked legal decision in such cases is *Burrage v. United States*. This case was adjudicated by the United States Supreme Court and the opinion of the court was written by Justice Antonin Scalia.[393,394] In his decision, Justice Scalia established a requirement of "but for" causation in the attribution of culpability for a drug-related death. By this, the court held that for a distributor of a drug to be held responsible for the intoxication fatality, the drug distributed needed to be lethal in and of itself, regardless of any other drugs included in a cause of death statement. This may complicate legal proceedings when there is more than one drug independently capable of causing death in a mixed drug intoxication (e.g., fentanyl and cocaine). In the prosecution of such cases, the prosecutor bears the burden of establishing that a defendant supplied any and all drugs capable of producing death. This burden goes beyond the demonstration of the distribution of a putative drug being capable of causing death and such a demonstration would be difficult to establish if other potentially lethal drugs (in isolation) were also present.

As noted previously, it is frequent for a death to be ascribed to multiple drugs and the exclusion of other drugs. The relative importance of each drug in a given death, however, often cannot be precisely assigned. For example, in deaths attributed to the combined effects of opioids and stimulants, such as cocaine and methamphetamine, it is inferred that the death has not been exclusively produced by either the opioids or the stimulants, largely because both of these classes of drugs have common and independent ("but for") lethal potential. By contrast, it is reasonable and possible to exclude certain classes of intoxicants as "but for" substances in causes of death, even when they are included with other drugs as part of the cause of death statement. This would be particularly true of benzodiazepines, given their high safety profile, or ethanol, given its relatively infrequent fatal role in isolation, particularly in habituated users. The consistent application of scientific principles in these cases is the best service to provide to the legal system.

The cause of death from inadvertent acute intoxications during drug misuse is certified as an acute intoxication. Historically, these have been called "narcotism" deaths, which indicates that a narcotic was the toxic agent. Today, toxicology laboratories are better at identifying the offending agent and so the specific intoxicant(s) can and should be listed.[157,158,395,396] This has important public health implications. It is important to know what drugs are causing health hazards, where they are doing so, and in whom.[9,153,157,158,397,398]

Medical terminology as it pertains to describing drug and alcohol "abuse" has also changed.[399–403] The current medically accepted diagnostic language found in the DSM-5 has shifted away from use of the word "abuse." Recently, many ME/C offices also have adopted this terminology for the certification of drug and alcohol "abuse" or "alcoholism" to the terms "use," "misuse," or "use disorder," in keeping with the addiction medicine terminology.[400–403]

Chronic substance use disorder is a disease, but, in contrast to alcohol use disorder (when alcohol is detected in isolation), acute exacerbations are chemical injuries that are directly responsible for death. Therefore, *acute drug intoxication deaths are not certified as natural deaths (see Chapter 3).* These deaths are certified as *accidents* if there is no intent to cause harm (either by the user or the supplier). Some certifiers are reluctant to certify these typical drug intoxication deaths as accidents and instead certify them as undetermined.[157,188] Such caution was more defensible in the past, when there was limited postmortem toxicological analysis. One reason given for the undetermined manner is that one cannot exclude that the supplier did not intentionally give the user a fatal dose (what was known in the old days as a "hot shot"[404]), and it also is a crime to sell these drugs. We disagree with this philosophy and believe that these deaths should be routinely certified as accidents. There are several reasons why this is the best action. The first is that our certifications have a primary public health function. The "undetermined" philosophy muddies the waters for public health experts. Even if it was later discovered that a person intentionally "poisoned" another, use of "accident" on the death certificate is not going to prevent prosecution, per se. In fact, by convention, most motor vehicle deaths also are certified as accidents, but deaths due to "vehicular manslaughter" are routinely prosecuted despite the ME/C determination of accident. In court, this disparity can be easily and simply explained. To take the undetermined intoxication paradigm to an extreme, one could say that all pedestrian deaths should be certified as undetermined in the off chance that the driver was intentionally trying to hit the pedestrian or that all apparent natural deaths may possibly represent an unrecognized smothering, a possibility that cannot be excluded with 100% certainty, but should not reasonably preclude the certification of hundreds of thousands of natural deaths *without cause*.

If the toxicological analyses disclose drug concentrations of one to two orders of magnitude greater than usual, it may give the investigator pause about considering other manners of death (i.e., suicide, homicide). In stark contrast to the Food and Drug Administration (FDA) dosing regulations of on-market prescription medications, the challenges of properly "dosing" an illicit pharmaceutical should not be forgotten. Most drug dealers are not pharmacists or chemists and are neither trained nor necessarily focused on careful drug dosing. But even if they had a Ph.D. in pharmacology, they still do not know what the "powders" consist of and at what amounts. The fentanyl crisis made this abundantly clear, as the difficulty to properly dose such a potent drug resulted an increasing inadvertent fentanyl fatalities. As a business model, it does not make sense for a proprietor to kill their customers. Therefore, when toxicological testing discloses substances in concentrations that are typically or commonly found with illicit drug use, they are best classified as accidents. Deaths due to the chronic effects of drug use are certified as natural, similar to the chronic effects of ethanol. The death of an intravenous user of drugs who dies from endocarditis or cirrhosis due to chronic hepatitis C would be classified as a natural death, assuming no acute intoxication contributed to the death.

The concept of **"efficient intervening cause"** in the setting of toxicological deaths (even if a remote back injury, for example, led to a drug/narcotic addiction) focuses on the notion that, if the death is from personal misuse, it is an accident by virtue of the drug use, not the original injury. No matter what the circumstances of that injury, the proximate cause is the drug intoxication, not what pain and/or injury the drugs were attempting to medicate, the latter having no place on the death certificate (note that this differs from a therapeutic complication (TC), where a medication that is taken "as directed" may lead to death, in which case the underlying indication for the medication administration must be listed as the proximate cause—a disease if a TC, an injury if an unnatural manner; see Chapters 3 and 8).

The conclusion that a death is from an acute intoxication requires three conditions: the toxicology results must be within the range typically encountered in such fatalities; the history and circumstances must be consistent with a fatal intoxication; and the autopsy must fail to disclose a disease or physical injury that has an extent or severity inconsistent with continued life.[3,327,405] For drug intoxication deaths with more than one drug in concentrations greater than trace amounts, *it is customary to include all of the identified drugs with potentiating and/or similar mechanisms of action in the cause of death statement.* The drug concentration may be helpful in determining the cause and manner of death, but it is not absolute. Various pre- and postmortem factors may affect the concentrations detected from autopsy samples. Even a very low concentration should not deter the certification of a death as an intoxication, given the appropriate circumstances and autopsy findings. This is particularly true for cocaine, which continues to "metabolize," through the action of blood esterases, well into the postmortem period. For this reason, in cocaine-related deaths, in isolation or in combination with other drugs, the concentration of the cocaine, per se, lends less forensic veracity, in most cases, than does the mere presence of parent cocaine. It often is reported with "trace" values, despite the fact that this may not accurately reflect concentration of cocaine in the blood at the moment of death. Postmortem intervals and redistribution, as well as survival intervals may also explain a "low" drug concentration in an otherwise clear-cut intoxication death.

Miscellaneous toxicological issues

Despite its rarity, ***homicidal*** **intoxications** do occur. It may be a challenge to the forensic pathologist when there is a baseless suspicion or it is erroneously invoked as the cause of some deaths. That said, unsuspected and/or undetected intoxication deaths undoubtedly occur despite having been ascribed to disease. A poison/intoxicant is defined as a substance that exerts a deleterious (and possibly fatal) effect by virtue of its *chemical* action, even when it is administered in relatively *small* amounts. *Allergic* or *atypical* responses, which occasionally follow administration of many drugs ("adverse reactions"), are not true intoxications. It is immaterial whether the harmful substance is swallowed (frequently mixed with food or drink), inhaled, injected, or absorbed. The following are some intoxicants that may be challenging for the toxicologist to identify and/or may be missed because they are not routinely screened for. If there is a concern for an unusual intoxicant, a discussion with the forensic toxicologist is warranted.

Nitrites and poppers

The nitrite ion has the chemical formula NO_2. Nitrite (mostly sodium nitrite) is widely used throughout chemical and pharmaceutical industries, and the nitrite anion is found normally in humans. Nitrate is a polyatomic ion with the chemical formula NO_3. Salts containing this ion are called nitrates. Nitrates are found in green, leafy vegetables and are common components of fertilizers and explosives. Sodium nitrite is used as a coloring agent or preservative in food and as an antimicrobial agent in meat products.[406] In adults, a dose of 2.6 g or higher may cause death.[381]

Sodium nitrite, which causes methemoglobinemia, has become a popular method of suicide. Detection and interpretation of nitrites in the body is difficult because it is a naturally occurring substance, and nitrites in blood samples are unstable at room temperature and are readily oxidized to nitrate.[407,408] Nitrites may be ordered on the internet, so computer searches or recent postal deliveries should be investigated.[380–386,409–412] Nitrites enter red blood cells and oxidize hemoglobin, forming methemoglobin. Methemoglobin is a hemoglobin in which the iron in the heme group is in the $Fe3+$ (ferric) state, not the $Fe2+$ (ferrous) state of normal hemoglobin, which results in its inability to bind oxygen. The blood becomes chocolate-brown and the person is cyanotic. It is important to identify the former at the time of autopsy; if sodium nitrite is the cause of death, the darkly discolored blood is usually grossly striking, even from across the autopsy suite. Lividity is also described as green-gray in sodium nitrite poisonings (Fig. 14.12).[412]

Normally, the percentage (divide the concentration of methemoglobin by the concentration of total hemoglobin.) of methemoglobin does not exceed 1%–2%, and cyanosis is seen when it reaches 10%–20%. Methemoglobins of 20%–50% may cause symptoms such as respiratory distress, dizziness, headache, and fatigue.[381] Loss of consciousness and death can occur at methemoglobin levels of 50%–70%.[381] Methylene blue is the first choice for treating acute methemoglobinemia. Therapeutic intervention with methylene blue may cause the brain to take on a bluish hue, something to be aware of during the autopsy if there was such a resuscitative effort.[413–415] The elevated methemoglobinemia is supportive evidence, in addition to the scene investigation; unfortunately, postmortem testing of methemoglobin (and nitrites, to an extent) are not as reliable in postmortem specimens.

Poppers is a term given to **alkyl nitrites** (e.g., isoamyl nitrite or isopentyl nitrite, and isopropyl nitrite), which are typically inhaled for recreational purposes. They are rapid-onset, short-acting, and potent vasodilators that produce a rush characterized by warm sensations and feelings of dizziness.[416] They may be sold as room deodorizers, leather polish, nail polish remover, or videotape head cleaner to evade antidrug laws. Poppers have a relaxation effect on involuntary smooth muscles and have a long history of being abused in social settings, particularly among homosexual and bisexual men.[416–419] The hypotensive effects in combination with underlying heart disease or prolonged use, with subsequent methemoglobinemia, may cause death. The manner in such deaths is typically accident.

Cyanide

Suspected suicidal ingestions in chemists and jewelers should make one consider cyanide intoxication. Both have access to cyanide (jewelers use cyanide to polish gold). As toxicology laboratories may not routinely test for cyanide, the forensic pathologist would need to have enough of an index of suspicion to request testing. Autopsy signs that may be supportive of such a death include corrosive changes of the oral cavity and gastrointestinal tract. Although cyanide causes death rapidly, a person may still have time to "clean up the scene" after ingestion. At autopsy, a

FIGURE 14.12 **Gray livor mortis with a suicidal sodium nitrite ingestion.**

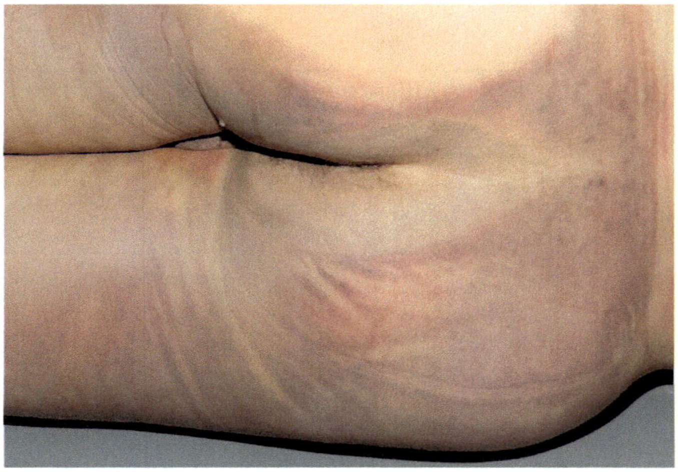

cherry-pink lividity has been described; however, it may not always be seen.[364–368,420–425] Cyanide has also been described in smoke inhalation deaths.[426–431] The ability to smell cyanide (described as bitter almond) is a genetic trait found only in some.[432] Caution should be exercised during autopsy, as it is possible for personnel involved in the postmortem examination of cases to inhale cyanide gas from the body.[433] Although cyanide can volatilize from autopsy tissues, the major risk occurs when the stomach is opened. The gastric acid converts cyanide salts to a highly volatile hydrocyanic gas. Prosectors risk inhaling this gas unless the stomach is opened in a biosafety chemical fume hood.[434]

Cyanide cannot be detected in an embalmed body because formalin breaks it down, but one may detect cyanide in specimens that have remained refrigerated for several weeks. That said, postmortem bacteria can produce cyanide.[423,435]

Antidotes of cyanide poisoning involve three strategies: binding of cyanide, induction of methemoglobinemia, and use of sulfur agents. Available antidotes are hydroxocobalamin (binds cyanide), sodium nitrite, and thiosulfate (detoxifies cyanide by transforming it to thiocyanate). Sodium nitrate works by converting hemoglobin to methemoglobin, which allows for the binding of cyanide (CN^-) anions by ferric ($Fe3+$) cations and the formation of cyanomethemoglobin. The immediate goal of forming this cyanide adduct is to prevent the binding of free cyanide to the cytochrome complex.[436]

Arsenic

Clinical manifestations of **acute arsenic poisoning** are predominantly gastrointestinal, cardiovascular, neurologic, and renal.[436–444] Some differences exist between the acute and chronic signs. Acute arsenic ingestion may result in rapid death. A *garlic-like odor* on the breath or in perspiration is suggestive of arsenic poisoning. Death from arsenic is generally caused by circulatory collapse, associated with intense gastroenteritis.

Arsenic combines with sulfhydral groups of mitochondrial enzymes and has endothelial affects. Repeated sublethal doses may cause Aldrich—Mees lines[445] (transverse white striae of finger nails), "raindrop"-shaped hyperpigmentation of skin, alopecia, stocking-glove paresthesias, keratoses, and a variety of cancers (lung and skin). Testing of bone, hair, and nails may show chronicity.[437,440,442,446,447]

Ethylene glycol

Ethylene glycol (e.g., antifreeze) testing may not be routine in most forensic toxicology laboratories, thus requiring a special request born out of an index of suspicion (it is not detected on routine gas chromatography testing).[377,378,448–451] Histologic detection of calcium oxylate crystals in the kidneys can be diagnostic. Diethylene glycol leads to renal necrosis and perilobular liver necrosis, but the deposition of few, if any, oxalate crystals.[448,449]

Anticholinesterases

Aldicarb, a carbamate insecticide/rodenticide, is a cholinesterase inhibitor that prevents the breakdown of synaptic acetylcholine, causing death by respiratory failure and a parasympathetic overdrive (SLUDE–salivation, lacrimation, urination, diarrhea, and emesis). *Tres Pasitos*, a mouse, rat, and roach killer contains high concentrations of aldicarb and has been sold illegally in the United States. It is highly toxic to animals and people.[452–454]

Acetaminophen

Acetaminophen (an active metabolite of phenacetin) is an over-the-counter pain reliever. Acetaminophen overdoses typically cause death through fulminant injury of the liver.[455–458] In contrast to many intoxicants that cause an acute death when intentionally or inadvertently taken in "overdose" quantities, the progression of the fatal sequelae is more protracted for acetaminophen, usually taking a few days to fully clinically manifest. Therefore, the acetaminophen concentration may no longer be markedly elevated at the time of presentation. There are graphs that can be used to estimate the initial concentration, based on the duration since ingestion.

Due to its nephrotoxicity (renal papillary necrosis), phenacetin has been largely replaced by acetaminophen. Four stages of toxicity have been described with acetaminophen. The first, 24 h after ingestion, may be asymptomatic or only mild anorexia, nausea, vomiting, and malaise. From 24 to 48 h, there is an elevation of the hepatic transaminases, with attendant abdominal pain and vomiting. Over the ensuing days, transaminases continue to rise, along with the international normalized ratio (INR), culminating in fulminant hepatic failure.[459]

Very large overdoses of acetaminophen are associated with early production of a toxic metabolite, called **oxoproline** (also called pyroglutamic acid). This metabolite appears early in the overdose, but it is clinically insignificant in most overdoses. With large overdoses, however, it may result in an oxoproline-induced anion gap metabolic acidosis. The oxoproline production appears before liver toxicity and may provide an alternate pathway to death.[455] Fatal acetaminophen toxicities may represent suicides or accidents, the latter less common, but encountered more commonly in those with substantial liver disease.

Sodium azide

Sodium azide is an inorganic compound (NaN_3). It is used as the gas-forming component of vehicular airbag systems. It is highly poisonous by ingestion or inhalation.[460–463] Sodium azide poisoning by ingestion presents challenges to the prosector and the funeral director alike. Due to its extreme lability, it may only be detected in the gastric contents, not the blood.[461] When it is mixed with water or an acid, sodium azide changes rapidly to a toxic gas with a pungent odor. It also changes into a toxic and explosive gas (hydrazoic acid) when it comes into contact with solid metals (for example, when it is poured into a drain pipe). Its toxicity may be linked to the inhibition of oxidative enzymes, such as catalases and cytochromes, and to a strong complex with hemoglobin that blocks oxygen transport. The direct vasodilatory properties of the salt probably induce hypotension.[461] Funeral director associations recommend that embalming should not be performed. As such, they should be notified if there is compelling reason to suspect a sodium azide-related death, even if not yet corroborated.

On occasion, an investigation may indicate specialized toxicology testing, resulting in samples needing to be sent to an "outside" laboratory.

Toxicological considerations in pediatric deaths

Depending on the circumstances of the death, various toxicological factors may be considered when evaluating the death of an infant or child.[71,464–471] Examples include a vitreous electrolyte analysis to exclude dehydration, basic testing to exclude drugs of abuse or medications of other family members in ambulatory young children who are suspected of a sudden chemical toxicity, analysis for medications used to treat seizure disorders, asthma, diabetes mellitus, and other conditions, particularly when there is concern for medical neglect, and the testing for common antipyretics and antihistamines, such as acetaminophen or diphenhydramine, when a caretaker reports having administered these to the child. The latter may be relevant in bathtub drownings, inflicted abdominal trauma with pain and fever sequelae, or other conditions that may trigger concern or interest for these medications. Some toxicology laboratories even offer a standard testing profile for pediatric deaths.

References

1. Warner M, Chen LH, Makuc DM, Anderson RN, Minino AM. Drug poisoning deaths in the United States, 1980–2008. *NCHS Data Brief.* 2011; 81:1–8.
2. CDC National Center for Health Statistics. *Drug Overdose Deaths in the U.S. Top 100,000 Annually*; November 17, 2021. www.cdc.gov/nchs/pressroom/nchs_press_releases/2021/20211117.htm.
3. Tardiff K, Marzuk PM, Leon AC, et al. Accidental fatal drug overdoses in New York City: 1990–1992. *Am J Drug Alcohol Abuse.* 1996;22(2): 135–146.
4. Krinsky CS, Lathrop SL, Brown P, Nolte KB. Drugs, detention, and death: a study of the mortality of recently released prisoners. *Am J Forensic Med Pathol.* 2009;30(1):6–9.
5. Binswanger IA, Stern MF, Deyo RA, et al. Release from prison–a high risk of death for former inmates. *N Engl J Med.* 2007;356(2):157–165.
6. Lozano JG, Healy NL, Kimberley Molina D. The prevalence of paraphernalia found at the scene of drug-related deaths. *J Forensic Sci.* 2016; 62(3):691–694.
7. Andrew TA, Duval JV. Confronting an upsurge in opiate deaths with limited resources. *Acad Forensic Pathol.* 2017;7(1):7–18.
8. Peterson GF, Clark SC. Forensic autopsy performance standards. *Am J Forensic Med Pathol.* 2020;27(3):200–225.
9. Davis GG, Cadwallader AB, Fligner CL, et al. Position paper: recommendations for the investigation, diagnosis, and certification of deaths related to opioid and other drugs. *Am J Forensic Med Pathol.* 2020;41(3):152–159.
10. Marzuk PM, Tardiff K, Leon AC, et al. Fatal injuries after cocaine use as a leading cause of death among young adults in New York City. *N Engl J Med.* 1995;332:1753–1757.
11. Tardiff K, Marzuk PM, Leon AC, et al. Cocaine, opiates, and ethanol in homicides in New York City: 1990 and 1991. *J Forensic Sci.* 1995;40(3): 387–390.
12. Tardiff K, Marzuk PM, Leon AC, et al. Homicide in New York City. Cocaine use and firearms. *JAMA.* 1994;272(1):43–46.

13. Helpern M. Fatalities from narcotic addiction in New York City. Incidence, circumstances, and pathologic findings. *Hum Pathol*. 1972;3(1): 13–21.
14. Siegel H, Helpern M, Ehrenreich T. The diagnosis of death from intravenous narcotism. With emphasis on the pathologic aspects. *J Forensic Sci*. 1966;11(1):1–16.
15. Helpern M, Rho YM. Deaths from narcotism in New York City. Incidence, circumstances, and postmortem findings. *N Y State J Med*. 1966; 66(18):2391–2408.
16. Chen HI, deJong J. Increased lung weights in drug-related fatalities. *J Forensic Sci*. 2017;62.
17. Molina DK, Vance K, Coleman ML, Hargrove VM. Testing an Age-old adage: can autopsy findings be of assistance in Differentiating opioid versus cardiac deaths? *J Forensic Sci*. 2020;65(1):112–116.
18. Pelletier DE, Andrew TA. Common findings and predictive measures of opioid overdoses. *Acad Forensic Pathol*. 2017;7(1):91–98.
19. Dolinak D. Opioid toxicity. *Acad Forensic Pathol*. 2017;7(1):19–35.
20. Szpilman D, Bierens JJ, Handley AJ, Orlowski JP. Drowning. *N Engl J Med*. 2012;366(22):2102–2110.
21. Lucas J, Goldfeder LB, Gill JR. Bodies found in the waterways of New York City. *J Forensic Sci*. 2002;47(1):137–141.
22. Girela-López E, Beltran-Aroca CM, Dye A, Gill JR. Epidemiology and autopsy findings of 500 drowning deaths. *Forensic Sci Int*. 2022;330: 111137.
23. Davis JH. Bodies found in the water. An investigative approach. *Am J Forensic Med Pathol*. 1986;7(4):291–297.
24. Terrence CF, Rao GR, Perper JA. Neurogenic pulmonary edema in unexpected, unexplained death of epileptic patients. *Ann Neurol*. 1981;9(5): 458–464.
25. Colice GL, Matthay MA, Bass E, Matthay RA. Neurogenic pulmonary edema. *Am Rev Respir Dis*. 1984;130(5):941–948.
26. JN K, JD H. Pulmonary edema after freebase cocaine smoking-not due to an adulterant. *Chest*. 1990;97:1009–1010.
27. Larkin SM, Williams DN, Osterholm MT, Tofte RW, Posalaky Z. Toxic shock syndrome: clinical, laboratory, and pathologic findings in nine fatal cases. *Ann Intern Med*. 1982;96(6 Pt 2):858–864.
28. Lindgren L, Yli-Hankala A, Halme L, Koskimies S, Orko R. Transfusion-related acute lung injury (TRALI) after fresh frozen plasma in a patient with coagulopathy. *Acta Anaesthesiol Scand*. 1996;40(5):641–644.
29. Chochinov AH, Baydock BM, Bristow GK, Giesbrecht GG. Recovery of a 62-year-old man from prolonged cold water submersion. *Ann Emerg Med*. 1998;31(1):127–131.
30. Durlacher SH, Banfield Jr WG, Bergner AD. Post-mortem pulmonary edema. *Am J Pathol*. 1951;27(4):688–689.
31. Kurtzman RA, Caruso JL. High-altitude illness death investigation. *Acad Forensic Pathol*. 2018;8(1):83–97.
32. Miller 3rd CC, Calder-Becker K, Modave F. Swimming-induced pulmonary edema in triathletes. *Am J Emerg Med*. 2010;28(8):941–946.
33. Karch SB, Stephens BG. Toxicology and pathology of deaths related to methadone: retrospective review. *West J Med*. 2000;172(1):11–14.
34. Henderson GL. Fentanyl-related deaths: demographics, circumstances, and toxicology of 112 cases. *J Forensic Sci*. 1991;36(2):422–433.
35. Wetli CV. Investigation of drug-related deaths. An overview. *Am J Forensic Med Pathol*. 1984;5(2):111–120.
36. Colak B, Baser L, Yayci N, Etiler N, Inanici MA. Deaths from drug overdose and toxicity in Turkey: 1997–2001. *Am J Forensic Med Pathol*. 2006; 27(1):50–54.
37. Graham JK, Hanzlick R. Accidental drug deaths in Fulton County, Georgia, 2002: characteristics, case management and certification issues. *Am J Forensic Med Pathol*. 2008;29(3):224–230.
38. Serinelli S, White S, Arunkumar P, Wang D, Gitto L. The outbreak of fentanyl-related deaths in Cook county, Illinois, between October 2015 and December 2017: a retrospective study and a comparison with previous data. *J Forensic Sci*. 2019;64(6):1735–1742.
39. Roizin L, Helpern M, Baden M, et al. Methadone fatalities in heroin addicts. *Psychiatr Q*. 1972;46(3):393–410.
40. Lapkus M, Prahlow JA. Infective endocarditis in an intravenous drug abuser. *Forensic Sci Med Pathol*. 2015;11(4):609–613.
41. Hejna P, Janik M, Dobias M. Tricuspid valve endocarditis complicated by septic pulmonary embolism in an intravenous drug user. *Forensic Sci Med Pathol*. 2014;10(1):126–129.
42. Weymann A, Schmack B, Rosendal C, et al. Tricuspid valve endocarditis with septic pulmonary emboli in a drug addict. *Ann Thorac Cardiovasc Surg*. 2012;18(5):481–484.
43. Skaria B, Kalra P, Simpson IA. Septic emboli from aortic valve endocarditis. *Heart*. 2004;90(8):865.
44. Ting W, Silverman NA, Arzouman DA, Levitsky S. Splenic septic emboli in endocarditis. *Circulation*. 1990;82(5 Suppl):IV105–109.
45. Chiu CL, Roelofs JD. Acute bacterial endocarditis with septic emboli. *J Iowa Med Soc*. 1974;64(10):436–437, 434.
46. Ambrosioni J, Hernandez-Meneses M, Tellez A, et al. The changing epidemiology of infective endocarditis in the twenty-first century. *Curr Infect Dis Rep*. 2017;19(5):21.
47. Miró JM, Moreno A, Mestres CA. Infective endocarditis in intravenous drug abusers. *Curr Infect Dis Rep*. 2003;5(4):307–316.
48. Basso C, Michaud K, d'Amati G, et al. Cardiac hypertrophy at autopsy. *Virchows Arch*. 2021;479.
49. Virmani R, Robinowitz M, Smialek JE, Smyth DF. Cardiovascular effects of cocaine: an autopsy study of 40 patients. *Am Heart J*. 1988;115(5): 1068–1076.
50. Karch SB, Billingham ME. The pathology and etiology of cocaine-induced heart disease. *Arch Pathol Lab Med*. 1988;112(3):225–230.
51. Kloner RA, Rezkalla SH. Cocaine and the heart. *N Engl J Med*. 2003;348(6):487–488.
52. Pilgrim JL, Woodford N, Drummer OH. Cocaine in sudden and unexpected death: a review of 49 post-mortem cases. *Forensic Sci Int*. 2013; 227(1–3):52–59.
53. Tomashefski Jr JF, Hirsch CS, Jolly PN. Microcrystalline cellulose pulmonary embolism and granulomatosis. A complication of illicit intravenous injections of pentazocine tablets. *Arch Pathol Lab Med*. 1981;105(2):89–93.
54. Siegel H. Human pulmonary pathology associated with narcotic and other addictive drugs. *Hum Pathol*. 1972;3(1):55–66.
55. Hopkins GB. Pulmonary angiothrombotic granulomatosis in drug offenders. *JAMA*. 1972;221(8):909–911.
56. Ganesan S, Felo J, Saldana M, Kalasinsky VF, Lewin-Smith MR, Tomashefski Jr JF. Embolized crospovidone (poly[N-vinyl-2-pyrrolidone]) in the lungs of intravenous drug users. *Mod Pathol*. 2003;16(4):286–292.
57. Lewin-Smith MR, Kalasinsky VF, Mullick FG. Histochemical identification of microcrystalline cellulose, calcium oxalate, and talc in tissue sections. *Arch Pathol Lab Med*. 2011;135(8):963. author reply 963-964.

58. Sigdel S, Gemind JT, Tomashefski Jr JF. The Movat pentachrome stain as a means of identifying microcrystalline cellulose among other particulates found in lung tissue. *Arch Pathol Lab Med*. 2011;135(2):249–254.
59. Nguyen VT, Chan ES, Chou SH, et al. Pulmonary effects of i.v. injection of crushed oral tablets: "excipient lung disease". *AJR Am J Roentgenol*. 2014;203(5):W506–W515.
60. Hossein-Zadeh Z, Shuman MJ, Rapkiewicz A. Fatal excipients: an autopsy case series of excipient lung disease. *Am J Forensic Med Pathol*. 2021;43.
61. Nogue S, Sanz P, Borondo JC, Picon M, de la Red G, Mestre G. Fatal lipoid pneumonia due to bronco-aspiration of isoparaffin after ingestion of an organophosphate insecticide. *Acta Anaesthesiol Scand*. 2003;47(6):777–779.
62. Pennza PT. Aspiration pneumonia, necrotizing pneumonia, and lung abscess. *Emerg Med Clin North Am*. 1989;7(2):279–307.
63. Disler D, Deluca SA. Necrotizing aspiration pneumonia. *Am Fam Physician*. 1991;44(5):1719–1721.
64. Naso-Kaspar CK, Herndon GW, Wyman JF, Felo JA, Lavins ES, Gilson TP. 'Lingering' opiate deaths? Concentration of opiates in medulla and femoral blood. *J Anal Toxicol*. 2013;37(8):507–511.
65. Vanbinst R, Koenig J, Di Fazio V, Hassoun A. Bile analysis of drugs in postmortem cases. *Forensic Sci Int*. 2002;128(1–2):35–40.
66. Moore C. Review of: hair in toxicology: an important bio-monitor by Tobin DJ. *J Forensic Sci*. 2006;51(4):954.
67. Stout PA, Simons KD, Kerrigan S. Quantitative analysis of gamma-hydroxybutyrate at endogenous concentrations in hair using liquid chromatography tandem mass spectrometry. *J Forensic Sci*. 2010;55(2):531–537.
68. Wang X, Johansen SS, Nielsen MKK, Linnet K. Segmental hair analysis-interpretation of the time of drug Intake in two patients undergoing drug treatment. *J Forensic Sci*. 2019;64(3):950–955.
69. Pascali JP, Palumbo D, Umani Ronchi F, Mari F, Fioravanti A, Bertol E. Ethyl glucuronide concentration in hair of detainees: a preliminary study. *J Forensic Sci*. 2020;65(1):189–192.
70. Zhuo Y, Wang X, Wu J, et al. Simultaneous quantitative determination of amphetamines, opiates, ketamine, cocaine and metabolites in human hair: application to forensic cases of drug abuse. *J Forensic Sci*. 2020;65(2):563–569.
71. Tournel G, Pollard J, Humbert L, Wiart JF, Hedouin V, Allorge D. Use of hair testing to determine methadone exposure in pediatric deaths. *J Forensic Sci*. 2014;59(5):1436–1440.
72. Chatterton C, Turner K, Klinger N, Etter M, Duez M, Cirimele V. Interpretation of pharmaceutical drug concentrations in young children's head hair. *J Forensic Sci*. 2014;59(1):281–286.
73. Garcia-Bournissen F, Rokach B, Karaskov T, Gareri J, Koren G. Detection of stimulant drugs of abuse in maternal and neonatal hair. *Forensic Sci Med Pathol*. 2007;3(2):115–118.
74. Gaulier JM, Maublanc J, Lamballais F, Bargel S, Lachatre G. LSD in pubic hair in a fatality. *Forensic Sci Int*. 2012;218(1–3):25–27.
75. Vignali C, Stramesi C, Vecchio M, Groppi A. Hair testing and self-report of cocaine use. *Forensic Sci Int*. 2012;215(1–3):77–80.
76. Kintz P. Segmental hair analysis can demonstrate external contamination in postmortem cases. *Forensic Sci Int*. 2012;215(1–3):73–76.
77. Kintz P. 2014 consensus for the use of alcohol markers in hair for assessment of both abstinence and chronic excessive alcohol consumption. *Forensic Sci Int*. 2015;249:A1–A2.
78. Alvarez JC, Etting I, Abe E, Villa A, Fabresse N. Identification and quantification of 4-methylethcathinone (4-MEC) and 3,4-methylenedioxypyrovalerone (MDPV) in hair by LC-MS/MS after chronic administration. *Forensic Sci Int*. 2017;270:39–45.
79. Cho HS, Cho B, Sim J, Baeck SK, In S, Kim E. Detection of 11-nor-9-carboxy-tetrahydrocannabinol in the hair of drug abusers by LC-MS/MS analysis. *Forensic Sci Int*. 2019;295:219–225.
80. Miyaguchi H, Takahashi H, Ohashi T, et al. Rapid analysis of methamphetamine in hair by micropulverized extraction and microchip-based competitive ELISA. *Forensic Sci Int*. 2009;184(1–3):1–5.
81. Lee S, Cordero R, Paterson S. Distribution of 6-monoacetylmorphine and morphine in head and pubic hair from heroin-related deaths. *Forensic Sci Int*. 2009;183(1–3):74–77.
82. Kalasinsky KS, Bosy TZ, Schmunk GA, et al. Regional distribution of cocaine in postmortem brain of chronic human cocaine users. *J Forensic Sci*. 2000;45(5):1041–1048.
83. Kalasinsky KS, Dixon MM, Schmunk GA, Kish SJ. Blood, brain, and hair GHB concentrations following fatal ingestion. *J Forensic Sci*. 2001;46(3):728–730.
84. Rosales CM, Young T, Laster MJ, Eger 2nd EI, Garg U. Sevoflurane concentrations in blood, brain, and lung after sevoflurane-induced death. *J Forensic Sci*. 2007;52(6):1408–1410.
85. Molina DK, Limfueco S, Hargrove VM. Brain: the best matrix for volatile testing? *Am J Forensic Med Pathol*. 2015;36(2):122–123.
86. Murrell MD, Cruz DA, Javors MA, Thompson PM. Distribution of venlafaxine, O-desmethylvenlafaxine, and O-desmethylvenlafaxine to venlafaxine ratio in postmortem human brain tissue. *J Forensic Sci*. 2014;59(3):683–689.
87. Mardal M, Johansen SS, Thomsen R, Linnet K. Advantages of analyzing postmortem brain samples in routine forensic drug screening-Case series of three non-natural deaths tested positive for lysergic acid diethylamide (LSD). *Forensic Sci Int*. 2017;278:e14–e18.
88. Chesser R, Pardi J, Concheiro M, Cooper G. Distribution of synthetic opioids in postmortem blood, vitreous humor and brain. *Forensic Sci Int*. 2019;305:109999.
89. Nedahl M, Johansen SS, Linnet K. Brain-blood ratio of morphine in heroin and morphine autopsy cases. *Forensic Sci Int*. 2019;301:388–393.
90. Rhee J, Jung J, Yeom H, et al. Distribution of cyanide in heart blood, peripheral blood and gastric contents in 21 cyanide related fatalities. *Forensic Sci Int*. 2011;210(1–3):e12–15.
91. Palamalai V, Olson KN, Kloss J, et al. Superiority of postmortem liver fentanyl concentrations over peripheral blood influenced by postmortem interval for determination of fentanyl toxicity. *Clin Biochem*. 2013;46(7–8):598–602.
92. McIntyre IM. Liver and peripheral blood concentration ratio (L/P) as a marker of postmortem drug redistribution: a literature review. *Forensic Sci Med Pathol*. 2014;10(1):91–96.
93. Gaillard Y, Breuil R, Doche C, et al. Detection of amitriptyline, nortriptyline and bromazepam in liver, CSF and hair in the homicidal poisoning of a one-month-old girl autopsied 8 months after death. *Forensic Sci Int*. 2011;207(1–3):e16–18.
94. Saitman A, Fitzgerald RL, McIntyre IM. Evaluation and comparison of postmortem hydrocodone concentrations in peripheral blood, central blood and liver specimens: a minimal potential for redistribution. *Forensic Sci Int*. 2015;247:36–40.

95. Wohlfarth A, Vikingsson S, Roman M, et al. Looking at flubromazolam metabolism from four different angles: metabolite profiling in human liver microsomes, human hepatocytes, mice and authentic human urine samples with liquid chromatography high-resolution mass spectrometry. *Forensic Sci Int*. 2017;274:55–63.
96. Kugelberg FC, Jones AW. Interpreting results of ethanol analysis in postmortem specimens: a review of the literature. *Forensic Sci Int*. 2007; 165(1):10–29.
97. Hirsch CS, Adelson L. Ethanol in sequestered hematomas. *Am J Clin Pathol*. 1973;59(3):429–433.
98. Aniline O, Pitts F. Phencyclidine (PCP): a review and perspective. *CRC Crit Rev Toxicol*. 1982;10:145–177.
99. Pestaner JP, Southall PE. Sudden death during arrest and phencyclidine intoxication. *Am J Forensic Med Pathol*. 2003;24(2):119–122.
100. deRoux SJ, Sgarlato A, Marker E. Phencyclidine: a 5-year retrospective review from the New York city medical examiner's office. *J Forensic Sci*. 2011;56(3):656–659.
101. Spiehler V, Sedgwick P. Radioimmunoassay Screening and GC/MS confirmation of whole blood samples for drugs of abuse. *J Anal Toxicol*. 1985;9:63–66.
102. Ameline A, Garnier D, Gheddar L, et al. Identification and analytical characterization of seven NPS, by combination of (1)H NMR spectroscopy, GC-MS and UPLC-MS/MS((R)), to resolve a complex toxicological fatal case. *Forensic Sci Int*. 2019;298:140–148.
103. Rengarajan A, Mullins ME. How often do false-positive phencyclidine urine screens occur with use of common medications? *Clin Toxicol*. 2013;51(6):493–496.
104. Leikin JB, Watson WA. Post-mortem toxicology: what the dead can and cannot tell us. *J Toxicol Clin Toxicol*. 2003;41(1):47–56.
105. Wennig R. Threshold values in toxicology – useful or not? *Forensic Sci Int*. 2000;113(1–3):323–330.
106. Drummer OH. Postmortem toxicology of drugs of abuse. *Forensic Sci Int*. 2004;142(2–3):101–113.
107. Gerostamoulos D, Beyer J, Staikos V, Tayler P, Woodford N, Drummer OH. The effect of the postmortem interval on the redistribution of drugs: a comparison of mortuary admission and autopsy blood specimens. *Forensic Sci Med Pathol*. 2012;8(4):373–379.
108. Ferrari LA, Triszcz JM, Giannuzzi L. Kinetics of ethanol degradation in forensic blood samples. *Forensic Sci Int*. 2006;161(2–3):144–150.
109. Musshoff F, Stamer UM, Madea B. Pharmacogenetics and forensic toxicology. *Forensic Sci Int*. 2010;203(1–3):53–62.
110. Jin M, Gock SB, Jannetto PJ, Jentzen JM, Wong SH. Pharmacogenomics as molecular autopsy for forensic toxicology: genotyping cytochrome P450 3A4*1B and 3A5*3 for 25 fentanyl cases. *J Anal Toxicol*. 2005;29(7):590–598.
111. Okie S. A flood of opioids, a rising tide of deaths. *N Engl J Med*. 2010;363(21):1981–1985.
112. Wong SC, Curtis JA, Wingert WE. Concurrent detection of heroin, fentanyl, and xylazine in seven drug-related deaths reported from the Philadelphia Medical Examiner's Office. *J Forensic Sci*. 2008;53(2):495–498.
113. Cicero TJ, Ellis MS, Surratt HL, Kurtz SP. The changing face of heroin use in the United States: a retrospective analysis of the past 50 years. *JAMA Psychiatr*. 2014;71(7):821–826.
114. Harruff R, FC C, Banta-Green C. Tracking the opioid drug overdose epidemic in king county, Washington using an improved methodology for certifying heroin related deaths. *Acad Forensic Pathol*. 2015;5(3):499–506.
115. Dart RC, Surratt HL, Cicero TJ, et al. Trends in opioid analgesic abuse and mortality in the United States. *N Engl J Med*. 2015;372(3):241–248.
116. Slavova S, O'Brien DB, Creppage K, et al. Drug overdose deaths: let's get specific. *Public Health Rep*. 2015;130(4):339–342.
117. CDC. *Rising Numbers of Deaths Involving Fentanyl and Fentanyl Analogs, Including Carfentanil, and Increased Usage and Mixing with Non-opioids*. 2018.
118. Jalal H, Buchanich JM, Roberts MS, Balmert LC, Zhang K, Burke DS. Changing dynamics of the drug overdose epidemic in the United States from 1979 through 2016. *Science*. 2018;361(6408).
119. Peterson BL, Schreiber S, Fumo N, Brooke Lerner E. Opioid deaths in Milwaukee county, Wisconsin 2013-2017: the primacy of heroin and fentanyl. *J Forensic Sci*. 2019;64(1):144–148.
120. Gill JR, DeJoseph ME. Death certification and investigation of opioid intoxication fatalities: the medical examiner perspective. *Conn Med*. 2019;83(5):203–209.
121. Karch SB, Stephens BG, Ho CH. Methamphetamine-related deaths in San Francisco: demographic, pathologic, and toxicologic profiles. *J Forensic Sci*. 1999;44(2):359–368.
122. Swalwell CI, Davis GG. Methamphetamine as a risk factor for acute aortic dissection. *J Forensic Sci*. 1999;44(1):23–26.
123. Byard RW, Gilbert J, James R, Lokan RJ. Amphetamine derivative fatalities in South Australia–is "Ecstasy" the culprit? *Am J Forensic Med Pathol*. 1998;19(3):261–265.
124. Rogers F, Byers. Cocaine-related deaths in Pima county Arizona 1982–4. *JFS*. 1986;31:1404–1408.
125. Lowenstein DH, Massa SM, Rowbotham MC, Collins SD, McKinney HE, Simon RP. Acute neurologic and psychiatric complications associated with cocaine abuse. *Am J Med*. 1987;83(5):841–846.
126. Nolte KB, Gelman BB. Intracerebral hemorrhage associated with cocaine abuse. *Arch Pathol Lab Med*. 1989;113(7):812–813.
127. Karch SB. The history of cocaine toxicity. *Hum Pathol*. 1989;20(11):1037–1039.
128. Kibayashi K, Mastri AR, Hirsch CS. Cocaine induced intracerebral hemorrhage: analysis of predisposing factors and mechanisms causing hemorrhagic strokes. *Hum Pathol*. 1995;26(6):659–663.
129. Kaufman R. Cocaine induced cerebral vasoconstriction in humans by MRI. *JAMA*. 1998;279:376–380.
130. Allred R, Ewer S. Fatal pulmonary edema following intravenous "freebase" cocaine use. *Ann Emerg Med*. 1981;10:441–442.
131. Lundberg G, Garriott J, Reynolds P, Cravey R, Shaw R. Cocaine-related deaths. *J Forensic Sci*. 1977;22(2):402–408.
132. Dressler FA, Malekzadeh S, Roberts WC. Quantitative analysis of amounts of coronary arterial narrowing in cocaine addicts. *Am J Cardiol*. 1990;65(5):303–308.
133. Kolodgie FD, Farb A, Virmani R. Pathobiological determinants of cocaine-associated cardiovascular syndromes. *Hum Pathol*. 1995;26(6): 583–586.
134. Hoang MP, Lee EL, Anand A. Histologic spectrum of arterial and arteriolar lesions in acute and chronic cocaine-induced mesenteric ischemia: report of three cases and literature review. *Am J Surg Pathol*. 1998;22(11):1404–1410.
135. Karch SB, Stephens B, Ho CH. Relating cocaine blood concentrations to toxicity—an autopsy study of 99 cases. *J Forensic Sci*. 1998;43(1):41–45.
136. Mittleman RE, Wetli CV. Cocaine and sudden "natural" death. *J Forensic Sci*. 1987;32(1):11–19.
137. Mittleman RE, Wetli CV. Death caused by recreational cocaine use. An update. *JAMA*. 1984;252(14):1889–1893.

138. Simpson RW, Edwards WD. Pathogenesis of cocaine-induced ischemic heart disease. Autopsy findings in a 21-year-old man. *Arch Pathol Lab Med*. 1986;110(6):479–484.
139. Daras M, Tuchman AJ, Koppel BS, Samkoff LM, Weitzner I, Marc J. Neurovascular complications of cocaine. *Acta Neurol Scand*. 1994;90(2): 124–129.
140. Levine SR, Brust JC, Futrell N, et al. Cerebrovascular complications of the use of the "crack" form of alkaloidal cocaine. *N Engl J Med*. 1990; 323(11):699–704.
141. Lange RA, Hillis LD. Cardiovascular complications of cocaine use. *N Engl J Med*. 2001;345(5):351–358.
142. Stephens BG, Jentzen JM, Karch S, Wetli CV, Mash DC. National Association of Medical Examiners position paper on the certification of cocaine-related deaths. *Am J Forensic Med Pathol*. 2004;25(1):11–13.
143. Roh LS, Hamele-Bena D. Cocaine-induced ischemic myocardial disease. *Am J Forensic Med Pathol*. 1990;11(2):130–135.
144. Marzuk PM, Tardiff K, Smyth D, Stajic M, Leon AC. Cocaine use, risk taking, and fatal Russian roulette. *JAMA*. 1992;267(19):2635–2637.
145. Lucena J, Blanco M, Jurado C, et al. Cocaine-related sudden death: a prospective investigation in south-west Spain. *Eur Heart J*. 2010;31(3): 318–329.
146. Wetli CV. Fatal cocaine intoxication. A review. *Am J Forensic Med Pathol*. 1987;8(1):1–2.
147. Stephens BG, Jentzen JM, Karch S, Mash DC, Wetli CV. Criteria for the interpretation of cocaine levels in human biological samples and their relation to the cause of death. *Am J Forensic Med Pathol*. 2004;25(1):1–10.
148. Spiehler VR, Reed D. Brain concentrations of cocaine and benzoylecgonine in fatal cases. *J Forensic Sci*. 1985;30(4):1003–1011.
149. Pitts WR, Lange RA, Cigarroa JE, Hillis LD. Cocaine-induced myocardial ischemia and infarction: pathophysiology, recognition, and management. *Prog Cardiovasc Dis*. 1997;40(1):65–76.
150. Buchanich JM, Balmert LC, Williams KE, Burke DS. The effect of incomplete death certificates on estimates of Unintentional opioid-related overdose deaths in the United States, 1999–2015. *Public Health Rep*. 2018;133(4):423–431.
151. Kircher T, Anderson R. Cause of death: proper completion of death certificate. *JAMA*. 1987;258:349–352.
152. Hanzlick R. Death certificates, natural death, and alcohol. The problem of underreporting. *Am J Forensic Med Pathol*. 1988;9(2):149–150.
153. Paulozzi LJ, Ibrahimova A, Rudd RA, Goldberger BA, Thogmartin JR, Shelton K. A comparison of Florida Medical Examiners' reports and death certificates for specific drug related overdose deaths. *Acad Forensic Pathol*. 2012;2(2):190–197.
154. Hanzlick R, Parrish RG. The role of medical examiners and coroners in public health surveillance and epidemiologic research. *Annu Rev Public Health*. 1996;17:383–409.
155. Brooks EG, Reed KD. Principles and Pitfalls: a guide to death certification. *Clin Med Res*. 2015;13(2):74–82.
156. Goldberger BA, Maxwell JC, Campbell A, Wilford BB. Uniform standards and case definitions for classifying opioid-related deaths: recommendations by a SAMHSA consensus panel. *J Addict Dis*. 2013;32(3):231–243.
157. Warner M, Paulozzi L, Nolte K, Davis G, Nelson L. State Variation in certifying manner of death and drugs involved 2013. *Acad Forensic Pathol*. 2013;3(2):231–237.
158. Gill J, Vincent G, Toriello A, Nelson L. An underestimation of heroin deaths due to the use of "acute opiate intoxication" on death certificates. *Acad Forensic Pathol*. 2016;6(1):114–121.
159. Ruhm CJ. Corrected US opioid-involved drug poisoning deaths and mortality rates, 1999–2015. *Addiction*. 2018;113.
160. Hurstak E, Rowe C, Turner C, et al. Using medical examiner case narratives to improve opioid overdose surveillance. *Int J Drug Policy*. 2018; 54:35–42.
161. Tatsumi K, Shapiro S, Bundock E. Death certificate surveillance: a component of death investigation. *Acad Forensic Pathol*. 2013;3(1):99–104.
162. McGivern L, Shulman L, Carney JK, Shapiro S, Bundock E. Death certification errors and the effect on mortality statistics. *Public Health Rep*. 2017;132(6):669–675.
163. US Department of Health and Human Services, Public Health Service. Physicians' handbook on medical certification of death. In: Hyattsville, MD: National Center for Health Statistics (US); 2003:1–57.
164. Greenblatt DJ, Allen MD, Noel BJ, Shader RI. Acute overdosage with benzodiazepine derivatives. *Clin Pharmacol Ther*. 1977;21(4):497–514.
165. Brett J, Wylie CE, Raubenheimer J, Isbister GK, Buckley NA. The relative lethal toxicity of pharmaceutical and illicit substances: a 16-year study of the Greater Newcastle Hunter Area, Australia. *Br J Clin Pharmacol*. 2019;85(9):2098–2107.
166. Drummer OH, Ranson DL. Sudden death and benzodiazepines. *Am J Forensic Med Pathol*. 1996;17(4):336–342.
167. Shields LB, Hunsaker DM, Hunsaker 3rd JC, Ward MK. Toxicologic findings in suicide: a 10-year retrospective review of Kentucky medical examiner cases. *Am J Forensic Med Pathol*. 2006;27(2):106–112.
168. Perret G, Abudureheman A, Perret-Catipovic M, Flomenbaum M, La Harpe R. Suicides in the young people of Geneva, Switzerland, from 1993 to 2002. *J Forensic Sci*. 2006;51(5):1169–1173.
169. Lapoint J, James LP, Moran CL, Nelson LS, Hoffman RS, Moran JH. Severe toxicity following synthetic cannabinoid ingestion. *Clin Toxicol*. 2011;49(8):760–764.
170. Labay LM, Caruso JL, Gilson TP, et al. Synthetic cannabinoid drug use as a cause or contributory cause of death. *Forensic Sci Int*. 2016;260: 31–39.
171. Boland DM, Reidy LJ, Seither JM, Radtke JM, Lew EO. Forty-three fatalities involving the synthetic cannabinoid, 5-fluoro-ADB: forensic pathology and toxicology implications. *J Forensic Sci*. 2020;65(1):170–182.
172. Paul ABM, Simms L, Amini S, Paul AE. Teens and spice: a review of adolescent fatalities associated with synthetic cannabinoid use. *J Forensic Sci*. 2018;63(4):1321–1324.
173. Patton AL, Chimalakonda KC, Moran CL, et al. K2 toxicity: fatal case of psychiatric complications following AM2201 exposure. *J Forensic Sci*. 2013;58(6):1676–1680.
174. Hess C, Stockhausen S, Kernbach-Wighton G, Madea B. Death due to diabetic ketoacidosis: induction by the consumption of synthetic cannabinoids? *Forensic Sci Int*. 2015;257:e6–e11.
175. Shanks KG, Winston D, Heidingsfelder J, Behonick G. Case reports of synthetic cannabinoid XLR-11 associated fatalities. *Forensic Sci Int*. 2015; 252:e6–9.
176. Angerer V, Jacobi S, Franz F, Auwarter V, Pietsch J. Three fatalities associated with the synthetic cannabinoids 5F-ADB, 5F-PB-22, and AB-CHMINACA. *Forensic Sci Int*. 2017;281:e9–e15.

177. Yeter O, Erol Ozturk Y. Detection and quantification of 5F-ADB and its methyl ester hydrolysis metabolite in fatal intoxication cases by liquid chromatography-high resolution mass spectrometry. *Forensic Sci Int*. 2019;302:109866.
178. Fels H, Lottner-Nau S, Sax T, et al. Postmortem concentrations of the synthetic opioid U-47700 in 26 fatalities associated with the drug. *Forensic Sci Int*. 2019;301:e20–e28.
179. Kraemer M, Boehmer A, Madea B, Maas A. Death cases involving certain new psychoactive substances: a review of the literature. *Forensic Sci Int*. 2019;298:186–267.
180. Ivanov ID, Stoykova S, Ivanova E, et al. A case of 5F-ADB/FUB-AMB abuse: drug-induced or drug-related death? *Forensic Sci Int*. 2019;297: 372–377.
181. Gladden RM, O'Donnell J, Mattson CL, Seth P. Changes in opioid-involved overdose deaths by opioid type and presence of benzodiazepines, cocaine, and methamphetamine – 25 states, July–December 2017 to January–June 2018. *MMWR Morb Mortal Wkly Rep*. 2019;68(34):737–744.
182. Edvardsen HE, Tverborgvik T, Frost J, et al. Differences in combinations and concentrations of drugs of abuse in fatal intoxication and driving under the influence cases. *Forensic Sci Int*. 2017;281:127–133.
183. Butler DC, Batalis NI. Opioid-associated deaths in South Carolina, 2013–2016: a retrospective review. *Acad Forensic Pathol*. 2017;7(4):640–648.
184. Hull MJ, Juhascik M, Mazur F, Flomenbaum MA, Behonick GS. Fatalities associated with fentanyl and co-administered cocaine or opiates. *J Forensic Sci*. 2007;52(6):1383–1388.
185. Mikolaenko I, Robinson Jr CA, Davis GG. A review of methadone deaths in Jefferson County, Alabama. *Am J Forensic Med Pathol*. 2002;23(3): 299–304.
186. Gill JR, Hayes JA, deSouza IS, Marker E, Stajic M. Ecstasy (MDMA) deaths in New York City: a case series and review of the literature. *J Forensic Sci*. 2002;47(1):121–126.
187. Hedegaard H, Bastian B, Trinidad J, Spencer M, Warner M. *Regional differences in the drugs most frequently involved in drug overdose deaths: United States, 2017. Natl Vital Stat Rep*. 2019;68(12):1–16.
188. Hedegaard HMA, Warner M. *Drug Overdose Deaths in the United States, 1999–2017. Drug Overdose Deaths in the United States, 1999–2017. NCHS Data Brief No. 329*. Hyattsville, MD: National Center for Health Statistics; 2018.
189. Hedegaard H, Warner M, Miniño A. *Drug Overdose Deaths in the United States, 1999–2016. NCHS Data Brief: Centers for Disease Control and Prevention*. 294. National Center for Health Statistics; 2017:1–7.
190. Varghese E, Abate M, Hu L, et al. Characterization of diphenhydramine-related accidental overdose deaths. *Acad Forensic Pathol*. 2015;Mar(5): 105–115.
191. Tharp AM, Hobron K, Wright T. Gabapentin-related deaths: patterns of abuse and postmortem levels. *J Forensic Sci*. 2019;64(4):1105–1111.
192. Middleton O. Suicide by gabapentin overdose. *J Forensic Sci*. 2011;56(5):1373–1375.
193. Finlayson G, Chavarria M, Chang S, et al. Gabapentin in mixed drug fatalities: does this frequent analyte deserve more attention? *Acad Forensic Pathol*. 2017;7(1):99–111.
194. Simpson KJ, Moran MT, McCall KL, et al. Increasing heroin, cocaine, and buprenorphine arrests reported to the Maine Diversion Alert Program. *Forensic Sci Int*. 2019;303:109924.
195. Gilson TP, Shannon H, Freiburger J. The evolution of the opiate/opioid crisis in Cuyahoga county. *Acad Forensic Pathol*. 2017;7(1):41–49.
196. Logan BK, Fligner CL, Haddix T. Cause and manner of death in fatalities involving methamphetamine. *J Forensic Sci*. 1998;43(1):28–34.
197. Won S, Hong RA, Shohet RV, Seto TB, Parikh NI. Methamphetamine-associated cardiomyopathy. *Clin Cardiol*. 2013;36(12):737–742.
198. Bin Abdul Rashid SN, Rahim AS, Thali MJ, Flach PM. Death by 'ice': fatal methamphetamine intoxication of a body packer case detected by postmortem computed tomography (PMCT) and validated by autopsy. *Forensic Sci Med Pathol*. 2013;9(1):82–87.
199. Darke S, Duflou J, Lappin J, Kaye S. Clinical and autopsy characteristics of fatal methamphetamine toxicity in Australia. *J Forensic Sci*. 2018; 63(5):1466–1471.
200. Halkitis PN, Fischgrund BN, Parsons JT. Explanations for methamphetamine use among gay and bisexual men in New York City. *Subst Use Misuse*. 2005;40(9–10):1331–1345.
201. Halkitis PN, Green KA, Mourgues P. Longitudinal investigation of methamphetamine use among gay and bisexual men in New York City: findings from Project BUMPS. *J Urban Health: Bulletin of the New York Academy of Medicine*. 2005;82(1 Suppl 1):i18–25.
202. Parsons JT, Kelly BC, Weiser JD. Initiation into methamphetamine use for young gay and bisexual men. *Drug Alcohol Depend*. 2007;90(2–3): 135–144.
203. Romberg RW, Needleman SB, Snyder JJ, Greedan A. Methamphetamine and amphetamine derived from the metabolism of selegiline. *J Forensic Sci*. 1995;40(6):1100–1102.
204. Leclercq M, Soichot M, Delhotal-Landes B, et al. False positive amphetamines and 3,4-methylenedioxymethamphetamine immunoassays in the presence of metoprolol-two cases reported in clinical toxicology. *J Anal Toxicol*. 2020;44(2):200–205.
205. Abbott DL, Limoges JF, Virkler KJ, Tracy SJ, Sarris GG. ELISA screens for fentanyl in urine are susceptible to false-positives in high concentration methamphetamine samples. *J Anal Toxicol*. 2021;46.
206. Felscher D, Schulz K. Screening of amphetamine/methamphetamine and their derivatives in urine using FPIA and Triage 8 and the Scope and limits of a subsequent identification by means of the REMEDi HS system. *J Forensic Sci*. 2000;45(6):1327–1331.
207. Anne S, Tse R, Cala AD. A fatal case of isolated methiopropamine (1-(thiophen-2-yl)-2-methylaminopropane) toxicity: a case report. *Am J Forensic Med Pathol*. 2015;36(3):205–206.
208. Rajotte JW, Palmentier JFP, Wallage HR. Drug recognition evaluation and chemical confirmation of a 25C-NBOMe-Impaired driver. *J Forensic Sci*. 2017;62(5):1410–1413.
209. Kriikku P, Pelander A, Rasanen I, Ojanpera I. Toxic lifespan of the synthetic opioid U-47,700 in Finland verified by re-analysis of UPLC-TOF-MS data. *Forensic Sci Int*. 2019;300:85–88.
210. Ferreira B, Dias da Silva D, Carvalho F, de Lourdes Bastos M, Carmo H. The novel psychoactive substance 3-methylmethcathinone (3-MMC or metaphedrone): a review. *Forensic Sci Int*. 2019;295:54–63.
211. Mueller F, Bogdal C, Pfeiffer B, et al. Isotonitazene: fatal intoxication in three cases involving this unreported novel psychoactive substance in Switzerland. *Forensic Sci Int*. 2021;320:110686.
212. Calle P, Maudens K, Lemoyne S, et al. Lessons to be learned from toxicological analyses in intoxicated patients and seized materials at an electronic music dance festival. *Forensic Sci Int*. 2019;299:174–179.

213. Busardo FP, Di Trana A, Montanari E, Mauloni S, Tagliabracci A, Giorgetti R. Is etizolam a safe medication? Effects on psychomotor perfomance at therapeutic dosages of a newly abused psychoactive substance. *Forensic Sci Int*. 2019;301:137–141.
214. Atherton D, Dye D, Robinson CA, Beck R. n-Ethyl pentylone-related deaths in Alabama. *J Forensic Sci*. 2019;64(1):304–308.
215. Adamowicz P, Malczyk A. Stability of synthetic cathinones in blood and urine. *Forensic Sci Int*. 2019;295:36–45.
216. Strehmel N, Dumpelmann D, Vejmelka E, et al. Another fatal case related to the recreational abuse of U-47700. *Forensic Sci Med Pathol*. 2018; 14(4):531–535.
217. Lee HH, Chen SC, Lee JF, Lin HY, Chen BH. Simultaneous drug identification in urine of sexual assault victims by using liquid chromatography tandem mass spectrometry. *Forensic Sci Int*. 2018;282:35–40.
218. Diestelmann M, Zangl A, Herrle I, Koch E, Graw M, Paul LD. MDPV in forensic routine cases: psychotic and aggressive behavior in relation to plasma concentrations. *Forensic Sci Int*. 2018;283:72–84.
219. Staeheli SN, Boxler MI, Oestreich A, et al. Postmortem distribution and redistribution of MDAI and 2-MAPB in blood and alternative matrices. *Forensic Sci Int*. 2017;279:83–87.
220. Peace MR, Krakowiak RI, Wolf CE, Poklis A, Poklis JL. Identification of MDMB-FUBINACA in commercially available e-liquid formulations sold for use in electronic cigarettes. *Forensic Sci Int*. 2017;271:92–97.
221. Machado Y, Coelho Neto J, Barbosa PEN, Lordeiro RA, Alves RB. Brephedrone: a new psychoactive substance seized in Brazil. *Forensic Sci Int*. 2017;275:302–307.
222. Lee J, Yang S, Kang Y, et al. Prevalence of new psychoactive substances in Northeast Asia from 2007 to 2015. *Forensic Sci Int*. 2017;272:1–9.
223. Karinen R, Hoiseth G. A literature review of blood concentrations of new psychoactive substances classified as phenethylamines, aminoindanes, arylalkylamines, arylcyclohexylamines, and indolalkylamines. *Forensic Sci Int*. 2017;276:120–125.
224. Hamnett HJ, Ilett M, Izzati F, Smith SS, Watson KH. Toxicological findings in driver and motorcyclist fatalities in Scotland 2012–2015. *Forensic Sci Int*. 2017;274:22–26.
225. Guirguis A, Girotto S, Berti B, Stair JL. Identification of new psychoactive substances (NPS) using handheld Raman spectroscopy employing both 785 and 1064 nm laser sources. *Forensic Sci Int*. 2017;273:113–123.
226. Guerrieri D, Rapp E, Roman M, Thelander G, Kronstrand R. Acrylfentanyl: another new psychoactive drug with fatal consequences. *Forensic Sci Int*. 2017;277:e21–e29.
227. Grapp M, Kaufmann C, Ebbecke M. Toxicological investigation of forensic cases related to the designer drug 3,4-methylenedioxypyrovalerone (MDPV): detection, quantification and studies on human metabolism by GC-MS. *Forensic Sci Int*. 2017;273:1–9.
228. Gerace E, Bovetto E, Corcia DD, Vincenti M, Salomone A. A case of nonfatal intoxication associated with the recreational use of diphenidine. *J Forensic Sci*. 2017;62(4):1107–1111.
229. Elliott SP, Burke T, Smith C. Determining the toxicological significance of pregabalin in fatalities. *J Forensic Sci*. 2017;62(1):169–173.
230. Brandt SD, Kavanagh PV, Westphal F, et al. Return of the lysergamides. Part II: analytical and behavioural characterization of N(6) -allyl-6-norlysergic acid diethylamide (AL-LAD) and (2'S,4'S)-lysergic acid 2,4-dimethylazetidide (LSZ). *Drug Test Anal*. 2017;9(1):38–50.
231. Bovens M, Bissig C, Staeheli SN, Poetzsch M, Pfeiffer B, Kraemer T. Structural characterization of the new synthetic cannabinoids CUMYL-PINACA, 5F-CUMYL-PINACA, CUMYL-4CN-BINACA, 5F-CUMYL-P7AICA and CUMYL-4CN-B7AICA. *Forensic Sci Int*. 2017;281:98–105.
232. Shevyrin V, Melkozerov V, Eltsov O, Shafran Y, Morzherin Y. Synthetic cannabinoid 3-benzyl-5-[1-(2-pyrrolidin-1-ylethyl)-1H-indol-3-yl]-1,2,4-oxadiazole. The first detection in illicit market of new psychoactive substances. *Forensic Sci Int*. 2016;259:95–100.
233. Gjerde H, Nordfjaern T, Bretteville-Jensen AL, et al. Comparison of drugs used by nightclub patrons and criminal offenders in Oslo, Norway. *Forensic Sci Int*. 2016;265:1–5.
234. Cunningham SM, Haikal NA, Kraner JC. Fatal intoxication with acetyl fentanyl. *J Forensic Sci*. 2016;61(Suppl 1):S276–S280.
235. Simonsen KW, Edvardsen HM, Thelander G, et al. Fatal poisoning in drug addicts in the Nordic countries in 2012. *Forensic Sci Int*. 2015;248: 172–180.
236. Kueppers VB, Cooke CT. 25I-NBOMe related death in Australia: a case report. *Forensic Sci Int*. 2015;249:e15–18.
237. Karinen R, Tuv SS, Oiestad EL, Vindenes V. Concentrations of APINACA, 5F-APINACA, UR-144 and its degradant product in blood samples from six impaired drivers compared to previous reported concentrations of other synthetic cannabinoids. *Forensic Sci Int*. 2015;246:98–103.
238. Institoris L, Arok Z, Seprenyi K, et al. Frequency and structure of stimulant designer drug consumption among suspected drug users in Budapest and South-East Hungary in 2012–2013. *Forensic Sci Int*. 2015;248:181–186.
239. Bertol E, Vaiano F, Di Milia MG, Mari F. In vivo detection of the new psychoactive substance AM-694 and its metabolites. *Forensic Sci Int*. 2015; 256:21–27.
240. Chen C, Kostakis C, Irvine RJ, White JM. Increases in use of novel synthetic stimulant are not directly linked to decreased use of 3,4-methylenedioxy-N-methylamphetamine (MDMA). *Forensic Sci Int*. 2013;231(1–3):278–283.
241. Kanamori T, Kuwayama K, Tsujikawa K, Miyaguchi H, Iwata YT, Inoue H. Synthesis and identification of urinary metabolites of 4-iodo-2,5-dimethoxyphenethylamine. *J Forensic Sci*. 2011;56(5):1319–1323.
242. Chiu SK, Li JF, Nolte KB. Evaluating the potential for unintentional occupational exposure to fentanyl and fentanyl analogues among medicolegal death investigators and autopsy technicians. *J Forensic Sci*. 2020;65(4):1324–1327.
243. Hikin L, Smith PR, Ringland E, Hudson S, Morley SR. Multiple fatalities in the North of England associated with synthetic fentanyl analogue exposure: detection and quantitation a case series from early 2017. *Forensic Sci Int*. 2018;282:179–183.
244. Rab E, Flanagan RJ, Hudson S. Detection of fentanyl and fentanyl analogues in biological samples using liquid chromatography-high resolution mass spectrometry. *Forensic Sci Int*. 2019;300:13–18.
245. Brockbals L, Staeheli SN, Gentile S, et al. Fatal poisoning involving cyclopropylfentanyl – investigation of time-dependent postmortem redistribution. *Forensic Sci Int*. 2019;294:80–85.
246. Sato S, Suzuki S, Lee XP, Sato K. Studies on 1-(2-phenethyl)-4-(N-propionylanilino)piperidine (fentanyl) and related compounds VII. Quantification of alpha-methylfentanyl metabolites excreted in rat urine. *Forensic Sci Int*. 2010;195(1–3):68–72.
247. Krotulski AJ, Papsun DM, Kacinko SL, Logan BK. Isotonitazene quantitation and metabolite discovery in authentic forensic casework. *J Anal Toxicol*. 2020;44(6):521–530.
248. Clinton HA, Thangada S, Gill JR, Mirizzi A, Logan SB. Improvements in toxicology testing to identify fentanyl analogs and other novel synthetic opioids in fatal drug overdoses, Connecticut, January 2016–June 2019. *Public Health Rep*. 2021;136(1_suppl):80S–86S.

249. Martucci HFH, Ingle EA, Hunter MD, Rodda LN. Distribution of furanyl fentanyl and 4-ANPP in an accidental acute death: a case report. *Forensic Sci Int*. 2018;283:e13–e17.
250. Freni F, Pezzella S, Vignali C, et al. A case report on potential postmortem redistribution of furanyl fentanyl and 4-ANPP. *Forensic Sci Int*. 2019; 304:109915.
251. Palmquist KB, Swortwood MJ. Data-independent screening method for 14 fentanyl analogs in whole blood and oral fluid using LC-QTOF-MS. *Forensic Sci Int*. 2019;297:189–197.
252. Somerville RF, Hassan VR, Kolbe E, et al. The identification and quantification of synthetic cannabinoids seized in New Zealand in 2017. *Forensic Sci Int*. 2019;300:19–27.
253. Ozturk HM, Yetkin E, Ozturk S. Synthetic cannabinoids and cardiac arrhythmia risk: review of the literature. *Cardiovasc Toxicol*. 2019;19(3): 191–197.
254. Krotulski AJ, Mohr ALA, Kacinko SL, et al. 4F-MDMB-BINACA: a new synthetic cannabinoid widely implicated in forensic casework. *J Forensic Sci*. 2019;64(5):1451–1461.
255. Ji Kwon N, Han E. A review of drug abuse in recently reported cases of driving under the influence of drugs (DUID) in Asia, USA, and Europe. *Forensic Sci Int*. 2019;302:109854.
256. Drummer OH, Gerostamoulos D, Woodford NW. Cannabis as a cause of death: a review. *Forensic Sci Int*. 2019;298:298–306.
257. Al-Matrouk A, Alqallaf M, AlShemmeri A, BoJbarah H. Identification of synthetic cannabinoids that were seized, consumed, or associated with deaths in Kuwait in 2018 using GC-MS and LC-MS-MS analysis. *Forensic Sci Int*. 2019;303:109960.
258. Pacher P, Steffens S, Hasko G, Schindler TH, Kunos G. Cardiovascular effects of marijuana and synthetic cannabinoids: the good, the bad, and the ugly. *Nat Rev Cardiol*. 2018;15(3):151–166.
259. Kelkar AH, Smith NA, Martial A, Moole H, Tarantino MD, Roberts JC. An outbreak of synthetic cannabinoid–associated coagulopathy in Illinois. *N Engl J Med*. 2018;379(13):1216–1223.
260. Tournebize J, Gibaja V, Kahn J-P. Acute effects of synthetic cannabinoids: update 2015. *Subst Abuse*. 2017;38(3):344–366.
261. Moore KN, Garvin D, Thomas BF, Grabenauer M. Identification of eight synthetic cannabinoids, including 5F-AKB48 in seized herbal products using DART-TOF-MS and LC-QTOF-MS as nontargeted screening methods. *J Forensic Sci*. 2017;62(5):1151–1158.
262. Liu X, Villamagna A, Yoo J. The importance of recognizing cannabinoid hyperemesis syndrome from synthetic marijuana use. *J Med Toxicol*. 2017;13(2):199–200.
263. Adamowicz P, Gieron J, Gil D, Lechowicz W, Skulska A, Tokarczyk B. The effects of synthetic cannabinoid UR-144 on the human body-A review of 39 cases. *Forensic Sci Int*. 2017;273:e18–e21.
264. Tait RJ, Caldicott D, Mountain D, Hill SL, Lenton S. A systematic review of adverse events arising from the use of synthetic cannabinoids and their associated treatment. *Clin Toxicol*. 2016;54(1):1–13.
265. Tyndall JA, Gerona R, De Portu G, et al. An outbreak of acute delirium from exposure to the synthetic cannabinoid AB-CHMINACA. *Clin Toxicol*. 2015;53(10):950–956.
266. Gerostamoulos D, Drummer OH, Woodford NW. Deaths linked to synthetic cannabinoids. *Forensic Sci Med Pathol*. 2015;11(3):478.
267. Seely KA, Patton AL, Moran CL, et al. Forensic investigation of K2, Spice, and "bath salt" commercial preparations: a three-year study of new designer drug products containing synthetic cannabinoid, stimulant, and hallucinogenic compounds. *Forensic Sci Int*. 2013;233(1–3):416–422.
268. Young AC, Schwarz E, Medina G, et al. Cardiotoxicity associated with the synthetic cannabinoid, K9, with laboratory confirmation. *Am J Emerg Med*. 2012;30(7):1320 e1325–1327.
269. Karch SB. Cannabis and cardiotoxicity. *Forensic Sci Med Pathol*. 2006;2(1):13–18.
270. Wolf BC, Lavezzi WA, Sullivan LM, Flannagan LM. Methadone-related deaths in Palm Beach county. *J Forensic Sci*. 2004;49(2):375–378.
271. Gagajewski A, Apple FS. Methadone-related deaths in Hennepin county, Minnesota: 1992–2002. *J Forensic Sci*. 2003;48(3):668–671.
272. Baden MM. Methadone related deaths in New York City. *Int J Addict*. 1970;5(3):489–498.
273. Vignali C, Stramesi C, Morini L, Pozzi F, Groppi A. Methadone-related deaths. A ten year overview. *Forensic Sci Int*. 2015;257:172–176.
274. Green H, James RA, Gilbert JD, Harpas P, Byard RW. Methadone maintenance programs—a two-edged sword? *Am J Forensic Med Pathol*. 2000; 21(4):359–361.
275. Letsky MC, Zumwalt RE, Seifert SA, Benson BE. Cause of death conundrum with methadone use: a case report. *Am J Forensic Med Pathol*. 2011;32(2):193–196.
276. Wong SH, Wagner MA, Jentzen JM, et al. Pharmacogenomics as an aspect of molecular autopsy for forensic pathology/toxicology: does genotyping CYP 2D6 serve as an adjunct for certifying methadone toxicity? *J Forensic Sci*. 2003;48(6):1406–1415.
277. Tarumi Y, Pereira J, Watanabe S. Methadone and fluconazole: respiratory depression by drug interaction. *J Pain Symptom Manage*. 2002;23(2): 148–153.
278. Cobb MN, Desai J, Brown Jr LS, Zannikos PN, Rainey PM. The effect of fluconazole on the clinical pharmacokinetics of methadone. *Clin Pharmacol Ther*. 1998;63(6):655–662.
279. Lev R, Petro S, Lee A, et al. Methadone related deaths compared to all prescription related deaths. *Forensic Sci Int*. 2015;257:347–352.
280. Mijatovic V, Samojlik I, Ajdukovic N, Durendic-Brenesel M, Petkovic S. Methadone-related deaths – epidemiological, pathohistological, and toxicological traits in 10-year retrospective study in Vojvodina, Serbia. *J Forensic Sci*. 2014;59(5):1280–1285.
281. Pilgrim JL, McDonough M, Drummer OH. A review of methadone deaths between 2001 and 2005 in Victoria, Australia. *Forensic Sci Int*. 2013; 226(1–3):216–222.
282. Jantos R, Skopp G. Postmortem blood and tissue concentrations of R- and S-enantiomers of methadone and its metabolite EDDP. *Forensic Sci Int*. 2013;226(1–3):254–260.
283. Bernard JP, Havnes I, Slordal L, Waal H, Morland J, Khiabani HZ. Methadone-related deaths in Norway. *Forensic Sci Int*. 2013;224(1–3): 111–116.
284. Buchard A, Linnet K, Johansen SS, Munkholm J, Fregerslev M, Morling N. Postmortem blood concentrations of R- and S-enantiomers of methadone and EDDP in drug users: influence of co-medication and p-glycoprotein genotype. *J Forensic Sci*. 2010;55(2):457–463.
285. Albion C, Shkrum M, Cairns J. Contributing factors to methadone-related deaths in Ontario. *Am J Forensic Med Pathol*. 2010;31(4):313–319.
286. Paulozzi LJ, Logan JE, Hall AJ, McKinstry E, Kaplan JA, Crosby AE. A comparison of drug overdose deaths involving methadone and other opioid analgesics in West Virginia. *Addiction*. 2009;104(9):1541–1548.

287. Danielson TJ, Mozayani A, Sanchez LA. Methadone and methadone metabolites in postmortem specimens. *Forensic Sci Med Pathol*. 2008;4(3): 170–174.
288. Shields LB, Hunsaker Iii JC, Corey TS, Ward MK, Stewart D. Methadone toxicity fatalities: a review of medical examiner cases in a large metropolitan area. *J Forensic Sci*. 2007;52(6):1389–1395.
289. Milroy CM, Forrest AR. Methadone deaths: a toxicological analysis. *J Clin Pathol*. 2000;53(4):277–281.
290. Rust KY, Baumgartner MR, Meggiolaro N, Kraemer T. Detection and validated quantification of 21 benzodiazepines and 3 "z-drugs" in human hair by LC-MS/MS. *Forensic Sci Int*. 2012;215(1–3):64–72.
291. Lee H, Park Y, Jo J, et al. Analysis of benzodiazepines and their metabolites using DBS cards and LC-MS/MS. *Forensic Sci Int*. 2015;255: 137–145.
292. Luk S, Atayee RS, Ma JD, Best BM. Urinary diazepam metabolite distribution in a chronic pain population. *J Anal Toxicol*. 2014;38(3):135–142.
293. Wyman J, Bultman S. Postmortem distribution of heroin metabolites in femoral blood, liver, cerebrospinal fluid, and vitreous humor. *J Anal Toxicol*. 2004;28(4):260–263.
294. Thaulow CH, Oiestad AML, Rogde S, et al. Metabolites of heroin in several different post-mortem matrices. *J Anal Toxicol*. 2018;42(5):311–320.
295. Anastos N, Lewis SW, Barnett NW, Pearson JR, Kirkbride KP. The rapid analysis of heroin drug seizures using micellar electrokinetic chromatography with short-end injection. *J Forensic Sci*. 2005;50(1):37–42.
296. Zhang D, Shi X, Yuan Z, Ju H. Component analysis of illicit heroin samples with GC/MS and its application in source identification. *J Forensic Sci*. 2004;49(1):81–86.
297. Moeller MR, Hammer K, Engel O. Poppy seed consumption and toxicological analysis of blood and urine samples. *Forensic Sci Int*. 2004; 143(2–3):183–186.
298. Rohrig TP, Moore C. The determination of morphine in urine and oral fluid following ingestion of poppy seeds. *J Anal Toxicol*. 2003;27(7): 449–452.
299. Ceder G, Jones AW. Concentrations of unconjugated morphine, codeine and 6-acetylmorphine in urine specimens from suspected drugged drivers. *J Forensic Sci*. 2002;47(2):366–368.
300. Fraser AD, Worth D. Experience with a urine opiate screening and confirmation cutoff of 2000 ng/mL. *J Anal Toxicol*. 1999;23(6):549–551.
301. Lo DS, Chua TH. Poppy seeds: implications of consumption. *Med Sci Law*. 1992;32(4):296–302.
302. Hayes LW, Krasselt WG, Mueggler PA. Concentrations of morphine and codeine in serum and urine after ingestion of poppy seeds. *Clin Chem*. 1987;33(6):806–808.
303. Boerner U. The metabolism of morphine and heroin in man. *Drug Metab Rev*. 1975;4(1):39–73.
304. Moriya F, Hashimoto Y. Redistribution of basic drugs into cardiac blood from surrounding tissues during early-stages postmortem. *J Forensic Sci*. 1999;44(1):10–16.
305. Robertson MD, Drummer OH. Postmortem distribution and redistribution of nitrobenzodiazepines in man. *J Forensic Sci*. 1998;43(1):9–13.
306. Pelissier-Alicot AL, Gaulier JM, Champsaur P, Marquet P. Mechanisms underlying postmortem redistribution of drugs: a review. *J Anal Toxicol*. 2003;27(8):533–544.
307. Olson KN, Luckenbill K, Thompson J, et al. Postmortem redistribution of fentanyl in blood. *Am J Clin Pathol*. 2010;133(3):447–453.
308. Cook DS, Braithwaite RA, Hale KA. Estimating antemortem drug concentrations from postmortem blood samples: the influence of postmortem redistribution. *J Clin Pathol*. 2000;53(4):282–285.
309. Yarema MC, Becker CE. Key concepts in postmortem drug redistribution. *Clin Toxicol*. 2005;43(4):235–241.
310. Anderson WH, Prouty RW. Postmortem redistribution of drugs. In: Baselt RC, ed. *Advances in Analytical Toxicology*. Vol. 2. Chicago, IL: Year-Book Medical; 1989:70–102 (Baselt, R.C., ed).
311. Hilberg T, Bugge A, Beylich KM, Ingum J, Bjorneboe A, Morland J. An animal model of postmortem amitriptyline redistribution. *J Forensic Sci*. 1993;38(1):81–90.
312. Koren G, MacLeod SM. Postmortem redistribution of digoxin in rats. *J Forensic Sci*. 1985;30(1):92–96.
313. Wu AH, Hamilton R, Middleberg R, Kacinko S, Kearney T. Investigation of postmortem absorption and redistribution after the application of a fentanyl patch. *J Forensic Sci*. 2015;60(2):532–535.
314. Holland MG, Schwope DM, Stoppacher R, Gillen SB, Huestis MA. Postmortem redistribution of delta9-tetrahydrocannabinol (THC), 11-hydroxy-THC (11-OH-THC), and 11-nor-9-carboxy-THC (THCCOOH). *Forensic Sci Int*. 2011;212(1–3):247–251.
315. Rodda LN, Gerostamoulos D, Drummer OH. The postmortem redistribution of iso-alpha-acids in postmortem specimens. *Forensic Sci Med Pathol*. 2014;10(4):550–556.
316. Hargrove VM, Molina DK. Peripheral postmortem redistribution of morphine. *Am J Forensic Med Pathol*. 2014;35(2):106–108.
317. Lemaire E, Schmidt C, Dubois N, Denooz R, Charlier C, Boxho P. Site-, technique-, and time-related aspects of the postmortem redistribution of diazepam, methadone, morphine, and their metabolites: interest of popliteal vein blood sampling. *J Forensic Sci*. 2017;62(6):1559–1574.
318. Lemaire E, Schmidt C, Denooz R, Charlier C, Boxho P. Popliteal vein blood sampling and the postmortem redistribution of diazepam, methadone, and morphine. *J Forensic Sci*. 2016;61(4):1017–1028.
319. Krinsky CS, Lathrop SL, Zumwalt R. An examination of the postmortem redistribution of fentanyl and interlaboratory variability. *J Forensic Sci*. 2014;59(5):1275–1279.
320. Johnson JT, Everly AG, Kpakima FEF, Detke HC. Postmortem redistribution of olanzapine following intramuscular administration of olanzapine pamoate in dogs. *Forensic Sci Int*. 2015;257:353–358.
321. Nagasawa S, Torimitsu S, Chiba F, Kubo Y, Yajima D, Iwase H. Donepezil distribution in postmortem cases and potential for redistribution. *Forensic Sci Int*. 2015;251:132–138.
322. Gill JR, Lin PT, Nelson L. Reliability of postmortem fentanyl concentrations in determining the cause of death. *J Med Toxicol*. 2012;9.
323. Robertson MD, Drummer OH. Postmortem drug metabolism by bacteria. *J Forensic Sci*. 1995;40(3):382–386.
324. Wad N. Degradation of clonazepam in serum by light confirmed by means of a high performance liquid chromatographic method. *Ther Drug Monit*. 1986;8(3):358–360.
325. Wyman JF, Dean DE, Yinger R, et al. The temporal fate of drugs in decomposing porcine tissue. *J Forensic Sci*. 2011;56(3):694–699.
326. Isenschmid DS, Levine BS, Caplan YH. A comprehensive study of the stability of cocaine and its metabolites. *J Anal Toxicol*. 1989;13(5): 250–256.

327. Gill J, Stajic M. Classical mistakes made in forensic toxicology by forensic pathologists. *Acad Forensic Pathol*. 2012;2(2) 228–234.
328. Gill J, Andrew T, Gilliland M, Love J, Matshes E, Reichard R. National association of medical examiners position paper: recommendations for the postmortem assessment of suspected head trauma in infant and young children. *Acad Forensic Pathol*. 2013;4(2):206–213.
329. Gill JR. From death to death certificate: what do the dead say? *J Med Toxicol*. 2016;13.
330. Adelson L. No anatomic cause of death; the enigma of the forensic pathologist. *Conn Med*. 1954;18(9):732–738.
331. Warner EA, Greene GS, Buchsbaum MS, Cooper DS, Robinson BE. Diabetic ketoacidosis associated with cocaine use. *Arch Intern Med*. 1998; 158(16):1799–1802.
332. Isidro ML, Jorge S. Recreational drug abuse in patients hospitalized for diabetic ketosis or diabetic ketoacidosis. *Acta Diabetol*. 2013;50(2): 183–187.
333. Nyenwe EA, Loganathan RS, Blum S, et al. Active use of cocaine: an independent risk factor for recurrent diabetic ketoacidosis in a city hospital. *Endocr Pract*. 2007;13(1):22–29.
334. Ely SF. Sudden death related to diabetes mellitus: current and emerging relevance to the forensic pathologist. *Acad Forensic Pathol*. 2016;6(2): 154–163.
335. Ng R, Darko D, Hillson R. Street drug use among young patients with Type 1 diabetes in the UK. *Diabet Med*. 2004;21(3):295–296.
336. Glasgow AM, Tynan D, Schwartz R, et al. Alcohol and drug use in teenagers with diabetes mellitus. *J Adolesc Health*. 1991;12(1):11–14.
337. Baudrie V, Chaouloff F. Mechanisms involved in the hyperglycemic effect of the 5-HT1C/5-HT2 receptor agonist, DOI. *Eur J Pharmacol*. 1992; 213(1):41–46.
338. Giugliano D. Morphine, opioid peptides, and pancreatic islet function. *Diabetes Care*. 1984;7(1):92–98.
339. Corry JE. A review. Possible sources of ethanol ante- and post-mortem: its relationship to the biochemistry and microbiology of decomposition. *J Appl Bacteriol*. 1978;44(1):1–56.
340. O'Neal CL, Poklis A. Postmortem production of ethanol and factors that influence interpretation: a critical review. *Am J Forensic Med Pathol*. 1996;17(1):8–20.
341. Hanzlick R. Postmortem ethanol concentrations and postmortem ethanol production. *Acad Forensic Pathol*. 2014;4(2):156–160.
342. Zumwalt RE, Bost RO, Sunshine I. Evaluation of ethanol concentrations in decomposed bodies. *J Forensic Sci*. 1982;27(3):549–554.
343. Jones AW. Forensic drug profile: cocaethylene. *J Anal Toxicol*. 2019;43(3):155–160.
344. Bailey D. Detection of isopropanol in acetonemic patients not exposed to isopropanol. *J Clin Tox*. 1990;28(4):459–466.
345. Jenkins AJ, Merrick TC, Oblock JM. Evaluation of isopropanol concentrations in the presence of acetone in postmortem biological fluids. *J Anal Toxicol*. 2008;32(8):719–720.
346. Denmark L. The investigation of beta-hydroxybutyrate as a marker for sudden death due to hypoglycemia in alcoholics. *For Sci Int*. 1993;62: 225–232.
347. Sadones N, Lambert WE, Stove CP. The (non)sense of routinely analysing beta-hydroxybutyric acid in forensic toxicology casework. *Forensic Sci Int*. 2017;274:38–43.
348. Palmiere C. The (non)sense of routinely analyzing beta-hydroxybutyrate in forensic toxicology casework. *Forensic Sci Int*. 2017;279:e18–e19.
349. Elliott S, Smith C, Cassidy D. The post-mortem relationship between beta-hydroxybutyrate (BHB), acetone and ethanol in ketoacidosis. *Forensic Sci Int*. 2010;198(1–3):53–57.
350. Heninger M. Postmortem vitreous beta-hydroxybutyrate: interpretation in a forensic setting. *J Forensic Sci*. 2012;57(5):1234–1240.
351. Klaric KA, Milroy CM, Parai JL. Utility of postmortem vitreous beta-hydroxybutyrate testing for distinguishing sudden from prolonged deaths and for diagnosing ketoacidosis. *J Forensic Sci*. 2020;65.
352. Wetli CV, Mittleman RE. The body packer syndrome: toxicity following ingestion of illicit drugs packaged for transportation. *J Forensic Sci*. 1981;26:492–500.
353. Reiber KB, McGee MP, Stajic M. *The Speeding Mule-Aan Unusual Fatality Associated with the Body Packer Syndrome*. Nashville, TN: American Academy of Forensic Sciences Platform Presentation; 1996. K17.
354. Klein C, Balash Y, Pollak L, Hiss J, Rabey MJ. Body packer: cocaine intoxication, causing death, masked by concomitant administration of major tranquilizers. *Eur J Neurol*. 2000;7(5):555–558.
355. Gill JR, Graham SM. Ten years of "body packers" in New York City: 50 deaths. *J Forensic Sci*. 2002;47(4):843–846.
356. Shields LB, Rolf CM, Hunsaker 3rd JC. Sudden death due to acute cocaine toxicity-excited delirium in a body packer. *J Forensic Sci*. 2015;60(6): 1647–1651.
357. Flach PM, Gascho D, Fader R, Martinez R, Thali MJ, Ebert LC. Death by "snow"! A fatal forensic case of cocaine leakage in a "drug mule" on postmortem computed and magnetic resonance tomography compared with autopsy. *Am J Forensic Med Pathol*. 2017;38(4):339–344.
358. Davidson LE, Berman AL, Murray D, et al. Operational criteria for determining suicide. *Mor Mortal Wkly Rep*. 1988;37:773–780.
359. Rosenberg ML, Davidson LE, Smith JC, et al. Operational criteria for the determination of suicide. *J Forensic Sci*. 1988;33(6):1445–1456.
360. Jonasson B, Jonasson U, Saldeen T. Suicides may be overreported and accidents underreported among fatalities due to dextropropoxyphene. *J Forensic Sci*. 1999;44(2):334–338.
361. Dierksen J, Gonsoulin M, Walterscheid JP. Poor man's methadone: a case report of loperamide toxicity. *Am J Forensic Med Pathol*. 2015;36(4): 268–270.
362. Powell JW, Presnell SE. Loperamide as a potential drug of abuse and misuse: fatal overdoses at the medical University of South Carolina. *J Forensic Sci*. 2019;64(6):1726–1730.
363. Eggleston W, Palmer R, Dube PA, et al. Loperamide toxicity: recommendations for patient monitoring and management. *Clin Toxicol*. 2020; 58(5):355–359.
364. Padwell A. Cyanide Poisoning. Case studies of one homicide and two suicides. *Am J Forensic Med Pathol*. 1997;18(2):185–188.
365. Vogel SN, Sultan TR, Ten Eyck RP. Cyanide poisoning. *Clin Toxicol*. 1981;18(3):367–383.
366. Musshoff F, Schmidt P, Daldrup T, Madea B. Cyanide fatalities: case studies of four suicides and one homicide. *Am J Forensic Med Pathol*. 2002; 23(4):315–320.
367. Cina SJ, Raso DS, Conradi SE. Suicidal cyanide ingestion as detailed in final exit. *J Forensic Sci*. 1994;39(6):1568–1570.
368. Gill JR, Marker E, Stajic M. Suicide by cyanide: 17 deaths. *J Forensic Sci*. 2004;49(4):826–828.
369. Jaffe FA. Pathogenicity of carbon monoxide. *Am J Forensic Med Pathol*. 1997;18(4):406–410.

370. Carson HJ, Esslinger K. Carbon monoxide poisoning without cherry-red livor. *Am J Forensic Med Pathol.* 2001;22(3):233–235.
371. Morgan DR, Poon P, Titley J, Jagger SF, Rutty GN. Unique case of fatal carbon monoxide poisoning in the absence of a combustible fossil fuel. *Am J Forensic Med Pathol.* 2001;22(3):220–224.
372. Fagin J, Bradley J, Williams D. Carbon monoxide poisoning secondary to inhaling methylene chloride. *Br Med J.* 1980;281(6253):1461.
373. deRoux SJ. Suicidal asphyxiation by inhalation of automobile emission without carbon monoxide poisoning. *J Forensic Sci.* 2006;51(5): 1158–1159.
374. Middleberg RA, Easterling DE, Zelonis SF, Rieders F, Rieders MF. Estimation of perimortal percent carboxy-heme in nonstandard postmortem specimens using analysis of carbon monoxide by GC/MS and iron by flame atomic absorption spectrophotometry. *J Anal Toxicol.* 1993; 17(1):11–13.
375. Lin PT, Dunn WA. Suicidal carbon monoxide poisoning by combining formic acid and sulfuric acid within a confined space. *J Forensic Sci.* 2014;59(1):271–273.
376. Hao H, Zhou H, Liu X, Zhang Z, Yu Z. An accurate method for microanalysis of carbon monoxide in putrid postmortem blood by head-space gas chromatography-mass spectrometry (HS/GC/MS). *Forensic Sci Int.* 2013;229(1–3):116–121.
377. deRoux SJ, Marker E, Stajic M. Fatalities by ingestion of propylene glycol. *J Forensic Sci.* 2005;50(4):939–941.
378. Leth PM, Gregersen M. Ethylene glycol poisoning. *Forensic Sci Int.* 2005;155(2–3):179–184.
379. Wittschieber D, Heuberger K, Schulz R, Kohler H, Varchmin-Schultheiss K. Fatal poisoning with diethylene glycol in an unusual setting. *Forensic Sci Med Pathol.* 2019;15(4):649–652.
380. Mudan A, Repplinger D, Lebin J, Lewis J, Vohra R, Smollin C. Severe methemoglobinemia and death from intentional sodium nitrite ingestions. *J Emerg Med.* 2020;59.
381. Katabami K, Hayakawa M, Gando S. Severe methemoglobinemia due to sodium nitrite poisoning. *Case Rep Emerg Med.* 2016;2016:9013816.
382. Greenberg M, Birnkrant WB, Schiftner JJ. Outbreak of sodium nitrite poisoning. *Am J Public Health Nation's Health.* 1945;35(11):1217–1220.
383. Gautami S, Rao RN, Raghuram TC, Rajagopalan S, Bhat RV. Accidental acute fatal sodium nitrite poisoning. *J Toxicol Clin Toxicol.* 1995;33(2): 131–133.
384. Harvey M, Cave G, Chanwai G. Fatal methaemoglobinaemia induced by self-poisoning with sodium nitrite. *Emerg Med Australas.* 2010;22(5): 463–465.
385. Gowans WJ. Fatal methaemoglobinaemia in a dental nurse. A case of sodium nitrite poisoning. *Br J Gen Pract.* 1990;40(340):470–471.
386. Durão C, Pedrosa F, Dinis-Oliveira RJ. Another suicide by sodium nitrite and multiple drugs: an alarming trend for "exit"? *Forensic Sci Med Pathol.* 2021;17(2):362–366.
387. Tang Y, Bieschke E, Labitzke E, et al. *Comprehensive Molecular Genetic Testing for the Cardiac Channelopathy Genes in 42 Cases of Sudden Infant Death and Sudden Unexplained Death in the City of New York Revealed High Mutation Rate.* Washington, DC: American Academy of Forensic Sciences; February 2008.
388. Wang D, Shah KR, Um SY, et al. Cardiac channelopathy testing in 274 ethnically diverse sudden unexplained deaths. *Forensic Sci Int.* 2014;237: 90–99.
389. Gando I, Yang HQ, Coetzee WA. Functional significance of channelopathy gene variants in unexplained death. *Forensic Sci Med Pathol.* 2019; 15(3):437–444.
390. Coe JI. Post-mortem chemistries on vitreous humor. *Am J Clin Pathol.* 1969;51:741–750.
391. Blumenfeld TA, Mantell CH, Catherman RL, Blanc WA. Postmortem vitreous humor chemistry in sudden infant death syndrome and in other causes of death in childhood. *Am J Clin Pathol.* 1979;71(2):219–223.
392. Chaturvedi AK, Botch SR, Canfield DV, Forster EM. Vitreous fluid and/or urine glucose concentrations in 1335 civil aviation accident pilot fatalities. *J Forensic Sci.* 2009;54(3):715–720.
393. Supreme-Court. *Burrage V. United States, 134 S.Ct.* 2014:881.
394. Gilson TP, Rendon C, Pinjuh J. Rules for establishing causation in opiate/opioid overdose prosecutions – the burrage decision. *Acad Forensic Pathol.* 2017;7(1):87–90.
395. Daula M, Hanzlick R. Evaluation of certifier practices regarding alcohol-related deaths: Fulton county medical examiner's center, Atlanta, Georgia, 2004. *Am J Forensic Med Pathol.* 2006;27(4):355–358.
396. Lahti RA, Sajantila A, Korpi H, Poikolainen K, Vuori E. Under-recording of ethanol intoxication and poisoning in cause-of-death data: causes and consequences. *Forensic Sci Int.* 2011;212(1–3):121–125.
397. Courts C, Madea B. Specific micro-RNA signatures for the detection of saliva and blood in forensic body-fluid identification. *J Forensic Sci.* 2011;56(6):1464–1470.
398. Svetla S, Daniella B. Drug overdose deaths: let's get specific. *Pub Health Rep.* 2015;130:339–342.
399. Hirsch CS, Flomenbaum M. *Problem-solving in Death Certification. American Society of Clinical Pathologists, Forensic Pathology Check Sample.* Vols. 95–1. 1995:1–31. FP.
400. Martin CS, Chung T, Langenbucher JW. How should we revise diagnostic criteria for substance use disorders in the DSM-V? *J Abnorm Psychol.* 2008;117(3):561.
401. Hasin DS, O'brien CP, Auriacombe M, et al. DSM-5 criteria for substance use disorders: recommendations and rationale. *Am J Psychiatr.* 2013; 170(8):834–851.
402. Hasin D, Hatzenbuehler ML, Keyes K, Ogburn E. Substance use disorders: diagnostic and statistical manual of mental disorders, (DSM-IV) and International Classification of Diseases, (ICD-10). *Addiction.* 2006;101:59–75.
403. O'Brien C. Addiction and dependence in DSM-V. *Addiction.* 2011;106(5):866–867.
404. Burroughs W. *Naked Lunch.* Paris: Olympia Press; 1959.
405. Gill JR. Practical toxicology for the forensic pathologist. In: Tsokos M, ed. *Forensic Pathology Reviews.* Vol. 2. Totowa, NJ: Humana Press; 2005: 243–272.
406. Roueché B. *Eleven Blue Men and Other Narratives of Medical Detection.* Boston, MA: Little Brown & Co; 1953.
407. Moshage H, Stegeman CA, Jansen PL. Determination of nitrite and nitrate in stored urine. *Clin Chem.* 1998;44(8 Pt 1):1780–1781.
408. Moshage H, Kok B, Huizenga JR, Jansen PL. Nitrite and nitrate determinations in plasma: a critical evaluation. *Clin Chem.* 1995;41(6 Pt 1): 892–896.

409. Yan H, Zhuo X, Shen B, Xiang P, Shen M. Determination of nitrite in whole blood by high-performance liquid chromatography with electrochemical detection and a case of nitrite poisoning. *J Forensic Sci*. 2016;61(1):254–258.
410. Cvetkovic D, Zivkovic V, Lukic V, Nikolic S. Sodium nitrite food poisoning in one family. *Forensic Sci Med Pathol*. 2019;15(1):102–105.
411. Neth MR, Love JS, Horowitz BZ, Shertz MD, Sahni R, Daya MR. Fatal sodium nitrite poisoning: key considerations for prehospital providers. *Prehosp Emerg Care*. 2021;25(6):844–850.
412. Hideyuki Nushida MN. An autopsy case of fatal methemoglobinemia due to ingestion of sodium nitrite. *J Forensic Res*. 2014;06(01).
413. Szymanski LJ, Hurth K. "Pistachio" and "Avatar" green-blue discoloration of the brain. *Acad Forensic Pathol*. 2016;6(1):150–152.
414. Flores M, Ogden E. Blue baby: an uncommon postmortem artifact due to perimortem methylene blue administration. In: *Paper Presented at: NAME Annual Meeting*. 2020. Kansas City, MO.
415. Durao C, Pedrosa F, Dinis-Oliveira RJ. Greenish-blue discoloration of the brain and heart after treatment with methylene blue. *Forensic Sci Med Pathol*. 2020;17.
416. Romanelli F, Smith KM, Thornton AC, Pomeroy C. Poppers: epidemiology and clinical management of inhaled nitrite abuse. *Pharmacotherapy*. 2004;24(1):69–78.
417. Plankey MW, Ostrow DG, Stall R, et al. The relationship between methamphetamine and popper use and risk of HIV seroconversion in the multicenter AIDS cohort study. *J Acquir Immune Defic Syndr*. 2007;45(1):85–92.
418. Hambrick HR, Park SH, Palamar JJ, Estreet A, Schneider JA, Duncan DT. Use of poppers and HIV risk behaviours among men who have sex with men in Paris, France: an observational study. *Sex Health*. 2018;15(4):370–373.
419. McManus TJ, Starrett LA, Harris JR. Amyl nitrite use by homosexuals. *Lancet*. 1982;1(8270):503.
420. Winek CL, Fusia E, Collom WD, Shanor SP. Cyanide poisoning as a mode of suicide. *Forensic Sci*. 1978;11(1):51–55.
421. Fernando GC, Busuttil A. Cyanide ingestion. Case studies of four suicides. *Am J Forensic Med Pathol*. 1991;12(3):241–246.
422. Ballantyne B. Autopsy findings following death by intramuscular hydrogen cyanide: an experimental study. *Med Sci Law*. 1970;10(3):171–174.
423. Curry AS, Price DE, Rutter ER. The production of cyanide in post mortem material. *Acta Pharmacol Toxicol*. 1967;25(3):339–344.
424. Le Garff E, Delannoy Y, Mesli V, Allorge D, Hedouin V, Tournel G. Cyanide suicide after deep web shopping: a case report. *Am J Forensic Med Pathol*. 2016;37(3):194–197.
425. Musshoff F, Kirschbaum KM, Madea B. An uncommon case of a suicide with inhalation of hydrogen cyanide. *Forensic Sci Int*. 2011;204(1–3): e4–7.
426. Moriya F, Hashimoto Y. Potential for error when assessing blood cyanide concentrations in fire victims. *J Forensic Sci*. 2001;46(6):1421–1425.
427. Lundquist P, Rammer L, Sorbo B. The role of hydrogen cyanide and carbon monoxide in fire casualties: a prospective study. *Forensic Sci Int*. 1989;43(1):9–14.
428. Noguchi TT, Eng JJ, Klatt EC. Significance of cyanide in medicolegal investigations involving fires. *Am J Forensic Med Pathol*. 1988;9(4): 304–309.
429. Wetherall HR. The occurrence of cyanide in the blood of fire victims. *J Forensic Sci*. 1966;11:167–173.
430. Ferrari LA, Giannuzzi L. Assessment of carboxyhemoglobin, hydrogen cyanide and methemoglobin in fire victims: a novel approach. *Forensic Sci Int*. 2015;256:46–52.
431. Birngruber CG, Veit F, Lang J, Verhoff MA. Inhaled cyanide poisoning as a vital sign in a room fire victim. *Forensic Sci Int*. 2017;281:e16–e18.
432. Brown KS, Robinette RR. No simple pattern of inheritance in ability to smell solutions of cyanide. *Nature*. 1967;215(99):406–408.
433. Andrews JM, Sweeney ES, Grey TC, Wetzel T. The biohazard potential of cyanide poisoning during postmortem examination. *J Forensic Sci*. 1989;34(5):1280–1284.
434. Nolte KB, Taylor DG, Richmond JY. Biosafety considerations for autopsy. *Am J Forensic Med Pathol*. 2002;23(2):107–122.
435. Lokan RJ, James RA, Dymock RB. Apparent post-mortem production of high levels of cyanide in blood. *J Forensic Sci Soc*. 1987;27(4):253–259.
436. Desai D, Su M. Cyanide poisonings. In: Waltham, MA: UpToDate; 2022 (Accessed 18 March 2022).
437. Weider B, Fournier JH. Activation analyses of authenticated hairs of Napoleon Bonaparte confirm arsenic poisoning. *Am J Forensic Med Pathol*. 1999;20(4):378–382.
438. Adelson L, George RA, Mandel A. Acute arsenic intoxication shown by roentgenograms. *Arch Intern Med*. 1961;107:401–404.
439. Tournel G, Houssaye C, Humbert L, et al. Acute arsenic poisoning: clinical, toxicological, histopathological, and forensic features. *J Forensic Sci*. 2011;56(Suppl 1):S275–S279.
440. Forney RB. Arsenic intoxication. *Postgrad Med*. 1954;15(6):A-26–38.
441. Zheng J, Zhang K, Liu Y, Wang Y. Fatal acute arsenic poisoning by external use of realgar: case report and 30 years literature retrospective study in China. *Forensic Sci Int*. 2019;300:e24–e30.
442. Lu P, Ma JQ, Li F, Xu GH, Guo W, Zhou HM. A fatal case of acute arsenic poisoning. *J Forensic Sci*. 2019;64(4):1271–1273.
443. Adelson L. Homicidal arsenic poisoning. Medicolegal investigation of four cases. 2. *Postgrad Med*. 1966;39(2):A-46–60.
444. Adelson L. Homicidal arsenic poisoning. Medicolegal investigation of four cases. I. *Postgrad Med*. 1966;39(1):A-72–86. contd).
445. Seavolt MB, Sarro RA, Levin K, Camisa C. Mees' lines in a patient following acute arsenic intoxication. *Int J Dermatol*. 2002;41(7):399–401.
446. Shapiro J. Arsenic and phosphate: measured by various techniques. *Science*. 1971;171(3968):234.
447. Shapiro HA. Arsenic content of human hair and nails: its interpretation. *J Forensic Med*. 1967;14(2):65–71.
448. Gupta A, Diaz FJ, Lal A, Sung L, Aaron CK. Basal ganglion hemorrhage as delayed complication of diethylene glycol ingestion. *Am J Forensic Med Pathol*. 2017;38(1):39–42.
449. O'Brien KL, Selanikio JD, Hecdivert C, et al. Epidemic of pediatric deaths from acute renal failure caused by diethylene glycol poisoning. Acute Renal Failure Investigation Team. *JAMA*. 1998;279(15):1175–1180.
450. Garg U, Frazee 3rd C, Johnson L, Turner JW. A fatal case involving extremely high levels of ethylene glycol without elevation of its metabolites or crystalluria. *Am J Forensic Med Pathol*. 2009;30(3):273–275.
451. Armstrong EJ, Engelhart DA, Jenkins AJ, Balraj EK. Homicidal ethylene glycol intoxication: a report of a case. *Am J Forensic Med Pathol*. 2006; 27(2):151–155.
452. Proenca P, Teixeira H, de Mendonca MC, et al. Aldicarb poisoning: one case report. *Forensic Sci Int*. 2004;146(Suppl):S79–S81.
453. Papoutsis I, Mendonis M, Nikolaou P, et al. Development and validation of a simple GC-MS method for the simultaneous determination of 11 anticholinesterase pesticides in blood–clinical and forensic toxicology applications. *J Forensic Sci*. 2012;57(3):806–812.

454. Sabino BD, Torraca TG, Moura CM, Rozenbaum HF, de Castro Faria MV. Development of a simple and low-cost enzymatic methodology for quantitative analysis of carbamates in meat samples of forensic interest. *J Forensic Sci*. 2010;55(3):808–812.
455. Singer PP, Jones GR, Bannach BG, Denmark L. Acute fatal acetaminophen overdose without liver necrosis. *J Forensic Sci*. 2007;52(4):992–994.
456. Bauer M, Babel B, Giesen H, Patzelt D. Fulminant liver failure in a young child following repeated acetaminophen overdosing. *J Forensic Sci*. 1999;44(6):1299–1303.
457. De-Giorgio F, Lodise M, Chiarotti M, d'Aloja E, Carbone A, Valerio L. Possible fatal acetaminophen intoxication with atypical clinical presentation. *J Forensic Sci*. 2013;58(5):1397–1400.
458. Gosselin M, Daze Y, Mireault P, Crahes M. Toxic myocarditis caused by acetaminophen in a multidrug overdose. *Am J Forensic Med Pathol*. 2017;38(4):349–352.
459. Bernal W, Auzinger G, Dhawan A, Wendon J. Acute liver failure. *Lancet*. 2010;376(9736):190–201.
460. Kruszyna R, Smith RP, Kruszyna H. Determining sodium azide concentration in blood by ion chromatography. *J Forensic Sci*. 1998;43(1): 200–202.
461. Le Blanc-Louvry I, Laburthe-Tolra P, Massol V, et al. Suicidal sodium azide intoxication: an analytical challenge based on a rare case. *Forensic Sci Int*. 2012;221(1–3):e17–20.
462. Chang S, Lamm SH. Human health effects of sodium azide exposure: a literature review and analysis. *Int J Toxicol*. 2003;22(3):175–186.
463. Marquet P, Clement S, Lotfi H, et al. Analytical findings in a suicide involving sodium azide. *J Anal Toxicol*. 1996;20(2):134–138.
464. Nine JS, Rund CR. Fatality from diphenhydramine monointoxication: a case report and review of the infant, pediatric, and adult literature. *Am J Forensic Med Pathol*. 2006;27(1):36–41.
465. Naso C, Jenkins AJ, Younger 3rd D. A study of drug detection in a postmortem pediatric population. *J Forensic Sci*. 2008;53(2):483–490.
466. Lowry JA, Fine JS, Calello DP, Marcus SM. Pediatric fatality review of the 2013 national poison database system (NPDS): focus on intent. *Clin Toxicol*. 2015;53(2):79–81.
467. Paul ABM, Simms L, Mahesan AM. The toxicology of methadone-related death in infants under 1 Year: three case series and review of the literature. *J Forensic Sci*. 2017;62(5):1414–1417.
468. Al-Abri SA, Yang HS, Olson KR. Unintentional pediatric ophthalmic tetrahydrozoline ingestion: case files of the medical toxicology fellowship at the University of California, San Francisco. *J Med Toxicol*. 2014;10(4):388–391.
469. Morley SR, Becker J, Al-Adnani M, Cohen MC. Drug- and alcohol-related deaths at a pediatric institution in the United Kingdom. *Am J Forensic Med Pathol*. 2012;33(4):390–394.
470. Madadi P, Kelly LE, Ross CJ, Kepron C, Edwards JN, Koren G. Forensic investigation of methadone concentrations in deceased breastfed infants. *J Forensic Sci*. 2016;61(2):576–580.
471. Allibe N, Eysseric-Guerin H, Kintz P, et al. Amitriptyline poisoning of a baby: how informative can hair analysis be? *Forensic Sci Int*. 2015;249: 53–58.

CHAPTER

15

Special topics: in-custody deaths, physical altercations, neglect (adult), approach to unexpected hemorrhage, and delayed and concealed homicides

Susan F. Ely and James R. Gill

At every homicide, there is one accurate witness without self-serving statements. The pathologist interrogates that witness.
-Dr. Charles S. Hirsch

Deaths in detention centers/correctional facilities/police custody

Incarcerated persons are at risk for deaths due to intentional self-harm (suicide), unintentional injury (accident), the hand of another (homicide), and disease (natural). Moreover, inmates (and ex-inmates) are at increased risk of death compared to the general public. This excess mortality has been attributed to an increased risk of unnatural death, particularly drug intoxications and suicide.[1–6]

In-custody deaths are typically investigated by the medical examiner or coroner (ME/C) and include those occurring in police custody or in a governmental detention facility (e.g., prison). In-custody fatalities are challenging because of the potential for death to occur at the hand of another (e.g., another inmate, corrections officer, or law enforcement agent) or due to neglect. When a government revokes a person's liberty, there is a duty to provide safe and adequate housing, food, and medical care. Failure to exercise proper care of a person in custody may affect the manner of death.

The investigation and certification of in-custody deaths involve medical, legal, and public health issues. Recent papers have reviewed the investigation of inmate deaths[7] and deaths associated with law enforcement apprehension.[8] Persons in custody do die from the same diseases and injuries that occur within the general public.[9] Inmates may commit suicide, overdose on drugs, or die of heart disease. Although these types of deaths are commonplace for the forensic pathologist (FP), they take on greater import when they occur in custody. Inmates live in close quarters, with people who may have been convicted of violent crimes. Is the hanging death a suicide or a ligature homicide made to appear as a suicide?[7] Even deaths due to natural disease may deserve ME/C scrutiny. In fact, some jurisdictions require an ME/C autopsy on all inmate deaths. Questions of inappropriate or lack of treatment may result in civil lawsuits and even criminal charges.

Graham categorized the causes of death into three stages: preapprehension, apprehension, and incarceration.[8] **Preapprehension deaths,** involving firearms by law enforcement, are relatively straightforward to certify.[10] The cause of death is clear and the generally accepted manner of death is homicide (see also "suicide by cop" below). During legal **apprehension**, an individual is at risk for death and/or injury from a variety of natural, chemical, and/or traumatic causes.[8,11,12] These include restraint asphyxia, blunt injury, electric shock, intoxication, and/or exacerbation of preexistent disease. *Nonfirearm* apprehension deaths that involve some combination of a physical altercation, restraint, pepper spray, and electromechanical disruption devices are challenging and potentially explosive deaths to investigate.[13–18] The addition of cocaine, psychosis, obesity, and other natural diseases potentially further

Principles of Forensic Pathology
https://doi.org/10.1016/B978-0-323-91796-4.00013-1

complicate these scenarios.[13,16,19–23] The determination of the manner of these deaths can be challenging and may vary by jurisdiction, depending upon local customs and the specific frame by frame details of the fatality.[8,24]

Deaths during apprehension may be due to an exacerbation of preexisting disease. Heart disease, sickle cell disease or trait, asthma, and cerebral artery aneurysms, in combination with stress, may be fatal. Stress increases epinephrine, causing the blood pressure and heart rate to increase, which, in turn, leads to greater demand on the cardiovascular system. Stress and/or prolonged physical altercations may cause arrhythmias and/or demand-related ischemia, exacerbate asthma, cause red blood cell sickling, or rupture preexisting arterial aneurysms.[25–34] If death at that moment was unlikely but for the stress/physical altercation of a hostile or criminal situation, it may qualify as a homicide by heart attack–type fatality (see below and Chapter 3).[27,35]

Deaths in the setting of police restraint

In 1992, Luke and Reay wrote about "The Perils of Investigating and Certifying Deaths in Police Custody."[36] They used the term "perilous" because in forensic pathology, "there is no more slippery slope than deaths in custody. Justice must be served, and served fairly, first and foremost. Judgments made have legal ramifications and are often rendered without recourse to legitimate second opinions. Political fallout may result, regardless of the conclusions reached." They continued: "An integral aspect of forensic pathology is the correlation between the circumstances of death and the pathologic and toxicologic findings of the postmortem examination. Deaths in police custody often demonstrate even less pathologic evidence than is the case with most violent asphyxial fatalities and, consequently, with fewer anatomic findings, circumstantial, historical, and scene investigative information becomes of the utmost importance."[36]

Deaths in the setting of police restraint exemplify the challenges of proper death investigation and certification.[13–16,18,23,26,36–86] As in other deaths, one first determines the proximate cause of death and the manner follows. Many of the possible etiologies and contributing conditions associated with these deaths, however, do not have anatomic correlates and/or demonstrable mechanisms of death. In addition, there may be competing or synergistic conditions that further complicate the determination. People may be forcibly restrained, have their neck or chest compressed, or have a primary cardiac arrhythmia without demonstration of any anatomic or microscopic evidence at autopsy.

These deaths may include *external factors* such as neck or chest compression, artifacts of resuscitative efforts, medication administration, exposure to noxious chemicals (e.g., pepper spray) and electromechanical devices, canine bites, blunt injury, smothering, placement of a spit-masks over the face, and potential positional compromise. *Internal factors* include comorbid conditions, such as, heart disease, obesity, hemoglobinopathies, and asthma, as well as drug intoxications, and psychiatric illness. It is the role of the FP to take this wide differential diagnosis and determine why the person died based on an investigation of the circumstances, the autopsy findings (or lack of findings), chemical and toxicological analyses, microscopic examination, etc.

There may be is no identifiable "smoking gun" with these fatalities, and the analysis will invariably entail the weighing of various factors to find a single or combination of causes that best explains the death. When facing competing causes of death in any fatality, the mechanism is usually a strong guide. In instances of restraint, for example, there often is no gross evidence of a mechanism (e.g., such as in the case of exsanguination, a subdural hemorrhage, or a fractured hyoid bone with petechiae) and so the FP must rely upon the circumstances. A review of witness statements and video recordings may allow for the "reconstruction" of events.[36] Determining when the person became unresponsive is the first step. The temporal relationship to unresponsiveness and restraint can help to include or exclude external factors. It is not unusual to find extensive superficial injuries (abrasions, contusions, lacerations), but linking them to the cause of death demands the identification of a mechanism (e.g., subdural hematoma, pneumothorax, functional lethal compromise, etc.).

Luke and Reay stated that in asphyxial and custody deaths in particular, "the pathology is often less important than are the investigative circumstances." In turn, both the pathologic evidence and the circumstances must be evaluated within the framework of, to paraphrase Dr. Lester Adelson, "that most precious (and all too uncommon) of human attributes, common sense."[36]

Asphyxia is described in great detail Chapter 11, but a few concepts would be both helpful and relevant to reiterate here. The term "asphyxia" or "restraint asphyxia" is often invoked in custody deaths. Asphyxia is a mechanism of death. Moreover, by definition, it arises as a result of an *injury,* structurally demonstrable or functional, not by natural causes. The problem is that there is no one diagnostic autopsy sign of asphyxia. This is because everyone ultimately dies from a breakdown in the body's oxygen cycle; in that sense, all deaths could be viewed as "asphyxial," but "asphyxia" as a pathophysiologic entity, per se, is not necessarily grossly or even microscopically identifiable. Rather, the question that demands an answer when someone asserts that the death was due to "asphyxia" or "restraint asphyxia" is, "how was the asphyxia brought about?" This is essentially asking, what is the proximate cause of death (the underlying cause of the asphyxia)? And then, secondarily, what evidence is there that supports this opinion and does it comport with our understanding of human physiology and the circumstances of this specific death? For example, the mechanism of death in neck compression is more complex than that of pure "asphyxia." The dominant mechanism is often a local cerebral ischemia due to a compromise of the cervical vasculature. There may or may not be a component of airway compromise depending on the nature and location of the application of force, but a person may or may not still move air and vocalize while sustaining a fatal compression of the neck. To make this diagnosis, one must consider all of the autopsy findings (which may be minimal to none) and the entirety of the circumstances/circumstantial data.

Also, when investigating these deaths, one must consider the interplay between disease and emotional stress.[25,28,87–98] Physical altercations are stressful. Some of these may fall into the realm of "homicide by heart attack," depending on the nature of the circumstances (see below).

Hyperactive delirium (so-called "excited delirium")

Excited delirium is not a cause of death, nor is it a recognized disease. It was originally introduced as a term to describe an acute psychosis caused by cocaine.[22,63,72,99–101] Many nonspecific medical terms, however, have worked their way into the medical lexicon (cerebral palsy is another example). Today, this clinical presentation would be diagnosed as a cocaine-induced hyperactive delirium, a well-recognized and accepted medical diagnosis.[102–104] A **hyperactive delirium** ("excited delirium") is defined as a hyperactive level of psychomotor activity that may be accompanied by mood lability, agitation, and/or refusal to cooperate with medical care.[105] There are five diagnostic criteria for a delirium (Table 15.1), and there are numerous etiologies, including a variety of drug intoxications (see Table 15.2).[106–108]

A common cause described as precipitating this syndrome is a *cocaine intoxication*. There is typically an abrupt onset of delirium with bizarre and often aggressive/destructive behavior. The person is agitated and uncooperative. In warm environments, hyperthermia is often present. Numerous studies have described its presentation and investigation.[14,16,22,23,43,47,58,63,67,72,77,78,80–83,100,101,106,109–135]

The term "excited delirium" was first invoked in published case reports by Wetli et al.[22,72,99] In one, entitled "Cocaine-Induced Psychosis and Sudden Death in Recreational Cocaine Users," the authors described the deaths of seven men, of which, five died in police custody.[72] They wrote, "Police, rescue personnel, and emergency room physicians should be aware that excited delirium may be the result of a potentially fatal cocaine intoxication;

TABLE 15.1 Definition of a hyperactive delirium.

1. A disturbance in attention (i.e., reduced ability to direct, focus, sustain, and shift attention) and awareness (reduced orientation to the environment).
2. An additional disturbance in cognition (e.g., memory deficit, disorientation, language, visuospatial ability, or perception).
3. The above disturbances are not better explained by another preexisting, established, or evolving neurocognitive disorder and do not occur in the context of a severely reduced level of arousal, such as coma.
4. The disturbance develops over a short period of time (usually hours to a few days), represents a change from baseline attention and awareness, and tends to fluctuate in severity during the course of a day.
5. There is evidence from the history, physical examination, or laboratory findings that the disturbance is a direct physiological consequence of another medical condition, substance intoxication or withdrawal (i.e., due to a drug of abuse or to a medication), or exposure to a toxin, or is due to multiple etiologies.

TABLE 15.2 Causes of delirium/psychosis

Delirium is an acute, confusional syndrome with a transient disturbance in consciousness and cognition that has a variety of causes:
Drug intoxications/withdrawals/noncompliance
Acute functional psychosis (schizophrenia, acute mania)
Endocrine/metabolic disease
Hyperthyroidism, hypothyroidism ("myxedema madness")
Hypoglycemia
Pituitary disease with secondary endocrine effects
Postpartum psychosis
Porphyria
Liver and kidney failure (encephalopathy, uremia)
Electrolyte disorders
Acid base imbalance
Nutritional disorders
Thiamin, vitamin B12 deficiencies ("megaloblastic madness")
Infections
Meningitis, encephalitis, sepsis
Neoplasia (brain tumors)
Seizure disorders
Complex partial seizures ("temporal lobe" or "psychomotor" epilepsy), postictal state
Vascular disorders
Hypertensive encephalopathy/cerebral infarcts
Systemic lupus erythematosus
Disseminated intravascular coagulopathy
Trauma
Concussion
Subdural, epidural, subarachnoid hematoma
Hypothermia/hyperthermia
Hypoxia
Factitious psychosis/malingering

its appearance should prompt immediate transport of the victim to a medical facility. Continuous monitoring, administration of appropriate cocaine antagonists, and respiratory support will hopefully avert a fatal outcome."

It is important that first responders recognize this entity, so they can treat it as a potentially fatal medical condition/presentation and not a criminal one. In fact, for many years, this was advocated and continues to be advocated for, having likely saved countless lives along the way.[77,135] Given its lethal potential, the American College of Emergency Physicians recognizes the importance of education regarding encounters with these patients. They recommend a coordinated training of law enforcement, emergency medical service, nursing, nurse practitioners, physician assistants, and physicians across the spectrum of medical specialties. The focus is on identification and the use of deescalation techniques.[136–144]

Unfortunately, the good intentions of getting this recognized as a medical condition may lead some to use it inappropriately or attempt to cover up wrongdoing.[134] Fortunately, law enforcement agents (and prosecutors and other first responders) are not the final arbiter of this diagnosis nor of its underlying cause. A hyperactive delirium is a medical diagnosis of a clinical presentation that is made by physicians. They are the diagnosticians who search for the *underlying cause* of this clinical presentation, which is a constellation of signs and symptoms, or for any potential *intervening* cause of death.

Instead of calling this clinical presentation "excited delirium," it should be called what it is known as medically when it is caused by a cocaine intoxication: a cocaine-induced hyperactive delirium. This is a recognized diagnosis in the DSM-V/ICD-10 (F14.221, *severe cocaine use disorder with cocaine intoxication delirium, acute, hyperactive*[105]); hyperactive delirium can also be caused by a variety of other conditions.[102,103,106,136] If a patient dies while in a hyperactive delirium (n.b., there are well-documented instances occurring outside of police presence[124,130]), then the cause of death would be the disease or injury that initiated the lethal chain of events (i.e., what caused the delirium). *The delirium is considered a symptom/mechanism/immediate cause, but it is not the underlying cause of death* (see Table 15.2).[106–108]

As evidenced by emergency physician's recognition of this clinical presentation in the living, it does exist; most people with this clinical presentation survive with proper treatment.[136] If first responders or correctional facility staff do not recognize this clinical presentation, however, then they may not initiate appropriate treatment.[135] Minimizing the existence of a diagnosis or stigmatizing it is not going to help a patient who requires urgent medical intervention.

It also must be noted that even though a person with a cocaine-induced hyperactive delirium may die while in the custody of the police, it does not necessarily mean that they died *from* it. People in a cocaine-induced hyperactive delirium can still be choked or die from traumatic asphyxia. It is the job of the FP, who is in the best position and has the necessary expertise, to make the proper diagnosis. The FP must determine if the person is dying from it, with it, or if it is even present. Unfortunately, with the national shortage of FPs, some of these deaths may not be properly investigated or certified.[145]

The reality is that people can and will continue to die from a hyperactive delirium, induced by cocaine or other stimulants, psychiatric conditions, or diseases. As Byard noted, it is important "to avoid becoming bogged down in the mire of semantic discussions—no matter what terminology is used, it is apparent that some individuals will collapse and die in an agitated state during encounters with police for reasons that are often far from clear."[100] Schizophrenics have an increased mortality and schizophrenia may cause death through a purported dysfunction in the activation of the autonomic nervous system and/or by a prolongation of the QT interval that is caused by psychotropic medications, but these are rare.[16,146] No matter what one calls this presentation, some may invoke it inappropriately and with ill intentions.[134,147,148] Some of these deaths occur at home, some in police custody, some in emergency departments, and some in intensive care units. If the death occurs in police custody, the medicolegal death investigator's (MDI's) job becomes much more challenging, as the differential diagnosis for the cause of death is greatly expanded.

There is no clear consensus for the manner of death for apprehension deaths involving a hyperactive delirium. DiMaio noted, "*The greater difficulty is designating a manner of death … Since a violent struggle has occurred with interaction between two or more individuals, the best classification of the manner of death is probably homicide. A good argument for an accident can be made, however. If the case is called a homicide, it is necessary to explain that the designation does not indicate that there was necessarily any criminal activity involved. The difference between homicide and murder should be carefully explained.*"[149] In the NAME position paper on police custody deaths, Mitchell et al. wrote, "*The more difficult cases are those where the individual is observed to be acting erratically due to a severe mental illness and/or acute drug intoxication. These cases have been defined in the literature as excited delirium and often result in a law enforcement response and the restraint of the decedent. It is not uncommon for the individual to die during or soon after restraint and/or an altercation with law enforcement. Manner of death in these instances can often be inconsistent from pathologist to pathologist and from office to office. Furthermore, manners of death in these cases have ranged from accidents due to the emphasis placed on drug toxicity, homicides due to the influence of the restraint and/or altercation, or undetermined due to the inability of the certifying physician to establish a definitive opinion.*"[24]

Graham writes, "*The certification of the cause and manner of death can be problematic in a death of an excited delirium subject lacking a discrete demonstrable nonexcited delirium cause of death, especially when forcible restraint has been employed.*"[8] Some certify these deaths based upon the underlying cause (e.g., cocaine-induced, as accident; due to schizophrenia, as natural) despite the restraint, as along as the restraint is not the cause of death (i.e., a neck restraint hold with a fractured larynx causing ventilatory failure).[8] On the death certificate in the former cases, some include the phrase "following forcible restraint" after the cause of death statement, the argument being that this acknowledges the restraint, even if the force of the restraint is not mechanistically viewed by the finder of fact as meeting the extent necessary to upgrade the manner to homicide.[8] Admittedly, this framing may cause confusion, as the cause of death statement and manner of death designation are, or at least appear to be, at odds, from a death certification standpoint (see further discussion in Chapter 3).

The fact is that the analysis of in-custody deaths, particularly those that occur during the apprehension phase, in which no structurally demonstrable fatal injury is identified, must carefully take into account many variables: circumstantial, physiologic, neuropsychiatric (e.g., delirium, psychiatric disorders), alcohol and/or drugs, comorbid conditions/baseline health, positional/mechanical factors (e.g., chest and/or neck compression), and applied factors (pepper spray, electromechanical disruption devices), all of which may have a potentially complex interplay and varied temporal relationships to the moment of death. There is no one size fits all algorithm to follow. Rather, each variable should be evaluated individually (and collectively) and on their own merits vis a vis their likely mechanistic contribution, as viewed through the lens of a frame by frame, well-documented sequence of events, if known. In the end, the cause(s) may be clearer in some cases than in others; in the latter, however, the most intellectually honest certification is often communicated in a *descriptive cause of death statement* in order to accurately encompass the complexity of the mechanism(s). Above all, it is imperative that the FP maintains their neutrality in the assessment of these deaths, following the facts and science at all times, irrespective of the opinions asserted and the pressures applied by sometimes competing stakeholders who are often in possession of an incomplete collection of facts and detail to which the FP is privy and painstakingly considers. The "distraction of the dramatic" should be assiduously avoided.

Luke and Reay concluded that "there is no need to rush to judgment. There will be ample time to discuss the case with a trusted colleague, and we would strongly encourage such interchange. The desired end is fairness, to the law enforcement officials involved and to the victim and his or her family. We certainly do not claim to be the ultimate authority or the oracle when it comes to asphyxia-related custody deaths, but we have learned over the years the value of training and experience, the risk of underestimating the importance of common sense, and the fact that there are no easy answers in such deaths."[36]

Suicidal behavior in law enforcement encounters

It is not unusual for some individuals to commit suicide (shoot themselves) when confronted by law enforcement, usually during an attempted arrest, a "police stand-off" involving a barricaded, armed individual, or during a perpetrated murder-suicide. The challenge for the FP is to determine if the death was caused by the decedent or the law enforcement agents. This entails witness information, as well as ballistics testing of the fatal bullet(s). If the person is shot by police in this scenario, the manner is typically certified as a homicide. Some suicidal individuals, however, may deliberately initiate a dangerous, public demonstration in order to invoke a fatal police encounter. This person may point a firearm (or a replica of a firearm) at civilians and/or at the police and intentionally escalate the encounter, ultimately resulting in the use of deadly force by law enforcement personnel. The phrase that has historically been used to describe this scenario is "suicide by cop," but the caveat is that this situation involves competing intentional acts that may result in dichotomous determinations of the manner of death among MEs/Cs.[150–156] One must weigh the absence of direct self-infliction with often overwhelming evidence that the individual intended to die. Regardless of the compelling nature of the person's intent to die, in some jurisdictions, these deaths are certified as homicides.[151,157] Since these deaths occur at the hand of another, the invoked reasoning is that homicide trumps other manners. The use of "suicide" in these deaths may be interpreted by some as an attempt to cover up police wrongdoing. An alternative approach might be to strike a compromise that would attempt to fairly and accurately reflect both aspects of the encounter: classify the death as a *homicide*, but describe the intentional escalation and self-destructive behavior in the *"how injury occurred" field* with a statement such as, "Deliberately provoked fatal shooting by police" (assuming this assertion meets the criteria outlined below).

In the general sense, there exists a slippery slope argument, as not everyone who brandishes a gun in the presence of a police officer is intending to commit suicide (by cop). Therefore, specific criteria should be met in order to invoke suicidal behavior anywhere on the death certificate, including a clear intent to die. This intent to die would not apply to "last-ditch" efforts or careless actions. A study by Hutson et al. proposed four criteria for the diagnosis of suicide by cop: (1) suicidal intent, (2) intent to be shot by law enforcement, (3) possession of a lethal weapon or facsimile, and (4) intentional escalation of the encounter.[151] A fifth criterion would require the *legal* use of force by law enforcement. The final criterion would be that the person is responsible for setting this suicidal process in motion prior to any involvement by police.

The investigation

In all custody fatalities, the investigative circumstances typically supply key information for their certification. *Disinterested investigators should go directly to primary sources of information* (witnesses or video footage) and not rely upon secondary reports from hospital staff. A step-by-step, freeze frame analysis of the events, as described by Luke and Reay, is performed.[36] What did witnesses see and hear? Was there neck or chest compression? If so, for how long? Was the person speaking, yelling, or making noises? Were the vocalizations continuous, and for how long? When did the person become unresponsive? The news, not the editorial should be primarily sought. This information, in combination with the autopsy findings and ancillary analyses, is needed for death certification, particularly when the mechanism of death is not structurally demonstrable.

Police investigations and the medical examiner/coroner

Although many ME/C offices have their own investigative staff, the ME/C still must rely upon the police for a variety of investigative material. The police and prosecutor's offices are the lead law enforcement agencies in the investigation of any suspected crime and ME/C regularly and appropriately avail themselves of their investigative information. That said, it is a two-way street, as information obtained from the FP's examination may also be important for the police investigation. The sooner investigators have information about the cause of death, the range of fire of a gunshot wound, the location of the entrance wound, the path of a bullet, the type of injury (should they be looking for a knife or a scissors?), or the identification of the decedent, the faster they can proceed with their investigation. It is not unusual (or inappropriate) for crime investigators to attend an autopsy. They want to "see for themselves" and not have to wait six weeks for an autopsy report, at which time the most fertile period for collecting investigative facts and locating witnesses may well have dried up. The most fruitful time for most criminal (or arguably any) investigations by law enforcement or other investigators occurs within days of injury occurrence (even if the death ensues at a later time).[158]

Besides needing to know the accurate time, date, and location of an injury occurrence (necessary for the death certificate), which is usually found embedded within the police record, it is also important to understand what the police specifically investigated, and of course, what they discovered as a result. That said, it important to point out that the information they have collected may or may not comport with the autopsy findings. If that is the case, they need to be made aware of this discordance, as the physical findings may take their investigation in an entirely different direction, and may well involve either the withdrawal of prior concerns for criminality or the reinforcement of them. Also, it is important to be mindful of the fact that the time of injury occurrence found within a police report may or may not necessarily represent the *actual* time the decedent was injured, depending on the case. When the body is simply discovered after having been injured, the "time of (injury) occurrence" documented in the police report may merely represent the first time that 911 was called following discovery of the body, not necessarily when the injury was actually witnessed or otherwise documented to have occurred. This is particularly well-illustrated with decomposed remains, discovered some time after having been killed or injured. In such instances, the "time of injury" is usually best indicated on the death certificate as "unknown," regardless of what the initial police records may indicate. A subsequent investigation may further reasonably elucidate the time of injury through witness statements, video footage, or other investigative data points, but it is often unknown on the day of the autopsy for death certification purposes.

With rare exceptions, a police investigation (with the attendant police report) is necessary for most all violent deaths, regardless of manner, or otherwise suspicious circumstances. Exceptions include well-documented/witnessed, nonsuspicious (accidental) falls of elderly and other persons in their homes or of patients sustaining accidental injuries in medical facilities. In those instances, there is rarely police involvement and a simple interview of the family and/or medical staff by the MDI or ME/C suffices to obtain the information needed to competently certify the death (time/place/circumstances of occurrence). That said, if a fatal injury is sustained in a medical facility (hospital, hospice, nursing home, rehabilitation center, etc.), it is best to request and obtain the **incident report,** if one was generated, before certifying the death. This can usually be quickly e-faxed to the MDI or FP while the autopsy is being performed and usually need not involve the police.

If the cause of death is traumatic, but the circumstances are as yet unclear, an **undetermined manner** should be deferred until there has been a police investigation. If no such police investigation has been started by the time of autopsy, it is best to initiate it by immediately speaking directly with the investigating detective/law enforcement officer (LEO) once one has been assigned. This police investigator should also be reminded that, because of *lucid interval* issues, the investigation of a subdural hemorrhage might need to be broader (temporally) than they might assume (i.e., not strictly limited to the location and moment of the terminal collapse/death). Regardless, it is best to provide very specific questions and requests to the LEO as early on in the investigation as possible. As pointed out above, beyond that time, witnesses may not be located and/or their memories may dim. Also, do not assume that the police will intuitively know the specific investigative information you seek. To that end, it is best to speak directly with the investigating LEO, not intermediaries, leaving nothing to chance and minimizing the risk of misinterpretation and the dissemination of misinformation. See Chapter 3 for further discussion regarding an undetermined manner.

It is also important to make sure the police report is not only received by the ME/C agency, but that it is actually *read* by the FP! One may be surprised at what is inadvertently omitted relative to what the LEO may have relayed over the phone. If that is the case, it is important to make sure the LEO updates their report to reflect and document all of the pertinent positives and negatives that they communicated in person or by phone. The police share the burden of an accurate medicolegal death investigation and communication with stakeholders, including families; for that reason, it is important to make sure their findings are *complete and in writing*. Remind them that you may refer the family members to them if they have further investigative questions, especially in homicides and suicides.

Finally, if the decedent is pronounced in a hospital, the police may or may not even be aware of the death or the injury event, depending on whether the hospital staff initiated communication with them. This is particularly true in noncriminal (suicidal and accidental) deaths, in which the injuries were sustained at home (as opposed to on the street). Emergency medical responders may respond to such events, with a police response not necessarily having been simultaneously triggered. Not only should the police be made aware of these deaths, but they also often need to initiate a potentially time-sensitive scene investigation before it has been altered substantially by family members, roommates, etc. This is particularly true in suicides, in terms of documenting open windows in descents from a height, pill counts in apparent intentional ingestions, reviewing video surveillance of rooftops, and in locating and securing suicides notes. Finally, in select cases (homicides), it is important to request that the police retrieve and deliver to the ME/C any clothing cut off during emergency medical management, before they are discarded in the hospital trash. This is particularly important in firearms deaths.

Physical struggles: contact, altercations, and the death certificate

The Constitution does not guarantee an assailant the right to a healthy victim -*Dr. Charles S. Hirsch*

Homicide-by-heart attack deaths[27,35] are those in which the cause of death is an acute exacerbation of underlying cardiac disease; however, the manner is homicide because a criminal act triggered the lethal pathologic cascade. The criminal activity results in a pathophysiologic stress response that leads to the acute fatal event in a predisposed person. A sudden cardiac arrest or event may occur during or shortly following a physical alteration that includes nonlife-threatening physical injury. In these complex cases, the pathologist must first determine if there is any *physical* injury and then, whether the injury caused or contributed to the death. If it did, then these deaths are usually certified as homicides.[28] In other instances, when there is either no physical injury or at least none that is life-threatening, a determination must be made as to whether the *events* precipitating the stress would be categorized as activities and stresses of daily living, or not. If not, then the event would be viewed as rising to the level of a **(functional) injury**, by definition, and in this instance, of a criminal nature. The latter might involve **a physical altercation or noncontact criminal menace** *(e.g., a robbery at gunpoint, being chased down the street and threatened, or physically threatened with a deadly weapon, such as a wielded knife).* In such circumstances, a homicidal manner would likely be appropriate.[27] This particular outcome is most commonly encountered in a person with advanced **heart disease (the proximate cause)** in which the stress of a **hostile altercation (the immediate cause)** results in a sudden cardiac event or arrest.[159] That said, one could equally view other acute noncardiac sequelae of such criminally stressful events similarly. *Ruptured cerebral aneurysms, hypertensive strokes, or ruptured aortic dissections* represent the most common examples. As such, one might reframe and broaden the "homicide-by-heart attack" designation to "homicide-by-heart attack or other acute vascular event."

When cardiac in nature (the majority are), *the most commonly encountered underlying disease is* **hypertensive and/or atherosclerotic cardiovascular disease;** however, other types may predispose to a cardiac dysrhythmia. Coronary arteries with aberrant origins, congenital valvular diseases, and other congenital cardiac anomalies may also be causally implicated in an assault.

Even without underlying cardiovascular disease, stress can result in considerable physiologic disturbances of the heart.[28,88] In these instances, an emotionally precipitated (catecholamine-driven) cardiac arrhythmia occurs and one may see acute myocyte injury.[27,28] In a study of victims who died as a direct result of a physical assault, without sustaining internal injuries, 11 of the 15 decedents had myofibrillar degeneration of the heart (**"stress cardiomyopathy"**).[28] A similar phenomenon has been described in the living, with so-called acute, stress-induced (*takotsubo*) cardiomyopathy.[88,160,161]

The history of an assault and/or exposure to an emotionally stressful criminal situation that precedes the cardiac event is usually obtained through a careful investigation. The autopsy is necessary to confirm or refute the substrate in which the acute life-threatening injury (criminal stress) was able to cause death: underlying cardiac disease, acute myocyte injury, and/or the presence of stimulant drugs. Cardiac disease or acute myocardial injury in five decedents who had a recent history of an assault or emotional stress associated with criminal activity was reported in an autopsy series.[35] The autopsy findings included **atherosclerotic cardiovascular disease, recent and remote myocardial infarcts, and acute cocaine intoxication**, the latter also capable of enhancing the risk of a cardiac arrhythmia.[35]

In some instances, the determination of the manner of death may be complicated if there is no life-threatening physical injury or any evidence of *physical contact* between the victim and the assailant.[27] For these deaths, any that involve unauthorized and offensive physical contact or a threatening action (e.g., pointing a gun at a person) would typically still result in a manner of homicide.[159] For example, one report describes a 67-year-old woman who was walking in a parking lot of a grocery store where she was stopped and threatened by a teenager demanding her purse. She refused and the boy struggled with her for the purse, threatening to kill her if she did not comply.[35] She was able to break free, run into the store, and tell others what happened before she collapsed. At autopsy, there was no evidence of acute injury, but there was hypertensive and atherosclerotic cardiovascular disease with cardiac hypertrophy, marked coronary artery atherosclerosis, and a remote myocardial infarct.[35] The cause of death was certified as: "cardiac arrhythmia due to hypertensive and atherosclerotic cardiovascular disease during an emotionally stressful event (victim of a robbery attempt)," the latter constituting a *functional* immediate injury. Given the criminal nature of this functional injury, the manner of death was classified as a homicide.[35]

Davis described **five criteria to evaluate cardiac-related homicides.**[27] These include: a criminal act; the decedent should recognize the threat to personal safety; highly emotional circumstances; collapse and death must occur during the emotional response period; and the demonstration of an organic disease.[27] The concept of the "emotional response period" is an important one. Most deaths falling into this category are characterized by *the onset of signs and symptoms of distress either during or not long after the criminal event.* The death, per se, may not occur until some later time, following potentially life-saving therapies or even a hospitalization involving life support, but, **importantly, the lethal sequence of events is set in motion reasonably proximate to the stressful event.** If, on the other hand, a man is involved in street fight, leaves the scene of the fight, goes home, has dinner with his wife, watches a movie, then is found dead in bed the following morning, the pathophysiologic link between the (criminally) stressful event and his death will have become substantially attenuated, particularly if he had suffered *no* appreciable signs or symptoms of cardiac distress during or close to the time of the fight or even in the intervening period, even if his ultimate mechanism of death is determined to be cardiac in nature. The link is further weakened by the high burden of proof one must meet in classifying homicides. The latter example introduces the reasonable possibility of an efficient intervening stressor or stressors due to the prolonged interval between the original altercation and his symptom onset and/or death (e.g., perhaps, he had an argument over finances with his wife before going to bed). Reasonable contingencies exist in this scenario; as such, the burden for a homicidal manner would not be met.

A revision of Davis's construct by Turner et al. focuses on physical contact and reworks the criteria.[35] They state that the *"action of the perpetrator toward the victim should be of such severity and have sufficient elements of intent to frighten, injure, or kill, either in fact or in statute, so as to lead to a charge of homicide in the event that death resulted from physical injury."*[35] The National Association of Medical Examiner's "A Guide for Manner of Death Classification" states that fatalities resulting from "fear/fright induced by verbal assault, threats of physical harm, or through acts of aggression intended to instill fear or fright may be classified as homicide if there is a close temporal relationship between the incident and death."[162] Pure verbal assaults represent a gray area with regard to manner of death. For example, a heated argument with a taxicab driver may be considered a stress of everyday living in New York City. By contrast, a verbal assault with the brandishing of a weapon or a threat of a physical assault may be sufficient for the determination of homicide. As such, the nuances of these noncontact circumstances demand guidance for classification purposes. To address that, those categorized as homicides are typically defined as being born out of events that are determined to rise above the stresses of "everyday living," especially when they occur during the commission of a crime. As such, the cause and manner of sudden deaths following such an encounter should take into consideration the nature of the putative event and whether it rises into the criminal realm and therefore constitutes a functional injury.

That said, a cause of death statement must contain in it a disease and/or an injury. The latter can be either structurally demonstrable/anatomic (i.e., subdural hemorrhage) or functional and descriptive (e.g., the pathophysiologic stress of a physical altercation or an assault). The latter (descriptive) type of death certification is more commonly used in "homicide-by-heart attack" fatalities (see Chapter 3 for further discussion regarding the certification of these deaths). Not all deaths classified as "homicide-by-heart attack" are prosecuted by law enforcement authorities, but this should have no bearing on manner classification decisions, per se, if the fatality clearly meets definitional criteria from a death certification standpoint.

With the foregoing in mind, it is important to recognize that **not all physical contact constitutes "death at the hand of another" (criminal contact)**; common sense must be applied. If someone dies from physical contact injury sustained during lawful police contact (handcuffing, even chasing someone who is shooting a loaded gun, etc.), or during a sanctioned sporting event (such as during football tackles) that is being played by the rules, we do not typically classify these as homicides. If, on the other hand, a fight breaks out on the field between two angry players, outside the confines of the sanctioned rules of the game, or during what is clearly the unlawful/excessive use of force by police, and someone dies as a result, that will likely be viewed altogether differently.

Equally important to consider is **the involvement of stimulant drugs and the possibility of more than one underlying or contributory cause of death in criminally stressful encounters.** Both natural disease and stimulant drugs may coexist in the same victim, acting in concert as the substrate on which the functional injury (criminal stress) exerts its lethal effects. The underlying cause isn't necessarily subject to an either/or criterion, nor is the immediate cause (stress injury) at odds with an underlying chemical substrate. Both heart disease *and* stimulant drugs can be implicated as underlying causes on the same death certificate (together or separated between parts 1 and 2), as can the "physical altercation" as the immediate cause, so long as the altercation, the acute drug intoxication, and the signs and symptoms of the cardiac event comport temporally. None of these entities is mutually exclusive; to the contrary, they are typically quite interrelated. But, the most important takeaway is that only one of these three variables informs the manner. To wit, any criminal injury cited on a death certificate will "trump" any noncriminal injury or natural disease process in terms of the manner designation. That said, the stress of the physical altercation, no matter how small its contribution, even with the involvement of one or more stimulant drugs, will inform the manner, not the drugs. As such, such a death would still be classified as a homicide. If one has determined that a physical altercation has occurred contemporaneously with the use of stimulant drugs (with or without coexistent heart disease), turning one's back on it and invoking only the drugs would head squarely into arbitrary and capricious territory.

Abuse and/or neglect in the debilitated adult or teen

Not uncommonly, concerns that a vulnerable adult has been fatally abused and/or neglected come to the attention of an ME/C office.[163–167] Such concerns are often alleged by family members about other family members or non-family members (e.g., a home health aide), but sometimes they come from outside sources, such as adult protective services or from physicians and hospital staff. Vulnerable adults are often elderly, but they also may be severely neurologically, psychiatrically, or cognitively impaired teens or young and middle-aged adults.

Elder abuse is an intentional act, or failure to act that, by definition, causes or creates a risk of harm to someone age 60 years or older.[165–167] The investigative concepts are similar to those applied in child abuse and neglect. Common types are physical, sexual, emotional, or financial. Because of underlying medical conditions and debility, these persons are more susceptible to fatal outcomes with abuse or neglect than would be a young, healthy person. Their susceptibility or vulnerability, however, does not have any bearing on the certification of death, nor does it absolve the assailant of criminal responsibility.

Elder abuse deaths from neglect often reveal evidence of *untreated disease, poor hygiene, decubitus ulcers, and malnutrition.* But, these findings alone are not diagnostic of neglect. The question is whether the decedent had the capacity to make an informed decision about their health and well-being and the agency to execute actions. The condition of a competent adult who decides to "die" at home, without medical care, if well documented and clear, would not usually constitute neglect, despite the sometimes horrendous condition of their body after death.

Regardless of the source or emotional charge of allegations that may come forward, FPs need to take any such allegation seriously and handle the case systematically, objectively, and thoroughly, with a consistent approach and checklist in their minds. If the allegations turn out to be unfounded, the FP will still have responsibly done their part to ascertain that. Uncovering abuse and neglect may be just as important as determining that there is no evidence of it. We are the neutral arbiters of the facts. *The approach can be viewed as* **two-tiered:** *the establishment of the* **decedent's dependence**, *followed by the* **gathering of evidence.**

The establishment of dependence

The concept of neglect is only relevant and proportionate to the extent that an individual relies upon another for nourishment, hydration, hygiene, bathroom assistance, heat, and safety. So, establishing and documenting this baseline level of functioning become central to the case investigation and must be addressed up front.

Infants, by virtue of their development, are dependent upon others/another for every part of their care and protection, by definition. For teens and adults, the degree of their dependence must be clearly assessed by accurate, thorough, and evidence-based documentation of their baseline level of cognitive, motor/neurologic, medical, and psychiatric functioning. Some of this documentation is independent of an autopsy assessment, but rather, is ascertained by gathering historical information, not just from family members, but from more trained and objective sources, such as primary care doctors, neurologists, psychiatrists, visiting nurses, and adult protective services staff.

Depending upon the circumstances and suspicion of the case, the prosecutor's office and police may assist with the investigation, through interviews, doctor's records, phone records, and subpoenas. But, more often than not, discussions with other physicians are more fruitful, efficient, and elucidating when done doctor to doctor. For this reason, we suggest the FP make those specific medical phone calls themselves, after which a thorough case note can be written to create a case file document. Be aware also that some prosecutor's offices have special elder abuse divisions, with a wealth of experienced prosecutors and investigators who are quite seasoned in these matters.

All that said, the bottom line is that if the baseline level of functioning does not support dependence on others (i.e., the decedent was fully capable of living independently and making their own decisions), the issue of "neglect" is moot, and the neglect work up should cease. ***Abuse***, *on the other hand, is not necessarily a function of dependency, per se, as fully functional and independent teens and adults can be abused emotionally and physically, depending on the nature of the relationships involved.*

This brings us to our *second tier:* the investigation and documentation of evidence. Once the decedent's baseline has reliably been established as "dependent," the physical and historical evidence of neglect and/or abuse should be sought in two ways. *First and foremost, it is best to approach a completely dependent teen or adult as you would an infant.*

Autopsy/physical evidence

Signs of both *neglect and physical abuse* can be explored and documented by conducting a thorough autopsy. A search for the former is much more involved and complex than the latter, and it will likely include nutritional and hydrational assessments and lab testing (body mass index and electrolyte analysis), toxicological analysis (were prescribed medications being given appropriately, excessively, or not at all?), and observations about hygiene (soiling of skin by dirt, feces, urine, extensive and/or infected diaper rashes), and/or extensive and uncared for decubitus ulcers, with maggots or superimposed infections (take sections for the stock jar and look at decubiti under the microscope).

Physical abuse is best documented at the autopsy table and through radiography. There may be both acute and chronic components. Besides the documentation and photography of demonstrable acute injuries, evidence of chronic patterns of abuse should also be sought and documented (contusions of varying ages, healed/healing fractures, especially spiral fractures of long bones, or fractures with uneven healing and large, irregular calluses, indicating no medical attention, unusual scars or healed burns, traumatic alopecia, etc.).

Police/district attorney/forensic investigator evidence

The investigative piece, including interviews of family members and potential suspects, should be conducted by the police, ideally, with the prosecutor's office also involved, if applicable. One doesn't want to later find themselves involved as a fact witness in a subsequent trial should the case end up being classified as a homicide. Avoid interviewing individuals who are possible perpetrators (this also applies to child abuse cases).

Beyond witness statements, patterns of abuse and neglect might also be supported by documentation of repeatedly missed doctor appointments, pharmacy records indicating unfilled, potentially life-saving prescriptions, interviews with neighbors, an empty refrigerator, despite cashed social security checks, or a pattern of domestic incident reports in police records. Clear communication with law enforcement and the prosecutor's office is often critical for the FP to ultimately render an opinion in these cases. Not uncommonly, the final manner of death opinion will be undetermined, but the ME/C must not arrive at that without due diligence. These investigations typically require a multidisciplinary team approach. The prosecutor, if involved, can be very helpful in that regard and subpoena whatever documents the FP may reasonably require to arrive at a determination.

As a final point, **the level of cognitive, psychiatric, or neurologic functioning of the caregiver is also potentially relevant.** As an example, if a moderately to severely demented husband had been entrusted with the care of his bed-bound and dependent wife and she dies related to myriad purported "neglect" issues, his ability to understand his role and have the volition to have reached out for help, if needed, is not an insubstantial consideration in the final manner of death determination. A wildly psychotic son, not oriented to place or time, having not taken his medications, may not even have the ability to acknowledge that his mother has died, much less to understand his potential responsibility in the death. These variables should reasonably be part of the ultimate death certification considerations. Because these are often domestic deaths, they can be complex, contentious, and difficult to adjudicate and defend in both criminal and family courts, so every effort should be made to arrive at well-documented, evidence-based conclusions, without capriciousness or arbitrariness.

As a conceptual extension of vulnerable persons, one might have a lower threshold for investigating **deaths at skilled nursing facilities** when they are reported to the ME/C.[164] Since these residents often have advanced disease and ready access to medical caregivers, these deaths usually are not "reportable" to the ME/C. In addition to occasional, if even rare, elder abuse and neglect, there are other unnatural deaths that may occur in skilled nursing facilities.[76,164–184] These include choking on food, neck compression due to entrapment in bed rails, and falls (both witnessed and unwitnessed).[164,174,185,186] The investigation can be further complicated because of advanced medical directives that include "do not resuscitate" or no "aggressive" measures or transport to a hospital. As a result, the 90-year-old woman who experiences a fall earlier in the evening and then is later found obtunded in bed, may not have undergone computed tomography (CT) imaging. This leaves the ME/C with the decision to take jurisdiction to exclude an occult injury, such as a subdural hematoma.[179,187–189.] In anticipation of likely objections to autopsy, there should also be a low threshold for having a conversation with families with these deaths, if an autopsy is felt to be necessary, taking their concerns and wishes into account, given the advanced ages of these decedents.

Nursing home deaths due to infectious disease (e.g., COVID-19) are an important public health issue. These deaths are best properly investigated and diagnosed. There is a risk that they may be underreported because of lack of testing, advanced medical directives preventing medical work-ups, or improperly completed death certificates.[190–194]

Approach to hemorrhage of unclear origin (unexpected hemorrhage found at autopsy)

Not uncommonly, during the performance of an autopsy, the FP encounters an unexpected, and often fatal hemorrhage of unclear origin and etiology. These include intra- and extra-axial intracranial hemorrhages, hemorrhages of the soft tissues of the anterior neck, hemothoraces, hemoperitoneum, gastrointestinal hemorrhages, and soft tissue hematomas of the retroperitoneum and rectus sheath. In the absence of structurally demonstrable injury, clear anatomic natural causes, or therapeutic interventions as explanations, one must have on hand a mental checklist of reasonable possibilities and probabilities within the context of each scenario.

Out of the gate and irrespective of the proximate cause, the most prevalent risk factors for unexplained hemorrhage are **disseminated intravascular coagulopathy (DIC) and other coagulopathies** of various causes, both iatrogenic and natural. Each must be considered, as they often set the stage for the so-called **"spontaneous" hemorrhage**. In order to consider the former (DIC), a clinicopathological correlation should be made, often with knowledge of hematologic laboratory indices and the specific clinical setting. In terms of other (non-DIC) coagulopathies, alcoholic cirrhosis, emergent thrombolytic therapy (TPA), and anticoagulation (and overanticoagulation) therapies represent the most common culprits, even if not the underlying causes, per se. Beyond coagulopathic risk, ***alcohol use disorder,*** *in general, presents the FP with the most frequent and varied mechanistic possibilities for the production of hemorrhage in all body compartments.*[195–202]

The first order of business after the initial discovery of substantial hemorrhage in *any* death, even when the etiology is clear, is to **assess its magnitude**, which, in turn, will indicate its relevance *as the mechanism of death (primary, contributory, or not at all).* If mechanistically significant, a search for its underlying etiology becomes all the more important, as it will likely determine the manner of death. As such, any sizable hemorrhage discovered at autopsy should first be quantified as accurately as possible. If a liquid hemorrhage, a careful collection into a measuring cup works best. If space-occupying (hematoma), it too can be either placed in a measuring cup or weighed in a scale pan (the latter works best for large soft tissue hematomas that require enucleation, such as those from the retroperitoneum or rectus sheath). Alternatively, space-occupying hematomas can be measured in three dimensions (centimeters) to facilitate the calculation of a volume. With any of these direct measurements, the percentage of blood loss relative to the decedent's total blood volume (body mass in kg x 70 mL/kg in adults) can be easily calculated. Beyond a direct measurement, **photographic evidence of exsanguination** (or lack thereof) should then be sought and documented. *Visceral (especially renal), muscular, and gingival pallor are excellent indicators of lethal blood loss (exsanguination), and, when present, support hemorrhage as the primary mechanism of death. Poorly formed or barely discernible lividity constitutes another commonly encountered sign of shock by exsanguination.* If the hemorrhage seems moderate, but evidence for a frank exsanguination does not exist, the blood loss itself or a comorbid contributory condition might be appropriately added in part 2 of the death certificate, once a cause for the hemorrhage has been identified (traumatic, natural, or iatrogenic), as part 2 conditions have the capacity to change the manner.

That said, once the pathophysiologic significance of the hemorrhage has been ascertained and documented, attention should then be turned to its **etiology**, *with exclusion of injury being a top priority.* If both trauma and iatrogenic causes are carefully sought and reasonably excluded, the FP backs into a natural manner, by exclusion, even in the absence of identifying a discrete and demonstrable pathologic vascular anatomic source (i.e., the so-called "spontaneous hemorrhage"). *Remember that the autopsy serves a critical purpose in what it excludes.* Even with its limitations, the exclusion of injury or a therapeutic intervention as the cause of a hemorrhage allows the FP to exclude unnatural manners and inform authorities that no crime was committed, then accurately classify the manner of death as natural, even if the cause is only descriptive (e.g., "Complications of nontraumatic retroperitoneal hemorrhage of undetermined etiology").

The following discussion outlines a systematic approach to the etiologies of hemorrhages by body region, moving from head to toe (see Table 15.1 through Table 15.6 for the differential diagnoses by location). In the interest of scope and for the purposes of this discussion, these sections will be limited to *adult decedents only,* as the contexts, diseases, injuries, and differential diagnoses in children differ substantially from their adult counterparts.

Head

So-called "intracranial" hemorrhages should be subclassified by location (i.e., epidural, subdural, subarachnoid, and intraparenchymal hemorrhage), each with their own attendant differential diagnoses (Table 15.3 and discussion in Chapter 7). In the absence of scalpular, facial, or other bodily injury, the most common explanation for an *intraparenchymal* hemorrhage is a **"stroke"** caused by natural cerebrovascular phenomena. *Basal ganglia, pontine, and cerebellar hemorrhagic strokes result from the vascular sequelae ("barotrauma") of hypertensive cardiovascular disease.* Ruptured saccular (berry) cerebral aneurysms present as either intracerebral or (often) basilar subarachnoid hemorrhages (SAHs), particularly in and around the circle of Willis, the latter typically producing a "tamponade" of the brainstem with a consequent sudden death. But, unlike hypertensive strokes, **berry aneurysms are etiologically specific, so do not require an underlying proximate cause** of HCVD to explain their formation, *although their rupture is typically associated with an acute surge in blood pressure in both hypertensive and nonhypertensive persons*, fueled by any number of circumstances, from consensual sexual intercourse to an assault (the latter would be appropriately included in the cause of death statement and the former, not). Decedents with both hypertensive cerebral hemorrhages and ruptured berry aneurysms typically present to the ME/C following a sudden, unexpected death that is sometimes preceded by a severe headache. Both also can be caused or exacerbated by *stimulant drugs*, most notably **cocaine.**[203–208] In the latter instance, the inclusion of an acute cocaine intoxication in parts 1 or 2 of the death certificate would inform an accidental, rather than a natural, manner. If related to criminal menace (see above), however, either may be classified as homicides.

A less common cause of an intracerebral hemorrhage involves the fatal rupture or leakage of an arteriovenous malformation (**AVM**).[209–211] The identification of this finding requires both a careful search and a high index of suspicion. Conversely, the likelihood of an expanding **contusion hematoma** as an explanation of an intracerebral hemorrhage is low if historical and demonstrable trauma have been excluded.[212–214]

TABLE 15.3 Differential diagnosis of intracranial hemorrhage.

Epidural
Trauma
Heat (artifact)
Subdural
Trauma (including occult trauma/"microtrauma")
"Burst" lobe from an intracerebral stroke
Disseminated intravascular coagulopathy (DIC) and other coagulopathies (e.g., renal dialysis, liver disease with chronic alcohol use)
Arteriovenous malformation (AVM)
Subarachnoid
Ruptured cerebral artery aneurysm or AVM
Blunt injury, including lacerated vertebral artery
Intracerebral
Hypertensive and embolic strokes (including cocaine-related)
Amyloid angiopathy
Injury (cortical contusions and contusion hematomas)
AVM
Cerebral artery aneurysm
Some malignancies and infectious foci
DIC and other coagulopathies

TABLE 15.4 Differential diagnosis of thoracic hemorrhage.

Trauma (including delayed rupture of an intercostal artery aneurysm)[289,290]
Ruptured aortic dissection
Spontaneous pneumothorax (e.g., from asthma, emphysema, drug insufflation)
Ruptured myocardial infarct (hemopericardium)
Malignancy
Fistulas
"Spontaneous"
Disseminated intravascular coagulopathy and other coagulopathies

Cerebral amyloid angiopathy (CAA) is an angiopathy in which amyloid beta peptide deposits in the walls of small- to medium-sized blood vessels of the central nervous system (CNS).[215,216] The amyloid material is only found in the brain and is not related to other forms of amyloidosis. CAA is associated with brain hemorrhages, particularly microhemorrhages. The bleeding is usually confined to a particular lobe, which may help in distinguishing it from hypertensive bleeds, which commonly involve the basal ganglia or pons.

The real challenge begins with unexpected and unexplained extra-axial hemorrhages, the most commonly encountered of which is a space-occupying **subdural hemorrhage or hematoma (SDH)**, not just a whisper of

TABLE 15.5 Differential diagnosis of abdominal hemorrhage.

Trauma (more commonly in those with alcohol use disorder, especially when involving cirrhosis and splenomegaly)
Ectopic pregnancy rupture[315]
Rupture of a liver hematoma (HELLP syndrome)[316–318]
Ruptutred aneurysm (aorta, spleen, liver, kidney)[220–224,319,320]
Malignancies[321]
Ruptured diaphragmatic varices[322,323] (only in the setting of cirrhosis)
Cirrhotic nodules[324]
Paracentesis
Trocar injury (laparoscopy, bone marrow biopsy)[325–327]
Anticoagulation therapy[328,329]
Surgery (e.g., lumbar disc surgery)[330]
Hemorrhagic pancreatitis
Resuscitation, with/without thrombolytic therapy and a liver laceration (a lesser volume is important, as is the absence of renal pallor); groin intravenous catheters with a high (peritoneal) advancement
Disseminated intravascular coagulopathy and other coagulopathies

TABLE 15.6 Differential diagnosis of upper gastrointestinal hemorrhage.

Bleeding gastric and/or duodenal ulcers (PUD, cancer, NSAID use)
Severe or erosive gastritis/duodenitis
Severe or erosive esophagitis
Ruptured esophagogastric varices (only in the setting of cirrhosis)
Portal hypertensive gastropathy (in cirrhotics)
Angiodysplasia/arteriovenous malformation (also known as vascular ectasia, gastric antral vascular ectasia)
Mallory–Weiss syndrome
Tumors (polyps/cancers)
Dieulafoy's lesion
Gastric antral vascular ectasia
Aortoenteric fistula
Ectopic varices
Iatrogenic bleeding after endoscopic interventions, including "trauma"/punctures
Disseminated intravascular coagulopathy and other coagulopathies (e.g., liver disease with chronic alcohol use)
No lesion identified (10% to 15% of patients)

subdural staining that is easily explained by a terminal collapse (i.e., consequence, not cause). The careful exclusion of historical and/or demonstrable trauma to the body and head (after a deliberate check behind the ears, and possibly, a posterior neck dissection) should prompt the FP to consider a **"spontaneous" subdural hemorrhage** due to or exacerbated by a coagulopathic state, once, of course, an AVM has been reasonably eliminated as the source. A similar "spontaneous" hemorrhage is occasionally described in **hemodialysis** (HD) patients, a phenomenon predicated at least in part on the combination of both renal failure and HD-related coagulopathies.[217–221] The incidence of spontaneous SDH in these patients has been estimated to be over 3% and is associated with a very high mortality.[217–222] In addition, **advanced age and dementia** are often associated with *cerebral atrophy*. This potentially enhances the risk in these patients to the vulnerable and tethered bridging veins in the setting of what is sometimes referred to as "microtrauma," injury to the head from such minor trauma that it defies detection and would not be considered an event that falls outside that viewed as the activities and stresses of daily living for most persons with normal cerebral anatomy.[223–227]

An **epidural hemorrhage (EDH)** typically accrues subjacent to a temporal skull fracture that injures the middle meningeal artery. It has rarely been described in the occipital fossa, related to a laceration of a venous sinus. Epidural hematomas (along with subdural hematomas) may present with a lucid interval until the hematomas expand and the intracranial pressure sufficiently increases to compromise the brain.[212,228–230] Admittedly, EDHs are not commonly encountered as intracranial hemorrhages of unclear etiology.

Neck

Nothing slows the pace of an autopsy faster than the detection of unexpected hemorrhage in the soft tissues of the anterior neck. As with all other findings identified in forensic pathology, these should be viewed not in a vacuum, but rather within the case context, both anatomical and circumstantial. In a clearly nonsuspicious death (by history and scene investigation), particularly with the identification of one or more compelling and competent causes of death and no concerning injury and/or facial/conjunctival petechiae, an alternate explanation for the hemorrhage should be sought. While the explanation still may be sinister, alarm bells should be muted until further warranted.

First, *the* **postmortem body position** *should be clarified*, including whether initially discovered prone or supine, and whether objects, jewelry, or upper extremities/hands may have been wedged against the neck in that position. **Resuscitative efforts** *should be described* (e.g., attempted placement of internal jugular vein catheters, difficult intubation with the application of cricoid pressure). A negative **posterior neck dissection** might further quell concerns for foul play. Also, one might ask if the pattern of the "hemorrhage" is actually more consistent with the ill-defined/"muddy" blood extravasation in the perivascular distribution (typically, the jugular veins) of the early **putrefactive process.** Context and further inquiry cannot be emphasized enough if the findings do not comport with the history and the rest of the autopsy.[231–245]

Finally, unlike fatal *hemorrhages* in other compartments of the body, those *in the neck do not typically represent an exsanguination mechanism* when of nonpenetrating etiology. Rather, when fatal (and traumatic), they most typically represent markers of forceful compression. *As such, documentation of their quality, rather than their quantity takes precedence.* That said, vertebral artery lacerations may occur with both neck manipulations and with blunt impacts[173,199,232,233,246–249]; when fatal, these soft tissue hemorrhages of the neck are usually seen in conjunction with an intracranial basilar SAH, very similar in appearance and distribution to the SAH associated with ruptured basilar saccular aneurysms within the circle of Willis.

Thorax

Aortic dissections and lacerations, as well as various **arterial fistulas**, may result in an unanticipated hemothorax.[250–277] Aside from a search for a coagulopathic cause, an unexpected, apparently nontraumatic and noniatrogenic hemothorax should prompt a search for **fistulas** involving the vasculature and pleura, a **spontaneous pneumothorax** in the setting of bronchial asthma, emphysema, or acute illicit drug intoxications with insufflation,[278–280] and **vasculoinvasive malignancies**, as all may lead to hemorrhage. Ruptured myocardial infarcts typically result in a hemopericardium.[281–284] Rib fractures can lead to injury of intercostal arteries that also may cause a **delayed hemorrhage from a traumatic aneurysm** formation.[285,286] There have been reports of "spontaneous" hemothoraces.[287] Of note, *cardiopulmonary resuscitative efforts rarely cause hemothoraces*, even in the setting of fairly extensive anterior chest wall rib fractures.[288] See Table 15.4.

Peritoneal cavity

A large and unexpected hemoperitoneum typically pours out of the peritoneal cavity at autopsy with the first deep abdominal wall incision. A measuring cup should be quickly located and the hemorrhage quantified as best as possible, along with a qualitative assessment (frank blood vs. watery, blood-tinged ascites). Then, before any further evisceration ensues, a careful inspection of the abdominal organs and tissues should be undertaken to determine whether any **lacerations** offer an explanation, the **spleen and liver** being the most commonly injured. If indeed identified, the injury should be photographed and any associated rib fractures and/or soft tissue contusions of the flanks, lateral trunk, back, and abdomen should be sought and documented. The specific size and location (lateral? anterior?) of the lacerations of the liver or spleen should also be made note of and correlated with the likelihood of having been caused by **resuscitative efforts**, as should the volume of blood loss (i.e., Is it too voluminous to be explained by resuscitative injury alone?).

Of note, **splenic lacerations** usually present fairly proximate to the traumatic event, but on occasion, the bleeding from the laceration is contained under the splenic capsule for some amount of time (hours to even days), effectively tamponading the defect and keeping further bleeding at bay, until, perhaps due to continued activity and blood pressure elevations, the capsule gives way under the pressure, allowing the blood to fatally spill into the peritoneal cavity, while the still unhealed laceration continues to bleed, no longer contained by the tamponade of the hematoma. This phenomenon is referred to as a **delayed splenic "rupture" (from a laceration),** but contrary to what its name may suggest, it is traumatic in etiology, rather than a spontaneous "rupture."[226,291–295] In the absence of injury (splenic laceration) and in the setting of marked splenomegaly, however, a true spontaneous rupture may also occur. That said, *splenic injury is more commonly encountered from relatively minor trauma when the spleen is enlarged for any number of reasons* (cirrhosis, intravenous drug use, hematologic disorders, etc.). To aggravate the situation even more, if the splenomegaly is due to cirrhosis or a hematologic disorder, it is likely that the patient will also be relatively coagulopathic, exacerbating their inadequate thrombophilic response to the injury and likely magnifying the blood loss.

If, on the other hand, no injury is identified once the hemoperitoneum has been evacuated, **the retroperitoneum should also be explored for any involvement.** If there is also retroperitoneal hemorrhage, it too should be quantitated and photographically documented in terms of both where it is most prominent and the overall area it spans (i.e., does it extend up to the diaphragm, into the groin, and cross into the anterior abdominal wall, or is it confined only to the retroperitoneum? See below for further discussion of the approach to retroperitoneal hematomas). *All major blood vessels (aorta, inferior vena cava, portal vein, and iliac and renal vasculature) and gastrointestinal structures and vasculature should be inspected for pathology/rupture (diverticula, gastric ulcers, pancreatitis, etc.) and associated surrounding hematomas.* If the patient received **TPA** for a presumed pulmonary thromboembolus (PTE), that too should at least be considered, particularly when the viscosity of the blood is thin/watery. If, indeed, a saddle (fatal) PTE is confirmed at autopsy, a Class I finding, it makes it all the easier to ignore the watery "hemoperitoneum" as a consequence of resuscitative efforts following the TPA therapy, rather than as the mechanism of death. The observation of the *absence* of visceral pallor is also helpful in further confirming this impression. Finally, on rare occasions, **an intravenous catheter of the groin** has been placed and used for the infusion of multiple units of packed red blood cells (depending on the details of the case). If the catheter is placed deeply up into the pelvic tissues, the tip may breach through the iliac vessels, allowing an excessive amount of blood products to pour into the peritoneal cavity and/or retroperitoneal space. For this reason, it is a good idea to inspect the location of the tips of catheters, the condition of the vessels traversed, inspect for the *absence* of expected visceral pallor, and inquire about what kinds of fluids had been infused when a hemorrhage is of still unclear origin.

Arguably, the most common scenario in which no source of the peritoneal bleeding is found involves an alcoholic with cirrhosis. Interestingly, these too are the same decedents for whom the FP may have heightened concern for nefarious encounters leading to fatal traumatic hemorrhagic complications in many body compartments. But, when all grossly *identifiable* vascular abnormalities and injuries have been excluded, one has to "believe what you see" (or don't see, as the case may be). The most reasonable conclusion as to the source of the hemoperitoneum under this particular set of circumstances is a **ruptured diaphragmatic varix**, a much under recognized entity.[296] This represents more of a diagnosis of exclusion based on context, pertinent negative findings, anatomy (cirrhosis), and the difficulty with which ruptured varices are detected (i.e., they tend to be quite grossly subtle, postrupture). **The most important conclusion drawn from such autopsies, of course, is the ruling out of injury**; the diaphragmatic variceal diagnosis is somewhat academic, but no less elegant to recognize.

A more exhaustive list of etiologies includes ruptured ectopic pregnancies,[297] the rupture of a liver hematoma (HELLP syndrome),[298–300] an aneurysmal rupture (aorta, spleen, liver, kidney),[301–306] malignancies,[307] (diaphragmatic varices),[195,196,296] cirrhotic nodules,[308] paracentesis, trocar injury (laparoscopy, bone marrow biopsy),[309–311] anticoagulation,[312,313] hemorrhagic pancreatitis, a ruptured gastric ulcer, and surgery (e.g., lumbar disc surgery).[314] See Table 15.5.

Gastrointestinal

Clearly lethal, as well as sublethal (contributory) gastrointestinal hemorrhage is not uncommonly encountered by the FP. More times than not, its source/etiology is easily identified. Common sources include ruptured esophageal varices, bleeding gastric ulcers of various etiologically specific causes (including peptic ulcer disease, gastric cancer, and chronic nonsteroidal antiinflammatory medication administration), hemorrhagic gastritis, and bleeding hemorrhoids (when the bleeding is isolated to the anorectal region).[331] As with the other body compartments, DIC or another coagulopathy, natural and/or iatrogenic, may serve as an exacerbating backdrop for such hemorrhages. On occasion, however, the body fails to cooperate with textbooks and leaves us scratching our heads when no clear source is discovered. That said, unlike the pathogenesis of a hemoperitoneum, a *gastrointestinal hemorrhage is rarely traumatic in nature*, unless one considers rare iatrogenic "trauma" from procedures such as percutaneous gastric tube placements, upper endoscopy, or colonoscopy. Rather, when the patient is *cirrhotic*, **ruptured esophageal varices** must be considered, even when they are grossly undetectable. Similarly, the complex hemophilia associated with chronic alcohol misuse must not be ignored. As with hemorrhagic complications in other body locations, **chronic alcohol misuse**, with its attendant hematologic, gastrointestinal, nutritional, and hepatic derangements (cirrhosis with portal hypertension, coagulation factor deficiencies), must always be considered as the proximate pathogenic link ("Common things are common because they are the things that occur with the greatest frequency").[202,332] Finally, intestinal **AVMs** are rarely grossly identifiable, but they must also be at least considered when the bleeding source remains unclear. The etiology of lower gastrointestinal bleeding may be categorized as anatomic, vascular, inflammatory, and neoplastic. *The most common causes for primary lower gastrointestinal hemorrhages in adults are: diverticulosis, ischemia, neoplasia, and angiodysplasia.*[333] See Table 15.6 and Table 15.7.

TABLE 15.7 Differential diagnosis of lower gastrointestinal hemorrhage.

Diverticulosis
Ischemia
Anorectal (hemorrhoids, anal fissures, rectal ulcers)
Neoplasia (polyps and cancers)
Angiodysplasia
Postpolypectomy (endoscopy-associated)
Inflammatory bowel disease
Radiation colitis
Other colitis (infectious, antibiotic associated, ischemic, colitis of unclear etiology)
Small bowel/upper gastrointestinal hemorrhage (see Table above)
Trauma/punctures complicating colonoscopy
Arteriovenous malformation
Disseminated intravascular coagulopathy and other coagulopathies
Unknown cause—6% to 23%

Retroperitoneal and rectus sheath (soft tissue)

The retroperitoneum is a large and highly vascular space (See Table 15.8). While not common, retroperitoneal and rectus sheath hematomas can be substantial, and therefore, fatal. Often, they are occult and go undiagnosed or misdiagnosed in the clinical setting, particularly when in the retroperitoneum, explaining their high mortality. As such, their discovery at autopsy is often unexpected. "Spontaneous" hemorrhages (nontraumatic, noniatrogenic, and no structurally demonstrable natural cause) in these compartments are well described in the clinical literature; when in the retroperitoneum, it is referred to as **spontaneous retroperitoneal hemorrhage (SRH),** or Wunderlich syndrome.[334–336] These hematomas are typically diagnosed in the setting of anticoagulation therapy and/or coagulopathies of other etiologies, including chronic alcohol use, with cirrhosis. In the latter scenario, is it postulated that these hematomas may result from ruptured soft tissue varices and/or "microtrauma"-induced shearing injury of the vulnerable epigastric artery as it penetrates the rectus sheath in a coagulopathic alcoholic decedent. Other difficult to detect vascular anomalies or vascular inflammatory conditions might also be at the root of these hemorrhages. Moreover, the retroperitoneal space can accommodate a prodigious volume of blood (up to a couple of liters). More commonly than not, this accrual of blood takes considerable time (hours to days), which must be taken into account in approaching the investigation. Another risk factor, especially for the "spontaneous" rectus sheath hemorrhage, is any condition leading to repeated Valsalva-type thoraco-abdominal contractions (coughing paroxysms in a patient with chronic obstructive pulmonary disease, tonic-clonic seizure activity, wheezing, etc.). In one series, the second most common cause of SRH were gynecologic and obstetrical complications.[337–339] Other risk factors include atherosclerosis, hypertension, and renal cancer.[337,340,341]

A separate setting in which retroperitoneal hematomas may occasionally be seen is in hospitalized **coagulopathic patients who have undergone procedures**, mainly central intravenous catheter placements with access through the groin vessels, which may produce small vascular punctures more proximally within the soft tissues of the groin and/or retroperitoneum. These often take time to reach clinical attention as the hemoglobin gradually trends down while blood slowly extravasates through one or more small venipuncture defects, only later reaching a critical volume that ends with the patient's sudden and often inexplicable shock and cardiac arrest. These, of course, would not, by definition, be classified as "spontaneous" hematomas; rather, they would be considered **therapeutic complications** in jurisdictions where that manner is available to use, even if the patient was at enhanced risk with cirrhosis or other coagulopathic risk factors (part 2). It is important to remember that the retroperitoneal space can accommodate a very large volume of blood without clinical impingement on other structures/organs, except for occasional femoral nerve compression, making it a relatively silent space for blood accumulation, until the patient suddenly goes into shock. That said, when they are symptomatic, the patient may complain of flank or abdominal pain, costovertebral angle tenderness, or even hematuria, but the "classic" presentation described as Lenk's triad (flank pain, flank mass, and hypovolemic shock) is not all that common.[336] **Dialysis** is another therapy-related risk factor for the formation of SRH.[342,343]

Occasionally, the volume of the retroperitoneal bleeding exceeds the capacity of that compartment and extends into the groin and thigh, and even into the **anterior abdominal wall (rectus sheath);** or, it might be isolated to the rectus sheath alone. Regardless, in many cases, marked visceral and muscular pallor would be expected and would support the mechanism of frank exsanguination; a lesser amount may cause or contribute to death in persons with comorbidities, particularly cardiac, who succumb to the stress of the blood loss sooner, before exsanguination, per se. Unlike the often asymptomatic hematomas in the larger retroperitoneal space, when in the rectus sheath, the

TABLE 15.8 Differential diagnosis of retroperitoneal hemorrhage.

Trauma to flank/kidney (penetrating or blunt)
Iatrogenic (deeply penetrating groin catheters, tissue/organ biopsies, especially of the kidney, hemodialysis)
Renal artery or other aneurysms or vascular malformations
Spontaneous retroperitoneal hemorrhage—SRH, ruptured varices or "microtrauma" in coagulopathic persons with alcohol use disorder, forceful and/or repeated thoracoabdominal contractions (Valsalva)
Disseminated intravascular coagulopathy and other coagulopathies

bleeding **commonly causes abdominal pain, especially with the patient's attempts abdominal flexion; even sitting up can prove exquisitely painful.** Occasionally, and importantly, *external signs of these hemorrhages might be confused with blunt force trauma/contusions.* Rather than trauma, however, these externally visible soft tissue hemorrhages often **represent hematomas (ecchymoses)** that have expanded and made their way from the deep tissues into the subcutaneous space close to skin surface, not the other way around, as would be the case with direct trauma to the skin surface (*contusions*).

As with all other body compartment hemorrhages, and especially in alcoholics, a *traumatic etiology must be assiduously sought and excluded before coming to the conclusion that the explanation lies in vascular and/or coagulopathic derangements.* The latter is viewed as a **diagnosis of exclusion**.

Delayed deaths due to remote injury: issues and concepts

Not uncommonly, a death occurs in a delayed fashion, long after a life-threatening injury has been sustained, rather than at the moment of injury occurrence. This time lapse might be relatively insignificant (hours) or it may be substantial (decades). The original circumstances can be accidental, suicidal, or homicidal in nature and the modality chemical or mechanical. Regardless of these differences, the common thread between these deaths is that, because there is **no efficient intervening cause**, the original injury is culpable for the ultimate fatality, irrespective of its delay.

That said, **the FP's burden is to prove the mechanistic link between the remote injury and the moment of death**, and when that death is a homicide, the burden is higher than it would be for any other manner. Because of the temporal lapse, these can be difficult to adjudicate and certify to a reasonable degree of medical certainty *if* the evidence and documentation have not been lined up correctly. By contrast, it is relatively straightforward for the FP to defend such a certification if they systematically gather the evidence that points only to one reasonable conclusion: but for the injury, the death would not have occurred because it was not interrupted by any efficient intervening cause. Even if the death ensues 30+ years after the injury was sustained, if properly thought through, worked up, and documented, the conclusion can be as clear as it would have been had the decedent died on the day of injury. When such a clear causal link exists, the amount of time that has lapsed between the injury and its fatal outcome in no way relieves the original traumatic event of its culpability.

If the remote injury results in some form of paralysis, a mental checklist should appear in the mind of the ME/C when the individual ultimately dies and the FP is presented with an investigation of their death.

For example, if the decedent had been rendered **paraplegic** as a result of a remote injury, what should the FP anticipate about how this type of injury might subsequently cause death? One mechanism is *deep vein thrombosis (DVT) with* ***PTE*** due to the associated relative immobility (being wheelchair bound), upper and lower urinary tract calculi and infections stemming from **neurogenic bladder-related stasis,** and occasionally, **infectious/septic decubitus ulcer complications,** the latter less so than in quadriplegic victims.

For a person rendered **quadriplegic**, a different, expanded set of complications should be considered, given the level and extent of the paralysis: more **respiratory difficulties** *and* **bronchopneumonia** are born out of the high cervical injury and complete immobility, **sepsis** *complicating decubitus ulcers,* **nutritional considerations and complications** (these patients are often tube fed), **PTE** (due to paralytic immobility/often bedbound and recumbent), and **gastrointestinal and urinary complications** born out of issues related to bowel motility dysfunction and a neurogenic bladder.

A forensic pathologist should have a broad understanding of the typical work-up and differential diagnoses for each type of paralytic injury and, as such, know how the autopsy should be approached (cultures, histology, retaining the brain/spinal cord, etc.). Beyond that, it is sometimes necessary to request confirmatory hospital and police records, if still available. Families can often be helpful resources in letting the FP know what hospital was involved on the day of the injury. They also occasionally (rarely) have a copy of the original police report (sometimes, the case has already been adjudicated by the time of the death, but the charges can still be upped by law enforcement from felony assault to murder, if the defendant is still alive and if the death is classified as a homicide). If the crime occurred in another state, local detectives can do their best to obtain those police records. Other jurisdictions are under no obligation to cooperate with an outside jurisdiction, but they usually do, if the records are not so old that they have been expunged. Refrain from asking for the entire, sometimes voluminous medical record on these deaths, some preceded by years of medical therapy, often involving multiple facilities. Rather, it is usually better to be more focused in your request, asking, for example, only for the admission note, the discharge summary, and perhaps some imaging and laboratory/microbiology studies, and operative notes from the original admission, depending on what is relevant to the case.

For all of the above reasons, it is more common than not that the death certificates for these fatalities are "pending" on the day of autopsy. That said, it is still important to clearly communicate to law enforcement officials about the likelihood of the death ultimately being certified as a homicide, depending upon what the FP knows and has concluded at that point in time, based on their sense of the likely mechanism of death. Many are clearly going to be certified as homicides later, but are made initially "pending" only to allow time for the gathering of confirmatory reports and the time and place of injury occurrence that will ultimately be included on the death certificate.

All this said, if a decedent was the victim of remote trauma (shooting, stabbing, etc.), but the injury has *not* led to sequelae that in turn led to death, there is no need to request the related police and medical records. In other words, if there is a clear and efficient intervening cause (such as a drug intoxication), avoid sending yourself or the police down unnecessary rabbit holes. Similarly, if an old bullet is incidentally found by radiography and/or during the performance of an autopsy, the FP need not pursue a police report about the associated circumstances of the remote shooting if it has no (causal) bearing on the death. That said, in some jurisdictions, there is a policy to retrieve and submit all bullets as evidence (as a matter of chain of custody practice), regardless of whether the firearms-related injury contributed in any way to the death.

Below is a summary of general categories of complications that may stem from remote injury of various trauma modalities of all manners, with their fatal/mechanistic links:

- Infectious Complications:
  - Related to immobility and recumbency (sepsis due to bronchopneumonia or decubitus ulcers)
  - Urinary tract infections, upper and/or lower, with/without urosepsis, from a neurogenic bladder related to CNS trauma
  - Pneumococcal sepsis due to a remote emergency/traumatic splenectomy (related to remote firearms or sharp/blunt force injuries)
- Thromboembolic Events: DVT/PTE (related to relative immobility/inactivity due to remote musculoskeletal blunt or penetrating trauma and/or neurotrauma—brain and/or spinal cord)
- Post-traumatic seizure disorder (with death related to seizure activity itself, or its sequelae: positional asphyxia, drowning, choking, motor vehicular collision, etc.)
- Sequelae of small bowel obstruction (e.g., perforation and acute peritonitis) as a delayed complication of adhesions, or incarcerated incisional hernias related directly to remote penetrating trauma (firearms, sharp, blunt) with/without associated remote emergency surgery
- Myriad toxic/metabolic/nutritional/deconditioning complications related to being chronically bedbound from remotely sustained neurotrauma (persistent vegetative state, para-/quadriplegia, etc.); complications of severe constipation and bowel dysmotility issues
- Complications of HIV/Hepatitis B or C acquired from emergency trauma-related blood product replacement for a variety of injuries involving substantial blood loss (much less common today, given stringent blood screening protocols).

Finally, besides locating police and (sometimes) medical/surgical records, it is not uncommon to delay certification of these cases while conducting analyses in neuropathology (to document remote brain and/or spinal cord trauma), histology (to document PTE/DVT/bronchopneumonia, etc.), microbiology (blood cultures to rule out sepsis), and toxicology (mainly for those persons dying at home, in order to exclude a drug intoxication as an efficient intervening cause; this in not typically necessary to perform in patients dying from a variety of complications in long-term care).

Approach to dismembered remains and other forms of postmortem concealment

The distraction of the dramatic.

Postmortem dismemberment, incineration, and burial are methods that perpetrators sometimes, if rarely, use to disfigure a body in an attempt to both conceal decedent identity and destroy evidence. The very act of going to this extreme lends strong support for the death being a homicide. Rarely, and in contrast, one may try to distance themselves from the victim of a drug overdose by removing the body from the place of death. This is often done in an attempt to avoid police involvement. In these instances, the body might be placed in a hallway or stairwell of an apartment building, dropped on the side of the road, or placed in an alley. While this postmortem activity is illegal and uncommon, the motivation appears to be driven by the need to simply distance oneself from the remains, rather than to cover up the death or its cause and circumstances, per se. Therefore, it would be extremely unusual for such a decedent, having died from an accidental cause (illicit drug intoxication), to be dismembered or set on fire.[344] There have also been descriptions of "dismemberments" involving drug traffickers (so-called "mules"), who ingest drug packets and subsequently die from an intoxication due to ruptured or leaking packets. Under these circumstances, there have been some reports describing postmortem incisions made to recover the drugs.[344] More accurately, these would be referred to as postmortem disembowelments rather than dismemberments, the latter not typically encountered in this setting.

The forensic anthropologist

Because of the condition of the remains in these deaths, the forensic anthropologist is not uncommonly called upon to assist in the effort to evaluate the decedent for the purposes of both identification and trauma analysis, primarily as it relates to the dismemberment.

Forensic anthropology is the science of applying anthropology to medicolegal investigations. The most typical tasks employ methods that estimate biological profiles (age, sex, ancestry, stature, and skeletal anomalies and pathologies),[345–351] particularly in skeletonized remains, as well as an assessment of skeletal trauma.[352–358] But, the practice of forensic anthropology has also experienced a considerable increase in sophistication, volume, and breadth of work, increasingly involving recovery of human remains, including those that are a part of mass fatality incidents, determination of human versus nonhuman origins, the assessment of forensic significance, trauma analysis, and identification (e.g., radiographic comparison or dental comparison, 3D imaging).[359–364] Many of these areas are discussed elsewhere in this text (Chapter 4).

Forensic anthropologists are important resources in certain medicolegal death investigations. Forensic anthropology is a board-accredited specialty that requires qualified practitioners to complete intensive and long-term postgraduate study. The American Board of Forensic Anthropology (ABFA) is the certification organization for forensic anthropologists. The ABFA offers certification programs at the MA/MS and Ph.D. levels.[359] *See* Box 15.1 *regarding the forensic anthropologist as a consultant in the investigation of deaths involving postmortem concealment.*

BOX 15.1
The anthropologic approach to dismemberment

Bradley Adams, Ph.D.

Cases of intentional postmortem dismemberment are not frequent, but may add a level of complexity to the autopsy and decedent identification.[365] Most commonly, intentional dismemberment occurs in a postmortem attempt to make the body more easily transportable, more easily disposable, and to hinder identification efforts, although other explanations exist.[366–371]

Postmortem dismemberment is accomplished by **disarticulation** (separation between joints) **or transection** (cutting through a bone). Disarticulation is usually performed with a knife, and transection occurs through sawing or chopping.[372] In some cases, a combination of disarticulation and transection is observed on the same body. Data derived from New York City show that 53% of their cases involved transection, 35% involved disarticulation, and 13% involved a combination of both.[344]

For the analysis of dismemberment cases within a medicolegal setting, a forensic anthropologist can provide valuable information regarding cuts to bone and/or cartilage. This can pertain to reassociation of body parts by conjoining along cut surfaces, documenting the anatomical distribution of cuts (aiding in the reunification of body parts), and conducting an analysis of the marks left in the bone/cartilage. In some instances, casts of the cut surfaces can assist with interpretation of the tool marks and may provide better resolution and be more easily assessed under a microscope.

Documentation of the tool marks associated with the dismemberment cuts can be indicative of the type of instrument(s) used in the dismemberment (of potential evidentiary value) and the directionality of the cuts. It is important to keep in mind that an anthropological assessment of the tool marks associated with postmortem dismemberment will focus on class characteristics. These will provide information about certain types of knives or saws, but are not individualizing to the specific tool that was used to make the cuts. That type of comparison may be possible by a trained tool mark examiner.

An analyst will examine the "kerf" to assess the type of instrument(s) used in a dismemberment. A **kerf** is a void in material (e.g., bone or cartilage) after it has been cut. An

BOX 15.1 (cont'd)

incomplete cut (often referred to as a "false start" kerf) results in both walls and the floor of the kerf being retained for evaluation. A complete transection will result in two kerf walls, but the floor will be missing. A kerf resulting from a tool with a beveled edge (e.g., knife blade) will be narrow and will have a V-shaped appearance in cross-section (Fig. 15.1). A saw will create a wider cut with a square-shaped or W-shaped kerf floor (Fig. 15.2).

Striations left on the kerf walls can provide information about the type of tool used. If the tool was a knife, striations may suggest whether or not the blade was serrated.[427,428] If the tool was a saw (Fig. 15.3), striations may indicate whether it was mechanical or hand-powered.[371,429]

For saws, the minimum kerf width can be extremely informative. The minimum width of the kerf will closely correlate with the width of the blade used to make the cut (Fig. 15.2). For example, a hacksaw has a very narrow blade and will result in a narrow kerf. A chainsaw, on the other hand, has a very wide blade, which will result in a very wide kerf. Studies provide data on the relationship between kerf widths and various types of saws.[371] While useful, only a small sampling of blade types is presented and should not be considered exhaustive.

Additional features associated with saw dismemberments include **breakaway spurs.** These vary in size, but are a shelf of bone that breaks away at the terminal end of a cut (Fig. 15.4). The presence of a breakaway spur (or the corresponding breakaway notch) indicates blade progress and is located opposite to where the cutting initiated. The shape, uniformity, and directionality of the striations on the kerf walls can provide information about the type of blade and whether it was hand-powered or mechanically powered (Figs. 15.3 and 15.4).[371,429]

Dismemberment via chopping will provide a distinctive signature in bone.[368,372,382,430,431] **Chopping or hacking** will result in a combination of blunt force and sharp force marks in the bone, resulting in an irregular appearance (Fig. 15.5).

Frequent anatomic locations for body dismemberment include the neck, shoulders, hips, and knees.[344] The number of body parts per case ranged from 2 to 14, with an average of 5.5 body parts per case, based on the New York City data. Most commonly, intentional dismemberment involves symmetric removal of body parts (e.g., both arms removed at the shoulders). It is uncommon to find cases that lack contralateral cuts between right and left locations.[344]

Dismemberment cases are almost exclusively associated with homicides, although rare exceptions do exist. Adams et al. presented two examples of nonhomicide dismemberments in which body packers accidentally died from a ruptured drug packet that had been ingested for smuggling.[366] In each case, an accomplice subsequently cut apart the decedent to retrieve the remaining drug packets and dispose of the body.

A study of New York City dismemberment cases between 1996 and 2017 showed that there was a total of 55 dismembered decedents, ranging from a high of eight bodies in 2005 and a low of 0 bodies in 2016.[344] On average, there were 2.5 annual dismemberment cases within the study period, which suggests one dismemberment occurs in every 224 homicides in New York City. The frequency seen in New York City is high and is comparable to counts reported for some countries.[366,367,369,373,432,433]

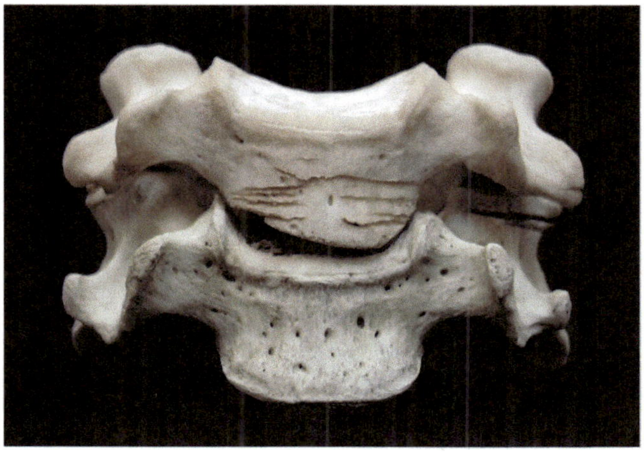

FIGURE 15.1 **Knife cut marks associated with removal of the head between the fourth and fifth cervical vertebrae.**

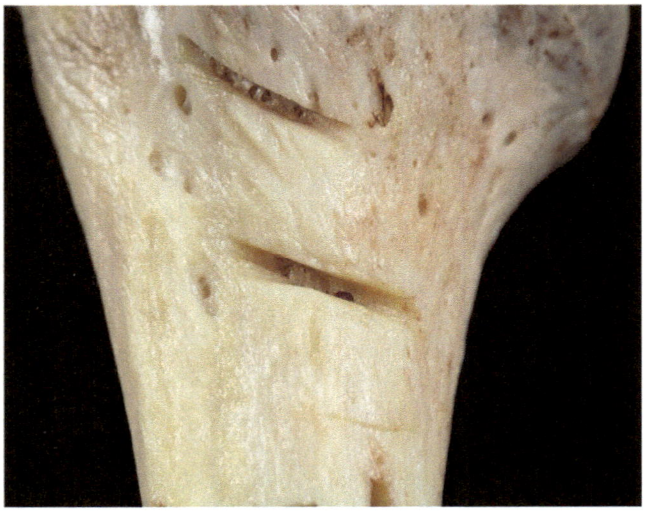

FIGURE 15.2 **Proximal humerus showing two incomplete "false start" kerfs from a saw.**

FIGURE 15.3 **Right femur cross-section.** Note cross-section showing striations from a hand-powered rip saw.

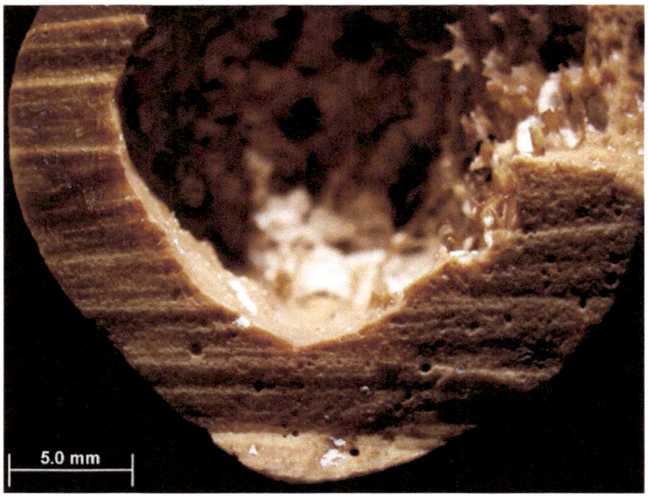

FIGURE 15.4 **Cross-section of a left femur.** Note prominent striations and a break-away spur (bottom of the image) indicating the terminal location of the blade progress.

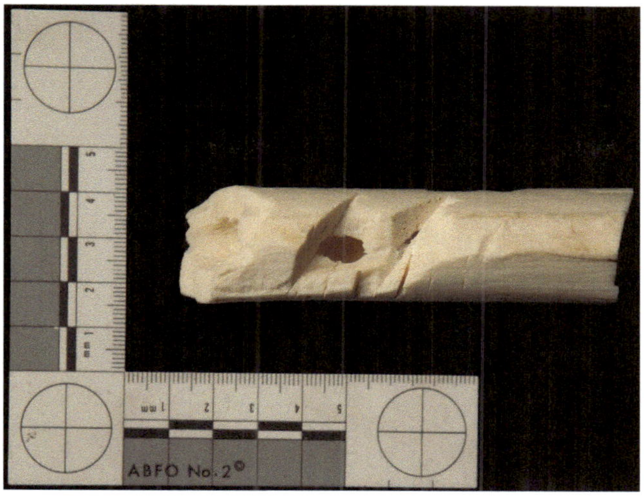

FIGURE 15.5 Chopping/hacking trauma to a femur.

Dismemberment

On rare occasion, the FP is faced with a body that has been dismembered.[344,373–377] *Overwhelmingly, a dismemberment represents a postmortem attempt at criminal concealment and disposal of remains rather than a cause of death,* per se. Regardless, it is no less important during the autopsy to carefully examine the remains and collect gross, trace, and DNA evidence with respect to the dismemberment activity, separate and apart from the focus on the cause and mechanism of death. With rare exception, the decedent has been the victim of homicidal violence prior to the concealment attempt.

Because the aim of the perpetrator in dismembering a body is presumably to dispose of it and/or to destroy evidence and attempts at decedent identity, the remains are not uncommonly deposited in different locations, typically left in trash bags or other forms of packaging, or sometimes thrown in public waterways or even buried. This can create substantial challenges for both law enforcement and the FP when, as a result, the remains are either not found in their entirety, putrefy or mummify at varying rates, are exposed to disparate environmental conditions, including animal predation and relocation, create multiple crime scenes that are separated in time and space and do not necessarily include the location where the homicide actually occurred, and sometimes span more than one jurisdiction. In New York City, for example, the Hudson River separates New York from New Jersey. If remains are deposited into this waterway, the tides will dictate which state ultimately takes jurisdiction, one not necessarily being aware of an active homicide investigation in the other. As such, it may take days to months for all or most of the remains to be discovered in series. For this reason, it is important that law enforcement authorities and ME/C offices clearly communicate with nearby jurisdictions if their search for further remains is ongoing.

Once remains are discovered, a crime scene investigation takes place for each location. This should also involve an ME/C death investigator and, if necessary, a forensic anthropologist, particularly when outdoors and the remains are skeletonized or partially skeletonized. Once the remains are at the morgue, the FP and forensic anthropologist carefully catalog the body parts and their relationships to one another i.e., when reunited, are the separate parts roughly consistent with one another in size, skin color, and sites of dismemberment? If the remains are completely skeletonized, the first question to answer is whether the remains are human. The second is whether there is evidence of more than one decedent (e.g., three femoral heads). Even when not skeletonized, the size and skin color of separate body parts of the same individual may take on dissimilar appearances if found in different locations and under varied environmental conditions. As such, in the end, anthropologic (evaluation of dismemberment disarticulation and transection sites), fingerprinting, when possible, and/or DNA matching can further scientifically confirm the identity of the remains and the accuracy of their reunification (i.e., that all remains represent one person, allowing them to take on the same assigned ME/C case number in the end).[378]

A standard autopsy is performed even if the remains are not available in their entirety. If only the torso is found, for example, the autopsy should be performed on that body part alone, with subsequent examinations of the head and limbs as they are found. On occasion, this may happen over a protracted period, but ideally, where feasible and for continuity of the examination and autopsy report, the same FP should handle the examination of the remains as they are discovered over time, with DNA being collected on each part. That said, depending on the mechanism(s) of death, the cause may not be elucidated or may only be partially elucidated early on, before all the remains are received. For this reason, the death certificate may need to be made "pending" if the discovered remains are initially incomplete.

As always, despite focusing on the cause and mechanism(s) of death, no finding should be considered unimportant. The FP should describe and document *all* that they see, as they would in any homicide, and, if available, work with the anthropologist/anthropology team during the autopsy examination. The anthropologist may guide the FP as to which joint and transection sites they need to have removed for separate processing and analysis, after they have been documented by the FP. Where anthropology support is not available, the FP has been adequately trained to make these decisions and retain samples for subsequent analysis by themselves or in consultation with an outside anthropologist. In the end, the careful analysis of the disarticulation and/or transection sites may be of great evidentiary value in terms of the tool(s) used, even if not in the determination of cause and manner of death, but with the former potentially aiding in the criminal investigation and its ultimate adjudication (see Box 15.1).[374,375,379–389]

Importantly, *in the ultimate certification of these deaths, mention of the dismemberment has no place in the cause, manner, or description of the circumstance of death* ("how the injury occurred") since, while it constitutes postmortem activity, albeit criminal, it is not relevant to the activities leading up to the moment of death. "The temptation to make an epicrisis of the death certificate must be resisted." That said, all of the anatomic and evidentiary details of the dismemberment injuries *do* belong both in the case file and in the autopsy report (in the final diagnoses *and* the body of the report), with a clear indication that they were sustained in the postmortem period, if, indeed, that is the opinion of the FP. Aside from the statistical unlikelihood of this category of injury having occurred antemortem, the absence of hemorrhage or any other vital reaction in these wounds, in combination with mortal injuries that clearly explain the death, support this conclusion. Regardless, the dismemberment injuries should still be included under the "Injuries" section of the autopsy report, but with the word "Postmortem" preceding the "Dismemberment" subheading; in other words, despite having been sustained after death, these injuries nonetheless constitute trauma inflicted on the remains and should be categorized as such.

The causes of death in these decedents are varied and no less likely to involve any one or more of the typical modalities seen in other homicides. On occasion, the remains are altered by such extensive postmortem change or postmortem injury that only some or none of the specific injuries leading to the homicidal death can be clearly identified. In such a case, *"Homicidal Violence, Type Undetermined," or "Homicidal Violence, including* whatever fatal injury was at least partially observed" can be used as cause of death statements (see the section on "Homicidal Violence and Homicidal Asphyxia" in Chapter 11 for a more granular discussion about the approach to this category of homicide).

Postmortem incineration

In an analogous fashion to postmortem dismemberment, the body of the rare victim of murder is set on fire after death in an attempt to destroy evidence or as an act of sadistic violence.[315–322,390–393] Also similar to dismemberment, this poses unique challenges to the finder of fact. As with dismemberment deaths, the identification of the decedent is key. While thermal injury can make this even more challenging, there are a variety of ways to make the identification.[323–328] Fortunately, the incineration attempt rarely, if ever, succeeds in disposing of the remains entirely since most criminals are unfamiliar with the exceedingly high and prolonged temperatures necessary for fire to incinerate the human body completely (adults more so than children).[289,290,329,330,394,395] As such, depending on whether an accelerant is employed and when the body is found relative to when it was set on fire, what the FP is left with are remains that may be only partially burned, allowing a complete autopsy of a minimally altered body on one end of the spectrum, to completely charred remains with exposure of bone and viscera, no obtainable blood, partially missing limbs, and external debris within the burned and exposed trachea on the other end.

Despite the considerable setback of the latter scenario on the postmortem examination, the FP can still determine a surprising amount. In many instances, firearms, sharp force trauma, and blunt force trauma wounds can be recognized on the skin once the charred debris is scraped away, and/or within the deeper tissues, even if not in the usual detail.[321,396,397] Similarly, sequelae such as hemothoraces, a hemoperitoneum, soft tissue hemorrhage, SDHs and SAHs, and fractures may still be detectable. Toxicological analysis is also possible in cases where viscera, soft tissue, and/or blood are available. At the extreme end of thermal injury, with partial skeletonization, certain discernments may not be possible. The most difficult examination to conduct in the setting of advanced destruction of tissue and bone is a thorough assessment of the neck. Physical findings in some homicidal strangulations can be subtle, even in totally intact remains, so extensive thermal destruction may make this determination impossible, depending on the case.[398] Attempts at postmortem incineration are rare outside the context of homicidal violence; this would constitute very unlikely activity following the "dumping" of a body after an accidental drug death, because it is unnecessary and only increases the risk of the perpetrator being seen and caught.

Regardless of the above limitations, the autopsy serves a central role in both identification efforts, as well as in what it can detect and rule out. When an otherwise healthy individual is found dead under obviously suspicious circumstances, including inflicted postmortem thermal injury, but has no structurally demonstrable cause of death, the diagnosis, by exclusion of other causes, is *homicidal violence*, even if the modality of the violence is obscured by the postmortem injury and tissue destruction. In other words, in many cases, the attempt to destroy the body fails to exclude a homicidal manner. Under a different set of circumstances, however, an undetermined cause and manner of death might be appropriate.

Importantly in these types of cases, in order to ascertain whether the decedent was alive or dead at the time of the fire setting, the upper and lower airways should be examined for evidence of smoke and heat inhalational injury (mucosal soot deposition and hyperemia, pulmonary edema, etc.) and the blood tested for an elevated carboxyhemoglobin saturation.[396,399–411] The absence of any/all of these markers of life at the time of the fire helps the FP to exclude thermal injury as a contributory cause of death. Finally, to the extent that it is possible, as in all homicides, every attempt should be made to collect any available trace evidence, despite the intentionally suboptimal condition of the body. Also important may be the submission to the appropriate laboratory of any remaining clothing or other fabric found with the remains for possible accelerant analysis.

Postmortem burial

Postmortem burial in the ground or inside walls or other hidden or sealed structures constitutes a third strategy that may rarely be used for body concealment.[412–426] As an aside, while postmortem concealment constitutes criminal activity (illegal disposal of a dead body) and these deaths are more commonly than not homicides, not all of them necessarily result from injury inflicted by another. A rare such scenario would be an elderly relative dying of natural causes, but where maintaining the appearance of her continued life benefits one or more survivors. For example, if a decedent stakes claim to a rent-controlled apartment (name on the lease) until the time of their death and also collects a social security check each month, her unemployed son who suffers from severe mental illness may have much to lose in reporting her death to authorities. Instead, he may feel compelled, cogently or psychotically, to conceal her body indefinitely while maintaining her residence and continuing to collect her social security income. As with other forms of concealment, putrefaction, mummification, and/or adipocere formation may pose considerable challenges to the forensic examiner in ascertaining the cause, manner, and time of death and in establishing identity (Box 15.1).

References

1. Wobeser WL, Datema J, Bechard B, Ford P. Causes of death among people in custody in Ontario, 1990–1999. *CMAJ*. 2002;167(10):1109–1113.
2. Kariminia A, Law MG, Butler TG, et al. Factors associated with mortality in a cohort of Australian prisoners. *Eur J Epidemiol*. 2007;22(7): 417–428.
3. Fazel S, Baillargeon J. The health of prisoners. *Lancet*. 2011;377(9769):956–965.
4. Binswanger IA, Stern MF, Deyo RA, et al. Release from prison–a high risk of death for former inmates. *N Engl J Med*. 2007;356(2):157–165.
5. Krinsky CS, Lathrop SL, Brown P, Nolte KB. Drugs, detention, and death: a study of the mortality of recently released prisoners. *Am J Forensic Med Pathol*. 2009;30(1):6–9.
6. Hser YI, Hoffman V, Grella CE, Anglin MD. A 33-year follow-up of narcotics addicts. *Arch Gen Psychiatr*. 2001;58(5):503–508.
7. Knight LD, Stoppacher R. Investigation of deaths and injury of prison inmates. *Acad For Path*. 2014;4(3):351–365.
8. Graham MA. Investigation of deaths temporally associated with law enforcement apprehension. *Acad For Path*. 2014;4(3):366–389.
9. Gill J, Girela-López E. Manner of death for in-custody deaths. *Acad For Path*. 2015;5(3):402–413.
10. Gill JR, Pasquale-Styles M. Firearm deaths by law enforcement. *J Forensic Sci*. 2009;54(1):185–188.
11. Bhana BD. Custody-related deaths in durban, South Africa 1998-2000. *Am J Forensic Med Pathol*. 2003;24(2):202–207.
12. Reay DT. Death in custody. *Clin Lab Med*. 1998;18(1):1–22.
13. O'Halloran RL, Frank JG. Asphyxial death during prone restraint revisited: a report of 21 cases. *Am J Forensic Med Pathol*. 2000;21(1):39–52.
14. O'Halloran RL, Lewman LV. Restraint asphyxiation in excited delirium. *Am J Forensic Med Pathol*. 1993;14(4):289–295.
15. Jauchem JR. Deaths in custody: are some due to electronic control devices (including TASER devices) or excited delirium? *J Forensic Leg Med*. 2010;17(1):1–7.

16. Ross DL. Factors associated with excited delirium deaths in police custody. *Mod Pathol*. 1998;11(11):1127–1137.
17. Study of deaths following electro muscular disruption: interim report. In: *NIJ*. Washington DC: National Institute of Justice; 2011:ed2011. http://www.ncjrs.gov/App/publications/abstract.aspx?ID=244890.
18. Steinberg A. Prone restraint cardiac arrest: a comprehensive review of the scientific literature and an explanation of the physiology. *Med Sci Law*. 2021;61(3):215–226.
19. Anastos N, Lewis SW, Barnett NW, Sims DN. The determination of psilocin and psilocybin in hallucinogenic mushrooms by HPLC utilizing a dual reagent acidic potassium permanganate and tris(2,2′-bipyridyl)ruthenium(II) chemiluminescence detection system. *J Forensic Sci*. 2006; 51(1):45–51.
20. Fuller RW, Dixon CM, Barnes PJ. Bronchoconstrictor response to inhaled capsaicin in humans. *J Appl Physiol*. 1985;58(4):1080–1084.
21. Chan TC, Vilke GM, Clausen J, et al. The effect of oleoresin capsicum "pepper" spray inhalation on respiratory function. *J Forensic Sci*. 2002; 47(2):299–304.
22. Raval MP, Welti CV. Sudden death from cocaine induced excited delirium: 45 cases (abstract). *Am J Clin Pathol*. 1995;104:329.
23. Stratton SJ, Rogers C, Brickett K, Gruzinski G. Factors associated with sudden death of individuals requiring restraint for excited delirium. *Am J Emerg Med*. 2001;19(3):187–191.
24. Mitchell RA, Diaz F, Goldfogel GA, et al. National association of medical examiners position paper: recommendations for the definition, investigation, postmortem examination, and reporting of deaths in custody. *Academic Forensic Pathology*. 2017;7(4):604–618.
25. Lecomte D, Fornes P, Nicolas G. Stressful events as a trigger of sudden death: a study of 43 medico-legal autopsy cases. *Forensic Sci Int*. 1996; 79(1):1–10.
26. Channa Perera SD, Pollanen MS. Sudden death due to sickle cell crisis during law enforcement restraint. *J Forensic Leg Med*. 2007;14(5): 297–300.
27. Davis JH. Can sudden cardiac death be murder? *J Forensic Sci*. 1978;23(2):384–387.
28. Cebelin M, Hirsch C. Human stress cardiomyopathy: myocardial lesions in victims of homicidal assaults without internal injuries. *Hum Pathol*. 1980;11(2):123–132.
29. Thogmartin JR. Sudden death in police pursuit. *J Forensic Sci*. 1998;43(6):1228–1231.
30. Scheinin L, Wetli CV. Sudden death and sickle cell trait: medicolegal considerations and implications. *Am J Forensic Med Pathol*. 2009;30(2): 204–208.
31. Prahlow JA. Homicidal cerebral artery aneurysm rupture. *J Forensic Sci*. 2004;49(5):1082–1085.
32. Gill JR, McCubbin KR, Landi KK. Homicide by asthma. *Acad For Path*. 2011;1(1):122–127.
33. Loerbroks A, Apfelbacher CJ, Thayer JF, Debling D, Sturmer T. Neuroticism, extraversion, stressful life events and asthma: a cohort study of middle-aged adults. *Allergy*. 2009;64(10):1444–1450.
34. Weiler JM. Exercise-induced asthma: a practical guide to definitions, diagnosis, prevalence, and treatment. *Allergy Asthma Proc*. 1996;17(6): 315–325.
35. Turner SA, Barnard JJ, Spotswood SD, Prahlow JA. "Homicide by heart attack" revisited. *J Forensic Sci*. 2004;49(3):598–600.
36. Luke JL, Reay DT. The perils of investigating and certifying deaths in police custody. *Am J Forensic Med Pathol*. 1992;13(2):98–100.
37. Reay DT, Howard JD, Fligner CL, Ward RJ. Effects of positional restraint on oxygen saturation and heart rate following exercise. *Am J Forensic Med Pathol*. 1988;9(1):16–18.
38. Reay DT, Howard JD. Restraint position and positional asphyxia. *Am J Forensic Med Pathol*. 1999;20(3):300–301.
39. Chan TC, Vilke GM, Neuman T. Reexamination of custody restraint position and positional asphyxia. *Am J Forensic Med Pathol*. 1998;19(3): 201–205.
40. Annas GJ. The last resort–the use of physical restraints in medical emergencies. *N Engl J Med*. 1999;341(18):1408–1412.
41. Hirsch CS. Restraint asphyxiation [letter]. *Am J Forensic Med Pathol*. 1994;15(3):266.
42. Siebert Jr CF, Thogmartin JR. Restraint-related fatalities in mental health facilities: report of two cases. *Am J Forensic Med Pathol*. 2000;21(3): 210–212.
43. Pounder D. Acute excited states and sudden death. Death after restraint can be avoided. *BMJ*. 1998;316(7138):1171.
44. Schmidt P, Snowden T. The effects of positional restraint on heart rate and oxygen saturation. *J Emerg Med*. 1999;17(5):777–782.
45. Chan TC, Vilke GM, Neuman T, Clausen JL. Restraint position and positional asphyxia. *Ann Emerg Med*. 1997;30(5):578–586.
46. Howard J. *Restraint Associated Deaths*. NAME Seattle Meeting (interim); 2001.
47. Pollanen MS, Chiasson DA, Cairns JT, Young JG. Unexpected death related to restraint for excited delirium: a retrospective study of deaths in police custody and in the community. *CMAJ (Can Med Assoc J)*. 1998;158(12):1603–1607.
48. Pudiak CM, Bozarth MA. Cocaine fatalities increased by restraint stress. *Life Sci*. 1994;55(19):PL379–382.
49. Hick JL, Smith SW, Lynch MT. Metabolic acidosis in restraint-associated cardiac arrest: a case series. *Acad Emerg Med*. 1999;6(3):239–243.
50. Miles SH, Irvine P. Deaths caused by physical restraints. *Gerontol*. 1992;32(6):762–766.
51. Fidone GS. Risks in physical restraint. *Hosp Community Psychiatry*. 1988;39(2):203.
52. Miles SH. Restraints and sudden death. *J Am Geriatr Soc*. 1993;41(9):1013.
53. Pare WP, Glavin GB. Restraint stress in biomedical research: a review. *Neurosci Biobehav Rev*. 1986;10(3):339–370.
54. Currier GW, Allen MH. Emergency psychiatry: physical and chemical restraint in the psychiatric emergency service. *Psychiatr Serv*. 2000; 51(6):717–719.
55. Mirchandani HG, Rorke LB, Sekula-Perlman A, Hood IC. Cocaine-induced agitated delirium, forceful struggle, and minor head injury. A further definition of sudden death during restraint. *Am J Forensic Med Pathol*. 1994;15(2):95–99.
56. O'Halloran RL. Reenactment of circumstances in deaths related to restraint. *Am J Forensic Med Pathol*. 2004;25(3):190–193.
57. Chan TC, Neuman T, Clausen J, Eisele J, Vilke GM. Weight force during prone restraint and respiratory function. *Am J Forensic Med Pathol*. 2004;25(3):185–189.
58. Stratton SJ, Rogers C, Green K. Sudden death in individuals in hobble restraints during paramedic transport. *Ann Emerg Med*. 1995;25(5): 710–712.
59. Zun LS. A prospective study of the complication rate of use of patient restraint in the emergency department. *J Emerg Med*. 2003;24(2): 119–124.

60. Michalewicz BA, Chan TC, Vilke GM, Levy SS, Neuman TS, Kolkhorst FW. Ventilatory and metabolic demands during aggressive physical restraint in healthy adults. *J Forensic Sci*. 2007;52(1):171–175.
61. Roeggla M, Wagner A, Muellner M, et al. Cardiorespiratory consequences to hobble restraint. *Wien Klin Wochenschr*. 1997;109(10):359–361.
62. Sathyavagiswaran L, Rogers C, Noguchi TT. Restraint asphyxia in in-custody deaths Medical examiner's role in prevention of deaths. *Leg Med*. 2007;9(2):88–93.
63. Otahbachi M, Cevik C, Bagdure S, Nugent K. Excited delirium, restraints, and unexpected death: a review of pathogenesis. *Am J Forensic Med Pathol*. 2010;31(2):107–112.
64. Cina SJ, Davis GJ. Cardiac dysrhythmia during restraint. *Am J Forensic Med Pathol*. 2010;31(4):e4.
65. Payne-James JJ, Smith G, Rivers E, O'Rourke S, Stark M, Sutcliffe N. Effects of incapacitant spray deployed in the restraint and arrest of detainees in the Metropolitan Police Service area, London, UK: a prospective study. *Forensic Sci Med Pathol*. 2014;10(1):62–68.
66. Kroll MW, Brave MA, Kleist SR, Ritter MB, Ross DL, Karch SB. Applied force during prone restraint: is officer weight a factor? *Am J Forensic Med Pathol*. 2019;40(1):1–7.
67. Strommer EMF, Leith W, Zeegers MP, Freeman MD. The role of restraint in fatal excited delirium: a research synthesis and pooled analysis. *Forensic Sci Med Pathol*. 2020;16(4):680–692.
68. Fernando T, Byard RW. Positional asphyxia without active restraint following an assault. *J Forensic Sci*. 2013;58(6):1633–1635.
69. Nissen T, Rorvik P, Haugslett L, Wynn R. Physical restraint and near death of a psychiatric patient. *J Forensic Sci*. 2013;58(1):259–262.
70. Hall C, Votova K, Heyd C, et al. Restraint in police use of force events: examining sudden in custody death for prone and not-prone positions. *J Forensic Leg Med*. 2015;31:29–35.
71. Woodford NWF. *Injuries and Death Resulting from Restraint*. 2006:171–188.
72. Wetli CV, Fishbain DA. Cocaine-induced psychosis and sudden death in recreational cocaine users. *J Forensic Sci*. 1985;30(3):873–880.
73. Reay DT, Fligner CL, Stilwell AD, Arnold J. Positional asphyxia during law enforcement transport. *Am J Forensic Med Pathol*. 1992;13(2):90–97.
74. Karch SB, Stephens BG. Acute excited states and sudden death. Acute excited states are not caused by high blood concentrations of cocaine. *BMJ*. 1998;316(7138):1171.
75. Karch SB, Wetli CV. Agitated delirium versus positional asphyxia. *Ann Emerg Med*. 1995;26(6):760–761.
76. Bell MD, Rao VJ, Wetli CV, Rodriguez RN. Positional asphyxiation in adults. A series of 30 cases from the dade and Broward county Florida medical examiner offices from 1982 to 1990. *Am J Forensic Med Pathol*. 1992;13(2):101–107.
77. DiMaio TG, DiMaio JM. *Excited Delirium Syndrome: Cause of Death and Prevention*. 1st ed. Boca Raton, FL: CRC; 2006.
78. Grant JR, Southall PE, Mealey J, Scott SR, Fowler DR. Excited delirium deaths in custody: past and present. *Am J Forensic Med Pathol*. 2009; 30(1):1–5.
79. Hall CA, McHale AM, Kader AS, Stewart LC, MacCarthy CS, Fick GH. Incidence and outcome of prone positioning following police use of force in a prospective, consecutive cohort of subjects. *J Forensic Leg Med*. 2011;19(2):83–89.
80. Maher PJ, Walsh M, Burns T, Strote J. Prehospital resuscitation of a man with excited delirium and cardiopulmonary arrest. *Cjem*. 2013;15(0): 1–4.
81. Hall CA, Kader AS, Danielle McHale AM, Stewart L, Fick GH, Vilke GM. Frequency of signs of excited delirium syndrome in subjects undergoing police use of force: descriptive evaluation of a prospective, consecutive cohort. *J Forensic Leg Med*. 2013;20(2):102–107.
82. Paquette M. Excited delirium: does it exist? *Psychiatr Care*. 2003;39(3):93–94.
83. Gill JR. The syndrome of excited delirium. *Forensic Sci Med Pathol*. 2014;10(2):223–228.
84. Rajagopalan A, Pollanen MS. Sudden death during struggle in the setting of heterozygosity for a mutation in calsequesterin 2. *Forensic Sci Med Pathol*. 2016;12(1):86–89.
85. Payne-James JJ, Rivers E, Green P, Johnston A. Trends in less-lethal use of force techniques by police services within England and Wales: 2007–2011. *Forensic Sci Med Pathol*. 2014;10(1):50–55.
86. Southall P, Grant J, Fowler D, Scott S. Police custody deaths in Maryland, USA: an examination of 45 cases. *J Forensic Leg Med*. 2008;15(4): 227–230.
87. Krantz DS, Sheps DS, Carney RM, Natelson BH. Effects of mental stress in patients with coronary artery disease: evidence and clinical implications. *JAMA*. 2000;283(14):1800–1802.
88. Wittstein IS, Thiemann DR, Lima JA, et al. Neurohumoral features of myocardial stunning due to sudden emotional stress. *N Engl J Med*. 2005; 352(6):539–548.
89. Malik MA. Emotional stress as a precipitating factor in sudden deaths due to coronary insufficiency. *J Forensic Sci*. 1973;18(1):47–52.
90. Chi JS, Kloner RA. Stress and myocardial infarction. *Heart*. 2003;89(5):475–476.
91. Moritz AR. Trauma, stress, and coronary thrombosis. *J Am Med Assoc*. 1954;156(14):1306–1309.
92. Lesnikova I, Leone L, Gilliland M. Manner of death certification after significant emotional stress: an inter-rater variability study and review of the literature. *Acad Forensic Pathol*. 2018;8(3):692–707.
93. Lyon AR, Rees PS, Prasad S, Poole-Wilson PA, Harding SE. Stress (Takotsubo) cardiomyopathy–a novel pathophysiological hypothesis to explain catecholamine-induced acute myocardial stunning. *Nat Clin Pract Cardiovasc Med*. 2008;5(1):22–29.
94. Boland TA, Lee VH, Bleck TP. Stress-induced cardiomyopathy. *Crit Care Med*. 2015;43(3):686–693.
95. Akashi YJ, Nef HM, Mollmann H, Ueyama T. Stress cardiomyopathy. *Annu Rev Med*. 2010;61:271–286.
96. Ramaraj R. Stress cardiomyopathy: aetiology and management. *Postgrad Med J*. 2007;83(982):543–546.
97. Berman M, Ali A, Ashley E, et al. Is stress cardiomyopathy the underlying cause of ventricular dysfunction associated with brain death? *J Heart Lung Transplant*. 2010;29(9):957–965.
98. Bybee KA, Prasad A. Stress-related cardiomyopathy syndromes. *Circulation*. 2008;118(4):397–409.
99. Wetli CV, Mittleman RE. The body packer syndrome: toxicity following ingestion of illicit drugs packaged for transportation. *J Forensic Sci*. 1981;26:492–500.
100. Byard RW. Ongoing issues with the diagnosis of excited delirium. *Forensic Sci Med Pathol*. 2018;14(2):149–151.
101. Kodikara S, Cunningham K, Pollanen MS. "Excited delirium syndrome": is it a cause of death? *Leg Med*. 2012;14(5):252–254.
102. Lee S, Gottlieb M, Mulhausen P, et al. Recognition, prevention, and treatment of delirium in emergency department: an evidence-based narrative review. *Am J Emerg Med*. 2020;38(2):349–357.

103. Bowman EML, Cunningham EL, Page VJ, McAuley DF. Phenotypes and subphenotypes of delirium: a review of current categorisations and suggestions for progression. *Crit Care*. 2021;25(1):334.
104. Cai A, Cai X. Toxin-induced acute delirium. *Neurol Clin*. 2020;38(4):781–798.
105. *Diagnostic and Statistical Manual of Mental Disorders (DSM-5)*. Washington, DC: American Psychiatric Association, American Psychiatric Publishing; 2013:598.
106. Hayes J. Sudden death during excited delirium, etiology unknown. *ASCP Check Sample*. 1991;91–5(FP-178):1–7.
107. Kasper D, Braunwald E, Fauci A, Hauser S, Longo D, Jameson J, eds. *Harrison's Principle of Medicine*. 16th ed. New York: McGraw-Hill Professional; 2004.
108. Brown TM, Boyle MF. Delirium. *BMJ*. 2002;325(7365):644–647.
109. Ruttenber AJ, McAnally HB, Wetli CV. Cocaine-associated rhabdomyolysis and excited delirium: different stages of the same syndrome. *Am J Forensic Med Pathol*. 1999;20(2):120–127.
110. Allam S, Noble JS. Cocaine-excited delirium and severe acidosis. *Anaesthesia*. 2001;56(4):385–386.
111. Ruttenber AJ, Lawler-Heavner J, Yin M, Wetli CV, Hearn WL, Mash DC. Fatal excited delirium following cocaine use: epidemiologic findings provide new evidence for mechanisms of cocaine toxicity. *J Forensic Sci*. 1997;42(1):25–31.
112. Wetli C. Excited delirium. In: Payne-James J, ed. *Encyclopedia of Forensic and Legal Medicine*. 1st ed. Vol. 2. Amsterdam: Elsevier Academic Press; 2005:276–281.
113. Gill JR. The medicolegal evaluation of excited delirium. In: Tsokos M, ed. *Forensic Pathology Reviews*. Vol. 5. Totowa, NJ: Humana Press; 2008: 91–114.
114. Physicians ACoE. *White Paper Report on Excited Delirium Syndrome. American College of Emergency Physicians; September 10, 2009*. 2009.
115. Mash DC, Duque L, Pablo J, et al. Brain biomarkers for identifying excited delirium as a cause of sudden death. *Forensic Sci Int*. 2009;190(1–3): e13–19.
116. Lucena J, Blanco M, Jurado C, et al. Cocaine-related sudden death: a prospective investigation in south-west Spain. *Eur Heart J*. 2010;31(3): 318–329.
117. Vilke GM, DeBard ML, Chan TC, et al. Excited Delirium Syndrome (ExDS): defining based on a review of the literature. *J Emerg Med*. 2012; 43(5):897–905.
118. Vilke GM, Payne-James J, Karch SB. Excited delirium syndrome (ExDS): redefining an old diagnosis. *J Forensic Leg Med*. 2011;19(1):7–11.
119. Penders TM, Gestring RE, Vilensky DA. Excited delirium following use of synthetic cathinones (bath salts). *Gen Hosp Psychiatr*. 2012;34(6): 647–650.
120. Johnson MM, David JA, Michelhaugh SK, Schmidt CJ, Bannon MJ. Increased heat shock protein 70 gene expression in the brains of cocaine-related fatalities may be reflective of postdrug survival and intervention rather than excited delirium. *J Forensic Sci*. 2012;57(6):1519–1523.
121. Takeuchi A, Ahern TL, Henderson SO. Excited delirium. *West J Emerg Med*. 2011;12(1):77–83.
122. Samuel E, Williams RB, Ferrell RB. Excited delirium: consideration of selected medical and psychiatric issues. *Neuropsychiatric Dis Treat*. 2009; 5:61–66.
123. Storey M. Explaining the unexplainable: excited delirium syndrome and its impact on the objective reasonableness standard for allegations of excessive force. *St Louis Univ Law J*. 2007;56:633–664.
124. Mileusnic-Polchan DLC, Garavan F, Oliver WR. *Excited Delirium Deaths Not Involving Police Custody*. Minneapolis, Minnesota: National Association of Medical Examiners Annual Meeting; 2016. Paper presented at.
125. Byard RW, Cox M, Stockham P. Blunt craniofacial trauma as a manifestation of excited delirium caused by new psychoactive substances. *J Forensic Sci*. 2016;61(6):1546–1548.
126. Shields LB, Rolf CM, Hunsaker 3rd JC. Sudden death due to acute cocaine toxicity-excited delirium in a body packer. *J Forensic Sci*. 2015;60(6): 1647–1651.
127. Kesha K, Boggs CL, Ripple MG, et al. Methylenedioxypyrovalerone ("bath salts"), related death: case report and review of the literature. *J Forensic Sci*. 2013;58(6):1654–1659.
128. Kroll MW, Hail SL, Kroll RM, Wetli CV, Criscione JC. Electrical weapons and excited delirium: shocks, stress, and serum serotonin. *Forensic Sci Med Pathol*. 2018;14(4):478–483.
129. Sn K, þ S, Jg J. Arrest-related death on the basis of a drug-induced excited delirium syndrome. *J Forensic Legal Med*. 2020:102091.
130. Maloney K. Excited delirium unassociated with police involvement (H37). In: *Paper Presented at: American Academy of Forensic Sciences 73rd Annual Scientific Meeting*. 2021 (Virtual).
131. DeBard ML. Identifying new disease as excited delirium syndrome rejects idea that police Brutality causes deaths. *Emerg Med News*. 2009; 31(11):3–5.
132. Pearson A. *Can Excited Delirium Get Cops off the Hook?* Elsevier; 2009.
133. Gonin P, Beysard N, Yersin B, Carron PN. Excited delirium: a systematic review. *Acad Emerg Med*. 2018;25(5):552–565.
134. Obasogie O. Excited delirium and police use of force. *Va Law Rev*. 2021;107(8):1545–1620.
135. Baker D. Making sense of 'excited delirium' in cases of death after police contact. *Policing: J Pol Pract*. 2018;12(4):361–371.
136. American College of Emergency Physicians (ACEP) Hyperactive Delirium Task Force. *ACEP Task Force Report on Hyperactive Delirium with Severe Agitation in Emergency Settings*; Jun. 23, 2021. www.acep.org/globalassets/new-pdfs/education/acep-task-force-report-on-hyperactive-delirium-final.pdf.
137. Harwood R. How to deal with violent and aggressive patients in acute medical settings. *J Roy Coll Phys Edinb*. 2017;47(2).
138. Patel MX, Sethi FN, Barnes TR, et al. Joint BAP NAPICU evidence-based consensus guidelines for the clinical management of acute disturbance: de-escalation and rapid tranquillisation. *J Psychiatr Intensive Care*. 2018;14(2):89–132.
139. Stevenson S. *Heading off Violence with Verbal De-escalation*. Vol. 29. NJ: SLACK Incorporated Thorofare; 1991:6–9.
140. Richmond JS, Berlin JS, Fishkind AB, et al. Verbal de-escalation of the agitated patient: consensus statement of the American association for emergency psychiatry project BETA de-escalation workgroup. *West J Emerg Med*. 2012;13(1):17.
141. Watson AC, Fulambarker AJ. The crisis intervention team model of police response to mental health crises. *Best Pract Mental Health*. 2012;8(2): 71–81.

142. Creed JO, Cyr JM, Owino H, et al. Acute crisis care for patients with mental health crises: initial assessment of an innovative prehospital alternative destination program in North Carolina. *Prehosp Emerg Care*. 2018;22(5):555–564.
143. Fernandez AR, Bourn SS, Crowe RP, et al. Out-of-hospital ketamine: indications for use, patient outcomes, and associated mortality. *Ann Emerg Med*. 2021;78(1):123–131.
144. Cole JB, Klein LR, Nystrom PC, et al. A prospective study of ketamine as primary therapy for prehospital profound agitation. *Am J Emerg Med*. 2018;36(5):789–796.
145. Weedn VW, Menendez MJ. Reclaiming the autopsy as the practice of medicine: a pathway to remediation of the forensic pathology workforce shortage? *Am J Forensic Med Pathol*. 2020;41(4):242–248.
146. Rosh A, Sampson BA, Hirsch CS. Schizophrenia as a cause of death. *J Forensic Sci*. 2003;48(1):164–167.
147. da Silva Bahatia B, Heisler M, Naples-Mitchell J, Saadi A. Excited delirium and Deaths in police custody. *Psychiatr Serv*. 2022;2:2022.
148. Saadi A, Naples-Mitchell J, da Silva Bhatia B, Heisler M. End the use of "excited delirium" as a cause of death in police custody. *Lancet*. 2022; 399(10329):1028–1030.
149. DiMaio V, DiMaio D. *Forensic Pathology*. 2nd ed. Boca Raton, FL: CRC Press; 2001.
150. Falk J, Riepert T, Rothschild MA. A case of suicide-by-cop. *Leg Med*. 2004;6(3):194–196.
151. Hutson HR, Anglin D, Yarbrough J, et al. Suicide by cop. *Ann Emerg Med*. 1998;32(6):665–669.
152. Van Zandt CR. Suicide by cop. *Police Chief*. 1993;July:24–30.
153. Geberth V. Suicide-by-cop. *Law Order*. 1993;July:105–108.
154. Mohandie K, Meloy JR, Collins PI. Suicide by cop among officer-involved shooting cases. *J Forensic Sci*. 2009;54(2):456–462.
155. Mohandie K, Meloy JR. Suicide by cop among female subjects in officer-involved shooting cases. *J Forensic Sci*. 2011;56(3):664–668.
156. Neitzel AR, Gill JR. Death certification of "suicide by cop". *J Forensic Sci*. 2011;56(6):1657–1660.
157. Wilson EF, Davis JH, Bloom JD, Batten PJ, Kamara SG. Homicide or suicide: the killing of suicidal persons by law enforcement officers. *J Forensic Sci*. 1998;43(1):46–52.
158. Geberth VJ. *Practical Homicide Investigation: Tactics, Procedures, and Forensic Techniques*. CRC Press; 2020.
159. Hirsch CS, Flomenbaum M. Problem-solving in death certification. *Am Soc Clin Pathol, Foren Pathol Check Sam*. 1995;95–1:1–31. FP.
160. Samuels MA. The brain-heart connection. *Circulation*. 2007;116(1):77–84.
161. Lindsay J, Paixao A, Chao T, Pichard AD. Pathogenesis of the Takotsubo syndrome: a unifying hypothesis. *Am J Cardiol*. 2010;106(9): 1360–1363.
162. Hanzlick R, Hunsaker JC, Davis GJ. *A Guide for the Manner of Death Classification*. Atlanta: National Association of Medical Examiners; 2002.
163. Collins KA, Presnell SE. Elder neglect and the pathophysiology of aging. *Am J Forensic Med Pathol*. June 2007;28(2):157–162.
164. Gruszecki AC, Edwards J, Powers RE, Davis GG. Investigation of elderly deaths in nursing homes by the medical examiner over a year. *Am J Forensic Med Pathol*. 2004;25(3):209–212.
165. Collins KA. Elder maltreatment: a review. *Arch Pathol Lab Med*. September 2006;130(9):1290–1296.
166. Lachs MS, Pillemer K. Abuse and neglect of elderly persons. *N Engl J Med*. 1995;332(7):437–443.
167. Julie AH, Collins KA, Pat SK, Yeoman C. Investigation of elder deaths. *Acad Forensic Pathol*. 2014;4(3):290–304.
168. Osculati A, Fassina G. Two cases of accidental asphyxia by neck compression between bed bars. *Am J Forensic Med Pathol*. 2000;21(3):217–219.
169. John SM, Koelmeyer TD. The forensic pathology of nonagenarians and centenarians: do they die of old age? (The Auckland experience). *Am J Forensic Med Pathol*. 2001;22(2):150–154.
170. Bennett AT, Collins KA. Elderly suicide: a 10-year retrospective study. *Am J Forensic Med Pathol*. 2001;22(2):169–172.
171. deRoux S. Suicidal incised wound of a fistula for hemodialysis access in an elderly woman. *Am J Forensic Med Pathol*. 2001;22(2):212–213.
172. Pounder DJ. The case of Dr. Shipman. *Am J Forensic Med Pathol*. 2003;24(3):219–226.
173. Collins KA, Sellars K. Vertebral artery laceration mimicking elder abuse. *Am J Forensic Med Pathol*. 2005;26(2):150–154.
174. Rothschild JM, Bates DW, Leape LL. Preventable medical injuries in older patients. *Arch Intern Med*. 2000;160(18):2717–2728.
175. Dolkas L, Stanley C, Smith AM, Vilke GM. Deaths associated with choking in San Diego county. *J Forensic Sci*. 2007;52(1):176–179.
176. Kibayashi K, Sumida T, Shojo H, Hanada M. Dementing diseases among elderly persons who suffered fatal accidents: a forensic autopsy study. *Am J Forensic Med Pathol*. 2007;28(1):73–79.
177. Abrams RC, Leon AC, Tardiff K, Marzuk PM, Sutherland K. "Gray murder": characteristics of elderly compared with nonelderly homicide victims in New York City. *Am J Publ Health*. 2007;97(9):1666–1670.
178. Dolinak D. Review of the significance of various low force fractures in the elderly. *Am J Forensic Med Pathol*. 2008;29(2):99–105.
179. Chisholm KM, Harruff RC. Elderly deaths due to ground-level falls. *Am J Forensic Med Pathol*. 2010;31(4):350–354.
180. Okamoto N, Tomioka K, Saeki K, et al. Relationship between swallowing problems and tooth loss in community-dwelling independent elderly adults: the Fujiwara-kyo study. *J Am Geriatr Soc*. 2012;60(5):849–853.
181. Berzlanovich AM, Fazeny-Dorner B, Waldhoer T, Fasching P, Keil W. Foreign body asphyxia: a preventable cause of death in the elderly. *Am J Prev Med*. 2005;28(1):65–69.
182. Drummer OH. Post-mortem toxicology in the elderly. *Forensic Sci Med Pathol*. 2013;9(2):258–259.
183. Carabellese F, Candelli C, Vinci F, Tamma M, Catanesi R. Elderly sexual offenders: two unusual cases. *J Forensic Sci*. 2012;57(5):1381–1383.
184. Biermann T, Dippel O, Bergner M, et al. Assaults in the elderly–a population-based study with victim and perpetrator characteristics. *J Forensic Sci*. 2011;56(3):669–673.
185. Di Nunno N, Vacca M, Costantinides F, Di Nunno C. Death following atypical compression of the neck. *Am J Forensic Med Pathol*. 2003;24(4): 364–368.
186. Byard RW, Gilbert JD. Sleeping accidents in the elderly. *J Forensic Sci*. 2011;56(6):1645–1647.
187. Hartshorne NJ, Harruff RC, Alvord Jr EC. Fatal head injuries in ground-level falls. *Am J Forensic Med Pathol*. 1997;18(3):258–264.
188. Mann RW, Kobayashi M, Schiller AL. Biparietal thinning: accidental death by a fall from standing height. *J Forensic Sci*. 2017;62(5):1406–1409.
189. Feghali J, Yang W, Huang J. Updates in chronic subdural hematoma: epidemiology, etiology, pathogenesis, treatment, and outcome. *World Neurosurg*. 2020;141:339–345.
190. Li Y, Temkin-Greener H, Shan G, Cai X. COVID-19 infections and deaths among Connecticut nursing home residents: facility correlates. *J Am Geriatr Soc*. 2020;68(9):1899–1906.

191. Su Z, McDonnell D, Li Y. Why is COVID-19 more deadly to nursing home residents? *QJM*. 2021;114(8):543–547.
192. Barnett ML, Grabowski DC. Nursing homes are ground zero for COVID-19 pandemic. *JAMA Health Forum*. 2020;1(3):e200369.
193. Bagchi S, Mak J, Li Q, et al. Rates of COVID-19 among residents and staff members in nursing homes—United States, May 25—November 22, 2020. *MMWR (Morb Mortal Wkly Rep)*. 2021;70(2):52.
194. Kim JJ, Coffey K, Morgan DJ, Roghmann M-C. Lessons learned—Outbreaks of COVID-19 in nursing homes. *Am J Infect Control*. 2020;48(10):1279.
195. Fox L, Crane SA, Bidari C, Jones A. Intra-abdominal hemorrhage from ruptured varices. *Arch Surg*. 1982;117(7):953–956.
196. Lyon DT, Mantia AG, Schubert TT. Hemoperitoneum from a ruptured varix in cirrhosis. Case report and literature review. *Am J Gastroenterol*. 1979;71(6):611–616.
197. Hamilton JR, Sunter JP, Cooper PN. Fatal hemorrhage from simple lacerations of the scalp. *Forensic Sci Med Pathol*. 2005;1(4):267–271.
198. Veevers AE, Lawler W, Rutty GN. Walk and die: an unusual presentation of head injury. *J Forensic Sci*. 2009;54(6):1466–1469.
199. Lee CK, Gray L, Maguire J. Traumatic vertebral artery injury: detailed clinicopathologic and morphometric analysis of 6 cases. *Am J Forensic Med Pathol*. 2009;30(2):134–136.
200. Hayashi T, Buschmann C, Matejic D, Ingold B, Tsokos M. Fatal intraabdominal bleeding from ectopic varices: report of two autopsy cases. *Forensic Sci Med Pathol*. 2013;9(1):96–99.
201. Simonsen J. Fatal subarachnoid haemorrhage in relation to minor head injuries. *J Forensic Med*. 1967;14(4):146–155.
202. Dedouit F, Piercecchi-Marti MD, Leonetti G, Rouge D, Telmon N. Unexpected natural death secondary to intra-abdominal bleeding: report of one idiopathic spontaneous intraperitoneal hemorrhage case. *Forensic Sci Int*. 2012;214(1–3):e43–46.
203. McIntyre IM, Hamm CE, Sherrard JL, Gary RD, Riley AC, Lucas JR. The analysis of an intracerebral hematoma for drugs of abuse. *J Forensic Sci*. 2003;48(3):680–682.
204. Kibayashi K, Mastri AR, Hirsch CS. Cocaine induced intracerebral hemorrhage: analysis of predisposing factors and mechanisms causing hemorrhagic strokes. *Hum Pathol*. 1995;26(6):659–663.
205. Shvartsbeyn M, Phillips DG, Markey MA, Morrison A, DeJong JL, Castellani RJ. Cocaine-induced intracerebral hemorrhage in a patient with cerebral amyloid angiopathy. *J Forensic Sci*. 2010;55(5):1389–1392.
206. Davis NL, Horan PM, Romich T, Roman JL, Lacy JM, Catanese CA. Giant berry aneurysms in a forensic setting: a series of 14 cases. *Am J Forensic Med Pathol*. 2010;31(4):330–334.
207. Nolte KB, Gelman BB. Intracerebral hemorrhage associated with cocaine abuse. *Arch Pathol Lab Med*. 1989;113(7):812–813.
208. Daras M, Tuchman AJ, Koppel BS, Samkoff LM, Weitzner I, Marc J. Neurovascular complications of cocaine. *Acta Neurol Scand*. 1994;90(2):124–129.
209. Mohr J. Arteriovenous malformations of the brain in adults. *N Engl J Med*. 1999;340(23):1812–1818.
210. Racette S, Sauvageau A. Brain arteriovenous malformation and its implication in forensic pathology. *J Forensic Sci*. 2007;52(1):189–191.
211. Matschke J, Lockemann U, Schulz F. Intracranial arteriovenous malformations presenting as sudden unexpected death: a report of 3 cases and review of the literature. *Am J Forensic Med Pathol*. 2007;28(2):173–176.
212. Lobato RD, Rivas JJ, Gomez PA, et al. Head-injured patients who talk and deteriorate into coma. Analysis of 211 cases studied with computerized tomography. *J Neurosurg*. 1991;75(2):256–261.
213. Reilly PL, Graham DI, Adams JH, Jennett B. Patients with head injury who talk and die. *Lancet*. 1975;2(7931):375–377.
214. Baratham G, Dennyson WG. Delayed traumatic intracerebral haemorrhage. *J Neurol Neurosurg Psychiat*. 1972;35(5):698–706.
215. Yamada M. Cerebral amyloid angiopathy: an overview. *Neuropathol: Offic J Japanese Soci Neuropathol*. 2000;20(1):8–22.
216. Pezzini A, Del Zotto E, Volonghi I, Giossi A, Costa P, Padovani A. Cerebral amyloid angiopathy: a common cause of cerebral hemorrhage. *Curr Med Chem*. 2009;16(20):2498–2513.
217. Del Greco F, Krumlovsky F. Subdural haematoma in the course of haemodialysis. *Lancet*. 1969;2(7628):1009–1010.
218. Leonard CD, Weil E, Scribner BH. Subdural haematomas in patients undergoing haemodialysis. *Lancet*. 1969;2(7614):239–240.
219. Brouns R, De Deyn PP. Neurological complications in renal failure: a review. *Clin Neurol Neurosurg*. 2004;107(1):1–16.
220. Bechar M, Lakke JP, van der Hem GK, Beks JW, Penning L. Subdural hematoma during long-term hemodialysis. *Arch Neurol*. 1972;26(6):513–516.
221. Sood P, Sinson GP, Cohen EP. Subdural hematomas in chronic dialysis patients: significant and increasing. *Clin J Am Soc Nephrol: CJASN*. 2007;2(5):956–959.
222. Feldman KW, Bethel R, Shugerman RP, Grossman DC, Grady MS, Ellenbogen RG. The cause of infant and toddler subdural hemorrhage: a prospective study. *Pediatrics*. 2001;108(3):636–646.
223. Bosma JJ, Miles JB, Shaw MD. Spontaneous chronic and subacute subdural haematoma in young adults. *Acta Neurochir*. 2000;142(11):1307–1310.
224. Buruma OJ, Sande JJ. Bilateral acute spontaneous subdural hematoma. A case report. *Clin Neurol Neurosurg*. 1976;79(3):211–214.
225. Burke MP, O'Donnell C, Opeskin K. Spontaneous acute subdural hematoma complicating arachnoid cyst. *Am J Forensic Med Pathol*. 2010;31(4):382–384.
226. Geisenberger D, Huppertz LM, Buchsel M, Kramer L, Pollak S, Grosse Perdekamp M. Non-traumatic subdural hematoma secondary to septic brain embolism: a rare cause of unexpected death in a drug addict suffering from undiagnosed bacterial endocarditis. *Forensic Sci Int*. 2015;257:e1–e5.
227. Freytag E. Fatal hypertensive intracerebral haematomas: a survey of the pathological anatomy of 393 cases. *J Neurol Neurosurg Psychiat*. 1968;31(6):616–620.
228. Bolliger SA, Thali MJ, Zollinger U. Nontraumatic intracranial epidural hematoma: a case report. *Am J Forensic Med Pathol*. 2007;28(3):227–229.
229. Aromatario M, Torsello A, D'Errico S, et al. Traumatic epidural and subdural hematoma: epidemiology, outcome, and dating. *Medicina)*. 2021;57(2).
230. Pearl GS. Traumatic neuropathology. *Clin Lab Med*. 1998;18(1):39–64.
231. Raven KP, Reay DT, Harruff RC. Artifactual injuries of the larynx produced by resuscitative intubation. *Am J Forensic Med Pathol*. 1999;20(1):31–36.
232. Opeskin K. Traumatic carotid artery dissection. *Am J Forensic Med Pathol*. 1997;18(3):251–257.

233. Sato Y, Kondo T, Ohshima T. Traumatic tear of the basilar artery associated with vertebral column injuries. *Am J Forensic Med Pathol.* 1997; 18(2):129–134.
234. Pollanen MS. A triad of laryngeal hemorrhages in strangulation: a report of eight cases. *J Forensic Sci.* 2000;45(3):614–618.
235. Plattner T, Bolliger S, Zollinger U. Forensic assessment of survived strangulation. *Forensic Sci Int.* 2005;153(2–3):202–207.
236. Pollanen MS, McAuliffe DN. Intra-cartilaginous laryngeal haemorrhages and strangulation. *Forensic Sci Int.* 1998;93(1):13–20.
237. Pollanen MS, Perera SD, Clutterbuck DJ. Hemorrhagic lividity of the neck: controlled induction of postmortem hypostatic hemorrhages. *Am J Forensic Med Pathol.* 2009;30(4):322–326.
238. Shields LB, Corey TS, Weakley-Jones B, Stewart D. Living victims of strangulation: a 10-year review of cases in a metropolitan community. *Am J Forensic Med Pathol.* 2010;31(4):320–325.
239. Betz P, Penning R, Keil W. The detection of petechial haemorrhages of the conjunctivae in dependency on the postmortem interval. *Forensic Sci Int.* 1994;64(1):61–67.
240. Pollanen MS. Subtle fatal manual neck compression. *Med Sci Law.* 2001;41(2):135–140.
241. Garland J, Tse R. Neck injury and conjunctival petechiae in a woman who underwent cardiopulmonary resuscitation and subsequently died from meningococcal sepsis. *Forensic Sci Med Pathol.* 2017;13(4):432–435.
242. Milling L, Leth PM, Astrup BS. Life-threatening and suspicious lesions caused by mechanical cardiopulmonary resuscitation. *Am J Forensic Med Pathol.* 2017;38(3):219–221.
243. Hejna P, Zatopkova L. Significance of hemorrhages at the origin of the sternocleidomastoid muscles in hanging. *Am J Forensic Med Pathol.* 2012;33(2):124–127.
244. Hejna P, Rejtarova O. Bleedings into the anterior aspect of the intervertebral disks in the lumbar region of the spine as a diagnostic sign of hanging. *J Forensic Sci.* 2010;55(2):428–431.
245. Pollanen MS. Pitfalls and artifacts in the neck at autopsy. *Acad Forensic Pathol.* 2016;6(1):45–62.
246. Contostavlos DL. Isolated basilar traumatic subarachnoid hemorrhage: an observer's 25 year re-evaluation of the pathogenetic possibilities. *Forensic Sci Int.* 1995;73(1):61–74.
247. Opeskin K, Burke MP. Vertebral artery trauma. *Am J Forensic Med Pathol.* 1998;19(3):206–217.
248. Melinek J, Hart AP. Vertebral artery dissection complicating occipital injection of heparin for treatment of thoracic outlet syndrome. *Am J Forensic Med Pathol.* 2012;33(1):76–79.
249. Galtes I, Borondo JC, Cos M, et al. Traumatic bilateral vertebral artery dissection. *Forensic Sci Int.* 2012;214(1–3):e12–15.
250. Prahlow JA, Barnard JJ, Milewicz DM. Familial thoracic aortic aneurysms and dissections. *J Forensic Sci.* 1998;43(6):1244–1249.
251. Bunai Y, Nagai A, Nakamura I, Ohya I. Traumatic rupture of an abdominal aortic aneurysm associated with the use of a seatbelt. *J Forensic Sci.* 1999;44(6):1304–1306.
252. Wilson SK, Hutchins GM. Aortic dissecting aneurysms: causative factors in 204 subjects. *Arch Pathol Lab Med.* 1982;106(4):175–180.
253. Duflou J, Mark A. Aortic dissection after ingestion of "ecstasy" (MDMA). *Am J Forensic Med Pathol.* 2000;21(3):261–263.
254. Lange RA, Hillis LD. Cardiovascular complications of cocaine use. *N Engl J Med.* 2001;345(5):351–358.
255. Quinton RA, McClain JL. Exsanguination by subclavian-esophageal fistula associated with a left subclavian aneurysm. *J Forensic Sci.* 2003; 48(3):643–645.
256. Weissmann-Brenner A, Schoen R, Divon MY. Aortic dissection in pregnancy. *Obstet Gynecol.* 2004;103(5 Pt 2):1110–1113.
257. Snir E, Levinsky L, Salomon J, Findler M, Levy MJ, Vidne BA. Dissecting aortic aneurysm in pregnant women without Marfan disease. *Surg Gynecol Obstet.* 1988;167(6):463–465.
258. Thiene G, Ho SY. Aortic root pathology and sudden death in youth: review of anatomical varieties. *Appl Pathol.* 1986;4(4):237–245.
259. Saint-Martin P, Rogers C, Carpenter E, Fishbein MC, Lau S, Sathyavagiswaran L. Subaortic pseudoaneurysm of the left ventricle complicating staphylococcal endocarditis. *J Forensic Sci.* 2009;54(4):930–932.
260. Shields LB, Rolf CM, Davis GJ, Hunsaker 3rd JC. Sudden and unexpected death in three cases of Ehlers-Danlos syndrome type IV. *J Forensic Sci.* 2010;55(6):1641–1645.
261. Monteiro FN, Bhagavath P, Rao L, et al. Descending thoracic aortic aneurysm rupture during postpartum period. *J Forensic Sci.* 2011;56(4): 1054–1057.
262. Fealey ME, Edwards WD, McMaster 3rd KR, Carter JB. Congenital aortic sinus aneurysm causing sudden unexpected death in a 56-year-old woman. *Am J Forensic Med Pathol.* 2009;30(2):195–197.
263. Williams C, Bland E. Syphilitic aortitis, with multiple fusiform and saccular aneurysms of thoracic aorta. *N Engl J Med.* 1946;235(20):730–733.
264. Tsokos M. Syphilitic aortic aneurysm rupture as cause of sudden death. *Forensic Sci Med Pathol.* 2012;8(3):325–326.
265. Chan D, O'Donnell C, Parsons S. Sudden cardiac death from aortoesophageal fistula: an autopsy case report of a rare complication of thoracic aortic aneurysm. *Forensic Sci Med Pathol.* 2017;13(4):504–507.
266. Shang EK, Nathan DP, Boonn WW, et al. A modern experience with saccular aortic aneurysms. *J Vasc Surg.* 2013;57(1):84–88.
267. Kallistratos MS, Pavlidis AN, Tsoukas AS, et al. Late traumatic aortic dissection. *Am J Emerg Med.* 2012;30(7):1328 e1325–1328.
268. Nath D, Vaideeswar P, Rojekar A, Vaz W. Spontaneous acute aortic dissection in pregnancy: report of a case devoid of known risk factors. *Am J Forensic Med Pathol.* 2013;34(2):103–106.
269. Ro A, Mori S, Kageyama N, Chiba S, Mukai T. Ruptured syphilitic aneurysm: a cause of sudden death in a man with human immunodeficiency coinfection. *J Forensic Sci.* 2019;64(5):1555–1558.
270. Srettabunjong S. Ascending aortic rupture through a penetrating atherosclerotic ulcer: a rare cause of sudden unexpected death. *J Forensic Sci.* 2018;63(2):608–610.
271. Rae G, Husain M, McGoey R, Swartz W. Postmortem aortic dissection: an artifact of the embalming process. *J Forensic Sci.* 2016;61(Suppl 1): S246–S249.
272. Hugar BS, Praveen S, Kainoor SK, Shetty AR. Sudden death in Marfan syndrome. *J Forensic Sci.* 2014;59(4):1126–1128.
273. Durmic T, Savic S, Bogdanovic M. Aortopulmonary fistula related death - forensic implications. *Forensic Sci Med Pathol.* 2019;15(3):485–487.
274. Hadzisejdic I, Mustac E, Krstulja M, Franjic N, Stimac D. Thoracic aortic aneurysm rupture into the esophagus. *Forensic Sci Med Pathol.* 2012; 8(3):327–329.

275. Bode-Janisch S, Schmidt A, Gunther D, Stuhrmann M, Fieguth A. Aortic dissecting aneurysms–histopathological findings. *Forensic Sci Int*. 2012;214(1–3):13–17.
276. Prijon T, Ermenc B. Classification of blunt aortic injuries a new systematic overview of aortic trauma. *Forensic Sci Int*. 2010;195(1–3):6–9.
277. Isselbacher E. Epidemiology of thoracic aortic aneurysms, aortic dissection, intramural hematoma, and penetrating atherosclerotic ulcers. In: Eagle K, Baliga R, Isselbacher E, eds. *Aortic Dissection and Related Syndromes*. Boston, MA: Springer; 2007:3–15.
278. Luque M, Cavallaro D, Torres M, Emanuel P, Hillman J. Pneumomediastinum, pneumothorax, and subcutaneous emphysema after alternate cocaine inhalation and marijuana smoking. *Pediatr Emerg Care*. 1987;3(2):107–109.
279. Douglass R, Levison M. Pneumothorax in drug abusers. An urban epidemic? *Am Surg*. 1986;52(7):377–380.
280. Bense L. Spontaneous pneumothorax. *Chest*. 1992;101(4):891–892.
281. Batts KP, Ackermann DM, Edwards WD. Postinfarction rupture of the left ventricular free wall: clinicopathologic correlates in 100 consecutive autopsy cases. *Hum Pathol*. 1990;21(5):530–535.
282. Hutchins KD, Skurnick J, Lavenhar M, Natarajan GA. Cardiac rupture in acute myocardial infarction: a reassessment. *Am J Forensic Med Pathol*. 2002;23(1):78–82.
283. Ampanozi G, Flach PM, Ruder TD, et al. Differentiation of hemopericardium due to ruptured myocardial infarction or aortic dissection on unenhanced postmortem computed tomography. *Forensic Sci Med Pathol*. 2017;13(2):170–176.
284. Shirani J, Yousefi J, Roberts WC. Major cardiac findings in necropsy in 366 American octogenarians. *Am J Cardiol*. 1995;75(2):151–156.
285. Takamure A, Nakagawa T, Kobayashi A, et al. Traumatic intercostal artery pseudoaneurysm following a bicycle accident. *Forensic Sci Med Pathol*. 2007;3(3):217–220.
286. Aoki T, Okada A, Tsuchida M, Hayashi J. Ruptured intercostal artery pseudoaneurysm after blunt thoracic trauma. *Thorac Cardiovasc Surg*. 2003;51(6):346–347.
287. Janik M, Straka L, Krajcovic J, Hejna P, Hamzik J, Novomesky F. Non-traumatic and spontaneous hemothorax in the setting of forensic medical examination: a systematic literature survey. *Forensic Sci Int*. 2014;236:22–29.
288. Plunkett J. Resuscitation injuries complicating the interpretation of premortem trauma and natural disease in children. *J Forensic Sci*. 2006; 51(1):127–130.
289. Moritz AR, Henriques FC. Studies of thermal injury: II. The relative importance of time and surface temperature in the causation of cutaneous burns. *Am J Pathol*. 1947;23(5):695–720.
290. Henriques FC, Moritz AR. Studies of thermal injury: I. The conduction of heat to and through skin and the temperatures attained therein. A theoretical and an experimental investigation. *Am J Pathol*. 1947;23(4):530–549.
291. Tsokos M, Puschel K. Isolated peliosis of the spleen: report of 2 autopsy cases. *Am J Forensic Med Pathol*. 2004;25(3):251–254.
292. Kodikara S. Death due to hemorrhagic shock after delayed rupture of spleen: a rare phenomenon. *Am J Forensic Med Pathol*. 2009;30(4): 382–383.
293. Riezzo I, Di Battista B, De Salvia A, et al. Delayed splenic rupture: dating the sub-capsular hemorrhage as a useful task to evaluate causal relationships with trauma. *Forensic Sci Int*. 2014;234:64–71.
294. Prats MM, Aronson JF. An autopsy case of spontaneous splenic rupture due to isolated splenic peliosis and review of the literature. *Acad Forensic Pathol*. 2016;6(2):331–337.
295. Buschmann CT, Tsokos M. Frequent and rare complications of resuscitation attempts. *Intensive Care Med*. 2009;35(3):397–404.
296. Bataille L, Baillieux J, Remy P, Gustin RM, Denie C. Spontaneous rupture of omental varices: an uncommon cause of hypovolemic shock in cirrhosis. *Acta gastro-enterologica Belgica*. 2004;67(4):351–354.
297. Hirose I, Harada K, Kuroda R, et al. An autopsy report on a ruptured rudimentary horn (uterine anomaly) with ectopic pregnancy. *Forensic Sci Int*. 2013;224(1–3):e4–6.
298. Sutton BC, Dunn ST, Landrum J, Mielke G. Fatal postpartum spontaneous liver rupture: case report and literature review. *J Forensic Sci*. 2008; 53(2):472–475.
299. Szeremeta M, Drobuliakova P, Przepiesc J, Zaluski J, Sackiewicz A, Niemcunowicz-Janica A. An unusual medicolegal case of 32-year-old mother and her 29-week fetus with hemolysis, elevated liver enzymes, and low platelet count syndrome: case report and short literature review. *Am J Forensic Med Pathol*. 2017;38(3):269–271.
300. Yajima D, Kondo F, Nakatani Y, et al. A fatal case of subcapsular liver hemorrhage in late pregnancy: a review of hemorrhages caused by hepatocellular hyperplastic nodules. *J Forensic Sci*. 2013;58(Suppl 1):S253–S257.
301. Stanley JC, Fry WJ. Pathogenesis and clinical significance of splenic artery aneurysms. *Surgery*. 1974;76(6):898–909.
302. Lamba M, Veinot JP, Acharya V, Moyana T. Fatal splenic arterial aneurysmal rupture associated with chronic pancreatitis. *Am J Forensic Med Pathol*. 2002;23(3):281–283.
303. Hunsaker DM, Turner S, Hunsaker 3rd JC. Sudden and unexpected death resulting from splenic artery aneurysm rupture: two case reports of pregnancy-related fatal rupture of splenic artery aneurysm. *Am J Forensic Med Pathol*. 2002;23(4):338–341.
304. Tsokos M, Nolting RO, Lockemann U. Sudden, unexpected death due to splenic artery aneurysm rupture. *Am J Forensic Med Pathol*. 2005; 26(1):83–85.
305. Barrett JM, Van Hooydonk JE, Boehm FH. Pregnancy-related rupture of arterial aneurysms. *Obstet Gynecol Surv*. 1982;37(9):557–566.
306. He MX, Zheng JM, Zhang SH, Wang JJ, Liu WQ, Zhu MH. Rupture of splenic artery aneurysm in pregnancy: a review of the literature and report of two cases. *Am J Forensic Med Pathol*. 2010;31(1):92–94.
307. Hansma P, Diaz FJ, Njiwaji C. Fatal liver cyst rupture due to anabolic steroid use: a case presentation. *Am J Forensic Med Pathol*. 2016;37(1): 21–22.
308. DiMaio VJ. Sudden, unexpected death due to massive, nontraumatic intra-abdominal hemorrhage in association with cirrhosis of the liver. *Am J Forensic Med Pathol*. 1987;8(3):266–268.
309. Mirtella D, Fedeli P, Cingolani M, Vacchiano G. Retroperitoneal hemorrhage caused by trephine biopsy. *Am J Forensic Med Pathol*. 2019;40(3): 289–292.
310. Ginelliova A, Farkas D, Farkasova Iannaccone S, Vyhnalkova V. Unexpected fatal outcome of laparoscopic inguinal hernia repair. *Forensic Sci Med Pathol*. 2016;12(2):178–180.

311. Romain N, Michaud K, Brandt-Casadevall C, Mangin P. Fatal aortic injury during laparoscopy: report of two cases. *Am J Forensic Med Pathol.* 2003;24(1):80–82.
312. Gonzalez C, Penado S, Llata L, Valero C, Riancho JA. The clinical spectrum of retroperitoneal hematoma in anticoagulated patients. *Medicine.* 2003;82(4):257–262.
313. Hosseini M, Hosseinzadeh A, Raufian K, Hedjazi A. Nontraumatic retroperitoneal hematoma after warfarin administration: fatal case report and review of the literature. *Am J Forensic Med Pathol.* 2015;36(4):245–248.
314. Busardo FP, Frati P, Carbone I, Pugnetti P, Fineschi V. Iatrogenic left common iliac artery and vein perforation during lumbar discectomy: a fatal case. *Forensic Sci Int.* 2015;246:e7–11.
315. Hejna P, Bohnert M, Janik M. Unique thermal destruction of the body following suicidal burning. *Forensic Sci Med Pathol.* 2019;15(2):262–266.
316. Byard RW. The autopsy evaluation of "straightforward" fire deaths. *Forensic Sci Med Pathol.* 2018;14(3):273–275.
317. Hill AJ, Lain R, Hewson I. Preservation of dental evidence following exposure to high temperatures. *Forensic Sci Int.* 2011;205(1–3):40–43.
318. Blau S, Briggs CA. The role of forensic anthropology in Disaster Victim Identification (DVI). *Forensic Sci Int.* 2011;205(1–3):29–35.
319. Matoso RI, Benedicto Ede N, de Lima SH, Prado FB, Daruge E, Daruge Junior E. Positive identification of a burned body using an implanted orthopedic plate. *Forensic Sci Int.* 2013;229(1–3):168 e161–165.
320. Keough N, L'Abbe EN, Steyn M, Pretorius S. Assessment of skeletal changes after post-mortem exposure to fire as an indicator of decomposition stage. *Forensic Sci Int.* 2015;246:17–24.
321. Macoveciuc I, Marquez-Grant N, Horsfall I, Zioupos P. Sharp and blunt force trauma concealment by thermal alteration in homicides: an in-vitro experiment for methodology and protocol development in forensic anthropological analysis of burnt bones. *Forensic Sci Int.* 2017;275: 260–271.
322. Ubelaker DH. The forensic evaluation of burned skeletal remains: a synthesis. *Forensic Sci Int.* 2009;183(1–3):1–5.
323. Martin-de las Heras S, Valenzuela A, Villanueva E, Marques T, Exposito N, Bohoyo JM. Methods for identification of 28 burn victims following a 1996 bus accident in Spain. *J Forensic Sci.* 1999;44(2):428–431.
324. Bassed R. Identification of severely incinerated human remains: the need for a cooperative approach between forensic specialities. A case report. *Med Sci Law.* 2003;43(4):356–361.
325. Berketa JW, Simpson E, Graves S, O'Donohue G, Liu YL. The utilization of incinerated hip and knee prostheses for identification. *Forensic Sci Med Pathol.* 2015;11(3):432–437.
326. Ellingham S, M AS. Determining volumetric shrinkage trends of burnt bone using micro-CT. *J Forensic Sci.* 2020;65(1):196–199.
327. Berketa J, James H, Marino V. Dental implant changes following incineration. *Forensic Sci Int.* 2011;207(1–3):50–54.
328. Palmiere C, Staub C, La Harpe R, Mangin P. Ignition of a human body by a modest external source: a case report. *Forensic Sci Int.* 2009; 188(1–3):e17–19.
329. Christensen AM. Experiments in the combustibility of the human body. *J Forensic Sci.* 2002;47(3):466–470.
330. Richards NF. Fire investigation–destruction of corpses. *Med Sci Law.* 1977;17(2):79–82.
331. Wuerth BA, Rockey DC. Changing epidemiology of upper gastrointestinal hemorrhage in the last decade: a nationwide analysis. *Dig Dis Sci.* 2018;63(5):1286–1293.
332. Unuma K, Harada K, Funakoshi T, Uemura K. Sudden death of an alcoholic elderly man with acute esophageal necrosis (black esophagus). *Forensic Sci Int.* 2011;212(1–3):e15–17.
333. Strate LL. Lower GI bleeding: epidemiology and diagnosis. *Gastroenterol Clin N Am.* 2005;34(4):643–664.
334. Parmar N, Langdon J, Kaliannan K, Mathur M, Guo Y, Mahalingam S. Wunderlich syndrome: wonder what it is. *Curr Probl Diagn Radiol.* 2021;51(2):270–281. https://doi.org/10.1067/j.cpradiol.2020.12.002.
335. Simkins A, Maiti A, Cherian SV. Wunderlich syndrome. *Am J Med.* 2017;130(5):e217–e218.
336. Kim JW, Kim JY, Ahn ST, et al. Spontaneous perirenal hemorrhage (Wunderlich syndrome): an analysis of 28 cases. *Am J Emerg Med.* 2019; 37(1):45–47.
337. Hatjipetrou A, Anyfantakis D, Kastanakis M. Rectus sheath hematoma: a review of the literature. *Int J Surg.* 2015;13:267–271.
338. Tolcher MC, Nitsche JF, Arendt KW, Rose CH. Spontaneous rectus sheath hematoma pregnancy: case report and review of the literature. *Obstet Gynecol Surv.* 2010;65(8):517–522.
339. Humphrey R, Carlan S, Greenbaum L. Rectus sheath hematoma in pregnancy. *J Clin Ultrasound.* 2001;29(5):306–311.
340. Alla VM, Karnam SM, Kaushik M, Porter J. Spontaneous rectus sheath hematoma. *West J Emerg Med.* 2010;11(1):76.
341. Smithson A, Ruiz J, Perello R, Valverde M, Ramos J, Garzo L. Diagnostic and management of spontaneous rectus sheath hematoma. *Eur J Intern Med.* 2013;24(6):579–582.
342. Sheth HS, Kumar R, DiNella J, Janov C, Kaldas H, Smith RE. Evaluation of risk factors for rectus sheath hematoma. *Clin Appl Thrombosis/Hemostasis.* 2016;22(3):292–296.
343. Jayawardene SA, Goldsmith DJ. Rectus sheath haematomata in patients with renal disease. *Nephrol Dial Transplant.* 2002;17(10):1832–1835.
344. Adams BJ, Rainwater CW, Yim AD, Alesbury HS. A retrospective study of intentional body dismemberment in New York city: 1996-2017. *J Forensic Sci.* 2019;64(4):1012–1016.
345. Adams BJ, Herrmann NP. Estimation of living stature from selected anthropometric (soft tissue) measurements: applications for forensic anthropology. *J Forensic Sci.* 2009;54(4):753–760.
346. Kanchan T, Rastogi P. Sex determination from hand dimensions of North and South Indians. *J Forensic Sci.* 2009;54(3):546–550.
347. Kanchan T, Menezes RG, Moudgil R, Kaur R, Kotian MS, Garg RK. Stature estimation from foot length using universal regression formula in a North Indian population. *J Forensic Sci.* 2010;55(1):163–166.
348. Sen J, Kanchan T, Ghosh S. Sex estimation from foot dimensions in an indigenous Indian population. *J Forensic Sci.* 2011;56(Suppl 1): S148–S153.
349. Uzun I, Daregenli O, Sirin G, Muslumanoglu O. Identification procedures as a part of death investigation in Turkey. *Am J Forensic Med Pathol.* 2012;33(1):1–3.
350. Gonzalez-Colmenares G, Medina CS, Baez LC. Estimation of stature by cephalometric facial dimensions in skeletonized bodies: study from a sample modern Colombians skeletal remains. *Forensic Sci Int.* 2016;258:101–106.
351. Albanese J, Tuck A, Gomes J, Cardoso HF. An alternative approach for estimating stature from long bones that is not population- or group-specific. *Forensic Sci Int.* 2016;259:59–68.

352. Walsh-Haney HA. Sharp-force trauma analysis and the forensic anthropologist: techniques advocated by William R. Maples, Ph.D. *J Forensic Sci*. 1999;44(4):720–723.
353. Thomas RM, Parks CL, Richard AH. Accuracy rates of ancestry estimation by forensic anthropologists using identified forensic cases. *J Forensic Sci*. 2017;62(4):971–974.
354. Thomas RM, Parks CL, Richard AH. Accuracy rates of sex estimation by forensic anthropologists through comparison with DNA typing results in forensic casework. *J Forensic Sci*. 2016;61(5):1307–1310.
355. Garvin HM, Passalacqua NV. Current practices by forensic anthropologists in adult skeletal age estimation. *J Forensic Sci*. 2012;57(2):427–433.
356. Ousley S, Jantz R, Freid D. Understanding race and human variation: why forensic anthropologists are good at identifying race. *Am J Phys Anthropol*. 2009;139(1):68–76.
357. Sauer NJ. Forensic anthropology and the concept of race: if races don't exist, why are forensic anthropologists so good at identifying them? *Soc Sci Med*. 1992;34(2):107–111.
358. Wiersema JM, Woody A. The forensic anthropologist in the mass fatality context. *Acad Forensic Pathol*. 2016;6(3):455–462.
359. Christensen AM, Passalacqua NV, Schmunk GA, et al. The value and availability of forensic anthropological consultation in medicolegal death investigations. *Forensic Sci Med Pathol*. 2015;11(3):438–441.
360. Adams BJ, Maves RC. Radiographic identification using the clavicle of an individual missing from the Vietnam conflict. *J Forensic Sci*. 2002; 47(2):369–373.
361. Adams BJ. Establishing personal identification based on specific patterns of missing, filled, and unrestored teeth. *J Forensic Sci*. 2003;48(3): 487–496.
362. Adams BJ, Aschheim KW. Computerized dental comparison: a critical review of dental coding and ranking algorithms used in victim identification. *J Forensic Sci*. 2016;61(1):76–86.
363. Adams BJ. The diversity of adult dental patterns in the United States and the implications for personal identification. *J Forensic Sci*. 2003;48(3): 497–503.
364. Love JC, Wiersema JM. Skeletal trauma: an anthropological review. *Acad Forensic Pathol*. 2016;6(3):463–477.
365. Conlon E. Mass fatality managment and the effects of commingling. In: Adams B, Byrd J, eds. *Commingled Human Remains*. Oxford: Academic Press; 2014.
366. Black S, Rutty G, Hainsworth S, Thomson MG. *Criminal Dismemberment*. CRC Press; 2017.
367. Hakkanen-Nyholm H, Weizmann-Henelius G, Salenius S, Lindberg N, Repo-Tiihonen E. Homicides with mutilation of the victim's body. *J Forensic Sci*. 2009;54(4):933–937.
368. Lynn KS, Fairgrieve SI. Microscopic indicators of axe and hatchet trauma in fleshed and defleshed mammalian long bones. *J Forensic Sci*. 2009; 54(4):793–797.
369. Rajs J, Lundstrom M, Broberg M, Lidberg L, Lindquist O. Criminal multilation of the human body in Sweden–a thirty-year medico-legal and forensic psychiatric study. *J Forensic Sci*. 1998;43(3):563–580.
370. Symes SA. *Morphology of Saw Marks in Human Bone: Identification of Class Characteristics*. The University of Tennessee; 1992.
371. Haglund WD, Sorg MH. *Advances in Forensic Taphonomy: Method, Theory, and Archaeological Perspectives*. CRC Press; 2001.
372. Passalacqua NV, Rainwater CW. *Skeletal Trauma Analysis: Case Studies in Context*. John Wiley & Sons; 2015.
373. Konopka T, Bolechala F, Strona M. An unusual case of corpse dismemberment. *Am J Forensic Med Pathol*. 2006;27(2):163–165.
374. Di Nunno N, Costantinides F, Vacca M, Di Nunno C. Dismemberment: a review of the literature and description of 3 cases. *Am J Forensic Med Pathol*. 2006;27(4):307–312.
375. Randall B. Blood and tissue spatter associated with chainsaw dismemberment. *J Forensic Sci*. 2009;54(6):1310–1314.
376. Dogan KH, Demirci S, Deniz I, Erkol Z. Decapitation and dismemberment of the corpse: a matricide case. *J Forensic Sci*. 2010;55(2):542–545.
377. Wilke-Schalhorst N, Schroder AS, Puschel K, Edler C. Criminal corpse dismemberment in Hamburg, Germany from 1959 to 2016. *Forensic Sci Int*. 2019;300:145–150.
378. Madea B, Saukko P, Oliva A, Musshoff F. Molecular pathology in forensic medicine–Introduction. *Forensic Sci Int*. 2010;203(1–3):3–14.
379. Cox M, Bell L. Recovery of human skeletal elements from a recent UK murder inquiry: preservational signatures. *J Forensic Sci*. 1999;44(5): 945–950.
380. Boglioli LR, Taff ML, Turkel SJ, Taylor JV, Peterson CD. Unusual infant death: dog attack or postmortem mutilation after child abuse? *Am J Forensic Med Pathol*. 2000;21(4):389–394.
381. Byard RW, James RA, Heath KJ. Recovery of human remains after shark attack. *Am J Forensic Med Pathol*. 2006;27(3):256–259.
382. Lynn KS, Fairgrieve SI. Macroscopic analysis of axe and hatchet trauma in fleshed and defleshed mammalian long bones. *J Forensic Sci*. 2009; 54(4):786–792.
383. Marciniak SM. A preliminary assessment of the identification of saw marks on burned bone. *J Forensic Sci*. 2009;54(4):779–785.
384. Freas LE. Assessment of wear-related features of the kerf wall from saw marks in bone. *J Forensic Sci*. 2010;55(6):1561–1569.
385. Gunawardena SA. Artefactual incised wounds due to postmortem predation by the Sri Lankan water monitor (kabaragoya). *Forensic Sci Med Pathol*. 2016;12(3):324–330.
386. Pettigrew M. Corpse dismemberment and a necrofetishist. *J Forensic Sci*. 2019;64(3):934–937.
387. Bailey JA, Wang Y, van de Goot FR, Gerretsen RR. Statistical analysis of kerf mark measurements in bone. *Forensic Sci Med Pathol*. 2011;7(1): 53–62.
388. Baier W, Norman DG, Warnett JM, et al. Novel application of three-dimensional technologies in a case of dismemberment. *Forensic Sci Int*. 2017;270:139–145.
389. Love JC. Sharp force trauma analysis in bone and cartilage: a literature review. *Forensic Sci Int*. 2019;299:119–127.
390. Galtes I, Scheirs S. Differentiation between perimortem trauma and heat-induced damage: the use of perimortem traits on burnt long bones. *Forensic Sci Med Pathol*. 2019;15(3):453–457.
391. Porta D, Poppa P, Regazzola V, et al. The importance of an anthropological scene of crime investigation in the case of burnt remains in vehicles: 3 case studies. *Am J Forensic Med Pathol*. 2013;34(3):195–200.
392. Gruenthal A, Moffatt C, Simmons T. Differential decomposition patterns in charred versus un-charred remains. *J Forensic Sci*. 2012;57(1): 12–18.

393. Masotti S, Pasini A, Gualdi-Russo E. Sex determination in cremated human remains using the lateral angle of the pars petrosa ossis temporalis: is old age a limiting factor? *Forensic Sci Med Pathol*. 2019;15(3):392–398.
394. Moritz AR. Studies of thermal injury: III. The pathology and pathogenesis of cutaneous burns. An experimental study. *Am J Pathol*. 1947;23(6): 915–941.
395. Shkrum MJ, Ramsay DA. *Thermal Injury. Forensic Pathology of Trauma: Common Problems for the Pathologist*. 2007:181–242.
396. Conway KS, Schmidt CJ, Brown TT. Medical examiner review of the characteristics of fire-related homicides. *Acad Forensic Pathol*. 2020;10(2): 87–93.
397. Pope EJ, Smith OC. Identification of traumatic injury in burned cranial bone: an experimental approach. *J Forensic Sci*. 2004;49(3):431–440.
398. Suarez-Penaranda JM, Munoz JI, Lopez de Abajo B, et al. Concealed homicidal strangulation by burning. *Am J Forensic Med Pathol*. 1999;20(2): 141–144.
399. Hirsch CS, Bost RO, Gerber SR, Cowan ME, Adelson L, Sunshine I. Carboxyhemoglobin concentrations in flash fire victims: report of six simultaneous fire fatalities without elevated carboxyhemoglobin. *Am J Clin Pathol*. 1977;68(3):317–320.
400. Iwasaki Y, Yashiki M, Kojima T, Miyazaki T. Interpretation of accelerants in blood of cadavers found in the wreckage after fire. *Am J Forensic Med Pathol*. 1998;19(1):80–86.
401. Jumbelic MI. Open air carbon monoxide poisoning. *J Forensic Sci*. 1998;43(1):228–230.
402. Gill JR, Goldfeder LB, Stajic M. The Happy Land homicides: 87 deaths due to smoke inhalation. *J Forensic Sci*. 2003;48(1):161–163.
403. Vreman HJ, Wong RJ, Stevenson DK, et al. Concentration of carbon monoxide (CO) in postmortem human tissues: effect of environmental CO exposure. *J Forensic Sci*. 2006;51(5):1182–1190.
404. Walter JE, Hirsch CS, Zumwalt RE. Never say never. Negligible carboxyhemoglobin in the victim of a smoldering mattress fire. *Am J Forensic Med Pathol*. 1984;5(3):239–244.
405. Gormsen H, Jeppesen N, Lund A. The causes of death in fire victims. *Forensic Sci Int*. 1984;24(2):107–111.
406. Middleberg RA, Easterling DE, Zelonis SF, Rieders F, Rieders MF. Estimation of perimortal percent carboxy-heme in nonstandard postmortem specimens using analysis of carbon monoxide by GC/MS and iron by flame atomic absorption spectrophotometry. *J Anal Toxicol*. 1993; 17(1):11–13.
407. Hirsch CS, Bost RO, Gerber SR, Cowan ME, Adelson L, Sunshine I. Carboxyhemoglobin concentrations in flash fire victims. *AJCP*. 1977;68: 317–320.
408. Stefanidou ME, Maravelias CP, Dona AA, Pistos CM, Spiliopoulou CA, Athanaselis SA. Carbon monoxide-related deaths in Greece: a 23-year survey. *Am J Forensic Med Pathol*. 2012;33(2):128–131.
409. Reumuth G, Alharbi Z, Houschyar KS, et al. Carbon monoxide intoxication: what we know. *Burns*. 2019;45(3):526–530.
410. Kinoshita H, Turkan H, Vucinic S, et al. Carbon monoxide poisoning. *Toxicol Rep*. 2020;7:169–173.
411. Gerostamoulos D, Beyer J, Wong K, Wort C, Drummer OH. Carbon monoxide concentrations in the 2009 Victorian Bushfire disaster victims. *Forensic Sci Int*. 2011;205(1–3):69–72.
412. VanLaerhoven SL, Anderson GS. Insect succession on buried carrion in two biogeoclimatic zones of British Columbia. *J Forensic Sci*. 1999; 44(1):32–43.
413. Pfeiffer S, Milne S, Stevenson RM. The natural decomposition of adipocere. *J Forensic Sci*. 1998;43(2):368–370.
414. Komar DA. Twenty-seven years of forensic anthropology casework in New Mexico. *J Forensic Sci*. 2003;48(3):521–524.
415. Vass AA, Smith RR, Thompson CV, et al. Decompositional odor analysis database. *J Forensic Sci*. 2004;49(4):760–769.
416. Schultz JJ, Collins ME, Falsetti AB. Sequential monitoring of burials containing large pig cadavers using ground-penetrating radar. *J Forensic Sci*. 2006;51(3):607–616.
417. Weitzel MA. A report of decomposition rates of a special burial type in Edmonton, Alberta from an experimental field study. *J Forensic Sci*. 2005;50(3):641–647.
418. Berryman HE, Bass WM, Symes SA, Smith OC. Recognition of cemetery remains in the forensic setting. *J Forensic Sci*. 1991;36(1):230–237.
419. Mann RW, Bass WM, Meadows L. Time since death and decomposition of the human body: variables and observations in case and experimental field studies. *J Forensic Sci*. 1990;35(1):103–111.
420. Bachmann J, Simmons T. The influence of preburial insect access on the decomposition rate. *J Forensic Sci*. 2010;55(4):893–900.
421. Gibelli D, Di Giancamillo A, Taborelli A, et al. Burial of piglet carcasses in cement: a study of macroscopic and microscopic alterations on an animal model. *Am J Forensic Med Pathol*. 2013;34(1):43–49.
422. Martin DC, Dabbs GR, Roberts LG, Cleary MK. The stone cold truth: the effect of concrete encasement on the rate and pattern of soft tissue decomposition. *J Forensic Sci*. 2016;61(2):302–308.
423. Fiedler S, Berns AE, Schwark L, Woelk AT, Graw M. The chemistry of death–Adipocere degradation in modern graveyards. *Forensic Sci Int*. 2015;257:320–328.
424. Salsarola D, Poppa P, Amadasi A, et al. The utility of ground-penetrating radar and its time-dependence in the discovery of clandestine burials. *Forensic Sci Int*. 2015;253:119–124.
425. Congram D, Kenyhercz M, Green AG. Grave mapping in support of the search for missing persons in conflict contexts. *Forensic Sci Int*. 2017; 278:260–268.
426. Ruffell A, Donnelly C, Carver N, Murphy E, Murray E, McCambridge J. Suspect burial excavation procedure: a cautionary tale. *Forensic Sci Int*. 2009;183(1–3):e11–16.
427. Crowder C, Rainwater CW, Fridie JS. Microscopic analysis of sharp force trauma in bone and cartilage: a validation study. *J Forensic Sci*. 2013; 58(5):1119–1126.
428. Steadman DW. *Hard Evidence. Case Studies in Forensic Anthropology*. Englewood Cliffs: Prentice Hall; 2002.
429. Reichs K, Bass WM. *Forensic Osteology: Advances in the Identification of Human Remains*. Charles C Thomas Pub Limited; 1998.
430. Humphrey JH, Hutchinson DL. Macroscopic characteristics of hacking trauma. *J Forensic Sci*. 2001;46(2):228–233.
431. Tucker BK, Hutchinson DL, Gilliland MF, Charles TM, Daniel HJ, Wolfe LD. Microscopic characteristics of hacking trauma. *J Forensic Sci*. 2001;46(2):234–240.
432. Püschel K, Koops E. Dismemberment and mutilation (2). *Archiv Fuur Kriminologie*. 1987;180(3–4):88–100.
433. Sea J, Beauregard E. Mutilation in Korean homicide: an exploratory study. *J Interpers Violence*. 2019;34(14):2863–2877.

CHAPTER

16

Communication

Susan F. Ely, James R. Gill, Jonathan Hayes and Jonathan Lucas

All our work, our whole life is a matter of semantics, because words are the tools with which we work, the material out of which laws are made, out of which the constitution was written. Everything depends on our understanding of them. **-*Hon. Felix Frankfurter, US Supreme Court***

Overview

We are in the business of communication and every word matters. At the center of a successful forensic pathology practice reside deliberate and finely honed communication skills. These skills include a mastery and conveyance of complex medical and scientific concepts, both written and verbal, to varied audiences, a reasonable familiarization with legal terminology as it pertains to forensic practice, the patience and willingness to teach both medical and nonmedical people one on one and in public forums, and a deftness in meeting the many raw iterations of grief and loss with compassion, equanimity, authenticity, and reflective listening. All of this must be balanced with the ability to provide the comfort of medical and scientific knowledge with the sensitive delivery of helpful and accurate information. Together and ideally, this skill set will produce clarity without condescension, while offering both honesty and neutrality. At its most elemental, however, it requires forensic pathologists to apply a degree of basic humanity to all that they do and say.

The reality of forensic practice is that our specific word choices in written communications are often consequential. At the crossroads of medicine and law, forensic pathologists must say what they mean and both know and mean what they say more than in nearly any other medical specialty. That said, despite our best intentions and responsible efforts, any word we utter or pen can and often will be dissected and not infrequently misunderstood, taken out of context, misused, or even weaponized, all outcomes ultimately beyond our control.

Still, as professionals, forensic pathologists are of limited use to anyone if they are unable to effectively communicate the data and opinions they have so assiduously gathered and formed. Our "audiences" are broad and varied, so our communications must be nimble enough to be specifically tailored to the needs and forensic aptitudes of each. These audiences include, but are not limited to, families and friends of the deceased, lawyers, jurors, clinicians, law enforcement, child and adult protective service workers, public health officials, myriad governmental officials, insurance claims adjusters, the media, medical students, residents, fellows, and the public at large. All benefit from the timely, accurate, and clear delivery of information surrounding highly complex and sometimes emotionally charged forensic matters. Similarly, the forensic pathologist often benefits greatly from detailed and accurate information communicated *in return* from direct, one-on-one conversations with law enforcement, physicians, public health officials, and the bereaved.

To that end, before doing anything else, the forensic pathology neophyte would be well advised to read Lester Adelson's *The Forensic Pathologist: "Family Physician" to the Bereaved*, published in JAMA in 1977, as applicable and timeless today as it was when it was written.[1] Immediately following that, they would move on to Charles Hirsch's *Talking to the Family After an Autopsy*.[2] Not only are each of these exquisitely written, broadly informative in defining the unique role of the forensic pathologist, and infinitely helpful in practice, they also prepare the reader, more than any other treatises, for what they will realistically encounter on a day to day basis as a professional in this field, allowing the novice to make a clear-eyed career choice, with no illusions. A perusal of these manuscripts will

Principles of Forensic Pathology
https://doi.org/10.1016/B978-0-323-91796-4.00021-0

either ignite a calling or have the reader turning to another profession. While the nobility of this profession is clearly conveyed in each, so is the fact that the practice of forensic pathology is not for the faint of heart, nor the emotionally imperceptive.

To wit, Dr. Adelson said the following,

> The queries put to the pathologist range widely. Based as they are on many complex combinations of grief, knowledge, guilt, ignorance, anger, animosity, avarice, and other personality traits of the profoundly disturbed questioners .... A straightforward question deserves a straightforward reply, even if it is only, 'I can't give you a good answer about that aspect of the death'. Under no circumstances should the postmortem family physician resort to untruths or half-truths. The best that he can do is to be gentle and circumspect in how he communicates unpleasant and disturbing information to his interrogator. Tact, compassion, patience, probity, sympathy, and even a little empathy are essential elements.[1]

Dr. Hirsch continued,

> Sensitized by a recognition of the pervasive guilt principle .... First and foremost, do not push. Most persons want to talk with a physician about the death of a loved one and will do so if provided an opportunity. However, not all persons are woven from the same emotional fabric, and some survivors do not want to talk about the death, at least not right away .... Sometimes, the survivors even wait for the anniversary of the death to ask questions about the autopsy findings, cause of death, and technical terms on the death certificate .... 'Doctor, I just couldn't talk about it until now'.[2]

The observations of Drs. Adelson and Hirsch represent the wholly realistic experiences of most practicing forensic pathologists today, as well as the sometimes surprising and varied manifestations of grief that we encounter nearly daily. In the authors' experiences, some of the most frequently asked questions from family members (excluding time of death) surround concerns of terminal suffering. Did he suffer? Was it fast? Did she know she was dying? Did anyone hurt her? Could I have saved her? It is often in hearing the answers to this specific group of questions that any fragile artifice of composure they may have been able to erect falls away. The responses to these sometimes desperate questions require nuance, but honesty. While we cannot answer all questions posed to us, on many occasions, mercifully, the forensic pathologist is in the unique position of being able to truthfully and clearly answer these specific inquiries, often relieving tremendous acute suffering in the process. Depending on the words we convey, an audible and emotional sigh of relief is often the result. There is likely a considerable degree of healing that accompanies knowledge and a comprehension of the facts. It is no wonder that the importance of good communication skills, of which listening is such a vital element, is central to any discussion about successful forensic pathology practice.

It is not unusual for a decedent's family to contact the forensic pathologist months or even years after a death. This may occur because the person was not ready to hear the details or it may be due to prolonged grief. Prolonged grief disorder, also known as complicated grief or persistent complex bereavement disorder, occurs when painful emotions are so long lasting and severe that the person has trouble recovering from the loss and resuming their own life. The person may experience intense longings for the decedent or a preoccupation with thoughts of the deceased. Prolonged grief disorder was recently added to the diagnostic and statistical manual of mental disorders (DSM).[3–6]

Another arena of communication that is important to mention in any discussion focused on forensic pathology practice is ***expert testimony*** (see ahead in this chapter). While careful and thorough preparation for expert testimony is essential, the real heavy lifting and important investment occur at the time of death certification. Lining up the necessary evidence, meeting the required burden of proof to match the case, carefully considering the mechanism through sound forensic thinking, and writing a clear and accurate opinion are the bigger hurdles to overcome in facing whatever might later be encountered on the stand. These up-front steps are the ones over which the forensic pathologist has actual control. By contrast, no amount of testimony preparation will extinguish the fact that there will be an adversarial legal proceeding in which the forensic pathologist's professional opinion may be questioned and tested. It is not personal. It is not about competence. It is by design and necessity, an inherent feature of the criminal justice system that, as a civilized society, we are all a part. It is a system of advocacy. Opposing attorneys must challenge the witness whose findings may support an allegation against their client. If they didn't, they would be abdicating their legal duty. That said, one will likely have a stronger stomach for having their opinions sometimes aggressively questioned if the foundation upon which those opinions are formed is evidence-based, consistent, and lacks capriciousness from the get-go. The hard work should happen up front, not at the time of the trial. By then, it is likely too late. No mastery of testifying will eclipse an ill-conceived and flawed conclusion, unsupported by science and the evidence. While nervousness is an unavoidable experience of early testimony, the anxiety will be substantially blunted by the confidence afforded from soundly formed opinions.

As outlined in Chapter 2, medical examiner/coroner (ME/C) communications can be categorized into one of four areas: **direct conversations** (telephone and in person discussions and meetings), **expert testimony/legal proceedings, public speaking** (including in-house case presentations, didactic lectures, hospital morbidity and mortality conferences, multidisciplinary conferences, presentations at professional meetings, and press conferences), and **written documents** (death certificates, autopsy reports, autopsy and case notes, press releases, emails, and social media). At the center of the forensic pathologist's practice, and desk, is *the telephone*, one of the most useful tools they will use in both data gathering and grief counseling. As a corollary to the use of the telephone, written documentation of selected communications is a natural compliment. In a nod to careful documentation and long before the digital age was in full gear, Dr. Hirsch was fond of saying, *"Memorialize that conversation on paper."*

Verbal: one-on-one communication

We solve more cases with our telephones than with our microscopes.
Information void is the mother of paranoia. -***Dr. Charles S. Hirsch***

The below first-person narrative details the personal reflections of **Dr. Jonathan Hayes**, a senior medical examiner who has practiced forensic pathology for over 30 years. In it, he shares his substantial insights and advice born out of a wealth of experience from both his conversations with families and his role as an expert witness.

On speaking with families and to juries

Although unexpectedness may or may not affect the investigation, it will certainly affect the next of kin. A death that is unexpected adds an additional element to the emotional burden of the family. As Adelson stated: "Even under the best of circumstances, death is almost always a traumatic experience to the survivors. The emotional impact of death is magnified enormously when it occurs unexpectedly."[1] Medical personnel who interact with families at this most emotional time of their lives need to be particularly sensitive to their questions and concerns.[2,7]

We all have our own ways of speaking with people, and the way we talk to others reflects both the context of the conversation and who we are as people. Speaking with families in the throes of their terrible loss can feel particularly daunting. Most are calm and lucid, but sometimes they are so distraught they can barely breathe, so angry they can barely speak, or so intoxicated that they are incoherent. In all of these situations, our goal is the same: we want to help them understand the death of a person they cherished, and guide them as they start their arrangements.

The sheer intensity of the situation makes many of us wary; medical examiners often put off talking with families, or delegate the responsibility to another team member. It is understandable, but unfortunate: your communication with the family builds a two-way bridge and will help you to manage your case better and to handle any curveballs that may arise down the line. Speaking to families is not just a responsibility, it's an opportunity and a privilege.

There is no one way to talk with people. My way of speaking with families will be different from yours, but I wanted to discuss some approaches that I have found helpful. What follows is a personal account of how I approach families; you should feel free to take what you like and reject what you do not as you build your skills in this sometimes uncomfortable arena. You will probably feel awkward at first, but with experience, you will find it easier to understand what people need from you, and how to say it to them in your own voice.

I would note that the manner in which you speak with the family is almost as important as the information you provide. Medical examiners often approach these conversations very gingerly, wary of awkwardness, worrying the family might get upset and start crying, or become enraged and start yelling. Bracing for that, we sometimes wrap ourselves in the cloak of our profession, stepping into the persona of the Doctor talking to the Patient, trying to avoid getting ensnarled in tricky emotions. Leaning on the distance of professionalism is not the best way to manage this interaction. It fosters a separation between you and the family member, and, even worse, a separation based on a hierarchy. Hiding behind your authority is a natural self-defense mechanism, but at the end of the day, it is helpful neither for you nor the family, as it is a less effective way of communicating and preparing them for the hard days that will follow. Instead of being a Doctor talking to a Patient, just be one human being talking to another. This is a moment to take off the white coat and embrace your own vulnerability. To make yourself truly available to a stranger who is suffering cruelly takes some courage, but it is the best way to work through this discussion, for both of you. It is a kindness that represents our profession at its best.

Newer forensic pathologists will feel that emotional pressure more intensely than veterans. As you dial the phone number or reach for the door handle of the Family Room, all of your insecurities—acute self-consciousness of your lack of experience, the sense that you are an impostor being sent out to handle a task that would be better handled by someone who actually *knows* what they're doing—may swell inside you. We have all been there. Just remember: *it is not about you.* The widow is the one who is suffering, the father who just lost his son is the one in agony, and the things you are going to say to them may ease their pain.

Instead of worrying about what you will say, or if the family will immediately sense that you are a beginner (*fraud,* your internal voice might be whispering), focus on the help you can give them, the relief you can provide. Your goal is to move them through the process as quickly and as painlessly as you can. You are in a unique position not only to give information, but also to help them work toward some form of acceptance as efficiently as possible. You are going to provide answers and guide them through the logistics of what happens next. Families are desperate for help: you are the person who is there to help them.

Again, it is not about you: it is about them. Their emotional state is going to play a large part in how your conversation or meeting goes. The exact depth of their need for information will shape the discussion you have—you will find that while some families only want the simplest of outlines, others will insist on every gory detail. Luckily, the families themselves will tell you what they need. The problem is, though they will tell you, they often do not say it explicitly with words, but through intonation, cadence, and silences. The only way you are going to understand what they need is if you really pay attention to what they say and how they are saying it. So, to be good at speaking with families, first and foremost, you have to be a good listener. Under these circumstances, active listening can be challenging, particularly when you are talking on the phone and are deprived of the visual cues we unconsciously rely upon when interacting with another person. What is more, there is often a language or cultural barrier; you may be talking with someone through an interpreter, every sentence in both directions mediated by a second party, unable to judge the intonations and actual words being shared. Just do your best to listen carefully to what they are saying. Occasionally repeating something they have just said to you can be a good way to clarify their meaning, but also shows them that you are listening carefully, and affirms that the things they have to say are important to you. For many people, speaking with a physician is intimidating. Indeed, I've met physicians who enjoy that little squirt of "power"—be that way at your peril. Accepting that not all conversations are going to go perfectly is helpful. Sometimes you will make blunders—an old colleague tells the story of how, speaking to a family whose son had hanged himself, somehow could not stop himself from repeatedly using the word "hang" in their conversation. "I don't want to keep you hanging around here," "hang on for a moment". I suspect that the discomfort was more acute for him than for the family. Just remember that you have information they want and need, they have information you want and need, and you will both do your best to contribute.

To get the best result from your interaction, it is important to go into the exchange focused on your desired outcome. Before dialing the number or walking into the meeting, ask yourself what you want to get from the conversation. Every discussion with the family is a two-way street, as much an opportunity to get information as to provide it. Before you speak with a family, make sure you know the case cold. Review the circumstantial information, the scene, the autopsy findings, and any available lab studies. Ask yourself what you need to finalize the report, and anticipate questions they might ask. Remember, families are usually eager to help to make a real contribution in sorting out what happened to the person they have lost. So, what is the first thing you say to someone who is grieving? What do you say to a man who has just lost his wife? It's a situation where we often feel the need to protect ourselves from their overwhelming anguish. We may slip into a sort of Death Response Theater, go through an empty ritual of saying what we think people expect us to say. I have strong opinions about this, and you may completely disagree, but for me, the worst to say to someone is the trite, "*I'm sorry for your loss.*" Have you ever spent time with someone who has just lost someone they loved? Almost every single person who approaches says the same thing: "Sorry for your loss." They hear it from the emergency medical technician (EMT) standing next to the body on the living room floor. Then the patrol officers. The detectives. Then neighbors, and friends. Then on social media. The same rote expression over and over again, from each friend or colleague or acquaintance seeing them for the first time since the death. What is more, when someone expresses their condolences, there is a social obligation to respond, to thank them from their expressions of concern. Saying "I'm sorry for your loss" makes a demand on the recipient for a reply. So, I never say that. Indeed, I have never said it. Instead, I focus on moving them through our interaction. They *know* you are the medical examiner, and what your role is in the miserable logistics of the death of their spouse. And every person, even in the depths of their grief, *knows* people feel badly for them for what they are going through—you really do not have to say it.

So, I do not. I introduce myself, and say, "My name is Dr. Hayes, and I'm with the Medical Examiner's Office. Do you have a few minutes to speak with me?" When they agree, I say, "I'm going to try and move you through this as quickly as I can. Do you mind if I ask you a few questions?" They always agree—again, they want to help. Being useful provides a concrete satisfaction to them, and for a few minutes at least, shifts the focus of their attention from their grief to understanding what happened.

I start the conversation by asking them what I want to know. I listen actively to their answers, and as I listen, I assess the person or people with whom I am speaking. I listen not just to what they are saying, but to how they are saying it, to the language and the tone of their speech. I try to gently gauge how upset they are, and to see how nuanced they are in their understanding of what might be going on. I use their language and vocabulary as a guide to the level at which I should pitch the information that I am sharing. Again, how you listen is every bit as important as how you express yourself.

Be wary of talking down to families. The gap in language, culture, education, and mood can be extreme, but it is still your responsibility to make sure that they understand what you are saying. When I first start talking, I speak as I would if I were speaking with a bright high schooler; if they are following me, fine, but if I sense that they are having problems understanding, I'll slow things down a bit until I know they are with me. Sometimes, the man or woman with whom you are speaking may be wrecked by anguish or alcohol, and may not be thinking as clearly as they do normally.

In your conversation, avoid medical terms, particularly abbreviations. Physicians have a tendency to throw out terms like PE or MI as if they were tossing beads from a Mardi Gras float—after all, they are the terms we use with other doctors to communicate quickly and efficiently, and we know precisely what we mean when we use them. But at another level, I think we sometimes use them because it makes us sound more authoritative, like we really know what we are talking about. Instead of ladling out medical alphabet soup, explain what happened simply and accessibly—"Because of the ankle fracture, he wasn't moving around very much. This made the blood flow through his legs slower, and eventually he developed a blood clot deep in his calf—that's why he was saying his leg hurt. And then the clot broke off and went to his lungs, and then his heart could no longer pump blood to the lungs, and he died. This is called a 'pulmonary embolism', when the clot breaks off and goes to the lung. And the clot in the leg is called a 'deep vein thrombosis'. So, on the death certificate, I'm going to say the cause of death was a "pulmonary embolism due to deep vein thrombosis in the leg due to immobilization following his ankle fracture." I will then ask if they understand what I am saying. Checking repeatedly that they are grasping what you're talking about helps cut down on follow-up calls, but remember: they are receiving important information at a time when they're stunned, and may not process it well, and may retain it very poorly. What is more, families will sometimes seize upon individual elements in what you are saying, and gradually manipulate their recollection until it fits with what they would prefer to be the truth. So occasionally I will get a call a week after the autopsy demanding the toxicology results, swearing I told them it would be ready next week. When that happens, I will always apologize for the misunderstanding, and explain how long those results will realistically take to come back. Even when you know you are right, it is both kindest and most efficient to apologize, offer the correction, and say you will call back to follow up when you have the results. And remember: anatomic terms are jargon! Consider using "Adam's apple" over "larynx," "windpipe" over "trachea," et cetera. And if you do use a technical term like "aorta," immediately clarify it by adding something like "That's the large artery that carries blood from the heart through to the rest of the body." The fact is, every autopsy finding, every cause of death, no matter how complex, can be simplified and explained. Sometimes that explanation may be lengthy, but it can always be explained—typically, the person with whom you are speaking will signal the level of detail they want. If a family member is clearly unable to follow along with an involved exegesis of the pathophysiology, sometimes I'll default to a shortcut, and distill it down to its essence—"he had a massive heart attack," perhaps, or "he had a stroke." The endless, quasi-Talmudic debates forensic pathologists get into over wording of the cause of death statement are rarely of major concern to families.

If a widow, on the other hand, shows an easy familiarity with her husband's medical history and prescriptions, I'll say something like, "You seem to know his health problems very well—are you in the medical field?" This is useful to know going in, as someone with medical training will quickly understand situations where lay people will need careful guidance—it is far faster to say "I think he had a fatal arrhythmia due to cardiac hypertrophy from his hypertension" than to explain the mechanism and underlying pathophysiology of an arrhythmic death to someone completely unversed in medicine. Furthermore, their familiarity with medical knowledge can streamline the conversation, and they may even direct the questions you ask when looking for more detail about the circumstances of death. At this point, I will step back for a second and note that in my experience, the first response a family member has to the death of a loved one is almost always guilt, no matter the cause of death, no matter how illogical it might seem to you. "I should've made him stop smoking!" or "I should've made him lose weight!" or "If I hadn't asked him to pick up a soda at the bodega, he'd still be alive."[2]

This guilt can be unbearable and can turn quickly into denial and anger. Nowhere is this truer than with suicides, where the family member may suddenly see themselves as having stood by doing nothing while someone they loved was suffering intensely right in front of them, suffering so badly that they chose to kill themselves rather than continue in agony. The tendency toward self-blame may be intense and can flip over into the most extreme anger and steadfast refusal to consider the possibility of suicide.[8]

In those circumstances, hear the anger, listen to the resistance and, where you can, avoid engaging in that battle, particularly in the period shortly after death. What I will do is say, "Well, I'm doing some further tests—I need to see if she was intoxicated at the time of her death, so I'm doing some routine testing to see if she had drugs or alcohol in her system. What I'm going to do for now is issue a temporary death certificate, which will give the cause of death as 'Pending Further Study'. When I have those test results, and the investigation is finished, I'll give you a call and we can discuss it then. I promise you I'm not leaping to any conclusions at this early stage, but I have to tell you that I do think there's a real possibility that she took her own life."

Even though they may adamantly deny that possibility, you have planted the seed. Over the following weeks, they will have time to reflect on it, and they may hear relevant stories from other family members or friends about how the decedent had been behaving strangely in the weeks before their death; it has been my experience that, over time, even the most resistant family members will likely come around to accepting a death as a suicide. After I have gotten the information I need, I thank them for their help, and then say something like, "Now, how can I help you?" and give them time to ask their questions, and answer as fully as I am able to—which might be limited both by the stage of the investigation or, in criminal cases, the office's practices about what information can be appropriately released.

I know what they want to know, but I let them ask, because I think it returns some control back to them. Their questions are fairly predictable, but sometimes the biggest question of all in a sudden death is hard for them to articulate directly. So, when I begin talking about my findings, I'll typically start off with, "First off, there was no foul play, no injuries. No one hurt her." That sentence can provide an immediate rush of relief, particularly where the decedent was estranged from his family, or when the family lives far away and is unfamiliar with the decedent's distant life.

Having said that I have ruled out foul play, I then move on to say what I *did* find. I'll talk first about any disease process I found, even if I suspect a drug intoxication death. "Now, he had some heart enlargement and scarring of the kidneys; we see these when people have high blood pressure. And as the heart becomes enlarged, it becomes weakened and electrically unstable—it can fail at any time." If I believe the death is an intoxication, I'll then gently segue into that by directly asking if they are aware of any alcohol or substance use. If they confirm the history, well and good, but if they do not know, or deny it, I will say something along the lines of "Well, I have to tell you that she had drugs on her bedside table" or "There were findings at the examination which make me suspect that he had been using drugs for some time." (I'm not big on euphemisms—they may soften the blow, but they also tend to obfuscate—but when talking with the family, I'll use "examination" over "autopsy." I try not to draw their attention to the more visceral aspects of the death).

I do not slam the door on explanations other than substance use at this point, but again, this is an idea I have implanted. I will say, "I didn't find any other medical cause of death, and she's young, so I think that substance use is the most likely explanation. I'm going to be doing routine testing to see if she has any drugs or alcohol in her system. If the testing comes back negative, I'll do further studies to see if I can find something else, but I have to tell you that at this point, I think it's very likely that she was using drugs."

I underplay my hand because I want to give the family time to come to the likely—or even obvious—conclusions themselves. Besides, on occasion the toxicology does, in fact, come back negative, and your investigation resets. I like to signal to them that I have not made up my mind yet, that I have not looked at the dead body of their daughter and judged her as a "drug addict." Sometimes, we have to disclose findings that are going to be very hard for a family to hear—signs of sexual assault, signs of torture, or dismemberment. Your every instinct may be to bend the truth to protect people's feelings, to sugarcoat things as much as possible; but while it is good to be sensitive and supportive, you have to communicate honestly what happened to the decedent. I have had situations where families have asked me if a man who burned to death suffered. In those cases, I've said that yes, he would have suffered, but that it was likely that he had lost consciousness quickly. Although difficult for the family to hear, this truth may turn out to have legal significance, for example, where a subsequent lawsuit is brought based on a claim of conscious pain and suffering. That said, you do not have to actively force information on a family. Listen to their wariness, and phrase things out of respect for that. And I will sometimes go with euphemisms in extreme cases—if someone in a car crash has been decapitated, I won't volunteer that (and I would never put that word on a death certificate). Instead, I will say "He sustained devastating injuries of his head and neck; he would have died instantly." That said, a decapitation is the sort of detail that some elements in the media just love, the sort of detail that has a way of being leaked. If the family were to directly ask me if he was decapitated, I would say that he was, and add that he would have died instantly; so far, a family has yet to ask me that.

Importantly, I have found that even when the truth is painful, the family is often still glad just to have that clarity, to know just what the person they loved went through. I am not sure that I truly believe in "closure," but I do believe that providing the family with the facts helps them come to terms with the death.

Finally, as noted above, sometimes the facts of the case, particularly when easily sensationalized, get out, often via the police (whether in formal press conference or in unauthorized leaks to the media). It is best that you handle informing the families yourself rather than that they hear it on the radio (similarly, if you have pended a case for further investigation, particularly a suicide, *always* alert the family before issuing the final death certificate, particularly if it's a case that has drawn media attention). Now, different offices have different policies for the timing of the release of information (see below), particularly with regard to homicides. In many jurisdictions, until a case has entered the legal system (by which I mean there has been an indictment), the information released to the families of homicide victims by the ME/C is strictly limited to what is actually written on the death certificate. It can be terribly frustrating for them, but until the case is fully investigated, we make sure we avoid potentially releasing information to the wrong person. When I explain that I cannot provide details, I apologize and tell them that as soon as the prosecutor's office authorizes it, I will be happy to discuss my findings with them at length. That said, once the body is released, the family will have access to it at the funeral home and will be able to see for themselves the condition of the body and any injuries. In some instances, you may be speaking with families who are objecting, whether for personal or religious reasons, to the autopsy you need to perform. Your response to their **objection** will very much be dictated by the policies of your jurisdiction. In some jurisdictions, families have very strong protection under the law for their religious objections: the medical examiner has to prove that there is a compelling medicolegal need for autopsy, typically either to determine or evaluate a homicide or to diagnose a potential public health issue, such as a major contagion.

You will handle their objection by balancing the public health aspect of the case with their personal or religious beliefs. If I am not firmly convinced of the importance of performing an autopsy in the case, but would prefer to do one, I will explain why it might be in the best interests of the family to have the autopsy. At the end of the day, if they persist, I will honor their objection. Furthermore, it is not my function to interpret the tenets of their faith for them, so if a Catholic family launches an objection "on religious grounds," I will treat it with the same gravity as I would if they were a practicing Muslim or Orthodox Jew. Now, if I feel an autopsy is critical, I will explain exactly why that is to the family. I find the most compelling argument is the honest one: "I need to make sure that no one hurt her." If a death is clearly a homicide, I'll say, "I'm sorry, but I need to document the injuries thoroughly so that whoever did this can be properly prosecuted. I've seen criminal cases fail because no autopsy was done, and I don't want to let her down." If I must do an autopsy, I'll do my best to make any accommodations the family might request. In Orthodox Jewish cases, I've often worked with a rabbi at my side, wiping my equipment and making sure any blood spilled stays with the body in accordance with the principles of Orthodox Jewish faith. The presence of the rabbi brings tremendous comfort to the families, and the involvement of a rabbi or imam is very common in cases of religious objection. Since they are usually not intimately involved with the decedent, they can be very helpful when managing the family's wishes and expectations; indeed, often the rabbi may be the family's primary point of contact with the ME. On occasion, I have encountered situations where an individual has killed themselves despite practicing a faith that proscribes suicide. Here, the family might be worried about being able to bury the decedent in a very restrictive religious cemetery (this concern, by the way, has been expressed to me by both Jewish and Christian families). I think that most religions have come round to viewing suicide as aberrant behavior that is the excusable product of mental illness, and few are so strict as to block a burial on the grounds of killing oneself, but in those circumstances, I have proposed waiting to issue the final certification until I have finished running the toxicology testing to determine if they were intoxicated at the time of the incident. By the time the results have returned and a final death certificate is issued, the person has long been interred—an ethical stretch for which I am able to forgive myself.

I will do my best to keep the family constantly updated with the state of the investigation, and more particularly, the stage of the body's progress through the system—there is often a religious imperative that the person be buried as quickly as possible. Sometimes the interment is to be in Israel, which adds extra pressure for rapid processing and release of the body.

I'll notify them when the procedure (another euphemism I like) is finished, when the identification has been completed, and when the body is ready for release. I will discuss the findings with them to the extent that I can, and let them know what the next stages of the investigation will involve. Keeping open lines of communication—letting the family know that you are sensitive to their needs and available for continued follow-up—is the best way to support them through what can be a nightmarishly painful ordeal.

At the end of my conversations with all families, I shift to the practicalities of handling the body. I explain when the autopsy report is likely to be ready, and how they can get a copy. I refer them to an appropriate administrator or department who can help them get access to the decedent's dwelling and possessions if, for example, their apartment has been sealed by the police. I tell them when the body can be released, and the protocol for the release, and I make sure they have all relevant telephone numbers for handling things going forward, including mine, for follow-up questions or should they learn something that might be of use to the investigation. As to the public health aspect of what we do, I am not shy about offering counsel. If I have a 42-year-old man who has collapsed while jogging and is found to have multifocal critical coronary artery stenoses at autopsy, I will warn his family that they need to take this finding seriously. I will explain that the death of a young man from this disease raises the possibility that there is some sort of genetic predisposition to atherosclerotic cardiovascular disease. I will encourage the family to see a physician, urging them to get their blood pressure, blood sugar, cholesterol, and lipids checked. I will advise them to focus on stopping smoking, and losing weight, and any other interventions I feel relevant to the circumstances of the death at hand.

Over the course of the conversation, families may get emotional. I will stop talking to let them grieve, staying quiet while they do. Sometimes, I feel their pain deeply myself—I have wept many times while speaking with families (Indeed, I have wept in court, on the stand, while testifying in child homicides; at that point, I have turned to the judge and asked to be excused for a couple of minutes). When I do feel intensely moved, I will say something to the family—"I'm so sorry", most commonly. That said, I will *never* trot out "Sorry for your loss". Now if *you* truly feel that expression, by all means, say it. Just do not say it because it's What People Say; say it because it is how you feel in the moment. Or even better, express your condolences in your own words. Occasionally, families will become enraged and start shouting. At this point, you are unlikely to be able to either elicit or share any further useful information, so it is best to extricate yourself from the conversation, while being clear that you are leaving the door open to further communication. I will say something like, "I think it's best that we speak again at another time. You have my number; please call me when you're ready." Remember that anger can be a part of the grieving process and you should not take it personally or respond in kind. If you are meeting with a family in person, and they become enraged, be very careful. I have had a father, his face three inches from mine, screaming at me that I was accusing him of shaking his son to death; it was disconcerting, and it made it hard to stay focused. This is a good time to remember what your role in the interaction is: you are the physician charged with investigating the case. You are in control of the meeting and can terminate it at any time; obviously, now would be the time. When you leave the room after an angry exchange, it is a good idea to have security respond for support. Indeed, if you are about to enter a possibly contentious interview, it is best to bring a colleague along, ideally a senior colleague. If you do not have security in your facility and feel truly threatened, you might decide to call 911. Just remember, no matter how much rage is directed against you, this is not about you. You're speaking with someone who has just suffered a devastating loss; you have not done anything wrong, you are not incompetent, and you are only there to help them understand. Your patience is your winning card, and when you call them or meet with them again later to follow up, they are generally apologetic, even a little sheepish. In closing, I will say that in the thousands of interactions I have had with families, I have pretty much always managed to get the information I needed, and to answer the family's questions—very few indeed have gone badly. The thing is, of those thousands of conversations, I remember the details of only a handful, but I know that for those families, my words have made an impression they will remember for years to come.

The principles of communicating clearly when speaking with families are equally applicable when testifying in court or giving depositions. Of course, in the court room, you don't usually have any direct feedback to confirm the level of understanding—although you'll sometimes see heads nodding in assent at something you have just said.

Even trickier, you are talking to a broad cross section of people, who are likely to have a range of levels of understanding in what you have to say. Aiming your vocabulary at about the level of a bright high schooler, avoiding medical jargon and abbreviations, and breaking pathophysiological processes down into their simplest form remain the essential skills. If you can do that—and really, it's not too hard if you yourself understand the physiology—you will be solid on the witness stand. Besides, the mechanics of most homicides are readily understood by lay people.

Court testimony can be tricky. While conversations with families involve an earnest exchange of information, in court, attorneys may attempt to manipulate or twist your words. Resist the urge to take that personally—they're just doing their jobs. Whereas you may feel like an impostor when you're just beginning to talk to families, the defense may actively try to undermine your credibility, emphasizing to the jury just how inexperienced you were when you did the autopsy. That's fine—it's to be expected. And again, they are only doing their jobs.

Just remember: you can wipe the floor with them when it comes to forensic knowledge. Juries will not base their opinion of your abilities on the defense attorney's posturing, but on the way you communicate your knowledge of the case to them. Even if you've only done four homicide autopsies, if you explain your case clearly, they'll recognize your authority. They may even feel sympathy toward you should the defense try to harass or bully you.

Which will likely happen sooner or later. Sometimes attorneys will deliberately try to fluster you. Just relax, and answer their questions calmly. You *know* the answers. One of the common techniques they may use is to ask compound questions. Teach them quickly that asking you compound questions is a mistake: I have no hesitation in saying, "I'm sorry, I don't understand the question, counselor. Would you mind repeating it?" This is a particularly effective technique when the defense has just poured out a long-winded compound question, stacked with subordinate clauses and conditional questions and inserted hypotheticals.

I pause for a second, look puzzled, and then ask them to repeat it. At this point, **they** may become flustered—the complexity of their question was such that it is difficult to reconstitute it, and generally they'll default to asking a simpler, more direct question. Sometimes, they will try to dodge that by having the court reporter read their bloated question back in its entirety. If it's the same question, with the same problematic construction, I'll just say, "I'm really sorry, but I'm just not clear what you're asking." At that point, the judge will typically instruct the attorney to rephrase the question; the more basic question then ensues.

At the end of the day, court testimony is straightforward—all you have to do is answer the questions honestly. Have a clear sense about what you believe to be the facts of the case, and focus on explaining them to the jury in a clear and comprehensible way. I largely ignore both the prosecuting and defense attorneys. When I step up to the stand, the first thing I do is turn my chair to face the jury. I make eye contact during my testimony, shifting focus from juror to juror, again trying to see if they are following me. I glance toward the prosecutor or defense attorney for their questions, but immediately turn back to the jury, often in mid question. My focus remains on the jury: they are the people I'm talking to, no matter who is asking the questions.

Do not assume that the members of the jury think you're omniscient—they do not. They assume you'll be scrupulously honest, and if you interfere with that assumption, your testimony will be weakened. Learn to say, "I don't know" or "I don't remember," and learn to say them easily, without an agonizing delay as you decide whether or not to admit your ignorance. If you do not know something, admit it immediately.

Similarly, if the defense presents you with a hypothetical that might hurt the prosecution's case, but which you consider reasonable, give it up immediately, without trying to weasel around it. "Yes, counselor, that's possible."

It is up to the prosecutor to come back to you and say, "Doctor, you said that the defense's hypothetical theory was 'possible'. Do you have an opinion as to how **likely** it was?" Then you can answer, "When I said that it was 'possible', I meant that I couldn't rule it out, but I think that that explanation is highly improbable." And then explain why you feel that way.

Sometimes an attorney will deliberately build a hypnotic rhythm to their questions, trying to lull you into placidity. They ask simple, straightforward questions to things that are quite evidently true and with which you agree, question after question. "Isn't it true, Doctor, that …" Yes. "Is it not also true that …" Yes.

And the questions will come one after another, and you find yourself agreeing with them, developing a kind of rhythm of affirmation—it is seductively easy to be able to just answer "Yes" without conflict, without even really having to think about it. But the questions will at one point start to shift, until an affirmative answer is *mostly* true, but not 100%; the expectation is that you will let it go because you have repeatedly answered "Yes" over and over again. But once you've accepted a less than 100% clear-cut assertion, the assertions become increasingly uneven, until you've been maneuvered into agreeing with something you really do not agree with. Watch out for that incremental drift from what is reasonable to what is not.

When an attorney says, "Wouldn't you agree, Doctor, that …," be on your guard. They are using you to lay the foundation for something you might strongly disagree with. You can answer something like, "In most cases, I think that's true," giving yourself an out should more specific and inaccurate assertions be made. Use a similar approach when the attorney produces a textbook and says, "Are you familiar with Spitz and Fisher? Wouldn't you agree, Doctor, that it's a standard reference in your field?" I say, "It's an excellent book, but I don't agree with every word of it. Forensic science is constantly changing, so it's very difficult for any one book to be authoritative."

Sometimes an attorney will produce a paper from a legitimate scientific journal. Expect a two-pronged attack when you hear "Doctor, are you familiar with this 2011 paper on Xerox machine homicides from Guy & Another Guy, from the *Journal of Forensic Sciences*"? There is an unspoken implication, if you're not familiar with the paper, that you aren't up to date in the scientific literature of your field.

Now, this paper has been produced with an agenda: the attorney is going to use it to buttress her or his rationale of the case. Rarely, the whole paper will support their position. More frequently, however, there will be something, sometimes as little as a single partial sentence, cherry-picked out of the context of the paper, on which they'll be hanging their theory. They want you to recognize the authority of that phrase, and they will hand the paper to you up on the witness stand for your scientific review.

Now whether or not that phrase is valid, perched up there at the witness stand, the eyes of the judge, jury, and courtroom upon you, the clock ticking noisily at the back of the courtroom, you will find it is not the easiest place in the world to do a little critical reading. What makes it worse is that the attorneys often hand you printouts of forensic articles that have been heavily marked up with highlighter and annotations. These are almost impossible to ignore as you try to read the article critically—your eyes go straight to the highlighted phrases, and your comprehension of the other elements in the paper is weakened. In these circumstances, particularly if the paper or textbook you have just been handed involves a lengthy passage for review, it is best to ask the judge if you can be excused for a few minutes to go through it. Instead of interrupting their flow, sometimes the attorney will urgently direct your attention to the passage on which she or he wants you to focus.

The key thing when testifying at trial is to recognize your function: you're not there to support the prosecution or to defeat the defense; you're there to testify in an unbiased way about the facts of the case, to share your knowledge of how and why the victim died. Avoid becoming invested in the outcome of the case: no matter how horrible the killing is, you've only seen one tiny corner of it. Murder cases are pursued aggressively in all jurisdictions, and there is a small, but profoundly significant incidence of false convictions. This means that the prosecution's case isn't always accurate; in some rare instances, an unscrupulous prosecutor may feed the forensic pathologist selective information to get the conclusions and testimony they want on the stand.

Just get up there and testify about what you found at autopsy. If you tell the truth straight down the line, you'll ensure that you haven't played a major role in the conviction of an innocent person. Personally, I don't follow up on the outcome of the cases in which I testify, although often the prosecutor will reach out to say thanks, or I'll learn of the verdict in the news media.

With practice, you'll feel more comfortable on the stand, and with continued experience, you'll feel more authoritative. Over time, you'll relax, and find trial testimony easier—believe it or not, even enjoyable!

Testifying in legal forums

No one is born a congenitally good witness. -*Dr. Lester Adelson*

The difference between an expert witness and an ordinary ("fact") witness is enormous. An ordinary witness is required to perform one simple task. They must relate to the jury what they remember of things they perceived with their senses. Whereas an ordinary witness is usually prohibited from offering any opinions, an expert witness is called for that very purpose. In general, a witness is allowed to give an opinion, if the subject matter is of a scientific or professional matter that is beyond the scope of the average juror and the witness has sufficient skill, knowledge, or expertise in that field to appear to be helpful to the jury in the search for the truth. Federal Rule 702 states that "If scientific, technical, or other specialized knowledge will assist the trier of fact to understand the evidence or to determine a fact in issue, a witness qualified as an expert by knowledge, skill, experience, training, or education may testify thereto in the form of an opinion or otherwise."[9]

Inquiries from attorneys

As a general guideline, aside from the prosecutor who is investigating the death, you should not discuss findings with an attorney before you have completed the final death certificate. One exception to this rule is when the decedent's family requests that you communicate directly with their attorney instead of with them, in which case you may speak with the attorney at any time to the extent that you would have spoken with the family.

In **criminal matters**, you should not discuss any details of a case with defense counsel until after the suspect has been charged with a crime; until they are formally charged, they are not a criminal defendant and, therefore, there is no criminal defense attorney who is privy to the case. Once a suspect has been charged with a crime (**indicted**), you generally are free to discuss the case with defense counsel, although you are not legally obligated to do so. If you have questions or concerns, do not hesitate to contact your agency's legal department. You may discuss a criminal matter with the prosecutor at any time.

In **civil matters**, you may need to discuss the case with the attorney for your jurisdiction, as the city, county, or state may be a party to the litigation. In some instances, it is unethical for an attorney to contact certain employees of the party being sued without first contacting the party's attorney. If you are contacted by a private attorney and asked about a matter that you believe is, or is likely to be, the subject of civil litigation involving your jurisdiction, politely tell the attorney that you will first need to speak with your agency's counsel. In all discussions with any attorney (e.g., an Assistant District/State's Attorney, a criminal defense attorney, a civil attorney), there are a number of points to remember:

1. What you say to the attorney may be used in court; be sure to discuss the case professionally and never "off the record."
2. You may freely and fully discuss your case, but you should not go beyond its parameters: limit your discussions to the facts of the case as you know them.
3. Do not speculate and do not discuss matters outside your areas of expertise. For example, if hospital or other records indicate how long the decedent lived after an injury, you can provide that information; you can also state, for example, that death would have been instantaneous, given the nature of the injuries (e.g., decapitation).
4. ME/C staff maintain objectivity and impartiality in criminal and civil matters, regardless of the identity of plaintiff and the defendant. They are neutral arbiters of the facts.
5. In general, ME/C do not provide free pathology advice to attorneys; if an inquiry does not involve an ME/C matter, the attorney may be told that private pathologists can be retained to provide that advice. Some jurisdictions require the ME/C to provide free pathology advice to other governmental agencies. If you have performed the autopsy, then you are a legitimate witness who may be called to testify. In civil cases, the attorneys may prefer to retain their own experts so as not to have to rely upon an independent medical examiner who may not be interested in facilitating a nuanced argument of the case.

On occasion, an outside attorney may ask you to save, or "preserve" documents, tissue, etc., in a case that you have handled. All such requests should be accommodated by having a "preservation order" (also known as a "litigation hold") placed on the items to prevent their destruction/discard while litigation is pending. You also may receive an email "preservation order" instructing you not to destroy or discard specific items related to a particular case; you must comply with this directive, while notifying the agency's appropriate department. It is best to put a time limit on these requests—such as a "five-year hold," which would require a renewal by the attorney. Otherwise, these specimens will be held for decades, even after the case has settled.

Testimony at depositions, grand juries, or trials

> If the law has made you a witness, remain a man of science. You have no victim to avenge, no guilty or innocent person to convict or save— you must bear testimony within the limits of science. -*Dr. P.C.H. Brouardel*

There are three forums during which formal testimony is taken of a witness. These are depositions, pretrial hearings/grand juries, and trials.

Depositions

Sworn testimony may be taken (and recorded by a court reporter or other recording device) at a deposition, *also known as an* **Examination Before Trial**. Depositions are pretrial proceedings, generally associated with civil litigation, used to discover facts (which may also lead to other evidence), preserve the testimony of a witness, and/or lock in a witness's version of events. **Depositions are less formal than trials (there is no judge or jury present), the scope of questioning may be broader, and trial-relevant objections may be waived (or "noted")**. Cases can be won or lost because of a witness's deposition testimony. One of the best ways to reduce the credibility of a trial witness, for example, is to show contradictions between trial and deposition testimony. So, it is wise to take depositions as seriously as you would your testimony at trial.

One's oral deposition may be preceded by a *subpoena* for the record. The subpoena may call for production of specific or all available records. The examining attorney may examine the witness as to all those documents. Documents are generally marked as exhibits in sequence before the witness is examined about the information contained in the documents, and the witness should have ample time to examine the document and ascertain whether or not it is the document the attorney is referring to in their questioning. *If you are not given ample time, take your time.* There is no written record of the amount of time you need to pause to think or review during the course of a deposition. Resist the temptation to please, to feel pressured, or to be rushed by anyone in the room. Your answers are being recorded, word for word. Take the time you need to consider them, and the specifics of the question you are being asked, before you speak.

A technique sometimes used by lawyers to attempt to prod a witness to volunteer additional information, particularly at a deposition, is the **"pregnant pause."** The deposing counsel asks a question, the witness responds, and then counsel simply stares at the witness in silence, waiting for some further response. The temptation for many witnesses is to fill the void with something, anything, even if the witness has nothing to say or has answered the question appropriately and thoroughly. Periods of silence are uncomfortable to the average person engaged in conversation, but the urge to speak must be modulated. When a witness has completed their answer, they should not speak again until another question is asked. It is also important to be aware of the fact that, since neither judge nor jury is in the room, behaviors are not being witnessed, nor do they necessarily translate in the written record of the exchange. Tone is not transcribed and may range from quite civil to unapologetically aggressive. Take that into account in the management of your expectations as you walk into the room, and **try to psychologically inoculate yourself against any potential bluster, sticking instead to the facts, the science, and, ultimately, the truth.**

Grand jury

Some jurisdictions use a grand jury system to indict a person for a crime. Others may use a *probable cause hearing* before a judge. A grand jury comprises residents from that jurisdiction who hear evidence (a grand jury is typically larger than a jury in a trial) and then decide whether or not to formally charge, or **indict**, a person for a crime. There is no judge in the grand jury room, nor is the defendant or their defense attorney. As such, there is no cross-examination; however, jurors may be allowed to ask questions that are usually filtered through a prosecutor. Grand jury testimony is recorded by a court reporter, who is in the grand jury room; therefore, your testimony becomes "discoverable" and available to the defense counsel if/when you are later called at trial as an expert witness. The grand jury inquiry is a closed proceeding. As such, you may not discuss your grand jury testimony with any lawyer except the assigned prosecutor. A suspect only becomes a defendant once indicted. Until that time, the forensic pathologist (FP) is unable to speak about the case to any defense attorneys.

Trial

Trials occur in civil, criminal, and family courts and may occur at the local, state, or federal levels. They usually involve a jury; however, some defendants may elect to have a bench (judge only) trial. Before testifying in court, most forensic pathologists anticipate their testimony and define the limits of their opinions. As such, it's important to spend a good deal of time reviewing all of the details of the case being adjudicated.

Steps in the preparation for testimony

Chance favors the prepared mind. *-Louis Pasteur*

After an attorney calls and/or you receive a subpoena:

1. *If you choose to bring anything to court, request a "true" or "certified" copy of the case file from your records department* (the original case file does not go to court). Be careful what you write on your copy of the autopsy report that you bring to court to testify, as the defense may ask to look at your entire file during the trial. There are two types of subpoenas: *Subpoena ad testificandum* orders a person to testify before the ordering authority or face punishment, and *Subpoena duces tecum* orders a person or organization to bring physical evidence before the ordering authority or face punishment. The latter is often used for requests to mail copies of documents to a requesting party or directly to court. Some prosecutor's offices do not summon forensic pathologists to testify by subpoena. Rather, they often do so by phone or email and organize any further scheduling that way. Others subpoena as a matter of policy and/or for the record. Civil attorneys more commonly subpoena neutral parties, such as expert witnesses from ME/C offices.
2. *It is best not to plan to go to court on the date listed on the subpoena without first speaking with the attorney.* Typically, that date is merely a place holder, an indicator of the day the trial starts, but they usually don't need the expert until some later time. That said, even though the date may change, the witness is still under subpoena to appear, whenever that may be.
3. *Fellows should review the case with a senior medical examiner* prior to their testimony, including the autopsy report, notes, diagrams, photographs, etc.
4. *It is important to meet with the attorney in advance, by phone, virtual meeting, or in person* (not in the courthouse hallway 10 minutes before testimony), to review the questions you will be asked (including the questions that they anticipate you may be asked on **cross-examination**—testimony directed by the opposing attorney). There should be no surprise questions on **direct examination** (testimony that is directed by the attorney calling the witness).
5. *What is the opposing counsel's "theory" of the case?* This may help you anticipate issues or questions that may arise on cross-examination. Not uncommonly, the issue at the center of the case is only one of *identity* (i.e., that the

defendant did not commit the crime, rather that someone else did). Rarely is the line of defense focused on issues related to the opinion of the ME/C (cause of death) in those instances. As such, cross-examination may be minimal in that scenario.

6. *Do not be humble; make certain to inform the attorney of all of your qualifications, educational background, certifications, etc.* It is often a good idea to provide the attorney with your curriculum vitae (use your office address and phone number).

In preparation for trial, it is best review the case file carefully, as well as any prior testimony pertaining to the case (e.g., from grand jury). There can be insubstantial (semantic) differences in testimony given on the same case by the same person, some of which may be focused on under the scrutiny of cross-examination. These typically trivial differences can be exploited by an attorney attempting to argue that you have changed your opinion. By reading prior testimony, it refreshes your recollection about the issues of the case and also reminds you of the opinions that were previously asserted, opinions that have not likely changed, but that are helpful to review. This will help prevent "innocent" contradictions or inconsistencies between deposition or grand jury and trial testimony. If their content is critical to your determination of cause and manner of death, one may review radiographs, microscopic slides, stock jars, etc. The autopsy images are invariably reviewed.

The courtroom

Once you are called into the courtroom (usually you are not allowed to sit in the courtroom when other witnesses are testifying), you will be immediately directed to the witness box by a court officer. Each of the legal parties will already be seated at separate tables, and the jury will already be seated in the jury box by the time the first witness is called. Once in the witness box, you will be asked to remain standing, raise your right hand, and then be sworn in (or affirmed) by the clerk. They will also typically ask you to state your full name (spell it), your place of employment, and, usually, your address (give your work address). The judge is usually the first to speak to the witness after they are sworn in. A cordial and respectful greeting from the judge is usually followed by some basic instructions. In terms of the etiquette in addressing and responding to a judge, "Yes, Your Honor. Thank you" is usually both sufficient and appropriate.

The four domains of credibility are: **trustworthiness, knowledge, confidence, and likability.**[10–17] These also are the attributes of good "witness-ship." Without credibility, the message of the witness, expert or fact, and the important information they need to convey may be lost entirely, no matter its veracity. **Likability** has been described as one of the most important attributes of the expert witness and is defined as the degree to which an expert is friendly, respectful, kind, well-mannered, and pleasant.[11] **Honesty and sincerity** constitute the first step in attaining credibility, and they are conveyed when the witness presents a modest demeanor and displays a sincere interest in the accuracy and truthfulness of their statements. Honesty is one of the first attributes that contribute to a witness's likability. The second is **brevity**. Most witnesses will not get flustered, confused, or embarrassed if they confine their statements to answering the question. **Be brief and to the point.** It is best to not volunteer information, argue, or make unneeded comments. An attorney once said that witnesses should be "truthful, but not candid." A third important quality of the likable and credible witness is **clarity**: errors, inconsistencies, and confusion undermine credibility. The fourth quality is **objectivity**. The ME/C is an impartial and conscientious public servant who is working for the interests of justice; they have no horse in this race. The "success" of an expert witness is not measured by whether the party having called them to court "wins" the case, but rather by whether they effectively communicated the facts and their opinions of the case from within the realm of their expertise, and have left the jurors with a better and more representative sense of the scientific basis of the case after the testimony has been put on record. The witness should be courteous, answer directly, and remain poised, irrespective of the quality, content, or tone of the questions asked of them. A final important quality is **equanimity** (emotional control), with which one ignores insults, badgering, and innuendoes. A display of anger results in a loss of credibility.[13,18]

Standard qualification questions (with some examples of responses):

1. What is your name and work address?
2. What is your occupation? (*A physician, a medical doctor*)
3. By whom are you employed?
4. How long have you been so employed?
5. What is the function of the ME/C?

To investigate all unnatural, suspicious, and unexpected deaths in this jurisdiction. The major function is to establish the cause and manner of death. This is usually done by performing an autopsy.

6. What are your duties as a ME/C?
7. What is pathology?

The area of medicine that specializes in the laboratory aspects of disease and injury; among other things, it includes performing autopsies.

8. What is forensic pathology?

A subspecialty of pathology that deals with the investigation of sudden, violent, or unexpected deaths. It is a subspecialty in pathology involved in the examination of living and dead persons to provide opinions concerning the cause, mechanism, and manner of disease, injury, or death. Among other areas of expertise, forensic pathologists are experts in patterns of injury.

9. What is your education and training in this field?
10. Are you licensed to practice medicine?
11. When and where did you receive your license (know the date)?
12. What is the American Board of Pathology?
13. Are you certified by the American Board of Pathology?
14. What is an autopsy?
15. Approximately how many autopsies have you performed or observed?

Approximately ______ and I have observed or assisted in two to three times as many.

16. Have you been qualified before as an expert in the field of forensic pathology in a court of law?
17. Approximately how often and in what courts?
18. Have you ever been denied qualification as an expert in forensic pathology?
19. Your Honor, at this time, I ask that Doctor X be qualified as an expert in the field of forensic pathology.

Typical direct examination questions:

Answer the questions as asked. It is best to try to get critical information out first—do not do a fancy windup—just throw it. One should answer clearly, concisely, and with conviction. Below are a few examples:

1. What is a cause of death?
2. What is a manner of death?
3. Did you perform an autopsy on Mr. X?
4. What is toxicology? *The science of drugs and poisons.*
5. Are you an expert in toxicology?

I am an expert in interpretive toxicology, but analytical (testing) toxicology falls within the realm of the forensic toxicologist.

6. What is sharp injury? What is a stab wound?
7. What is an incision?
8. What is blunt injury? What is a contusion?
9. What is a laceration?
10. What is an abrasion?
11. What did you find on the external examination, on the internal examination?
12. What is your opinion of the cause of death?

Cross-examination questions:

Cross-examination questions often attempt to limit the witness's responses to simple "yes" or "no" answers, some of which may be used later during closing arguments to support a theory of the case. The questions may contain subtle differences that attempt to extract a useful response. The structure of the question (e.g., compound questions) may be used to try to achieve this result. One of the most important answers is, "I don't know." Many experts are reluctant to admit to not knowing an answer to a question. As Dr. Hirsch was fond of pointing out, "Someone who has all the answers all the time is either a scoundrel or a fool." Saying "I don't know" has two benefits. First, it is truthful. Second, it will increase your credibility with the triers of fact. Do not be afraid to say it, Such an unapologetically honest answer puts both your humanity and your integrity on display.

The following are examples of issues and questions that may be encountered in cross-examination exchanges, followed by potential answers that represent a larger illustration. In all circumstances, the assumption is that any responses that you offer in court will be truthful, first and foremost. That said, the illustrative exchanges below are not intended to offer a script to be memorized, but rather represent examples of appropriate and respectful responses that reveal a broader point: how to field sometimes intentionally difficult questions truthfully, respectfully, and artfully, without falling into traps that may obfuscate the facts, and potentially derail your duty as an expert witness. The goal of the proceeding, therefore, is to help maximize the conveyance of facts and fact-based opinions, while minimizing the effects of strategies that may distract or misrepresent the truth (to "Slay a beautiful theory with an ugly fact.").

Potentially challenging or otherwise tricky questions

1. If given multiple questions in one (compound question):

If you are asked several questions rolled into one, it will usually be impossible to answer accurately unless you break them down. In such an instance, it is appropriate to say:

"Your question contains several components, which I'll try to answer one by one."

If you are confused about how to break the questions down, simply say to the questioner: "That question seems to contain a number of separate concerns. Can you break them down for me and ask me the questions one at a time?" For example, a question such as, "Did you find the gun in the alley and take it with you?" should be asked as, "Did you find the gun in the alley?" and "Did you take it with you?"

2. Beware of leading questions (i.e., which contain the answer within the question):

These questions often begin with phrases such as, "Isn't it a fact that…?" or "The fact is, isn't it, doctor, that…" These are usually leading questions containing implications or assumptions that may be untrue or only partially true. Some questioners will put their leading questions in another form by presenting a hypothesis and following it with, "Isn't that correct?" or "Isn't that true?" Remember: you know the answer; the questioner doesn't. The court is most interested in your answer, not the foregone conclusion of the questioner. In each instance of this category of query, the questioner is trying to lead you to give an answer that may or may not be the truth.

Witnesses are frequently asked leading questions that suggest information which is either half true or contains facts not within the witness's knowledge. These questions may sound plausible at first blush, but it is the expert witness's responsibility to look at the entire question and determine whether or not they know enough to give a valid answer. For example, one may be asked: "This is a contact gunshot wound, correct?" The proper question is: "What is the range of fire of this gunshot wound?" Leading questions are permitted under certain circumstances (e.g., cross-examinations, expert witnesses).

3. Beware of exact numbers: In a humorous, if still illustrative example of how far he would go in allowing a questioner to pressure him into opining in a narrowly quantitative fashion, Dr. Hirsch suggested answering the following to the question, "how many does 'a few' mean?": *"A few is more than a couple and less than several."* The point of his sharing this was not at all to encourage petulance, but to point out that we sometimes simply cannot accurately quantitate. Exact numbers are, at times and not uncommonly in forensic pathology practice, unknowable, at least with the kind of precision that may be asked of us. "Speculation is built on a foundation of assumption."

If pressed about an exact number (e.g., how many hours had the decedent been dead when discovered?), some reasonable responses may include:

"In order to answer any more specifically than that, I would have to speculate."

This response will likely compel the judge to intervene, as a witness is never allowed to speculate.

4. If you are asked an apparent "yes" or "no" question that cannot be answered simply "yes" or "no" without misleading the jury (e.g., do you still cheat on your taxes?):

Common responses include:

"I can't truthfully answer the question with a yes or no; may I be allowed to explain?"

"I am afraid that if I try to answer that with a simple yes or no, I may mislead the jury."

5. If asked if a certain textbook is an authoritative or definitive text:

Authoritative means that you have read the entire book and agree with everything in it. If this is not the case, the answer should be "no." That also applies to this book! That said, be careful to not contradict your own writing.

6. If you have been read a line from book:

"I cannot answer the question without seeing the quote in context."
(Besides the context, also check the year and the edition)

7. If asked if something is "possible":

One must say "yes" if it is possible, that is, if it does not defy the laws of physics/nature. There do exist, however, reasonable and unreasonable possibilities. Often these answers get clarified on redirect. If presented with an *unreasonable* possibility, one might clarify with some version of:

"Yes, it is possible because it doesn't defy the laws of nature, but it is extremely unlikely, and therefore, represents an unreasonable possibility."

Dr. Hirsch once memorably responded, on redirect, to further qualify an unreasonable hypothetical possibility that had been presented to him under cross-examination:

"While I can't say that's impossible, it is so highly unlikely that it stands shoulder to shoulder with impossibility."

8. If a mistake in the autopsy report is asked about (comma, spelling mistake):

"Thank you for pointing that out to me; I hope it did not cause any confusion."

9. If asked to criticize another person's autopsy report:

"The report is not optimal, but it has sufficient information. When I look back at some of my own reports, they are not always optimal."

10. If interrupted:

"Excuse me counsel, but I would like the opportunity to finish my answer to the question that you already asked me."

11. If an attorney gets overly aggressive:

"Your honor, he is distracting me by speaking so loudly and standing so closely."

"There is no need to yell, I can hear you quite well." This may be useful in depositions to put into the record that an attorney is yelling. One usually cannot discern from a transcribed deposition that a person was yelling.

12. Be careful of the questioner's attempts to conflate "necessary" for "important." For example, if some areas are not filled out on an autopsy worksheet:

"I did not feel it was necessary. Those areas and systems were normal."

"The issue is not how I fill out forms but, whether I documented the injuries of the decedent. As you can see, my notes and records in that regard are quite thorough and complete."

13. "Have you ever made a mistake?"

"Yes."

14. "What do you do when you make a mistake?"

"I utter the two most difficult sentences to say: 'I was wrong' and 'You were right.'"

15. "Why did it take so long to determine the cause of death?"

"I wanted to be fair to everyone involved and make sure that we reviewed all the evidence before making a final decision."

16. If asked if you will agree to answer the following questions with a "yes" or "no" answer, you may respond: "I will try."

17. **Having a thorough understanding of degrees of certainty** is one of the many responsibilities of individuals afforded the privilege of expert witness status. Being asked about your degree of certainty is fair game in courts of law. See Chapters 2 and 3 for a more granular discussions about this.

Time of death estimates and postmortem changes: caveats and approach

Estimation of the postmortem interval is a slippery slope and merely an estimate, becoming increasingly inaccurate as that interval widens (see also Chapter 4). For that reason, forensic pathologists typically speak in ranges. Try to avoid playing the numbers game. Also, resist the urge to over explain how you arrived at your conclusion, unless specifically asked to do so, of course. Opine and answer the questions as best you can. Usually, the best way to handle this issue is to suggest that the attorney ask a hypothetical question, to which a simple "yes or no" may be the response. In response, one may be able to say whether a hypothetical time and date are "consistent" (reasonably possible) or "inconsistent" (not reasonably possible) with the physical findings (i.e., after careful consideration of the changes in the body) and the documented circumstances. "Everything is consistent; nothing is inconsistent."

For example, one may be asked, "Are the postmortem findings and circumstances *consistent with* Mr. Smith dying at around noon on July 1st?" as opposed to, "When did Mr. Smith die?" The latter is not likely to be answered with specificity or veracity, unless the death was witnessed by medical professionals and/or other reliable witnesses. The opposing counsel may also ask their own hypotheticals to which you may answer in kind with the same responses.

Objections raised while on the stand

Objections may be raised during your testimony by either counsel. The common objections surround: leading questions, hearsay, badgering, relevance, vagueness, "asked and answered," speculation, argumentativeness, or a nonresponsive answer.

In general, *hearsay*, which is testifying about the observations of another, is not admissible testimony. The medical chart is a common exception. *Surrogate testimony*, in which a forensic pathologist testifies on the report of another, may be allowed with the record entered into evidence as a business record and when the expert's opinions are based on their own *independent* review of all the evidence related to the case.[19] *Badgering the witness* is based on an *argumentative* question in which the attorney states a conclusion and then asks the witness to argue with it, often in an attempt to get the witness to change their mind. An objection based on *foundation* involves a question or response in which the attorneys failed to explain the background circumstances of how they know the information about which their witness is testifying or the attorney is inquiring.

Testimony tips

In their intended role in supplying a critical public service, ME/C offices are ideally objective, neutral, and transparent agencies by design, and therefore have no agenda, take no sides, and support no individuals or political interests. Our job as ME/Cs is to determine the cause and manner of death and to identify decedents, regardless of who is helped or harmed (in the litigation context) by these determinations. To that end, it is best too not volunteer information or to make suggestions. This is particularly true during cross-examination at a trial. Be careful if you are aware of a previous trial for the same defendant—you should not refer to any previous trial (e.g., one ending with a hung jury) in front of the jury, as this may result in a mistrial. If required, you may refer to prior "testimony," but do not use the word "trial."

If an objection to a question is interposed, or attorneys start to argue about the question, do not answer the question. At trial, wait until the judge has ruled on the objection; if you are unsure of what to do, ask the judge if you should answer the question. The term **"overruled"** (open your mouth) means that you may answer the question, while **"sustained"** (silence) means that you may *not* answer the question. If you get confused, it is easiest to just not say anything. At that point, you will either be asked a new question or the judge will say "you may answer." You also may find that, by the time the dispute is resolved, you have forgotten the question. If so, simply ask that it be read back by the court reporter before responding.

Pay no attention to the demeanor of the questioner. The attorney may be particularly charming or notably confrontational; either style may be a way of distracting you from listening to the question and providing an accurate answer. It is best to keep your demeanor professional and calm. It enhances your credibility and also makes the experience of testifying considerably less stressful.

Unlike the death certification opinions asserted in **criminal proceedings** (determined to a reasonable degree of medical/scientific certainty), those in **civil matters** require a lower burden: preponderance of evidence (more likely than not). This is predicated on the fact that, in the adjudication of criminal matters, the *civil liberties* of one or more persons are at stake, whereas, in civil proceedings, a *monetary reward* for damages is being argued. In another divergence from most criminal matters, issues of **pain and suffering** are common, even central, to many **civil trials**, as they may lead to a verdict with a larger monetary award. One may be able to confidently and medically testify that a particular injury would not be likely to have caused immediate death, but you may (or may not) be able to opine as to whether and/or for how long the decedent had likely remained conscious after that injury. This is relevant since, logically, and legally, *issues of pain and suffering apply only to conscious individuals, as these constitute perceptive experiences.* Unconscious persons do not feel pain—that is why surgery is performed on unconscious people. As a forensic pathologist, by the time you encounter these decedents, they are no longer suffering and, depending on the circumstances of the injury and the specific case, statements about pain and suffering may be speculative. That said, in other instances, as a physician and forensic pathologist, you may be able to offer a reasonable opinion that a certain injury would more likely than not (burden of proof in civil proceedings) be associated with the experience of pain and, further, that there is no good reason to believe that the injury would have been associated with unconsciousness (based on both the evaluation of all injuries and the toxicology testing, for example). As an expert witness, the forensic pathologist may not only give their opinion regarding the cause and manner of death, but also about other inquiries to which they are confident in their responses based on their education, training, and experience. These opinions often may involve a review of witness statements and first responder findings (e.g., emergency medical service records). Without that information and depending on the case details, one may be on shaky ground, absent a clearly immediately incapacitating injury at autopsy (most forensic pathologists have autopsied a person with a gunshot through the heart who was still able to ambulate for several yards before collapsing).[20–23]

Being called to testify in a former jurisdiction

Forensic pathologists may not work in the same office in which they completed their fellowship and, in fact, may end up working in multiple offices during the course of their careers. A request that they testify in a legal proceeding involving an autopsy they performed in a jurisdiction in which they no longer work does occur. Due to various Supreme Court decisions (Melendez-Diaz v. Massachusetts, Bullcoming v. New Mexico, Williams v. Illinois),[24–26] more prosecutors are seeking testimony from the original pathologist who performed the autopsy. In the past, surrogate testimony was widely accepted. This allowed another pathologist to review the autopsy report and images and provide their *own opinion* about the death in court. That is, another forensic pathologist was allowed to testify in place of the original prosector. In these instances, the substituting pathologist would review all of the case material and, typically, the autopsy report would be submitted as a business record. A detailed, well-written autopsy report that is adequately supplemented with photographs usually suffices. The body of such an autopsy report is an objective description of the findings,[27] free of speculation and opinion. As such, any forensic pathologist would be able to review it and offer their own opinion. In addition, if this requirement was absolute, the statute of limitations of a homicide would then be the life span of the forensic pathologist.

Appropriate photographic and radiologic studies are important in providing redundant documentation in the event that autopsy notes or transcriptions are lost. As Mortiz said, "If a negative or positive postmortem finding is so important that it may make the difference between the freedom or imprisonment, or the life and death of someone, every attempt should be made to protect, preserve, and record it for others to see and evaluate."[28] Autopsy images also may be particularly useful at trial, particularly when a surrogate witness is called. For example, a defense attorney may object to the admission of an autopsy report at trial because the original pathologist is not testifying. If, in response, the judge offered that, in lieu of the autopsy report, the court would allow the surrogate forensic pathologist to use all 53 autopsy photographs to demonstrate the multiple gunshot wounds, the defense attorney may stipulate to the admission of the autopsy report, after further consideration.

Although there may be legal reasons to have the original pathologist testify, does it make a practical difference? Any forensic pathologist who practices in an office with a large number of homicides is unlikely to have an independent recollection of the details of a specific autopsy performed two or more years earlier. But, even if they were able to recall the autopsy, how reliable would that recollection be? If the original forensic pathologist essentially relies upon the report and images when they testify, what would prevent someone else from doing the same? In fact, defense and plaintiff experts are routinely allowed to give opinions regarding autopsy reports and autopsy findings in which they had no part in writing or documenting, respectively.

Live video conferencing from a remote jurisdiction may be an alternative that would reduce the time and travel issues. Having multiple forensic pathologists cosign each autopsy report may also be sufficient to allow testimony by one of the signatories in place of the actual prosector. Some forensic pathologists have been pressured to return to their primary jurisdiction to testify on an old homicide case without receiving remuneration of missed work at their current job. Attorneys may threaten subpoenas (interstate state subpoenas can be a challenge to obtain), but subpoenas cannot compel a person to give an opinion. Therefore, the attorney may end up with a witness who refuses to provide any expert testimony.[19]

Miscellaneous

1. Bring your badge/official identification when testifying. It may be of assistance in getting you through the courthouse security checkpoints.
2. Keep a record of the times that you have testified (grand jury vs. trial, county, attorney, case name, type of death, etc.). You should be prepared to state during testimony how many times you have testified/been qualified prior to the current case.
3. You also may be asked how many autopsies you have performed, how many autopsies are performed annually by your agency, and how many homicides there are annually in your jurisdiction. Be prepared to answer these informational/background questions as well. If asked to estimate the number of firearm deaths that you have evaluated, the estimate should include all manners of death, not just homicides.

Top 10 testimony tips

1. TELL THE TRUTH.

You will be put under oath at the beginning of the testimony. The oath means that you have sworn to answer every question honestly. Falsehoods on the most minor or irrelevant answers may be used to destroy your credibility on the more important issues. Because you should be on your guard to confine your answers to the demonstrable truth, it is wise to avoid volunteering information not expressly called for by the questions asked. As everyone involved in the litigation will know, intentionally false answers may constitute the crime of perjury. As Dr. Hirsch was known to say with considerable frequency, "When all else fails, tell the truth."

2. LISTEN CAREFULLY TO EACH QUESTION.

Be sure you understand the entire question completely. Do not mentally interpret the question to assist the questioner or mentally supply definitions for vague terms. It is the questioner's job to make the questions clear. Do not hesitate to ask for clarification of a vague phrase or an ambiguous term. If the question was long and complex, if your mind wandered because of outside distraction, if it makes assumptions with which you do not agree, or if you simply did not understand the question, do not hesitate to ask for a rereading of the question. Beware of leading and compound questions.

3. PAUSE BEFORE BEGINNING EACH ANSWER.

Answer slowly. This gives you the time to reflect on the question. The written transcript will not reflect how long you take to answer a question or how long you pause. Remember, nothing a lawyer says is evidence of anything unless it is answered affirmatively by the witness. Pausing for a moment also gives either attorney the chance to object to the question, after which you must remain silent until a ruling has been made by the judge (sustained or overruled).

4. DO NOT VOLUNTEER INFORMATION YOU ARE NOT ASKED.

Answer only the question that is asked. Do not anticipate questions. Keep your answers short and to the point. Do not promise to look up information or to obtain materials, or to conduct further calculations or tests.

5. READ A DOCUMENT BEFORE YOU TESTIFY CONCERNING ITS CONTENTS.

If the document is not available and you cannot recall it exactly, say so. Testifying is not meant to be a test of memory. You may always qualify your answers with phrases such as, "to the best of my recollection."

6. DON'T ENGAGE IN CONJECTURE (DO *NOT* SPECULATE).

Answer each question, as asked. If you do not know an answer, say so. Do not assume that the examiner has accurately described your prior testimony or that of another witness when he/she/they purport to do so. Never guess if you don't know the answer. Don't guess on the basis of facts that may not be reliable. Know the facts first-hand before you are willing to give an opinion. You may ask to see documents to which references are being made. You may refer to documents contained in the file present at a deposition or the trial. Avoid being drawn into areas beyond your expertise. If you are asked an opinion concerning something outside your expertise, do not hesitate to say that you do not have the expertise to form an opinion and that anything you offered would be speculative.

7. BE CAREFUL WITH YOUR DEMEANOR.

Your position is one of responsibility and your demeanor should reflect that. Speak slowly, clearly, and audibly, so that the court reporter will be able to take down every word. Keep your dignity and avoid joking with the attorneys or the court reporter during the examination. Be courteous to the attorneys and to the court reporter. Avoid losing your temper. Some attorneys may not be conscious of an unpleasant manner which may grate on your nerves; others may try to provoke you as a stratagem. Whatever type of attorney you are dealing with, sparring with that attorney should not obscure the purpose of the proceeding, which is to provide a true distillation of the facts as you know them. Keep your concentration on that purpose, not on the personalities. Attorneys sometimes attempt to wear witnesses down or rile them up in order to provoke an inappropriate response. A witness should remain calm, take a break if they need to (particularly during a deposition), and otherwise remember that controlled anger may be a tool used by lawyers and should not be mistaken for actual anger.

8. IF YOU HAVE DISCUSSED THE CASE PREVIOUSLY, ADMIT IT.

You may be asked if you have discussed your testimony with anyone. It is perfectly acceptable and expected to have discussed your testimony prior to your appearance in a deposition or courtroom. Simply say so. It would actually be an outlier situation for a conscientious and responsible professional *not* to prepare. An attorney may attempt to insinuate that you were influenced about how to testify in a previous conversation with the other attorney. If that did not occur, say so as well. You will be able to get through this tricky area by listening carefully to the question and answering only the question that is asked. One way to avoid insinuations of this nature is to arrange to discuss the case with both or all sides present at the same time prior to trial, or at least make the offer if you are contacted by the opposing attorney. The offer may not be accepted.

9. IF THE ATTORNEYS BEGIN TO ARGUE, SIT SILENTLY.

Particularly in a deposition, one attorney may dispute whether another attorney's question is proper. If they strongly disagree, the recourse is to call the judge and obtain a ruling. The trend in some jurisdictions and in federal discovery is to allow for broad discovery, so attorneys are motivated to find a way for the deposition to proceed. Let them do their job. In most cases, you will be a disinterested third-party witness, meaning you have no stake in the outcome. Just as physicians have their own medical jargon, so do attorneys (legalese). See Table 16.1.

TABLE 16.1 Common legal terms with definitions.

Battery: Unauthorized and offensive touching.
Assault (legal): Act that *threatens* physical harm. Definition: Physical *or* verbal attack. Intent: Purpose or aim behind commission of act, i.e., hoped for or desired outcome. Motive: Instigating force responsible for initiating or carrying out the deed.
Corpus delicti: The facts constituting or proving a crime. Material substance or foundation of a crime. It is not just the body in the murder case, but the fact that he has been murdered.
Habeas corpus: "You have the body." An order requiring that a detained person be brought before a court at a stated time and place to decide the legality of his imprisonment.
Arbitrary: Not fixed by rules, left to one's judgment. Based on one's whim.
Capricious: Erratic, tending to change abruptly without apparent reason.
Writ of mandamus: A court order commanding that a specified thing be done.
Stipulate: Factual statement accepted as true.
Hearsay: Testifying about the observations of another (medical chart is an exception)
Continuance: Legal proceeding moved to a later date.
Speculate: Guess (never do it).
Res ipsa loquitur: The thing speaks for itself.

High-profile deaths, media inquiries, and press offices

Distribution of partial information guarantees complete misunderstanding.
On sensationalism: "It makes heat without shedding light." -*Dr. Charles S. Hirsch*

The general public often has great interest in the work of the forensic pathologist.[29] That interest may range from simple curiosity in a particularly odd or unusual death, a desire to know details of the death of a well-known celebrity or a multiple fatality incident, or interest in a death that stems from having a nexus with complex, emotional and political issues affecting the community. Communication with the public occurs by way of both reputable and non-reputable media outlets through a number of methods, including print, television, radio, and the internet. In addition, today, social media channels can be an effective and far-reaching medium for news stories.

The frontline forensic pathologist may have little opportunity to be "in front of the cameras" and may believe their presence in the public space is marked by anonymity. However, the words in the autopsy report and any associated courtroom or deposition testimony may be directly quoted in news stories or headlines. Facts and opinions that are clear to the forensic pathologist may be reported inaccurately, or they can be intentionally distorted or taken out of context to support an "interesting" article. Indeed, forensic pathology is not for the faint of heart.

In cases for which there is sufficient public interest to warrant a more public communication or response, the ME/C office will need to decide how to address this. There are a number of avenues to take.

Silence or nonresponsiveness

Although it is an option, nonresponsiveness to the media is generally a poor approach and will likely create animosity and suspicion. As public servants, we have an obligation to communicate with and speak to the public. "Information void is the mother of paranoia." Avoid using the words, "no comment."

Handling individual inquiries from the media

If possible, the best model is to have as few people as possible or, ideally, a dedicated public relations professions (PRPs) officer speak with media or provide updates on behalf of the office. This provides consistency, allows the PRP to build relationships with reporters, and also allows you to be nimble when new information or changes in strategy arise.

When sharing information about a case or incident to different inquirers, it is paramount to remain consistent and to not show favoritism. Provide the same information to each caller. Some offices may have a "press" email distribution list.

Press releases

The written press release is a reliable and efficient means by which to share information. It can take the form of an updatable web page, emails to known media outlets, or common social media platforms. A press release has several advantages. The same information is provided to everyone at the same time and at a time of your choosing. The information being shared is well-defined, unlike a verbal conversation, which can vary in nuance as time goes on. Moreover, something in writing can, if desired, be reviewed by various people prior to release. Invariably, media will follow up with additional questions to the release (as they may have a desire to get "exclusive" information). One can then either respond as needed or edit/update the press release to ensure consistency.

Press releases are particularly useful when a matter is complicated or very careful messaging is necessary.

Press conferences

Under certain circumstances, usually when a situation is extremely acute and there is heightened public interest, a press conference is in order. A press conference is suitable when release of information must be coordinated with other agencies such as law enforcement, government officials, or public health officials, such as in a mass fatality incident.

A press conference usually has two or three parts: the statements by those who are speaking, which are usually followed by questions from the press, while still on camera, sometimes followed by requests for interviews. As much as possible, preparation is extremely important and can make the difference between a success or a failure. It is helpful to have written speaking points prepared, either verbatim or bulleted, using larger text with generous spacing. If possible, do not go to the press conference alone. Bring either a PRP or another staff member who can monitor questions, take notes, and keep an eye on interviews. Know the order of the speakers, and be prepared to potentially introduce the speaker following you, if necessary. Introduce yourself. Look professional. Speak clearly and slowly. Do everything possible to consider what you say, given the circumstances—don't be tone deaf. You may be asked to spell your name.

When speaking in any crisis situation, the following is a quick set of items to address. Being familiar with this list can help you, particularly when you might be caught off guard or be asked to speak with little warning.

- Condolences
- What we know
- How we know it
- What is being done
- What you can do
- Next planned update

Example: "Good afternoon. My name is Dr. ______, Chief/Deputy Medical Examiner of _____. On behalf of my department, I would like to extend our deepest condolences to the families affected by this tragedy. As of 15 minutes ago, our staff have recovered eight of the 10 bodies from the scene, and we have positively[30] identified seven of them. We are in the process of locating and notifying next of kin. Names will be provided as soon as the families are notified. We will continue to work to recover the remaining bodes and to work quickly to establish identification. If you believe someone you know may have been involved in this tragic incident, please call 1800-xxx-xxxx. Again, I wish to extend my condolences to all those affected. We will be providing another update tomorrow morning at 9:00 a.m."

Interviews

Interviews can happen suddenly and catch you off guard, or they may be planned and, as such, allow time to prepare. When preparing for an interview, always ask for the topic or even the questions—never walk in cold. Give yourself time to review the issue and prepare the points you may want to make.

Or,

"That information cannot be released at the present time because of the ongoing police investigation, but we will keep you apprised as soon as it can be released."

Some additional points to consider when being interviewed in person.

- Look and act professional.
- Watch your facial expressions and reactions. If asked an obnoxious or an inappropriate question, take care not to overreact. Those moments tend to make their way to the broadcast.
- Never speculate—It can be interpreted as fact.
- Never say "no comment"—This may sound like you are hiding something. Instead, say "our findings are not yet complete" or "we are not yet prepared to discuss our findings."
- Never lie—This will destroy your credibility.
- Never speak off the record (or "on background")—the reporter may misinterpret the statement as being on the record. It is best to take the approach that there is no such thing as "off the record."
- The incident is under investigation, and we are taking this incident very seriously.
- Do not play favorites—give everyone the same information and access.
- If approached and not ready: Smile—"I would love to speak to you about this issue, but I need a few minutes first."
- We reviewed the autopsy findings in light of additional information that was not available on the day of the autopsy. This additional information raised a reasonable possibility that this was not a homicide.

Some quotes that may be of help:

"Decisions cannot transcend the information upon which they are based." (Dr. Charles S. Hirsch)

"We render a diagnosis on everything we see and not on isolated findings." (Dr. Charles S. Hirsch)

"Death certificates are written on paper and not engraved in stone. If we obtain additional information that may affect our opinion, we will consider it." (Dr. Charles S. Hirsch)

"Everyone is entitled to his opinion, but not his own facts." (Daniel Patrick Moynihan)

In an in-person interview, if possible, bringing a PRP can again be extremely helpful. Because these professionals tend to have experience in the journalism industry and may very well have worked for the very media company with whom you are speaking, they have special insight about where reporters might be going or can even intervene if they feel the interview is becoming problematic. Importantly, they can also provide a third-party record of what happened.

If the interview is for print, you may be asked if the conversation can be recorded. If it is not brought up by the reporter, it is advisable to ask at the outset of the interview whether they are recording you. You should make a decision prior to the interview on this topic. Recordings can assist in providing an accurate record of what was asked and what the responses were. If asked whether they can record, you might ask if you also can record.

If the issue is particularly complicated or sensitive, one alternative to the face-to-face interview is obtaining questions in writing and then providing a written response. This has several advantages. First, responses can be carefully crafted with consideration of risk. Second, there is written documentation of what was asked and what was answered. Typically, this approach may lead to additional questions, and it may be less desirable to certain reporters.

Relationship with the media

Just like any relationship, yours with the media will take time to build and nurture. Make it clear that their access to you is based upon their fairness to you. There is a balance or natural tension between how you understand information you wish to provide and the reporter's need to write an article that will be interesting and read as many times as possible, by as many people as possible. It can seem confrontational or adversarial at times, depending on the "story."

Release of information

The type of investigative information that can be released to the public varies from jurisdiction to jurisdiction. Some offices cannot generally release autopsy and investigative reports, while others must if they are requested. Even if reports must be released upon request, redaction of certain information in the documents may be required or preferable. Death certificate information is public. In an active, ongoing investigation, sharing certain information with the public can adversely affect a law enforcement investigation. One must familiarize themselves with the local laws that govern what can and cannot be shared and the acute investigative sensitivities at play prior to speaking with or releasing information to the public or the media.

For example, homicides and deaths of civilians involving law enforcement actions may constitute high-profile deaths. The family and press want to know what happened. Families may even obtain their own second autopsies. The risk of releasing information too soon may compromise the investigation. The released information may allow a suspect to tailor their testimony to fit the findings. For example, a suspect may intend to say that the gun inadvertently discharged and struck the decedent, who was sitting several feet away. If the suspect learns that the autopsy documented a close-range gunshot wound, the suspect could take that into account and avoid giving an inconsistent statement.

Consideration of families and the media

The ME/C office is not alone in having the attention of the media and the public. The families of the decedent(s) have it as well. They may very well have reporters calling them, showing up at their homes, or intercepting them as they go about their day. For that reason, families may often have an intermediary to deal with government agencies and the media. Usually, these are attorneys, but they may be family friends or others.

As in all cases, it is important to keep open lines of communication with the family. However, if a family tends to spend time giving press conferences or sharing information with the public, one must be cautious or at least crystal clear with what is shared and what cannot be made public. Letting details get "in the wind" can cause problems later at trial.

Prior to releasing a high-profile report or case file to the public, share the completed reports with the family in private, answer their questions, and address their concerns. This should be done when everything is queued up (i.e., the press release page or other releases are vetted and written, but require only "the push of a button" to have them go live). Coordinate the timing of the release with the family. They may have their own public relations representative or desire to send out their own press release, and it is beneficial for everyone to work together. What should be avoided is the family releasing a report prior to your office.

Tone

Under no circumstances should you embrace sensationalism and hyperbole. Avoid leaning toward either a prosecutorial or defensive stance—the ME/C is not the trier of fact. Stick with the facts, what you know, and what you can share. The role of the forensic pathologist is to be neutral and to provide reasonable opinions as to the cause and manner of death, not to be entertaining or to create drama.

Mistakes

Everyone makes mistakes. The old joke is, "I thought I had made a mistake once, but I was wrong. My spouse has certainly believed so on several occasions." Charles Hirsch was known to say that he was "glad my job description doesn't contain the adjective infallible." When mistakes are made, they should be acknowledged and corrected. Initially, it may not be clear what happened and it may be best to describe them as "cases" and not mistakes.

Public speaking: case presentations, lectures, and teaching

Medical education has awakened to the realization that trauma deserves and demands serious study, and forensic pathologists have important contributions to make to this educational effort.[31] ***Dr. Charles S. Hirsch***

Case presentations

The level of detail best used as part of in-agency case presentations should be inversely proportionate to that in our reports. The former is focused on pertinent negatives and positives, and the other is exhaustive.

Although the tendency is to face and speak to the chief or deputy chief, you must remember that a case conference includes everyone in the room. Make believe you're talking to a jury in court (case presentations are excellent practice for this): make each person around the table feel as if you're speaking directly to them individually, make roving eye contact, speak up, and be professional. Avoid reading from your notes; be familiar with your cases before you walk into the room.

Think about your cases before you present. Anticipate the "what if" questions. Anticipate relevant questions and answer them before they are asked. Start with the obvious things your audience is wondering about; don't leave them in suspense. If you have just described the sudden death of a morbidly obese 36-year-old woman who experienced precipitous dyspnea prior to a cardiac arrest, the first thing you should be communicating is not the length of her hair or the presence of gallstones, but rather the presence or absence of a pulmonary thromboembolism!

In your preparation, as in your certifications, it is important and helpful in your focus to start with the mechanism of death. You should never leave the autopsy table without doing so, and certainly should not present a case without having considered and landed on a mechanistic theory. If you fail to share your conclusion or current theory, you should not be surprised by that question. It is one of the most important and predictable questions you will be asked. Not asking it before determining cause and manner puts the cart well ahead of the horse.

Case presentations should be succinct. Start with the age and sex of the decedent (25-year-old *man*, not male). Give a brief synopsis of the circumstances, followed by the pertinent negative and positive external and internal examination findings. In trauma deaths, summarize the overall picture, rather than presenting the granular details that would be found in an autopsy report, with measurements and tracks. Where there is extensive and expected trauma, from a descent from height, for example, you would be well-advised to make a broader statement, as opposed to a comprehensive cataloging of every mark on the body. Something like the following would suffice: "There was extensive/devastating blunt force trauma of the body with multiple fractures and lacerations of nearly every organ." If, on the other hand, a particular injury is unusual, dramatic, vexing, or otherwise worth discussing, by all means, share it in more detail and/or ask the group questions about it.

State the cause and manner of death, if determined, or, alternatively, your theory/differential diagnoses for cause of death and your focused plan to sort it out, if pending further study. If you certified a mechanical injury, share the wording of your "how injury occurred" section. If a student or resident plans to present one of your cases, take the time to prepare them by reviewing with them some of the above concepts.

Formal lectures

The brain can absorb no more than the seat can endure. -*Dr. Charles S. Hirsch*

Not surprisingly, PowerPoint-type presentations are the mainstay of forensic pathology lectures and presentations. They are even occasionally used in court. Therefore, familiarity with this software, as well as virtual presentation platforms, is key. Font size for titles should be between 36–44 and 20–32 for the text. Avoid using more than one or two different font types. For the introduction, it can be quite effective to open with a "hook," such as a quote, a question, a dramatic statement, or a scenario ("Imagine ..."). It is best to then preview key points and let the audience know what is in it for *them*. The body of the talk should aim to engage the audience roughly every 10 min, using a variety of techniques such as questions, sharing a story, polls, etc. The conclusion should summarize the key points, restate the benefits to them, an action step/key takeaway, and then present a closing hook. Graphic images may have important implications for a given presentation, but keeping such images frozen and projected on the screen for a prolonged period, while continuing to talk, serves as a sometimes uncomfortable and unnecessary distraction, even in a room filled with forensic pathologists. Similarly, efforts at deidentification of the decedents are a basic and reasonable expectation for any formal presentation.

Teaching

There are a wide range of professionals and medical students and residents who rotate at ME/C offices. These may include EMT students, pathologists' assistant students, medical students, and pathology residents. Due to the various levels of medical knowledge, it is helpful to tailor your teaching to the visitor. EMT students are particularly interested in the anatomy of the neck ("here is the cricoid cartilage"); it is helpful to clarify the concepts surrounding terminal aspiration versus choking.

Medical students will like everything—a review of anatomy always works. Some worthwhile topics include:

- Explaining cause versus manner (ask them how to fill out the death certificate on the case)
- The differences between and definitions of hypertensive and atherosclerotic heart disease; the definition of cardiac hypertrophy
- The differences between blunt and sharp injuries
- The power of vitreous chemistry
- Toxicology testing (the value of a peripheral blood sample, a sample that supports an acute intoxication, etc.)
- In situ/organ by organ evisceration and dissection tips
- Cutting through rib cartilage, not bone.
- The cardiac conduction system.
- Spinal cord removal.
- Stripping of the dura, pleura, and psoas muscle to look for fractures.
- Dissection and microscopic sections in pulmonary embolism cases.
- Removing the anterior neck structures after the brain and thoracic organs have been removed.

During a fellow's training year, a lot of time is spent in the autopsy room, working up cases and, hopefully, taking advantage of the opportunity to observe interesting findings in other cases. In trying to complete one's own cases, it is easy to miss out on these other examinations. It also is easy to understand why one would not be concerned with anything else but the matter "at your own hands." Being prepared to give lectures, sometimes at the spur of the moment, may feel like even less of a priority.

But if a fellow waits for a less harried time (i.e., after the fellowship year) to start collecting images as teaching material, one will have passed up a golden opportunity. The FP training year is a singular time, during which the fellow has exposure to so many cases. Between their assigned cases and those of colleagues, they will have seen enough examples of injuries and disease to establish a solid collection. During one's fellowship year, one will be exposed to many typical, as well as atypical forensic injuries, and the basic concepts they represent. If documented photographically, the training year can provide the fellow with an excellent start to a comprehensive personal collection. An organized photo collection will be of tremendous help for future talks and presentations.

Written communication

Memorialize that conversation on paper. *-Dr. Charles S. Hirsch*

While well-cultivated observational and investigative acumen yield the most accurate and comprehensive fact gathering and form the basis for asserting sound forensic opinions, the momentum of their benefit is slowed or nearly halted by poor execution in translating and organizing those observations and opinions into the words that comprise a clear and communicative autopsy report. The same can be said of death certificates (covered in Chapter 3). A bad report is only marginally helpful.

At the center of the transition from hospital to forensic autopsy practice is the synthesis of findings into a written narrative that reflects the opinions for which we must take increasing ownership. A natural extension of this transition then is the development of more declaratively stated observations and conclusions and a resistance to hedge or hide behind disclaimers and the passive voice. In essence, this represents a shift toward getting comfortable with the notion that a defense of opinions will be a focal point of this particular chosen career and practice.

In terms of enhancing our written communication, the best forensic autopsy report is predicated on equal parts high quality content and deliberate organization. One page (sometimes two), first or last, should be dedicated to the synthesis of findings, anatomic and historical, organized in decreasing order of importance to reflect the FP's specific impressions of the case, as well as of the final cause and manner of death determination. This final presentation of information (**the Final Diagnoses**) requires thoughtful analysis, mechanistic thinking, and sound forensic knowledge, representing **"the editorial"** segment of the report. In general, the main headings on this page (often Roman numerals) should reflect primary diseases or injuries that are typically etiologically specific, with nonetiologically specific subheadings and sometimes subsubheadings that reflect their anatomic and/or clinical sequelae. The inverse organization is more typical of the death certificate, where the immediate causes and mechanisms (sequelae), if cited, precede the proximate/underlying cause (i.e., the "bottom line"), the former, however, not being requisite in death certification, per se (see Chapter 3 for examples). With regard to the latter, clear, deliberate, and cogent cause of death statement wording is of the utmost importance; word choice seemingly as minor as "of" versus "to," "including," "with," or "following" may have specific relevance to stakeholders and potential future analytic and adjudicative issues and to our very framing of the case, depending on the details.

In terms of the rest of the report (**the body**), its primary purpose is to establish clear and thorough *documentation*, with emphasis on "clear and thorough"! The body of the report constitutes **"the news,"** the *objective* documentation of findings that support the ultimate synthesis and conclusions represented by the "Final Diagnoses." The objective, comprehensive, and descriptive features of this part of the report are critical for the forensic pathologist in training to grasp early and ultimately master in their practice. It also allows third parties to review the report, along with associated imaging, such that independent conclusions may be drawn, forming the basis of both internal quality assurance efforts and our ability to testify in courts of law in lieu of the original medical examiner, when they are unavailable. In his now famous (and timeless) treatise, "Classical Mistakes in Forensic Pathology," Dr. Alan Moritz cautioned the following,

> It is as surprising as it is distressing to note how frequently pathologists include statements of opinion and interpretation in the part of the protocol that is supposed to be objective and factual. The purpose of the protocol is twofold. One is to record a sufficiently detailed, factual, and non-interpretive description of the observed conditions, in order that a competent reader may form his own opinions in regard to the significance of the changes that were observed and described. The other is to interpret the significance of the changes described.[28]

Put more plainly, Dr. Hirsch simply suggested, "Describe what you see" *vis a vis* the **body** of the autopsy report. See also Appendices A–C.

On occasion, an inadvertent written error, substantial or minor, will be discovered in an autopsy report at a later time, sometimes even in court. The best way to respond to this discovery, whether in your own report or that of another, is to correct it (if before adjudication), and/or own it and humbly apologize for the mistake, if discovered too late, by a defense attorney, for example. Dr. Hirsch was fond of saying that you will never be fired for making a mistake, but you will be in deep trouble for trying to cover it up: "I'm glad my job description doesn't contain the adjective 'infallible'," he would say. Human error is inevitable, but when discovered, it must be corrected and/or recognized on record. Full disclosure is always the rule. Admitting that you're human will go a long way in maintaining your credibility.

While it is important to choose the most accurate and descriptive words and strike a balance in the amount of illustrative detail used to succinctly characterize findings, the overall report organization and sentence syntax can distinguish an excellent and highly communicative document from one that confuses more than it clarifies. The devil, as they say, is in the details. See Table 16.2.

TABLE 16.2 Terminology tips.

Penetrating: Enters, but does not exit
Perforating: Enters and exits (organ or body region).
Tract: System of body parts that serves a combined anatomic purpose (e.g., gastrointestinal tract)
Track: Detectable evidence that something has passed; the course along which something has moved (the path of a bullet is a "track," not a "tract").
"On" versus **"of"**: "Bracelet *on* the wrist" versus "tattoo *of* the forearm"; **"to"** versus **"of"**: ("Gunshot wound *of* the torso and neck"; "gunshot *to* the left buttock"); see Chapter 9
Hanged: For a suicidal hanging, the past tense of hang is hanged (pictures are hung).
Rupture: Use only with natural disease processes (e.g., ruptured cerebral artery aneurysm)
Use "slight," "moderate," "marked" (avoid: "mild," "significant," "massive").
Use "recent," "organizing," and "remote" to date a pathologic finding (e.g., an acute myocardial infarction).
Use "approximately" for volumes, unless measured.
Use "dead by neurological criteria" (do not use "brain dead").
Use "cerebral infarct or hemorrhage" or "cerebral vascular event" (rather than "cerebral vascular accident").
Use "stenosis" to describe the percent narrowing of blood vessels and "occlusion" for complete blockage.
The phrases "are present," "in shape," "in color," and "evidence of" are often unnecessary and/or redundant, with some exceptions.
Format and sequence for locations of findings/modifiers: surface, side (R or L), body part. Example: "anterior left forearm"
The word "ANAMNESTIC" is used commonly in the final diagnoses outlines when referring to a diagnosis listed as historical/by history, not by autopsy, per se. "Diabetes mellitus" or "schizophrenia" or other psychiatric diagnoses fall into this category.

Two reports documenting the same careful attention to the findings might communicate them in vastly disparate ways, depending on the construction of the report and the quality of the writing. Clumsily described, overly wordy, and poorly organized injury descriptions, with ill-conceived laundry lists masquerading as final diagnoses, often create a "more is less" scenario, leading the reader, particularly a nonforensic one, into a morass of confusing detail that may be evocative of little. Ideally, the FP should be able to review a thorough, but succinct description of an injury or natural finding, close their eyes, and then see it clearly on the decedent's body, in size, morphology, and location, the image in their mind's eye nearly identically matching the one that will ultimately be confirmed with a look at the autopsy photographs. This should be a central writing goal for trainees.

The specific injury distribution pattern characterized, for example, by 10 abrasions and a brow laceration isolated to the prominences of one side of the body, most likely representing a single impact of the body to a flat, resistant surface (planar injury) during a terminal fall, will be completely lost on the reader if the granular detail of the 10 independent abrasions are listed without further describing and highlighting their relationship to one another, of describing "the company they keep." Going one step further and carefully describing a specific and well-recognized injury distribution pattern (i.e., planar injury) will convey an interpretation of what is being observed. Both approaches are comprehensive in their documentation of the actual injuries, but one totally ignores the contextual pattern, or at least fails to communicate that pattern, therefore missing the boat entirely in highlighting the mechanics of these injuries as a whole, instead, giving the impression to an inexperienced reader that there is perhaps a concerning multiplicity of wounds (i.e., multiple impact sites).

In this vein, **the approach to the trauma autopsy report** *requires a specific and well thought out approach to best communicate injuries.* As such, it is imperative to have a systematic approach in both the content (e.g., a mental checklist of what to include in the description of all firearms wounds) and in the reproducible organization of these wound descriptions (standardized headings, describing external followed by internal findings, organizing the findings by body region, and including standard features by wound type, with pertinent positive and/or negative findings, for example).

Appendices A-C of this book include autopsy report suggestions for the content and organization of common final diagnoses lists (Appendix A), a basic autopsy organizational format (Appendix B), and general autopsy templates (adult and pediatric), as well as specific templates for various injury modalities (Appendix C). As is the case for all templates, it is important to point out that they serve only as scaffolding, a start point from which the report should be tailored to the specifics of the case and the injuries involved, rather than as a rigid dictum to which an adherence may well communicate less. Templates are helpful in serving as mental reminders of what to look for and document (pertinent positives and negatives), how to best organize, and how to create a reproducible report structure for training and quality assurance purposes, but they are not and should not be a substitute for common sense and the careful documentation of unique or unusual findings. Never are they intended to encourage a mindless recitation or "canned" text. "Routine is never a substitute for judgment."

Tips on crafting a forensic autopsy report

1. *Avoid using the passive voice:* It may convey an unnecessary and vague uncertainty and/or unwillingness to commit. While sometimes necessary, it is more often a somewhat cumbersome and less communicative writing style. Instead, try to get in the habit of making concise, *declarative* statements wherever possible.

Avoid: "The gallbladder is without gallstones." Instead, write: "There are no gallstones." Or, "The gallbladder has no gallstones."

2. *Avoid wordiness:* One usually does not need to use phrases such as "is seen," "in color," "in shape," etc. Attempt to be direct and concise.
3. *Avoid redundancy:* Examples: "posterior back," "nape of the neck."
4. *Don't be afraid to use a string of adjectives that modify (and precede) a noun/finding:* Avoid placing these descriptors after the fact in parentheses or in a dependent clause; this creates choppiness that potentially interferes with flow and reading comprehension.
5. *Try to avoid the monotony of a paragraph composed of a string several short sentences that begin with "There is ..."* Instead, where logical, attempt to consolidate a series of related findings (e.g., abrasions/contusions), separated by commas, in a clear sentence that flows. For example, instead of several consecutive "there is" sentences, one might write: "There are three 1/2–3/4" irregular, dried abrasions of the lateral left knee, 1/4"–3/8" focally abraded contusions of the right orbital rim, lateral right zygoma, and right chin, and a 6" × 4", vertically oriented, brush-type abrasion of the left scapular back."
6. If you are unsure about what you are seeing, describe it as clearly and objectively as possible, *without* labeling/naming it; but, if you *do* know what you're looking at, so name it, followed by its description—avoiding unnecessary hedging (e.g., if you know you've identified a contusion, write "a 3/4", roughly circular, purple-red *contusion*," not "a 3/4", roughly circular, purple-red *discoloration*").

If you are clearly looking at a stab wound and certifying the death as "Stab Wound of Chest", then, in the body of the report, identify the putative wound clearly as a "stab wound," followed by a description of its features. Describing it as a "slit-like defect," without further qualification, despite knowing what it is, is both illogical and confusing, given the opinion offered in the cause of death. Once again, we must not "deny reality for the sake of objectivity." It is best to name it, specifically, etiologically, if you know what it is, *then* go on to describe it.

When a finding being described reflects an historical event or therapy, as opposed to a diagnosis, a parenthetical factual comment is often helpful and clarifying when placed after the description, such as in, "there are symmetrical, triangular 0.5 × 0.3 cm ulcers of the posterior true vocal cords and submucosa (comment: long-term endotracheal intubation)." This leaves nothing to the imagination and prevents the reader from remaining in a sustained state of suspense.

That said, when a finding is equivocal/unclear/modified by artifacts, etc., one might use such language as "consistent with" or "apparent" or "discernible," to intentionally communicate that uncertainty or ambiguity, and/or use only descriptive language without further offering a diagnosis.

Two apt "Hirschisms" come to mind for such a situation: "Just describe what you see" and "Sometimes, ambiguity is your best friend." While it's very important to take ownership of the things we *can* confidently diagnose, it is equally important, and often unavoidable, to occasionally describe findings, but not name them, recognizing our limitations when we are unsure of what we are observing.

7. We suggest "slight, moderate, marked" rather than "mild, significant, massive" ("Salsa is mild; atherosclerosis is slight"–Charles S. Hirsch). We also suggest "yellow metal" rather than "gold" and "white metal" rather than "silver."
8. For consistency and to avoid confusion for any lawyers/MEs, etc. reading the report at a later time, we advise making an effort to ensure that the letters/numbers that are designated for wounds on the autopsy diagram match the letter/number headings of the same wounds described in the body of the report.
9. *General guidelines for constructing Final Diagnoses lists:* These are the final bullet point diagnoses, followed by associated findings/sequelae. Lengthier and more detailed anatomic descriptions are more appropriately placed in the body of the report, however.

In general, use **Roman numeral (or lettered) headings to organize etiologically specific findings: injuries** grouped by body region in trauma cases and by **disease categories** in natural deaths, followed by their associated sequelae (subheadings), and sub-subheadings, where relevant. Aim to list the etiologically specific **diagnoses in decreasing order of pathophysiologic importance** (common examples of findings not relevant to death, but that should be found "lower down" on the diagnoses list often include such entities as benign prostatic hypertrophy, diverticulosis, uterine leiomyomata, and status post remote appendectomy). Anatomic findings are usually listed before historical or clinical ("anamnestic"/by history) diagnoses, unless the latter are primary or contributory causes or are otherwise important to highlight. Avoid *exhaustive* scene information in the Final Diagnoses lists.

10. *Write the body of the report in paragraphs, not in phrases, bullet points, lists, or outlines.* Outlines are reserved for the consolidation of conclusive information on the first (or last) page of the report (the "Final Diagnoses" list). Admittedly, certain lengthier categories in the body of the report (e.g., clothing, tattoos) lend themselves nicely to itemized lists and may be easier to read in that format. "Let common sense prevail."
11. *All injuries and their direct sequelae, external and internal, are best separated from the rest of the nontraumatic findings and therapies and systematically described* (external findings first, followed by internal) under a single "Injury" heading, further organized by injury modality, then further by body region (head, torso/trunk, and extremities), followed with a statement at the end of this section, such as, *"These injuries, having been described, will not be repeated,"*

    Avoid "mixing metaphors" such as, placing postmortem changes under the general external or internal examination sections, etc. (see **Appendix B** for suggested report organization). For example, if the aorta is appropriately described as having a red-orange intima under the "postmortem changes" section, but is otherwise unremarkable, it should be described as "normal" or "unremarkable" under the "internal examination" section, the postmortem discoloration notwithstanding (and already described). The inference with this framing is that the aorta is *otherwise* unremarkable. All noninjurious headings, including "Therapeutic Interventions," provide a space for clear descriptions of both objective observations and pathologic processes that are not directly caused by or otherwise resulting from the injuries or postmortem changes.
12. *Check all names and dates in the report:* Check the name against the officially identified one, especially in homicides, where aliases are occasionally encountered.
13. **With the extremities**, it is best to speak in terms of **proximal and distal**, not superior and inferior, as they can move in space and make the latter designation uninterpretable. If you lift your arm or leg up, what had been "inferior" may now be "superior." By contrast, proximal and distal never change with movement because they are terms that are used in relation to the fixed torso. Do not measure extremity wounds from the midline for the same reason; it's neither relevant nor reliable as a measurement, and doing so may well misrepresent the location of the limb relative to the body's midline at the moment of injury occurrence. For the supinating and pronating **forearm**, we suggest designating **dorsal/volar** and **ulnar/radial** as landmark descriptors. For the **arm** (upper arm), **lateral and medial** suffice, as they are fixed (no supination/pronation). Landmarks such as the axilla, the antecubital fossa, the elbow, triceps/biceps/deltoid regions, the groins, the knees, and the ankles may be additionally helpful visual points of reference when describing findings of the extremities. Finally, palms and soles should be specified.

With respect to the order of descriptors, we suggest the following sequence: *vertical location on the body, surface (anterior/volar, lateral, posterior/dorsal, or medial), side (right or left), then body part.* Example: Distal lateral right thigh.

14. The word "Anamnestic" (or equivalent, such as "by history," "clinical," etc.) can be used in the final diagnoses outlines when referring to a diagnosis listed as historical/by history, not by autopsy, per se. "Diabetes mellitus" or "schizophrenia," or other psychiatric diagnoses, fall into this category.
15. *Attempt to remain nimble in report writing:* Create extra headings and qualifying statements where needed (funeral home activity, autopsy modified for religious purposes/minimally invasive autopsy, organ and tissue donation/procurement, and previous hospital autopsies). This is where thinking outside of the template is best illustrated (see also Chapter 5, The Forensic Autopsy, for suggested additional headings and phraseology).
16. While more a matter of style than substance, if a complex cause of death statement is particularly long and cumbersome, by necessity and despite one's best efforts, it may well require several "due to's" and become, at best, monotonous to read, but, at worst, confusing. *Keep in mind that there are alternatives to the standard "due to" parlance on death certificates. Phrases such as "related to," "complicating," or "following" offer just a few examples.*

Beyond the communicative information contained in autopsy reports and death certificates, written correspondence and documentative case notes require similar thoughtfulness and succinctness, keeping in mind that all are potentially discoverable (i.e., for legal purposes). Memorializing important conversations in a case note, particularly those that include relevant additional facts or processes relating to the case, may prove crucial for any number of reasons down the line. The forensic pathologist will field thousands of phone calls throughout the course of a career; remembering the details of all represents a virtual impossibility. It is for this reason that documentation is so important. While the adage, "if it's not written down, it didn't happen" isn't actually true, the limitations of the human memory can make it seem that way. That said, and importantly, substantive email discussions regarding the details and theories of a death, particularly a homicide, should be expressly avoided for a variety of reasons, not the least of which is the time-consuming burden of having to later locate and retrieve all of those emails for discovery requests. While testimony and meeting scheduling is a perfectly innocuous topic to communicate through email, a phone call or an in-person meeting is a much more advisable communication forum for all other case discussions. Finally, and needless to say, professionalism in all emails and public posts is requisite and expected, but this is not unique to the practice of forensic pathology. Let common sense prevail.

References

1. Adelson L. The forensic pathologist. "Family physician" to the bereaved. *JAMA*. 1977;237(15):1585–1588.
2. Hirsch CS. Talking to the family after an autopsy. *Arch Pathol Lab Med*. 1984;108(6):513–514.
3. Prigerson HG, Horowitz MJ, Jacobs SC, et al. Prolonged grief disorder: psychometric validation of criteria proposed for DSM-V and ICD-11. *PLoS Med*. 2009;6(8):e1000121.
4. Jordan AH, Litz BT. Prolonged grief disorder: diagnostic, assessment, and treatment considerations. *Prof Psychol Res Pract*. 2014;45(3):180.
5. Maercker A, Lalor J. Diagnostic and clinical considerations in prolonged grief disorder. *Dialogues Clin Neurosci*. 2012;14(2):167.
6. Maciejewski PK, Maercker A, Boelen PA, Prigerson HG. "Prolonged grief disorder" and "persistent complex bereavement disorder", but not "complicated grief", are one and the same diagnostic entity: an analysis of data from the Yale Bereavement Study. *World Psychiatr*. 2016;15(3):266–275.
7. Haglund WD, Reay DT, Fligner CL. Death notification. *Am J Forensic Med Pathol*. 1990;11(4):342–347.
8. Kõlves K, Zhao Q, Ross V, Hawgood J, Spence SH, De Leo D. Suicide and other sudden death bereavement of immediate family members: an analysis of grief reactions six-months after death. *J Affect Disord*. 2019;243:96–102.
9. United States Code. *Edition, Supplement 4, Title 28—JUDICIARY AND JUDICIAL PROCEDURE*. 2006.
10. Adelson L, Dewitt C, Gerber SR, Moritz AR. *Physician in the Courtroom*. Press of Western Reserve University; 1957.
11. Brodsky S, Neal T, Cramer R, Ziemke M. Credibility in the courtroom: how likeable should an expert witness be? *J Am Acad Psychiatry Law*. 2009;37:525–532.
12. Cramer R, Brodsky S, DeCoster J. Expert witness confidence and juror personality: their impact on credibility and persuasion in the courtroom. *J Am Acad Psychiatry Law*. 2009;37:63–74.
13. Neal T, Guadagno R, Eno C, Brodsky S. Warmth and competence on the witness stand: implications for the credibility of male and female expert witnesses. *J Am Acad Psychiatry Law*. 2012;40:488–497.
14. Brodsky SL, Dvoskin JA, Neal TMS. Temptations for the expert witness. *J Am Acad Psychiatry Law*. 2017;45(4):460–463.
15. Parrott CT, Neal TM, Wilson JK, Brodsky SL. Differences in expert witness knowledge: do mock jurors notice and does it matter? *J Am Acad Psychiatry Law*. 2015;43(1):69–81.
16. Nagle JE, Brodsky SL, Weeter K. Gender, smiling, and witness credibility in actual trials. *Behav Sci Law*. 2014;32(2):195–206.
17. Brodsky SL, Griffin MP, Cramer RJ. The witness credibility scale: an outcome measure for expert witness research. *Behav Sci Law*. 2010;28(6):892–907.
18. Cramer RJ, Neal TM, DeCoster J, Brodsky SL. Witness self-efficacy: development and validation of the construct. *Behav Sci Law*. 2010;28(6):784–800.
19. Gill J. The dilemma of surrogate testimony for the forensic pathologist. *Acad Forensic Pathol*. 2012;2(2):284–289.
20. Levy V, Rao VJ. Survival time in gunshot and stab wound victims. *Am J Forensic Med Pathol*. 1988;9(3):215–217.

21. Gonsoulin M, Barnard JJ, Prahlow JA. Death resulting from ruptured cerebral artery aneurysm: 219 cases. *Am J Forensic Med Pathol*. 2002;23(1): 5–14.
22. Spitz W, Petty C, Fisher R. Physical Activity until collapse following fatal injury by firearms and sharp pointed weapons. *J Forensic Sci*. 1961; 6(3):290–300.
23. Gilbert JD, Jensen L, Byard RW. Further observations on the speed of death in hanging. *J Forensic Sci*. 2008;53(5):1204–1205.
24. Melendez-Diaz v. Massachusetts, 557 U.S. U.S. 305(Supreme Court of the United States 2009).
25. Bullcoming v. New Mexico, 564 U.S. (Supreme Court of the United States 2011).
26. Williams v. Illinois, 567 U.S. U.S. (Supreme Court of the United States 2012).
27. Adams VI. *Guidelines for Reports by Autopsy Pathologists*. New Jersey: Humana Press; 2008.
28. Moritz AR. Classical mistakes in forensic path. *Am J Clin Pathol*. 1956;26(12):1383–1397.
29. Perper JA, Juste GM, Schueler HE, Motte RW, Cina SJ. Suggested guidelines for the management of high-profile fatality cases. *Arch Pathol Lab Med*. 2008;132(10):1630–1634.
30. Kelly CC. Dealing with the news media: effective communication strategies for medical examiners. *Am J Forensic Med Pathol*. 1998;19(2): 181–185.
31. Hirsch C. Anniversary discourse: the office of chief medical examiner. *Bull NY Acad Med*. 1989;65:822–827.

CHAPTER

17

The forensic pathologist's public health role

Erin G. Brooks

Any government with an interest in the lives of its citizens must have a compelling interest in their deaths. -***Drs. Charles S. Hirsch and Ross Zumwalt***

Introduction

While historically some have viewed the work of medical examiners/coroners (ME/C) and death investigators as predominantly serving the criminal justice system, in fact, the act of certifying a death is—in and of itself—a public health endeavor.[1] At its most fundamental level, the purpose of the death certificate is to track how and why individuals die as a means of helping others survive.[2] Death certificates do save lives. In addition to generating these vital statistics, which guide public health policy, ME/C also perform a public health service in a variety of other ways. For instance, autopsy pathologists report unsafe consumer products[3–5] (e.g., the association of infant suffocation to soft sleeping environments, such as commercial loungers), warn of recreational hazards[6–8] (e.g., the dangers of snowmobile or all-terrain vehicle use), publicize and report occupational hazards[9,10] (e.g., falls and electrocution risks), identify toxic exposures (e.g., carbon monoxide from defective heating units), and evaluate the safety of medical therapies. At autopsy, ME/C perform infectious disease surveillance, diagnosing and reporting communicable diseases such as bacterial meningitis, tuberculosis, and Legionella, thereby helping to mitigate disease spread in the community.[1,11–14] They identify and characterize emerging infectious threats such as COVID-19.[15,16] They track violent deaths, including homicides, suicides, accidents, child and domestic abuse deaths, maternal mortality, and overdose deaths—to name a few—and participate as members of death review teams (DRTs) attempting to identify systemic issues and prevent further such deaths.[17] In one way or another (i.e., at the county, regional, or state level), ME/C are employed by the government, and maintaining public health is a basic governmental responsibility.[1]

Background

In the United States, ME/C jurisdictions are defined geographically, which makes them ideal for population-based surveillance, as infectious disease outbreaks and violent death trends often begin locally, prior to spreading.[1,11] Additionally, death investigations performed by ME/C offices typically begin as soon as death notification occurs, providing for timely tracking of causes of death. Finally, in the process of death investigation, ME/C offices compile substantial data that is of public health interest, including scene investigative findings, reports of interviews, medical records, and ancillary testing (e.g., toxicology, microbiology, and radiology) results.[1,11,18] These data are distilled into a final autopsy report that contains a succinct cause and manner of death statement. In approximately 50% of cases, the manner of death certified by ME/C will be natural—with the remainder of cases representing some combination of unnatural manners (e.g., suicide, homicide, and accident.)[11] This percentage, however, may vary depending upon the jurisdiction.

The proportion of medical examiner cases having some relation to public health is 100%. If nothing else, ME/C issue a death certificate with an etiologically specific underlying cause of death. This properly completed death certificate is an extremely important document for the decedent's family. It allows for funeral arrangements to proceed, the collection of life insurance, and the settlement of estates. Beyond the family, vital statistical data from death certificates help to define public health priorities and allocation of grant funds.

Principles of Forensic Pathology
https://doi.org/10.1016/B978-0-323-91796-4.00023-4

While statewide medical examiner systems arguably provide for the most reliable and efficient public health surveillance, only 25 states are currently organized in this manner.[19] In the remainder, ME/C death investigative jurisdictions are divided along county, district, or parish lines;[19] such jurisdictions operate largely independently of each other—including their performance of mortality surveillance. As of 2006, there were roughly 2185 jurisdictions in the United States.[1] Each jurisdiction is overseen by either a coroner or medical examiner. Typically, coroners are elected officials who are not required to be physicians, whereas medical examiners are appointed by a governing authority and are physicians; however, there are exceptions to these rules on a state-by-state basis.[1,11,18,19] While there are far more coroner than medical examiner jurisdictions, more than half of the US population reside in medical examiner jurisdictions, as these tend to represent more densely populated urban areas.[1] ME/C offices are charged with investigating sudden, unexpected, and unnatural deaths; overall, ME/C's certify approximately 20% of US deaths.[1,18]

Passive surveillance

Data generated on death certificates and autopsy reports are utilized by a variety of governmental public health agencies to assess trends and determine where and how to best allocate public health resources.[1,11,18] There are a variety of well-characterized surveillance programs, in addition to various independent research groups that rely upon ME/C data. These are considered "passive" public health surveillance systems, meaning ME/C are not responsible for actively reporting information directly to the system. Three of the best characterized and longer standing passive surveillance programs include those tracking US vehicular fatalities, occupational fatalities, and violent deaths.

Vehicular fatalities

In 2020, there were 38,824 fatalities due to US traffic collisions, the highest number of deaths since 2007 (Fig. 17.1).[20] The US Department of Transportation maintains a Fatality Analysis Reporting System (FARS), which tracks roadway fatalities nationwide and annually releases a summary report.[21] A crash fatality is defined as a motor vehicle crash occurring while traveling on a public traffic roadway and that results in death within 30 days of the crash.[1] The FARS database was established in 1975 and is searchable online.[21] The intent of FARS is to "provide an overall measure of highway safety, to help identify traffic safety problems, to suggest solutions, and to help provide an objective basis to evaluate the effectiveness of motor vehicle safety standards and highway safety programs."[22] The location of the collision, intoxication status of the driver, road conditions, vehicle type, and many other findings can all be used to focus attention on ways to decrease future motor vehicle deaths.

FIGURE 17.1 **Motor vehicle collision.**

Occupational fatalities

Beginning in 1992, the US Bureau of Labor began publishing an annual Census of Fatal Occupational Injuries (CFOI) in an effort to help improve workplace safety.[23] The census includes traumatic work-related fatalities that occurred in the reference year, regardless of when the initial workplace traumatic injury occurred.[24] The CFOI database is searchable online. The National Institute for Occupational Safety and Health (NIOSH) likewise incorporates traumatic occupational fatality data in their Traumatic Injury Prevention Program.[25] Note that nontraumatic occupational illnesses are tracked annually via the Survey of Occupational Injuries and Illnesses (SOII) data by the US Bureau of Labor, but this survey only considers nonfatal cases.[26] ME/C may work with the Occupational Safety and Health Administration (OSHA) in certain workplace deaths. The standard US death certificate has a checkbox to designate injuries that have occurred at work, whether work-related per se or not (see Chapter 3 for definitional criteria).

Violent deaths

On an annual basis, roughly seven US residents die a violent death each hour. In 2019, there were approximately 47,500 deaths due to suicide and 19,100 due to homicide.[27] Violent deaths are all theoretically preventable, and the intent of the National Violent Death Reporting System (NVDRS), established in 2002 by the Centers for Disease Control and Prevention (CDC), is to utilize violent death data to help develop, implement, and evaluate violence prevention strategies.[28] While initially it collected data from only six states, currently NVDRS collates data from all 50 states, the District of Columbia, and Puerto Rico. A violent death is defined as one resulting from the "intentional use of physical force or power, threatened or actual, against oneself, another person, or a group or community."[27] Thus, suicides, homicides, certain in-custody deaths, and violent deaths where intent is unclear are all included.[27]

Another recently established surveillance system is the State Unintentional Drug Overdose Reporting System (SUDORS).[29] Instituted by the CDC in 2016, SUDORS provides a comprehensive review of drug intoxication deaths by abstracting data from ME/C autopsy reports, death certificates, and toxicology reports. The opioid crisis has only accelerated during the recent COVID-19 pandemic, with over 100,000 overdose deaths between April 2020 and 2021.[30] The intent of SUDORS is to provide timely drug intoxication data—including identification of emerging substance or polysubstance intoxication trends— and generally provide an improved understanding of the circumstances surrounding overdose fatalities in order to better prevent and respond to them.[29]

Active surveillance

In some instances, data cannot be easily extracted from death certificates and/or autopsy reports, and so the ME/C must directly report adverse events to sentinel surveillance systems. These would thus be considered "active" surveillance programs. Typically, ME/C reporting to such systems is strongly encouraged but not mandated. There are well-established active surveillance reporting systems for multiple case types that an ME/C might encounter, including cases of adverse medication/medical device events, adverse vaccination events, and unsafe consumer product events.

Adverse medication/medical device events

The US Food and Drug Administration (FDA) maintains MedWatch, an online site in which adverse events (including deaths) that arise as a result of medication/medical device use can be reported.[31] In addition to prescription and over-the-counter medications, adverse events related to dietary supplements, medical foods/beverages, biologics (e.g., blood components, blood/plasma derivatives, and gene therapies), and cosmetics may also be reported by the ME/C.[31] As a result of such reports, there may be modification in the product design, safety alerts issued, or even a recall of the product from the market in order to protect public safety.[32]

Adverse vaccination events

The CDC and FDA comanage the Vaccine Adverse Event Reporting System (VAERS), which was established in 1990 with the intent of assessing and improving vaccine safety.[33] ME/C are strongly encouraged to report potential vaccine-associated deaths to VAERS. Guidance regarding which vaccine-associated events are reportable, the time frame following the vaccination from which such events typically arise, and the particular vaccines these events have been associated with can be found on the VAERS site.[34] More recently, myocarditis and pericarditis have been reported as rare potential side effects following mRNA-based COVID-19 vaccination (i.e., Pfizer-BioNTech or Moderna vaccines).[35] It is important to point out that, in order to maximize the capture of events that may have time-sensitive public health importance, the putative events reported to the VAERS site represent a very large aggregate of reporting by a variety of people with disparate expertise, so are not necessarily confirmed vaccination-associated events. The vetting and analysis of these reports is conducted subsequently, with any trends having scientific veracity being later confirmed and publicly announced. The CDC and National Association of Medical Examiners (NAME) have collaborated on a recommended ME/C protocol for deaths occurring after COVID-19 vaccinations.[36]

Unsafe consumer product events

The Medical Examiners and Coroners Alert Project (MECAP) was established in 1976 by the US Consumer Product Safety Commission (CPSC) as a quick way for ME/C to report deaths in which a consumer product appears to have played a significant role in causing the death.[37] The CPSC is charged with protecting the public from injury/death resulting from products posing a fire, chemical, or mechanical hazard, and/or that can injure children—and ME/C reports are integral to this mission. Among previous issues brought to light by MECAP reports were the hazards of infants drowning in five gallon buckets, infant suffocations due to infant cushion loungers, and chain saw-related fatalities due to lack of adequate safety standards and devices.[37]

Death review teams

DRTs represent a more collaborative and multidisciplinary approach to fatality review and prevention.[11,17] Typically such teams are composed of death investigative professionals with a diverse array of expertise, such as that related to social services, law enforcement, medicolegal death investigation/forensic pathology, mental health, and other areas.[17] Review teams discuss circumstances leading up to and surrounding the deaths, which often consist of a complex series of events, with the goal of identifying steps that could have been taken to prevent the death and what systemic changes could be implemented to prevent other such deaths.[17] DRTs have historically been subdivided by age and demographic group, i.e., maternal mortality review committees,[38] fetal and infant mortality review,[39] child death review (CDR),[40] domestic violence fatality review,[41] elder abuse review,[42] and prison death review.[43] Certain categories of DRTs are more widespread than others. For instance, currently, CDR teams exist in all 50 states, active domestic violence fatality review teams in 42 states, and fetal and infant mortality review in 28 states.[39–41] Death reviews also occur at the level of local trauma centers, with the ME/C providing follow-up, by telephone or through formal morbidity and mortality conferences, with autopsy findings from trauma patients who had received medical/surgical treatment. This information may provide the trauma and other physicians with critical information that may benefit future patients.

Identification of missing and unidentified persons

The National Institute of Justice has characterized the numbers of active missing persons and unidentified remains as representing our "nation's silent mass disaster," with up to 100,000 missing persons and 40,000 collections of human remains waiting to be sorted out at any given time.[18,44] ME/C play a key role in this important public health endeavor. Via pursuit of positive identification through modalities such as fingerprinting, dental analysis, comparative radiologic analyses, DNA and anthropologic studies—as well as voluntarily contributing information to the National Missing and Unidentified Persons System (NamUs)—ME/C directly assist in the effort to match unidentified remains with missing persons.[18,45] NamUs is a free online system, established by the National Institute of

Justice, in which ME/C may enter information regarding unidentified remains. Families and law enforcement may also enter information on missing persons, and information may be entered regarding unclaimed persons who have been identified but lack family.[18,45] The database attempts to connect missing persons with unidentified remains. The other main database providing assistance in this endeavor is the National Crime Information Center (NCIC) Missing Person and Unidentified Person Files; however, this system is accessible only by law enforcement.[18,46]

Detection and characterization of infectious disease

One of the more traditional and better-publicized public health roles of ME/C has been their service as infectious disease sentinels. Not only do ME/C offices identify notifiable communicable diseases of public health interest but also they help detect and characterize novel emerging infections.[12–14] As a result of the rapidity with which some infections progress, diagnosis may not be made until the time of postmortem examination.[14] ME/C offices regularly report communicable diseases of public health interest to their respective state or local departments of health, who in turn notify the CDC. In this way, at-risk contacts of the infected decedent can be informed, and epidemiologic investigations attempting to mitigate disease spread can be initiated. Each state's department of health maintains an online listing of infectious diseases which ME/C are mandated to report, as well as the time frame within which such reports should be made. Certain bacterial meningitis deaths, namely when due to *Neisseria meningitidis*, represent an urgent public health matter in terms of the need to rapidly prophylax close contacts (Fig. 17.2). These communicable diseases typically parallel those designated as "notifiable" diseases/conditions by the CDC; currently, there are approximately 120 such diseases/conditions.[47] The CDC's Infectious Diseases Pathology Branch (IDPB) section offers advanced immunohistochemical and molecular testing for ME/C offices.

Autopsy tissue, serum, and/or swab samples have proven to be an effective means of performing infectious disease surveillance in a variety of communities—for instance, postmortem samples have been utilized for Nipah viral encephalitis surveillance in Malaysia and Bangladesh, dengue virus surveillance in Puerto Rico, and pediatric influenza mortality surveillance.[14] More recently, a protocol for SARS-CoV-2 decedent surveillance has been piloted in an effort to provide an improved understanding of asymptomatic/atypical disease presentations, severe disease outcomes, and to strengthen death certification (Fig. 17.3).[15] Some ME/C offices employ autopsy protocols with defined symptom criteria to facilitate a more systematic approach to postmortem infectious disease surveillance overall (e.g., the Medical Examiner Syndromic Surveillance System [Med-X] and Surveillance of Probable Infectious Etiology for Unexplained Death [UNEX]).[12,13]

As ME/C offices are charged with investigating sudden and unexpected deaths, novel emerging infections are very likely to fall under their jurisdiction. Among the emerging infectious outbreaks in which autopsy has played a key role in identifying and characterizing the etiologic agent are hantavirus pulmonary syndrome due to the Sin Nombre virus, West Nile viral encephalitis, and pulmonary hemorrhage due to leptospirosis.[12,14,48] Since the COVID-19 pandemic, autopsy studies have played a major role in helping to characterize the disease spread, which

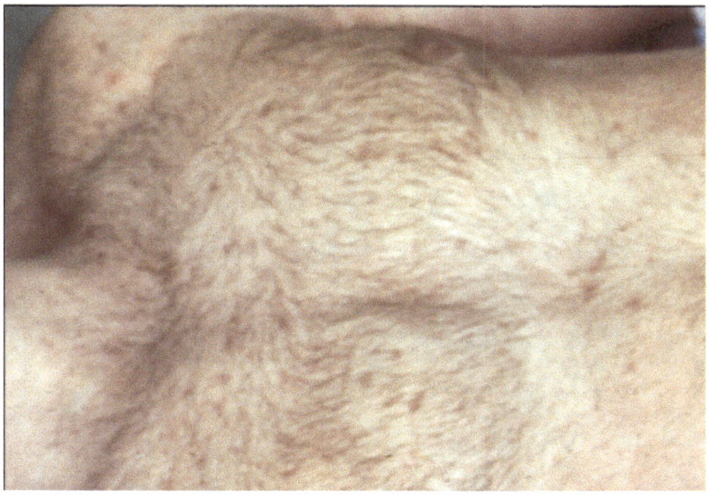

FIGURE 17.2 **Petechial rash of meningococcemia.**

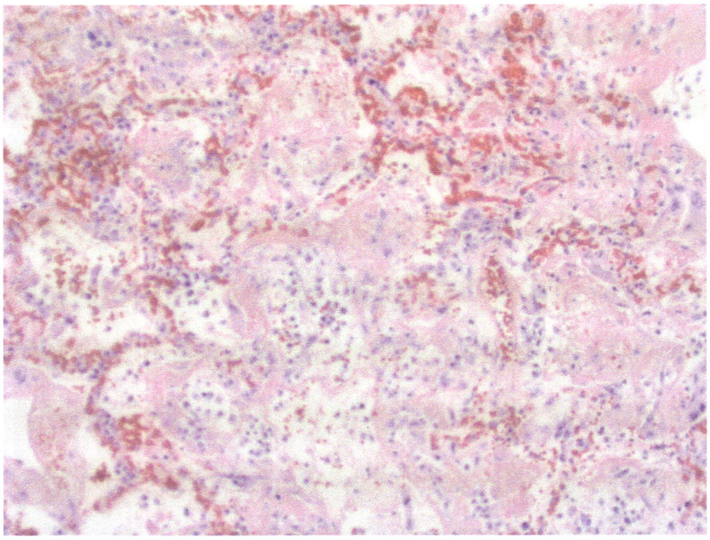

FIGURE 17.3 **Diffuse alveolar damage in COVID-19 death.**

individuals are particularly at risk of fatal outcomes, and what organ systems may be most affected.[16] Autopsy pathologists have procured and contributed tissue samples utilized by researchers to advance understanding of the pathogenesis of SARS-CoV-2. In addition, ME/C have played a major role in spearheading efforts to ensure that deaths attributed to COVID-19 are being accurately tracked and certified.[49]

Conclusion

ME/C have been, and continue to be, integrally involved in public health endeavors in the United States. Participation in DRTs, in multiple active and passive electronic mortality surveillance systems, in ongoing efforts to identify missing persons and unidentified remains, and in identifying and characterizing infectious disease are but a few of the myriad ways death investigative offices contribute to maintaining the public health.

References

1. Hanzlick R. Medical examiners, coroners, and public health: a review and update. *Arch Pathol Lab Med*. September 2006;130(9):1274–1282.
2. Schulz K. *The History of Final Forms. The New Yorker*; April 7, 2014. www.newyorker.com/magazine/2014/04/07/final-forms. [Accessed March 27, 2022].
3. Peachman RR. *While They Were Sleeping. Consumer Reports*; December 30, 2019. www.consumerreports.org/child-safety/while-they-were-sleeping/. [Accessed March 27, 2022].
4. United States Consumer Product Safety Commission. *The Boppy Company Recalls Over 3 Million Original Newborn Loungers, Boppy Preferred Newborn Loungers and Pottery Barn Kids Boppy Newborn Loungers After 8 Infant Deaths; Suffocation Risk*. 2021. [Accessed March 27, 2022].
5. Parks SE, Erck Lambert AB, Hauck FR, Cottengim CR, Faulkner M, Shapiro-Mendoza CK. Explaining sudden unexpected infant deaths, 2011–2017. *Pediatrics*. May 2021;147(5):e2020035873.
6. Lin PT, Blessing MM. The characteristics of all-terrain vehicle (ATV)-related deaths: a forensic autopsy data-based study. *Forensic Sci Med Pathol*. December 2018;14(4):509–514.
7. Hall AJ, Bixler D, Helmkamp JC, Kraner JC, Kaplan JA. Fatal all-terrain vehicle crashes: injury types and alcohol use. *Am J Prev Med*. April 2009;36(4):311–316.
8. Gustafsson T, Eriksson A. Off-road vehicle fatalities: a comparison of all-terrain vehicle and snowmobile accidents in Sweden. *IATSS Res*. July 2013;31(1):12–15.
9. Taylor AJ, McGwin Jr G, Valent F, Rue 3rd LW. Fatal occupational electrocutions in the United States. *Inj Prev*. 2002;8(4):306–312.
10. Copeland AR. Accidental death due to falls at work. *Am J Forensic Med Pathol*. March 1989;10(1):17–20.
11. Hanzlick R, Parrish RG. The role of medical examiners and coroners in public health surveillance and epidemiologic research. *Annu Rev Publ Health*. 1996;17:383–409.
12. Zaki S, Blau DM, Hughes JM, et al. CDC Grand Rounds: discovering new diseases via enhanced partnership between public health and pathology experts. *MMWR Morb Mortal Wkly Rep*. February 14, 2014;63(6):121–126.

13. Blau DM, Clark SC, Nolte KB. National association of medical examiners ad-hoc committee for bioterrorism and infectious diseases. Infectious disease surveillance by medical examiners and coroners. *Emerg Infect Dis.* 2013;19(5):821–822.
14. Singh VD, Lathrop SL. Role of the medical examiner in zika virus and other emerging infections. *Arch Pathol Lab Med.* January 2017;141(1): 82–84.
15. Brouwer AF, Myers JL, Martin ET, et al. Severe acute respiratory syndrome coronavirus 2 surveillance in decedents in a large, urban medical examiner's office. *Clin Infect Dis.* May 18, 2021;72(10):e580–e585.
16. Maiese A, Manetti AC, La Russa R, et al. Autopsy findings in COVID-19-related deaths: a literature review. *Forensic Sci Med Pathol.* 2021;17(2): 279–296.
17. *National Center for the Review and Prevention of Child Deaths at the Michigan Public Health Institute. The Coordination and Integration of Fatality Reviews: Improving Health and Safety Outcomes across the Life Course*; December 2011. www.ncfrp.org/wp-content/uploads/NCRPCD-Docs/CoordinatedReviews.pdf. [Accessed March 17, 2022].
18. Levy B. The need for informatics to support forensic pathology and death investigation. *J Pathol Inf.* June 23, 2015;6:32.
19. Centers for Disease Control and Prevention (CDC). Death Investigation Systems. www.cdc.gov/phlp/publications/coroner/death.html. [Accessed March 28, 2022].
20. US Department of Transportation. NHTSA Releases 2020 Traffic Crash Data. www.nhtsa.gov/press-releases/2020-traffic-crash-data-fatalities#:~:text=The%20U.S.%20Department%20of%20Transportation's,number%20of%20fatalities%20since%202007. [Accessed March 28, 2022].
21. US Department of Transportation. Fatality Analysis Reporting System (FARS). www.nhtsa.gov/research-data/fatality-analysis-reporting-system-fars. [Accessed March 28, 2022].
22. US Department of Transportation. Fatality Analysis Reporting System (FARS) Brochure. https://crashstats.nhtsa.dot.gov/Api/Public/ViewPublication/811992#:~:text=Some%20specific%20policy%20and%20research%20uses%20of%20FARS%20data%20include%3A&text=Alcohol%2Drelated%20legislation%2C%20%E2%80%A2%20Motorcycle,and%20%E2%80%A2%20Air%20bag%20effectiveness. [Accessed March 28, 2022].
23. US Bureau of Labor Statistics. Census of Fatal Occupational Injuries (CFOI). www.bls.gov/iif/oshcfoi1.htm. [Accessed March 28, 2022].
24. US Bureau of Labor Statistics. Scope of the Census of Fatal Occupational Injuries (CFOI). www.bls.gov/iif/cfoiscope.htm. [Accessed March 28, 2022].
25. Centers for Disease Control and Prevention (CDC). Traumatic Injury Prevention Program PPOP. www.cdc.gov/niosh/programs/ppops/tipp.html. [Accessed March 28, 2022].
26. US Bureau of Labor Statistics. Survey of Occupational Injuries and Illnesses Data. www.bls.gov/iif/soii-data.htm. [Accessed March 28, 2022].
27. Centers for Disease Control and Prevention (CDC). NVDRS Frequently Asked Questions. www.cdc.gov/violenceprevention/datasources/nvdrs/faqs.html. [Accessed March 28, 2022].
28. Centers for Disease Control and Prevention (CDC). National Violent Death Reporting System (NVDRS). www.cdc.gov/violenceprevention/datasources/nvdrs/index.html. [Accessed March 28, 2022].
29. Centers for Disease Control and Prevention (CDC). State Unintentional Drug Overdose Reporting System (SUDORS). www.cdc.gov/drugoverdose/od2a/pdf/SUDORS-Fact-Sheet.pdf. [Accessed March 29, 2022].
30. Centers for Disease Control and Prevention (CDC). Drug Overdose Deaths in the U.S. Top 100,000 Annually. November 17, 2021. www.cdc.gov/nchs/pressroom/nchs_press_releases/2021/20211117.htm. [Accessed March 29, 2022].
31. US Food and Drug Administration (FDA). MedWatch: The FDA Safety Information and Adverse Event Reporting Program. www.fda.gov/safety/medwatch-fda-safety-information-and-adverse-event-reporting-program. [Accessed March 28, 2022].
32. US Food and Drug Administration (FDA). Recalls, Market Withdrawals, & Safety Alerts. www.fda.gov/safety/recalls-market-withdrawals-safety-alerts. [Accessed March 28, 2022].
33. Vaccine Adverse Event Reporting System (VAERS). About VAERS. https://vaers.hhs.gov/about.html. [Accessed March 28, 2022].
34. Vaccine Adverse Event Reporting System (VAERS). Information for Healthcare Providers: Guidance on Reportable Events. https://vaers.hhs.gov/resources/infoproviders.html. [Accessed March 28, 2022].
35. Centers for Disease Control and Prevention (CDC). Myocarditis and Pericarditis after mRNA COVID-19 Vaccination. www.cdc.gov/coronavirus/2019-ncov/vaccines/safety/myocarditis.html?s_cid=11374:cdc%20covid%20vaccine%20heart%20inflammation:sem.ga:p:RG:GM:gen:PTN:FY21. [Accessed March 28, 2022].
36. National Association of Medical Examiners (NAME). Protocol for Deaths Occurring after COVID-19 Vaccination. March 17, 2021. www.thename.org/covid-19-resources. [Accessed March 28, 2022].
37. US Consumer Product Safety Commission (CPSC). MECAP News General Informational Flier. www.cpsc.gov/s3fs-public/MECAP_News_General_Info.pdf?VersionId=grjjdXGfGLlqkqKneCoOfqXZrDoMbK1P. [Accessed March 28, 2022].
38. Centers for Disease Control and Prevention (CDC). Enhancing Reviews and Surveillance to Eliminate Maternal Mortality. www.cdc.gov/reproductivehealth/maternal-mortality/erase-mm/index.html. [Accessed March 28, 2022].
39. National Center for Fatality Review and Prevention. Fetal & Infant Mortality Review. https://ncfrp.org/fimr/. [Accessed March 28, 2022].
40. National Center for Fatality Review and Prevention. CDR Map. https://ncfrp.org/cdr-map/. [Accessed March 28, 2022].
41. National Domestic Violence Fatality Review Initiative. About NVDFRI. https://ndvfri.org/about/. [Accessed March 28, 2022].
42. American Bar Association Commission on Law and Aging. Elder Abuse Fatality Review Teams. [Accessed March 28, 2022].
43. California Correctional Healthcare System. Analysis of 2019 Inmate Death Reviews in the California Correctional Healthcare System. https://cchcs.ca.gov/wp-content/uploads/sites/60/MS/2019-Inmate-Death-Reviews.pdf. [Accessed March 28, 2022].
44. Ritter N. Missing persons and unidentified remains: the nation's silent mass disaster (NCJ number 216523). *National Institute of Justice Journal.* 2007;256:2–7.
45. National Missing and Unidentified Persons System (NamUs). The NamUs Mission. https://namus.nij.ojp.gov/about. [Accessed March 28, 2022].

46. Federal Bureau of Investigation (FBI). 2020 NCIC Missing Person and Unidentified Person Statistics. www.fbi.gov/file-repository/2020-ncic-missing-person-and-unidentified-person-statistics.pdf/view. [Accessed March 28, 2022].
47. Centers for Disease Control and Prevention (CDC). Nationally Notifiable Diseases Surveillance System (NNDSS). https://ndc.services.cdc.gov/search-results-year/. [Accessed March 29, 2022].
48. Sampson BA, Ambrosi C, Charlot A, Reiber K, Veress JF, Armbrustmacher V. The pathology of human West Nile Virus infection. *Hum Pathol.* May 2000;31(5):527–531.
49. Gill JR, DeJoseph ME. The importance of proper death certification during the COVID-19 pandemic. *JAMA.* 2020;324(1):27–28.

CHAPTER

18

Developments in postmortem imaging

Jamie Elifritz and Heather S. Jarrell

A difference, to be a difference, must make a difference. -*Gertrude Stein*

Introduction

Forensic imaging (FI) and clinical imaging are like the two heads of Janus. Depending on the situation and timing, any imaging study can be useful for forensic purposes. The infancy of all imaging of the human body dates back to Wilhelm Conrad Rontgen's discovery of the "X-ray." His wife, Bertha Rontgen, was his first human subject in November of 1895, when he acquired an exposure of her hand.[1] The first known court case in North America involving radiographs occurred in December of 1895, involving an X-ray for localization of a bullet in a man's leg and the ensuing conviction for attempted murder.[2,3]

Rapid technological advancements have propelled imaging in clinical medicine. Similarly, the utility of imaging in forensic pathology has been astutely recognized. The Virtopsy Project was the first to amalgamate forensic pathology and radiology.[4–9] Launched in 2000, at the Institute of Legal Medicine at the University of Bern in Switzerland, the Virtopsy Project developed benchmarks in FI. The collaborators, led by Dr. Michael J. Thali, were the first to establish routine imaging practices in the setting of medicolegal death investigation.

FI has been blooming on a global stage, with numerous burgeoning centers established all over the world from the Department of Legal Medicine of Chiba University, where Japan's first installment of a computed tomography (CT) scanner took place in 2006, to the Victorian Institute of Forensic Medicine where the routine use of **CT scanning** in Australia occurred in 2005. Currently, formal FI courses (Virtopsy) are offered by the University of Zurich, where one can learn from the original innovators in the field.

CT basics

In 1967, Sir Godfrey Hounsfield, a British engineer, invented the CT scanner. The first CT scanners were installed in the United States in 1973.[10] Since that time, there has been rapid technological advancement leading to increasingly sophisticated imaging, as well as greater availability of CT scanners.

CT scanners utilize X-rays to make images. The gantry, sometimes compared to a doughnut, houses both the source of X-rays and the detector. The table that the patient/decedent lies prone on advances horizontally through the gantry as the gantry spins. All the while, the X-rays are emitted through the body and detected at 180° from their source (Fig. 18.1). This information is transmitted to the computer, which applies a mathematical model to formulate an image. The thickness of the tissue imaged at each step varies and is referred to as the slice thickness and depends on the specific parameters of the individual machine. Slice thickness typically varies between 1 and 10 mm.

Image quality depends on numerous factors, many of which are beyond the scope of this discussion. However, it is important to be familiar with a few fundamental principles. The term quantum noise (noise) refers to the graininess of an image that results in poorer image quality. One key concept is that image quality improves when more X-rays reach the detector. It is important to know these parameters when discussing imaging artifacts.

Principles of Forensic Pathology
https://doi.org/10.1016/B978-0-323-91796-4.00002-7

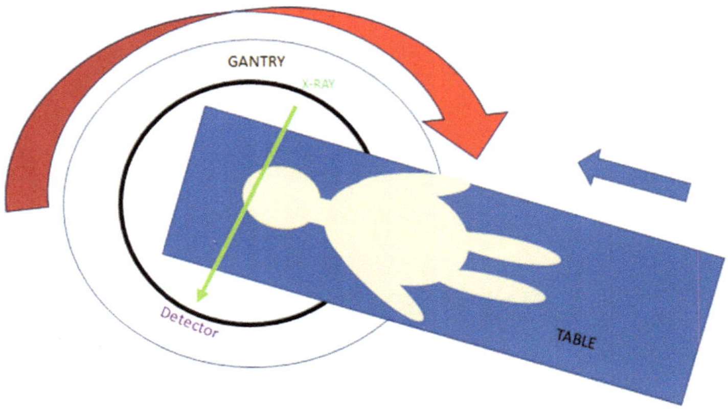

FIGURE 18.1 **Computed tomography schematic.** The Gantry rotates around the table as the table advances through the gantry. X-rays are emitted from one side of the gantry and detected at 180°.

It is also helpful to be familiar with certain specifications when choosing a CT scanner, as competing concepts must be balanced for image optimization. The main point regarding the choice of a scanner is that while having a 64-slice scanner (or higher) is optimal for clinical interpretation (often obtained in conjunction with the administration of intravenous and/or intraoral contrast agents), *a 16-slice scanner is appropriate for postmortem imaging*. At the New Mexico Office of the medical investigator, decedents are imaged with a 16-slice CT scanner with an 85 cm bore and a flat bariatric table, with Big Bore version 4.2 software.

For the standard nondecomposed adult, the scanning protocol is performed over two volumes: (1) vertex of cranium to just below the upper extremities (with arms crossed over the chest/abdomen) and (2) sternal notch to just below the toes (with arms raised over the head). *The total time required to scan one decedent (including transport time) averages 16 min*. Axial slices are reconstructed from both volumes with a 3 mm slice thickness (3 mm spacing) and a 1 mm slice thickness (0.5 mm spacing), using both a soft tissue algorithm and a bone algorithm. Other reconstructions include sagittal and coronal reformats (bone, soft tissue, and brain), axial brain, dental, and coronal and axial lung views. When interpreting a CT scan, it is important to determine the orientation. Images are most commonly viewed in the transverse plane and are orientated so that one is looking up the body from the patient's toes. A helpful way to get your bearings is the acronym **RALP**. Starting at the 9 o'clock position and moving clockwise in 90° intervals, we are looking at the **R**ight, **A**nterior, **L**eft and **P**osterior aspects of the patient.

Hounsfield units

The mechanism for categorizing the shades of gray seen on CT is named after its inventor, Sir Godfrey Hounsfield, the Hounsfield Unit (HU). A region of interest can be assessed by selected to reveal the tissue's HU which provides important information regarding tissue content. Centered around the image attenuation of water (HU = 0), there is a scale of positive and negative values ranging from +1000 for dense tissue found in bone (appears white on CT) to −1000 for lucent structures, such as air in the lungs (appears black on CT). Other tissue components, such as fat, blood, muscle, and brain fall in the spectrum (Fig. 18.2). When a tissue is *bright or white on CT*, it is described at **hyperattenuating**. Likewise, when a tissue is *dark on CT*, it is described as **hypoattenuating**.

HUs provide important information about the makeup of tissues that can be used to diagnose pathology. For instance, a normal liver should measure approximately +60 HU and hyperattenuates (appears brighter) relative to the spleen.[11,12] As the liver becomes fatty (steatosis), it appears darker than the spleen (hypoattenuating) and as a result, the HUs decrease to less than +40.[13] Likewise, blood can be differentiated from simple fluid based on the HU scale, with a simple fluid collection measuring 0 to +20 HU and a hemorrhagic collection measuring in the range of +65−95 HU.

Artifacts

Artifacts are undesired imaging findings that compromise image quality. The most common artifact in clinical imaging is motion, which is negated by virtue of the reference population in the postmortem setting. *The most common artifacts encountered with postmortem CT (PMCT) are beam-hardening, metal, streak, quantum mottle, and ring.*

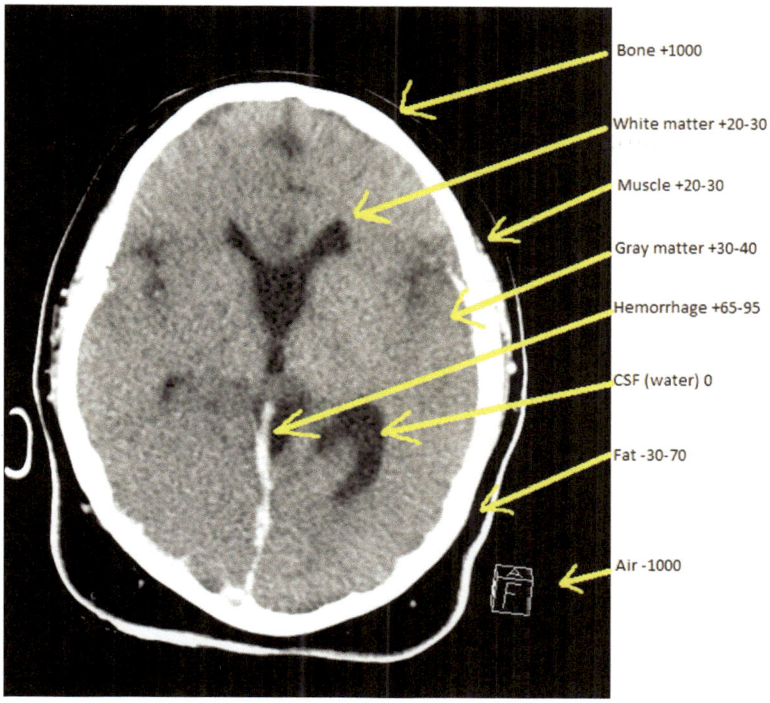

FIGURE 18.2 **Hounsfield units.** CT unit to standardize relative densities of tissue.

Beam-hardening artifact occurs when the X-rays pass through a dense (or hard) tissue, such as bone. One can think of it as runners racing through a muddy field. The lower energy X-rays (or less strong runners) will become attenuated (stuck in the mud) and will not make it to the detector (finish line). Thus, the detector only "sees" the high energy X-rays and interprets this as a dark streak in the image. This commonly occurs at the skull base, as the lower energy X-rays become attenuated by the petrous portion of the temporal bone (Fig. 18.3).

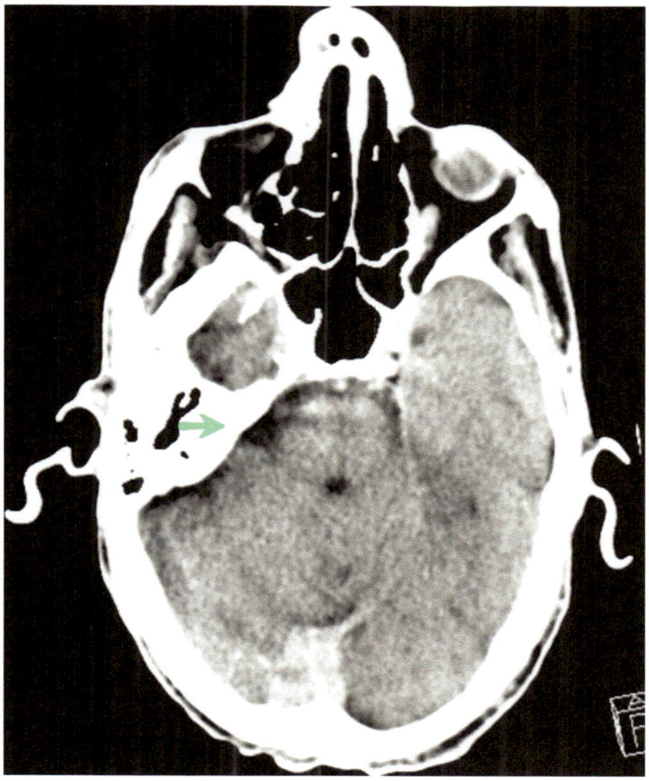

FIGURE 18.3 **Beam-hardening artifact.** Axial PMCT brain with brain window. Note the dark streak through the pons.

Metal artifact is caused partially by beam hardening, however, it also involves scatter, Poisson noise, and edge effects. The result is streaks of bright and black lines through tissue adjacent to metal (Fig. 18.4). This is most troublesome in the pelvis, where hip arthroplasties wreak havoc on the evaluation of the pelvic contents. There are scanning techniques that can reduce metal artifact, but this is typically not needed in standard postmortem imaging.

Streak artifact and **quantum mottle** result from insufficient X-rays reaching the detector. In both cases, there is a component of the X-rays passing through tissue that causes a "scatter" of the X-rays, such that they are redirected through the tissue and do not follow the anticipated 180-degree path to the detector. Streak artifact is commonly seen in the chest and abdomen when the arms are down by the sides, causing streak artifact (dark lines) through the adjacent thoracoabdominal structures (Fig. 18.5). Quantum mottle is seen in decedents with large body habitus, causing scatter of the X-rays (related to the volume of soft tissue that the X-rays must traverse) and resultant poor image quality (Fig. 18.6).

Streak artifact can be fixed by positioning the arms over the head for imaging of the chest and abdomen. Quantum mottle can be alleviated by "turning up the juice" on the CT scanner, otherwise known as, increasing the

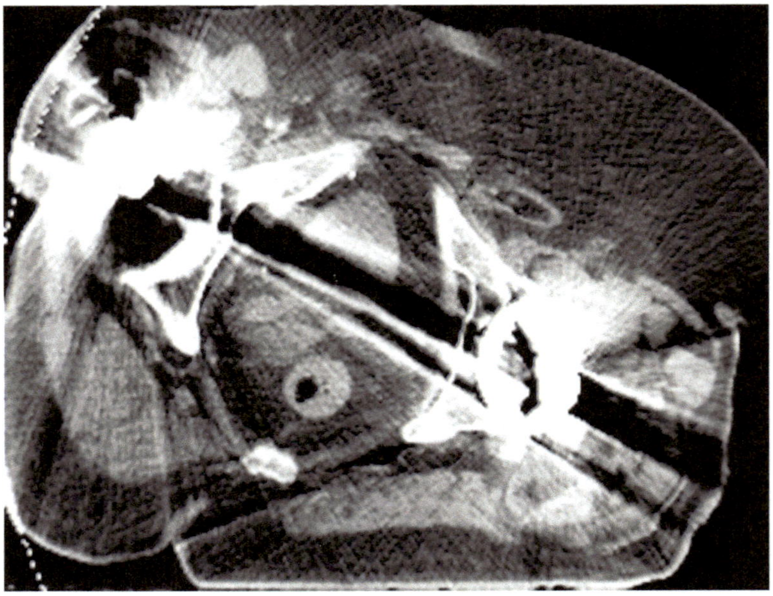

FIGURE 18.4 **Metal artifact.** Axial PMCT pelvis with soft tissue window. Metal artifact limits evaluation of adjacent structures.

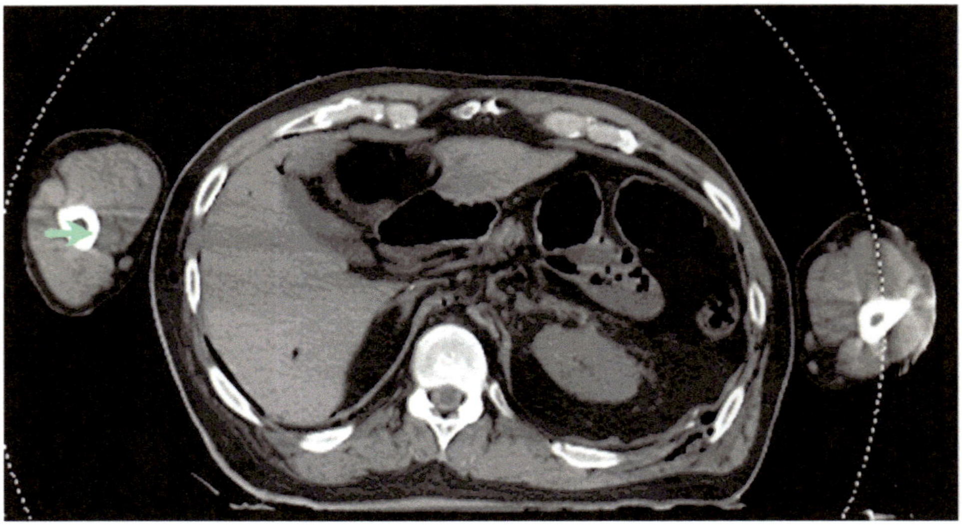

FIGURE 18.5 **Streak artifact.** Axial PMCT abdomen with soft tissue window. Streaking can mask underlying pathology.

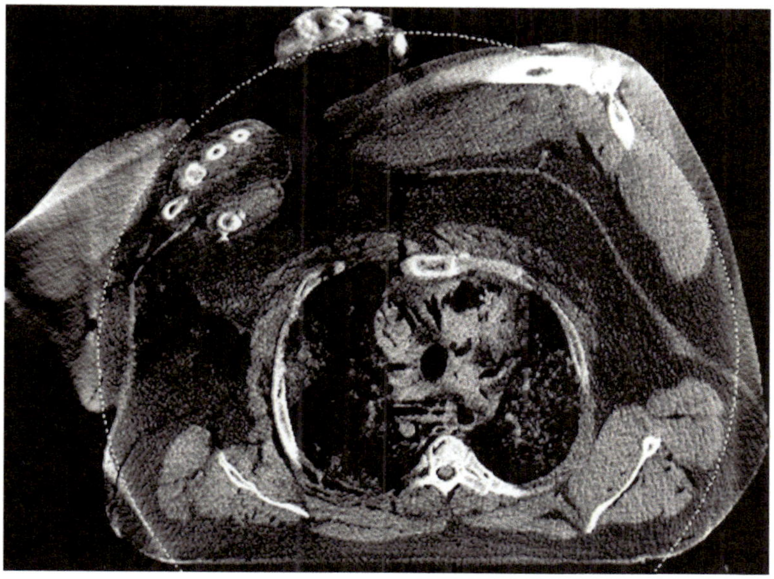

FIGURE 18.6 **Scatter artifact/quantum mottle.** Axial PMCT chest with soft tissue window. Increases the noise, causing grainy images.

milliamperage (mA) and/or peak tube voltage (kVp) to ensure that more X-rays are sent through the tissue so more X-rays will make it to the detector.

Ring artifact most commonly occurs as a result of miscalibration or failure of a detector element. It results in a ring near the isocenter of the image which is typically present throughout numerous slices at that same location on numerous images (Fig. 18.7). The remedy is to let your CT technologist know about the artifact so they can figure out if the machine needs to be recalibrated or if a detector needs to be replaced.

Approach to PMCT interpretation

The first step to CT interpretation is to evaluate the scout views. This easy step is often forgotten and can lead you to miss otherwise obvious findings, such as a gun in the decedent's pocket (Fig. 18.8).

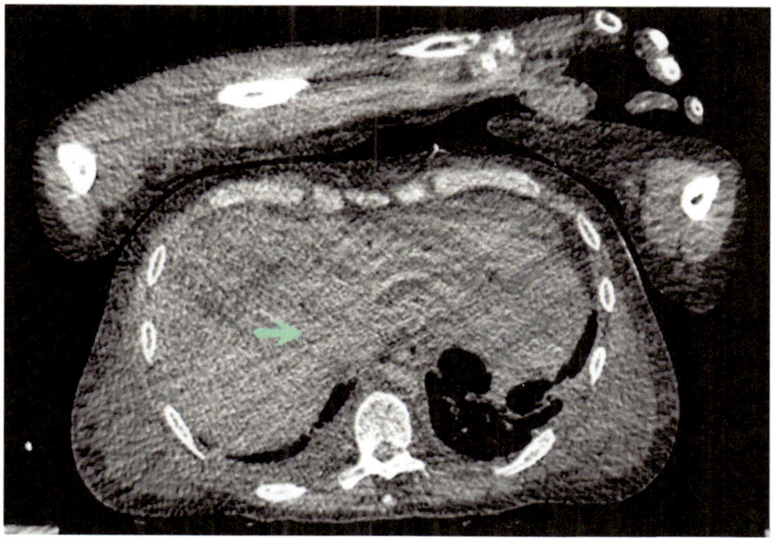

FIGURE 18.7 **Ring artifact.** Axial PMCT abdomen with soft tissue window. Distorts images on multiple slices in the same location.

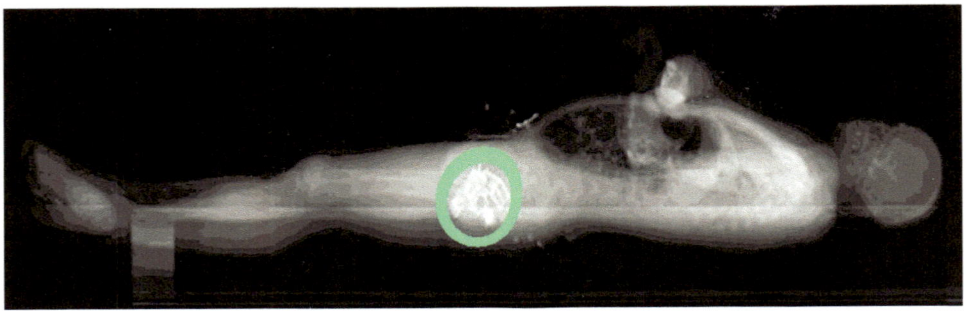

FIGURE 18.8 **Scout view.** Note gun-shaped lighter in pants pocket.

CT images are first viewed in the axial plane, ideally with the body in anatomic position. All search patterns are repeated in the sagittal and coronal planes with reformatted images. A systematic approach is recommended with a checklist of key anatomical structures.

Head and neck

Always assess the brain using the brain window and evaluate for symmetry, edema and/or atrophy, intraaxial or extraaxial blood, and parenchymal mass(es) or foreign bodies.

The imaging window is changed to evaluate the soft tissues. Again, symmetry is our friend when looking for soft tissue injuries, masses, and lymphadenopathy. Subsequently, the bone window is used to identify and categorize fractures of the cranium, face, and cervical spine (c-spine) in three planes. Special attention is required at the skull base

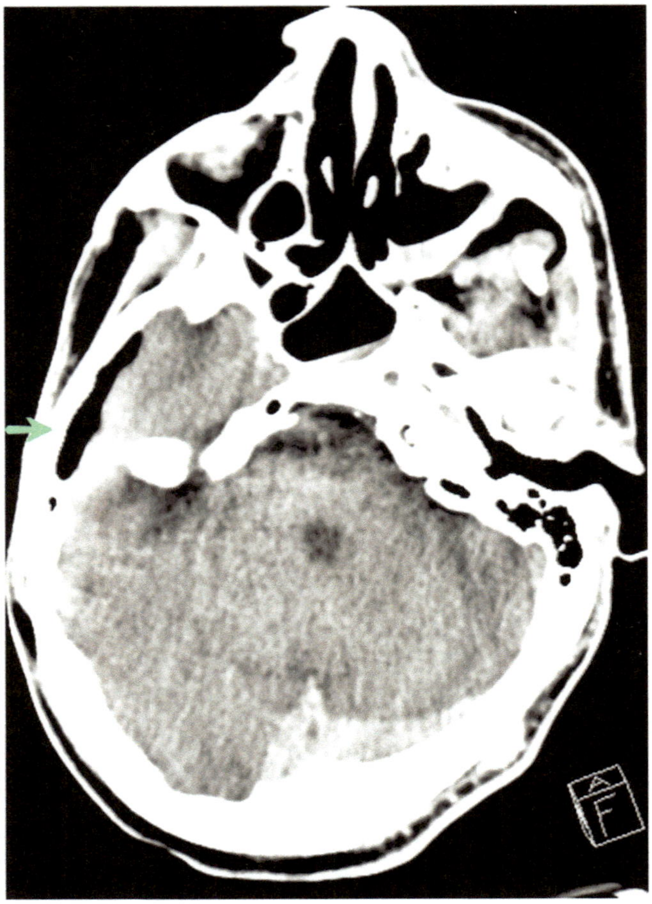

FIGURE 18.9 **Pneumocephalus.** Pneumocephalus on soft tissue and lung windows. Intracranial air is better depicted on the lung window.

and upper c-spine to evaluate the atlanto-occipital and c-spine alignment. The spectrum of injury ranges from severe, with transection of the spinal cord to subtle, with minimally displaced avulsion fractures of the occipital condyle. Sometimes, the lung window can be helpful to identify small volume pneumocephalus (Figs. 18.9 and 18.10).

The bone window is employed for interrogation of possible cranial, facial, or c-spine fractures. Recognition of the normal sutures and vascular grooves can be helpful to contrast with fractures. Careful attention should be directed at the skull base and the atlanto-occipital junction, where injuries may be subtle or may be best appreciated in orthogonal planes (c-spine fractures may be easier to see with sagittal than axial views).

Chest

Starting with the mediastinum, the great vessels are evaluated for signs of natural disease, including aortic and coronary artery disease. The size and attenuation of the aorta and main pulmonary artery can provide information about aortic ectasia, dissection, and pulmonary artery hypertension. A pulmonary thromboembolism is difficult to assess on postmortem imaging; however, this finding is reported to be smooth in contour and low in density. In practice, imaging findings in the pulmonary artery do not always align with autopsy results as lividity findings are frequently confounding. The trachea and esophagus are assessed on sequential images. Fluid is often in one or both structures. Sometimes, a food bolus (in the setting of choking) or a tooth (in the setting of trauma) or foreign body can be found.

Next, the lung windows are assessed for lividity, noting the position of the body from the scene images. Patterns of opacification can be seen with infection or trauma and pulmonary nodules or masses. This also is the perfect time to identify a pneumothorax or pneumomediastinum, as well as gas in the more superficial soft tissues. The pleural space is best evaluated for fluid, masses, or a calcified plaque on the soft tissue window. Last, but not least, the muscles and superficial soft tissues on the soft tissue window are examined. Soft tissue findings can be subtle and are often overlooked. Evaluation of the osseous structures are deferred until after evaluation of the abdominal and pelvic structures.

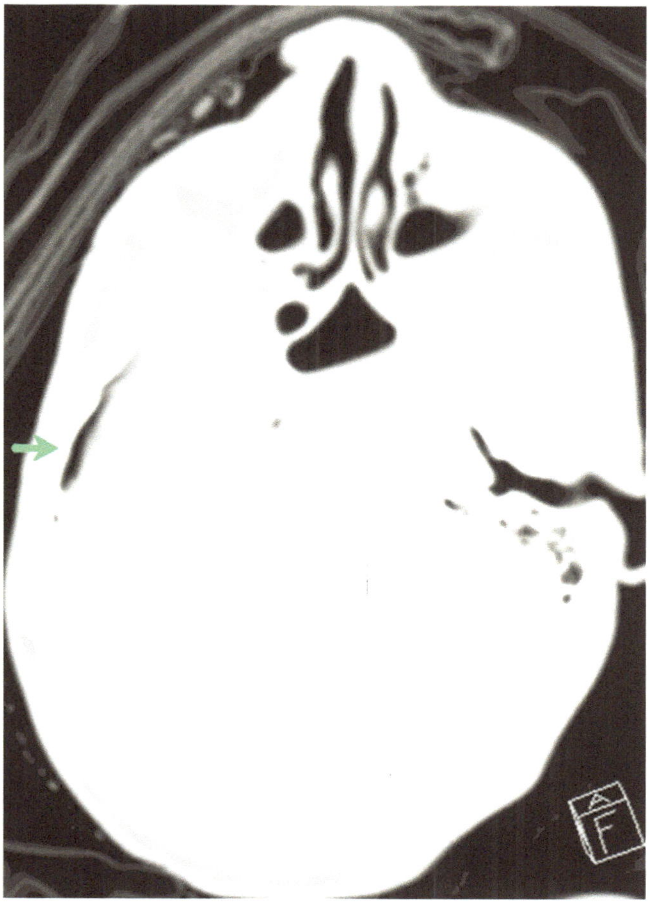

FIGURE 18.10 **Pneumocephalus.** Small volume pneumocephalus seen better on lung window.

Abdomen and pelvis

The organs and remaining soft tissues are evaluated in the same order on every exam. First, start with the liver (Fig. 18.11) and evaluate the density for steatosis (Fig. 18.12), fibrosis (Fig. 18.13), or masses (Fig. 18.14), the size, and contour for hepatomegaly and cirrhotic features. Hepatic steatosis is diagnosed if the liver is at least 10HU less than spleen.[14] Gallstones are common. If there is focal wall thickening or associated fat stranding, it may indicate underlying cholecystitis. Be careful when evaluating wall thickening in the setting of ascites or stigmata of cirrhosis because this pathology can cause a similar appearance. If the gallbladder is surgically absent, surgical clips are frequently seen in its place. The spleen is evaluated for density and size. For the pancreas, evaluate for mass, atrophy, duct abnormalities, and regional mesenteric stranding. The adrenal glands should look like inverted Y's and have smooth thin limbs. The most common abnormality is a fat density mass, consistent with an adenoma. Renal inspection focuses on parenchymal attenuation and contours which, when abnormal, may indicate masses and/or scarring. Renal stones are common, and any obstruction of the collecting system should be documented. The ureters can be difficult to assess when there is little intraabdominal fat or with extensive bowel gas; however, one should attempt to follow them throughout their course. The bladder wall is typically not thicker than 3 mm, however, can appear thick related to decompression. Bladder wall thickening can be seen in men with enlarged prostates

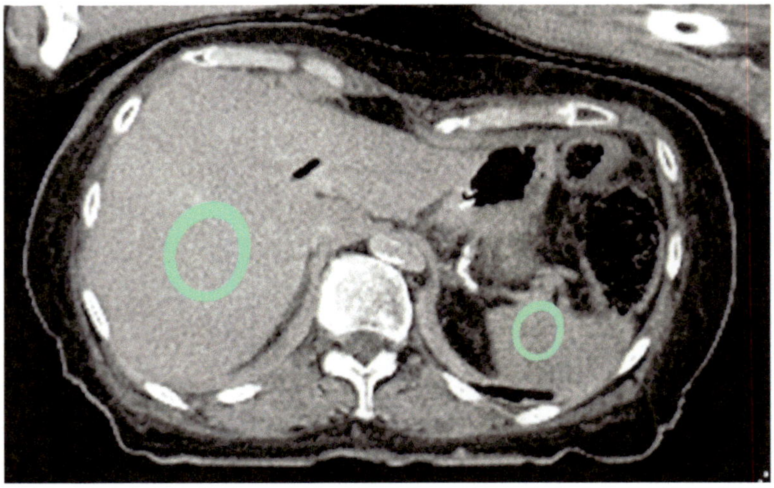

FIGURE 18.11 **Normal liver on axial PMCT abdomen with soft tissue window.** Normal liver is hyperdense (brighter) to the spleen.

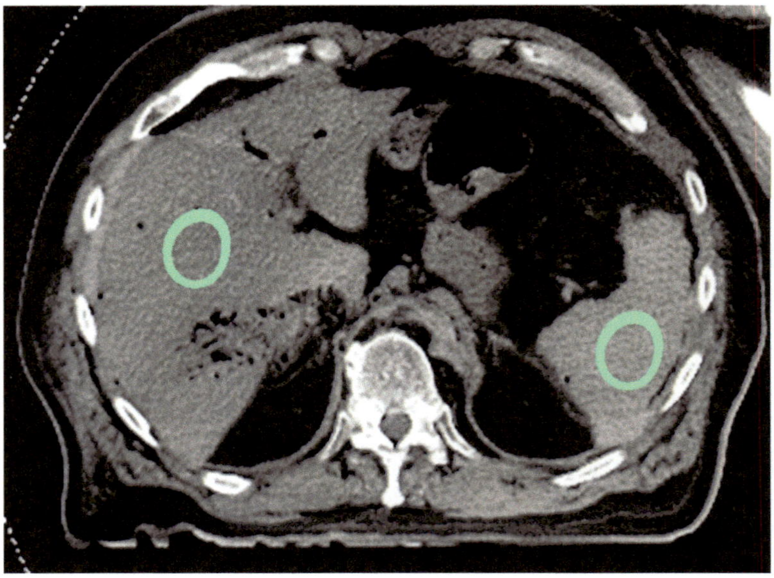

FIGURE 18.12 **Liver with steatosis.** Fatty liver is hypodense (darker) to spleen.

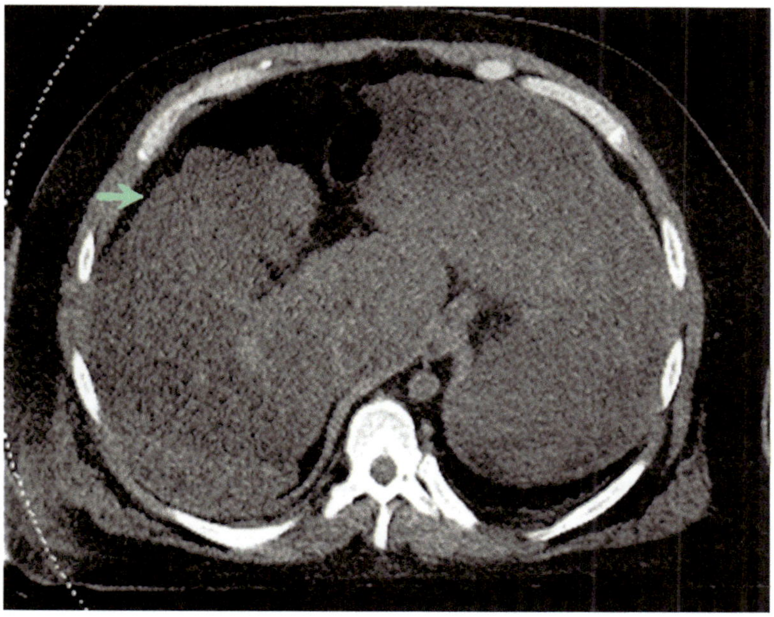

FIGURE 18.13 **Cirrhosis of liver.** Cirrhosis with enlarged, nodular, hypodense liver.

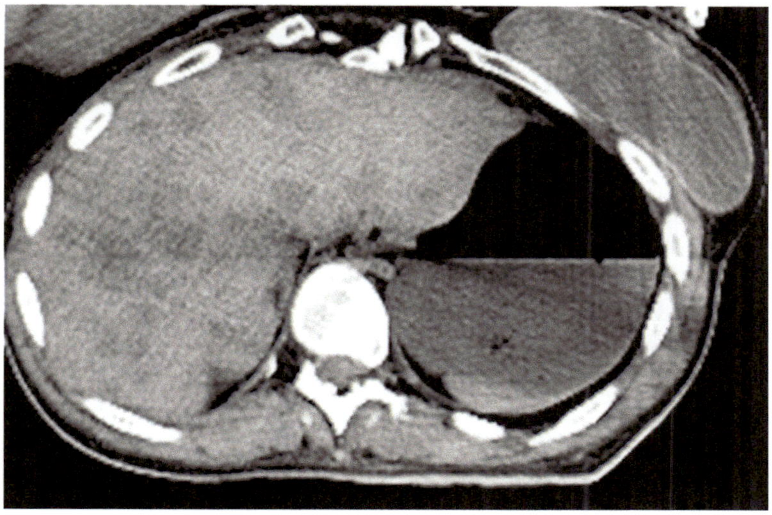

FIGURE 18.14 **Liver with masses.** Liver metastases. Note left breast implant.

in the setting of chronic outlet obstruction, as well as in either sex with cystitis. Bladder wall diverticula can be encountered and cause stasis that is responsible for infection and stone formation (Fig. 18.15).

Once the solid organs and renal collecting systems are analyzed, the retroperitoneum is evaluated. The aortic contour, size, and calcifications should be assessed. The abdominal aorta is typically collapsed in postmortem imaging and has been compared to a collapsed rubber band in cases of ectasia or aneurysm (Fig. 18.16). On noncontrast CT, aortic dissections can appear as a smooth and sharply delineated attenuation differential; however, this also can be seen with postmortem change. *If there is central calcification associated with the attenuation differential, it is diagnostic of a dissection.*[15,16] Injury to the aorta will be associated with secondary findings of retroperitoneal hemorrhage even if the exact site of injury is unclear (Fig. 18.17). The inferior vena cava (IVC) is usually collapsed and can be followed superiorly to the right atrium and superior vena cava (SVC).

Soft tissue injuries can be subtle on imaging, partly due to lack of intravenous contrast and partly due to the cessation of bleeding after death. Solid organs may demonstrate very subtle attenuation differences in the setting of lacerations. Additionally, a small amount of free fluid or fat stranding adjacent to an organ may be the only indication of underlying injury (Figs. 18.18 and 18.19).

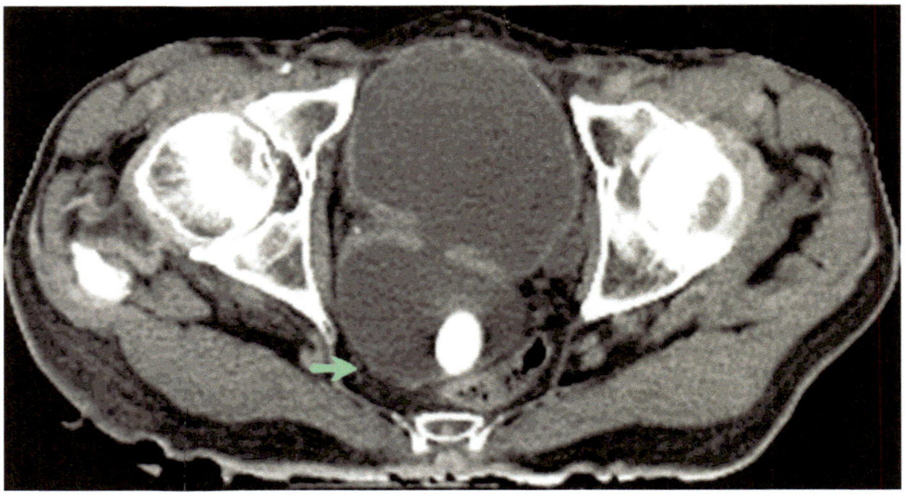

FIGURE 18.15 Bladder wall diverticula. Posterior bladder diverticulum with wall thickening and large stone.

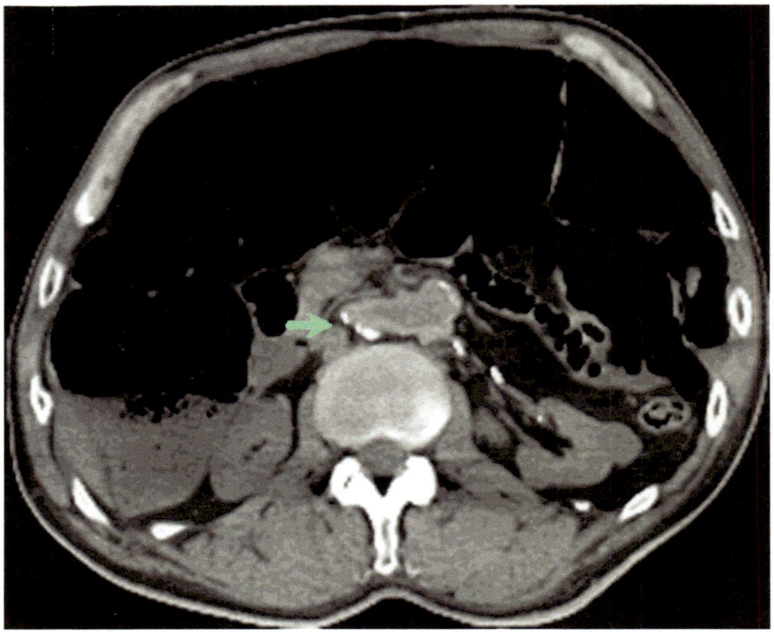

FIGURE 18.16 Abdominal aorta. Abdominal aortic aneurysm with "rubber band" appearance.

The stomach is often distended with an air-fluid level and a varying amount of intraluminal contents. The pylorus can be followed to the duodenum, which can be followed across the midline to the ligament of Treitz. This aids in evaluation of duodenal ulcers and free air due to gastric or duodenal perforation. The small bowel is often difficult to follow sequentially throughout its course and is frequently dilated and air-filled secondary to putrefaction. The large bowel can almost always be followed in entirety. It is easiest to start at the rectum and follow the large bowel proximally to the cecum. The appendix, if present, will come off the cecum near or at the same level at the ileocecal valve and is usually air-filled. If absent, surgical clips or suture material may be evident.

After the bowel is analyzed, the mesentery is reviewed for stranding, lymphadenopathy, and masses. Free fluid, as well as the attenuation of the fluid (which will be 0–20 HU with simple fluid and higher in attenuation, such as +60 HU with blood) are noted (Figs. 18.20 and 18.21). The lung windows are helpful for the evaluation of small amounts of free air in the abdomen and pelvis (Fig. 18.22).

The pelvis is a common location for small amounts of free fluid, which can be physiologic in females and can be related to decomposition in either sex. If present, the prostate, uterus, and/or ovaries should be noted. If there is pelvic or hip surgical hardware, the pelvic contents may be obscured by streak artifact.

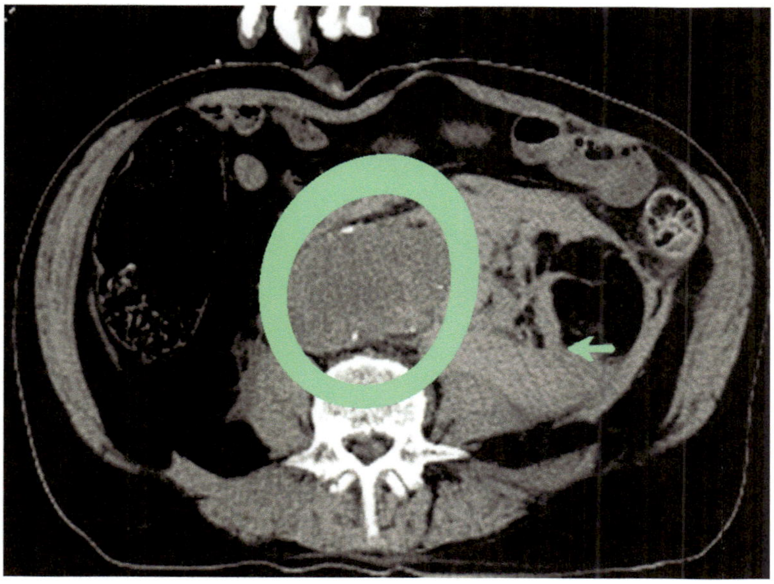

FIGURE 18.17 **Ruptured abdominal aortic aneurysm.** Ruptured abdominal aortic aneurysm with hemorrhage in the left retroperitoneum.

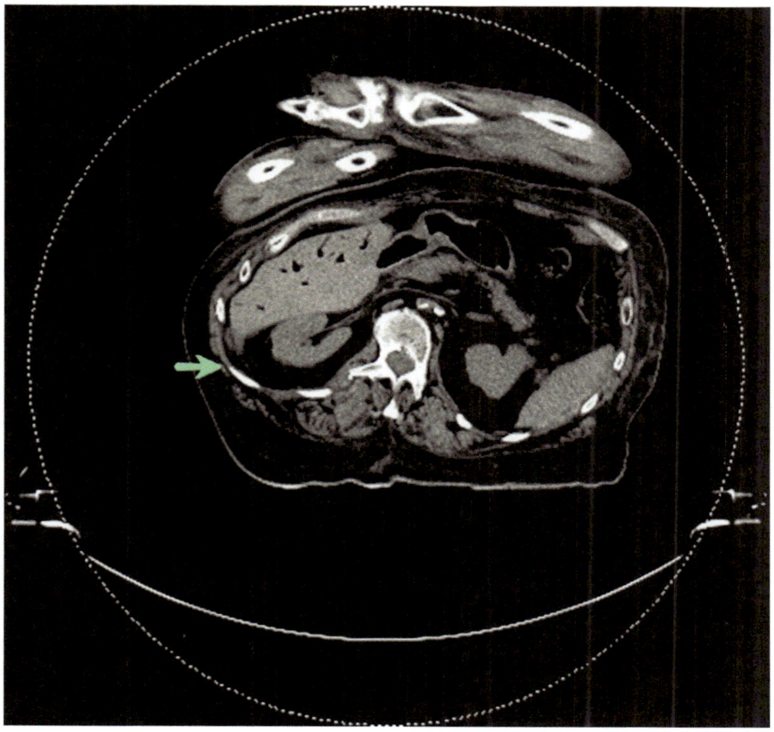

FIGURE 18.18 **Liver laceration.** Subtle hepatic laceration with stranding and fluid around the posterior segment of the right lobe.

As with the head, neck, and chest, the muscles and superficial soft tissues should be examined on soft tissue windows. Doing so often yields additional information (Fig. 18.23).

Bones

A standard sequential analysis of the osseous structures is advised. For the abdomen and pelvis, bone windows in the axial plane are used to evaluate the spine and shoulder girdle, to be subsequently cross-referenced on the sagittal and coronal sequences. Care should be taken to evaluate the alignment of the spine, which is best viewed on the sagittal sequence by looking at the right and left facet joint alignment at each level, as well as the vertebral body

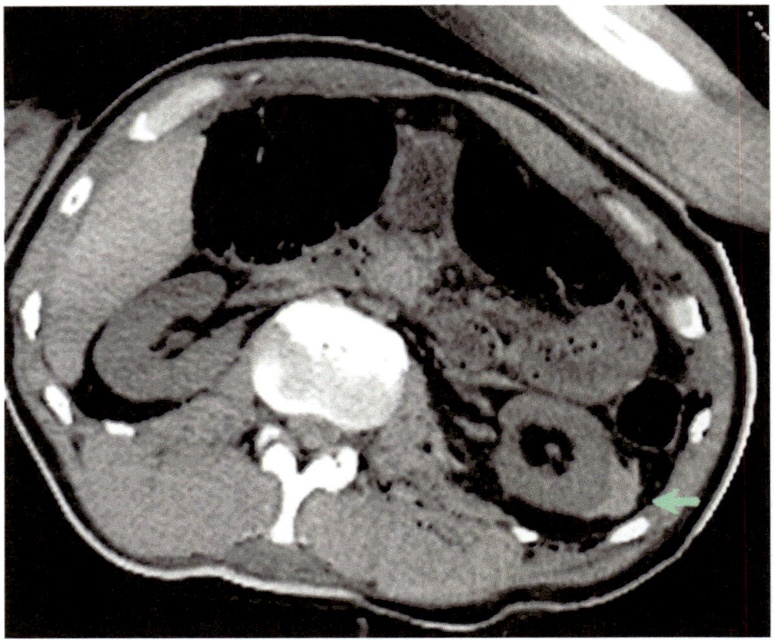

FIGURE 18.19 **Kidney laceration.** Left renal laceration with adjacent hemorrhage.

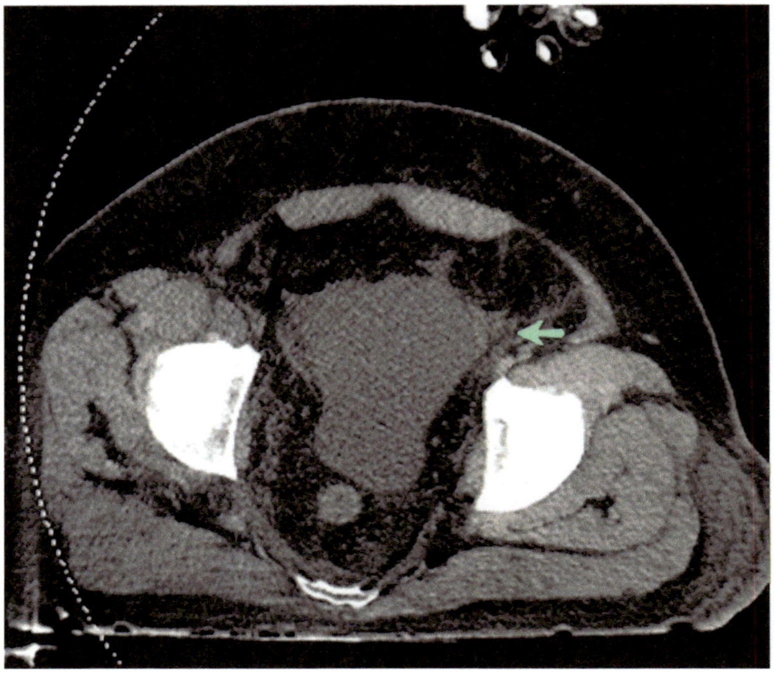

FIGURE 18.20 **Fluid with cirrhosis.** Free fluid associated with ascites. HU near that for water.

alignment and vertebral body integrity at each level, systematically, in a craniocaudal fashion (Figs. 18.24–18.26). It is best to focus on two vertebral bodies and facet joints at a time in this sequence (think of "two by two" like the animals on Noah's Ark). Going back to the cranial portion of the thorax, look specifically at the sternum in the axial plane and cross-reference it primarily with the sagittal plane, then turn your attention to the ribcage, starting with the left ribs and counting each rib as it is viewed, starting with the first anterior rib and moving craniocaudally. Follow a similar pattern in reverse order to assess the posterior ribs on the left from the twelfth rib to the first rib in a caudocranial fashion, then repeat this sequence for the right ribs.

The upper extremities are evaluated in their entirety in the axial plane and are cross-referenced in the sagittal and coronal planes. The pelvis is carefully inspected with attention to common injuries which can be subtle in the form of

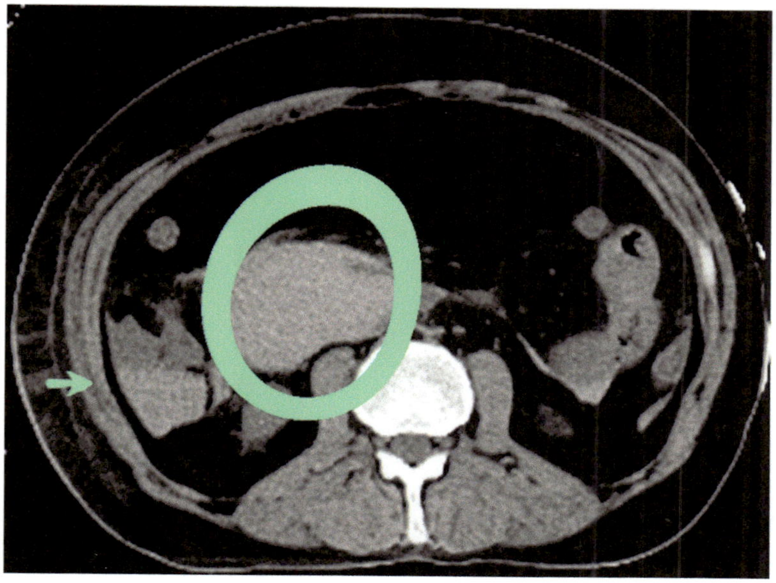

FIGURE 18.21 **Blood from pancreatitis.** Hemorrhage from hemorrhagic pancreatitis. HU +60–90.

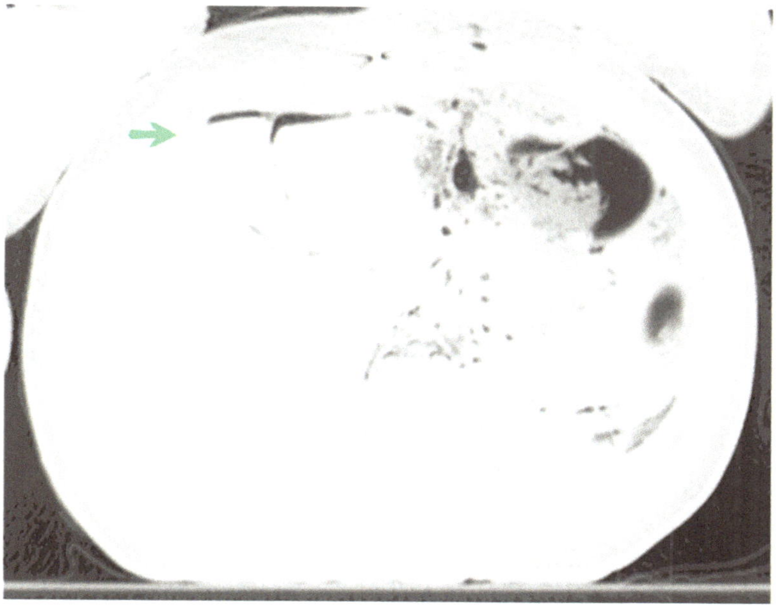

FIGURE 18.22 **Pneumoperitoneum.** Small volume pneumoperitoneum, best visualized on lung window.

femoral neck fractures, acetabular fractures, and associated pelvic ring disruption (Figs. 18.27 and 18.28). Pubic symphysis and sacroiliac alignment are assessed. Finally, the lower extremities should be evaluated in their entirety in the axial plane and are cross-referenced in the sagittal and coronal planes.

The ability to date fractures based on imaging features has been debated.[17–20] When looking at fractures, it is important to look for new periosteal bone, which over time, creates a healed callus. In this manner, *fractures can be classified as acute, subacute, or healed*. Acute fractures (Fig. 18.29) demonstrate sharp margins and lack new periosteal bone formation. Subacute or healing fractures (Fig. 18.30) are evident based on the presence of new periosteal bone which translates to blurring of the fracture margins during the early healing phase that progresses over time. Remote fractures exhibit a mature bridging callus (Fig. 18.31).

This classification scheme gives a rough idea of the age of the fracture. It does not pin down an exact date or range of dates during which the trauma occurred. Many factors influence the healing process, such as age, bone mineral

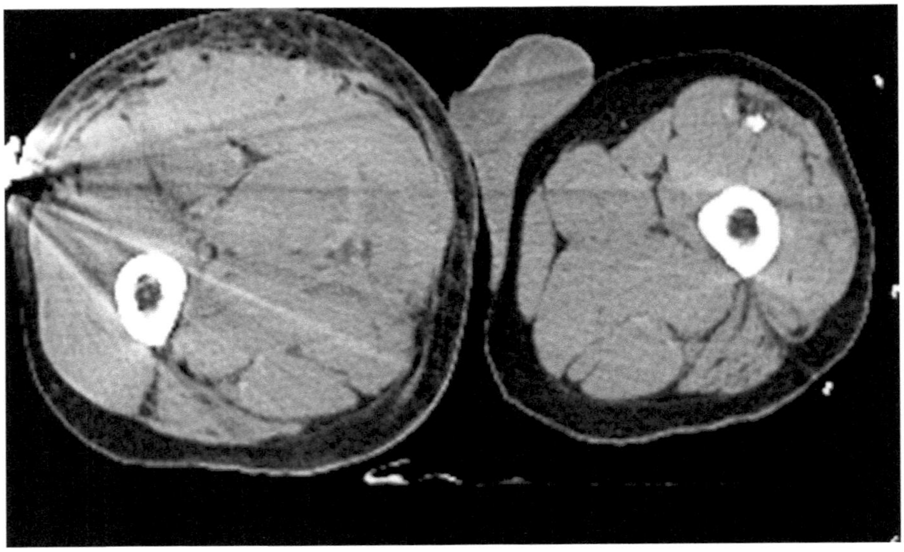

FIGURE 18.23 **Right thigh with hemorrhage.** Right thigh enlargement with hemorrhage. Decedent kicked by horse.

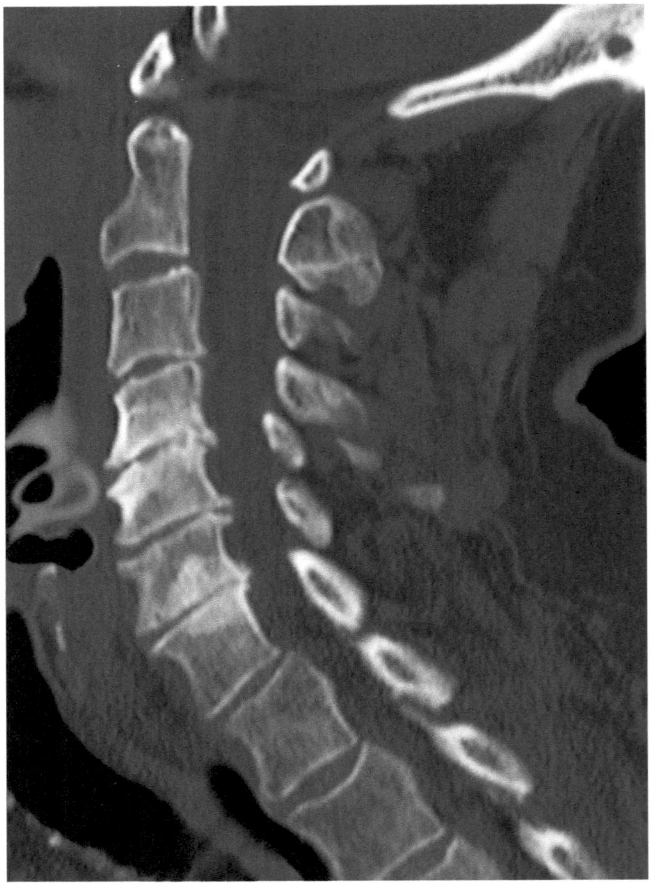

FIGURE 18.24 **C-spine, midline.** Sagittal images of the cervical spine with bone window. Approach to spine interpretation. Evaluate two vertebral bodies at a time working cranial to caudal. Left and right facet joints normal alignment.

density, weight bearing status, nutrition, smoking, underlying disease, and inflammatory states.[21–25] One must be mindful about these factors and recognize the limitations in more exact dating. Delineating antemortem from postmortem fractures is similarly fraught with controversy and limitations.[26]

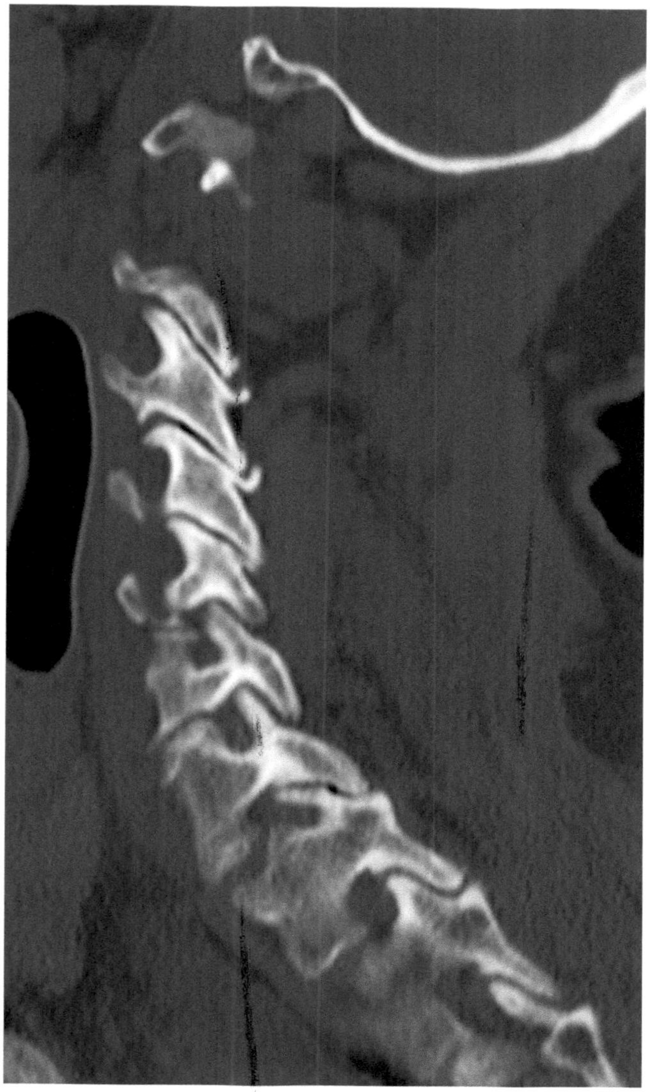

FIGURE 18.25 C-spine, left.

Support devices

Referred to in radiology as "lines and tubes," support devices should be evaluated for their presence and proper placement. This information is monumental for providing feedback to our colleagues who are providing emergency care in our communities. While support devices may not have the ultimate power to revive a dead person, improper placement makes all further resuscitation efforts futile and may, in some cases, cause harm.[27]

One of the most commonly placed airway support devices is **the endotracheal tube (ET tube)**. ET tubes should be placed in the trachea, with correct termination of the tube below the level of the clavicular heads (below the larynx to avoid injury) and, ideally, 2 cm above the carina. ET tube misplacements commonly involve the right main stem bronchus in the clinical setting, which can lead to collapse of the left lung, hyperinflation of the right lung, and a possible pneumothorax (Fig. 18.32). There are often helpful clinical symptoms that indicate such a misplacement, as well as the ubiquitous chest X-ray, which is commonly obtained immediately following placement in the clinical setting. Unfortunately, the emergency setting out in the field constitutes a less controlled environment, lacking many of the conveniences of the emergency room or clinic. It is common to discover esophageal placement of an ET tube placement at PMCT (Fig. 18.33). But, to be clear, when it occurs during resuscitative efforts, these misplacements do not constitute causes of death (a common misconceptions of families); rather, they simply represent a failure of the intervention to resuscitate an already dying or dead patient. This is an important distinction (see Chapter 3).

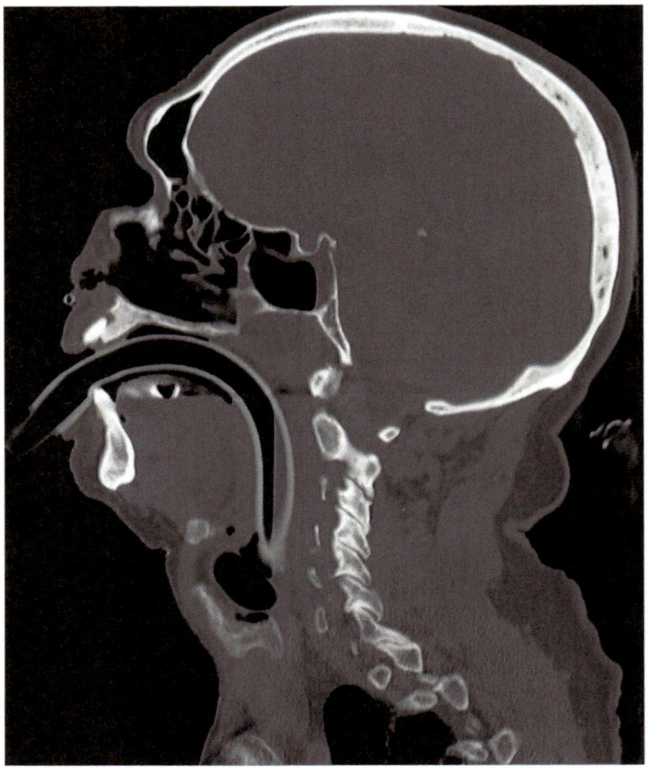

FIGURE 18.26 C-spine, right.

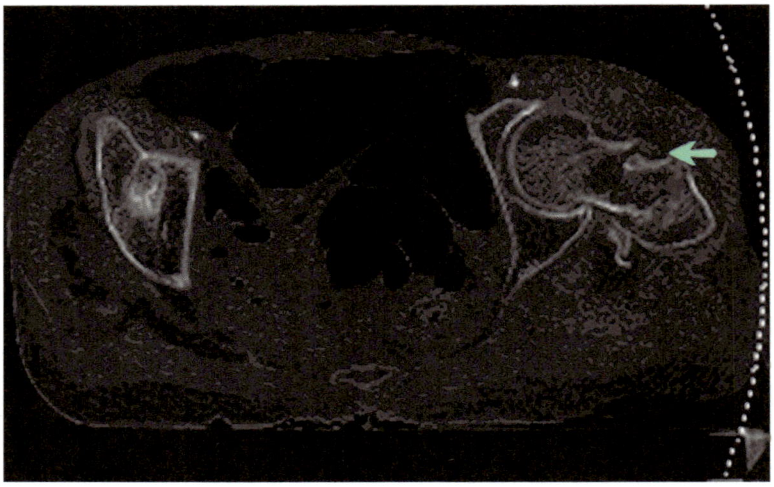

FIGURE 18.27 Left femur fracture.

Nasoenteric tubes should terminate in the stomach, preferably advanced multiple centimeters beyond the gastroesophageal junction. These devices are used to decompress the stomach and, therefore, reduce the risk of aspiration. They can also be used to administer medications. Suboptimal shallow placement can increase the risk of aspiration. Placement in the setting of facial trauma may exacerbate existing trauma. Additionally, spurious placement into the airway can lead to direct pulmonary damage.

Central venous catheters are sometimes placed in the field. Common locations include the groin (terminating in the IVC), as well as the neck and subclavian access sites (terminating in the SVC). When misplaced, they can have an arterial course, or in the case of the neck and subclavian access sites, they may take a wrong turn cephalad and terminate in the jugular veins. Even if placed appropriately in the vein, a mural puncture can result in substantial, even fatal hemorrhage into the thorax (neck and/or subclavian placements) or the retroperitoneum or peritoneal cavity (groin placements), particularly in anticoagulated patients.

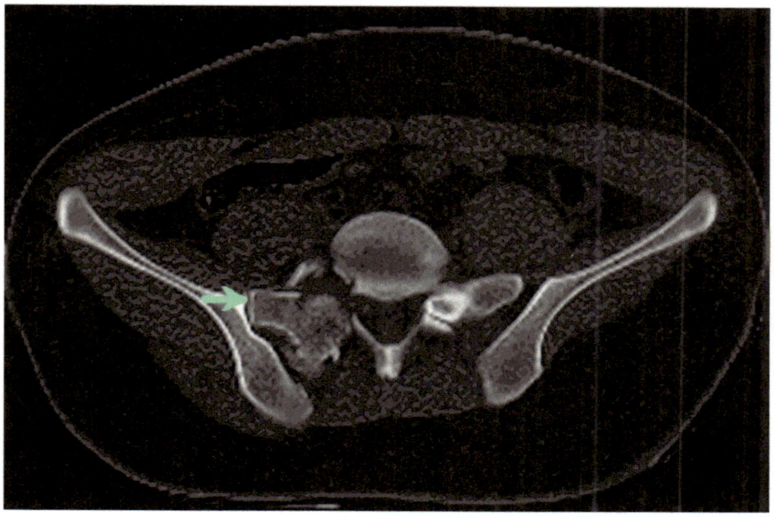

FIGURE 18.28 Right sacral intraarticular fracture.

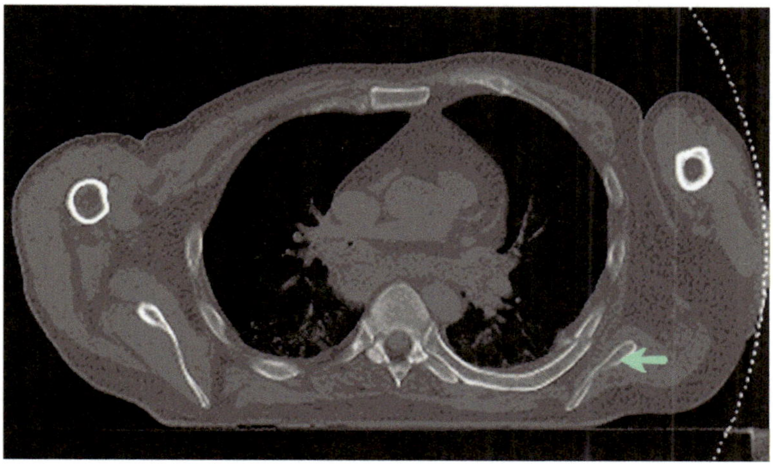

FIGURE 18.29 **Acute rib fracture.** Note the sharp margination of the left rib fracture.

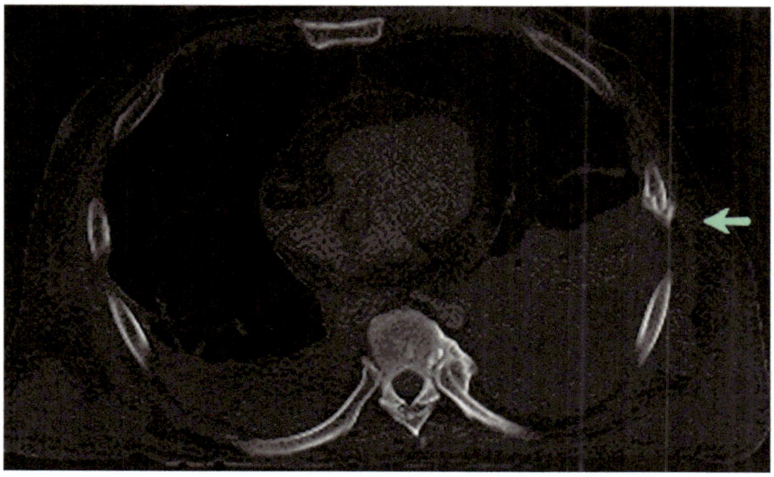

FIGURE 18.30 **Subacute rib fracture with callus.** The left rib fracture margins are blurred, secondary to new periosteal bone/callus formation.

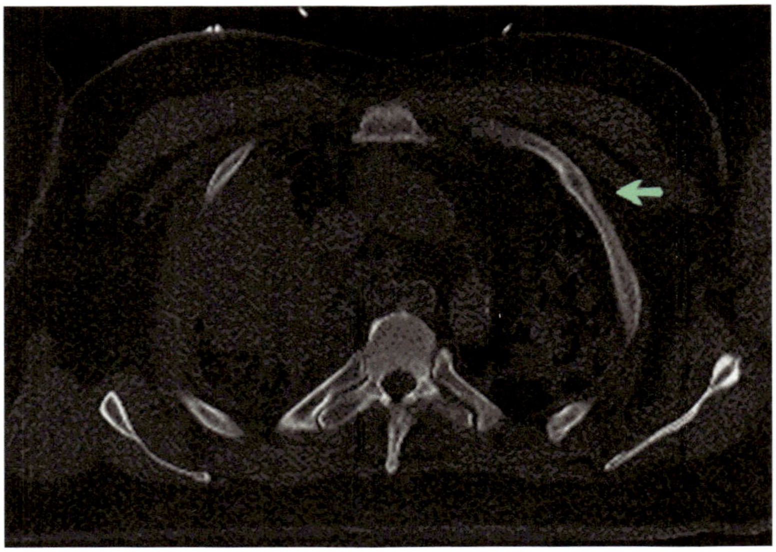

FIGURE 18.31 **Healed rib fracture.** Chronic/healed. The left rib is healed in deformity.

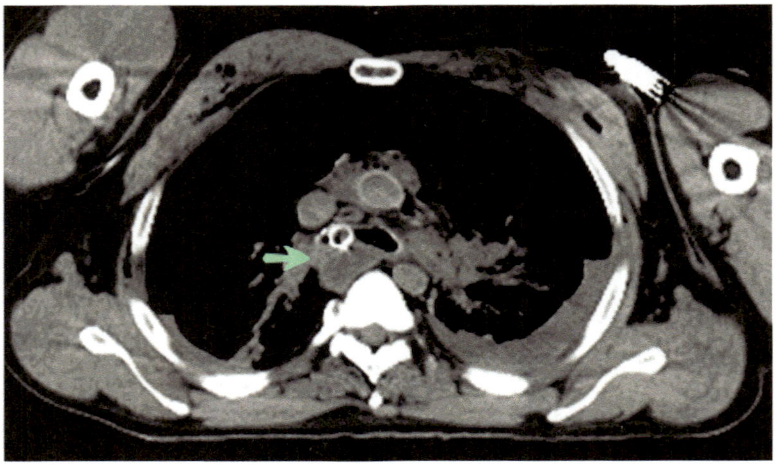

FIGURE 18.32 Endotracheal tube in right mainstem bronchus.

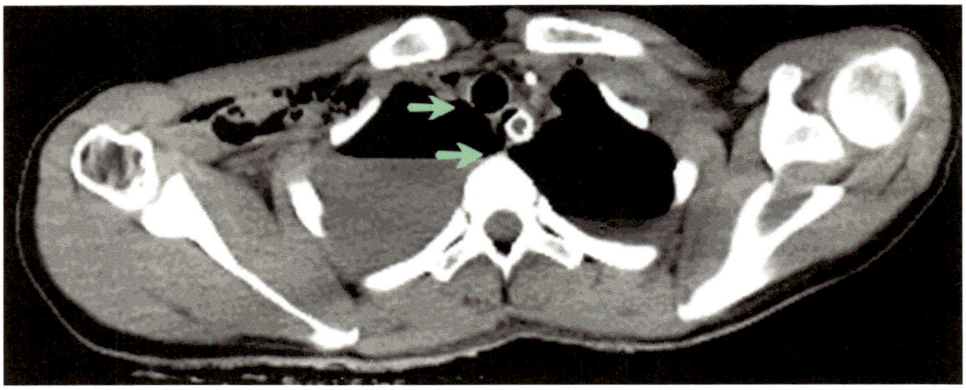

FIGURE 18.33 **Endotracheal tube in esophagus.** Note the trachea anterior to the support tube.

Intraosseous catheters are very commonly placed in the emergency setting. They should terminate in the marrow cavity (Fig. 18.34). The most frequent complication is shallow placement in the soft tissues (Fig. 18.35).

Tube and needle thoracostomies are two support devices that may cause the most harm when misplaced. Ideally, these devices will terminate in the pleural space to relieve a pneumothorax and/or hemothorax. Problems arise when they are placed unnecessarily (in a patient lacking a pneumothorax), placed too shallowly, or placed too deeply (causing end-organ trauma that may or may not contribute to the death, Figs. 18.36 and 18.37).

Pacemakers are common support devices. While not placed in the field under emergent conditions, the electrodes can displace, leading to events such as lead malposition, lead fracture, and end-organ injury resulting in cardiac rupture, tamponade, hemorrhage, or an arrhythmia. The classic device is composed of a pulse generator (located in the subcutaneous tissues of the chest wall) and one or more electrode leads. The prototypical single lead device should terminate with the tip in the apex of the right ventricle. The more common dual lead device contains one lead that terminates in the right atrium and the other, in the right ventricular apex. Leads are not placed in the left side of

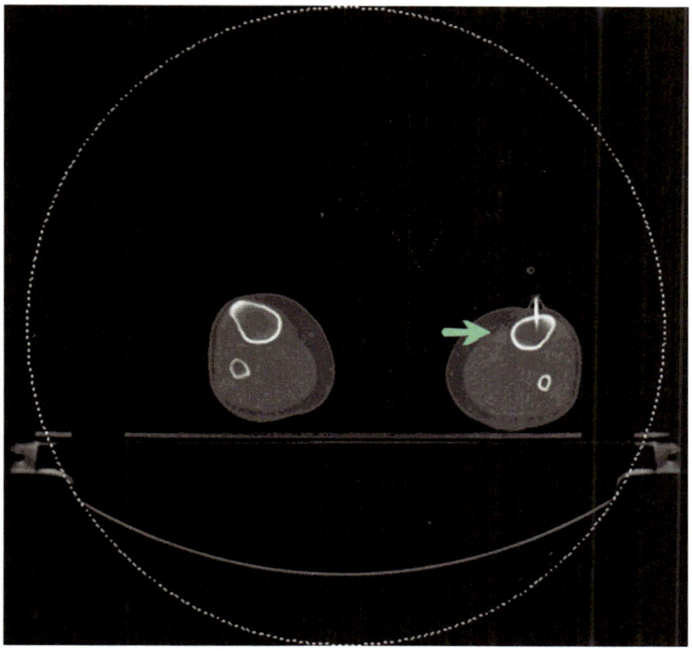

FIGURE 18.34 **Intraosseous line.** Proper termination of IO in medullary cavity of adult left tibia.

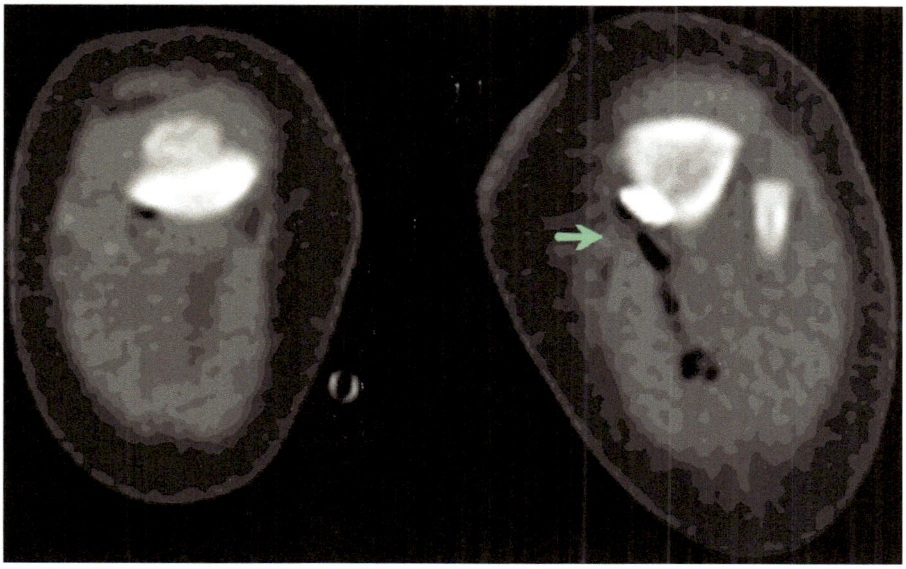

FIGURE 18.35 **Misplaced intraosseous line.** Termination of IO in the soft tissues posterior to the pediatric left tibia with air track.

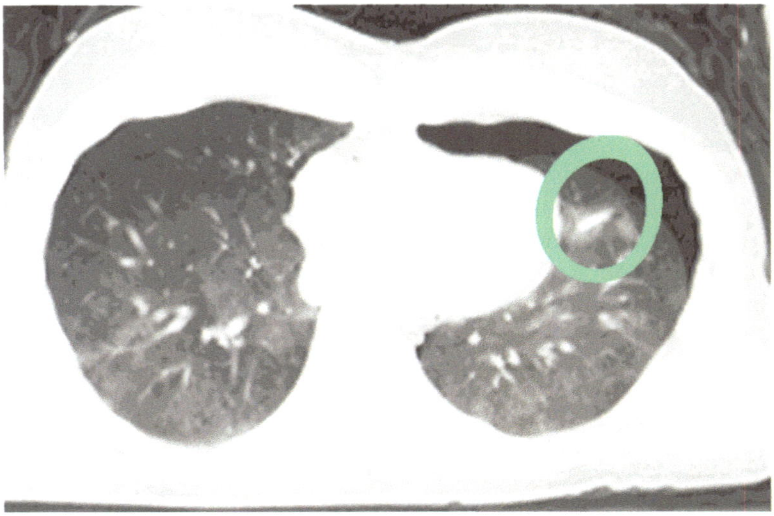

FIGURE 18.36 **Needle thoracostomy, lacerates lung.** Deep location of thoracostomy needle terminating in the left lung with resultant pulmonary laceration.

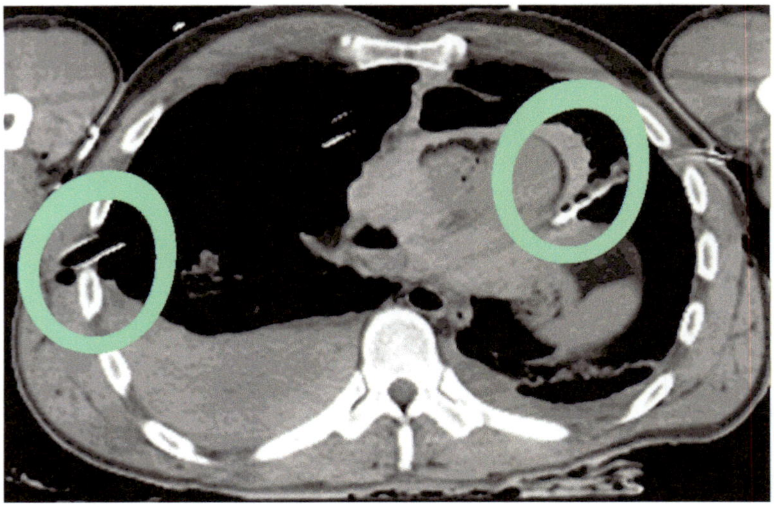

FIGURE 18.37 **Needle thoracostomy in pericardium.** Deep location of thoracostomy needle terminating in the heart with hemopericardium. Note right chest tube.

the heart owing to the high pressures in these chambers. Numerous other devices are implanted for cardiac pacing and defibrillation. When encountering an unknown device, it is best to do a quick literature search to confirm proper placement.

CT reporting

Currently, there is no standard for reporting PMCT studies. In many cases, the CT is used for triage purposes, and a full report is never rendered.

Using clinical imaging CT reporting and autopsy reporting as reference standards, a complete PMCT report should mention all organ systems. Pertinent findings from each organ system should be mentioned, and the report should be easy to read and standardized for quick reference. At the end of the report, an impression or key findings section is necessary to highlight the findings of greatest importance.

Example CT reporting template

Full report:
Degree of postmortem change:

Head, neck, and brain:
Head—Brain is symmetric without mass or midline shift. No intra- or extraaxial blood or fluid collection. The calvarium is intact.
Neck osseous structures—No displaced fracture or osseous lesion.
Neck soft tissues—No focal abnormality.
Chest:
Lungs—No focal abnormality.
Mediastinum—No focal lesions or evidence of trauma.
Osseous structures—No displaced fracture or osseous lesion.
Soft tissues—No focal abnormality.
Abdomen and pelvis:
Liver—No focal abnormality.
Gallbladder—No focal abnormality.
Spleen—No focal abnormality.
Pancreas—No focal abnormality.
Adrenal glands—No focal abnormality.
Kidneys—No focal abnormality.
Stomach—No focal lesion abnormality.
Large and small bowel—No focal abnormality.
Retroperitoneum—No focal abnormality.
Urinary bladder—No focal abnormality.
Other pelvic contents—Unremarkable.
Osseous structures—No displaced fracture or osseous lesions.
Soft tissues—No focal abnormality.
Imaged extremities:
Osseous structures—No displaced fracture or osseous lesions.
Soft tissues—No focal abnormality.
Key findings:
Insert clinically relevant information here.

Radiologic identification

Radiologic Identification **(RadID)** is one of the many tools at the medical investigator's disposal for decedent identification. Visual review, fingerprints, and dental records, as well as genetic and anthropologic examinations are valuable instruments as well. At the most basic level, all methods of identification are anchored to pattern recognition.

Today, we all benefit from the fact that clinical imaging is so widely applied.

CT offers the additional benefit of optimizing the imaging plane and rotation for an improved "apples to apples" assessment. Ideally, numerous points of reference will "match" between antemortem and postmortem images, facilitating radiological identification. That said, there is no agreed upon standard anatomical landmark or number of reference points that define a "positive" identification. Two classic anatomical points of reference include the **frontal sinuses and the sternal notch**.[28–39] In practice, all the available information, including provided history, scene photographs, prior medical records (including imaging), and autopsy/external examination results are aggregated to make the best assessment for identification purposes.

The more "matching" features, the better, with "unique" matching features favored over the more common imaging findings. For example, in a case involving a severely decomposed body, a "unique" partially calcified left upper quadrant mass was utilized to help make a definitive identification (in light of other known information, including imaging of degenerative joint disease changes) (Fig. 18.38). In some cases, the decedent has a constellation of common imaging findings/anatomical variants to support a positive identification (e.g., a right toe deformity, left hip hardware, and right iliac stent) and was also found in a secured residence, enhancing the confidence of the identification by context (Figs. 18.39–18.44).

In practice, the most efficient approach is to accumulate all relevant information from the case file and examine the antemortem imaging first, after which we create a list of antemortem imaging findings that are available to compare to PMCT. If possible, frontal sinuses from a head CT or the sternal notch from a chest CT will be the

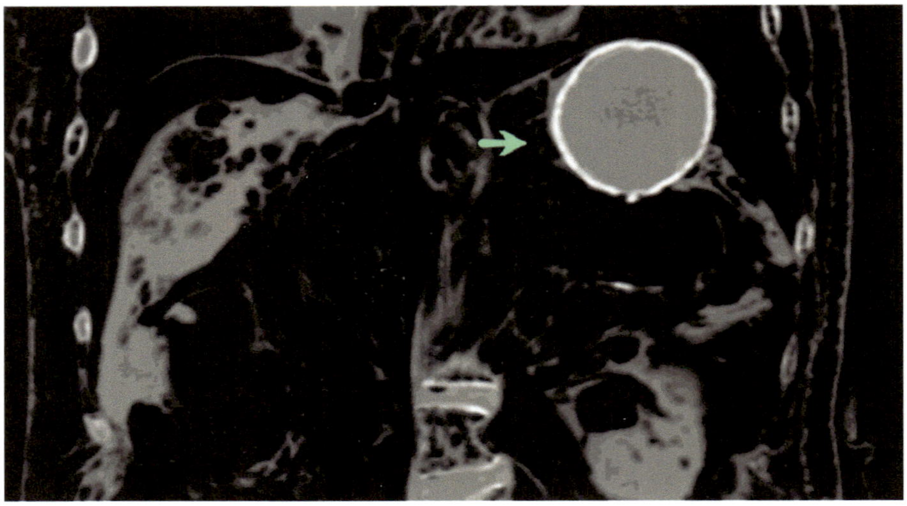

FIGURE 18.38 **Calcified mass in left upper quadrant of abdomen.** This match was supported by the left upper quadrant peripherally calcified mass seen on PMCT and documented on antemortem imaging.

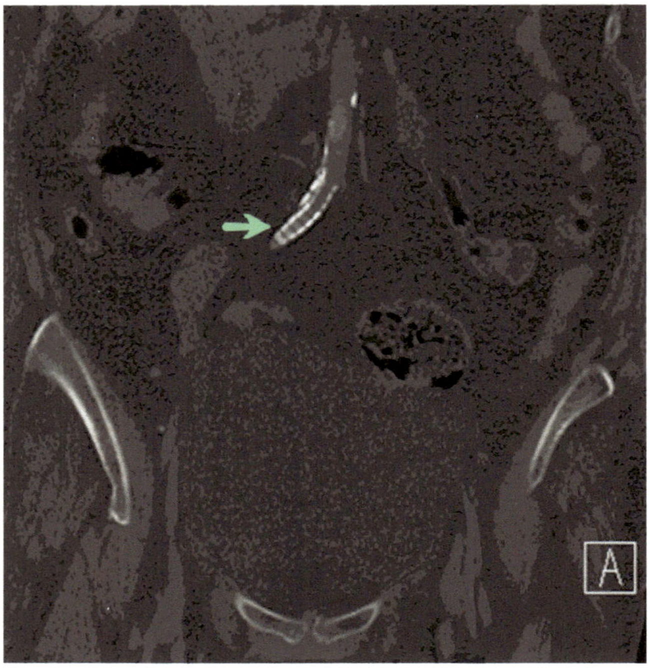

FIGURE 18.39 Antemortem right iliac stent.

most valuable sites for comparison. Next, any acute or healed fractures/deformities are noted. Patterns of degenerative change and anatomic variants are also helpful (such as cervical ribs, bone islands, and accessory ossicles). Even common pathology, such as rotator cuff calcific tendonitis, may be helpful. If the soft tissues are preserved, the size, location, and number of findings, such as hepatic lesions, renal cysts, and splenules should be noted. Create a list of antemortem findings as long as possible and use that to compare to the PMCT. If everything looks similar and the case file supports a match, the medical examiner can use this information to confirm the official identification.

Postmortem CT angiography

A major drawback to conventional PMCT is the lack of soft tissue contrast, especially when it comes to the identification of vascular injury. Postmortem change poses a significant impedance to this process. At its core, the PMCT

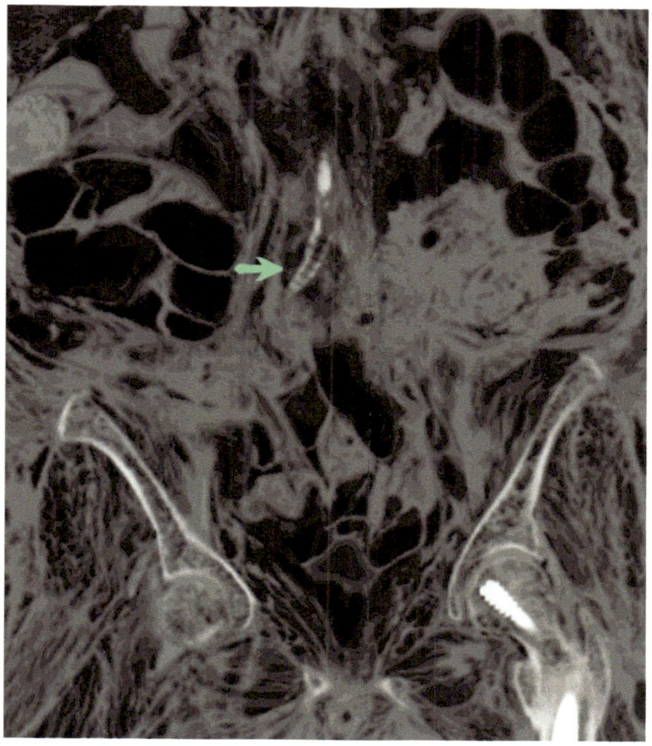

FIGURE 18.40 Postmortem right iliac stent.

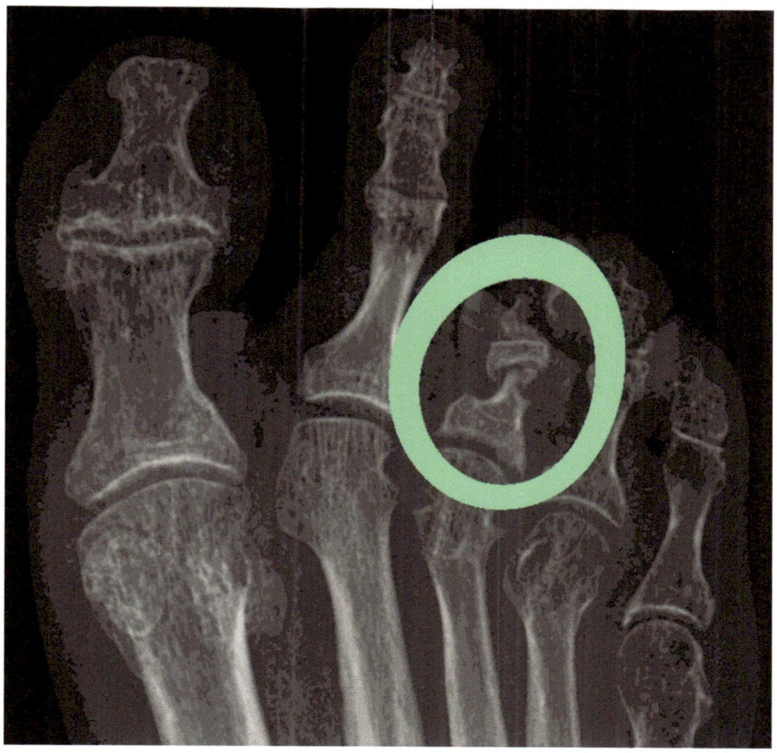

FIGURE 18.41 Antemortem right third distal toe deformity.

angiography (PMCTA) process involves a device with enough pressure to push contrast media through postmortem vessels that lack blood pressure. Multiple techniques have been developed with an array of injection pumps, contrast media, and protocols.[40–42]

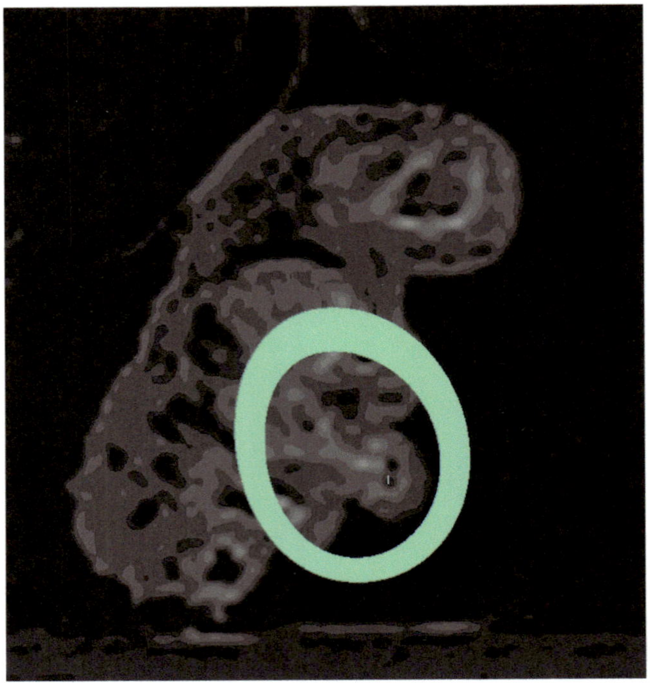

FIGURE 18.42 Postmortem right third distal toe deformity.

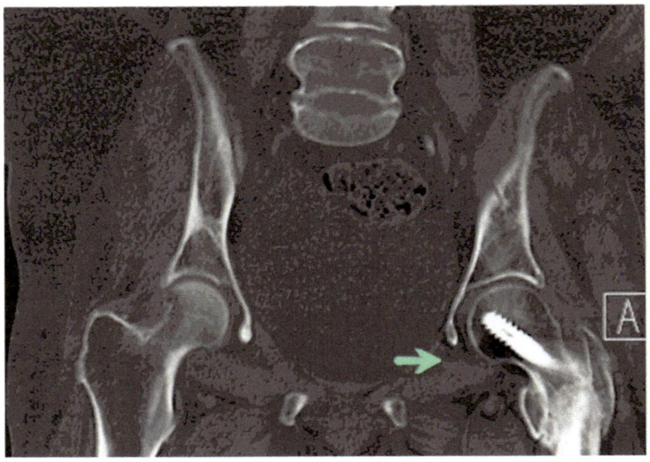

FIGURE 18.43 Antemortem hip arthroplasty.

The *femoral vein* is the ideal cannulation site, both related to its size and its inconspicuous location (this may be of particular concern when a viewing is planned as a part of funeral services). *Bodily fluids are procured for toxicological analysis prior to injection.* Care should be taken to close the loop for any possible extracorporeal fluid escape during the exam with a leak-proof body bag. Ross et al. recommend a control volume of up to 600 mL for optimal vascular opacification and to avoid pressure-related intravascular injuries.[43]

PMCTA can also be executed via a directed approach, with hand-injected contrast media for focused visualization of specific anatomic structures. In our office, we have benefited from this directed approach (Figs. 18.45–18.47).

Regardless of the approach, it is well known that CT and autopsy are complementary procedures, the latter being more sensitive for the identification of soft tissue findings. PMCTA can optimize soft tissue analysis to bridge this gap. *It has been reported that PMCTA increases PMCT sensitivity for detecting pathologic findings from 64% to 81%.*[44] The same study found that the sensitivity at conventional autopsy was 83%. It is important to note, however, that is not clear what importance *all* of PMCTA findings offer. In other words, it is unclear how many of those PMCTA-enhanced findings have any relevance to the mechanism and cause of death, and to what extent, if any, they would move the certification of death needle.

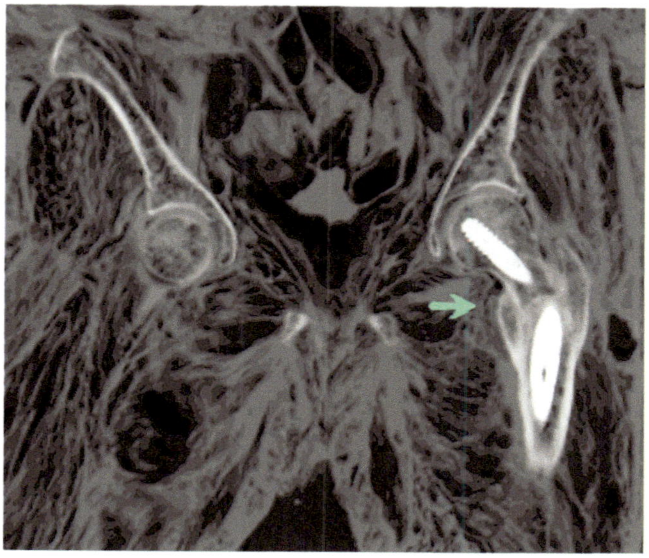

FIGURE 18.44 Postmortem hip arthroplasty.

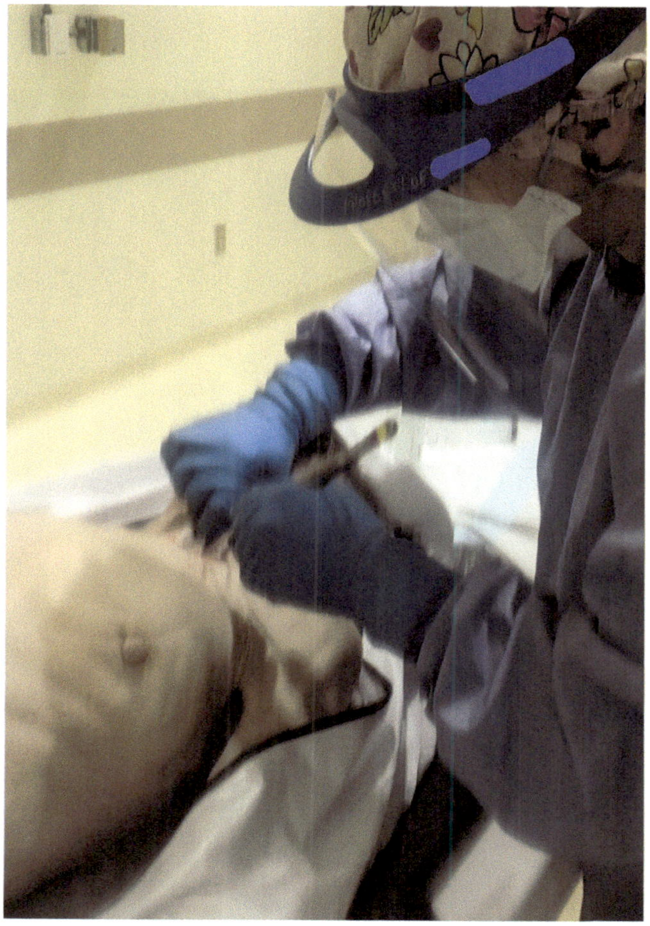

FIGURE 18.45 Carotid artery dissection for contrast media.

A subsequent multicenter study comparing conventional PMCT to PMCTA found that the latter detected over half of the important findings.[44] In addition to its strength at revealing skeletal trauma, PMCTA detected many vascular abnormalities that were not evident at autopsy. This is not to say that PMCTA is superior to autopsy; however, it highlights that it is a valuable complement to autopsy and could be used in specific cases.

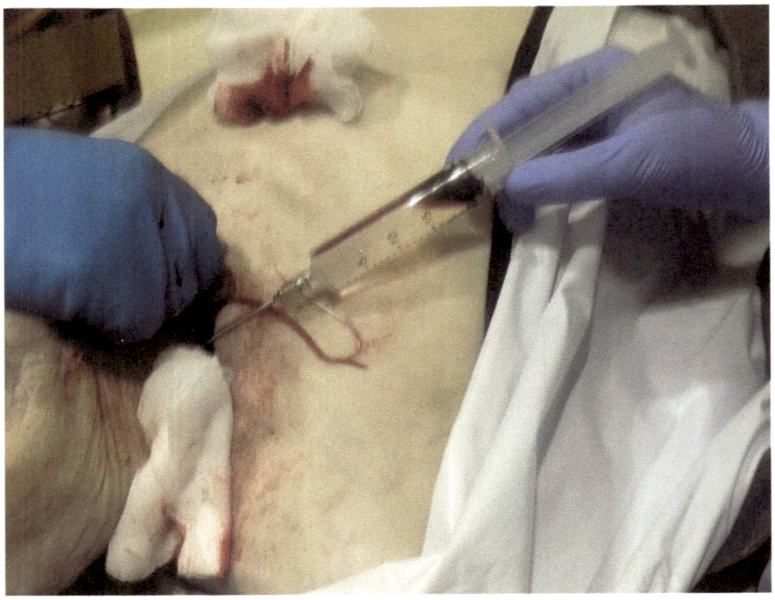

FIGURE 18.46 Carotid artery injection of contrast media.

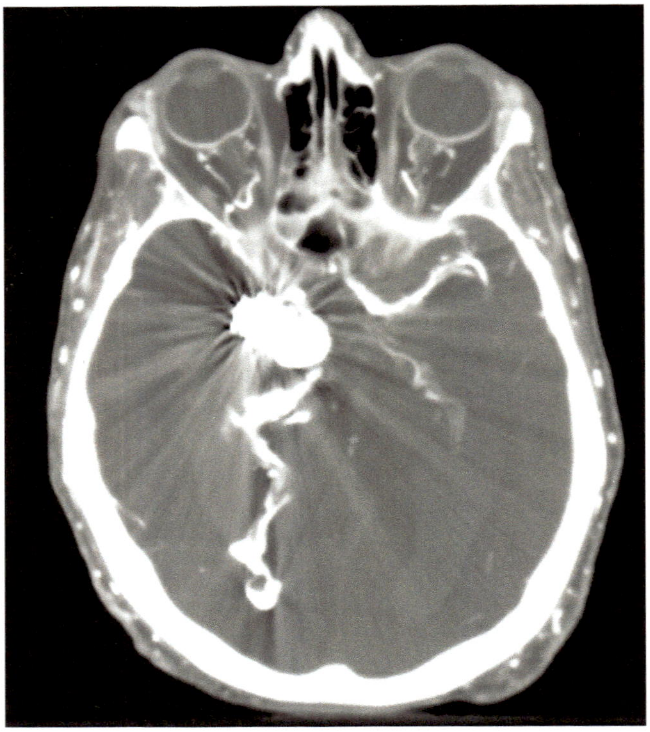

FIGURE 18.47 **Carotid artery dissection.** Axial CTA intermediate bone window. Hand injected CT angiogram (CTA).

For example, PMCTA can be employed to improve delineation of gray and white brain matter and to map out the cerebral vasculature.[43,45] In this way, even subtle findings such as small aneurysms can be detected.[46] Furthermore, the precise location of intimal injury can be detected by the identification and pattern of the extravasated contrast medium. In the neck, the difficult to reach vertebral arteries can be easily evaluated and traumatic brainstem injuries are readily delineated.[46]

Coronary artery stenosis can be quantified on PMCTA, and congenital anomalies can be nicely illustrated. Cardiac rupture and tamponade are clearly depicted as well as penetrating injuries. Pulmonary arterial filling defects are

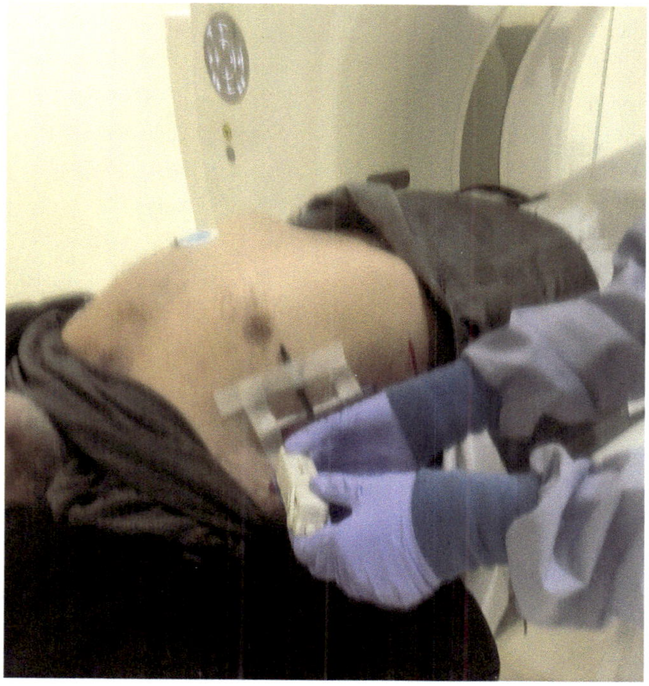

FIGURE 18.48 **Abdominal mass biopsy.** Picture of CT-guided biopsy.

well demonstrated; however, there should be a word of caution given the overlap of imaging findings with postmortem clot. As such, gross and histopathology remains the gold standard for the diagnosis of pulmonary thromboembolism.

The abdominal aorta is typically decompressed on PMCT; therefore, PMCTA can be helpful to characterize aneurysms, dissections, and the location of ruptures. Likewise, solid organ blunt force trauma and penetrating injury can be directly examined. Extremity vasculature may also demonstrate the location of injury of a major blood vessel.

Image-guided procedures

Clinically, if you can visualize a finding on imaging, you can sample it with image guidance. The same is true for postmortem examinations. *Both PMCT and ultrasound are powerful tools for fluid and tissue sampling.*

PMCT is best employed for solid tissue sampling. CT is a great tool for localizing a lesion or mass (including osseous lesions) and for guiding needle placement. First, the anatomic structure is localized on PMCT. Next, a radiopaque marker can be placed on the skin to be used as a reference for skin entry. Subsequently, measurements can be made to guide the appropriate tissue depth. Standard needle sizes include 14 and 18 gauge. Numerous specimens should be obtained to diminish the risk of sampling error. After the needle is advanced, the decedent can be placed back in the scanner to confirm needle placement (Fig. 18.49). Once, properly positioned, the samples can be obtained and stored for future analysis. Images can be stored on the Picture Archiving and Communication System (PACS) to document appropriate sampling as a reference if needed.

Similarly, PMCT can be used to localize foreign bodies for extrication. Most commonly, this involves the removal of a retained bullet. The original decedent scan is often sufficient to guide a directed search for a ballistic item. If necessary, a radiopaque marker, as employed with CT-guided tissue sampling, can help narrow the search by allowing the practitioner to dynamically visualize the most direct approach.

Ultrasound is an exceptionally good tool for localizing and sampling fluid collections. Some basic ultrasound techniques need to considered, such as artifacts and transducer/needle positioning. There are artifacts of acoustic enhancement and acoustic shadowing. Acoustic enhancement is sometimes referred to as "increased through transmission," and is seen as brightness deep/posterior to a structure that transmits sound well (such as fluid). It is a useful artifact that helps discern fluid-containing structures from nonfluid-containing structures. Acoustic shadowing is a signal void, or darkness, behind a structure that does not transmit sound well (like bone and gallstones). The biopsy/aspiration

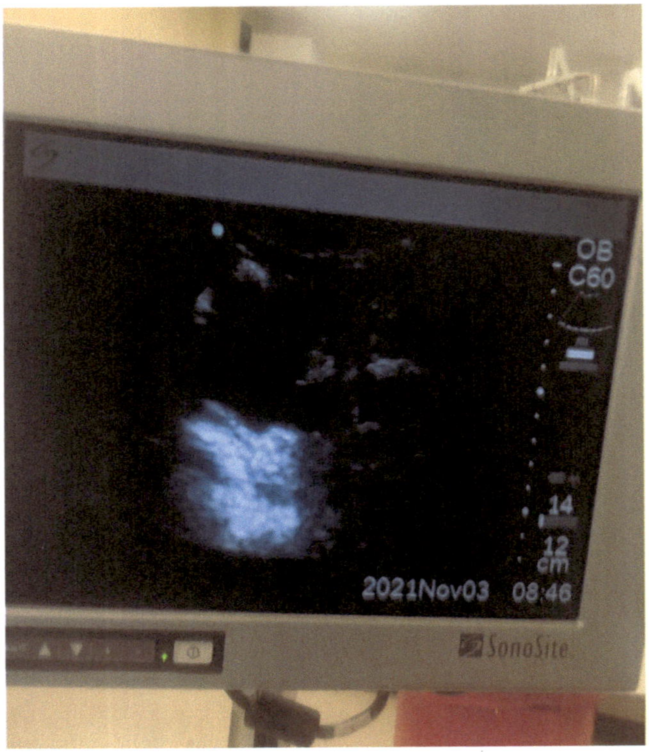

FIGURE 18.49 Ultrasound image of the bladder.

needle will also cause shadowing, requiring proper technique to allow for constant visualization of your target during the procedure.

The in-plane technique is the most commonly used for sampling tissue and fluid with ultrasound guidance. It involves holding the transducer such that the short axis is held in the direction where you plan to direct you needle into the skin. Using this method, you can see your needle in the fluid collection (Figs. 18.50 and 18.51). The needle is well visualized when the ultrasound beam is perpendicular to it. It is advisable to localize the largest portion of your target directly under the transducer (aim for the broad side of the barn) and move your needle into plane with the target. Moving the transducer and the needle at the same time will result in frustration due to the difficultly in

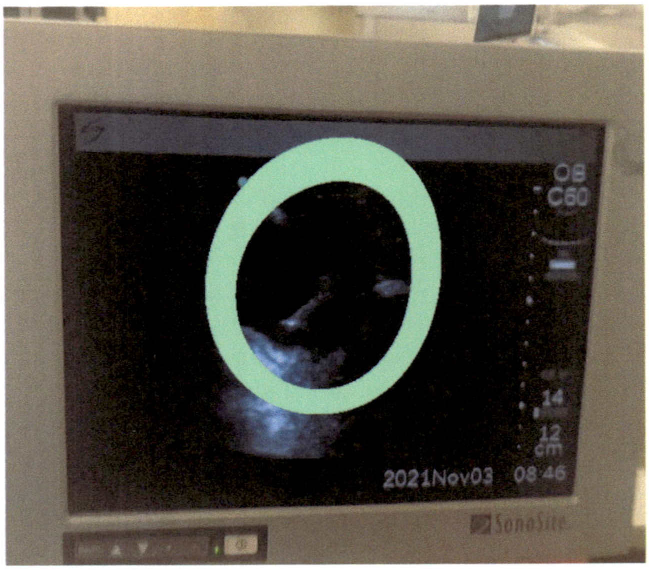

FIGURE 18.50 **Ultrasound image of the bladder with aspiration needle inserted.** Note posterior acoustic enhancement deep to the bladder.

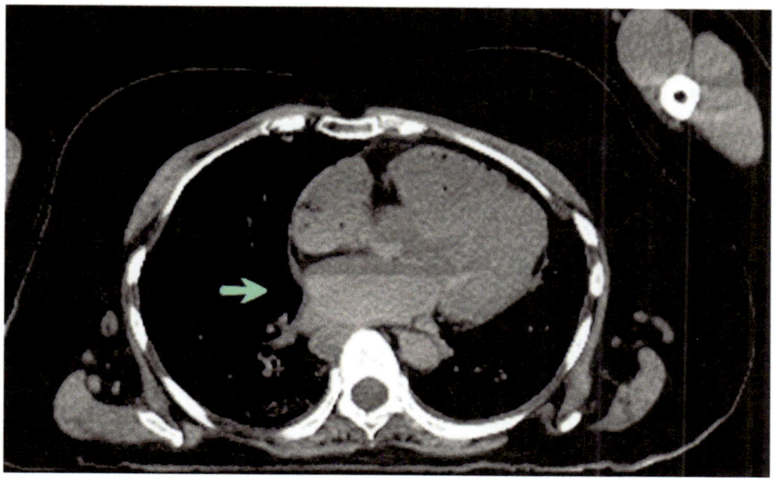

FIGURE 18.51 Axial CT thorax with soft tissue window. Livor in the heart.

maintaining visualization of one's target. A good problem-solving tip is to gently rock the transducer back and forth along the long axis (centered over your target) until you see both the target and the needle aligned. You can "cheat" and line up your needle with the long axis of the transducer at the skin surface to improve alignment with the transducer. Once appropriately positioned, the needle is advanced under direct visualization (think of jaws approaching a victim underwater). In terms of aspiration, imaging can capture the real-time decrease in size of the fluid collection. Large bore needles work best for quick access, especially for viscous fluids.

In the case of an *ultrasound-guided soft tissue biopsy,* the needle can be aligned adjacent to the target for a spring-loaded core needle biopsy and can be subsequently deployed under direct visualization. Other devices can be advanced into the lesion with the cutting apparatus in the "open" position and closed within the center of the target for sampling.

Fine needle aspiration can also be easily navigated with ultrasound guidance. The same approach is used as described above. A helpful trick it to attach a syringe to the needle and create a small vacuum by gently pulling back the plunger at a constant volume of air while entering and exiting the target. Allow the plunger to return to a neutral position, and then remove the needle to assess your sample.

Postmortem and decomposition changes on imaging

Imaging findings related to postmortem decomposition are dictated by the same processes as those seen at autopsy: livor mortis, rigor mortis, and decomposition. The degree of postmortem change is influenced by the position of the body (gravity), as well as other factors, such as the postmortem interval, temperature, and humidity.

Livor mortis is the first postmortem change that is readily apparent upon visual inspection. Complementary to visual inspection, dependent skin thickening and increased attenuation of the dependent skin and subcutaneous fat are appreciated on CT. Livor mortis is nicely manifested in the great vessels and the heart by way of the hematocrit effect, with layering of the dependent cells (Fig. 18.52).

The lungs are also a characteristic location for findings of lividity. For example, dependent ground glass opacities are the hallmark of lividity in the lungs; the ground glass opacities do not cross the fissures (Fig. 18.53). These findings are positionally dependent, such that a body that is found prone will have a different pattern of lividity compared to that of a body found supine.

Rigor mortis typically does not result in significant imaging findings aside from those related to the difficulty of positioning the body for imaging purposes.[47–50]

Changes related to postmortem tissue breakdown are characterized by autolysis, **putrefaction**, and liquefaction. Autolysis (nonbacterially mediated tissue breakdown) is characteristically seen in the pancreas, where digestive enzymes begin to break down pancreatic tissue in the postmortem period. This process is typically superimposed by the development of gas within the pancreas (putrefaction) given its proximity to the digestive tract (Fig. 18.54).

Putrefactive findings often dominate decomposition changes on imaging and can be found throughout all tissues of the body, even in the early postmortem period. Typically, soft tissue gas is first seen in the gastrointestinal track,

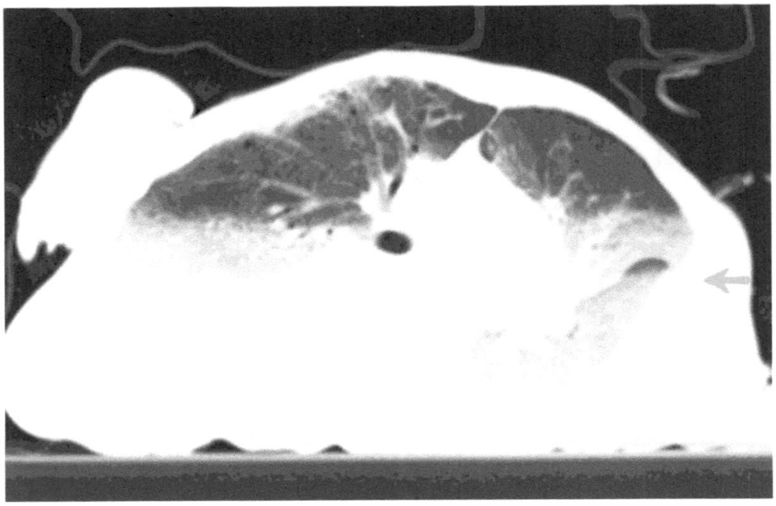

FIGURE 18.52 **Axial CT thorax with lung window. Lividity in the lungs.** Note that lividity does not cross the fissures.

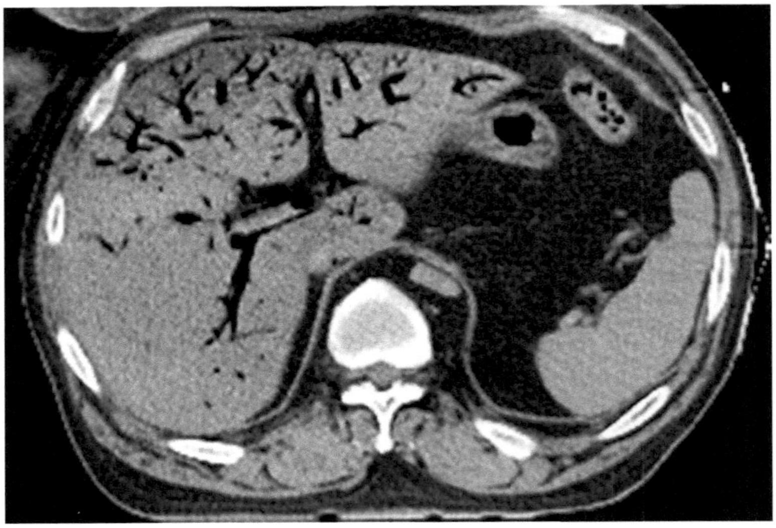

FIGURE 18.53 Early decomposition with gas in the biliary system.

fueled by local flora. In early cases of decomposition, gas is frequently seen in the vessels, the heart, and the liver (Fig. 18.55). As putrefaction progresses, (either related to the postmortem interval, natural disease, body temperature, environmental factors, or medical intervention), gas will be seen throughout all soft tissues, solid organs, and even in the bones (Fig. 18.54). Cardiopulmonary resuscitative efforts may be related to more extensive soft tissue gas.[51,52] At moderate to advanced states of decomposition, definitive assessments of pneumocephalus, pneumothorax, pneumomediastinum, and pneumoperitoneum becomes impractical (Fig. 18.56). In advanced putrefaction, the bones become diffusely saturated with gas, complicating the evaluation of nondisplaced fractures and subtle osseous lesions (Fig. 18.57).

Autolysis is well appreciated on imaging of the brain. Early effects of autolysis manifest as cerebral edema with poor gray–white differentiation, with variable effacement of the ventricles. In moderate to advanced states of putrefaction, the brain liquefies and a meniscus is present in the cranium (Fig. 18.58). Similarly, fluid is released from other tissues and blood vessels, causing purged pleural fluid, pericardial fluid, and peritoneal fluid which can imitate pathologic fluid accumulations.

Findings of atlanto-occipital subluxation can be confounding with postmortem imaging. Lack of muscle tone and the high variability of head positioning add to the inherent challenges of the evaluation of subtle craniocervical injury. That said, some cases demonstrate clear malalignment and/or characteristic avulsed fractures (Figs. 18.59 and 18.60). At our institution, we carefully evaluate atlanto-occipital alignment in three planes and cases with

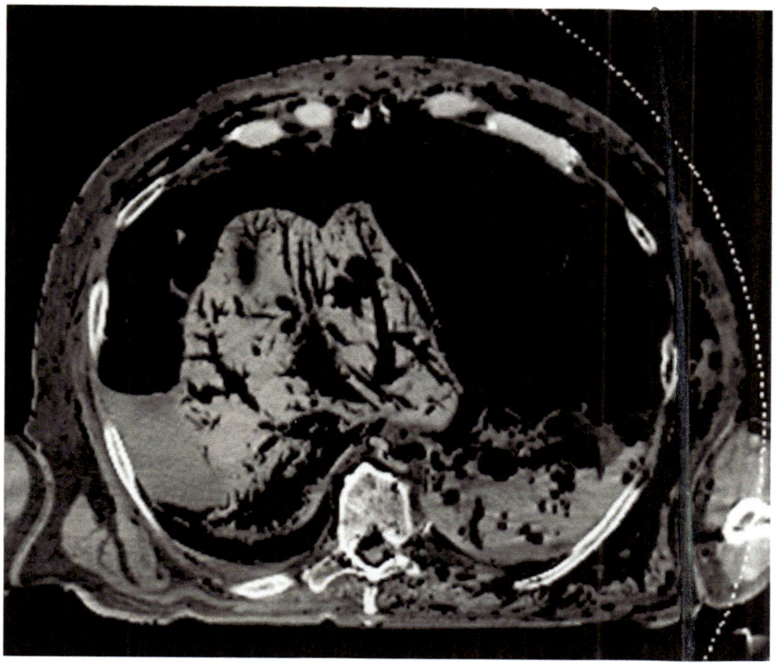

FIGURE 18.54 **Moderate decomposition.** Progressive putrefactive gas throughout the liver as well as the spleen and other soft tissues. Free fluid is present from autolysis.

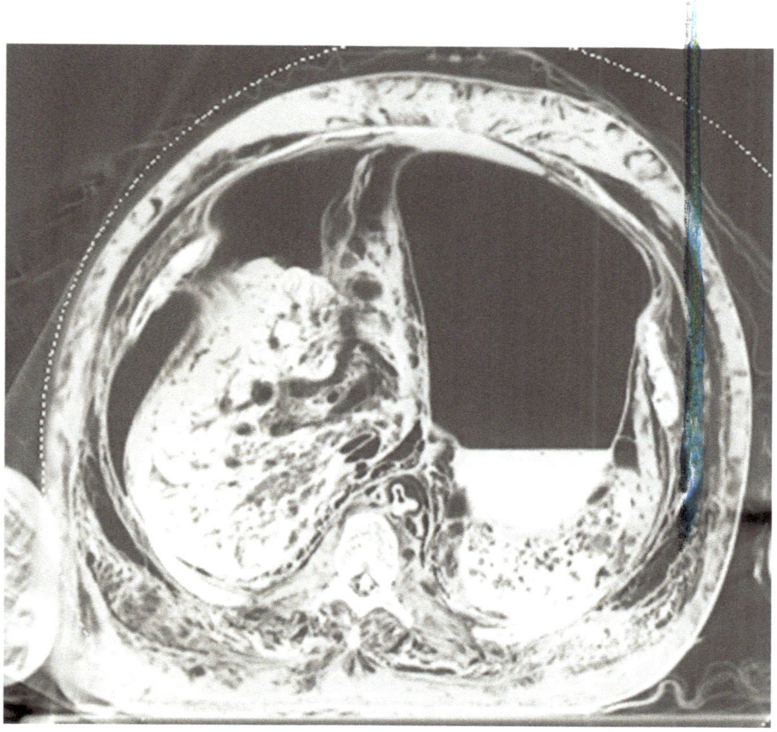

FIGURE 18.55 **Advanced decomposition.** Progressive air limits evaluation for pathologic gas collections.

unequivocal CT findings and compelling histories are further evaluated with a posterior neck dissection for radiological and autopsy correlation.

Once familiar with the expected decomposition changes, PMCT is helpful in a multitude of cases. In the majority of cases, PMCT can either exclude or shed light on underlying skeletal trauma. However, there are those exceptional cases where soft tissue injuries remain evident and not only elucidate, but also help to document the cause of death (Figs. 18.61 and 18.62).

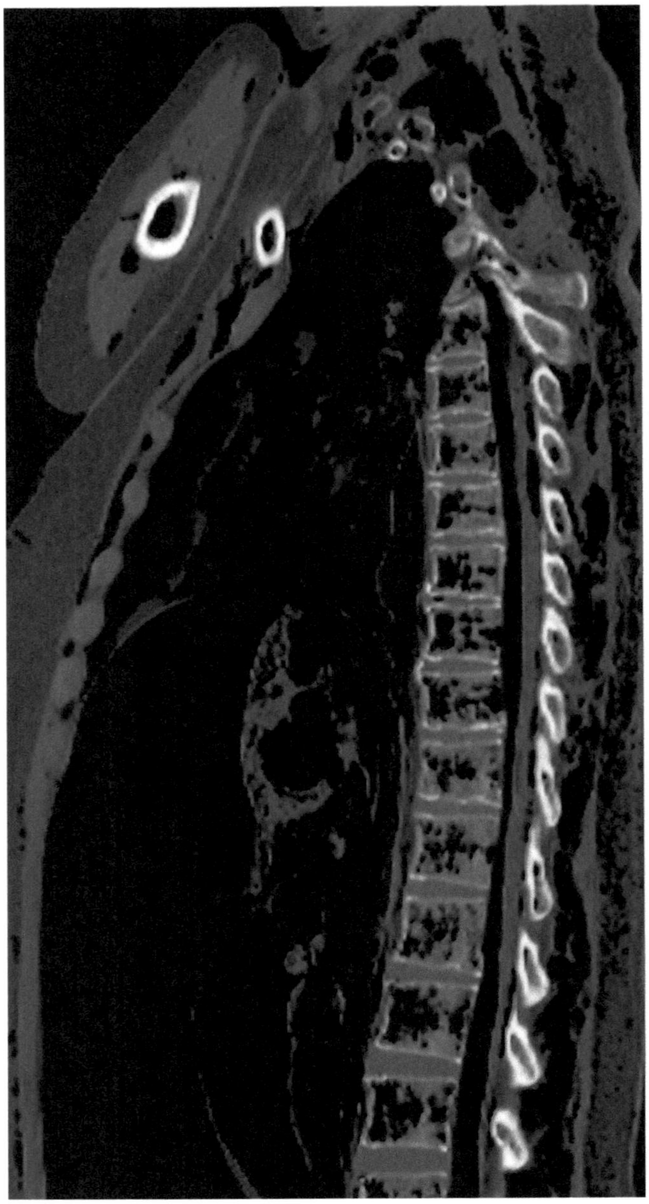

FIGURE 18.56 Gas throughout the bones limits evaluation for subtle fractures.

Mummification, skeletonization, predation, and charred remains

In cases of Mummification, Skeletonization, and Predation, the greatest strength of PMCT is its exquisite ability to delineate skeletal anatomy. As detailed under RadID, patterns of skeletal findings can be used for identification purposes. Additionally, even under extreme circumstances, skeletal injuries may provide critical information about underlying injuries *relating to the cause of death.*

Mummification and skeletonization

The first thing that comes to mind with mummified remains is mummies. But, mummification and skeletonization also occur frequently in contemporary investigations (especially in arid dessert climates of the Southwest). In mummified remains, the skeleton is protected by the outer envelope of desiccated soft tissue. Skeletal remains can provide valuable information about the decedent's sex, age, and prior trauma. Every so often, a surprise may be hidden for you to find (Figs. 18.63 and 18.64).

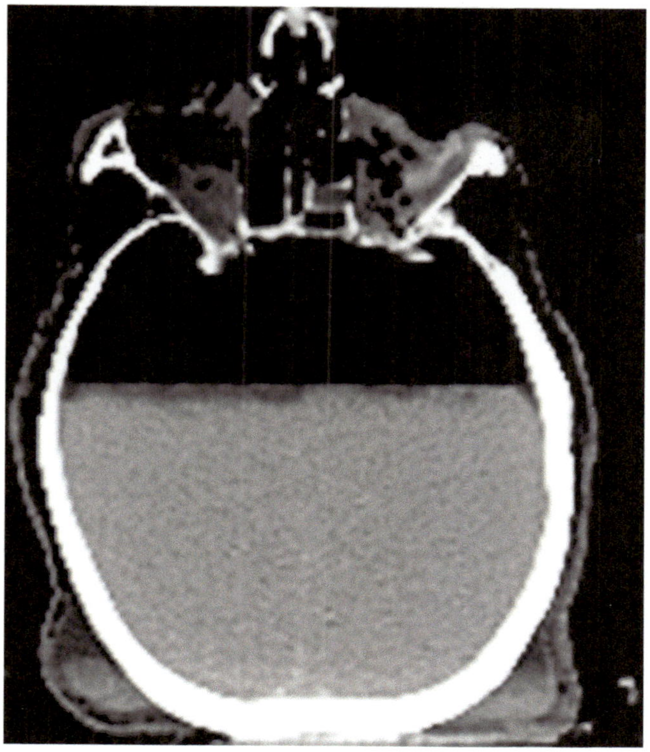

FIGURE 18.57 Axial CT brain with brain window. Advanced brain decomposition with fluid level.

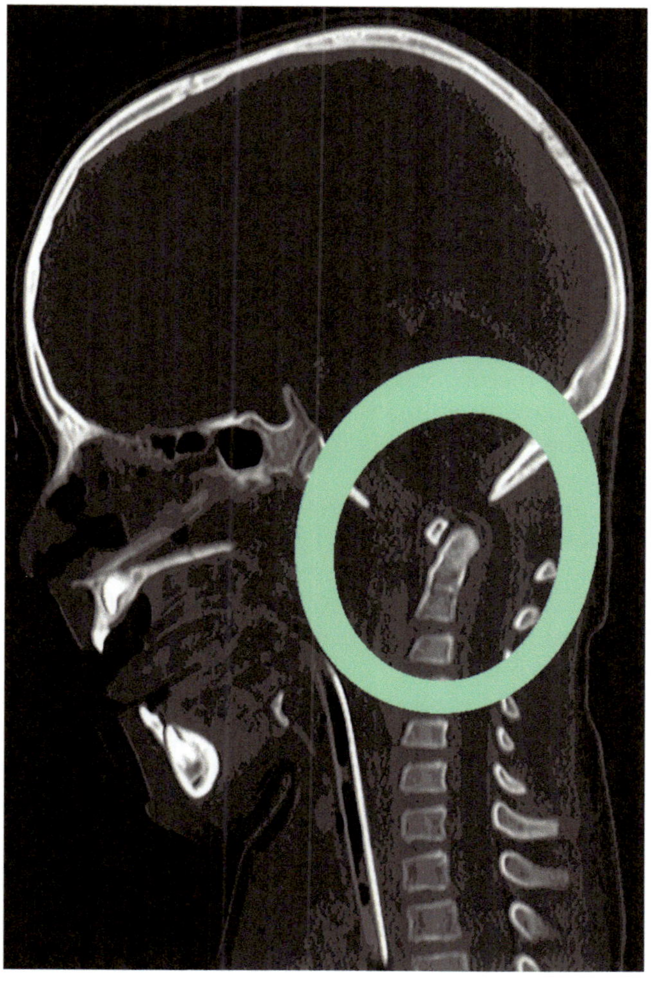

FIGURE 18.58 **Atlantooccipital injuries.** Severely displaced AOD.

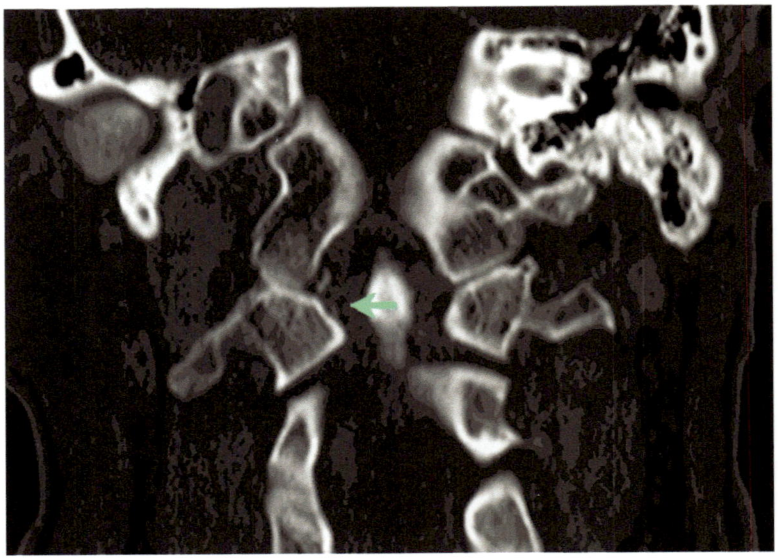

FIGURE 18.59 Atlantooccipital injuries. Subtle right occipital avulsion fracture.

Predation

Predation findings can be confused with antemortem injury. Exposed anatomy is typically targeted in the outdoor setting; however, any body part can be affected and may create further difficulty when making an identification.[53] Unfortunately, imaging is limited in its ability to differentiate predation changes from antemortem injury. These cases require careful assessment by the medical examiner in the autopsy suite (Figs. 18.65–18.67).

Charred remains

The tissue destruction inherent to charred remains can add considerable complexity to the task of determining cause and manner of death. Precious fluids and soft tissues required for sampling can be nonexistent. Furthermore, the skeletal examination may be thwarted by the difficultly in removing charred soft tissues and also in recognizing thermal-related fractures.[54,55]

In such remains, the skin and immediately underlying soft tissues may be consumed, exposing muscle. The thinner the soft tissue layer, the more rapid the destruction of that tissue and the more likely that thermal osseous injury occurs. The exposed muscle shrinks/contracts, leading to flexion deformities of the "pugilistic" posture, with the long bones undergoing consequent thermal fractures. In contradistinction to antemortem trauma, thermal fractures are smoothly marginated and have been described having a "flute mouthpiece or stair step" morphology (Fig. 18.68). Additionally, thermal fractures are more likely to be bilateral and symmetric as compared to antemortem injuries.

Despite extensive burn injuries, the skull may still provide valuable information relating to identification. The paranasal sinuses are extremely useful in that regard and are relatively specific for identification purposes. Additionally, the teeth are typically shielded from thermal injury by the overlying buccal mucosa. Native teeth, restorations, and prosthetics are often very well preserved. PMCT is exceptional for comparisons to antemortem CT and dental imaging and records. A "pseudoradiograph" can be reconstructed from the CT images of the mandible and maxilla (Fig. 18.69) for comparison purposes.

One of the pitfalls in the evaluation of a skull subjected to thermal injuries is an incomplete thermal fracture, which can mimic antemortem trauma. It is important to closely examine areas of the skull that lack overlying soft tissues, as these locations are at greatest risk for these linear fractures of the outer table (the inner table integrity remains preserved). Conversely, antemortem injury-related fractures are perpendicular to bone. Additionally, a thermal-related epidural hematoma (EDH) detected by PMCT can mimic antemortem injury. In both instances, the collection will be low density and crescentic in shape; however, thermal-related collections will often cross the midline and be associated with detachment of the venous sinus secondary to exudation of blood from it.[56] That said, if subsequently performed, the autopsy identification of such epidural hemorrhages will easily discern

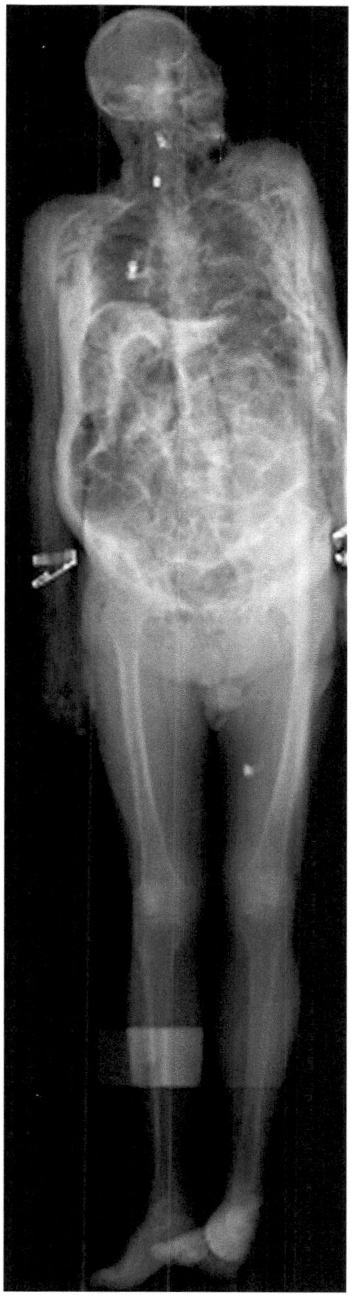

FIGURE 18.60 Scout view demonstrates findings associated with decomposition. PMCT reveals ruptured aortic aneurysm with left retroperitoneal hemorrhage (Fig. 18.62).

a thermally induced EDH from one that is trauma-related by its gross appearance alone (the former having a dark maroon color and a granular consistency).

Natural disease

PMCT can provide an excellent picture of the overall health of an individual. Findings of atherosclerosis, hypertensive organ effects, fatty changes of the liver, bacterial and viral pneumonia, and fatty changes of the liver are well visualized on PMCT.

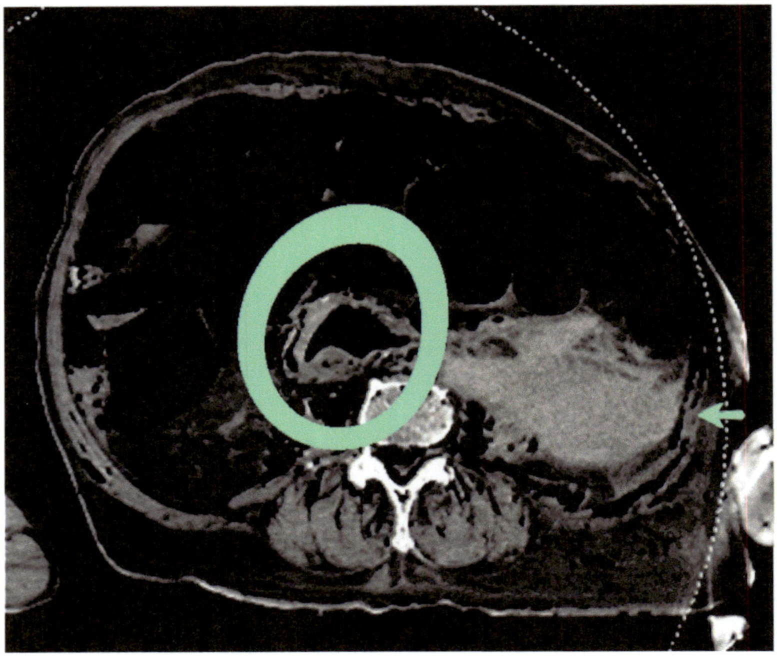

FIGURE 18.61 **Decomposed decadent with ruptured aortic aneurysm.** PMCT reveals ruptured aortic aneurysm with left retroperitoneal hemorrhage.

Brain

Cerebral atrophy, demonstrated by loss of brain parenchyma, is apparent on PMCT on coronal sections as enlargement of the cerebral ventricles. While this feature is nonspecific, as many different types of neurodegenerative diseases and chronic alcohol use may be the cause, neuropathologic examination is the gold standard for further elucidation of these diseases (Fig. 18.70). Unilateral ventricular enlargement is less likely to be associated with neurodegenerative disease, and one should divert attention to carefully evaluate for focal parenchymal loss, such as a remote cerebrovascular event (Fig. 18.71).

Nonenhanced PMCT has a high sensitivity for the detection of **hemorrhagic stroke**, while ischemic strokes are not well visualized. If the survival interval following an acute stroke is long enough (usually longer than 2–3 h), edema will be present (Fig. 18.72). Clinically, perfusion CT scans are useful tools in the evaluation of acute strokes, but they are not able to be performed reliably in a postmortem setting. Hemorrhagic stroke is evident in 10%–15% of all strokes and is more common in the Western part of the world, and have a poorer prognosis than ischemic strokes.[12] The majority of cerebral hemorrhages (approximately 80%) are primary in nature, with chronic hypertension as the underlying cause (Fig. 18.73).[13] That said, it is worth bearing in mind that superficial lobar hemorrhages in an elderly decedent should warrant the consideration of cerebral amyloid angiopathy. Common causes for other intracerebral hemorrhages include vascular malformations, illicit stimulant drugs, and trauma. Aside from hemorrhage arising in the classic locations for hypertensive hemorrhage (basal ganglia, pons, cerebellum), autopsy is recommended to better discern the cause of a hemorrhage and classify the most appropriate manner of death. **Basilar subarachnoid hemorrhage** is most often the harbinger of a ruptured aneurysm involving the circle of Willis, which is discussed in the angiography section of this chapter.

Heart

The evaluation of hypertensive and atherosclerotic cardiovascular disease constitutes a less obvious use of CT evaluation clinically, but it is an extremely useful tool in a postmortem setting, when considering natural disease. The constellation of numerous findings, in the proper context, can suggest the cause of death as being due to natural causes.

While CT examination of the heart shows objective findings, clinical judgment is necessary to decide whether these findings are the cause of death. The most helpful findings in the evaluation of **hypertensive and atherosclerotic cardiovascular disease** are cardiomegaly, vascular calcifications of the aorta, coronary arteries, and peripheral arteries, as well as chronic kidney disease. Discerning the presence of cardiomegaly can be difficult and takes

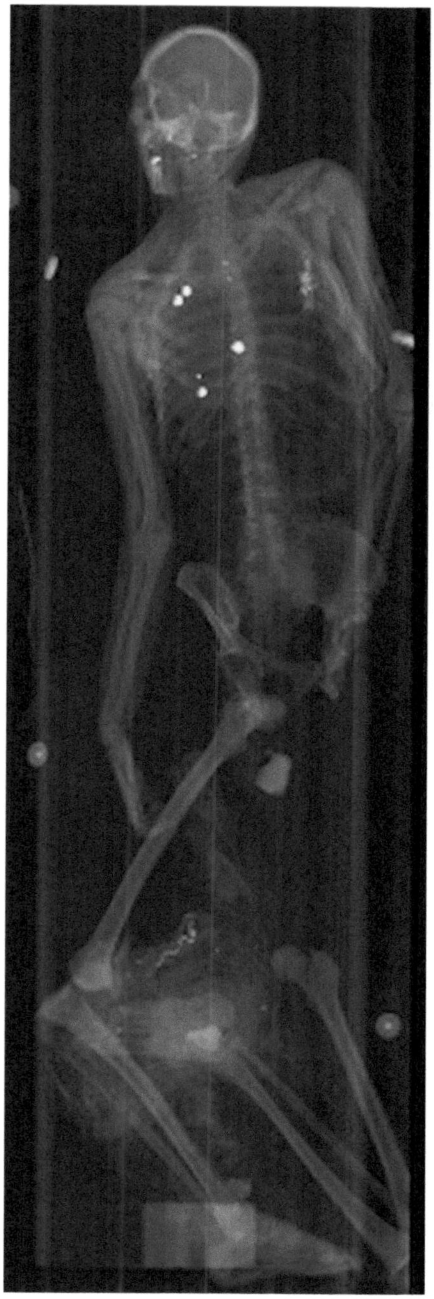

FIGURE 18.62 **Skeletal remains with multiple retained ballistics.** Scout view demonstrates multiple retained ballistics.

practice by subjective observation. If in doubt, the cardiothoracic ratio is a crude measurement of the widest aspect of the heart compared to the widest aspect of the chest cavity. If greater than 50%, cardiomegaly is likely present (Fig. 18.74).[14] When present, it is also an indicator of the decedent's antemortem path to congestive heart failure. Calcifications are easily identified on PMCT, but the degree of stenosis cannot be ascertained and even coronary CT angiography has been well known to overestimate the degree of stenosis.[15,16] While coronary artery calcifications may be diffuse, one cannot immediately conclude that the degree of stenosis is clinically significant and clinical judgment must be used to decide if the findings likely caused death given the specific circumstances. Vessel calcifications are seen as hyperintense lesions along the vascular wall, such that, in severe cases, the entire vessel outline is readily apparent (Fig. 18.75).

Hemopericardium (blood within the pericardial cavity) is most often the result of an aortic dissection, ruptured myocardial infarction, or trauma, and is best visualized as high-density fluid with an HU measurement of 35 units or

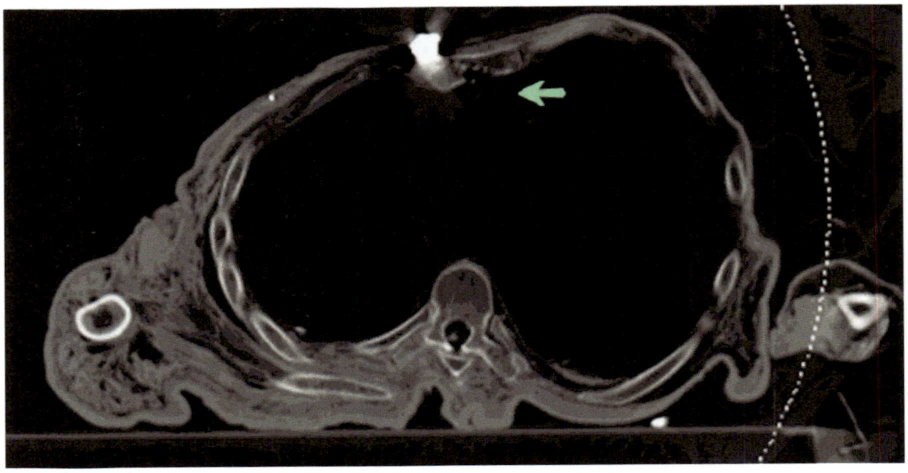

FIGURE 18.63 **Skeletal remains with multiple retained ballistics.** Axial PMCT thorax with bone window reveals acute fracture of the sternum associated with a retained ballistic.

greater (Fig. 18.76). When cardiopulmonary resuscitation (CPR)-related injuries are present, particularly sternal fractures, one must also consider whether the hemopericardium could be a result of penetrating injury due to CPR.[17]

An **aortic aneurysm**, another consequence of hypertensive and atherosclerotic cardiovascular disease, is seen as focal dilation of the aorta. Rupture of the abdominal aorta is associated with retroperitoneal and/or free peritoneal hemorrhage with the occasional visual identification of the discontinuity of the aortic wall, which signifies the point of aortic rupture.[18]

Chronic hypertensive changes in the kidneys are evident on PMCT, with PMCT being useful for the detection of cysts, neoplasms, hemorrhage, and calcifications. Chronic hypertensive changes are most evident by thinning of the renal cortex (Fig. 18.78).

Pulmonary

Viruses are the most common culprits of upper respiratory infection, and while their features might overlap somewhat on PMCT, subtle changes may be present based on the activity of the virus. The value of PMCT in the evaluation of pneumonia proved to be a very beneficial tool for case triage during the COVID-19 pandemic. As such, we have devoted a section within this chapter on COVID-19.

The **viral pattern** of lung infection on PMCT are diffuse ground glass opacities.[19] Additionally, as all viruses are not created equal, their effects on the lungs are not equal. For example, respiratory syncytial virus and human metapneumovirus cause patchy consolidation with ground glass opacities, with centrilobular nodules; influenza causes reticulonodular opacities, with or without focal consolidation, that usually involve the lower lobes (Fig. 18.79).[20] In this author's opinion, **differentiating between CPR-related changes and viral pneumonia** can be challenging. However, CPR-related changes tend to be centrally located, without opacities that extend to the periphery of the lung and often have associated anterolateral rib fractures. An added benefit of using CT as a triage tool, especially in the setting of COVID-19, is in the evaluation of a superimposed bacterial pneumonia (Fig. 18.80). In offices where technology affords the opportunity, CT-guided needle biopsies can be utilized to make a diagnosis.

Bacterial pneumonia can often be discerned from viral pneumonia with CT examination. Considerations must be made when triaging for autopsy versus external examination, reconciling that identification of the bacteria is not possible with external examination and weighing the need to make such an identification as part of the medical examiner's role in tracking emerging infectious diseases. Secondly, careful evaluation for the presence of rib fractures should be made, which could feasibly change the manner of death from natural to accident. This consideration should be made in the presence of an acute or subacute rib fracture that may have caused subsequent immobilization of the rib cage, resulting in a mechanical decrease in inspiration, with the subsequent development of pneumonia.

Previously, pneumonia was categorized clinically into lobar and aspiration pneumonia, but this terminology has been largely discouraged, with the subsequent classifications of community-acquired pneumonia (CAP), nosocomial pneumonia, and aspiration pneumonia. CAP most closely correlates to lobar pneumonia, which is limited to one lobe or segment of a lung. Aspiration pneumonia is demonstrated by centralized opacities that correlate to aspirated material within the bronchioles and is usually limited to the lower lobes.[21] Nosocomial pneumonia is a term

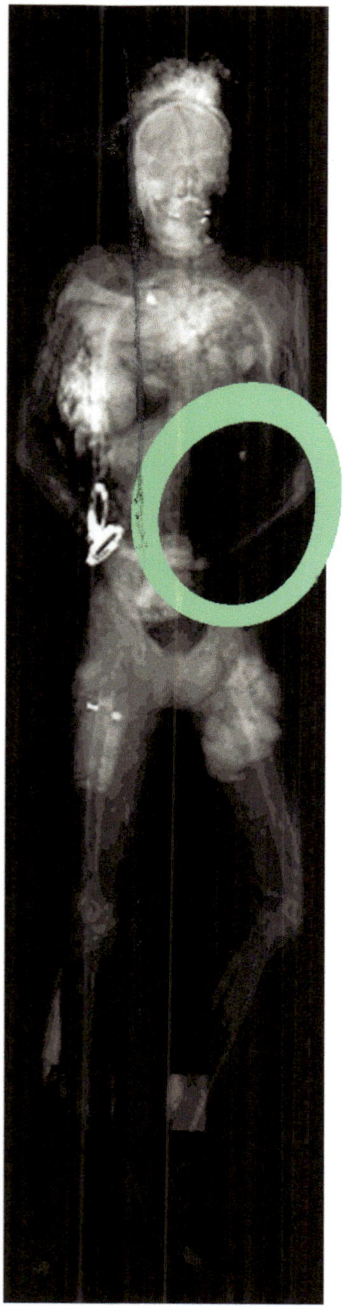

FIGURE 18.64 Scout view demonstrates missing tissue from predation.

that is used for pneumonia that develops while hospitalized, and the role of imaging in this area is limited. Nosocomial pneumonias are associated with higher mortality, but the pathologist is often able to request culture results prior to antibiotic administration and autopsy may not be warranted from a forensics standpoint.

Pleural collections are easily seen on PMCT and can often be distinguished between hydrothorax (effusion) and hemothorax by to the presence of sedimentation, which is seen in hemothorax. However, sedimentation can sometimes be seen in decomposition changes. When questionable, needle aspiration of the fluid is warranted during external examination. Correlation of a pleural effusion without sedimentation with other findings of atherosclerotic and hypertensive cardiovascular disease, such as cardiomegaly, atrophic kidneys, and coronary artery atherosclerosis, supports pleural effusions associated with congestive heart failure. Pleural nodules associated with pleural effusions should raise the possibility of a malignant pleural effusion.[22] Bilateral large pleural effusions in a person

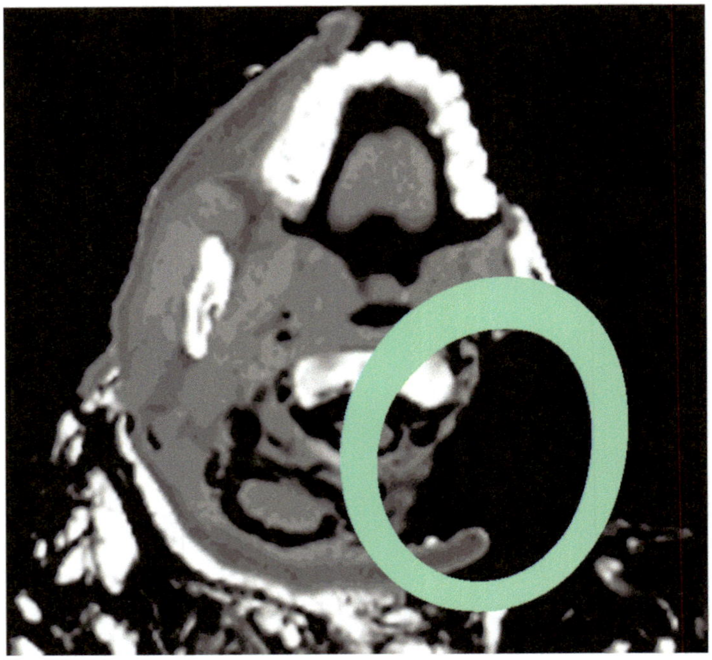

FIGURE 18.65 Predation of head.

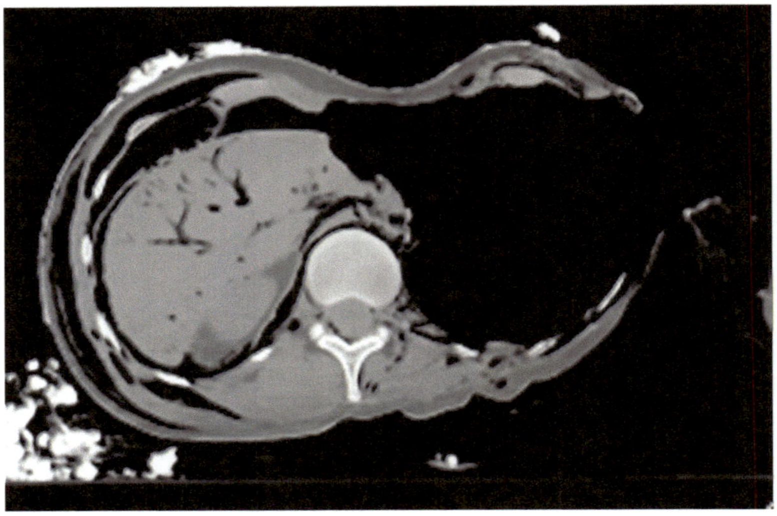

FIGURE 18.66 Predation with missing organs (including the spleen).

with a history of cough, fatigue, and fever in rural regions of the Western United States should always raise a mental red flag for the differential diagnosis of *Hantavirus pulmonary syndrome.*

Emphysematous changes occur in smokers and nonsmokers, but most commonly in the former. They are demonstrated by low-density spaces within the lungs secondary to loss of the alveolar septae.[23]

Liver

The evaluation of **cirrhosis** in PMCT is challenging, and in this author's experience, as a pathologist, is only made when obvious and in the context of significant chronic alcohol use. PMCT findings suggestive of cirrhosis include a nodular liver contour, an enlarged caudate and left lateral lobe segment, and widening of the fissures and porta hepatis.[24] Probably the least challenging assessment for liver disease is the evaluation of **fatty infiltration of the liver**, which is demonstrated by low attenuation, such that the *HU is less than +40.*[25] While fatty changes of the liver on PMCT are not significant enough to account for death alone, its finding supports significant disease processes and organ damage which is often not isolated, and in the context of proper history, one can often reasonably conclude the cause of death. The presence of ascites, along with PMCT findings suggestive of cirrhosis, is also

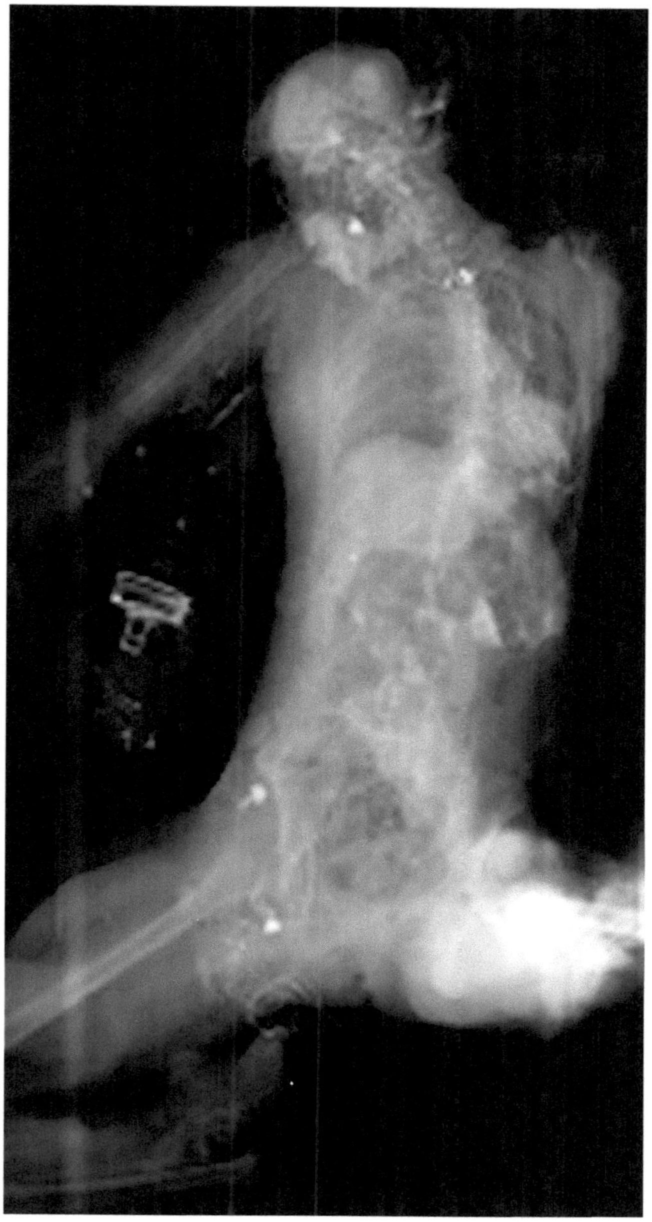

FIGURE 18.67 **Charred remains.** Scout view with contraction deformities of the extremities

supportive of end-stage liver disease and/or heart failure. **Ascites** appears as fluid collections in gravity-dependent regions, such as the pelvic cul-de-sac, perihepatic, and perisplenic regions, with *HUs of less than +20 (correlates with protein and albumin concentrations), while HUs associated with hemoperitoneum are +40 or greater.*[26]

SARS-CoV-2 relating findings

Pulmonary CT findings in SARS-CoV-2 infection have been extensively studied.[57–68] While there are no pathognomonic imaging findings for SARS-CoV-2, bilateral multifocal ground glass opacities tend to predominate (Figs. 18.81–18.83). Less common findings, such as, "crazy paving "and the "halo sign" have been also been reported, as well as uncommon findings (suggestive of an alternative diagnosis), including but not limited to, pleural effusion, "tree-in-bud" sign, lymphadenopathy, and cavitating lesions.[69,70]

Postmortem imaging introduces challenges to this heterogeneous compilation of potential findings with superimposed lividity. As previously discussed, lividity is a gravity-dependent process that is nearly always seen to

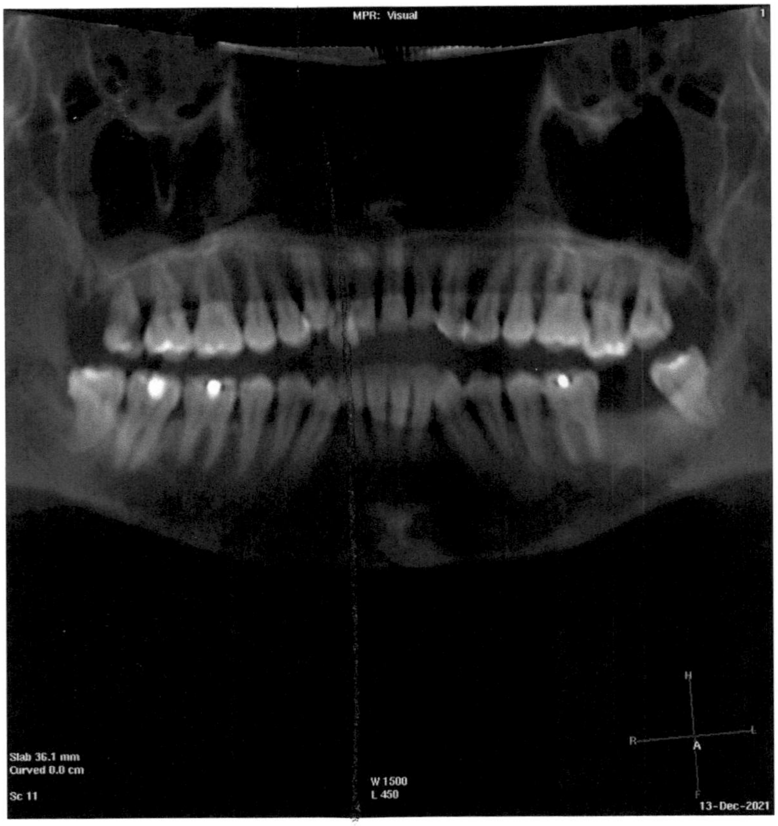

FIGURE 18.68 **Pseudoradiograph.** PMCT of the face is reconstructed to make a pseudoradiograph for comparison to antemortem panorex.

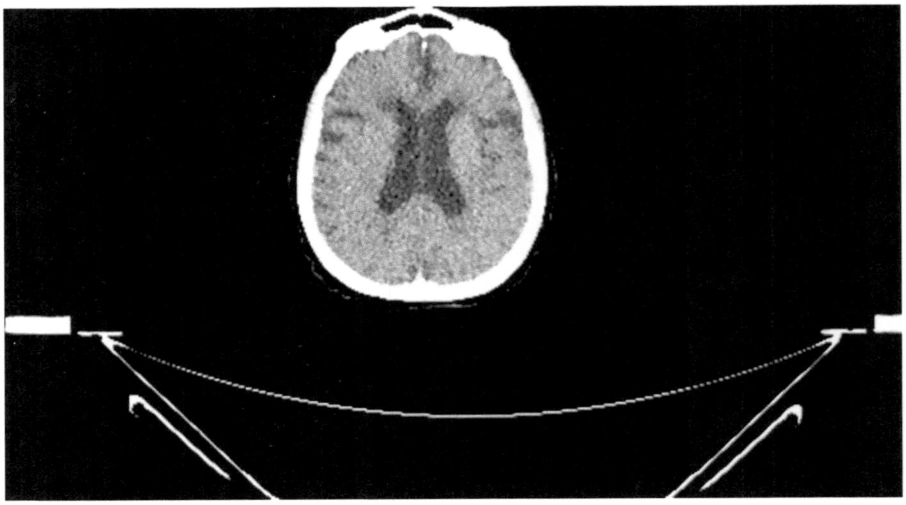

FIGURE 18.69 Cerebral atrophy.

some degree in the lungs. On PMCT, lividity is represented by dependent ground glass opacities (dependent with respect to the position of the body and not necessarily the position the body is placed in to obtain CT images). As such, it can be challenging to determine pathological versus nonpathological pulmonary findings on PMCT.[71,72] There is a spectrum of PMCT imaging findings that supports the diagnosis of SARS-CoV-2.[73–81] A radiologic–histopathological correlation of lung findings further supports the notion that CT observations correlate with multiple histopathological changes including diffuse alveolar damage, interstitial edema, capillary dilatation and congestion.[82]

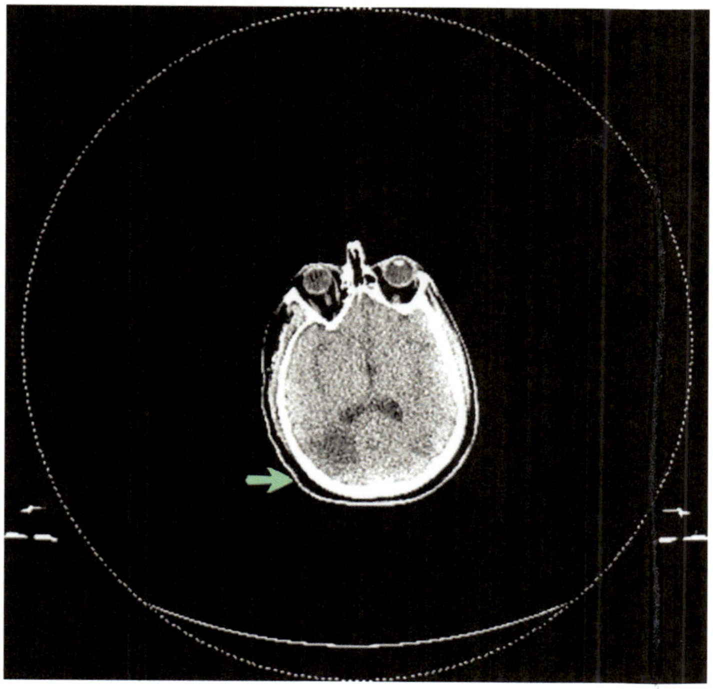

FIGURE 18.70 Remote cerebral infarct.

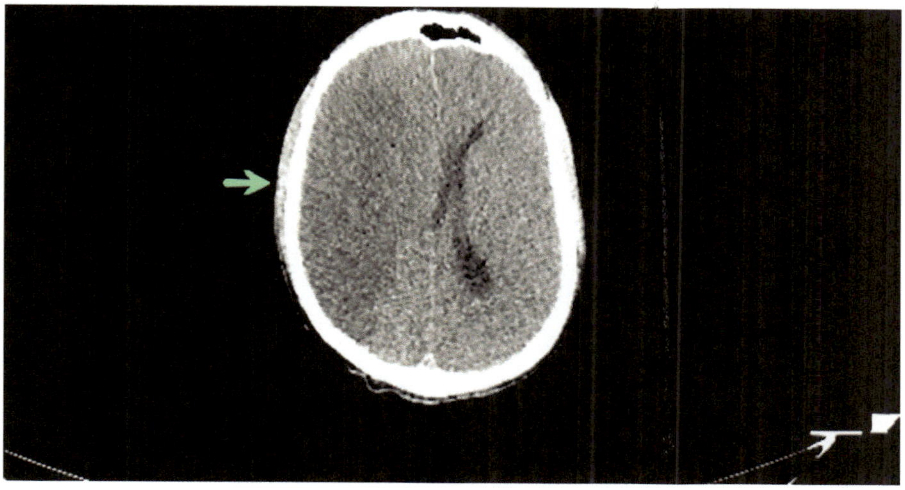

FIGURE 18.71 Acute cerebral infarct.

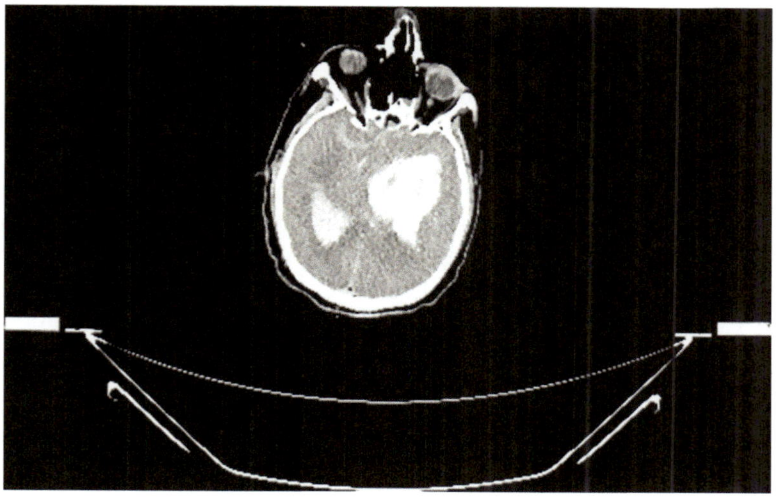

FIGURE 18.72 Hypertensive intracerebral hemorrhage.

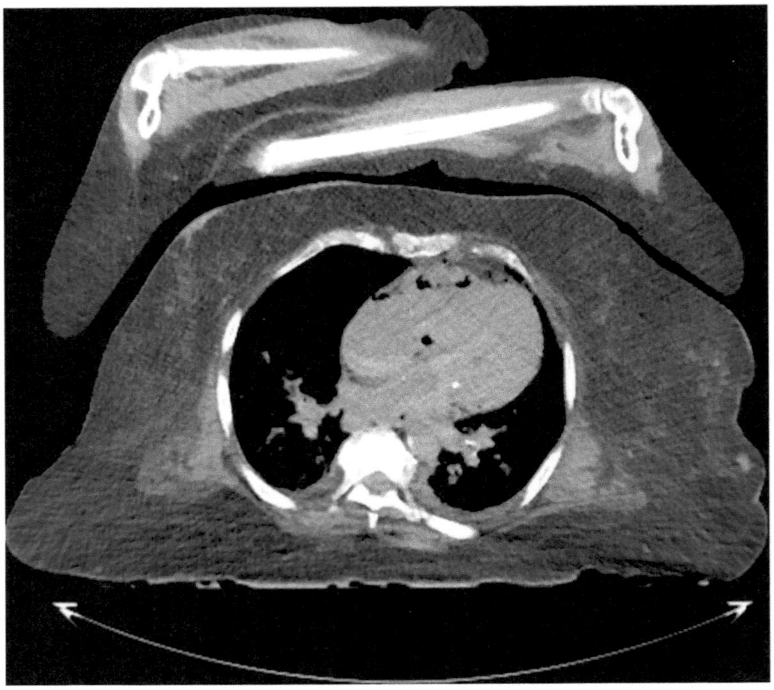

FIGURE 18.73 Cardiac hypertrophy.

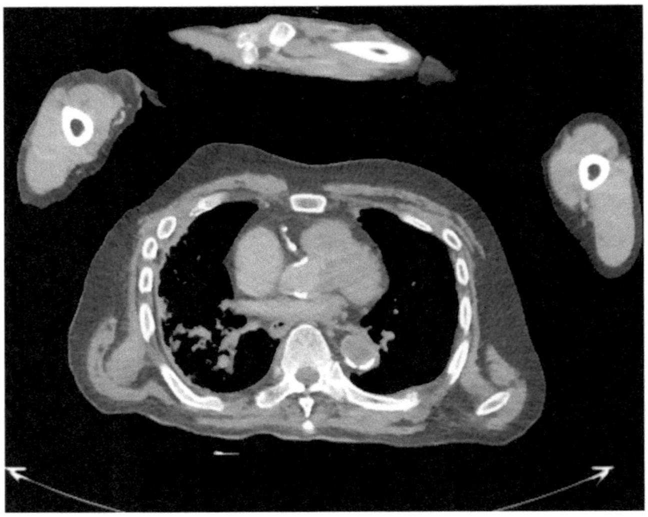

FIGURE 18.74 Coronary artery disease.

SARS-CoV-2 infection also has a high association with clotting events. It is unknown at this time whether this might be related to the direct effects of the virus or perhaps an immune modulation event. There are numerous permutations of imaging findings associated with clotting disorders in living patients, with pulmonary embolism and ischemic stroke being the most common.[83–89] Unfortunately, PMCT has a limited dynamic range for soft tissue contrast and is, therefore, not sensitive for these findings. Furthermore, postmortem changes, primarily related to stasis and gravity, confound the specificity of some observations. For instance, nonhemorrhagic strokes are unlikely to be evident in many cases of living patients and even less apparent with PMCT. Postmortem changes evolve rapidly in the brain with an end result of poor gray–white imaging resolution and limited detailed visualization of anatomical structures. As previously discussed, livor mortis results in a hematocrit effect (well depicted in the great vessels) resulting in findings that mimic pathological clotting.

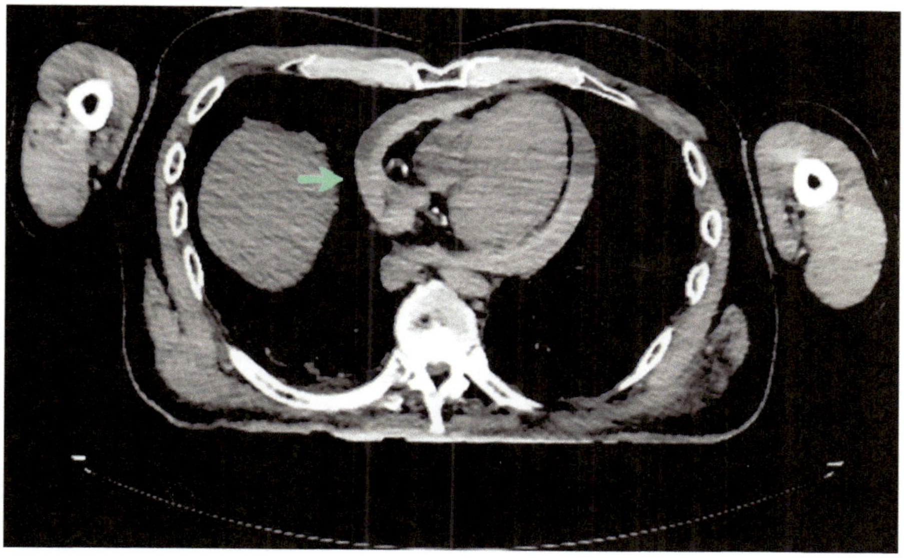

FIGURE 18.75 Hemopericardium.

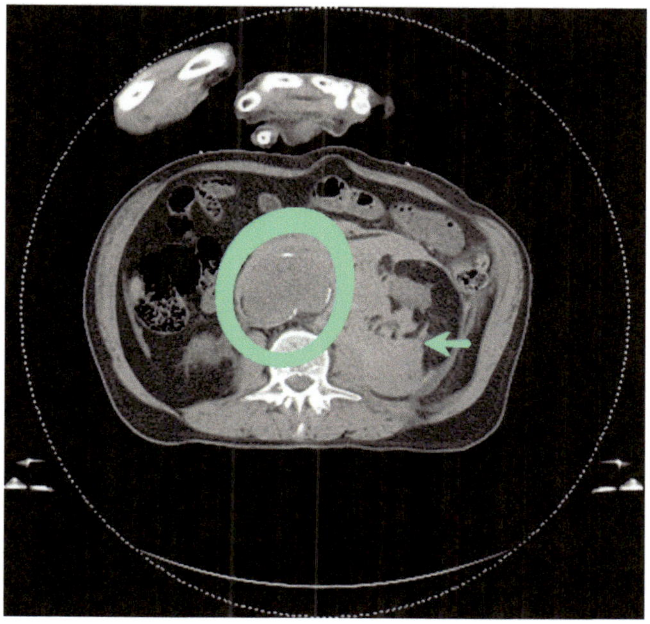

FIGURE 18.76 Ruptured abdominal aortic aneurysm.

Much more research is needed to further define the radiologic and pathologic findings of SARS-CoV-2 infection. The search for answers will undoubtedly lead to an ongoing exploratory effort for years, if not decades to come.

Blunt force trauma

The evaluation of blunt force trauma is one of the most effective applications of PMCT in supplanting an autopsy. Moreover, PMCT can provide additional data, such as skeletal trauma (scapular fractures, pelvic fractures, extremity fractures) that might otherwise be overlooked or not evaluated by a standard autopsy and conventional radiography. Careful evaluation of PMCT also often reveals organ trauma, including lacerations of the lungs, liver, and less often, the spleen and kidneys.

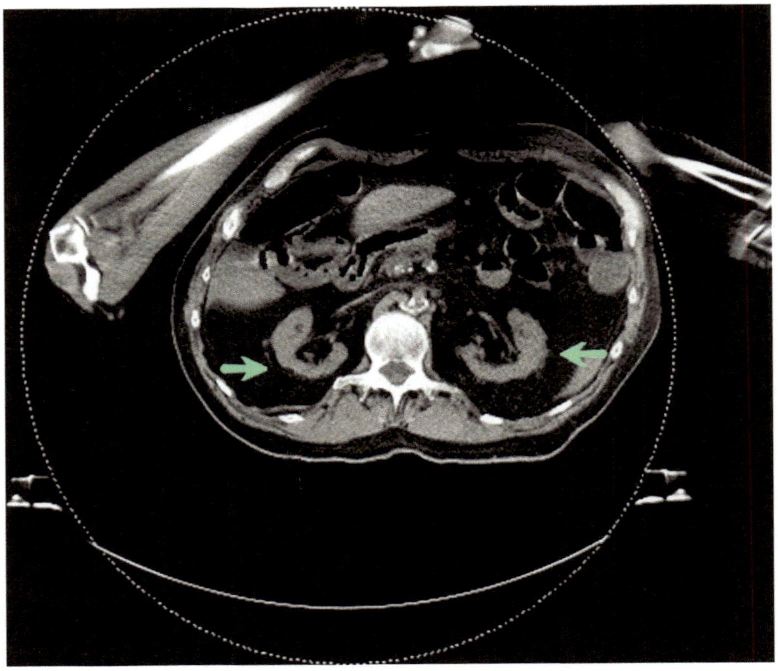

FIGURE 18.77 Hypertensive kidney disease.

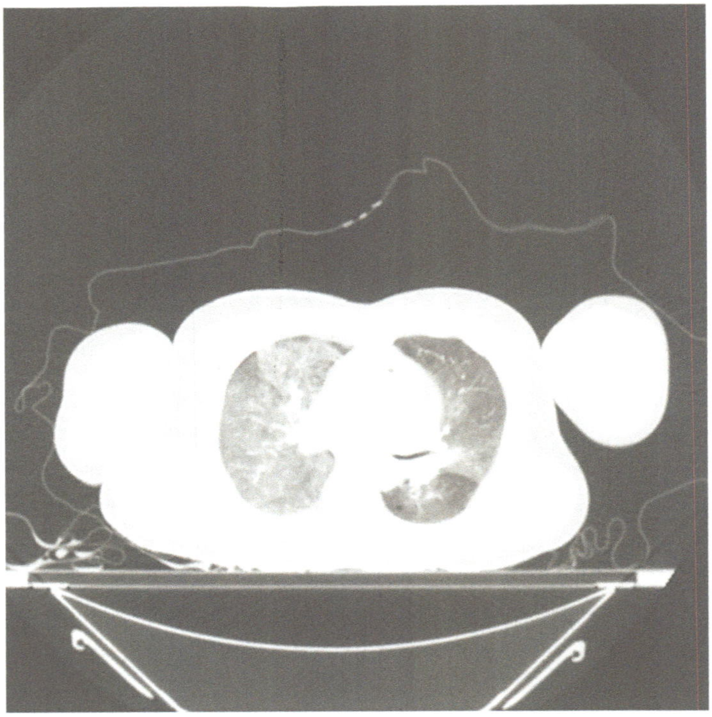

FIGURE 18.78 Influenza (H1N1).

Blunt head trauma

The most obvious lethal trauma depicted on PMCT is blunt head trauma, which often consists of skull fractures with intracranial hemorrhage. In the absence of decomposition changes, pneumocephalus is a harbinger of a skull fracture (Fig. 18.84) or severe spinal fracture (with an accompanying pneumorrhachis, intraspinal air), so one should be on alert for fractures when pneumocephalus is present. Localization of basilar skull fractures is often challenging and requires knowledge in the manipulation of the software that "removes" the skull cap and allows an axial view

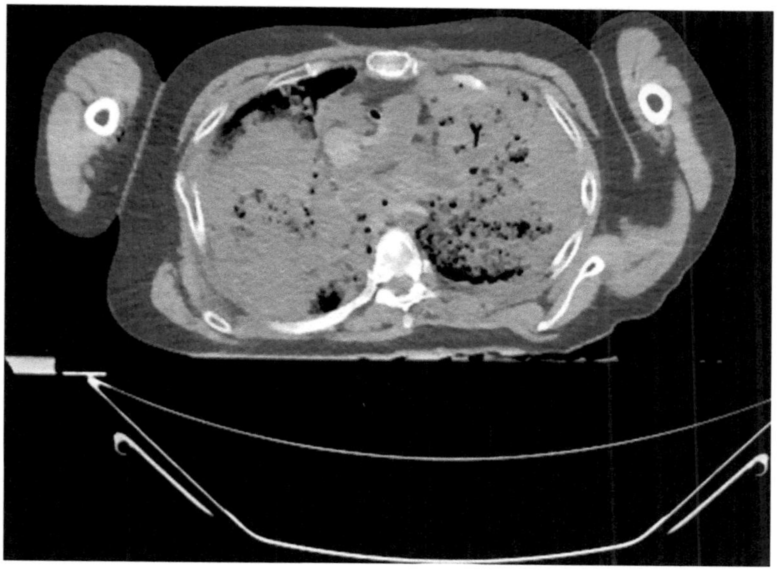

FIGURE 18.79 Superimposed bacterial pneumonia on viral infection.

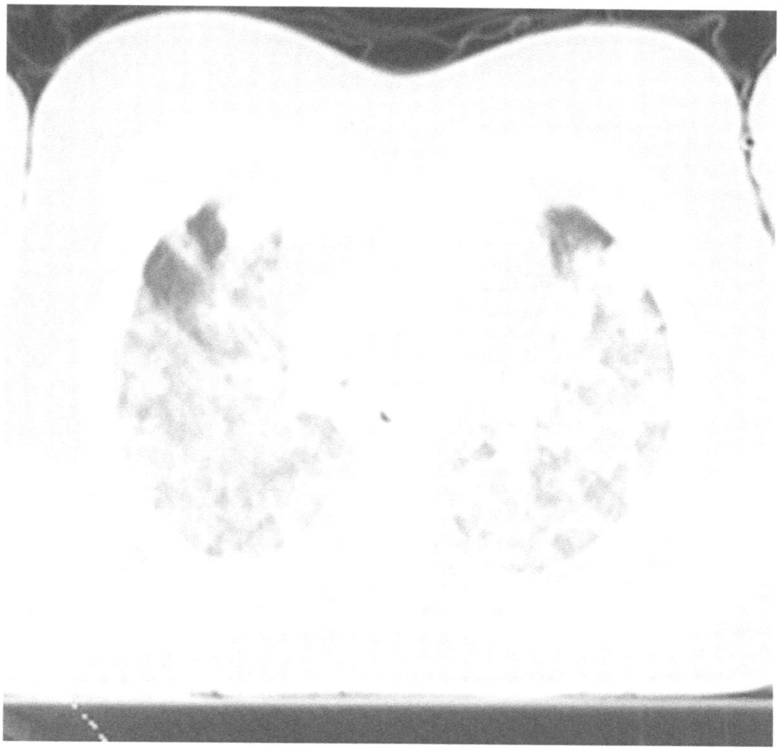

FIGURE 18.80 **PMCT SARS-CoV-2 pneumonia.** Axial CT thorax with lung window PMCT SARS-CoV-2 pneumonia. Note the bilateral, multifocal ground glass opacities with nondependent and noncentral distribution.

(Fig. 18.85). As discussed in Chapter 19, 3-D, reconstruction may be valuable in visualizing fracture patterns and can also be helpful as a visual aid in court testimony, although 3D reconstruction may not display some basilar skull fractures adequately (Fig. 18.86). Lastly, review of the axial, coronal, and sagittal views is strongly recommended for the confirmation of fractures.

Subdural hematomas are well visualized on PMCT and are demonstrated by blood accumulation in the potential space between the dura mater and the arachnoid mater (Fig. 18.87). They are most often caused by blunt force head trauma, but nontraumatic causes such as vascular malformations should be considered (see Chapter 10). Subdural hematomas are caused by tearing of the bridging veins as they cross into the dural venous sinus (Fig. 18.88). They appear as crescent-shaped collections of hemorrhage and are unilateral in 85% of adults, but bilateral in

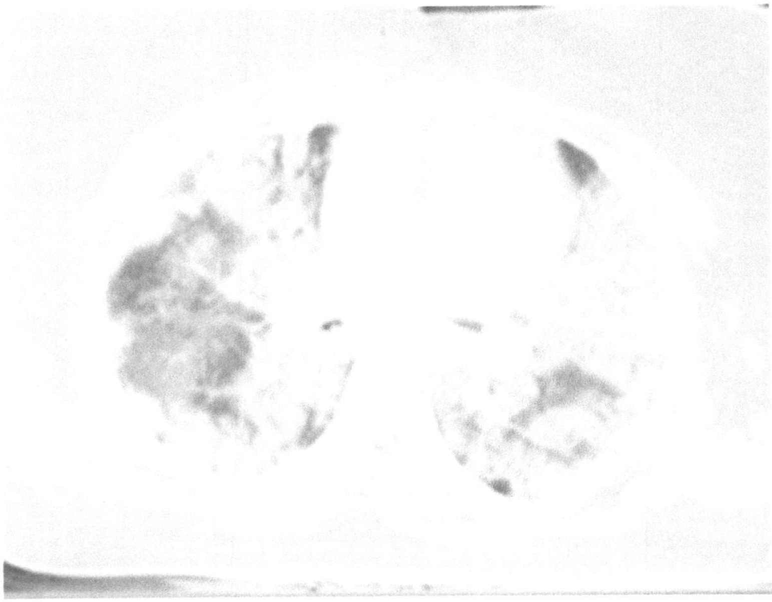

FIGURE 18.81 **PMCT SARS-CoV-2 pneumonia.** Axial CT thorax with lung window PMCT SARS-CoV-2 pneumonia. Note the bilateral, multifocal ground glass opacities with nondependent and noncentral distribution.

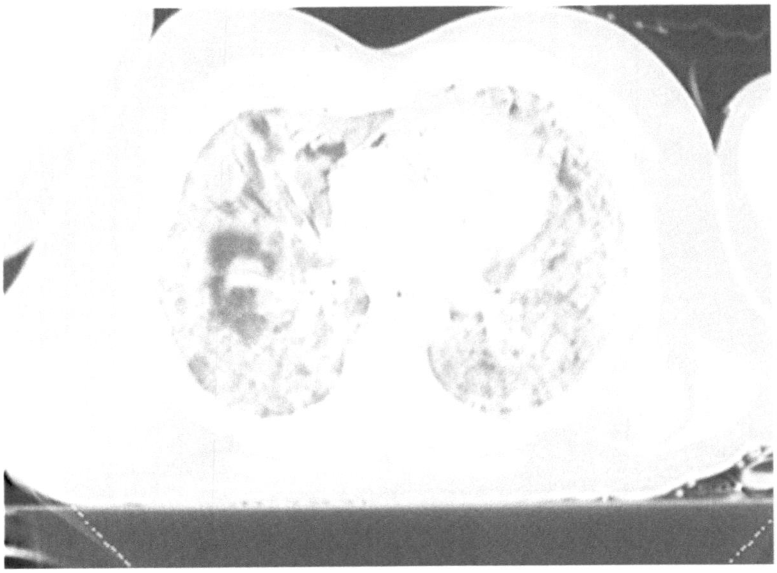

FIGURE 18.82 **PMCT SARS-CoV-2 pneumonia.** Axial CT thorax with lung window PMCT SARS-CoV-2 pneumonia. Note the bilateral, multifocal ground glass opacities with nondependent and noncentral distribution.

approximately 75% of infants. *Acute subdural hemorrhage appears as a crescentic, homogenous, hyperdense lesion that is typically greater than +50 HU.* As the hemorrhage begins to organize and enters the subacute phase, typically over a week, the density lowers to around +40 HU and may be less obvious. *Chronic subdural hematomas are typically around 21 days of age and appear as a hypodense lesion with a 0 HU.* The evaluation of acute subdural hematomas should involve consideration of mass effects, which include midline shift and various herniations, if the pathologist feels comfortable evaluating these changes on PMCT.

EDHs are usually associated with a skull fracture that produces injury of the associated vasculature, such as the middle meningeal artery/vein in temporal fractures, and do not cross suture lines (Fig. 18.89). However, associated fractures are not invariable in EDHs in pediatric decedents due to the malleability of bone. The classic presentation is a lucid interval following blunt head trauma, with subsequent loss of consciousness due to the mass effect and consequent herniation. The lucid interval may last longer in pediatric patients because the dura mater is particularly adherent to the bone. EDHs are biconvex in appearance, since their origin is actually subperiosteal (see figure).

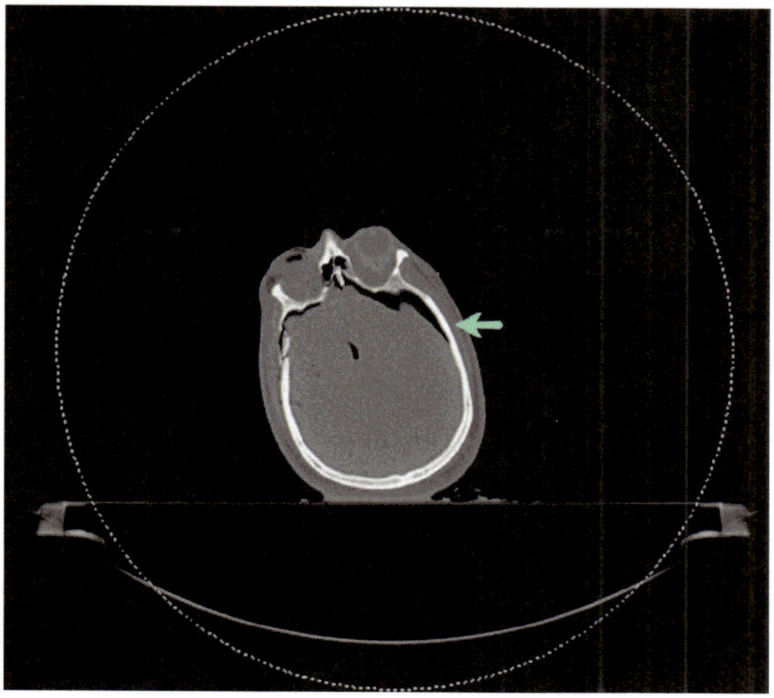

FIGURE 18.83 **Pneumocephalus.** A harbinger of a skull fracture.

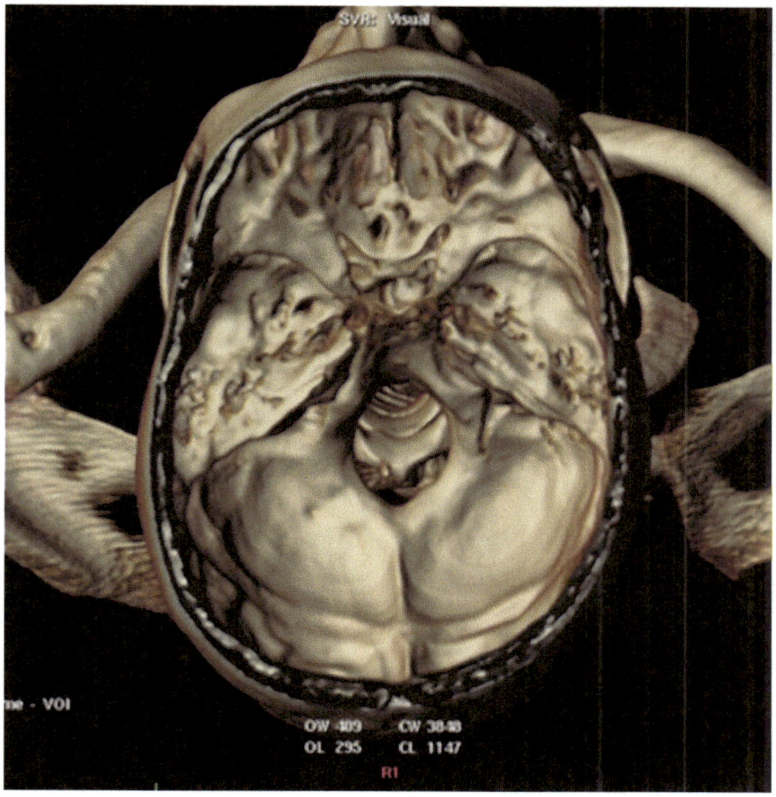

FIGURE 18.84 **Basilar skull fracture.** Manipulation of the software "removes" the skull cap and allows an axial view.

Subarachnoid hemorrhage is characterized by hemorrhage under the arachnoid mater and, although commonly caused by blunt head trauma, is also commonly caused by aneurysmal rupture, processes that causes intraventricular hemorrhage, cerebral amyloid angiopathy, and vascular malformations. Subarachnoid hemorrhage is

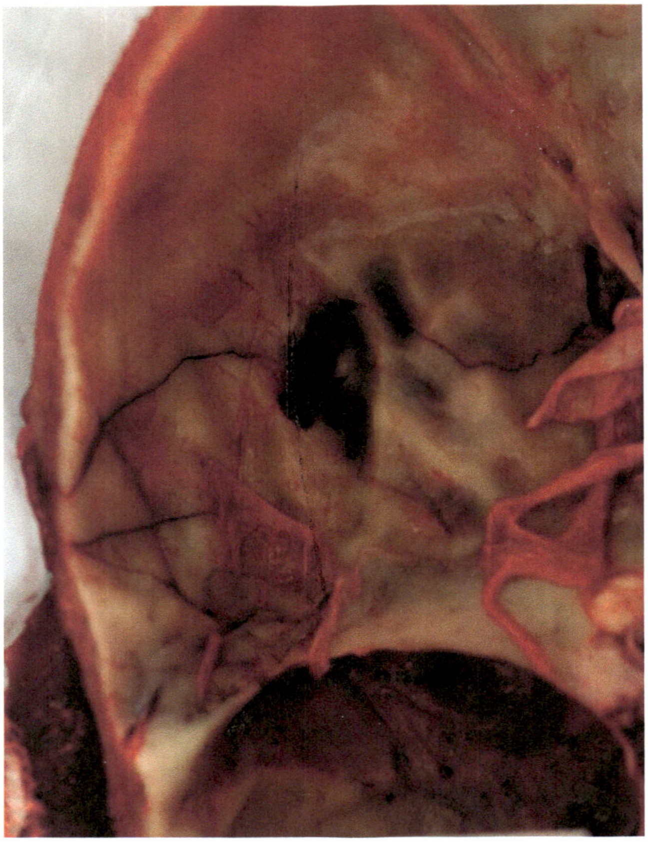

FIGURE 18.85 Basilar skull fracture at autopsy.

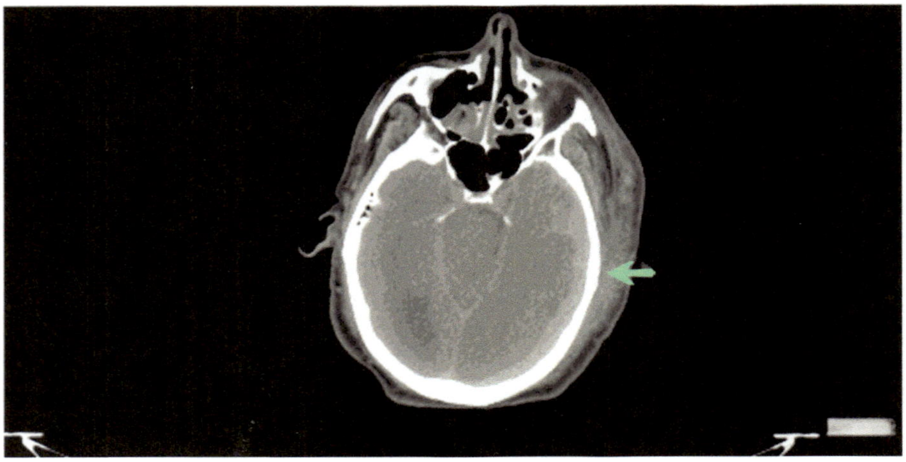

FIGURE 18.86 Subdural hematoma.

visualized as hyperdensity in the subarachnoid spaces (Fig. 18.90). Accumulation of subarachnoid hemorrhage within an isolated region, such as the brain base, may more strongly suggest a specific etiology, such as a ruptured cerebral artery aneurysm. If vascular injury (e.g., vertebral artery) is suspected based on the location of the hemorrhage, the use of contrast may be helpful in localizing the rupture (see postmortem angiography section).[46,90,91]

Blunt neck trauma

C-spine fractures are most often caused by axial loading, during which a force is transmitted through the top of the head down into the c-spine, causing hyperflexion of the head, with resultant cervical fracture. PMCT can be a

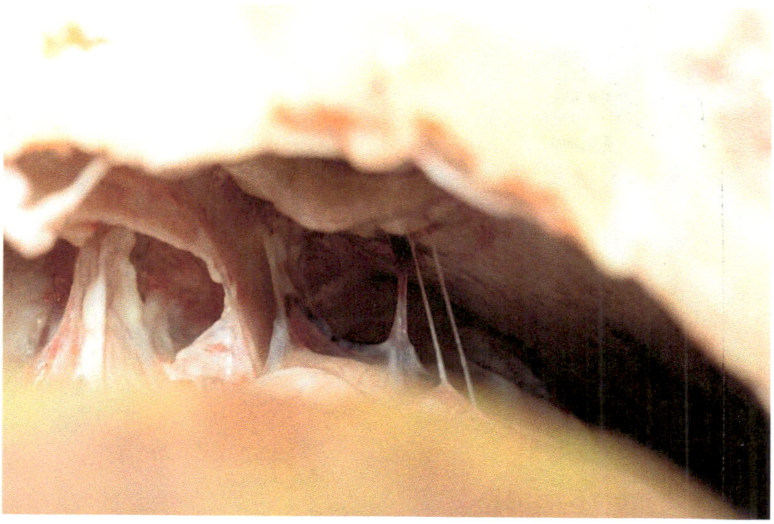

FIGURE 18.87 Bridging veins.

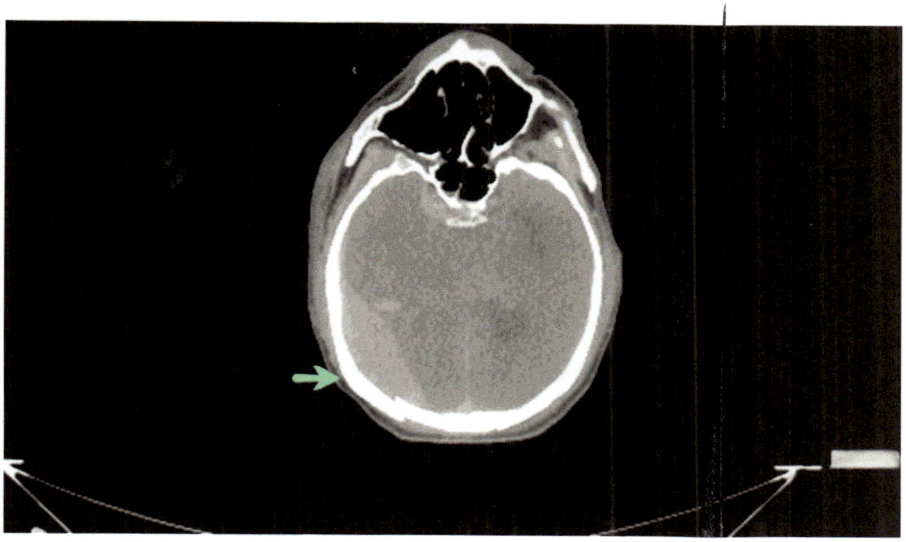

FIGURE 18.88 Epidural hematoma.

useful tool in the evaluation of blunt neck trauma, with a sensitivity for c-spine injury of approximately 70%, and a specificity and accuracy of approximately 90%, while autopsy has yielded a sensitivity of 23%, a specificity of 98%, and an accuracy of 89%.[92] Neurologic sequelae are estimated to result from as many as 50% of cases with c-spine fracture, which implies that its presence likely represents a significant neuropathologic finding in the right setting (i.e., in a decedent without other significant trauma or an alternate explanation of death).

Knowing the strengths and weaknesses of PMCT in the evaluation of cervical trauma is valuable when triaging these cases. Atlas and axis fractures are well evaluated with imaging, whereas craniocervical fracture–dislocation is easily identified by autopsy. Additionally, decomposition changes may demonstrate gas patterns that can be confused with fractures, something to be kept in mind during the interpretation of the imaging. Establishing a methodical approach when evaluating for the presence of cervical fractures is helpful, keeping in mind that different views allow better representations of certain cervical elements; certain views are also more (or less) beneficial for different disease processes.

During triage, decisions regarding external examinations versus autopsy examination (excluding cases where autopsies are mandatory) are often based on the presence or absence of fatal trauma. Concurrent knowledge of the circumstances surrounding death are helpful when making these decisions. For example, motor vehicle collisions with a side impact (T-bone impact) should warrant careful evaluation of the axial, sagittal, and coronal views of the c-

FIGURE 18.89 Subarachnoid hemorrhage.

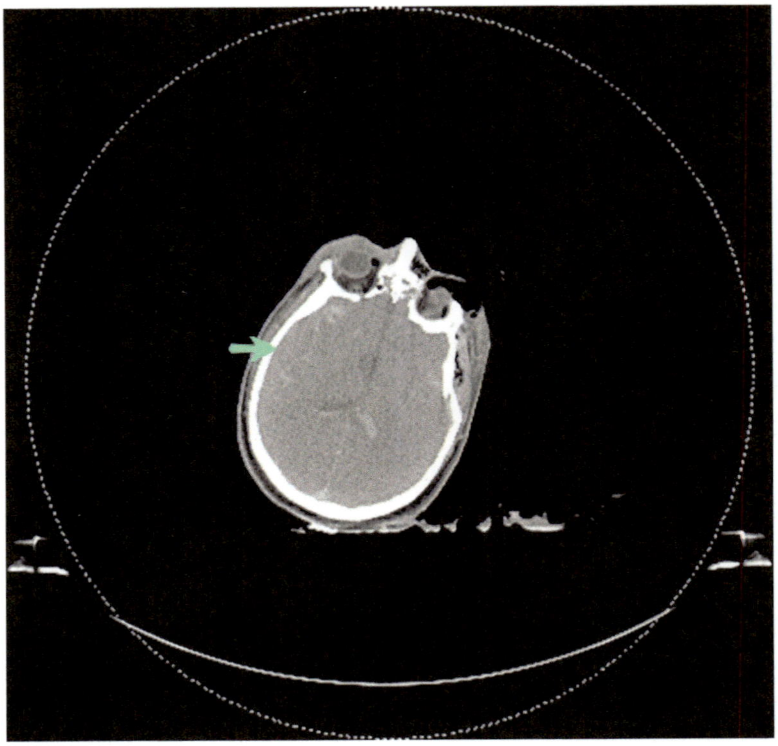

FIGURE 18.90 **Basion–axial interval.** See marker in image.

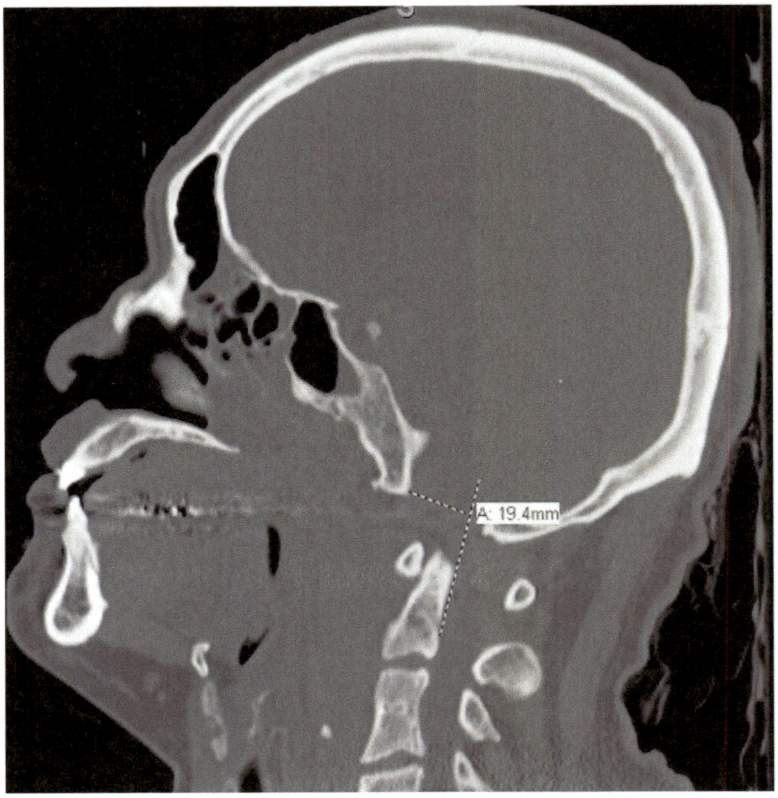

spine. If fatal injury is not identified, one should carefully weigh the possibility of a c-spine fracture versus a natural event that precipitated the crash, and an autopsy with a posterior neck dissection may be indicated.

Measurement tools within the radiologic software are helpful when evaluating for certain types of fractures, and they are specific and accurate when used. For example, measurement of the basion–axial interval (BAI), which is the horizontal distance between the basion and the posterior cortex of the axis, can detect craniocervical dissociation when the distance is greater than 12 mm in adults, and is best evaluated on the sagittal view (Fig. 18.91). A second helpful measurement is the basion–dens interval, which is the distance between the basion and the tip of the dens, also best seen on the sagittal view. A distance greater than 10 mm indicates craniocervical dissociation (Fig. 18.92).

In the postmortem setting, C1 and C2 fractures indicate likely mechanistically significant blunt trauma, given their proximity to vital neurologic and vascular structures. A burst fracture of C1, known commonly as a Jefferson fracture, represents an axial load force and often does not necessarily signify a neurologic deficit, but is most commonly associated with vertebral artery injury (Figs. 18.93 and 18.94). The odontoid is the most commonly fractured element of the second cervical vertebra, and fractures are classified into three types. The type 1 fracture, which involves the tip of the odontoid, is a rare fracture, but is clinically stable. Type 2 involves the base of the odontoid and is most commonly seen in elderly people, including people with osteoporosis. Type 3 involves the body of C2 and typically requires internal fixation if the person survives the initial injury (Fig. 18.95). C2 fractures from traumatic spondylolisthesis/pars interarticularis are famously known as *hangman's fractures*, which formerly occurred as a result of judicial execution hangings, are caused by cervical *hyperextension, are typically lethal, and are most often seen in fatalities related to traffic accidents, sports accidents, and falls.*

Hyperflexion/hyperextension injuries of the c-spine (the head violently and/or excessively bending forward or backward, respectively) are commonly known as **whiplash injuries**. When the load exceeds the physiologic ability of the spine to bend beyond the limit of its angle, consequent injury occurs. Hyperflexion/hyperextension injuries may be caused by contact forces, such as a head strike on a dashboard or steering wheel, or noncontact, such as a rear-end collision. These injuries *may produce a teardrop fracture of the anterior endplate of the vertebral body by tearing the anterior longitudinal ligament*, with neurologic symptoms of central cord syndrome in the vast majority of

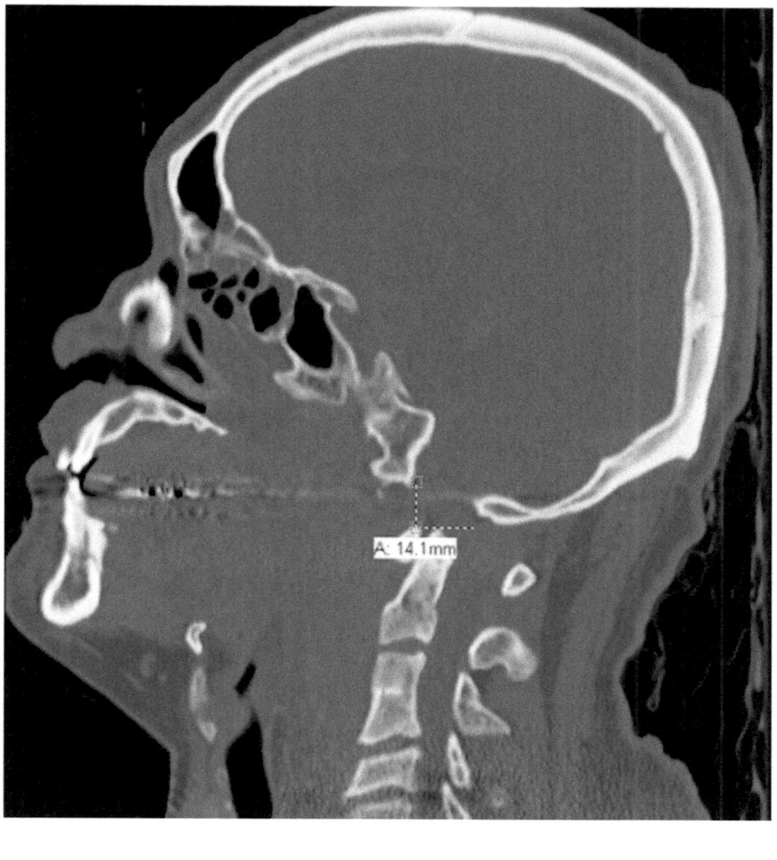

FIGURE 18.91 **Basion–dens interval.** See marker in image.

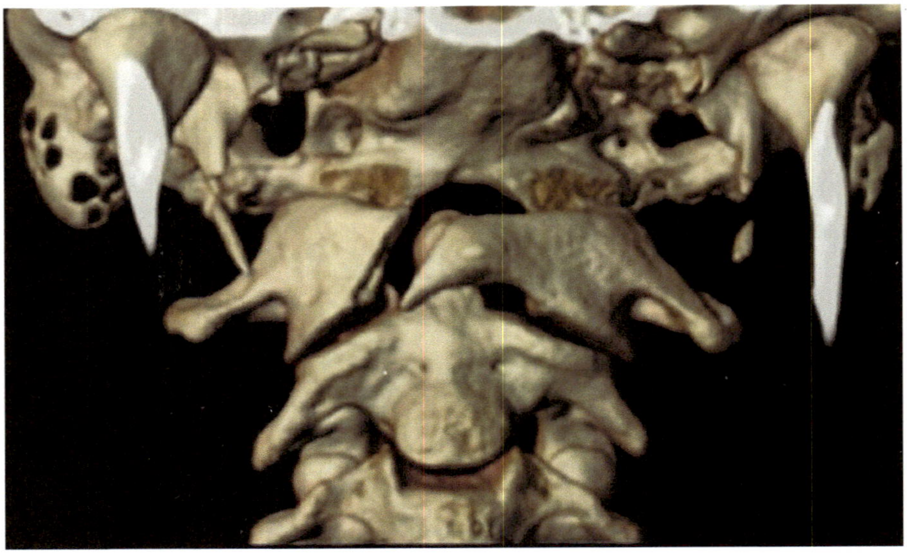

FIGURE 18.92 A burst fracture of C1, known commonly as a Jefferson fracture.

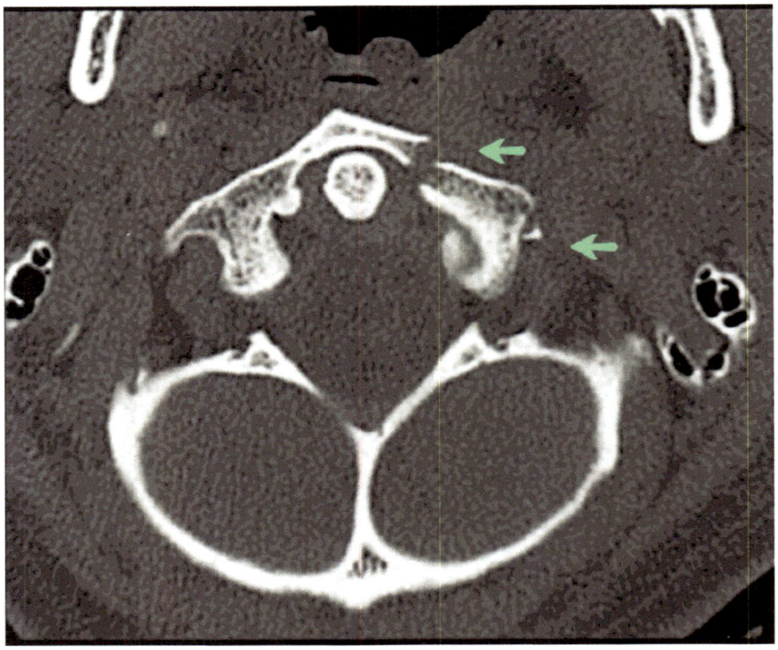

FIGURE 18.93 A burst fracture of C1, known commonly as a Jefferson fracture.

survivors. If the injury is significant enough to cause posterior displacement, injuries to the anterior and posterior ligaments, annulus, and intervertebral disc occur.

Blunt chest trauma

Blunt chest trauma carries a high morbidity and mortality for trauma-related deaths.[93,94] The evaluation of blunt force trauma of the chest is perhaps one of the most practical uses of PMCT and is also one of the areas in which those new to PMCT evaluations feel most comfortable. The most obvious findings include pneumothoraces, hemothoraces, and rib fractures, but caution must be exercised when deciding whether the injuries are significant enough to account for death.

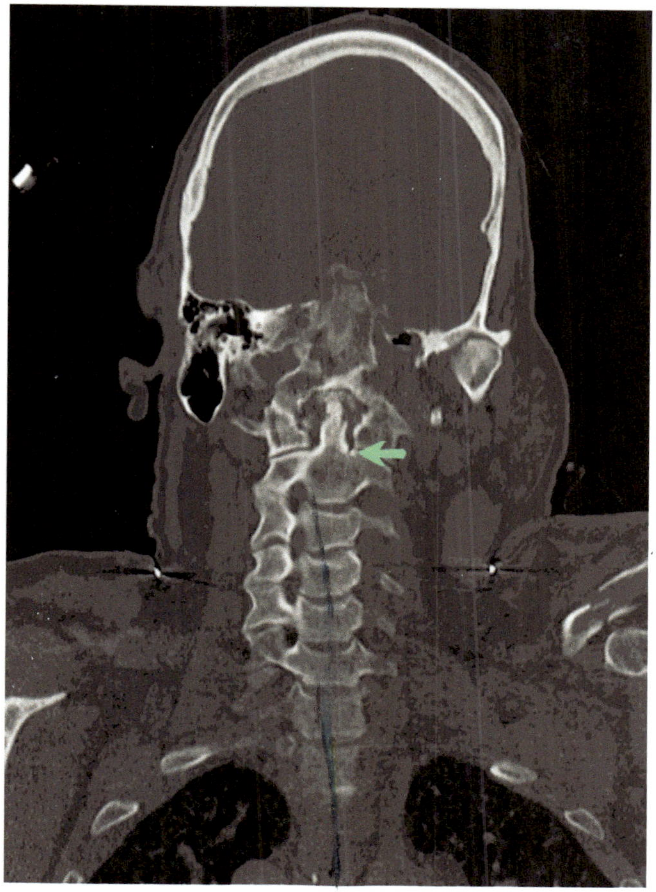

FIGURE 18.94 Odontoid fracture, type 3.

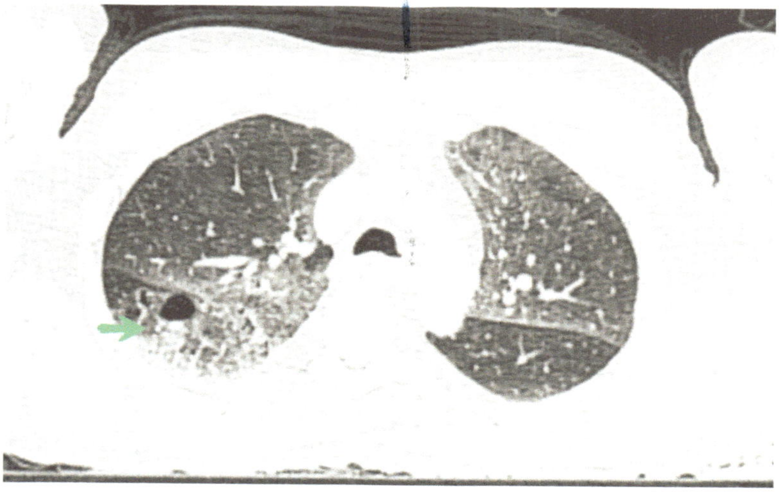

FIGURE 18.95 Pneumatocele.

A pneumothorax is demonstrated by air surrounding the lungs in the pleural cavity, which inhibits the ability of the lung to expand and fill with air. **Pneumothoraces** are commonly caused by blunt chest trauma in which rib fractures cause pulmonary lacerations, penetrating and perforating chest cavity trauma, with or without direct lung injury, and iatrogenic injury. The detection of a pneumothorax is particularly useful before an autopsy is performed, and the thoracic space is breached or otherwise manipulated. Small pneumothoraces may not be significant enough to account for death, so if the constellation of findings isn't significant enough to account for death, other causes and mechanisms should be considered. The presence of thoracostomy needles, chest tubes, and/or CPR-related anterolateral rib fractures may be the sole cause of a pneumothorax; review of emergency medical service run sheets and other medical records is warranted to search for the documentation of audible air escape with needle insertion. A mediastinal shift indicates the presence of a tension pneumothorax and is a clear-cut sole or contributory mechanism of death.

A characteristic finding associated with **lung lacerations** on PMCT is the pneumatocele, which appears as a "Swiss cheese" density on the lung window or a lucent defect. Pulmonary contusions can be inconspicuous on PMCT, but they appear as nodular opacities with a surrounding ground glass appearance, which is due to the leakage of blood into the adjacent lung parenchyma (Fig. 18.96).

Hemothoraces are demonstrated by an accumulation of blood in the pleural cavity surrounding the lungs and are caused by injury to the chest wall, lung parenchyma, heart, or great vessels. A hemothorax can be differentiated from a hydrothorax (effusion) by the presence of sedimentation and through the use of HUs, which are typically +30–45 HU for liquid blood and +50–90 for clotted blood (Fig. 18.97).

Rib fractures should be evaluated for their location (anterior, lateral, posterior), as well as their age (acute vs. healing). *Anterolateral rib fractures are likely CPR-related in cases where medical intervention occurred.* **Flail chest**, defined as

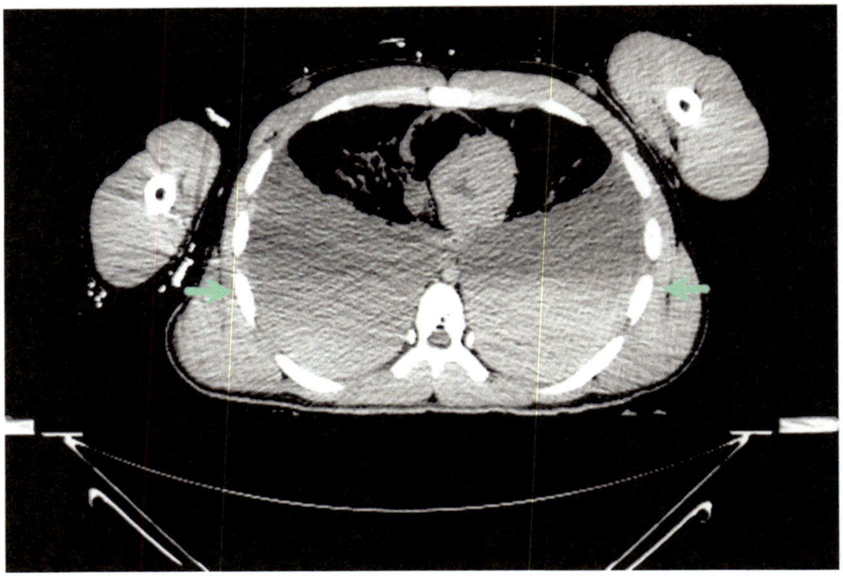

FIGURE 18.96 Hemothoraces.

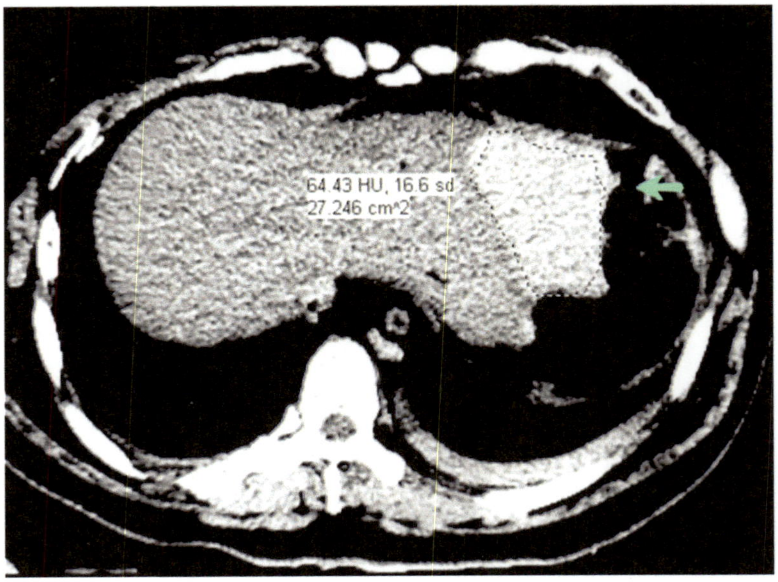

FIGURE 18.97 Liver lacerations.

two or more contiguous rib fractures with two or more fractures per rib, is easily identified on PMCT, and death may be attributed to it, depending on the context of the case and even if it exists in isolation, due to the mechanical failure of the rib cage to expand sufficiently for adequate breathing.

Less commonly, representing less than 1% of all traumatic injuries, is a **traumatic diaphragmatic rupture/laceration**, which typically signifies *severe, and often compressive blunt force trauma* and almost always coexists with other similarly severe injuries. Diaphragmatic defects are best seen on the coronal view or sagittal of the torso and most often occur on the *left side*, as opposed to the right side, where the liver offers the diaphragm protection at its own expense.

Blunt abdominal trauma

The evaluation of solid organ injury is one of the trickier aspects of PMCT, particularly for the untrained pathologist. The liver and spleen are the most commonly injured abdominal organs.[95] Liver, spleen, and kidney injuries represent severe blunt force trauma and often coexist with other significant injuries. These injuries are best viewed on the soft tissue window, with HU measurements being the best assistant for discernment between blood and natural disease.

Liver lacerations appear as dense lesions within the liver, with an HU of +30–45 for liquid blood and +50–90 for clotted blood. Fluid from natural disease demonstrates a much lower HU (Figs. 18.98 and 18.99). The same principles apply to the spleen, as far fluid collections are concerned, but one should keep in mind that splenic clefts can be misinterpreted as lacerations.

Blunt extremity and pelvic trauma

A search for extremity fractures is often an afterthought in the evaluation of blunt trauma, but it should be documented in an effort to provide a complete representation of all injuries. That said, the presence of **femur fractures** can result in significant blood loss in the thigh, which can also be visualized on PMCT. **Pelvic fractures**, particularly the so-called "open book" fractures (typically caused by severe compressive forces or a descent from a substantial height), can also result in significant blood loss due to the associated injury of major pelvic blood vessels (primarily iliac and femoral).

PMCT can provide valuable information in pedestrian–motor vehicle deaths, including those involving a "hit-and-run" scenario, in that it facilitates point of impact documentation and measurements. Depending on how the bone fractures, the direction of vehicular travel may be determined. A wedge-shaped fracture is known as a

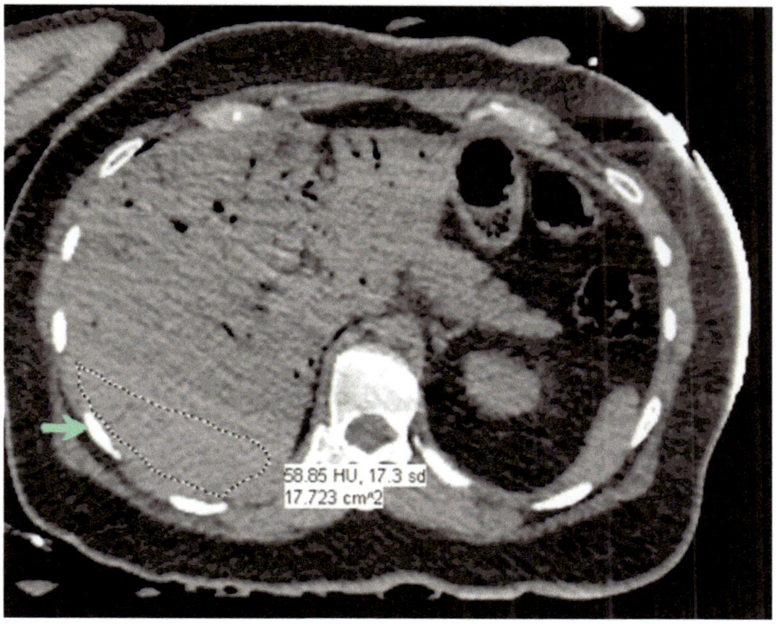

FIGURE 18.98 Liver lacerations.

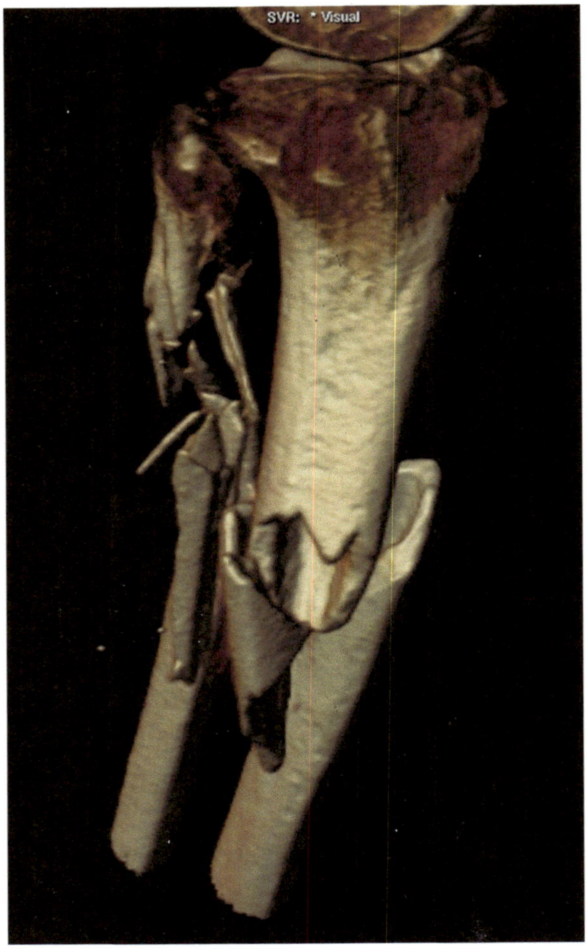

FIGURE 18.99 Messerer fracture.

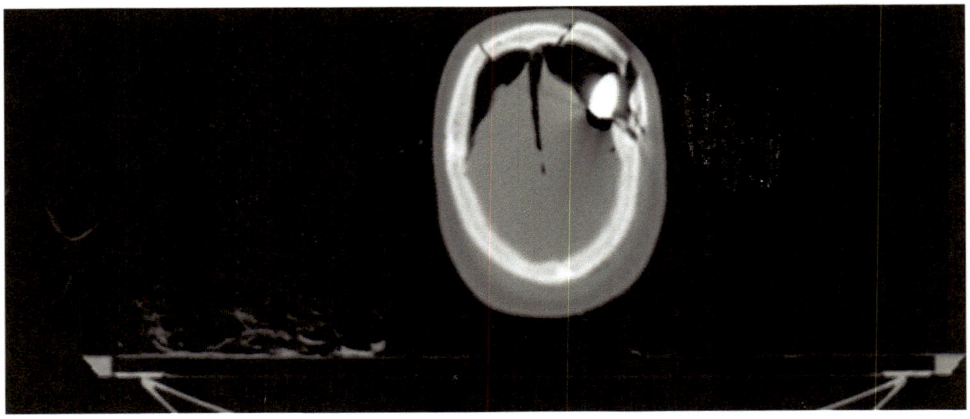

FIGURE 18.100 Retained projectile.

Messerer fracture (Fig. 18.100) in which the apex of the wedge points in the direction opposite the direction of travel of the vehicle (i.e., points at the vehicle itself).

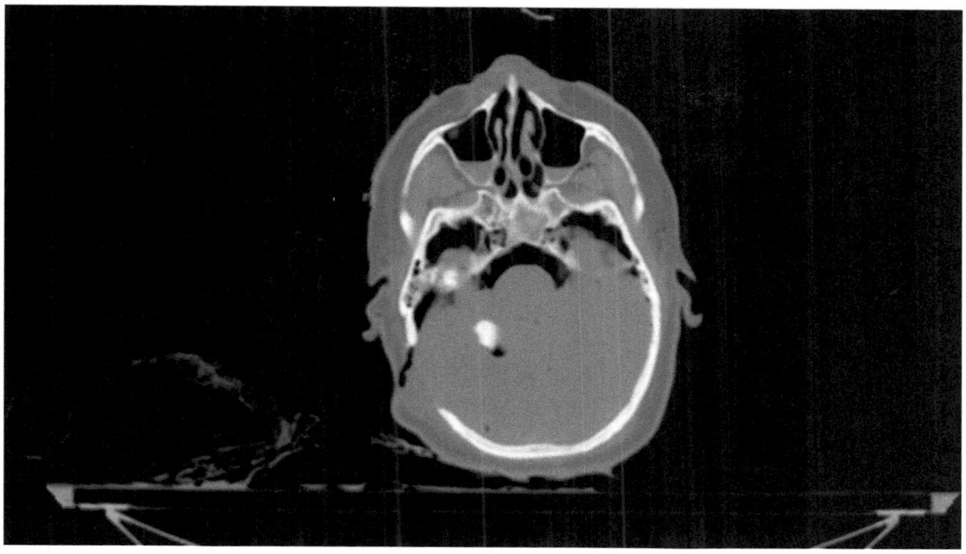

FIGURE 18.101 Gunshot wound with beveling of skull.

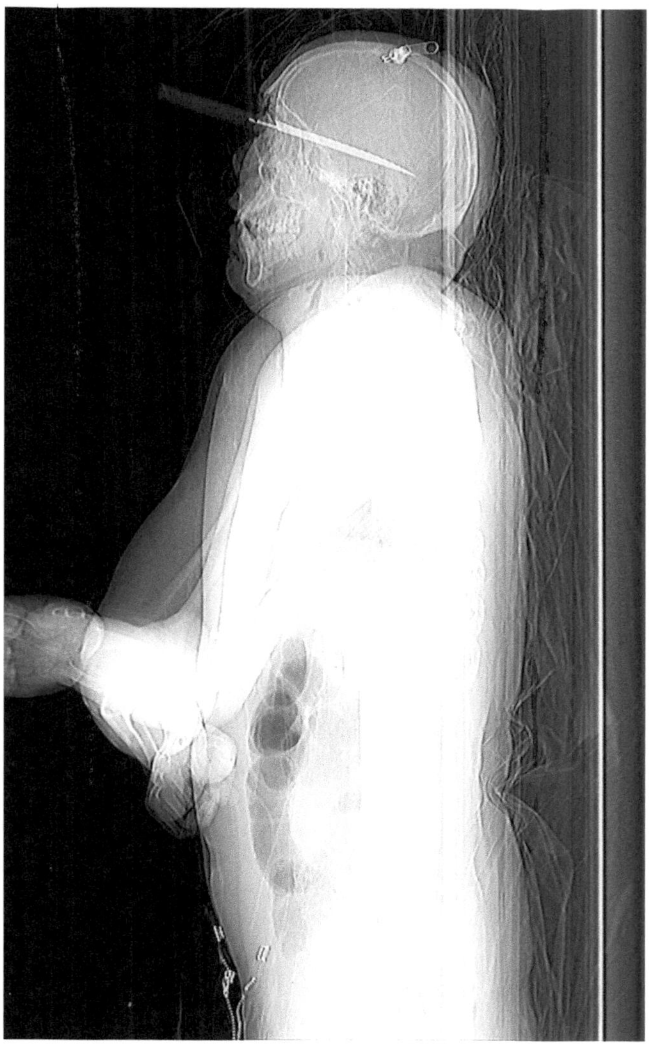

FIGURE 18.102 Knife in head.

Penetrating injury

Penetrating injury and its relevance to imaging studies primarily involves **sharp and firearms** injuries. The use of PMCT for these injuries may determine what body cavities and organs have been injured, whether there are blood and/or air collections, and the location of any retained **projectiles** (Fig. 18.101) or broken sharp objects.

Most often, the wound tract can be discerned by PMCT examination, but it can be made difficult or impossible to ascertain when more than four or five gunshot wounds involve the same cavity. The thinner the slices are, the better the wound tracks can be discretely visualized. However, thinner slices require more time to scan and the benefit must be weighed against efficiency. With regard to gunshot wounds, the bullet trajectory can usually be determined (for example, front to back, left to right, and downward) and beveled bone defects can also be detected, aiding in discriminating entrance from exit wounds (internally beveled vs. externally beveled, respectively, Fig. 18.102).

Stab wounds can be well visualized on PMCT examination when they involve the soft tissue, but they may not be obvious within body cavities. However, the sequelae of the injury, such as hemopneumothoraces, a hemopericardium, or a cardiac air embolism can often provide clues to the organs involved. Thus, PMCT examination provides an overall picture of what one might expect before performing the autopsy. Additionally, one must be mindful to evaluate for a retained fragmented sharp object within the wound track, which, when retrieved, has evidentiary

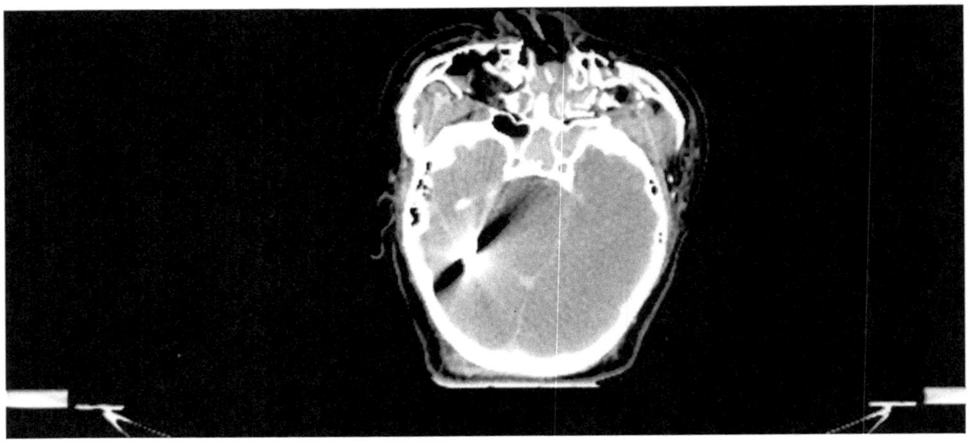

FIGURE 18.103 Axial view of stab wound track of head.

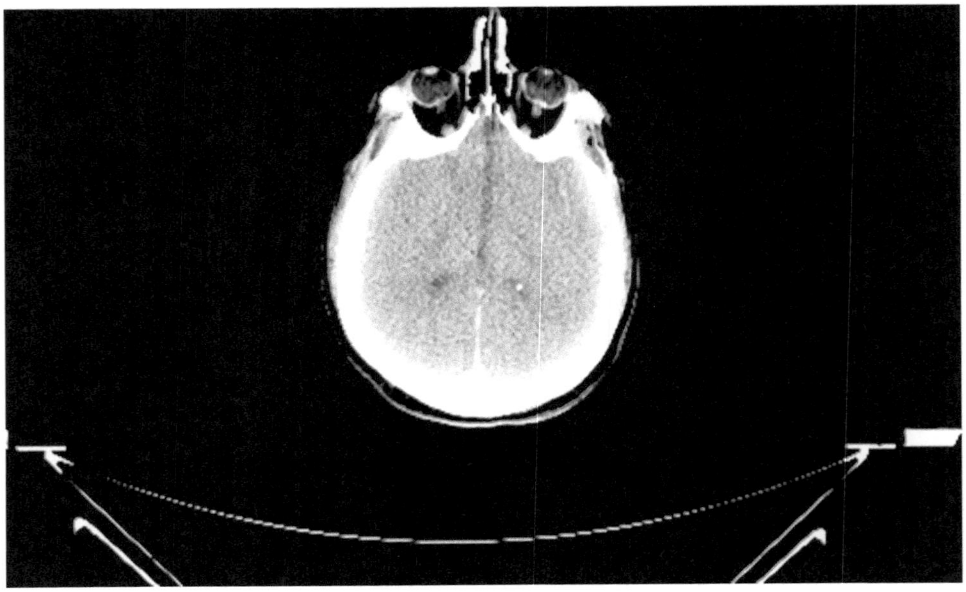

FIGURE 18.104 Cerebral edema.

value for tool comparisons. These objects may be seen on plain film radiographs as well; however, the PMCT allows better localization, which might improve the efficiency of recovery (Figs. 18.103 and 18.104).

Drug intoxication

The benefits of PMCT in drug intoxication deaths are less obvious, but a constellation of findings can provide subtle hints that indicate a drug-related cause of death for certain drug classes. When urine drug screening is used in combination with PMCT, the cause of death is often apparent before the autopsy is performed. PMCT also can help

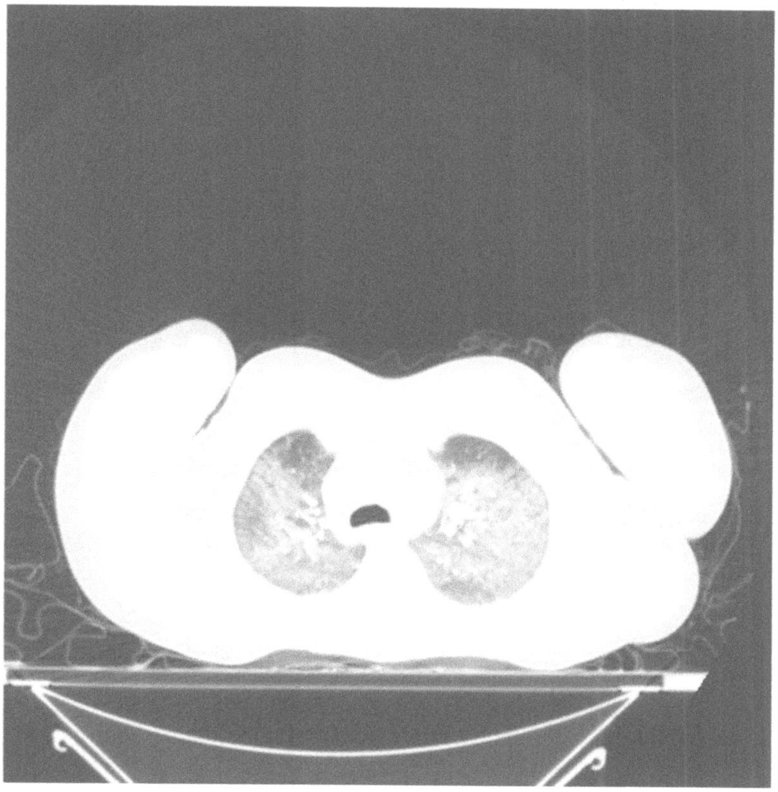

FIGURE 18.105 Pulmonary edema.

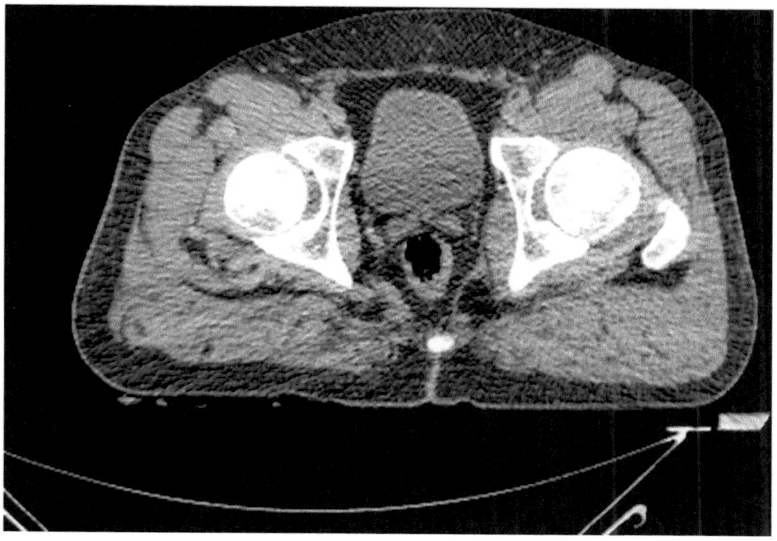

FIGURE 18.106 Urinary bladder fluid retention.

exclude injury or disease to the extent that it could explain the death and possibly trump a drug intoxication diagnosis. Autopsy, however, is the most accurate method for determining the cause and manner of death.

The most common autopsy findings in opiate-caused deaths are cerebral edema, pulmonary edema, and distension of the urinary bladder. Hence, there should be no surprise that the same triad of findings is visible on PMCT. When the triad is present, the findings are reported as having a high specificity for an opioid or depressant drug fatality.

Cerebral edema is best visualized on PCMT as compression of the cerebral ventricles (Fig. 18.105). **Pulmonary edema** results when fluid accumulates in the extravascular compartments of the lung and may be caused by both cardiogenic (heart failure) and noncardiogenic factors (CPR, fluid overload, drug overdose/neurogenic, etc.). Pulmonary edema is best visualized on PMCT as ground glass opacification and thickening of the bronchovascular bundles (Fig. 18.106) Not to be confused with pulmonary livor mortis, the ground glass opacities are not restricted to positional settling of fluid, but rather are diffuse in nature. Opiates, aside from respiratory and central nervous system depression, also cause sympathetic overstimulation and thus, increase the sphincter tone of the urinary bladder, inhibiting the parasympathetic nerves, which, in turn, reduces the sensation of bladder fullness and causes acute urinary bladder retention (Fig. 18.107). This finding may occur in acute ethanol intoxication.

The toxic, often functional (e.g., arrhythmia) effects of acute stimulant use are not visualized on PMCT. However, structurally demonstrable sequelae, such as a hemopericardium secondary to an aortic dissection or ruptured myocardial infarction, as well as hypertensive brain hemorrhages or ruptured saccular cerebral aneurysms, should alert the pathologist to consider the potential involvement of illicit stimulant drugs, such as methamphetamine or cocaine, which would change the manner of death from natural to accident. Hypertensive hemorrhages are well documented on PMCT and are characterized by hemorrhage within the basal ganglia, pons, and/or cerebellum, and result from the rupture of end perforating blood vessels in these areas. Long-term effects of stimulant use are reflected by chronic hypertensive changes, such as cardiac hypertrophy and renal atrophy with perinephric fat stranding.

Pediatric imaging

While cross-sectional imaging provides a great deal of information about both the soft tissues and osseous structures, the *skeletal survey remains the gold standard imaging examination in cases of known or suspected* **Nonaccidental Trauma** (NAT). As per the American College of Radiology (ACR), the Society of Pediatric Imaging (SPR), and the American Association of Pediatrics (AAP), a complete skeletal survey is recommended on all children younger than the age of two years old for whom abuse is suspected. Additionally, a skeletal survey is advised in select children aged 2–5 years of age or any child who is unable to communicate and for whom abuse is suspected (ACR/SPI guidelines).[96]

The skeletal survey includes separate radiographic exposures of specific anatomical parts to include the entire axial and appendicular skeleton, while ensuring maximal image uniformity and sharpness. This is in contradistinction to the "babygram," which is a single radiograph of the entire infant.

FIGURE 18.107 Skeletal survey.

Anatomical Site	Number of Images	Projection
Skull	2	Frontal, Lateral
Cervical Spine	1	Lateral
Thorax	4	AP, Lateral, Bilateral Obliques
Lumbar Spine	1	Lateral
Pelvis	1	AP
Humeri	2	AP
Forearms	2	AP
Hands	2	PA
Femurs	2	AP
Lower legs	2	AP
Feet	2	AP

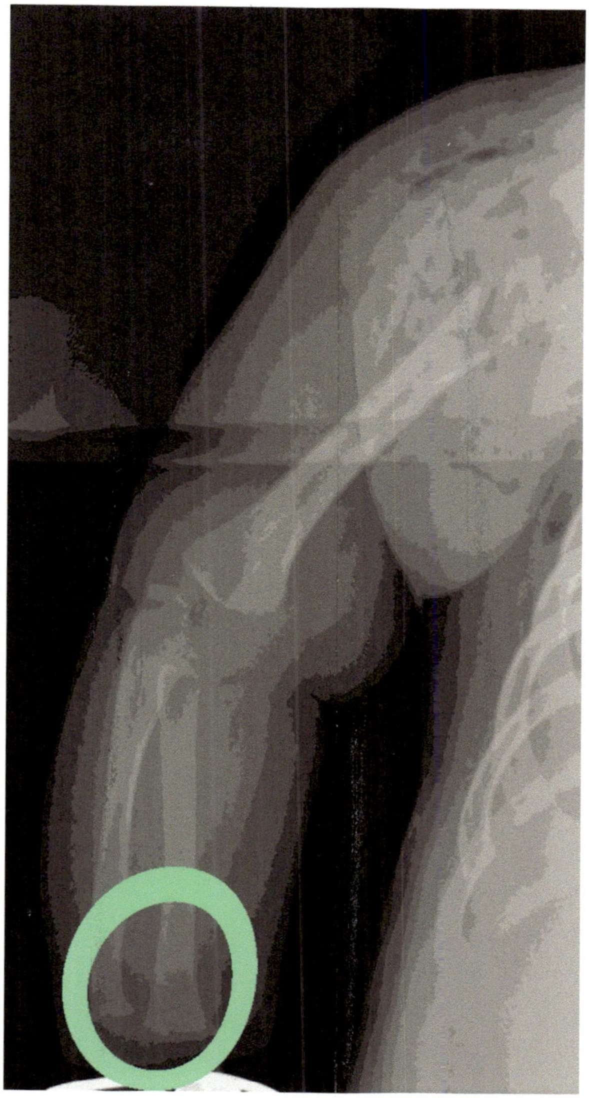

FIGURE 18.108 **Pediatric buckle fracture.** Distal radius and ulna fractures are poorly demonstrated on the dedicated right upper extremity image.

The axial skeleton should be imaged in two orthogonal planes: frontal and lateral. Bilateral oblique views of the ribs are important to increase the sensitivity of detecting rib fractures. Likewise, the skull will be imaged in two orthogonal planes. The appendicular skeleton is imaged in the following projections: bilateral humerii (AP), bilateral ulna and radius (AP), bilateral hands (PA), bilateral femurs (AP), bilateral tibia and fibula (AP), and bilateral feet (AP) (Fig. 18.108).

Additional views may be needed to fully document suspected abnormalities. For instance, orthogonal views of the extremities can be obtained to further evaluate or to document suspected or demonstrated fractures, respectively. The grandfather of FI, Dr. Gil Brogdon, once said that "one view is no view." Simply explained, findings can be missed on a single view. In fact, all anatomical parts should be evaluated in every image because an unexpected abnormal finding may be revealed when moving from a dedicated image to a separate image that was obtained for evaluation of a different anatomic site (Figs. 18.109 and 18.110).

Findings on the skeletal survey may prompt additional imaging studies for complementary analyses. For instance, brain and/or ocular magnetic resonance imaging may be helpful to document associated soft tissue injuries such as subdural hematoma and/or retinal hemorrhage in a child with history and findings suggestive of abusive head trauma. At some institutions, all decedents undergo PMCT. The skeletal survey and the PMCT are often complimentary (Fig. 18.111).

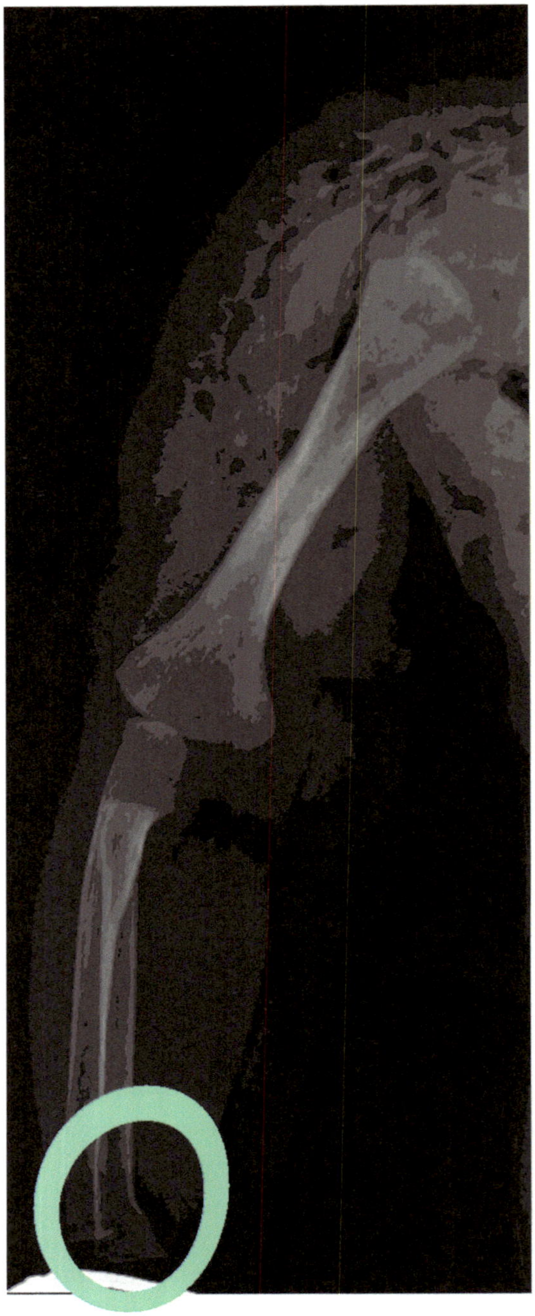

FIGURE 18.109 **Pediatric buckle fracture.** Distal radius and ulna fractures are more clearly demonstrated on the frontal chest radiograph. These fractures were not seen on CT, likely related, in part, to the right wrist band with associated streak artifact. Note gas in the right shoulder was related to attempted IO catheter placement.

Certain skeletal findings are highly concerning for NAT, such as the *classic metaphyseal lesion (CML) in long bones, posterior rib fractures, scapular fractures, as well as spinous process and sternal fractures* (Fig. 18.112). In addition to these locations, *fractures of different ages* are highly concerning for NAT.

Quality assurance is important during every portion of the skeletal survey examination. This starts with adapting the recommended imaging protocol and acquiring the essential training for the imaging technologist. The technologist should be familiar with the proper positioning and other technical parameters of the examination. Next, the physician interpreting the exam should have the appropriate training and experience. When possible, it is best to have a second opinion from a subspecialized pediatric radiologist.

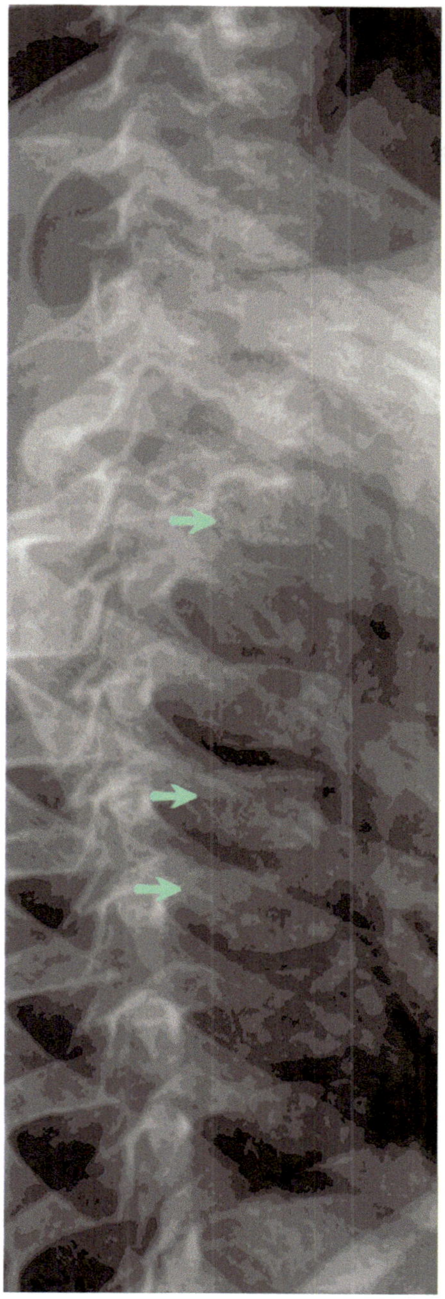

FIGURE 18.110 **Pediatric spine healing compression fractures.** Lateral view of the chest with superior endplate sclerosis of T2, T5, and T6.

Appendicular fractures

Long bone fractures can be extremely subtle in the pediatric population. When in doubt, request orthogonal views. *The CML has been described as one of the most specific imaging findings in abuse.* However, it is important to keep in mind that there is controversy about the histological validity of this presumed specificity.[97] The mechanism of the CML is related to stress across the metaphysis that results in a fracture plane perpendicular to the long axis of the bone. On imaging, the appearance may be subtle and is dependent upon the projection/patient positioning. Typically, it appears as though a triangular "corner" of the metaphysis has been broken off (Fig. 18.113).

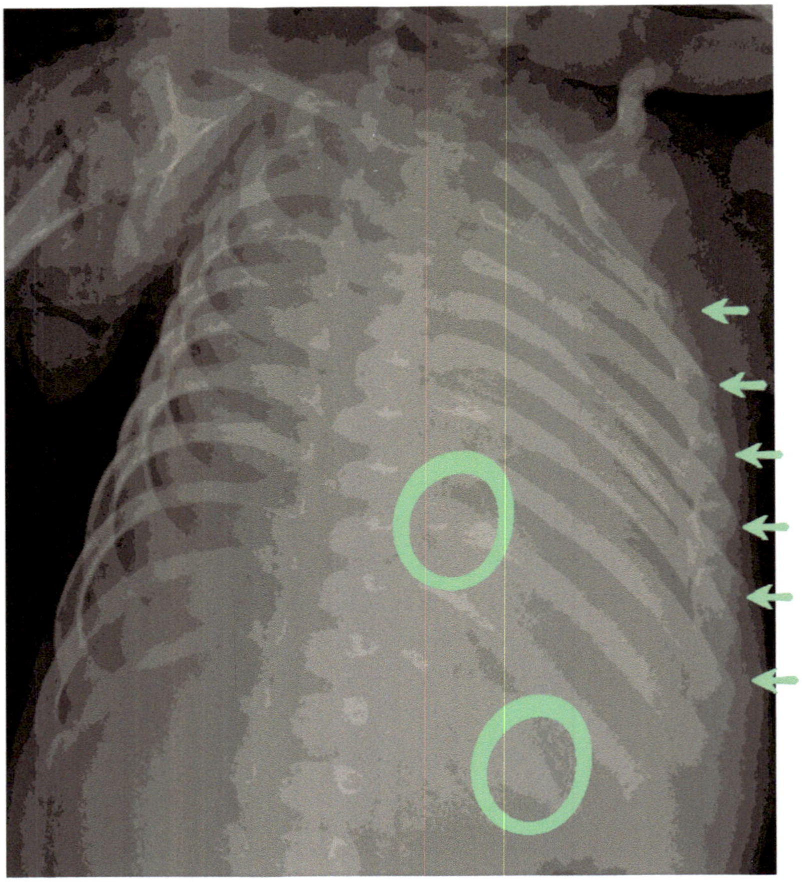

FIGURE 18.111 **Rib fractures.** Pediatric rib fractures suspicious for Nonaccidental Trauma. Acute to subacute lateral left sided rib fractures and healed left ninth posterior rib fracture as well as healed left lateral 11th rib fracture.

Foreign bodies

The most common foreign bodies encountered by imaging are implanted medical devices, although unexpected foreign bodies can be seen in the setting of body packers, legacy ballistics, body piercings, and mental illnesses. These objects can be readily apparent through multiple imaging modalities including radiographs, CT, and ultrasound.[98,99] Overwhelmingly, however, the most forensically important foreign bodies a medical examiner encounters are projectiles of various sorts in the setting of (acute) firearms death.

PMCT can be extremely useful for the identification and localization of foreign bodies. Often, as in cases of legacy bullets, the cause of death is unrelated (Fig. 18.114). However, in cases of fatal body packing, imaging can be used not only to localize drug packet shadows but also can demonstrate bowel obstruction or perforation.[100,101]

As one might expect, the composition of the foreign object and its anatomic location will influence the perceptibility of the object. As described under *CT Basics* below, different tissues have inherently varied densities (measured on the HU scale). For instance, bone is very dense and is seen as white on the gray scale continuum. Likewise, air is lucent and is seen as black. The appearance of foreign objects on CT will depend partially on their density. Dense objects, such as ballistic fragments, will appear densely white and air-containing objects will appear black. Items composed of plastic typically have an intermediate density, and items made of wood have a more dynamic range of densities based on wood type, as well as air and water content.[102–105]

When dealing with shades of gray, items of varied densities are easier to differentiate than items of similar density. This is why, without intravenous contrast material, solid organ and vascular pathology may be occult on PMCT.

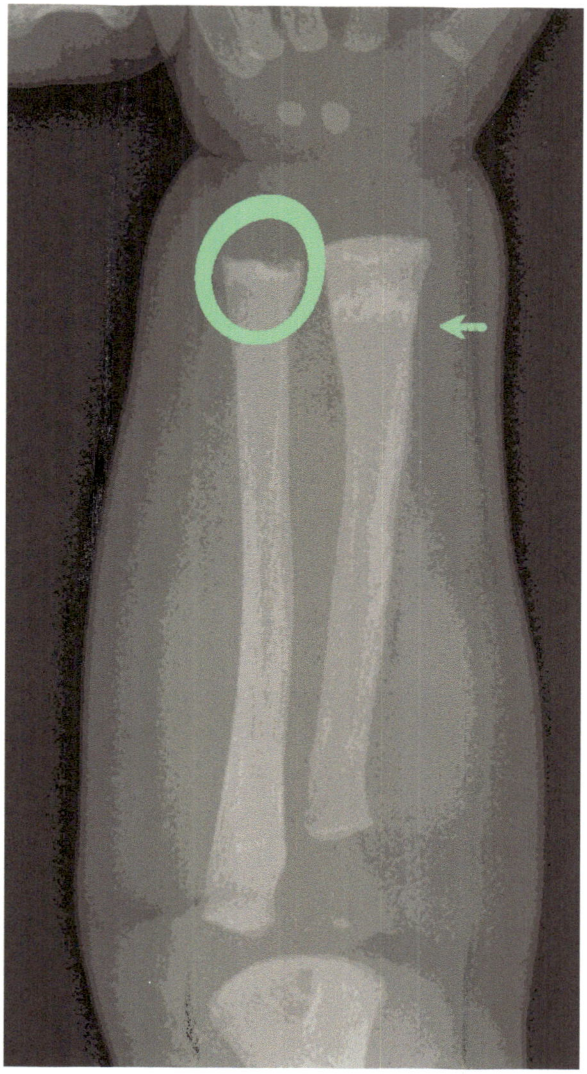

FIGURE 18.112 **Healing CML.** AP radiograph of the right forearm. Subacute distal ulna classic metaphyseal fracture. Note band of sclerosis in the distal radius from subacute buckle fracture.

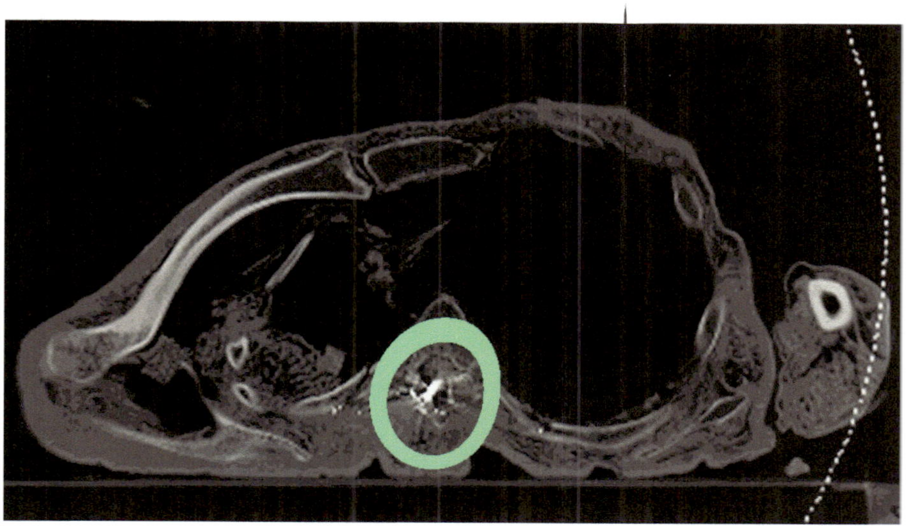

FIGURE 18.113 Retained projectile in spine.

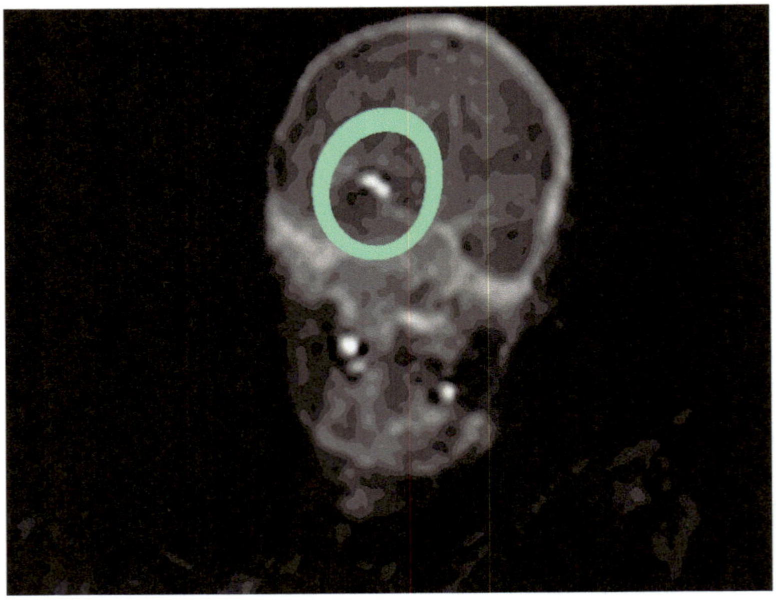

FIGURE 18.114 **Implant for seventh nerve palsy.** An internet search revealed the identity of the implant.

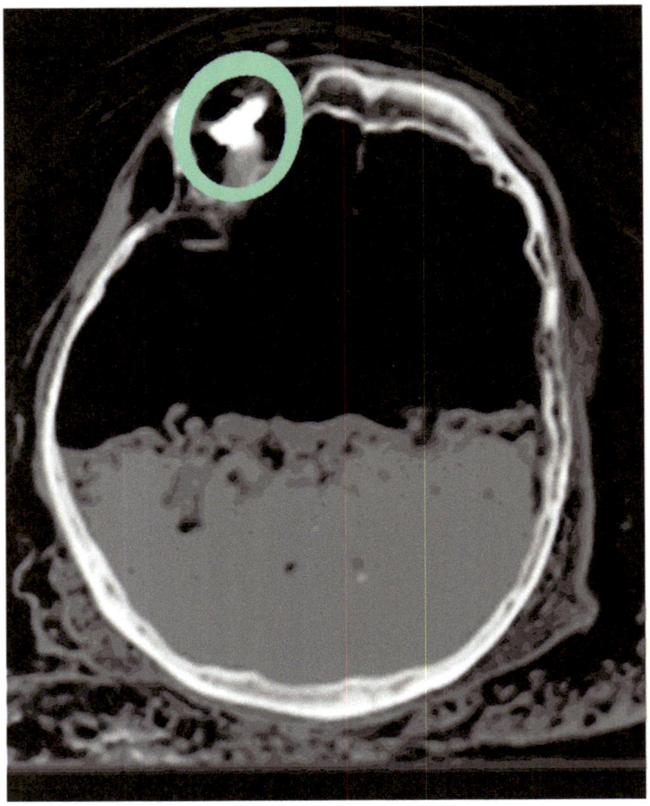

FIGURE 18.115 Implant for seventh nerve palsy.

Similarly, if a foreign body has a similar density to the tissue surrounding it, visualization may not be possible. For example, a food bolus in the larynx or trachea may not be readily apparent without pertinent history, and in most cases, requires confirmation with autopsy. However, a metallic implant for seventh nerve palsy is easily discernible compared to the soft tissues in the orbit (Figs. 18.115 and 18.116).

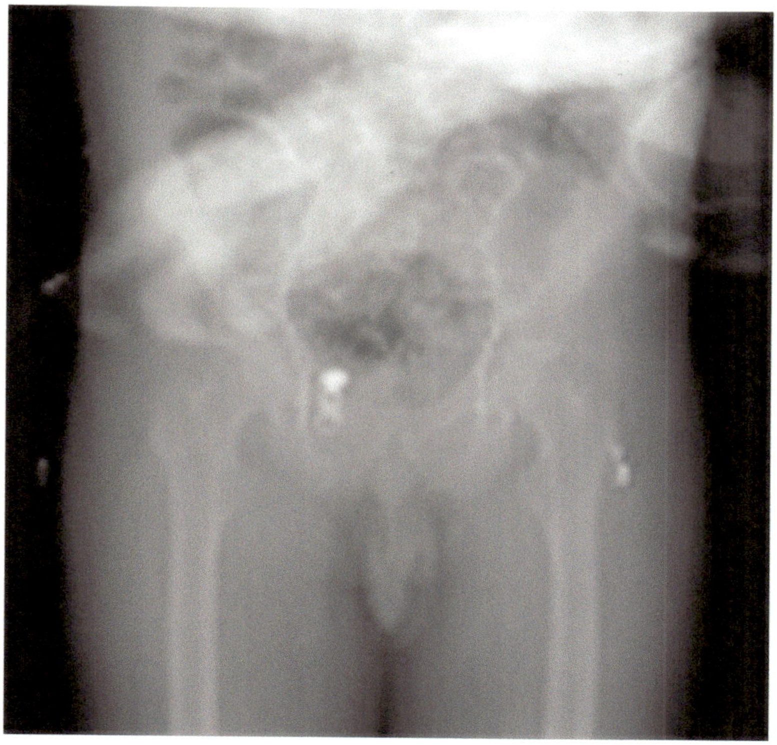

FIGURE 18.116 **Body packer.** Scout view is unremarkable.

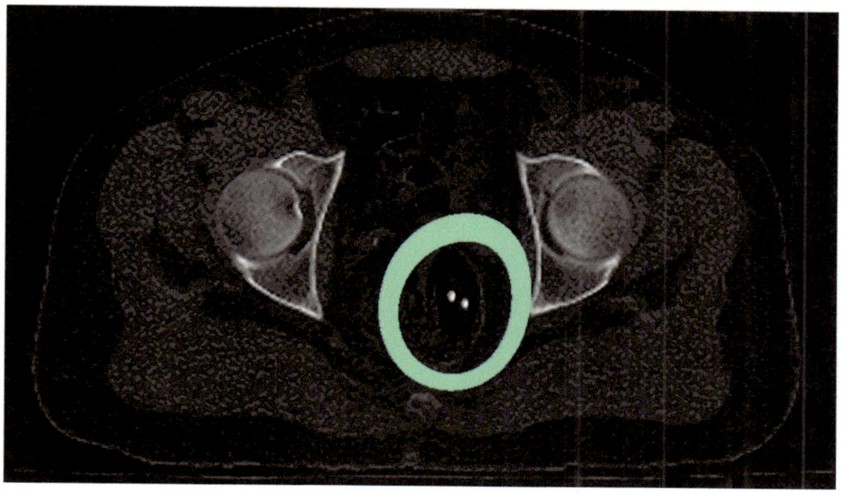

FIGURE 18.117 **Body packer.** Axial CT bone windows reveal a foreign body in the rectum.

The location, number, and dimensions of foreign bodies can offer significant information prior to autopsy. This is especially helpful in instances of unsuspected body packing deaths in which an autopsy may not have been initially indicated (Figs. 18.117–18.119).

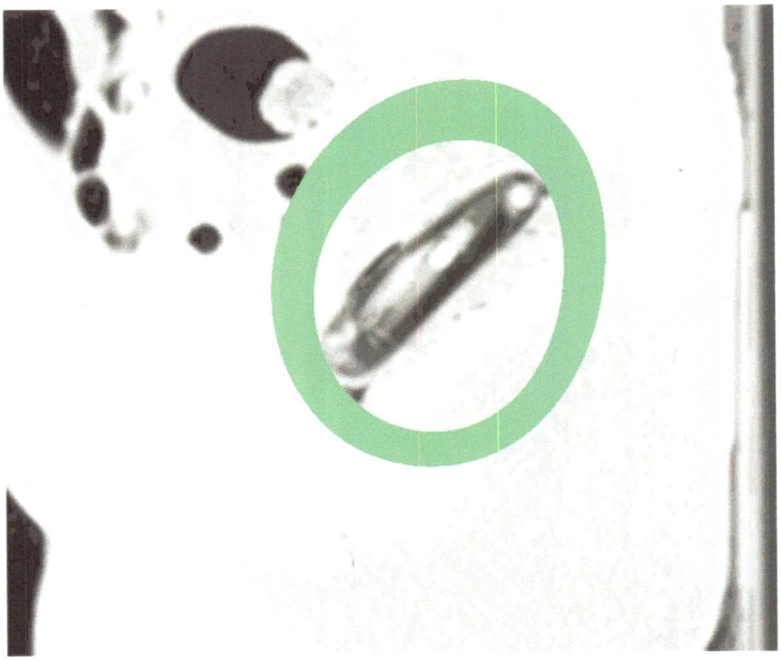

FIGURE 18.118 **Body packer.** Sagittal lung window.

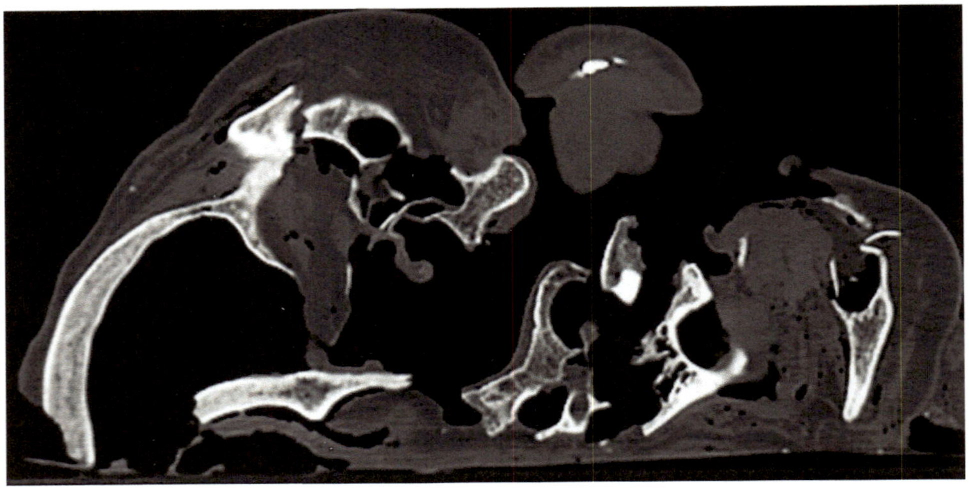

FIGURE 18.119 Axial CT head with bone window. Marked cranial disruption.

Pro tips/pearls

The best "Pro tip" is to get to know your PACS and learn how to maximize your efficiency when interpreting PMCT. Learn how to use "hot keys" if they are available, for quick switching of the imaging window (bone, brain, lung, soft tissue). Knowing how to zoom and pan images aids in the evaluation of subtle findings and small structures. Scout lines and localizers help cross-reference findings in orthogonal planes. Always look at the scout images prior to CT interpretation; you will be surprised about the skeletal injuries that are easily overlooked on cross-sectional imaging, but that are readily apparent on the scout. It can be very helpful to pull up the axial plane next to the sagittal or coronal plane with scout lines. Also, when using CT to triage cases for autopsy, consider the pertinent history to quickly hone into and assess critical structures that are likely to exhibit pathology and that could influence decisions to autopsy, such as marked cranial disruption in the case of a motor vehicle collision (Fig. 18.120).

Forensic pathology postmortem imaging training curriculum: the New Mexico experience

Given the wide range of applicability to forensic investigations and the rapidly expanding access to CT (and MR), FI instruction of some sort should be taught in forensic pathology fellowship training programs. As this is an evolving field, there is no standard curriculum. The Office of the Medical Examiner (OMI) in Albuquerque, New Mexico has developed a curriculum that can be used as a framework for other programs.

For example, at the New Mexico OMI, all decedents are imaged with whole body CT (including scout views) at intake. The CT images are reviewed at case triage, along with the scene photos and the investigator's report at the morning report case conference. This information is synthesized and cases are selected for autopsy, partial autopsy, or external exam based on **PMCT** observations. Any case can be sent to a forensic radiologist for consult.

A one-year didactic curriculum was developed to cover basic FI topics with formal lectures presented by the forensic radiologist, clinical radiology faculty (with specific subspecialty expertise in pediatric imaging and neuroradiology) as well as with a clinical forensic pediatrician.

Future directions

It is the sincere belief of the authors that education is the future of FI. PMCT is becoming more accessible and has proven strengths that can be taken advantage of to compliment autopsy findings; in some select instances, PMCT can be used to supplant autopsy. This might be particularly useful in the setting of religious or other objections to autopsy. At the New Mexico OMI, PMCT is a valuable tool for triaging cases prior to autopsy.

Few forensic pathology fellowships offer a dedicated FI curriculum. New trainees (and some practicing forensic pathologists) would benefit from both formal and informal instruction.

References

1. Glasser O. The life of Wilhelm Conrad Rontgen as revealed in his letters. *Sci Mon.* 1937;45:193–206.
2. Brecher R, Brecher EM. *The Rays: A History of Radiology in the United States and Canada.* Baltimore: Williams and Wilkins; 1969.
3. Cox JaK RC. The new photography with report of a case in which a bullet was photographed in the leg. *Montreal Med J.* 1984;24:661.
4. Bolliger SA, Thali MJ, Ross S, Buck U, Naether S, Vock P. Virtual autopsy using imaging: bridging radiologic and forensic sciences. A review of the Virtopsy and similar projects. *Eur Radiol.* 2008;18(2):273–282.
5. Dirnhofer R, Jackowski C, Vock P, Potter K, Thali MJ. VIRTOPSY: minimally invasive, imaging-guided virtual autopsy. *Radiographics.* 2006; 26(5):1305–1333.
6. Germerott T, Preiss US, Ebert LC, et al. A new approach in virtopsy: postmortem ventilation in multislice computed tomography. *Leg Med (Tokyo).* 2010;12(6):276–279.
7. Thali M. Virtual autopsy (virtopsy) in forensic science: from the scalpel to the scanner. *Pathologe.* 2011;32(Suppl 2):292–295.
8. Thali MJ, Jackowski C, Oesterhelweg L, Ross SG, Dirnhofer R. VIRTOPSY - the Swiss virtual autopsy approach. *Leg Med (Tokyo).* 2007;9(2): 100–104.
9. Thali MJ, Yen K, Schweitzer W, et al. Virtopsy, a new imaging horizon in forensic pathology: virtual autopsy by postmortem multislice computed tomography (MSCT) and magnetic resonance imaging (MRI)—a feasibility study. *J Forensic Sci.* 2003;48(2):386–403.
10. Stockburger WT. CT imaging, then and now: a 30-year review of the economics of computed tomography. *Radiol Manag.* 2004;26(6):20–22, 24-27; quiz 28-30.
11. Piekarski J, Goldberg HI, Royal SA, Axel L, Moss AA. Difference between liver and spleen CT numbers in the normal adult: its usefulness in predicting the presence of diffuse liver disease. *Radiology.* 1980;137(3):727–729.
12. Boyce CJ, Pickhardt PJ, Kim DH, et al. Hepatic steatosis (fatty liver disease) in asymptomatic adults identified by unenhanced low-dose CT. *AJR Am J Roentgenol.* 2010;194(3):623–628.
13. Bohte AE, van Werven JR, Bipat S, Stoker J. The diagnostic accuracy of US, CT, MRI and 1H-MRS for the evaluation of hepatic steatosis compared with liver biopsy: a meta-analysis. *Eur Radiol.* 2011;21(1):87–97.
14. Hamer OW, Aguirre DA, Casola G, Lavine JE, Woenckhaus M, Sirlin CB. Fatty liver: imaging patterns and pitfalls. *Radiographics.* 2006;26(6): 1637–1653.
15. de Jong PA, Hellings WE, Takx RA, Išgum I, van Herwaarden JA, Mali WPTM. Computed tomography of aortic wall calcifications in aortic dissection patients. *PLoS One.* 2014;9(7):e102036.
16. LePage MA, Quint LE, Sonnad SS, Deeb GM, Williams DM. Aortic dissection: CT features that distinguish true lumen from false lumen. *Am J Roentgenol.* 2001;177(1):207–211.
17. Prosser I, Maguire S, Harrison S, et al. Dating of fractures: an analysis of key radiological features in children aged 5 years and under. *Arch Dis Child.* 2010;95(Suppl 1):A3.
18. Halliday KE, Broderick NJ, Somers JM, Hawkes R. Dating fractures in infants. *Clin Radiol.* 2011;66(11):1049–1054.
19. Prosser I, Maguire S, Harrison SK, Mann M, Sibert JR, Kemp AM. How old is this fracture? Radiologic dating of fractures in children: a systematic review. *Am J Roentgenol.* 2005;184(4):1282–1286.

20. Prosser I, Lawson Z, Evans A, et al. A timetable for the radiologic features of fracture healing in young children. *AJR Am J Roentgenol*. 2012; 198(5):1014–1020.
21. Marsell R, Einhorn TA. The biology of fracture healing. *Injury*. 2011;42(6):551–555.
22. Claes LE, Heigele CA, Neidlinger-Wilke C, et al. Effects of mechanical factors on the fracture healing process. *Clin Orthop Relat Res*. 1998;355.
23. Frost HM. The biology of fracture healing. An overview for clinicians. Part I. *Clin Orthop Relat Res*. 1989;248:283–293.
24. Cornell CN, Lane JM. Newest factors in fracture healing. *Clin Orthop Relat Res*. 1992;277:297–311.
25. Claes L, Recknagel S, Ignatius A. Fracture healing under healthy and inflammatory conditions. *Nat Rev Rheumatol*. 2012;8(3):133–143.
26. Cappella A, Amadasi A, Castoldi E, Mazzarelli D, Gaudio D, Cattaneo C. The difficult task of assessing perimortem and postmortem fractures on the skeleton: a blind text on 210 fractures of known origin. *J Forensic Sci*. 2014;59(6):1598–1601.
27. Gupta PK, Gupta K, Jain M, Garg T. Postprocedural chest radiograph: impact on the management in critical care unit. *Anesth Essays Res*. 2014; 8(2):139–144.
28. Brun CN, Christensen AM, Kravarski M, et al. Comparative radiologic identification with standardized single CT images of the paranasal sinuses-Evaluation of inter-rater reliability. *Forensic Sci Int*. 2017;280:81–86.
29. Deloire L, Diallo I, Cadieu R, et al. Post-mortem X-ray computed tomography (PMCT) identification using ante-mortem CT-scan of the sphenoid sinus. *J Neuroradiol*. 2019;46(4):248–255.
30. Gascho D, Philipp H, Flach PM, Thali MJ, Kottner S. Standardized medical image registration for radiological identification of decedents based on paranasal sinuses. *J Forensic Leg Med*. 2018;54:96–101.
31. Gibney B, Ryan JW, MacMahon PJ, O'Connor GS, Bolster F. Assessment of RAdiopaque Patient IDentification Stickers (RAPIDS) for patient-scan correlation in a mass casualty incident. *Emerg Radiol*. 2020;27(3):293–301.
32. Marek Z, Kuśmiderski J, Lisowski Z. [Radiograms of the paranasal sinuses as a principle of identifying catastrophe victims and unknown skeletons]. *Arch Kriminol*. 1983;172(1–2):1–6.
33. Pereira JGD, Santos JBS, Sousa SP, Franco A, Silva RHA. Frontal sinuses as tools for human identification: a systematic review of imaging methods. *Dentomaxillofacial Radiol*. 2021;50(5):20200599.
34. Pfaeffli M, Vock P, Dirnhofer R, Braun M, Bolliger SA, Thali MJ. Post-mortem radiological CT identification based on classical ante-mortem X-ray examinations. *Forensic Sci Int*. 2007;171(2–3):111–117.
35. Ruder TD, Kraehenbuehl M, Gotsmy WF, et al. Radiologic identification of disaster victims: a simple and reliable method using CT of the paranasal sinuses. *Eur J Radiol*. 2012;81(2):e132–138.
36. Scendoni R, Kelmendi J, Cossellu G, et al. Comparison of frontal sinuses for personal identification in 3 populations using Cameriere's code number. *Am J Forensic Med Pathol*. 2021;42(1):42–45.
37. Weiss G, Wittig H, Scheurer E, Ruprecht R, Lenz C. Identification of deceased based on sternal bone computed tomography features. *Forensic Sci Int*. 2018;286:233–238.
38. Wood RE. Forensic aspects of maxillofacial radiology. *Forensic Sci Int*. 2006;159(Suppl 1):S47–S55.
39. Wu W, Li Y, Fan F, Zhang K, Deng ZH. Research progress on individual identification by frontal sinus imaging. *Fa Yi Xue Za Zhi*. 2021;37(1): 81–86.
40. Busardò FP, Frati P, Guglielmi G, et al. Postmortem-computed tomography and postmortem-computed tomography-angiography: a focused update. *Radiol Med*. 2015;120(9):810–823.
41. Filograna L, Pugliese L, Muto M, et al. A practical guide to virtual autopsy: why, when and how. *Semin Ultrasound CT MR*. 2019;40(1):56–66.
42. Flach PM, Gascho D, Schweitzer W, et al. Imaging in forensic radiology: an illustrated guide for postmortem computed tomography technique and protocols. *Forensic Sci Med Pathol*. 2014;10(4):583–606.
43. Ross SG, Bolliger SA, Ampanozi G, Oesterhelweg L, Thali MJ, Flach PM. Postmortem CT angiography: capabilities and limitations in traumatic and natural causes of death. *Radiographics*. 2014;34(3):830–846.
44. Chevallier C, Doenz F, Vaucher P, et al. Postmortem computed tomography angiography vs. conventional autopsy: advantages and inconveniences of each method. *Int J Leg Med*. 2013;127(5):981–989.
45. van Eijk RP, van der Zwan A, Bleys RL, Regli L, Esposito G. Novel application of postmortem CT angiography for evaluation of the intracranial vascular anatomy in cadaver heads. *AJR Am J Roentgenol*. 2015;205(6):1276–1280.
46. Inokuchi G, Yajima D, Hayakawa M, et al. Postmortem dynamic cerebral angiography for detecting aneurysm and bleeding sites in cases of subarachnoid hemorrhage. *Forensic Sci Med Pathol*. 2014;10(4):487–495.
47. Cadaveric spasm. J Am Med Assoc. 1898;XXX(16):923–924.
48. Knüsel CJ, Janaway RC, King SE. Death, decay, and ritual reconstruction: archaeological evidence of cadaveric spasm. *Oxf J Archaeol*. 1996; 15(2):121–128.
49. Chauhan M, Behera C, Bodwal J, Dikshit P. Rare gripping object in electrocution: cadaveric spasm at its best. *Med Leg J*. 2017;85(3):155–158.
50. Gupta B, Jani C, Singh G. Cadaveric spasm. *Indian Internet J Forensic Med Toxicol*. 2003;1(3).
51. Okuda T, Shiotani S, Kobayashi T, et al. Immediate non-traumatic postmortem computed tomographic demonstration of myocardial intravascular gas of the left ventricle: effects from cardiopulmonary resuscitation. *SpringerPlus*. 2013;2(1):86.
52. Shiotani S, Kohno M, Ohashi N, Atake S, Yamazaki K, Nakayama H. Cardiovascular gas on non-traumatic postmortem computed tomography (PMCT): the influence of cardiopulmonary resuscitation. *Radiat Med*. 2005;23(4):225–229.
53. Buschmann CT, Wrobel D, Tsokos M. [Post-mortem animal predation of the genital region caused by a half-breed dog]. *Arch Kriminol*. 2008; 222(5–6):182–186.
54. Hammarlebiod S, Farrugia A, Bierry G, Raul JS, Willaume T. Thermal bone injuries: postmortem computed tomography findings in 25 cases. *Int J Leg Med*. 2021;136:219–227.
55. Thali MJ, Yen K, Plattner T, et al. Charred body: virtual autopsy with multi-slice computed tomography and magnetic resonance imaging. *J Forensic Sci*. 2002;47(6):1326–1331.
56. Coty JB, Nedelcu C, Yahya S, et al. Burned bodies: post-mortem computed tomography, an essential tool for modern forensic medicine. *Insights Imaging*. 2018;9(5):731–743.
57. Roberts IS, Traill ZC. Use of post-mortem computed tomography during the COVID-19 pandemic. *Diagn Histopathol*. 2021;27(10):418–421.

58. O'Donnell C, Iles L, Woodford N. Post-mortem CT lung findings at a medicolegal institute in SARS-CoV-2 RT-PCR positive cases with autopsy correlation. *Forensic Sci Med Pathol*. 2021:1–10.
59. Leth PM, Rasmussen CH, Pagh M. Findings in post-mortem CT and autopsy in a 53-year-old-man with COVID-19. *Ugeskr Laeger*. 2020; 182(28).
60. Kniep I, Heinemann A, Edler C, et al. COVID-19 lungs in post-mortem computed tomography. *Rechtsmedizin*. 2021:1–3.
61. De-Giorgio F, Cittadini F, Cina A, et al. Use of post-mortem chest computed tomography in Covid-19 pneumonia. *Forensic Sci Int*. 2021;325: 110851.
62. Frija-Masson J, Debray MP, Boussouar S, et al. Residual ground glass opacities three months after Covid-19 pneumonia correlate to alteration of respiratory function: the post Covid M3 study. *Respir Med*. 2021;184:106435.
63. Zheng Y, Wang L, Ben S. Meta-analysis of chest CT features of patients with COVID-19 pneumonia. *J Med Virol*. 2021;93(1):241–249.
64. Ufuk F, Savas R. Chest CT features of the novel coronavirus disease (COVID-19). *Turk J Med Sci*. 2020;50(4):664–678.
65. Kovács A, Palásti P, Veréb D, Bozsik B, Palkó A, Kincses ZT. The sensitivity and specificity of chest CT in the diagnosis of COVID-19. *Eur Radiol*. 2021;31(5):2819–2824.
66. Chaimayo C, Kaewnaphan B, Tanlieng N, et al. Rapid SARS-CoV-2 antigen detection assay in comparison with real-time RT-PCR assay for laboratory diagnosis of COVID-19 in Thailand. *Virol J*. 2020;17(1):177.
67. Ceylan N, Çinkooğlu A, Bayraktaroğlu S, Savas R. Atypical chest CT findings of COVID-19 pneumonia: a pictorial review. *Diagn Interv Radiol*. 2021;27(3):344–349.
68. Carpenter CR, Mudd PA, West CP, Wilber E, Wilber ST. Diagnosing COVID-19 in the emergency department: a scoping review of clinical examinations, laboratory tests, imaging accuracy, and biases. *Acad Emerg Med*. 2020;27(8):653–670.
69. Kwee TC, Kwee RM. Chest CT in COVID-19: what the radiologist needs to know. *Radiographics*. 2020;40(7):1848–1865.
70. Helmrich E, Decker L, Adolphi N, Makino Y. Postmortem CT lung findings in decedents with Covid-19: a review of 14 decedents and potential triage implications. *Forensic Imaging*. 2020;23, 200419-200419.
71. Ishida M, Gonoi W, Okuma H, et al. Common postmortem computed tomography findings following atraumatic death: differentiation between normal postmortem changes and pathologic lesions. *Korean J Radiol*. 2015;16(4):798–809.
72. Filograna L, Thali MJ. Post-mortem CT imaging of the lungs: pathological versus non-pathological findings. *Radiol Med*. 2017;122(12): 902–908.
73. Sushentsev N, Bura V, Kotnik M, et al. A head-to-head comparison of the intra- and interobserver agreement of COVID-RADS and CO-RADS grading systems in a population with high estimated prevalence of COVID-19. *BJR Open*. 2020;2(1):20200053.
74. O'Neill SB, Byrne D, Müller NL, et al. Radiological Society of North America (RSNA) expert consensus statement related to chest CT findings in COVID-19 versus CO-RADS: comparison of reporting system performance among chest radiologists and end-user preference. *Can Assoc Radiol J*. 2021;72(4):806–813.
75. Prokop M, van Everdingen W, van Rees Vellinga T, et al. CO-RADS: a categorical CT assessment scheme for patients suspected of having COVID-19-definition and evaluation. *Radiology*. 2020;296(2):E97–e104.
76. Penha D, Pinto EG, Matos F, et al. CO-RADS: coronavirus classification review. *J Clin Imaging Sci*. 2021;11:9.
77. Lieveld AWE, Azijli K, Teunissen BP, et al. Chest CT in COVID-19 at the ED: validation of the COVID-19 reporting and data system (CO-RADS) and CT severity score: a prospective, multicenter, observational study. *Chest*. 2021;159(3):1126–1135.
78. Fujioka T, Takahashi M, Mori M, et al. Evaluation of the usefulness of CO-RADS for chest CT in patients suspected of having COVID-19. *Diagnostics*. 2020;10(9).
79. De Smet K, De Smet D, Ryckaert T, et al. Diagnostic performance of chest CT for SARS-CoV-2 infection in individuals with or without COVID-19 symptoms. *Radiology*. 2021;298(1):E30–e37.
80. Çomoğlu S, Öztürk S, Topçu A, et al. The role of CO-RADS scoring system in the diagnosis of COVID-19 infection and its correlation with clinical signs. *Curr Med Imaging*. 2021;18:381–386.
81. Bellini D, Panvini N, Rengo M, et al. Diagnostic accuracy and interobserver variability of CO-RADS in patients with suspected coronavirus disease-2019: a multireader validation study. *Eur Radiol*. 2021;31(4):1932–1940.
82. Henkel M, Weikert T, Marston K, et al. Lethal COVID-19: radiologic-pathologic correlation of the lungs. *Radiol Cardiothorac Imaging*. 2020;2(6): e200406.
83. Poh KC, Tay VYJ, Lin SH, Chee HL, Suhitharan T. A review of COVID-19-related thrombosis and anticoagulation strategies specific to the Asian population. *Singapore Med J*. 2020 Dec. https://doi.org/10.11622/smedj.2020174. PMID: 33264831.
84. Paterson RW, Brown RL, Benjamin L, et al. The emerging spectrum of COVID-19 neurology: clinical, radiological and laboratory findings. Brain. 2020;143(10):3104–3120.
85. Lorenzo C, Francesca B, Francesco P, Elena C, Luca S, Paolo S. Acute pulmonary embolism in COVID-19 related hypercoagulability. *J Thromb Thrombolysis*. 2020;50(1):223–226.
86. Kamel MH, Yin W, Zavaro C, Francis JM, Chitalia VC. Hyperthrombotic Milieu in COVID-19 patients. *Cells*. 2020;9(11).
87. Jain R, Young M, Dogra S, et al. COVID-19 related neuroimaging findings: a signal of thromboembolic complications and a strong prognostic marker of poor patient outcome. *J Neurol Sci*. 2020;414:116923.
88. Izzedine H, Jhaveri KD, Perazella MA. Vascular injury and COVID-19-related mortality: what lies below the tip of the iceberg? *Clin Nephrol*. 2020;94(1):11–13.
89. Brüggemann R, Gietema H, Jallah B, Ten Cate H, Stehouwer C, Spaetgens B. Arterial and venous thromboembolic disease in a patient with COVID-19: a case report. *Thromb Res*. 2020;191:153–155.
90. Opeskin K, Burke MP. Vertebral artery trauma. *Am J Forensic Med Pathol*. 1998;19(3):206–217.
91. Savall F, Dedouit F, Mokrane FZ, Rouge D, Saint-Martin P, Telmon N. An unusual homicidal stab wound of the cervical spinal cord: a single case examined by post-mortem computed tomography angiography (PMCTA). *Forensic Sci Int*. 2015;254:e18–e21.
92. Makino Y, Yokota H, Nakatani E, et al. Differences between postmortem CT and autopsy in death investigation of cervical spine injuries. *Forensic Sci Int*. 2017;281:44–51.
93. Battle CE, Hutchings H, Evans PA. Risk factors that predict mortality in patients with blunt chest wall trauma: a systematic review and meta-analysis. *Injury*. 2012;43(1):8–17.

94. Veysi VT, Nikolaou VS, Paliobeis C, Efstathopoulos N, Giannoudis PV. Prevalence of chest trauma, associated injuries and mortality: a level I trauma centre experience. *Int Orthop*. 2009;33(5):1425–1433.
95. Arumugam S, Al-Hassani A, El-Menyar A, et al. Frequency, causes and pattern of abdominal trauma: a 4-year descriptive analysis. *J Emerg Trauma Shock*. 2015;8(4):193–198.
96. Radiology ACo. *ACR–SPR practice parameter for the performance and interpretation of skeletal surveys in children*. Reston: American College of Radiology ACR Standards American College of Radiology; 2016:1–9.
97. Ayoub DM, Hyman C, Cohen M, Miller M. A critical review of the classic metaphyseal lesion: traumatic or metabolic? *AJR Am J Roentgenol*. 2014;202(1):185–196.
98. Grassi R, Faggian A, Somma F, De Cecco CN, Laghi A, Caseiro-Alves F. Application of imaging guidelines in patients with foreign body ingestion or inhalation: literature review. *Semin Ultrasound CT MR*. 2015;36(1):48–56.
99. Laya BF, Restrepo R, Lee EY. Practical imaging evaluation of foreign bodies in children: an update. *Radiol Clin North Am*. 2017;55(4):845–867.
100. Heinemann A, Miyaishi S, Iwersen S, Schmoldt A, Püschel K. Body-packing as cause of unexpected sudden death. *Forensic Sci Int*. 1998;92(1): 1–10.
101. Joynt BP, Mikhael NZ. Sudden death of a heroin body packer. *J Anal Toxicol*. 1985;9(5):238–240.
102. Pinto A, Brunese L, Daniele S, et al. Role of computed tomography in the assessment of intraorbital foreign bodies. *Semin Ultrasound CT MR*. 2012;33(5):392–395.
103. Tseng HJ, Hanna TN, Shuaib W, Aized M, Khosa F, Linnau KF. Imaging foreign bodies: ingested, aspirated, and inserted. *Ann Emerg Med*. 2015;66(6):570–582. e575.
104. You YY, Shi BJ, Wang XY, et al. Intraorbital wooden foreign bodies: case series and literature review. *Int J Ophthalmol*. 2021;14(10):1619–1627.
105. Rong AJ, Fan KC, Golshani B, et al. Multimodal imaging features of intraocular foreign bodies. *Semin Ophthalmol*. 2019;34(7–8):518–532.

CHAPTER

19

Recent and ongoing technological advancements in forensic pathology practice

Michelle N. Stram, James R. Gill and Susan F. Ely

The role of molecular genetics in sudden and unexpected deaths

Molecular genetics plays a critical role in the autopsy-negative sudden deaths of apparently healthy individuals and can provide key familial diagnostic information in children and young adults with a history of epilepsy, findings of cardiomyopathy, fatal venous thromboembolic events, and aortic dissection. In the forensic setting, this companion diagnostic information enables the identification of etiologically specific cardiac causes of death in grossly and microscopically normal hearts. These typically involve sudden, unexpected deaths, but they also can diagnose an underlying genetic abnormality in nontraumatic seizure disorders. Molecular testing also can help to identify the underlying cause of an aortic dissection and a pulmonary thromboembolism (PTE) in instances of genetic aortopathies and hereditary thrombophilias, respectively. It is therefore important for medical examiners to include molecular diagnostics testing in select cases, as some detected mutations may be central to the cause of death, and the results also may have important implications for surviving family members in terms of genetic counseling, testing, and clinical follow-up.[1] Conditions encountered in medicolegal death investigations that may have an important genetic component are outlined in Table 19.1. The genes and conditions listed are not exhaustive, but provide a representative sampling of the mutations and diseases that maybe be included in a molecular testing panel. Detection of a mutation in a decedent also can focus testing for surviving family members by directing them to be screened for that specific mutation instead of the more costly broad panel of tests.

A thorough medicolegal investigation is critical in identifying deaths that may benefit from a molecular diagnostic work up. We recommend a comprehensive investigation of the circumstances surrounding the death, including a thorough scene investigation, interviews with family and friends, family and social histories, obtaining a medical history from treating physicians, and a review of any medical and/or hospital records. For apparent sudden cardiac deaths, the inclusion of specific questions about dyspnea, palpitations, syncope, and the relationship of symptoms to physical activity are especially important, as well as inquiring about a family history of sudden death at a young age or known cardiac conditions manifesting before middle age.[2] A study of referred medical examiner sudden undetermined deaths found that one-third of decedents harbored a putative cardiac channel mutation: seven previously reported to host mutations in the RyR2-encoded calcium release channel and 10 with LQTS susceptibility mutations.[3]

For epilepsy and nontraumatic seizure disorders, it is likewise important to obtain a specific seizure history, including the age of onset, a history of childhood febrile seizures, seizure frequency (presence/frequency of breakthrough seizures), seizure type (posttraumatic vs. natural), medication and compliance (drugs, dosage, date of refill, pill count, prescribing physician), surgical interventions, if any, comorbid conditions (including cardiac findings such as syncope and any psychiatric history and/or cognitive impairment), and attention to specific details of the scene investigation (death during sleep, prone positioning, placement of the nose and mouth, etc.).[4]

After a comprehensive investigative and autopsy examination, including gross and microscopic evaluations, appropriate laboratory tests, such as toxicology, vitreous chemistry, microbiology tests, metabolic screening (in

Principles of Forensic Pathology
https://doi.org/10.1016/B978-0-323-91796-4.00007-6

TABLE 19.1 Forensically important conditions with mutations: cardiac channelopathies, cardiomyopathies, aortopathies, thrombophilias, and epilepsy.

Diseases	Genes*	Comments
Cardiac channelopathies		
Long QT syndromes	*KCNQ1 (LQT1), KCNH2 (LQT2), SCN5A (LQT3), CACNA1C (LQT8), CALM1, CALM2, CALM3, TRDN*	Predominantly cardiac ion channels (Na+, K+, Ca2+) or regulatory proteins
Brugada syndrome/ J-wave syndromes	*SCN5A*	Mediation of voltage gated ion channels
Short QT syndromes	*KCNH2 (SQT1), KCNQ1 (SQT2), KCNJ2 (SQT3), SLC4A3*	Cardiac potassium channels or ion exchange
Cardiomyopathies		
Hypertrophic cardiomyopathy	*MYBPC3, TPM1, MYL2, ACTC1, MYL3, MYH7, TNNT2, TNNI3, PRKAG2, ALPK3, JPH2, TNNC1, CSRP3*	Components of sarcomere, e.g., myosin heavy and light chains
Arrhythmogenic right ventricular cardiomyopathy	*PKP2, DSP, DSG2, DSC2, TGFB3, JUP, RyR2, TMEM43*	Includes components of desmosome and Ca2+ regulation
Idiopathic dilated cardiomyopathy	*TTN, SCN5A, MYH7, LMNA, FLNC, BAG3, TNNT2, DES, TNNC1, RBM20, DSP, NEXN, TPM1, TNNI3, JPH2, VCL, ACTC1*	Components of sarcomere, Z-disk, nuclear lamina proteins, and intermediate filaments, RNA-binding proteins, molecular chaperone regulatory proteins
Left ventricular noncompaction cardiomyopathy	*ACTC1, LDB3, LMNA MYBPC3, MYH7, TNNT2*	Components of sarcomere and nuclear lamina proteins
Aortopathies		
Marfan syndrome	*FBN-1*	Fibrillin-1 (extracellular matrix proteins)
Type IV (vascular) Ehlers–Danlos syndrome	*COL3A1*	Collagen, type III (extracellular matrix proteins)
Loeys–Dietz syndrome	*TGFBR1, TGFBR2, TGFB3, TGFB2, SMAD3*	Transforming growth factor-beta (TGF-β) pathway, transmitting chemical signals from the cell surface to the nucleus
Familial thoracic aortic aneurysm	*TGFB2, MYH11, SMAD3, FBN1, ACTA2, TGFBR2, TGFBR1, PRKG1, MYLK, LOX, EFEMP2, SMAD2, MFAP5, FOXE3*	Includes smooth muscle contraction proteins
Thrombophilias		
Antithrombin deficiency	*SERPINC1*	Antithrombin is the major plasma inhibitor of blood coagulation by inactivating thrombin and factor Xa
Protein C deficiency	*PROC*	Activated protein C cleaves and inactivates procoagulant factors Va and VIIIa
Protein S deficiency	*PROS1*	Protein S is a cofactor or protein C–mediated inactivation of procoagulant factors Va and VIIIa
Factor V Leiden	*F5*	Activated protein C resistance due to the factor V Leiden variant (F5 c.1601G > A; p.Arg534Gln, historically known as R506Q)
Prothrombin 20210A mutation	*F2*	Prothrombin F2 c.*97G > A variant (historically known as G20210A) leads to increased mRNA and protein expression of prothrombin

TABLE 19.1 Forensically important conditions with mutations: cardiac channelopathies, cardiomyopathies, aortopathies, thrombophilias, and epilepsy.—cont'd

Diseases	Genes*	Comments
Epilepsy		
Dravet syndrome	*SCN1A*	Severe epilepsy syndrome with features including prolonged seizures often triggered by high body temperature, developmental delay, speech impairment, ataxia, hypotonia, and sleep disturbances
Lennox–Gastaut syndrome	*CHD2, FOXG1 SCN1A, SCN8A, STXBP1*	Pediatric epilepsy syndrome characterized by multiple seizure types; mental retardation or regression; and abnormal findings on electroencephalogram (EEG)
Epilepsy–aphasia spectrum	*GRIN2A*	Severe epilepsy syndrome with a characteristic electroencephalogram (EEG) pattern and developmental regression particularly impacting language
Epilepsy with myoclonic–atonic seizures	*SCN1A, SCN1B, GABRG2, CHD2, SLC6A1*	Epilepsy syndrome of early childhood characterized by seizures of many different types, most often myoclonic-atonic, astatic, or generalized tonic-clonic seizures
West syndrome (infantile spasms)	*ARX1, CDKL5, PAFAH1B1/LIS1, DCX, TUBA1A, STXBP1, KCNQ2, GRIN2B, GRIN2A, MAGI2, SPTAN, FOXG1, NSD1*	Triad of (1) infantile spasms, (2) hypsarrhythmia, and (3) developmental arrest or regression

** These gene/disease associations are continually being updated[8] (see https://clinicalgenome.org/).*

infants), and neuropathology/cardiac pathology consultations, targeted molecular testing is warranted if these examinations fail to diagnose the cause of death. Detected mutations in specific genes associated with cardiac and/or epilepsy-focused sudden death can provide critical information about the cause of death. In selected cases, such molecular analyses can reveal pathogenic or likely pathogenic mutations that may be causally implicated in the death. In order to allow for future molecular genetic testing, we recommend the collection of blood (at a minimum, a bloodspot card) in all autopsies and fresh/nonfixed tissue placed in a nucleic acid stabilization and storage solution (RNA later) for all pediatric deaths or for those where the need for future molecular testing is reasonably anticipated. A purple (or lavender) top K2 EDTA tube is routinely collected by some offices. The recommended fresh tissue source is myocardium, as that also can be used for molecular virology testing in instances of myocarditis (tissues in descending order of preference: heart, liver, spleen, skeletal muscle). Fresh tissue in RNA later solution can be stored for a week at room temperature, a month at 4°C, or indefinitely at −20°C.

A bloodspot card is routinely prepared and stored and can be used for testing when fresh tissue is unavailable. Currently, formalin-fixed tissue is suboptimal and may be insufficient, as genetic testing may require analysis of large segments of DNA or whole gene sequencing, rather than the smaller fragments that are useful for identification testing. As technology advances, formalin-fixed tissue may become more useful. Typically, one first reviews the toxicological results, histopathological slides, neuropathology and cardiac pathology consultation reports, and gathers additional investigative, medical, and social history, as available, before pursuing genetic testing. Instances in which all initial studies are negative and/or the history/autopsy findings suggest a sudden cardiac event, seizure, aortic dissection, or thromboembolic event in a young person, without the identification of sufficient risk factors, derive the greatest benefit and produce the highest yield from the inclusion of molecular diagnostics. If possible, either a partnership with an academic center or other institution or the inclusion of a genetic counselor in the molecular laboratory can provide the critical service of counseling the surviving family regarding the potential implications of the decedent's molecular results.

The process of molecular testing includes preanalytical, analytical, and postanalytical components. Commonly used molecular diagnostic techniques in the forensic setting include sequencing methods, assays to identify copy number variants, and targeted variant identification methods. Next-generation sequencing (NGS), or massively parallel sequencing, is a high-throughput approach that can interrogate multiple regions of genomic DNA in a single assay, and NGS panels for sudden cardiac death and epilepsy are becoming an increasingly important tool for incorporation into forensic practice.[5,6] Automated, capillary-based, fluorescent analysis of polymerase chain reaction (PCR) amplicons (Sanger sequencing) can be used for identification of single nucleotide polymorphisms and small insertions/deletions, and is helpful for confirming variants identified by NGS. Sanger sequencing also provides a means to "patch" the coverage of regions that are poorly covered by NGS. Multiplex ligation-dependent probe amplification is a multiplex assay to detect copy number variations of genomic DNA sequences. Targeted techniques usually start with nucleic acid target amplification, such as by PCR, followed by an analytical method. For example, DNA fragment analysis uses automated, capillary-based, multicolor, fluorescent analysis of PCR-based DNA fragments to identify point mutations within a decedent's DNA. RNA- or DNA-based real-time PCR using fluorescent probes also can be employed to detect point mutations. A detailed review of the above methodologies is beyond the scope of this chapter. The clinical relevance of detected sequence variants should be evaluated in accordance with the standards and guidelines for the interpretation of sequencing variants established by the American College of Medical Genetics and Genomics (ACMG) and the Association for Molecular Pathology (AMP).[7] All clinically relevant variants, including pathogenic, likely pathogenic, and variants of uncertain significance (VUS) should be reported. Interpretation of the VUS should be made in the context of the autopsy findings, investigation, and case history and include a review of clinical databases (e.g., ClinGen, OMIM, ClinVar, HGMD, etc.) and publications, evaluation of minor allele frequency (e.g., ESP6500, gnomAD), and in silico predictions (e.g., MutationTaster, SIFT, PROVEAN, PolyPhen2, etc.).[8,9] Identification of VUS, the nuances of which may take some time and expertise to explain, have been called "genetic purgatory"; although identification of a VUS can be unsettling for the family, the results are important to understand in the context of a dynamic field where new information is being continually added and the possibility of greater clarity on the ramifications of a given mutation may occur with time.

Cardiac-focused

Broadly, a cardiac-focused panel can be divided into (1) cardiac channelopathies (e.g., long QT syndrome, Brugada syndrome, short QT syndromes, etc.); (2) noncardiac channelopathies (e.g., Bartter syndrome, familial hyperinsulinemia, etc.); and (3) cardiomyopathies (e.g., hypertrophic cardiomyopathy, arrhythmogenic cardiomyopathy, dilated cardiomyopathy, left ventricular noncompaction cardiomyopathy, etc.). Although a few are inherited in an autosomal recessive or x-linked pattern, most channelopathies and cardiomyopathies are inherited in an autosomal dominant pattern; they can, however, exhibit both reduced penetrance (a carrier of a disease-causing variant may not show the characteristic phenotype) and variable expressivity (different carriers of the same disease-causing variant manifest different symptoms with a wide spectrum of disease severity).

Aortopathies

Inherited thoracic aortopathies are conditions that can predispose affected individuals to a mural weakening and dilation of the thoracic aorta, and although the diseases may be silent, they can also result in sudden and unexpected death due to dissection or rupture.[10] The thoracic aorta, as the most proximal portion of the aorta to the heart, withstands the greatest wall stresses and the highest pressures. The ascending aorta and aortic root are embryologically distinct from the remainder of the aorta (descending thoracic and abdominal portions), with the former arising from the neural crest and the latter arising from the mesoderm, leading to differences in gene expression.[11] For these developmental and physiologic reasons, the mechanisms of aneurysm formation, age of presentation, and the risks of dissection and rupture differ greatly in thoracic and abdominal aortic disease. Abdominal aortic aneurysms (AAAs) are the most commonly identified aneurysm of the aorta, and the vast majority of AAAs are related to atherosclerotic risk factors. By contrast, thoracic aortic aneurysms (TAAs) are more likely to be associated with a genetic cause with the majority of mutations occurring in genes associated with the development of the thoracic aorta (e.g., vascular smooth muscle cell contraction or metabolism, the transforming growth factor-beta (TGF-β) signaling pathway, and extracellular matrix proteins).[12] TAAs most commonly involve the ascending aorta (60%) followed by the descending aorta (40%), with 10% involving the aortic arch.[13] TAAs are commonly divided into either syndromic disorders, with extravascular organ involvement, or nonsyndromic disorders, where the aortic abnormality is seemingly isolated. Familial TAA is a nonsyndromic disorder. Syndromic disorders include Marfan syndrome, Type IV (vascular) Ehlers–Danlos syndrome, and Loeys–Dietz syndrome.

Thrombophilias

Hereditary thrombophilias are a group of inherited conditions which predispose individuals to form thrombi inappropriately. The genetic mutations, in addition to other biologic and environmental factors, result in variably increased risks for venous thromboses, where in some instances, the sentinel event can be fatal (e.g., bilateral saddle PTE). Mutations can impact different parts of the clotting and fibrinolysis pathways. For example, genetic mutations can involve natural inhibitors of coagulation (e.g., inherited deficiencies in antithrombin, protein C, and protein S) or can render clotting factors less susceptible to inactivation (e.g., factor V gene mutation, which renders it less susceptible to inactivation by activated protein C). The relative risk for a first venous thromboembolic event in a person with a heterozygous mutation is 5–10 for antithrombin deficiency, 4–6.5 for protein C deficiency, 1–10 for protein S deficiency, 3-5 for Factor V Leiden mutation, and 2–3 for the prothrombin 20210A mutation. In general, the risk for venous thromboembolic events is markedly increased in individuals with homozygous or compound heterozygous mutations.[14] Although homozygous c.665C→T (p.Ala222Val) polymorphisms in methylenetetrahydrofolate reductase (MTHFR) may mildly increase the risk of venous thromboembolism (odds ratio 1.27) when associated with elevated plasma homocysteine levels, testing for MTHFR polymorphisms is no longer recommended as part of the routine clinical evaluation for thrombophilia or recurrent pregnancy loss.[15]

In some instances, no other risk factors, natural, therapeutic, or injurious, are identified in a death caused by a PTE. In such individuals, genetic testing may afford the only opportunity to provide a proximate cause and, if not, at least confidently conclude that the cause of death is "PTE of undetermined etiology," by exclusion. In other instances, "soft" proximate causes are identified (slight obesity, slight dehydration, an ankle sprain reducing mobility, but not preventing it, etc.), leaving the forensic pathologist with the nagging feeling that the underlying risk has not been definitively or thoroughly uncovered. It is important to recognize that there are acquired (nongenetic) hypercoagulable states, such as the antiphospholipid antibody syndrome (APS), in which the primary presenting feature may be a venous, arterial, or small vessel thromboembolic event and/or pregnancy morbidity accompanied by the persistence presence of antiphospholipid antibodies (aPL).[16,17] More than half of patients with this disease have primary APS, and by definition do *not* have a concomitant systemic autoimmune disease (e.g., systemic lupus erythematosus). In a large retrospective study of patients without autoimmune disease, aPLs were present in up to 20% of patients with DVTs[18] (the same prevalence as a heterozygous mutation in Factor V Leiden in patients with VTE)[19] and it is estimated that aPLs are responsible for at least 30,000 deep venous thromboses (DVTs) annually in the United States.[20,21] Therefore, it is possible that a significant proportion of unexplained, unprovoked fatal VTEs may in fact be caused by pathogenic aPLs.[16,22] However, making a postmortem diagnosis of APS is challenging due to the pre-analytical considerations for a clot-based detection of a lupus anticoagulant, and the requirement in consensus guidelines to demonstrate persistent positivity of aPLs (usually measured 12 weeks apart), which has clear limitations in cases where the first thromboembolic event is fatal. Cases of unprovoked DVTs may well benefit from a thrombophilia work-up, if available and fiscally feasible, with the consideration that this testing is an important, but limited part of a thrombophilia workup.

Epilepsy-focused

Epilepsy is a common, heterogeneous neurological disorder with many complex genetic and developmental underpinnings. Epilepsy is diagnostically challenging in the forensic setting as a potential cause of death, as it requires a detailed scene investigation, medical history, and autopsy protocol, and, because it kills through a functional mechanism (with microscopic ± gross findings that support the diagnosis of the underlying seizure disorder), it is most often a diagnosis of exclusion, particularly if the death is unwitnessed (see Chapter 7).

Although there are many underlying causes for epilepsy, several epilepsy syndromes are caused by mutations in single genes. An epilepsy-focused panel can be divided into the genes implicated in (1) epilepsy syndromes (e.g., Dravet syndrome, West syndrome, Lennox–Gastaut syndrome, etc.) and (2) conditions that feature seizures as a symptom (e.g., cardiac channelopathies, tuberous sclerosis, etc.). These conditions follow several inheritance patterns such as autosomal dominant, x-linked, and autosomal recessive. Patients with epilepsy are at risk of sudden unexpected death, a condition referred to as sudden unexpected death in epilepsy (SUDEP). SUDEP is defined as a "sudden, unexpected, witnessed or unwitnessed, nontraumatic and nondrowning death, in an individual with epilepsy, with or without evidence for a seizure, and excluding documented terminal status epilepticus, in which the postmortem examination does not reveal a cause of death."[4] The potential for genetic causes of SUDEP has also been raised by the identification of ion channel mutations that can be associated with both cardiac arrhythmias and some forms of epilepsy. In addition to a thorough history, complete autopsy, and neuropathologic examination, molecular genetics can potentially provide additional etiologically specific molecular genetic information about the underlying cause of a seizure disorder.

Hemoglobinopathies

Hemoglobinopathies encompass all genetic diseases of hemoglobin and can be broadly divided into two main categories: thalassemia syndromes (e.g., α- and β-thalassemia) and structural hemoglobin variants (e.g., HbS and HbC). There are many subtypes and combined types of hemoglobinopathies, and the severity of disease can be highly variable depending on the genotype, ranging from mild hypochromic anemia to severe hematologic disease. Molecular testing is used for the diagnosis of hemoglobinopathies (e.g., sickle cell disease) in the forensic setting. It is important to recognize the potential limitations of sequencing methodologies in differentiating homozygous structural variants (e.g., HbS/HbS) and hemizygous states (e.g., HbS/β-thal^{0}), as this may have implications for targeted testing of surviving relatives. Additionally, it is important to note that newborn screening for sickle cell disease is mandated in all US states, so obtaining those results, along with any follow-up testing/consultation reports can also help in clarifying key information about a decedent's hemoglobinopathy diagnosis.[23] That said, in medical examiner/coroner (ME/C) offices without access to genetic diagnostics, hemoglobin electrophoresis offers another potential methodology.[24]

Importantly and to reemphasize, for all of the above types of diagnostic testing in which a pathogenic mutation is identified, the potential implications for surviving family members cannot be overstated. The identification of a pathogenic mutation may provide important information in elucidating the proximate cause of death, but simultaneously opens up a host of additional questions for the survivors, not all of which the ME/C has the bandwidth or expertise to address. Some ME/C offices may have the luxury of an on-site genetic counselor, but this is largely an outlier situation. As such, equally important to careful case selection and high-quality methodologies in molecular testing is access to partners in clinical genetic medicine for both family counseling and diagnostics. Ideally, a relationship with a local academic center with a multidisciplinary team that offers both genetic counseling and targeted testing can provide a ready referral for next of kin, sometimes even with dispensation packages for families experiencing financial hardship. These relationships benefit all involved and expand even further the tremendous public service of an ME/C office.

With careful selection and access to experienced counseling services, molecular testing can yield very satisfying results and succeed in providing clarity and closure for families, sometimes even preventing premature deaths in survivors. Common mutations are listed in Table 19.1.[1,25,26]

Rapid DNA

Rapid DNA testing describes the fully automated process of developing a CODIS Core STR profile from a reference sample, such as a buccal swab. The "swab in—profile out" process consists of automated extraction, amplification, separation, detection, and allele calling, without human intervention. They have been used in mass disaster events on a variety of biological samples.[27–31] Rapid DNA testing was useful in quickly identifying decedents in a boat fire that involved 34 people.[32] In the boat fire investigation, the first rapid DNA identification was completed within 4 h. The limiting step was obtaining the family reference samples. All 34 decedents were identified within one day of discovery.

The instrument, about the size of a desktop printer, identifies samples in as little as 90 min and can test up to 2000 samples in 24 h. Use of different chips allows for testing of a variety of samples including buccal swabs, tissue (e.g., liver, muscle), blood, and even bone. These device applications hold decedents' and relatives' information, so geographically distant relatives can provide samples. The Federal government's Disaster Mortuary Operational Response Teams have this capability.

Digital fingerprinting

Largely gone are the days of "Book him, Dano," with ink fingerprinting of arrestees or of decedents.[33] Digital scanning technology can capture rolls and tap fingerprints, which can be easily sent to be searched in a variety of fingerprint databases, including state, Federal Bureau of Investigation (FBI), military, and Homeland Security. Devices can even capture latent fingerprints off of a variety of surfaces.[34–36] These devices allow for the rapid identification of decedents.[37] Special techniques also allow for fingerprinting of decomposed or mummified remains.[33,38,39]

There are four types of fingerprint scanners: optical (visual image), capacitance/complementary metal—oxide semiconductor (capacitors and current form an image), ultrasonic (high frequency sound waves), and thermal (temperature differences). The basic function is to obtain an image of a fingerprint and then find its match in a database.[40–43] The measure of the fingerprint image quality is in dots per inch.

Exemplar prints, or known prints, are fingerprints deliberately collected from a subject, whether for enrollment in a system or when under arrest. A partial fingerprint obtained ("lifted") from a surface is called a latent fingerprint. Moisture and grease on fingers may result in latent fingerprints on certain surfaces such as glass. Because they are not clearly visible, their detection may require powder dusting, spraying with ninhydrin, iodine fuming, or soaking in silver nitrate.[44] The Integrated Automated Fingerprint Identification System (IAFIS) is a national computerized automated fingerprint identification and criminal history system maintained by the FBI. IAFIS provides automated fingerprint searching, latent searching, electronic image storage, and electronic exchange of fingerprints and responses. IAFIS houses the fingerprints of 70 million people in the criminal master file and 31 million civilian prints.

3D printing and anthropology

The advent of 3D printing, or additive manufacturing, has introduced the technology of constructing a three-dimensional object from a computer-aided design (CAD) model or a digital 3D model. The 3D printable models can be created with a CAD package using a 3D scanner or by a plain digital camera and photogrammetry software. The term "3D printing" can refer to a variety of processes in which material is deposited, joined, or solidified under computer control to create a three-dimensional object. The material is typically added together (the fusion of plastics, liquids, or powder grains), layer by layer.[45–52]

It is particularly useful for forensic anthropology (Fig. 19.1).[53] Digital Imaging and Communications in Medicine (DICOM) is the standard for the communication and management of medical imaging information. DICOM is most commonly used for storing and transmitting medical images, enabling the integration of medical imaging devices such as scanners, servers, workstations, printers, network hardware, and picture archiving. It has been widely adopted by hospitals and is making inroads into smaller applications, such as in dentists' and doctors' offices. Computed tomography scan DICOM files can be converted to 3D print formats. With 3D printing technology, one can create exact replicas that can be useful for court testimony (e.g., showing skeletal trauma) via a 3D replica (Fig. 19.2). There is also a potential training application in the use of 3D prints generated from scans of cases, in lieu of the actual human remains, for broader educational purposes.

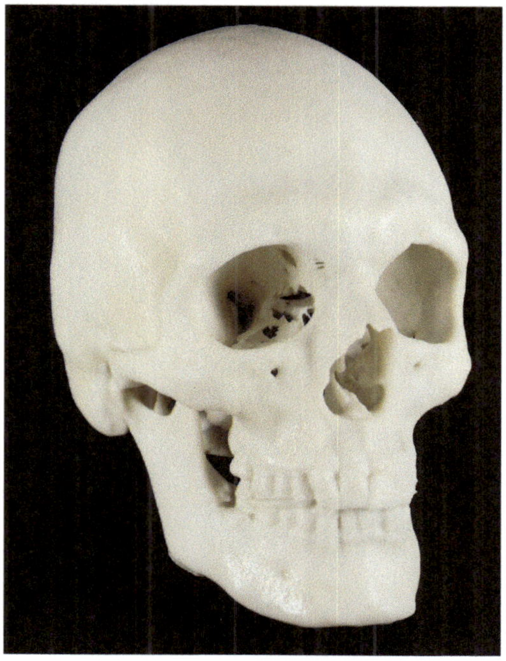

FIGURE 19.1 3D print of a skull.

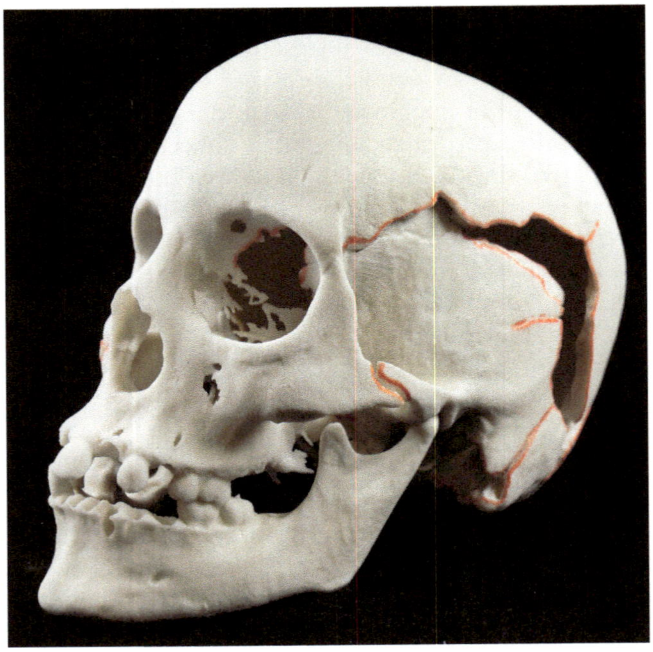

FIGURE 19.2 3D print of a fractured skull.

Facial reconstruction and recognition

The skull provides clues to the facial appearance. The brow ridge, the interorbital distance, the shapes of the nasal bones and chin, and the overall profile of the bones determine facial features. In **facial reconstruction**, a sculptor and a forensic anthropologist work together to interpret features of a skull to create a facial rendering that approximates the actual face of an unknown individual (Figs. 19.3–19.5). A **facial recognition system** is capable of matching a human face from a digital image or video frame against a databases of known faces. It works by pinpointing and measuring facial features from a given image. Some facial recognition algorithms identify and extract landmarks or features from an image of the subject's face. For example, an algorithm may analyze the relative position, size, and/or shape of the eyes, nose, cheekbones, and jaw. These features are then used to search for other images with the same features.[54,55]

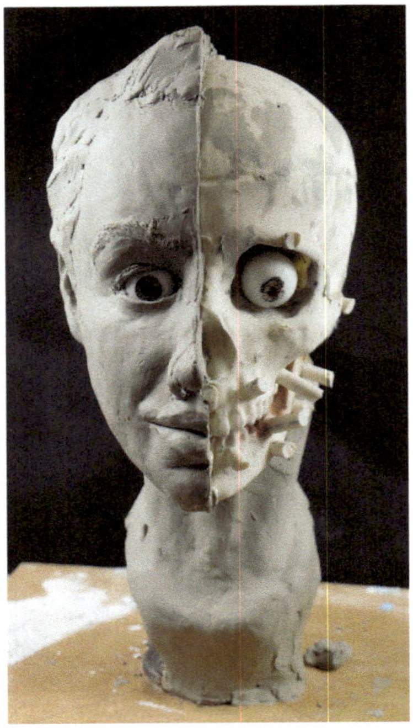

FIGURE 19.3 Facial reconstruction 1.

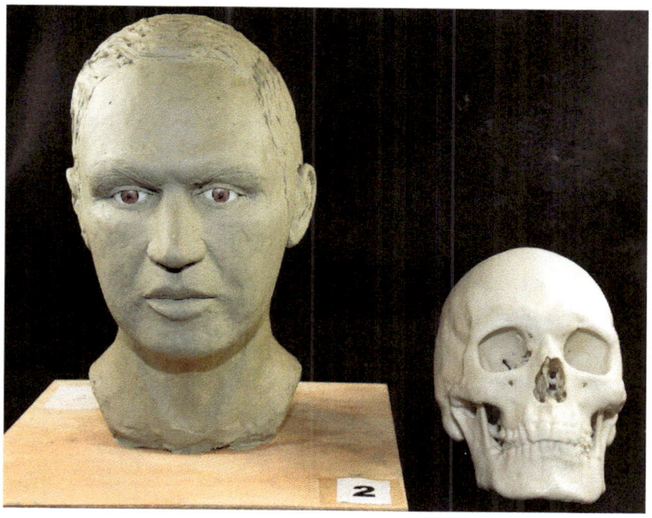

FIGURE 19.4 Facial reconstruction 2.

FIGURE 19.5 Facial reconstruction 3.

Some algorithms normalize a gallery of face images and then compress the face data, only saving the data in the image that is useful for facial recognition. Recognition algorithms are divided into two main approaches: geometric (distinguishing features) and photometric (a statistical approach that distills an image into values and compares the values with templates to eliminate variances). Some classify these algorithms into holistic (former attempts to recognize the face in its entirety) and feature-based models, which subdivide it into components, such as according to features, and analyze each with its spatial location with respect to other features.

References

1. Tang Y, Stahl-Herz J, Sampson BA. Molecular diagnostics of cardiovascular diseases in sudden unexplained death. *Cardiovasc Pathol*. 2014; 23(1):1–4.
2. Koene RJ, Adkisson WO, Benditt DG. Syncope and the risk of sudden cardiac death: evaluation, management, and prevention. *J Arrhythm*. 2017;33(6):533–544.

3. Tester DJ, Ackerman MJ. Postmortem long QT syndrome genetic testing for sudden unexplained death in the young. *J Am Coll Cardiol*. 2007;49(2):240–246.
4. Middleton OL, Atherton DS, Bundock EA, et al. National association of medical examiners position paper: recommendations for the investigation and certification of deaths in people with epilepsy. *Acadc Foren Pathol*. 2018;8(1):119–135.
5. Behjati S, Tarpey PS. What is next generation sequencing? *Arch Dis Child Educ Pract Ed*. 2013;98(6):236–238.
6. Yohe S, Thyagarajan B. Review of clinical next-generation sequencing. *Arch Pathol Lab Med*. 2017;141(11):1544–1557.
7. Richards S, Aziz N, Bale S, et al. Standards and guidelines for the interpretation of sequence variants: a joint consensus recommendation of the American College of medical genetics and genomics and the association for molecular pathology. *Genet Med*. 2015;17(5):405–424.
8. Ghosh R, Harrison SM, Rehm HL, Plon SE, Biesecker LG. ClinGen sequence variant interpretation working G. Updated recommendation for the benign stand-alone ACMG/AMP criterion. *Hum Mutat*. 2018;39(11):1525–1530.
9. Leong IU, Stuckey A, Lai D, Skinner JR, Love DR. Assessment of the predictive accuracy of five in silico prediction tools, alone or in combination, and two metaservers to classify long QT syndrome gene mutations. *BMC Med Genet*. 2015;16:34.
10. Fletcher AJ, Syed MBJ, Aitman TJ, Newby DE, Walker NL. Inherited thoracic aortic disease: new insights and translational targets. *Circulation*. 2020;141(19):1570–1587.
11. Pfaltzgraff ER, Shelton EL, Galindo CL, et al. Embryonic domains of the aorta derived from diverse origins exhibit distinct properties that converge into a common phenotype in the adult. *J Mol Cell Cardiol*. 2014;69:88–96.
12. Bhandari R, Aatre RD, Kanthi Y. Diagnostic approach and management of genetic aortopathies. *Vasc Med*. 2020;25(1):63–77.
13. Isselbacher E. Epidemiology of thoracic aortic aneurysms, aortic dissection, intramural hematoma, and penetrating atherosclerotic ulcers. In: Eagle K, Baliga R, Isselbacher E, eds. *Aortic Dissection and Related Syndromes*. Boston, MA: Springer; 2007:3–15.
14. Middeldorp S. Inherited thrombophilia: a double-edged sword. *Hematol Am Soc Hematol Educ Program*. 2016;2016(1):1–9.
15. Hickey SE, Curry CJ, Toriello HV. ACMG Practice Guideline: lack of evidence for MTHFR polymorphism testing. *Genet Med*. 2013;15(2):153–156.
16. Miyakis S, Lockshin MD, Atsumi T, et al. International consensus statement on an update of the classification criteria for definite antiphospholipid syndrome (APS). *J Thromb Haemostasis*. 2006;4(2):295–306.
17. Duarte-Garcia A, Pham MM, Crowson CS, et al. The epidemiology of antiphospholipid syndrome: a population-based study. *Arthritis Rheumatol*. 2019;71(9):1545–1552.
18. Farmer-Boatwright MK, Roubey RA. Venous thrombosis in the antiphospholipid syndrome. *Arterioscler Thromb Vasc Biol*. 2009;29(3):321–325.
19. Mannucci PM, Franchini M. Classic thrombophilic gene variants. *Thromb Haemostasis*. 2015;114(5):885–889.
20. Andreoli L, Chighizola CB, Banzato A, Pons-Estel GJ, Ramire de Jesus G, Erkan D. Estimated frequency of antiphospholipid antibodies in patients with pregnancy morbidity, stroke, myocardial infarction, and deep vein thrombosis: a critical review of the literature. *Arthritis Care Res*. 2013;65(11):1869–1873.
21. White RH. The epidemiology of venous thromboembolism. *Circulation*. 2003;107(23 Suppl 1):I4–I8.
22. Hudzik B, Szkodzinski J, Polonski L. Testing for antiphospholipid antibodies at autopsy. *Forensic Sci Med Pathol*. 2014;10(2):288–289.
23. Force USPST. Screening for sickle cell disease in newborns: recommendation statement. *Am Fam Physician*. 2008;77(9):1300–1302.
24. Hughes RL, Feig J. Sickle cell trait-related exertional deaths: observations at autopsy and review of the literature. *Mil Med*. 2015;180(8):e929–932.
25. McTague A, Howell KB, Cross JH, Kurian MA, Scheffer IE. The genetic landscape of the epileptic encephalopathies of infancy and childhood. *Lancet Neurol*. 2016;15(3):304–316.
26. Pavone P, Polizzi A, Marino SD, et al. West syndrome: a comprehensive review. *Neurol Sci*. 2020;41(12):3547–3562.
27. Amick GD, Swiger RR. Internal validation of RapidHIT((R)) ID ACE sample cartridge and assessment of the EXT sample cartridge*(dagger). *J Forensic Sci*. 2019;64(3):857–868.
28. Laurin N, Fregeau CJ. An accelerated analytical process for the development of STR profiles for casework samples. *J Forensic Sci*. 2015;60(4):983–989.
29. Lounsbury JA, Landers JP. Ultrafast amplification of DNA on plastic microdevices for forensic short tandem repeat analysis. *J Forensic Sci*. 2013;58(4):866–874.
30. Graham EA. Lab-on-a-chip technology. *Forensic Sci Med Pathol*. 2005;1(3):221–223.
31. Chong KWY, Wong Y, Ng BK, Lim WSH, Rosli AR, Syn CKC. A practical study on direct PCR amplification using the GlobalFiler PCR Amplification Kit on human bloodstains collected with microFLOQ Direct swabs. *Forensic Sci Int*. 2019;300:43–50.
32. Gin K, Tovar J, Bartelink EJ, et al. The 2018 California wildfires: integration of rapid DNA to dramatically accelerate victim identification. *J Forensic Sci*. 2020;65(3):791–799.
33. Morgan LO, Johnson M, Cornelison JB, Isaac CV, deJong JL, Prahlow JA. Autopsy fingerprint technique using fingerprint powder. *J Forensic Sci*. 2018;63(1):262–265.
34. Williams G, McMurray HN, Worsley DA. Latent fingerprint detection using a scanning Kelvin microprobe. *J Forensic Sci*. 2001;46(5):1085–1092.
35. Connatser RM, Prokes SM, Glembocki OJ, et al. Toward surface-enhanced Raman imaging of latent fingerprints. *J Forensic Sci*. 2010;55(6):1462–1470.
36. Goddard AJ, Hillman AR, Bond JW. High resolution imaging of latent fingerprints by localized corrosion on brass surfaces. *J Forensic Sci*. 2010;55(1):58–65.
37. G R. Printing decamps: livescan and digital fingerprint systems streamline identifying the deceased. *Law Enforc Technol*. 2006;33(22–33).
38. Morgan LO, Johnson M, Cornelison J, Isaac C, deJong J, Prahlow JA. Two novel methods for enhancing postmortem fingerprint recovery from mummified remains. *J Forensic Sci*. 2019;64(2):602–606.
39. Schmidt C, Nawrocki S, Williamson M, Marlin D. Obtaining fingerprints from mummified fingers: a method for tissue rehydration adapted from the archeological literature. *J Forensic Sci*. 2000;45(4):874–875.
40. de Jongh A, Rodriguez CM. Performance evaluation of automated fingerprint identification systems for specific conditions observed in casework using simulated fingermarks. *J Forensic Sci*. 2012;57(4):1075–1081.
41. Langenburg G, Hall C, Rosemarie Q. Utilizing AFIS searching tools to reduce errors in fingerprint casework. *Forensic Sci Int*. 2015;257:123–133.

42. Ratha NK, Karu K, Chen S, Jain AK. A real-time matching system for large fingerprint databases. *IEEE Trans Pattern Anal Mach Intell.* 1996; 18(8):799–813.
43. de Boer J, Bazen AM, Gerez SH. Indexing fingerprint databases based on multiple features. In: *Paper Presented at: ProRISC the 12th Annual Workshop on Circuits, Systems and Signal Processing*. 2001.
44. De Alcaraz-Fossoul J, Barrot-Feixat C, Zapico SC, et al. Ridge width correlations between inked prints and powdered latent fingerprints. *J Forensic Sci.* 2018;63(4):1085–1091.
45. Martin-de las Heras S, Valenzuela A, Ogayar C, Valverde AJ, Torres JC. Computer-based production of comparison overlays from 3D-scanned dental casts for bite mark analysis. *J Forensic Sci.* 2005;50(1):127–133.
46. Baier W, Warnett JM, Payne M, Williams MA. Introducing 3D printed models as demonstrative evidence at criminal trials. *J Forensic Sci.* 2018; 63(4):1298–1302.
47. Edwards J, Rogers T. The accuracy and applicability of 3D modeling and printing blunt force cranial injuries. *J Forensic Sci.* 2018;63(3):683–691.
48. Newcomb TL, Bruhn AM, Giles B, Garcia HM, Diawara N. Testing a novel 3D printed radiographic imaging device for use in forensic odontology. *J Forensic Sci.* 2017;62(1):223–228.
49. Molina A, Martin-de-las-Heras S. Accuracy of 3D scanners in tooth mark analysis. *J Forensic Sci.* 2015;60(Suppl 1):S222–S226.
50. Brough AL, Rutty GN, Black S, Morgan B. Post-mortem computed tomography and 3D imaging: anthropological applications for juvenile remains. *Forensic Sci Med Pathol.* 2012;8(3):270–279.
51. Ebert LC, Thali MJ, Ross S. Getting in touch–3D printing in forensic imaging. *Forensic Sci Int.* 2011;211(1–3):e1–6.
52. Eckhardt M, Shah K, Bois M, Maleszewski J, Moore K, Lin P. Healed fracture of superior horn of thyroid cartilage in autoerotic asphyxia: an indication of prior activity? A case report utilizing 3D scanning and printing of the larynx. *Acad Forensic Pathol.* 2018;8(1):170–179.
53. Colman KL, de Boer HH, Dobbe JGG, et al. Virtual forensic anthropology: the accuracy of osteometric analysis of 3D bone models derived from clinical computed tomography (CT) scans. *Forensic Sci Int.* 2019;304:109963.
54. Tome P, Vera-Rodriguez R, Fierrez J, Ortega-Garcia J. Facial soft biometric features for forensic face recognition. *Forensic Sci Int.* 2015;257: 271–284.
55. Stephan CN, Cicolini J. Measuring the accuracy of facial approximations: a comparative study of resemblance rating and face array methods. *J Forensic Sci.* 2008;53(1):58–64.

Appendices

APPENDIX

A

Formats of autopsy report final diagnoses

General formats for final diagnoses lists

The proceeding outlines should correspond to the details in the autopsy notes and the organization of the body of the autopsy report. By following the suggested format, one should be able to look at a report, analyze the fatal injury, and reconstruct lethal mechanisms in a few minutes.

The underlying *etiologically specific diseases and/or injuries* are listed as the *headings* and their *sequelae as subunits* (see examples below). Unless functional and/or historical entities represent the primary mechanism of death (e.g., epilepsy, diabetes mellitus), the diagnoses are usually organized with gross findings toward the top of the list and historical ("anamnestic") entities listed lower down (Boxes A.1 and A.2).[1]

Look for these entities anatomically or historically to support an acute and chronic substance use diagnosis. If you identify them, incorporate them into your diagnoses. It's often helpful to organize subheadings from acute to chronic (i.e., acute findings first, chronic, second), from outside to inside, and from anatomical to historical. So, for example, even though "track marks" are a chronic sequela, you may put them as "A" under the "Acute and Chronic Substance Use Disorder" heading because they are part of the external findings. There is no right or wrong way, but, whatever you do, try to be consistent and have a thoughtful process, one that makes sense to you. If you do, you are much less likely to forget to document or fail to observe findings. Having a consistent and systematic approach also enhances your likelihood of observing findings through the lens of a unified diagnostic theory (Box A.2).

Gunshot wounds

Note that **nonspecific sequelae can be listed under more than one etiologically specific heading,** as they may develop as a result of more than one disease/injury process (e.g., "visceral pallor" below). Using this technique in natural deaths is exemplified by placing "cardiac hypertrophy" under hypertensive cardiovascular disease, obesity, and chronic alcohol use disorder. One disease/injury needn't/shouldn't be arbitrarily chosen for being solely responsible for a finding that likely has more than one contributory variable (disease or injury).

Note also that some forensic pathologists list the anatomic locations of entrance and exit wounds under each gunshot wound heading. This is yet another perfectly reasonable approach that allows for a glance at the front page of the autopsy report for the reader to quickly and easily extract pertinent details for each wound. More important than strictly adhering to the exact formats suggested below, attempt to incorporate the concepts and organizational logic. Above all, remain consistent and clear in your approach (Box A.4).

Box A.1 Natural

I. Atherosclerotic cardiovascular disease, with:
- **A.** Marked coronary artery atherosclerosis with multifocal stenoses.
- **B.** Recent thrombus of left anterior descending coronary artery.
  - **1.** Recent infarct, myocardium of left ventricle, and interventricular septum, with rupture.
    - **a.** Hemopericardium (450 mL) and cardiac tamponade.
- **C.** Remote thrombus of right coronary artery.
- **D.** Remote infarct of myocardium of left ventricle.
- **E.** See microscopic examination report.

II. Diabetes mellitus, anamnestic.

Box A.2 Substance use*

I. Acute and chronic intravenous drug use (heroin, fentanyl, cocaine), with:

- **A.** Linear scars overlying subcutaneous veins of upper extremities.
- **B.** Marked pulmonary edema (combined lung weight, 2200 g).
- **C.** Full/distended urinary bladder.
- **D.** Cardiac hypertrophy (480 g).
- **E.** Chronic viral hepatitis, anamnestic.
- **F.** In methadone maintenance treatment program, anamnestic; see toxicology report.

In terms of the* **"Final Diagnoses" in drug deaths, *be mindful of some of the frequently encountered findings. If these findings are documented in the body of the report, they should find their way to the front page. Specifically:*

Possible acute findings:

- **Frothy or watery pulmonary edema** (the lungs are not just heavy, but they exude fluid, often frothy, when expressed)
- **Acute bronchopneumonia**

Possible/frequent chronic findings:

- **Track marks** (described instead as "linear scars overlying subcutaneous veins of left/right/bilateral upper extremity/upper extremities/antecubital fossae")
- **Hepatosplenomegaly** (with weights parenthetically)
- **Hepatic steatosis with/without sclerosis or cirrhosis and splenomegaly**
- **Hepatic portal lymphadenopathy**
- **Perforated nasal septum**
- **Cardiac hypertrophy** (especially with chronic cocaine and/or ethanol use), with/without dilation
- **Hepatitis B or C, anamnestic** (with/without cirrhosis as a sequela)

Box A.3 Thermal and smoke/heat inhalation injuries

I. Complications of thermal injuries, with:

- **A.** Full thickness burns of head, torso, and extremities, involving approximately 65% of body surface area.
- **B.** Anasarca

II. Complications of Smoke inhalation

- **A.** Mucosal hyperemia and soot deposition in larynx and trachea.
- **B.** Carboxyhemoglobinemia (55%, anamnestic), with:
  - **1.** Hypoxic encephalopathy.
- **C.** Acute, bilateral bronchopneumonia with diffuse alveolar damage (requiring ventilatory support, anamnestic).
- **D.** Septicemia and disseminated intravascular coagulation, on pressor support, anamnestic.

III. Atherosclerosis of coronary arteries, slight.

Box A.4 Gunshot wounds (with no therapy)

I. Perforating gunshot wound of chest, with:

- **A.** Perforations of sternum, heart, left main stem bronchus, left lung, and left 6th rib.
- **B.** Hemopericardium (approximately 300 mL), left hemothorax (approximately 1200 mL), and bilateral hemoaspiration.
- **C.** Visceral pallor.

II. Penetrating gunshot wound of abdomen, with:

- **A.** Perforations of left lobe of liver, stomach, pancreas, and spleen.
- **B.** Hemoperitoneum (approximately 1000 mL).
- **C.** Visceral pallor.
- **D.** One bullet recovered and submitted.

III. Perforating gunshot wound of left upper extremity with fracture of humerus.

Gunshot wounds with subsequent therapy

With injuries followed by complex complications and treatment, write the diagnoses as a chronological sequence under the heading of the underlying cause: (Box A.5)

Box A.5 Gunshot wound post complex therapy

I. Complications of perforating gunshot wound of abdomen, with:
- **A.** Perforations of stomach and colon.
- **B.** Status post emergency laparotomy with colostomy and repair of perforated stomach (February 2, 2022), anamnestic.
- **C.** Status post laparotomy for drainage of subphrenic abscess (February 26, 2022), anamnestic.
- **D.** Status post laparotomy for drainage of multiple abdominal abscesses (March 18, 2022), anamnestic.
- **E.** Status post laparotomy for lysis of adhesions and resection of necrotic small intestine (April 7, 2022), anamnestic
- **F.** Chronic peritonitis.
- **G.** Sepsis and disseminated intravascular coagulation, anamnestic.
- **H.** See microscopic examination report.

Blunt trauma

List diagnoses by body regions and outline in layers, from external to internal (Boxes A.6—A.10)

Box A.6 Fatal blunt head injury (assault)

I. Six blunt impacts of head, with:
- **A.** Abrasions and lacerations, face and scalp.
- **B.** Basilar and vertex comminuted skull fractures, extensive.
- **C.** Bilateral hemispheric and basilar subdural hemorrhage (approximately 70 ml).
- **D.** Cerebral cortical contusions, bilateral temporal and frontal poles.
- **E.** Cerebral swelling and diffuse necrosis (comment: hypoxic/ischemic encephalopathy).
- **F.** Bilateral bronchopneumonia; see microscopic examination report.

Box A.7 Extensive blunt injuries (motor vehicle collision)

I. Blunt force trauma of head, with:

(A—F, as above)

II. Blunt force trauma of torso, including crush-type injuries, with:
- **A.** Focally patterned abraded contusions of chest and abdomen (tire tread marks).
- **B.** Multiple, bilateral displaced rib fractures, fracture of sternum, and extensive comminuted ("open book") fractures of pelvis.
- **C.** Lacerations of aorta and punctures (from rib fractures) of lungs, with:
  1. Bilateral hemothoraces (right: 800 ml; left: 1400 ml)
- **D.** Laceration of liver, with:
  1. Hemoperitoneum (300 ml)
- **E.** Visceral and muscular pallor.

III. Blunt force trauma of extremities, with:
- **A.** Abrasions and lacerations of upper and lower extremities with open, displaced fractures of left humerus and right femur.

Box A.8 Stab and incised wounds

I. Stab wound of chest, with:
 A. Penetration of right lung and superior vena cava.
 B. Right hemothorax (1500 mL)
 C. Visceral and muscular pallor.
II. Incised wounds of left hand (2) and right forearm (4); no vital injury.

Box A.9 Hanging

I. Hanging, with:
 A. Circumferential patterned cervical ligature abrasion.
 B. Periorbital, conjunctival, and oral mucosal petechiae.
 C. Visceral congestion.
 D. Ligature (belt) recovered and submitted to evidence.

Box A.10 Manual strangulation

I. Manual strangulation, with:
 A. Multiple contusions and abrasions of skin of anterior and posterior neck.
 B. Facial and conjunctival petechiae and confluent hemorrhages, florid.
 C. Multifocal cervical strap muscle hemorrhages.
 D. Fractured hyoid bone.

Reference

1. Hirsch CS. The format of the medicolegal autopsy protocol. *Am J Clin Pathol*. 1971;55(4):407–409.

APPENDIX B

Format of the autopsy report

Final Diagnoses (listed in outline form; see Appendix A for examples)
Cause of Death
[Contributing Condition (Part 2)—if applicable]
Manner of Death

External Examination:

- Age (appears older or younger than identified age), sex, height, weight (do not accept blindly).
- Scalp hair (color, length, texture, male pattern baldness, alopecia), eyes (color, arcus senilis, corneal clouding), conjunctiva/sclera (petechiae, hemorrhage, jaundice).
- Ears (piercings—how many?), nose (septum intact?), oral cavity (injury, petechiae, blood pooling, coffee ground emesis), teeth (condition, missing maxillary incisors?, edentulous/predominantly edentulous), dentures (full/partial, maxillary/mandibular).
- Neck, chest, abdomen, back, external genitalia (atraumatic? circumcised?), extremities (track marks, cutting scars, pitting pretibial/pedal edema).
- Scars: Location, shape, length, color (there is no need to get exquisitely granular or detailed with their descriptions, particularly when they are numerous. Unless particularly germane to the case, these can be described in series in one sentence).[1]
- Identifying tags/bands: list location of all appropriately labeled hospital tags, handwritten and digital (check for accuracy)

Postmortem changes: Rigor, livor, and algor mortis; other postmortem changes (degree of putrefaction/mummification or adipocere) with specific descriptions.

Tattoos:

- Professional or amateur, color, location:
- If one or two or if special significance, describe.
- If the identity of the decedent is unknown, describe and photograph.

Therapeutic interventions/procedures:

- *Describe*: Endotracheal and nasogastric tubes, tracheostomies/gastrostomies/colostomies, cervical collars, needle punctures, taped gauze dressings, adhesive echocardiogram or defibrillator pads, sutured versus open incisions (include internal alterations), devices, arteriovenous shunts, (sutured) intravenous/urinary catheters/intracavitary tubing extending from the chest or abdomen, quantity of blood-soaked laparotomy pads.
- Avoid acronyms.

[1] If there are surgical scars, do not keep the reader in suspense. For example, "There is a 4″ well-healed, obliquely-oriented scar of the right lower abdominal quadrant. Internally, the appendix is absent."

Postmortem tissue/organ procurement: Describe incisions made, corneae/tissues/organs procured, locations of skin harvest sites, and a list tissues and/or organs removed prior to autopsy.

Clothing:

- Describe color, type of garment, bloodstained? blood-soaked? Perforations/defects ("roughly corresponding to gunshot wounds A-D described below"), and whether cut by emergency services prior to arrival.
- Record details of sizes and brands (when an unknown and putrefied decedent, found in a public waterway, for example).

Injuries, External and Internal: (SEE "INJURY" TEMPLATES, Appendix C).

All injuries, if any, and no matter how minor, are described in this section and this section only. If there are no injuries, this section may be omitted from the report, *or* some prefer to include the "Injuries" heading with the word "none" next to it to emphasize the absence of injury. That's fine too. **See the injury autopsy templates** (Appendix C) **for the suggested format to describe the various main types of injuries (blunt, sharp, firearms, asphyxial, and thermal).** Adding the below declaration as the last statement in the "Injuries" section precludes the necessity of alluding to the injuries again in the "Internal Examination" section of the report:

These injuries, having been described, will not be repeated.

Internal examination:

Head:
Scalp, skull, subdural or epidural hemorrhage
General configuration of brain, arachnoid, arteries, nerves, cut surfaces
Spinal cord (if indicated)

Neck: larynx, hyoid bone, soft tissues, trachea, tongue, and cervical vertebrae; posterior neck, if warranted.

Body cavities: normal situs, effusions or hemorrhage (measured volumes), fibrous adhesions.
Thickness of abdominal wall pannus (reported in inches and measured 3 cm below umbilicus to standardize it)

Cardiovascular system:
Aorta, major branches, and great veins (thromboemboli)
Weight (heart)
Coronary arteries—origins/ostia location and shape/morphology, distribution, % atheromatous stenoses in major arteries (proximal, distal, both), recent or old thrombi.
Pericardium, epicardium, and endocardium
Myocardium—left ventricular and right ventricular thickness, interventricular septum, myocardial lesions (size and location of infarcts)
Valves: chordae, calcifications, vegetations, number of cusps

Respiratory system: Lungs—weights, pleural surfaces, major bronchi, pulmonary arteries and veins, and parenchyma. Consolidations, emphysema blebs/bullae (where?), masses, vascular thrombi (where?) and describe quality and extent, and intrabronchial mucus plugging or inspissation

Hepatobiliary system: Liver (consistency, color, nodularity), gallbladder (gallstones), extrahepatic bile ducts, pancreas

Hemic and lymphatic systems: spleen, thymus, lymph nodes, prominent white pulp? Shriveled splenic capsule? (usually indicative of some form of shock: septic, hypovolemic, neurogenic, or cardiogenic)

Genitourinary system:
Kidneys—surface, architecture, *color* (normal versus slight/mod/marked pallor?), ureters (caliber), calculi
Urinary bladder (including contents—quality and quantity), internal genitalia

Endocrine system: pituitary, thyroid (parathyroid glands, only if appropriate), adrenal glands

Digestive system: include gastroesophageal and gastric mucosal descriptions, stomach contents-quality/quantity, small and large intestine, vermiform appendix

Musculoskeletal system: general statement about skeletal musculature and axial skeletal (WNL, kyphosis, scoliosis, osteophytes/evidence of degenerative joint disease, brittle), muscular pallor

Formats for the "Injury" section

When measuring gunshot and sharp force trauma wounds from the top of the head and from the left or right of midline, do not use a soft/bendable measuring tape and wrap it around the contours of the body. Rather, use a rigid ruler and measure from the midline with it straight.

Gunshot wounds

Begin with the following summary statement and clarification:

There is/are (#) gunshot wound(s) of the body. All directions and descriptions are stated with reference to the standard anatomic planes with the body measured in the horizontal position. These injuries are labeled 'A' through 'Z' for descriptive purposes only; no sequence is implied. (The last sentence is not relevant if there is only one gunshot wound).

Penetrating/perforating gunshot wound of[2] ...

1. **Entrance wound:** Anatomic location (*preceded by* anterior/posterior, medial/lateral, superior/inferior for the head and torso, proximal/distal for the extremities, left/right. for all), preferably including a reference to a local landmark (e.g., nipple, umbilicus, axilla).
   - Distance below the top of the head and to the right or left of midline (use inches, not metric, for all injuries, external and internal).
   - Characteristics (without and with release of skin tension): Size and shape of perforation, margin of abrasion—symmetry/asymmetry/radius/atypicality (irregularity), fouling and/or stippling of the surrounding skin—symmetry/asymmetry, radius, density, and any well-demarcated areas of sparing (may be seen with hats, eyeglasses, skin folds, blocking hands, etc.).

If the entrance wound is on the lateral left flank and the exit wound is on the posterior right neck, the heading would state, "Perforating Gunshot Wound **of** the Torso and Neck" to reflect the entirety of the wound. If wishing to refer only to the entrance site (commonly preferred when there are a multiplicity of wounds entering the same body region, such as the torso), one could instead indicate "Perforating Gunshot Wound **to** Lateral Left Flank." It is preferable to pick one or the other per case (depending on the need and number of wounds), remaining consistent for all wounds in one report.

2. **Path of bullet and resultant injuries/sequelae:**
   - Describe **track** (not tract) with the *sequence* of structures perforated ("After perforation of the skin, the bullet/projectile sequentially perforates..."); state whether it is deflected by bone.
   - General character of internal injuries (slight, moderate, marked tissue destruction); this is sometimes quite remarkable with high-velocity projectiles or magnum loads.
   - Hemorrhages (soft tissue—slight, moderate, marked; if free blood in cavities, measure; clotted? liquid? bright red?), hemoaspiration (patchy, extensive, bilateral, etc.), associated visceral/muscular pallor (slight, moderate, marked), shriveled splenic capsule?
3. **Exit wound or site of lodgment:**
   - Anatomic location (*preceded by* anterior/posterior, medial/lateral, superior/inferior for the head and torso, proximal/distal for the extremities, left/right for all bod regions), preferably including reference to a local landmark (e.g., nipple, umbilicus, axilla).
   - Distance below the top of head and to the right/left of midline.
   - Wound description (or bullet—see below): Nonabraded, slit-like, microlacerated, etc. If not atypical or shored, it is often best to measure in one slit-like linear dimension when reapposed.

OR,

[2] In the heading, **"of"** refers to the entirety of the wound, from entrance to exit/lodgment. **"To"** refers only to the entrance site location.

4. **Bullet:** Deformed? If so how?/where? (nose, base, side, all), fully/partially jacketed? caliber (small, medium, large), and is submitted to evidence. Verify the site of lodgment or perforated bones with preautopsy X-rays.

5. **Course/direction:**
   - Downward/upward, left/right, front/back (or, "without discernible vertical/horizontal/anteroposterior deviation").

Multiple gunshot wounds

For each gunshot wound with a distinct track and exit/bullet, use the format for a single gunshot wound (above) in the body of the report. Give each wound its own correlated *Roman numeral* on the "Final Diagnoses" list. That said, if there are *many* gunshot wounds overall, including multiple in one region and in close proximity to each other, you may consider grouping them in your diagnoses headings only, by body region—"multiple perforating (2) and penetrating (4) gunshot wounds of head, with," for example, but with *separate* designators and descriptions of *each* individual wound in the subsequent *body* of the report ("Injury" section).

If several entrance wounds are so closely approximated that distinct tracks cannot be discerned, so state. In this scenario, you can describe entrance wounds as a group, exit wounds or bullets recovered as a group, internal perforated structures and their sequelae (combined effects), and a general statement about the overall course and direction (this should be a very *rare* occurrence).

Shotgun wounds

Begin with a similar, if slightly modified introductory statement (see above).

1. **Entrance wound**
   - Anatomic location: centered below the top of the head and to the right or left of midline; central defect (size, shape, margins), approximate number and a description of the spread of satellite perforations, if any. Fouling, stippling, (wad) abrasion, or filler.
2. **Path of bullet and associated injuries/sequelae:** Internal injuries, combined effects ("sequential" perforations more difficult), hemothorax, visceral pallor, etc.
3. **Exit wound or wads and shot recovered** (there will be no single lodgment site as in gunshot wounds, if there are shot, not a slug). If there's an exit wound, it may be gaping/devastating.
4. **Projectiles submitted to evidence:** Wad (if relevant), *representative* pellets (buckshot, birdshot) or slug.
5. **Course and direction:** Downward/upward, left/right, front/back.

Blunt force trauma

The descriptions should follow the same sequence as the final diagnoses list (i.e., organize injuries by body region and describe in layers, starting from the outside [skin] and moving inward [soft tissues, then bone, organ, cavity]):

> There are blunt injuries of the head, trunk, and extremities. These injuries are described by body region.

Head and neck

- Palpable facial fractures?
- Externally visible injuries: Lacerations, contusions, abrasions—describe their size, color, and shape, making note of any specific **patterned injuries** (i.e., characteristics of a wound caused by contact with a specific surface/impact by a specific object) or **patterns of injury** (e.g., planar distribution of multiple injuries)
- Subscalpular hemorrhages, strap muscle hemorrhages
- Skull (fractures: linear, depressed/comminuted, size, location, basilar/vertex, shatter type)
- Subdural or epidural hemorrhage (hemorrhage vs. hematoma; weigh and/or measure; save for toxicology, as needed); subarachnoid hemorrhage (diffuse, patchy/multifocal, hemispheric/vertex, basilar)
- Brain (same scheme as natural, primary traumatic lesions, secondary effects of trauma, swelling—gyri flattened and widened/sulcal effacement)
- (Retention for neuropathologic examination, as needed)

- Neck and spinal cord (epidural hematoma, subdural hematoma, softening, contusions)
- (Facial soft tissue and skeletal dissection, as needed)

Trunk/Torso

- Externally visible injuries
- Subjacent soft tissue injury
- Avulsions/avulsion pockets from shearing forces (common in pedestrians struck and/or run over by a motor vehicle, forceful grabbing, squeezing, and twisting of limbs). Make a note of the quality and quantity of the soft tissue hemorrhage. A *gelatinous appearance* is common with *shearing* injury. Multiple avulsion pockets that are filled with blood or soft tissue hematoma may represent a pathophysiologic/mechanistic contribution to death, depending on the case. Look for characteristic "crush" or compressive type injuries
- Fractures—clavicles, sternum, ribs (rib number, left/right, linear/displaced, parasternal/anterior/anterolateral/lateral/posterolateral/posterior/paravertebral), vertebral column, pelvis (linear, slightly displaced, comminuted and widely separated and flattened—"open book")
- Visceral/organ injuries
- Body cavities—hemorrhages, abdominal organs displaced into thorax through diaphragmatic lacerations.

Extremities

With the extremities, speak in terms of **proximal and distal**, *as opposed to superior and inferior,* as they can move in space and make the latter designation uninterpretable. If one lifts their arm or leg up, what had been "inferior" may now be "superior." By contrast, proximal and distal never change despite limb movement because they are terms that are used in relation to the fixed torso. There is no need to measure wounds from the midline for the same reason (unclear location in space at the moment of injury occurrence).

- Make a note as to whether a lower extremity is externally rotated and/or foreshortened. If so, consider a pelvic/hip or femoral fracture and dissect accordingly to rule this in or out (or ask for an X-ray to exclude if an external examination).
- Upper extremities—look for externally visible injuries/deformities/fractures; dissect as needed.
- For pedestrian fatalities, remember to document points of impact (external/internal skin and/or soft tissue injuries and/or fractures) above the level of heel(s); if the injury is large (shearing/hematoma), report the span above the heel (e.g., "a large, gelatinous soft tissue hemorrhage/contusion hematoma of the posterolateral left calf and popliteal fossa spans an area 7–11 inches above the heel").

When describing blunt force trauma injuries, remember that while contusions and lacerations can be seen both externally (skin/soft tissue) and viscerally, **abrasions, by definition, are strictly cutaneous injuries** *(stripping away of epidermis by frictional forces); the word "abrasion" is not relevant in the description of internal blunt trauma.*

Sharp force trauma

There are (#) sharp force injuries of the head, trunk, and extremities: # stab wounds, # incised wounds, and # puncture wounds. All directions and descriptions are stated with reference to the standard anatomic planes with the body measured in the horizontal position. (when multiple) These injuries are labeled "A" through "Z" for descriptive purposes only; no sequence is implied.

Categorize by injury modality (stab, incised), within which the injuries should be organized by body region, then further described from superficial (outer) to deep (inner).

Stab wounds

Anatomic location (*preceded by* anterior/posterior, medial/lateral, superior/inferior for the head and torso, proximal/distal for the extremities, left/right for all body regions), preferably including reference to a local landmark (e.g., nipple, umbilicus, axilla).

Wound appearance

- Description of appearance and size (i.e., "gaping," swallow-tail, slit-like, etc.), but measure **length only after skin tension has been released** (i.e., relaxed length, not gaping width)
- Orientation (horizontal, vertical, oblique)
- Angles: locations of blunt and sharp ends/angles, if a single edged knife (most); for a blunt end, measure its width (usually 1/16″)
- Any surrounding abrasion/contusion: describe the size and pattern (consistent with "hilt mark"?)
- Track (bone or cartilage involved?): organs and tissues *sequentially* penetrated; # and specific locations of rib/intercostal space(s) penetrated (parasternal, anterior, anterolateral, lateral, posterolateral, posterior, paravertebral), if relevant; if bone is injured-save it; give the *direction* of penetration
- *Approximate* depth of penetration (give an estimated depth as a range in inches, which accounts for the nonstatic space and moving organs within body cavities); the only time a more precise measurement is given is when there is penetration through the soft tissues/muscle only (i.e., no body cavity penetration)
- Associated injuries and body cavity and soft tissue hemorrhage, hemopneumothoraces, hemoaspiration, visceral and muscular pallor/shriveled splenic capsule, etc. **Quantitate** body cavity accumulations.

Note: *If there are an excessive number of wounds, they can be carefully grouped by region/distribution, giving ranges of location from the top of the head/shoulder and from the midline, and ranges of sizes, orientation, and depth. Outliers/unusual wounds and/or wounds with a more lethal contribution should be further described separately/highlighted. No matter what,* **the total number of wounds should always be determined, photographed, and reported as part of the total tally,** *even if not every single one is has an individual heading in the autopsy report.*

Incised wounds

- Describe by anatomic location (as with stab wounds—see above), length, and depth (note: there is *NO* directionality description for incised wounds). These wounds typically have *two sharp ends* (i.e., caused by the long, sharp cutting edge of a knife or other implement, such as a box cutter or razor blade).
- Describe the structures penetrated (when nonfatal, this is often just soft tissue—skin and dermis, with or without involvement of the subcutis, muscle, or substantial vasculature).
- Often not fatal unless deeply penetrating on the neck (homicides, suicides) or wrists (suicides).
- In homicides, always look for incised wounds on the hands (including webs!), forearms, and lower extremities. Even if not mechanistically contributory to death, these wounds may have import in later framing what may have occurred during a violent encounter, including evidence of the likely consciousness of the decedent at the time that the injuries were sustained.

Asphyxia

- **Findings will vary by modality** [hanging (see template), homicidal strangulations, smothering-inflicted versus accidental, suicidal covering of the head with a plastic bag, chest compression, etc.]
- **External injuries:** Note quantity and location of conjunctival (bulbar/palpebral, superior/inferior, lateral/medial), oral, and facial petechiae (+ or −), neck contusions, and/or ligature furrow (trench) versus abrasion (no trench) versus "impression"—mark, but no injury, per se; describe the ligature mark in width, single/double, horizontal versus courses upwards and disappears, specific pattern or texture. Facial pallor or plethora, protruding tongue, stocking/glove lividity. Peri- and/or intraoral injury? Frenulum tears? (the latter two findings are most important in a suspected smothering or gagging).
- *If there is* a **ligature**: Is it tight/loose, single/multiple loops, color, length and width, texture (abrasive versus smooth), location, and character of knot (do not untie the ligature knot, if fixed—not a slipknot; rather, cut through it on the opposite side to the knot, label it, and photograph).
- **Internal injuries:** Fractures, soft tissue hemorrhage/contusions, upper chest injury, including rib fractures, petechiae, visceral congestion (nonspecific), pulmonary edema (nonspecific). Only with the stripping off of the soft tissues of the larynx can one truly assess the bony and cartilaginous structures.
- Dissect the neck after removal of the brain and thoracic organs. Perform both anterior and posterior neck dissections, as deemed necessary.

Thermal

A. **Smoke inhalation**—soot in the nares, oral cavity, larynx, trachea, and bronchi? Hyperemia of the larynx/trachea/bronchi? Frothy airway and pulmonary edema? Cherry red lividity and/or bright red blood/tissues? **Photograph.**

B. **Thermal injury**—partial versus full thickness thermal injury (first/second/and third-degree thermal burns) of the (head, trunk, extremities), involving approximately ____% body surface area? Use the rule of nines. Describe charring (4th-degree thermal burn injury).
- *Describe artifacts of "4th-degree burns":* pugilistic posture, longitudinal muscle splitting, long bone fractures, partial traumatic amputations of limbs, (granular) epidural hemorrhage.
- *Scald injury* (distribution, inflicted/"submersion" pattern—discrete line of demarcation—with flexion sparing, versus an accidental splash pattern); postmortem versus antemortem submersion/hot water contact.

APPENDIX

C

Autopsy report templates

Autopsy templates by type

Autopsy templates are merely guideposts. They should never be used as a substitute for thoughtful and careful observation, documentation, and thorough, clear, and objective descriptions. These are not cut and paste documents, but rather they are a start point for dictations, a framework. They should be used with sound judgment in order to reflect the specificity of each individual case and its unique findings, not as a means to cut corners or to force descriptions to fit within strict confines. The template is written by default as nonpathologic; findings/abnormalities should be added as identified by case (examples are offered).

I. General Template
II. Pediatric Template
(Injury Section Templates)
III. Firearms
Handgun/Rifle
Shotgun
IV. Blunt Injuries
V. Sharp Injuries
Stab
Cut
Chop
VI. Hanging
VII. Thermal/Inhalation Injury

General/generic dictation template

I hereby certify that I, **John Doe**, *M.D./D.O., Medical Examiner/Coroner, have performed an autopsy on the body of ___, on* **April 4, 2022**, *starting at* **9:30AM** *in the mortuary of the Office of Medical Examiner/Coroner of (jurisdiction). This autopsy was performed in the presence of Dr(s). ______.*

External examination

The body is received in the supine position in a plastic body bag. A handwritten tag encircles the left great toe; a handwritten band is on the left wrist. Both appropriately reflect the above-named case number. (E.g., if from a hospital, "A digital hospital tag encircles the right great toe and a digital hospital band is on the right wrist.")

The body is that of a well-developed, well-nourished, *average-framed/large-framed/markedly obese/cachectic, etc., ___' ___", ___* lb. (BMI), *light brown-skinned/medium brown-skinned/lightly pigmented (transgender?)* man/woman, whose appearance is consistent with the given age of ___ years [*VERSUS* "whose age cannot be discerned due to moderate/advanced putrefactive change (documented age, 52 years)"].

The (coarse, curly, fine/thick, straight/wavy, color) scalp hair measures up to ___". There is a ___" mustache and ___" beard (goatee? unkempt?/neither mustache nor beard vs. dense/slight stubble in the distribution of the mustache and beard). There are no palpable facial fractures. The eyes have (clouded) (brown/blue/green/hazel) irides (with arcus senilis) and *clear/slightly or moderately congested* conjunctivae without petechiae, hemorrhage, or jaundice (if petechiae or hemorrhage, where? left/right, bulbar, inferior/superior palpebral? etc.). The oral cavity has ____ *(describe teeth: natural, natural, but missing many, edentulous with full dentures/partial dentures in place, in disrepair, edentulous in the maxilla with natural teeth in the mandible, etc.)* and an atraumatic mucosa. The anterior and posterior aspects of the torso are unremarkable. The extremities are free of edema or linear scars overlying subcutaneous veins (or there are linear scars overlying subcutaneous veins of the upper extremities/left/right antecubital fossa/volar forearm, etc.). The atraumatic fingernails are well-manicured/painted and extend just to/slightly beyond/well beyond the fingertips. The atraumatic, unremarkable external genitalia are those of an adult man/woman; (the testes are descended/undescended and free of masses, penis circumcised/uncircumcised). The anus is atraumatic. There are no discernible scars or tattoos *(if tattoos are present, generally describe and give locations in a separate heading; see below).*

[Note: If the individual is transgender and status post surgery, consider writing something such as the following:

- *The breasts are status post remote breast augmentation with intact saline implants.*
- *The atraumatic genitalia are status post remote (male-to-female OR female-to-male) gender confirmation/reassignment surgery.*]

Postmortem changes

There is (no/easily broken/moderate/strong) symmetric rigor mortis of the upper and lower extremities, neck, and jaw. Livor mortis is *fixed/unfixed/predominantly fixed* and *posterior, sparing pressure points (or, with irregular blanching consistent with contact with bed linens, etc.).* The body is cool.

If putrefaction, mummification, adipocere, *describe*:

Example:

(Early visceral/moderate/advanced putrefactive change is characterized by diffuse green-red skin and tissue discoloration, focal skin slippage, abdominal, scrotal, and facial bloating, marbling, vitreous humor volume loss, purged thoracic cavity blood, and visceral softening and sponginess)-or any combination thereof.

Tattoos

There are multiple monochromatic/polychromatic tattoos on the (torso/chest/upper/lower extremities) as follows: Location followed by parenthetical general description ("clown face," "crucifix," "skull design," etc.).

Therapeutic/diagnostic procedures

The following are in place: [E.g.: taped endotracheal and nasogastric tubes, adhesive defibrillator and echocardiogram patches on the anterior torso, intravenous catheters in the antecubital fossae and distributions of the right internal jugular and left subclavian veins, an intra-osseous needle in the proximal left tibia, a sutured thoracostomy tube extending from the lateral left chest (fourth intercostal space), etc. There is a dried, presternal abrasion and anterior/anterolateral rib fractures, status postresuscitative efforts].

Clothing

The body is received clad in the following items of clothing: (list with details; pockets empty? bloodstained?). See injury templates below for instructions in describing clothing with defects, blood, debris, etc.

versus,

The body is received unclad.

Injuries, external and internal

***(ALL injuries belong in this section, and no other, no matter how minor; see specific injury templates below).

End this section with:

These injuries, having been described, will not be repeated.

Internal examination

HEAD: The scalp has no contusion. The skull has no fracture. There is no epidural, subdural, or subarachnoid hemorrhage. The brain weighs _____ gm and is normal in size and shape. The cerebral hemispheres are approximately symmetric and have the usual pattern of sulci and gyri *(versus: the gyri are widened and flattened, etc.)*. The leptomeninges are thin and translucent. The cerebral vessels are normally distributed and free of atherosclerosis or aneurysms. The cranial nerves are normally distributed. The white and gray matter, deep nuclei, brainstem, cerebellum, and ventricles are unremarkable and free of focal lesions. ***(If submitting the brain and/or spinal cord for neuropathology consultation, describe the scalp, skull, and extra-axial spaces, but largely leave the remaining brain description to the neuropathologist and state, "The brain/brain and spinal cord are retained for later neuropathologic evaluation.").*

NECK: The cervical vertebrae, hyoid bone, tracheal and laryngeal cartilages, and paratracheal soft tissues have no trauma. The upper airway is unobstructed. The tongue is unremarkable.

BODY CAVITIES AND BREASTS: The organs are in their normal situs with neither fibrous adhesions nor abnormal fluid accumulations. The abdominal wall pannus is ___″ in thickness. The breasts are symmetrical and free of focal lesions, purulence, or masses.

(OR, there is gynecomastia).

***Fibrous adhesions and nontraumatic fluid accumulations (exudative collections, effusions, and hemorrhages resulting from natural disease should be described here, but if hemorrhagic accumulations in the cavities or soft tissues, liquid or clotted, are the result of trauma, they should be described and quantitated under the "Injury" section, as sequelae. If they are caused by a therapeutic procedure, they would best be placed under the "Therapeutic/Diagnostic Procedures" section.*

CARDIOVASCULAR SYSTEM: The aorta *has no/moderate/marked/focally calcified/ulcerated* atherosclerosis. The venae cavae and pulmonary arteries have no thrombus or embolus.

The ____ gm heart has right predominant coronary arteries without atherosclerotic stenosis/with atherosclerotic stenoses of the epicardial vessels as follows: ___% eccentric/concentric stenosis of the proximal/mid/distal left main artery, ___% eccentric/concentric stenosis of the proximal/mid/distal left anterior descending artery, ___% eccentric/concentric stenosis of the proximal/mid/distal right coronary artery, and ___% eccentric/concentric stenosis of the proximal/mid/distal circumflex artery. There are no recent thrombi. The myocardium is homogeneous, dark red, and firm, and free of discernible pallor, hemorrhage, softening, or fibrosis (versus a description of focal lesions). The left and right ventricular walls are concentrically ___ cm and ___ cm in thickness, respectively. The endocardial surfaces and four cardiac valves are unremarkable and free of vegetations or thrombi. The coronary ostia are normal in location and morphology.

RESPIRATORY SYSTEM: The right lung weighs ___ gm; the left lung weighs ___ gm. The parenchyma is free of masses, hemorrhage, consolidation, or other focal lesions. The bronchial tree and vasculature are unremarkable and free of obstruction or thrombi.

LIVER, GALLBLADDER, AND PANCREAS: The liver weighs ___ gm and has an intact capsule. The parenchyma is (red-brown, soft, and free of focal lesions or nodularity). The gallbladder contains approximately ___ mL of dark green bile fluid and no gallstones (or, with multiple 0.7 to 1.2 cm light green, faceted gallstones, etc.). The pancreas is unremarkable in lobulation, color, and texture.

HEMIC AND LYMPHATIC SYSTEMS: The ____ gm spleen has an intact, smooth/shriveled capsule and soft, maroon parenchyma with/without prominent white pulp. There is no lymphadenopathy.

GENITOURINARY SYSTEM: The right kidney weighs ____ gm; the left kidney weighs ___ gm. Each kidney has a *smooth/slightly/moderately/markedly granular* subcapsular surface and is maroon-brown (versus tan, moderately/markedly pale or mottled, etc.) with unremarkable architecture and vasculature. The ureters maintain uniform caliber into an unremarkable bladder containing approximately ___ mL of *(clear/cloudy, light/dark yellow/amber, blood-tinged, etc.) urine (or into an empty bladder)*. The prostate gland is unremarkable and consistent with age *(or symmetrically enlarged and nodular, but free of focal of focal lesions)*. The testes are descended and unremarkable.

[versus, "The ovaries, fallopian tubes, uterus, cervix, and atraumatic vaginal vault are unremarkable and consistent with age. The vaginal vault is free of discharge and atraumatic. The cervical os is slit-like/or circular" OR "are unremarkable with the exception of 0.5 to 2.0 cm intramural and subserosal uterine leiomymata," etc.]

ENDOCRINE SYSTEM: The pituitary, thyroid, and adrenal glands are each normal in color, size, and consistency.

DIGESTIVE SYSTEM: The esophagus and gastroesophageal junction are unremarkable. The stomach contains approximately ___ mL of _____ (describe pills or food material: e.g., nondescript? green-tan? viscid? pasty? with recognizable food fragments, including green vegetables and white meat, etc.). The gastric mucosa, small and large intestine, and appendix are unremarkable (versus *have scattered noninflamed diverticulae of the left colon, and/or the appendix has been remotely surgically removed/is absent,* etc.).

MUSCULOSKELETAL SYSTEM: The vertebrae, clavicles, sternum, ribs, and pelvis have no fracture. The musculature is normal in distribution and appearance.

Microscopic tissue examination

Tissue or Organ x # of fragments (# = block number)
HEART x (#):
LUNG x (#):
LIVER x (#):
KIDNEY x (#):
One or more of the following, depending on the case:

Toxicology

Samples of femoral (or cavity? right or pooled heart?) blood, urine, and vitreous humor are submitted for toxicological analysis; a separate report will be issued.

OR (if no testing requested):

Samples of femoral blood, urine, and vitreous humor are submitted to the toxicology laboratory for storage.

Neuropathology/cardiac pathology

The brain/spinal cord/heart is/are retained for neuropathologic/cardiac pathology consultation; a separate report will be issued.

Microbiology

A swab of the nasopharynx (or upper and lower respiratory tract, where applicable) is submitted for viral analysis; a separate report will be issued.

and/or:

A sample(s)/swab(s) of (blood/tissue/peritoneal exudates) is/are submitted for bacterial culture; a separate report(s) will be issued.

Molecular genetics

A sample of heart tissue is submitted/held for possible future molecular genetic studies. (A separate report will be issued, if applicable).

Forensic biology

A blood spot card (bone, fingernails) is/are submitted to the forensic biology laboratory.
(A sexual assault evidence collection kit is submitted to the forensic biology laboratory).

Forensic imaging

Postmortem radiographs/postmortem computed tomography (PMCT) are taken and retained.
Digital photographs are taken and retained.

Evidence: (*examples*)

Three bullet envelopes and the bags removed from the hands are submitted to the Evidence Unit.
The ligature is submitted to the Evidence Unit.
All clothing is submitted to the Evidence Unit.

Pediatric dictation

Generic pediatric autopsy dictation template

I hereby certify that I, ________, M.D./D.O., Medical Examiner/Coroner, have performed an autopsy on the body of (name) on (date), starting at (time) in the mortuary of the Office of the Medical Examiner/Coroner of (jurisdiction). This autopsy was performed in the presence of Dr(s). ______.

External examination

The body is received in a small white plastic body bag. (A handwritten Medical Examiner/Coroner tag encircles the left ankle/a digital hospital tag encircles the left ankle, etc.). All tags and bands appropriately reflect the above-named case number.

Comment: expected weights and measurements, where available, are listed parenthetically following the autopsy weights and measurements, correspond to a corrected gestational age of _____, and are taken from (cite pediatric pathology reference of your choice)

The body is that of a normally developed, well/poorly nourished, well/poorly hydrated, _____ lb./____ gm, [female/male infant, boy/girl child, female/male adolescent], whose appearance is consistent with the given age of [_____ years/months/weeks/days].

(For fetus and neonate) The crown-heel length is ___ cm/___"; crown-rump length is___ cm/___"; the head circumference is ___ cm/___"; and the foot length is ___ cm/___." The fetal measurements are consistent with a gestational age of ______ *[other measurements that may sometimes be relevant: outer canthal distance, inner canthal distance]*. (For infant) The body length is ___"; the body weight is___ lb.; the head circumference is ___." The height and weight are the ___ percentile and ___ percentile for age, respectively.

The fine, curly/straight (color) hair measures up to (inches) in length. The nose and facial bones are palpably intact. The eyes have (brown, blue, hazel, gray, etc.) irides and conjunctivae that are free of hemorrhage, petechiae, or jaundice. There are no epicanthal folds. The ears are normally set. The choanae are patent and the palate is intact. (If infant) The anterior and posterior fontanelles are (open and sunken, closed, bulging, etc.). The oral cavity is edentulous/has natural deciduous anterior incisors of the maxilla and mandible. The lips, gums, and frenulae are atraumatic. The chest is/is not symmetrical. The abdomen is (flat/protuberant and soft/hard); the umbilicus is unremarkable (the umbilical stump is frayed/fragmented/dried brown/well-healed). The posterior torso is unremarkable. The upper and lower extremities are unremarkable. There are normal numbers and configurations of fingers and toes.

(There is a 4" hyperpigmented macule overlying the sacral region).

The atraumatic external genitalia are free of rashes and are those of a normal (un/circumcised) male/female infant/boy/girl child. (The testes are descended/undescended). The anus is patent and atraumatic. The anus contains scant/abundant meconium.

[Add any anatomical external anomalies, if noted: ears, jaw size, palmar creases, scars, facies, club feet, etc.]

If applicable, describe the placenta and cord, with a separate heading:

Placenta and umbilical cord

Received separately, but with the infant/fetus, in a labeled hospital container (or attached to the infant/fetus…) is a (formalin-fixed) placenta and attached umbilical cord, together weighing ____ gm *(normal for gestational age, ____ gm). The placenta is roughly circular/ovoid in shape and measures ____× ___×___ cm.* The pink-beige umbilical cord is peripherally clamped, with a central/eccentric or membranous attachment to the placenta, has normal coiling without knots, and has three patent blood vessels. The amniotic membranes are translucent/free of purulence or meconium; the cotyledons are maroon-red and free of grossly discernible focal abnormalities or attached blood clot(s). The maternal surface is/is not disrupted.

Postmortem changes

There is (no/slight/moderate/strong symmetric) rigor mortis of the upper and lower extremities, neck, and jaw. Lividity is pink/purple, (non)fixed, anterior/posterior/lateral on the torso, patchy/has a confluent distribution/has a sharp line of demarcation over anterior torso and/or face, with blanching of abdomen/central face. The body is cool. *(If stillborn, comment on presence or absence of maceration; if present, describe: diffuse red skin discoloration, overriding cranial sutures, etc.).*

CLOTHING: (may be retained in: homicides, leaving the scene motor vehicle collisions, otherwise suspicious deaths with an active/ongoing investigation)

The body is received unclad.

Or,

The body is received clad only in a clean/wet/feces-stained/urine-soaked, etc. diaper and a hospital blanket or gown, etc.

Or,

The body is received clad in the following items of clothing: (list)

The garments are intact/cut/torn/clean/soiled with blood/urine/feces/gastric contents, etc.

Therapeutic interventions

The decedent is received with the following in place: *(Example: Endotracheal and nasogastric tubes, a proximal left tibial intraosseous catheter, and pacemaker patches on the anterior torso and extremities).*

Injuries, external and internal

***(ALL injuries belong in this section, and no other, no matter how minor; see specific injury templates below).

End this section with,

"These injuries, having been described, will not be repeated."

Internal examination

HEAD: The scalp has no contusion. The skull has no fracture. The cranial sutures are/are not fused. There is no subgaleal, epidural, subdural, or subarachnoid hemorrhage. The brain weighs _____ gm *(normal range for age, ____ gm)*. The brain and spinal cord are retained in formalin for later neuropathologic examination. [There are/are not hemorrhages of the retrobulbar soft tissues or optic nerve sheaths.]

OR (if brain not retained—unusual),

[The cerebral hemispheres, brainstem, and cerebellum are roughly symmetric with the usual pattern of sulci and gyri. The leptomeninges are smooth and glistening. The cerebral vessels are free of aneurysms. The cranial nerves are normally distributed. The white and gray matter, deep nuclei, and ventricles are unremarkable. There are no focal lesions.]

BODY CAVITIES: The organs are in their normal situs. The pleural and peritoneal cavities contain no fluid, hemorrhage, or adhesions. The diaphragm is intact.

NECK: The cervical vertebrae, hyoid bone, and tracheal and laryngeal cartilages are intact. The subcutaneous, skeletal musculature, and paratracheal soft tissues have no trauma. The upper airway is unobstructed; there are no posterior pharyngeal abscesses, masses, or foreign bodies. The tongue is normal in size and otherwise unremarkable. The posterior neck is unremarkable.

CARDIOVASCULAR SYSTEM: The heart weighs ____ gm *(normal range for age, ____ gm)*. The coronary arteries pursue normal courses; the coronary ostia are normal in location and morphology. The myocardium is homogenous, dark-red, firm, and free of pallor, hemorrhage, softening, or fibrosis. The left and right ventricular walls are ____ cm and ____ cm in thickness, respectively. The endocardial surfaces and four cardiac valves are unremarkable. The foramen ovale is probe patent; there are no other interatrial or interventricular septal defects. The ductus arteriosus is closed. The aorta and great vessels, venae cavae, and pulmonary arteries and veins arise and branch normally.

RESPIRATORY SYSTEM: The right lung weighs ____ gm; the left weighs ____ gm *(normal range for age, ___ gm, combined)*. The pleural surfaces are smooth, with normal lobar divisions and no petechiae. The *(pink/gray, fluffy/firm, hyperaerated, etc.)* parenchyma is free of *(masses, hemorrhages, consolidation, or frothy, sanguinous fluid)*. The bronchial distribution and vasculature are unremarkable and unobstructed (versus, "contain multifocal tenacious mucus plugs"). There are no tracheoesophageal fistulas.

LIVER, GALLBLADDER, AND PANCREAS: The ____ gm liver *(normal range for age, ____ gm)* has an intact capsule and brown/tan/yellow parenchyma that is free of a slippery or fibrous texture or focal lesions. The gallbladder contains approximately ____ mL of green fluid/thick bile fluid with(out) gallstones. The pancreas is unremarkable in lobulation, color, and texture.

HEMIC AND LYMPHATIC SYSTEMS: The spleen weighs _____ gm *(normal range for age, ____ gm)*, has an intact capsule, and unremarkable maroon parenchyma. There is no lymphadenopathy (versus there is shotty mesenteric lymphadenopathy, etc.). The thymus weighs ____ gm (normal range for age, ____ gm), is in its normal mediastinal location, and is unremarkable [versus has scattered petechiae/is barely discernible/involuted, etc.].

GENITOURINARY SYSTEM: The right kidney weighs _____ gm *(normal range for age, ____ gm)*; the left weighs ____ gm *(normal range for age, ____ gm)*. Each kidney has a (smooth), red-brown subcapsular surface [with persistent fetal lobulations], and unremarkable architecture and vasculature. The ureters maintain uniform caliber into an unremarkable (empty) bladder/bladder containing approximately ____ ml of (color, clarity) urine. The prostate gland is unremarkable for age and development. The testes are (descended/undescended) and unremarkable.

OR,

The ovaries, fallopian tubes, and uterus are unremarkable and consistent with age and development. The vagina is atraumatic. The hymen is intact (if not, describe).

ENDOCRINE SYSTEM: The pituitary, thyroid, and adrenal glands are normal in size, color, and consistency. The combined adrenal gland weight is _____ gm *(normal range for age, ____ gm)*.

DIGESTIVE SYSTEM: The esophagus and gastroesophageal junction are unremarkable. The stomach contains approximately ____ ml of nondescript, beige fluid/partially digested material/fluid. The gastric mucosa, small intestine, and large intestine are unremarkable *(no anatomic anomalies, intussusception, volvulus, hemorrhages, obstruction, etc.)*. The small intestine contains *(scant/abundant), (tan/yellow/green/seedy)* material throughout; the large intestine contains *(scant/copious), (liquid/semisolid/softened/formed), (green/brown)* feces/meconium throughout. The appendix is present and unremarkable (versus absent).

MUSCULOSKELETAL SYSTEM: The vertebrae, clavicles, sternum, ribs, and pelvis have no fracture. (Or, if there are healing and/or acute fractures, describe in the injury section). The musculature is normally distributed. [If a flay dissection was performed, "There is no hemorrhage of the soft tissues of the anterior and posterior torso and upper and lower extremities." Or, describe positive findings.].

Microscopic tissue examination

Tissue or Organ x # of fragments (# = block number)
HEART x (#):
LUNGS AND BRONCHI x (#):
LIVER x (#):
KIDNEY x (#):
PANCREAS x (#):
BONE MARROW x (#):
THYROID GLAND x (#):
ADRENAL GLANDS x (#)
THYMUS x (#):
SPLEEN x (#):
(PLACENTA/UMBILICAL CORD)
One or more of the following:

Toxicology

Samples of blood, urine, and vitreous humor (bile, liver, brain, and gastric contents, if relevant) are submitted for toxicological analysis; a separate report will be issued.

Neuropathology/cardiac pathology

The brain/spinal cord/heart is/are retained for neuropathologic/cardiac pathology consultation; a separate report will be issued.

Microbiology

Specimens are submitted for viral studies; a separate report will be issued.

A sample(s)/swab(s) of (blood/tissue/peritoneal exudates/cerebrospinal fluid) is/are submitted for bacterial culture; a separate report will be issued.

Molecular genetics

A sample of heart tissue is retained for possible future genetic studies.

Metabolic studies

Blood and bile spots are submitted for metabolic analysis; a separate report will be issued.

Forensic biology

A blood spot card (or bone) is submitted to the forensic biology laboratory.

Forensic imaging

Full-body radiographs/PMCT are taken and retained.
Digital photographs are taken and retained.

Evidence

All clothing is submitted to the Evidence Unit.

Gunshot/shotgun wound templates

Handgun/rifle wounds

Injuries, external and internal

There is/are (#) gunshot wound(s) (GSW(s)) of the body: # is/are penetrating; # is/are perforating/all are penetrating/perforating, etc. All descriptions and directions are stated with reference to the standard anatomical planes with the body measured in the horizontal position. (the next line *only* when there is more than one wound) These injuries are labeled "A" through "X" for descriptive purposes only; no sequence is implied.

A. Penetrating/Perforating Gunshot Wound of... [description of span/location of entire wound—e.g., "...of left forearm and chest"]:

Entrance wound

"A gunshot entrance wound of the (e.g., "inferomedial left mammary chest," "left preauricular face," see Figure C.1 for anatomic nomenclature) is centered ____″ below the top of the head and ____″ to the left/right of midline (you can also give additional landmarks as needed, such as "at the mid axillary line, 3″ inferior to the axilla," or "at the level of the nipple," etc.). It consists of a _____" circular perforation with a symmetrical _____" (often 1/16″)/asymmetrical surrounding margin of (pink, dried/maroon) abrasion [if asymmetrical: e.g., 1/8″ on the inferolateral aspect and 1/16″ on the remaining circumference]. There is neither fouling nor stippling of the adjacent skin/hair-bearing skin" (if either is present, measure pattern, density, radius, and symmetry).

Notes

- Using the glabella or forehead in GSWs of the head: allows a measurement in the anteroposterior dimension as well; either can be used as a point of reference. This gives a third dimension on a more spherical part of the body. For example: 2″ below the top of the head, 2–1/2″ to the right of midline, and 2″ posterior to the plane of the forehead/glabella.

1. Supraclavicular
2. Infraclavicular/Supramammary
3. Mammary (L/R, superomedial/inferomedial, superolateral/inferolateral)
4. Inframammary
5. Parasternal
6. Sternal
7. Axillary and Infra-axillary
8. Costal margin
9. Abdominal Quadrants (right lower)
10. Periumbilical
11. Suprapubic
12. Groin
13. Arm (anterior/posterior/medial/lateral); 13+14 = upper extremity
14. Forearm (volar/dorsal)
15. Thigh (15+16 = lower extremity)
16. Leg (shin, calf, ankle)
17. Suprascapular
18. Scapular
19. Infrascapular
20. Lateral scapular
21. Interscapular
22. Lumbar
23. Paravertebral
24. Flank (anterior/mid/posterior axillary line)
25. Sacral
26. Gluteal

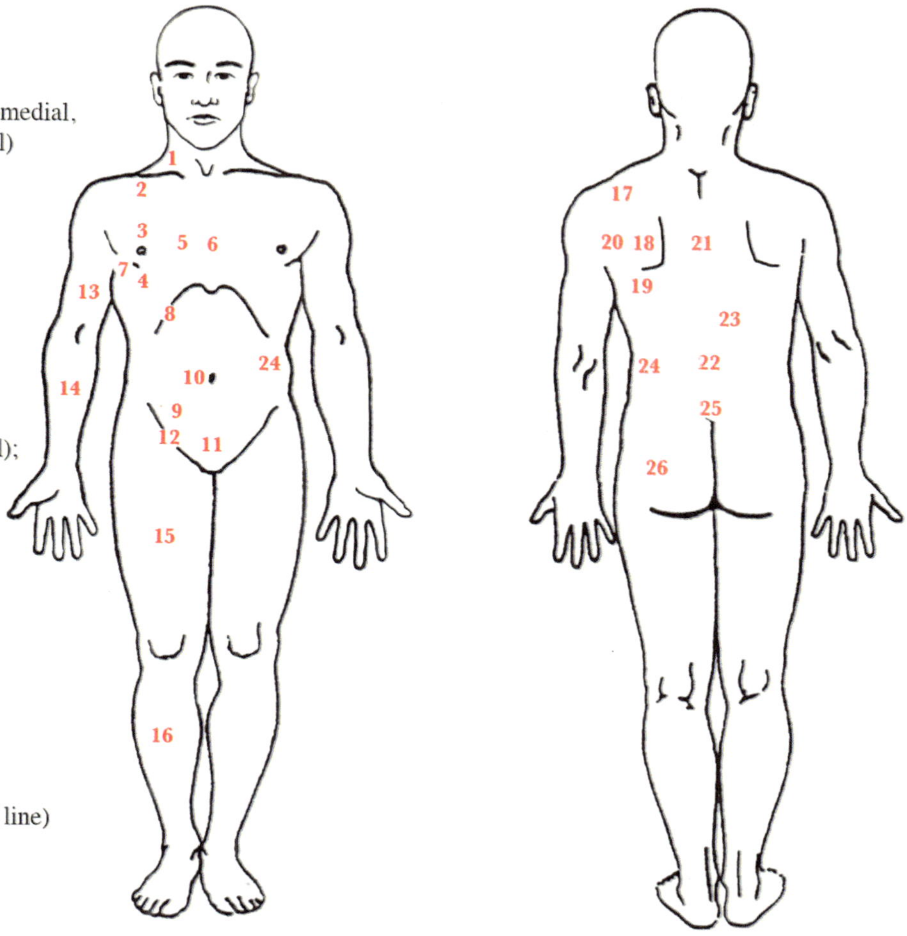

FIGURE C.1 Anatomic landmark nomenclature.

- If referring to the **upper extremities**, measure wounds from below the top of the shoulder, in lieu of or in addition to a separate measurement from the top of the head; extremity wounds should have no "from the midline" measurement, as they are variably mobile in space relative to the midline, giving midline references no relevance/interpretive value.
- If an entrance wound has a surrounding abrasion that is irregular/"atypical," so state, and describe how it's atypical (e.g., "broad, dried surrounding margin of abrasion with irregular contours, measuring up to 1/2″ inferiorly, 3/8″ laterally, and 1/8″ superiorly"). This is commonly found in intermediate target, reentrance wounds, and shored wounds.

Track with associated injury/sequelae

"After perforation of the skin and soft tissues of the (anatomic location, as stated under "entrance wound" above), there is *sequential* perforation of ______ (e.g., "the anterolateral left sixth rib/ICS, the medial aspect of the upper lobe of the left lung, the anterior thoracic aorta, and the right parasternal third rib, before becoming lodged in the soft tissues of the right parasternal chest wall," OR, "before exiting the body.")."

"Associated injuries include _______." (Example: "an 1800 mL left hemopneumothorax, moderate soft tissue hemorrhage, visceral pallor, and a shriveled splenic capsule.").

Exit wound (or site of lodgment)

"The site of lodgment (or gunshot exit wound) on the (*location with anatomic description; see anatomic nomenclature diagram below*) is centered ______″ below the top of the head and ______″ to the left/right of midline. It consists of an (irregular/stellate/microlacerated, nonabraded, slit-like, etc.) ____″ defect," OR "is lodged in the musculature of the right parasternal chest."

Course and direction

"The direction this bullet travels is: front to back/back to front, left to right/right to left, and downwards/upwards."

Notes

- *If there is indiscernible/insignificant deviation in any of the three planes, say "without significant/discernible (vertical/horizontal/anteroposterior) deviation."*
- If remarkably steep in vertical course, use "steeply" as a modifier (e.g., "steeply upwards"); this might be important in later testimony/questions related to relative body positions.

Bullet (only for penetrating wounds or for bullet fragments left behind in perforating wounds)

"The bullet is nondeformed (OR) slightly/moderately/markedly deformed (e.g., at the nose, flattened on the side, etc.), fully jacketed/unjacketed, gray metal, and of apparent (small/medium/large) caliber. It is submitted to the Evidence Unit without inscriptions."

"These injuries, having been described, will not be repeated."

In a separate section of the autopsy report, before the "Injuries" section:

Clothing

"The decedent is received clad in the following items of focally/extensively bloodstained/blood-soaked clothing, most/all of which has/have been cut prior to arrival (status post resuscitative efforts) and described as worn from outer to innermost layers."

Or,

"The decedent is received unclad. The following items of clothing are received separately from police personnel through the Evidence Unit."

Example:

"One black, nylon-type, hooded winter jacket; one blue, plaid, button down shirt; one gray T-shirt; one white tank top; one pair of khaki pants with an attached black, cloth belt; one pair of beige thermal pants; one pair black boxer briefs; one pair of white athletic socks; one pair of yellow and black, high top Nike sneakers; and one knit cap. All pockets are empty. These items have multiple defects *(don't call these "wounds"!)* roughly corresponding to GSWs A through D described below. No gunpowder residue is **grossly** discernible."

"These items are labeled and submitted as evidence after a period of drying."

NOTES: *If fouling or apparent gunpowder flakes are seen on the clothing around any entrance defect, photograph and describe what you see, objectively, in quality, radius, and distribution/symmetry. Most ballistics labs will more definitively analyze and confirm with chemical testing (if tested/relevant to case, as deemed by law enforcement authorities). This is just a gross description.*

For example, "there is smoky, black-gray residue/discoloration symmetrically surrounding the defect on the front of the white tank top that corresponds to gunshot entrance wound B on the body; it measures up to radius of 1–1/4″."

And/or,

"There are multiple minute, flake-like particles asymmetrically surrounding the defect on the white tank top corresponding to gunshot entrance wound C on the body (up to ¾″ inferiorly and laterally and 1–1/4″ superiorly)."

NOTE: never call these particles on the clothing "stippling"; stippling is an injury.

And/or,

"There is a thin, dense, discrete rim of black residue on the margins of the defect on the front of the white tank top that corresponds to gunshot entrance wound A on the body (comment: bullet wipe)."

Shotgun wounds

Injuries, external and internal

"There is/are (#) shotgun wounds(s) of the body: # is/are penetrating; # is/are perforating/all are penetrating/perforating, etc. All descriptions and directions are stated with reference to the standard anatomical planes with the body measured in the horizontal position." (Offer the following only when there is more than one wound) "These injuries are labeled 'A' through 'X' for descriptive purposes only; no sequence is implied."

A. Penetrating Shotgun Wound of ___ [description of span/location of entire wound; e.g., "...of left forearm and chest"]: (most shotgun injury is *penetrating*).

Entrance wound

A shotgun entrance wound of the (e.g., "inferomedial left mammary chest," "right preauricular face," "posterior hard palate"; *see Figure C.1, anatomic landmark diagram, for nomenclature*) is centered _____" below the top of the head and _____" to the left/right of midline (you can also give additional landmarks as needed, such as "at the mid axillary line, 3" inferior to the axilla," or "at the level of the nipple," etc.). It consists of a large _____" circular perforation with a symmetrical/asymmetrical _____" (usually 1/16") margin(s) of (pink, dried/maroon) abrasion. There is neither fouling, stippling, nor wad abrasion of the adjacent skin/hair-bearing skin (if any *is* present, measure size, pattern, density, radius, symmetry, etc.).

Modifications

"The internal margins of the entrance wound are focally scalloped, the scallops averaging ½" diameter.

Central wound with pellet spread: "A shotgun entrance wound is centered _____" from the top of the head and ____" to the left/right of midline. It consists of a ¾" roughly circular defect and is surrounded by ____ (#) smaller individual pellet perforations, each averaging 1/8–1/4" in diameter and distributed in an average spread of ___" in diameter, with rare outlier/more distant perforations.

Versus only pellet perforations: "There are approximately ____ (#) circular pellet perforations averaging ___" in diameter, each with symmetrical 1/16" surrounding pink margin of abrasion."

Track

"After perforation of the skin and soft tissues of the (anatomic location, as above), together the pellets sequentially perforate ______. (e.g., "the anterolateral left sixth rib/ICS, the medial aspect of the upper lobe of the left lung, the anterior thoracic aorta, and the right parasternal third rib, before becoming lodged in the soft tissues of the right parasternal chest wall, the right lung parenchyma, and the right hemidiaphragm.").

"As a result of this shotgun wound, there is____." (Example: "an 1800 mL left hemopneumothorax, moderate soft tissue hemorrhage, visceral pallor, and a shriveled splenic capsule").

Exit wound/site of lodgment

"The site of exit on the (*location with anatomic description; see Figure C.1*) is centered ____" below the top of the head and ______" to the left/right of midline. It consists of a (? large, gaping, irregular/stellate/lacerated, non-abraded, etc.) ____" wound."

Or,

Example,

"There are multiple sites of pellet lodgment distributed within the right parasternal chest wall, the right lung parenchyma, and the right hemidiaphragm."

Course and direction

"The direction these pellets travel is: front to back/back to front, left to right/right to left, and downwards/upwards." [OR, "front to back, and without discernible horizontal or vertical deviation," etc.].

Notes

- These three dimensions may not be applicable/discernible when multiple pellets are lodged in the body ("billiard ball effect"). Say only what you can. Sometimes, that's only front to back or back to front.
- If there is indiscernible/insignificant deviation in any of the three planes, say "without significant/discernible (vertical/horizontal/anteroposterior) deviation."

Shot and wadding: *(for penetrating shotgun injuries only)*
"There are numerous (if possible, "approximately ___) bird/buckshot in the body. On average, they measure ___" in diameter; some are moderately deformed. (There is also a ___" plastic/cardboard wadding with ____ (#) open petals). *Representative* bird/buckshot pellets (and the wadding) are recovered and submitted to the Evidence Unit."

"These injuries, having been described, will not be repeated."

In a separate section of the autopsy report, before the "Injuries" section:

Clothing

"The decedent is received clad in the following items of focally/extensively bloodstained/blood-soaked clothing, most/all of which has/have been cut prior to arrival (status post resuscitative efforts) and described as worn from outer to innermost layers."

(*Or,* "The decedent is received unclad. The following items of clothing are received separately from police personnel through the evidence unit.")

Example,

"One black, nylon-type, hooded winter jacket; one blue, plaid, button down shirt; one gray T-shirt; one white tank top; one pair of khaki pants with an attached black, cloth belt; one pair of beige thermal pants; one pair black boxer briefs; one pair of white athletic socks; one pair of yellow and black, high top Nike sneakers; and one knit hat. All pockets are empty. These items have multiple defects *(don't call these "wounds"!)* roughly corresponding to shotgun wound A below. No gunpowder residue is **grossly** discernible."

"These items are labeled and submitted as evidence after a period of drying."

NOTE: *If fouling or apparent gunpowder flakes are seen on the clothing around any entrance defect, photograph it and describe what you see objectively, in quality, radius, and distribution/symmetry.*

Blunt force trauma template

Blunt Force Trauma

Injuries, external and internal

There is blunt force trauma of the (head and neck/torso/extremities).

All descriptions and directions are stated with reference to the standard anatomical planes with the body measured in the horizontal position. The injuries are separated and described by body region.

Blunt force trauma of______ (Head/Head and Neck/Torso/Upper Extremities/Lower Extremities):

(In general, describe external first, followed by internal findings; headings organized by body region):

There is a (_____ × _____"), (color: _____) (oval/roughly circular/irregular/patterned) contusion/focally abraded contusion of the (*see Figure C.1 for anatomic nomenclature*). There are/is a (stellate, Y-shaped/linear/oblong) laceration of the (*see Figure C.1 for anatomic nomenclature*).

Abrasion examples:

There is a broad 10" × 6" area of dried, orange-red, vertically-oriented, brush-type abrasion of the anterolateral right flank.

There is a cluster of ½–3/4" dried, pink-purple abrasions of the inferomedial left knee.

Laceration example:

There is a Y-shaped laceration of the posterior left parietal scalp. It is vertically-oriented with branching at its inferior aspect, where it is associated with inferiorly oriented undermining measuring up to ¾". It has a 1/8" irregular surrounding abrasion and tissue bridging within the wound base. (if overlying bone) There is moderate subjacent subscalpular hemorrhage, but no skull fracture. *Or,* there is a subjacent linear/slightly displaced/depressed, comminuted skull fracture.

Example of internal injury descriptions/sequelae:

"Internally, associated injuries include…slight/moderate/marked/gelatinous subjacent soft tissue hemorrhage, multiple linear/displaced/shatter-type, vertex/basilar, left parietal/right temporal/etc., skull fractures, hinge/

ring/longitudinal basilar fractures, 70 mL of liquid/clotted/focally adherent subdural hemorrhage/hematoma with associated compression of the brain, patchy hemispheric subarachnoid hemorrhage, displaced fractures of right anterolateral right ribs #2–8, gaping anteroposterior hepatic lacerations at the falciform ligament, 1000 of hemoperitoneum, etc."

Notes

- *Describe fractures:* anatomic location, linear/comminuted, depressed/nondepressed, open/closed, patterned? stellate, radiating (if skull), basilar/vertex.
- *Organ injuries:* contusions/lacerations (not ruptures), hemoaspiration (airways, lungs: describe pattern: bilateral, patchy, diffuse).
- *Body cavity hemorrhages:* hemopericardium (not tamponade), hemothorax/hemothoraces, hemoperitoneum (always measure these).
- *Associated findings:* empty heart, visceral pallor, etc.
- *Patterned injuries:* Describe in detail.
  ["The brain is retained for neuropathologic evaluation."]

Sharp force trauma templates

Stab wounds

Injuries, external and internal

There is/are (a/multiple) stab (and incised) wound(s) of the (head and neck/torso/extremities).

All descriptions and directions are stated with reference to the standard anatomical planes with the body measured in the horizontal position. [*For multiple:* These injuries are labeled "A" through "___" for descriptive purposes only; no sequence is implied.] The injuries are described by body region.

Stab wound of________ (Head/Neck/Torso/Upper Extremities/Lower Extremities; *separate and organize each heading by body region; letter/number, if multiple):*

Example:

"A stab wound of the (*see Figure C.1 for anatomic nomenclature*) is centered ____" below the top of the head and ____" to the right/left of midline. It consists of a slit-like (gaping? V-shaped?) wound measuring ____" in length on the surface of the skin after the skin tension is released. It is horizontally/vertically/obliquely oriented with an (inferomedial/inferolateral/superolateral/superomedial) sharp angle and a ____" (inferomedial/inferolateral/superolateral/superomedial) blunt end. There is/is no abrasion or contusion of the surrounding skin. (Describe details—size, shape, color—of any surrounding abrasion or contusion).

"After the skin and subcutaneous soft tissues, there is sequential penetration of…"

"There is an associated ____ mL of (left/right liquid/clotted hemothorax/hemopericardium, etc., hemoaspiration, and marked visceral pallor)."

"The direction of penetration is: front to back/back to front, left to right/right to left, and downwards/upwards."

"The **estimated** depth of penetration is _____" "*(can give a range here: e.g., 4–1/2 to 6"; this is especially relevant when the knife/sharp object traverses an area that has both a soft surface and mobile tissue within cavity space, such as the abdomen/abdominal cavity or thorax).*

"These injuries, having been described, will not be repeated."

Notes

- **Common associated findings to look for and document in sharp injury***: empty heart, visceral/muscular pallor, shriveled splenic capsule, etc.*
- **Patterned injuries in stabbings:** From ice picks, forks, barbeque forks, screwdrivers (Phillips head, flathead), broken glass shards/bottles, scissors (closed and open to varying degrees), gaping wounds/swallow tail wounds indicating relative movement of body to weapon/weapon to body, often during some form of struggle.

- **Sometimes, the sharp/blunt angles are ambiguous/not discernibly distinct on the skin.** If this is the case, even with release of skin tension and rehydrating, call the angles **"ambiguous"** in terms of blunt or sharp. Know, however, that sometimes, wound ambiguity on the skin surface is clarified by subjacent bone or firm tissue like the myocardium or cartilage, where blunt and sharp angles can be better discerned. If that's the case, so describe and state in the autopsy report.
- **Stab wounds are only measured in LENGTH, and that, only after skin tension is released.** Don't bother measuring "width" of stab wounds that are gaping prior to relaxation of the skin; this has no real relevance in terms of weapon matching or wound pattern analysis. The only width that should be referred to is the "width" of the blunt angle of the wound (usually 1/16").
- ****Do not forget to assess/look for and document visceral and muscular pallor at autopsy, as exsanguination is the most common mechanism of sharp force deaths.**

NOTE: If there are an **excessive number of wounds, they can be carefully grouped/clustered by region/distribution, giving ranges of location from the top of the head/shoulder and from the midline, and ranges of sizes, orientation, and depths. Outliers/unusual wounds and/or wounds with a more lethal contribution should be further described separately/highlighted.

Describing injuries in clusters is sometimes not only necessary, but more communicative (again, especially when the number of wounds is particularly complex and/or excessive with a confluence of wound tracks). There is an art to creating a narrative that draws a visual picture, as opposed to just methodically describing wound after wound, without taking the further step of contextualizing them. That is the art of crafting a report, of creating a document that evokes a clear visual, not just of individual wounds, but of the entirety of the wounds on the body.

In the end, **all wounds** and their features/measurements must be documented photographically and in the autopsy notes, but depending on their number and distribution, they may lend themselves better to this "cluster" description in the final report. **Even if clustered, the total number of wounds identified on the body should still be noted in the report; there should be no approximations in the number of wounds.** It is sometimes helpful with an excessive wounding pattern to summarize the total wound tally at the end of the injury section (i.e., total number of stab wounds and total number of incised wounds).

Incised wounds

Incised wounds of_______ (Head/Neck/Torso/Upper Extremities/Lower Extremities):

Example:

"There is a (____") in length × (_____") depth, vertically/horizontally/obliquely oriented incised wound of the ___ (e.g., distal volar right forearm, proximal, dorsal right second finger, etc.) and a (____") in length × (_____") depth, vertically/horizontally/obliquely oriented incised wound of the the ______ (site).

["None of the incised wounds listed above injure major blood vessels"].

OR (if on the neck and penetrating vital structures):

"After incision of the skin, there is transection of the left jugular vein, left carotid artery, and left larynx. Associated injuries include marked associated soft tissue hemorrhage, extensive, bilateral hemoaspiration (diffuse leopard skin pattern of pulmonary hemorrhage), and marked visceral and muscular pallor. There is (no) intracardiac air embolism."

"These injuries, having been described, will not be repeated."

Notes

- **There are no blunt and sharp angles of incised wounds,** *as they are caused by the long sharp edge of the knife, not the tip first.*
- **There is no directionality statement for incised wounds** *(i.e., front to back, left to right, etc.).*
- **Potential patterned injuries to consider in incised wounds:** hesitation marks (suicides), serrated knife pattern from the serrated knife edge scraping across the skin (multiple evenly spaced, parallel linear abrasions: measure the distance between these parallel abrasions and describe in detail, as a serrated patterned abrasion), pattern of disposable/metal handheld razors.

- **The location on the body** of incised wounds is important for later interpretation of a struggle and likely movement of the upper extremities/consciousness of the victim during a violent encounter with a sharp instrument (forearms/hands are the most commonly involved).
- **Superficial incised wounds can look like linear abrasions:** in sharp force trauma cases, place these wounds under the "Incised Wound" heading and describe them as "superficial incised wounds/linear abrasions." These are often caused by the knife tip scraping across the skin during an altercation involving a sharp object/knife.
- *"There is/is no* **intracardiac air embolism***": this is a pertinent (positive or) negative comment helpful primarily in deep incised wound(s) of the neck.*

Chop wounds

Injuries, external and internal

Chop wounds of ____ (Head/Neck/Torso/Upper Extremities/Lower Extremities; *separate and organize each heading by body region; letter, if multiple):*

"There is a/are/multiple (____") in length × (_____") depth, vertically/horizontally/obliquely oriented, gaping wound(s) of the ___(anatomic location/s)___, exposing avulsed muscular soft tissues. There is an associated (linear chip/impression, depressed fracture) of the subjacent bone (name bone)."

"These injuries, having been described, will not be repeated."

Notes

- *Seen with often dense/heavy weapons with relatively sharp edges, such as meat cleavers, hatchets, machetes, and axes.*
- *If no crush injuries, check the underlying bones for grooves/divots, etc.*
- *A gaping incised wound of the skin with an underlying fracture or deep groove in the bone indicates that one is likely dealing with a chopping weapon.*
- *They are nearly uniformly fatal.*
- *A chop weapon is usually unwieldy and dense/heavy, but sharp. As such, the wounds they produce are characteristically devastating, gaping, and when of skin and underlying soft tissue, may have large tissue flaps hanging off from one side. The heavy and large nature of the weapon, combined with its sharpness creates wounds that are a hybrid or have features at the* **intersection of sharp and blunt force injuries***, and when of the skull, have characteristic crush-type, depressed, and comminuted underlying fractures that often become confluent. The blunt/sharp hybrid is also often evidenced by a thin rim of abrasion around the wounds because of the girth/thickness of the weapon (not as sleek as a knife, so can abrade the skin as it makes contact).*

Hanging template

Injuries, external and internal

The decedent is received with a ligature (type) (tightly/loosely) wrapped around the neck (once/twice, etc.), with a single (double), fixed (slip) knot located (posteriorly/anteriorly/laterally, etc.). It consists of a (e.g., coarse, 1–1/4" in width/diameter, tan rope) with (one/two) free ends (e.g., 7" and 24", respectively) extending from the knot.

Underlying the ligature, there is a circumferential/partially circumferential ligature abrasion (furrow, if trenched) of the neck. It ranges from ____ to ____" in width, is (color/dried and parchment-like), and is (e.g., most prominent, dried, leathery, and trenched anteriorly/L/R), with a (e.g., rope weave pattern impressed into the skin). At the (front, left, right, and back) of the neck, the abrasion is horizontally oriented, just (at/inferior to/superior to) the thyroid prominence. It courses upwards and posteriorly along the lateral aspects of the neck and toward the ears, until it disappears entirely at the posterior neck, (the location of the knot and the point of ligature suspension).

Internally, there are (no) fractures of the neck (hyoid, thyroid, cricoid cartilage, vertebrae, L/R), and no/scant hemorrhage of the subcutaneous and muscular soft tissues immediately subjacent to the deepest segment of the ligature furrow. There are (no) petechial hemorrhages of the conjunctivae (if present, indicate where: L/R, medial/lateral, superior/inferior, bulbar, palpebral), face, larynx, and oral mucosa (comment on facial pallor or plethora, protruding tongue, etc.).

Thermal burn injury template

Injuries, external and internal

Head and neck: There is [no/slight/moderate/marked] swelling of the soft tissues of the head and neck. There is singeing of the nasal hairs, eyebrows, eyelashes, or [frontal] scalp hair. There are (no) [first/second/third degree] burns of the head and neck. The edges are [rolled and sloughing/charred/well-demarcated with a bright red, hyperemic rim]. There are (no/few small foci/multiple foci] of fluid-filled blisters. The exposed dermis is pink/red/white/glistening. There is (no) charring with exposure of the underlying subcutaneous fatty tissue or muscle. Describe % involvement, anteriorly and posteriorly.

Trunk: There is [no/slight/moderate/marked] swelling of the soft tissues of the [anterior/posterior] torso. [The (anterior/posterior) torso is spared.] There is (no) singeing of the body hair. There are (no) [first/second/third degree] burns of the (anterior/posterior) torso (give distribution and % BSA). The edges are [slightly rolled and sloughing/charred/well-demarcated with a bright red, hyperemic rim]. There are (no/few small foci/multiple foci] of fluid-filled blisters. The exposed dermis is pink/red/white. There is (no) charring with exposure of the underlying subcutaneous fatty tissue or muscle.

Extremities: There is [no/slight/moderate/marked] swelling of the soft tissues of the [RUE/LUE/RLE/LLE]. [The buttocks/anterior/posterior thighs are spared]. There is (no) singeing of the body hair. There are (no) [first/second/third degree] burns of the [RUE/LUE/RLE/LLE]—give distribution and % BSA. The edges are [slightly rolled and sloughing/charred/well-demarcated with a bright red, hyperemic rim]. There are (no/few small foci/multiple foci] of fluid-filled blisters. The exposed dermis is pink/red/white. There is degloving injury. There is (no) charring with exposure of the underlying subcutaneous fatty tissue or muscle.

General description of artifacts of thermal burns/heat: "The body has a pugilistic posture and is extensively (near totally) charred, with multiple areas of longitudinal muscle splitting, displaced long bone (heat) fractures, traumatic amputations of the hands, and a granular, dark maroon left hemispheric epidural blood collection that crosses the midline."

APPENDIX

D

Library for forensic pathology fellows

The forensic pathology fellow faces numerous print and electronic sources of educational material. The following selection (Table D.1) has expanded on recommendations by Hanzlick, Molina, and Quinton.[1]

TABLE D.1 Educational material for forensic pathology fellows.

Required reading	
NAME	Guide for Manner of Death Classification
NAME	Forensic Autopsy Performance Standards
DiMaio, VJM	Gunshot Wounds
Moritz, A	Classic Mistakes in Forensic Pathology[2]
Spitz and Fisher et al	Medicolegal Investigation of Death
Other reading	
Adelson L	The Pathology of Homicide
Byard RW, Krous HF	Sudden Infant Death Syndrome
CAP, Randall B, et al.	Practice Guidelines for Forensic Pathology
CAP/Forensic Pathology Committee/National Association of Medical Examiners	Basic Competencies in Forensic Pathology: A Primer
College of American Pathologists (CAP)	Handbook of Forensic Pathology
Collins K, Byard R	Forensic Pathology of Infancy and Childhood
Cumming PM, Trelka DP, Springer KM	Atlas of Forensic Histopathology
Dettmeyer R	Forensic Histopathology
DiMaio VJM, Dana SE	Handbook of Forensic Pathology
DiMaio VJM, Molina DK	Forensic Pathology
Dolinak D, Matshes E, Lew E	Forensic Pathology Principles and Practice
Feliciano DV, Mattox KL, Moore EE	Trauma (2nd Ed)
Gonzales TA, Vance M, Helpern M, Umberger CJ	Legal Medicine: Pathology and Toxicology
Gordon I, Shapiro HA	Forensic Medicine: A Guide to Principles
Hanzlick R, Graham M	Forensic Pathology in Criminal Cases
Henssge C et al.	The Estimation of Time Since Death in the Early Postmortem Period
Knight B	Forensic Pathology
Perper JA, Wecht CH	Microscopic Diagnosis in Forensic Pathology

Continued

TABLE D.1 Educational material for forensic pathology fellows.—cont'd

Reece RM	Child Abuse
Sens, MA, Hughes, R	Diagnostic Pathology: Forensic Autopsy
Simpson K, Knight B	Forensic Medicine
Valdes-Dapena M et al.	Histopathology Atlas for the Sudden Infant Death Syndrome
Yeh I	Atlas of Microscopic Artifacts and Foreign Materials
Zumwalt, RE	Zumwalt's Forensic Pathology
General	
Berton Roueche	Series of articles on medical detection published in the New Yorker (e.g., The Case of the Eleven Blue Men)
Blum D	The Poisoner's Handbook
Hanzlick RL	Death Investigation: Systems and Procedures
Jentzen JM	Death Investigation in America
Nordby JJ	Dead Reckoning: The Art of Forensic Detection
Death scene investigation	
Fisher BAJ	Techniques of Crime Scene Investigation
Horswell J	The Practice of Crime Scene Investigation
Jentzen J et al.	Medicolegal Death Investigator
Lee HC et al.	Henry Lee's Crime Scene Handbook
National Institute of Justice (NIJ)	Guide for the Death Scene Investigator
Rebman A, David E, Sorg MH	Cadaver Dog Handbook
U.S. Department of Justice	Crime Scene Investigation: A Guide for Law Enforcement
Toxicology	
Baselt R	Disposition of Toxins and Drugs in Man
Cupp MJ	Toxicology and Pharmacology of Herbal Products
Dolinak D	Forensic Toxicology: A Physiologic Perspective
Ellenhorn MJ	Ellenhorn's Medical Toxicology
Karch SB	Drug Abuse Handbook
Karch SB	Pathology of Drug Abuse
Kintz P	Analytical and Practical Aspects of Drug Testing in Hair
Levine B	Principles of Forensic Toxicology
Molina DK	Handbook of Forensic Toxicology for Medical Examiners
Trestrail JH	Criminal Poisoning
Yudko E, Hall H, McPherson S	Methamphetamine Use: Clinical and Forensic Aspects
Neuropathology	
Dolinak D, Matshes E	Medicolegal Neuropathology: A Color Atlas
Itabashi HH, et al.	Forensic Neuropathology: A Practical Review of the Fundamentals
Leestma J	Forensic Neuropathology
Oehmichen M, Auer RN, Konig HG	Forensic Neuropathology and Neurology
Rengachary SS, Ellenbogen RG	Principles of Neurosurgery
Tronscoso, Rubio, Fowler	Essential Forensic Neuropathology

TABLE D.1 Educational material for forensic pathology fellows.—cont'd

Forensic cardiac pathology	
Finsechi V, Baroldi G, Silver M	Pathology of the Heart and Sudden Death in Forensic Medicine
Autopsy-related	
Adams V	Guidelines for Reports by Autopsy Pathologists
College of American Pathologists (CAP)/Collins, K	Special Autopsy Dissection Techniques
Multiple authors	Multiple topics related to autopsy (emedicine.medscape.com)
Death certification	
CAP	Cause of Death and the Death Certificate
CDC	CDC Sudden Infant Death Training Materials
Gill JR, Goldfeder LB, Hirsch CS.	Use of "therapeutic complication" as a manner of death[3]
Hirsch CS, Flomenbaum M.	Problem-solving in death certification. ASCP, Forensic Pathology Check Sample. 1995; FP 95-1:1–31.
Rosenberg ML, et al.	Operational Criteria for the Determination of Suicide[4]
Decedent identification	
CAP (Fierro MF)	CAP Handbook for Postmortem Examination of Unidentified Remains
Champod C, Lennard, C, Margot, P, Stoilovic, M	Fingerprints and Other Ridge Skin Impressions
Thompson T, Black S	Forensic Human Identification: An Introduction
Organ/tissue procurement	
NAME	Medical Examiner Release of Organs and Tissue for Transplantation
Ethics	
Barnett PD	Ethics in Forensic Science: Professional Standards for the Practice of Criminalistics
Multiple Authors	Academic Forensic Pathology Ethics Issue
NAME	Code of Ethics and Conduct
Communication with families of deceased	
Adelson L	Forensic Pathologist. "Family Physician" to the Bereaved[5]
Hirsch CS	Talking with the Family After an Autopsy[6]
Testimony and Law	
Adelson L, Dewitt C, Gerber SR, Moritz, AR	Physician in the Courtroom (1954)
Bronstein DA	Law for the Expert Witness
CAP	The Pathologist in Court
Davis GG	Pathology and Law – A Practical Guide for the Pathologist
Feder H	Succeeding as an Expert Witness
Matson J	Effective Expert Witnessing
Volkell R	Quick Legal Terminology
Wecht CH, Rago JT	Forensic Science and Law
Galloway A, Hartnett-McCann K, Fulginiti LC	Forensic Anthropology and the United States Judicial System
Special types of death	
Armstrong EJ, Erskine KL	Water-Related Death Investigation
Auerbach PS	Wilderness Medicine

Continued

TABLE D.1 Educational material for forensic pathology fellows.—cont'd

CAP/Langston C, et al.	Practice Guideline for Examination of the Placenta
Davis GG, Cadwallader AB, Fligner CL et al.	Position Paper: Recommendations for the Investigation, Diagnosis, and Certification of Deaths Related to Opioid and Other Drugs. Am J Forensic Med Pathol. 2020
DiMaio TG, DiMaio VJM	Excited Delirium
Dix J, Graham M, Hanzlick R	Asphyxia and Drowning: An Atlas
Ely SF, Hirsch CS.	Asphyxial deaths and petechiae: a review.[7]
Gebgerth VJ	Sex-Related Homicide and Death Investigation
Geddes LA	Medical Device Accidents
Lathers CM et al	Sudden Death in Epilepsy
Middleton OL, Atherton DS, Bundock EA, et al.	NAME Position Paper: Recommendations for the Investigation and Certification of Deaths in People with Epilepsy. Acad Forensic Pathol. 2018
Reay DT, et al.	A Procedure for the Investigation of Anesthetic/Surgical Deaths
Sauvageau A, Gerberth VJ	Autoerotic Deaths
Criminalistics	
Bevel T, Gard- ner R	Bloodstain Pattern Analysis
Bodziak WJ	Footwear Impression Evidence: Detection, Recovery, and Examination
James SH, Kish PE, Sutton TP	Principles of Bloodstain Pattern Analysis
James SH, Nordby JJ	Forensic Science: An Introduction to Scientific and Investigative Techniques
Kelly SJ, Lindblom BS	Scientific Examination of Questioned Documents
Lewis PR, Reynolds K, Gagg C	Forensic Materials Engineering Case Studies
Poorman K	Common Objects and their Impressions
Saferstein R	Criminalistics: An introduction to Forensic Science
Siegal JA	Forensic Science: The Basics
Anthropology/Osteology/Radiology	
Adams BJ, Byrd, JE	Recovery, Analysis, and Identification of Commingled Human Remains
Adams, BJ	Forensic Anthropology
Bass WM	Human Osteology: A laboratory and Field Manual (3rd Ed)
Brogdon BG, Vogel H, Mc- Dowell JD	A Radiologic Atlas of Abuse, Torture, Terrorism, and Inflicted Trauma
Langley NR, Tersigni-Tarrant MA	Forensic Anthropology: A Comprehensive Introduction
Levy A, Harcke HT	Essentials of Forensic Imaging: A Text-Atlas
Matshes E, et al	Human Osteology and Skeletal Radiology: An Atlas and Guide
Thali MJ, Viner MD, Brogdon BG	Brogdon's Forensic Radiology (2nd Ed)
Ubelaker D	Human Skeletal Remains: Excavation, Analysis, Interpretation
Wedel, VL, Galloway, A	Broken Bones: Anthropologic Analysis of Blunt Force Trauma
Odontology	
American Society of Forensic Odontology (ASFO), Senn DR, Weems DA	Manual of Forensic Odontology
Dorion BJ	Bitemark Evidence
Entomology	
Byrd JH, Cast- ner JL	Forensic Entomology

TABLE D.1 Educational material for forensic pathology fellows.—cont'd

Goddard J	Physician's Guide to Arthropods of Medical Importance
Stewart A	Wicked Bugs: The Louse that Conquered Napoleon's Army and Other Diabolical Insects
Pediatric pathology	
Bundock, E, Corey, T	Unexplained Pediatric Deaths: Investigation, Certification and Family Needs
Case, ME	Forensic Pathology of Child Death Assessment: For Autopsy Results and Diagnoses
Collins, K and Byard, RW	Forensic Pathology of Infancy and Childhood
Husain A, Stocker JT, Dehner LP.	Pediatric Pathology
Botany	
American Medical Association (AMA)	AMA Handbook of Poisonous and Injurious Plants
Coyle HM	Forensic Botany
Stewart A	Wicked Plants: The Weed That Killed Lincoln's Mother and Other Botanical Atrocities
Terrorism and mass fatalities	
Carroll E, Johnson A, DePaolo F, Adams BJ, Mazone D, and Sampson B.	Trends in United States mass fatality incidents and recommendations for medical examiners and coroners.[8]
CDC	Guidelines for Handling Decedents Contaminated with Radioactive Materials
Centers for Disease Control and Prevention (CDC)	Medical Examiners, Coroners, and Biologic Terrorism: A guidebook for surveillance and case management
Ellison HD	Handbook of Chemical and Biological Warfare Agents
Hanzlick R, Nolte K, deJong J, et al.	The Medical Examiner/Coroners Guide for Contaminated Deceased Body Management
Jensen RA	Mass Fatality and Casualty Incidents: A Field Guide
Motor vehicle collision investigation	
Burke MP	Forensic Medical Investigation of Motor Vehicle Incidents
Dix J, Graham M, Hanzlick R	Investigation of Road Traffic Fatalities: An Atlas
Limpert R	Motor Vehicle Accident Reconstruction and Cause Analysis (4th Ed)
Firearms and ballistics	
Dodd MJ	Terminal Ballistics: A Text and Atlas of Gunshot Wounds
Hatch DE, Dickson R	Officer Involved Shootings and the Use of Force: Practical Investigative Techniques
Hueske EE	Practical Analysis and Reconstruction of Shooting Incidents
Fire/explosion investigation	
Crippin JB	Explosive and Chemical Weapons Detection
Drysdale D	An Introduction to Fire Dynamics
Fairgrieve S	Forensic Cremation: Recovery and Analysis
Lentini JJ	Scientific Protocols for Fire Investigation
U.S. Department of Justice	Fire and Arson Scene Evidence: A Guide for Public Safety Personnel
U.S. Department of Justice	A Guide for Explosion and Bombing Scene Investigation

Continued

TABLE D.1 Educational material for forensic pathology fellows.—cont'd

Journal	*Publisher*
Academic Forensic Pathology	Academic Forensic Pathology (The Official Publication of NAME Foundation)
American Journal of Forensic Medicine and Pathology	Lippincott Williams and Wilkins (The Official Publication of NAME)
Forensic Science, Medicine, and Pathology	Springer
Forensic Science International	Elsevier
Medicine, Science, and the Law	Royal Society of Medicine (Official Journal of the British Academy of Forensic Sciences)

References

1. Hanzlick R, Molina DK, Quinton R. Recommended reading and reference bibliography for forensic pathology fellows. *Acad For Path*. 2014;4(4): 463–473.
2. Moritz AR. Classical mistakes in forensic pathology. *Am J Clin Pathol*. 1956;26(12):1383–1397.
3. Gill JR, Goldfeder LB, Hirsch CS. Use of "therapeutic complication" as a manner of death. *J Forensic Sci*. 2006;51(5):1127–1133.
4. Rosenberg ML, Davidson LE, Smith JC, et al. Operational criteria for the determination of suicide. *J Forensic Sci*. 1988;33(6):1445–1456.
5. Adelson L. The forensic pathologist. "Family physician" to the bereaved. *JAMA*. 1977;237(15):1585–1588.
6. Hirsch CS. Talking to the family after an autopsy. *Arch Pathol Lab Med*. 1984;108(6):513–514.
7. Ely SF, Hirsch CS. Asphyxial deaths and petechiae: a review. *J Forensic Sci*. 2000;45(6):1274–1277.
8. Carroll E, Johnson A, DePaolo F, Adams BJ, Mazone D, Sampson B. Trends in United States mass fatality incidents and recommendations for medical examiners and coroners. *Academic forensic pathology*. 2017 Sep;7(3):318–329.

APPENDIX

E

Hirschisms and Adelsonisms

"The old adage that dead men [and women] tell no tales does not hold true as far as the medical profession is concerned. Dead men [and women] have for centuries been telling doctors the story of *how* they got sick and *why* they died, and doctors have made use of this information very effectively in the treatment of other sick people…The public owes to the dissectors of the dead a debt that it has seldom recognized or acknowledged"[1] SL Wilens

Hirschisms

"Nothing can equal the thoughtful medicolegal autopsy as a teacher of the amazing variability of human responses to disease and injury."

"We never mix public responsibility with private gain."

"At every homicide, there is one accurate witness without self-serving statements. The pathologist interrogates that witness."

"Our science and art are useful tools in the larger patchwork of investigation that ultimately is stitched into a quilt of truth."

"Absence of evidence is not evidence of absence."

"The plural of anecdote is not data."

"A foolish consistency is the hobgoblin of little minds..." (RW Emerson)

"I'd rather autopsy six people shot once than one person shot six times."

"It doesn't matter what a reporter tells you, you're *never* off the record."

"We solve more cases with the telephone than with the microscope."

"You can call someone any name in the book. Just put the telephone receiver down first."

"Reality always meets the paranoid halfway."

"I wouldn't want to lean on a broken crutch."

"The Crown never wins and the Crown never loses, let justice be done." (Great Britain)

"Beware of the person who has seen one case." (Sir William Osler)

"Once a suicide, always a suicide."

"Partial autopsies give partial answers."

"I would rather explain why I did something than why I did not."

"Common things are common because they are the things that occur with the greatest frequency."

"Information void is the mother of paranoia."

"If you think of it, do it."

"The temptation to make an epicrisis of the death certificate must be resisted."

"Someone who has all the answers all the time is either a scoundrel or a fool."

"When all else fails, tell the truth."

"She didn't know whether she was on horse or on foot."

"Memorialize that conversation on paper."

"A nickel's worth of pneumonia."

"Many a beautiful idea has been slain by ugly fact." William Boyd[2]

"The distraction of the dramatic."

"As nervous as the devil in cold water."

"As Dave Barry says, 'I swear I am not making this up'..."
"There are none so blind as those who refuse to see."
"The south end of a northbound horse."
"Distribution of partial information guarantees complete misunderstanding."
"There is an infinite gap between the lab and the street."
"I have to think with a pencil when I do one of these."
"That must be rare squared."
"Funny way to run a railroad."
"Inspired, but wrong."
"Speculation is built on a firm foundation of assumption."
Hirsch's law: "When you weigh less than your age, the likelihood of mortality is very high."
"If you do not believe a witness, you better have a compelling reason to doubt him/her."
"A difference, to be a difference, must make a difference." (Gertrude Stein)
"If I had to check one box or the other, I would check both."
"Minor surgery: Surgery that someone else has."
About a mistake: "Glad my job description doesn't contain the adjective 'infallible'."
"When emotion collides with intellect, intellect always loses."
"Merchants of Doubt"
"Hunches are not evidence."
"Sequence doesn't make a consequence."
"It is much easier to be a historian than a prophet."
"A negative is never proof."
"The tyranny of routine"
"Routine is never a substitute for judgment."
"Sometimes ambiguity is your best friend."
"It takes circulation to get inflammation."
"Once your standards go, your principles won't be far behind."
"It makes heat without shedding light."
"He is somewhere in centerfield waiting for a bunt."
"Omentum: The policeman of the abdomen."
"The brain can absorb no more than the seat can endure."
"Toxicology gives us information, but not conclusions."
"Do not confuse correlation with causation."
"Render a diagnosis on everything we see and not on isolated findings."
"Wrestling with 1000 devils."
"Let us agree to disagree, but not be disagreeable."
"The only time they let loose is to get a tighter grip."
"You don't want to drown in a drop of water."
"She woke up dead."
"Everything is consistent, nothing is inconsistent."
"Let common sense prevail."
"While I can't say that's impossible, it's so highly unlikely that it's standing shoulder to shoulder with impossibility."
"Describe what you see."
"You've got to believe what you see."
"You can't see what you can't believe."
"Bacchus has drowned more men than Neptune." (Thomas Fuller)
"You can drown in a teacup."
"I don't know, but it's an excellent example of whatever it is." (upon being presented with an interesting patterned injury at the autopsy table)
"If government has an interest in the lives of its citizens, it must also have an interest in their deaths."
"Film is cheap."
"Think dirty."
"Refusal to consider circumstantial information can lead to a denial of reality in the name of objectivity."
"One cause of death per customer."

"*A* cause of death is not necessarily *the* cause of death."

"We're in the business of communication."

"We're not in the business of making assumptions." (or of "rushing to judgment").

"No problem was ever solved by moving papers to the bottom of the pile."

"You should never walk away from the autopsy table without first considering the *mechanism* of death."

"I'm glad we have an excellent medical examiner on the case."

"Just do the autopsy; you'll sleep better at night."

"A few: more than a couple and less than several."

"We want the news, not the editorial."

"Always seek primary sources of information."

Sign at a deli: "We have an agreement with the bank. They don't sell pastrami and we don't cash checks."

"Salsa is mild; atherosclerosis is slight."

"Honesty is the only stigma that attaches to a pathologist who leaves an occasional case undetermined."

"The absolute dependence of infants and children upon their parents or adult custodians renders them susceptible to a range of fatal maltreatment and neglect that defies the imagination of a thousand nightmares."[3]

"…the autopsy findings in this group range from nothing at all to blatantly grotesque. However, regardless of the nature of the autopsy findings, no genus of homicide victims presents a more complex or difficult challenge to the forensic pathologist and other investigative authorities."

"We strongly emphasize the need for investigative teamwork and the imperative to interpret autopsy findings in the light of the entire case study. Narrowly focused tunnel vision, restricted exclusively to autopsy findings taken out of context of the investigative universe, will see only part of the truth. Errors of interpretation resulting from such defective perception will leave some instances of fatal maltreatment unrecognized and, in other instances, will not provide truthful, nonsinister explanations for injuries. None of us can rest comfortably with the notion that we may have overlooked a homicide or misinterpreted homicidal injuries as an accident. However, any discomfort resulting from such false conclusions is luxury compared to our dread of the converse."

"In reality, pathologic processes seldom cooperate fully with attempts to date or age them with precise reliability."

The two most difficult sentences to say: "I was wrong." and "You were right."

"You have to know the rules well enough to break them."

"A biased investigator is a useless investigator."[4]

"The truism that 'The rare things are rare, and the common things are common' does not exclude the occurrence of the most infrequent events. 'Once in a lifetime' events do occur in one's lifetime."[4]

"The body doesn't lie."

"The fundamental difference between living and dead organisms is functional, not structural; therefore, the pathologist who focuses exclusively on structurally demonstrable, anatomic causes of death is doomed to fail."

"At last, I have seen a purple cow."[4]

"Any old business, new business, monkey business?"

"Why then, I think I'll go home and have a double..."

Adelsonisms

"The pituitary is the conductor of the endocrine symphony."

"Death is democratic with a small 'd.' You see your colleague, the postman, and the kid next door in the autopsy room and it gets you where you live. You don't become hardened to death; you become sensitized."

"In the autopsy room, we see people with their naked bodies. When families see us, we see guilt, grief, and greed. We see families with their naked souls."

"Autopsies are for facts. Hunches have no place in autopsies. If you want hunches, try the lottery."

"The most important part of the autopsy is what you do with your head, not with your hands."

"Ischemia is a two-headed dragon. You can't get what you need and you can't get rid of what you don't need."

"Nature never lies."

"A question is halfway to wisdom."

"No one is born a congenitally good witness."

"When the blind lead the blind, they both fall into a ditch."

"Don't waste time. If you waste anything, waste money. You can get more money, but you can't get more time. Every day is a day closer to the grave whether you live to be 50 or 150. There are no blank checks. Our earthly existence is finite."

"Everyone looks through to the world with individual eyes. Some people see the donut and some people see the hole."

"No one dies suddenly and unexpectedly at the age of 100."

"The most uncommon commodity, common sense, is soluble in alcohol."[5]

References

1. Wilens S. *My Friends the Doctors*. New York: Atheneum; 1961.
2. Boyd W. *A textbook of Pathology*. 8th edition. Philadelphia: Lea and Febiger; 1970.
3. Zumwalt RE, Hirsch CS. Pathology of Fatal Child Abuse. In: Helfer RE, Kempe RS, eds. *The Battered Child*. 4th ed. Chicago, IL: The University of Chicago Press; 1987:247–285.
4. Hirsch CS, Adelson L. A suicidal gunshot wound of the back. *J Forensic Sci*. 1976;21(3):659–666.
5. Adelson L. *The Pathology of Homicide*. Springfield: Charles C. Thomas; 1974.

Index

Note: 'Page numbers followed by "*f*" indicate figures, and "*t*" indicate tables and "*b*" indicate boxes.'

D

T

U

V

W

Y